D1640102

33 UAB 1150-2
DeInventarisiert

UB Essen

VOLUME II

BUILDING AND RUNNING THE LABORATORY,
1954 – 1965

HISTORY OF CERN

ARMIN HERMANN

JOHN KRIGE

ULRIKE MERSITS

DOMINIQUE PESTRE

and with a contribution by

LAURA WEISS

Study Team for CERN History

NORTH-HOLLAND

1990   Amsterdam · Oxford · New York · Tokyo

Elsevier Science Publishers, B.V. 1990

Alle rights reserved. No part of this publication may be reproduced, stored in a retrieval system, or transmitted, in any form or by any means, electronic, mechanical, photocopying, recording or otherwise, without the prior permission of the publisher, Elsevier Science Publishers B.V., P.O. Box 211, 1000 AE Amsterdam, The Netherlands.

Special regulations for readers in the USA: This publication has been registered with the Copyright Clearance Center Inc. (CCC), Salem, Massachusetts. Information can be obtained from the CCC about conditions under which photocopies of parts of this publication may be made in the USA.

All other copyright questions, including photocopying outside of the USA, should be referred to the publisher.

No responsibility is assumed by the Publisher for any injury and/or damage to persons or property as a matter of products liability, negligence or otherwise, or from any use or operation of any methods, products, instructions or ideas contained in the material herein.

ISBN 0 444 88207 3

Published by:

North-Holland
Elsevier Science Publishers B.V.
P.O. Box 211
1000 AE Amsterdam
The Netherlands

Sole distributors for the USA and Canada:

Elsevier Science Publishing Company Inc.
655 Avenue of the Americas
New York, N.Y. 10017
USA

Printed in The Netherlands

Preface

The first volume of the History of CERN dealt with the launching of the European Organization for Nuclear Research. As we explained in that book, the process, initiated around the end of 1949 and the beginning of 1950, assumed a more precise shape in December 1950 during a meeting held at the Centre Européen de la Culture in Geneva. In 1951 the project was transferred to UNESCO by Pierre Auger, a determined partisan of the scheme, and it became a proposal to build two accelerators, one of them intended to be the biggest in the world. In February 1952 a dozen European states agreed to finance a provisional organization. Its role was to refine the ideas put forward by Auger's 'consultants', to draft an intergovernmental convention and to present more detailed budget estimates. In October 1952 the provisional Council of delegates accepted that the centrepiece of the new laboratory be a strong-focusing proton synchrotron of some 30 GeV, and agreed to locate it in Geneva. In June 1953 the Council got the text of the Convention accepted by twelve governments. Fifteen months later, the text having been ratified by a sufficient number of parliaments, the Convention entered into force. On 7-8 October 1954 the Council of the new organization met for the first time.

The present volume starts at this point and takes the history through to the mid-1960s, when it was decided to equip the laboratory with a second generation of accelerators, and a new Director-General was nominated. It covers the building and the running of the laboratory during these dozen years, it studies the construction and exploitation of the 600 MeV Synchro-cyclotron and the 28 GeV Proton Synchrotron, it considers the setting up of the material and organizational infrastructure which made this possible, and it covers the reigns of four Director-Generals, Felix Bloch, Cornelis Bakker, John Adams, and Victor Weisskopf.

The two volumes are organized in quite different ways. For the years 1949-1954 the object of our study led us to put the emphasis on the *political* dimension of the *process* of creation. Two organizational principles followed from this. The first was that a chronological approach should be dominant, since it alone would allow us to follow the complex path of development, since it alone would enable us to grasp the character of the 'final product' that was CERN in 1954. The second was that it seemed preferable to present events successively from two points of view, that of the principal actors on the one hand, and that of the state apparatuses on the other. The main reason for this was that a group of some 30 to 40 people managed to act more or less autonomously between 1950 and 1953, and that European governments were placed in a reactive position, were somewhat marginalized as far as key decisions were concerned. Parts II and III of Volume I thus treated the creation of CERN through the activities of this core group—and described the split that divided the physics community of the day on the question of setting up a European 'big science' laboratory. And Part IV was devoted to case studies dealing with the decision-making processes in the four 'major' member states.

The material to be covered in this book has demanded a different treatment. Three considerations are relevant here. Firstly, the political dimension, in the broad sense of the term, was no longer omnipresent as during the process of creation. Scientific and technical determinations were at work along with it. This is so, for example, when one looks at the choice of certain equipment or at the drawing up of the experimental programme. The second consideration is that the organization itself was now established, that the activity in Geneva was circumscribed by various rules, that a number of constraints shaped its development—in short, that the institutional dimension was also inescapably present. In 1965, for example, CERN employed more than 2000 people with very diverse qualifications—experimentalists, machine designers, engineers, technicians, administrators...—and its successful functioning called for a proper integration of tasks. Finally, there was no longer one dominant process in the organization's life but several, and it was no longer possible to tell just one story—at least not without producing a text so dense as to be illegible. Since anyway no point of view ever allows one to say everything, since no study can hope to deal simultaneously with all the legitimate questions that can be asked of a complex ensemble, we decided to focus attention successively on particular aspects of CERN's life, to tell several parallel stories, to refine them so as to grasp their individual logics—recombining their threads at a different, 'coarser' level.

The plan of the book flows from these remarks. Part I attempts to hold the threads in one hand, as it were, to describe as concisely as possible the various strands which together constitute the history of CERN—it aims to offer a synchronic panorama year by year of CERN's many activities. Written by John Krige and Dominique Pestre, this part is deliberately descriptive and stays close to the events. Its purpose is to provide, in an unfussy, 'neutral' tone, a chronological overview of what happened in and around the organization at Geneva. The two chapters treat the accelerator building years (1954-1959) and then the period during which the first experimental programme was brought into operation (1960-1965). Essentially an introduction to the book, they help the reader get a first feel for the developments, the debates, and the issues at stake inside the organization.

Similarly, part IV, the conclusion to the book, tries to be synthetic, tries to make a cut through the particular. However, unlike the opening chapters, it is not chronological, aiming instead to capture the whole period in a single movement. Its more precise purpose is to bring out the specificity of CERN, to identify the ways in which it differed from other big science laboratories in the 1950s and 1960s, to understand wherein lay its uniqueness, its originality.

In contrast, the two other parts of the book—which are also the longest—are deliberately analytical. The individual chapters composing them are intended to follow a particular path, to probe deeply into a problem, or to describe a segment of the organization's activities or of what we call the CERN 'system'. The ten chapters have been put into two blocks, each dealt with by two different groups of authors and each characterized by its own dominant angle of approach, by its own main preoccupation. The overriding concern in part II is that which we often find among scientists: for them the essential thing in the practice of science is the result and its temporal unfolding—in other words the accumulation and progressive development of knowledge and technique. Accordingly this section of the book deals primarily with technological achievements and scientific results, and it includes the most technical chapters in the volume, chapters using as main sources publications in the open literature, internal reports, and minutes of specialized committees or of divisional meetings.

Part III is more in line with a classical historical approach. Focussed on human actors, their ways of being, their modes of working, it studies groups and their relations, it aims to define how this science-based organization worked, how, at one pole, it chose, planned, and concretely realized its experimental programme on the shop-floor, how, at another, it identified the equipment it would need in the long term and organized its relations with the outside world, notably the political world.

Have we then divided the book on the lines of a good old internal history/external history? We do not think so, we do not think that our approach has been shaped by this dichotomy. What we have sought to do in each case is to grasp historically an organization-dedicated-to-science, to see how several kinds of practices interleaved, how multiple determinations entangled with each other in one and the same event or process. In part II, chapter 5, for example, a chapter dedicated to the building and exploitation of the 28 GeV PS, there is an analysis of the important set of experiments done with a neutrino beam which takes into account scientific research programmes, practical considerations, the rush to succeed before the rival Brookhaven laboratory, and so on. Similarly in part III, chapter 12, which deals with the decision-making process which in 1965 led to the 'choice' to build a new accelerator, one finds a technical history of the new acceleration principles discussed inside the scientific community between 1956 and 1963, along with a description of mental phenomena bordering on the limits of consciousness – the fear of 'technological monsters', for example – and an account of the 'perverse effects' of the ways in which the CERN Council functioned. In short, once the question to be explored has been chosen—and as always it is the historian who delimits his object—we kept, or at least we tried to keep an open mind about the kind of material relevant to answering that question. Of course when one tackles an issue one has to decide what angles to take, how deep to delve, but this does not determine the nature of the 'explanatory' elements to be considered. One may put them into two broad categories labelled 'internal' or 'external' if one wants to, but we did not, nor do we think that it would have been helpful to do so.

Let us now give a little more detailed information on the contents of parts II and III. The chapters in part II were written by Armin Hermann, Ulrike Mersits, and Laura Weiss. The first (chapter 3) is of a general nature. Written by Armin Hermann it gives an overview of a number of social and technological developments in high-energy physics during our period. Chapters 4 and 5, written by Ulrike Mersits, describe the construction of CERN's two accelerators in the 1950s, and the scientific work done with them between 1957, when the Synchro-cyclotron was commissioned, and 1965. The first contains a survey of the particle physics done with the smaller machine along with case studies of four notable experiments. The physics results discussed in chapter 5, by contrast, are divided according to detection technique (bubble chambers, electronic methods) and the all-important CERN neutrino experiments are given separate treatment. Finally chapter 6, written by Laura Weiss, describes the construction of CERN's first three hydrogen bubble chambers. It is included because it was these heavy and complex pieces of equipment that transformed high-energy physics between 1950 and 1970 into big science. Originally we planned to write separate chapters on the construction of the two accelerators. This plan had to be modified due to inadequate resources, and the work done on the building of the machines was folded into the chapters dealing with their scientific exploitation.

John Krige and Dominique Pestre are jointly responsible for the six chapters in part III. The first describes what is called the CERN system, that is the unity constituted by the laboratory as such and the Council of representatives of member states. The aim of this chapter is to define the concrete rules whereby this entity functioned, to explain both the nature of its relationship towards the states and the distribution of powers inside the executive. The next chapter looks at how experimental work was organized in Geneva—an essential task in a big science laboratory whose facilities have to be 'shared' between dozens of research teams. It describes the various solutions tried in the early years and the nature of the compromise finally arrived at, showing how it differed from American ways of dealing with the problem. While this chapter is concerned with the debate on how to organize the work around the equipment CERN had, the next, chapter 9, studies the process of deciding what experimental material the senior staff thought the laboratory ought to acquire. It concentrates on the needs of the PS experimental programme in the late 1950s and offers some insight into why the laboratory was not able to capitalize on the early lead it had over BNL by virtue of the commissioning of the CERN PS some nine months ahead of the equivalent Brookhaven AGS. The focus in chapter 10 is on the most basic need of all in any expanding scientific laboratory—money. It studies the estimates laid by the executive before the Finance Committee, and it shows how its delegates' responses were initially shaped by different national policies vis-à-vis CERN—different policies which led to an explosive and highly- politicized debate in 1961 and an 'agreement' that CERN's budgets should grow annually at a rate comparable to national and American facilities. While chapter 10 deals with finance policy in general, Chapter 11 deals with a small but important aspect of it: CERN's contract policy with industry. The approach here is thematic rather than historical, if only because the policy, which holds that CERN award its contracts competitively, has not undergone any major changes. The chapter reveals the gap between the competitive 'ideal' and everyday practice, it discusses some of the pitfalls of the Finance Committee's way of monitoring the CERN executive's behaviour, and it describes the advantages accruing to the host state when no attempt is made to ensure that there is a 'fair' distribution of contracts between the member states. Finally, we come to chapter 12 which discusses in detail the process of decision for the accelerators which were to succeed the 28 GeV PS. Its main findings are that, in its initial phases, that process was dominated by two principal and opposed groups of actors, the engineer-builders of the machines and the physicists, and that it subsequently took a decisive turn as the debate was carried into the sphere of the Council and of national bureaucracies in the member states—leaving the European physics community in disarray.

A project of this magnitude and scope would not have been possible without the active co-operation and help of many people. For this second volume we would like to begin by again thanking the members of the CERN History Advisory Committee, Paul Levaux (chairman), Alfred Gunther (secretary), Edoardo Amaldi[†], Jan Bannier, Michel Crozon, Olle Edqvist, Axel Horstmann, Geoff Oldham, Michel Paty, Wolfgang Paul, Olaf Pedersen, Herbert Pietschmann, Giorgio Salvini, Roman Sexl[†], and John Ziman, who have continued to give us their support and encouragement. Thanks to them it has been possible to consider launching an additional programme to take the history of CERN up the late 1970s. John Krige is coordinating the work, which is already well under way.

All of those who sponsored the project are listed on p.XX. Special mention should again be made of the Stiftung Volkswagenwerk in the Federal Republic of Germany, the Joint Committee of the Science and Engineering Research Council and the Social Science Research Council in the United Kingdom, the Centre National de la Recherche Scientifique in France, the Fonds zur Förderung der Wissenschaftlichen Forschung in Austria, and the Netherlands Organization for the Advancement of Pure Research.

As with volume I, John Krige translated Dominique Pestre's and Laura Weiss's chapters from French. We also circulated preliminary versions of most of the chapters widely among the scientific community and our professional colleagues. Leon Van Hove showed an ongoing interest in our work, and made many helpful comments and criticisms. We are also indebted to E. Amaldi†, R. Armenteros, J. Bannier, R. Budde, G. Fidecaro, J. Freeman, P. Germain†, Y. Goldschmidt-Clermont†, M. Hine, K. Johnsen, G. Lemaine, G. Munday, G. Moorhead, D.R.O. Morrison, C. Peyrou, B. Powell, H.-P. Reinhard, G. Salvini, P. Standley, V. Weisskopf, K. Winter, D. Wiskott, and C. Zilverschoon for information and for suggestions on how to improve our texts. Finally we benefited from discussions of our results with the participants at the Stanford Workshop on the History of Big Science organized by Peter Galison and Bruce Hevly in August 1988.

CERN has continued to support our work by giving us effectively unlimited access to the documents in the CERN archive, and we must again thank Roswitha Rahmy, the archivist, and Yves Felt, her assistant, for putting up with our repeated demands for files and papers. We have felt at home in the Theory division, first under Maurice Jacob and then under John Ellis, and lately in Stefan Schwarz's TH/SIS group. We have always been able to rely on David Stungo and his pool of typists to help us in the painstaking task of producing camera-ready text, and we would particularly like to thank Rosmarie Meyer for her outstanding contribution. A final word of thanks is due to our publishers, and to Joost Kircz, Jan Visser, and Jane Kuurman in particular, for their continued interest in the project.

John Krige
Dominique Pestre

This project has been supported financially by the following institutions:

Austrian Foundation for the Advancement of Scientific Research
Centre National de la Recherche Scientifique, France
Consiglio Nazionale delle Ricerche, Italy
Danish Natural Science Research Council
Economic and Social Research Council, United Kingdom
Federal Ministry for Science and Research, Austria
Institut Interuniversitaire des Sciences Nucléaires, Bruxelles
Istituto Nazionale di Fisica Nucleare, Italy
Joint Committee of the Science and Engineering Research Council and the
   Social Science Research Council, United Kingdom
LK-NES Foundation, Denmark (private industry)
Netherlands Organization for the Advancement of Pure Research (Z.W.O.)
Norwegian Research Council
République et Canton de Genève, Département de l'Instruction Publique
Science and Engineering Research Council, United Kingdom
University of Milan, Italy
Volkswagen Foundation, Federal Republic of Germany

Contents

Preface   v
List of supporting institutions   xi
Contents   xiii
Remarks on the notes and bibliographies   xix
List of archival sources consulted   xx
List of abbreviations   xxi

PART I.   AN OVERVIEW OF THE FIRST DECADE OF CERN

1   Building the laboratory, 1954–1959   3
John KRIGE

1.1   1954/55, the year when the magnitude of the task that lay ahead began to emerge   4
1.2   1956, the year when CERN gained international recognition, and took the first steps needed to retain it   7
1.3   1957, the year when experimental work first became possible, and when financial constraints were first imposed   9
1.4   1958, the year when the big equipment needed to exploit the PS was settled, the year when the SC yielded its first important results   12
1.5   1959, the year when the PS gave its first beam, the year when the future of CERN was no longer in doubt   15
Notes   17
Bibliography   809

2   Running the laboratory, 1960–1965   19
Dominique PESTRE

2.1   1960, the year when all seemed possible, the year of reorganization and of large-scale projects   20
2.2   1961, the year of doubts and anxieties, the year of tensions   23
2.3   1962, the year when trends were reversed, the year of selfcriticism and of new hopes   27
2.4   1963, the first year in which tasks were satisfactorily integrated   29
2.5   1964, the year of the debate on the medium- and long-term programmes   32
2.6   1965, the year of consolidation   34
Notes   37
Bibliography   809

PART II. PHYSICS AND ENGINEERING AT CERN

3 Some aspects of the history of high-energy physics, 1952–66   41
Armin HERMANN

3.1 Introduction   42
3.2 Social aspects   43
3.3 Scientific developments   73
3.4 HEP and society   79
Notes   87
Bibliography   92

4 The 600 MeV synchro-cyclotron: constructing CERN's smaller accelerator and the machine's role in the scientific programme   95
Ulrike MERSITS

4.1 Introduction   96
4.2 Design and construction of the 600 MeV SC   97
4.3 Particle physics with the SC from 1958 to 1965   109
4.4 Concluding remarks   126
Notes   131
Bibliography   135

5 The construction of the 28 GeV Proton Synchrotron and the first six years of its scientific exploitation   139
Ulrike MERSITS

5.1 Introductory remarks   141
5.2 Planning and constructing Europe's biggest proton accelerator   145
5.3 High-energy physics using bubble chambers   162
5.4 The use of electronic detectors to study elementary particles   190
5.5 The neutrino experiments at CERN: confronting the reality of big science (1960–1965)   215
5.6 Conclusion   234
Appendices   240
Notes   243
Bibliography   254

6 The construction of CERN's first hydrogen bubble chambers   269
Laura WEISS

6.1 Historical overview   271
6.2 The principle of the bubble chamber   275
6.3 CERN's bubble chambers   284
6.4 Technical elements and construction   297
6.5 Conclusion   321
Notes   323
Bibliography   335

PART III. PLANNING AND MANAGING RESEARCH AT CERN

7 The CERN system, its deliberative and executive arms and some global statistics on how it functioned 341
Dominique PESTRE

7.1 The member states, the Council, and the executive: relationships strongly mediated by a stable group of people in the Council and in the Scientific Policy Committee 345
7.2 The CERN executive: an ensemble strongly influenced by the scientific senior staff 370
7.3 CERN by numbers: some statistical data on the personnel and its activities 393
Notes 410
Bibliography 809

7A Another aspect of CERN's European dimension: the 'European Study Group on Fusion', 1958-1964 416
Dominique PESTRE

7A1. The origin of the proposal: stakes and motives 418
7A2. The June 1958 Council session, and the political refusal to collaborate with Euratom 421
7A3. The activities of the study group, 1958-1964 422
7A4. Conclusions to be drawn from this episode on the nature and place of CERN in Europe at the end of the 1950s 424
Notes 426
Bibliography 809

8 The organization of the experimental work around the Proton Synchrotron, 1960-1965: the learning phase 429
Dominique PESTRE

8.1 Setting up research around the Synchro-cyclotron, 1957-1960 432
8.2 The arrangements concerning the loan of big bubble chambers to CERN; 1957-1959 436
8.3 The debate in 1959 around 'National Participation in Research at CERN': towards the disappearance of truck teams 444
8.4 The choice of experiments: the laborious setting up of the first committee system, 1959-1960 450
8.5 The changes made to Adams' decision-making system for experiments, 1961-1962 465
8.6 Experimenting with bubble chambers: the strength of the system of collaborations, 1960-1965 473
8.7 Electronic experimentation: the struggle for access to the PS, 1960-1965 480
8.8 Conclusion 489
Notes 494
Bibliography 809

Collection of photographs and documents of historic interest

9 Planning the infrastructure for the PS experimental programme 503
John KRIGE

9.1 A few technical essentials 506
9.2 Some reflections on the SC bubble chamber programme (1955-1956) 508
9.3 Defining big equipment for the PS experimental programme 511
9.4 Acquiring the material for the PS experimental programme 530
9.5 The beam transport crisis in 1961 and its causes 535
9.6 Data-handling facilities for track-chamber pictures 545
Notes 561
Bibliography 809

10 Finance policy: the debates in the Finance Committee and the Council over the level of the CERN budget 571
John KRIGE

10.1 The emergence of the idea that there be a ceiling imposed on CERN's expenditure 573
10.2 1958 to 1960: the continuation of the ceiling policy and its use as a means to stabilize the CERN budget- 586
10.3 1961: the three-year ceiling is broken and the foundations are laid for a new policy 596
10.4 The Bannier Report and the budget debates from 1962 to 1965 604
10.5 Concluding remarks: the significance of the Bannier Report 620
Notes 623
Bibliography 809

11 The contract policy with industry 637
John KRIGE

11.1 Laying the foundations of CERN's contract policy: the gap between principle and practice 640
11.2 The competitive award of technical contracts: a critique of the 'rational model' 643
11.3 The award of building contracts: the pitfalls of *a posteriori* control 650
11.4 Some global statistics on the distribution of contracts awarded by CERN between 1955 and 1965: the advantages of being the host state 664
11.5 Final remarks: the tenacity of CERN's contract policy 671
Notes 671
Bibliography 809

12 The second generation of accelerators for CERN, 1956—1965: the decision-making process 679
Dominique PESTRE

12.1 New acceleration principles and future accelerators: a chronological account of the evolution of the field, 1956–65 683
12.2 The time of new ideas: the CERN Research Group on Accelerators, 1957–1960 699

12.3 The tempestuous period: the first important year of the debate
     between scientists, 1961   706
12.4 The calm after the storm: the emergence of CERN's
     'New Study Group on Accelerators', December 1961–January 1963   729
12.5 The European high-energy project: ECFA's first report,
     January–June 1963   736
12.6 The 'selling' of the ISR to the Member States and the postponement
     of the 300 GeV project by the Council, June 1963–December 1965   745
12.7 Conclusions   759
Annex 1   763
Annex 2   764
Notes   765
Bibliography   809

PART IV. CONCLUDING REMARKS

13 Some characteristic features of CERN in the 1950s and 1960s   783
Dominique PESTRE

13.1 CERN in the mid-60s: cruising steadily ahead   784
13.2 The Council and the member states: an unusual relationship   786
13.3 The 'corporate spirit' of CERN: an important element making for cohesion   789
13.4 Learning to work together: a problem more or less solved around 1965   792
13.5 Learning to do 'big physics': a process still at work in the mid-1960s   794
13.6 American and European physicists: an historical difference   799
13.7 CERN and the history project: some lessons   803
Notes   806
Bibliography   809

Bibliography for parts I, III and IV   809
John KRIGE and Dominique PESTRE

Appendix 1   826
   Organigramme of CERN, 1955–1960, CERN divisions and their leaders
Appendix 2   828
   CERN Director-Generals and Members of the Directorate
Appendix 3   828
   Member States and their percentage contributions to the CERN basic programme,
   at three-year intervals
Appendix 4   829
   Senior Office Bearers in the Council and its Committees
Appendix 5   829
   Members of the SPC, up to 1965
Appendix 6   830
   V. Weisskopf's table of some important research results obtained at CERN or elsewhere
   in Europe in connection with CERN within the period 1960–1965

Who's who in the history of CERN 833
    Armin HERMANN

Chronology of events 847
    John KRIGE and Dominique PESTRE

Name Index 869
Thematic Subject Index 874

Remarks on the notes and bibliographies

The text of each chapter is immediately followed by its own notes.

There is a bibliography at the end of each chapter in part II and a collective bibliography at the end of part III for the chapters in parts I, III, and IV.

The following reference system has been used:

- Official CERN documents are referred to as CERN/..., and are in CERN's archives. Further information is only given when we have reason to believe that it is needed to retrieve the document easily. Note that the date given is that of the document, not of the meeting.

- The location of other documents in the CERN archives is specified in brackets using the file name. Thus (DG20328) means box 20328 in the series DG (Director-General) in CERN's archives. For documents in other archives the format (archive-file name) is used. Thus (AEF-80) means box 80 in the series Affaires Atomiques of the French Ministry of Foreign Affairs, Paris. A list of archival sources consulted follows these remarks.

- The notes refer to items in the bibliographies using the following format: Author or subject (year). Thus Adams (1965) or Schweber (1985). The full reference is given in the associated bibliography.

List of archival sources consulted

In compiling the list we have included the abbreviations used in the notes and references. In general, the abbreviation identifies a specific collection of documents in an archive, which is italicized in the text below.

AA	–	Auswärtiges Amt, Politisches Archiv, Bonn.
AEF	–	Ministère des Affaires Etrangères, Paris: *Affaires Atomiques* file.
AF	–	Archives de France, Paris: AF–F17bis: *Ministère de l'Education Nationale* file AF–307AP: *Archives privées, Raoul Dautry* file AF–CNRS: *'Versement effectué par le CNRS le 22/8/80, cote 80/284'*.
AIP	–	American Institute of Physics, New York City.
BNL	–	Brookhaven National Laboratory, Long Island (NY).
CNR	–	Consiglio Nazionale delle Ricerche, Ufficio di Presidenza, Rome.
DFG	–	Deutsche Forschungsgemeinschaft, Bonn-Bad Godesberg.
DM	–	Deutsches Museum, Sondersammlungen der Bibliothek, Munich: mainly *Max von Laue's* papers.
LBL	–	Lawrence Berkeley Laboratory, Berkeley, California.
MPG	–	Bibliothek und Archiv zur Geschichte der Max-Planck-Gesellschaft, Berlin: mainly *Walther Bothe's* papers.
MPI/K	–	Max-Planck-Institut für Kernphysik, Heidelberg: mainly *Wolfgang Gentner's* papers.
MPI/P	–	Max-Planck-Institut für Physik und Astrophysik, Munich: mainly *Werner Heisenberg's* papers.
PRO	–	Public Record Office, Kew, London: PRO-AB: *UK Atomic Energy Authority* file PRO-CAB: *Cabinet Office* file PRO-DSIR: *Department of Scientific and Industrial Research* file PRO-ED: *Department of Education and Science* file PRO-FO: *Foreign Office* file
SERC	–	Science and Engineering Research Council, Hayes, Middlesex: *NP* file.
CERN	–	CERN, Geneva.

List of abbreviations[*]

Organizations and Commitees

CEA	Commissariat à l'Energie Atomique (F)
CNES	Centre National d'Etudes Spatiales
CNR	Consiglio Nazionale delle Ricerche (I)
CNRN	Comitato Nazionale per le Ricerche Nucleari (I)
CNRS	Centre National de la Recherche Scientifique (F)
DFG	Deutsche Forschungsgemeinschaft (FRG)
DSIR	Department of Scientific and Industrial Research (UK)
ECFA	European Committee for Future Accelerators
ECSC	European Coal and Steel Community
EDC	European Defence Community
EEC	European Economic Community
ELDO	European Space Vehicle Launcher Development Organization
ESRO	European Space Research Organization
EURATOM	European Atomic Energy Community
FOM	Institute for Fundamental Research into Matter (NL)
IIPN	Institut Interuniversitaire de Physique Nucléaire (B)
IISN	Institut Interuniversitaire des Sciences Nucléaires (B)
INFN	Istituto Nazionale di Fisica Nucleare (I)
IUPAP	International Union of Pure and Applied Physics
MPG	Max-Planck-Gesellschaft (FRG)
NASA	National Aeronautics and Space Administration
NATO	North Atlantic Treaty Organization
OECD	Organization for European Co-operation and Development
OEEC	Organization for European Economic Cooperation
OPEC	Organization of Petroleum Exporting Countries
UNESCO	United Nations Educational, Scientific and Cultural Organization
UNO	United Nations Organization
ZWO	Organization for the Advancement of Pure Research (NL)

[*] The internal organization of CERN, and the abbreviations used, can be found in appendix 1.

Countries

A	Austria
B	Belgium
CH	Switzerland
DK	Denmark
E	Spain
F	France
FRG	Federal Republic of Germany
G	Greece
GDR	German Democratic Republic
I	Italy
L	Luxembourg
N	Norway
NL	The Netherlands
S	Sweden
UK	United Kingdom
USA	United States of America
USSR	Union of Soviet Socialist Republics
Y	Yugoslavia

Currencies

FF	French francs
SF	Swiss francs
MSF	*Million* Swiss francs (and other currencies)

PART I

An overview of the first decade of CERN

CHAPTER 1

Building the laboratory, 1954–1959[1]

John KRIGE

Contents

1.1 1954/55, the year when the magnitude of the task that lay ahead began to emerge	4
1.2 1956, the year when CERN gained international recognition, and took the first steps needed to retain it	7
1.3 1957, the year when experimental work first became possible, and when financial constraints were first imposed	9
1.4 1958, the year when the big equipment needed to exploit the PS was settled, the year when the SC yielded its first important results	12
1.5 1959, the year when the PS gave its first beam, the year when the future of CERN was no longer in doubt	15
Notes	17
Bibliography	809

1.1 1954/55, the year when the magnitude of the task that lay ahead began to emerge

On 29 September 1954 CERN, or rather the Conseil Européen de la Recherche Nucleaire, officially ceased to exist. On that day representatives from France and Germany deposited their instruments of ratification at UNESCO House in Paris, and the Convention setting up the European Organization for Nuclear Research came into force. The first session of the new CERN Council—the acronym was kept for convenience—was held on 7 and 8 October 1954 in the presence of Pierre Auger, Denis de Rougemont, and Isidore I. Rabi, all of whom had played an important role in the prehistory of 'the first scientific organization of its kind in the world'. On 10 June 1955 the foundation stone of the laboratory was laid at the site in Meyrin on the outskirts of Geneva. The following day an agreement defining the legal status of the Organization in the host state was signed by representatives from the Swiss Federal Council and from CERN.

The ratification of the Convention lifted the formal barrier to the implementation of the plans made and decisions taken during the preceding two years. The machines the laboratory would have—a 600 MeV synchro-cyclotron of standard design and a 25 GeV proton synchroton embodying the newly-discovered strong-focusing principle—had been settled and some of their basic parameters had been fixed. Now they had to be built. The internal organization of the laboratory and the committee structure that would watch over it had been drawn up. Now they had to function together as a system.

The supreme governing body of CERN is the Council, which is made up of two representatives from each of its member states, one a leading scientist, the other a senior administrator or diplomat. It is aided in its work by three committees—a Committee of Council which is smaller, and which prepares the ground for policy decisions that the Council has to take, a Finance Committee, comprising one science administrator from each member state, and which deals with all matters having important financial consequences, notably the budget, and a Scientific Policy Committee, where scientific eminence, not nationality, is the main criterion for membership. In fact four of the eight members of CERN's first SPC (Blackett, Bohr, Cockcroft, and Heisenberg) were Nobel prizewinners. A fifth, Alfvén, was awarded the Nobel prize for physics in 1970.

The first organigramme for the laboratory foresaw six divisions, one each to build the machines (PS and SC), one to prepare instruments for their exploitation (STS), a Theoretical Study Division, a division to care for the site and its buildings (SB), and an administrative division which was broken down into Finance, Personnel, and

Purchasing. Above this there was a Directorate, made up of the Director-General, who was the head of the laboratory, the Deputy Director-General, and a third Member of the Directorate. For the first three or four years of the laboratory's life the personnel in these divisions were geographically dispersed. The staff of the PS Division was situated in offices and barracks at the Institute of Physics of the University of Geneva, the division devoted to theory was located in Copenhagen at Niels Bohr's Institute, and the people in the remaining divisions were housed in buildings, barracks, and hangars around Geneva airport.

This internal structure more or less mirrored that put in place during the planning phase of the laboratory, and the bulk of the senior posts in it were filled by men who had already been active in the so-called 'Executive Group' set up in 1952. The PS Division was put in the hands of John Adams, who had been serving as Odd Dahl's deputy. Cornelis Bakker remained the head of SC and was the third member of the Directorate. Lew Kowarski and his deputy in the earlier Laboratory Group, Peter Preiswerk, became respectively the heads of STS and SB. Bohr asked Christian Møller to take over responsibility for Theoretical Studies in Denmark. Sam Dakin, who had earlier been loaned to CERN by the British Board of Trade, was left temporarily in charge of Administration. Finally, Edoardo Amaldi, the Secretary General during the planning phase, was nominated Deputy to the only new man at the top, Felix Bloch, CERN's first Director-General and a Nobel prizewinner for physics in 1952.[2]

As the European laboratory project was put into practice a number of unrealistic, unworkable ideas which had informed its planning soon began to surface. We shall concentrate on three in particular in this quick survey.

Firstly, much to the disappointment of the senior members of the SPC, it proved impossible to persuade leading physicists from abroad to spend time in Geneva before the experimental work with the SC got under way. Concerned about maintaining a 'scientific spirit' in the laboratory during the long years of construction, some of the SPC's Nobel laureates had insisted the year before on having a really eminent scientist to head the Organization. During 1955 the Committee continued to try to build up a small nucleus of theoreticians in Geneva who could help shape the experimental programme. They also proposed instituting a Guest Professorship. Their efforts were in vain. All of those approached—Wick, Dalitz, Panofsky—declined. More dramatically, their choice for Director-General, Felix Bloch, resigned and went back to Stanford as soon as he reasonably could.

Bloch had always been reluctant to take on the most senior post in the laboratory. He was not a high-energy physicist—his main interest was nuclear magnetic resonance, and he had insisted on bringing some of his equipment and two of his assistants with him from California. More importantly, he was not willing to handle the administrative duties that were required of a person in his position. Attracted by the prestige of taking charge of CERN, and assuming that Amaldi would bear the

Notes: p. 17

administrative burden, Bloch was lured to Geneva by Heisenberg and, above all, by Bohr. No sooner had he arrived than Amaldi informed him that he wanted to leave Switzerland. Six weeks in office was enough to convince Bloch that he had made a mistake. In mid-November 1954 he wrote to Council President Lockspeiser informing him that he wished to be released from his post by autumn 1955.

The Council accepted Bloch's resignation early in 1955. Amaldi left, and Bakker, the man whom the Executive Group had favoured for the post, was made 'Deputy' Director-General until the end of August, whereupon he officially became DG. Wolfgang Gentner took charge of the construction of the SC. Bakker remained head of the Organization throughout the period covered in this chapter. In December 1959 he was appointed to serve as Director-General for a further five years as from 1 September 1960. He never did so; he was killed in a flying accident in April the same year.[3]

1954/55 also witnessed a slight, and unanticipated extension of the CERN site to ensure that the proton synchrotron was built on suitably stable foundations. The new strong-focusing principle reduced the cost per GeV by drastically reducing the cross-section of the vacuum tank: the aperture in the PS was 7 cm × 14 cm, that in its 'predecessor', the Bevatron at Berkeley, was 30 cm × 120 cm. Concerned to eliminate every possible factor which might lead to a loss of the primary circulating proton beam, Adams' group made an extensive survey of ground structure and ground movements in the area set aside for their new accelerator. Expert consultants were called in, boreholes were dug, and a complete map of the underlying sandstone layers was made. It was concluded that there was no suitable location for the PS within the existing boundaries of CERN's land, and a further strip on the south side of the site was bought. Even then, Adams told the Council, it was 'very lucky that it had been possible to find a position for the machine on the site' at all. And to be quite sure, the PS group drove eight cylindrical pillars deep into the rock around the 100m radius circle of the machine, and inserted flexible couplings between them and the concrete ring supporting the magnets guiding the proton beam.[4]

A third area in which adjustments quickly had to be made was that of contract policy. The basic principles guiding the purchase of goods, equipment and supplies had been laid down in 1953. The Council then agreed that, while CERN should make every effort to place its orders in the member states, it need not feel constrained to distribute its expenditure according to a principle of just return: there was to be no *a priori* correlation between the percentage a country contributed to the CERN budget and the percentage of its purchases CERN placed in that member state. Quality and price, not 'political' considerations should determine where CERN placed its orders, and to secure the best prices the laboratory should call for tenders from suitable firms in as many members states as possible.

This competitive policy was challenged by the CERN management from the start.

Calling for tenders, the staff argued, was completely uneconomic for small purchases, was far less effective than informal bargaining for some materials, could amount to little more than a 'costly ritual' for specialized equipment (since the researcher needed to discuss the detailed design and specifications with a particular supplier), and so on. The tendering procedure, they concluded, should be implemented 'where practicable', rather than being the mechanism 'normally' invoked to secure the best value for money.

These ideas were debated by a working group set up by the Finance Committee at the end of 1954, and a 'compromise' was reached early in the new year. The Finance Committee emphasized that competitive tendering was to be used wherever possible, and that at least three competitive tenders should 'normally' be obtained for the purchase of material by CERN. At the same time the Committee recognized that this procedure could be modified in practice to satisfy technical and administrative requirements. These policies had two striking consequences during the first decade of the laboratory's life. Firstly, CERN made about one third of its purchases (by value) in the host state, Switzerland, which contributed some 3.5% to its budget. Secondly, the percentage of contracts (by value) placed 'competitively' (i.e. by full and formal tendering) decreased steadily as the laboratory moved from the phase of constructing its accelerators to exploiting them, dropping to less than 15% in 1964.[5]

1.2 1956, the year when CERN gained international recognition, and took the first steps needed to retain it

Events at CERN in 1956 were dominated by the symposium on 'High Energy Accelerators and Pion Physics'. The meeting was held from 11 to 23 June at the Institute of Physics of the University of Geneva, and was attended by over two hundred physicists and engineers representing nearly all the high-energy nuclear physics centres in the world. Adams was responsible for organizing the first week, which dealt mainly with new ideas on accelerators. During the second week, organized by Gentner, attention was focussed on some advanced experimental techniques to be used with high-energy accelerators (bubble chambers, fast counters, etc.) and on pion physics. As a result of discussions held during the symposium it was agreed that CERN would regularly play host to the now famous Rochester conferences on high-energy physics. A year later, at the IUPAP General Assembly, it was decided that henceforth the venue of the meetings would rotate automatically between Europe, the Soviet Union, and the United States. The next in the series was held at CERN in 1958, the ninth took place in Kiev, and the tenth conference in 1960 was again in Rochester.[6]

In 1956 there was another important sign that CERN, and Europe, were becoming recognized as players on the world stage of high-energy physics. At the December Council session the Director-General announced that the Ford Foundation had

Notes: p. 17

agreed to grant CERN $400,000 to be spent over a period of five years. The money was to be used for fellowships enabling scientists from non-member states to work at the Geneva laboratory. The duration of the grant was subsequently reduced to four years, and during 1959 another award for $500,000 for a further four years was negotiated with the Ford Foundation.[7]

After the June symposium the CERN staff concluded that a number of initiatives had to be taken immediately if they were to keep up with developments in their field. In December the Council agreed to set up a small research group to study new acceleration principles. It also pondered a revised Capital Investment Programme for 1957-60 which showed major increases in the expected costs of research at the Synchro-cyclotron.

The leader of the PS Division came away from the CERN symposium feeling that the organization had been so busy constructing its machines that it had lost touch with new ideas on accelerators being put forward elsewhere. CERN staff not only had very little of their own to contribute to the meeting, he told the SPC in October, but also found that they had a hard time following the work that other groups had done. To ensure that the laboratory was on a par with centres like Brookhaven and Berkeley, Adams suggested that an accelerator research group, the bulk of them visitors from Europe and the United States, be set up inside his division. The SPC agreed, the Council did not object, and in 1957 the AR Group under A. Schoch began its work. The Group, which remained rather small and marginal in the organization for the first few years, reached a plateau of about 15 people in mid-1958. During this period the majority of the full-time personnel worked on plasma accelerators, the remainder studying intersecting beam machines. The situation changed dramatically at the end of 1959 when the Group was joined by a dozen of the best engineers released from Adams' division, and its energies were redirected to the study of storage rings.[8]

Senior staff in the SC division were also struck by what they learnt at the June symposium. Various steps had already been taken in anticipation of commissioning the small machine in 1957. A group in STS was building a double cloud chamber. A young physicist had been sent to the USA to study bubble chamber technique and, in November 1955, an ad-hoc advisory committee had met in Geneva to decide on the most suitable experiments for the machine. Their views on the SC research programme for 1956 and 1957 were laid before the SPC just before the June symposium. In the light of what they learnt there, and a subsequent visit to the Soviet Union, Bakker and Gentner concluded that most of the research they were planning to do was already being taken care of elsewhere. CERN, Gentner told the Finance Committee, had to think in terms of doing more complicated experiments with more sophisticated equipment than originally foreseen. With that the SC Division immediately increased its estimated annual expenditure for 1957-1960 by between 75 and 100%, almost doubling its staff estimates and planning to have 25 fellows regularly associated with its experimental activities.⁹

The most significant event in 1956 as far as the construction of the PS was concerned was the placing of the order for the 100 magnet units needed to guide the primary circulating proton beam. Each unit consisted of ten C-shaped blocks of laminated steel, and for two years engineers in the PS Division had laboured to find a steel which would satisfy the stringent conditions on magnetic field performance which they believed the strong-focusing principle demanded. Originally the group had planned to minimize fluctuations in the magnetic properties of the steel by mixing the 1000 or so magnet blocks before tying them together into groups of ten. They soon found that this was not enough to compensate for inhomogeneities in normally produced commercial steels. Instead they decided that it was necessary to mix, not the 1000 magnet blocks, but the 250,000 laminations making up the blocks. A major diagnostic programme was launched on three new models in 1955 to study different steels on the market, and to explore the effects of the methods used by different suppliers for ageing steel and for producing blocks. A scheme was worked out for collecting and storing steel plates and for feeding them to the manufacturing process. On 9 March 1956 invitations to tender were sent to eight European blockmakers, four of whom submitted offers. After further discussion, the four were invited to submit a second tender by mid-July. These were gone over thoroughly with the firms, and in October John Adams invited the Finance Committee to award the contract for manufacturing the magnet blocks to Ansaldo San Giorgio in Genoa, Italy. A few weeks later the contract for manufacturing the steel was awarded to Societa Italiana Acciaierie Cornigliano, a steelmaker conveniently situated in the same city. The total contract was worth 8 MSF.¹⁰

1.3 1957, the year when experimental work first became possible, and when financial constraints were first imposed

On 1 August 1957 the first proton beam was accelerated in the CERN synchro-cyclotron and immediately reached its full energy of 600 MeV. Though the machine was of standard design and of relatively low energy, and though everyone realized that the future of CERN depended on the PS, the achievement was not without significance. Scientifically speaking, the SC was now the biggest accelerator on the continent, and was not only well-suited to doing meson physics, but was also invaluable as a means of training physicists, often of different nationalities, to work together, often for the first time, around a shared piece of big equipment. Politically speaking, it reassured the member states that the fledgling laboratory was able to achieve the goals that it had set, and in this case at least do so on time and on budget. Personally speaking, it gave an enormous boost to the confidence of the CERN staff,

Notes: pp. 17 ff.

and especially to those who had been responsible for constructing it, many of them relatively inexperienced in building heavy equipment of this type.[11]

The variable condenser in the high frequency system was the only component which caused major difficulties during the construction of the SC. This system was used to produce a high frequency voltage in the regions where the protons were to be accelerated. Models of the system and of its components, including a variable condenser of the tuning fork type, were built and tested, and an order was placed in September 1955. A year later, and in the light of serious doubts about the performance of the tuning fork, it was decided to construct a rather more conventional rotating condenser as a possible alternative. In the event, by the end of 1957 the tuning fork assembly was working adequately: unexpected difficulties in running the device in the SC's magnetic field had been overcome days before the official inauguration of the machine, and the fork was being pulsed to avoid overheating. The Synchro-cyclotron was routinely available for physics by April 1958.[12]

Along with the commissioning of the SC a number of organizational measures were taken in 1957 in anticipation of experimental work getting underway at CERN. The Director-General proposed a new scheme, which he never managed to implement, whereby the laboratory would be internally split along functional lines, one main division being dedicated to experimental research, the other to applied physics and engineering. In September Gilberto Bernardini joined the SC division as Director of Research. On 1 October Bruno Ferretti took up his post as Leader of the Theoretical Study Division in Geneva, with the specific task of co-operating in the preparation of experimental research. At the same time the CERN TH division in Copenhagen was closed down. Originally installed at the Institute for Theoretical Physics to appease Bohr, the group had devoted most of its energies to quantum field theory and the theory of nuclear constitution, as well as arranging for the training of young European physicists on the synchro-cyclotrons at Liverpool (400 MeV) and at Uppsala (200 MeV). Finally, in August a special Advisory Committee comprising ten CERN staff and seven visitors' representatives met for the first time to discuss, and to weigh, experimental proposals for the SC.

CERN was a laboratory constructed by and for the Organization's member states, and they expected their national scientific communities to have privileged access to its facilities. To satisfy this requirement, the laboratory management, in consultation with outside users, had to lay down ground rules and devise mechanisms for regulating access to its machines, and it had to do so long before national laboratories in the United States were called upon to deal formally with 'the user problem'. The system put into practice from 1957 onwards at the SC had one marked characteristic: a sharp divide was maintained between CERN groups and national groups (at least until 1960).[13]

There was a divide, firstly, in the organization of research: we find very few mixed teams of CERN staff and visitors formed. The outside users who came to CERN to work at the SC at the end of the 1950s did so almost exclusively as members of so-called truck teams. This meant that they tended to come as a group, rather than as individuals, that they were from one, or perhaps two national institutions, and that they brought their experimental equipment, a small bubble chamber perhaps, with them. CERN provided infrastructural support (secondary beam magnets, power supplies, liquid nitrogen, etc.), but otherwise the in-house staff did not participate in the experiment. Their work done, the truck teams left taking their equipment, and their results, with them.

There was also a divide at the level of the decision-making structures. As conceived, the Advisory Committee's task was to choose the scientifically best experimental proposals laid before it by CERN staff and outside users, referring to the DG for a final decision when conflicts arose. In practice the CERN direction decided on its own what experiments would be done by CERN teams (whether or not they included the odd visitor), while the Advisory Committee restricted its attention to proposals for experiments from visiting teams. If this system worked smoothly, as it seems to have done, it was basically because visitors were allocated 30% of the machine time and apparently did not need more.

1957 was also the year in which a new financial policy was accepted by the CERN Council. There had been grumblings for several years about the rising estimates of constructing CERN, which were originally set at 120 MSF. In 1955 the Norwegian delegation was unhappy to see the revised estimate for expenditure to 1960 climb to 197 MSF. The Swedish delegation voiced its concern a year later, when the figure had increased to 219 MSF, while the French toyed with the idea of changing the voting rules for the budget, which could be passed by simple majority, to give more weight to the major contributors. But it was the British delegation that took the bull by the horns in 1957 and suggested that a new policy was needed to deal adequately with the relentless upward pressure on expenditure from the rising costs of research.

What was the prevailing policy in the Finance Commitee *vis-à-vis* CERN budgets? Essentially, one of *laisser-aller*. The management simply told the delegates what it thought was needed for the following year, and the delegates took it upon themselves to raise the money at home. The British delegation felt that this could not go on, that some form of external control had to be imposed on CERN's expenditure. CERN, it stressed, just like any national research facility, had to realize that funds for the laboratory were limited and that estimates, once accepted, had to be adhered to. With these considerations in mind, in summer 1957 the UK proposed, and the CERN management 'accepted', that a ceiling be imposed on CERN's expenditure, a global envelope below which research projects would have to be accommodated by a scheme of priorities.

Many FC delegates were unhapy about this idea when it was first put forward, believing that CERN could not hope to remain competitive if it did not invest heavily

Notes: p. 18

in new, expensive experimental equipment, and that this was a singularly inopportune moment to be imposing 'external' constraints on expenditure. Notwithstanding their doubts, they had little choice but to go along with the British suggestion later in the year. At the end of 1957 the CERN management put forward a budget for 1958 which, at 59 MSF, was about 50% higher than the figure foreseen the year before (40 MSF). However, most domestic financial authorities had already set aside their funds for CERN for 1958 on the basis of the 40 MSF estimate. To meet CERN's 'needs' and at the same time to overcome the resulting procedural difficulties at home, Jean Willems (B) and Jan Bannier (NL) proposed that the laboratory be given 56 MSF for 1958, that the budgets for 1958 and 1959 together be voted at 100 MSF, and that the member states make up the shortfall in their 1958 allocations when setting aside money for 1959. The British, backed by the French and the Scandinavians, agreed—on condition that the 100 MSF be regarded as a ceiling not to be broken.[14]

1.4 1958, the year when the big equipment needed to exploit the PS was settled, the year when the SC yielded its first important results

The ceiling policy adopted for the first time at the end of 1957 was imposed again by the Council in 1958. Towards the end of that year the Director-General indicated that CERN would like some 55 MSF for 1959—11 MSF above the limit imposed the year before—, and a further 69 MSF for 1960, i.e. a total of 124 MSF for the two years. After long and difficult negotiations, in December 1958 the Council accepted, against the vote of Sweden, to break the 1959 ceiling it had laid down only a year before. This 'concession', though, was conditional on another ceiling being imposed, and in May 1959 the terms of the 'deal' were settled. CERN would be given more or less what it wanted for 1959 and 1960 (i.e. 120 MSF), but expenditure for 1960, 1961 and 1962 was to be stabilized at 65 MSF per annum. This attempt to impose a level ceiling on CERN led to an explosive Council debate at the end of 1961, and to a fundamental revision of the UK's position on financing the laboratory.[15]

The most important item pushing up CERN's budget estimates in 1958 was the cost of equipment for the PS experimental programme. The first serious steps to plan this programme were taken in mid-1957, and the ensuing debate, which lasted almost a year, was dominated by concerns about the kind and size of big bubble chamber that CERN should build. Although a big hydrogen chamber was an attractive option, the staff initially thought that it might be better to build a large propane chamber instead: it could do useful physics, it would tie up fewer human and material resources, it was not dangerous, and there was far less to lose if new developments in detectors should render bubble chambers obsolete in a few years time. The uncertainty on this score was dispelled by the end of 1957: persuaded by Luis

Alvarez's argument that the hydrogen bubble chamber was likely to be a kind of universal detector at high energies, and aware that at least Berkeley and a British group planned to have big (i.e. 1.5-2 m) hydrogen bubble chambers operating in the early 1960s, the CERN staff decided that they too should build one. But how big should it be? One school of thought, dominated by experimental physicists, felt that it was crucial that the detector be ready when the PS first worked. To satisfy this time contraint they concluded that CERN should build a 1m chamber. Against them an engineer like Adams argued that it was better to build the biggest chamber that was technically possible even if it was ready 'late'. The latter view prevailed. Experimentalists recognized that, scientifically speaking, the bigger the chamber was the better, they appreciated that only something comparable to what was being built in the United States could ensure their competitiveness in the medium term, they were comforted by the UK's offer to loan its 1.5 m chamber to CERN for a couple of years, beginning around 1961, and they were pleased to have Adams suggest that CERN also build a propane chamber to be ready 'in time' for the start of research at the PS. In May 1958 the SPC was asked to endorse the construction of a 2 m hydrogen and 1 m propane bubble chamber at CERN, and in June the Finance Committee accepted to raise the money for them and their ancillary equipment.[16]

Various other needs of the PS experimental programme were attended to in 1958. A number of secondary beams sufficient to fill the South experimental hall were laid out by CERN's propane bubble chamber and counter groups, and in spring the next year a standard set of 18 bending magnets and 26 quadrupole lenses were agreed on and ordered. European emulsion physicists met with the CERN staff, first in November 1958, then in January 1959, to discuss what they would do with the new machine. Plans were made to build a number of electrostatic separators suitably long to 'purify' high-energy secondary beams. Kowarski's STS group pressed ahead with building instruments for automatically digitizing the coordinates of tracks in bubble chamber pictures, and had an exploratory meeting with potential users to discuss the possibility of standardizing this kind of equipment in Europe. CERN's first computer, a Ferranti Mercury with a cycle time of 60 microseconds and a 'fast memory' capacity of '1000 words', passed its acceptance tests. And on 25 September 1958 the Executive Committee for the PS Experimental Programme, the body whose task it was to coordinate all these diverse activities, met for the first time.[17]

For many of CERN's experimental physicists the need to plan the PS experimental programme was something of a nuisance, a distraction from the research opportunities provided by the SC. Indeed, as far as doing physics was concerned, this machine exceeded all expectations, its first outstanding result being announced at the Atoms for Peace conference in Geneva in September 1958. It was the discovery of a rare decay mode of the pion.

For some years physicists had believed that, besides its ordinary decay into a muon

Notes: p. 18

and a neutrino, a pion should also decay into an electron and a neutrino. Although a number of experiments had already been done elsewhere to search for this 'rare' decay, no evidence for it had been found. Its 'absence' became all that more problematic when it could not be reconciled with new developments in the theory of weak interactions, and a systematic search for it was started at CERN early in 1958. The experiment was difficult because the electrons produced directly from pion decay had to be detected against a high background of electrons from the decay of muons resulting from normal pion decay. By using a range telescope to discrimate between electrons coming from the two processes, and by using a photographic system of recording, a team around Merrison and Fidecaro successfully established the existence of the rare decay mode, and set the lower limit for the branching ratio of the two modes at 4×10^{-5}, a figure that they later refined. This was the first step in what was to become a major research programme at CERN in weak interaction physics.[18]

A number of other important scientific results were to be achieved at the SC during the next few years. For example, the most precise measurement yet of the anomalous magnetic moment of the muon ($g-2$) was first made in 1960. A valuable research programme was undertaken with a 'pure' muon beam formed from π-mesons decaying in a long strong-focusing channel. The beta decay mode of the pion was observed unambiguously for the first time in 1962, providing confirmation for the theory of a conserved vector current. All the same, and its significant contributions to π- and μ-physics notwithstanding, the SC's role as a 'high-energy' research tool obviously declined with the successful commissioning of the PS. In 1964 a proposal was put forward to couple an on-line isotope separator to its beam (the ISOLDE project), signalling a shift in emphasis to nuclear spectroscopy and nuclear structure studies.[19]

Scientific results of a very different kind to those obtained at the SC, and coming from a very different source, also had an important impact at CERN in 1958. Towards the end of 1957 scientists at Harwell believed that they may have produced thermonuclear neutrons in their experimental torus ZETA, suggesting that a fusion reactor was, in fact, a practical possibility. In the wake of this discovery a number of European specialists on fusion met at CERN in March 1958 to discuss the new results. At about the same time CERN Council President de Rose, Director-General Bakker, and the Chief Scientific Adviser to Euratom, Guéron, looked into the possibility of CERN carrying out fundamental research into plasma physics on behalf of Euratom. At the end of May Adams, Bakker, and Guéron met, and proposed that a 'Euratom-CERN Joint Study Group for Fusion Research' be established quickly and, in particular, early enough to reap the benefits of the forthcoming Atoms for Peace conference in Geneva. This proposal was laid before the CERN Council on 19 June 1958 where it provoked such resistance from the delegates from the non-member states of Euratom that another Council meeting was called a week later to iron out the differences. The opposition feared that CERN's autonomy would be compromised by linking it formally to another organization, particularly one whose work was of direct industrial and military significance. And while not unwilling to sanction the formation of a study group to evaluate plasma research programmes aimed at fusion, they were emphatic that its membership should be restricted to CERN staff, and that other organizations like Euratom and the OEEC should do no more than send observers to the group's meetings. Its goals and composition being thus defined, the 'CERN Study Group on Fusion' was set up in summer 1958, and met three times before submitting its report to the Council in March 1959. The context in which it deliberated had been changed dramatically by the discovery, in April 1958, that the Harwell results were false, and by the revelations in September 1958, following the lifting of security restrictions, that no group in the world was anywhere near producing a working fusion reactor. At the Council meeting in May 1959 the Study Group asked for, and was accorded, CERN's patronage to continue its work, which was essentially technical, and dedicated to exchanging information and results on various European national programmes. It seems to have had its last meeting in Aachen in Germany in May 1964.[20]

1.5 1959, the year when the PS gave its first beam, the year when the future of CERN was no longer in doubt

The build-up to achieving a circulating primary beam at the design energy of the PS began in May 1959. The linear injector produced its first beam of 10 MeV protons on the 22nd; a week later a second tank was brought into operation, and the beam energy increased to 20 MeV. By the end of August the linac was producing 50 MeV protons, and injection studies could begin. Particles went around once inside the ring in mid-September, to coincide with the international accelerator conference being held in Geneva at the time. A month later RF acceleration was applied to the beam for the first time, and on 13 November the machine was regularly accelerating protons up to a few GeV. Finally, on the evening of 24 November protons were successfully brought up to full energy: 24 GeV. A few weeks later the poleface windings were energized and the proton beam reached an energy of just over 28 GeV. The most powerful accelerator in the world was now at CERN, and in Europe.[21]

No one doubted that the machine would work: orbit calculations by the PS group and model studies at both CERN and BNL had long since confirmed the viability of strong focusing. But this in no way diluted the mixture of joy and relief of those who, gazing intently at their oscilloscopes, saw the beam survive the transition energy which had seemed such a formidable barrier during the design stage, or noted that there was no loss of particles when control of the accelerating frequency switched from a programmed RF system to the beam itself. Each component of the PS had

Notes: p. 18

been tested individually. On the evening of 24 November 1959 they worked *together* for the first time, and they worked *at once*. The future of CERN was assured.

With the PS about to come on stream there were protracted negotiations and several, sometimes bruising, debates in 1959 on how to organize the experimental work around it. Underpinning them lay one key consideration: that the large hydrogen bubble chambers being built by the British (1.5 m), the French (80 cm), and CERN (2 m) were going to be installed at the laboratory. These detectors were costly, dangerous, immobile, and immobilized space and ancillary equipment for others in the experimental halls. This meant that visitors could no longer come as members of truck teams, as they were doing at the SC. An alternative arrangement had to be found, an arrangment which had to reconcile the rights of those who had invested time and money in building 'their' equipment, with the rights of those who had joined and paid for the European laboratory so as to have access to facilities they regarded as communal. The search for a compromise was complicated by mutual mistrust between the national groups, by the fact that all parties, including CERN itself, had launched independent construction programmes rather on the truck team model and 'automatically' assumed that they would reap the first benefits from 'their' equipment, by the international character of CERN itself, with its inbuilt tendency to manage conflict formally by drafting texts and laying down abstract principles, and by the fact that the debate on the use of the visitors' chambers became intertwined with a more general discussion on how teamwork at the PS should be organized. In the event, though nothing was unambiguously settled, it was 'agreed' that in the short term some privileges would be granted to the builders, though by and large the use of the three chambers was to be as communal as possible, that in the long term some effort should be made to coordinate plans for the provision of big detectors at CERN, and that the basic experimental unit for exploiting the equipment would be a 'mixed team' of CERN staff and national groups.[22]

There were several signs of confidence in the future of CERN in 1959. For one thing, the complex of offices and meeting halls known as the Main Building was brought into use. In May the Finance Committee and the Council met for the first time in the new auditorium. In June the Director-General and the staff of the Administration division moved into their offices. And in December the Council used the Council Chamber for the first time. The Main Building had been completed as planned thanks to a special financial contribution by the Swiss Confederation, and a number of national and local dignitaries were present at its inauguration at the start of the December Council session. So too were representatives of CERN's thirteenth member state, Austria, which was officially admitted to the organization on 1 July.

The whole management of the forthcoming site and buildings programme was also put on a new footing with the nomination of Charles Mallet, a civil engineer, as Divisional Director of SB. This sector had previously been something of an organizational anomaly in the CERN structure. Its first head, Peter Preiswerk, was essentially a nuclear physicist, and had been given SB not in recognition of his expertise in the area but because he had taken some responsibility for this aspect of the laboratory in the planning phase. Obviously he needed help, and help came in the form of Chief Architect Steiger and his Zurich firm, who were effectively awarded a contract to build CERN. This was not an ideal arrangement. It was not that anyone was unhappy with the quality or aesthetics of the buildings— on the contrary. But Steiger felt unduly constrained at having to work within the framework of CERN's contract policy and purchasing procedures—after all, he was not a member of the organization—while, for the very same reason, the Finance Committee found it difficult to monitor what he was doing. This arrangement was effectively abandoned with Mallet's arrival: from henceforth the SB Division would be in charge of drawing up plans and supervising the technical installations of buildings on the CERN site.[23]

Finally, in 1959, a worked out policy for awarding indefinite contracts to CERN staff was adopted for the first time. For two or three years John Adams, in particular, had been stressing that it was essential for CERN to provide challenging work and a career structure for the engineers and applied physicists in the PS group who would otherwise be forced to leave the organization even before the machine was commissioned. This is one of the reasons why he strongly supported the construction of a big hydrogen bubble chamber—it was an engineering project which could absorb his staff as they were released from building the accelerator. It was also the main reason which both he and Bakker gave for setting up a study group on fusion at CERN.[24] In 1957 the Council had already agreed on the principle of indefinite appointments. Now they accepted a policy which entailed that 15% of the scientific staff would be eligible for such contracts in January 1961 and a further 15-20% in January 1964. At the same time, in 1959, the Personnel Office launched the largest recruitment campaign in CERN's short history— 343 appointments were made, bringing the total staff at the end of the year, including fellows, to almost 900 people.[25]

CERN was here to stay.

Notes

1. Only summary references are given. The chapter is based on CERN Annual Reports, Council, FC and SPC minutes, and the results presented in chapters 3 to 13 in this volume.
2. See chapter 7 and Appendices 1-5.
3. See History of CERN (1987), chapter 8.4. The full reference is in the bibliography to parts III and IV of this book.
4. See Council, minutes, 7-8/10/54, 15, 24/2/55, 32, Annual Report, 1955, 18.
5. See chapter 11, sections 11.1 and 11.4.
6. See chapter 3, section 3.2.3.
7. See Council, minutes, 14/12/56, 16, Annual Report, 1959, 17.

Notes: p. 18

8. See chapter 12, section 12.2.
9. See chapter 6, section 6.3, chapter 9, section 9.2, chapter 10, section 10.1.1.
10. See chapter 11, section 11.2.1.
11. See chapter 4, section 4.4.
12. See chapter 4, sections 4.2.2, and 4.2.3.
13. For this paragraph and the next two see chapter 8, sections 8.1.1 and 8.1.2.
14. See chapter 10, sections 10.1.2 and 10.1.3.
15. See chapter 10, sections 10.2.2 and 10.2.3.
16. See chapter 9, section 9.3.
17. See chapter 9, sections 9.4 and 9.6.
18. See chapter 4, section 4.3.2.1 and Annual Report, 1958, 46-7.
19. See chapter 4, section 4.3.
20. See chapter 7, Annex, section 7A1.
21. See Council, minutes, 2/12/59, 43-4, and chapter 5, section 5.1.
22. See chapter 8, sections 8.2, 8.3 and 8.8.
23. See chapter 11, section 11.3.
24. See chapter 7, Annex, section 7A1, and chapter 9, sections 9.3.2 and 9.3.3.
25. See CERN/325, 29/10/59, and Annual Report, 1959, 90-1.

CHAPTER 2

Running the laboratory, 1960–1965[1]

Dominique PESTRE

Contents

2.1 1960, the year when all seemed possible, the year of reorganization and of large-scale projects	20
2.2 1961, the year of doubts and anxieties, the year of tensions	23
2.3 1962, the year when trends were reversed, the year of selfcriticism and of new hopes	27
2.4 1963, the first year in which tasks were satisfactorily integrated	29
2.5 1964, the year of the debate on the medium- and long-term programmes	32
2.6 1965, the year of consolidation	34
Notes	37
Bibliography	809

2.1 1960. The year when all seemed possible, the year of reorganization and of large-scale projects

Within the timespan of the first decade of CERN's life, 1960 emerges as the end of an epoch, the year that everybody had been waiting for. The giant synchrotron envisaged nine years before had accelerated its first beam of protons to 24 GeV on 24 November 1959 and, for the first time since the war, European physicists were able seriously to pit themselves against their American colleagues in the prestigious field of high-energy physics. A ring with a circumference of more than 600m, and comprising 16 accelerating cavities and 100 magnets each weighing 38 tons and of maximum power input of 27 Megawatts, the PS became the focal point of life at CERN. There was, then, an understandable euphoria early in 1960, the kind of feeling one has as new horizons are opened up in the aftermath of a long, difficult, but successful preparatory phase.[2]

Contrary to what one might perhaps think, a new machine on this scale was not 'available' on the day it started to work. It had, as it were, to be run in, to be optimized, and the people who were to keep it working routinely had to be trained. It was necessary, for example, to instal different target systems in the vacuum chamber— striking these, the primary accelerated protons created secondary particles which were transferred out of the main ring for the use of physicists—, and to increase the intensity of the beam circulating in the ring as much as possible. Thus, in February 1960, only one afternoon per week was set aside for physicists, the rest of the time being taken by the engineers. By summer 1960 some 25% of the machine time (about 19 hours) was available for doing physics, while by November, one year after the PS was 'commissioned', it was agreed to reserve about 75% of the time for experiments, the remaining quarter being at the disposal of those whose task it was to keep the machine running and to improve its performance.[3]

Just as 1960 was the year in which the PS was run in, so too was it the year in which the physicists cut their teeth. With little experience of working at energies as high as those reached by the big machine, they had to become used to working on what was for them an unprecedented scale (to give an order of magnitude, note simply that the shielding for the first neutrino experiment was calculated in September 1960 to require 4000 tons of cement blocks and 650 tons of steel). In 1960 only one experimental hall was ready, the South hall, of some 3000m^2. By the end of the year, and in anticipation of opening up the North hall, it was cluttered with men and with equipment. Two bubble chambers were in action, a third in assembly, several liquid hydrogen targets were installed in special booths, and ethylene

Cerenkov counters were in operation — all bulky items carrying risks of explosion (see the photographs in the centre of the book). In the midst of them, and from the autumn onwards, the bulk of the beam focusing and deflecting magnets ordered 18 months before were to be found, along with their batteries of generators. These magnets were needed to 'build' the secondary beams, i.e. magnetically to capture the particles emitted at the targets and to guide them to the detectors. Only one kind of beam element was lacking at this time, the particle separators, a component which was essential for bubble chamber physics. Separators were used to 'purify' secondary beams and to focus only certain particles having particular momenta. Decided on early in 1959, CERN's electrostatic separators were still under construction at the time, and were only expected to be available during the first half of 1961. In the interim a small electrostatic separator (two elements each 3.5 m long) was borrowed from Cresti's group in Padua. Arriving at the end of the year, it was to be used in conjunction with the 80cm hydrogen chamber on loan from Saclay, a chamber which was also to start to work in Geneva at the beginning of 1961.[4]

Not surprisingly hopes were running high for the year to come. Early in 1961 the 80 cm chamber would be working, and everyone expected that the huge 1.5 m British chamber would be on the site eight or nine months later. Include the big separators being built at CERN and one looked forward to working conditions better than those at Berkeley — where beam energies were lower — or even at Brookhaven where, at the end of 1960, the 30 GeV accelerator had only just been commissioned, and the big 2 m chamber was only expected to be ready in a couple of years. If one remembers also that seven Ieps (instruments for the evaluation of bubble chamber photographs) were to be available at CERN in 1961, and that the Organization took delivery of an IBM 709 computer in November 1960, one can understand why people at CERN felt relatively optimistic about the future.[5]

Groups using electronic means of detection shared these sentiments. Throughout 1960 they familiarized themselves with the machine (and their equipment) by studying the composition of the secondary beams produced by different targets and at various angles, and began measuring cross-sections. Growing in confidence, they envisaged for 1961 and 1962 a number of specific experiments centred on particular questions, and calling for more complex apparatus. It is nevertheless the preparations for the 'neutrino' experiment which strike the eye. Interest in this experiment was aroused early in 1960 when Gilberto Bernardini proposed to use the PS to confirm whether two kinds of neutrinos existed (ν_e and ν_μ). This experiment was worth trying, not only because the question was essential for theoreticians, but above all because of the difficulty, the sophistication, and the scale of the venture — which was probably worth a Nobel prize. In September 1960, after similar proposals had been made at Brookhaven, the decision was taken at CERN to dedicate the entire organization to the project. It was given priority vis-à-vis other experiments in November, and became the direct responsibility of the directorate (notably that of Bernardini and of Weisskopf). Scheduled for the first quarter of

Notes: p. 37

1961, the experiment called on the services of 40 physicists, two heavy liquid bubble chambers, the cloud chamber, and a large complex of electronic detectors. The final lay-out was settled on 12 December and preparations got under way.[6]

We find this same confidence about the years that lay ahead in the Council. Not only because its members were proud of CERN's technological (the PS) and scientific (the work done at the SC) successes, but also because the organization, as institution, was growing in renown. It had become *the* European laboratory capable of competing with its American counterparts, and it was surrounded by an aura of glory—which was not without importance at a time when many European governments were becoming increasingly willing to invest in 'science', seen as the new key to economic progress. As proof of this let us bear in mind that the first intergovernmental conference called to set up a European space research organization was held at CERN in December 1960, and that in that same month the Council discussed the possibility of transforming itself into 'a kind of European council for scientific research'. The idea here was that the body would become the 'head' of several laboratories constructed on the CERN model and each devoted to its own particular research field: high energy, controlled fusion, space research, etc.[7]

In 1960 the Council also appointed a new Director-General. After the death of Cornelis J. Bakker in April in an accident, the man who had been DG since 1955, the Council approved the nomination of John Adams, the undisputed leader of the PS Division, and the most powerful man in Geneva at the time. This arrangement was at best temporary as Adams was already committed to take on the directorship of Britain's new Culham laboratory on controlled fusion. At its session in December 1960 the Council thus appointed a new Director-General, who officially took up office on 1 August 1961. He was Victor Weisskopf, a renowned theoretician of Austrian origin, and a professor at MIT at the time.[8]

To experiment at 25 GeV, particularly with bubble chambers, entailed a break with previous experimental practices. To keep the PS running, to improve its performance, to instal the secondary beams and their shielding, to deal with bubble chambers, with their power supplies, with liquid hydrogen, etc, all this called for the participation of hundreds of people having different skills. In consequence a *reorganization* was called for.

Up until 1957 CERN comprised six divisions, two to build the machines, one to prepare equipment for experiments, and the Theory, Site and Buildings, and Administration Divisions. With the coming into operation of the SC in 1957 the question arose as to whether one should set up a division for experimental research as such. Despite his efforts Director-General Bakker could not get this arrangement accepted inside the organization and CERN continued with its six divisions until 1959. The imminent functioning of the PS reactivated the problem, and a consensus was reached on an organigramme of twelve divisions in 1960. There were two major complexes set up alongside the Administration, Finance, and Site and Buildings Divisions, one essentially staffed by engineers and builders, the other by physicists. In the former there were four divisions: Accelerator Research (reponsible for new projects and for fundamental modifications to the PS), PS Machine and SC Machine (operation, maintenance, and development of the two accelerators), and Engineering (responsible for the execution of large mechanical and electrical projects). In the second group we find the Theory Division, the Data-Handling or 'Données et Documents' Division, the Nuclear Physics Division (responsible for experiments with electronic detectors) and two divisions dealing with bubble chambers. Their task was both to build the chambers and to get them to work. One, the Track Chamber Division, was responsible for hydrogen chambers (including CERN's 2 m chamber), while the other, the Nuclear Physics Apparatus Division, dealt with propane chambers and secondary beam equipment.[9]

To use properly one of the biggest accelerators in the world also meant *choosing* the experiments to be done, installing some beams and not others, reserving machine-time, equipment, and space for some groups and denying it to others. A decision-making system for the experimental programme thus also had to be set up—and this too was done, by stages, in 1960.

In January it was Adams, PS Division leader, and Cocconi, representing the CERN physicists, who organized access to the machine. In March European bubble chamber specialists put forward a new idea, that of integrating CERN physicists and visitors into 'collaborations'. This was because a good deal of the equipment was supplied by non-CERN teams (Padua, Saclay, Ecole Polytechnique...), because the know-how and experience of the personnel in Geneva was indispensable (they knew the PS and understood how the laboratory functioned organizationally), and because the photographs taken with the chambers could easily be shared. The new DG was attracted by this conception of how to organize experimental work, all the more so since, during the summer, leading figures in the European physics establishment complained that CERN was too closed in upon itself. In September Adams thus proposed to generalize the new system: experimental teams would be mixed teams and Emulsion and Electronic Experiments Committees would be established alongside the Track Chamber Committee. The decisions taken by these three committees were to be coordinated by a 'central' committee, the Nuclear Physics Research Committee, chaired by the Director of Research. The NPRC met for the first time in November with Weisskopf as Chairman—and the whole complex was to enter into operation in January 1961.[10]

2.2 1961, the year of doubts and anxieties, the year of tensions

If 1960 was the year rich with potential in which all seemed possible, 1961 was the

Notes: p. 37

year in which one difficulty or even setback seemed to follow hot on the heels of another. In 1960 one was organizing, planning, projects; in 1961 the time had come to implement them concretely, the machine was fully operational — and things turned out to be more complex than anyone had expected.

As far as equipment was concerned the persistent leitmotiv throughout 1961 was that CERN lacked adequate particle separators, the first two built in the laboratory only being available in *December*. Without them the number of quality beams for the bubble chambers was highly restricted — only Cresti's much smaller separator was at hand — and the accessible purified beam energies were far lower than physicists would have liked. The second source of anxiety was the ever-increasing delay in the completion of the British bubble chamber — in fact it only became operational in 1964. Fortunately the Saclay 80cm chamber functioned smoothly, and CERN was able to have at least one medium-sized hydrogen chamber available for physics. Finally, as a number of experiments were set up in parallel, it became clear that CERN was short of standard beam equipment (bending magnets and quadrupole lenses) and of space: certainly the North hall was opened in August (which is not to say that it was equipped), but the big East hall was not yet ready and the South hall was saturated.[11]

As far as experimentation as such was concerned, the most painful event was the decision, taken in summer 1961, temporarily to stop work on what had been CERN's great hope, the neutrino experiment. In April, with the installation complete and the experiment ready to run, a senior CERN physicist raised serious doubts about the possible interaction rates. Having measured the π^+ flux which was to serve as the primary beam for the experiment, he announced that it was an order of magnitude below that which had been calculated, and that the experiment was not materially feasible. After discussing several ways of saving the experiment, it was regretfully, if ineluctably, decided to suspend the work, to dismantle the beam, and to release the bubble chambers for other purposes. Psychologically the blow was very hard, touching as it did the confidence in the ability of the Europeans. Indeed it seemed as though CERN had been unable to take any real advantage of its eight month lead over Brookhaven, and it was the American competitor which was then in the position to run successfully the 'crucial' two neutrinos experiment — which it did in 1962. And as Weisskopf liked to say, it is no good in physics to be always late.[12]

In this somewhat delicate atmosphere tensions began to emerge between bubble chamber physicists and physicists using electronic means of detection. The root of the problem was simple: to the extent that the period of apprenticeship was partly over, more and more people wanted to work with sophisticated and complex tools. Rivalries were inevitable since machine-time and other resources remained limited. They were reinforced by the system of parallel experiments committees based on the mode of detection used, each committee trying to impose its priorities, its choices of beams, its needs for equipment and space. To resolve the problem, towards the end of 1961, there was a growing disposition to divide *a priori* the machine-time between bubble chambers and electronic detectors (emulsion work being mainly done parasitically). This 'arithmetic' solution to what was really a question of scientific choice did not, however, resolve the disputes afflicting the different electronic groups. This type of experimentation was by nature very individualistic, even idiosynchratic, it called for a more ongoing 'interaction' with the machine, and the battles for access to the accelerator were waged daily. Pressured by the CERN physicists who claimed that visitors were less efficient (allegedly because they had less know-how on the experimental floor), the new Director-General admitted, at the end of 1961, that it was necessary to reconsider and to simplify the decision-making process for experiments. In the following years he let the CERN staff organize electronics experiments more or less as they liked.[13]

In short doing physics at 28 GeV was turning out to be far more difficult than expected. It called for the integration of very different kinds of elements, each having its own time-scale, but each being essential at the crucial moment of experimentation. A good example of this difficulty of getting the timing right is that of the separators: since they were not ready when they were needed—and since CERN, a new laboratory, had not built up a stock of equipment on which it could draw—the entire experimental programme suffered.

The running of the experimental programme was not the only major scientific concern at CERN in 1961. The other important subject widely debated was the *second generation of accelerators* which was to replace the SC and the PS within the next decade. We have seen that an accelerator research group was set up in 1957 inside the division building the PS. In 1960 it was strengthened by some of the best specialists in the PS, then 'without work'. Wanting to commit the organization to a new project (one has to allow for 8 to 12 years between the first discussions about an accelerator and its coming into operation), in December 1960 they proposed that CERN construct colliding rings linked to the PS, the so-called Intersecting Storage Rings. This involved extracting the proton beam from the accelerator, storing it in two rings of the same size, and doing physics in the zones where the two beams crossed over and intersected more or less head-on. In March 1961 the directorate (in fact Adams and Hine, the previous leaders of the PS Division) suggested to the Council that it budget for this machine. A Homeric debate between physicists and engineer-builders ensued, two groups clearly demarcated at CERN and, in Europe at least, immersed in very different traditions. In April the physicists succeeded in stopping the project coming before the Scientific Policy Committee, insisting that its merits first be compared with more classical, fixed target machines like the PS. In their view only this type of machine would enable them to do physics 'properly' by virtue of the secondary beams it produced. In June these same physicists organized an international conference which concluded that it was of little interest to have only the ISR. During the summer and the autumn, the debate continued to rage, at Berkeley, at Brookhaven, and in Europe. Granted the clear-cut differences of

Notes: p. 37

opinion involved, a 'compromise' was accepted in November: since money was not likely to be forthcoming in the short term, the choice should be postponed and all options kept open for the time being. In consequence in 1962 and 1963 both types of machines were to be studied in parallel.[14]

In 1961, then, the European high-energy physics community split. It was divided concerning very long-term projects, and it was divided concerning the everyday arrangements for experimental work which was plagued by setbacks—the neutrino experiment—and by the lack of certain essential items of equipment. The last great debate of 1961, however, occurred in the Council. At issue was the rhythm of development of the organization, and in particular the level of the *budgets* in the years to come.

In 1959 the British delegation got the Council to accept that CERN's budget would be stabilized (65 MSF for the three years 1960, 1961, and 1962). In December 1960 this ceiling was more or less respected (the 1961 budget was voted at 67.7 MSF), though the CERN management drew the delegates' attention to the negative consequences of a budget as low as that favoured by the UK. In March 1961 Adams insisted that the budget for 1962 be revised drastically. Suggesting various alternatives (69, 75 and 83 MSF), he asked the SPC in April to look into the matter. The committee found the situation to be serious—one reason for the relative lack of equipment at CERN was said to be the inadequate level of financing in the previous years—and suggested that the 1962 budget be of the order of 78 MSF. During the Finance Committee meetings in June and October the majority of the delegations recognized that an extra effort was necessary and accepted the principle that budgets should not be stabilized but should grow substantially from one year to the next. The British delegation, for its part, remained determined to fix a three year limit which, while not completely stabilizing annual expenditure, allowed for only very low rates of growth.

In November the British government decided to strike a major blow. Dissatisfied with the Council's behaviour, which it thought to be too subservient to the ever-increasing demands laid before it by the CERN management, it chose to contact directly the governments of the other member states asking them to fix, by direct arrangement, financial envelopes to be imposed on CERN.

The tension at CERN and in the Council then reached a level never attained before or in the following years. For Germany, France, and the bulk of the delegations, Britain's attitude threatened the very future of CERN itself. Not only because a policy of sustained growth seemed to them to be the only way of keeping CERN competitive with the best American laboratories, but also because the UK government's wish to by-pass the Council seemed to its delegates to be a direct attack on their prerogatives and on what constituted the uniqueness and dynamism of 'their' laboratory. The December Council session was thus explosively tense, with threats and accusations flying back and forth—and it concluded with the total defeat of the British. On 18 December the budget for 1962 was voted at 78 MSF by 10 votes to 2 without any official protest from the UK delegation. At the same time a working group was set up under one of the 'founding fathers' of CERN, J.H. Bannier, and was asked to propose a framework for financing the laboratory in the years ahead. The group met for the first time in February 1962.[15]

2.3 1962, the year when trends were reversed, the year of selfcriticism and of new hopes

The Bannier working party on the CERN programme and budget met four times between February and early April 1962. In its report laid before the Council in June it proposed that the budgets of the organization grow annually and suggested that the rates vary between 14 and 10% per annum for the next four years. These rates were established primarily by studying the current rates of growth in comparable American facilities and in certain European countries like France and Germany. To assist the negotiations between Council delegates and their domestic financial authorities the working party also put forward a four-year planning procedure, in which the rate of growth for the first two years was fixed along with a provisional estimate of the laboratory's needs for the next two years.

In June the Council took note of these proposals and agreed to discuss them in October. It also encouraged the CERN management to base their forward estimates on them (it is worth noting that there was no *a priori* opposition to the proposals, notably concerning the rates of growth of the budgets). In autumn the administration duly put forward its estimates for the forthcoming years—which called for rates of expansion considerably in excess of the figures suggested by the working party! To justify this they argued that an 'equipment gap' had built up at CERN the extent of which was only now becoming clear. Disturbed, the Council nevertheless refused to give in to the Director-General on this point, declaring that it would remain within the limits to growth set in the Bannier report. In December, after a long debate, it fixed the rates of growth for 1963 and 1964 at 13% and 11% respectively, but could not agree on the provisional figures for 1965 and 1966. All the same a new atmosphere reigned in the Council for the first time since 1957. Despite the fact that the management's request for even greater sums of money had been rejected, a certain satisfaction was perceptible, a satisfaction because everyone had now accepted that growth was essential—and because the rates of growth just agreed on were substantial.[16]

1962 was also a watershed from the scientific and technical point of view: the organization pulled itself together and felt sufficiently confident to cast a critical eye over its activities during the preceding two or three years. One symptom of this is the report submitted by Victor Weisskopf to the Council in June 1962, a report

Notes: p. 37

reproduced *verbatim* in the minutes and which stimulated a lively (self-)critical debate on CERN's strengths and weaknesses. The theme developed by the Director-General was that the relative setbacks in experimentation around the PS could be attributed to just those factors which had made of research at the SC a brilliant success. He reminded his audience that CERN's reputation still rested on the results obtained with the small machine, notably the study of various disintegrations of pions and the measurement of $g-2$ of the muon (the magnetic moment of the 'heavy electron'). What Weisskopf was at pains to point out (and which Gilberto Bernardini took up with greater finesse) was that the SC had been in service far longer than the PS—and one needed time to adapt to it—, that the small machine called for less organization and planning than the big one—as well as less heavy experimental equipment—, and that at 600 MeV the still inexperienced Europeans were in a world that was far more familiar than that of 25 GeV—and so better able to assess what was feasible and scientifically important.[17]

Certainly the atmosphere was far more encouraging at the end of 1962. Technically, first of all, the 10m electrostatic separators had functioned satisfactorily throughout the year at voltages of 700-800 kV. Relatively sophisticated in comparison with American separators— which partially explains why they took that much longer to build—they enabled physicists in 1962 to construct a separated antiproton beam of 3 to 4 GeV/c momentum, as well as a separated K-meson beam which was expected to reach several GeV/c early in 1963. It also looked this time as if the British bubble chamber was about to arrive. Its magnet was installed at the end of 1962 and the chamber was expected in the East hall in mid-1963. This hall, whose construction was started early in 1962, was 117 meters long and covered a uniform area of over 5000 m^2. The generators were installed during the year, making 12,300 kW of power (double the previous capacity) available for beams and various experimental set-ups. Finally, radio frequency separators were expected to be available in the short- to medium- term. Tested bit by bit at the end of 1962 by the Accelerator Research Division, they would permit the separation of beams of up to 15 GeV/c, a figure never reached before.[18]

Marked improvement was also noticeable as far as scientific results were concerned. Naturally one must not exaggerate the point: it was still the scientists at Berkeley and at Brookhaven who were the undisputed leaders in the field. Resonances, for example, were still very largely discovered in California and in Long Island—and it was of course Brookhaven who announced, in June 1962, the existence of two neutrinos, one associated with the electron, the other with the muon. All the same CERN made a noteworthy announcement in February 1962, that of the discovery of the antiparticle of the χ hyperon made with its separated antiproton beam (the announcement was made simultaneously by Brookhaven). As for future research the most noteworthy feature in 1962 was the systematic investment made by the laboratory in neutrino physics. Technically difficult—let us not forget that neutrinos have an extremely low interaction rate with matter—the domain was novel and potentially very rich (study of weak interactions).[19]

Early in 1962 Guy von Dardel was nominated the interdisciplinary group leader for a long-running neutrino experiment aimed at getting very precise quantitative results. To this end it was essential to have a flexible and sufficiently intense neutrino beam (30 to 100 times greater than before) as well as detection equipment superior to that used in CERN's first neutrino experiment. Three elements were at the heart of the scheme, a neutrino horn, a system for the rapid ejection of the primary proton beam out of the accelerator, and a group of spark chambers of 30 tons.

A neutrino horn is the magnetic equivalent of a convex lens intended to increase the intensity of the neutrino flux. Conceived by Simon van der Meer in 1961, its construction, as well as its cooling system and its 180kW power supply, were completed at the end of 1962. The fast ejector was at this time being assembled and the beam transport system being installed. As for the set of spark chambers, the elements were tested at the end of the year. It thus seemed likely that CERN's second neutrino experiment would get under way somewhere around mid-1963.[20]

2.4 1963, the first year in which tasks were satisfactorily integrated

The first half of 1963 was dominated, in Europe, by two events, the setting up of the neutrino experiment and the definition of the future accelerator construction programme. The debate on what accelerators to build at CERN in the following years was reopened in response to pressure from the French physicists. Their main spokesmen, Francis Perrin and Louis Leprince-Ringuet, were strong supporters of CERN. In contrast to their British colleagues, they wanted national programmes to be subservient to the European programme, and not vice versa. Thus they asked, insisted, that the 'central' programme be fixed before accepting that money be spent in their own country. In November 1962, aware of the risks encumbent on any further delay, the CERN management decided to act and to formalize the process by which the choice would be made. After consulting the most influential members of the Council and the SPC, they called a meeting in January 1963 of the European high-energy physics élite (about 50 people). On 17 and 18 January the group met, constituted itself as the European Committee for Future Accelerators (ECFA), and set up a working group chaired by Edoardo Amaldi. From January to June this working group met nine times, while ECFA met three times. On 12 June a report was submitted to the Council which it discussed at its session on the 20th.

The report concluded that in the 1970s Europe should be furnished with a pyramid of accelerators. Two European machines were proposed for the summit, intersecting rings fed by the 25 GeV PS (the ISR) and a high energy (300 GeV) fixed target proton synchrotron. At the base ECFA suggested that European states contruct a coordinated range of machines of different types. Financially, it estimated that annual expenditure would be of the order of 500 MSF around 1977 for the apex of the pyramid, and of some 1600 MSF for the entire European high-energy physics effort. In the short term the report proposed that a supplementary programme of 3-4 MSF be voted at the end of the year for the continuation of studies at CERN needed for the construction of the ISR and the 300 GeV. (Let us remember that CERN was financed by programmes; the basic programme included the construction and operation of the SC and the PS; the building of any other accelerator, by contrast, had to be treated as an independent—or supplementary—programme, in which only those states who so chose needed to participate.)

At its session on 20 June the Council took note of these proposals and charged its delegates to prepare the ground in their respective countries. It also decided to meet again in October so as to set up the supplementary programme for preparatory studies. On 11 October the Council unanimously approved the creation of the new programme and asked that the states wishing to contribute financially to it declare their intentions in December. At its meeting on 17 December 1963 all member states bar Spain and Greece announced that they would participate in the supplementary study programme. Its budget for 1964 was 3.8 MSF.[21]

The introduction of a new supplementary programme was not without effects on the determination of the level of the basic budget. At the Council meeting in December 1962, the United Kingdom opposed the majority on the rate of increase of the budget. The British proposed to increase the budget for 1963 by +13% with respect to that for 1962, by 11% for 1964, and by +10% and +9% for 1965 and 1966. Insisting that the recommendations of the Bannier Report were not being adhered to, the CERN management asked for +13% and +13% for 1963 and 1964, and +12% and +11% for the next two years. And although a period of intense negotiations followed in June and autumn 1963, the figures finally agreed on in December were close to those advocated by the UK delegation. The main reason for this was that no one wanted to put the British delegates in an embarrassing and difficult situation just when they had got their government to agree to the supplementary programme, and when the process of parliamentary ratification of the conventions establishing the European Space Research Organization (ESRO) and the European Launcher Development Organization (ELDO) were proceeding favourably.[22]

The other item of importance in 1963 was the second neutrino experiment. From January to April the different components of the experimental set-up were installed around the PS. There was a world first on 12 May when a proton beam was extracted from the PS and guided into the neutrino horn—and all functioned perfectly. Experiments as such got under way in June with a neutrino beam 50 times more intense than that used at Brookhaven a year before. And in a little more than 30 days about 360 candidate events were registered by the CERN heavy liquid chamber while the spark chamber complex recorded 4000. In 1962 Brookhaven had drawn its conclusions from 56 events.[23]

Alongside this experiment whose analysis continued throughout 1963 and 1964 and in which the results achieved the year before at Brookhaven were refined, another important experiment in the field of weak interactions was carried out at CERN. Using two scintillation counters, a Cerenkov counter and a spark chamber, the beta decay of the neutral lambda hyperon (a rare phenomenon) was studied; it led to the first measurement of the angular distribution of the disintegration products. Thanks too to spark chambers and to electromagnetic detectors another type of rare event was dealt with at CERN in 1963, that of the production of electron and muon pairs by p-$\bar{p}$ (proton–antiproton) annihilation. As for work with bubble chambers, it is worth mentioning the determination of the relative parity of the sigma and lambda hyperons using slow antiprotons and the Saclay 80 cm chamber, and the finding that the diffraction peak shrinked with increasing energy in p-p, but not π-p, collisions. From 1963 onwards then CERN seemed to be in a position to take a less marginal role in work of an international standard— something that we can also see from the study of resonances, in which the Geneva laboratory was now a fully-fledged participant.[24]

This is not to say that all the bottlenecks had been cleared at CERN. In the process that is an experiment (and which goes from the initial idea to publication via the assembly of the equipment and the obtaining of beam time), it is essential that the whole not be slowed down at any particular stage. Delays occurred at numerous points in 1961 and 1962; in 1963 their number was reduced—though some still remained. The most striking were the analysis of photographic data (the millions of pictures taken by bubble and spark chambers), and the computing capacity. In June, for example, the Director-General raised a cry of alarm in the Council because the IBM 709 was no longer adequate and CERN was having to rent more and more computer time from outside. The consequences were the decision to reduce the tasks of DD (to leave it free to concentrate on programming), the arrival at the end of the year of an IBM 7090, four times more powerful than the 709, and the decision to acquire a CDC 6600 for the start of 1965.[25]

1963 emerges then as the first year in which matters were more or less under control, a year in which the different aspects of experimental work seem to have been integrated, a year in which the last remaining bottlenecks tended to be eliminated. This state of affairs was symptomized by a rationalization of the institutional structure. Firstly, a new post was created in the Directorate, that of Technical Management, filled by Pierre Germain. Its purpose was to reduce the load on Mervyn Hine, Weisskopf's right-hand man for medium-term planning. The historic leader of the DD (Data-Handling) Division, Lew Kowarski, left temporarily for the United States to be replaced by the man responsible for computing, Ross Macleod (who was officially called the Deputy Director). As for the Engineering division, it

Notes: p. 37

was abolished and its members distributed between the divisions for which they had worked. Finally a new policy for employment contracts was instituted. Until 1963 CERN physicists and engineers had been given short-term contracts so as to facilitate the rotation of staff with European universities. In June indefinite contracts were awarded to those who were willing to stop doing research on their own behalves, and to form instead the infrastructure needed by researchers the bulk of whom would now be coming from outside the laboratory.[26]

2.5 1964, the year of the debate on the medium- and long-term programmes

Without a doubt 1964 was the year in which the needs of the future were the dominating concern in the organization. More precisely, it was the year in which there was a clear willingness to decide soon on the second generation of accelerators, as well as on certain other large equipment required for the end of the 60s. In these two debates it is difficult to distinguish, or to estimate the relative weights of, technical and financial considerations: at this stage judgements on whether it was opportune, scientifically or politically, to put proposals were intimately mixed one with the other.

The debate on the accelerator programme once more came to the fore after six to eight months of reflection in the member states since certain countries maintained their pressure to have decisions taken soon, and since the idea of building accelerators to form 'the base of the pyramid' was abandoned by the small countries (Scandinavia, Netherlands,...). At the same time everyone recognized that ECFA's programme was *very* expensive (even if restricted to the two machines at the summit) and that the international context was less than perfect: in the United States, in particular, it seemed as though the era of lavish funding which followed on the scare created by the Soviet Sputnik was coming to an end—and this drew the sting somewhat from the competitive argument. In this contradictory situation the Director-General took a decision of the first importance in May: he changed CERN 'selling strategy', if one might call it that, and agreed to stagger in time the decisions on the ISR (the less expensive machine) and the 300 GeV. His aim was to break the deadlock, to get the decision-making process moving again.

The majority of the physicists (in the strict sense of the term) reacted negatively to this move, fearing that the member states would feel that they had discharged their responsibility once the ISR was financed. They stuck to the line they had adopted in 1961 and had reiterated in 1963—the 300 GeV was the more important machine—, and most of them insisted that the ISR be withdrawn from the programme. The management and the SPC did not follow them on this latter point, recommending instead to the Council that *both* machines be built, with the decisions *separated* in time. The Council accepted this policy in June 1964, and asked its delegates to get a favourable decision from their governments during the next six to twelve months on the financing of the ISR, the decision on the 300 GeV being postponed for a further two or three years.

Once triggered, the process followed its course in the various state apparatuses—which in practice meant discussing only the ISR seriously and pushing the other decision to an indeterminate future date. Only a few countries were willing to commit themselves in December (France, Germany, The Netherlands, Switzerland), and an exceptional Council session was scheduled for March 1965. In fact the decision was only taken in June: on the 16th all member states bar Greece declared themselves to be in favour of the ISR—and no reference was made to a decision in the short- to medium-term on the 300 GeV project.[27]

To all intents and purposes, then, the 300 GeV project was (temporarily) eliminated even if no one had explicitly 'decided' so, and even if the physicists continued to protest vociferously. It was along with the ISR only that the management—with the support of the physicists this time—tried to raise the money needed for what was called a PS improvement programme. Spelt out in spring 1964, it involved a number of elements which were collectively intended to make experimentation around the big synchrotron competitive between 1968 and 1972, the year when the ISR was to come on line. According to the management, this programme, unforeseen and unforeseeable in 1962, was to be treated as falling outside the framework of the Bannier Report's recommendations. In essence they asked for a new linac injector for the PS (so as not to be outstripped by Brookhaven who wanted to improve the intensity of their AGS), a new experimental area with its beam equipment (indispensable if the number of visitors from the member states continued to grow), and a new batch of detectors (including a 5 m bubble chamber). For the last France and Germany announced that they were ready to share in the construction and its costs.

The cost of the improvement programme was some 130 MSF over six years. The proposal made little impact on the national delegates during the course of 1964, who refused to break the limits proposed by the Bannier working party in 1962. Which is not to say that they did not make a 'special effort' to meet the administration's request: indeed it was the maximum percentage growth proposed by the Bannier Report which was voted in December 1964 for 1966. The Scandinavian countries' delegates played a decisive role here. Convinced by the arguments of the management, they left the United Kingdom alone in defending a lower figure. If one adds that 6.8 MSF were voted for a supplementary programme to go deeper into the ISR and 300 GeV projects, one sees that the overall growth for the next two years agreed to by the Council was appreciable. On the other hand, no decision could be reached on the provisional estimates for 1967 and 1968.[28]

The two big debates we have just described were of marginal relevance to the daily

Notes: p. 37

functioning of the laboratory in 1964. The short-term development worth noting in this year was the commissioning of new equipment. Some of this can simply be seen as of a 'routine' or 'unexceptional' nature—even though it significantly improved the situation on the experimental floor. An example here would be the polarized proton target developed by Saclay which was first to be used to determine the parity of the Ξ^- particle. Other equipment was of wider significance. Firstly, in the area of hydrogen bubble chambers where CERN was falling further and further behind Brookhaven whose 2 m chamber had gone into service in June 1963. In January 1964 a 6 GeV/c separated kaon beam using three CERN-built 10 m long electrostatic separators was ready in the East hall for the British bubble chamber. This was unfortunately not yet operational, and the 80 cm Saclay chamber was used with the beam. The 1.5 m chamber was finally ready in June, and 700,000 photographs were taken during the rest of the year. In December it was the turn of the CERN 2 m chamber to pass its first trials. Continually delayed during its construction for lack of staff (and because the British chamber was expected to be used in the interim), the CERN device was ready for installation in a 5.6 GeV/c K$^+$ and $\bar{\text{p}}$ beam—an experiment which was done in the first quarter of 1965.

The second area in which important improvements were made was that of data-handling. In terms of computing the use of the IBM 7090 at maximum capacity (24 hours a day 7 days a week) is worth noting; all the same it was quickly saturated. Preparations were made in parallel for the arrival of the CDC 6600, 20 times more powerful than the IBM, and considerable time was spent in 1964 on making the programmes compatible. As for picture scanning, the development worth noting is the operation of the first flying spot digitizer (HPD) which could treat hundreds of spark chamber events per hour, and which analyzed 200,000 pictures in 1964. Its rate of handling bubble chamber pictures remained somewhat slow, however, and only a few thousand photographs were dealt with in this way in 1964. A cathode ray digitizer (Luciole) was being developed for the future; in 1965 it was used primarily for an experiment on the missing mass of the hyperon.

Finally it is noteworthy that in 1964 the first attempts were made to bypass the use of film with spark chambers and to instal small computers on-line to experiments. (In March a conference on this topic was organized at CERN.) The idea was to exploit to the maximum the fact that spark chambers were electronic devices, that one could directly digitalize all the information received, and that an on-line computer accelerated data-taking and handling (work done in real time). Two experiments were done in this way in 1964.[29]

2.6 1965, the year of consolidation

As this survey ends in 1965 let us quickly look at some of the lines of research followed by the laboratory in 1964-65—recognizing that it is difficult to limit oneself to a two-year period since a research topic often persists over three to four years, if not more. With bubble chambers the major area of investigation remained the properties of already-identified as well as new resonances. In 1964 and 1965 teams working at CERN announced several discoveries of such quantum states and developed a theory which enabled them to account for much of their data (so-called peripheral collisions). It was, however, a team at Brookhaven which first announced the discovery judged to be the most significant, that of the Ω^- predicted by the SU_3 classification in 1962. Quark hunters were also at work after Gell-Mann and Zweig had postulated the existence of these entities at the beginning of 1964. The main conclusions CERN physicists drew from their experiments were that if isolated quarks existed their mass was not inferior to 2.3 times that of the proton. Several of the experiments we mentioned in our survey of 1963 were also continued; this was so, for example, with the study of the annihilation of p-$\bar{\text{p}}$ into lepton pairs. With electronic means of detection the analysis of the scattering of different high-energy particles remained essential; it led, amongst other things, to interesting results on resonant states. After 1963, however, and thanks to the availability of energetic and purified beams, the study of low cross-section processes (rare events) tended to move centre stage; we could mention as example the investigation of large momentum transfer and charge exchange scattering.

As for weak interactions, the analysis of data taken in the neutrino experiments in mid-1963 and at the beginning of 1964 (6000 events in 1964) led physicists to accept the conservation of the μ-lepton number and above all to draw the conclusion that, if the W boson responsible for the weak interaction existed, its mass was above 1.8 GeV. In 1965 it was further decided to launch a third neutrino experiment; important investments in beams for it (notably of anti-neutrinos) were made, and it was planned to dedicate an entire experimental area to this kind of physics. The second set of experiments associated with weak interactions which took shape in 1964 was related to the violation of *CP* parity announced for the first time by researchers at Brookhaven. Several teams were working at CERN on this topic in 1964 and 1965, the most noteworthy being that studying the disintegration of K^0 into two pions at 10 GeV/c, which allowed for a comparison with the work done at 1.1 GeV/c at BNL. To give one more indication of the increasing importance of the study of weak interactions at CERN as from 1964, we could also mention the remeasurement of $g-2$ of the muon, this time with the PS. A muon storage ring was installed at CERN for this purpose in 1965.[30]

Three remarks to conclude this quick summary. Firstly, experimentation at the SC struck out in a new direction in 1964, when the work was deliberately oriented towards the study of nuclear structure and the physics of elementary particles progressively decreased. Secondly, there was no decrease in the number of emulsion experiments—unlike what was happening in the US—and important results were obtained (the study of the magnetic moment of the Λ^0 hyperon, for example). Thirdly, the nature of the teams working at CERN changed markedly between 1964 and 1966:

Notes: p. 38

whereas before nine out of ten electronic experiments were organized by, or around, a CERN core, we now find a perfect balance between visitors and CERN physicists.[31]

Taking stock, we can say that in the mid-1960s the CERN PS came of age as far as physics results were concerned, the apprenticeship was over. The 25 GeV machine in Geneva produced about 30% of all experimental high-energy physics papers published, these articles received the same number of citations as those produced at the Brookhaven AGS in 1965 and 1966, and, on the whole, the community deemed the precision of the data taken at CERN to be excellent or even superior to that obtained in the United States. This is not to ignore that some of the experiments judged by physicists to be crucial—like the discoveries of the Ω^- or of the violation of *CP* parity—were performed at the rival Brookhaven laboratory.[32]

1965 was a promising year for another reason: the ISR project was approved at the June Council session. Formally the vote only authorized the programme as such (to cost about 330 MSF over six years), and CERN had to wait until December for those states wishing to participate financially to announce their intentions. All the same, after the favourable statement by the British delegate on 16 June everyone knew that all states bar Greece would contribute to the project—as was confirmed in 1965 and 1966.

As for the laboratory's basic budget—including the improvement programme asked for by the management in 1964—here too CERN could be satisfied. In March 1965, after having again explained to the Council various aspects of the PS improvements envisaged, the management once more put the figures it had proposed in 1964 on the table, and asked for 166 and 186 MSF for 1967 and 1968 (of which 19 and 32 MSF were for improvements to the synchrotron). These figures corresponded to growth rates of 17 and 12%, and were well above the rates the Council had in mind. Despite considerable pressure from the SPC the figures favoured in the spring by the most 'generous' countries were below those the management wanted (France agreed to increase its offer for 1967 from 158 to 159.5 MSF, so from +11% to +12% with respect to the previous year). No decision was then taken by the Council in June, and during the summer the management persisted with its demands. In November the Director-General asked for 163 MSF for 1967, 181 and 200 MSF for 1968 and 1969. France accepted the sequence 162, 180, and 197, while Britain preferred 160, 180, 198, and in December it was the British combination that was adopted—which corresponded to a rate of growth of 12.5% for the first two years. At this same Council session 20.7 MSF were also voted for the ISR programme in 1966, 68.6 MSF for 1967 and some 73-74 MSF for 1968 and 1969. Finally 4 MSF were set aside for continuing studies on the 300 GeV project. In short, the constant pressure applied by Weisskopf and Hine over several years had borne fruit, and CERN's member states had shown their willingness to make a major investment in the Geneva laboratory. After all the budget for 1961 was only 67.7 MSF—while an expenditure of something like 230 MSF was foreseen for 1967.[33]

One outstanding problem remained in 1965, the nomination of a successor to Victor Weisskopf who, having served as CERN's Director-General for four years, now wanted to go back to MIT. This matter was settled without undue tension in June, when Bernard Gregory was appointed to serve for five years. Having arrived in Geneva in 1960 with the Saclay 80 cm chamber, first chairman of the Track Chamber Committee and then Director of Research, his nomination was a tribute to a man and a country which were totally dedicated to the organization.

It was a new era which started at CERN on 1 January 1966.

Translated by John Krige

Notes

1. As for the previous chapter only summary references are given. We have primarily used the CERN Annual Reports, Council and SPC minutes, and the results presented in chapters 3 to 13 of this volume.
2. For the PS see E. Regenstreif, *Le synchrotron à protons du CERN,* 3 volumes (Geneva: CERN 58-6, 59-26, 61-9, 1958-61).
3. See chapter 8, section 8.4.2.
4. Annual Report (1960); Council, minutes, 14/6/60, 29-38; 8-9/12/60, 59-64.
5. See chapter 9, section 9.6.
6. See chapter 5, section 5.5.
7. Annual report (1960); Council, minutes, 8-9/12/60, 59-60.
8. See chapter 7, section 7.2.4.1.
9. See chapter 7, sections 7.2.1 to 7.2.3.
10. See chapter 8, sections 8.4.3 to 8.4.5.
11. See chapter 9, section 9.5.2 and chapter 13, sections 13.5 and 13.6.
12. See chapter 5, section 5.5.
13. See chapter 8, section 8.5.
14. See chapter 12, section 12.3.
15. See chapter 10, section 10.3.
16. See chapter 10, sections 10.4.1 and 10.4.2.1.
17. Council, minutes, 13/6/62, 14-19.
18. Annual Report (1962); Council, minutes, 13/6/62, 14-19; 19/12/62, 42.
19. See chapter 5, and sources cited in note 18.
20. See chapter 5, section 5.5.
21. See chapter 12, sections 12.4 and 12.5.
22. See chapter 10, section 10.4.2.
23. See chapter 5, section 5.5.
24. See chapter 5, sections 5.3 and 5.4.
25. See chapter 9, section 9.6.3.
26. See chapter 7, section 7.2.4.
27. See chapter 12, section 12.6.
28. See chapter 10, section 10.4.2.

29. Annual Report (1964), (1965); Council, minutes, 18-19/6/64, 9-31; 15-16/12/64, 39-42.
30. See chapter 5, sections 5.3 to 5.5.
31. See chapter 8, section 8.7.
32. See chapter 13, sections 13.5 and 13.6, and the references there to Irvine and Martin (1984).
33. See chapter 10, section 10.4.2.2.

PART II

Physics and engineering at CERN

CHAPTER 3

Some aspects of the history of high-energy physics, 1952–66

Armin HERMANN

Contents

3.1 Introduction	42
3.2 Social aspects	43
3.2.1 The scientific community	44
3.2.2 Communication	54
3.2.3 Conferences	61
3.2.4 Summer schools	69
3.3 Scientific developments	73
3.3.1 Accelerators	73
3.3.2 Particle detection	78
3.4 HEP and society	79
3.4.1 The intellectual and cultural importance	80
3.4.2 The economic value	83
3.4.3 The importance for military technology	86
Notes	87
Bibliography	92

3.1 Introduction

In the period under review, that is roughly the years from the foundation of CERN up to the mid-sixties, elementary-particle (or high-energy) physics was a young field of human activity. It is the purpose of this chapter to describe in general patterns the development of the scientific community and of high-energy physics itself.

For high-energy physics (or HEP for short) to take form, two processes were essential: separation from nuclear physics and from cosmic-ray research. This double emancipation has already been described in Volume I;[1] nevertheless, a few words of repetition seem appropriate.

When the 'International Physics Conference'[2] sponsored by the provisional CERN Council was held at the Institute for Theoretical Physics in Copenhagen from 3 to 17 June 1952, its purpose was threefold: to promote contacts between European physicists,[3] to provide a survey of the 'various domains of nuclear physics', and to receive guidance in the planning of the European laboratory.[4]

Here we are interested in the second item, the survey of the field. The conference was concerned with nuclear physics in general. A report, drafted after the conference by Werner Heisenberg and adopted by the Council, clearly distinguished HEP from nuclear physics and emphasized the problems of high-energy physics to be 'more fundamental and therefore more interesting' than those remaining at lower energies.[5]

Such a statement did not need to be approved by every physicist. For the formation of HEP it was sufficient that a group of adequate size shared this opinion. As it turned out, the group was rather strong: a large number of famous physicists felt attracted by the emerging field and gave inspiring examples to the younger generation. Among these was Enrico Fermi, who believed, at the end of the war, 'that nuclear physics was reaching a stage of maturity and that the future fundamental developments would be in the study of elementary particles'.[6] Robert Wilson also wanted, after the war, 'to go back and scatter protons from protons. I felt that there was a fundamental problem there'.[7]

Edoardo Amaldi deemed it certain, in 1949, 'that the solution of the fundamental problems of elementary particle interactions ... will have decisive influence ... on all research into nuclear physics and engineering'.[8]

In 1950 Robert E. Marshak, the newly-installed Chairman of the Rochester Physics Department, which operated one of the few meson-producing machines, decided 'to organize an invitational conference that would thoroughly discuss the latest developments in high energy physics'.[9] Indeed a series of annual Conferences, exclusively devoted to high-energy physics, was founded. The organizers tried to keep the number of participants small; nevertheless, the conference grew in size and scope, 'reflecting the increase in world activity and interest in particle physics'.[10] The Rochester conferences not only reflected the increase in the HEP community, but themselves played an important role in the development of the field.[11]

The separation from cosmic-ray research took place parallel to the emancipation of HEP from nuclear physics. In his report of June 1952 Heisenberg stated that 'the big accelerators are ... an indispensable tool for investigations ... of elementary particles', but he admitted that 'cosmic ray work ... seems to be superior to the use of big machines, since it can easily be extended to extremely high energies'.[12]

The idea of the Cosmotron in Brookhaven was explicitly to work with cosmic energies in the laboratory. When, after the Cosmotron (completed in 1952 and brought to 3 GeV in 1954) the Bevatron came into operation at 5 GeV in 1954 and reached 6.3 GeV in 1956, a stream of new results came in. At the Sixth Rochester Conference, held from 3 to 7 April 1956, Robert B. Leighton made the remark that 'next year those people still studying strange particles using cosmic rays had better hold a rump session of the Rochester Conference somewhere else'.[13]

In the same year, 1956, the 'Annual Review of Nuclear Science' explained that cosmic ray physicists had been displaced from the study of elementary particles by their more successful colleagues: 'Cosmic ray studies ... now seem to have returned to the basic problem of identifying the origin and sequence of events which take place before terrestrial detection, leaving the apparently endless task of finding new particles to those associated with the newer high-energy accelerators'.[14]

An illustration of this is the cosmic-ray work done at CERN. When the organization was founded in 1954, two cosmic-ray groups were set up. The Geneva group constructed two multiplate cloud-chambers to measure the lifetime of K-mesons. The experiment was stopped in 1957, and the members of the group transferred to other divisions within CERN. The Jungfraujoch group, working with equipment built at Manchester University, collected about 16 000 photographs on nuclear interactions produced in paraffin, of which some 200 were at energies higher than 25 GeV. The work came to an end in April 1958. The members of the team were engaged in the measurements of 50 events up to July, after which the group started to disperse.[15]

3.2 Social aspects

In the development of a discipline historians generally distinguish between the social and the cognitive level. At the social level one deals with the community of scholars, its rules and taboos, hierarchies and structures, and at the cognitive level with the scientific problems defined by the community, the methods of solving them, the theories and the arguments for and against.

We shall follow these generally adopted patterns and deal first with the social aspects, beginning with the development of the community and its structure.

Notes: p. 87

3.2.1 THE SCIENTIFIC COMMUNITY

For every discipline whether it is able to attract young talents is a question of 'to be or not to be'. Too many young people are nearly as bad as too few, the ideal being a steady flow of highly trained and creative new researchers into the field.

As is well known (see section 3.3.1), the planning of a new generation of accelerators must be commenced ten years in advance. One has to be sure that the manpower for the construction of the accelerators and for the future research programmes will be available, and this too implies planning ten or more years ahead. The manpower situation has therefore perhaps been subject to longer and more detailed investigation in high-energy physics than in any other academic field.

In this respect the historian is in an easy position. All he has to do is to collect the empirical data already evaluated.

Manpower situation

At the time in question, investigations of the manpower situation were always carried out for a national or regional scientific community. Since the United States was ahead in HEP in every respect (discoveries, planning of scientific programmes and development of the community) the U.S.A. deserves special interest; more data are available than for any other country or region. Apart from the U.S.A. we shall study the development of the HEP community in Europe, and restrict ourselves to these two cases.[16]

Most manpower investigations concentrated on one class of worker, namely the Ph.D. physicist, as the most essential member of the labour pool and the one requiring the longest training. In the following figures we also restrict ourselves to the Ph.D. physicist.

a. United States

Table 3.1
The Ph.D. population in HEP

	Total	Exp.	Theory	Engineering
1959/60[17]	718	386	229	103
1963 (fiscal year)[18]	880			
1965 (fiscal year)[19]	1060			
1973 (January)[20]	1681	1072	587	
1975 (January)[21]	1732	1107	609	
1981[22]	2000			

Another important figure is the production of PhDs.[23]

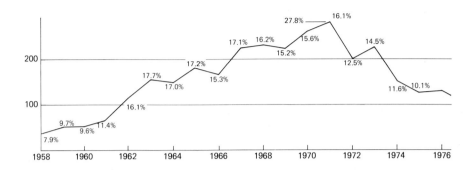

Fig. 3.1. HEP Ph.D. production (in total numbers and as % of all physics Ph.D.s), theory plus experiment

According to another source[24] the figure for 1982 is about 120.

On the basis of these figures we are able to produce Table 3.2, which presents a detailed picture of the year-to-year development of the community and the newcomers' prospects of employment. Our calculations are based on the following plausible assumptions:

(1) The figures for the development of the community for the years 1960, 1963, 1965, 1973, 1975 and 1981 are taken from empirical data. We accept these figures as they are and have made no attempt to check the validity of each census. Figures for 1959/60 were based on a survey made 'in early 1960'. For the sake of simplicity we use these figures for January 1960.

The report from which we have taken the figure for 1963 states: 'One is led to a figure between 800 and 900 [...] now [April 1963] active in the field'. The calculation was based on the assumption of 880 for (the beginning of) the fiscal year 1963. As the fiscal year started on 1 July in the 1960s we have to reduce the number for January accordingly. The same holds for the 1965 figures.

(2) To fill in the gaps between these years we assumed (a) an equal increase annually from 1960 to 1963, which gives 5.4% every year; (b) a decline of the increase from the 10% in 1963/64 by 1% each year (1965:10%), (1966:9%, up to 1974 with 1%); and (c) a steady increase from 1975 to 81 by 2.5% each year.

(3) New jobs (the second column) arise from two processes. One is attrition. For our purposes, it seems unnecessary to find out the real numbers and we therefore assume the 2% attrition rate on which all forecasts in the U.S.A. were based. The other process is the increase in the number of posts, which in retrospect is simply the increase in the number of Ph.D.s in the community.

Notes: p. 88

(4) The numbers of new Ph.D.s (third column) are taken from Fig. 3.1 and are therefore (except for the peak) only approximations (±5).

Table 3.2
US HEP community (Actual figures)

Year	Numbers (January)	New jobs	New Ph.D.s	Job prospects	Increase in community
1960	718	53	50	106%	5.4%
1961	757	56	65	86%	5.4%
1962	798	58	115	50%	5.4%
1963	840	102	155	66%	10%
1964	925	113	150	75%	10%
1965	1020	122	180	68%	10%
1966	1122	123	165	75%	9%
1967	1223	122	220	55%	8%
1968	1321	118	230	57%	7%
1969	1413	113	225	50%	6%
1970	1498	105	260	40%	5%
1971	1573	94	278	34%	4%
1972	1636	82	200	41%	3%
1973	1685	68	225	30%	2%
1974	1719	47	150	31%	1%
1975	1732	78	125	62%	2.5%
1976	1775	81	130	62%	2.5%
1977	1820	81	100	68%	2.5%
1978	1865	84	120	70%	2.5%
1979	1912	86	120	72%	2.5%
1980	1960	88	120	73%	2.5%
1981	2009	90	120	75%	2.5%
1982	2100	91	120	75%	2.5%

We are mainly concerned with the period 1952–66. One important element of the 'spirit of a period' is expectations for the future. With respect to manpower development these expectations had been formulated in 1965 with considerable precision when a 'Policy for National Action in the Field of High-Energy Physics', covering the period 1965–81 was prepared by the Joint Committee on Atomic Energy.[25]

We are now able, from the high of the year 1986, to compare the expectations of the mid-sixties with reality. The comparison reveals that in the middle of the sixties the community was far too optimistic about its future. Nobody envisaged the job crisis of the early seventies, which brought stagnation of the community with an increase of only 1.5% annually in 1973/74.

To explain this more explicitly, we have to go into the details of the 1964 calculations about the development of the community. These calculations were based on the following assumptions:[26]

(1) An existing number of 1060 Ph.D.s in fiscal 1965
(2) An attrition rate of 2% on the existing number of scientists
(3) With respect to the last of the relevant figures, the number of new Ph.D.s entering the field, the basis was somewhat more complex:
(3a) It was well known that the increase in the number of Ph.D.s for the last three years had 'averaged 28 percent per year'. With this figure, however, the HEP Community would have developed rather like a rabbit population. Therefore an annual increase of only 7% was assumed, commencing with a number of 130 new Ph.D.s in fiscal 1963 (which was also below the actual figure of ~ 150).
(3b) For the numbers of new Ph.D.s *integrated* into the community three different cases were calculated: (a) assumed, that the community retains 100% of the new Ph.D.s, (b) assumed 75% and (c), the most pessimistic, only 50%.

We have repeated all the calculations from year to year, and give the results in Table 3.3. Except where the figures are rounded off, the table is in full agreement with the figures given in the report of the Joint Committee on Atomic Energy.

Let us now compare this forecast with the actual figures given in Table 3.2. To do so correctly, it is necessary to make the following changes in Table 3.3:

(1) For fiscal year 1964 (running from 1 July 1964 to 30 June 1965) simply read January 1965, and do the same for all other years.
(2) The basic assumption of 1060 Ph.D.s in fiscal 1964 differs from the Joint Committee's statement that: 'During fiscal year 1965 there are about 1060 Ph.D.s in the field.' Since this is only a slight difference, simply subtract 40 from each figure relating to the size of the community (the first three columns).

It is now easy to see how close the calculations of 1964 came to reality:

(1) The calculations extend up to fiscal 1981 (or January 1982). The figures given for this date are, for the three different cases, 4.537, 3.592 and 2.664, from each of which 40 should be subtracted in order to make the comparison fair.

All the figures are remote from reality. In January 1981 the HEP community was fairly close to 2100. We conclude that nobody expected the community to increase so slowly. *The expectations in the mid-sixties were over-optimistic.*

(2) Over all the years the most conservative assumption (Case c) was closest to reality. Around 1976 the calculated figures met reality; prior to that the calculations were below and afterwards above it. In this case too, a few years after 1976 the calculated figures were much higher than reality.

Assumption (b) was in very good agreement with reality for the first few years (up to 1970). Table 3.2 indicates a real figure of 1498 for January 1970. For the same date, Table 3.3 gives a value of 1496 after the deduction of 40. Subsequently the calculated values rose far above the real ones.

It is misleading to speak, as the 1965 Report does, of Case (b) as assuming a 75% chance of employment for newcomers. The real figures of Ph.D. production were, in the years up to 1971 (Table 3.2, Column 3), much higher than the fictive numbers on

Table 3.3
US HEP community forecast in 1964

Our repetition of the 1964 calculations. The years given are fiscal years. To get ordinary years simply read every year as the January of the following year: the 1970 figures, for example, are actually the figures for January 1971. To compare with Table 3.2 simply subtract 40 from every figure in the first three columns.

	community			newcomers		
	(a)	(b)	(c)	(a)	(b)	(c)
1963				130	97.5	65
1964	1060	1060	1060	139	104	70
1965	1178	1143	1109	149	112	74
1966	1303	1232	1161	159	119	80
1967	1436	1328	1217	170	128	85
1968	1577	1428	1302	182	137	91
1969	1728	1536	1367	195	146	98
1970	1888	1651	1438	209	157	104
1971	2060	1775	1513	223	168	112
1972	2241	1908	1595	239	179	119
1973	2436	2049	1682	256	192	128
1974	2643	2200	1776	274	205	137
1975	2864	2361	1878	292	220	146
1976	3099	2534	1986	313	235	157
1977	3350	2718	2104	335	251	168
1978	3618	2915	2230	359	269	179
1979	3905	3125	2364	384	288	192
1980	4211	3351	2509	411	308	205
1981	4537	3592	2664	440	330	220

which the calculation was based. Therefore, the real chances dropped drastically from around 75% in 1964 to 50% in 1968/69 and to 30% in 1973/74. The accordance of Case (b) with reality for the years up to 1970 is due to the assumption of 98 new Ph.D.s entering the field in 1963 with an annual and steady increase of 7%. In fact, the increase was 11% in 1964, zero in 1966, and −5% at the end of the sixties.

(3) The basic assumption for the growth of the community was, in each of the three cases, an increase of 7% every year. The asymptotic behaviour of the community is then also given by a 7% increase. Thus in Case (a) the calculated increase in the first year is about 11%, going down to 7.74% in 1981. In Case (b) the calculated increase from 1964 to 1965 (the first year) is 7.8%, declining slowly to 7.2% in 1981. In Case (c) the increase in the first year is 4.6%, rising to 6.2% in 1981. The values slowly approach 7% in all three cases.

In reality, the community's rate of increase was around 10% in 1965 and diminished

Social aspects

drastically to about 1.5% annually in 1973/75. *In 1965 nobody expected a job crisis and such a period of stagnation in the growth of the HEP community (and in all other branches of physics).*

(4) If the employment chances of new Ph.D.s entering the field (Table 3.2) is examined, it will be seen that the most pessimistic assumption, i.e. 50%, was the nearest to reality. The actual chances went down from 100% in 1960 to around 50% at the end of the sixties, with a minimum of 30% in 1973 and a rise ~ 75% in 1980.

As a first reaction to the job crisis an abrupt decline in Ph.D. production can be seen immediately after the peak (1971: 278, 1972: 200, 1973: 220). The still high production of Ph.D.s coincides with the stagnation of 1973–75, when the growth of the HEP community was only 51 Ph.D.s or 1.5% annually. The opportunities for newcomers changed drastically for the worse.

(5) That all the calculations in 1964 were remote from reality is also revealed if the number of new Ph.D.s in the early eighties is checked. For fiscal 1981 Table 3.3 shows 440 new Ph.D.s, with 440, 330, or, in the worst case, 220 of them actually entering the field. In reality, there were only 120 new Ph.D.s, of whom probably around 90 entered the field.[27]

b. Europe[28]

Thorough investigations into HEP manpower were made for the years 1962/63 and 1978/80.[29] These studies are due to the European Committee on Future Accelerators (ECFA), and both were undertaken when a new high-energy accelerator programme was being developed. The figures are:

	Total	Experiment	Theory
1962	685		
1978 (1st November)	2257	1517	740

The famous Amaldi Report, from which the 1962 figures are taken, made some predictions up to 1977, namely 'about 1800–2000 being available in the CERN Member States by 1972, compared with an estimated need of 1500.' The predicted figure of 1800–2000 is in excellent agreement with the real figure (~ 2100).

The figures show an average increase of 7.72% per year, or a doubling time of 9.3 years. If an exponential increase is assumed (which holds with more or less accuracy) it can be seen that Europe caught up with the U.S.A. during the latter's period of stagnation in 1973–75. In 1973 there were around 1560 HEP postdoctoral researchers in Europe and 1681 in the U.S.A. Whereas in Europe the figure rose by 7.72%

annually to reach 1806 in 1975, in the U.S.A. annual growth was only 1.5%, resulting in 1732 HEP Ph.D.s in the same year.

Comparison with earlier times

If the foregoing figures are compared with the corresponding ones half a century ago, it will be realized that a historical process is taking place with vast consequences for science and society. In the twenties, it was a community of some 300 people who solved the riddle of the atomic structure. Taking up a famous phrase of Churchill's, Weisskopf used to say, that maybe never before in the history of the human mind had so few achieved so much in such a short time.[30]

With increasing numbers in the community, interest shifted to the nucleus and further on to the elementary particles. The situation in the year 1981 was characterized by a population of 2000 postdoctoral high-energy physicists in the United States and about 2800 in Europe. The nuclear physics community was a little smaller, with about 1450 Ph.D.s in the U.S.A. and 2450 in Europe.[31]

For our purposes it is not necessary to check the corresponding figures for China, Japan and the U.S.S.R. It is sufficient to assume a total population of 6–7000 high energy- and 5–6000 nuclear physicists of Ph.D. (or comparable) level in the world as a whole in 1981.

Similarly, it is not necessary to analyse to what extent other branches (especially solid-state research) had also developed from atomic physics. Compared with the 300 or 400 people half a century ago, the 1981 figures indicate the tremendous expansion clearly enough.

The structure of the community

In the 18th century, most scientific research was done at home and as a private hobby. Many people were able to buy or to build the apparatus needed for the experiments on their own. At the end of the 18th and during the first decades of the 19th century professors of physics applied for funds, and at every university a 'cabinet de physique' was established. This 'cabinet' was simply a collection of apparatus, stored in a single room, and mainly used for demonstrations in the courses. Research was generally done with these state-owned instruments, but mostly still in the private home of the professor.

By the middle of the 19th century physicists argued strongly for a building of their own, the physics institute. It housed the collection of apparatus, had rooms for the research of the professor and the small staff, one or two others for the practical student courses and (rather often) a lecture hall and appartments for the professor and assistant professor, the mechanics and the servant.

From a private hobby, research had become a public matter. Rules had to be established regarding admission to the state-owned institute and the use of the state-owned instruments. Conflicts were frequent between the director of the institute and its other users such as associate professors and research students.[32]

By around the turn of the last century, research in physics was concentrated in university institutes and a few other institutions such as the academies. In the first decades of the 20th century the physics institutes specialized in subfields, for example in spectroscopy, and it became a problem to change the field when a new director was appointed. In addition to those of the universities, pure research institutes such as the Kaiser Wilhelm Institutes were founded for special activities, for example radioactive research. They generally held a leading position in their field of activity and were in competition with a few university institutes.

The next and final step to the present-day situation was the creation of big research establishments during and shortly after World War II such as the Los Alamos Scientific Laboratories, Oak Ridge National Laboratory, Brookhaven National Laboratory (BNL) and the Lawrence Radiation Laboratory (LRL).

Special instrumentation such as reactors for neutron research and accelerators for nuclear-collision experiments was available only in these central facilities. While a few of the larger universities acquired reactors and accelerators of their own, the central facilities generally possessed much bigger ones. At least in the field we are interested in here, high-energy physics, the central facilities always had a monopoly. In 1954, for example, inelastic proton–proton collisions above the threshold energy of proton–antiproton pair-production could be produced nowhere but at Berkeley.

It is well known, and does not need to be repeated here, that the introduction of increasingly large accelerators was inevitable once the interest of research shifted to smaller and smaller structures. To resolve the structure of an atom, which is of the order of 10^{-8} cm, it was sufficient to have particle beams with an energy of a few eV. To resolve nuclear structures in the region of 10^{-12} cm one has to go into the MeV range, and to 'see' the electromagnetic structure of the proton it is necessary to have electron beams up into the GeV range.

Of course, physicists were aware that with the end of World War II a new development had started, and from the late 40s articles appeared on the new structure of physical research, which was described as 'large-scale science' or 'big science'.[33] It was probably Alvin M. Weinberg, Director of Oak Ridge National Laboratory, who in 1961 first used this expression.[34] The new term was made popular by Derek J. de Solla Price. His Columbia lectures were widely read when they appeared in 1963 as a book under the title 'Little Science, Big Science'.[35]

There were fundamental differences between the old type of research institute and the new facilities, of which the most obvious was the scale. In manpower, instrumentation and size, the new installations were larger by one, two or even more orders of magnitude. In 1906, when he had just received the Nobel Prize, Philipp Lenard proposed a new 'Institute for Physical Research'; he thought that all the instrumentation might cost up to 100 000 Marks and that he would need a handful of men to assist him. Today, CERN employs about 3 500 people and has an annual

Notes: p. 88

budget of about 800 Million Francs.

Another difference is the relationship with the universities. The fin-de-siècle research institutes supplemented university research, whereas the national laboratories such as BNL and LRL left no chance to compete except by setting up a research centre of a similar kind. Nearly all significant experiments in HEP can only be done at big and specialized laboratories with their accelerators and their instrumentation.

When the old type research institutes were founded the universities expressed some opposition, until they finally accepted them as fair competition. However, it was never the idea that national laboratories such as BNL and LRL should also be 'fair competitors' to the university institutes, but rather that they should do all the research in HEP alone. One might expect, therefore, that the universities would have been outraged when the monopolists (or rather oligopolists) appeared on the scene.

This, however, was not the case. The universities themselves created the accelerator laboratories, sometimes one university for itself, as in the case of LRL, sometimes in collaboration with government (Argonne, SLAC), sometimes a small number of universities (CEA, PPA) or a regional group (BNL, Fermi Accelerator Laboratory).

The universities had realized that the new fields of science (and nuclear physics in particular) required more money and manpower than any research hitherto, and that the funds needed for nuclear science alone exceeded what a big university would have spent in the past for all its faculties combined.

Faced with this challenge, the universities decided to create independent research facilities.[36] They understood perfectly that if they wanted to maintain themselves as sources of ideas and action, and not just become depositories and dispensers of knowledge, they had to be heavily involved in the research at these centres.[37]

Conditions were favourable for co-operation between the universities and the new centres. Experiments with accelerators are much more complex than any in earlier times and call for more skill and manpower. This is why the physics departments of the universities did not become obsolete, but found a new role in collaboration with the research teams of these centres and generally also with teams from other universities.

Right from the beginning, the HEP community (or rather the experimentalists) therefore consisted of two groups, the outside users (i.e. the university physicists) and the inside users (the staff members of the accelerator centres).

In 1962, out of the 685 HEP post-doctoral researchers in Europe, 550 worked for existing national projects (at the universities and old type research institutes) and 135 (=20%) for CERN.[38] In 1978 CERN and DESY together represented 15% of the total manpower in experimental (229 out of 1517) and 8% in theoretical high-energy physics (59 out of 740). In the U.S.A. the figures for 1975 show 'that about 25 percent of the HEP manpower is employed at the laboratory centres'.[39]

Collaboration did not become more difficult with each new generation of accelerators. On the contrary, the larger the accelerator, said Val Fitch, the more accessible it is to outside users:

> It has been the experience of the Brookhaven Laboratory that the AGS accelerator is a far more suitable accelerator for handling groups of researchers from universities than is the Cosmotron. Because of the AGS design, coupled with the greater space that is naturally available around the machine, many more experiments can be handled simultaneously, as well as many more secondary beams from the same target, and many more targets. The cost of each experiment is substantially less than that of the Cosmotron.[40]

As mentioned above, the admission of users to laboratories had already raised problems 100 years ago. Now 100 times more people took part in the enterprise:

> High-energy physics is a complex and involved field in which successful advances require a close partnership among experimentalists, theorists, accelerator physicists, engineers, data analysts, computer programmers, and highly skilled specialists and technicians. The research requires not only large accelerators with energetic and intense particle beams but also well-equipped laboratories with large staffs of technical specialists.
> The accelerator laboratory staff is responsible for the efficient operation, maintenance, and improvement of the accelerator; the organization, planning, and management of experimental areas with their shielded secondary particle beams; and the development, construction, and operation of large experimental pieces of apparatus such as bubble chambers, spark chambers, beam transport equipment, and data analysis facilities. Furthermore, the local staff plays a particularly important role in the study of new principles, the development of new ideas and the invention of new devices, both in the areas of accelerator design and improvement, and in particle detection and analysis.
> There is great diversity in the types of experiment undertaken in high-energy physics and in the methods employed in carrying them out. Some experiments are performed entirely by members of the local research staff. However, a large portion of the facilities of the accelerator laboratory are provided to accommodate broadly based university participation in elementary-particle physics research. The initial planning for university user experiments, the preparation of specialized equipment for a particular experiment, and the final analysis of the data are usually carried out by the user group at its home institution.[41]

This statement of the Atomic Energy Commission of early 1965 'is to be understood', Lew Kowarski commented, 'as an endeavor to see things as they are, rather than an expression of somebody's ideal'.[42] Kowarski had been requested by the CERN directorate 'to take a look at the present state of the accelerator/user relations in the U.S.A.', and we follow his description.

The accelerator side is a clearly structured organization, and therefore forums had to be developed, so that people on 'the other side of the fence' (as Kowarski put it) could also articulate their wishes. This was done by the so-called 'Users' Group', to which membership was in some cases rather precisely defined: 'Experimental physicists holding active guest appointments at BNL and having tenure appointments in their universities', and 'BNL scientists of tenure rank who are active in experimental h.e. physics'.[43]

To enable members of the Users' Group to express their views there is at every central HEP laboratory in the U.S.A. a kind of a general assembly, usually held two or three times a year and lasting one or two days. An attendance of about 70 is typical, but figures may go up according to mood and topics of the agenda:

> A typical agenda contains reports from the leading officials (with an informal discussion from the floor); reports from the standing committees of any, and technical papers of more or less general interest. Progress reports are given concerning the construction and operation of facilities offered for common use, and an exchange of users' opinions is invited (this is supposed to be one of the main purposes of the Meetings). Questions of machine scheduling and of specific priorities are rigorously excluded. Keeping and circulation of the minutes is either required (BNL) or considered desirable (Argonne).[44]

3.2.2 COMMUNICATION

In previous sections we have given some figures about manpower in HEP which indicate tremendous expansion. Expansion always creates all sorts of problems as explained generally (maybe more for amusement than for instruction) by Ortega y Gasset in his 'Revolt of the masses'.

One of the most serious problems, as far as science is concerned, is communication.

The 'second Malthusian dilemma'

As the number of scientists working on a special field increases, so does the quantity of information which a single scholar has to absorb and work with. If n stands for the number of members of a scientific community, the number of contacts (where everyone exchanges ideas with everyone) is, in principle, $(1/2)n(n-1)$. The increase of information therefore goes proportionally with n^2, i.e. geometrical progression. Alvin M. Weinberg spoke of the 'second Malthusian dilemma'.[45]

Eugene P. Wigner was one of the first to ask how the n^2-law would affect science.[46] In retrospect we can easily recognize some effects.

The increasing difficulty of communication clearly enforces specialization or, in other words, the establishment and finally the separation of subgroups. Thus the community of atomic scientists split up into elementary-particle, nuclear and solid-state physicists in the late forties and early fifties. The HEP community, which was then formed, already revealed a tendency to specialization by the mid-sixties.

Another effect is the constant feeling of probably a majority of physicists that 'we should know more about what these people at X and Y are doing'. This feeling undoubtedly stimulates the organization of all sorts of meetings and, in many cases, frequent travels to find out what is going on. Val Fitch spoke of the 'Bedouins' of HEP.

There is another important effect on the communication process itself. In the small scientific community that established quantum theory, the exchange of letters played an important role in the communication process.[47] Leading scientists in particular, such as Niels Bohr, Werner Heisenberg, Wolfgang Pauli, Arnold Sommerfeld and others drew on letters to a very considerable extent for their information. 'You wretched eggheads. What with your chats and letter-writing, you know what's wrong long before it's published', as Paul Ehrenfest once grumbled jokingly to Wolfgang Pauli.[48]

Where correspondence is concerned the $n(n-1)$ configuration obviously applies. Clearly, the more scientists there are, the less adequate this form of communication within the scientific community becomes. As A.J. Meadows says, in terms of information transfer private correspondence is much less important nowadays.[49]

As the number of scientists increases, other forms necessarily gain importance over letters. This is particularly true in the case of periodicals, which, as we know, started to appear in the last third of the 17th century as a substitute for correspondence between scientists. Nor should the importance of preprints and reports in HEP be underestimated.

When measuring the relative importance of the various types of literature, social scientists make quotation counts. Thus, it has been found that, in the mid-sixties, scientists took 82% of the information relevant for their research from periodicals, 12% from books and only 6% from other publications (including reports).[50] We discuss in the following two sections the importance of periodicals and of reports and preprints.

Periodicals

The following table for 1969 puts physics periodicals into order according to how often they are quoted.[51]

Rank	Journal	Country	Citations	Rank 1954
1	Phys. Rev.	U.S.A.	1401	1
2	Phys. Rev. Lett.	U.S.A.	483	–
3	J. Appl. Physics	U.S.A.	412	6
4	J. Chem. Physics	U.S.A.	361	16
5	Nucl. Physics	Netherlands	294	–

Rather than discussing the existing periodicals, we will restrict ourselves here to considering The Physical Review and Physical Review Letters.

The pre-eminence of The Physical Review can be seen from the fact that authors of overview papers endeavour to deal with all the papers published in this periodical. Blatt and Weisskopf, for instance, in their foreword to 'Theoretical Nuclear Physics', June 1952, note that they did not look at all the literature systematically: 'This applies particularly to papers which appeared in periodicals other than The Physical Review'.[52]

In 1950, The Physical Review totalled 3 290 pages; by 1980 there were 29 544 pages.

To enable important news to be published quickly, Physical Review Letters was founded in 1958. This too has never stopped growing. In 1960 it comprised 1 257 pages, by 1980 it had reached 4 615.

A page count gives the following results:

The Physical Review and Physical Review Letters

Year	Phys. Rev. (pages)	Phys. Rev. Lett. (pages)	Total (pages)
1950	3.920		3.920
1955	7.558		7.558
1960	7.758	1.257	9.015
1965	14.600	2.184	16.784
1970	25.612	3.356	28.968
1975	28.021	3.525	31.546
1980	29.544	3.987	33.531
1983	33.927	4.615	38.542

Until 1960 the growth of Physical Review paralleled that of the membership of the American Physical Society (APS), doubling every ten years. From 1960 to 1965 Physical Review doubled its size. and between 1965 and 1975 it doubled again, an increase greater than that of APS members.

This indicates that the number of published pages goes up more steeply than the number of scholars. One must keep in mind, however, that a lot of physicists outside the APS community (especially the Europeans) also publish in Physical Review, and their contribution may distort the picture. To check the relation we have to take the overall production and compare it with the size of the community.

It is much easier to register the overall production of articles in HEP than to find out the overall production of printed journal pages. We are aware that there may also have been a general increase in the number of pages per article. In Physical Review Letters, despite repeated warnings that papers should be kept brief, in eight and a half years the average length of a paper grew from approximately 1.8 printed pages in Volume 1 (1958) to 2.9 pages in Volume 16 (1966).[53]

Physics Abstracts

	All fields	Elementary particles	Increase annually
1953	8829		
1958	9201	713	13% annually
1963	26000	1313	19% annually
1968	50477	3152	4.5% annually
1973	81352	3939	

Nuclear Science Abstracts

	All fields	High-energy physics	Increase annually
1967	47055	5.411	5.4%
1972	60848	7.044	

In the early sixties the HEP community in the U.S.A. increased by 5.4% annually and in Europe by 7.7%, so the worldwide increase is also around these figures. The figures above show that the number of articles published went up more steeply than the increase of the HEP community.

The editorials reveal despair at the inexorably rising tide. In 1965 the Physical Review was facing a crisis:

> It is rapidly growing so large as to jeopardize one of its major virtues—its wide availability in the laboratories and offices of research physicists as a working tool for immediate reference and detailed study.[54]

Beginning in January 1964, the Review was devided into two sections, Section A being devoted to the physics of atoms, molecules and condensed matter, Section B to the physics of nuclei and elementary particles.[55] In 1966 the two sections were each again divided into two, and Physical Review was then published in four parts entitled:

Notes: p. 89

A General physics
B Condensed matter
C Nuclear physics
D Particles and fields

Of course growth such as that shown in Table 3.4 is a problem not only for any one journal, but to some degree for nearly every scientific periodical. 'We receive over three times as many manuscripts and publish twice as many letters as we believe to be right',[56] said Samuel Goudsmit in an editorial in Physical Review Letters in 1961.

Table 3.4
Pages published in Physical Review according to field

	1960	1965	1970	1975	1980	1983
General physics			4427	4871	5112	6764
Condensed matter			9939	11232	12404	15312
General physics and condensed matter		7992	14366	16103	17516	22076
Nuclear physics			4626	4256	5379	5619
Particles and fields			6620	7663	6649	6232
Nuclear physics and particles and fields		6608	11246	11919	12028	11851
Total	7758	14600	25612	28022	29544	33927

Many scientists were driven (not to say obsessed) by the urge to ensure that their latest research was given top priority. Competition led to the most bizarre abuses. Goudsmit gave some examples:

> An author who gets an interesting 'Letter' published and now believes that all subsequent results of his work must be published in a series of 'Letters'; another author arriving at a later result in the same subject who demands the right to have his work also published as a 'Letter'; an author who uses the 'Letters' merely to announce a later paper and whose Letter is incomprehensible by itself; an author who submits a 'Letter' which is merely an amplification of a previously published meeting abstract; an author who submits many Letters hoping that statistics rather than quality will cause one to be accepted; an author who carries a chip on his shoulder and casts aspersions on the motives and integrity of a referee who gives an adverse report on his paper; an author who tries to sneak a Letter in to 'scoop' a competitor who has already submitted a full article; an author who fails to make clear in the introduction the scope and significance of his paper; an author who has so little regard for his paper that he doesn't check it for typographical errors and omissions; an author who pays no attention to Physical Review Letters style; etc., etc.[57]

With their sometimes pathological desire to get their work published, many authors disregarded all established standards:

> There appeared in Nature Physical Science an article of some importance which was described by the principal author as 'an extremely important breakthrough into the new field of ...'. With all despatch, the article was sent to a referee and published in the usual way. Only afterwards did it become known that the same article had been submitted to two other journals [...]. It seems that the article was rejected by one journal but published, in a format substantially the same as that which appeared in Nature, by the second.[58]

In order to beat real or supposed competitors, many scientists adopted the habit of sending material first to newspapers and magazines. Publication is much quicker in the mass media, because they don't send their articles beforehand to a referee.

Reports and preprints

We have already spoken about the relative importance of the various types of publication. In HEP, besides books and periodicals, 'other publications (including reports)' definitely play a greater role than the 6% which is typical of science as a whole.

In order to show the sort of literature involved, a description of what CERN produces in this context is needed. To date, some 10 000 documents have been produced. They fall into three categories: yellow reports, which bibliographically speaking are independent publications; internal papers, which are not intended for outside use; and preprints.

Let us consider the yellow reports first. Bibliographically they constitute a series of monographs. They are referred to by the name of the author(s), title and serial number, e.g. Yves Goldschmidt-Clermont, The analysis of nuclear particle tracks by digital computer. CERN 57-29. Each year numbering starts again. In 1955, the first year, 31 monographs appeared; in 1956, again 31; in 1957, 56 and so on.

However, these yellow reports, which, as we have said, are bibliographically speaking independent publications, are only a small proportion of the papers produced at CERN. The large majority are internal papers or preprints, although it is not always clear to which of the two categories papers should be assigned.

The CERN Archives have drawn up an (incomplete) list of all the papers produced by CERN, recording 123 series of documents issued within CERN by all its committees, divisions, groups, etc. The first four series are quite clearly internal papers. Their references are CERN/..., CERN/CC/..., CERN/FC/..., and CERN/SPC/..., (the dots = the number within the series) and are issued by the Council, Committee of Council, Finance Committee and the Scientific Policy Committee respectively. Taking these four series together, a total of 7 332 documents have been issued from the establishment of the Organization up to the present

Notes: p. 89

(February 1986): 4015 by the Finance Committee and 797 by the Scientific Policy Committee.

Other series include 'divisional reports' (or 'internal reports'). They are used to inform about results or to propose solutions to problems, whether of a scientific, technical or administrative nature. One example is the reports issued by the Accelerator Division under the reference CERN/AR/..., of which a total of 274 appeared between 1960 and 1966. (In 1960 what was to become the Accelerator Division was a group in the PS Division. The papers issued in 1960 bear the reference PS/AR/...).

Another example is the approximately 250 reports issued by the Nuclear Physics Apparatus Division between 1961 and 1969. These include the 1961 'Report about the breakage of the main glass window in the 500 liter heavy liquid bubble chamber' under reference NPA/Int. 61-1 and 'The work of the N.P.A. division', reference NPA/Int. 61-2. The former is marked: 'Distribution: (open)', the latter 'Distribution: (closed). Copy for each member of scientific staff of NPA + 20 copies'.

Thus some of these divisional reports are available to outsiders, like the preprints, whereas others are treated like the numerous internal papers indispensable for the functioning of a big international organization.

We come to the third category of CERN literature, the preprints. The typical preprint is a scientific paper destined for publication, either in its present form or after some reworking, and usually in a periodical. The preprint's historical antecedent is the correspondence between scientists.

When quantum theory was being worked out in the twenties, scientific correspondence still had an important part to play. Letters from Bohr, Pauli, Heisenberg, Sommerfeld and many others were decisive to exchanges within the scientific community. Although sent to a particular person, they were often intended to be read by a wider circle (e.g. those to Bohr and his colleagues in Copenhagen). They were passed around at the Institute.[59]

As the number of scientists increased, the system ceased to work (cf. sect. 3.1). Letters were increasingly replaced by typed manuscripts, which generally had the character of a draft for an article. These preprints were sent out to a dozen people or more, the number of addressees increasing considerably with time. Just as, in the past, only those who were part of the network of letter exchange could contribute to scientific progress, now it has become vital to obtain preprints written by the leading figures in the field. The number of copies produced is nowadays considerably high, in the case of the CERN theory division 900.

According to A.J. Meadows, it has frequently been proposed 'that the distribution of preprints should be regularized':

> An example of such a proposal is provided by the discussions in the 1960s concerning a Physics Information Exchange for research workers in theoretical

high-energy physics [...]. It was suggested that a new exchange system should be set up so that accredited researchers throughout the world would have access to all important preprints. A central preprint office would be established together with local preprint libraries.[60]

The proposed system never got under way. Instead, the libraries of the large research centres also try to collect preprints, at least those written by staff members.

As a general rule, the corresponding periodical article is quoted rather than the preprint itself. What if a preprint is not to appear or has not yet appeared as an article? If the article is expected soon, then it is quoted by its title with the note 'to be published'. However, if the paper is rejected or withdrawn—this does not happen all that often—then it is quoted as 'G. Zweig, CERN Preprints TH 401 and TH 402 (unpublished)'.

3.2.3 CONFERENCES

Comparing HEP with other fields of scientific activity, we find a particularly large number of meetings of all sorts, from huge and well-planned international conferences covering more or less the whole field to ad hoc workshops on special questions with the participation of a few people from a handful of institutions. It would be difficult or even impossible to find a topic of some relevance which had not been thoroughly discussed at such occasions.

We shall concentrate here on the big, institutionalized congresses, beginning with the most general ones and then passing on to conferences on special topics and limited in their audience.

Conferences covering the whole field

High-Energy Physics originally developed in the United States, and in the postwar years most of the congresses dealing with HEP were held there. The first meetings devoted exclusively to this field, the famous Rochester conferences, changed their character from informal gatherings of a small group of friends to big, institutionalized, international conferences as HEP expanded.[61]

This development reflects 'the increase in world activity and interest in particle physics'.[62] When writing these words for the Proceedings of the 1966 International Conference on High-Energy Physics, McMillan thought that the Rochester conferences were initiated in 1952. Apparently he was not aware that it was generally accepted to count in retrospect as the 'First Annual Conference on High-Energy Nuclear Physics', a small meeting at the University of Rochester on 16 December 1950, with a total attendence of about 50 people, most of them from Rochester, with a few visitors from neighboring universities such as Cornell and Columbia.

In 1952 there were two conferences at the University of Rochester, one on 11–12

Notes: p. 89

January, and another from 18 to 20 December. The January meeting, called the 'Rochester Conference on Meson Physics', was 'modest in scope, of short duration, supported locally and still not truly international'.[63] Robert E. Marshak, the father of the Rochester conferences, gave this description 20 years later. 'Still not truly international' was somewhat euphemistic. The proceedings show that only 23 members of the local faculty, 15 local graduate students and 52 'out-of-town guests' participated, all of them from the United States except Rudolf Peierls.[64]

The December 1952 meeting was held under the title 'Third Annual Conference on High-Energy Nuclear Physics'. Among the 138 participants, Edoardo Amaldi, Cornelis J. Bakker, Louis Leprince-Ringuet, Donald H. Perkins and L. Riddiford were officially listed as coming from Europe.[65]

Marshak, the conference chairman, undoubtedly himself responsible for the 'pretentious title', was obviously convinced that from then on the conference could be held annually. The National Science Foundation, which part-sponsored the December meeting, had taken a favorite position towards future financial support.

Whereas the topic of these conferences was always the same, elementary-particle physics, the title shifted from Meson Physics[66] to High-Energy Nuclear Physics and (from the VIIIth Conference 1958 onwards) to High-Energy Physics, thus documenting the early stage of a field in which an accepted nomenclature does not exist.

Aware of the stimulating effect of an exchange of results and ideas, the Europeans also organized meetings when the field developed. An 'International Nuclear Physics Conference' was held at Harwell in September 1950. Of the ten main sessions, four were devoted to high-energy physics.[67] In June 1952, as already mentioned in the introduction, the provisional CERN Council sponsored the 'International Physics Conference' in Copenhagen. The topics were what we would now call nuclear physics and elementary-particle physics. In July 1953 there was the 'International Conference on Cosmic Rays' in Bagnères-de-Bigorre, remembered as 'one of the most remarkable of the century'.[68] A year later, in July 1954, the 'Conference on Nuclear and Meson Physics' was held in Glasgow, and in June 1955 there was the 'Conferenza Internationale sulle particelle elementari' in Pisa.

These were only the major conferences in Europe; there had been many other meetings, colloquia and summer courses. The Soviet Union and the socialist countries also had their big scientific congresses, garnished with smaller meetings.

Embarras de richesse: the year 1956 started with the 'Sixth Annual Rochester Conference' (3–7 April), followed by the 'Conference of Physics in High-Energy Particles' (14–20 May) in Moscow, and by the 'CERN Symposium on High-Energy Accelerators and Pion Physics' (11–23 June) in Geneva.

The physicists, who believed so deeply in the fundamental order of nature, felt that they should bring somewhat more order into the scheme of their conferences. At the IUPAP General Assembly in September 1957, Marshak proposed international co-ordination and, for this purpose, the creation of the High-Energy Physics

Social aspects

Commission. The first commission consisted of two Americans (Marshak and Panofsky), two Europeans (Bakker and Peierls) and two Soviets (Tamm and Veksler). Bakker, the Director General of CERN, acted as chairman and Marshak as secretary:

> At our first meeting, it was decided to have an automatic three-way rotation of the 'Rochester' conference starting with CERN in 1958, Kiev in 1959 and back to Rochester in 1960. The 1958 conference in Geneva became the Eighth Annual International Conference on High-Energy Physics, taking on the sequential ordering of the first seven Rochester conferences.[69]

The 'Tenth Annual International Conference on High-Energy Physics' (25 August–1 September 1960) was the last 'Rochester Conference' held in Rochester, and the last 'annual' conference as it was decided to go over to biennial conferences. So the 'Eleventh International Conference on High-Energy Physics' was held at CERN in 1962, the Twelfth at Dubna in 1964 and the Thirteenth at Berkeley in 1966.

In 1966, altogether 480 'invited delegates and observers' were listed and more than 500 papers, 'were received from workers in the field of high-energy physics all over the world.' Edwin M. McMillan spoke, as already quoted, about the growth of these conferences in size and scope, 'reflecting the increase in world activity and interest in particle physics' during the previous 15 years.[70]

International Conferences on High Energy Physics

(titles according to official proceedings)

1	First Annual Conference on High Energy Nuclear Physics* 16 December 1950	Rochester
2	Rochester Conference on Meson Physics 11–12 January 1952	Rochester
3	Third Annual Conference on High Energy Nuclear Physics 18–20 December 1952	Rochester
4	Annual Conference on High Energy Nuclear Physics 25–27 January 1954	Rochester
5	Fifth Rochester Conference on High Energy Nuclear Physics 31 January–2 February 1955	Rochester
6	Sixth Annual Rochester Conference on High Energy Nuclear Physics 3–7 April 1956	Rochester
7	Seventh Annual Rochester Conference on High Energy Nuclear Physics 15–19 April 1957	Rochester

8	1958 Annual International Conference on High Energy Physics 30 June–5 July 1958	Geneva
9	Ninth International Annual Conference on High Energy Physics 15–20 July 1959	Kiev
10	1960 Annual International Conference on High Energy Physics 25 August–1 September 1960	Rochester
11	International Conference on High Energy Physics** 4–11 July 1962	Geneva
12	5–15 August 1964	Dubna, USSR
13	31 August–7 September 1966	Berkeley, CA
14	28 August–5 September 1968	Vienna
15	26 August–4 September 1970	Kiev
16	6–13 September 1972	Chicago and Batavia
17	1–10 July 1974	London
18	15–21 July 1976	Tbilisi, USSR
19	23–30 August 1978	Tokyo
20	17–23 July 1980	Madison, WI
21	26–31 July 1982	Paris
22	19–25 July 1984	Leipzig
23	16–23 July 1986	Berkeley, CA
24	4–10 August 1988	Munich

* No proceedings. Title given in retrospect.
** All following conferences under this title.

When the Rochester conferences changed from an annual to a biennial sequence, it became possible to have another biennial meeting in the years between. The new series was called 'International Conference on Elementary Particles' and started 1961 in Aix-en-Provence, went to Sienna in 1963, to Oxford in 1965 and to Heidelberg in 1967. The idea was to give 'in particular young physicists the opportunity to attend and present their scientific contributions at a major international meeting'.[71]

'These biennial conferences are now well established', stated the chairman of the 1983 meeting. 'Although European in organization and location, the conference is as broadly international as is our subject'.[72] To give one example: the number of participants at the Heidelberg conference 1967 was about 700, among them about

Social aspects

500 from CERN Member States, 78 from the U.S.A. and Canada, 27 from the U.S.S.R., 22 from other European countries and 22 from the rest of the world.

After the creation of the European Physical Society (EPS) in 1968 it became a regular sponsor, thus helping to ensure continuity, and it became customary to call the series the 'Europhysics Conferences on High-Energy Physics'.

International Conferences on Elementary Particles

1	14–20 September 1961	Aix-en-Provence
2	30 September–5 October 1963	Sienna
3	19–25 September 1965	Oxford
4	20–27 September 1967	Heidelberg
5	25 June–1 July 1969	Lund
6	30 June–6 July 1971	Amsterdam
7	6–12 September 1973	Aix-en-Provence
8	23–28 June 1975	Palermo
9	4–9 July 1977	Budapest
10	27 June–4 July 1979	Geneva
11	9–15 July 1981	Lisbon
12	20–27 July 1983	Brighton

Besides these regular conferences, there were always other meetings. We have already mentioned a few for the years 1950–55. Other examples of important international conferences on HEP held in Europe include the 'International Conference on Mesons and Recently Discovered Particles',[73] which took place in Padua and Venice 22–28 September 1957, and the Twelfth Solvay Conference on Physics, held in Brussels, 9–14 October 1961.[74]

Topical Conferences

One of the most important features of science is specialization. When high-energy physics became established as a branch of physics, the process went on. Many physicists of the HEP community concentrated their work, at least for some years, on special topics, and correspondingly an interest for specialized conferences developed. One example of a topical conference is the Biennial Symposium on Electron-Photon Interactions, which was inaugurated in February 1963 when the 6 GeV Cambridge Electron Accelerator (CEA) had just come into operation and the 6 GeV DESY Synchrotron was nearing completion.

International Biennial Symposia on Electron and Photon Interactions at High Energies[*]

1	26–30 January 1963	Cambridge
2	8–12 June 1965	Hamburg
3	5–9 September 1967	Stanford
4	14–20 September 1969	Liverpool
5	23–27 August 1971	Ithaca
6	27–31 August 1973	Bonn
7	21–27 August 1975	Stanford
8	25–29 August 1977	Hamburg
9	23–29 August 1979	Batavia

[*] From 1975 onwards called the International Symposia on Lepton and Photon Interactions at High Energies

Another example of a topical conference is the 'International Symposium on Multiparticle Hadrodynamics' which was held annually in the seventies. In the preface of the proceedings of the Fourth Symposium, held in Pavia in 1973, we read about what we regard as the 'spirit of HEP':

> Some 125 physicists from 17 countries met and discuss[ed] the main experimental and theoretical problems in high multiplicity hadronic interactions. The importance of this topic increased very rapidly during the past years. (A long trip has been done since 1970.) From the presentation and the first understanding of the general gross features of multiparticle reactions, through a personal confrontation of different ideas, one has now reached the stage of 'studying details'. The few partisans who met first in Paris (1970), and grew in Helsinki (1971) and Zakopane (1972), are now almost an army.[75]

Just substitute 'multiparticle hadronic interaction' by some other topic, and Paris (1970), Helsinki (1971) and Zakopane (1972) by other places and years, and you have the physicists' description of any other series of topical conferences.

Accelerator conferences

All the conferences that have been mentioned so far were (and are) devoted to theory and experiment in HEP. Accelerator design is another side of the field, and accordingly there was also an intense exchange of ideas.

From 11 to 23 June 1956 a CERN Symposium was held at the Geneva University Institut de Physique, since the main building and conference halls at CERN were still under construction. The first week was devoted to accelerators.[76] 'The philosophy behind this was', Bakker explained, 'to provide an opportunity for those more particularly interested in the technical side.' Accelerator physicists and engineers did not feel less inspired than the elementary particle physicists. 'If any conclusion can be drawn from the conference', John Adams said, 'it must be that accelerating machine projects are in no way dying from lack of nourishment. More money seems to be available to build machines nowadays than ever before'.

This CERN Symposium was later counted as the first of a series of 'International Conferences on High-Energy Accelerators'. The second was held from 14 to 19 September 1959, this time at CERN, with 245 physicists and engineers from 28 countries. The topic was high-energy accelerators and instrumentation.[77] The first session was devoted to the need for new particle accelerators. 'Although some of the traditional fields ... may be less productive as the next range in energy and intensity is explored, there are very strong reasons', Wolfgang Panofsky believed, 'that other, less explored areas will be the source of important results'.[78]

From this 1959 conference onwards, 'high-energy accelerators' and 'instrumentation' each got its own series of international conferences. Both series were co-sponsored by IUPAP, and the same principle was adopted as with the Rochester conferences, i.e. to hold them in rotation in the Soviet Union, the United States and Western Europe.

International Conferences on High-Energy Accelerators

1	CERN Symposium on High-Energy Accelerators and Pion Physics	
	11–23 June 1956	Geneva
2	International Conference on High-Energy Accelerators and Instrumentation	
	11–19 September 1959	Geneva
3	International Conference on High-Energy Accelerators*	
	6–12 September 1961	Brookhaven
4	21–27 August 1963	Dubna
5	9–16 September 1965	Frascati
6	11–15 September 1967	Cambridge, Mass.
7	27 August–2 September 1969	Yerevan-Tsahkadzor
8	20–24 September 1971	Geneva
9	2–7 May 1974	Stanford
10	11–17 July 1977	Serpukhov
11	7–11 July 1980	Geneva
12	11–16 August 1983	Fermilab

* All following conferences under this title.

Notes: p. 89

International Conferences on Instrumentation for HEP

1	12–14 September 1960	Berkeley
2	16–18 July 1962	Geneva
3	9–10 September 1966	Stanford
4	8–12 September 1970	Dubna
5	8–12 May 1973	Frascati

Both series, the accelerator as well as the instrumentation conferences, were supplemented by regional and national meetings. Here we shall consider only the accelerator conferences.

In 1965 the Nuclear Science Group of the Institute of Electrical and Electronic Engineers started a series of 'National Particle Accelerator Conferences' which were held every second year. At the first meeting 760 participants were registered, at the second 812, with roughly ten percent from other countries. In the Soviet Union a similar biennial 'All-Union Conference on Charged Particle Accelerators' was organized starting in 1968.

We may mention, as a special type of regular meeting in the field of accelerator development, what are called study groups. A study group is given a special task, and to this extent it is similar to a subdivision or group of an institution like CERN. Generally, however, members of a study group (or working party) belong to different institutions and limited time is available for the work. In order to exchange ideas, the members hold regular meetings, usually closed.

One example is the study group which met at Brookhaven National Laboratory in August 1961 to consider the design and use of accelerators with energies from 300 to 1000 GeV.[79] In 1961, also, a 'European Accelerator Group' (or 'European Accelerator Study Group') was set up and had its first meeting on 30–31 May, 30 participants being invited. The idea was that 'in planning and designing the next generation of accelerators a reasonable amount of co-ordination' was needed 'to avoid too much overlapping by allowing the new equipment to become complementary rather than competitive'.[80] When in January 1963 the Director-General of CERN and the chairman of the Science Policy Committee convened a meeting of leading high-energy physicists from CERN member states, they constituted the European Committee on Future Accelerators (ECFA).[81] ECFA was to play a decisive role in the development of a HEP policy for Europe.

We have spoken of the international accelerator conferences and their complementary meetings. Another form of complementarity is the conferences on related fields and those more specialized in character. An example of a conference on a related field is one held in Cambridge, Mass., from 14 to 16 October 1958. It was the begin of a series of international conferences on low-energy accelerators, the second of which was held in Amsterdam, 4–6 October 1960.

Social aspects

An example of more specialized meetings is the series of Linear Accelerator Conferences, mostly devoted to proton linacs, which started in April 1961 in Upton N.Y. Another example is the 'International Conference on the Application of Accelerators in Research and Industry', started in Oak Ridge, Tennessee, in 1968, and currently (1986) held biennially in even years, usually in early November.[82] A third example is the International Cyclotron Conference.

International Cyclotron Conferences
(title according to official proceedings

1	Informal Conference on Sector-Focused Cyclotron	
	2–4 February 1959	Sea Island, Georgia
2	International Conference on Sector-Focused Cyclotron	
	17–20 April 1962	Los Angeles, Cal.
3	International Conference on Sector-Focused Cyclotrons and Meson Factories	
	23–25 April 1963	Geneva
4	International Conference on Isochronous Cyclotrons	
	2–5 May 1966	Gatlinburg, Tenn.
5	Fifth International Cyclotron Conference	
	17–20 September 1969	Oxford
6	Sixth International Cyclotron Conference	
	18–21 July 1972	Vancouver
7	Seventh International Conference on Cyclotrons and their Applications*	
	19–22 August 1975	Zurich
8	18–21 September 1978	Bloomington, Ind.
9	7–10 September 1981	Caen
10	30 April–3 May 1984	East Lansing, Mich.

* All further conferences under this title.

3.2.4 SUMMER SCHOOLS

Part of the ensemble of conferences and meetings described in the previous section are the summer schools. They generally deal with special topics and may therefore be listed among the topical conferences. But this is only one aspect of them. The unique value of a summer school lies in the fact that it is especially meant for beginners, forming an important part of their scientific education.

At the turn of the century a student could learn his physics at a single university. Since then physics has developed into a number of branches, each a profession in itself. The courses offered by a physics department are generally not sufficient for the education of a solid-state physicist or a high-energy physicist, who urgently needs further instruction in a great number of topics. This gap is filled by special courses offered by summer schools.

The final decision of a young man to devote his life to physics is often strongly influenced by attendance at such a school. The basis of a long scientific co-operation may be laid by the contacts made there, and even the initial ideas for new experiments and theories may be conceived on such an occasion.

The first in Europe was the 'Summer School of Theoretical Physics' of the University of Grenoble in Les Houches at the foot of Mont Blanc. This school came into existence in close context with the foundation of CERN.

Early in 1950 a young French physicist, Mlle Cécile Morette, had the idea of creating a summer school of theoretical physics in France. She herself had learnt meson physics not during her studies in Caen and Paris, but only when she worked first as a graduate student, then as postdoctoral fellow, with Walter Heitler in Dublin, Christian Møller in Copenhagen and J. Robert Oppenheimer at Princeton.[83] She was convinced that new methods must be developed to implement modern theoretical physics in France. Since she had been particularly inspired by the courses of Richard Feynman during a Michigan Summer Symposium in Ann Arbor, the value of such an institution was obvious to her.

In the autumn of 1950 she met Pierre Auger, who advised her to present her project to the Commission of Scientific Co-operation of the Centre Européen de la Culture. At the meeting on 12 December 1950 which played such an important role in the history of CERN,[84] Mlle Morette explained her 'Projet de création d'une école d'été'.[85]

Some 20 of the most brilliant young physicists should be made familiar with new developments in high-energy physics. Two-month courses in quantum mechanics and quantum field theory were envisaged. They should be given by four eminent scholars supported by some visiting professors.[86] Even the place had been found: Les Houches near Chamonix in Haute Savoie.[87] The idea was to avoid distractions but to offer possibilities of recreation.

The project was well received. The Commission recommended 'la création dès à présent d'un Centre de formation de physiciens théoriciens, destinée à constituer la Section théoretique indispensable au Laboratoire' and asked Raoul Dautry and Pierre Auger to raise the necessary funds.

The Les Houches summer school was however established separately from the European Laboratory project. Financial support for the summer school was granted by the Higher Education Department of the French Ministry of Education and the money was given via the University of Grenoble.

The first course was held in the summer of 1951 and was devoted mainly to quantum mechanics and quantum field theory. Thirty students attended, 13 from France and 12 from other European countries. Among the lecturers were Léon van Hove, Res Jost, Wolfgang Pauli, Emilio Segré and Victor F. Weisskopf.[88] In 1952 the topics were quantum mechanics, statistical mechanics and nuclear physics; in 1953 quantum mechanics, solid-state physics, statistical mechanics and elementary-particle physics.

During the first few years the lecture notes were reproduced only for the participants and some other students, but later about 100 copies were made and distributed to various laboratories and institutions. Since 1958 they have been published in a series of monographs.

After the creation of a new graduate programme for higher education in France, the 'Troisième cycle', the school had become something like a 'école de quatrième cycle'.[89] A determining factor in the success of the school was, according to a later summary, 'the assembly of students and teachers in such an isolated environment':

> A list of previous students reveals the scientific quality and originality of the atmosphere: a good number of them became renowned researchers. More unexpectedly, certain professors returned as students in later sessions.[90]

In 1953 the Italian Physical Society created an International School for Physics, and since then courses have been held regularly in the Villa Monastero at Varenna, situated beside Lake Como in wonderful countryside. The first course took place from 18 August to 12 September 1953 and was devoted primarily to experimental methods in particle physics. In the inaugural session Madame Cécile de Witt-Morette brought 'les vœux de vos camarades de l'Ecole d'été de physique théoretique de l'Université de Grenoble'.[91]

From 18 July to 7 August 1954 the second course was held with the participation of Edoardo Amaldi, Gilberto Bernardini, Enrico Fermi, Werner Heisenberg, Bruno Rossi and other outstanding physicists. When the papers were prepared for publication a few months later (as the proceedings of the first course in the 'Supplemento al Nuovo Cimento') the world of physics was shocked by the untimely death of Enrico Fermi on 28 November 1954. B.T. Feld edited the 'Lectures of Enrico Fermi on Pions and Nucleons'. Working with tape recordings, he evoked a lively picture of Fermi 'as he lectured during those lovely mornings in the beautiful setting of the Villa Monastero on Lake Como.'[92] Later on the summer school in Varenna was officially named 'Scuola Internazione di Fisica «Enrico Fermi»'.

In the following years the courses dealt with a wide range of topics. For example the eighth with 'Mathematical Problems of the Quantum Theory of Particles and Fields' (21 July to 9 August 1958), the ninth with 'Physics of the Pion' (18 to 30 August 1958), the tenth with 'Thermodynamics of Irreversible Processes' (15 to 27 June 1959) and the eleventh with 'Weak Interactions' (29 June to 11 July 1959).

The number of schools increased considerably during the fifties. 'It is difficult to count the summer schools', Cécile De Witt wrote in 1959, 'the organization being fundamentally different':

Notes: pp. 89 ff.

On peut citer, entre autres, celles d'Italie, de Yougoslavie, du Japon, de Suède, des Etats-Unies, de Grèce, une école 'ambulante' consacrée à l'état solide (Paris 1958; Cambridge 1959), une école en Amérique latine (Mexique 1959; Brazil 1960). Le Comité scientifique de l'OTAN a prévu depuis 1959 un programme permettant chaque année l'organisation d'un certain nombre d'écoles d'été (5 écoles subventionnées en 1959; 14 en 1960).[93]

Here are a few examples from among this plethora.

In 1958 the Air Force Office of Scientific Research (ARDC) and the University of Colorado jointly sponsored the Summer Institute for Theoretical Physics in Boulder, Colorado. The function of the Institute was 'to bring together scientists and students with the view of providing a means for exchange of ideas and information'.[94] As always lectures, seminars and informal discussions made up the activities of the institute. Courses were generally held from mid-June to the end of August at the Department of Physics of the University of Colorado.

At the instigation of Antonino Zichichi, the 'International School of Physics «Ettore Majorana»' was founded in Erice (Sicily) in 1963, with the participation of 120 young physicists from 23 countries.[95]

Every year thereafter around 100 beginners came for two weeks to hear lectures by distinguished scientists. A few years later other schools started courses, and finally, under the directorship of A. Zichichi, the 'Ettore Majorana Centre for Scientific Culture' was founded. Its aim is 'to create in Europe a cultural forum of high scientific standard, which will allow experienced research workers to study and discuss with their more qualified colleagues the result of their research'.[96]

Today the centre consists of more than 80 schools, among them the 'International School of Subnuclear Physics' which held its 24th course from 7 to 15 August 1986; thus it had not missed a single year since its foundation.

In 1961 the theoretical physicist Paul Urban of Graz University had the idea (not astonishing for an Austrian) of organizing ski holidays with his collaborators.[97] One of his pupils proposed to combine this with physics, thus creating a winter school. On the Planer Alm that same winter the 20 participants discussed Non-Abelian gauge theory (the SU(2) Yang-Mills-theory). The following year many students attended. The topic was vector mesons and the place Schladming, where the Socialist Youth Hostel had a lecture hall holding 200–300 participants. In 1965 static SU(6) properties in the quark model were discussed. The meeting became famous: Keith Johnson and Gunnar Källen had a hard duel on the question of whether QED was a finite theory.[98]

Finally, the 'Internationale Universitätswochen für Kernphysik in Schladming', abbreviated IUKS, developed into a permanent institution of the University of Graz. The topics are always special high-energy problems. The name of the school (Kernphysik = nuclear physics) is reminiscent of its foundation: at that time particle physics was still part of nuclear physics. 'Since Austrians stick to tradition, we have kept the name'.

The courses are held every year for two weeks late in February or early March. After 20 years Urban asked his pupil and successor Heinrich Mitter to organize the school. In 1986 the first 25 years of the Schladming Winter School were celebrated, and in the Proceedings the editors and organizers Heimo Latal and Heinrich Mitter explained that they always had participants from the east, west, north and south, and that they successfully tried to bridge the political borders.[99]

Starting with the 'Easter School for Physicists using the Nuclear Emulsion Technique in Conjunction with the CERN Proton Synchrotron and Synchro-Cyclotron', held at St. Cergue, 8–12 April 1962, CERN has organized a series of Summer Schools dealing with various topics, predominantly theoretical, in HEP. Most specular was the creation of a school, organized by CERN and the Joint Institute for Nuclear Research (JINR), at Dubna. The first courses were held in Loma-Koli (North Karelia, Finland) from 21 June to 5 July 1970 and attended by 86 young physicists.[100]

3.3 Scientific developments

In the preceding section we have dealt extensively with the social aspects of HEP. We now have to give a review about the development of accelerators and detection devices. We will not speak, however, about experiments and their interpretation, nor about the development of concepts and theories. A detailed survey is scheduled for Volume III of this 'History of CERN'.

For all physicists engaged in the study of the fundamental constituents of matter and their interactions, the existence and availability of particle accelerators producing empirical data on a 'frontier of science' is a sine qua non. For this reason, we shall start with the development of accelerators.

3.3.1 ACCELERATORS

'Why does particle physics require high-energy accelerators?', Léon van Hove asked at the 1961 International Accelerator Conference, and gave the familiar answer: 'Most elementary particles are unstable and, to be studied at all, must be created in high-energy collisions; furthermore, the most interesting properties of these particles manifest themselves at short distances, i.e. in collisions with large momentum transfers'.[101]

The construction and operation of accelerators has become a major concern since the end of World War II. In the HEP community a considerable percentage of scientists devote most of their time to accelerator design; a figure given for early 1960 in the U.S.A. shows that, out of 718 physicists of PhD level, no less than 103 specialized in 'accelerator design and construction', having spent a total of 86 man years on these tasks in 1959.[102]

Notes: p. 90

In the scientific community, the 'excellent engineering work that had gone into the construction' of the big machines was often praised,[103] but nevertheless the performance of an experiment and its theoretical interpretation was generally held in higher esteem than the work done by accelerator physicists. In the GAC–PSAC Panel Report of 1963, the universities, the supporting government agencies and the community were asked to give more recognition to accelerator design and construction: 'The intellectual effort involved in solving a difficult problem in accelerator design, or in evolving the design of a new detection device, is often considerably greater than that required in the actual performance and analysis of an experiment'.[104]

The accelerator literature increased considerably in the fifties and sixties. Numerous articles and books described the various types of machine and their special components.[105] The papers in which the fundamental principles of particle acceleration were laid down by their inventors were already regarded as 'classical', and edited by M. Stanley Livingston in 1966.[106]

In the period under review, many different types of accelerators were used. Well over 50% of all machines applied the simplest and oldest method, direct-voltage acceleration. However, the picture changes totally if we concentrate on high-energy accelerators. In the following two tables we give a list for 1955 and another for 1965.

In 1957, out of all operating accelerators above 1 GeV, there were six synchrotrons (all weak focusing) and one linac. During the next decade (up to 1967) 21 new accelerators above 1 GeV came into operation, of which 12 were strong focusing synchrotrons, six other synchrotrons and three linacs. This marks the success of the alternating gradient focusing principle invented in 1952 by E.D. Courant, M. Stanley Livingston and H.S. Snyder.[107]

Accelerators with energies of 1 GeV and above operating in 1955
(machines near completion marked with an asterisk)

Location and name		Type	Particle	Energy (GeV)	Completion
BNL	Cosmotron	w.f.s.	p	3	1952
LRL	Bevatron	w.f.s.	p	6.2	1954
Stanford	Mark III	Linac	e	1.2	1950
CalTech		w.f.s.	e	1.5	1952
Cornell		w.f.s.	e	1.3	1955
Birmingham		w.f.s.	p	1	1953
			d	0.65	
Dubna*		w.f.s.	p	10	1957

w.f.s. stands for weak focusing Synchrotron.

Accelerators with energies above 1 GeV built in the decade 1955-65
(machines near completion marked with asterisk)

Location and name		Type	Particle	Energy	Completion
Princeton	PPA	w.f.s.	p	3	1963
Argonne	ZGS	w.f.s.	p	12.5	1963
BNL	AGS	s.f.s.	p	33	1960
Cambridge	CEA	s.f.s.	e	6	1962
Cornell*		s.f.s.	e	10	1968
Stanford	SLAC	Linac	e	20	1966
Saclay	Saturne	w.f.s.	p	3	1958
Harwell	Nimrod	w.f.s.	p	8	1963
CERN	PS	s.f.s.	p	28	1959
Frascati		w.f.s.	e	1.1	1959
Lund		s.f.s.	e	1.2	1962
Hamburg	DESY	s.f.s.	e	6.5	1964
Bonn		s.f.s.	e	2.3	1966
Daresbury	U.K.	s.f.s.	e	4.5	1965
Orsay		Linac	e	1.3	1961
Serpukhov*		s.f.s.	p	70	1967/68
Moscow		s.f.s.	p	7.3	1961
Yerevan		s.f.s.	e	6	1965
Kharkov		Linac	e	2	1965
Tokyo	INS	s.f.s.	e	1.3	1961
Canberra	ANU	w.f.s.	p	10.6	1962/63

w.f.s.: weak focusing Synchrotron
s.f.s.: strong focusing Synchrotron.

Up to 1960 physicists generally believed that it was necessary to choose between intensity and energy. The new AG focusing synchrotrons however, with the CERN PS as the first example, produced remarkably high intensities at high energies. In a review article written in 1967, M. Hildred Blewett praised the 'outstanding success of these machines in producing intensities more than an order of magnitude above what had been hoped for, their reliability of operation, and the ease with which many types of beams can be provided for experiments'.[108]

In the early sixties physicists wished to reach far higher energies and the feasibility of a new generation of machines was discussed. It was realized that radically new technologies and different basic principles were not yet promising, but that the AG focusing concept could be extended into the 10^2 GeV region,[109] whereas the highest energy ever reached by a weak-focusing synchrotron was the 12.5 GeV of the Argonne ZGS.

Apart from the AG synchrotron, the only alternative was the linear accelerator, and only for electrons. The advantage of a linac is, first, that the particle beam can easily be extracted at its high-energy end; and second, that its energy can be increased by adding extra sections or more power sources.

In accelerator development the main desire was always to reach higher energies, of course with sufficient intensity. From the very beginning there was always an increase by a factor of ten about every six or seven years. This can easily be proved by listing the milestones of accelerator development:

1932	Cockcroft-Walton generator	710 keV	(p)
1939	Berkeley 60" Cyclotron	22 MeV	(d)
		44 MeV	(α)
1946	Berkeley 184" Cyclotron	190 MeV	(d)
		380 MeV	(α)
1952	Cosmotron	3 GeV	
1957	Dubna Synchro-Phasotron	10 GeV	
1959	CERN PS	28 GeV	
1967	Serpukhov	70 GeV	
1974	NAL Accelerator (Fermilab)	600 GeV	

This development from one generation of accelerators to the next was stimulated by the hope for important new discoveries in the new energy range and by national prestige. However, before dealing with these motivations, let us concentrate on another determinant: the manpower factor.

When a big accelerator is completed, accelerator physicists always feel a strong desire for another attractive project. The case of CERN provides an illustration of this phenomenon. In the Scientific Policy Committee in 1957 John Adams explained that his 'PS group was working on an extremely interesting object, but saw no future once the machine had been build'.[110] He asked for 'a new construction programme for CERN'.[111]

Indeed, a 'European high-energy accelerator programme' was developed up to 1963 and 'a pair of storage rings' as well as 'a new proton accelerator of a very high energy' proposed in the famous Amaldi Report.[112] The first studies for the intersecting storage rings were made in 1962, the decision to construct taken in 1965, and the first beam produced in 1970, whereas the decision for the 300 GeV accelerator was postponed until February 1971 and the SPS completed in 1976.

In the case of Brookhaven National Laboratory there was no gap between the completion of the Cosmotron and the start of the work on the AGS. The Cosmotron began operating in 1952 and the design of the AGS commenced one year later. The eight years between the completion of the Cosmotron (1952) and that of the AGS (July 1960) is roughly the shortest period between two generations of accelerators that one laboratory at the forefront of technological development could achieve.[113]

Apart from the manpower factor, the two other determinants of accelerator development were scientific interest and national prestige, both mixed, as always, with personal ambition.

The scientific interest was enormous. The HEP community was well aware, as Robert Serber put it, 'of the dependence of particle physics on the great accelerators, since the day when Gardner and Lattes observed the first scrawny machine-made meson'.[114]

There had been some people in the early fifties (Niels Bohr, for example), and quite a few in the early sixties (this time Heisenberg among them) who felt that it might not be worth spending a lot of money and manpower to attain higher energy levels. They thought that the 'asymptotic region', where no new effects could occur, was already reached. However, such people were definitely outsiders in the HEP community.

The abundance of new results that came in when the Cosmotron and the Bevatron started operation deeply impressed the vast majority of physicists.[115] The next generation of accelerators, the CERN PS and the Brookhaven AGS, again fulfilled expectations, not only in engineering but especially with respect to the experimental results. 'Every time in the past', Robert Serber said in 1961, 'when we have entered a new energy range we have struck gold'.[116]

When the Bevatron was conceived, the idea was to produce antiprotons. There were strong arguments that such a particle should exist, but, as Pauli put it, 'there is a big difference between believing and knowing'.[117]

The first design studies for the European 'big machine' envisaged a synchrotron with an energy of 10 GeV. The basic philosophy was that such a big accelerator in the vanguard of science would be so expensive that no European country could afford to build it on its own, and that the project 'virtually demanded European co-operation'.[118] When the AG principle was invented a 10 GeV machine was simply too cheap and the energy was raised to 25 GeV.

In the early sixties, the same situation existed in principle. Before the decision was taken to build a new generation of accelerators, quite a number of conferences and all sorts of meetings were held to specify the scientific need more explicitly. Nevertheless, expectations and the possible scientific programme remained rather vague. In this connection a short discussion that took place at the 1961 International Conference on High-Energy Accelerators is revealing. M. Hildred Blewett asked whether accelerator builders should think about '100 GeV, 500 GeV, 1,000 GeV or what?'

> Robert Serber: I think the question becomes an economic one rather than a physical one. We'd like to go up as far as is technically feasible and still not ask for something too unreasonable from the government.
> Léon van Hove: If I try to give an answer it will be even vaguer.[119]

Notes: p. 90

Coming now to national prestige, we will restrict ourselves to a few remarks. This topic is dealt with in more detail in section 3.4, which is about HEP and society.

Since a big accelerator is such an expensive and complex thing, and since these accelerators could be compared with each other even by the layman simply by using the yardstick of the energy of the particles produced, they became objects of prestige in their own right. Up to 1957, the region above 1 GeV was purely an American domain. In the early fifties a decision was taken in the Soviet Union to build the most powerful machine in the world, the weak-focusing 10 GeV Synchrotron at Dubna. In January 1956 it was decided that the energy of the ZGS to be built at Argonne would be 12.5 GeV 'to just beat the 10 GeV Dubna synchro-phasotron' in energy, and a crash programme was also envisaged in order 'to beat' the competitor to the post.[120]

3.3.2 PARTICLE DETECTION

Some devices for the detection and identification of charged particles, such as the Wilson cloud chamber, the Cerenkov and scintillation counters, invented decades ago, were still widely used in the fifties and sixties.

In 1946 Powell and his collaborators succeeded in producing a new photographic emulsion with 'greatly improved properties'. In their famous paper the Powell group predicted 'a widespread adoption for work in nuclear physics'.[121] This was indeed the case when Powell was able, in 1953, to pack together stripped emulsions 'to form, in effect, a solid, sensitive mass'.[122]

For the exploitation of an experiment the whole stack of emulsions had to be examined. Powell himself organized a collaboration between six European laboratories in the study of unstable particles, which became known as the 'G-Stack Collaboration'. The results were published in a number of papers with altogether 31 authors in Nuovo Cimento 4, Supp. 2 (1956).[123]

One of the last great experiments of this kind was the 'International Cooperative Emulsion Flight', abbreviated I.C.E.F., which took place in 1960, and was headed by Marcel Schein and later by Maratoshi Koshiba. For the exploitation 32 groups from 15 countries comprising 98 physicists collaborated.[124]

A new development began in 1952 with the invention of the bubble chamber. In his Nobel lecture Donald A. Glaser described how he systematically reviewed all possible amplifying mechanisms to achieve a sufficient sensitivity to the minute amounts of energy deposited by a fast charged particle: 'At the time that I was studying these instabilities, I knew that the large proton synchrotrons in the few GeV energy range would come into operation in the early 1950's and that they would have pulse repetition times of a few seconds. It was therefore important that the new detector be able to cycle in a few seconds'.[125]

All hopes for accelerator experiments were fulfilled, and large quantities of data on elementary particles and their properties were produced by bubble chambers. The result of an experiment is, as with emulsions, given in a great number of photographs.

Another important advance in particle detection occurring during the years under review was the electrically sensitized track detectors (flash, spark and streamer chambers). The principle of the spark chamber had been known since 1949, but it became a useful device only ten years later. Pioneer work was done by Marcello Conversi with Adriano Gozzini at Pisa,[126] and by Shuji Fukui with Sigenori Miyamoto at the newly founded Institute for Nuclear Study, University of Tokyo.[127]

The results of the experiments with spark chambers were again photographs in huge quantities.

Techniques had to be developed to gain the relevant information from these photographs. A decisive step was the digitalization and automatic handling of the data. Since the mid-fifties computing machines became available in the laboratories.

The exploitation of the bubble chamber was greatly improved by the use of computing machines. In 1964 the measuring devices on-line to computers reached a capacity of about 10000 events per week and a successful development to reach 100000 events per week was expected. The computer however, which developed the bubble chamber to its highest efficiency, finally caused its demise. The digitalization of the location of the spark is much easier to achieve than is the track of a charged particle in a bubble chamber. Therefore spark chambers are superior to bubble chambers. This was even more the case with the wire spark chamber. When its development started in 1964, it promised 'even more useful applications', because the information obtained 'is already digitized and can be easily fed into a digital computer'.[128]

3.4 HEP and society

One of the most characteristic features of modern society is the rapid growth of science. Whereas 150 years ago one professor worked with a small collection of instruments, we now generally have a large department and big laboratories.

Where, previously, there used to be a small department in a ministry responsible for universities, nowadays whole ministries and specialized agencies, such as the C.N.R.S. in France or the Z.W.O. in the Netherlands, are concerned with the promotion of research.

The costs of scientific research have been rising all the time and scientists have been continuously faced with the task of asking for more funds from the state. In so doing they have been forced not only to justify their particular project but to make statements about the value of their research for human society. Thus applications for

Notes: p. 91

government funding and the written expert opinions marshalled in support of them provide a very accurate picture of science's needs and its role in the state.

The arguments advanced in 1860 were still being used with very little modification in 1960. Though one new aspect has been added: the significance of research for arms and warfare. The important contributions of scientists and engineers, especially chemists, to the war effort 1914–18 and the more important contributions during World War II, with the Anglo-American physicists as the heros, considerably strenghtened the self-confidence of scientists and honed their arguments.

Below we give an overview of the arguments categorized according to science's intellectual and cultural, economic and military importance for state and society.

3.4.1 THE INTELLECTUAL AND CULTURAL IMPORTANCE

This may be divided into three aspects: creative cultural achievement, broad effects and national prestige (including intellectual influence on other countries).

Creative cultural achievements

As Francis Bacon put it, science is 'illuminating and fruitful'.[129] In the 17th and 18th centuries when the arbor scientiarum flowered but had not yet borne fruit, scientists emphasized the importance of science in the furtherance of knowledge. Leibniz said: 'Every truth, every experiment or theorem must be seen as a newly discovered reflection of God's glory, even though it does not immediately profit us but merely brings us light'.[130]

Since the middle of the 19th century, by which time science had produced important technical applications, scientists stressed the importance of science for trade and industry. As Justus von Liebig put it in 1840, while he was loath to judge the worth of a scientific descipline exclusively according to its usefulness, the government had a duty to make particular provision for those subjects, such as chemistry, which were of material benefit.[131]

In many areas of science, such as solid-state physics, it is still argued that the value of research is proved by its technical applications. In high-energy physics the arguments used in the 18th century have been revived (though without any awareness of their historical context). Eugene W. Wigner, for example, put it 1960 like this:

> We do not expect at present that high-energy physics will lead to 'practical results' in the near future. Its aim is a deeper understanding of nature, not the increase of our power to accomplish something tangible.[132]

Similarly, in the Report on Policy for National Action in the Field of High-Energy Physics from January 1965, it is stated: 'Despite the usefulness of the technological

Broad effects

Quantum theory—created in the twenties by a relatively small circle of physicists—has had an immensely profound and wide-ranging impact on the whole of our intellectual life. Following in the physicists' footsteps, chemists, biologists, philosophers, science historians and teachers have studied and continue to study quantum theory. Tens of thousands of books have been written about the subject.

It is a moot point whether present-day high-energy physics which, like quantum theory, devotes itself to the smallest (or at least what are currently thought to be the smallest) building blocks of matter can ever achieve again such importance for human thought. It is undoubtedly true, as the Amaldi Report said in 1963, that 'the subject of high-energy physics possesses very great intrinsic cultural value.'[134] There is a whole series of text books on the subject which the general public can understand, indicating that interest in HEP is alive or at least dormant.

Nonetheless, it may well be that the much more abstract and therefore more difficult concepts will not have any broad influence. It is also possible that educated people will turn away from elementary-particle physics, an idea canvassed by Jakob von Uexküll, the founder of the alternative Nobel Prize, since it is becoming 'an increasingly dubious science'.[135]

What is certain is that high-energy physics has a very important role to play in higher education. In most industrialized countries, what in Germany is known as the 'Humboldt principle of the unity between research and teaching' is now being practised. Universities take part in experiments at the big laboratories, so their physics departments learn about research results straightaway and this in turn forms the basis of scientific work of professors, postdoctoral fellows and those working towards their doctorates:

> Experiments are planned in advance at the university, equipment is set up and tested in local laboratories, and at the appropriate time the scientist with his equipment goes to the accelerator centre where the basic data are obtained. Later, analysis and interpretation of the data are performed at the university. Sixty to seventy percent of the research with the Alternating Gradient Synchrotron and the Cosmotron at Brookhaven National Laboratory is conducted by university users. Fifty to seventy percent of the Research of the Zero Gradient Synchrotron at Argonne is done by users and a similar situation is expected at the Stanford Linear Accelerator Center.[136]

National Prestige

Science is always both international and national. It is international in so far as its

Notes: p. 91

methods and results are accessible to everybody, independent of faith, race or language. It is often a great experience for a travelling scholar to meet 'on the other side of the globe, a man of very different cultural and ethnic associations, busy with the same problem and using the same technique, so that conversation immediately becomes easy, cordial, and absorbing'.[137]

Pierre Auger, whom we have quoted, regarded 'science as a force for unity among men'. A convinced European, he believed with Denis de Rougemont and Raoul Dautry, that there is no better field than science in which to stimulate European co-operation.

Indeed, CERN has influenced political co-operation in Europe, but international collaboration in HEP to an even greater extent. The first informal contacts between Soviet and CERN scientists were established during the first United Nations Atoms for Peace Conference, held in Geneva in August 1955.[138] In the following year three CERN scientists and 14 Americans attended the Conference of Physics of High-Energy Particles in Moscow (May 14–20, 1956), and a group of Soviet scientists took part in the CERN Symposium in High-Energy Accelerator and Pion Physics and in the Sixth Rochester Conference. In 1957 a High-Energy Physics Commission was created, consisting of two Americans, two Soviet physicists and two Europeans, responsible for the organization of the International Conference on HEP (see Section 2.3).

Other examples of international collaboration are an exchange of scientists between CERN and the Joint Institute for Nuclear Research in Dubna (JINR) which started in 1960, and the joint JINR–CERN summer school founded in 1969. The detailed history of the collaboration between CERN and JINR (and other Soviet research institutes) has been described by W.O. Lock.[139]

Physicists are convinced (and they are probably right) that collaboration in HEP 'has contributed its part toward lessening of world tension'.[140] In a period of political thaw, plans for much closer co-operation were developed by Victor F. Weisskopf, such as the construction of a common East-West accelerator. However, the project came to nothing.

But science also always sails under a national flag. Its achievements primarily serve domestic industry and sometimes national defence. A major discovery not only confers prestige on the fortunate discoverer but also on the nation (or social system) to which the discoverer belongs. Eminence in science has not only an idealistic importance but it also is of prime political significance.

Thus the French preponderance in the planned European Laboratory was an additional argument for approval by the French government. The laboratory would foster 'the development of French in these circles' and contribute to the country's intellectual influence.[141]

The preservation of U.S. leadership in elementary particle physics was a major concern to the 1960 PSAC–GAC Panel (Priore Panel) on high-energy accelerator physics. In his commentary, Eugene P. Wigner wrote:

HEP and society

> In the past, the Unites States played a leading role in high-energy physics and we do not hesitate to claim the discovery of the majority of new important phenomena and concepts in this field by scientists of our Nation. It is natural for every American physicist to wish to maintain the U.S. leading position in the high-energy field and the present writer shares this wish with the writers of the majority report.[142]

The Ramsey Panel of 1963 again stressed the fact that 'the United States has maintained is leadership in high-energy research':

> Over the last decade, most of the major inventions and discoveries in high energy physics have been made in U.S. laboratories [...] We believe that, apart from the intrinsic scientific interest ... it is essential that the U.S. maintain its leading position in an area of research which ranks among our most prominent scientific undertakings.[143]

It was difficult for the layman to evaluate the many different discoveries and inventions in the field or to distinguish the relative rank of Japan, the U.S.A., the U.S.S.R. and Western Europe. To the public, by far the most important discoveries were those rewarded by the Nobel prize, and the most important technical achievements the particle accelerators, or rather the energy of the particles produced.

Inventories of the Nobel prizewinners are published regularly by the Royal Swedish Academy of Sciences. If we extract from all the physics prizes from 1953 to 1970 those awarded for HEP and elementary particle physics, the following list emerges (an asterix indicates that the second half of the prize was given for another branch of physics).

1955:	W.E. Lamb, P. Kusch
1957:	T.D. Lee, C.N. Yang
1959:	E. Segré, O. Chamberlain
1960:	D.A. Glaser
1961:	R. Hofstadter*
1963:	E.P. Wigner
1965:	S.-I. Tomonaga, J. Schwinger, R.P. Feynman
1968:	L.W. Alvarez
1969:	M. Gell-Mann

Out of 13 prizewinners, 12 were American.

3.4.2 THE ECONOMIC VALUE

Generally three different aspects are mentioned:
a. scientific discoveries, such as the deflection of the magnetic needle by the electric current, may constitute a basic innovation for a new branch of industry

Notes: p. 91

b. scientific equipment and techniques may stimulate industry to develop new products and methods
c. the HEP field provides excellent training for physicists, engineers and technicians, which is of great value to other scientific disciplines, industry and administration.

The importance of scientific advance for technology

Man invented the wheel because he had to carry heavy loads. But Ørsted in 1820 did not discover that a magnetic needle is deflected by an electric current because he wanted to telegraph or build a motor. No, he was convinced that all forces were fundamentally related and that they could be converted one into the other. He therefore attempted to convert these forces. Not till he had made his discovery did the idea for its potential technical application take shape.

In 1869 Hermann Helmholtz, like Francis Bacon and Gottfried Wilhelm Leibniz before him, declared that a physicist should strive only for pure knowledge. In the course of time the natural sciences had transformed 'the whole life of modern man by the practical application of their achievements' and Helmholtz was convinced research would always result in important applications. But that would generally be the case 'where applications were least expected; to chase after them usually leads nowhere'.[144]

In 1869 Helmholtz said that recent inventions had 'mostly' derived from science. By the beginning of the 20th century it had become clear that science was the sole source of major innovations.

The natural sciences have therefore fulfilled their role of acting as an impetus for technology magnificently. But it has to be asked whether technical applications are contributed by every single field of research and more particularly by high-energy physics. End 1951 Heisenberg wrote in his report for the Foreign Office in Bonn: 'In the long run this new branch of nuclear physics may of course also open a field of application of similar importance as that of current nuclear physics'.[145]

In 1976 Victor F. Weisskopf said:

> History has shown that every advance in physics, however esoteric it may seem, will at some time in the future find important technical applications [...] It may take decades, but applications there will be.[146]

Other physicists, more cautiously, have said that a very long time may elapse between a particular scientific discovery and its technical application. Such statements cannot be verified and it would be preferable to admit straight away that there may be fields of fundamental research for which there are no applications. High-energy physics is one of these — for the time being at least.

To put it another way, the scientific achievements of high-energy physics — i.e. the discoveries about the behaviour of elementary particles and the understanding of their structural characteristics as such — have not led to any technical applications. Of course, the development and building of accelerators and elementary particle detection devices is quite a different matter. This side of high-energy physics has acted as a very powerful stimulant on technical developments.

Stimulating industry

When a new accelerator or an expensive new detection device is under construction, the accelerator centre offers contracts to industry. What the physicists want to achieve generally lies at the forefront of existing technological capability. The accelerator centre does not spend its money for fully commercially developed products but for prototypes, thus stimulating industry in the development of new technology. There is also a further aspect, which makes the collaboration even more valuable to industry: the HEP physicist is often well acquainted with the physical and technological principles of the new device: 'Therefore, the researcher can often provide information back to the manufacturer'.[147]

Well-known examples of the stimulation of industry are high-vacuum technology, new devices in electronics and computers, and super-conducting magnet technology. In 1963 the Ramsey Panel Report spoke of 'technological by-products':

> Accelerator physics, beginning with the earliest cyclotrons, has produced some ideas and stimulated inventions useful in other areas of science and industry. The development of high power transmitting tubes in the late thirties was stimulated by the needs of the cyclotron. The early impetus to the development of the klystron arose from the desire to make a linear accelerator [...] The alternating gradient principle, discovered by high energy accelerator physicists, has been applied in electron tubes in the communication industry.[148]

The educational values of HEP

Already in the early fifties, when CERN was being planned, it was repeatedly stated that the European laboratory would be 'a training centre' for other disciplines such as chemistry and medicine, and would create 'a type of research worker adaptable to industrial research in the home countries'.[149] Heisenberg recommended German participation, and one of his arguments was that it would provide urgently needed scientific training for younger physicists, especially with respect to the future economic utilization of nuclear energy in Germany.[150]

In a very recent review on 'physics through the 1990s' it was again stressed that young scholars working in HEP 'develop a general problem-solving ability of a high order', and therefore: 'Those who leave the field find excellent uses for their training in other areas'.[151]

The educational value is not restricted to high-energy physicists. Engineers and

Notes: p. 91

technicians working for industry and collaborating with the staff of the accelerator centres benefit enormously.[152]

3.4.3 THE IMPORTANCE FOR MILITARY TECHNOLOGY

When Galileo founded modern physics, he intended his new science to have some effect on everyday life. In 1609 he succeeded in reinventing the telescope: turning it to the heavens he made some wonderful scientific discoveries. At the same time he gained approval for it from the Signoria in Venice by pointing out how the telescope would allow the Venetian fleet to detect enemy ships much earlier.

Wolfgang Pauli spoke of the 'böse Hinterseite der Physik', the unfortunate side of physics, which is an integral part of physics and cannot be separated.[153]

As in the previous section, where we spoke about the use of science for peaceful economic purposes, we may distinguish (like John Krige and Dominique Pestre in Vol. I) three different aspects: applications of scientific results, the military use of scientific equipment and techniques, and the training of scholars.

Applications of scientific results

Up to the present time, as already stated, there have been no applications of scientific results and in particular no military applications. Neither the observed decay-modes of elementary particles nor the great discovery of the quarks as fundamental constituents of matter, nor indeed any other result, can be used for military purposes. Indeed, in some people's view, the fact that there have so far been no military applications has been decisive for the survival of mankind.

If we accept that throughout the ages, poets and writers have articulated man's secret fears and hopes, then it is obvious that man has always feared the apocalypse. Towards the end of the 20th century he fears nothing so much as the achievements of science. In Heinar Kipphardt's play 'In der Sache J. Robert Oppenheimer', Oppenheimer says: 'It is not the fault of physicists that nowadays brilliant ideas are always being turned into bombs'.[154] In Friedrich Dürrenmatt's 'Die Physiker' a brilliant physicist discovers the unified field theory, the universal formula and (out of curiosity, as he says), 'the system behind all potential inventions'. However, if it were put into effect it would have such appalling consequences that to save mankind, he hides himself away in a lunatic asylum. In vain. He is run to earth because what he knows makes him too important. Fate must run its course. The play's last words are: 'Once an idea is thought up, it can never be thought away'.[155]

Military use of scientific techniques

Even if so far there is no military application of man's insight into the structure of matter, HEP does not entirely escape the interest of the military. Second only to laser beams, particle beams play an important role in the U.S. Strategic Defense Initiative. There are, of course, big differences between operating a particle beam in an accelerator or storage ring and doing so in space. It is obvious, however, that the equipment, techniques and know-how developed by accelerator physicists and engineers form (with respect to particle beams as a weapon) the basis on which the SDI project has been started.

The problem of how to avoid as far as possible the damage caused by high-energy particle beams when they hit targets, has been thoroughly studied at every accelerator laboratory including, of course, CERN. The purpose of the SDI programme is different: in this case, the researchers try to achieve maximum destruction. However, the basis of their work is the experience of the radiation-protection teams.

Training of scholars

During World War II the United States and the United Kingdom realized that the military strength of a nation depended not only on purely military factors such as a corps of highly trained fighter pilots. Another important factor is the abilities of the scientific community. Ever since, the U.S. Department of Defense and its specialized agencies have invested heavily in fundamental research (including HEP) and given support to numerous scholars.

Notes

1. Mersits (1987).
2. Report of the International Physics Conference sponsored by the Council of Representatives of European States ... Edited by O. Kofoed-Hansen, P. Kristensen, M. Scharff, A. Winter.
3. Ibid., preface by N. Bohr.
4. Provisional Council. First session. Minutes, p. 4. CERN/Gen/1.
5. 'The centre of interest ... has changed from the nucleus to the elementary particles. Many of the questions which are still left open on the structure of atomic nuclei ... will probably find their solution only when more is known about the properties of elementary particles.' Provisional Council. Second session. Minutes, p. 7. CERN/Gen/2.
6. E. Segré, *Enrico Fermi*. Dictionary of Scientific Biography. Vol. IV (New York: Charles Scribner's Sons, 1971), 581.
7. R.E. Wilson, in: Weiner (1972), 74f.
8. Amaldi (1949).
9. Marshak (1985).
10. E. McMillan, Foreword. Proceedings of the XIIIth International Conference on High Energy Physics. Berkeley 1966.
11. See section 2.3
12. Note 5, 7.
13. Marshak (1970).
14. Annual Review of Nuclear Science. Vol. 6 (1956), p. V (preface).
15. First Annual Report of the European Organization for Nuclear Research (1955), 11; European

Organization for Nuclear Research. Second Annual Report (1956), 13; Annual Report 1957, 17; Annual Report 1958, 19.
16. The U.S.S.R. and the Asian countries deserve special interest; we feel however that a review would go beyond the limits of this study.
17. Braden, Fregeau & Frye (1960). The survey was made 'in early 1960'. For simplicity we take these figures for January 1960.
18. AEC (1963). - The report states: 'One is led to a figure between 800 and 900 ... now [April 1963] active in the field.' It based its calculation on the assumption of 880 for (the beginning of) the fiscal year 1963. Since in the 1960's the fiscal year started on 1 July, we have to reduce the numbers accordingly.
19. HEP Program (1965), 27. It says: 'During fiscal year 1965 there are about 1.060 Ph.D.s in the field.' We assume that this figure was valid rather (like the 1963 figures) for the beginning of the fiscal year, that is July 1965. To get the January figure, we subtracted half the increase for the year.
20. Report (1978), 6. - To the numbers of experimentalists (1072) and theorists (587) altogether 22 unclassified must be added.
21. Report (1978). Add 16 unclassified to arrive at the total number.
22. DPF Summer Study (1982), 651.
23. Report (1978).
24. DPF Summer Study (1982), 651.
25. Joint Committee (1965).
26. Joint Committee (1965), 28.
27. We have no empirical data. We have assumed, as always, an attrition rate of 2%, and simply calculated the figure of 90 entering the field.
28. Here we understand Europe as the community of the CERN member states.
29. Report of the Working Party on the European High Energy Accelerator Programme. FA/WP/23/Rev. 3, 74.
30. Pauli (1979), V.
31. Nuclear Physics in Europe. Present state and outlook. Strasbourg 1983, 37.
32. Scheuch & Alemann (1978), 95ff.
33. Kowarski (1955).
34. Weinberg (1961).
35. de Solla Price (1963).
36. Ramsey (1966), Mersits (1987).
37. Round-Table (1964).
38. Report of the Working Party on the European High-Energy Accelerator Programme. CERN FA/WP/23/Rev. 3, 74.
39. Report (1978), 5.
40. Round-Table (1964).
41. HEP Program (1965), 21.
42. Kowarski (1967).
43. Kowarski (1967), 5.
44. Ibid.
45. Weinberg (1967), 3f.
46. Wigner (1950).
47. Hermann (1979).
48. Pauli (1985), 136.
49. Meadows (1974), 118. Meadows quotes a survey of U.K. astronomers and space scientists, which indicated that less than 20% obtained regular information relevant to research from their correspondence.
50. Ibid., 90.
51. Dierks (1972), 26.

Notes

52. Blatt & Weisskopf (1952), preface. The quote has been retranslated from the German edition.
53. S.A. Goudsmit & G.L. Trigy, 'Editorial. Growth of Physical Review Letters', Phys. Rev. Lett. **18** (1967), 1f.
54. S. Pasternack and A. Herschman, 'Editorial: A Proposal for Changing The Physical Review', Phys. Rev. **137** (1965), AB1.
55. Important Announcement, Phys. Rev. **132** (1964), 1.
56. S.A. Goudsmit, 'Editorial', Phys. Rev. Lett. **6** (1961), 587.
57. Ibid.
58. 'Publish and be damned a second time', Nature **233** (1971), 294.
59. Hermann (1979), XX.
60. Meadows (1974), 114.
61. Mersits (1987), 38.
62. E. McMillan, 'Foreword', Proceedings of the XIIIth International Conference on High-Energy Physics, Berkeley, 1966.
63. R.E. Marshak, 'The Rochester Conferences: The Rise of International Cooperation in High-Energy Physics', Bulletin of the Atomic Scientists, **26** (1970), 6, 92–98.
64. Proceedings of the Rochester Conference on Meson Physics. January 11–12, 1952, Appendix I. Including the three representatives from the sponsoring local industry there were altogether 93 participants. The first Rochester Conference in 1950 had about 50, the third about 140 participants.
65. High Energy Nuclear Physics, Proceedings of the Third Annual Rochester Conference, December 18–20, 1952, 109f.
66. Unfortunately there are no proceedings of the Rochester meeting on 16 December 1950, and we do not know how this conference was called officially.
67. International Nuclear Physics Conference. September 1950, Harwell. Proceedings ... Edited by E.W. Titterton. AERE-9/M/68.
68. Ceallaight (1985), 188.
69. Marshak (1985).
70. E. McMillan, cf. 62.
71. Proceedings of the Lund International Conference on Elementary Particle Physics. Held at Lund, Sweden, June 25–July 1, 1969. Preface.
72. Proceedings of HEP. International Europhysics Conference on High Energy Physics, Brighton (UK), 20–27 July, 1983. Foreword.
73. International Conference on Mesons and Recently Discovered Particles. The 43° Congresso Nazionale di Fisica. Communcazioni. Padova 1957.
74. Mehra (1975).
75. The IV International Symposium on Multiparticle Hadrodynamics ... Proceedings, p. V.
76. CERN Symposium on High Energy Accelerators and Pion Physics ... Proceedings. Vol. 1. High energy accelerators. Geneva 1956.
77. International Conference on High-Energy Accelerators and Instrumentation. Geneva 1959.
78. W. Panofsky, The future of high energy accelerators in physics. Ibid. p. 6.
79. M.H. Blewett, Preface. International Conference on High Energy Accelerators. Proceedings. Brookhaven 1961, iii.
80. European Accelerator Group. First Meeting. Paris (Saclay) 30–31 May 1961. CERN/AG/1.
81. Report of the working party on the European high energy accelerator programme. FA/WP/23/Rev. 3.
82. Proceedings of the Eighth International Conference on the Application of Accelerators in Research and Industry held in Denton, Texas, 12–14 November 1984. Amsterdam 1985.
83. Information is mainly based on a interview with Madame DeWitt-Morette made by Pomian and Pestre around 1980 and a letter from Madame DeWitt to A. Hermann 29/9/1986.
84. See Volume I, p. 109.

85. A dossier 'Projet de Création d'une Ecole d'Eté' was sent by Cécile Morette to the C.E.C. in January 1951. Cf. letter of Raymond Silva to Cécile Morette, 29 January 1951. Archives C.E.C.
86. Compte-rendu analytique de la réunion du 12 décembre 1950.
87. Not mentioned in the official compte-rendu, but in Rollier's report.
88. Université de Grenoble. Ecole d'été de physique théorique. Rapport d'activité 1951-1959.
89. Université scientifique et médicale de Grenoble and Institut national polytechnique de Grenoble. Les Houches Summer School of Theoretical Physics. Centre of Physics [Information Booklet, 1983].
90. Ibid.
91. Cécile DeWitt, Voeux exprimés au nom de l'Ecole d'Eté de Physique Théorique Les Houches. In: Supplemento al Volume XI, Serie IX (1954) del Nuovo Cimento, p. 146.
92. Supplemento al Volume II, Serie X (1955) del Nuovo Cimento, p. 18.
93. Université de Grenoble. Ecole d'été de physique théorique. Rapport d'activité 1951-1959.
94. Summer Institute for Theoretical Physics. Boulder, Col. 1958.
95. Antonino Zichichi (ed.). Strong, Electromagnetic and Weak Interactions (New York: W.A. Benjamin, Inc. 1964), Foreword.
96. Antonino Zichichi (ed.), Progress in Scientific Culture. The Interdisciplinary Journal of the Ettore Majorana Centre. The quoted text is on the cover of every number.
97. P. Urban, Erinnerungen. Graz 1985, p. 55.
98. E. Guth, 'Schladming School Examines Special High-Energy Problems', Physics Today, July 1967, p. 109-111.
99. H. Latal & H. Mitter (eds.), Concepts and Trends in Particle Physics. Proceedings of the XXV. Int. Universitätswochen für Kernphysik, Schladming, Austria, February 19-27, 1986 (Berlin: Springer 1986), p. VI.
100. Lock (1975), 10f.
101. L. van Hove, 'The role of high-energy accelerators in particle physics', International Conference on High Energy Accelerators, Brookhaven 1961, p. 6.
102. Braden, Fregeau & Frye (1960). - In Vol. I of this 'History of CERN' we have given short biographies of some of these 'accelerator physicists' (Adams, J. Blewett, H. Blewett, Goldschmidt-Clermont, Goward, Fry, Hine, Schmelzer, Wideroe).
103. Our quote is from Herbert W.B. Skinner, who praised the Cosmotron in a closed meeting on 7 June 1952 (Conference on High Energy Accelerators for Nuclear Research; Files of Churchill College Cambridge). Another example is M. Hildred Blewett's judgement about the AG-machines, Blewett (1967).
104. HEP Program (1965), p. 92.
105. Blewett (1967). - On page 468 a review is given about the accelerator literature.
106. Livingston (1966), 5.
107. Courant, Livingston & Snyder (1952). - It is a well known story, that the idea was conceived independently and at an earlier date by Nicholas Christofilos.
108. Blewett (1967), 460.
109. A. Schoch, Prospects and problems of future accelerator projects, International Confererence on theoretical aspects of very high-energy phenomena (Geneva: CERN 61-22, 1961), 390.
110. Scientific Policy Committee. Seventh Meeting, 29 October 1957. CERN/SPC/54.
113. We assume of course that the accelerator physicists were not working on two generations of machines simultaneously, which would be possible in principle by an overexpansion of staff.
114. International Conference on High Energy Accelerators. Brookhaven 1961, p. 3.
115. Marshak (1970).
116. Item 14, 3.
117. Pauli (1979), XXVI.
118. History of CERN (1987), 407.
119. Item 114, 11.

Notes

120. Symposium ZGS (1980), 11. - Actually the Dubna machine (for which the work had begun much earlier) was ready in 1957, the ZGS only in 1963. In this year the CERN PS and the AGS were already operating at much higher energies (28 and 33 GeV respectively). One record was acquired: The ZGS went into history 'as the highest energy weak focusing synchrotron'.
121. Powell et al. (1946), 102.
122. Powell (1953), 220.
123. Powell (1956).
124. Waloschek (1986), 71 ff.
125. Glaser (1960), 74.
126. Conversi (1982).
127. Fukui (1985).
128. HEP Program (1965), 76.
129. F. Bacon, Neues Organ der Wissenschaften [Novum Organum] (Darmstadt: Wissenschaftliche Buchgesellschaft, 1974), 51.
130. G.W. Leibniz, Auswahl und Einleitung von F. Heer (Frankfurt: Fischer Bücherei, 1958), 90.
131. J. v. Liebig, Reden und Abhandlungen (Wiesbaden: Dr. Martin Sändig, 1965), 9.
132. HEP Program (1965), 132.
133. Ibid., 19.
134. Amaldi Report (1963), 9.
135. Zeit-Magazin, No. 44 (1987), 89.
136. HEP Program (1965), 23.
137. Auger (1965), 209.
138. Lock (1975), 2.
139. Ibid.
140. HEP Program (1965), 29.
141. History of CERN (1987), 324.
142. HEP Program (1965), 132.
143. AEC (1963), 3.
144. Hermann von Helmholtz, Vorträge und Reden. Vol. I (Braunschweig: Vieweg 1896), 372.
145. History of CERN (1987), 402.
146. 'Physik wohin?' Panel Discussion at the 40th Congress of the German Physical Society on September 15, 1976. Bild der Wissenschaft, 14 (1977), No. 4, 168–181.
147. Physics Through the 1990s: Elementary Particle Physics (Washington D.C.: National Academy Press, 1986), 171.
148. AEC (1963), 3.
149. History of CERN (1987), 542.
150. Ibid., 402.
151. See note 147, 182.
152. HEP Program (1965), 25.
153. Pauli (1964), 1295.
154. Heinar Kipphardt, 'In der Sache J. Robert Oppenheimer'. Spectaculum, VII (1965), 247.
155. Friedrich Dürrenmatt, 'Die Physiker'. Spectaculum, VII (1965), 146.

Bibliography

AEC (1963) — United States Atomic Energy Commission, *Report of the Panel on High-Energy Accelerator Physics of the General Advisory Committee to the Atomic Energy Commission and the President's Science Advisory Committee.* April 26, 1963.

Amaldi (1949) — E. Amaldi, 'Recenti progressi e prospettive nello sviluppo delle applicazioni dell'energia atomica', *Scienza e Tecnica*, **10-12** (1949), 240-264.

Amaldi Report (1963) — Report of the Working Party on the European High Energy Accelerator Programme, CERN/FA/WP/23/Rev.3.

Auger (1965) — Pierre Auger, 'Science as a Force for Unity Among Men', *Bulletin of the Atomic Scientists*, 12 (1965), 208-210.

Blatt & Weisskopf (1952) — J.M. Blatt and V.F. Weisskopf, *Theoretical Nuclear Physics* (New York: John Wiley & Sons, 1952).

Blewett (1967) — Hildred Blewett, 'Characteristics of typical accelerators', *Annual Review of Nuclear Science,* 17 (1967), 427-468.

Braden, Fregeau & Frye (1960) — Charles H. Braden, Jerome H. Fergeau and Glenn M. Frye, 'Manpower in high-energy physics', *Physics Today,* Dec. 1960, 42-46.

Brown & Hoddeson (1983) — L.M. Brown and L. Hoddeson (eds.), *The birth of particle physics* (Cambridge: Cambridge University Press, 1983).

Ceallaight (1982) — C.O. Ceallaight, *'A Contribution to the History of C.F. Powell's Group in the University of Bristol 1949-65',* Colloquium-Paris (1982), 185-188.

Colloquium-Paris (1982) — *International Colloquium on the History of Particle Physics: Some Discoveries, Concepts, Institutions from the Thirties to the Fifties,* 21-23/7/1982, Paris (Paris: Edition de Physique, supplement *Journal de Physique,* Tome 43, 1982).

Conversi (1982) — M. Conversi, 'The development of the Flash and Spark Chambers in the 1950's', Colloquium (Paris), 91-99.

Courant (1968) — E.D. Courant, 'Accelerators for high intensities and high energies', *Annual Review of Nuclear Science,* 18 (1968), 435-464.

Courant, Livingston & Snyder (1952) — E.D. Courant, M.S. Livingston and H.S. Snyder, 'The strong-focusing synchrotron ... a new high energy accelerator', *Phys. Rev.* 88 (1952), 1190-1196.

Dierks (1972) — H. Dierks, *Über de Zitierhäufigkeit von Zeitschriften auf dem Gebiete der Physik* (Köln: Greven, 1972).

DPF Summer Study (1982) — R. Donaldson, R. Gustafson and F. Paige (eds.), *Proceedings of the 1982 DPF Summer Study on Elementary Particle Physics and Future Facilities.*

Doncel (1987) — M.G. Doncel, A. Hermann, L. Michel and A. Pais, *Symmetries in Physics (1600-1980).* Proceedings ... (Bellaterra: Seminari d'Història de les Ciències, 1987).

Fukui (1985) — S. Fukui, 'Development of discharge (spark) chamber in Japan in the 1950s'. Preprint. To appear in the Proceedings of the 'International Symposium on Particle Physics in the 1950s: Pions to Quarks'. Fermilab, 1-4 May 1985.

Glaser (1960) — D.A. Glaser, 'Elementary Particles and Bubble Chambers'. Nobel Lecture, December 12, 1960. Les Prix Nobel en 1960. Stockholm 1961, 72-94.

HEP Program (1965) — High-Energy Physics Program: Report on National Policy and Background Information. Joint Committee on Atomic Energy (Washington: U.S. Government 1965).

Bibliography

Hermann (1979)	Armin Hermann, 'Die Funktion und Bedeutung von Briefen', Pauli (1979), XI–XLVII.
History of CERN (1987)	A. Hermann, J. Krige, U. Mersits & D. Pestre, *History of CERN*. Vol. I (Amsterdam: North-Holland, 1987).
Hoddeson (1983)	Lillian Hoddeson, 'Establishing KEK in Japan and Fermilab in the U.S.: Internationalism, Nationalism and High-Energy Accelerators'. *Social Studies in Science,* 13 (1983), 1–48.
Joint Committee (1965)	*High Energy Physics Program: Report on National Policy and Background Information*. Joint Committee on Atomic Energy. Congress of the United States. Washington 1965.
Van Hove & Jacob (1980)	L. Van Hove and M. Jacob, *Highlights of 25 years of physics at CERN* (Amsterdam: North-Holland Publishing Company, 1980).
Kowarski (1955)	L. Kowarski, 'The Making of CERN. An Experiment in Cooperation', *Bulletin of the Atomic Scientists,* 11 (1955), 354–357.
Kowarski (1967)	L. Kowarski, *An observer's account of user relations in the U.S. accelerator laboratories* (Geneva: CERN 67-4, 1967).
Livingston (1966)	M.S. Livingston (ed.), *The Development of High-Energy Accelerators* (New York: Dover Publications, 1966).
Lock (1975)	W.O. Lock, *A History of the Collaboration between the European Organization for Nuclear Research (CERN) and the Joint Institute for Nuclear Research (JINR), and with Soviet Research Institutes in the USSR 1955–1970* (Geneva: CERN 75-7, 1975).
Marshak (1970)	R.E. Marshak, 'The Rochester Conferences: The rise of international cooperation in high-energy physics', *Bulletin of the Atomic Scientists,* 26 (1970), 92–98.
Marshak (1985)	R.E. Marshak, '*Scientific impact of the first decade of the Rochester Conferences (1950–1960).*' Preprint. To appear in the Proceedings of the 'International Symposium on Particle Physics in the 1950s: Pions to Quarks'. Fermilab, 1–4 May 1985.
Meadows (1974)	A.J. Meadows, *Communication in Science* (London: Butterworths, 1974).
Mehra (1975)	J. Mehra, *The Solvay Conferences on Physics. Aspects of the Development of Physics since 1911* (Dordrecht-Holland: D. Reidel, 1975).
Mersits (1987)	U. Mersits, 'From cosmic-ray and nuclear physics to high-energy physics', History of CERN. Vol. I (Amsterdam: North-Holland, 1987), 3–60.
Pauli (1964)	Wolfgang Pauli, *Collected Scientific Papers*. 2 Vols. (New York: Interscience Publishers, 1964).
Pauli (1979)	Wolfgang Pauli, *Scientific Correspondence with Bohr, Einstein, Heisenberg a.o.* Vol. I (New York: Springer Verlag, 1979).
Pauli (1985)	Wolfgang Pauli, *Scientific Correspondence with Bohr, Einstein, Heisenberg a.o.* Vol. II (Berlin: Springer Verlag, 1985).
Pickering (1984)	Andrew Pickering, *Constructing Quarks. A sociological History of Particle Physics* (Chicago: The University of Chicago Press, 1984).
Powell et al. (1946)	C.F. Powell, G.P.S. Occhialini, D.L. Livesey, and V.F. Chilton, 'A New Photographic Emulsion for the Detection of Fast Charged Particles', *Journal of Scientific Instruments,* 23 (1946), 102–106.
Powell (1953)	C.F. Powell, 'The Use of Stripped Emulsions for Recording the Track of Charged Particles', *Philosophical Magazine,* 44 (1953), 219–222.
Powell (1956)	C.F. Powell, 'Introduction – G stock collaboration', *Nuovo Cimento,* 4 Supp. 2 (1956), 399–401.

Ramsey (1966)	N.F. Ramsey, *Early History of Associated Universities and Brookhaven National Laboratory,* Brookhaven Lecture Series, **55** (Brookhaven: BNL 992, 1966).
Report (1978)	Report of the Subpanel on High-Energy Physics Manpower of the High-Energy Physics Advisory Panel. Washington 1978.
Round-Table (1964)	'High-Energy Physics Round-Table Discussion', *Physics Today,* Nov. 1964, 50–57.
Scheuch & Alemann (1978)	E.K. Scheuch and H.v. Alemann (eds.), *Das Forschungsinstitut. Formen der Institutionalisierung von Wissenschaft* (Erlangen: Institut für Gesellschaft und Wissenschaft, 1978).
Scuola Fermi (1977)	*Rendiconti della Scuola Internazionale di Fisica Enrico Fermi,* LVII Corso (New York: Academic Press, 1977).
de Solla Price (1963)	Derek J. de Solla Price, *Little science, big science* (New York: Columbia University Press, 1963).
Symposium ZGS (1980)	*Symposium on the History of the ZGS* (New York: American Institute of Physics, 1980).
Waloschek (1986)	P. Waloschek, *Der Multimensch. Forscherteams auf den Spuren der Quarks und Leptonen* (Düsseldorf: Econ, 1986).
Weinberg (1961)	A.M. Weinberg, 'The impact of large-scale science on the United States', *Science,* **134** (1961), 161–164.
Weinberg (1967)	A.M. Weinberg, *Reflexions on big science* (Oxford, London, etc.: Pergamon Press, 1967).
Weiner (1972)	C. Weiner, *Exploring the History of Nuclear Physics* (New York: American Institute of Physics, 1972).
Wigner (1950)	E.P. Wigner, 'The limits of science', *Proceedings of the American Philosophical Society,* **94** (1950), 422–427.
Wilson (1980)	Robert R. Wilson, 'The next generation of particle accelerators', *Scientific American,* 242 (1980), 1, 42–57.

CHAPTER 4

The 600 MeV synchro-cyclotron: constructing CERN's smaller accelerator and the machine's role in the scientific programme[1]

Ulrike MERSITS

Contents

4.1. Introduction	96
4.2 Design and construction of the 600 MeV SC	97
4.2.1 Decision on a medium-sized accelerator for the European Laboratory	97
4.2.2 Fixing the design parameters (1952–1955)	98
4.2.3 The construction period (1955–1957)	106
4.3 Particle physics with the SC from 1958 to 1965	109
4.3.1 A survey of activities around the machine	109
4.3.2 Four case studies on the main research topics carried out with the SC	116
4.3.2.1 The electron decay mode of the pion ($\pi \to e\nu$)	116
4.3.2.2 A precision measurement of the anomalous magnetic moment of the muon: the $(g-2)$ experiment	119
4.3.2.3 The beta decay of the pion ($\pi^+ \to \pi^0 e\nu$)	122
4.3.2.4 The μ-channel and the scientific programme carried out with it	123
4.4 Concluding remarks	126
Notes	131
Bibliography	135

4.1 Introduction

In June 1952 it was decided that the future European Laboratory should be equipped with a gigantic proton synchrotron of 28 GeV along with a far smaller machine, a synchro-cyclotron of 600 MeV (SC). On 1 August 1957 this smaller accelerator first reached its full design energy. In a certain sense this date can thus be regarded as the starting point of the active scientific life of the organization. The arguments underlying the choice of this particular kind of accelerator were that it should be of a standard kind, technically not too novel, and mainly calling for the scale-up of similar existing and reliable machines. It was also to be relatively fast to build and not too costly, and was primarily intended to bridge the gap until the 28 GeV proton synchrotron was ready in the early 60s. The SC was thought of as a tool for training European physicists who were unaccustomed to working with accelerators and for adapting the functioning of the organization to the real needs of every-day scientific life.

Originally it was believed that the SC would play—scientifically speaking—only a subordinate role. However, when one looks at the first eight years (1958–1965) of experimental work with the machine the amount and the variety of good results produced by groups working with the SC exceeded all expectations and was indeed impressive. The accelerator was used for experiments stretching from particle physics to nuclear physics and chemistry, and a large number of teams from in and outside CERN performed their first experiments here. That said, an all-embracing description of this scientific programme is not possible here; instead we have decided to restrict our study to the work done on particle physics. This choice was based on two criteria. Firstly, the central research aim of CERN and thus consequently also of this accelerator was particle physics, and nuclear physics and chemistry were only regarded as by-products. We are aware that this fact changed drastically in the mid-60s for the SC but we feel that this aspect should form part of a history describing the period from 1965 onwards. Secondly, it was primarily particle-physics experiments that constituted the first real scientific successes for CERN, and which were of crucial importance for a smooth start of this European venture.

Our chapter has three main parts. In the first we offer a short account of the design and construction of the accelerator. We will explain how and why the decision to build such a machine was taken, describe the main steps in its evolution, and survey briefly the building period starting in 1955.

In the second part we will give the reader a general impression of what happened in the eight years after the completion of the SC. On the one hand we will describe the technical evolution of the accelerator, i.e. the problems arising and the improvements carried out, as this had important implications for what could be done scientifically. On the other hand we will present the kinds of physics done, underlining trends or specialities which appeared during this period.

Thereafter we discuss three particularly well-known experiments, i.e. the rare decay mode of the pion $\pi \to e\nu$, the precision measurement of the anomalous magnetic moment of the muon $(g-2)$ and the beta decay of the pion $\pi^+ \to \pi^0 e^+ \nu$, and the subsequent research of the teams doing this work, as well as the setting up of a novel facility—the μ-channel—and the kind of physics performed with it. This emphasis on a few selected experiments should not be seen as a judgement on the large number of other experiments which are hardly mentioned, if at all. In our case studies we try to give an impression of the experiment as a whole, describing the scientific environment which motivated its choice, the concept of the experiment, the major difficulties involved, and finally the result and its place in the theoretical context. We also describe the continuation of the group's activity in order to see how it evolved at CERN.

In the conclusion to the chapter we will try to grasp the spirit in this period, will summarize the central points of the scientific programme, and will illustrate the development of the SC from a particle-physics instrument to a central tool for nuclear physics and chemistry at CERN.

4.2 Design and construction of the 600 MeV SC

4.2.1 DECISION ON A MEDIUM-SIZED ACCELERATOR FOR THE EUROPEAN LABORATORY[2]

In September 1951 the idea that the European Laboratory should have two accelerators was proposed for the first time. During the three months that followed it took further shape, and by December 1951 at the UNESCO meeting in Paris the synchro-cyclotron was stressed as an ideal solution for a medium-energy accelerator. The reasons given were that it would be relatively easy and fast to build, would nevertheless offer interesting physics to do, and that a number of machines of this type were existing and functioning smoothly all over the world. Having thus in principle agreed on the kind of accelerator to build, it remained to choose the energy. The lower limit was given by the biggest existing synchro-cyclotron, namely the Chicago machine of 450 GeV. In the upper direction there was an economic limit imposed by the nonlinear rise in magnet size—and therefore costs—with energy. Thus in a first draft programme of the SC Study Group presented at the first session of the Council in May 1952 a 600 MeV machine was suggested. This choice seemed to be dictated by a scientific argument resulting from the work of the Chicago group around Fermi on the interactions of pions with hydrogen. These results—the finding of the first excited proton state (Δ resonance)—had been reported at the second

Notes: p. 131

Rochester Conference in January 1952, and it seemed fascinating to investigate this point at higher energies.[3]

As a consequence, the Council decided to call an international physics conference in June 1952 in Copenhagen to permit a broad exchange of views and thus provide a survey of the present situation in nuclear physics. Such a discussion between European and American experts should help to fix the energies of the particle accelerators. Immediately after a report was to be drawn up 'from which the Council might take guidance to choose the type and range of energies of the machines to be studied'. In this Council session also Cornelis Bakker was appointed head of the group in charge of studying the possible construction of a synchro-cyclotron for energies of at least 500 MeV.[4]

The conference was held from 3-17 June 1952, just before the second session of the Council in Copenhagen. The conclusion presented by Werner Heisenberg was that 'the synchro-cyclotron for protons of about 600 MeV could be built more readily than any other accelerator type in this energy range and would give the best results'.[5]

And thus the design of a 600 MeV SC was officially sanctioned.

4.2.2 FIXING THE DESIGN PARAMETERS (1952-1955)

The first meeting of the SC Group took place between the scientific conference and the Council session from 15 to 19 June 1952.[6] The aim was to discuss the preliminary work done so far even without a definite decision being taken, and to programme the further steps needed. The design work was divided into five specific topics each requiring different skills and experiences from scientists and technicians: magnet, high-frequency system, vacuum system, shielding and control, and the general layout. Each subgroup had a certain autonomy, being stationed generally in the home institute of the respective leader. However they also had to accept a large part of responsibility for their task. To exchange views, coordinate the ongoing work, and discuss about how to continue, a meeting of all subgroups, headed by Bakker, was held roughly every two months. This was kept up until the groups started moving to Geneva late in 1954.

The *magnet*, which represented the central part of the SC, was of particular importance as its shape influenced other parts like the vacuum chamber, for example. Thus a rough design had to be submitted as soon as possible. Further, as the magnet and the coils would consume more than one third of the entire SC budget, a lot of effort was spent on working out its specifications. Bengt Hedin, a Swedish specialist in cyclotrons, who had also taken part in the construction of the 200 MeV synchro-cyclotron at Uppsala, was chosen to lead this subgroup. Thus from June 1952 on he spent half of his time, assisted by an electrical engineer O. Fredriksson, on designing the SC magnet.[7]

As a synchro-cyclotron was at that time already a standard machine, they first

Design and construction

investigated and compared the features of those already existing. On the basis of this they could decide on suitable parameters for the CERN machine and could then go over to construct a model to learn about details of the design and its feasibility. At the first meeting of the SC Group it was already agreed that the magnet design would be similar to the one of the Pittsburgh 440 MeV SC and a scale model was constructed. As far as the coils were concerned, the choice was between copper and aluminium as material, and water or oil for the cooling system. Here a balance had to be found between the price of the raw material used, the amount needed, the power consumption involved, and other technical features. In this early stage the choice fell on copper coils and water cooling. Oil cooling had shown serious problems on existing US machines of this type. Guided by these first choices, negotiations regarding manufacturing possibilities began, and by the end of 1952 it was clear that European firms had the capacity to manufacture both the magnet and the coils.[8]

Meanwhile the above-mentioned model with a pole diameter of 28 cm (scaling factor 0.056) had been built and on the basis of measurements carried out with it in January 1953 some of the final parameters could be regarded as fixed: the size of the magnet frame, of the poles, and of the coils.[9] (See fig. 4.1.)

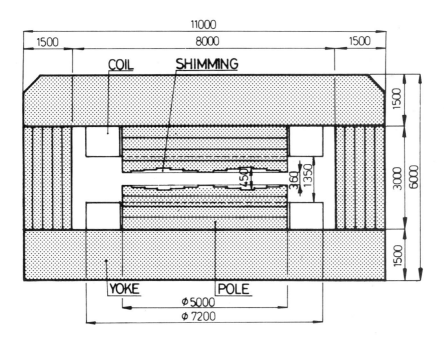

Fig. 4.1. Cross-section of the CERN SC magnet[10]

Notes: pp. 131 ff.

The rough layout being settled, the details of the coil structure, the number of Ampere turns needed, the quality of the steel, the plans for manufacturing the 2500 tons of steel for the yoke and the poles in sections, and similar questions had to be dealt with. For this reason it was decided in March 1953 that another model was required, this time on a 1:10 scale. It was to be an exact scale model of the final component. In this way the shimming needed to ensure a contour for the pole faces guaranteeing a decrease of 5% of the magnetic field with increasing radius (to provide focusing forces for the beam), could be determined.[11]

The extraction radius was to be 2.25 m (finally increased to 2.27 m), at a field gradient of $n = 0.2$. The magnetic flux density in the centre was to be 1.88 Wb/m^2 and 1.78 Wb/m^2 at the exit radius. Only one last change was made to the coil dimensions, namely their distance was decreased from 1.35 to 1.2 m. However the coil material was rediscussed in June 1953 and finally the original decision was reversed in favour of aluminium. The advantages were that the price would be considerably lower, that the coils would be easier to manufacture, and that the aluminium would have lower weight, facilitating transport and assembly.[12]

By October 1953 the design of the magnet and the coils was essentially settled, and besides finalizing the shimming of the poles the group's main task for the next half year was to prepare the documents with precise specifications to be sent out with the invitations for tender in July 1954. From then onwards much effort had to be devoted to dealing with the manufacturers. Visits were arranged to those firms under serious consideration so that sufficient information could be gathered to be able to take a well-informed decision.

On 6 October 1954 the Interim Finance Committee discussed this subject and it was decided to recommend the following choice to the Council: Schneider-Creusot (F) for the yoke, RDM Rotterdam (NL) for the pole pieces, and ACEC Liege (B) for the Al-coils. Furthermore, the firm Lloyds (St. Etienne) was contracted to survey the taking of magnetic samples, which then had to be tested by the 'Eidgenössische Amt für Maß und Gewicht' in Bern to ensure the high magnetic quality of the material for any part of the magnet.[13]

In late 1954/early 1955 Hedin and Fredriksson moved to Geneva. The magnet model was shipped from Uppsala to Geneva and a motor generator was ordered so that the model could be operated.

The subgroup in charge of the *high-frequency system* was headed by Frank Krienen who was joined by M. Morpurgo at the beginning of 1954. The former had been chosen to guide the development of this component as it was regarded as 'the most difficult and elaborate part', and he could draw on his experience gained at Philips on radar and related topics. With the change from a non-relativistic to a relativistic energy range, and thus from a cyclotron to a synchro-cyclotron, it had become necessary that the accelerating H.F. voltage between the dee and the dummy dee (a copper lining of the pole faces of the magnet) should vary during the period of acceleration. The higher the energy of an SC the larger this frequency sweep had to be, and the more difficult the H.F. system became. Given the field strength in the centre of magnet and at the exit radius, and the desired energy of 600 MeV, the frequency variation in CERN's case had to be roughly from 32 MHz to 16 MHz.

Such an H.F. system consisted in fact of two main parts: the dee system (resonator) and the H.F. supply and modulator. The dee system was built of three components: the dee itself, a hollow, light-alloy construction with copper lining, the stem, which was needed to produce the necessary electrical length for the dee which was limited to less than 2.5 m (the pole face radius), and the stub, an inductance added in series to the resonant circuit in order to obtain a greater frequency sweep. The values of these parts were first calculated on a theoretical basis and then designed along more or less empirical lines with the help of a 1:10 and a full-scale model (see fig. 4.2). The variation in frequency of the resonator system was then obtained by a variable capacitor, the construction of which became the central task of this subgroup. This part was completely novel and, as we shall see later, proved to be highly complex and difficult to build, and for a while even threatened the success of the SC.[14]

Although in the June 1952 meeting a rotating condenser system for frequency modulation seemed appropriate, doubts soon arose. Firstly, the existing rotating condensers were causing a lot of problems and secondly, different systems were also planned for the new machines in America. In August Bakker visited Berkeley and learned that for its renewed 184" synchro-cyclotron two different systems were envisaged: a vibrating reed system or a ferrite tuning element. In comparison with a rotating condenser model, these new devices allowed a higher rate of frequency modulation and as a result a correspondingly higher average beam intensity.

By the end of 1952, after extensive studies on these new systems, a first choice was made. A vibrating reed system was to be used incorporating a variable capacitor of tuning fork type, which distinguished it from the model envisaged by the Berkeley team. This form for the vibrating system was chosen because most of the forces between the prongs were counterbalanced in it. Excitation would be at the bottom of the tuning fork using an electro-mechanical transducer. Aluminium was thought to be an ideal material. However several difficulties had to be taken into account in the design such as material stresses proportional to the frequency of free vibration and the fatigue limit of the material, which would have to withstand some 10^9 vibrations. Further the weight of such equipment, the heat generated, and the problems of manufacturing a huge tuning fork 200 cm long, 54 cm high and 9 cm between the tips had to be kept in mind.[15]

Further, at the end of 1952 it was decided that two models should be constructed: one to a quarter scale, the other full scale with a moderate width (distance between the prongs). This was a rather complex venture and on 15 March 1953 an agreement was signed between CERN and Philips for research and development of this component.[16] The CERN subgroup was expected to collaborate with their colleagues at Philips and submit their results regularly for approval. When the description and

Notes: p. 132

specification of the H.F. system were determined to CERN's satisfaction, this agreement was to be regarded as terminated and the final manufacturing contract could then be put out to tender. Thus the first half of the year was mainly devoted to constructing the two models and to developing a driving source which would give sufficient amplitude to the fork tips. The full-scale model was especially important because of the problems posed by the production of such a large device with the desired quality and precision. Throughout the second half of 1953 live tests were made on the material of the tuning fork. The insulators for the suspension system were in preparation and problems like the damping of the vibrating system in the stray field of the magnet were considered. As was realized much later, the latter problem turned out to be a major one. It was also necessary to develop and test detectors to determine the amplitude and position of the tuning fork. The next stage was then to construct a vacuum test bench to provide necessary information on the full-scale model under running conditions and on the achievable frequency range. In addition a new quarter-scale model was needed incorporating some new features.

In February 1954, Bakker and Krienen travelled to the United States again, visiting Pittsburgh, Chicago, Berkeley, and Stanford. In Berkeley their main interest was to exchange information on the use of the vibrating reed H.F. system. Though it was the most risky part of the CERN design for the SC, no points arose which called for any changes. It appeared that the development of this part of the machine had proceeded satisfactorily and one began to think that the mechanical parts of the design would be more difficult than the electrical ones.[17]

The vacuum test bench was finally ready in June 1954 after 6 months delay, and the new quarter-scale model was prepared for final tests. By January 1955 these experiments had given enough data to justify ordering a full-sized tuning fork. Time began to run short and a tight schedule had to be set up for the first half of 1955 if the full-size model of the H.F. system was to work by May.

With development nearing completion, the contract with Philips was discussed at the fourth meeting of the Finance Committee on 8 June 1955. It was stated that the stage had been reached where the manufacturing contract had to be placed if the high-frequency system was to be ready in time. In any case it would be difficult to follow the usual tender procedure since, having done all the development work, Philips would be in the strongest position to win the contract. As a result the manufacturing of the frequency modulating system started in September 1955 at Philips, with R.D.M. Rotterdam providing the aluminium tuning forks.[18]

The development of the third component, the *vacuum system*,[20] took a somewhat different path to that of the two just described. First of all its design depended on how the magnet and the H.F. system would look, and secondly only standard techniques were to be applied. The leader of this subgroup was M.J. Moore, working in co-operation with F. Bonaudi. Early in 1954 the team was joined by F. Blythe as draughtsman. Moore was at that time essentially the chief engineer at the Nuclear Physics Institute at the University of Liverpool. Bonaudi had just taken his

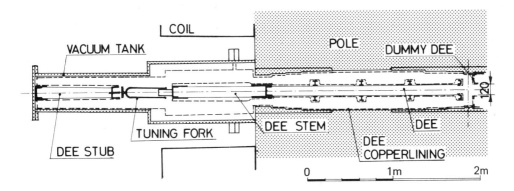

Fig. 4.2. Cross-section of the high-frequency system[19]

post-graduate diploma in engineering at Turin.

The first rough ideas can already be found in the August 1952 report. One would need 2–3 oil diffusion pumps of 80 cm diameter, they should be placed on the side of the H.F. system to provide enough place for the experiments, there was a preference for stainless steel as material, and the tank should be made in one piece (contrary to Pittsburgh). However during the following months very little work could be done, as the design of the magnet and the H.F. system was still too vague.[21]

At the beginning of 1953 the subgroup obtained information on these two components and could therefore prepare first drawings. In their March progress report it was clearly stated that the basic design would follow that of other cyclotrons and that no major differences were envisaged. The total volume to be evacuated would be 23 000 litres and the expected presure in the tank 3×10^{-6} mm Hg (5×10^{-6} mm Hg with a beam of particles). It was to take less than 40 minutes to reach a pressure of 10^{-2} mm Hg with rotary pumps, at which point the two diffusion pumps would be put into operation; the desired pressure would be reached after about 70 minutes. Thus except for details the design was already clear.[22]

From then on, a major part of the subgroup's work consisted in collaborating with the Control subgroup to develop a good vacuum control system. In addition, contacts were established with firms able to produce the vacuum equipment so that no time would be lost when the design was finalized. This was the case by the end of 1953, whereupon the drawings giving all the necessary information to the manufacturers could be made. However, it was not until February 1955 that these were actually mailed. It had been necessary to wait for decisions on the contracts for the manufacturing of the magnet frame and the coils since changes in these two components would have had consequences for the vacuum system.

Eight firms were invited to tender, but only four of them replied, and in fact only two of them were competitive from the point of view of price. Although Leybold

Notes: p. 132

Hochvakuum-Anlagen (FRG) was not the lowest in price, this firm got the contract for the pumping system. Reasons for this choice were discussed in the meeting of the Finance Committee of 8 June 1955. They lay mainly in the lower overhaul time per year, due to the design of the diffusion pumps and to the rapid cooling device, which would, if necessary prevent the SC from being down for four working days. The contract for the vacuum chamber was given to the Swedish firm Avesta.[23]

A fourth subgroup dealt with the *shielding and control system*[24] of the accelerator, both important parts in the design of such a machine. The layout of the shielding, for example, is closely connected to the design of the building as a whole and especially to the design of the experimental rooms. A badly conceived shielding system would thus not only be dangerous but background radiation would also be a major handicap for the physicists doing experiments. The control system ensured smooth functioning of the accelerator and, in the event of a problem in one part of the machine, it provided security for the other parts. The subgroup's headquarters were in Paris. The leader was P. Debrain, who was joined in November 1952 by M. Lazanski, a Yugoslavian engineer.

As far as the control system was concerned, there was no existing model for it. Each machine had to have its own, meeting particular needs. The final layout depended thus on the design specifications of the various main components and on the way it was planned to operate them. The first step Debraine took was to clarify policy questions in building-up the control system and in organizing and delineating the various types of equipment. By August 1952 it was already generally agreed that there should be one central interconnection point in the control room so that all the switching could be done from there.[25]

While waiting for the preliminary design, and a sketch of the operating procedure, of every piece of equipment developed by the other subgroups, various control circuits and their running conditions were studied on the assumption that the SC would be of the standard type (with rotating condenser). The idea was to work out something like a purchasing list with prices and a corresponding complete design record.

In the next step the proposal for the control system was discussed component by component with the subgroup concerned and the manufacturers. However it seems that this was not without its difficulties, for Debraine and Lazanski felt it necessary to spell out clearly the task of their subgroup in September 1953. 'The building up of such a group from the very beginning implies that the SC Group realized the necessity for the control system to be designed from a centralized point of view and built as homogeneously as possible.' They continued by explaining that they would take responsibility only for a system that they had put together themselves and that they would not assemble pre-existing equipment chosen by somebody else.[26]

From then onwards the work continued gradually in constant interaction with the technicians and scientists involved, up to the ordering of the standardized alarm and control units early 1955.

The second task, namely, to design the shielding was handled by three different groups, the Laboratory group, the subgroups in charge of the general layout and the Control subgroup. Indeed each time the energy of an accelerator was increased it was difficult to estimate the final size of the shielding, hence difficult to predict exactly the type and intensity of radiation that would occur. So as a first step an attempt was made to evaluate realistically the expected neutron-flux density all over the SC building. This was done by Pickavance and Cassels in October 1952 on the basis of the Harwell SC and gave values between 10^6 and 10^7 neutrons/cm^2/sec.[27]

Having settled the question of intensity, the time came to make estimates for the wall thickness for given concrete densities. Experiments with baryte-loaded concrete were started in order to learn about the stopping power of such a material. It was also agreed that for health reasons there should be no regularly occupied rooms within 50 meters of the centre of the SC. The handling of the huge shielding walls which appeared to be necessary would be the responsibility of the subgroup in charge of the layout.

In the final version the walls facing the experimental rooms were supposed to be 5.8 m thick, that of the wall towards the equipment room 5.1 m. For these three walls baryte-loaded concrete of density 3.6 was to be used. The fourth wall 4 m thick would consist of normal concrete of rather low density 2.4 (see fig. 4.3). Altogether some 22 000 tons of concrete were to be used in constructing the shielding. For the neighbouring walls of the experimental rooms, large lifting platforms were to be installed to facilitate the handling of the concrete blocks out of which they were constructed. Heavy sliding doors were envisaged to optimize the protection of the entrances to the cyclotron hall.

The last subgroup was in charge of the *general layout*. This meant coordinating the wishes and needs of physicists and technicians with the boundary conditions imposed by the design of the machine. The three people in charge, all working as consultants, were T.G. Pickavance, J.M. Cassels, and M.J. Moore. The two first were from the accelerator group at Harwell, and Moore we have already met: thus all three were accustomed to the needs and practical problems involved in the running of such an accelerator. Their task was to determine the size and form of the SC building, to arrange the different rooms needed, to organize the building up and the handling of the shielding, and to design the crane and lifting platform structure.

The first preliminary proposals were presented in October 1952. A two-storey building was foreseen including already two experimental halls, one on each side of the machine. The proton beam was to be extracted from the vacuum chamber on the opposite side of the neutron and meson beams. The mesons were to be produced at an internal target which would be movable from outside. On the 'proton-side' of the SC a concrete wall was planned with slits through which the protons would enter the experimental room. On the 'neutron-meson side' the shielding was of concrete blocks

Notes: pp. 132 ff.

rather close to the magnet to take account of the decay in flight of the mesons.[28]

In January 1953 the initial plan for a two-storey building was abandoned for financial reasons, and attention was then directed towards the exact layout of the neutron and proton walls. The idea was to construct them in such a way that they could be relatively easily rearranged in order to allow different particle beams for the experiments. After lengthy discussions between the architect Steiger and the Laboratory group detailed though not final drawings were ready by the end of 1953. The crane structure for the whole building was also specified. As for moving the shielding blocks, a lifting platform above the 17 m long neutron wall was advocated, as well as a smaller device above a part (6.5 m) of the proton wall.[29]

On June 1954, the final contract between CERN and the architect was signed. This meant that he was entrusted with the plans for the construction of all the buildings envisaged at that time on the Meyrin site. In September 1954 the design was essentially frozen. The lifting platforms were then ordered from Ruhrstahl, the cranes from Mohr and Federhaff, and the steel construction of the building from Ateliers de Construction Metallique (CH).[30]

4.2.3 THE CONSTRUCTION PERIOD (1955–1957)[31]

Work on the site in Meyrin began in the middle of June 1954, so already three months before the Convention entered into force; the excavations were completed by the end of the year. In October a Site and Buildings Division whose members had been formerly attached to the Laboratory group was set up with responsibility for all the construction on the site. By May 1955 the building of the SC was finished up to ground level. This was followed by the construction of the shielding and the roofing which was completed in September. During the last months of 1955, the internal walls were erected, the cranes which had been delivered were assembled and tested, the lifting platforms were built, and the work on air-conditioning equipment, plumbing, etc., was started.

To give an idea of the huge work the SB Division had to organize all over the site we wish to give the following figures: 115,000 m³ of earth was excavated, 4.6 km of pipes laid, 11,000 m³ of reinforced concrete was used requiring a total of 428,000 man hours.

One design feature of the SC building is worth pointing out. The entire complex was constructed like a boat with the SC hall in the middle and the experimental rooms to the left and to the right. Thus, if the weight in one of the experimental rooms became too high owing to heavy equipment used for experiments—thus threatening the stability of the whole construction—this could be counterbalanced by filling the foundation of the experimental room on the other side. In this way the whole building could float on the layer of sand on which it was constructed.

In August 1955 Wolfgang Gentner became leader of the SC Group, thus replacing Cornelis Bakker, who took over from Felix Bloch as Director General. During the last years of World War II, Gentner had built the 12 MeV cyclotron in Heidelberg. After the war he was Professor in Freiburg, where he worked on natural radioactivity and cosmic radiation. From the very beginning, in 1952, he had also worked for CERN as a consultant to the PS group. However for him the decision to lead the SC Group was not easy. First of all he was more of a physicist than an engineer, and not that interested in the technical work still to be done in 1955. Secondly he did not want to leave his institute completely, where he had a number of different projects running. Nevertheless he decided to join CERN on a full-time basis. Since he was mainly interested in the physics one would be able to do with the SC, he started to build up experimental teams in 1955 and also to prepare the necessary experimental equipment to be ready in time.[32]

At the end of 1955 the SC Group had acquired a staff of 32. The yoke and one coil arrived and the assembly of the magnet could be started. Further the control and alarm circuits that had been delivered were put through their acceptance tests.

Early in 1956 the low-voltage supply was available for the group. The magnet had been assembled and the early measurements showed that the eccentricity was only 0.11 mm. Thus a start was made on placing the shim plates in position. The motor-generator set, which was to be commissioned in August, still gave some problems which had to be solved by the manufacturer BBC. As a result the initial plan for testing the coils before assembly of the vacuum tank had to be abandoned because of the tight schedule. At the end of August, after a few days of vacuum tests, the diffusion pumps were installed. The supply system for the cooling water had been completed in the first half of the year. The control system was also well under way. Thus by this stage, no major delay had occurred to disrupt the overall schedule.

Indeed the only device causing serious problems was the H.F. modulation system. The measurements on the tuning fork showed a permanent distortion of the prongs. In particular, there was a dynamic distortion due to unwanted modes and the temperature gradient. The deadline for its delivery had originally been December 1956, but it was clear that the solution of these problems would still take some time, and that delays were inevitable. Thus it was decided that the construction of a rotating condenser system should start in parallel in case these technical difficulties could not be overcome.

During the year the staff had nearly doubled and was 61 at the end of 1956. The SC building was completed, the magnet construction finished, the control system was largely working, the vacuum system was assembled, and the cooling system was in place. Only the H.F. system was still not ready. Meanwhile the last stage of preparation got under way—the extraction channel for the proton beam, the ion source, the targets, bending and focusing magnets, and so on, were made ready.

In the first months of 1957, all parts of the accelerator were finally tested. In March the H.F. system was delivered even though there were still problems with the tuning fork. It was decided that it should run *in situ* under actual operating conditions. The dee and the dee liner were installed followed by all other necessary

Notes: p. 133

parts like the control and supply apparatus. Many adjustments had to be made until the installation was in full operation. However there were considerable difficulties in operating the tuning fork in the high magnetic field, difficulties which had never come to light in the model tests. The seriousness of the situation was well expressed by Bonaudi: 'Still, we fixed the date of inauguration of the cyclotron. We wrote an invitation to everybody, including ministers, and governments and so on, city officials, universities [...]. And then the H.F. system was not working, and it was only one week away'. However the problem was solved at the last minute by using an additional coil to compensate the magnetic field outside the gap. Moreover the difficulties of having parasitic modes were overcome by mounting stiffening ribs on the prongs of the tuning fork.[33]

On 1 August 1957, the CERN synchro-cyclotron produced its first beam. Neutrons from an internal target were detected and their energy was measured. The full energy of 600 MeV had been reached.

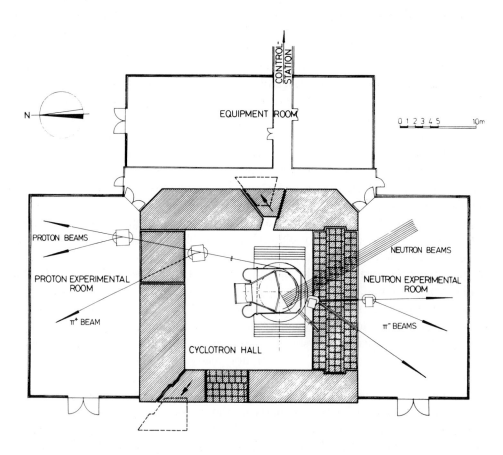

Fig. 4.3. General layout showing the various beams leaving the SC[34]

4.3 Particle physics with the SC from 1958 to 1965

4.3.1 A SURVEY OF ACTIVITIES AROUND THE MACHINE

The months following the first operation of CERN's small accelerator were entirely devoted to finalizing and testing parts of the machine and preparing the necessary infrastructure to allow for a start of the experimental programme by early 1958.[35] Thus the SC Division under Wolfgang Gentner (Director) and Gilberto Bernardini (Director of Research from September 1957) spent most of the time on technical and organizational tasks. Considerable effort was devoted to the design and building of instruments and apparatus to be used with the accelerator, while physicists and technicians were recruited to build up the research teams themselves, both in their internal structure and in their contacts with the scientific community outside. At the same time the properties of the internal proton beam had to be studied in order to learn about its energy, focusing, and intensity at different radii. Two types of ion sources were constructed and tested, the bending magnets which had been ordered for setting up beams were controlled for field strength and homogeneity, and the shielding walls between the machine and the experimental halls were erected.

To prepare for the future experiments it was crucially important to investigate the energy-intensity-angle correlations of the secondary beams. Generally these beams were produced either by an internal target and then deflected into the experimental hall, or via a proton beam directed onto an external target. To begin with, the internally produced beams were studied using a target positioned at different radii. The negative pions emitted in the forward direction were deflected in the fringing field of the cyclotron and could leave the machine if their energy was greater than 100 MeV. The same was true for positive pions, which were produced in the backward direction relative to the incoming protons. The mesons leaving the machine were then collected by a magnet immediately outside the machine. (For a layout of the experimental halls see fig. 4.3)

Apart from these different pion beams A. Citron and H. Øverås pointed out that the accelerator could serve as a tool for extensive studies in muon physics. These muons arise as decay products of the pions and the idea was to construct a focusing channel to obtain a high-intensity purified beam. The great scientific interest of such an installation was widely agreed on, and a detailed study of the channel got under way.

As far as the planned pion experiments were concerned, the preparation of the equipment started. This stretched from building a variety of electronic parts, two hydrogen targets, and a lead glass Cerenkov counter to the setting up of the first negative pion beam roughly 200 m in length.

By the end of 1957 the SC Division had grown to 109 members including 26 physicists and 19 fellows and research associates. It comprised eight groups in all:

Notes: p. 133

three counter teams (under G. Bernardini, A. Lundby and A.W. Merrison), a spallation group (K. Goebel), a polarized proton group (R. Keller), a mu meson group (A. Citron), an emulsion group (W. Gibson) and an electronics group (H.I. Pizer). The latter had moved during the year from the Scientific and Technical Services (STS) Division and was responsible for giving support to the SC users concerning electronic parts.

By spring 1958 the real experimental work began. The accelerator operated for a total of 3200 hours during this year, in the second half, for as much as 24 hours a day, six days a week. As for beams, the physicists had access to an internal p-beam of 0.3 μA flux, a far less intense external p-beam of roughly 0.02 μA (corresponding to 1.1×10^{11} protons/sec), and both positive and negative pion beams with energies ranging from 70 to 320 MeV with a maximal intensity of 4×10^5 pions/sec (for π^+ at 150 MeV) and 10^4 pions/sec (for π^- at 70 MeV). In addition decay muons could be used for research.[36]

Among the particle-physics experiments three gave results during the year. Two of them were devoted to studying aspects in weak interactions. This had been triggered partly by the new situation in this field—the discovery of parity violation which opened the way to the V–A theory, and the strong impact this had on the scientific community. The first, by the Fidecaro–Merrison group, was a search for the rare decay mode of the pion into an electron. Within a few months they managed to prove quantitatively the existence of this reaction and made a first measurement of the branching ratio to test theory. This was in a way CERN's first big scientific success and so contributed essentially to the good start of the scientific programme of the small accelerator and of CERN's scientific life in general. We will therefore discuss it in more detail in section 4.3.2.1. The second weak interaction experiment was the one done by Lundby's team. Its aim was to see if muons and electrons were equally coupled to nuclei. This was achieved by measuring the relative rate of muon capture in C^{12} to the beta disintegration rate of the so formed B^{12} nucleus. Their result also confirmed the theoretical prediction within the experimental limits.[37] The third experiment, by Dick Harting's group, dealt with charge independence in strong interactions. They measured the ratio of the cross sections for p + d $\rightarrow$ He^3 + π^0 and p + d $\rightarrow$ H^3 + π^+ and found a value very close to the theoretical one (2.2 $\pm$ 0.07).[38]

Apart from that we should also mention Citron's team who continued their preparation of the muon channel. Further the 10 cm hydrogen bubble chamber, constructed by the STS and SC Divisions, was exposed to a 270 MeV pion beam. This was CERN's first bubble chamber and was primarily constructed to serve as model for the 30 cm chamber. About 6000 photographs were analyzed for scattering events.[39]

Three other teams were working on the SC along with the above. We shall only briefly touch on their work here, and will not return to them again. The first we wish to mention were the physicists using emulsions, a technique mainly applied in cosmic-ray research. Their way of experimenting, i.e. by exposing photographic plates to a beam and then distributing them and investigating them from various scientific points of view, makes it difficult to follow up all the investigations performed with this detection method. In addition it played only a rather small role in the particle-physics programme. We thus decided not to describe this aspect. However it is worthwhile stressing that at an organizational level this 'subcommunity' of experimental physicists was the first at CERN to plan their work around the accelerators in the long-term. They started to discuss the kind of emulsion experiments possible in 1958, as well as the type of collaboration which would be needed in the future. As a result the emulsion physicists were the first group to be organized for the work with the PS, and they as well as cosmic-ray physicists served as an example for bubble-chamber physicists when organizing collaborations.[40]

The two other groups we want to touch on here were the spallation group and the nuclear chemistry group. The first started working at Geneva University and moved to Meyrin in the course of 1958. Their speciality was the study of the tritium content in meteorites. Together with the knowledge of the absolute tritium production cross section in different materials which they had determined with the help of the SC, they were able to draw conclusions on the intensity of cosmic rays in outer space. Their experiments, illustrated that an accelerator like the SC enabled one to do 'laboratory experiments' which contributed to the understanding of astrophysical problems.[41]

The nuclear chemistry group was founded in spring 1958 and was initially mainly concerned with studies such as the fission of uranium or the spallation of arsenic. However as the number of particle-physics experiments decreased so the activity of this group around the SC diversified and grew in importance. In the second half of the 60s nuclear chemistry was actually to become one of the main activities at the SC.[42]

In addition to CERN groups it was foreseen that the SC would be used by physicists from institutions outside the laboratory—the so-called truck teams. Such teams began to arrive with their equipment in 1958 and their experiments started from mid-1958 onwards. The first came from Harwell, Liverpool, Padua, Saclay, Trieste, U.C. London, and Utrecht.[43]

The first year of full operation of the synchro-cyclotron was 1959. The Annual Report identified the work of no less than 13 groups from inside CERN and of 6 outside teams. The schedule was tight and the number of users had soon reached a limit above which effective working was complicated or even impossible. The staff of the SC Division had grown to 184 not counting the visitors, an increase of 25%, mostly due to the large number of technical staff necessary to run and improve the machine. As far as the latter was concerned, two groups had been set up, one to study stochastic acceleration in order to increase the beam intensity, the other to investigate the feasibility of a polarized ion source.[44]

Seven of the experiments carried out dealt with topics in particle physics. Four groups, more or less continued or extended the investigations they had undertaken in

1958: Citron's team started setting up the muon beam and preparing the layout of their first experiment. Fidecaro and Merrison extended the $\pi \to e\nu$ experiment to a search for the $\mu \to e\gamma$ decay and began working on a precision measurement of the Panofsky ratio. Harting continued improving the charge independence investigations and Lundby chose to determine helicity and lifetime of the muon in order to test the validity of the conservation of the weak-coupling vector current (CVC hypothesis) in the presence of strong interactions.[45]

One of the three new CERN groups working at the SC was that led by Francis Farley. They planned a precision measurement of the anomalous magnetic moment $(g-2)$ of the muon, a crucial test for the predictive power of QED and a way of indicating if there was a fundamental difference between the muon and the electron. Because of its important role we will describe this experiment in detail in section 4.3.2.2. The second group led by E. Zavattini investigated the inverse photoproduction reaction $\pi^- p \to \gamma n$. Before this the cross-section for the photoproduction of pions off free neutrons had been extracted from photoproduction off deuterons, a procedure which called for making many assumptions.[46] The third group around Bernard Hyams tried to determine precisely the experimental limit to the stability of the nucleon.[47]

In November 1959 CERN's 30 cm hydrogen bubble chamber was tested for the first time at the SC. 100,000 photographs were taken of positive pions of 265 and 330 MeV looking for reactions $\pi^+ p \to \pi^+\pi^- n$ and $\pi^+ p \to \pi^+\pi^0 p$. This was meant to serve as an extensive test of the chamber, which was later transferred to the PS and was the first bubble chamber taking pictures of high-energy interactions at CERN.[48]

Among the truck teams coming from institutions in the Member States we should like to mention a group from Utrecht investigating pion scattering by nuclei, a Harwell–UCL collaboration studying the parity non-conservation in π^0 decay, a team from Harwell working on polarization in n–p scattering, and a group from Padua looking at π–p elastic scattering in their 2 litre propane bubble chamber. The SC was also used to test the fluoro-carbon bubble chamber from Fribourg, and a team from Bologna brought their 20 cm HBC and exposed it to a π^+-beam of 120 MeV to study $\pi^+ p$ scattering.[49]

At the end of 1959 the 28 GeV proton synchrotron operated for the first time and the focus of interest at CERN shifted to this new and fascinating tool. Concretely this meant that many of the CERN teams finished off their experiments at the SC and went over to work at the PS in 1960. The SC was shut down from November 1959 until April 1960 in order to carry out major improvements. It was modified so that several experiments with secondary particles of different energies could be carried out simultaneously. The neutron room was arranged in such a way that one experiment using neutrons and three counter experiments working with muons at 280, 125 and 80 MeV could be done at the same time. As a result only 500 effective machine hours and 600 h of parasiting could be allocated. As for the attempts to increase substantially the beam intensity of the SC by using stochastic acceleration, they were of rather limited success, leading to an improvement by a factor 2 — only an increase of one or two orders of magnitude would have allowed one to open up a new field of research in pion and muon physics. Meanwhile the size of the SC Division had reached 223 by the end of 1960, an expansion which was again due to taking on 25% more technical staff.[50]

1960 was also the first year in which we see an equilibrium in the participation of outside teams relative to CERN groups. Thus we find records on the work of eight CERN teams, five of which were carrying out particle-physics experiments, and seven outside groups. Among the results obtained the most outstanding was the precision measurement of the anomalous magnetic moment of the muon. In the first run the error margin of $(g-2)$ was decreased from about 10% to 1.9% and the value found was in excellent agreement with the predictions of QED. Other work contributing to the understanding of pion and muon physics were an experiment on the Panofsky ratio, the spectra of decay electrons emitted by a captured muon, and the inverse photo-mesic effect. In addition Citron's μ-channel was completed during the year and gave a μ-beam far superior in quality to what had been expected. As a result the study of muons and their properties was to become an important part in the SC research programme. A second new line of research explored was the use of polarized beams, the feasibility of which was investigated in depth by a CERN group.[51]

As far as the visiting teams were concerned four of them — from Utrecht, Harwell, UCL, and Bologna — continued their work. New topics were the polarization of recoil protons in π-p scattering and the study of pp interactions at 591 MeV with the 20 cm bubble chamber. Three other groups arrived from Rome, Trieste, and Darmstadt. Among their experiments one rather interesting idea was that put forward by the Rome group under M. Conversi. They suggested using Citron's μ-beam to look for the neutrinoless conversion of a muon into an electron, i.e. $N + \mu^- \rightarrow N + e$. This would serve as a crucial test for the understanding of the relation muon-electron. This question was treated extensively in the ν-experiment carried out with the PS in the early 60s when the non-existence of such a reaction was clearly proven. The two other groups mentioned above investigated mu-mesic atoms and the forward scattering amplitude in the charge-exchange process of π^- in hydrogen.[52]

At the start of 1961 the new internal organization of CERN came into operation, along with a restructuring of the staff into new divisions.[53] Thus an MSC Division was set up for improving the machine, for guaranteeing its optimal functioning for experiments, and for developing it further. All those actually doing experiments (with the exception of the bubble-chamber physicists) were gathered in the so-called Nuclear Physics (NP) Division. Furthermore the clear separation between CERN groups and visiting teams was abandoned, a fact which is reflected in the presentation of the Annual Report. As for the synchro-cyclotron itself, it did experiments for no less than 6285 h with 1407 h of parasiting. At the same time beam intensity was increased from 0.3 μA to 0.8 μA in two steps, firstly by mixing a small

Notes: p. 133

amount of argon with the hydrogen supply of the ion source and secondly by improving the electrostatic focusing near the ion source.[54]

Despite the good working climate which had been built up around the SC, the future of the small accelerator was put in doubt at the end of 1961. The reason for this was the threatened limitations to CERN's budget for 1962; one way of guaranteeing the unimpeded continuation of the PS was to close down the SC and to transfer staff and equipment to the bigger accelerator. And although this suggestion of abandoning the smaller accelerator was probably more or less a political weapon in a budgetary discussion — and indeed the budget voted for 1962 was considerably higher than that for 1961 —, it caused a rather uneasy feeling inside the division for a short while.[55]

Scientifically speaking, as we have said, the centre of interest in the SC research programme had now moved to muon physics. Here we should mention the scattering of muons by various elements, muon capture in hydrogen, the neutrinoless conversion of muons into electrons, and the decay of muons from K-orbits around Fe. All of these experiments used the Citron beam. Another attempt was made to improve the accuracy in the determination of $(g-2)$, and the error was further reduced by a factor 4. The experiments on μ-mesic atoms, on charge exchange scattering, and on inverse photoproduction were continued and one new experiment searching for the ω^0 produced in the reaction $\pi^- p \to \omega^0 n$ and having a mass of 305 MeV was started. Work at Berkeley hinted at this diboson, but in the CERN experiment no evidence for such a particle was found.[56]

Throughout 1962 a big effort was made to improve further the machine performance in three particular ways. Firstly, an attempt was made to increase the machine intensity. In the long term this required modifying the RF system and studying the conditions in the centre of the machine; however as an immediate step the mean magnetic field was increased, so that by June 1962 the beam intensity had reached 1.2 μA. Secondly, the extraction and the optics of the secondary beams were studied carefully. And lastly, the possible installation of a polarized proton source, which would open up completely new experimental possibilities, was looked into.

As parts of the accelerator (such as rubber joints) showed ageing problems due to the strong irradiation, and as faults further increased as machine performance was pushed to the limit, the time lost due to breakdowns was double that in 1961. Nevertheless this could be compensated for by organizing the maintenance more efficiently so that the machine time for experiments decreased by only 10% to about 5610 hours. What is more the improvements allowed the parasiting time to double to 3070 hours.[57]

On the scientific side muon physics was still the dominant subject. These experiments embraced the scattering of muons on carbon (carried out by Citron's group) to test if the muon would behave like a heavy electron, the capture of μ in liquid hydrogen using the 20 cm Italian bubble chamber, the radiative capture of μ^- which had hitherto been unobserved, a precision measurement of the muon lifetime, the formation of mu-mesic molecules and the study of muon depolarization in a magnetic field.

Finally two other experiments, both carried out by a group around Heintze, Rubbia, and Soergel, should be mentioned. The first was a study of the β-decay of a charged pion, i.e. $\pi^+ \to \pi^0 e\nu$. As it represented an essential test for the validity of weak interaction theory it will be discussed in section 4.3.2.3. The second experiment carried out in parallel investigated the radiative decay of positive pions ($\pi^+ \to e^+ \nu\gamma$), a process containing information on the non-locality of the π-decay.[58]

In 1963 the proportion of breakdown increased again by nearly a factor 2 illustrating clearly the seriousness of the problem. Nevertheless with a lot of effort it was possible to offer roughly the same amount of machine time to experimenters as in 1962. As far as the improvement programme was concerned the study of a new RF acceleration installation, the extraction problem for secondary beams, and the conditions at the centre of the machine were still studied. An important step was taken in January when a new RF amplitude modulator was put into operation, again increasing the machine intensity to $1.6\ \mu A$.[59]

Speaking generally, nuclear chemistry and physics began to play an increasingly important role in the experimental programme. Among the particle-physics experiments four dealt with topics in weak interactions: the radiative muon capture in Ca^{40} was studied, a process which allowed the determination of the pseudoscalar coupling strength; the measurements on the reactions $\pi \to e\nu\gamma$ and $\pi^+ \to \pi^0 e^+ \nu$ were continued and gave numerical values for vector and axial vector form factors; the e^+ polarization in μ-decay was measured by two different methods, and the results were more or less in agreement with the values predicted by V–A theory. In addition an experiment on the charge exchange reaction of π^- in hydrogenic substances, a search for the so-called ABC anomaly in $\pi\pi$ interactions at low energy, and a study of μ capture in O^{16} were carried out.[60]

After the shut-down in January and February 1964 the SC could be used very intensively and nearly 6000 hours of machine time and 4000 hours of parasiting was available to physicists. By now the machine was able to deliver beams to one main user and to three parasiting groups. Furthermore the utilization of the μ-channel was improved by extending the neutron room.[61]

At the same time the increasing success of the PS in the field of particle physics had as a consequence that the SC programme shifted more and more towards nuclear physics and chemistry topics. Indeed only three of the experiments reported in the Annual Report for 1964 could strictly speaking be classified as particle-physics experiments. Two of them were a continuation of experiments, already started, namely the π^+ beta decay and the search for the ABC resonance. The third was on μ capture in hydrogen gas. The nuclear structure programme contained experiments such as inelastic scattering of pions on light nuclei, pion double charge exchange in nuclei, muonic X-ray transitions and many more.[62]

This trend was even more accentuated in 1965, the end of the period covered in this

Notes: pp. 133 ff.

chapter. The importance of nuclear physics work was further underlined by the construction of the on-line isotope separator ISOLDE which had been approved by the Nuclear Physics Research Committee in December 1964. During the shut-down of the machine in January and February improvements were again made as a result of which the external proton beam intensity increased by a factor 2.[63]

During the 5311 hours of machine time only three particle-physics experiments were done: a continuation of the study of μ capture in H_2 gas, the search for charge conjugation violation in $\pi^0 \to 3\gamma$ decay, and the investigation of pp scattering at 600 MeV on a polarized target. Finally it is interesting to see that this shift in the central scientific topic in the SC research programme was formalized in the Annual Report by clearly separating for the first time 'Elementary particle physics' from the 'Nuclear Structure Programme'.[64]

4.3.2 FOUR CASE STUDIES ON THE MAIN RESEARCH TOPICS CARRIED OUT WITH THE SC

4.3.2.1 The electron decay mode of the pion ($\pi \to e\nu$)

The long tradition of weak interaction research for which CERN is well known began as early as 1958 with an experiment on the newly completed SC. Remembering the scientific situation, we realize that this was only a few months after the suggestion of V–A theory by Sudarshan and Marshak, Feynman and Gell-Mann, and Sakurai which caused a breakthrough in weak interaction physics. Nevertheless although this theory at first seemed to be the solution, several experimental results could not be reconciled with this hypothesis. These were the $e\nu$ angular correlation in ^{6}He, the sign of electron polarization from μ decay, the frequency of the electron mode in π-decay, and the absence of reactions like $\mu \to e\gamma$ or $\mu \to 3e$ although theoretically no rule excluded them. This of course made weak interaction physics an attractive and challenging subject to study particularly with respect to testing V–A theory.[65]

At CERN in early 1958 a small group around Alec Merrison and Giuseppe Fidecaro suggested using the relatively high intensity pion beam to carry out a systematic search for a rare decay mode of the pion, $\pi \to e\nu$. This reaction had been proposed by H. Yukawa in 1935 to explain β-decay but—although searched for several times since the discovery of the π-meson in 1947—no one had yet observed it. Now, however, using V–A theory the branching ratio of this rare mode relative to the muon decay mode was calculated to be

$$R = \frac{\pi \to e\nu}{\pi \to \mu\nu} = 1.2 \times 10^{-4},$$

thus confirming the rate predicted on the basis of electron–muon universality.

There was a major experimental difficulty however: the large background of electrons coming from the $\pi \to \mu \to e$ chain. Thus the electron energy spectrum had to be determined as precisely as possible in order to distinguish the interesting events from the rest. This was done using the set-up shown in fig. 4.4.

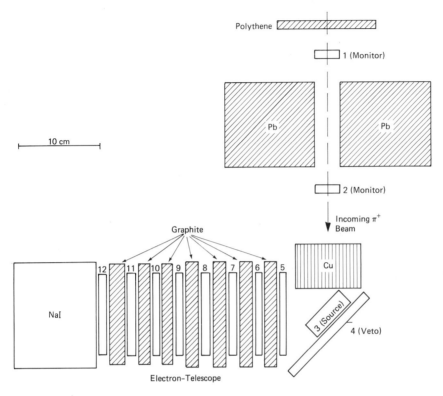

Fig. 4.4. Experimental set-up for searching for $\pi \to e\nu$[66]

The π-beam, filtered from protons of the same energy, then collimated and slowed down in a copper block, was brought to rest in a scintillation counter. The energetic particles coming from this counter were examined in a range telescope consisting of 8 counters with graphite in between them which allowed one to discriminate the electrons coming directly from pion decay from the background. Complete time information was recorded by photographing the pulses on a fast travelling wave oscilloscope. A large sodium iodide crystal (NaI) was situated behind the range-telescope, delivering additional information, namely no, or a very small, pulse in the $\pi \to \mu \to e$ case, and generally a large pulse when it was a $\pi \to e$ event.

By September 1958 the existence of this decay mode could already be regarded as established. Some 40 examples of the rare mode $\pi \to e$ had been observed allowing one to make a first estimate to the lower limit for the branching ratio of

$$R = \frac{\pi \to e\mu}{\pi \to \mu\nu} > 4 \times 10^{-5}.$$

Notes: p. 134

This result stimulated the interest and admiration of the international physics community. However for CERN—as we can see from the press conference called immediately afterwards—it was much more than a successful experiment; it was regarded as a first landmark for the Organization in the field of elementary-particle physics, and it created an enthusiastic atmosphere in the laboratory.[67]

In order to correct this first rough value for the detection efficiency of the telescope the Fidecaro–Merrison group made use of the new Mercury computer which had arrived in the summer of 1958. More precisely they applied the so-called Monte Carlo Method simulating a series of statistical experiments on a random basis. It was also the first scientific use of this computer. Thus by mid 1959 a final result for the branching ratio of this rare decay mode could be given, based on some 73 events accumulated before the end of 1958, namely

$$R = (1.22 \pm 0.3) \times 10^{-4},$$

in excellent agreement with the predictions of V–A theory.[68]

Having described this particular successful experiment let us now quickly summarize the continuation of the research of the Fidecaro–Merrison group. As a first step they used the same experimental set-up to make a precision measurement of the mean life of π^+. The value at the time was $\tau_\pi = (25.6 \pm 0.5)$ ns, an average over seven experiments based on about 1100 samples of pion decays. At CERN it was possible to photograph in less than one hour of machine time the pulses of about 8000 $\pi\mu$ events. The analysis was then done using a so-called IEP, an instrument designed to measure bubble-chamber photographs, and the Mercury computer. The result was

$$\tau_\pi = (25.46 \pm 0.32) \text{ ns}$$

a figure having the highest statistical accuracy yet reached.[69]

The team stayed in the same line of research by starting a search for the $\mu \to e\gamma$ decay, or at least determining its degree of forbiddenness, in 1959. As we have already said, the absence of this process was one of the most interesting problems in weak interactions. The experimental set-up consisted of an electron range telescope, behind which an NaI counter was placed. The result was a negative one within the limit of sensitivity of the apparatus, thus giving

$$\frac{\mu \to e\gamma}{\mu \to e\nu\bar{\nu}} < (1.2 \pm 1.5) \times 10^{-6}.$$

In addition the decay $\mu \to e\nu\bar{\nu}\gamma$ was observed directly for the first time and its rate was consistent with theoretical expectations.[70]

Having an NaI counter of excellent resolution at hand, the next and last experiment they performed with the SC, was a precision measurement of the Panofsky ratio

$$P = \frac{\pi^- p \to n\pi^0}{\pi^- p \to n\gamma}.$$

This work began in 1959 and was finished by 1960. The correlation between low energy pion–proton scattering and pion photoproduction had been shown in 1952 by Anderson and Fermi. The early discrepancies having been removed in the middle of the fifties by two experiments in Liverpool (both Fidecaro and Merrison had been involved), the group felt that a precise redefinition of the Panofsky ratio and more accurate measurements were worthwhile. The group found

$$P = (1.533 \pm 0.021),$$

the most precise value at that time.[71]

In 1960 the Fidecaro-Merrison team finished their experimental work at the SC and were among the first to start working with the PS.

4.3.2.2 A precision measurement of the anomalous magnetic moment of the muon: the (g-2) experiment

One of the most interesting puzzles in particle physics was, and still is, the relation between the electron and the muon. Here physicists had two particles apparently behaving in exactly the same way, though differing in mass by no less than a factor 200. Hence the wish to determine a distinguishing feature of these particles.

Muons, as well as electrons, have no interaction, other than electromagnetic and weak. The former, which we want to focus on here, was described by quantum electrodynamics (QED), an extremely valuable theory developed in the late 40s. In the framework of this theory muons and electrons take part in virtual interactions with the electromagnetic field, represented by the following Feynman diagrams in fig. 4.5.

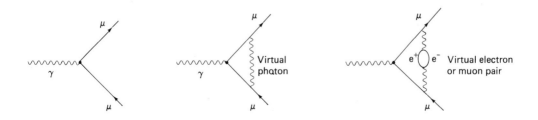

Fig. 4.5. Feynman diagrams for radiative corrections to the muon vertex

Because of these interactions the gyromagnetic factor g which, according to Dirac should be 2 for spin $\frac{1}{2}$ particles, differed from this value. By measuring this difference for the muon and comparing it with the value calculated on the basis of QED, one thus had a perfect test for the predictive power of the theory. What is more, any clear deviation from the QED value of $\frac{1}{2}(g-2) = 0.001165$ would imply that there was a fundamental difference between the electron and the muon.

This then was the general idea behind the experiment suggested by F. Farley and R. Garwin.[72] More precisely they thought of determining $(g-2)$ by letting circling polarized muons, which were naturally produced in π-decay, drift slowly in a large magnet. If g differed from 2 then the direction of polarization of the muon would gradually change as it turned in the magnetic field. Thus by measuring the spin direction of the muons before and after they had circulated in a magnetic field for as many turns as possible, one would be able to calculate $g-2$ precisely. Until this time $(g-2)$ had always been determined indirectly, and to an accuracy of only about 10%. Three difficulties had to be overcome in practice: the muons had to be injected and extracted from the magnet without producing disturbing effects, they had to be stored for as many turns as possible in a magnet, and the initial and final polarization angles had to be measured as precisely as possible. To get started the group borrowed a magnet from the University of Liverpool in 1959. It was 83 cm long, 52 cm wide, and had a gap of 10 cm. They managed to inject a muon beam in such a way that the particles made 30 turns up to the end of the magnet.[73] The intensity of trapped particles at the end of the magnet was 20 μ/sec. Extrapolating from this first test they decided that a magnet of about 6 m long with the same width was needed in order to reach 500–1000 turns—the number estimated to obtain the desired accuracy of 1% in $(g-2)$—and a magnet of this size was ordered.

Meanwhile some changes were made on the 'Liverpool' magnet resulting in about 100 turns, though with a lower intensity and without the possibility of extracting the muons properly. Nevertheless a first rough estimate of $(g-2)$ was made, and a value 1.8 times the theoretical value was obtained. While waiting for the big magnet to arrive, the final details of the experimental arrangement were settled. A large amount of auxiliary equipment was designed and constructed such as the vacuum chamber, the magnet support, the injection magnetic channel, a variety of counters and, in particular, the system to measure the polarization of the ejected particles.

By July 1960 the 6 m magnet with a gap of 14 cm had arrived and the final preparations for the experiment could start (for the set-up see fig. 4.6).

Having assembled all the individually tested components of the arrangement, the $(g-2)$ team managed that same year to store muons and to eject them with good intensity after they had made up to 1100 turns. This first published results gave

$$\tfrac{1}{2}(g-2) = 0.001145 \pm 0.000022,$$

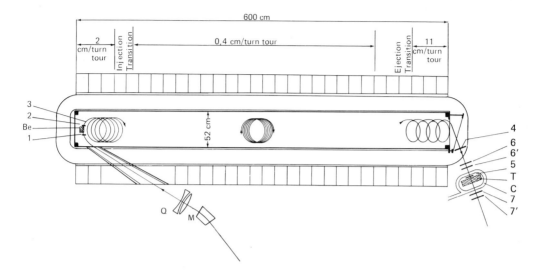

Fig. 4.6. Set-up of the $(g-2)$ experiment[74]

in good agreement with theory and with an accuracy of 1.9%, the best value yet obtained.[75]

Following this immediate success, their next aim was to improve the set-up to achieve the utmost precision in the measurement. This was done during 1961 and involved having a higher trapping efficiency for muons, reshimming the 6 m magnet, using a magnetic channel for ejection, improving the polarization analyzer and so on. Indeed the result obtained after these modifications was excellent, i.e.

$$\tfrac{1}{2}(g-2) = 0.001162 \pm 0.000005 .$$

The error was decreased to 0.42%,[76] and compared to the theoretical value gave

$$\tfrac{1}{2}(g-2) = [\tfrac{1}{2}(g-2)]_{\text{theor}} (0.9974 \pm 0.0042)$$

which underlined in a rather impressive way the strength of QED as a theory for electromagnetic interactions.

After completing this experiment at the SC Farley's team suggested constructing a muon storage ring at the PS to increase the accuracy further. This project was agreed on and by 1966 the first measurements were carried out with great success bringing the experimental value even closer to the theoretical predictions (see chapter 5.4).

Notes: p. 134

4.3.2.3 The beta decay of the pion ($\pi^+ \to \pi^0 e\nu$)

The idea of investigating the beta decay of the positive pion was put forward early in 1961 by the Citron group together with Carlo Rubbia. The background to the proposal was the theory of universal Fermi interaction of Richard Feynman and Murray Gell-Mann which led to a pion–lepton vector interaction of strength equal to the ordinary beta-decay vector interaction, under the so-called conserved vector current hypothesis (CVC). This permitted definite predictions for the branching ratio, namely

$$R_{\text{theor}} = \frac{\pi^+ \to \pi^0 e^+ \nu}{\pi \to \mu\nu} = (1.06 \pm 0.02) \cdot 10^{-8}.$$

Thus a precision measurement of this ratio would help clarify an additional point in weak interactions.

Having settled the final composition of the team by the end of 1961 (Depommier, Heintze, Mukhin, Rubbia, Soergel, Winter) the experiment got under way in 1962.[77] The set-up consisted of an NaI crystal and a lead glass Cerenkov counter, both for detecting the decay photons of the π^0, and a large plastic scintillator in which the π beam was brought to rest and the decay electron was detected. The pulses of the counters were vizualized on two oscilloscopes and photographed. (For the set-up see fig. 4.7.) In a first run 16 such beta decay events were found, leading to a branching ratio of $R_{\text{exp}} = (1.7 \pm 0.5) \times 10^{-8}$.[78]

The experiment was continued in 1963 with a view to improving the statistics and in particular taking into account the detection efficiencies. This time a branching ratio of

$$R = (1.15 \pm 0.22) \times 10^{-8}$$

was found based on 52 events. Thus there was already rather good agreement with the predictions of CVC theory.[79]

With a new γ detector consisting of eight lead glass counters which together formed a cylinder, the experiment was redone for a last time in 1964, confirming the earlier results.[80]

In parallel this team used the same equipment to investigate the radiative beta decay $\pi^+ \to e^+ \nu\gamma$. This process was of interest as it could possibly give some information on non-local effects in pion decay due to strongly interacting particles or perhaps due to an intermediate boson transmitting weak interactions. In two experiments in 1962 and 1963, 143 such events were observed, a number far too high if π-e interaction was to be point-like. It was however not possible to draw any more definite conclusions.[81]

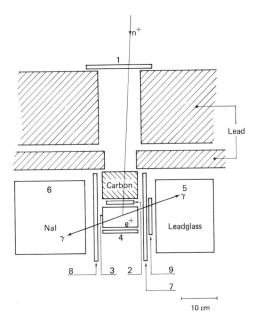

Fig. 4.7. The set-up for the pion beta decay experiment[82]

4.3.2.4 The μ-channel and the scientific programme carried out with it

The idea of constructing a device for collecting μ-mesons originating from π-decay in flight was suggested by A. Citron and H. Øverås early in 1957.[83] Such a focusing channel would turn the SC into a facility for doing experiments in which high-intensity μ-beams were required and for which the muons always present as a contaminant in pion beams were insufficient. In other words, muon experiments with virtually no disturbing pion background would be possible. If one simply tried to filter out the pions the transmitted muons would only be of low energy.

More precisely, the idea was to capture a π-beam in a focusing device and to send it through a magnetic analyzer which would eliminate all mesons of momentum smaller than a given one. Subsequently the muons were to pass through a channel of a length of the order of a π-μ decay length. At the end of it the beam would again be momentum analyzed and only those particles having lost momentum in the channel would be selected. In this way a relatively pure μ-beam could be obtained.

At the first meeting of the Advisory Committee responsible for the scientific programme of the SC, held in September 1957, Citron expressed the hope that with such a procedure one would be able to produce μ-beams with a total intensity of 3×10^3 μ/sec and with a contamination of only 1%. He estimated that the channel would cost of the order of 400 000 SF. His ideas were received with great interest,

Notes: p. 134

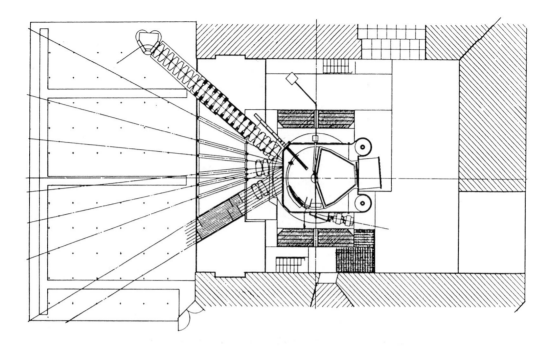

Fig. 4.8. The layout of the μ-beam[87]

particularly since with such a beam 'Hofstadter scattering experiments' could be done with muons, so throwing considerable light on the electron–muon relation.[84]

1958 was spent designing in detail the strong-focusing channel and the necessary quadrupole lenses, and orders were placed for the parts of the channel and the bending magnet for analyzing the output. During 1959 these parts were delivered and tested carefully. In addition the properties of the initial π-beam were investigated. By the end of the year, the position of the channel and the layout for a muon scattering experiment were fixed. Installation work started during the shut-down to overhaul the SC.

What were the exact features of this channel? (see also fig. 4.8). The pions generated when 600 MeV protons fall on a Be target were momentum analyzed by the magnetic field of the SC. At this point one could choose between high-energy and low-energy beams depending on the position and inclination of the decay channel with respect to the target. The decay channel itself had a total length of 13 m, and consisted of twenty four equally spaced quadrupole lenses which provided the focusing. About 20–30% of the μ-mesons generated in the channel emerged at the end, where they were momentum analyzed a second time. To cause as little loss as possible a 70 degree alternating-gradient magnet comprising three sections was designed for passing on the beam directly from the channel to a detector without having too much drift space in between.[85]

The result obtained by 1960 was a high-energy muon beam of 270 MeV/c momentum containing 6×10^3 μ/sec/100 cm^2. The pion contamination integrated over all momenta was roughly 1.5%. Furthermore with the low-energy beam one would stop 10 μ/sec per gram of C-absorber over at least 100 cm^2. By 1961, when the intensity of the SC internal beam was increased from 0.3 to 0.8 μA, the μ-beam intensity was raised considerably to about 2.5×10^4 μ$^-$/sec/100 cm^2 at high energy.

In 1961 the Citron group also performed their first experiment, the scattering of μ$^-$ on carbon. The idea here was to check whether the muon and the electron showed any difference of behaviour other than that due to their different rest masses when scattered by nucleons. The target material carbon was chosen since electron scattering had already been thoroughly investigated for this case and because the excited states were well spaced. Muons of 180 and 240 MeV/c incident momenta were scattered in C through angles up to 72° corresponding to a four-momentum transfer up to 250 MeV/c. The conclusion drawn was that there appeared to be no deviation in the scattering cross-section from that expected for heavy electrons within the limits of error ($\pm 7\%$). In addition if one assumed that the muon's charge was not point-like as the electron, but smeared out over a certain volume in space an upper limit of 0.64×10^{-13} cm could be given with 95% confidence on the radius of such a charge distribution.[88]

Many other groups apart from that led by Citron made use of this new facility. The first to mention was a group from Rome around M. Conversi.[89] In 1960 they had already started a systematic search for a reaction in which a muon was converted in the Coulomb field of the nucleus into an electron without the emission of neutrinos. $\mu^- + N \rightarrow N + e^+$ belonged to those processes which, though compatible with the conservation laws established at the time, nevertheless did not occur. Thus they aimed to determine as precisely as possible the branching ratio

$$R = \frac{\mu + Cu \rightarrow Cu + e}{\mu + Cu \rightarrow Ni + \nu}.$$

However this group did not find one event showing an electron and thus, taking into account the detection efficiency, they could only give an upper limit for this branching ratio, namely 6.6×10^{-6}.

As the μ-beam intensity was improved in 1961 Conversi and his team decided to repeat their measurements—a decision which was triggered by the world-wide increase in interest in weak-interaction physics due to the two neutrino experiments and related topics. On the other hand an experiment at Berkeley had not ruled out completely the existence of a neutrinoless conversion of the muon into an electron. However, even with improved detection equipment and the more intense beam the earlier result was confirmed, the new limit being 2.2×10^{-7}. Thus one speculated that there was some selection rule separating muons from electrons. And indeed this result fitted perfectly with the two kinds of neutrinos then recently discovered and with the independent conservation of lepton number for electrons and muons.

In 1962 the same group used the μ-beam to search for a rare process, namely the radiative μ-capture $\mu^- + N \to N' + \nu + \gamma$ (N = Fe nuclei), which was as yet unobserved. The branching ratio was indeed found to be 1×10^{-4} relative to ordinary μ-capture in medium nuclei, thus agreeing well with theoretical estimates.[90]

This kind of experiment was continued in 1963, this time using Ca^{40} to capture the muons. Here it was interesting that the rate of this process, $\mu^- p \to n\nu_\mu\gamma$ was sensitive to the strength of the pseudoscalar coupling which was induced in weak processes by the strong interaction. A measurement of the capture rate thus allowed a determination of the strength of the coupling constant. However for unknown reasons the result they obtained did not fit with theoretical predictions.[91]

A second team which made extensive use of the μ-beam from 1961 onwards was the Rome–Bologna–CERN collaboration exposing the 20 cm Italian hydrogen bubble chamber.[92] The beam was brought to rest taking 260 000 pictures showing 3.5 million muon stops. The main question of interest was how much time the muon would spend as a (μp) atom. This process $\mu^- p \to n\nu$ had first been detected by Hildebrand in 1961 and was characterized by the absence of an electron and by a neutron of unique energy (5.3 MeV). The result for the capture rate was $R = (450 \pm 50)$ sec^{-1} for the CERN experiment, in good agreement with Hildebrand's results at Chicago. Later the experiment was also extended to measure μ-capture in hydrogen gas.

4.4 Concluding remarks

In reading this chapter it may seem — particularly to those who have been involved — that we have not adequately captured the spirit and the human element of all the work done on the SC. The technological steps in the development of the machine and the general lines of the experimental programme are — we hope — well represented, but only a fraction of what happened around its construction and exploitation, which was in fact the source of the enthusiasm of the people involved, has been recorded. Questions like: What attracted engineers and physicists at the time?, What made them join the enterprise, and leave their home countries?, What did they expect from this small machine, from the research laboratory as a whole?, should be asked and their answers would give colour to the picture we have drawn of this period.

But it is precisely at this point that our difficulties begin. The sources available reflect only a little of the spirit of this early period. As the reader can see from the notes, for the construction period a complete set of documents of the SC Group was at our disposal, containing reports of meetings which took place every two months, progress reports of the different subgroups and of the group as a whole, travel reports, letters to the group, etc. All these documents are of a highly technical nature, containing details of the different parts of the SC, of the considerations which dictated a particular design, and of plans of how to proceed with the work.

They enabled us to gain an insight into the technical part of the enterprise, but did not provide us with any other sort of information. There are no reflections of the personalities who played a key role, of the development of ideas, or of the atmosphere in the group at different stages of the work. And as far as the scientific life around the SC is concerned, the documentation inside CERN is even poorer. Besides the published articles, which reflect nothing of the surrounding situation in which the experiment was performed, we have only a few documents describing the scientific work of the SC. No trace of discussions of the experimental programme, no proposals, only a little correspondence and a few punctual press releases or articles in the CERN Courier were found.

Of course, besides these documents we always have another source of information to consult, namely, interviews with people who had experienced all this at first hand. Here we could use interviews already made, or we had personal discussions with Blythe, Bonaudi, Gentner, Hedin, Krienen and Lazanski on the more technical side and Citron, Fidecaro, Goebel, Harting, Lundby, and Merrison on the experimental side. We learned a good deal about the life at CERN in these early days, of the small problems they encountered from day to day during the process of organization and also, of considerable importance to us, their attitude towards the SC, their expectations, and how the machine lived up to them. But we should not forget that these interviews were made 20 years and more after the design and construction period. Consequently the information is more general and seldom related to a special moment in the development or to a particular event. Thus they do not allow us to draw a balanced and rather complete picture. However these interviews enable us to derive something of a feeling for the spirit of the time in general, for the enthusiasm of the engineers and physicists who worked together and for the broad hopes and expectations they had invested into this first machine of the first big European venture.

Thus, to conclude this article we will try to do two things. In the first place let us try to grasp this spirit or at least get a feeling for what happened alongside the purely technical and scientific work. We will surely not be able to answer the questions we posed at the beginning of the conclusion, but we may get an idea of what the answer could look like. Secondly, having presented in this chapter the scientific work around the SC in a rather general way and having elaborated the most substantial experiments in more detail, we will try to identify the essential points to be retained of the evolution up to 1965.

Let us start by considering this small machine—a political one, as Blythe referred to it in an interview—from an objective point of view. This accelerator was of standard type to be built using more or less existing techniques. The projected energy of 600 MeV thus represented nothing more than a step up in the energy scale similar to that already taken several times in the development of every type of accelerator. Furthermore, the SC would in any case remain in the shadow of the proton synchrotron which was something absolutely new and which could reach energies of

Notes: p. 135

28 GeV. That granted, one might well ask: Why should the construction and the subsequent use of this accelerator have been exciting?

Nevertheless, the fact remains that this excitement and enthusiasm were felt by physicists and engineers alike. To understand these complex feelings, we must look at the whole 'SC enterprise' from three main points of view. First of all, What was its significance from a more technical standpoint? secondly, What did it represent for the physicists? and finally, What place would it have in the CERN institution as a whole?

Let us summarize the situation in post-war Europe. America was well advanced in the building of accelerators. In Europe, only two 'big' accelerators of the synchro-cyclotron type existed in 1952: the 90″ (200 MeV) SC in Uppsala just a bit too small to do π physics and the 150″ (400 MeV) SC in Liverpool which was under construction. This was not all, however, since there was also a lack of experts for big accelerators and the team chosen for the SC had more or less only worked in energy regions well below the planned 600 MeV. For them even this 'small' machine meant entering new territory in a technological sense.

Then there were the physicists who had found Europe in a rather difficult situation after the war. For financial and technical reasons it had not been possible for many of them to keep pace with the developments in America and work in the forefront of this new field of high-energy physics. Thus, the European laboratory was a big hope for them. It would enable them to participate in research in the field, and to take an active role in its development. The CERN synchro-cyclotron provided them with the possibility of doing experiments in 1958, two long years before the big accelerator was ready.

Last but not least, the SC also played a very important role in the development of the research establishment as a whole. Its success strengthened confidence outside and inside CERN in the future of this new form of international collaboration, affecting not only the SC Group but also the other groups and perhaps even the governments of the countries involved in the enterprise. The first task, that of building and running the SC, was a somewhat 'routine' one, and so a kind of objective test of how well the whole establishment would function. Correlatively, failure would probably have had negative consequences. In fact, this small machine was easily conceived in the sense that models for it already existed and, moreover, a realistic budget could be drawn up from the outset. The accelerator was predestined to be the test piece. The masterpiece would be the PS.

In addition to these points, there were other feelings of a rather more emotional kind in the minds of the people taking part in this enterprise. There was the fact that everything was exceptional, promising, dynamic, which made the laboratory in some way attractive for its own sake. Lazanski described this feeling very well in saying, 'it was a completely new undertaking, everyone was young, the organization was young ... we were frequently together. It contributed to create a team spirit.'

With all this in mind, it becomes understandable how great the excitement among those early scientists was, why they subjectively felt they were pioneers, why the expectations both inside and outside the group were so high and why the enthusiasm was so great. Objectively, it may be true that the SC was a small machine—which nevertheless gave outstanding results—, but it was also the nucleus of a whole organization, the success of which contributed a great deal to the future of high-energy physics in Europe.

Let us now turn to the scientific achievements.

For the time being looking only at the particle physics aspect of the programme, we find a clear dominance of experiments dealing with topics in weak interactions. Two reasons can be given for this. Firstly a more technical one, related to the kind of particle beams and their relatively low energies which were at the physicists' disposition. For example, the construction of a highly successful μ-channel in the early 60s led to a concentration on studies related to this particle. Secondly, and this was surely the more important reason, a particular scientific situation in this field prevailed. With the detection of parity violation in weak interactions and the establishment of the so-called $V-A$ theory, both in 1957, physicists found themselves confronted with a completely new landscape to explore. A large number of questions remained unanswered and theory had to be carefully tested. These experiments on weak interactions did not merely lead to the production of a rich stock of valuable information, but were also the start of a long tradition in weak interaction physics at CERN.

When looking at the scientific programme of the SC as a whole, and at the place this small accelerator took in the life of CERN, we can see a drastic change taking place between 1958 and 1965. Indeed this change occurred in two clear steps. The starting point was the period 1957 up to 1960, thus from the coming into operation of the SC up to the completion of the PS. In this period the SC was CERN's only accelerator and therefore the centre of its scientific life. It not only provided the first opportunity for physicists to gather experience but also served as a test for the whole organization. One had to 'learn' how to conceive an experimental programme, thus choosing the topics and teams that would take part, dealing with the difficulties between CERN teams and groups coming from outside, and handling the accelerator as such in an optimal way. Scientifically the aim—as defined from the start—was to do particle physics, and nuclear physics or chemistry were only regarded as side aspects. And indeed the good scientific results obtained with this small accelerator ensured that the whole organization got off to a fine start.

The first real changes in this perception occurred in 1960. The gigantic accelerator of 28 GeV energy which put European physicists on an equal footing with the USA for the first time, at least in terms of machine energy, logically became *the* attraction at CERN. As a result the role of the SC as an instrument for particle physics diminished slowly but steadily. Indeed by using the SC one was limited to low energies and to only four kinds of particles, namely p, n, π, and μ. Many of the groups interested in particle physics thus moved to the PS and as a consequence the time spent for nuclear physics and chemistry increased.

However putting the PS into operation did not only bring a change in the scientific work with the SC, but also caused a change in the structure of the experimental teams. While CERN teams and truck teams were strictly separated in the early years, from 1961 onwards many mixed groups were working at CERN.

All the same in the early 60s the SC still contributed important results to the understanding of particle physics. Let us not forget the $g-2$ experiment, the beta decay of the pion, and the large amount of μ physics carried out with the μ-channel, a unique device of its kind.

This brings us to the last step which took place in the period 1963 to 1965. In fact the turning point was the discussions about the future of the SC,[93] which finally led to a more or less explicit redefinition of the accelerator as a nuclear physics instrument. Indeed it had become clear that this small machine and the physics done with it should not be regarded as competing with what one could do at the PS, but rather as complementing it. This new role for the SC, and its ready acceptance inside CERN, owed much to the acting Director General Victor Weisskopf, who, being a nuclear physicist himself, had always insisted on the crucial importance of this kind of physics for a global understanding of the way matter was built. On the occasion of the 1963 cyclotron conference at CERN he expressed this very clearly, remarking that the

> "most important contributions to physics here in Geneva were, in fact, done at the small machine [...]. It was possible because of an unusual good and intense beam of pions, [and] in spite of the fact that new fields of qualitative discovery are open only at very high energies. Much can and should be done at lower energy with good intensities and manageable beams."[94]

Thus the SC had changed its initial role and had become the centre of a nuclear physics and chemistry programme, a programme which had been marginal in the beginning.

The SC is still working today after more than 30 years and one might say that the decision to follow this line of research has guaranteed its survival. It has undergone many improvements and changes and has ensured, after a splendid start in particle physics, a continuous tradition in nuclear physics and chemistry at CERN.

Notes: p. 135

Notes

1. The following interviews have been used as background for writing this chapter. Transcripts of them are available in the CERN archives.
   - G. Bernardini by L. Kowarski and S. Newman, 3/7/74
   - F. Blythe by C. Taylor, 8/9/76 (no transcript)
   - F. Bonaudi by C. Taylor, 5/10/76
   - A. Citron by L. Kowarski and S. Newman, 12/8/74
         by U. Mersits and D. Pestre, 18/11/87 (no transcript)
   - G. Fidecaro by L. Kowarski and S. Newman, 2 parts, 8/1/75, 27/1/75
   - W. Gentner by S. Newman, 2/10/74
   - K. Goebel by L. Kowarski and S. Newman, 13/8/74
   - D. Harting by L. Kowarski and S. Newman, 2 parts, 27/1/75, 13/2/75
   - B. Hedin by C. Taylor, 10/8/76
   - I.F. Krienen by L. Kowarski, no date, no transcript
   - M. Lazanski by C. Taylor, 14/9/76
   - A. Lundby by L. Kowarski and S. Newman, 21/5/74, 22/5/74
   - A.W. Merrison by M. Gowing and L. Kowarski, 29, 30/5/75

   The documents referred to in the following as CERN/SC are to be found in the boxes SC22836 and SC22837 in the CERN archives.
2. For a more detailed account on the decision process of the SC see History of CERN (1987).
3. Synchro-cyclotrons existing or under construction by 1952 [see Accelerators (1951, 1954)]:

	Pole diameter	Proton-energy [MeV]
America		
University of California, Berkeley	184″	350
Carnegie Institute of Technology, Pittsburgh	145″	440
University of Chicago	170″	450
Harvard University	95″	125
Columbia University	164″	384
University of Rochester	130″	250
Canada		
McGill University	82″	100
Great Britain		
Atomic Energy Research Establishment, Harwell	110″	175
University of Liverpool	150″	400
Netherlands		
Institute for Nuclear Research, Amsterdam	71″	56 α-particles
Sweden		
University of Uppsala	90″	200

See 'Draft Programme of Synchro-Cyclotron Study Group', CERN/GEN/1 (May 1952), 31-32 and Meson Conf. (1952), 19 ff.
4. See CERN/GEN/1 (May 1952).
5. W. Heisenberg, 'Report on the Scientific Conference held in Copenhagen from 3 to 17 June 1952', CERN/GEN/2 (June 1952), pp. 9-11.
6. *Report on the activities of the Synchro-Cyclotron Group during the third week of the Physics Conference held in Copenhagen, June 15-19, 1952*, CERN/SC 6, 26/6/52.
7. For further technical information see Hedin (1955). See also interview with B. Hedin, *op. cit.*, note 1.
8. For the comparison of the existing SCs see CERN/SC 3, 6/6/52.
   C.J. Bakker, B. Hedin, *Report on Journey in USA July 15th-30th 1952*, CERN/SC 11, 8/52. From the synchro-cyclotron teams visited, much information on their machines could be gained. The CERN team was provided with drawings and technical specifications of the main parts of the different SCs they saw. For the less experienced European team this sort of help was of the utmost importance.
9. *Report on the Synchro-Cyclotron meeting at Brussels, January 15, 16 and 17, 1953*, CERN/SC 38, 19/1/53.
10. This drawing is a simplified version made of Fig. 3. *CERN/Synchro-cyclotron Magnet*, to be found in CERN/GEN/5 (5/5/53).
11. *Report on the Synchro-Cyclotron group-meeting at Geneva, March 16, 17, 1953*, CERN/SC 50, 23/3/53.
12. *Report on the Amsterdam meeting of SC group, June 4, 5 and 6, 1953*, CERN/SC 64, 10/6/53.
13. See CERN/IFC/51, 20/10/54.
14. For more technical information on the H.F. system see Krienen (1958), Schmitter (1959); K. Schmitter and S. Kortleven, 'The Radio-Frequency System'; B. Bolée and F. Krienen, 'The Tuning-Fork Modulator', both in SC (1960/61).
    The quote is taken from CERN/SC 1.
15. Most of the information on the H.F. system in the early period can be found in the following four reports: F. Krienen, *Dee and R.F. system for 600 MeV Synchro-Cyclotron*, CERN/SC 13, 18/08/52; C.J. Bakker, *Report on visit to Radiation Laboratory, Berkeley August 1952*, CERN/SC 18, 11/09/52; F. Krienen, *Report on vibrating reed for R.F. system of 600 MeV Synchro-Cyclotron*, CERN/SC 20, 9/10/52; F. Krienen, *Report on High Frequency system of the 600 MeV Synchro-Cyclotron*, CERN/SC 34, 12/52.
16. Information on the contract between CERN and Philips can be found in Box *SC-RF System (1952-55)* (SC22806).
17. *Report on visit U.S.A. by C.J. Bakker and F. Krienen, January 30-February 16, 1954;* CERN/SC 97, 18/2/54. The main difference between the vibrating reed system developed at Berkeley and at CERN was, that in Berkeley, instead of a tuning fork, a system consisting of four pairs of blades was used which was to move with equal amplitude and phase.
18. *Fourth meeting* of the Finance Committee, 8 June, 1955; CERN/IFC/55*.
19. This drawing is a simplified version we made of drawing RF009 to be found in CERN/SC 45, 5/3/53.
20. Plans, tender documents, some correspondence, etc. can be found in Box *Vacuum System* (SC22807).
21. See J.M. Cassels, I.F. Krienen, M.J. Moore, T.G. Pickavance, *Experimental layout of the 600 MeV Cyclotron*, CERN/SC 12, 11/08/52.
22. M.J. Moore, F. Bonaudi, *Progress report on vacuum tank and pumping system*, CERN/SC 47, 13/03/53.
23. See CERN/FC/49, May 1955 and CERN/FC/50, June 1955.
24. For details on the control system see Debraine, Lazanski & Boyadjian (1956).
25. See *Report on the Synchro-Cyclotron meeting at Amsterdam, August 28-29, 1952*, CERN/SC 17, 1/09/52.
26. See P. Debraine, M. Lazanski, *Note on Responsibilities of Subgroups as to local control equipment*, CERN/SC 62, 30/5/53 and P. Debraine, M. Lazanski, *Report on Synchro-Cyclotron Control System*, CERN/SC 75, 7/9/53.
27. T.G. Pickavance, J.M. Cassels, *Notes on the Shielding of the 600 MeV Cyclotron*, CERN/SC 24, 10/52.

Notes

28. C.J. Bakker, *Progress Report of the Synchro-Cyclotron Group, October 1952,* CERN/GEN/4, 4-7/10/52, p. 10-11.
29. See Progress reports of the SC group CERN/GEN/10 (1 April-15 June 1953) and CERN/GEN/13 (15 October-30 December 1953).
30. See CERN/IFC/33, 18 March 1954 and CERN/IFC/43, 17 June 1954.
31. This part was mainly based on the Annual Reports 1955-1957.
32. For information on W. Gentner see 'Who is Who at CERN', *CERN Courier, 1960,* no. 12, p. 2 and an interview with him, *op. cit.,* note 1.
33. Interview with F. Bonaudi, *op. cit.,* note 1.
34. This drawing is a simplified version we made of 'General Layout showing the extracted beams', *CERN Annual Report* (1956), 37.
35. See *CERN Annual Report,* 1957, 41-55 and CERN/SPC/37, 22/2/1957.
    A tentative experimental programme for the start of the SC can be found in CERN/SPC/28, 1/10/56 and CERN/AC/1, 26/9/57.
    During 1957 only one experiment, which did not use the accelerator, however, was carried out by the SC division. The group around Lundby made a check of the results on parity violation in weak interactions recently obtained in the United States. They measured the circular polarization of the photons after beta decay and confirmed these findings.
36. The information is taken from *CERN Annual Report,* 1958, 45-46, *SC research programme and activities,* CERN/SPC/60, 18/2/58, and *Research programme of the synchro cyclotron,* CERN/SPC/79, 14/11/58.
37. For the results see Burgman et al. (1958).
38. The results were published as Harting et al. (1959, 1960).
39. This information is taken from *CERN Annual Report,* 1958, 34.
40. For a general description of emulsion work at CERN and in particular also for the experiments carried out at the SC between 60 and 63, see Combe & Lock (1963).
41. To get an impression of the work performed by this group, see for example Interview with K. Goebel, *op. cit.,* note 1.
42. For a short summary of the work of the nuclear chemistry group in the early 60's see G. Rudstam, 'Nuclear Chemistry at CERN', *CERN Courier,* 1963, 31-34.
43. A short description of the experiments performed by the visiting teams is in *CERN Annual Report,* 1959, 67-68.
44. The information for this paragraph is taken from *CERN Annual Report,* 1959, 55-68 and 71, and *Twelfth Meeting of the Scientific Policy Committee,* CERN/SPC/92, 17/4/59, p. 4.
45. For publications see Harting et al. (1959, 1960), Fischer et al. (1959) and Lundby et al. (1959).
46. See Gatti et al. (1961).
47. See Backenstoss et al. (1960).
48. See *CERN Annual Report,* 1959, 47-48.
49. Publications on these experiments: Davis et al. (1960), Loria et al. (1961).
50. See *CERN Annual Report,* 1960, 53, 56, 78-79, 98.
51. See *CERN Annual Report,* 1960, 54-62.
52. For the abstracts of the experiments of the visiting teams see *CERN Annual Report,* 1960, 63-66.
    The neutrino experiment and the scientific motives for it have been treated extensively in Mersits (1987b).
53. A detailed description of the new organizational structure is given in chapter 9.
54. This information is taken from the report of the MSC Division in *CERN Annual Report,* 1961, 35-38.
    On the occasion of the increase of beam intensity a press release was sent out underlining its importance (DIRADM20400).
55. The shut-down of the SC was mentioned at the *Twentieth Session of Council,* 19/12/61, CERN/429, p. 3. This was a consequence of a discussion held at the *Twenty second Meeting of the Scientific Policy Committee,* 25/11/61, CERN/SPC/151.

For the reaction from the SC Division see Memo Lapostolle to Weisskopf, 25/1/62 and the answer Memo DG to Lapostolle, 29/1/61, both in (DIRADM20267).

56. See *CERN Annual Report,* 1961, 53-60. Attention: The ω^0 mentioned in this experiment is not to be confused with the $\omega(783)$ resonance discovered at Berkeley.
57. The information is taken from the report of the MSC Division for the *CERN Annual Report,* 1962, 44-47.
58. See *CERN Annual Report,* 1962, 64-69.
59. A summary of the improvements carried out on the SC during 1961 and 1963 can be found in P. Lapostolle, *Three years of development at the CERN synchro-cyclotron,* January 1964 (MGNH22004).
60. Some of the published results can be found in: Blair et al. (1964), Buhler et al. (1963).
61. See 'Shut-down at the SC', *CERN Courier,* April 1964, 44-48.
62. By mid-1964 a so-called 'Nuclear Structure Physics Committee' was established, dealing in particular with nuclear physics experiments at the SC.
See also 'Nuclear physics at the SC', *CERN Courier,* Jan. 1965, 7.
63. See letter Weisskopf to Bohr, Nielsen, Taillac, Gentner, Tangen, Andersson, Bergström, 17/12/64 (DG20587). Attached one finds the proposal for ISOLDE. This isotope separator had been approved at the NPRC meeting on 2/12/64.
64. Publications: Duclos et al. (1965).
65. E.C.G. Sudarshan and R.E. Marshak, 'The nature of the four-fermion interaction', in Meson Conf. (1957), V-13; Feynman & Gell-Mann (1958); Sakurai (1958).
66. This figure is reproduced from Ashkin et al. (1959), 1245.
67. See Fazzini et al. (1958). On 11/9/58 a press release with the title 'Crucial experiment with the CERN 600 MeV synchro-cyclotron' was sent out (DIRADM20400).
68. See Ashkin et al. (1959a).
69. See Ashkin et al. (1960).
70. See Ashkin et al. (1959b).
71. See Anderson & Fermi (1952). The result of the CERN experiment was published as Cocconi V.I. et al. (1961).
72. In the 1958 conference on high-energy physics at CERN, W. Panofsky reported on a number of experiments undertaken at Columbia University, the Joint Institute for Nuclear Research and the University of Chicago to measure precisely the $(g-2)$ of the muon in order to yield the limits of QED (HEP Conf. (1958)).
For a presentation of the members of the $g-2$ group see 'The $(g-2)$ team', *CERN Courier,* 4 (1962), 8-9.
73. See *CERN Annual Report,* 1959, 63 and Charpak et al. (1969) for technical details. The set-up to measure (g-2) was also used to redetermine the electric dipole moment: Charpak et al. (1960, 1961b).
74. This figure is reproduced from Charpak et al. (1961a).
75. See Charpak et al. (1961a).
Press release: *New light on the muon,* 17/1/61 (DIRADM20400).
76. See Charpak et al. (1962) and Charpak et al. (1965). For further information see: 'g minus two', *CERN Courier,* 4 (1962), 3-7, Farley (1964).
77. For the proposal see Citron Group and C. Rubbia, *A search for the $\pi^\pm \to \pi^0 e^\pm \nu$ decay mode* (no date, Fidecaro papers). It was submitted to the EEC Meeting of 11 April 1961 and declared of great interest. However many of the members were still involved in the muon experiment and so some delay occurred. See Memo Citron, Heintze, Rubbia to Preiswerk, 7/8/61 (DG20888).
78. First publication: Depommier et al. (1962).
79. See Depommier et al. (1963a).
80. See *CERN Annual Report,* 1964, 57-58.
81. See Depommier et al. (1963b).
82. This figure is reproduced from *CERN Annual Report,* 1962, 68.
83. Proposal: A. Citron and H. Øveras, *On a focusing channel for collecting µ-mesons from π-μ decay in*

flight, SC/143, 27/3/57. For a general impression on the way this project developed see Interview with A. Citron, *op. cit.,* note 1.

84. See Minutes of the *First Meeting of the Advisory Committee on the SC Experimental Programme,* 26/9/57, CERN/AC/1.
85. For a description of the beam see Citron et al. (1960).
   A method to test the beam purity was published as Citron et al. (1962b).
86. See *CERN Annual Report,* 1960, 58.
87. This figure was reproduced from Citron et al. (1960), 287.
88. See Citron et al. (1962a).
89. See Conversi et al. (1960, 1962) and Conforto et al. (1962).
90. See Conforto, Conversi & DiLella (1962).
91. See Conversi, Diebold & DiLella (1964).
92. For the results see Bertolini et al. (1962); a more general description of the scientific situation can be found in Rubbia (1963).
93. In 1961 a new group was created to investigate the possibility of converting the SC into a high-intensity accelerator (meson factory). This was thought to be possible by modifying the actual SC into a spiral isochronous cyclotron, or by building a new isochronous cyclotron or a cryogenic linac. However by 1962 this idea was abandoned so as not to hinder the decisions to be taken on a 300 GeV accelerator or the PS storage rings. Instead all effort was concentrated on optimizing the existing machine.
94. The quotation is taken from 'Opening Remarks' by Victor Weisskopf given on the occasion of the conference on sector-focused cyclotrons and meson factories (Cyclotron Conf. (1963)).

Bibliography

Accelerators (1951)	B.E. Cusham, *Bibliography of Particle Accelerators, July 1948 to December 1950,* University of California, UCRL-1238, March 1951.
Accelerators (1954)	F.E. Frost and J.M. Putman, *Particle Accelerators,* University of California, UCRL-2672, 16 November 1954.
Anderson & Fermi (1952)	H.L. Anderson and E. Fermi, 'Scattering and capture of pions by hydrogen', *Physical Review,* **86** (1952), 794.
Ashkin et al. (1959a)	J. Ashkin et al., 'The electron decay mode of the pion', *Nuovo Cimento,* **13** (1959), 1240–1262.
Ashkin et al. (1959b)	J. Ashkin et al., 'Search for the decay $\mu \to e + \gamma$ and observation of the decay $\mu \to e + \nu + \bar{\nu} + \gamma$', *Nuovo Cimento,* **14** (1959), 1266–1281.
Ashkin et al. (1960)	J. Ashkin et al., 'A new measurement of the mean life of the positive pion', *Nuovo Cimento,* **16** (1960), 490–504.
Backenstoss et al. (1960)	G.K. Backenstoss et al., 'An investigation of the stability of nucleons', *Nuovo Cimento,* **16** (1960), 749–755.
Bertolini et al. (1962)	E. Bertolini et al., 'Determination of the μ^- total capture rate in liquid hydrogen', in *1962 International Conference on high energy physics at CERN, 4-11/7/62.*
Blair et al. (1964)	I.M. Blair et al., 'Results of a search for $\pi^+ - \pi^-$ resonance in the dipion mass region 280–350 MeV', *Physics Letters,* **11** (1964), 79–82.
Buehler et al. (1963)	A. Buehler et al., 'A measurement of the e^+ polarisation in muon decay: the e^+ annihilation method', *Physics Letters,* **7** (1963), 368–371.
Burgman et al. (1958)	J.O. Burgman et al., 'Muon K-capture compared to β decay for $C^{12} \leftrightarrow B^{12}$', *Physical Review Letters,* **1** (1958), 469.

Charpak et al. (1960)	G. Charpak et al., 'A method for trapping muons in magnetic fields, and its application to a redetermination of the EDM of the muon', *Nuovo Cimento,* **17** (1960), 288–303.
Charpak et al. (1961a)	G. Charpak et al., 'Measurement of the anomalous magnetic moment of the muon', *Physical Review Letters,* **6** (1961), 128–132.
Charpak et al. (1961b)	G. Charpak et al., 'A new limit to the electric dipole moment of the muon', *Nuovo Cimento,* **22** (1961), 1043–1050.
Charpak et al. (1962)	G. Charpak et al., 'A new measurement of the anomalous magnetic moment of the muon', *Physics Letters,* **1** (1962), 16–20.
Charpak et al. (1965)	G. Charpak et al., 'The anomalous magnetic moment of the muon', *Nuovo Cimento,* **37** (1965), 1241–1361.
Citron et al. (1960)	A. Citron et al., 'A high-intensity μ-meson beam from the 600 MeV CERN synchro-cyclotron', in *Proceedings of the Conference on instrumentation for high-energy physics,* Sept. 1960, 286–288.
Citron et al. (1962a)	A. Citron et al., 'Scattering of μ-mesons by carbon', *Physics Letters,* **1** (1962), 175–178.
Citron et al. (1962b)	A. Citron et al., 'An investigation of π^- contamination in μ^- beams', *Nuclear Instruments and Methods,* **15** (1962), 121–128.
Cocconi et al. (1961)	V.T. Cocconi et al., 'A redetermination of the Panofsky ratio', *Nuovo Cimento,* **22** (1961), 494–518.
Combe & Lock (1963)	J.C. Combe and W.O. Lock, *Nuclear Emulsion work using the CERN accelerators, December 1959–April 1963* (Geneva: CERN 63-22, 1963).
Conforto, Conversi & DiLella (1962)	G. Conforto, M. Conversi and L. DiLella, 'Observation of radiative capture of negative muons in iron', *Physical Review Letters,* **9** (1962), 22–25.
Conforto et al. (1962)	G. Conforto et al., 'Search for neutrinoless coherent nuclear capture of μ^- mesons', *Nuovo Cimento,* **26** (1962), 261–282.
Conversi et al. (1960)	M. Conversi et al., 'Search for electrons from muon capture', *Nuovo Cimento,* **18** (1960), 1283–1286.
Conversi et al. (1962)	M. Conversi et al., 'Search for conversion of muons into electrons', *Physical Review Letters,* **8** (1962), 125–127.
Conversi, Diebold & DiLella (1964)	M. Conversi, R. Diebold, and L. DiLella, 'Radiative muon capture in Ca^{40} and the induced pseudoscalar coupling constant', *Physical Review,* **136** (1964), B1077–B1091.
Cyclotron Conf. (1963)	*International conference on sector-focused cyclotrons and meson factories,* 23–26/4/64 (Geneva: CERN 63-19, 1963).
Davis et al. (1960)	D.G. Davis et al., 'An experimental test of parity conservation in π^0-meson production by neutrons', *Nuovo Cimento,* **15** (1960) 641–651.
Debraine, Lazanski & Boyadjian (1956)	P. Debraine, M. Lazanski and G. Boyadjian, *The design of the CERN synchro-cyclotron control system,* (Geneva: CERN 56-1, 1956).
Depommier et al. (1962)	P. Depommier et al., 'Determination of the $\pi^+ \to \pi^0 + e^+ + \nu$ decay rate', *Physics Letters,* **2** (1962), 23–26.
Depommier et al. (1963a)	P. Depommier et al., 'Further measurements of the $\pi^+ \to \pi^0 + e^+ + \nu$ decay rate', *Physics Letters,* **5** (1963), 61–63.
Depommier et al. (1963b)	P. Depommier et al., 'Further measurements on the decay $\pi^+ \to e^+ + \nu + \gamma$', *Physics Letters,* **7** (1962), 285–287.
Duclos et al. (1965)	J. Duclos et al., 'A search for the decay $\pi^0 \to 3\gamma$', *Physics Letters,* **19** (1965), 253–255.
Farley (1964)	F.J.M. Farley, 'Electromagnetic properties of the muon', *Progress in*

	Nuclear Physics, **9** (1964), 257-294.
Fazzini et al. (1958)	T. Fazzini et al., 'Electron decay of the pion', *Physical Review Letters,* **1** (1958), 247-249.
Feynman & Gell-Mann (1958)	R.P. Feynman and M. Gell-Mann, 'Theory of the Fermi interaction', *Physical Review,* **109** (1958), 193-198.
Fischer et al. (1959)	J. Fischer et al., 'Measurement of μ^+ lifetime', *Physical Review Letters,* **3** (1959), 349-350.
Gatti et al. (1961)	G. Gatti et al., 'Inverse photoproduction reaction $\pi^- + p \rightarrow \gamma + n$ in flight', *Physical Review Letters,* **6** (1961), 706-708.
Harting et al. (1959)	D. Harting et al., 'Experiment on charge independence in pion interactions', *Physical Review Letters,* **3** (1959), 52-54.
Harting et al. (1960)	D. Harting et al., 'Experiment on charge independence in interactions of nucleons and pions', *Physical Review,* **119** (1960), 1716-1725.
Hedin (1955)	B. Hedin, *Design of CERN synchro-cyclotron magnet* (Geneva: CERN 55-3, 1955).
Hedin (1959)	B. Hedin, 'The construction of the 600 MeV synchro-cyclotron at CERN', *Ned. T. Nathurk,* **25** (1959), 61-75.
HEP Conf. (1958)	*1958 annual internation conference on high-energy physics at CERN,* 30/6-5/7/58.
History of CERN (1987)	A. Hermann, J. Krige, U. Mersits and D. Pestre, *History of CERN, Vol. I, Launching the European Organisation for Nuclear Research* (Amsterdam North Holland, 1987).
Int. Physics Conf. (1952)	O. Kofoed-Hansen et al. (eds.), *Report of the International Physics Conference,* Copenhagen, 3-17 June 1952.
Krienen (1958)	F. Krienen, *Modulator CERN synchro-cyclotron* (Geneva: CERN 58-8, 1958).
Loria et al. (1961)	A. Loria et al., 'The scattering of positive 120 MeV pions on protons', *Nuovo Cimento,* **22** (1961), 820.
Lundby (1958)	A. Lundby, 'The CERN 600 MeV synchro-cyclotron', *Discovery,* **19** (1958), 56-59.
Lundby et al. (1959)	A. Lundby et al., *Attempt to determine muon helicity by measuring polarization of B^{12} formed in the reaction $C^{12} + \mu^- \rightarrow B^{12} + \nu$* (Geneva: CERN 59-5, 1959).
Meson Conf. (1952)	A.M.I. Messiah and N.P. Noyes (eds.) *Proceedings of the Rochester Conference on Meson Physics,* 11-12 January 1952.
Meson Conf. (1957)	*International conference on mesons and recently discovered particles,* Padua-Venice, 22-28 September 1957.
Rubbia (1963)	C. Rubbia, 'Muon capture in hydrogen' in *International conference on fundamental aspects of weak interactions,* Brookhaven 9-11/9/63, 278-291.
Sakurai (1958)	J.J. Sakurai, 'Mass reversal and weak interactions', *Nuovo Cimento,* **7** (1958), 649-660.
SC (1960/61)	'The CERN 600 MeV synchro-cyclotron at Geneva', *Philips Technical Review,* **22** (1960/61), 141-180.
Schmitter (1959)	K. Schmitter, *Le système haute fréquence du synchro-cyclotron du CERN* (Geneva: CERN 59-33, 1959).

CHAPTER 5

The construction of the 28 GeV Proton Synchrotron and the first six years of its scientific exploitation[1]

Ulrike MERSITS

Contents

5.1 Introductory remarks	141
5.2 Planning and constructing Europe's biggest proton accelerator	145
5.2.1 From a 10 GeV constant gradient to a 30 GeV alternating gradient synchrotron: the period up to October 1952	145
5.2.2 The feasibility study and the fixing of the main machine parameters: November 1952 to October 1953	147
5.2.3 The period of engineering design studies, manufacturing, and construction (1954 to 1959)	150
5.2.3.1 An overview on the development of the PS group	150
5.2.3.2 The PS magnet	152
5.2.3.3 The radio-frequency system	154
5.2.3.4 The injection linear accelerator	156
5.2.3.5 Some remarks on other major activities around the PS	159
5.3 High-energy physics using bubble chambers	162
5.3.1 Bubble chambers, beams, and data handling—the three central parts of a bubble chamber experiment	162
5.3.1.1 The bubble chambers in operation at CERN between 1960 and 1965	162
5.3.1.2 The major bubble chamber beams	165
5.3.1.3 From the photographs to scientific results	168
5.3.2 The scientific contributions made with bubble chambers	169
5.3.2.1 The interactions of pions with nucleons	169
5.3.2.2 Investigating proton–proton collisions	177
5.3.2.3 Antiproton–proton interactions at low energy—a rich source of resonances	178
5.3.2.4 Experiments with high-energy antiprotons between 3 and 5.6 GeV/c	180
5.3.2.5 Kaon–proton interactions at energies up to 1.5 GeV/c	182
5.3.2.6 Kaon–proton interactions at the highest available energy—from 3 to 10 GeV/c	185

5.4 The use of electronic detectors to study elementary particles	190
5.4.1 Strong interaction physics	190
5.4.1.1 Beam survey experiments—the production of secondaries	190
5.4.1.2 Investigating the fundamental total cross-sections	193
5.4.1.3 Scattering experiments	198
5.4.1.4 Resonances: their production and decay	202
5.4.1.5 Some tests for the unitary symmetry model and the search for quarks	206
5.4.2 Electromagnetic interactions	207
5.4.2.1 Purely quantum electrodynamical problems	207
5.4.2.2 Electromagnetic interaction and properties of hadrons	208
5.4.2.3 Searching for magnetic monopoles	210
5.4.3 Weak interaction processes	211
5.4.3.1 The study of K-mesons, *CP* violation and related questions	211
5.4.3.2 The helicity of the muon	214
5.4.3.3 The beta decay of Λ^0	214
5.4.4 A test of the validity of relativity theory	215
5.5 The neutrino experiments at CERN: confronting the reality of big science (1960–1965)	215
5.5.1 The state of the art in neutrino physics (1960)	215
5.5.2 The first neutrino experiment at CERN	218
5.5.2.1 Discussing the feasibility—the period of scepticism (1960)	218
5.5.2.2 The period of optimism: September 1960 – May 1961	219
5.5.2.3 The first setback: von Dardel's measurements and its consequences (May 1961)	220
5.5.3 Conceiving a completely new neutrino programme (mid-61 to mid-63)	221
5.5.3.1 Two new techniques to improve the beam performance	222
5.5.3.2 The new detector: a multi-ton spark chamber	225
5.5.4 Results from the 1963 and 1964 experiments	226
5.5.5 Why Brookhaven and not CERN?	229
5.6 Conclusion	234
Appendices	240
Notes	243
Bibliography	254

5.1 Introductory remarks

With the coming into operation of the 28 GeV proton accelerator at CERN late in 1959, after nearly eight years of design and construction work, European high-energy physics stood at the dawn of a new era. Until then scientists who wanted to contribute actively to this rather prestigious new field of physics had had only two alternatives. Either they made use of one of the fairly small machines which had been constructed in the post-war period in a few places in Europe knowing, however, that from a technical point of view they had little chance to compete with the well equipped American laboratories; or they had to establish a collaboration with American physicists, so having access to one of the 'big' machines at Berkeley, or Brookhaven. Now the situation had suddenly changed.

As already described in the previous chapter, the first step made at CERN towards becoming a leading laboratory had been taken two years earlier when the smaller machine, the 600 MeV synchro-cyclotron, started operating. Despite the success of this accelerator, physicists' interest was focused on the new 'gigantic' synchrotron for which they had so many hopes. It was not only the fact that with 28 GeV a completely new energy range would be opened up; it was also that CERN was to become **the** place where European high-energy physics would be done in the years to come. What was more, for the first time scientists would work under conditions comparable to those in the United States. Half a year's advance of the CERN PS over its American counterpart—the Brookhaven AGS—would further offer a small advantage at the beginning. One hoped that in this way the bulk of the first harvest of results could be reaped by CERN. In consequence European physics, which had suffered so much due to World War II, would recapture its old position in the world wide scientific context.

However physicists were not only confronted with a completely new machine and all the consequences of this, but they had to learn that the way of doing particle physics was also about to undergo drastic changes. The time when physics could be decided from day to day, where the scientists worked on their own or with a few others using humble means, was definitely gone. Organizational structures had to be found satisfying the scientists and administrators, at the same time allowing for efficient work. Long-term planning and nearly perfect timing started to play a more and more decisive role for the progress of high-energy physics and therefore it was of utmost importance for a smooth development to agree upon a common scientific policy. This embraced not only the choice of what kind of physics should be done, but also what means (such as beams, detectors, etc.) had to be put at physicists' disposition.

Another major factor to be taken into consideration when writing the history of the physics done with the PS is of course the world-wide state-of-the-art in the field. In the late 50s/early 60s numerous new findings drastically changed the landscape in elementary-particles physics. The discovery of a large variety of new particles and resonances made it a particularly puzzling time. New results arrived steadily, notably from the USA. At Berkeley Luis Alvarez had built up a strong bubble-chamber group working with a 15″ (38 cm) and a 72″ (182 cm) hydrogen chamber from 1958 and 1959 respectively, and using the secondary beams of the 6 GeV Bevatron. This combination of bubble chambers with good beams bore immediate success. In 1960, the detection of the first strange resonance $Y_1^*(1385)$ was reported. 'Although the famous Fermi 3,3 resonance had been known for years, and although other resonances in the π^- nucleon system had since then shown up in total cross-section experiments ... the impact of the Y_1^* resonance on the thinking of particle physicists was quite different—the Y_1^* really acted like a new particle, and not simply as a resonance in a cross-section'. And after the announcement at the Rochester conference in 1960, 'the hunt for more short-lived particles began in earnest'. The $K^*(890)$ and the 3π resonances ω^0 and η^0 had to be added to the list of particles in 1961. Scientists at Brookhaven also succeeded in finding the resonance ϱ.[2] At the same time a number of new theories also enriched the scientific world, trying to explain the way strong interactions took place, and attempting to impose an order on the sudden chaos. Here we would just like to mention names such as Regge, Mandelstam, Chew, Low, Pomeranchuk, Gell-Mann and Ne'eman. In the late fifties/early sixties two theories had been put forward to describe strong-interaction processes: Regge theory and the bootstrap theory by Chew and Frautschi. Both were a step away from field theory, which had not been in a position to successfully describe, for example, pion–nucleon interactions, towards a more phenomenological S-matrix theory. The theory of unitary symmetry or 'eightfold way' put forward by Gell-Mann and Ne'eman in 1961 was then an attempt to find again a field-theoretical description. It sought to impose an order on this large number of particles by attributing all particles having the same quantum numbers except hypercharge Y and isospin I to the same multiplet. Not only did it give a place to the already known particles and resonances, but it pointed the way to the discovery of further new ones to complete certain multiplets. What is more, by the middle of 1964 the notion of what should be regarded as 'elementary' began to be challenged with the proposal—by Gell-Mann and Zweig—of a new kind of subparticle—QUARK—which would build up all known particles. And not only strong interaction physics saw tremendous changes in the period under consideration. The domain of weak interactions was also characterized by fantastic new discoveries such as the finding of two distinct neutrinos or *CP* violation.[3]

A lot of changes were also taking place in detection techniques. The invention of the bubble chamber in 1953 made a major impact, as did its further development in which Luis Alvarez at Berkeley played the decisive role. By the late 50s this detector had become one of the most promising devices for carrying out high-energy physics experiments. Photographing an interaction between particles (a so-called event) in its entirety, it was an ideal tool to look for the unknown, for example hunting new particles or resonances. It offered universality in taking a large number of pictures, but it had at the same time the disadvantage of low statistics when searching for a special kind of event. Electronic detectors were therefore an ideal counterpart, and they too underwent a tremendous improvement and diversification. In particular, they became a powerful tool with the increase in speed in electronic circuits and the later use of computers on-line. This way of working enabled the physicist to pursue a rather specific aim and to choose directly (via triggers) the kind of events to be studied, thus offering larger statistical accuracy. In parallel to these two methods a third kind of technique was still used, namely photographic emulsions. However, this method 'borrowed' from cosmic-ray research was used only marginally compared to the two others.[4]

From an organizatorial point of view it was soon realized at CERN that these three detection methods were not only technically distinct, but called for research teams which were differently structured and clearly separated on the individual level, demanded other skills, needed different timing and infrastructure, and much more. More explicitly, this meant that the bubble chamber groups were huge (containing a number of teams from the Member States), only a small part of them was stationed at CERN, and in the majority of cases data evaluation was done at the home institutes. In this case the major concern was to bring those physicists together who were interested in working on interactions of a certain secondary particle at a given energy (even with different aims) and then to organize the distribution of photographs. To a certain extent the same was true for emulsion experiments though on a far smaller scale. The electronic groups, in contrast, were far smaller, generally consisting of some staff members of CERN and incorporating individuals from outside institutions. Here the whole team had to be based at CERN because setting-up, data taking, and the output were one big unit, and not separable as in the case of bubble-chamber experiments.[5]

With that in mind a new organizational structure was put into operation on January 1961 having three different committees dealing with the problems inside each community: the Track Chamber Committee (TCC), the Electronic Experiments Committee (EEC) and the Emulsion Committee (EmC). The three were under the Nuclear Physics Research Committee (NPRC) which decided on the overall structure of the scientific programme.

When describing in what follows the construction of the CERN proton synchrotron and the scientific work performed with it, we will make use, for the latter, of this division into distinct communities according to the detection method. This choice was guided by several considerations. Firstly, the above-mentioned three communities were clearly distinct from the point of view of persons involved. One could therefore say that each of them had something like an internal history. Then, when placed in the general scientific context it is interesting to see how the different

Notes: pp. 243 ff.

instrumental conditions, with its advantages and constraints, led to a particular kind of research programme, having different strengths. Lastly, the policy decisions were generally first formed inside each subcommittee and only then globally arranged by the NPRC. We are aware that such a division into parts has its weaknesses—for example it misses to a large extent the interactions between the communities. Further we will restrict ourselves in the account to the two major communities, namely those using bubble chambers and electronic methods respectively. This choice was mainly dictated by the time at our disposition, and is not intended to diminish the role which the emulsion technique played, particularly in the early days at CERN.

The chapter thus falls into four main parts. The first will describe the design and construction of the accelerator, beginning with the implementation of the alternating-gradient focusing principle and all the difficulties this entailed. Then, after some remarks on the PS Group as such, we continue to handle the three major parts of the machine, the magnet, the radio-frequency system, and the injection linear accelerator. To close we touch shortly on the development of some of the other main parts of the machine.

The second part deals with research using bubble chambers as detectors. In a first section we describe what such an experiment involved, stretching from more technical aspects, i.e. the bubble chambers and the beams, to the extraction of scientific results out of the 'raw material'—the photographs. Thus we will refer to the diverse bubble chambers in use at CERN, we will give an overview on the evolution of the particle beams involved and lastly give an idea about the path followed by the thousands of photographs before physicists could draw a scientific conclusion. As there is a chapter dealing with the functioning of bubble chambers in general and in particular with the hydrogen chambers built at CERN, we do not go into very much detail here. Further there is also a section in this book dealing with the question of evaluating photographs where the reader will find detail on the problems involved and on the techniques developed to overcome them. In the second section the main lines of research with bubble chambers are followed, documented by tables giving an overview on all experiments made. More precisely, this means the study of interactions of protons, antiprotons, and both positive and negative kaons and pions with nucleons.

The third part will then describe the electronics experiments. As here the experiments are generally designed to answer one rather specific scientific question, we arranged them in four sections according to the interaction mechanism involved: strong, weak, electromagnetic, or gravitational.

The fourth and last part is a case study on the neutrino experiments. Here we will not restrict ourselves to a simple description of the experiment but will try to use it as an example of a very big experiment carried out at CERN to show the complexity and the difficulties of experimental work on this scale. This has become of immediate interest as the Nobel Prize in Physics for 1988 went to the Brookhaven team for their 1962 experiment which was carried out in competition with CERN.

5.2 Planning and constructing Europe's biggest proton accelerator[6]

5.2.1 FROM A 10 GeV CONSTANT GRADIENT TO A 30 GeV ALTERNATING GRADIENT SYNCHROTRON: THE PERIOD UP TO OCTOBER 1952

When early in the 1950s the plans for the European Laboratory began to take shape two different accelerator projects emerged. One machine was to be of standard type, easy and relatively fast to build, and of reasonable cost. Thus, as we have seen, the decision was taken to build a 600 MeV synchro-cyclotron. The second device, however, was to be a much more ambitious undertaking — an accelerator bigger than any other then existing. This would not only allow the European physics community to readjust their position *vis-à-vis* their American competitors, but also to break new ground in physics and to keep CERN in the forefront of high-energy physics.[7]

In May 1952 at the first meeting of the provisional Council in Paris, Odd Dahl, Norwegian specialist in accelerators, was appointed as 'Head of the Study Group in charge of studies and investigations regarding accelerators of particles for energies higher than 1 BeV'. The other members of the preliminary group were H. Alfen, W. Gentner, F.K. Goward (Deputy Director), F. Regenstreif, and R. Wideröe. The energy of the machine was not fixed at the time, a fact underlined by the vague statement in the minutes that proton energies between 1 and 20 GeV would be 'most useful from a scientific point of view' and the task of Dahl's group was defined as 'to indicate the scope of the undertaking without going into technical details'. A first report was to give an idea about a provisional design, manufacturing possibilities, delivery times, and above all the costs, and was to be submitted to the Council by the end of 1952. So as to base the choice of the energy of this big accelerator on good scientific and technical grounds, it was decided to organize an international conference in Copenhagen: this took place from 3 to 19 June 1952.[8]

When Odd Dahl presented his report on what machine should be built to the Council at its second meeting right after the conference, ideas had already taken a clearer shape. A proton synchrotron with energies between 10 and 20 GeV promised the most interesting and important scientific results and, what is more, its construction was regarded as less risky than any other type of machine, as one could draw on the experience of American accelerator builders. The competing accelerator types, namely electron and proton linear accelerators, both had clear disadvantages. The high cost and the technical difficulties of the former would cause a limitation in energy to about 3 GeV and doubts were expressed if really new results could be obtained in this energy range. The second possibility, a proton linear accelerator, would also be rather expensive and this device was technologically at a rather early stage of development. Anyway it would be fairly difficult to design a proton linac reaching energies of 30 MeV which would be needed as injector to the proton synchrotron in order to obtain beam intensities comparable to the Brookhaven machine.[9] Thus, it was decided that a proton synchrotron was preferable, and the

Notes: p. 244

energy, led mainly by budgetary considerations, was fixed at 10 GeV, as this was in any case sufficiently bigger than the US machines.[10]

Having taken this decision, it was now important to define the scientific and technical problems in order to determine the number of staff needed in this first stage and in particular the skills and experience they should have. This done, a permanent group was formed, starting the work on 1 July. New collaborators were D.W. Fry (part-time consultant on accelerator physics), K. Johnsen (accelerator theory) and Chr. Schmelzer (radio-frequency applications).[11]

As the CERN proton synchrotron was to be an extrapolated version of the Cosmotron Dahl, Goward, and Wideröe decided to make a visit to Brookhaven in August 1952 in order to discuss with the staff and to learn as much as possible about the machine. However, the discussions on what accelerator to construct in Europe had also stimulated the American specialists to think about the problems of building bigger and better machines. And indeed this had born fruit and when the visitors from CERN arrived E.D. Courant, M.S. Livingston, J.P. Blewett, H.S. Snyder and other members of the BNL accelerator group could report on a completely new, revolutionary concept—which came to be known as the alternating-gradient or strong-focusing principle. It should however be mentioned here that the same idea had already been elaborated and published by a Greek engineer N.C. Christofilos in 1950, but nobody had taken note of it.[12]

The way of making the synchrotron financially more attractive while raising the energy was to decrease the magnet cross-sections substantially: this was again linked to the size of the vacuum chamber and thus to the dimension of the beam. However field indexes below 1, as used in the conventional type of machine would never allow one to confine the beam further. Thus the decisive idea was to use magnet sectors with a very high field gradient arranged in the following way. In the first section the magnet field dropped with increasing radius (positive n) focusing the beam vertically while defocusing it radially at the same time. The next section had then a magnet field increasing with increasing radius (negative n) and exactly the inverse happened. Keeping up this sequence over the whole ring resulted in an overall focusing of the beam, notably reducing the oscillation amplitude and thus reducing the beam cross-section. Now, on the basis of these considerations, it suddenly seemed possible to construct a 30 GeV alternating-gradient synchrotron for the same price as the 10 GeV conventional one.[13]

Odd Dahl was fascinated and challenged by these completely new possibilities and on the occasion of the third Council meeting in October 1952 he already presented two different design projects. The first, titled 'Project 1', fixed the basic specifications of a conventional 10 GeV proton synchrotron, whereas the second was based on the 'Brookhaven scheme' and described a 30 GeV accelerator incorporating this completely novel technique—the AG principle. Indeed in this new project the vacuum chamber cross-section was reduced from 45×15 cm^2 for the 10 GeV machine to only 4×5 cm^2, entailing a decrease of the magnet weight from 6000 tons to 700 tons although at the same time the orbit radius had increased from 29 to 100 m. Further the injection energy was lowered to 3.5 MeV compared to the 30 MeV envisaged for the conventional machine, thus not calling for the development of a proton linac. The field index for these new magnets was to be several thousand.

The arguments brought forward in favour of this new project were an energy three times higher for the same price, that the 10 GeV machine 'would stretch cost and engineering to practical limits', and that one had here a project which would give CERN the opportunity also to make a contribution to technical developments in experimental high-energy physics. And indeed the Council decided to authorize a more detailed study on such a 30 GeV machine, and to make a final choice later.[14]

5.2.2 THE FEASIBILITY STUDY AND THE FIXING OF THE MAIN MACHINE PARAMETERS: NOVEMBER 1952 TO OCTOBER 1953

Since the 30 GeV accelerator was based on a completely new principle for confining protons in a path under acceleration, it was clear that no concrete planning could be undertaken as long as the theoretical basis for such a design was not clarified. Therefore in December 1952 the PS group was formally established augmenting the original staff. Four subgroups were formed dictated by geographical considerations. They were to move to Geneva at a later stage of the project. Theoretical and computational work on orbit stability was done at Harwell by a group around D.W. Fry and F.K. Goward and partly also at Copenhagen by G. Luders; magnet studies were concentrated in Grivet's laboratory in Paris under the guidance of E. Regenstreif; problems connected with the radio-frequency system were investigated by Chr. Schmelzer together with A. Schnell at the Bothe Institute in Heidelberg; and lastly the headquarters of this undertaking were at the Chr. Michelsens Institute in Bergen where Kjell Johnson and Odd Dahl were working. Wolfgang Gentner and Wideröe as permanent consultants evaluated building requirements and radiation protection measures.[15]

The idea of 'a machine three times bigger for the same money' was really seductive. The principle as such seemed unrefutable, the technological novelty of the project was a real challenge for the physicists and engineers involved. However, reality soon showed major imperfections. Indeed magnets with a field gradient of several thousand required virtually perfect mechanical alignment maintainable over a long period. Any small displacement would correspond to a large field change and therefore cause a serious distortion of the beam stability. Furthermore, inhomogenieties in the magnetic field and its gradient caused by remanent magnetism, eddy currents, saturation effects, or imperfections in the steel quality would make the beam hit the walls of the vacuum chamber.[16]

Hence the criticism that the 30 GeV project was far too ambitious, that the complexity of such an undertaking was too great, and the doubts that the technological know-how and the experience of the PS staff was sufficient. But the physicists and engineers involved were not ready to abandon the scheme, and began to study minutely all possible problems and to discuss them with their American collegues. And by March 1953, when they presented a kind of progress report in the 'Second Report to the Member States' the central problems linked to this new kind of accelerator were clearly formulated and some first solutions already presented.[17]

To begin with, it was proposed to undertake a second design study with changed parameters. Based on calculations of the orbit group it seemed more realistic to decrease the field gradient to 900; however, this entailed an increase of the magnet aperture by a factor 3 to 4. Therewith one had to give up one of the basic advantages of such an AG focusing machine, i.e. the low magnet weight. It was estimated that in this new version the magnets would weigh no less than 10,000 tons — even more than a conventional 10 GeV synchrotron. Whilst the problem of azimuthal inhomogeneities was resolved by lowering the field gradient, a completely new problem had been created, which caused quite some concern. Under these changed conditions the machine would have to operate through the so-called 'transition energy' — an energy where the frequency of rotation of the particles is independent of radius. This would necessitate the development of rather sophisticated frequency and phase control systems. Further, the initially planned low-energy injection (3 MeV) had to be given up in favour of a 50 MeV linear accelerator as injector in order to take account of field inhomogeneities at injection.[18]

Concerning the RF system physicists were conscious from the start that for such a big accelerator using the strong-focusing principle it would differ in many respects from that used in existing machines. Thus the RF group in Heidelberg started a programme studying a variety of different systems. In general three components had to be investigated: the RF programming system, a suitable form of accelerating units, and a system to control the beam. The main difficulty to overcome was the perfect synchronisation between the motion of the particles conditioned by the magnetic field and the frequency of the accelerating voltage. The tolerable frequency error had to be below 0.01% at the end of the accelerating cycle and only one part in 10^6 at the transition energy. Further this precision had to reproducible over a long period in time.[19]

On the magnet side it was clear that progress could only be made by using a rather empirical approach, i.e. learning by constructing a series of magnet models. Two such models were planned for 1953, both to learn about remanent field behaviour, the saturation region and in particular also to develop and test techniques to measure the magnets.[20]

At this point also the shielding was discussed. Indeed information on interaction processes in this range of energy were only known from cosmic-ray experiments at high altitudes. When the protons lose their energy by nuclear collision, mesons and nucleons are created and it is therefore important to study the absorption, angular distribution and spectrum of the secondaries and tertiaries. As these secondaries would be produced all over the circumference of the ring the ideal solution seemed, 'a trench type building with the machine below ground level and with a thick roof to attenuate scattering'. Exact numbers could of course not be given for the time being, but globally such a design was later adopted.[21]

The more the project progressed, the more staff was needed in order to tackle all important problems connected to the accelerator. But for the time being one had to draw on the collaboration of the staff available at the laboratories in which the different subgroups were working. Further it was clear that nothing would change fundamentally before a final decision was taken by the Council. Support, however, was largely given by the USA. In spring 1953 Hildred and John Blewett, two accelerator specialists from the Brookhaven staff, joined the PS group for some months. This meant not only assistance from the United States which went far beyond the normal, but also that during these crucial months of 1953, which would be the basis for a definite decision, the PS group could draw on the expertise and the professional authority of their American colleagues.[22]

October 1953 was then the decisive moment in this early phase of the PS's development. It started with the moving of the PS group to Geneva on 5 October, where they were hosted in the Institute of Physics at the University. The groups which had been spread all over Europe were now gathered in one place, and this considerably simplified efficient coordination. Further the number of scientists, engineers, and technicians had increased to about 20, and from then onwards they would meet every week to discuss the progress made, the difficulties encountered, and in particular the further steps to take.[23]

It was also by this time that the understanding of this new kind of machine was judged as sufficiently advanced to organize a conference on that topic in Geneva. It was meant to present the results of CERN's design work, to exchange views on them with American experts, and to hear about American projects—in particular, about the one at Brookhaven, where a machine virtually identical to the one at CERN was discussed. As the Council would meet right after this conference, any decision would be largely based on the outcome of these discussions. The presentations of the CERN PS were structured in three big parts. It started with a review of the project, the machine parameters and the leading design principles by O. Dahl and F. Goward. This was then followed by theoretical design considerations presented by G. Luders, M.G.N. Hine (who had joined CERN in September 1953), K. Johnsen, and J.B. Adams. Lastly, the component design of magnet, RF system and shielding was described by Chr. Schmelzer, P. Grivet, J.C. Jacobsen, and A. Citron.

The main parameters for the 30 GeV AG synchrotron given by Dahl can be summarized as follows: The machine would have a maximum magnetic field of 12,000 Gauss and thus a radius of 112 meters. 114 magnet sectors, each having half a focusing and half a defocusing sector, weighing altogether 4000 tons and with a field gradient of 392 would be installed in the ring. The inside dimensions of the vacuum system would be 8 cm × 12 cm. 50 MeV was given as injection energy. Further a

Notes: p. 244

more detailed cost and manpower analysis was presented. The former had risen to 69.2 MSF (including 17.2 MSF for buildings) and regarding the latter 700 man-years were foreseen, rising from 53 in the first year to about 140.[24]

Thus when the Council met for the 7th time from 29 to 31 October 1953 Odd Dahl—strengthened by the conference in his belief of the feasibility of the project—presented the basic design of the machine and the rough budget plan. Indeed the conference had also convinced the Council of the capability and the seriousness of the PS group and as John Cockcroft stated 'it had demonstrated the excellent work of the Group'. Now it was up to the representatives of the Member States to choose in particular the exact size of the accelerator. After a lengthy discussion balancing the higher costs and the scientific merits of such a big machine, a final decision was taken: the PS group should design and construct an AG proton synchrotron of 25 GeV energy with a magnetic field of 12,000 Gauss, thus with a radius of 72 meters.[25]

5.2.3 THE PERIOD OF ENGINEERING DESIGN STUDIES, MANUFACTURING, AND CONSTRUCTION (1954 TO 1959)

5.2.3.1 *An overview on the development of the PS group*

With the moving of the PS group to Geneva and the final decision of the Council to construct a 25 GeV alternating-gradient proton synchrotron—both in October 1953—a new phase started. So far the task had been to make a feasibility study, which was obviously of a rather general nature; now the detailed design started. Thus in many senses 1954 was to become a very important year in the construction of this huge European accelerator. It was not only the year during which the design of the final machine was fixed, but it meant serious changes in the organizational structure and size of the PS group. The number of physicists, engineers, and technicians was increased steadily reaching the number of 50 by September 1954. Each of them had a precisely defined task in the design study which had to be tackled in every detail. The existing subgroups were restructured and new subgroups were added. Also the positions of the leader of the subgroups were reconsidered and—with the exception of the RF group which underwent no changes—, all other posts were newly filled. Thus the magnet group was now led by C.A. Ramm; a mechanical engineering group was set up, embracing design and drawing offices and a workshop, under C.J. Zilverschoon (it included also work on the vacuum system which was later guided by G.L. Munday); the theory group was restructured, dealing now mainly with specific orbit problems of the CERN machine and was led by A. Schoch who was later joined by M.G.N. Hine; and finally a linear accelerator group was founded guided first by H.G. Hereward and in the final phase by P. Lapostolle. At a later point in the development an electrical engineering group under F. Grütter was also set up to deal with the power supply and the control system. The problem of surveying the ground in order to decide where exactly to build the PS was entrusted to a local firm and the buildings were designed in cooperation with the architect Steiger.[26]

1954 was also a year of dramatic changes among the leadership of the PS group. In February Frank Goward, the Deputy Director of the group, fell ill and died a month later. John Adams, who had joined CERN in September 1953, took over his position. However, Odd Dahl had by that time already decided to leave CERN and had taken up other committments, and so he resigned when the permanent organization came into being in September 1954, and John Adams became leader of the PS group. Luckily, both these personnel changes and the transition from the interim to the final organization took place smoothly and without any abrupt changes. Thus the PS project which was well on its way despite the numerous organizational changes was not touched by these events.[27]

When speaking about the PS group we should of course also say some words about their relation with their American colleagues. In the previous section we remarked that the links were extremely good, or perhaps we should say extraordinary, as BNL went so far as to send staff for some months to CERN to help in the early phase. Similarly later on, with the exception of the first half of 1954 (when there was no exchange at all of information), staff from CERN regularly visited Brookhaven, where a similar project on a 25 GeV accelerator had been launched early in 1954 and *vice versa*. This was crucial for a direct exchange of views and to compare different approaches to problems. And although at later stages CERN was absolutely on an equal footing with Brookhaven in the practical development of this new machine, and a process of mutual learning from each other had been established, one was always conscious that the support from the American side in the very beginning of the project had been indispensable for a smooth advancement of the European project, and a particularly close relationship was established between these two labs.[28]

There are a few things to be said concerning staff and activities within the PS Division during the period up to 1959. Firstly with the advancement of the project the staff increased—as foreseen—rapidly from 75 in January 1955, to 120 in November 1955, to 143 in November 1956, to 163 in October 1957 (when the group moved on to the Meyrin site), and reached 216 by the end of the construction period—thus needing more personnel than the planned 700 man-years. It should, however, be said that during the late 50s the PS Division had also taken over other tasks not directly related to the construction of the PS. In the beginning of 1957 a group led by A. Schoch and containing at first two staff members and seven CERN and Ford Associates started to investigate new ideas for accelerators. The idea was to build up quickly a strong group of specialists in the design of new high-energy accelerators in order to ensure CERN a place among the best in this domain. The second activity was started early 1958 in cooperation with the STS Division, namely the work on bubble chambers, beam transport systems, and the services required by such a device. Charles Peyrou with a staff of 26 physicists and engineers was to take over the task of constructing the three hydrogen bubble chambers (10 cm, 30 cm,

Notes: p. 244

2 m), whereas the envisaged 1 meter propane bubble chamber and the beam transport equipment were to be handled by C. Ramm and the staff of the PS magnet group.[29]

5.2.3.2 The PS magnet[30]

The studies carried out late in 1952 and in 1953 had revealed that the magnet was an extremely sensitive part of the new accelerator and that its design would also largely affect the overall budget. Thus it was decided that the way to learn more about the exact shape of the magnet units and the effect of the magnetic parameters — permeability of the steel, magnetic remanence and effects of eddy currents — on the structure of the field, was to construct a series of magnet models. This was a slow and costly method — no less than 10 models were constructed and tested throughout the design period —, but an alternating-gradient synchrotron was a completely new type of accelerator and such a procedure permitted the group to take a well-based final decision.

The first two models constructed during 1953 were one with very high field gradient and a 1/5 scale model for an $n = 100$ magnet. Both had been ordered when an extremely high field gradient magnet had been envisaged for the PS magnet. The first to take account of the changed situation was model 3, a 1/3 scale model of an $n = 400$ machine. It was used during 1954 to study the field on the equilibrium orbit as a function of the existing current, the flux densities in different sections, the non-linearities, the field behaviour in the air-gap, and many more properties. Similar investigations were then also carried out on model 4, which was the first full-scale $n = 400, 0, +400$ model, and by December 1954 after close collaboration with the theoretical group, the main parameters of the PS magnet were fixed, and did not change during the following year, apart from slight changes in the field index. There would be 100 magnet units, with a field index of about 282 around the circumference. The operating mode μ (the phase shift per magnet unit) was $\pi/4$, and the number of betatron oscillations per revolution Q was 6.25. The magnet units, weighing altogether 3200 tons (far less than feared originally), would usually be 4.4 meters long with a field-free gap of 1.6 meters in between, except for ten 3 meter long straight sections.[31]

All that the magnet group had learned to date was incorporated in the models 5, 6 and 7, constructed during 1955. The first two were intended to decide what thickness — 1 mm or 1 cm — the laminations of the magnet should be. Model seven was identical to number five except with a pole profile reversed with respect to the yoke. After detailed studies and comparisons between these three, sufficient information had been collected to decide on a satisfactory construction of the magnet. One would use 1 mm thick precision stamped steel laminations interleaved with paper. In this way effects due to eddy currents in the body of the yoke would be negligible. Further one had learned that the magnetic characteristics of the steel to be used seriously affected the field distribution. As uniformity and stability of the magnet were key concerns, investigations on the available steel qualities in Europe were started, and soon showed that it would be nearly impossible to come up to the severe specifications demanded by CERN. What is more, one was not clear about the influence of the method of magnet production, of steel ageing and of non-uniformity in the magnetic properties of the steel supply—all peculiar to the individual block makers and steel suppliers. As a consequence the decision was taken to ask each of the three firms which had come into closer considerations to produce a model as close as possible to the final version, and incorporating the available results of the recent pole profile studies.[32]

Early in 1956 the exact magnet specifications were communicated to the firms interested, first replies were received by June and in October the contract was finally placed with the Italian firm Ansaldo San Giorgio. The whole magnet was to have a hundred 4.4 m long units each composed of two half units of 5 blocks. The latter would be made from 1.5 mm precision stamped steel sheets. The first blocks were to be delivered by September 1957, the last arriving before May 1958 to keep to the overall schedule. Further, a way of dealing with the fluctuations in steel quality was found. The 250 000 laminations would be suitably mixed before putting them together in order to obtain overall uniformity of the magnet units.[33]

By this time the question of coils—copper versus aluminium as basic material—had also been resolved. Aluminium was to be used for reasons of lower costs, less weight, and ease of manufacture. They would be water cooled, the conductor having a cross-section of 55 × 88 mm with a hole of 12 mm to let the cooling water through.[34]

At the beginning of 1957 a first prototype block was delivered and the controls were satisfactory. The design was thus frozen and all attention was turned to the smooth execution of the construction. Two representatives of the magnet group were stationed at the manufacturing firm to survey the control of the quality of the incoming steel. By mid-1957 the production started and as foreseen all blocks had arrived at CERN a year later.[35] (For the magnet cross-section see Fig. 5.1.)

The time had now come to test all these blocks and, to the satisfaction of everybody, the results showed that the effort put into the initial development of the production technique had resulted in a quality better than anticipated. Thus during 1958 the blocks were assembled into units. At this stage, however, a delay occurred, as the pole face windings—made to correct the magnetic field in the gap—were delivered late and had to be mounted before installing the magnet units in the ring.[37]

Nevertheless, by January 1959 all 100 magnet units were ready for testing and a very tight schedule was drawn up to make up the time lost. After six months of mechanical, electrical, and magnetic testing and measuring, all components of the magnet system were installed in the ring without major problems. It had been possible to reach a precision in the alignment of no less than $1/10^6$. Further the diverse magnetic lenses needed for corrections were positioned and on 27 July 1959 the main power supply was connected without difficulties.[38]

Notes: pp. 244 ff.

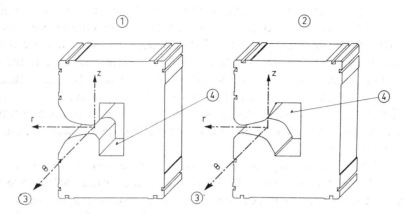

Fig. 5.1 Cross-section of a PS magnet unit[36]

5.2.3.3 The radio-frequency system[39]

As already mentioned before, the tasks of this subgroup were concentrated on three main items: the design and development of the accelerating cavities, the setting up of a frequency programming system embracing the so-called Hall computer and the RF generator, and lastly a beam control system.

Concerning the accelerating cavities, the basic design which was chosen from various possibilities made use of two quarter-wave co-axial resonators, where the protons passed the hollow inner conductor and were accelerated when crossing the gap in the middle of the resonator. Owing to the lack of space between the magnet units in the PS ring the necessary electrical length of the resonator had to be achieved by loading with ferrite, thus also allowing the tuning by means of an auxiliary magnetic field acting on the permeability of the ferrite. The accelerating cycle was to cover the frequency band between 2 and 10 Mc/sec. By the end of 1954 this part of the RF system had reached a stage where a mechanical design could be made and it was planned to start more detailed model tests.[40]

The frequency programming system, which would ensure that the frequency kept in step with the rise of the magnetic field, consisted basically of a computer linked to a master oscillator generating the radio frequency. More precisely, the Hall effect would be used to determine the magnetic field intensity at the equilibrium orbit and these data were converted by a computer according to the theoretical law (field-frequency) into a signal steering the RF generator which again fed the accelerating cavities (see schematic drawing in Fig. 5.2). However, both the 'Hall computer' and the RF generator made only a little progress during 1954.[41]

Since the accelerating cavities were the central and most expensive part of the RF system, two models with scaled cross-section were built and tested during 1955. The first was mainly used to test the properties of ferrites, whereas the second was

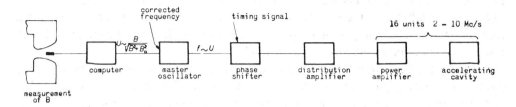

Fig. 5.2 Schematic drawing of the RF system[42]

connected to an amplifier to make dynamic tests. Indeed the results showed good agreement with the values predicted theoretically. Therefore a full-scale model was built and was ready for extensive tests by November 1955. The results obtained turned out to be most encouraging and the final design for the cavities and the RF power amplifiers seemed virtually settled. Now it was time to prepare the specifications for the manufacturers, in time to call for tenders in May 1956. In any case the final technological problems would be solved in collaboration with the manufacturer, which was chosen to be Magneti-Marelli in Milan.[43]

On the side of the frequency control system, work concentrated on the computer — an analog computer pushed to its extreme limits — which was in the stage of component design. By 1956 a refined model of the Hall computer was working successfully and was tested for accuracy and reproducibility. Also the spin generator — the type of master oscillator initially envisaged — was further developed during 1955, and in 1956 a first version of a beat frequency generator using electron spin resonance was built. However the frequency stability could only be reached in the short time range.[44]

In the course of 1957 Magneti-Marelli delivered a prototype of an accelerating unit to CERN where it underwent thorough tests, which led to changes in the design. Meanwhile, the delivery of the ferrite cores, a contract that had been allocated to Philips, was well on its way, the quality being even better than foreseen. Also major progress was made with the frequency programming system. As far as the Hall computer was concerned, 'Mark I' was constructed and underwent a series of performance tests. Drawing on the experience thus gained the 'Mark II' computer was designed, reaching a reproducibility of 0.5% over 8 hours. Concerning the RF generator, however, a completely new track was followed, with effort now being placed on the development of a servo-controlled generator as an alternative to the spin generator, the signal to noise ratio of which had been too low. A first model of such a generator was built giving satisfactory results.[45]

As the project was advancing well it was also time to start investigating a beam control system. The idea of having a 'phase-lock' system — proposed at the 1956 CERN Symposium — was mainly followed here. In such a system a phase discriminator measured the phase between the accelerating gap and a pick-up electrode. If a difference between measured and desired value was recorded, an error signal was produced, which acted on a frequency-modulated oscillator. A radial control device containing a pick-up electrode was further envisaged in order to correct any transverse deviation of the beam.[46]

In spring 1958 the improved model of the accelerating unit was received and tests showed that the design could now be frozen. It was therefore brought to the tunnel to undergo life-tests. The development of the Hall computer had also reached its final stage and Mark III was being set up in the Computer Room, where reproducibility tests gave a satisfactory value as low as $\pm 7 \times 10^{-4}$. As for the frequency generator, a servo master oscillator had been built, successfully tested and installed in the Computer Room. With this device the group had reached a relative frequency stability of 3×10^{-4} over ten hours and any substantial improvements of these performances seemed impossible. Anyway as the Hall computer also had certain limits of accuracy it was of no interest to push this device too far. Thus, all parts being virtually ready, a testing programme of the whole installation could be started.[47]

By 1958 the RF group, now comprising a total staff of 19, had reached a phase where they were mainly testing and installing the arriving parts. This was true both for the accelerating units which arrived from beginning September onwards at a rate of about one unit per week, as well as for the whole frequency programming system which underwent a series of tests for long-term stability. The beam control system was also nearly completely designed and prototypes of all major parts had been constructed and tested. The same could be said about the pick-up electrode system.[48]

During 1959 the entire RF system was installed, and tests on the accelerating units and the frequency programming system were carried out giving satisfactory results. The beam control system was also set up and it was decided that its automatic phase control system should be used during programmed acceleration thus raising the overall performances of the RF system. Further the pick-up electrode system was installed and tested. And on 1 Ocober 1959 the RF system was ready for acceleration tests.[49]

5.2.3.4 The injection linear accelerator[50]

While it had been assumed in the beginning that an AG focusing synchrotron would only demand a rather low injection energy, i.e. 3 MeV, it soon turned out that an energy of about 50 MeV was needed to obtain overall satisfactory results; otherwise field inhomogeneities would play a substantial role. This meant that physicists at CERN had to design a proton linear accelerator as injector for the PS.[51]

During 1954 mainly two different types of linacs—the helix type and the Alvarez type—were closely investigated by H.G. Hereward and K. Johnsen in order to balance advantages and drawbacks and to take a final decision. This choice was however simplified, as meanwhile a project on a proton linear accelerator of no less than 600 MeV based on Alvarez' ideas was started at Harwell, and the PS group was able to draw on both the design and the manufacturing experience gained there. Therefore it was clear by the end of the year that at CERN one should follow a similar design incorporating some minor changes.[52]

The linac was to function as follows. There were three cylindrical capacity resonators arranged in series, each containing a set of drift tubes along the axis. The complex was enclosed in a vacuum tank and connected to an RF source as power supply. The protons were produced by an ion source, accelerated to 500 kV by a cascade generator, and injected into the linac, where they were further accelerated at each gap between the drift tubes. At the end, a system of lenses and bending magnets guided these protons into the synchrotron. In the three cavities the protons would be accelerated from 0.5 to 10 MeV, from 10 to 30 MeV, and from 30 to 50 MeV respectively. The total length of the device was to be about 30 meters. Before the protons were injected into the linac they would pass a 'buncher', whose task it was to modulate the particle velocity and to group them inside the boundaries of the trapping region. At the exit of the linac debunching was then necessary to reduce the energy spread in the beam. As for focusing—a rather central question—it was decided to use grid focusing in tank I, whereas alternating gradient quadrupole magnets were to be installed in tanks II and III.[53] (See Fig. 5.3.)

In late 1954/early 1955 the PS Division started to take up contacts with the firm Metropolitan Vickers, which was already manufacturing the Harwell linac. The contract later signed embraced the three tanks and the vacuum system. The 500 kV generator for pre-acceleration of the protons was standard and could be ordered without further development work from Haefely in Switzerland. As a result the design work was mainly concentrated on the linac RF system (which proved rather complicated as it was a multi-cavity accelerator), the buncher and debuncher, and the inflector. Further it was decided that in parallel to the grid focusing foreseen in tank I pulsed quadrupole magnets should be designed, built, and incorporated at a later stage.[55]

Due to unforeseen problems with the supply of stainless steel the delivery of the first part of the linac was seriously delayed and thus the first vacuum tank arrived only by the end of 1956, was thoroughly tested and leaks repaired. Meanwhile a prototype of the RF supplies was tested and after initial difficulties produced (in 1957) the 2 Megawatts peak power needed. The definite order could thus be placed. The pulsed ion source had also been completed by June 1956 and underwent tests resulting in beams of 10 and 40 mA. Indeed, all major parts of the linac were well advanced with the exception of the accelerating structure. Although tanks II and III had been scheduled to arrive at CERN by mid-1957, during the year only tank II could be installed together with its pumping system.[56]

In spring 1958, the ion source, the accelerating column, all the associated electronics, the vacuum and the RF system as well as the liner and the grid focusing drift tubes of tank I were put in place, so that the first accelerated proton beam could

Notes: p. 245

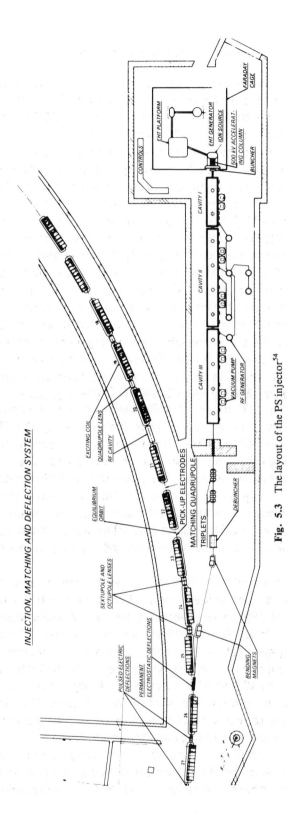

Fig. 5.3 The layout of the PS injector[54]

be produced in the first section. Liner and drift tubes were then installed in tank II. Further satisfactory progress on the buncher, debuncher and the inflection system could be reported. During the rest of the year, with the help of a staff of 27, the final adjustments were carried out: the current of accelerated protons in the first linac was increased, the ion source was replaced by a more stable one, the second and third cavities were installed together with their necessary RF equipment, and a start was made to set-up the inflector. Considerable delays occurred when aligning the drift tubes in the last two linac sections, as specified manufacturing tolerances had not been satisfied and numerous vacuum leaks had to be repaired.[57]

In May 1959 protons were accelerated by tanks I and II giving 30 MeV with a current of 200 μA. This value was then improved by July to 3 mA. Then the grid focusing in tank I was replaced by quadrupole focusing, raising the current to 5 mA. Finally, on 24 August the first protons were accelerated by the third tank and gave as expected 50 MeV and a current of 500 μA, which was increased to 4.8 mA during the first days of running. And after having tested different components of the inflector, the linac, and the synchrotron a 50 MeV beam made its first orbit in the proton synchrotron on 16 September 1959.[58]

5.2.3.5 Some remarks on other major activities around the PS[59]

So far we have discussed in some detail three major parts of the proton synchrotron—the magnet, the RF system, and the linear accelerator injector. Now we want to touch quickly on the general power distribution, the control, and the vacuum system as well as on the geodetic survey programme and the linked special construction of the PS ring.

Starting with the general power supply, the CERN site was connected by 3 power cables able to carry 21 MW to a power station in the Canton of Geneva. The power was needed for three purposes: to run the machine, to give the necessary power for experiments, and to assure the general services. In the first category by far the most power was needed by the magnet, followed by the accelerating system, and the linac. The second group covers mainly the detection devices (bubble chambers and counters) and beam analysis equipment (separators and magnets). However by far the biggest consumption was registered for general services such as lighting, heating or machine driving. To ensure the supply of all this one main substation and two further substations were installed on site. The detailed design of the power distribution system was started only on 1957 by the electrical engineering subgroup under F. Grütter, when exact specifications on the main components of the accelerator were known. By 1958 no less than 10 staff members were working on this task and no major problems were reported.[60]

The setting up of the control system, which was also carried out by the same subgroup, could also start only rather late due to the novelty of the machine design and the complexity of the equipment. To guarantee optimal flexibility it was decided

Notes: p. 245

to have one main control room and four ancillary centres. The first would be in a sense the brain of the machine getting all information on the linac, the internal proton beam and its transverse position, the RF system, the correcting lenses and pole face windings, the magnets, the targets, and the vacuum, as well as all kinds of safety information such as radiation measures. The other four would be interconnected and control individually important parts of the whole accelerator. In general, the system was built up as the relevant parts of the PS were delivered, using in general prefabricated standard units.[61]

As for the vacuum system it was handled by the mechanical subgroup, and in fact there was not a lot of development work needed. Owing to the high beam energy at injection the problem of scattering by residual gas molecules became far less important and therefore a pressure of 10^{-5} torr was sufficient and one did not have to worry very much about pressure effects. The choice of the material fell, after

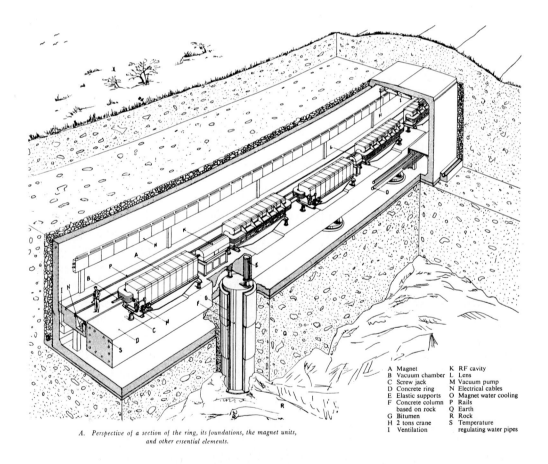

A. Perspective of a section of the ring, its foundations, the magnet units, and other essential elements.

A Magnet   K RF cavity
B Vacuum chamber   L Lens
C Screw jack   M Vacuum pump
D Concrete ring   N Electrical cables
E Elastic supports   O Magnet water cooling
F Concrete column   P Rails
   based on rock   Q Earth
G Bitumen   R Rock
H 2 tons crane   S Temperature
I Ventilation   regulating water pipes

Fig. 5.4 Perspective of a section of the ring[64]

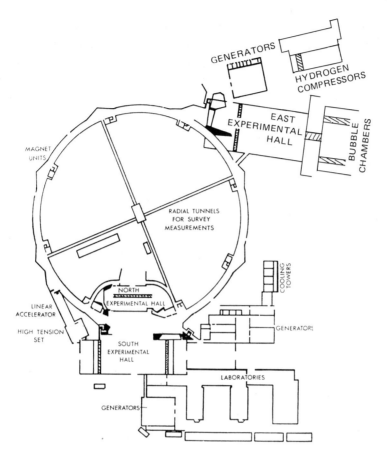

Fig. 5.5 The PS and its experimental areas[65]

having made a series of tests, on metal 2 mm thick, as this gave satisfactory mechanical rigidity and took minimum space from the useful aperture for the beam. The whole system was to consist of 100 elliptical tubes of 7 × 14 cm cross-section placed between the poles of the magnet units, twenty 3 meter long and eighty 1 meter long tubes. To reduce the shut-downs to a minimum in case of maintenance work or insertion of targets, the chamber was divided into 10 equal sections separated by plate valves in the field-free sections. The pumping was done by 50 oil-diffusion pumps equally spaced over the whole ring with additional units at crucial points such as the inflector or the target area. Early in 1958 the contracted firm Usines Jean Gallay in Geneva delivered the parts of the vacuum system and they were tested mechanically, magnetically and for vacuum tightness. This took until mid-1959. The results were rather satisfying, though some problems with the alignment occurred when the installation started. They could be solved fairly quickly and by September

Notes: p. 245

1959 the whole ring was put under vacuum for the first time.[62]

The last point to mention here was the choice of an ideal emplacement for the accelerator on the Meyrin site, crucial as the machine would be extremely sensitive to ground movements. Thus a detailed site survey programme was undertaken in cooperation with a local firm and a complete map of the ground conditions was drawn. On the basis of this it was possible by late 1954 to identify a satisfactory position for the machine, necessitating however asking for new ground on the south border of the site. A number of survey markers were then placed in the morraine layer there, and an accurate study of the ground movements started. To give additional security it was important to have a structure for the magnet foundations which was insensitive to temperature changes and ground movements. This problem was solved by embedding eighty pillars firmly into the molasse rock around the circle of the machine. The magnets themselves were then placed on a heavy concrete ring beam, 100 m in radius and 2 m square cross-section which rested on the pillars. Between ring and pillars a flexible coupling allowed free movement of the ring with respect to the pillars. Cooling water was circulating in the ring beam to protect it from local temperature gradients and the building was air conditioned. The whole was incorporated in a ring shaped underground building to keep harmful radiation at a low level.[63] (See Fig. 5.4.)

On 24 November for the first time a beam of 24 GeV protons was obtained, which was increased to as much as 28.3 GeV on 8 December. And as no major problems occurred a regular experimental programme could already be started three months later in spring 1960.

5.3 High-energy physics using bubble chambers[66]

5.3.1 BUBBLE CHAMBERS, BEAMS, AND DATA HANDLING—THE THREE CENTRAL PARTS OF A BUBBLE CHAMBER EXPERIMENT

5.3.1.1 The bubble chambers in operation at CERN between 1960 and 1965

Six bubble chambers were used with the PS for carrying out research in the field of particle physics during our period. They stayed in the laboratory for different lengths of time, and had rather different technical features, facts which are reflected in the physics programme and in the role each of them played in it. Four of them were hydrogen bubble chambers ranging in size from 30 cm to 2 meters. The other two were both of the heavy liquid type and 1 meter in size. Three had been constructed in laboratories in the member states, two in France and one by a British team.

Which liquid to choose for what kind of experiment? The interest in using liquid hydrogen, although much more difficult to handle, is the fact that the hydrogen nucleus consists of protons and thus the interaction of the incoming beam takes place with a free proton at rest. This creates ideal conditions as the dynamics of the interaction is much easier to determine. Heavy liquids, on the contrary, are a very complex filling material, as the protons are bound and secondary interactions occur in the complex nucleus. However, if one wants to study a reaction in its entirety, one also has to detect the directly invisible neutral particles involved via their decay products. This is only possible by using a liquid with high density and thus higher stopping power. In other words, a heavy liquid chamber is ideal for studying rare events, electron-decay modes of strange particles, and many other processes.

- The *30 cm CERN hydrogen bubble chamber* (HBC30): Originally intended for use with the synchro-cyclotron, the design of this relatively small device started early in 1957 under the leadership of Charles Peyrou at CERN. It was assembled during 1958 and was CERN's first bubble chamber to be equipped with a magnetic field of 17,500 Gauss. It was filled with hydrogen in May 1959 and was used for the first time in summer of that year at the SC, though not without major complications. Being the only bubble chamber at hand when the PS started working it was moved to the South Hall, where it took photographs on high-energy pion and proton interactions from March 1960 onwards. It operated for two years until it was moved away in March 1962. During this period nearly half a million pictures of p, π, and K interactions had been taken. Later it was transported to Frascati where a new electron accelerator had been put into operation.

- The *81 cm Saclay hydrogen bubble chamber* (HBC81):[67] Without any doubt this device played the central role in CERN's bubble chamber programme from 1960 to 1965. Constructed by a team of scientists and technicians at Saclay in collaboration with the Ecole Polytechnique under the guidance of Bernard Gregory, it was transported to CERN in early 1961 and installed in the bubble chamber area of the South Hall. Its size was chosen so as to be big enough to have clear advantages over the chambers at hand (30 cm CERN and 50 cm Saclay chambers), but at the same time small enough to be ready in time to bridge the gap until the 152 cm British and the 200 cm CERN chamber were completed. Thus it was the first big hydrogen bubble chamber put into operation at CERN and it remained in this role far longer than foreseen, indeed until mid-1964. Over the period described by us some 6.5 million photographs, representing over half of those taken at CERN, were made with this device, which corresponded to a length of film of some 2000 km. During 1961 and 1962 the chamber ran exclusively with hydrogen while in May 1963 it was successfully filled with deuterium for the first time. For this purpose it was equipped with a temperature control system designed and manufactured originally for the 10 and 30 cm CERN bubble chambers. This material which was relatively easy to handle, but rather expensive and difficult to purchase, also allowed the study of interactions with neutrons. In the first two years 15% of the time was already devoted to taking photographs with deuterium in the chamber. This increased to about 80% (corresponding to 1.6 million pictures) by 1965, when both of the bigger hydrogen bubble chambers were at hand.

Notes: p. 245

– The *152 cm British National hydrogen bubble chamber* (BNHBC152):[68] The construction of this chamber began early in 1959 by a British collaboration, containing groups from Birmingham, Liverpool, Imperial College, and NIRNS. Originally it was planned that this chamber should arrive at CERN by the end of 1961 and be installed in a building specially constructed for it in the East Hall. This chamber was intended to bridge the time until the CERN 2 meter chamber came into operation, whereupon it was to be transported back to Britain to be used with the 7 GeV proton synchrotron 'Nimrod' built at the Rutherford High Energy Laboratory. However in 1961 only the magnet was ready and transported to CERN. The chamber as such followed as late as 1963 and by October of this year its first cool down for a technical run was undertaken. But there were still a lot of 'teething troubles' and so it was decided early in 1964 that the 81 cm Saclay chamber would take over the planned programme until the British chamber functioned without difficulties. By June 1964 the device passed all the necessary tests and finally, after a delay of more than two years, experimental work could start. Returning to Great Britain in September 1965, this bubble chamber thus only stayed for one year at CERN. During this period 1.3 million photographs were taken, the majority (64%) of which were on Kp interactions between 5 and 10 GeV/c, 23% with pions and only a very small number with protons.

– The *2 meter CERN hydrogen bubble chamber* (HBC200):[69] The decision that CERN should construct a huge hydrogen bubble chamber to be used with the PS was taken in the late 1950s (see chapter 9). In 1959 the team was essentially recruited and the design and construction phase could start. It went through the stage of numerous models and tests, and the first successful trials were only made with the chamber as late as December 1964. However when in March the chamber went finally into operation it showed the first problems after 200 000 photographs. Thus it was dismantled and underwent a complete overhaul during the summer. Then it worked without difficulties, taking about 1 million photographs from September until December 1965, so only at the very end of the period covered by us.

– The *1 meter Ecole Polytechnique heavy liquid bubble chamber* (EPHLC):[70] The decision to investigate the construction of a big heavy liquid bubble chamber was taken in July 1957, so as to be ready when the PS started operating. The choice fell on a heavy liquid filling as at that time the bubble chamber technique still posed a number of problems and this seemed a less risky option. To gain experience two smaller chambers, one of 1.5 and one with 40 litres were first constructed, but this could nevertheless not avoid problems arising with a 300 litre chamber. A. Lagarrigue was named project leader and they managed to start assembly of the parts by September 1959. In the first four months of 1960 the chamber was tested at the 3 GeV accelerator Saturne and indeed, as originally planned, it arrived by mid-1960 at CERN ready for experimental work.

– The *1 meter CERN heavy liquid bubble chamber (HLBC):*[71] Formed by members of the PS magnet group, a team under C.A. Ramm started to work on the propane bubble chamber during 1958. However the bubble chamber project advanced only slowly as long as there was still work to be done for the PS. Nevertheless the design reached a stage that at least parts of the chamber could be ordered and delivered during 1959. By December 1960 this chamber containing 500 litres of heavy liquid underwent first operational trials, but due to a leak of freon the main window broke and the chamber had to be repaired. Foreseen as a major component in the neutrino experiment the chamber was used during May and July 1961 for preliminary background runs and removed again to the NPA building in January 1962 for further modifications. From then on—with one exception—this chamber was entirely used for the neutrino experiment. By mid-1964 the chamber was again transported to the NPA building, as an enlargemnent of the chamber had been envisaged. The new version ready by May 1965 was twice as big as the old one containing 1200 liters of heavy liquid and had thus become the world's biggest heavy liquid bubble chamber.

5.3.1.2 The major bubble chamber beams

As far as the setting-up of beams was concerned physicists were rather involved, and in fact this should be regarded as a major preparative activity for their experiments. In terms of time this could take several weeks or even months. The more the bubble chambers grew in size and became instruments installed in a particular place, the more important it became to provide a variety of rather complex and at the same time flexible beams to use it. Advances in beam technology started to become an important factor influencing the speed of progress in particle physics. And as the setting up of beams became an increasingly complex and technically sophisticated venture, specialists were largely occupied with this task.

What was needed for a secondary beam? When a bunch of particles, about 10^{11} in the case of the PS, hits a target put into the vacuum chamber of the machine, secondary particles of different kinds and in different numbers are produced. These particles are guided by bending magnets, and focused by quadrupole lenses until they reach the place where their interaction with other particles will be studied. The result is a beam of one given momentum containing a variety of particles. In a next step the aim is to 'filter out' one sort of particle contaminated as little as possible by the others. This was of particular importance for the study of rare particles like kaons and antiprotons which constituted only a small percentage of all secondaries produced. Particle separation can be achieved by several methods. In the simple *electrostatic separator* all the particles of a given momentum pass through a strong electrostatic field created by applying a high voltage between two parallel metal plates. Having the same momentum, particles of different mass move with different velocities and are therefore deflected by different amounts. At very high energies this technique called for various independent stages. A further improvement was the construction of the so-called *radio-frequency separator*. This consists in principle of

Notes: pp. 245 ff.

two or more radio-frequency cavities with a certain distance between them. The particles arrive at the first cavity with the same momentum, which they pass at the same speed as the electromagnetic wave. Some 50 m further there is a second cavity guiding an identical wave. However since the particles in the beam have different masses they travel between the two cavities at different speeds and arrive therefore shifted in time. The system has then to be adjusted so that a chosen particle arrives exactly at the same phase of the wave as in the first cavity and thus continues in the direction of the initial beam. All other particles arriving sooner or later are deflected at different angles and caught in a 'beam stopper'. This new kind of separator was only of use for bubble chamber experiments as at that time it could only deliver a short pulse length, ideal for this detection technique.[72]

In 1960, when the PS started to function it was soon realized that beam construction had not been given sufficient weight in the scale of priorities for allocation of resources. Too few magnets and separators were at hand and the construction of a new beam often meant waiting for the possibility to reuse parts of other beams, or to borrow them from laboratories in the member states. Consequently, an optimal exploitation of the machine was severely hampered.[73]

Having given a general picture of what was involved in such a secondary beam, we would like to give now a short overview of the kind of beams constructed. Here we should say first some words about the nomenclature used. At CERN the beams were identified by a letter followed by a number indicating how many times this beam had been modified. Among the separated beams used for bubble chamber work we find f, m, k, o, and u, of which we shortly describe the history.

The *f-beam*, or low-energy, antiproton beam, was designed and constructed by a group consisting of members of the NP Division (Fidecaro group) and of the Ecole Polytechnique early in 1961. Because of the above-mentioned lack of bending magnets, quadrupoles, and separators, this beam had to go through several trial stages. The final version incorporated two electrostatic separators lent by the Padua group and after no less than 30 shifts of tuning, this beam line could be regarded as ready by mid-61. It was the first well-collimated secondary beam built at the PS, delivering antiprotons with a momentum up to 1 GeV/c to the bubble chamber area in the South Hall and up to 4 GeV/c to the counters. It was replaced in September of the same year by the first member of a series of *m-beams* set up by the NPA Division. This was, in contrast, the first fast antiproton beam envisaged to transport these particles up to a maximum energy of 6 GeV/c, as well as kaons or pions. However in the initial stage only 3 GeV/c were reached, increasing to 3.6 GeV/c when the second of CERN's electrostatic separators was installed. During September 1962 this beam-line was rebuilt ($m_{2,3}$) including then two 10 meter separators built at CERN and reaching thus 3.5 GeV/c for kaons and no less than 5.7 GeV/c for antiprotons. These were the highest values reached until then for kaons and antiprotons. The m_4 being a counter beam, only the m_5 was used again for bubble chambers, this time being located in the North Hall however. It included the new 6 meter separator of the

NPA Division and transported 3–3.5 GeV/c kaons, pions, protons, and antiprotons of less than 6 GeV/c.[74]

The first member of the family of *K-beams* was set up in August 1961 in the newly commissioned North Hall. Including a 10 meter CERN separator it transported K-mesons of momentum between 1 and 1.5 GeV/c. In 1962 when higher intensity was needed for experimental reasons, the beam-line was modified to include two borrowed separators from Saclay and Padua. The problem with slow kaon beams was that high losses occurred due to decay, and thus the beam had to accept, on the one hand, a rather large number of particles, but had then to guarantee a good separation. The same year the Padua separator was replaced by a 3 meter separator built by the NPA Division, giving a beam of separated K-mesons (k_3) up to 800 MeV/c as well as low-energy antiprotons. A new K-beam (k_4) was only set up in 1964. It was a short two-stage electrostatically separated beam (3 and 6 meter separators) yielding kaons and antiprotons between 600 and 1200 MeV/c. The novelty was that it was designed to need only 25% of the circulating beam to provide the wanted intensity. In 1965 a similar beam was rebuilt coming from a different target to be used for the CERN heavy liquid bubble chamber.[75]

In 1963 the third experimental area, the East Hall was inaugurated. There extremely long beams (more than 100 meters) would reach a building where the two biggest hydrogen bubble chambers were installed, the 152 and the 200 cm chambers. The setting up of the *o-beams* started in summer 1963 so as to be ready for the arrival of the British chamber. Being 150 m long new problems had to be faced in building this beam line. A novel method was used for the alignment of the magnets leading to a precision of 0.2 mm, and the curvature of the earth and its magnetic field had to be taken into account. The beam containing three 10 meter separators was designed to transport high-energy pions, protons, antiprotons, and kaons (o_2). The o_3 beam, sharing the first part of o_2, then branched off before the separators towards the EPHLC where high-energy pions were needed. In January 1965 most of the operating time in the o_2 beam line was devoted to test the new radio-frequency separator system. These separators, as described above, demanded a very short burst (2 μs) on the internal target, which was produced by giving a 'kick' to the beam, by means of the kicker magnet normally used for fast ejection. This kick caused the beam to oscillate slightly about its normal path in such a way that it struck the target after a further $1\frac{2}{3}$ circuits of the machine, i.e. after roughly 1000 m. The adjustment of such a system and the co-ordination between its various components, was rather complex. On the 25 January the full system was operated for the first time when the o_2 beam line, equipped with two RF separator cavities, provided a beam of negative kaons at a momentum of 10 GeV/c. Apart from the beams described which were set up for the British chamber, two beams were also built for the 2 m chamber. The first was the o_8 with three electrostatic separators providing kaons up to 5 GeV/c, pions up to 12 GeV/c, and antiprotons up to 8 GeV/c. In addition, it was equipped with a shutter magnet which would cut the beam off if more than 20 particles arrived at

Notes: p. 246

once in the chamber. The second and last beam to be mentioned here was the u_1 which was constructed in August and September 1965, using a rapid ejection system and an external target. The beam contained two RF separators, permitting physicists to obtain kaons of 10 and 14 GeV/c and antiprotons of 18 GeV/c, as well as protons and pions up to 20 GeV/c.[76]

5.3.1.3 From the photographs to scientific results[77]

When looking at the tables presented in the following subsection on the bubble chamber experiments it is interesting to see that the number of photographs taken for one particular experiment varied from some tens of thousands to more than half a million. Some of them were taken in one go, others spread over years. How to explain this? Indeed there are two important factors influencing this. The first of course is the kind of physics which is studied. From the cross-sections of a particular interaction physicists could estimate how many pictures were needed to obtain a reasonable amount of information. But the way in which these data had to be treated in order to extract scientific results also played a decisive role. As we will sketch here very shortly this was a rather tiresome and complex procedure and needed both technical equipment and large amounts to manpower. Therefore it was also important to judge—particularly in the beginning, when machine time was very scarce and the infrastructure for bubble chamber work was just being set-up in many European laboratories—how fast a collaboration could treat their photographs in order to ask for a reasonable and justifiable number of photographs.

How to extract the interesting physics results out of many thousands of photographs taken in the course of an experiment? Indeed for this purpose each picture had to pass through five stages: scanning, measuring, geometrical reconstruction, kinematic analysis, and lastly scientific interpretation. The scanning was a first selection done manually by the so-called scanning girls to sort out the events regarded *a priori* as interesting according to criteria defined by physicists. The measuring procedure which then followed, was also done manually in the beginning of bubble chamber work in the 50s. However, it was soon shown to be the bottle neck in the course of the experiment and people realized that the further success of this detector was strongly linked to an automatization of data handling. This meant automatic measuring devices on the one hand, and computers for the efficient and fast manipulation of these data on the other. Thus from the late 50s onwards at CERN as well as in the United States the evolution went from manual measuring tables to semiautomatized digitized projectors like the socalled IEPs (Instrument for the Evaluation of Photographs) developed and built at CERN (group around Y. Goldschmidt-Clermont). Simultaneously, it was central to treat this increasing amount of information by computers, and therefore in the early sixties adequate soft-ware (named REAP) was developed at CERN for converting the information received on punched papertape from the measuring projectors into a suitable form for the computer. In a second step the event was reconstructed geometrically in three dimensions by converting the measurements made on each of the two or three stereotopic photographs into space coordinates (THRESH). Finally, another programme (GRIND) made it possible for physicists to test certain kinematic hypotheses, by using the 3-momenta of each particle (reconstructed by the geometric programme) in the equation defining the momentum-energy balance. The last part in the chain was then the scientific interpretation. Making use of refined statistical analysis techniques based on samples as large as possible one tried either to situate the results in the frame of already existing theories, or to use them together with others to form the basis of a new theoretical hypothesis.[78]

We should however mention here that CERN did not stop with the development of semi-automatized IEPs but also played an important role in developing an automatic measuring device for track chamber photographs—the Hough-Powell device (HPD). This device was based on the use of a 'flying spot' which scanned the whole photograph, measuring the coordinates of all the images of the bubbles and feeding them straight into a computer. From these data the computer programme, instructed by a special input resulting from rough measurements of the wanted events, could then select the events to be analyzed. The HPD first went into production on bubble chamber pictures in 1965 and was therefore less important for the period described here.[79]

5.3.2 THE SCIENTIFIC CONTRIBUTIONS MADE WITH BUBBLE CHAMBERS

5.3.2.1 *The interactions of pions with nucleons*

When the PS was ready for scientific exploitation by early 1960 the study of high energy pion–proton and proton–proton interactions, at 16 GeV/c and 24 GeV/c respectively, were the first two bubble-chamber experiments scheduled. Both pions and protons were produced abundantly when the circulating beam hit the target and therefore no complicated set-up for particle separation was necessary. The choice of the bubble chamber was obvious as the 30 cm chamber was the only one existing at that time at CERN. As this was the first investigation of these fundamental interactions in a high-energy range it is not surprising that virtually everybody involved so far with bubble chambers at CERN, i.e. the hydrogen bubble chamber and the IEP groups, took part in this experiment. Thus we find for example R. Budde, Y. Goldschmidt-Clermont, L. Montanet, D.R.O. Morrison and C. Peyrou in one group who would all later specialize on different aspects of the study of particle interactions with the help of bubble chambers. Further participants were two Italian Universities, a mixed British team, and marginally also Swiss and German universities.

Restricting our discussion here to pion–proton interactions, and—as the

Notes: p. 246

designation of the experiment said—in particular to strange particle production, the most interesting result concerned the interaction mechanism as such. Indeed, studying the angular distribution of hyperons in the pion–proton centre-of-mass-system a clear backward peak showed up whereas the neutral kaons which were also produced showed a preference for forward emission. These two facts allowed one to conclude that non-statistical processes played an important role here and hinted towards an interpretation of the production mechanism as being of a peripheral nature via single meson exchange. And indeed these findings opened up a new line of studies concerning the 'reaction mechanism' which was pursued extensively at CERN over the following years. In addition, these peripheral or quasi-elastic events could be used—following the idea of Chew and Low—to gain information on the pion–pion cross-section. Other interesting findings were the striking low production cross-section of the xi-particle and anti-hyperons, the small transverse momentum of the generated secondaries—also linked to the peripheral nature of the process—, and the preponderance of positive hyperons.[80]

Meanwhile the Ecole Polytechnique heavy liquid bubble chamber had arrived at CERN and by November 1960 the first experiment could be carried out investigating hydrogen-like interactions produced by pions with energies of *6, 11, and 18 GeV/c*. It was thus in some sense a continuation, or rather an extension, of the pioneering work with the 30 cm HBC. The advantage of this chamber was, however, its much greater size and stopping power, which meant an increased detection efficiency for both neutral particles in a given reaction. The collaboration contained groups from the Ecole Polytechnique—the constructors of the chamber—, Milan, Padua, and Turin.

Among the variety of results we want to mention, firstly, a study on the correlation in strange-particle pair production, in particular on those events with two visible neutral strange particles. Further the behaviour of K_1^0's in $\pi^- N \to K^0 \bar{K}^0 (n\pi) N$ and $\pi^- N \to K^0 \Lambda (\Sigma^0) n\pi$ was investigated. In all cases full agreement with the earlier results was found. The photographs at the lowest energy—6 GeV/c—also gave interesting information on resonances. The $K^0 \Lambda$ channel was dominated by $K^*(890)$ production, and there was evidence for the $Y^*(1385)$. Studying the mass distribution of the dipion system physicists saw besides a peak at 800 MeV due to the ϱ resonance, an enhancement at 1250 MeV. It was interpreted as a further hint towards the so-called f-meson, which had been suggested by a team at Berkeley. Soon after this particle was established definitely in the 4 GeV/c π^- experiment performed with the 81 cm hydrogen bubble chamber at CERN. Finally, the validity of the peripheral collision model in the case of low multiplicity pion–proton collisions was also investigated. Indeed, the observations were consistent with the assumption that the one-pion exchange (OPE) mechanism was responsible for the majority of the events studied.[81]

Early in 1961 the third bubble chamber—the 81 cm Saclay chamber—was put into operation and among the first proposals was again a study of high-energy $\pi^- p$

interactions. Thus by June of that year the chamber was exposed to a *10 GeV/c* π beam, an energy chosen so low for reasons of intensity.

Again the production of strange particles and in particular KK and YK pairs were studied, this time by a CERN team under Charles Peyrou together with physicists from Warsaw. The results showed that 50% of the events were peripheral, i.e. the hyperons were emitted strongly backwards while the accompanying kaon closely followed the direction of the incident pion. Assuming the validity of a single meson exchange peripheral model for πp interactions this meant that the exchange particle would rather be a kaon than a pion, a fact of particular interest in the light of the Regge model discussed at the time. The second interesting feature observed for the first time was an apparent dependence of the transverse momentum p_T of the secondaries on their mass. They tried to account for this fact by assuming that intermediate bodies were formed in the interaction which broke up afterwards in the particles actually observed. Such two-intermediate-states models were not completely new as they had already been used for the interpretation of interactions at cosmic-ray energies. Lastly, investigating K-pair production—the most frequent type of associate production—they found that this process could most likely be explained in terms of a peripheral model with pion exchange. Besides strong forward emission of the Ks, there appeared also a strong correlation between the two Ks in a pair, representing the first hint towards a resonant state in the KK system—a point which was clarified only much later.[82]

A second collaboration between a CERN group around D.R.O. Morrison and physicists from the Ecole Polytechnique reused the same photographs, this time to study elastic pion–proton scattering, notably for comparison with the findings in the pp case. Indeed the results showed that, in contrast to the case of pp scattering, the diffraction peak in the pion–proton case did not shrink, a fact which was confirmed in counter experiments at CERN and Brookhaven. Further they found that the differential cross-section did not depend on the momentum transfer according to a simple exponential law, but showed strong deviations at small angles.[83]

Using the same bubble chamber an experiment with *4 GeV/c* pions of both signs was started a few months later in November 1961. This comparatively low energy was chosen as the investigations planned demanded a complete analysis of the event and 4 GeV/c was thought to be the upper limit for a 81 cm chamber. The collaboration—also referred to as the Jet collaboration—consisted of no less than five German teams from Aachen, Bonn, Berlin, Hamburg, and Munich and two British teams from Birmingham and London. Indeed it was not only the first time that so many groups worked together, but also that the participants neither came from CERN nor were members of the group who had constructed the chamber.[84]

The results extracted from these photographs were of particular interest with regard to resonances. Firstly, as mentioned above, the two-pion resonance f⁰ was definitely confirmed in the π^- data. Its mass was established to be 1260 MeV, and the spin and isospin were determined as 2 and 0 respectively. Secondly, investigating

Notes: p. 246

the reaction channel $\pi^+ p \to p\pi^+\pi^+\pi^-$ and plotting the three-pion effective mass distribution, restricted to those events where at last one two-pion combination was in the ϱ region, they saw two clearly separated peaks at 1080 and 1320 MeV which they interpreted as $(\varrho\pi)$ resonances A_1 and A_2. A first hint in this direction had already come from Goldhaber and his team at Berkeley who reported such an enhancement spreading from 1 to 1.45 GeV with a minimum at 1.2 GeV when studying the same reaction at a slightly lower energy.[85]

Concerning the $\pi^- p$ data a systematic analysis of all four- and two-prong events was made. In the former case this embraced the determination of cross-sections for N^{*++}, ϱ^0 and ω production, the reactions were discussed in the light of the OPE model, it was determined that the three above-mentioned resonances were produced in peripheral interactions, and studies were undertaken on a connection between multiplicity and angular distribution. The same cross-sections, angular and momentum distributions for a number of reaction channels were also measured for two-prong events. The results showed that the single pion production channels were largely dominated by pion resonances ϱ and f. For the data with positive pions, elastic scattering, (3/2,3/2) isobar production, and many more were investigated. Lastly, studying the reaction $\pi^- p \to \pi^- p \omega$ and assuming that it was dominated by the OPE diagram, physicists tried to determine the cross-section of the reaction $\pi^- \pi^0 \to \pi^- \omega^0$ via the Chew–Low model. For comparison they also used a model adapted by Selleri, without decisive results however.[87]

In spring 1962 the 81 HBC was used again for the study of *pion–proton interactions* this time at *2.75 GeV/c*. The collaboration consisted of groups from Saclay, Orsay, Bari, and Bologna and had worked previously on this subject at Saclay, exposing a 50 cm HBC to a π beam of 1.59 GeV/c and studying elastic

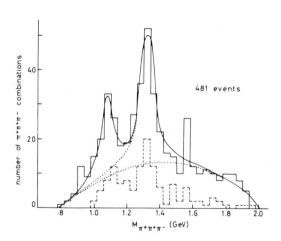

Fig. 5.6 $M(\pi^+ \varrho^0)$ distribution showing the two peaks A_1 and A_2[86]

scattering, inelastic scattering accompanied by a single pion production or by more than one pion in four-prong events. Their initial results obtained had agreed rather well with the OPE model, and thus it seemed interesting to investigate the dependence of the dynamics of pion-production and decay on the value of the incident π momentum. With this in mind a new experiment at slightly higher energy and involving a bigger chamber was proposed. Altogether they took no less than 255 000 photographs — the biggest number so far on this type of interaction.[88]

Summarizing the main results from the negative pion data, we should start with the studies of the interaction mechanism. Firstly, in the case of single pion production the OPE mechanism could be considered as dominant. Secondly, the pion–pion elastic scattering cross-section was found to be independent of the incident π energy — at least in this energy range. Then, the four-prong events of the kind $\pi^- p \to p (3\pi)$ showed important production of N^{*++} (3/2,3/2) mainly according to an OPE mechanism. Lastly, the channels giving five body final states ($\pi^- p \to \pi^- p\, 3\pi$) were dominated by quasi-two body states $\omega N^{*0}_{33}(1236)$ and $N^{*0}_{13}(1518)$. As for resonances, an extensive study on the A_1 and A_2 was made, with particular attention to a determination of the spin and parity. For the A_1 evidence was found for $J^P = 1^+$, although 2^- could not be excluded, for the A_2 $J^P = 2^+$ was clearly favoured. Further they studied their decay modes in particular into $\varrho\pi$, but also made a search for the eta-pi mode, though without success.[89]

The π^+ data were investigated along similar lines. Looking at the $N^{*++}(3/2,3/2)$ production in $\pi^+ p \to \pi^+ p \pi^0$ they learned that there was good agreement of the angular distribution in the isobar c.m.s. with theoretical predictions for a peripheral process with ϱ exchange. In the case of five-body final states again particular attention was drawn to quasi-two-body processes, namely

$$\pi^+ p \to \left\{ \begin{array}{c} N^{*++}\omega \\ p(\pi^+\omega) \end{array} \right\} \to p\pi^+\pi^+\pi^-\pi^0 \; .$$

The first was of particular interest as several theories attempted to explain this quasi-two-body associated production by a peripheral mechanism involving one meson exchange — in this case fairly good agreement was reached for a ϱ-exchange model. The second channel was used to make a search for the B meson, though without success. The four-prong reaction $\pi^+ p \to \pi^+ p\pi^+\pi^-$ was also studied with respect to the production mechanism in the channel $N^{*++}\varrho^0$. Although the angular distributions of the two resonances indicated a production dominated by the OPE model, the differential cross-section was found not to agree with theoretical predictions.[90]

Filled with deuterium for the first time in spring 1963, *two experiments in pion beams* were scheduled for the 81 cm chamber. The first collaboration from Saclay, Florence, Orsay, Bari, and Bologna made an exposure at *4.5 GeV/c*, which was extended in mid-65 to 5 GeV/c. Besides a study of the neutral decay modes of the f-meson, the double pion production was looked at, in particular the reaction $\pi^+ d \to d\pi^+\pi^+\pi^-$. There were two aims: a study of the 3-pion system in a pure $I=1$ state and the examination of the $d\pi$ system. Further one hoped to learn about the production mechanism for the A_2 resonance. What is more, a new resonance had shown up at 1.32 GeV in the (3π) mass spectrum of the reaction $\pi^+ d \to p_s p \pi^+ \pi^- \pi^0$. It had a $\varrho\pi$ structure, corresponding to a $I=1$, $J^P = 2^+$ state and was thus supposed to be the neutral counterpart to the A_2 resonance.[91]

The second pion experiment with deuterium in the chamber this time at *6 GeV/c*, was carried out by two teams from CERN and the Ecole Polytechnique. When studying the neutral particles produced in $\pi^+ d$ interactions, they saw an enhancement at 1250 MeV in the mass plot, which was interpreted as due mainly to the neutral decay of the f-meson. The meson's isospin was confirmed to be even, thus in agreement with the results obtained in the 4 GeV/c π^+ experiment. A second rather interesting result arose from the study of the reaction $\pi^+ d \to pp\pi^+\pi^-$. Indeed combining the results with those of the Orsay-Milan-Saclay collaboration on the reaction $\pi^- p \to n\pi^-\pi^+$ at 8 GeV/c (see below) physicists managed to demonstrate the existence of a new resonance in the $\pi^-\pi^+$ mass plot at 1670 MeV, which they named the g-meson. None of the two sets of data on their own would have been sufficient to draw this conclusion.[92]

To continue let us say a few more words about the above mentioned 8 GeV/c π experiment. It was part of a programme taking pictures of negative pions at 8, 16, and 18 GeV/c in the East Hall (o_3 beam) using the Ecole Polytechnique heavy liquid bubble chamber. Many members of the collaboration had already studied pion-proton interactions between 6 and 18 GeV/c earlier. This time, however, they took nearly half a million photographs—six times more than before. Some of them were taken with a hydrogen target inside the chamber, combining thus the advantages of hydrogen with those of heavy liquids. Further one of the 16 GeV/c runs was made with a lead target.

The 8 GeV/c pictures were used to study pion-proton interactions giving two to four pions and the resonance formation showing up in such channels. Besides the detection of the g-meson, they established that the $N\pi^+\pi^-$ channel was dominated by ϱ and f production, the $\pi^+\pi^-\pi^- p$ channel by N^{*++} and ϱ^0, and a small number of ω^0 was seen in the $\pi^+\pi^-\pi^0 n$ channel and ϱ^0 in the $\pi^+\pi^-\pi^+\pi^- n$ channel. It was also interesting that neither A_1 nor A_2 states were present in the three-charged pion production channel, and that only rare cases of quasi two-body final states were observed.[93]

The results at 16 GeV/c provided confirmation of the phenomenon of 'diffraction dissociation'. This meant that the pion gave the appearance of changing into three pions in the nucleon field. These three pions were present in the form of an A_1 resonance and the results for spin and parity were identical to the prior findings, namely 1^+ or 2^-. The data taken with a lead target at this energy had as purpose to study the $\varrho \to \pi\gamma$ decay by observing the inverse phenomenon $\pi\gamma \to \varrho$. In using a lead target and comparing the results with those obtained in the liquid alone one could give a clear separation of electromagnetically produced $\pi^-\pi^0$ in lead. The $\varrho \to \pi\gamma$ width could be determined in that way.[94]

Using the 18 GeV/c data physicists studied the mechanism of those reactions containing $\pi^+\pi^-$ pairs. In the channel $\pi^-p \to \pi^+\pi^-n$ the mass distribution was found to be very spread out and there was only evidence for ϱ production. However, the production cross-section had decreased sharply, much more pronounced than the total cross-section from 18 to 6 GeV/c.[95]

Refilled with hydrogen, the 81 cm chamber was scheduled during February and March 1964 for *a positive pion run at 8 GeV/c*. The collaboration contained one group from CERN, which had already participated in the 10 GeV/c experiment, two German groups (Aachen and Berlin) which had taken part in the 4 GeV/c experiment, and two Polish groups from Warsaw and Cracow. The designation was 'Study of the mechanism of two body production', the idea being that in fact little had been published on resonance production (other than ϱ and N^{*++}) at energies bigger than 6 GeV/c. A strict kinetic analysis could be applied for the first time to interactions at such high energies with this relatively small chamber as the initial system (π^+p) had charge 2 and consequently the production of neutral particles was much lower than in the π^-p case.

Indeed it turned out that resonances were produced abundantly in two- and four-prong interactions making the reaction often appear as a two-body process. More explicit this meant reactions such as

$$\pi^+p \to \left\{ \begin{array}{c} N^{*++}\pi^0 \\ p\varrho^+ \\ N^{*+}\pi^+ \end{array} \right\} \to p\pi^+\pi^0$$

in the two-prong case, and

$$\pi^+p \to \left\{ \begin{array}{c} N^{*++}\varrho^0 \\ N^{*++}f^0 \end{array} \right\} \to p + 3\pi$$

in the four-prong case. Further these reactions were found to be of peripheral nature, which made the separation of the resonances from background clearer than at lower energies (3–4 GeV).[96]

Looking at the slope of differential cross-sections in quasi-two-body reactions as a function of four-momentum transfer, it was found to be exponential and about the same as that for elastic scattering. This could, however, not be reproduced by calculations based on the absorption model suggesting that this model was unsatisfactory. Indeed it only worked well for describing reactions with pion exchange, disagreeing however completely in the case of vector-meson exchange.[97]

Notes: p. 246 ff.

Further this collaboration gave important contributions to the understanding of the A-mesons. Besides the quantum numbers of A_2 which were settled as $J^P = 2^+$, the A_1 was definitely confirmed as a resonance, ruling out the possibility that it was just a kinematic effect. A determination of branching ratios of rare f, A_1, and A_2 decays was also carried out.[98]

One general conclusion could, however, be drawn from this experiment: contrary to what was often thought, high-energy primaries were very useful for the study of resonances. However as so many reaction channels were possible, systematic investigations with rather good statistics needed both a larger chamber and extremely fast measuring instruments to 'digest' the high number of photographs. With this in mind the collaboration continued this research topic using the 2 m HBC towards the end of 1965.

Early in 1964, a fascinating new theoretical idea was put forward by Gell-Mann and Zweig independently. Assuming that strong interactions of baryons and mesons were correctly described in terms of a broken SU(3) symmetry, they suggested that all strongly interacting particles were just different combinations of three basic units. The fundamental triplet of particles would be similar to proton, neutron, and lambda, except for their baryon number 1/3 and their electrical charge (2/3, −1/3, −1/3). They would be named u,d,s. If these 'quarks' or 'aces' were produced in high-energy interactions, physicists supposed that they would behave like the other charged particles, though having an ionizing power of 1/9 due to their charge of 1/3.

The idea proposed by D.R.O. Morrison was that 26 GeV/c protons in the PS hit a target and that a beam line selecting particles of charge one with 21 GeV/c be set-up. If charge 1/3 particles passed this beam line they would have 7 GeV/c and particle tracks in the BC with 1/9 bubble density would show the existence of these quarks.

As a first step, and to be as quick as possible, the photographs taken with 16 GeV/c pions in the 30 cm bubble chamber in 1960 were reexamined, as well as 100 000 pictures taken with the EPHLC. Both attempts were unsuccessful. The same was true for the above mentioned experiment, from which it could only be concluded that if these particles existed they were either too heavy ($m > 2.3$ times the proton mass) to be produced with the energies available, or they were strongly bound within the particle.[99]

Finally, there are two last experiments we want to touch on here. The first, started late in 1964, was carried out by groups from Bonn, Durham, Nijmegen, Ecole Polytechnique, Strasbourg, and Turin with the 152 cm British chamber in *a beam of 5 GeV/c positive pions*. They studied six prong events, a rather rare subject, as physicists had generally focused on low multiplicity phenomena. Three characteristic features were observed: the target proton retained its charge in most cases, rarely more than one neutral pion was produced, and, lastly, no 'glancing collisions' were observed. In general this was more or less descriptive work which fitted into no existing theoretical frame.[100]

The last pion exposure, and at the same time also the last run of the British bubble chamber at CERN, was done in summer 1965 at an energy of 11 GeV/c. Later in the year the data were completed using the 2 m chamber. Let us shortly sketch three of the results found by the Hamburg, Genova, Milan, and Saclay collaboration. Firstly, the 4π final system was investigated for any apparent structure. Indeed one enhancement was found at 1.68 GeV, which was interpreted as the g-meson, and one at 1.3 GeV which could however not be clearly attributed to the B-meson. Secondly, the production cross-sections and their t-dependence was studied. The channel $\pi^- p \rightarrow \pi^- p \pi^+ \pi^-$ was shown to be strongly dominated by $N^{*++}(1238)$, ϱ and f production and so permitted the study of the t-dependence of quasi-two-body reactions. The general trend established was that the slope of the distribution in the momentum transfer to the target proton decreased with increasing mass of the 3π system. Lastly, the production of both A mesons was studied in the above mentioned (3π) channel.[101]

5.3.2.2 Investigating proton–proton collisions

Only a very small part of CERN's early bubble chamber programme, as we will see below, was devoted to the study of p–p interactions. Indeed, most of the experiments on this subject were performed with counter techniques which were used to study rather specific features of the interaction. Further, it should be said that apart from the first proton exposure carried out with the PS in 1960, no CERN group got involved in this line of research.

The first of the four experiments which will be mentioned here was made with the 30 cm bubble chamber exposed to a *24 GeV/c* elastically scattered proton beam in October 1960. As was the case in the first pion experiment, here too the members of the CERN IEP and HBC groups took part. The interest was twofold: to gain information on p–p interactions in the new-high energy range as well as to have comparative data to the π results. Indeed, confirmation was found for the low transverse momentum of all secondaries with the exception of positive hyperons, for the small production cross-section of xi-particles and antihyperons, and for the preponderance of positive hyperons. However no asymmetry in the angular distribution of kaons and hyperons was observed. Finally, the peripheral nature of the p–p interaction was also clearly seen, an interesting phenomenon already known from pp interactions at 20 GeV/c using counter techniques and from the 16 GeV/c π bubble chamber experiment.[102]

It took nearly two years until the next proton exposure was carried out, this time using *protons of 5.5 GeV/c* together with the 81 cm HBC. An Israeli team investigated neutral strange particle production and found that the dominant process of Λ production was associated with the formation of $Y_1^*(1385)$. Further the absence of K^* resonances together with the small KK production cross-section, indicated that the formation of mesonic resonances in pp interaction at high energy was

Notes: p. 247

unimportant. The peripheral production of $N^*(1518)$ and $N^*(1688)$ was also studied, in particular so as to compare total and differential cross-sections calculated both in the unmodified OPE model, as well as taking into account absorptive corrections to the experimental results. Good agreement was, however, only obtained for the total cross-section in the case of $N^*(1688)$ production. The shape of the angular distribution was found to be more or less reproduced by the OPE model with absorptive modifications. Similar investigations were also done for production and decay of $N^{*++}(1238)$ in the reaction $pp \rightarrow nN^{*++}(1238)$ where the absorption model gave satisfactory results.[103]

The third proton experiment—*at 10 GeV/c* using the Saclay and later the far bigger British bubble chamber—was scheduled in 1964 for a collaboration of physicists from Cambridge, Hamburg, Stockholm, and Vienna. Their primary concern was four-prong events such as $pp \rightarrow pp\pi^+\pi^-$ with respect to double isobar production in the final state, and to determine if the process involved a peripheral interaction. And indeed, evidence for this was found, namely $pp \rightarrow N^{*++}(1238) + N^{*0}(1688)$. Secondly, when studying in detail resonance production in the mass region between 1400 and 1700 MeV/c^2—a region when already counter and bubble chamber experiments had shown significant peaks—they saw indications for $N^*(1450)$ in the $p\pi^+\pi^-$ and $n\pi^+$ mass distribution. Lastly, the reaction $pp \rightarrow N^{*++}(1238)n \rightarrow pn\pi^+$ was used to make a comparative study between the one-pion exchange model with form factors or with absorptive corrections.[104]

To terminate this short section on pp interactions we want to mention an experiment carried out in mid-1965 with the British bubble chamber in a beam of 6 GeV/c protons, proposed by groups from Genova, Milan, and Oxford. The aim was to produce data on pp collisions in the multi GeV region in order to check the validity of different theoretical models over a wider energy range, but no definite conclusions for the production of various final states were drawn from this experiment.

5.3.2.3 Antiproton–proton interactions at low energy—a rich source of resonances

Among the first experiments to be proposed by R. Armenteros and his group for 'their' new 81 cm hydrogen bubble chamber to be installed at CERN, was a study of $p\bar{p}$ interactions at low energy. This choice was, on the one hand, triggered by a theoretical idea suggesting that a proper interpretation of the nucleon form-factor and of NN and πN scattering required the introduction of a two-pion resonance with isospin $I=1$ and spin $J=1$ and a three-pion resonance with $I=0$ and $J=1$. Their expected masses would be four to six times that of the pion. On the other hand, results from Berkeley had shown that $p\bar{p}$ annihilation led to an abundant production of pions and therefore offered ideal conditions for a search and study of those phenomena. However, when the EP chamber was in place and the experiment went into the phase of taking photographs in June 1961 the annoucements of the first American results on such new resonances arrived at CERN. Physicists were thus confronted with the unpleasant reality that the first rich harvest had been reaped by teams at Berkeley and at Brookhaven. This, however, did not remove the interest of European high-energy physicists in the experiment, as people believed that they had only seen the 'tip of the iceberg', and that many more interesting things were still waiting to be discovered.[105]

Accordingly one collaboration under the leadership of R. Armenteros containing groups from CERN, the Ecole Polytechnique, and the Collège de France investigated the aspect of p$\bar{\text{p}}$ annihilation at rest into K pairs,[106] those without visible K production being studied by physicists from Padua, Cambridge, and Oxford.[107] Finally, the aspect of p$\bar{\text{p}}$ interaction in flight was treated by the Trieste–Rome–CERN (Fidecaro group) collaboration. This would, on top of the phenomena also observed at rest, allow a study of elastic collisions and charge-exchange processes. Detailed investigations of this kind could deliver information about the nature of p$\bar{\text{p}}$ interaction.[108]

As the results of this first data taking phase proved to be very interesting, further exposures were made by the CERN–Collège de France collaboration in December 1962, summer 1964, and at the beginning of 1965.

What were the main results?

The first set of photographs allowed the observation of a significant production of p$\bar{\text{p}} \to K_S^0 K_L^0$ and the non-observation of p$\bar{\text{p}} \to K_S^0 K_S^0$ which led to the conclusion that the annihilation was produced in the S-state. Further the presence of the first reaction contradicted the Sakata model of unitary symmetry, whereas in the eightfold way of Gell-Mann and Ne'eman this was possible. Secondly, the detection of p$\bar{\text{p}} \to K_S^0 K^*(840) \to K_S^0 K_L^0 \pi^0$ showed that the spin of $K^*(840)$ was not 0, as one had thought on the basis of the first results collected at Berkeley,—a result which marked a turning point in the understanding of SU(3) for mesons. Lastly, due to the absence of any disturbing background, one had nearly perfect conditions for the study of the ω^0 meson in the annihilation p$\bar{\text{p}} \to K^+ K^- \pi^+ \pi^- \pi^0$. It allowed for a determination of the width and branching ratio into neutral particles, and the quantum numbers were confirmed to be $J^{PS} = 1^{--}$.[109]

The photographs taken in 1962, 1964, and 1965 in particular enabled physicists to contribute in a rather impressive way to the field of resonances. The first resonance detected, called the E-meson (for it was found in a European laboratory) manifested its existence as enhancement in the $(K^0 K^\pm \pi^\mp)$ system produced in the reaction p$\bar{\text{p}} \to \pi^+ \pi^- (K^0 K^\pm \pi^\mp)$. Its mass was measured as 1410 MeV, but due to insufficient statistics no spin-parity assignment was initially possible. The second resonance, the so-called C-meson, was seen when studying p$\bar{\text{p}}$ annihilation at rest into two neutral kaons and two charged pions. They found a pronounced peak around 1250 MeV in the $(K\pi\pi)$ mass distribution. Thirdly, a (KK) resonance was proposed with $I = 1$ and mass 1003 MeV in the annihilation p$\bar{\text{p}} \to K_S^0 K^\pm \pi^\mp$. It was later confirmed, being named the δ-meson. Lastly, the D-meson, a (KKπ) resonance at 1290 MeV, was announced when working on p$\bar{\text{p}}$ at 1.2 GeV simultaneously with Berkeley, where it had been found in πp interactions at 4 GeV/c.[110]

Notes: p. 247

5.3.2.4 Experiments with high-energy antiprotons between 3 and 5.6 GeV/c

With the first fast anti-proton beam ready by autumn 1961 (see section 5.3.1.2) a series of exposures were started with the 81 cm chamber stretching from November 1961 to April 1962. Half a million pictures divided equally over *three energies 3, 3.6 and 4 GeV/c* were taken and their evaluation involved no less than eight groups from CERN, and British, French, and German laboratories.

The central interest of this experiment was a search for the anti-cascade particle $\bar{\Xi}^-$ by the CERN–Ecole Polytechnique–Saclay collaboration. The xi was the heaviest elementary particle known at the time, belonging to the family of strange particles. Convinced by the idea that each particle should have its anti-particle, physicists had so far looked without success for the anti-xi. One was, however, aware that the sought after event would have a complicated structure and a low cross-section, and one neither knew the life-time nor was one sure if it would always decay into lambda-pion or if there were other decay modes. Accordingly physicists had proposed to look for this particle in antiproton–proton interactions at 3 GeV/c, as this energy lay just above the calculated threshold for xi-antixi production, where only a few disturbing pions would be produced.

The same idea was also followed independently on the other side of the ocean, where at Brookhaven physicists were exposing their 20″ bubble chamber to an anti-proton beam of 3.3 GeV/c. Indeed, both laboratories—in each case one event was measured completely—were successful and after a series of rather complex negotiations the European and the American results were published simultaneously in the March 1962 issue of *Physical Review Letters*. Further it is worth remarking that in this case for the first time only the collaboration signed the publication and not the individual authors.[111]

The detection of this particle, predicted by theory, was not only a scientific success, but also a nice demonstration how powerful the bubble chamber technique was for the detection of new particles and for precise measurements on a single event.

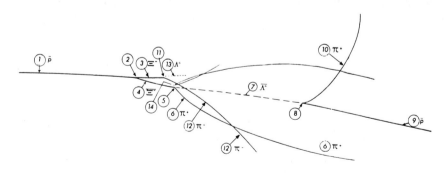

Fig. 5.7 Production of $\bar{\Xi}^-$ by $\bar{p}$ of 3 GeV/c[112]

This rather spectacular result was not the only important outcome of this experiment. Firstly, a collaboration of physicists from Birmingham, CERN (under B. French), Imperial College, and the Ecole Polytechnique undertook a comprehensive study of hyperon–antihyperon production in two- and four-body final states for all three energies. The result was that the cross-section for two-body final states decreased with increasing antiproton momentum, whereas in the three- and four-body final states it remained constant. The angular distributions showed a strong forward peaking of the anti-hyperon in the c.m. system of the reaction, suggesting that a peripheral mechanism also dominated this process. Further in three- and four-body reactions the pion–hyperon resonances $Y_1^*(1385)$, $Y_0^*(1405)$, and $Y_0^*(1520)$ and their charge conjugate were produced abundantly and gave mass values, width, and decay branching ratios confirming those found in other experiments. Finally, also the C- and E-meson observed earlier in low energy $p\bar{p}$ interactions were confirmed here.[113]

Secondly, a CERN group around Y. Goldschmidt-Clermont and D.R.O. Morrison investigated closely the aspect of $p\bar{p}$ elastic scattering and charge exchange reactions at this high energy. Much work had already been done on πp and pp elastic scattering, both theoretically and experimentally, and it seemed important also to investigate the $p\bar{p}$ data. In fact a stunning result was reported, namely the 'antishrinking' of the $p\bar{p}$ diffraction peak with increasing energy. What is more, the so-called $\bar{N}^*(3/2,3/2)$ resonance was also detected here for the first time.[114]

Thirdly, a team from Hamburg investigated $p\bar{p}$ interactions at 3.6 GeV/c in order to compare the results with the one-pion-exchange model. Further studies were made on pion production without annihilation, a channel which increased with rising energy. Single-pion production was found to agree with the predictions of the OPE model for small values of four-momentum transfer, double pion production agreed for all values.[115]

The next step to even higher energies was taken with the installation of the m_2 beam during late 1962, which delivered separated anti-protons of no less than 5.7 GeV/c. Thus during October and December 1963 photographs were taken with the 81 cm chamber for groups from Bonn, CERN (under B. French), Hamburg, Milan, and Saclay. In the first place the groups from Bonn, Hamburg and Milan investigated the reaction $p\bar{p} \to N^*\bar{N}^*(3/2,3/2)$ using both the data taken at 3.6 and 5.7 GeV/c and a comparison was made with theoretical predictions for the peripheral model including form factors. Further they studied elastic scattering, pion production and annihilation into pions.[116]

The second collaboration—CERN and Saclay—was particularly interested in heavy hyperon resonances decaying into (KN) and (KNπ). Indeed besides $Y_0^*(1520)$, $Y_1^*(1660)$, and $Y_0^*(1815)$ resonances they found peaks at 1940, 2100 and 2300 MeV. And studying the effective mass distribution of $(\Sigma\pi)^0$, along with the abundant appearance of light resonances in channels accessible to them, only the 2100 MeV peak turned up again.[117]

Notes: p. 247

This set of data was then increased during 1965 using the biggest bubble chamber available—the 2 m chamber—, giving rise to a large number of publications in the late 60s which will not be discussed here.

The last high-energy antiproton exposure in the period discussed by us was then late in 1965 with the 2 m chamber. A collaboration from the Ecole Polytechnique and Liverpool planned to study $\Lambda\bar{\Lambda}$ production and annihilation into Ks of $p\bar{p}$ at 2.5 GeV/c.

5.3.2.5 Kaon-proton interactions at energies up to 1.5 GeV/c

Physics with kaons began at CERN in December 1961 when the first separated K-beam with an energy of 1.5 GeV/c was installed in the North Hall. Two collaborations, one from Amsterdam, CERN, and Glasgow, the other from Oxford and Padua, took part, respectively studying K^-p and K^+p interactions with the 30 cm bubble chamber. The experiment's designation was 'Ξ production and K, Y and π resonances'.

The comprehensive effort to study strange baryonic resonances was started in the most frequent channels such as $\Lambda 2\pi$, $\Sigma 2\pi$, $\Lambda 3\pi$, and $\bar{K}^0\pi^-p$. By that time three such resonances were already known—$Y_1^*(1385)$, $Y_0^*(1405)$, and $Y_0^*(1520)$. Indeed in the $\Sigma 2\pi$ reaction two $I=0$ pion-hyperon resonant states were dominant, $Y_0^*(1405)$ and $Y_0^*(1520)$, both already known from experiments at Berkeley, whereas in the $\Lambda 2\pi$ channel $Y_1^*(1385)$ was produced in large amounts and could thus be investigated in detail. However no really new findings were reported in this direction. Nevertheless there was a hint of an as yet unknown resonance at 1660 MeV. But although two different experiments at Berkeley showed the same effect, for the time being no decision could be taken about the existence of this particle—later referred to as $Y_1^*(1660)$. Looking at the third channel—$\Lambda 3\pi$—abundant production of the vector meson $\omega^0(784)$ was found, but only a few pseudoscalar mesons $\eta^0(548)$, an effect which could explain the relatively high cross-section for this channel. The last channel, $\bar{K}^0\pi^-p$, went entirely via the pK^{*-} state, $K^*(890)$ being a resonance already established. Finally, besides these studies of resonances the mass and decay parameters of the rare Ξ particles were investigated. 62 examples of Ξ^- and 4 Ξ^0 were found in the photographs and the mass values were determined as 1321.1 and 1325 MeV/c^2 respectively. The decay time was measured to be $(1.55 \pm 0.31) \times 10^{-10}$ sec for the negative particle.[118]

Concerning the K^+p run, the aim was mainly to investigate single production of pions by positive kaons with special reference to the OPE model. The K^+p system was 'particularly suitable for a study of reaction mechanisms, because there were no known baryon states with strangeness plus one'. As $K^*(890)$ and $N^*(1220)$ were strongly produced their differential production cross-section could be measured. The conclusion drawn was that if the OPE model were to be the right one to describe the production of these particles then the form factors could not only be a function of four-momentum transfer, but also of the total c.m.s. energy.[119]

Replacing the 30 cm chamber the Ecole Polytechnique heavy liquid bubble chamber was moved into the same K-beam in spring 1962. As heavy liquid chambers were particularly useful for investigating neutral particles due to their high stopping power two experiments were made on that topic during the year. Firstly, a collaboration including physicists from the Ecole Polytechnique, University College, Rutherford Lab., Bergen, and CERN made a study of Λ and Σ particles. They investigated elastic scattering of Λ, the muonic decay mode of Λ and the beta-decay of Λ and Σ. The decay mode of the ω^0 into two electrons was also looked at.[120]

Secondly, groups from the Ecole Polytechnique, Orsay, and Milan had designed an experiment to test if the $\Delta S = \Delta Q$ selection rule, which was a consequence of the theory of particle currents used to describe weak interactions, would hold in the leptonic decay of neutral kaons. Put differently: when a strongly interacting particle decays into another, together with two further weakly interacting particles, does the difference in strangeness have to be equal to the difference in charge, or are other decay channels also possible?

In practice the criterion for deciding on the validity of the $\Delta S = \Delta Q$ rule was to measure the decay rates of K^0_1 and K^0_2 into the leptonic channel $e^\pm \pi^\mp \gamma$. If the ratio of these two was unity, the selection rule would be strictly valid, a fact which was indeed confirmed in this experiment within the limits of precision. Further the photographs were used to study the validity of $\Delta I = 1/2$ leptonic rule in K^0 decay, the decay rates of various modes of K^0, to measure the mean life of K^0_2, and many more.[121]

In September 1962 the 81 cm HBC was engaged for the first time in a low energy kaon experiment. This experiment, carried out by CERN, the University of Maryland and the Naval Research Lab., had one precise aim: determine the relative parity of lambda and sigma.[122]

According to the idea of Ne'eman and Gell-Mann on the classification of elementary particles in multiplets, lambda and sigma would be assigned to the same octet of unitary symmetry, if they had even parity. Measuring this quantity would therefore be a crucial test for SU(3) theory. The proposal put forward as early as October 1961 suggested a detailed study of the invariant mass spectrum of the electron-positron pairs (Dalitz pairs) coming from the electromagnetic decay

$$K^- p \rightarrow \Sigma^0 + \pi^0$$
$$\hookrightarrow \Lambda^0 + e^+ + e^-.$$

If the relative parity was odd, then the matrix element would be independent of the pair momentum, which meant that more Dalitz pairs exhibiting large invariant mass would occur than in the case of even parity.

Scanning and measuring of these photographs was given high priority as the result was of utmost theoretical interest and three months after the last pictures were taken

Notes: pp. 247 ff.

the results were ready for publication. The CERN data favoured even lambda–sigma parity thus clearly supporting the theory of unitary symmetry, and marking a great scientific success for CERN.[123]

Besides this rather sound result the leptonic decays of the $\Sigma^\pm$ were studied, and the results compared with Cabibbo's theory of weak interactions, which had been published in 1963. Indeed from these decays the Cabibbo angle was calculated to be about 0.26, in good agreement with theory. Further they showed that the $\Delta S = \Delta Q$ rule held with an accuracy of 12%.[124]

The next experiment using low energetic kaons was carried out by a CERN group under R. Armenteros, together with physicists from Heidelberg and Saclay—a collaboration who became known as CHS. It got started in mid-1964 and was the first step in a comprehensive study of hyperon resonances which would continue for years at CERN. At that time various hyperon resonances were already known, but their confirmation through a determination of their quantum numbers and branching ratios was rather limited. It was therefore of the utmost importance to learn more about them in order to test the SU(3) model. Further there were two rather clear hints from other experiments that this energy region was interesting from the point of view of resonances. On the one hand a determination of the total cross-section in counter experiments had shown a broad bump in $I=0$ centered at 1.8 GeV/c c.m.s. energy and on the other hand experiments at Berkeley had hinted at a new resonance at 1765 MeV in the reaction $K^-n \to K^-p\pi^-$ at incident K momentum of 1.5 GeV/c.[125]

The CHS collaboration thus suggested performing a so-called K^-p formation experiment. This meant that, in contrast to what had been done so far (i.e. to remain at one particular energy), photographs were taken of primary kaons at 20 different, regularly spaced momenta between 0.8 and 1.2 GeV/c. Thus one studied the evolution of the total and differential cross-section as a function of momentum of the elastic, charge exchange, and inelastic channels with visible Λ and Σ decays. In this way physicists could not only observe a resonance as a bump in the cross-section, but it was also possible to deduce the angular distribution, spin, and parity of this particle. Further K^-n photographs were taken for reasons of comparison.[126]

Let us summarize here some of the major results. The existence of the reaction sequence

$$K^-p \to Y^*(1760) \to Y^*(1520) + \pi^0$$
$$\hookrightarrow \Sigma^\pm \pi^\mp$$
$$\hookrightarrow \Lambda \pi^+ \pi^-$$

allowed the determination of both the isospin of this new resonance $Y^*(1760)$ as 1, and via the production angular distribution of $Y^*(1520)$, the spin-parity assignment as $5/2^-$.

An analysis of the elastic and charge exchange K⁻p channels showed a behaviour suggestive of two resonances with spin 5/2 with opposite parities and masses near 1760 and 1815 MeV. With the result $J^P = 5/2^-$ for the Y*(1760) it could be concluded that Y*(1820) had the spin parity assignment $5/2^+$.

The branching ratio $\Sigma\pi$ of the Y*(1385) was also determined as 0.14, so in good agreement with the value expected in the framework of SU(3).

There are still two experiments we shortly want to touch on here. The first was carried out with the 81 cm chamber filled with deuterium by an Italian collaboration. It resulted in two publications on the longitudinal polarization of low energy muons from $K_{\mu 3}$ decay and on the existence of pions with spin.

The second experiment, and the last with low-energy kaons in the period up to 1965, was made with the enlarged version of the CERN heavy liquid bubble chamber. Two collaborations were involved in this so-called X2 experiment, namely one with physicists from Aachen, Bari, Bergen, Brussels, CERN, Nijmegen, and the Ecole Polytechnique, the other from Berkeley, University College and Wisconsin.[127] This bubble chamber was the biggest instrument of the kind in the world and represented the enormous advantage of being able to reconstruct complete events. Further it has a very strong magnetic field which allowed precise momentum determination. The central aim was to investigate K⁺ decays in particular with respect to testing the $\Delta S = \Delta Q$ selection rule, the exact determination of branching ratios, and the measurement of decay spectrum and rare decay modes. All this had become particularly important as, in 1964, CP violation had been found in weak interactions.

5.3.2.6 Kaon-proton interactions at the highest available energy—from 3 to 10 GeV/c

A comprehensive effort to study Kp interactions at ever higher energies began in late 1962/early 1963 and occupied an increasingly important place in CERN's research programme during the following years. With every technical improvement in setting up beams the energy could be raised, starting from 3 GeV/c and reaching 10 GeV/c by 1965. However the bubble chambers involved also grew bigger and bigger starting with the 81 cm chamber and ending with the 2 m. Altogether no less than 1.7 million photographs were taken up to 1965 to study this topic.

The first exposure to a high-energy K-beam of 3.5 GeV/c was made late in 1962 with the 1 meter CERN heavy liquid bubble chamber by a collaboration who had already worked on Kp interactions at 1.5 GeV/c the same year. Besides the aim of studying the rare cascade particle xi and searching for new resonances, this energy opened a completely new possibility, the search for the Ω^-. Murray Gell-Mann had predicted such a particle on the basis of his unitary symmetry model and therefore the proof of its existence or nonexistence was very important for the evolution of this theory. More explicitly, physicists were looking for a resonance with strangeness -3 and isospin 3/2 which should be detectable above the threshold of 3.2 GeV/c in the reaction $K^-p \to K^+K^0\Omega^-$. This first attempt was however unsuccessful.[128]

During 1963 an intensive K-programme was initiated with the 81 cm hydrogen bubble chamber, taking pictures of positive and negative kaons at 3 and 3.5 GeV/c. The negative kaon–proton interactions at 3 GeV/c were analyzed by a collaboration from Amsterdam, the Ecole Polytechnique, and Saclay, those at 3.5 GeV/c by an association of British Universities and finally the positive kaon pictures at both energies by a group from CERN under Yves Goldschmidt-Clermont together with physicists from Brussels and Stockholm. Further during late 1965 more than half a million photographs were taken with deuterium filling for reasons of comparison.

Interesting new results in the domain of resonances were revealed from the K^-p data. Investigating the reaction $K^-p \to \bar{K}^0\pi^-p$, i.e. drawing its Dalitz plot and the $\bar{K}^0\pi^-$ effective mass distribution, showed not only that this channel was dominated by peripheral $K^*(890)$ production, but also an enhancement at 1400 MeV. A detailed analysis of the decay distribution of this newly detected resonance $K^*(1400)$ based on

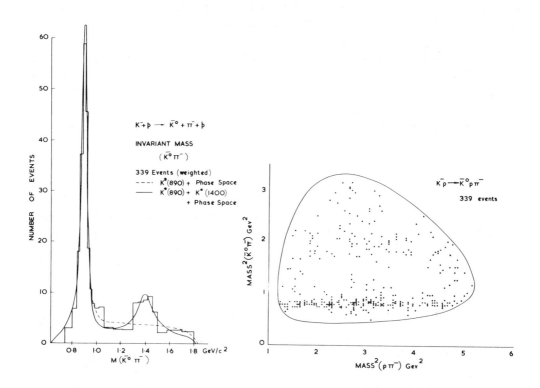

Fig. 5.8 $M(\bar{K}^0\pi^-)$ distribution showing the production of $K^*(1400)$[129]

the Gottfried–Jackson model of meson exchange favoured the quantum numbers $J^P = 2^+$. Further, in the reaction $K^-p \to \Xi^- K^0 \pi^+ \pi^0$ the existence of a baryonic resonance with strangeness -2 at 1820 MeV ($\Xi^*(1820)$)—earlier suggested by Berkeley—was confirmed, and there as well first indications of a $\Xi^*(1935)$ were seen. Thirdly, spin and parity of the already known $Y^*(1660)$ was determined as $(3/2)^+$ via the reaction

$$K^-p \to Y_1^*(1660) + \pi^-$$
$$\to \Sigma^- \pi^+ \pi^- .$$

This was not in agreement with the Berkeley results and as both suffered from rather poor statistics no conclusion could be drawn. Lastly, these photographs were also used to make a search for the Ω^- but again without success.[130]

As far as the K^+p interactions were concerned the first interesting result was the non-observation of direct production of resonant strangeness $+1$ states. However, the central point investigated was the production mechanism of K^* and N^*, a topic of considerable theoretical interest in the light of the absorption model of Gottfried and Jackson and the Regge pole model. About 50% of the reaction channels proceeded through two-body or quasi-two-body intermediate states and when looking in detail at the reaction $K^+p \to K^0 \pi^+ p$ there were two equally dominant channels: $K^+p \to K^{*+}p$ and $K^+p \to N^{*++}K^0$. Indeed here one observed for the first time for strange particles a predominant vector-meson exchange, only a small percentage going via pions for the K^*, and only a vector meson exchange in the N^* case (see Fig. 5.9). In addition, together with the results of the study of K^+p at 1.5 GeV/c, there was some evidence that the vector meson exchange became more dominant as the energy increased.[131] However, on the field of resonances this experiment caused quite some confusion. Indeed the data seemed to show two KK resonances M_1 and M_2, one $K\pi\pi$ resonance at 1270 MeV and a so-called $\varkappa$ resonance at 735 MeV, all of which were never confirmed.[132]

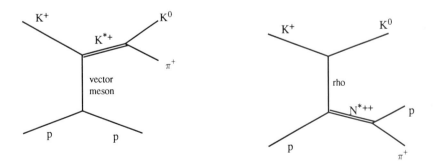

Fig. 5.9 Feynman diagrams illustrating the exchange mechanism

Notes: p. 248

In February 1964 experiments with negative kaons were taken up again by the British collaboration, this time including also one German group. In view of using 'their' 152 cm chamber for the same purpose later that year they exposed the HBC 81 to a beam of 6 GeV/c negative kaons. This time, they hoped, the energy would be high enough to be more successful in the search for Ω^-. But again they found no sign of the particle in the photographs. Meanwhile, however, the Brookhaven group was reporting a first success. Working in a mass separated beam of 5 GeV/c negative kaons and using the new 80 inch hydrogen bubble chamber—the biggest chamber of its kind—they saw the first of the rare examples of the Ω^- by the end of January 1964. This was not only a demonstration of the strength of bubble chambers for studying a single event, but it was another important 'victory' for Gell-Mann's theory.[133]

When the British bubble chamber started operating by mid-1964 the study of $K^- p$ interactions was continued as foreseen. The aim was to find more examples of the Ω^- particles and in particular to determine cross-section, mass, and life-time and more than half a million pictures were taken at 5 and 6 GeV/c. Indeed in the 6 GeV/c photographs one event was found rather quickly corresponding to the production of an Ω^-. It decayed into a neutral xi and a negative pion, the mass was determined as (1666 ± 8) MeV and the lifetime as 1.85×10^{-10} sec (see Fig. 5.10). It added one further example to the four others found meanwhile at Brookhaven. Finally, altogether nine examples were found in these photographs, and were collectively published in 1968. They showed that two decay modes were possible for this particle, one via $\Xi\pi$, the other via ΛK.[134]

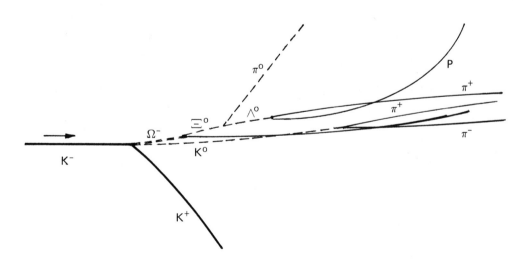

Fig. 5.10 Decay of the Ω^- meson in the British bubble chamber[135]

$$K^- + p \to K^0 + K^+ + \Omega^-$$
$$ \hookrightarrow \pi^+\pi^- \ \ \hookrightarrow \Xi^0 + \pi^-$$
$$ \hookrightarrow \Lambda^0 + \pi^0$$
$$ \hookrightarrow p + \pi^+$$

The K^+p investigations were also continued by the CERN–Brussels collaboration, this time at 5 GeV/c using first the 81 cm chamber, then the British one and finally the 2 meter CERN chamber. Two detailed studies, one on K^+p elastic scattering and the other on two-body channels in K^+p interaction were made. In the first total and differential elastic cross-sections were determined and examined in the light of models for high-energy scattering such as the Regge-pole model. Both cross-sections agreed rather well for small t which gave support to the presence of a real part of the forward scattering amplitude. Further the diffraction peak showed definite shrinking with increasing momenta. In the second paper a systematic study of two-body and quasi-two-body reactions and their change with increasing energy was published. The general conclusion drawn was that all reactions were characterized by strongly peripheral production and that they exhibited a shrinking which could qualitatively be described by the simplest Regge-pole model.[136]

These two publications were followed by a multitude of others from 1967 onwards using in particular the photographs of the 150 and 200 cm chambers, and dealing also with the search and study of resonances. We will, however, not study them in detail here.

To close this part on high-energy kaon–proton interactions we want to touch on two experiments shortly as they only started at the end of the period discussed by us. It seems however important to us to mention them (without discussing any of their results) as they show the continuation of the K-physics programme to ever higher energies.

The first started in April 1965 by exposing the 152 cm British chamber to a beam of 10 GeV/c negative kaons. It was carried out by a CERN group under D.R.O. Morrison together with physicists from Aachen, Berlin, Imperial College, and Vienna. The same energy—10 GeV/c—was also attained for positive kaons by December of the year and thus the second experiment carried out by the British collaboration, who had previously worked on K^-p interactions between 3.5 and 6 GeV/c, was scheduled for the last month of 1965.

Notes: p. 248

5.4 The use of electronic detectors to study elementary particles[137]

5.4.1 STRONG INTERACTION PHYSICS

5.4.1.1 Beam survey experiments — the production of secondaries

The first step to take whenever an accelerator opens up a new energy region is to investigate the secondary particle intensity as a function of angle, momentum, and primary beam energy. This knowledge is essential, enabling the physicists to judge which experiments are feasible, which separated beams they can exploit, and where to install them. Further a beam survey provides data allowing both phenomenological and theoretical analyses of the basic high-energy particle production mechanisms. This was particularly important in the early 60's, since no theory existed which was capable to completely describing the production process of elementary particles. Only statistical theories, such as the one elaborated by Rolf Hagedorn at CERN, could be used as a guideline for estimating the energy distribution and composition of the secondaries produced when the accelerated protons hit the target. It was, however, not possible to make predictions on angular distributions and energy–angle correlations in the centre-of-mass system.[138]

The pressing need for an adequate beam survey, and the fact that only one small bubble chamber was at hand in 1960 explains why the major part of the machine time allocated during this year was for counter groups doing this kind of research. Five teams, led by senior staff of the SC (later NP) Division, took part in this survey programme. The first was gathered around Guy von Dardel. Their experimental set-up had already been prepared in 1959 and in January 1960 they started to investigate the *secondary beams at 3° and 6°,* using a 2.4 m bending magnet to analyze the momentum and a large velocity-selective gas Čerenkov counter to determine the speed of the beam components.[139] If physicists wanted to do experiments with rare particles such as kaons and antiprotons, it was important to know if they would be detectable against the background of abundantly produced pions. Thus the numbers of protons, antiprotons and kaons were measured relative to the pions. The results were then compared to the predictions of statistical theory.

Three main conclusions were drawn from this experiment: that the p–π ratio was not well reproduced by statistical theories, showing a large excess at higher energies (> 14 GeV/c) and a deficiency below; that the ratio of positive kaons to pions stayed constant over the energy range of 5–18 GeV, whereas the negative ones were less abundant and diminished with increasing energy; and that the antiproton–pion ratio was far lower than predicted and decreased by a factor 5 between 6 and 16 GeV/c. These discrepancies lay in the fact that theory assumed a proton target (and not Al) and isotropy for the energy spectrum as a function of the c.m. angle.[140]

The second experimental group was led by Giuseppe Fidecaro and Alec Merrison. They studied the *secondaries* at larger angles, namely *at 15.9° produced by a 25 GeV p-beam* hitting the target. The detection and measurement was done using a high resolution time-of-flight system (TOF), consisting of three scintillation counters. It was designed to separate, after momentum selection by a magnet, particles having a difference in velocity of $\Delta\beta/\beta \simeq 0.015$. The intensities for $(K^-, \bar{p})$ and (K^+, p, d) were here also measured relative to the produced pions. Three interesting points were established. First, the measured kaon intensities showed that the PS could efficiently be used for low-energy K-meson physics and consequently Fidecaro and Merrison proposed that a K-beam should be set up soon.[141] Secondly, with their high precision TOF measuring system a new determination of the mass ratio of protons to antiprotons was made, leading to the most precise value at the time, namely (1.008 ± 0.005). The last and most unexpected result was however a large number of deuterons which were produced with high momentum and at large angles.[142]

Covering a similar subject a third team with Arne Lundby as leader started to work with the PS in spring 1960. They investigated in particular *long-lived* (i.e. $\tau > 10^{-8}$ sec) *charged particles produced in the forward direction* when 24 GeV protons hit a target. The mass of the particles was identified by measuring both its momentum and its velocity. At the centre of the set-up were two differential, isochronous, self-collimating Čerenkov counters—called DISC[143]—developed specially for this purpose by the team. Besides additional information on the intensity of already known secondaries and their energy dependence, they could report two new findings among the particles produced, namely H^3 and He^3.[144]

Whereas the three experiments mentioned so far all studied the charged components of the secondary beams, the two following investigated the neutral particles produced by a high-energy proton synchrotron. The first group, gathered around Giuseppe Cocconi, measured the *spectrum of photons* emitted at 180° to the beam using a total absorption Čerenkov counter. They learned that the photon spectrum extended far beyond the energies which would be expected if they came from neutral pions emitted in nucleon–nucleon collisions. This could then only be explained by taking into account secondary interactions in the nuclei. One noteworthy practical spin-off was that these backwards photons could be used to monitor the target activity.[145]

The second experiment was carried out by a team led by G. Giacomelli and W. Middelkoop. They had constructed two total absorption lead glass Čerenkov counters in order to measure the *photon flux originating from the decay of neutral pions* coming from the target. They used both Al and Be targets and chose two angles, 3.2° and 15.9°. Also here, they showed clearly that the statistical model was a poor approximation to the results. Indeed whereas at 3.2°, the minimum angle at which the neutral particles could escape the production region, the number of photons was a factor three higher than predicted, at 15.9°—corresponding to roughly 90° in the c.m. system—the photon flux was an order of magnitude too small. These deviations from theory could be interpreted as indicating anisotropy in the angular distribution in the centre-of-mass system. Secondly, they found that

Notes: pp. 248 ff.

above 2 GeV the number of neutrons was roughly a factor two lower than the number of photons. Lastly, a Be target would yield far more photons than an Al one over the whole spectrum.[146]

As a by-product of this experiment the absorption coefficients of photons at 10 GeV in CH_2 and C and at 13.5 GeV in Li, C, Cu and Pb were measured. The results proved to be in good agreement with those calculated theoretically on the basis of three absorption mechanisms, namely electron–positron pair production, Compton scattering, and the photonuclear effect.[147]

This team followed the same line of research in 1961, investigating *photon production* this time *in proton–proton collisions at 23.1 GeV*. As in the previous experiment the energy spectra of the photons produced were measured between 1.75° and 27° by means of a lead glass total absorption Čerenkov counter, and four conclusions were deduced from the data. Firstly, there was a strong energy-dependent anisotropy in the c.m. system. Secondly, as the γ's were mainly produced by decay of the π^0 mesons, the π^0 spectra could be derived from the photon spectra. Indeed they agreed reasonably well with the spectra of the charged pions measured at CERN and at BNL. Thirdly, comparing the experimental values with a statistical model showed that the total cross-section was reasonably well predicted, whereas this was not the case for the angular distribution. Lastly, looking at photon production in p-Be and p-Al collisions the cross-sections turned out to be inferior by a factor 1.4 and 1.9 respectively to the free nucleon case.[148]

After these first overview experiments carried out in 1960/61, more specific investigations on the production of secondary particles were undertaken. Here we would like to mention the experiments on deuteron production carried out by Fidecaro's group in 1961, by Taylor in 1962, and taken up again by Cocconi's team in 1963. Then a CERN–Orsay collaboration worked throughout 1963 on the production of pions, kaons, protons and antiprotons in proton collisions with various materials. Lastly, in 1965 the antideuteron production in p-Be collisions was studied by Zichichi and his group.

Let us say a few words about each of these experiments. Fidecaro's idea of studying *deuteron production*, born from the detection of this particle in his beam survey, created considerable interest inside the scientific community. Its aim was to establish whether the predominant production mechanism involved only a single nucleon–nucleon collision, or if the observed deuterons had their main origin, as various theoretical models suggested, in more complicated processes requiring the presence of complex nuclei. For this purpose two targets, CH_2 and C, were flipped alternatively into the circulating beam of 26.6 GeV, and the relative yields were measured. However only a rather vague conclusion could be drawn, namely that the amount of deuterons observed could not be explained without admitting a nuclear effect, though which kind of nuclear mechanism was involved was not clear. Similar studies were made by A. Taylor and his team (Cocconi group and physicists from Rutherford Lab.) in the course of a p-p scattering experiment. They looked for d-production in pure hydrogen collisions at 19 and 24 GeV/c and detected those particles with momenta up to 9.2 GeV/c, but with low production cross-section. Again no definite conclusion could be drawn. Cocconi's group continued to study the pp → dπ process along the same lines as Fidecaro, but using incident proton energies between 2.1 and 8.9 GeV/c. The data revealed one novelty: a non-explicable resonance-like behaviour of the cross-section at a diproton mass of 2.9 GeV. It was clear that further investigations would be needed to understand this reaction.[149]

The experiment carried out by the CERN (K. Winter and Mermod) – Orsay (Vivargent) collaboration, was originally undertaken to provide information on *pion and kaon production* for an improved calculation of the ν spectrum. It was set up in the first beam available in the newly constructed east experimental area in August 1963. This was a scattered-out proton beam between 19 and 23 GeV/c which hit targets of different materials. The secondaries produced were investigated with three threshold and one differential Čerenkov counter. A novelty was that, due to the scattered-out beam, absolute cross-sections for the production of pions, kaons, protons, and antiprotons in pp collisions could also be obtained at 0°. The results were consistent with the assumption that isobar excitation played an important role in pp collisions.[150]

The last experiment to be mentioned was the *antideuteron experiment,* which had a relatively long history. At the first meeting of the Electronics Experiments Committee held in March 1961 von Dardel suggested investigating seriously antideuteron production in p–p collisions. Although preliminary measurements had already started it was nevertheless decided not to follow this research topic as no immediate success was expected and the number of experiments waiting for machine time was very high. Fidecaro also looked for antideuterons when setting up the antiproton beam, though without success. Finally, Antonio Zichichi apparently got interested in this question, and used the PAPEP experiment (see 5.4.2.2) in which he participated to make a short search for antideuterons. No results were published, though he took up the question later and found a feeble production of these particles with $\bar{d}/\pi = (1.2 \pm 0.2) \times 10^{-8}$ in an experiment done in 1965.[151]

5.4.1.2 Investigating the fundamental total cross-sections

The measuring of total cross-sections is one of the most straightforward and basic things to do whenever a new energy range is entered. The interest in this topic was stimulated in the late fifties by Pomeranchuk's theoretical prediction that the total cross-section for a particle and its antiparticle on hydrogen approached equal values at high energies.[152] With the aim of proving or disproving this theorem three groups performed experiments between 1960 and 1962 measuring various total cross-sections. They were led by Guy von Dardel, Giuseppe Fidecaro, and Giuseppe Cocconi.

Von Dardel's team started the first real experiment early in 1960, after the beam survey measurements, to investigate the *total cross-sections of p, p̄, K⁺ and π⁺ on hydrogen* in the energy range *between 3 and 10 GeV/c*. For this purpose they installed their velocity-selective gas Čerenkov counter in a beam leaving the internal target at 6°. With the exception of the $\pi^- p$ cross-section, which was measured in more detail, only values at three or four different energies were taken for each kind of interaction. The results were then merged with those already found in experiments elsewhere. The conclusion drawn was that for all cases of interaction the cross-sections for both particles and antiparticles tended to approach a constant value at high energy. However, for kaons and nucleons the difference between particle and antiparticle cross-section on hydrogen, together with their weak dependence on energy, suggested that the limiting equality—predicted by Pomeranchuk—would be reached, if at all, only at energies considerably higher than 10 GeV/c. These measurements also hinted that the Pomeranchuk theorem could possibly be satisfied at lower energies for pions because of their smaller mass. This latter finding led this group to continue their research in this particular direction.[153]

Indeed in 1961 an extensive study of the *pion–proton cross-sections* was started. It was carried out in two stages, first improving the values up to 10 GeV/c, and then increasing the energy *up to 20 GeV/c*. In the second stage a ten-meter threshold gas Čerenkov counter was added. It was found that, unlike the proton cross-section which Cocconi's group had found to be constant between 5 and 28 GeV/c (see next paragraph), both $\sigma_{tot}(\pi^+ p)$ and $\sigma_{tot}(\pi^- p)$ decreased monotonically with increasing energy up to 10 GeV/c. On the other hand, in the energy range between 10 and 20 GeV/c, a $1/E$ function gave a good fit to the data (see Fig. 5.11), thus strongly supporting Pomeranchuk's theorem.[154]

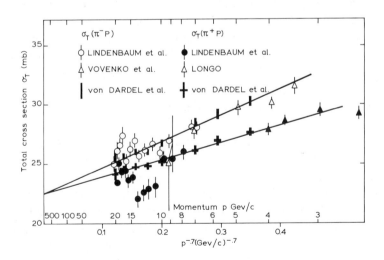

Fig. 5.11 Total $(\pi^+ p)$ and $(\pi^- p)$ cross-section[155] versus $p^{-0.7}$

The second group interested in the energy behaviour of cross-sections was the one led by Giuseppe Cocconi. During summer 1960 they extended von Dardel's measurements on the *p-p cross-section up to maximum momentum of 28 GeV/c*, and also studied protons on neutrons and complex nuclei. The values found at 10 GeV/c were in good agreement with those of von Dardel and at 24 GeV/c with those obtained by the 30 cm hydrogen bubble chamber group.[156] The most striking result however was the apparent constancy of σ_{tot} over the whole energy region investigated (see Fig. 5.12).

Further the measurements of the neutron–proton total cross-section were made at three momenta, and gave — within experimental limits — the same results as in the proton–proton case. This was confirmed in 1962, in the context of a scattering experiment, where a precision measurement of the total cross-section of protons on deuterons and protons was made. As far as the scattering on complex nuclei was concerned, six materials were used: Be, C, Al, Cu, Cd and Pb. The data were found to fit an $A^{2/3}$ law rather well, A being the mass number.[157]

The last group to mention was headed by Giuseppe Fidecaro. They were particularly interested in the behaviour of the *p$\bar{p}$ cross-section* at energies not investigated. The study of antiprotons had only become possible in the mid-fifties when the Bevatron at Berkeley came into operation. However the available intensity at that time was very low, making precision measurements impossible. Indeed quite a

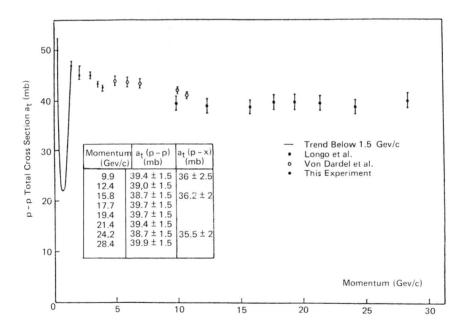

Fig. 5.12 p-p and p-n total cross-section as a function of proton momentum[158]

Notes: p. 249

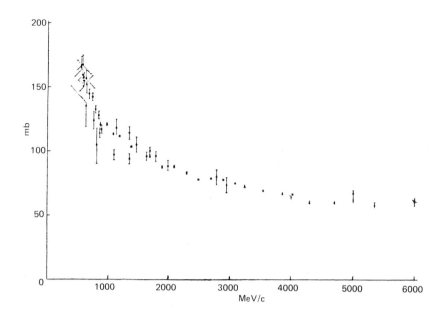

Fig. 5.13 Total p$\bar{\text{p}}$ cross-section as a function of the antiproton momentum[159]

number of properties of the antiproton were still unknown such as spin, magnetic moment, details on the annihilation process, etc. Accordingly as early as May 1959 at a meeting on the future experimental programme of the PS, it was clearly recognized that CERN should investigate this area in more detail.[160]

In a first experiment in 1960 the transmission of a p̄ beam through a liquid hydrogen target was measured covering the energy region between 1.4 and 5.5 GeV/c. For values bigger than 2.7 GeV/c no previous data existed, and at low energies the data available were somethat inconsistent. However when Fidecaro wanted to go below 1.4 GeV he was immediately confronted with the problem of building such a low energy p̄ beam with reasonable intensity and purity. By spring 1961 they had managed to set up and tune such a beam, transporting antiprotons of 1 GeV/c maximum energy to the 81 cm hydrogen bubble chamber and up to 4 GeV/c to a liquid hydrogen target and a counter system. This beam was the first well collimated secondary beam at the PS.[161]

The runs made in this beam between June and August 1961 delivered data on the pp̄ total cross-section from 0.58 up to 5.35 GeV/c. It showed a function decreasing smoothly with increasing energy, with no structure in the energy region studied (see Fig. 5.13).[162]

5.4.1.3 Scattering experiments

CERN was very active in this first half of the sixties in the field of scattering experiments and various groups undertook investigations producing interesting results. The first to be mentioned was that of Cocconi. In a sequence of experiments covering the period 1961 to 1965 they carefully studied proton–proton scattering. The second group was Dick Harting's which specialized during 1962/63 on pion–proton scattering. Then around 1963 high intensity, pure beams began to be produced at the PS, allowing one to investigate a new set of reactions with low cross-sections, such as large momentum transfer and charge-exchange scattering. This was done by the groups of Bernard Hyams, Arne Lundby, Paul Falk-Vairant (Saclay) and Peter Astbury (visitor from Imperial College).

Let us give some more details about these experiments. As we have seen in the previous parts Cocconi started to investigate the total cross-section of p–p interactions in 1960. He continued this research line in 1961, though concentrating on the aspect of *elastic and near-elastic processes between 9 and 25 GeV/c*. The aim of this experiment was to momentum-analyze the protons coming from a Be target put in the internal beam. This was done with the help of a 4 m long bending magnet and a telescope made of two scintillators. The most stunning observation was a so-called quasi-elastic peak (later renamed inelastic peak) in the momentum spectrum. Indeed cosmic-ray experiments had already shown some evidence for quasi-elastic collisions, which meant that particles emerging from collisions against nucleons at rest had lost only a little of their initial energy. This peak was present whenever and only when the elastic peak was present. Further it was noticed that the momentum difference between the two peaks stayed constant, i.e. 1 GeV, over the whole region independent of the scattering angle. None of this could be explained in the framework of Fermi's statistical theory.[163]

In the course of 1961 Cocconi's group was joined by physicists from Rutherford Laboratory led by A. Taylor. They continued to investigate these interesting new phenomena, this time using the CH_2–C difference method, and covering the energy range between 12 and 27 GeV/c and 10–60 mrad. Again they observed an elastic and an inelastic peak, the latter however showing a fine structure. Though several theoretical explanations for the origin of this inelastic bump were put forward, none of them was capable of reproducing all the features of the experimental data.[164]

In summer 1961 Giuseppe Cocconi left CERN to spend a sabbatical year in the United States and Taylor took over the role of group leader. The p–p programme continued, now measuring the *elastic scattering between 12 and 26.2 GeV/c* between 10 and 50 mrad. This was later extended to larger values of momentum transfer, i.e. the scattering angle was changed to 110 mrad.

One new phenomenon was revealed in this study, namely the 'shrinking of the diffraction peak' with increasing energy. This was important as the measurement of

Notes: pp. 249 ff.

diffraction scattering allowed a determination of the effective proton radius, a fundamental quantity to know. Until the late fifties physicists believed that at energies bigger than 10 GeV the effective radius of the proton would approach a constant and the effective density of nucleon matter within the proton would also be approximately constant. In 1958 Regge pole theory completely changed these assumptions. It predicted that the size of the proton grew logarithmically with energy, but that at the same time the density of nuclear matter decreased. This had as a consequence a logarithmic shrinkage of the width of the diffraction peak with increasing energy: a prediction clearly confirmed in the data taken (see Fig. 5.14).

A second observation to be mentioned was a fine structure in the inelastically scattered proton spectrum which had already appeared in the 1961 experiment. This effect seemed to depend on the four-momentum transfer; as its values increased the structure appeared more clearly.[166]

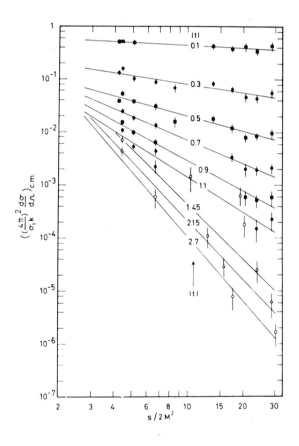

Fig. 5.14 Normalized elastic differential cross-section as a function of s/M^2
(s is the square of the c.m. total energy, M is the nucleon mass)[165]

After Cocconi's return to CERN and the departure of the Rutherford group, the team started a study of *inelastic p–p scattering and nuclear isobar production in p–p collision* in 1963. They measured the momentum spectra of positively charged particles emitted at a fixed laboratory angle of 60 mrad for incident proton momenta between 3.6 and 11.8 GeV/c. The features of these spectra were then discussed in terms of nucleon isobar production

$$p + p \to N + N^*$$
$$\hookrightarrow N + \text{pions}$$

where N^* was a resonant pion–nucleon state. It showed that the production cross-section for the (3/2, 3/2) state decreased much more rapidly with increasing momentum transfer than the cross-section for production of more massive isotopic spin 1/2 states. The data further indicated the possible existence of a new excited nucleon at 1.4 GeV.[167]

Starting in 1962 and throughout 1963 Bogdan Maglic together with the Cocconi group investigated closely the feasibility of using sonic spark chambers as a detector in high-energy experiments. Conventional spark chambers had already two striking advantages—a large solid angle and an improved spatial resolution—over scintillation counter systems. The new idea was to use the sonic method to locate sparks rather than photographing them, thus permitting an immediate digitization of the information and reducing the time needed for data analysis. The first tests were carried out with a system designed to measure angular and momentum distribution of protons elastically and inelastically scattered on hydrogen.[168]

Using this newly-constructed equipment Cocconi's team undertook an experiment on *small angle scattering of protons in hydrogen and deuterium between 10 and 26 GeV/c*. And indeed they were able to record an appreciable 20 000 events per day on magnetic tape. In September 1964 the system was extended by the on-line use of an SDS 920 computer which controlled the operation of the detector. In this way the rate of data collection was increased by a factor five to 100 000 events per day. The full computation of the spark coordinates, momenta, and scattering angles was performed, off-line, from the magnetic information using an IBM 7090 computer. The measurements continued throughout 1965.

Four results are worth stressing. First, they investigated the differential cross-section for elastic proton–proton scattering at 10, 19 and 26 GeV/c in the range of very small angles, extending into the Coulomb scattering region. In this region the assumption that diffraction scattering arose from a pure imaginary scattering amplitude, as believed in the fifties, could be tested best. Indeed, the data showed that the shape of the angular distribution could only be adequately explained by the interference of a substantial, repulsive, *real* part of the nuclear scattering amplitude with the Coulomb amplitude. Secondly, the momentum spectra of protons scattered

Notes: p. 250

with small inelasticity were measured at the energies mentioned above. As the structure observed in the inelastic spectra was attributed to the excitation of the target nucleon by the process p + p → p + X, a study on the missing mass (m_X) spectrum was carried out. Besides various pion–nucleon resonances already known between 1.24 and 1.92 GeV, the resonance at 1.4 GeV, already observed in the 1963 experiment, showed up again. It was assumed to be a nucleon isobar with quantum numbers $(1/2, 1/2)^+$, manifesting itself preponderantly at small momentum transfers. As a third aspect proton–deuteron scattering was studied at 19.3 GeV/c to obtain information on n-p interaction. At small angles the analysis showed that p-n and p-p scattering amplitudes were very similar and contained approximately equal real parts of the same sign. Lastly, we should mention a systematic study of differential cross-sections of 20 GeV protons scattered elastically and quasi-elastically by a series of nuclei, ranging from Li to U. The group found that the light nuclei at small angles showed the characteristic central diffraction peak produced by an absorbing disk. At larger angles however the quasi-elastic scattering produced by single nucleons predominated. The interest was to compare these results to the predictions of the optical theorem, which relates the total cross-section of a reaction to the imaginary part of the scattering amplitude in the forward direction. Indeed if one limited oneself to the region where slighly inelastic and quasi-elastic scattering was absent, the optical theorem gave a valid picture of the interaction of 20 GeV protons with nuclei.[169]

The second fundamental process, namely *pion–proton scattering,* was investigated in 1962/63 by a collaboration between CERN, Bologna, Liverpool and Michigan University, with Dick Harting as group leader. Their aim was to measure a large number of elastic π-p and, for reasons of comparison, also p–p scattering events behind the diffraction peak, since little was known about angular dependence, magnitude and even existence of this very large momentum transfer part of the elastic scattering. The energies chosen were 8, 12 and 18 GeV. The results would be particularly interesting in the light of Regge theory.

With a detector consisting of two magnets and nine spark chambers viewed by a single fast cycling camera, 450 000 events were registered. Whilst they confirmed the shrinking of the diffraction peak with increasing incident energy in p-p collisions, they also showed the absence of this effect in π-p scattering. This very interesting finding had already been made at Brookhaven a bit earlier, where the more efficient data handling installations had allowed quicker production of results.[170]

The next experiment studying *large momentum transfer pion–proton scattering* was carried out by Arne Lundby's group together with physicists from Ivry. They measured the differential cross-section from the diffraction peak up to 180°. Previous experiments had given evidence for a peak in the elastic scattering cross-section in the backward direction, but the relatively small number of events observed did not allow one to gain any detailed knowledge. The rather complex set-up consisted of six spark chambers arranged around a 60 cm long liquid hydrogen target, two spectrometer magnets and a number of Čerenkov and scintillation counters. This was installed in a high-intensity pion-beam of 3.5 GeV/c and 225 000 spark chamber photographs were taken.[171]

The analysis of the photographs showed a steeply rising backward peak of the positive pions, while the shape was less clear for negative ones. This backward peak had been predicted on the basis of an optical model analysis of the scattering from a disc, and the results were in general agreement. There was however no clear explanation for the marked difference between the behaviour of positive and negative pions.

The last three experiments we will discuss in this part all dealt with *charge exchange processes*. A study of the *pion–proton* case had been suggested early 1963 by a team from Saclay and Orsay under Paul Falk-Vairant.[172] This was one of the rare examples where only outside physicists participated in a counter experiment at the PS. While work on this subject had already been carried out at Saturne in the momentum range up to 2 GeV, at CERN much higher momenta, namely between 6 and 18 GeV/c, were investigated. The study of this reaction provided a critical test of the currently popular exchange models in high-energy interactions.

The experimental set-up used consisted of a liquid hydrogen target surrounded by a set of lead converters and counters. Spark chambers were added to register the π^0n as two showers. The data collected showed, firstly, that the cross-section bent downwards towards zero momentum transfer, having its maximum at $-t = 0.04$, a behaviour which was different from the elastic scattering case. (It could however be explained by applying a ϱ-exchange model.) Secondly, the cross-section and the real part of the scattering amplitude at $t = 0$ found experimentally was in reasonably good agreement with the theoretical calculations, and thirdly, the differential cross-section for $-t < 0.5$ $(\text{GeV}/c)^2$ showed a $1/p$ dependence together with evidence for a shrinkage of the diffraction peak, thus confirming again the Regge pole model.[173]

A second study of the *pion–proton* case was undertaken by Bernard Hyams during 1963/64. In 1962 he started an experiment to study the peripheral production of photons in π-p interactions. Using a total absorption scintillation detector as a first step, the energy and angular distribution of all γ rays produced in the forward direction was measured. The following year investigations on both the peripheral photon production and the elastic pion–proton charge exchange reaction occurring simultaneously were made in a beam of 10 GeV/c π^-. Whereas we found no publication concerning the first reaction, the results of the second were similar to those of the Falk-Vairant experiment, though with a 30% lower differential cross-section. This was explained as being due to possible systematic and statistical errors.[174]

The collaboration studying K^-p *charge exchange scattering* was formed by physicists from the ETH Zurich and from CERN with Peter Astbury as group leader, who had constructed a large magnetic spark chamber for this purpose. During 1964 data on the total and differential cross-sections of the reaction $\pi^- p \to \bar{K}^0 n$ were taken at 9.5 GeV/c. It was again, as in the πp case, noticed that the momentum transfer distribution was quite different to the one for elastic scattering. It was found to be flat for $-t$ between 0 and 0.15 (GeV/c)^2 and fell off rapidly thereafter. The results were compared to different theoretical models, such as the optical model and the Regge pole model, but due to the large experimental errors no definite conclusion in favour of one or the other could be drawn. The investigations were extended in 1965 to other reactions[175] such as $p\bar{p} \to n\bar{n}$, $\pi^- p \to \Lambda\bar{\Lambda}n$ or $\pi^- p \to K_0^1 \bar{K}_0^1 n$.

5.4.1.4 Resonances: their production and decay

Whereas the subjects described so far had been mainly studied using counter techniques, the field of resonance research was dominated worldwide by bubble chamber experiments throughout the late fifties/early sixties. The strength of the bubble chamber lay in the completeness of the information contained in a single picture, which made it an ideal tool to search for the unknown. It is thus not surprising to find very little activity devoted to this topic among the counter groups.

Resonances, being very short-lived and consequently not directly detectable, could however be observed indirectly in various manners. The most frequent method was to look for structure in the total or differential cross-section versus energy curve. In addition, CERN took up a new and very promising approach to this problem from 1963 onwards, namely the so-called missing mass spectroscopy.

The search for *resonances in the hyperon system* was the first experiment in this category, and was undertaken in 1961 by Arne Lundby's group. Installing a 2 m hydrogen target in the newly set up high intensity π-beam in the North hall they looked at the reaction $\pi^- p \to K^+ Y^-$. The outcoming K's were analyzed by a bending magnet and a differential Čerenkov counter (DISC). Plotting the positive kaon intensity as a function of the π^- momentum, physicists saw the production of Σ^-(1196), Y_1^*(1380) and hints of a Y(1550). In parallel to the detection of the K^+ a second DISC was installed to measure the proton intensity coming from the reaction $\pi^- p \to B + p$. However there was no sign of 'excited pions' B. This was explained by the fact that here one studied high momentum transfer reactions and it had been suggested already by other experiments that pion resonances were produced in peripheral interactions.[176]

In 1963 Lundby's team continued the research on production and decay of hyperon resonances, this time using a spark chamber set-up in the north hall. The results of the first experiment were confirmed, showing in addition two already known resonances at 1660 and 1765 MeV and a new one at 1455 MeV.[177]

Another group working on *resonances*, particularly *in the diboson system* was the one of Dick Harting. In collaboration with the University of Michigan they worked on a detection device using thin-plate spark chambers in order to improve the precision in the determination of the angle of tracks. Further, to present all data on a single 35 mm film frame, they elaborated a system of mirrors folding the images of the several spark chambers and their stereoscopic views.[178] This new set up was first used in a short prototype experiment by Caldwell et al. studying the ϱ^0 production at 12 and 17 GeV in πN interactions. Glancing collisions were sought by sending a high-energy pion beam onto nucleons. More precisely, the equipment was designed to observe the reaction presented in Fig. 5.15.

As expected, the ϱ peak dominated the mass histogram for the dipion system. Besides that a second bump at 885 MeV was registered, but needed more confirmation. A quantitative comparison of the data with predictions from the one-pion exchange model was impossible due to systematic uncertainties; however, the shape of the momentum transfer spectrum of the ϱ's at 17 GeV agreed rather well with the calculated one.[179]

The actual diboson experiment started in the second half of 1962. The incident pions of 12 and 18 GeV/c traversing a CH$_2$ target, produced sometimes (in a peripheral interaction) a short-lived neutral particle decaying into $\pi^+\pi^-$, K^+K^-, $K^+\pi^-$ and π^+K^- pairs. By measuring the angles and momenta of the incoming pion and the decay particles, the mass of this short-lived parent particle and the four-momentum transfer could be calculated. The information on the nature of the decay particles was gained by two large gas threshold Čerenkov counters.

Since the group separated physically after the experiment the extraction of results was very slow. Indeed first preliminary conclusions were presented only at the Dubna conference in 1964, so almost two years after the end of the data taking phase. The final publication continued up to 1968. Peripheral production of nearly 10 000 pion pairs was reported and analyzed in terms of a one-pion-exchange model. It was shown that there was fair agreement between the data and the calculated values of the ϱ^0 production cross-section.[180]

The idea for the next experiment was born from a number of preceding experiments, among them one carried out by Cocconi's team in 1963 on the reaction pp → π^+d. They had found a narrow peak in the differential cross-section around 2.9 GeV/c suggesting a resonance behaviour due to an excited deuteron state.[181] The

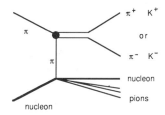

Fig. 5.15 π^0 production in πN interactions

Notes: p. 250

question was: would diprotons (dibosons) exist, a fact not foreseen in the framework of SU(3). With that in mind a team led by Klaus Winter together with physicists from Orsay proposed an investigation of the *angular distribution and total cross-section* of the inverse reaction $\pi^+ d \to pp$ between 0.65 and 1.95 GeV/c.

As far as the angular distribution was concerned its shape changed rapidly between 0.85 and 1 GeV/c. Up to 0.85 GeV/c the differential cross-section had its maximum between $\cos\theta = 0.7$ and 1, whereas above 1 GeV/c it was strongly peaked at $\cos\theta = 1$. The results obtained for the total cross-section were compared to theoretical calculations assuming the intermediate formation of the $N^*_{3/2,3/2}$ isobar.[182]

The most extensive study, however, of *the production of heavy charged bosons* was carried out by Bogdan Maglic and his team. They proposed to use the missing-mass method in order to scan the mass spectrum of both boson and nucleon isobars. More precisely, the aim was to observe *all* resonances produced with similar cross-section, with high statistical significance and good resolution. This knowledge on the complete mass spectrum was of key importance for theoreticians investigating an underlying law. We have to keep in mind that until then systematic research in this direction had been very difficult for a number of reasons. Firstly, kinematic fitting could only be obtained if there was not more than one neutral decay product. Secondly, even in the case of charged decay products the number of meson combinations was often too high for an effective mass computation. Lastly, as most of the information was gained by bubble-chamber experiments the statistics were rather poor.

Maglic and Gosta elaborated a new approach to this problem, namely the 'Jacobian peak' method. Its novelty was that in fact only one single quantity, namely the angle of the final state proton in the reaction $\pi^- p \to pX^-$, needed to be measured. The existence of a discrete mass would manifest itself as a peak in the angular distribution of protons of all momenta in a given momentum band.[183] To turn this idea into practice they constructed a missing mass spectrometer. It consisted of acoustic spark chambers and scintillation counter hodoscopes connected on-line to the Mercury computer. This allowed them not only to monitor the technical performance but also to sample the high-statistics physics data taken. The work with this apparatus started late in 1964 and continued up to 1966.[184]

The spectrum of X^- was investigated up to a mass of 2.5 GeV at incident pion momenta varying between 3 and 12 GeV/c. In this region two already known resonances, namely the ϱ^- and the A_2^- were found. They were produced with high statistics and excellent mass resolution allowing detailed studies of their properties. Apart from that, evidence for a new boson was found at 962 MeV with a rather narrow width. The existence of this particle, called $\delta(962)$, was then confirmed in 1966 at Saclay. In much higher mass regions four other bosons were registered, the R(1675), S(1929), T(2195) and the U(2382). All these data were taken in May and November 1965 and in January 1966 (see Fig. 5.16). We should however add here that this experiment and the results it delivered caused quite some confusion in the

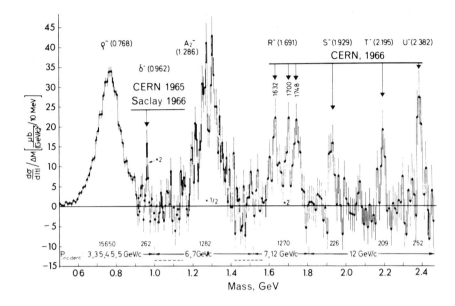

Fig. 5.16 Spectrum of bosons X^- of isospin $I = 1$ or 2 produced in the reaction[185] $\pi^- p \to pX^-$

scientific community. The most important point was the so-called splitting of the A_2. Indeed, as we can see in Fig. 5.16 the A_2 peak seemed to consist of two peaks. This observation incited a lot of theoretical speculations and a large number of papers were written in the following years on this subject. However any attempt to confirm this effect experimentally using an invariant mass method failed. It emerged later that Maglic's method of searching for resonances was very sensitive to the incident pion moment, which apparently simulated wrong results. This was also the case for the resonances in the higher mass regions.[186]

So far we have described experiments studying the creation of resonances in various reactions. Now let us turn to those investigating specific decay modes of resonances, namely ϱ, η and ω. Two groups were working on this topic. The first was that of Giuseppe Fidecaro studying the $\varrho \to \pi \gamma$ *decay*. Fidecaro had started to work on spark chamber techniques in the early 60's with the aim of measuring the $\Lambda\Sigma$ parity. However, unfortunately the first set-up built by his group was not sophisticated enough and the second, improved one, was ready too late, as the experiment had been carried out successfully in the meantime by a bubble chamber group at CERN. Due to the lack of machine time the counter experiment was stopped immediately, and the group decided to use their skills gained in the spark chamber technique to investigate a rare decay mode of the ϱ-meson. Their strength was that such an experiment could not be carried out with bubble chambers due to the low statistics of the process. In fact this decay mode was unknown until then, but on theoretical grounds there were good reasons for believing in its existence. The result obtained by Fidecaro's team was an upper limit for the partial width, namely $\Gamma(\varrho\pi\gamma) < 0.6\,\text{MeV}$.[187]

The last experiment we want to mention was carried out by a group from Frascati, Naples, and Trieste investigating *η and ω decays*. The η meson was a particularly astonishing object as it underwent various decay modes with very similar probabilities. To chose between the available theoretical explanations, one needed, however, more concrete information. Further the dynamics of the decay was connected to a variety of other interesting questions such as the validity of the unitary symmetry model, determination of coupling constants ($\eta\omega\omega$, $\eta\varrho\varrho$,...) and many more. Here again bubble chambers would be inefficient for the study of neutral η decays. To study the decay modes of the η in $\gamma\gamma$, $3\pi^0$ and $\pi^0\gamma\gamma$ a lead glass Čerenkov counter was used and they were found to be 41.6, 20.9 and 37.5% respectively of all neutral decays. Measuring the neutral ω decays relative to the $\omega \to \pi^+\pi^-\pi^0$ the figure found was 13.4%, so slightly higher than the prevailing average of 10.6%.[188]

5.4.1.5 Some tests for the unitary symmetry model and the search for quarks

After Gell-Mann and Ne'eman had suggested independently in 1961 a unitary symmetry model for the treatment of strongly interacting particles, a number of experiments were started at CERN to test its validity. However, as far as the counter groups were concerned they did not obtain as many results as initially hoped. Indeed in the period 1960 to 65 the two most striking proofs sustaining this theory both resulted from bubble chamber experiments. The first was the measurement of the relative $\Lambda\Sigma$ parity at CERN in 1962 and the second the finding of the Ω^- at Brookhaven in 1964. Nevertheless we shall mention the two experiments undertaken by the NP Division.

The first guided by Giuseppe Fidecaro was designed to measure the $\Lambda\Sigma$ *parity*. However for a number of reasons mentioned earlier they had to abandon the study in the middle of the experiment.[189]

A second rather fundamental check for testing the eightfold way was to measure the *parity of the Ξ hyperon* relative to the proton. This experiment proposed in 1963 by Carlo Rubbia intended to look at the reaction $K^-p \to \Xi K^+$, where the protons would be part of a polarized target. The (Ξp) parity was now related to the dependence of the differential cross-section on the orientation of the protons in the target. More precisely, if the relative parity was even (odd) the K^+ tend to be emitted on the right (left) of an observer oriented as the proton spin and looking along the direction of the incident beam. Throughout 1964 and 1965 several thousand events were recorded in a large spark chamber set-up, using a polarized target built at Saclay. However, no final publication of the results is known to us.[190]

Under this heading we want to add one further experiment namely the *search for quarks*. Although this was not an experiment carried out with the PS it was nevertheless a counter experiment within the NP Division. In 1964 Gell-Mann and Zweig had suggested the existence of particles having fractional electric charge and baryonic numbers and which were to be the building blocks of all mesons and baryons. This led to an immediate search for these new constituents of matter. This experiment, carried out by the group of A. Zichichi, never achieved positive results. Only upper limits for the fluxes of $(2/3)\,e$ and $(1/3)\,e$ were found, namely $5 \times 10^{-8}\,\text{cm}^2\,\text{sr}^{-1}\,\text{s}^{-1}$ and $1.5 \times 10^{-7}\,\text{cm}^2\,\text{sr}^{-1}\,\text{s}^{-1}$ respectively.[191]

5.4.2 ELECTROMAGNETIC INTERACTIONS

5.4.2.1 *Purely quantum electrodynamical problems*

Quantum electrodynamics (QED) created in the late forties, was and still is the best physical theory describing electromagnetic interactions.[192] However two points needed further clarification. The first concerned the relation between the electron and the muon. More precisely, in order to solve the puzzle as to why there were two particles behaving in exactly the same way and differing only very much in mass, physicists were trying to find a dissimilarity in their properties. The second point to mention was that, although the predictions of QED agreed remarkably well with the experimental results obtained at the time, physicists nevertheless felt that the occurence of infinite renormalization constants might itself reflect a failure of QED to describe correctly processes in which very small distances and high momentum transfers were involved. Several authors had tried to introduce non-locality in the QED, which led to a theory in which at least one of the fundamental postulates proper to QED was not valid, and to results differing by an order of magnitude according to the way this non-locality was introduced. Thus there was some interest in testing this theory at higher energies and two experiments were undertaken on the subject in the first half of the sixties at CERN.

We should first mention the group gathered around G. Giacomelli and D. Harting studying *electron–positron annihilation in the multi-GeV region*. The total cross-section of the reaction $e^+e^- \to 2\gamma$ was a good test of the validity of QED at small distances, and only electron machines and intersecting beams would offer better conditions than the PS for such an experiment. Thus using a scintillation counter and a lead glass total absorption Čerenkov counter the total cross-section of this reaction was measured at laboratory energies between 1.94 and 9.64 GeV. The values found agreed rather well with the theoretical calculations including radiative corrections. Further it was considered whether these results could be used to give a limit for the distance down to which QED seemed to be valid. However only a rather general conclusion could be drawn: within the limits of experimental accuracy the validity of QED was confirmed.[193]

Notes: pp. 250 ff.

However the biggest, the longest, and the most costly project in this direction proposed by Francis Farley was started in collaboration between the Nuclear Physics and Nuclear Physics Apparatus Divisions in 1963. The aim was to carry out a *precision measurement of the anomalous magnetic moment of the muon* and test in this way the validity of the predictions made by QED. It was meant to improve the results of the successful $(g-2)$ experiments performed with the synchro-cyclotron (for details see chapter 4.3.2.2). The new idea Farley suggested was to construct a muon storage ring consisting of a weak focusing ring magnet. It would have an equilibrium orbit of 5 m diameter near which the muons would be trapped and stored until their decay. Further a magnetic horn was constructed in order to focus the pion parents, as in the case of the neutrino experiment, and thus increase the number of muons in the ring. The energy of the stored polarized muons would be 1.3 GeV. In 1965 the magnet was delivered and installed in the South hall in the former bubble chamber area. It was fed via a PS beam extracted by a fast ejection system and guided by a beamline to the horn target. The experiments carried out in 1966 and 1967 delivered the most precise measurement of the $(g-2)$ factor of the negative muon, namely $(11\,666 \pm 5) \times 10^{-7}$. This was impressively consistent with the value of $11\,656 \times 10^{-7}$ calculated in QED.[194]

5.4.2.2 *Electromagnetic interaction and properties of hadrons*

Quantum electrodynamics concerns the interaction between the electromagnetic field and electrons and muons. Since many hadrons carry electric charge or magnetic moment, they also interact with the electromagnetic field—a fact which cannot be described in the framework of QED. However at the same time no quantitatively precise theory existed for the strong interaction. Thus, in order to learn about the electromagnetic properties of the proton M. Conversi and Antonio Zichichi suggested an experiment with the designation *PAPEP (PAPLEP)*. The precise aim was to study the *p*roton-*a*ntiproton annihilation into *e*lectron *p*airs (*lep*ton pairs) in order to collect information on the charge structure of the nucleons. The electromagnetic form factors, which furnish this information, could be investigated in three ways: by electron scattering as Hofstaetter had done in the USA, by p-$\bar{\text{p}}$ annihilation into electron–positron pairs, and by the inverse process. The electron scattering experiments measure the form factors for space-like momentum transfer, whereas the annihilation experiments measure it for time-like momentum transfer (see Fig. 5.17). However the last possibility mentioned, namely electron-positron annihilation was ruled out, as no operating storage rings existed with an energy sufficiently large to study this process. Thus clearly at CERN one could best study the second process (p$\bar{\text{p}}$ → e^+e^-) and the experiment was started in 1962.

Three ingredients determine the cross-section of the reaction p$\bar{\text{p}}$ → e^+e^-: the eγ

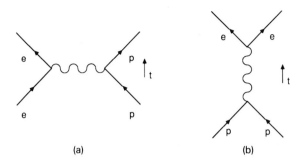

Fig. 5.17 a) electron–proton scattering, b) proton–antiproton annihilation

vertex, the photon propagator, and the pγ vertex. Thus measuring the cross-section and knowing rather well the first two items it would be possible to gain a lot of information on the nucleon–photon vertex. The fact that a nucleon's behaviour is not point-like is expressed by two functions, called Dirac and Pauli form factors, which would both be 1 in the case of point-like behaviour.[195]

The detector, a combination of spark chambers, scintillation and Čerenkov counters, was installed in a high-intensity, partially separated antiproton beam of 2.5 GeV/c. At this energy and under the assumption of a point-like proton the cross-section was calculated to be 240 nanobarn (1 nanobarn = 10^{-33} cm^2). If the number of detected e$^+$e$^-$ pairs was far lower than expected on the basis of this value, this could be interpreted as evidence against a point-like structure of the nucleon. Unluckily, however, the results did not reach the expected quality. The data taken during 1962 and 1963 only enabled the group to give an upper limit to the cross-section, though it did show some evidence for an electromagnetic structure of the proton in the time-like region.[196]

During 1964 this experiment was extended to a study of proton–antiproton annihilation into lepton pairs in general. The purpose here was not only to measure the form factor, but also to check the equivalence of electromagnetic interactions of the electron and muon for time-like momentum transfer. For this purpose a new electron telescope was constructed, consisting of five layers of plastic scintillator plus lead plus some spark chambers. Again no precise study of the processes was possible and as a result only an upper limit for the annihilation cross-section into lepton pairs could be given, namely

$$\sigma_{p\bar{p} \to \ell\bar{\ell}} < 0.54 \text{ nanobarn} .$$

This was 500 times smaller than the value expected in the case of point-like behaviour cited above, and therefore it was regarded as a demonstration that nucleons were far from being point-like particles.[197]

Notes: p. 251

The second experiment on this topic which should be mentioned was carried out by Guy von Dardel and his group and aimed at precision measurement of the *lifetime of the neutral pion*. At Caltech, Berkeley, and Frascati recent measurements on this property had been carried out, however with relatively high errors. Von Dardel's proposal said that if the lifetime was of the order of 2×10^{-16} sec then he could possibly reach a precision of 5%. The new method he suggested for this purpose would take advantage of the relativistic dilatation of the decay path. The neutral pions, produced in platinum foils of 3-60 microns exposed to the internal PS beam, would decay into photons partially producing electrons by conversion. The number of these converted electrons was thus a function of the π^0 lifetime and was measured with varying thickness of the foils with two specially developed 10 m long high resolution gas threshold Čerenkov counters.[198]

After a first run in January 1962, where quite a number of unexpected difficulties arose, the data-taking phase of the experiment was completed by August of the same year. It resulted in the most precise direct measurement of the neutral pion lifetime known until then and it remained so for many years. With a value of $(1.05 \pm 0.18) \times 10^{-16}$ sec it was about a factor two smaller than expected.[199]

5.4.2.3 Search for magnetic monopoles

To end the description of the study of electromagnetic interactions we would like to mention an experiment carried out to search for magnetic monopoles. Historically speaking, they had been predicted in 1931 by Dirac,[200] who had shown that one could construct a first quantization theory which contained as sources of the magnetic field, point magnetic poles besides electric poles. The values of these magnetic poles did not appear as a new universal constant, but were — if they existed — an integral multiple of $g = (137/2)\,e$. Although causing theoretical problems on the level of second quantization, seen from a more practical point of view these magnetic poles were not in contradiction with any fundamental law of nature.

The main properties of such *magnetic monopoles* were the following: high ionization energy loss, gain of 20.5 MeV/cm by an accelerating force of a kilo-oersted field, repelled by diamagnetic substances and attracted by paramagnetic ones. The hope of the CERN experiment, carried out by Giacomelli and his group, was to produce monopole-antimonopole pairs up to a mass of 2.8 GeV in nucleon-nucleon collisions. For this purpose, two different types of experiments were performed. In the first the monopoles were to be detected directly by means of scintillation counters after having been accelerated by solenoids. In the second, a counter telescope was used to detect the high-energy induced activity. Both of them delivered negative results. The only conclusion which could be drawn was that the upper limit for the production cross-section was $(10^{-36}-10^{-35})$ cm^2.[201]

5.4.3 WEAK INTERACTION PROCESSES

5.4.3.1 The study of K-mesons, CP violation and related questions

The main focus in weak interaction experiments with the PS was on the study of K mesons. Before entering this subject let us summarize shortly the state of the art in this part of elementary particle physics. Indeed the kaon occupied a rather peculiar position among elementary particles. The neutral kaon seemed to be capable of transforming itself into a neutral antikaon even though the latter had quite different strong interactions with other particles. This paradox was resolved by distinguishing between the objects formed by the strong interaction K^0 and $\bar{K}^0$, whilst those undergoing the weak decay were K_1^0 (K_S) and K_2^0 (K_L). When formed the K^0 was a certain combination of K_1^0 and K_2^0, whereas $\bar{K}^0$ was a different combination of the same particles. Now K_1^0 and K_2^0 differ in lifetime (10^{-10} and 10^{-8} sec respectively), decay mode and also very slightly in mass. As a consequence, the composition of a mixture of these two particles will also change in time and therefore also change the proportion of K^0 to $\bar{K}^0$.

Another striking observation was the so-called 'regeneration' of the K_1^0. When a beam of neutral kaons travelled for a long distance it soon contained only K_2^0's, owing to the decay of the K_1^0 component. However, as the K_2^0 consisted of K^0 and $\bar{K}^0$, and the latter was intensely absorbed in colliding with nuclei, the proportion of mixing between K^0 and $\bar{K}^0$ changed when passing through matter. In this way K_1^0 were 'put back' into the beam. This sequence of transformations could be written as[202]

$$K^0 = 1/\sqrt{2}\,(K_1^0 + K_2^0) \xrightarrow{\text{decay } K_1^0} 1/\sqrt{2}\,K_2^0 = 1/2\,(\bar{K}^0 + K^0) \xrightarrow{\bar{K}^0 \text{ absorption in matter}} 1/2\,K^0$$

In the period 1960–1963 only two experiments studying K's were reported. The first was undertaken late 1962/early 1963 by a CERN-ETH (Zurich) collaboration with Peter Astbury as group leader, who had constructed a large cloud-chamber of $170 \times 60 \times 40$ cm^3 in the early 60's.[203] The aim was to *study the K_2^0*. For this purpose the chamber was installed at the end of a high-intensity pion beam (a3) in the north hall. The K's were produced in the reaction

$$\pi^- p \rightarrow \Lambda^0 K^0$$
$$\rightarrow \Sigma^0 K^0$$

and their decay was then studied in the cloud chamber. In particular three decay channels were investigated, namely:

$$K_2^0 \rightarrow \pi^+ \pi^- \pi^0$$
$$\rightarrow \pi^\pm \mu^\mp \nu$$
$$\rightarrow \pi^\pm e^\mp \nu$$

Notes: p. 251

giving a branching ratio relative to all charged decays of 15:39:46.[204]

A second experiment, this time concerning the positive kaon decays, had been suggested by a group from Argonne led by Arthur Roberts, whose speciality it was to construct spark chambers for use in high-energy experiments. However no concrete results were reached.[205]

Thus the field of K physics at CERN was far from being a central research topic until 1964, when news from Brookhaven suddenly changed the whole situation. To study closely K_2^0 decays and interactions with the help of spark chambers and counters, a group from Princeton under the leadership of V.L. Fitch had proposed an experiment in April 1963 at the Brookhaven AGS. It started in June of the same year. A beryllium target was bombarded by the internal proton beam, giving K^0's among other particles. After a long travel time the beam would consist only of K_2^0 and neutrons. The assumption was that whilst K_1^0 nearly always decayed into two pions, the K_2^0 would have a three-body decay mode. A decay into two pions would be strictly forbidden by the fundamental principle of *CP* (charge conjugation and parity) invariance.

After nearly one year of data taking and evaluation, the result was definite. Christenson, Cronin, Fitch and Turley had observed the 2π decay of the K_2^0 with a rate of 2 in every 100 decays. This process thus violated *CP* invariance and strongly implied that time reversal invariance also did not hold. *CP* and *T* were interdependent through the invariance of *CPT*, which seemed firmly established for all interactions.[206]

As these new facts touched the most fundamental belief of physicists, other explanations were immediately searched for. The most simple proposal was that the measurements were wrong, which was ruled out rather quickly by an experiment at CERN. Another argued that *CP* would hold in some cases and not others. The third idea was that an unknown effect had transformed K_2^0 in K_1^0 mesons and the latter were decaying normally into 2π. This, however, had as a consequence the suggestion of a fifth extremely weak but long-range force acting differently on particles and antiparticles.[207]

These new facts stimulated both theoreticians and experimentalists also at CERN. Thus in 1964 a collaboration between a group around Klaus Winter and physicists from Orsay (Vivargent) decided to investigate closely the *2π decay of the K_2^0*. They were trying to find a possible energy dependence of this effect which might have as a source the 'fifth force'. More precisely, if physicists could find such an energy square dependence the theory of a new force would be proven and *CP* would stay invariant in K^0 decay. At Brookhaven the decay rate had been measured at 1.1 GeV/c, at CERN results at 10 GeV/c would be obtained.

The detector used for this purpose was a combination of spark chambers, a threshold gas Čerenkov counter, a 1 m bending magnet, a 1 m iron absorber and a number of scintillation counters to trigger the whole apparatus. From the results obtained three points should be retained. First, the results of the Brookhaven experiment were confirmed. Secondly, at 10.7 GeV/c the branching ratio for the two pion decay as compared to all charged modes was 3.5×10^{-3}, thus a factor 200 smaller than expected if a fifth force existed, i.e. one could not reconcile *CP* invariance and the 2π decay of K_2^0 with the help of an additional force. Lastly, a $K_2^0 \to \mu\mu$ decay had been observed and an upper limit of 2×10^{-4} for the branching as compared to all charged decay modes was given.[208]

In 1965 a programme on the $K_L \to \pi^+\pi^-$ decay was started by two groups with the aim of studying the interference between K_L and K_S amplitudes of the $\pi^+\pi^-$ decay. The interest in these experiments lay, on the one hand, in their being a sensitive check of the validity of the interpretation of the Christenson et al. experiment in terms of *CP* violation. On the other hand, if such an interference was found, it would permit a determination of the K_L-K_S *mass difference* as well as the *relative phase of the $\pi^+\pi^-$ decay amplitudes*.

One of the experiments carried out by the same group who had already studied the 2π decay[209] of the K_L was designed to measure the decay law of a coherent mixture of K_L and K_S in the $\pi^+\pi^-$ decay mode behind a regenerator. The second group lead by Carlo Rubbia[210] investigated the $\pi^+\pi^-$ decay as a function of distance from a plate of copper placed in a neutral beam at 33 m. The conclusion reached from both of these experiments was that there existed an interference term of K_L and K_S amplitudes in the $\pi^+\pi^-$ decay. Thus the interpretation of $K_L \to \pi^+\pi^-$ decay by *CP* violation was supported strongly. The mass difference between these two neutral kaons was measured at

$$\Delta m/\Gamma_S = 0.480 \pm 0.024 \quad \text{and} \quad 0.44 \pm 0.06$$

respectively. Further it was stated here that no strangeness changing neutral currents existed.

The same year the ETH–CERN collaboration undertook a cloud chamber experiment determining further features of the K_2^0 such as its *lifetime* and its $K_2^0 \to \pi^+\pi^-\pi^0$ decay.[211] The lifetime of $\tau = 61$ ns was found to be in agreement with the other values obtained until then. The same could be said about the branching ratio $\Gamma(K_2^0 \to \pi^+\pi^-\pi^0)/\Gamma(K_2^0 \to \text{all charged})$ which was measured to be 0.159.

Two further experiments were also started in 1965. The first, carried out by a collaboration between CERN and Rutherford Laboratory, investigated the $K_L \to 2\pi^0$ decay. They used parts of the ν set-up, i.e. counters and spark chambers, adding some 50 large scintillation counters. The second experiment which had also been stimulated by the apparent *CP* violation in K decay was done by a CERN–ETH–Saclay group. They studied the Dalitz plot of the $\eta \to \pi^+\pi^-\pi^0$ decay with the idea of checking the existence of a semi-strong or electromagnetic *C*-non-conserving interaction which would allow one to forget about *CP* violation.[212]

5.4.3.2 The helicity of the muon

An experiment to investigate the aspect of μ-e high momentum transfer collision, more precisely, determining the *helicity of the muon,* was undertaken by Bernard Hyams' group. Current theory of weak interactions with a single two-component neutrino and lepton conservation predicted a positive helicity for the negative muon observed in the rest frame of the parent pion. The method used here was to measure the spin dependence of the differential cross-section for the scattering of polarized muons on polarized electrons, by reversing the magnetization in the iron sheet to change the spin orientation of the electron. In February 1960 the group started with calculations on the feasibility and design of a high-energy, high-flux, polarized μ-beam, and in autumn the first beam tests were made. As far as the detector was concerned a sandwich counter was envisaged consisting of iron plates and scintillators, measuring electron and γ ray energies by total absorption. By April 1960 the design was settled and the construction started. The actual measurements carried out early in 1961 gave a value of $+1.17 \pm 0.32$ for the helicity and were therefore in good agreement with the theoretical predictions[213] of $+1$.

5.4.3.3 The beta decay of Λ^0

In 1962 Carlo Rubbia proposed an experiment measuring the *beta decay of the neutral lambda hyperon,* i.e. $\Lambda^0 \rightarrow p + e^- + \nu$, a strangeness violating process. All observed strangeness-conserving weak interactions could be explained in terms of a Universal Fermi Interaction based on a V-A current. While the conserved vector current predicted that there existed one universal value for the vector coupling constant c_V, the axial vector coupling constant was slightly modified from its 'universal' value due to renormalization effects. As far as strangeness non-conserving interactions were concerned, the case was not so well understood. For this reason the group intended to investigate the asymmetry in the angular distribution of the decay electron relative to the direction of the polarized Λ. In the context of V-A theory this asymmetry parameter α of the electron angular distribution was sensitive to the sign and to the magnitude in the ratio of axial (c_A) and polar (c_V) vector coupling coefficients. For that reason it could be used to derive information on the ratio c_A/c_V in strangeness violating beta decay.

The Λ's were produced in a Be target via the reaction $\pi^+ N \rightarrow K^+ \Lambda^0$ with a separated π^+ beam of 1 GeV/c. As the rate of beta decay was a factor 1000 lower than the decay into pions and nucleons, a high discrimination factor had to be reached by combining two dE/dx scintillation counters, a water Čerenkov counter and a thick-plate spark chamber for shower production and observation. With this set-up some 350 000 pictures were taken during 1963 and early 1964. The asymmetry parameter was determined to be 0.06 ± 0.19, in agreement with predictions of Cabibbo's theory which gave 0.02. As far as the coefficients c_A and c_V were concerned, the results of this experiment were combined with those found in the heavy liquid bubble chamber of the Ecole Polytechnique and led to a value of

$$c_A/c_V = -(0.9^{+0.25}_{-0.3})$$

also in good agreement with Cabibbo who predicted[214] -0.65 ± 0.1.

5.4.4 A TEST OF THE VALIDITY OF RELATIVITY THEORY

This experiment had a somewhat unique role in the experimental programme of the PS. It was meant to *verify the second postulate of special relativity* in the GeV region. This postulate stated that the velocity of electromagnetic radiation was independent of the motion of the source. Thus if the velocity of radiation $c' = c + k_v$, where v is the source velocity, then k should be zero if the second postulate holds. Until the early 60's mainly astronomical tests had been carried out and the terrestrial tests reported were non-conclusive. Then in 1963 T. Alvager, Nilsson and Kjellmann did an experiment measuring the velocity difference of gamma radiation emitted from a moving source C^{16*}. They came to the conclusion that $k = 0.1$.[215]

Now at CERN again Alvager and Kjellmann, together with Farley and Wallin, measured the velocity of γ rays emitted by rapidly moving π^0 mesons produced in the PS target. The result for the velocity measurement of γ rays of energy bigger than 6 GeV from a source moving with $\beta = 0.99975$ lead to a k value of $(-3 \pm 13) \times 10^{-5}$, which was a clear confirmation of the theory of relativity.[216]

5.5 The neutrino experiments at CERN: confronting the reality of big science (1960–1965)

5.5.1 THE STATE OF THE ART IN NEUTRINO PHYSICS (1960)[217]

The neutrino was postulated by Wolfgang Pauli in 1930 in order to explain the apparent non-conservation of energy in the beta decay of the neutron. This new particle had then to be electrically neutral, to have a very small mass (or even be massless), and to have spin 1/2. It had, however, a 'disturbing' feature: due to the small cross-section for inverse β-decay ($< 10^{-44}$ cm^2) it was then thought to be undetectable.

Pauli's idea was taken up by Enrico Fermi who formulated the first theory of β-decay in 1933—thus the first theory of weak interactions.[218] By analogy to photon–emission by an excited atom in the normal radiation process, he stated that e and v were created in the decay process, assuming that the interaction took place in one space-time point. Fermi's theory won an immediate success; it was adapted and extended in the following years in the light of new results, but remained untouched in its central core. A fundamentally new element appeared with the detection of the muon in 1946 as an additional member of the family of weakly interacting particles.

Notes: pp. 251 ff.

Further various experiments carried out between 1947 and 1949 showed that apparently all weak processes investigated to date were due to the same coupling constant g (a measure of the strength of coupling)—a very nice analogy with electromagnetic interaction. This led to the formulation of the so-called *Universal Fermi Interaction (UFI).*[219]

The next important step in the evolution of the understanding of the neutrino occurred years later in 1956 and 1957. Cowan and Reines, working at the Savannah River reactors, succeeded in demonstrating the existence of the neutrino via the inverse β-decay reaction $\bar{\nu}_e + p \rightarrow n + e^+$. Thus this particle could in no way be regarded any longer merely as a central ingredient in a well-functioning theory but had become a physical reality. What is more, these experiments using reactor neutrinos allowed a clear proof that the neutrino and the antineutrino were distinct particles.[220]

At the same time, new experimental results from cosmic-ray research put the theory of weak interactions, which had been applied so far satisfactorily, in front of a problem. Two particles, the θ^+ ($K^+_{\pi2}$) and the τ^+ ($K^+_{\pi3}$), had been detected, equal in mass and lifetime, though having different modes of decay (2π and 3π respectively). Now the question was: Were these two particles identical having 'simply' two different decay modes or were they distinct having different features only when decaying? The first case would mean that parity was not conserved, a rather 'strange' thought in a period where parity conservation was generally taken for granted.[221] However, the second possibility also did not seem very plausible. A solution to this dilemna was proposed by T.D. Lee and C.N. Yang in the same year. They stressed that, although until then parity conservation had been accepted generally, *de facto* it had only been proven for strong and electromagnetic interactions. Thus they suggested that parity conservation should be rigorously tested by experiments for the case of weak interactions.[222] And indeed in 1957 C.S. Wu and her collaborators succeeded in demonstrating the parity violation of weak interactions in their famous Co-60 experiment. Thus it had also become clear that θ and τ were the same particle.[223]

With all this new information it was nevertheless still not possible to lay down exactly the form of the 4-fermion interaction, a situation which would change in late 1957. In their talk on the 'Nature of the four-fermion interaction' at the Padua–Venice conference, Sudarshan and Marshak came to the conclusion that 'the only possible universal four-fermion interaction [was] an equal mixture of vector and axial vector interaction'. Thus the 'V–A theory' of weak interactions was born. We should add, that this same conclusion was also reached by Feynman and Gell-Mann, and Sakurai independently, although with a different approach.[224]

By 1958, after two rather turbulent years, a solution to the theoretical treatment of weak interactions seemed to be at hand. However, the confrontation with experiments soon revealed some major imperfections, of which two concerned directly the neutrino. Firstly, there was the reaction $\mu \rightarrow e\gamma$ which was consistent with all known selection rules, but nevertheless had so far not been observed. This was in contradiction to the 'philosophy' generally adopted by physicists, that anything which was 'allowed' to exist would do so. The second puzzling feature was the energy behaviour of the neutrino cross-section. It was assumed that the cross-section would rise with the square of the momentum, so that from a given momentum upwards the unitarity of the S-matrix would not be guaranteed anymore (implying a violation of the law of probability). Consequently, the theory of the universal Fermi interaction as it stood would hardly be applicable to high-energy reactions.

To explain the apparent non-existence of the reaction $\mu \to e\gamma$ theorists proposed to introduce two distinct neutrinos. One would be emitted in β-decay, thus being ν_e, the other would be produced in π-decay, therefore being ν_μ. Such a theory would also be attractive from the point of view of symmetry and systematics of particles, and would help to explain the difference in the nature of the muon and the electron. The problem of the ν cross-section was 'solved' by abandoning the idea of a point-like weak interaction. Indeed, in analogy to the electromagnetic interaction, where QED was applied with great success, weak interaction could be transmitted by an intermediate vector boson W. Further, on a more abstract level, one hoped to construct in this way a renormalizable theory—which Fermi's was not.[225]

The idea of using the new generation of accelerators at CERN and at Brookhaven as a source for high-energy neutrinos was proposed by Bruno Pontecorvo in mid-1959 and independently by Melvin Schwartz early 1960.[226] They explained that the pions produced when accelerated protons hit a target decay into muons and neutrinos, and as muons could be stopped by using massive shielding, only high-energy neutrinos would enter the experimental area. In this way both the two-neutrino question and the idea of the W boson could be tested experimentally. More precisely, if the electron and the muon–neutrino were different, physicists would only detect the reaction

$$\nu_\mu + p \to \mu^- + n$$

but not

$$\nu_\mu + p \to e^- + n \ .$$

As far as the W boson was concerned it could—if it existed—be produced via the reaction

$$\nu + Z \to W^+ + \ell^- + Z$$
$$\big|_{\to e + \nu}$$
$$\big|_{\to \mu + \nu}$$

and the lepton pair would be observed.

Notes: p. 252

5.5.2 THE FIRST NEUTRINO EXPERIMENT AT CERN

5.5.2.1 Discussing the feasibility—the period of scepticism (1960)

At CERN Pontecorvo's idea was taken up by Gilberto Bernardini, the Director of the SC Division at the time, and the man responsible for the preparation of the PS experimental programme. Fascinated by the possibility of using the proton synchrotron for studying ν interactions, early in 1960 he started to investigate the feasibility and the needs of such an experiment. At first sight the project depended on three key elements: the magnitude of the neutrino flux, the shieldability of the background coming from the machine and from cosmic radiation, and the size of the detector.

However, the first rough estimates were rather discouraging: the neutrino flux available would be very low, the shielding would need thousands of tons of iron and concrete, costing about 1 MSF, and only a detector of some 10 tons would permit a decent detection rate. Nevertheless, for Bernardini and the physicists around him there were a number of good reasons for pressing on. One main factor was the fascination not only of going to a higher energy, but of entering completely unknown territory in particle physics. Secondly, the community of theoretical physicists became increasingly interested in this experiment. Thirdly, American physicists were also seriously investigating the possibility of doing an experiment of this type with the AGS, their analogon to the CERN PS. Lastly, and apart from all this, the neutrino experiment was a rather prestigious scientific enterprise and the hope that, as had been the case for the antiproton experiment at the Bevatron, it would be a good candidate for a future Nobel prize was clearly present.[227]

Estimating the neutrino flux to be roughly 1 event/ton of detector/day (based on Hagedorn's calculations), and knowing that the shielding required more detailed background measurements, Bernardini's primary concern at this point turned to the detector. After investigating a multitude of possibilities he came to the conclusion that the ideal solution would be the construction of a huge spark chamber and as he had no staff with the necessary know-how at CERN, he took up contacts with experts in the United States and Great Britain. However at this stage his plans for such a device did not bear fruit, for a complex set of reasons that we will discuss later. Instead, in the second half of 1960, it was decided that for a quick experiment, detection devices which were at hand or under construction had to be used, i.e. the 1 m Ecole Polytechnique heavy-liquid bubble chamber of A. Lagarrigue and (or) the 1 m CERN heavy-liquid bubble chamber of C. Ramm. The project was also joined by a counter group under H. Faissner.[228]

On 27 May 1960 the experiment was presented for the first time to the Scientific Policy Committee (SPC) where it was referred to as 'very promising'. However, though it had gained official status, it was not supported enthusiastically by the organization at all. It was in a sense regarded by many physicists as a crazily ambitious venture, and on the administrative side one was reluctant to tie up a lot of money, manpower, and machine time in one single rather risky project.[229]

5.5.2.2 The period of optimism: September 1960 – May 1961

A change in the general attitude vis-à-vis the neutrino experiment took place in the last third of 1960. It was triggered by an ensemble of independent events, which forced physicists as well as administrators to take a clear position in the neutrino discussion. Firstly, in the autumn of 1960 Gilberto Bernardini had presented CERN's preliminary neutrino programme at the Rochester Conference, where it was regarded with utmost interest and enthusiasm by the American physics community. Then in the summer of 1960 the Brookhaven AGS was completed, which meant that CERN was no longer in the position of having the biggest accelerator in the world and shortly afterwards at the Berkeley Instrumentation Conference Lederman, Schwartz, and Gaillard presented their project for a 10 ton spark chamber for a neutrino experiment with this new accelerator. Finally in September 1960 Krienen, Steinberger, and Salmeron published the first detailed calculations on the neutrino flux to be expected at the PS and the shielding necessary. They reached the conclusion that *1 event/ton of detector/day* would be the expected value, depending very little on the target position. As far as the shielding was concerned they calculated 4000 t of heavy concrete and 650 t of steel.[230]

With people convinced of the importance of the experiment *and* of its feasibility, a new element—the timing—entered into consideration. If one wanted to use the small advance in time over Brookhaven, one had to react quickly. The interest lay, as Leprince-Ringuet explained, in 'a quick experiment using all possibilities', and he added that CERN was in a 'privileged situation and should profit from it'.[231] In this wave of optimism the first neutrino experiment was planned for February/March 61—a date already revised by November 1960. By then it was decided that the neutrino experiment should be carried out in two stages. The first would be a combined bubble chamber-counter experiment. It would take place in about June 61, and should give in 2–3 weeks a first rough answer to questions about the existence of the two neutrinos and the W-boson. Based on these results a long-term ν research programme would be built up. It would involve the construction of a huge neutrino detector, for example a spark chamber.[232]

In the meantime, in the course of the reorganization of CERN, a body named the Directorate was founded to assist the Director-General. It started functioning late in 1960 and initially had comprised the following members: John Adams (DG), Gilberto Bernardini and Victor Weisskopf for Research, Samuel Dakin for Administration and Mervin Hine for Applied Physics. And as a lot of importance was attached to the success of the ν experiment, it was, contrary to all other experiments, put directly under the supervision of this body, Bernardini and Weisskopf being responsible for the physics, Hine for the more technical aspects.

Notes: p. 252

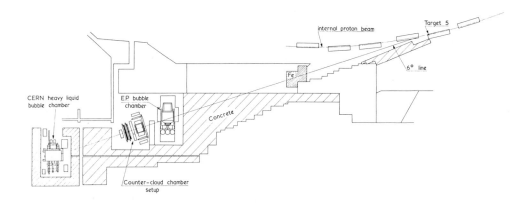

Fig. 5.18 The layout of the first neutrino experiment[234]

The last two months of 1960 were marked by vigorous discussions on the exact layout. We have to keep in mind that no less than two bubble chamber groups and one counter team were now involved, all *a priori* interested in optimal conditions for their respective group. This resulted in a decision taken by the Directorate on 12 December 1960 containing four major points: The first concerned the layout. The solution adopted foresaw Lagarrigue's chamber close to the target, followed by the counter-cloud chamber set-up of Faissner's group and lastly Ramm's bubble chamber (see Fig. 5.18). This promised the best overall conditions for such an experiment. The three other points stated were more of a policy character. They underlined that the experiment should start as early as possible, that it should aim at statistics and not at simply recording a few events, and lastly, the most important point, the neutrino experiment would have *priority over all other experiments* in the first half of 1961.[233]

In January 1961 busy preparations got under way. A very tight time schedule was worked out and the importance attached to this experiment made things which had been impossible until then now seem possible.

5.5.2.3 *The first setback: von Dardel's measurements and its consequences (May 1961)*

In March and April 1961 the first background and engineering preruns for the Ecole Polytechnique chamber were made. The building of the huge shielding was well under way, and last efforts were undertaken to finish the multiplate cloud chamber and the CERN heavy-liquid bubble chamber. Nothing seemed to be able to hinder the success of this experiment.[235]

However in the midst of all these preparations, just one month before the actual start of the experiment, bad news spread rapidly through the organization. Guy von Dardel, at that time not a member of the neutrino group, had measured the π^+-flux at 6° coming from the target placed in the middle of straight section 1 — thus simulating the conditions for the neutrino experiment. His results were alarming: apparently the π-flux was an order of magnitude smaller than calculated. He concluded that 'the experiment [was] not feasible under those conditions with present machine intensity and with detectors containing only a few tons of matter'.[236]

The shock inside the organization was tremendous; the neutrino experiment had become unfeasible under these conditions. A full year of hard work, of devotion, of fighting for manpower, space and machine time, and of dreaming of a really great success had vanished into nothing. At the same time the confidence of the administration in the 'neutrino people', built up with so much difficulty, had been shaken. Von Dardel described his feelings in a letter to Weisskopf about the situation using rather drastic terms: 'My first reaction was one of complete unbelief and shock that the top priority experiment of CERN, one of the biggest and most complicated and expensive, and important experiments ever done in nuclear physics, which involv(ed) three physics groups with two bubble chambers, one cloud chamber, and enormous counters, which (had) taxed to the utmost a big organization's resources in money, manpower, shielding blocks and installations, and which is directly under two of CERN's directors, should be so badly prepared that a complete outsider like me, can (...) show an error of almost an order of magnitude in the intensity, which is the most vital parameter in the experiment'.[237]

Some last efforts to save the experiment were suggested, but none of them offered a real solution. The experiment as it stood had to be abandoned and it was clear that now Brookhaven would be first.

5.5.3 CONCEIVING A COMPLETELY NEW NEUTRINO PROGRAMME (MID-61 TO MID-63)

The race with Brookhaven concerning the neutrino experiment being definitely lost, one realized that the only sensible thing to do — being convinced of the importance — was to start again from the very beginning and draw up a completely new programme. Thus on 21 July 1961, at the 20th meeting of the Scientific Policy Committee, this experiment was widely discussed. It was decided that the future programme should contain two crucial improvements: a beam of higher neutrino flux reached by a focusing device for pions and an ejected proton beam, and a new multi-ton detector. Further the shielding should be optimized by using steel and cast-iron blocks. Reviewing the situation at Brookhaven, it was stated that they 'had put practically all their resources into the qualitative part (...), intended to obtain a quick answer' to the two-neutrino question. CERN's strength, by contrast, should be a very sophisticated, elaborate, and efficient set-up, providing optimal working conditions for a whole series of experiments to be performed with neutrinos. To guarantee the smooth organization of this new programme which would involve a large number of different groups from different divisions the Directorate decided early in 1962 to appoint Guy von Dardel as interdivisional group leader for the neutrino experiment.[238]

In spring 1962 the first data were collected at Brookhaven and indeed they suggested that there existed two different neutrinos as expected and by June of the same year the final results were submitted for publication.[239] Thus when the CERN Council met on 13 June 1962, it was confronted with the reality that Brookhaven had indeed succeeded in solving this fundamental question. After discussion on the why and the how of this failure of CERN, it was agreed that it was now important for the organization to be offensive and to try to succeed in the further steps. CERN's neutrino programme would be a long-range programme supposed to deliver results far superior in quality to the American ones.[240]

These general ideas about the future were put down in a memo by Victor Weisskopf, acting Director General, to the neutrino staff. He stressed that it had now become obvious that neutrino physics was an essential part of high-energy physics and that there were still a number of very important questions, in particular the existence of the intermediate vector boson W, to be answered. Entering into more detail, he underlined that CERN should aim at having the most intense and flexible neutrino beam together with the optimal detection equipment. However he warned that all this should be done as quickly as possible, as it was of 'no use either to perform a good experiment after it has been done elsewhere'.[241] It was in this spirit that the preparations were carried out.

5.5.3.1 Two new techniques to improve the beam performance

The idea of constructing a device to increase the neutrino beam intensity was formulated by Simon van der Meer early in 1961 and published as a CERN Report under the title 'A directive device for charged particles and its use in an enhanced neutrino beam'.[242] The author had realized that the average production angle of pions and kaons in the laboratory would be greater than the decay angle and that therefore the intensity could be increased by focusing these particles before their decay. This *'neutrino horn'*, as it was called later, was based on an analogy to an internally reflecting conical surface in geometrical optics. One knew that here, after each reflection, the angle between the light ray and the axis decreased by an amount equal to the opening angle of the cone. Similar effects could now be reached with charged particles in a magnetic field of the shape presented in Fig. 5.19.

The horn consisted thus of two co-axial cones (2.5 degrees semi-angle) made out of conducting material through which a current passed (300 kA in the case of CERN). Between the two surfaces a magnetic field was built up. Since its strength was inversely proportional to the distance from the axis, the central cone was free from any magnetic field. The target, positioned in the neck of the horn, had to be several

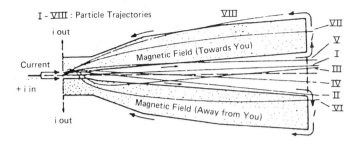

Fig. 5.19 Principle of a magnetic horn (cross-section)[244]

interaction lengths long but very narrow, in order to avoid secondary interactions of particles produced. Thus a particle coming from the target (mostly pions and a small mixture of kaons) would enter the magnetic field between the cones and be deflected back into the field free region, though with an angle to the axis smaller than before. Summarizing one could thus say that the magnetic horn 'transforms a real particle source of small spatial dimension and large angular spread out into a virtual one of large dimensions and smaller angular spread'.[243]

This device would improve the neutrino experiment in two ways. Firstly, the concentration of pions in the direction of the detector, and therefore also the neutrino flux, would increase by an order of magnitude. Secondly, the fact that the horn focused either positive or negative pions (or kaons) and removed those of opposite sign nearly completely, made it possible to choose between a neutrino and an antineutrino beam by simply reversing the excitation current. The pureness of these beams was very high though contamination of a few percent was produced by rare decay modes of the K-mesons.

Two neutrino horns were constructed in the period covered in this section. One gave a maximum energy of 9 GeV, the second of 11 GeV. The latter was built with the view to extending upwards the lower limit of the intermediate vector boson mass.[245]

The second innovation — of crucial importance as the magnetic horn could only work with an external beam — was a *fast ejection system* proposed already in December 1959 by B. Kuiper and G. Plass.[246] By fast ejection was understood the extraction of particles from the accelerator in a time interval of the order of or shorter than the revolution time of the protons in the machine (in the case of the CERN PS 2.1 μs), as distinct from slow extraction, which would provide particles over time intervals of the order of milliseconds.

To summarize the main idea: Due to phase-focusing, the proton beam in the CERN synchrotron was in fact split into 20 so-called bunches along the 628 m circumference of the ring. Each of them had a length in time of roughly 10 ns

Notes: p. 253

(roughly 3 m) and 100 ns of space in between. For extracting these bunches of particles out of the PS one had to place in a straight section between two synchrotron-magnets an additional magnet unit, the kicker-magnet. A quick rise (< 100 nsec) of the magnetic field in this unit would allow a deflection of the passing bunch. The latter would oscillate back performing betatron oscillations. Thus a further magnet unit, called a bending magnet, had to be placed after a quarter wave-length of this oscillation. It was designed in such a way that only the deflected bunch was influenced by its magnetic field and was finally guided out of the synchrotron field (see Fig. 5.20).

Both kicker and bending magnet were moved into their place by hydraulic actuators just before the moment of ejection and therefore very precise timing was of the utmost importance. Once the bunch of protons had left the synchrotron it was guided by quadrupoles and bending magnets to the target. Since little space was available near the synchrotron these magnets had to be of specially small size, and in order to hit the very narrow target the beam had to have a spot size of a few mm, which was reached with quadrupoles of high field gradient (6 kG/cm).

This method of fast ejection would have two major advantages. First, it allowed for a high-intensity external proton beam, so that all problems of small angle scattering vanished. Secondly, it made it possible to install a beam-sharing system as

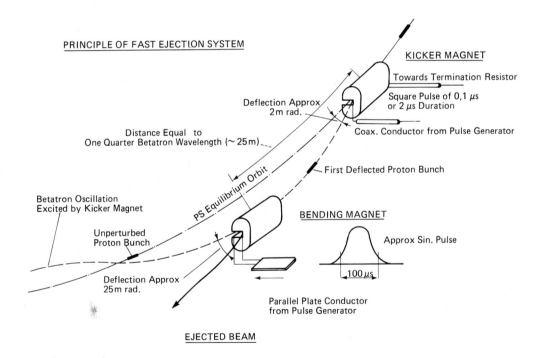

Fig. 5.20 Schematic drawing of the fast ejection system[247]

one could extract the desired amount of bunches of the circulating beam at different positions.

On 12 May 1963 the full PS beam was ejected for the first time and virtually no deviations from the predicted trajectory were seen. By the middle of 1963 the fast ejection system worked with an extraction efficiency of 95% in a time of 2.1 μsec and at the end of the pulsed beam transport system the beam had a focus spot of 2 mm diameter.[248]

5.5.3.2 The new detector: a multi-ton spark chamber

Having achieved in this way the world's most intense beam of neutrinos, CERN also decided to invest in the construction of a massive detector, namely a spark chamber. The choice was based on the fact that such a device was relatively easy to build in any size, it would provide high statistics, and it could be triggered by counters, thus eliminating unwanted events right from the beginning. The only disadvantage compared to bubble chambers was that less precise information about one single event could be gained. Thus it was decided that Ramm's bubble chamber would be run in parallel to the spark chamber to study the details of neutrino events. It would be filled with freon and have an effective weight of 0.75 t.

As we have mentioned, before Bernardini had already suggested the use of spark chambers in early 1960. However first tests were only made by mid-61 and then discussion on the detailed design began. As the central idea was to develop a device as flexible as possible, it was decided to construct modular sub-units, which could rather easily be arranged in different ways. By end-61 the construction of these sub-units, i.e. three-plate spark chambers, started. They were either made of aluminium or brass, filled with a mixture of 30% neon and 70% helium, and closed by plexiglass optical windows. The dimension of each module was 1 m $\times$ 1.6 m $\times$ 0.035 m. By mid-62 the first batch was tested satisfactorily.[249]

Regarding the final detector as a whole it consisted of three regions. The first was a low-density 'production region', built of aluminium and brass modules with counters in between. The aluminium chambers were meant to give detailed information on complex events, the brass chambers were to ensure the conversion of electrons and photons due to their higher density. This section was followed by a magnet, a Helmholtz-type pair of coils able to produce a field of 4 kG on a gap 1.7 $\times$ 1.1 $\times$ 1.2 m^3, with 7 aluminium spark chamber units sampling the paths of the particles. It enabled the determination of sign and momentum of particles (up to 10 GeV) leaving the first region. The third and last part was a thick walled 'range region', made out of spark chamber modules with lead or iron walls of increasing thickness (5–20 cm) between them. It allowed discrimination of muons from strongly interacting particles. Since each spark chamber module was only 1 m wide, two rows of them were mounted side by side in both the production and the range region. The set up was triggered by counters and photographed from the side by pairs of stereo- cameras.[250]

Notes: p. 253

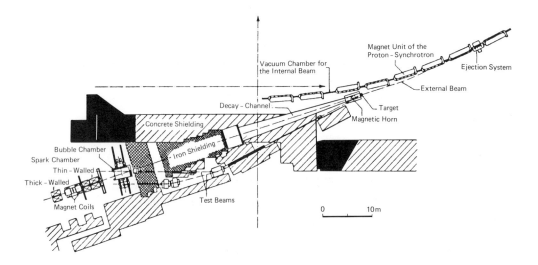

Fig. 5.21 The layout of the 1963 experiment[251]

After the first series of experiments performed in 1963 the detector was redesigned for the search of the intermediate boson. The main improvement for the 1964 experiment was a magnetized iron spark chamber to determine the spin of the muons replacing the magnet in the previous set-up. This part of the detector would therefore be both analyzing the particles with energies up to 15 GeV emerging from the first region and a region where very penetrating particles were produced. Secondly, two slabs of magnetized iron were placed at the end of the range region, permitting the measurement of the charge of particles leaving the detector. Lastly, the arrangement of the spark chamber modules in the production region was changed.[252]

5.5.4 RESULTS FROM THE 1963 AND 1964 EXPERIMENTS[253]

Two runs were undertaken for the neutrino experiment at CERN: one from June to August 1963 and the other from February to May 1964, each lasting about 30 days.

The first question investigated at CERN was to what level the assumption that there were *two distinct neutrinos* held. The Columbia-Brookhaven group had based their conclusions on only 29 muon events versus 6 electron events and had worked with a rather high background. At CERN the American result was confirmed with a rather impressive statistical accuracy. In the case of the bubble chamber 459 events were analyzed, of which 454 contained a negative muon and only 5 a negative electron with energy higher than 400 MeV. That meant that

$$R = \frac{\nu_e + n \to p + e^- + \text{anything}}{\nu_\mu + n \to p + \mu^- + \text{anything}} < 1/100 .$$

The number of detected electrons was consistent with the ν_e contamination of the beam coming from K_{e3} decays. The spark chamber results were less good, as discrimination of confusing events was rather difficult.

The next point to clarify was *to what limit ν_μ acted differently from ν_e*. Indeed, assuming that UFI holds, the same phenomenology was expected for reactions due to electron neutrinos as for those due to muon neutrinos; the ratio of elastic electron to muon production was thus measured. Altogether, 39 elastic electron and 418 elastic muon production events were registered in the spark chamber. With the given beam-composition and the respective detection probabilities in the spark chamber, the following prediction could be made assuming that there are two different kinds of neutrinos and that all electron neutrinos in the beam came from electronic K decays:

$$R = \frac{\nu_e + n \to p + e^-}{\nu_\mu + n \to p + \mu^-} = 0.6\% .$$

If there were only one neutrino this ratio should be roughly 1. Assuming a neutrino-flip hypothesis, i.e. that there are two distinct neutrinos but $K_{\mu 2} \to \nu_e$, the result should be 8%. Experimentally the value was found to be $(1.7 \pm 0.5)\%$. This virtually ruled out the neutrino flip hypothesis, and it was concluded that the theory of UFI held up to a momentum transfer of 1 GeV, and that a new kind of selection rule forbids an e lepton–μ lepton transition.

This brings us to the next topic, namely the test on *lepton conservation*. Until the Brookhaven experiment it was supposed that muonic and electronic leptons would carry the same lepton number, namely +1 for leptons and −1 for antileptons, obeying the additive selection rule, i.e. the number of leptons minus the number of antileptons was always constant. Now if $\nu_e \neq \nu_\mu$ it was found that this selection rule was limited to the e-leptons and therefore it was assumed that there were two distinct leptonic quantum numbers for e-leptons and μ-leptons each independently obeying an additive conservation law. This fact was tested at CERN by determining the sign of the muons produced by a ν_μ beam. If they were all negative this would show that the lepton number was conserved. In practice this meant measuring the sign of all charged particles with momentum bigger than 600 MeV/c which penetrated the magnetized iron slabs at the end of the set-up. Only $0.027 \pm 0.006\%$ of positive tracks were observed which is within the purity of the beam—a rather clear proof for the conservation of μ-lepton number.

Notes: p. 253

A fourth center of investigation was the *form factors due to strong interaction*. The form factor is of the type

$$F_i(q^2) = (1 + q^2/M_i^2)^{-2}$$

where q^2 is the four momentum transfer and M_i is the cut-off mass. The cut-off mass in the isovector form factor M_V was known as 0.84 GeV from electron scattering experiments. Fitting the histogram for the angular distribution of elastically produced muons with various values for the cut-off in the axial vector form factor (under the assumption of point-like interaction ($M_w = \infty$)), the spark chamber results indicated that M_A should be between 0.5 and 1.5 GeV. Bubble chamber picture analysis permitted a more precise result namely

$$M_A = (1.0^{+0.5}_{-0.3}) \text{ GeV}.$$

Further, by measuring the energy and emission angle in elastic electron reactions it was possible to obtain a q^2 distribution. Fitting this histogram with various values for the axial vector form factors physicists found that M_A for electrons and muons were roughly equal, which showed again that the μ-e universality held at least up to a momentum transfer of ~ 1 GeV.

By far the most interesting subject was however the *search for the W-boson*. As this particle would have a life-time shorter than 10^{-18} sec, physicists had to look for the decay products. In fact one could consider charged lepton pairs ($\mu^-\mu^+$ and μ^-e^+) as a signature of the production and leptonic decay of the intermediate vector boson (see the equation in section 5.5.1). In the first case one should have two visible tracks longer than 1.5 and 0.8 Λ_0 (interaction length) respectively, and the projected angles of the two tracks with the neutrino direction should be smaller than 45° in both stereo views. 350 events fell into this category. Now assuming that one of the tracks could be a strongly interacting particle (p, π, ...), and knowing the interaction length in each material, it is possible to calculate the expected interaction rate. Comparing it with the number actually observed they were found to be equivalent, permitting the conclusion that none of the observed events contained two muons and therefore no W boson had been observed. However it was not possible to determine a lower limit to the W-mass like that.

Another possibility would be to look for events which showed two non-interacting tracks, one with positive sign longer than 7 Λ_0 and one with negative sign longer than 2.4 Λ_0. No such event was found. The expected number depended on the production cross-section (function of M_w), the detection efficiency, which was roughly 11%, the neutrino spectrum at high energy, and on the branching ratio between leptonic and non-leptonic decays R which was assumed as 1. The number of expected events calculated in this way lay between 4 and 51 events.

Looking for muon–electron pairs one would expect, for $M_w = 1.8$ GeV and

$R = 1$, between 6 and 16 events. Less than 3 were observed. This led to the conclusion that $M_w \geq 1.8$ GeV if $R \geq 1$.

While the spark chamber was particularly good for detecting muonic decay of the W, the bubble chamber delivered better results for the electron decay channel. Some 900 selected events were analyzed, and only one candidate for an e-decay, which was assumed to come from the $\bar{\nu}_e$ background was found. If $M_w = 1.8$ GeV and the branching ratio for e decay was 50% one would have expected some 2.5 events. From these results it was concluded that the W mass should be larger than 1.8 GeV unless the pionic decay mode was dominant.

Lastly, we would like to mention the investigations on the existence of *neutral lepton currents*. Although one had detected no signs for neutral leptonic currents in the low energy region, there was no plausible reason why they should not exist. This meant that one was looking for reactions of the type

$$\nu + p \rightarrow \nu + p$$

or

$$\nu + p \rightarrow \nu + n + \pi^-.$$

However this proved very difficult as there was a high background of fast secondary neutrons. The only possible result which could thus be obtained was a relatively poor limit of the ratio neutral to charged current

$$\frac{\nu_\mu + p \rightarrow \nu_\mu + p}{\nu_\mu + n \rightarrow \mu^+ + p} < 0.03 \text{ (or 0.2 taking into account the scanning efficiency)}$$

5.5.5 WHY BROOKHAVEN AND NOT CERN?

On the occasion of the 21st Session of Council on 13 June 1962 a rather disturbing question was spoken out officially: 'Why did we (CERN) not succeed when (Brookhaven) succeeded so impressively?'.[254] The interest here is of course not to pass a judgement on the people involved in this experiment, but rather to try to understand the conditions which led to this, and to throw light on the more general kind of difficulties CERN had to face when it first started. What is more we should say that the ν experiments done from 1963 onwards were rather successful, delivering interesting and good results. Now concerning this initial 'failure' we identified three main forces: the first, and perhaps the most important one, a complete lack of experience of European physicists in big science research; secondly, a very complex relationship to the Unites States, leading to a fierce competition, and hindering up to a certain extent a 'normal' development of this experiment; and lastly, CERN being an intergovernmental organization and having Europe's biggest accelerator in addition weighed heavily on the way of carrying out research.

Notes: p. 253

Let us explain in some detail what we mean exactly by *lack of experience*. For most of the scientists, technicians, and administrators involved in this venture CERN, the main aim until 1960 had been the completion of the big accelerator, and that, if possible, before the competitor, the Brookhaven AGS. They did not fully realize that 'big science' was not just the construction of bigger and bigger machines, was not just a scaling-up process, but would create different needs, demanded a different timing, needed different organizational procedures, and implied huge amounts of money for exploiting these big machines. Virtually none of the physicists involved had any real experience in this direction. For many of them the biggest machine they had worked on was the synchro-cyclotron of 600 MeV, where most of the experiments were small compared to those at the PS.

The consequences of this were manifold. Firstly, the size of this experiment, the time it would need for preparation, and particularly the load of work it involved was grossly underestimated in the beginning. Bernardini, in his enthusiasm, explained in various letters early 1960, that for him the neutrino experiment seemed absolutely feasible, he estimated that it could start late in the year and that results could be obtained in a few weeks of running. He, and many others with him, were not worried about the fact that even the simplest set-up, would mean moving the detector to its place and testing it, purchasing and erecting the shielding, and adapting it according to test measurements, and finally establishing the optimal tuning for the accelerator. All this required in reality more than a year and large amounts of manpower from outside the neutrino group, which obviously were missing elsewhere during this time and paralyzed the preparation of other experiments.

Secondly this unexperience led to problems in organizing the experiment. To allow an immediate start it was decided to use a sequence of no less than three detectors — two bubble chambers and a counter-controlled cloud chamber — indeed a rather complex set-up. It meant that the timing had to be perfect to meet the same deadlines in all three groups, as none of the equipment was ready in 1960, and that the probability for difficulties was multiplied by three. But much more important than this, it meant merging three groups into one team of altogether 30 scientists. Compared to the Brookhaven team of 7 physicists this was enormous. It involved a heavy organization and also entailed various problems of sociological nature. Structural tensions appeared, which were due to the fact, that although the team had to succeed above all, nevertheless within the team each group tried to reach optimal conditions for themselves. This is rather well illustrated by the spirited and lengthy discussion on the layout of the experiment, where the choice of the arrangement of the detectors was an overall optimization, which meant in this particular case a disadvantage for one of the bubble-chamber groups. Another serious consequence of putting together such a big group of rather unexperienced physicists was pointed out by von Dardel. There is 'a tendency' he explained, 'for everyone to concentrate on his little part of the total experience, and to assume that somebody else cares about the overall planning'.[255] Lastly, in a collaboration of that size there was little possibility of following an individual approach to any problem, which was not appreciated by many physicists who were not used to working that way. As John Adams explained in an interview in the early 60s, 'organizational procedures, which are absolutely essential for experiments on these machines, were abhorred by many European physicists, and so a lot of time was wasted on discussions of the significance and working methods of physics'.[256]

After the problems of timing and organizing the experiment we have just mentioned, the third important point was the decision-process on the large scale equipment that was needed. In Brookhaven, when it had become clear that neutrino physics with accelerators would be a central research topic, the construction of a 10 ton spark chamber was launched in mid-60 and was finished a year after, just in time for the experiment. At CERN nothing of this kind was started before the end of 1961. The reasons are twofold. First, the construction of the PS had costed much more than foreseen and the Member States had been promised that after completion the costs would not increase. Unfortunately, though, no money had been budgeted for such a big project. Secondly, as in the case of bubble chambers, the construction of which was started at CERN only after long struggles and much too late, there was a certain reluctance to enter this completely new technology. The spark chamber technique was still regarded as too unreliable, too little experience had been gained with it, and there was unfortunately also not enough confidence in the know-how of CERN staff in this field. Even as late as January 1961 John Adams insisted that there 'is no group with sufficient staff for such a project and there is no budget in CERN allocated for such a major piece of equipment'.[257] In fact, one waited at CERN for the results of the American project and only if it was a success, would one be ready to really invest in the same direction. Although this was the only promising technique for very big detectors, although some influential physicists at CERN tried to push this programme, and although several groups were building very small spark chambers, CERN was not ready to take a risk, and preferred to stay in a waiting position.

So much for the lack of experience and its consequences on the neutrino experiment. The second determining factor in the failure of this experiment was the *complex relation to the United States*. With the construction of the PS, European high-energy physics was for the first time on an equal level with the USA, at least in terms of machine energy. Now it was up to the European physicists to prove that they were capable of contributing to the forefront of research. Further they could not disregard the fact that in 1959 Segre and Chamberlain had been awarded with the Nobel Prize for the detection of the anti-proton and some physicists had the more or less explicit hope that this could also be the case for the detection of the second kind of neutrino or of the W. This made the neutrino experiment an excellent possibility for competition. The race between Brookhaven and CERN was further underlined by the fact that CERN gave a special status to this experiment in putting it directly under

Notes: p. 253

the Directorate, and in attributing top priority to it. The pressure to succeed became enormous and caused a certain blindness vis-à-vis many problems. As Helmut Faissner put it in 1964, a kind of 'neurosis had seized us: a 'big discovery' complex'.[258]

However CERN's relationship with the USA was twofold. Brookhaven did not only play the role of a competitor, but also that of reference frame, and source of expertise. As we have seen in the case of the spark chambers one was watching the developments in the USA before launching a similar effort, simply being afraid to invest in the wrong direction, to be misoriented right from the beginning. Bernardini expressed this in a rather bitter way: 'I consider this attitude as regrettable in so far as ideas are only taken up when they have received their blessing at Rochester. I had hoped that CERN would become a place where Europe would develop its own style in accelerator physics in accordance with its tradition and culture, emphasizing originality as distinct from the development of physics towards mass production'.[259]

As a result a strong will to win the race goes along with a conservatism which has just the opposite effect.

This leads us to the last point, namely the *way CERN was organized* and its influence on this experiment. CERN's multinational character puts the organization inevitably under great pressure to provide a service that would be, above all, reliable, and accessible to all countries. To a certain extent this caused a kind of conservatism hindersome to spontaneous developments, and often quoted as the reason why CERN on many occasions was late compared to American labs. The need of balancing national interests combined with a lack of experience on this terrain, resulted in a tendency to a rather heavy committee structure, and meant being overcautious in commitments to very new ideas, foreseeing rather slow and complicated decision processes, staying too long in a waiting position, and trying to proceed by the most precise and secure way, which was often far from being the most efficient and quick one. On the other hand we have to keep in mind—and this should in no way be understood as an excuse—that both scientists and administrators had to build up CERN from nothing. There was no tradition in experimental high-energy physics and thus they had to learn to draw up a scientific programme, set aside enough money for carrying it out, and provide sufficient manpower.

Having given some reasons why this first attempt at CERN failed, let us speak about the successful experiments which followed and which started a fruitful area of neutrino research. Indeed the results obtained in the 1963 and 1964 experiments were of excellent quality and of far superior statistical accuracy than the Brookhaven results. This, together with the fact that a variety of highly interesting open questions remained, and that the experimental conditions could still be improved, led to the decision that at CERN neutrino physics should continue to be a central part in the research programme of the PS.[260]

When in January 1965 Victor Weisskopf opened the first conference on experimental neutrino physics held at CERN he called it 'an historic event'. Indeed

Neutrino experiments

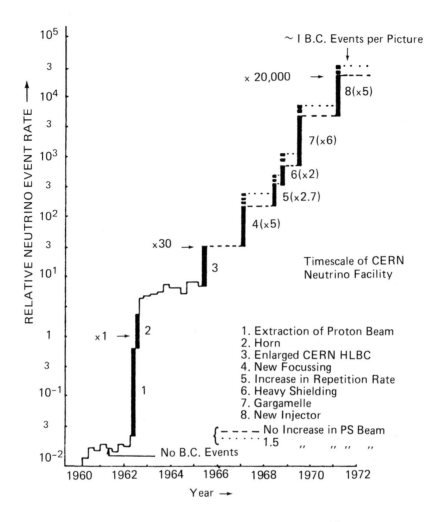

Fig. 5.22 Past and future of the CERN neutrino programme[262] in 1965

this kind of experiment had been suggested only five years before, it had developed at great speed, and such a conference signalled that neutrino physics was about to become an independent research field within high-energy physics.[261]

Further, during the experiments in 1963 and 1964, it had been proved that bubble chambers would deliver results superior in quality to spark chambers, so that the immediate future in neutrino physics at CERN should be based on such a device. As a first step the 500 liter HLBC which had been increased to 1200 liter early in 1965 was to be used in an antineutrino experiment. It had also been decided that a new neutrino area should be constructed and major improvements concerning the beam intensity should be carried out (see Fig. 5.22). The work on this started in 1965.

Notes: pp. 253 ff.

However the most important improvement for the future was the construction of a huge heavy liquid bubble chamber called Gargamelle, 6–7 times bigger than the CERN chamber. The convention for this purpose was signed between CERN and the Commissariat français à l'Energie Atomique on 2 December 1965. It was with this chamber that the famous detection of neutral currents was made at CERN in the early 70s.[263]

5.6 Conclusion

Now that we have described the construction of CERN's—and Europe's—first really big accelerator and the first years of its scientific exploitation, we want to conclude by discussing more general points concerning this period. We would like to outline the emergence of a new form of cooperation in high-energy physics and the consequences this entailed, to touch briefly on the problems related to the 'human aspects' in historico-scientific accounts of our kind, to tackle some points around 'CERN being late' in this early period, and lastly to discuss the general lines in the scientific programme carried out in these first six years with the PS.

With the invention of the bubble chamber in 1952 by Glaser and the further development first in the USA and then also in Europe, not only was a splendid new kind of particle detector added to the list of those already existing, but a device was put at scientists' disposition which would completely change the existing ways of experimenting and collaborating within this field of physics. Indeed it was this technique which triggered the setting up of ever larger collaborations—an aspect which is now inseparately connected to high-energy physics. Thus it was not only the need for huge equipment—accelerators—, but in particular this new way of working which caused the definite transition from 'little' to 'big science'. It was here that researchers had to learn to cooperate, to set up and organize big teams based in various countries, and of course also to cope with all the negative side-effects this entailed. Indeed, seen from outside, the field of high-energy physics started to become anonymous, to be done by 'nameless' actors, as thirty scientists and more—not to speak about the several hundreds who form a collaboration today—signed a publication. When one spoke about an experiment its label was generally used, and it was thus not connected to a name, to a personality anymore. Physicists faced this situation with rather ambiguous feelings. On the one hand they felt the enthusiasm of collaborating on a supra-national level and at being part of this prestigious European venture, of reconstructing physics in Europe, of competing against the Americans. On the other hand they had lost strictly speaking their individual role which many of them had been used to playing before. Added to this with the electronics experiments the 'old' way of doing physics was still present, thus a strong contrast. Here only a few physicists were working together, most of them stationed at CERN, and the experiment was clearly correlated with one person—the

Conclusion

team leader. And although it was to a certain extent a conscious decision of bubble chamber physicists to put their spirit of collaboration and internationalism ahead of any personal 'glory', this difference was present. We should, however, say here that these differences in the way of experimenting disappeared over the years. Today all experiments in high-energy physics are based on huge collaborations and in a sense the early bubble chamber work formed the social basis for them.

However bubble chamber physics not only had a strong social impact on the scientific community, but through the way it was carried out it also had an important place in the effort to reconstruct important centres in physics after the war. Bubble chamber pictures could be easily transported so permitting nearly all labs, even those with few means, to participate if they so chose. It thus also provided an excellent training ground for young physicists in all the Member States. To be more precise in the period discussed by us some 12.5 million pictures were taken allowing physicists from about 50 institutions to participate and deliver results in high-energy physics research.

Let us now come to the human side of this history. We have often been confronted with the reproach that our accounts failed to describe the atmosphere that prevailed. We did not describe the enthusiasm, 'the hopes as the experiments were designed, the frustrations when delays occurred, the efforts and pains during the data taking, the patience, meticulosity and inventiveness during analysis, the elation when new results came out, the disappointments when hopeful preliminary indications were not confirmed'. Yves Goldschmidt-Clermont used a very sensitive comparison for this. 'The chronological survey is perhaps a useful skeleton. Could it be endowed with muscles and nerves? Stomach? Even intelligence? Perhaps emotions?' — A justified claim and the expression of a certain expectation and disappointment of those who have been involved in the enterprise that is CERN from the very beginning and who find their work, their thoughts, their ideas, their whole intellectual life reduced to a few lines — if mentioned at all — in a historical account, an account which reflects perhaps the scientific core but surely not 'the experiment' in all that it meant for them. They had come to CERN, left their home countries, they had been drawn by the fascination of this new organization and this huge unique accelerator, to do *the* physics they thought should be done out of their deepest conviction. Thousands of physicists — playing more or less crucial roles in the life of CERN — have worked in our period at CERN carrying out hundreds of experiments and delivering results that had their importance and their place at that time, even if today they are not regarded as 'milestones' in the history of the Organization. They — all of them — collectively laid the foundation for what CERN is now. It is, however, impossible and would go far beyond the scope of such a history, to describe all of them in detail. One way to draw a more colourful picture of the whole situation surrounding an experiment in this period would be to pick out one or other collaboration which left its trace in the historical landscape of CERN and of elementary particle physics. With this in mind we chose the neutrino experiment for several reasons. It embraced major factors such

Notes: p. 254

as its importance for the Organization, the fundamental nature of the physics tackled and, last but not least, the completeness of the material in the archives at our disposal. Here we have encountered all the feelings mentioned above such as enthusiasm, disappointment, we have seen the human difficulties, organizational frontiers, conflicting interests, limits of timing, unforeseen difficulties, problems of resources, and many more. Though this is only one experiment it could however stand in a sense for the many others.

This brings us to the third point we would like to touch on here. 'CERN was born a little too late', Victor Weisskopf said in his report in 1965 when he left CERN as Director General. But was CERN really born too late or was it not rather the inexperience and the lack of foresight which caused major delays in the start of the experimental programme. Many physicists, in particular the bubble chamber physicists, complained that the first rich harvest of resonances and new particles was made at American accelerator centres, although CERN had a unique facility at hand—the PS. Indeed one should say that the bubble chamber programme was marked by noticeable delays as far as the commissioning of the chambers was concerned. Seen from the point of view of size and number of devices, CERN was on an equal footing with Brookhaven. However the schedule drawn up around 1960 for the completion of them was rather unrealistic. More explicitly, this meant that in 1960 a 30 cm hydrogen chamber and a 1 m propane chamber constructed by the Ecole Polytechnique were at hand for the exploitation of the world's biggest accelerator. The 1 m CERN heavy liquid chamber was ready by end-1960, though almost entirely tied up in the neutrino programme. Finally, by mid-1961 the 81 cm Saclay chamber, the first hydrogen chamber of sensible size in relation to the energy of the accelerator, started operating. The 152 cm British National Bubble Chamber scheduled first for 1961 and intended to bridge the gap until the completion of the 2 m CERN chamber, was more than two years late starting to work only by mid-1964. Similarly, the 2 meter chamber, scheduled for 1963–64 was only functioning by early 1965. In comparison, at BNL besides the 14", 20", and 30" chambers, the big 80" hydrogen chamber was already at hand by mid-1963 for use with the AGS.

But the bubble chambers were not the only problem. CERN was also late with the equipment for exploiting the machine, such as beam transport equipment, i.e. quadrupoles, bending magnets and separators. This obviously restricted the beams that could be built. Material had to be borrowed from outside and the construction of a new beam had to be carefully planned, in order to recuperate the needed material from other beam-lines in time.

Furthermore we should mention here a constant lack of space and machine time. We have to keep in mind that the PS was a unique machine for European physicists and everyone was keen, using bubble chambers, counters or emulsions, to perform his experiment as quickly as possible. In addition there was the neutrino experiment which CERN decided as early as 1960 should take a special place in the

Organization's scientific programme. And this meant of course that other experiments had to wait. Further, the PS experimental area was only extended in 1961 by the north hall and as late as 1963 by the east hall. CERN—as well as Brookhaven—had underestimated the space needed for an appropriate exploitation of such an accelerator.

To close this chapter let us say a few words about the scientific results of this experimental programme. As far as bubble chambers were concerned and despite the difficulties encountered, the results were nevertheless numerous and of rather high quality. We should mention here the outstanding contributions in the field of particle spectroscopy, such as the detection of the anti-xi, first evidence for the excited anti-nucleon N^* (3/2,3/2), the discovery of the mesons E(1420), C(1215), and K^*(1400) work contributing to the understanding of the A_1 and A_2 mesons, and the measurement of the properties of the numerous new particles found. Nor should we forget important successes such as the measurement of the lambda-sigma parity, which was an important step forward for the general acceptance of SU(3) theory. Further also diverse studies on the interaction mechanism—in particular peripheral processes—were carried out and allowed scientists to gain a better understanding of what was going on in 'strong interaction processes'. Problems of weak interaction physics were only marginally tackled, and that only in the later part of the period covered by us.

Turning now to the electronic experiments, this part of CERN's scientific programme with the PS was also dominated by investigations on strong interaction processes. This was for two reasons. The first lies in the nature of the machine which physicists had at their disposition, namely a high-energy *proton* accelerator. The second was that the period from the late fifties up to the mid sixties was marked by a flood of new, stimulating ideas for understanding this interaction. More precisely, there was Pomeranchuk's theorem and the Regge theory in 1958 and 1959 which attracted physicists' attention. It was followed by the famous idea of Gell-Mann and Ne'eman who suggested the SU(3) classification of strongly interacting particles in 1961. And last but not least in 1964 Gell-Mann and Zweig put forward their proposal that there existed a still deeper level in the hadron structure, namely the quarks (aces). Thus a large number of very valuable results coming from CERN experiments must be remembered, among them for example the completely unexpected constancy of the p-p cross-section at high energies, the appearance of a quasi-elastic peak and the logarithmic shrinkage of the diffraction peak.

A far less central role was occupied by weak interaction physics—with one big exception, the neutrino experiment. For the period 1960 to 1963 one could even go so far as to speak of neglect. It is not easy to give a clear explanation for such a choice, even if two main factors come to mind. The first is that the existing theory of weak interactions was quite satisfying and nothing indicating a really new direction had been suggested. Indeed the general attitude adopted by the Electronics Experiments Committee was not to choose equally out of all sections of high-energy physics because the chances of making appreciable contributions were not equally high. If one concentrated on one field the probability was higher of obtaining valuable results, reasonably complete and consistent as to choice of energy. Thus for 1962–63 for example the EEC had chosen as the main subject the measurement of differential cross-sections for high-energy elastic scattering. The second factor was that these first years at the PS were marked — as already said — by a lack of beam transport equipment, space for setting up the experiments, and machine time. The last factor was particularly important, as the bubble chamber and the neutrino experiments already took much of the time, so that the competition between the different counter experiments was very high. This 'underestimation' of the importance of weak interactions showed up rather clearly in 1964. Whereas only two experiments had been prepared in this field up to 1963, six were scheduled for 1964/65. As we have shown, this change occurred because of the results obtained at Brookhaven about the possible non-validity of *CP* invariance. However the bitter consequence for CERN people was that again an American team had found the key result which first caused the shift.

Concerning electromagnetic processes we find the same phenomenon as in the case of weak interactions, namely that physicists had a theory (here QED) which reproduced rather perfectly the measured data. Obviously no outstanding results could be expected for the near future. Indeed this was even more accentuated at Brookhaven where no real research project in this direction was undertaken. Nevertheless a small number of experiments were carried out testing the boundary of the validity of QED, trying to answer questions concerning the electromagnetic properties of hadrons (are they point-like or not?) and searching for magnetic monopoles.

After these purely scientific aspects, we would like to add a few words on the tremendous technical changes which took place in this period. The first occurred in detection devices, not only in bubble chambers but also for electronic devices. CERN physicists took a very active role in developing Čerenkov and scintillation counters to a high grade of reliability and sophistication. We should also mention the spark chamber technique which evolved incredibly rapidly at the beginning of the sixties, and in which CERN, after a relatively slow start, contributed rather actively. Different structures adapted to experimental needs were designed, they were combined with magnetic fields and they were changed to sonic spark chambers. Secondly, enormous progress was made in both design and use of computers in the early sixties. In particular bubble chambers and spark chambers which delivered their results in the form of many thousands of photographs, made computers indispensable if one was to evaluate the data within reasonable time limits. However this was only one aspect. The second and most important development of data-processing methods for counter techniques was the so-called 'on-line' use of computers which started in about 1964. This meant that the data were transmitted directly to a computer, and the acquired information was fed back to the detector in order to modify and control its functioning. This not only simplified and rationalized the process of data taking and evaluation; it also meant an enormous increase in speed. Indeed this was a key point opening the area where computers became a central tool in high-energy physics experiments.

Thus, despite all the initial difficulties, physicists at CERN delivered a number of excellent results up to 1965. During this period everyone had gained experience 'in doing things, running things, in thinking about things, in acting together'. Now, as Weisskopf said, the European Organization for Nuclear Research was well on its way, thinking carefully about the future and planning 'for the golden age' which still lay ahead.

APPENDIX I

Pion exposures between 1960 and 1965

beam energy [GeV/c]	date of exposure	number of photographs taken	bubble chamber	collaboration
16	March/Oct. 60	64,000 (π^-)	HBC 30	CERN (IEP & HBC groups)
6/11/18	Nov. 60	76,000 (π^-)	EPHLC	EP, Milan, Padua, Saclay
10	June 61	92,000 (π^-)	HBC 81	– CERN, Warsaw – CERN, EP
4	late 61/ early 62	65,000 (π^-) 73,000 (π^+)	HBC 81	Aachen, Bonn, Berlin, Hamburg, Munich, Birmingham, London
2.75	1962 Nov. 65	145,000 (π^-) 110,000 (π^+) 144,000 (π^-)	HBC 81 HBC 200	Saclay, Orsay, Bari, Bologna
4.5/5	1963/1965	310,000/310,000 (π^+)	DBC 81	Saclay, Florence, Orsay, Bari, Bologna
6	mid-63	100,000 (π^+)	DBC 81	CERN, EP
8/16/18	Dec. 63/Oct. 64	470,000 (π^-)	EPHLC	Orsay, Milan, Saclay
8	Feb./March 64	141,000 (π^+) 70,000	HBC 81 HBC 200	Aachen, Berlin, Crakow, Warsaw
20	March 64	14,000	HBC 81	CERN
5	Oct/Dec 64	130,000 (π^+)	BNHBC 152	Bonn, Durham, Nijmegen, EP, Strasbourg, Turin
11	1965	135,000/236,000 (π^-)	BNHBC 152/ HBC 200	Hamburg, Genova, Milan, Saclay

APPENDIX II
Proton exposures between 1960 and 1965

beam energy [GeV/c]	date of exposure	number of photographs taken	bubble chamber	collaboration
24	Oct. 60	78,000	HBC 30	CERN (IEP & HBC groups)
5.5	mid-63	60,000	HBC 81	Israel
10	Feb/March 64	84,000	HBC 81	Cambridge, Hamburg, Stockholm, Vienna
	Dec. 64	50,000	BNHBC 151	
6	Apr. 65	110,000	BNHBC 152	Genova, Milan, Oxford
8	Nov./Dec. 65	63,000	HBC 200	Saclay, Orsay

APPENDIX III
Antiproton exposures between 1960 and 1965

beam energy [GeV/c]	date of exposure	number of photographs taken	bubble chamber	collaboration
at rest		170,000	June 61	– CERN, EP, Coll. de France
	June 61			
slow		90,000	HBC 81	– Padua, Cambridge, Oxford
				– Trieste, Rome, CERN
at rest	Dec. 62	120,000		
1.2	June/July 64	300,000		CERN, Collège de France
stop	Febr. 65	190,000		
3, 3.6, 4	Nov. 61/Apr 62	500,000	HBC 81	– CERN-EP-Saclay
				– Birmingham,
3.6	Late 65	209,000	HBC200	– CERN, IC, EP, Saclay
				– CERN
				– Hamburg
5.7	Oct./Dec. 1963	310,000	HBC 81	CERN, Hamburg, Milan, Saclay
	March/Nov. 1965	220,000	HBC 200	
2,5	1965	182,000	HBC 200	Saclay
stop	Oct. 1964	70,000	DBC 81	Padua, Pisa
flight	Oct. 1964	80,000	DBC 81	Rome, Trieste

APPENDIX IV
Kaon exposures between 1960 and 1965

beam energy [GeV/c]	date of exposure	number of photographs taken	bubble chamber	collaboration
1.5	Dec. 61/Jan, 62	190,000 (K⁻) 50,000 (K⁺)	HBC 30	Amsterdam, CERN, Glasgow, Oxford, Padua
1.5	March 62	250,000 (K⁻)	EPHLC	EP, UC, Rutherford Lab,, Bergen, CERN
0.8	Sept. 62	300,000 (K⁻)	HBC 81	CERN, Uni Maryland, Naval Res, Lab,
3.5	Nov/Dec 62	188,000 (K⁻)	HLBC 100	EP, UC, Rutherford Lab,, Bergen, CERN
3/3.5	1963 Aug-Dec 65	646,000 (K⁻) 330,000 (K⁺) 687,000 (K⁻)	HBC 81 DBC 81	– Amsterdam, EP, Saclay – Birmingham, Glasgow, IC, Oxford, Ruth, Lab, – CERN, Brussels, Stockholm
0.8–1.2	mid 1964 Dec. 64 & Apr. 65 Jan 65	250,000 380,000 (K⁻) 220,000	HBC 81 DBC 81	CERN, Heidelberg Saclay
6 5, 6 5	Feb 64 June-Oct, 64 Feb/March 64 June/July 64 mid 65	70,000 (K⁻) 538,000 90,000 60,000 (K⁺) 87,000	HBC 81 BNHBC 152 HBC 81 BNHBC 152 HBC 200	British Collab, + 2 German groups CERN, Cambridge, Brussels
stop	Oct 64	120,000 (K⁺)	DBC 81	Bari, Bern, Genova, Turin
10	April 65 Dec. 65	232,000 (K⁻) 30,000 (K⁺)	BNHBC 152	Aachen, Berlin, CERN, IC, Vienna
stop	May/June 65	1,350,000 (K⁺)	HLBC	X₂ collaboration

Notes

1. Several interviews were made and copies of the transcripts are available in the CERN archives:
   Gilberto Bernardini by L. Kowarski and S. Newman, 3 July 1974.
   H. Blewett, M.G.N. Hine and K. Johnsen by M. Gowing, L. Kowarski and S. Newman, 3 July 1975 and 20 November 1975.
   Anselm Citron by L. Kowarski and S. Newman, 12 August 1974.
   Giuseppe Cocconi by L. Kowarski and S. Newman, 30 July 1974.
   Giuseppe Fidecaro by L. Kowarski and S. Newman, 8 and 27 January 1975.
   Dirk Harting by L. Kowarski and S. Newman, 27 January and 13 February 1975.
   Arne Lundby by L. Kowarski and S. Newman, 21 and 22 May 1974.
   Guy von Dardel by L. Kowarski and S. Newman, 24 March 1975.
   Antonino Zichichi by L. Kowarski and S. Newman, 7 March 1975.
   Leon Lederman by S. Newman, 14 October 1974.
   Jack Steinberger by L. Kowarski and S. Newman, 15 August 1974.
   In September 1987 an interview with V. Weisskopf was carried out. Further discussions with R. Armenteros, Y. Goldschmidt-Clermont, K. Johnson and D.R.O. Morrison helped in writing parts of this chapter.
   Further a large number of conference proceedings were used to follow the development in high-energy physics: HEP Conf. (1960, 1962, 1964), Elementary Particle Conf. (1961, 1963, 1965), Weak Interaction Conf. (1963, 1965), Instr. Conf. (1960), Theory Conf. (1961).
   For a history on the high-lights in the 25 years of CERN see Van Hove & Jacob (1980).
   The information on the developments at the Brookhaven National Laboratory was mainly taken from the BNL Annual Reports.
   For background literature on particle physics in the 60s see for example: Chew, Gell-Mann & Rosenfeld (1964), Fermi School (1964a,b), Tassie (1973), Lichtenberg (1965), Puppi (1963), Dalitz (1963).
2. The publications of the new resonances:
   The $Y_1^*(1385)$ had been found in the 15 inch hydrogen bubble chamber studying K^-p interactions at 1.15 GeV/c (Alston et al. (1960)). The $K^*(890)$ was detected in the same experiment a year later (Alston et al. (1961a)). For the quotation see Alvarez (1968).
   The ω^0 had already been predicted by theory, i.e. by Nambu to explain the electromagnetic form factors of p and n and by Chew on dynamical grounds as the 3π resonance or bound state. It was found in the 72 inch hydrogen chamber when studying the annihilation of proton–antiproton into five pions at 1.61 GeV/c (Maglic et al. (1961)).
   The η^0 was observed when investigating π^+d interactions at 1.23 GeV/c with the 72 inch chamber (Pevsner et al. (1961)).
   The ϱ meson was detected at Brookhaven during studies on collisions of pions with target pions furnished in a virtual state by the nucleon using the 14 inch hydrogen bubble chamber (Erwin et al. (1961)).
3. Some of the major theoretical papers of the described period:
   Chew & Frautschi (1961), Regge (1959, 1960), Pomeranchuk (1958).
   Ne'eman (1961), Gell-Mann's work was first published as an internal report: Gell-Mann, Internal Report of California Institute of Technology, CTSL-20, March 1961, published as Gell-Mann (1962); see also Gell-Mann & Ne'eman (1964).
   Gell-Mann (1964) and G. Zweig, 'An SU(3) model for strong interaction symmetry and its breaking'. *CERN Preprint* 8419/TH.412 (Feb. 1964, the paper has never been published).
   See also *CERN Courier,* **3** (1964), 26–27.
   For the weak interactions see Danby et al. (1962) and Christenson et al. (1964).
4. As we will not deal in this chapter with nuclear emulsion physics let us quote here some summary reports written about this period: E.H.S. Burhop, 'Recent experiments with nuclear emulsions', *Elementary Particle Discussion (1963),* 350–364. Herz & Lock (1966).

5. For details on the organizational structure of CERN see chapter 8.
   See also Morrison (1978).
6. For a very detailed technical description of the CERN PS see: Regenstreif (1959, 1960, 1962).
   Overview articles: Adams (1955, 1957, 1960).
   J.B. Adams, 'Nuclear Particle Accelerators', in *Elementary Particle Discussion* (1963), 303–323.
   Also a complete set of the Minutes of the CERN-PS Staff Meetings starting on 2/11/53 up to the end of 1959 were the source for many details mentioned in this part. They can be found in the CERN archives and are referred to, in what follows as (PSM-...).
7. For a detailed discussion of the decision process on the two CERN accelerators see CERN History (1987).
8. See First session of the provisional CERN Council held in Paris 5–8 May 1952, CERN/GEN/1, 31–33.
9. O. Dahl, *Report on the proton synchrotron project*, CERN/GEN/2, Copenhagen 20–21 June 1952, p. 12–13.
10. See Second Session of the prov. CERN Council, CERN/GEN/2, 20–21 June 1952.
11. See O. Dahl, *Progress report of the proton synchrotron group*, in CERN/GEN/4, 4–7 October 1952, p. 7–8.
12. See O. Dahl, 'Report to the Secretary, coverning visit to USA by Goward, Wideröe and Dahl in August 1952', CERN-PS/S4, 25/8/52 (CERN-DG20551).
    Courant, Livingston & Snyder (1952), Christofilos (1950).
13. See the tentative design worked out by Livingston in 'Trip report', *op. cit.*, note 12, p. 5.
14. See CERN/GEN/4.
15. See CERN/GEN/6, 12–14/1/53.
16. J.D. Lawson, *The Effect of Magnet Inhomogeneities on the Performance of the Strong Focusing Synchrotron*, Report CERN/PS/JDL1, 2/12/52 (rev. 10/12/52) both in (CC-CHADI,1/2).
17. For a description of the prevailing situation see the interview with H. Blewett, M.G.N. Hine and K. Johnsen, *op. cit.*, note 1.
18. See *Progress to orbits work*, Annex I – Part 1 of the Second Report to Member States, CERN/GEN/5, 5/5/53. The problem of the 'transition energy' is described in Adams (1960).
19. *Progress of the RF work*, Annex I – Part 2, CERN/GEN/5. See also C. Schmelzer, 'Radio-Frequency System for the CERN Proton Synchrotron', in *Accel-Conf. (1953)*, 20–21.
20. *Progress of magnet design work*, Annex I – Part 3, CERN/GEN/5.
21. *Building and shielding status*, Annex I – Part 4, CERN/GEN/5.
22. For the description of the developments see Interview with H. Blewett, M.G.N. Hine and K. Johnsen, *op. cit.*, note 1.
23. See Seventh Session of the Council, 29–31/10/53, CERN/GEN/12.
24. See Accel-Conf. (1953).
25. See CERN/GEN/12, 29–31/10/53; for the quotation see p. 7.
26. See changing structure in the progress reports to the Council from 1955 onwards.
27. See short biographies in A. Hermann, Who's who in the foundation of CERN, in *CERN History (1987)*, 545–565.
28. See *CERN Annual Report 1955*, p. 25. See also, for example, K. Johnsen, Report from my visit to Harwell and Brookhaven, 30/11–22/12/52, CERN-PS/KJ13.
29. The figures are taken from CERN/123, 9/12/55, p. 1, CERN/148(A), 25/11/55, p. 1, CERN/197(A), 14/12/56, p. 7, CERN/249, 19/12/57, p. 7. See also *CERN Annual Report 1958*, 34 and 43.
30. The magnet design and construction is described in detail in Regenstreif (1959), 70–158.
31. See PSM-1, 2/11/53 and PSM-2, 11/11/53, *Progress Report of the Proton Synchrotron Group*, 28/12/53, CERN/GEN/13, p. 21–22 and *PS Division Progress Report*, CERN/123(A), 9/2/55, p. 2.
32. J.B. Adams, *PS Division progress reports*, Oct. 1954–Jan. 1955, CERN/123(A), p. 4–5 and February–November 1955, CERN/148(A), p. 4–5.
33. J.B. Adams, *Progress report*, 1 May–31 October 1956, CERN/197(A), p. 8–9.

34. See Regenstreif (1959), 114.
35. J.B. Adams, *Progress Report,* 1 November 56–30 April 57, CERN/225(A), p. 14-15.
36. The figure is reproduced from *CERN Courier*, 6-7, January-February 1960, p. 14.
37. J.B. Adams, *Progress Report,* 1 November 57–30 April 58, CERN/268(A), p. 9-11 and 1 May-31 October 58, CERN/291(A), p. 11-12.
38. See *Progress report of the Director-General and Divisional Directors (1959),* CERN/319, 18/11/59, p. 12.
39. For a detailed technical description see Regenstreif (1959), 159-224.
40. See PSM-7, 16/12/53.
41. 9th Session of Council, 8-9 April 1954, CERN/GEN/14, p. 31.
42. The figure is reproduced from *Regenstreif (1959),* 161.
43. See CERN/148(A), 25/11/55, p. 5-6 and CERN/174(A), 1/6/56, p. 2.
44. See CERN/197(A), 14/12/56, p. 10.
45. J.B. Adams, *Progress Report,* 1 May-31 October 57, CERN/249(A), p. 11-12.
46. See J.B. Adams, *op. cit.,* note 45 and CERN/268(A), 5/6/58, p. 12-13.
47. See *Progress Report Proton Synchrotron Division* (Nov–April 1958), CERN/268(A), 5/06/58, p. 11-13.
48. See *Progress report of the Director-General and Divisional Directors* (May–October 1959), CERN/291(A), 14/11/59, p. 13-14.
49. See *Progress report of the Director-General and Divisional Directors* (1959), CERN/319(A), 18/11/59, p. 11-12.
50. For a detailed description of the injection system see Regenstreif (1960).
52. See PSM-25, 19/5/54 (discussion of the two types of linacs under consideration); PSM-40, 1/9/54 (description of the Harwell linac).
51. See *Second Report to Member States,* CERN/GEN/5, p. 33.
53. Regenstreif (1960), 1-2.
54. This figure is reproduced from Regenstreif (1960), Fig. V.1.
55. See Report of the Director-General and Divisional Directors, CERN/148A, 25/11/55, p. 6-8.
56. For the progress of the linac system up to 1956 see PSM-81, 2/10/56.
57. See CERN/268(A), 5/6/58, p. 13-14 and CERN/291(A), 14/11/58, p. 14-16.
58. See *Progress report of the Director-General and Divisional Directors* (1959), CERN/319(A), 18/11/59, p. 13-15.
59. For technical details on these parts of the PS see Regenstreif (1962).
60. See, for example, CERN/291(A), 14/11/58, p. 16. Regenstreif (1962), 1-90.
61. Regenstreif (1962), 119-196.
62. See CERN/148(A), 25/11/55, p. 12; CERN/197(A), 14/12/56, p. 12; CERN/249(A), 19/12/57, p. 14; CERN/291(A), 14/11/58, p. 18-19 and CERN/319(A), 18/11/59, p. 19-20.
63. In the final report to the Member States (CERN/GEN/15, 29/09/54) O. Dahl and J.B. Adams described the problem of the stability of the ground as one of the three major factors that may prejudice the successful operation of the machine. See also CERN/123(A), 9/2/55, p. 7, CERN/148(A), 25/11/55, p. 10-11, CERN/197(A), 14/12/56, p. 14-16, CERN/268(A), 5/6/58, p. 17-18 and Adams (1956).
64. This figure is reproduced from *CERN Annual Report 1955,* p. 22.
65. This figure is reproduced from Elementary Particle Discussion (1963), 318.
66. For a compilation of all publications covering bubble chamber experiments at CERN during the period 1960-1974 see Steel (1976). Further see Krige & Pestre (1986). For a compilation of the bubble chamber exposures with regard to the beam particles see Annexes I, II, III and IV.
67. For a description of this chamber see HBC81 (1961).
68. See 'Test of British National Hydrogen Bubble chamber', *CERN Courier,* **3** (1963), 37; '60-inch bubble chamber at CERN', *CERN Courier,* **9** (1964), 120-123; 'Nimrod', *CERN Courier,* **9** (1963), 113 and **6** (1964), 77; Nimrod (1979).
69. For a detailed discussion on the construction of the 2 m chamber see chapter 6.

70. For a short description see EPHLC (1961) and Bloch et al. (1961).
71. See C.A. Ramm and L. Resegotti, 'The principles and the design of the CERN propane bubble chamber', in *Instr. Conf. (1960)*, 127-132; Shutt (1967), 181.
    See also CERN/SPC/90, 31/3/59 (The CERN propane bubble chamber project, CERN/PS/EP/5).
72. Considerations about the beams to construct around the PS started in 1959, see CERN/SPC/90, 31/3/59 (CERN/PS/EP/3, p. 4-6). See *CERN Courier*, **1** (1965), 4 and 'Sifting high-energy particles', *CERN Courier*, **3** (1965), 35-37.
73. This problem is dealt in some length in chapter 9.5.
74. For the f-beam see *CERN Annual Report 1961*, p. 42-43, Gregory (1961). The information on the following series of m-beams is taken from the *CERN Annual Report* and the *CERN Courier*. See for example: *CERN Courier*, **10** (1962), 2.
75. For a description of the k_1 beam see Amato et al. (1963). Further Aubert et al. (1963), *CERN Courier*, **10** (1962), 2 and 'Beam layout in the experimental areas in September 1964', *CERN Annual Report, 1964*, p. 33.
76. For the o_2 beam see *CERN Courier*, **9** (1963), 111 and **10** (1963), 127 and *CERN Courier*, **1** (1965), 4. For the o_8 and the u_1 beam see *CERN Annual Report 1965*, 83.
77. For a detailed description of the developments taking place at CERN see chapter 9.6.
78. For a short overview see L. Kowarski, 'Bubble chamber data reduction: Leave it to machines', *CERN Courier, Winter* (1960), 27-29, Benot et al. (1963).
79. To gain a short insight in the development of this device see *CERN Annual Reports, 1961*, p. 75; *1962*, p. 88-89; *1963*, p. 93; *1964*, p. 88-89.
80. The results were presented extensively at the two big conferences Elementary Particle Conf. (1961) and HEP Conf. (1962). See also Bartke et al. (1962a) and CERN/SPC/128, 29/4/61, p. 2.
    For the study of the aspect of peripheral collision see D.R.O. Morrison, 'Peripheral effects of 16 GeV/c negative pions and the $\pi\pi$ cross-section', in *HEP Conf. (1962)*, 606-609. See Chew & Low (1959) for the method of extracting $\pi\pi$ cross-sections from πp data.
81. See Bingham et al., 'V^0 pairs produced in the Ecole Polytechnique 1 m propane bubble chamber by π^- of 6.1, 11.6, and 18.1 GeV/c', in *Elementary Particle Conf. (1961)*, 101-110, and some publications: Baglin et al. (1963), Bellini et al. (1963, 1964, 1965), Veillet et al. (1963). The f-meson was first mentioned in Selove et al. (1962).
82. Main results were published as A. Bigi et al., 'Strange particle production in 10 GeV/c π^- interactions in hydrogen bubble chambers', *HEP Conf. (1962)*, 247-251; Bigi et al. (1964a,b); Brandt et al. (1963); P. Fleury et al., 'The study of two-prong events produced by 10 GeV/c π^- in hydrogen', *HEP Conf. (1962)* 597-602. See also Gregory (1961).
83. See *CERN Annual Report, 1963*, 76.
84. Presentation of the experiment: CERN/TC/COM 62-1, 8/1/62, App. A.
85. See Aderholz (1964a,c).
86. This figure is reproduced from Aderholz et al. (1964c).
87. See Bondar (1963a,b, 1964a,b), Aderholz (1964b, 1965), Bartsch (1964).
88. See Alitti et al. (1963).
89. Publications: Alitti et al. (1964, 1965, 1966) and Baton et al. (1964).
90. See Armenise et al. (1964, 1965a,b, 1966).
91. See CERN/TC/COM63-21, 27/2/63; CERN/TC/COM64-7, 9/1/64.
    Publication of results: Forino et al. (1964, 1965), Armenise et al. (1967).
92. See Bruyant et al. (1964a,b) and Goldberg et al. (1965).
93. See Goldberg et al. (1965) and Allard et al. (1967). For a progress report see CERN/TC/COM65-40.
94. Publication: Allard et al. (1964).
95. See Allard et al. (1965).
96. Deutschmann et al. (1964, 1965a).

97. Deutschmann et al. (1965b).
98. Deutschmann et al. (1966) and Bartsch et al. (1967).
99. For the proposal see CERN/TC/COM64-14, 17/2/64.
 Publications: Morrison (1964), Bingham et al. (1964) and Blum et al. (1964).
100. See, for example, Drevermann et al. (1967).
101. Proposal: CERN/TC/COM64-70, 20/10/64.
102. See Bartke et al. (1963); D.R.O. Morrison, 'Peripheral or quasi elastic interactions of 24 GeV/c protons and 16 GeV/c negative pions in hydrogen', in *Elementary Particle Conf. (1961)*, 407-417.
103. Publications: Alexander et al. (1964, 1965).
104. See Almeida et al. (1965a), CERN/TC/COM66-1, 3/1/66.
105. See CERN/SPC/90, 31/03/59 (CERN/PS/EP/8).
106. Presentation of the experiment at a Track Chamber Committee Meeting CERN/TC/COM61-27, 18/08/61.
107. Publication of this collaboration: G.B. Chadwick et al., A study of the annihilation of stopped antiprotons in hydrogen: the reaction $\bar{p}+p \to \pi^+\pi^-\pi^0$' and 'Production and decay of the ϱ and the ω mesons in antiproton annihilation', both in *HEP Conf. (1962)*, 69-72 and 73-75; Chadwick et al. (1963).
108. See U. Amaldi et al., 'Antiproton-proton low energy cross-sections', in *Elementary Particle Conf. (1963)*, 243-248.
109. This paragraph is based on the following publications: R. Armenteros et al., 'Study of the ω^0 meson in annihilations $p+\bar{p} \to K+\bar{K}+\omega^0$ at rest', Study of the K^* resonance in (p$\bar{p}$) annihilation at rest', and 'p-$\bar{p}$ annihilation at rest into two mesons', all in *HEP Conf. (1962)*, 90-95, 295-297 and 351-355; D'Andlau et al. (1963).
110. For the E-meson see: R. Armenteros et al., Evidence for a $(K_1^0 K^{\pm} \pi^{\mp})$ resonance near 1410 MeV in $p+\bar{p} \to K\bar{K} 3\pi$ annihilations at rest and study of the $p\bar{p} \to K\bar{K}\pi$ annihilations at rest', in *Elementary Particle Conf. (1963)*, 287-291; R. Armenteros et al., 'Study of a $(K\pi\pi)$ enhancement in $K\bar{K} 3\pi$ annihilations of antiprotons at rest', in *HEP Conf. (1964)*, 467-468; Baillon et al. (1967).
 For the C-meson see: Armenteros et al. (1964).
 For the D-meson see: D'Andlau et al. (1965).
 See also Armenteros et al. (1965a,b).
111. See CERN-Saclay (1962) and Brown et al. (1962). For a narrative account of this experiment, see 'New fundamental particle discovered, the anti-xi-minus', *CERN Courier*, **3** (1962), 4-5.
 The one remaining undiscovered quasi-stable elementary particle the anti-xi-zero was detected in 1963 at BNL in studying 3.69 GeV/c anti-protons with the 20 inch HBC. In 300,000 photographs, three examples of this particle were found. A very detailed kinematic analysis of all decay products is necessary (see *CERN Courier*, **9** (1963), 113).
112. This figure is reproduced from *CERN Courier*, **3** (1962), 5.
113. See Musgrave et al. (1964) and Böck et al. (1964).
114. See Czyzewski et al. (1965, 1966).
115. See Dehne et al. (1964a,b).
116. See Boeckmann et al. (1965, 1966).
117. See Böck et al. (1965) and Accensi et al. (1966).
118. Preliminary results were presented in W.A. Cooper et al., 'Interactions of 1.47 GeV/c negative K-mesons in hydrogen', *HEP Conf. (1962)*, 298-302. See also Cooper et al. (1964).
 The resonances $Y_0^*(1405)$ and (1520) were first mentioned in Alston et al. (1961b) (Berkeley).
 For the 1660 resonance see W.A. Cooper et al., 'The 1660 MeV/c^2 and Y_0^* resonances', in *Elementary Particle Conf. (1963)*, 154-160 for the CERN results and G. Alexander et al., 'Study of strange particle resonant states produced in 1.89-2.24 GeV/c π^-p interactions', in *HEP Conf. (1962)*, 320-324 for the American results. Final confirmation for the existence of $Y_1^*(1660)$ was given by Alvarez et al. (1963).
 See also B. Buschbeck-Czapp et al., 'Measurement of the ω^0 and η^0 branching ratios and a study of the reaction $K^-p \to K^0n$', in *Elementary Particle Conf. (1963)*, 166-170 and Schneider (1963).

119. See Chadwick et al., 'A study of the reaction $K^+p \to K^0\pi^+p$', in *HEP Conf. (1962)*, 378-379 and Bettini et al. (1965).
120. See Bezaguet et al. (1964a,b), Baglin et al. (1964).
121. See Aubert et al. (1964a,b).
122. For the proposal see CERN/TC/COM61-45, 26/10/61.
123. Courant et al. (1963).
124. See Cabibbo (1963) and N. Cabibbo, 'Weak interactions and the unitary symmetry', in *Fermi School (1964a)*, 299-310. Willis et al. (1964).
125. Proposal: CERN/TC/COM64-18, 24/02/64.
126. The results are published in: Armenteros et al. (1965c,d, 1967).
127. For the proposals see CERN/TC/COM64-1, 6/1/64; CERN/TC/COM64-41, 2/6/64; CERN/TC/COM64-46, 9/7/64; CERN/TC/COM64-57, 7/9/64 and CERN/TC/COM65-51, 17/8/65.
128. See discussion to the talk of G.A. Snow, 'Strong interactions of strange particles', in *HEP Conf. (1962)*, 805.
129. This figure is reproduced from Haque et al. (1965).
130. The paragraph is based on the publications: Focardi et al. (1965), Haque et al. (1965), Badier et al. (1965c); J. Badier et al., '$B = 1$, $S = -2$ states in K^-p interactions at 3 GeV/c', in *HEP Conf. (1964)*, 593-596, Badier et al. (1965a,b). The $\Xi^*(1820)$ had already been reported before by Smith et al. (1964). For $Y^*(1660)$ see Léveque et al. (1965).
131. See Ferro-Luzzi et al. (1965b) and Lynch et al. (1964).
132. See Ferro-Luzzi et al. (1964, 1965a).
133. Barnes et al. (1964); for a narrative account on the experiment see Fowler & Samios (1964). Although the finding of the Ω was regarded as a great success, there were nevertheless anxieties about the fact that physicists had only seen one such example. See letter Citron to Weisskopf, 7/6/64 (DIRADM-20278).
134. See Colley et al. (1965) and Scotter et al. (1968).
135. The figure is reproduced in a modified version from Colley et al. (1965).
136. See DeBaere et al. (1966), Goldschmidt-Clermont et al. (1966), Almeida et al. (1965b).
137. Articles dealing with counter experiments carried out at CERN: J.M. Cassels, 'Recent experiments with counters at the CERN proton synchrotron', *Elementary Particle Discussion (1963)*, 340-350; A. Citron, *Comparison of counter experiments at the Brookhaven AGS and the CERN PS*, 11/3/65 (CHIP 10004); van Hove & Jacob (1980).
138. Hagedorn's statistical theory and in particular his detailed calculations on the number of each kind of secondary particles produced at a certain angle and with a certain momentum were of great help for setting up the first beams. See Hagedorn (1960a) and von Behr & Hagedorn (1960).
139. For a description of the detector and a general survey on the development of Čerenkov counters see R. Mermod et al., 'A velocity-selective gas Čerenkov counter', and G. von Dardel, 'Čerenkov counters', both in *Instr. Conf. (1960)*, 172-173 and 166-171.
140. The results were presented at the 1960 Rochester Conference: G. von Dardel, 'Particle abundance in interactions with 25 GeV protons', *HEP Conf. (1960)*, 837-839; for the comparison with statistical theory see R. Hagedorn, 'Limitation of the description of strong interactions by central collisions and the statistical model', in *Theory Conf. (1961)*, 183-203.
141. G. Fidecaro and A.W. Merrison, *Proposal for a low-momentum K-meson beam* (priv. comm.).
142. For the results see Cocconi V.T. (1960). The numerical calculations of deuteron production in nucleon-nucleon collision were made by Hagedorn (1960b).
143. The detector is described in detail in L. Gilly et al., 'DISC: A differential isochronous self-collimating Čerenkov counter', in *Instr. Conf. (1960)*, 87-88.
144. The results are reported in L. Gilly et al., 'Particle production by 24 GeV/c protons', in *HEP Conf. (1960)*, 808-809. For a more detailed presentation of these results see L. Gilly et al., *Search for unusual particles II and III,* Internal Report SC60-3 and 60-4.

145. G. Cocconi et al., *Proposal for an experiment to study the γ rays produced by strange particle decay in the neighbourhood of the PS target,* 7/7/59 (KHR 22141). Further information is taken from *Annual Report 1960,* p. 68.
146. For the detector see Gatti et al. (1961); for the presentation of the results Fidecaro M. et al. (1961).
147. See Fidecaro M. et al. (1962).
148. Proposal by W. Middlekoop, *Measurement of the photon spectra in p-p collisions,* 7/2/61 (priv. comm.); the results are presented in Fidecaro M. et al. (1962).
149. The importance of such an experiment was underlined by van Hove in a Memo to Harting, 3/3/61 (attached to the Minutes of the 1st EEC Meeting). See Amaldi U. et al. (1962, 1963a).

    Proposal: A.E. Taylor, *Proposal for the study of production of deuterons in p-p collisions,* 20/11/61 (LK 22465). The results of Taylor and his group were only referred to in the *Annual Report 1962,* p. 55.

    See Cocconi G. et al. (1963).
150. Let us say a few words about the von Dardel group. Von Dardel left his team early in 1962 when he was appointed by the Directorate as interdivisional group leader for the neutrino experiment, 14/2/62 (DIRADM 20278). However the other members of the team continued to work together.

    For the results see Dekkers et al., 'Production cross-section of pions, kaons, and protons in proton-proton collision', *Elementary Particle Conf. (1963),* 641-643 and Dekkers et al. (1965).

    The remark on the importance of isobar excitation in pp collision has to be seen in the light of Cocconi's experiment on isobar production carried out in 1963.
151. See discussion at the 1st EEC Meeting 1-2/3/61, p. 9-11 and Interview with Zichichi, *op. cit.* note 1. The results of the 1965 experiment are presented in Massam et al. (1965). See also Pilkuhn (1964) for a theoretical calculation of deuteron production.
152. See Pomeranchuk (1958).
153. von Dardel et al. (1960). See also G. von Dardel, *Experimental programme of the von Dardel group for 1961,* 7/2/61 (KHR 22141).
154. Proposal: G. von Dardel, *Measurement of total cross-sections in hydrogen,* CERN/Ex.C/72, 26/2/60; For the results see von Dardel et al. (1961, 1962).

    When measuring the charged pion flux coming from the proton target von Dardel realized that a serious mistake had been made in the flux calculations for the neutrino experiment. His report *Counting rate in the planned experiment to detect neutrino induced reactions,* NP Int-Rep. 61-5, in May 61 meant the end for the first neutrino experiment at CERN.
155. This figure is reproduced from *Annual Report 1961,* 41.
156. The bubble chamber results were reported in Bartke et al. (1963). The determination of $\sigma(pp)$ in the case of bubble chambers was far less precise than with counter techniques.
157. Ashmore et al. (1960).
158. The figure is reproduced from Ashmore et al. (1960).
159. The figure is reproduced from Amaldi U et al. (1964).
160. See *Experiments for studying properties of antiprotons,* CERN/PS/EP/8, in CERN/309, 1/5/59.
161. For a detailed description of the beam see Amaldi U. (1963b).
162. The results are published in Amaldi U. et al. (1964). See also *Experimental programme of the 'Fidecaro' group for 1961,* 13/2/61 (KHR 22141).

    The diagram presented in Fig. 5.13 contains also values from other experiments having measured the $p\bar{p}$ cross-section.
163. Cocconi G. et al. (1961a). For more general information of the activities on the group see: G. Cocconi, *Experimental programme Cocconi group for the next 6 months' period,* 3/2/61 (KHR 22141).
164. The results of the second experiment were published in Cocconi G. et al. (1961b).

    Two examples of explanations for the origin of the inelastic bump: Drell & Hiida (1961); Van Hove suggested that it may be possible to explain this effect in terms of diffraction disintegration of the target nucleons (Note 10 of Cocconi G. et al. (1961b)).

    For an overview on the strong interaction experiments, see Wetherell (1961).

165. This figure is reproduced from *Annual Report 1962*, 55.
166. See Diddens et al. (1962a,b); for a summary on all data of elastic proton-proton scattering see Cocconi G. (1964).
167. See Cocconi G. et al. (1964).
168. The work of B. Maglic and the Cocconi group is published as E. Lillethun et al., 'Operation of a sonic spark chamber system', in *Spark chamber Meeting (1964)*, 157-169. It had been published as a CERN preprint in December 1963.
169. For a detailed description of the sonic spark chamber used in the Cocconi experiment see Bellettini et al. (1965c). The results are published in Bellettini et al. (1965a,b,c,d, 1966a,b).
170. See Caldwell et al. (1964) and Harting et al. (1965).
171. Proposal: CERN-Ivry group, *π-p large-angle elastic scattering*, 10/9/63 (WOL 23232); for the detector see CERN-Ivry group, *An apparatus for the study of elastic scattering at large momentum transfer*, 8/63, NP Int.-Rep. 63-8; results in Baker et al. (1966).
172. Proposal: P. Falk-Vairant, Study of $\pi^- p \to \pi^0 n$ *charge exchange in forward direction at high energy*, 25/3/63 (KHR 22141).
173. See Stirling et al. (1965).
174. Proposal: Hyams and Stirling, *Measurement on peripheral collisions and the π form factor*, 18/10/61 (WOL 23232) and Backenstoss et al., $\pi^- + P \to \pi^0 + N$, 12/7/63 (WOL 23232). Results are presented in Backenstoss et al. (1964, 1966).
175. See Astbury (1966a,b).
176. A. Lundby, *Experimental programme of 'Lundby' group for 1961*, 7/2/61 (KHR 22141); for the results see, Dowell et al., *Resonances in the pion hyperon system*, 2/8/61, NP Int. Rep. 61-9 and Dowell et al. (1962).
177. See Blackall et al. (1965).
178. For the spark chamber construction see G. Culligan, D. Harting and N.H. Lipman, *A spark chamber hodoscope* (Geneva: CERN 61-25, 1961) and Bleuler et al. (1963).
179. See Caldwell et al. (1962).
180. See L.W. Jones et al., 'Diboson production by pions of 12 and 18 GeV/c', in *HEP Conf. (1964)*, 481-484; see also Jones et al. (1966a,b, 1968).
181. For Cocconi's work see note 149.
182. Proposal: C.C. Ting et al., *Proposal for measuring the differential and total cross-section of the reaction* $\pi d \to pp$, 10/10/63 (KHR 22141); for results see Dekkers et al. (1964).
183. Proposal: Maglic and Gosta, *Missing mass spectrometer for heavy mesons*, 30/5/63 (WOL 23232); Maglic & Gosta (1965).
184. For the detector see H. Blieden et al., 'System consisting of sonic spark chambers, time-of-flight and pulse-height counters ('Missing-Mass spectrometer') with on-line computer', in *Spark Chamber Meeting (1964)*, 49-56.
185. This figure is reproduced from *Annual Report 1966*, 25.
186. The results were presented in numerous papers of which we want to quote some: Blieden et al. (1965, 1966a,b), Seguinot et al. (1966), Kienzle et al. (1965), Lefebvres et al. (1965), Chikovani et al. (1966), Levrat et al. (1966), Focacci et al. (1966).
187. Proposal: U. Amaldi et al., *Missing mass spectrometry for the study of multiparticles resonances: - a first application to the measurement of the branching ratio* $\varrho^\pm \to \pi^\pm \gamma / \varrho^\pm \to \pi^\pm \pi^0$, no date (1963) (KHR 22141); Fidecaro G. et al. (1966).
188. Proposal: R. Querzoli, G. di Giugno and V. Silvestrini, *Proposal for a measurement of the branching ratio* $\eta \to \gamma\gamma/\eta \to$ *other neutrals*, May 1964 (WOL 23233). The results are published in Di Giugno et al. (1966a,b).
189. Proposal: B. Cork and G. Fidecaro, Σ^0 *polarization and measurement of the relative* Σ-Λ *parity*, 31/10/61 (rev. version) (LK 22465).

190. Rubbia sent his letter of intention to Preiswerk (Chairman of the EEC) in June 1963; see Memo Rubbia to Preiswerk, 10/6/63 (WOL 23232) Information taken from *Annual Reports 1964* p. 50 and *1965* p. 54.
191. Buhler-Broglin et al. (1965, 1966). Massam and Zichichi had occupied themselves rather intensively with various kinds of questions related to quarks; see Massam & Zichichi (1966). Indeed in parallel also Cocconi had suggested an experiment searching for quarks. We found however no trace of the experiment. Proposal: G. Cocconi et al., *An experiment to search for the production of chage 1/3 e particles at the CERN proton synchrotron*, 3/2/64 (KHR 22141).
192. For a short overview on the history of QED see CERN History (1987), p. 32–41.
193. Proposal for this experiment was made by G. Giacomelli, *Positron-electron annihilation in the multi GeV region*, 7/2/61 (KHR 22141); Results: Fabiani et al. (1962).
194. Information taken from *Annual Reports 1964,* p. 54 and 110; *1965*, p. 55 and 110–111.
195. On 27 October 1961 the PAPEP group advanced their first proposal: Conversi et al., *Antiproton annihilation to electron pair,* 27/10/61 (LK 22465). This proposal was however regarded as not detailed enough and after various discussion the EEC asked for a further more elaborate one. This was done by the end of 1962 and the tests for the experiment started. See also Zichichi et al. (1962).
196. See Conversi et al. (1963).
197. See Conversi et al. (1965).
198. The experiment has been discussed at length in the 1st Meeting of the Electronics Experiments Committee held on 1–2/3/61, p. 16–17.
199. In a Memorandum to the DG von Dardel explains the difficulties but also the merits of this experiment. Memo von Dardel to Weisskopf, 9/3/62 (DG 20888). Indeed the experiment was rather controverted by the community of theorists, as this measurement could not be compared to any theoretical calculations. Von Dardel et al. (1963).
200. See Dirac (1931).
201. Proposal: G. Giacomelli, *Search for magnetic monopoles,* no date (roughly early 1961) (KHR 22141). See Fidecaro M., Finocchiaro & Giacomelli (1961).
202. Two examples for a textbook and a lecture note dealing with the complex question of neutral kaons: Okun (1965), Physics School (1962).
203. CERN-ETH Cloud chamber group, *Proposal for an experiment on K_2^0 decay,* 18/10/61 (DG 20877).
204. Results were presented first by A. Zichichi, 'Branching ratios for K_2^0 decay', in *Elementary Particle Conf. (1963),* 23–29; final publication Astbury et al. (1965c).
205. Proposal by A. Robert, *Study of the leptonic decay modes of K^+*: K_{e3} and $K_{\mu3}$, 2/3/61 (KHR 22141).
206. Christenson et al. (1964). For a more narrative account of this experiment see Val Fitch, 'A discovery of inner Mongolia', in N.V. Baggett (ed.), *AGS 20th Anniversary celebration,* BNL 51377 UC-28, 22/5/80.
207. At CERN quite some effort was devoted to the theoretical investigation of a fifth force.
208. De Bouard et al. (1965).
209. Proposal: M. Bott-Bodenhausen et al., *Proposal to search for interference in $K_2^0 \to 2\pi$ and $K_1^0 \to 2\pi$*, no date (WOL 23233). Results are published as Bott-Bodenhausen et al. (1966).
210. See Alff-Steinberger et al. (1966a,b).
211. See Astbury et al. (1965a,b).
212. See Cnops et al. (1966).
213. *Experimental programme of the Hyams group for 1961,* no date (KHR 22141). The results are presented in Backenstoss et al. (1961); for the detector see Backenstoss et al. (1963).
214. Proposal: J. Barlow et al., *A proposed experiment to measure the angular correlation of electron relative to hyperon spin direction in the β-decay of the Λ^0*, 3/63 (KHR 22141), Barlow et al. (1965).
215. See Alvager, Nilsson & Kjellmann (1964).
216. Proposal: T. Alvager, F.J.M. Farley and J. Kjellmann, *Proposal for a test of the second postulate of special relativity in the GeV region,* 3/4/64 (WOL 23233); Alvager et al. (1964).
217. Some articles and books written on neutrino physics up to the sixties: Gell-Mann (1959), Lederman (1963, 1967, 1970), Fermi School (1964a), Jungk (1968), Mglic (1972), Reines (1960).

218. See Fermi (1934).
219. A short description is given in CERN History (1987), 37.
220. See Cowan et al. (1956).
221. In 1953 Dalitz and Fabri had pointed out that from the decay of theta and tau information about spin and parity of these particles could be gained. Research on these particles was carried out in many laboratories and by 1956 the data indicated that they would have different parity. This topic was discussed at length at the 1956 Rochester Conference (see Introduction by C.N. Yang in *HEP Conf. (1956),* VIII).
222. See Lee & Yang (1956). Lee and Yang were awarded the Nobel Prize in 1957 for the finding of parity violation in weak interactions.
223. See Wu et al. (1957).
224. E.C.G. Sudarshan and R.E. Marshak, 'The nature of the four-fermion interaction', in *Meson Conf. (1957),* VC-13; Sudarshan & Marshak (1958); Feynman & Gell-Mann (1958); Sakurai (1958).
225. See Nishijima (1957).
226. See Pontecorvo (1959) and comment by B. Pontecorvo to a talk by A.I. Alikhanov in Session VIII on 'Weak Interactions', *HEP Conf. (1959),* 233-236; Schwartz (1960); for a theoretical discussion of the subject see Lee & Yang (1960a,b), Cabibbo & Gatto (1960) and Yamagucchi (1960).
227. The last two paragraphs were based on: Memo Bernardini to Frisch, 11/3/60 (SC 22822); Dakin, *Note for a file PS Research Programme,* 29/3/60 (DIRADM 20262); Letter Bernardini to Lee, 31/3/60; Letter Bernardini to Steinberger, 13/4/60; Letter Bernardini to Zavattini, 23/5/60 all in (NP 22822).
228. Letter Bernardini to Lee, 31/3/60, Letter Bernardini to Cranshaw, 4/4/60 both in (NP 22822); Only the Ramm bubble chamber is foreseen in the proposal made by Krienen, Steinberger and Salmeron (see note 230).
    See draft of a *Proposal for a 6.5 ton neutrino detector* by H. Faissner 12/6/60 (DG 20890).
229. See CERN/SPC/112, 27/5/60.
230. G. Bernardini, 'The programme of 'neutrino experiments' at CERN', in *HEP Conf. (1960),* 581-585.
    L. Lederman, M. Schwartz and J.M. Gaillard, 'A neutrino detector for use at the Brookhaven AGS', in *Instr. Conf. (1960);* see also C.A. Ramm, 'The possibility of the detection of neutrino interactions in the CERN heavy-liquid bubble chamber', in *Instrum. Conf. (1960).*
    F. Krienen, J. Steinberger and R.A. Salmeron, *Proposal for an experiment to detect neutrino induced reactions,* PS/Int. EA 60-10, 12/9/60 (KHR 22139).
231. Letter Leprince-Ringuet to Adams, 26/9/60 (DG 20890); the quotation was translated by the author from the French original.
232. G. Bernardini, *Progress report on the status of the neutrino experiment projects at CERN,* 4/11/60, CERN/SPC/121.
233. See M.G.N. Hine, *Neutrino experiment: Layout and shielding,* 1/12/60 (DG 20890); Memo Bernardini to Hine, 9/12/60 (DIRADM 20278); Memo DG to Division Leaders and Members of the Directorate, 12/12/60 attached to CERN/DIR/15. A more personal impression on these discussions can be gained from two letters: Faissner to Weisskopf, 30/11/60 and 21/12/60 (DG 20890).
234. Reproduced from CERN/SPC/127/Rev., 18/4/61.
235. For a description of the situation and the progress made see for example *Minutes of meeting on the neutrino experiment held at CERN on February 2nd, 1961,* in (DG 20890) and G. Bernardini & M.G.N. Hine, *Progress report on the status of the neutrino experiment at CERN,* 18/4/61, CERN/SPC/127/Rev. The Ecole Polytechnique chamber carried out the first background and engineering preruns in March and April 1961.
    T.D. Lee gave an extensive lecture on neutrino questions in 1961 at CERN. See Seminar-CERN (1961).
236. G. von Dardel, *Counting rate in the planned experiment to detect neutrino induced reactions,* NP Int. Rep. 61-5, May 1961.
    He illustrates rather clearly the impossibility of this experiment by using the example of the Ecole Polytechnique chamber. Having a weight of 0.4 t it would register one event every 8 months running 75 hours per week.

Notes

237. Letter von Dardel to Weisskopf, 3/6/61 (DG 20889).
A series of letters of von Dardel to Weisskopf draw a rather colourful picture of the situation, 3/6/61, 10/6/61, 17/6/61 (DG 20889); see also letter Bernardini to Weisskopf, 16/6/61 (DG 20889).
238. See G. Bernardini, *The neutrino experiment,* CERN/SPC/138, 20/7/61; for the quotation see CERN/SPC/140, 27/9/61, p. 14.
Memo DG to Division Leaders, 14/2/62 (DIRADM 20278).
239. See Danby et al. (1962).
240. 21 Session of Council, 13/6/62, CERN/457, p. 16.
The future of the physics programme with the PS and the AGS were discussed at length in a common conference in September 1962. Among many other topics also neutrino physics was dealt with at length. See Meeting-Brookhaven (1962).
241. Memo DG to Staff members concerned, 3/7/62, CERN/DGO/Memo/1269.
242. See Van der Meer (1961). Van der Meer and de Raad suggested immediately the construction of an enhanced neutrino beam; see NPA/Int. 61-3, 28/2/61 (DG 20890).
243. See Faissner (1964), 201.
244. Reproduced from Fermi-School (1964a), 18.
245. Giesch et al. (1963); M. Giesch et al., 'Magnetic horn and neutrino flux calculations', in *Elementary Particle Conf. (1963),* 536–545.
246. See Kuiper & Plass (1959).
247. Reproduced from Faissner (1964), 194.
248. R. Bertolotto et al., 'The extracted 25 GeV/c proton beam for the CERN neutrino experiment', in *Elementary Particle Conf. (1963),* 523–535; A.G.H. (1963); Memo Weisskopf to all staff members, 21/6/63 (KHR 22139).
249. See *Discussion on spark chambers for the neutrino experiment,* 27/6 and 29/6/61 (DG 20889); R.A. Salmeron, *Comments on spark chambers for the neutrino experiment,* NP Int. Rep. 61-12; *Proposal for a 30-ton neutrino spark chamber* (Faissner group), 20/11/61 (GP 22566).
250. H. Faissner et al., 'Design and performance of the large spark chambers used in the high-energy neutrino experiment at CERN', in *Elementary Particle Conf. (1963),* 546–554.
251. Reproduced from Faissner (1964), 203.
252. For a description of the 1964 spark chamber see G. Bernardini, 'Neutrino Physics', in *HEP Conf. (1964),* 40–41.
253. This chapter was written on the basis of the following publications:
H.H. Bingham et al., 'CERN neutrino experiment – preliminary bubble chamber results', in *Elementary Particle Conf. (1963),* 555–570; G. Bernardini et al., 'CERN neutrino spark chamber experiment', in *Elementary Particle Conf. (1963),* 571–584; J.S. Bell, J. Lovseth and M. Veltman, 'CERN neutrino experiment: conclusion', in *Elementary Particle Conf. (1963),* 584–590; M.M. Block et al., 'Progress report on experimental study of neutrino interactions in the CERN heavy liquid bubble chamber', in *HEP Conf. (1964),* 7–15; G. Bernardini et al., 'Spark chamber study on the elastic production of muons and electrons by high-energy neutrinos', in *HEP Conf. (1964),* 16–23; G. Bernardini et al., 'Search for charged lepton pairs in high energy neutrino interactions', in *HEP Conf. (1964),* 24–28; Block et al. (1964), Bienlein et al. (1964), Bernardini et al. (1964, 1965). See also G. Bernardini, 'Weak interactions', in *Fermi School (1964a),* 1–51, H. Faissner (1964), Neutrino-results (1965), Franzinetti (1963), Weak Interaction Conf. (1963), Paty (1965), von Dardel (1964), Zichichi (1964).
254. 21 Session of Council, 13/6/1962, CERN/457, p. 16.
255. Letter von Dardel to Weisskopf, 3/6/61 (DG 20889).
256. See Jungk (1968), 102.
257. Memo Adams to Preiswerk, 11/1/61 (DG 20861).
258. See Jungk (1968), 111.
259. Letter Bernardini to Leprince-Ringuet, 31/10/60 (DG 20890).

260. Bernardini (1964).
261. See Neutrino Conf. (1965), Plass (1965).
262. The figure is reproduced from Weak Interaction Conf. (1965), 283.
263. See *Future of bubble chambers at CERN and in Europe,* CERN/SPC/194, 10/2/65. P. Germain, *Nouveau faisceau de neutrinos et l'utilisation des chambres à liquide lourds,* CERN/SPC/208, 24/9/65. A. Lagariggue, 'The 'Gargamelle' project', in *Neutrino Conf. (1965),* 263-266.

Bibliography

Accel.-Conf. (1953)   M.H. Blewett, *Notes from the conference on the alternating-gradient proton synchrotron,* Geneva, 26-28/10/53.

Accensi et al. (1966)   A. Accensi et al., 'Annihilation of 5.7 GeV/c $\bar{p}$ in hydrogen into four charged pions', *Physics Letters,* **20** (1966), 557-560.

Adams (1955)   J.B. Adams, 'The Alternating Gradient Proton Synchrotron', *Suppl. Nuovo Cimento,* **10** (1955), 355-374.

Adams (1956)   J.B. Adams, The Design of the Foundations for the Magnet of the CERN Alternating Gradient Proton Synchrotron (Geneva: CERN 56-21, 1956).

Adams (1957)   J.B. Adams, 'Some Engineering Problems of the CERN Proton Synchrotron', *Discovery* (July 1957), 286-292.

Adams (1960)   J.B. Adams, 'The CERN proton synchrotron', *Nature,* **185** (1960), 568-572.

Aderholz et al. (1964a)   M. Aderholz et al., 'The isospin and neutral decay of the f-meson', *Physics Letters,* **10** (1964), 240-242.

Aderholz et al. (1964b)   M. Aderholz et al., 'Elastic $\pi^+ p$ scattering at 4 GeV/c', *Physics Letters,* **10** (1964), 248-249.

Aderholz et al. (1964c)   M. Aderholz et al., 'The observation of two $\varrho\pi$ resonances', *Physics Letters,* **10** (1964), 226-228.

Aderholz et al. (1965)   M. Aderholz et al., 'Investigation of exchange mechanism in the reaction $\pi^+ p \to p\pi^+ \pi^0$ at 4 GeV/c', *Nuovo Cimento,* **35** (1965), 659-663.

A.G.H. (1963)   A.G.H., 'Fast ejection of protons from the CERN proton synchrotron', *CERN Courier,* **6** (1963), 79-82.

Alexander et al. (1964)   G. Alexander et al., 'Neutral strange particle production and $Y_1^*(1385)$ formation in pp collision at 5.5 GeV/c', *Physical Review Letters,* **13** (1964), 355-358.

Alexander et al. (1965)   G. Alexander et al., 'Peripheral production of $N^*(1518) \to p\pi^+\pi^-$, $N^*(1688) \to p\pi^+\pi^-$ resonances in pp $\to$ pp$\pi^+\pi^-$ reactions at 5.5 GeV/c', *Nuovo Cimento,* **40** (1965), 839.

Alff-Steinberger et al. (1966a)   C. Alff-Steinberger et al., 'K_S and K_L interference in the $\pi^+\pi^-$ decay mode, CP invariance and the K_S-K_L mass difference', *Physics Letters,* **20** (1966), 207-211.

Alff-Steinberger et al. (1966a)   C. Alff-Steinberger et al., 'Further results from the interference of K_S and K_L in the $\pi^+\pi^-$ decay modes', *Physics Letters,* **21** (1966), 595-597.

Alitti et al. (1963)   J.-Alitti et al., '$\pi^- p$ interactions at 1.59 GeV/c', *Nuovo Cimento,* **29** (1963), 515.

Alitti et al. (1964)   J. Alitti et al., 'Study of the reaction $\pi^- p \to \pi_{33}^{*++} \pi^- \pi^-$ and $\pi\pi$ elastic scattering', *Nuovo Cimento,* **35** (1964), 1.

Alitti et al. (1965)   J. Alitti et al., 'A_1 and A_2 resonances in $\pi^- p$ interactions at 2.75 GeV/c', *Physics Letters,* **15** (1965), 69-72.

Bibliography

Alitti et al. (1966) — J. Alitti et al., 'The reaction $\pi^- p \to \omega N^{*0}$ at 2.75 GeV/c', *Physics Letters*, **21B** (1966), 354.

Allard et al. (1964) — Allard et al., 'Properties of the A1 studied in 16 GeV/c π^- interactions on nuclei', *Physics Letters*, **12** (1964), 143–146.

Allard et al. (1965) — J.F. Allard et al., 'Coherent production of multipion final states in 6 to 18 GeV/c π^- interactions with nuclei', *Physics Letters*, **19** (1965), 431.

Allard et al. (1967) — J.F. Allard et al., 'Experimental results on 8 GeV/c $\pi^- p$ interactions with two, three and four pions', *Nuovo Cimento*, **50** (1967), 106.

Almeida et al. (1965a) — S.P. Almeida et al., 'Double isobar production in 10 GeV/c proton-proton collision', *Physics Letters*, **14** (1965), 240–242.

Almeida et al. (1965b) — S.P. Almeida et al., 'Resonant states in the $K^+ \pi^+ \pi^- p$ system', *Physics Letters*, **16** (1965), 184–186.

Alston et al. (1960) — M.H. Alston et al., 'Resonance in the $\Lambda\pi$ system', *Physical Review Letters*, **5** (1960), 520–524.

Alston et al. (1961a) — M.H. Alston et al., 'Resonance in the K-π system', *Physical Review Letters*, **6** (1961), 300–302.

Alston et al. (1961b) — M.H. Alston et al., 'Study of resonances in the Σ-π system', *Physical Review Letters*, **6** (1961), 698–702.

Alvager, Nilsson & Kjellmann (1964) — T. Alvager, A. Nilsson and J. Kjellmann, 'On the independence of the velocity of light of the motion of the light source', *Arkiv for Fysik*, **26** (1964), 209–221.

Alvager et al. (1964) — T. Alvager et al., 'Tests of the second postulate of special relativity in the GeV region', *Physics Letters*, **12** (1964), 260–262.

Alvarez et al. (1963) — L.W. Alvarez et al., '1660 MeV Y_1^*', *Physical Review Letters*, **10** (1963), 184–188.

Alvarez (1968) — L.W. Alvarez, 'Recent developments in particle physics', Nobel Lecture, 11/12/68.

Amaldi U. et al. (1962) — U. Amaldi et al., 'Deuteron production in p-p collisions at 26 GeV', *Bulletin of the American Physical Society*, **7** (1962), 348.

Amaldi U. et al. (1963a) — U. Amaldi et al., 'Deuteron and α-particle production in carbon and polyethylene by 26.6 GeV/c protons', *Nuovo Cimento*, **29** (1963), 476–486.

Amaldi U. et al. (1963b) — U. Amaldi et al., 'The CERN separated low-momentum antiproton beam', *Nuovo Cimento*, **30** (1963), 973.

Amaldi U. et al. (1964) — U. Amaldi et al., 'Antiproton-proton total cross-sections between 0.575 and 5.35 GeV/c', *Nuovo Cimento*, **34** (1964), 825–853.

Amato et al. (1963) — G. Amato et al., 'A one stage separated K-meson beam of 1.5 GeV/c momentum at the CPS', *Nuclear Instruments and Methods*, **20** (1963), 47–50.

Armenise et al. (1964) — N. Armenise et al., 'On the production of N^{++} 3/2 3/2 isobar in $\pi^+ p$ interactions at 2.75 GeV/c', *Physics Letters*, **13** (1964), 341–343.

Armenise et al. (1965a) — N. Armenise et al., '$\pi\pi$ interactions in $\pi^+ p \to n\pi\pi$ reactions at 2.75 GeV/c', *Nuovo Cimento*, **37** (1965), 361.

Armenise et al. (1965b) — N. Armenise et al., 'Quasi-two-body processes in the reaction $p\pi^+ \pi^+ \pi^- \pi^0$ at 2.75 GeV/c', *Nuovo Cimento*, **40A** (1965), 273.

Armenise et al. (1966) — N. Armenise et al., 'Exchange mechanism for $\pi^+ p \to N^{*++} \varrho^0$ reaction at 2.75 GeV/c', *Nuovo Dimento*, **41A** (1966), 159.

Armenise et al. (1967) — N. Armenise et al., 'A_2^0 production in $\pi^+ d$ interaction', *Physics Letters*, **25B** (1967), 53–56.

Armenteros et al. (1964) — R. Armenteros et al., 'Evidence for a (K$\pi\pi$) resonance with a mass of 1230 MeV/c^2', *Physics Letters*, **9** (1964), 207–209.

Armenteros et al. (1965a)   R. Armenteros et al., 'Experimental results on the annihilation $p\bar{p} \to \bar{K} K\pi$ at rest: K^* production', *Physics Letters*, **17** (1965), 170–171.

Armenteros et al. (1965b)   R. Armenteros et al., 'Experimental results on the annihilation $p\bar{p} \to \bar{K} K\pi$ at rest: non-K^* resonating events', *Physics Letters*, **17** (1965), 344–347.

Armenteros et al. (1965c)   R. Armenteros et al., 'Mass, width and branching fraction of $Y^*(1385)$', *Physics Letters*, **19** (1965), 75.

Armenteros et al. (1965d)   R. Armenteros et al., 'Spin, Parity and Isospin of $Y^*(1760)$', *Physics Letters*, **19** (1965), 338.

Armenteros et al. (1967)   R. Armenteros et al., 'Formation of resonant states in $K^-p \to KN$ between 780 and 1220 MeV/c', *Nuclear Physics*, **B3** (1967), 592–600.

Ashmore et al. (1960)   A. Ashmore et al., 'Total cross-section of protons with momentum between 10 and 28 GeV/c', *Physical Review Letters*, **5** (1960), 576–578.

Astbury et al. (1965a)   P. Astbury et al., '$K_2^0 \to \pi^+\pi^-\pi^0$ decay: branching ratio and π^0 energy spectrum', *Physics Letters*, **18** (1965), 175–178.

Astbury et al. (1965b)   P. Astbury et al., 'A measurement of the mean life of the K_2^0', *Physics Letters*, **18** (1965), 178–181.

Astbury et al. (1965c)   P. Astbury et al., 'Branching ratios of K_2^0 decays', *Physics Letters*, **16** (1965), 80–83.

Astbury et al. (1966a)   P. Astbury et al., 'The charge exchange $K^- + p \to \bar{K}^0 + n$ at 9.50 GeV/c', *Physics Letters*, **16** (1966), 328–331.

Astbury et al. (1966b)   P. Astbury et al., 'The charge exchange $\bar{p}p \to \bar{n}n$ at 5, 6, 7 and 9 GeV/c', *Physics Letters*, **22** (1966), 537–539.

Aubert et al. (1963)   B. Aubert et al., 'Low energy separated beam at the CERN PS', *Nuclear Instruments and Methods*, **20** (1963), 51–54.

Aubert et al. (1964a)   B. Aubert et al., 'Direct observation of the decay mode $K_2^0 \to \pi^0\pi^0\pi^0$', *Physics Letters*, **10** (1964), 250–251.

Aubert et al. (1964b)   B. Aubert et al., 'Further results on the $\Delta S = \Delta Q$ rule in $K^0\beta$ decay and K_1^0-K_2^0 mass', *Physics Letters*, **10** (1964), 215–219.

Backenstoss et al. (1961)   G. Backenstoss et al., 'Helicity of μ^- mesons from π-meson decay', *Physical Review Letters*, **6** (1961), 415–416.

Backenstoss et al. (1963)   G. Backenstoss et al., 'A total absorption scintillation detector for electrons, photons and other particles in the GeV region', *Nuclear Instruments and Methods*, **21** (1963), 155–160.

Backenstoss et al. (1964)   G. Backenstoss et al., 'Forward production of photons from 9 GeV π-p interactions', *Nuovo Cimento*, **23** (1964), 14–17.

Backenstoss et al. (1966)   G. Backenstoss et al., 'π-p charge exchange scattering at 10 GeV/c', *Nuovo Cimento*, **42** (1966), 814–821.

Badier et al. (1965a)   J. Badier et al., 'Baryonic states of strangeness-2 produced in K^-p interaction at 3 GeV/c', *Physics Letters*, **16** (1965), 171–174.

Badier et al. (1965b)   J. Badier et al., 'On the decay of the χ and ϕ mesons', *Physics Letters*, **17** (1965), 337–340.

Badier et al. (1965c)   J. Badier et al., 'Spin, isospin and decay rates of $K^*(1400)$', *Physics Letters*, **19** (1965), 612.

Baglin et al. (1963)   Baglin et al., 'Correlations in strange-particle pairs produced by π^- of 6, 11 and 18 GeV/c', *Nuovo Cimento*, **29** (1963), 339–370.

Baglin et al. (1964)   C. Baglin et al., 'Branching ratio for the muonic decay mode of the lambda hyperon', *Physics Letters*, **11** (1964), 357.

Baillon et al. (1967)   P. Baillon et al., 'Further study of the E-meson in antiproton annihilation at rest', *Nuovo Cimento*, **50A** (1967), 393–421.

Baker et al. (1966)	W.F. Baker et al., 'The angular distribution of backward elastically scattered pions at 3.55 GeV/c', *Physics Letters,* **23** (1966), 605-609.
Barlow et al. (1965)	J. Barlow et al., 'Measurement of the electron asymmetry in the leptonic decay of polarized Λ^0', *Physics Letters,* **18** (1965), 64-68.
Barnes et al. (1964)	V.E. Barnes et al., 'Observation of a hyperon with strangeness minus three', *Physical Review Letters,* **12** (1964), 204-206.
Bartke et al. (1962a)	J. Bartke et al., 'Hyperon and kaon production by 16 GeV/c negative pions on protons', *Nuovo Cimento,* **24** (1962), 876-895.
Bartke et al. (1962b)	J. Bartke et al., 'Hyperon and kaon production by 24.5 GeV/c protons on protons', *Nuovo Cimento,* **29** (1963), 8-18.
Bartsch et al. (1964)	J. Bartsch et al., 'On the production of the 3/2 3/2 isobar in 4 GeV/c π^+p interactions', *Physics Letters,* **10** (1964), 229-231.
Bartsch et al. (1967)	J. Bartsch et al., 'Quantum number assignment for the A_2^0 meson', *Physics Letters,* **25B** (1967), 48-52.
Baton et al. (1964)	J.P. Baton et al., 'Single pion production in π^-p interactions at 2.75 GeV/c', *Nuovo Cimento,* **35** (1964), 713.
Bellettini et al. (1965a)	G. Bellettini et al., 'Absolute measurements of proton-proton small-angle elastic scattering and total cross-sections at 10, 19 and 26 GeV/c', *Physics Letters,* **14** (1965), 164-168.
Bellettini et al. (1965b)	G. Bellettini et al., 'The real part of the proton-neutron scattering amplitude at 19.3 GeV/c', *Physics Letters,* **19** (1965), 341-345.
Bellettini et al. (1965c)	G. Bellettini et al., *A sonic spark chamber system* (Geneva: CERN 65-33, 1965).
Bellettini et al. (1965d)	G. Bellettini et al., 'Evidence for a 1.40 GeV nucleon resonance from high energy inelastic proton scattering in hydrogen and deuterium', *Physics Letters,* **18** (1965), 167-171.
Bellettini et al. (1966a)	G. Bellettini et al., 'Proton-proton small angle scattering and total cross-section at 10.0 GeV/c', *Physics Letters,* **19** (1966), 705.
Bellettini et al. (1966b)	G. Bellettini et al., 'Proton-nuclei cross-sections at 20 GeV', *Nuclear Physics,* **79** (1966), 609-624.
Bellini et al. (1963)	G. Bellini et al., 'Glancing collisions of pions with protons at 6.1 and 18.1 GeV/c', *Nuovo Cimento,* **27** (1963), 816-833.
Bellini et al. (1964)	G. Bellini et al., 'Experimental results on $\pi^+\pi^-$ production by 18.1 GeV/c π^-', *Physics Letters,* **10** (1964), 126-128.
Bellini et al. (1965)	G. Bellini et al., 'A study of 6.1 GeV/c π^- interactions in a heavy liquid bubble chamber', *Nuovo Cimento,* **40** (1965), 948-978.
Benot et al. (1963)	M. Benot et al., 'La mesure des photographies de chambres à bulles au CERN', *Industries Atomiques,* **7/8** (1963), 54-69.
Bernardini (1964)	G. Bernardini, 'Some motivations for a very costly long-range programme on high-energy neutrino physics at C.E.R.N.', *Proceedings of the Royal Society A,* **278** (1964), 369-390.
Bernardini et al. (1964)	G. Bernardini et al., 'Search for intermediate boson production in high-energy neutrino interactions', *Physics Letters,* **13** (1964), 86-91.
Bernardini et al. (1965)	G. Bernardini et al., 'Lower Limit for the Mass of the Intermediate Boson', *Nuovo Cimento,* **38** (1965), 608-622.
Bettini et al. (1965)	A. Bettini et al., 'The K^+ proton interactions at 1.45 GeV/c', *Physics Letters,* **16** (1965), 83-86.
Bezaguet et al. (1964a)	A. Bezaguet et al., 'A study of the beta-decay of $\Sigma^\pm$ hyperons', *Physics Letters,* **11** (1964), 262.

Bezaguet et al. (1964b)	A. Bezaguet et al., 'On the upper limit of the decay $\omega^0 \to e^+e^-$', *Physics Letters*, **12** (1964), 70.
Bienlein et al. (1964)	J.K. Bienlein et al., 'Spark chamber study of high-energy neutrino interactions', *Physics Letters*, **13** (1964), 80-86.
Bigi et al. (1964a)	A. Bigi et al., 'The production o K^0K^0 pairs by 10 GeV/c negative pions on protons', *Nuovo Cimento*, **33** (1964), 1249.
Bigi et al. (1964b)	A. Bigi et al., 'On the associated production of K-Y pairs by 10 GeV/c π^- in the hydrogen bubble chamber', *Nuovo Cimento*, **34** (1964), 1122.
Bingham et al. (1964)	H.H. Bingham et al., 'An unsuccessful search for fractionally charged particles with mass ≤ 2.2 GeV', *Physics Letters*, **9** (1964), 201-203.
Blackall et al. (1965)	P. Blackall et al., 'Search for hyperon resonances with spark chambers', *Physics Letters*, **16** (1965), 336-338.
Bleuler et al. (1963)	E. Bleuler et al., 'The application of thin-plate spark chambers to high-energy π-p experiments', *Nuclear Instruments and Methods*, **20** (1963), 208-212.
Blieden et al. (1965)	H.R. Blieden et al., 'Observation of ϱ^- meson with a missing mass spectrometer operating in region of Jacobian peaks', *Physics Letters*, **19** (1965), 444-448.
Blieden et al. (1966a)	H.R. Blieden et al., 'Evidence for deviation from one-pion exchange mechanism in ϱ^- production', *Physics Letters*, **19** (1966), 708-712.
Blieden et al. (1966b)	H.R. Blieden et al., 'Observation of charged ϱ-meson and its properties using a missing mass spectrometer', *Nuovo Cimento*, **43** (1966), 71-83.
Bloch et al. (1961)	Bloch et al., 'A 300-liter heavy liquid bubble chamber', *Review of Scientific Instruments*, **32** (1961), 1302-1307.
Block et al. (1964)	M.M. Block et al., 'Neutrino interactions in the CERN heavy liquid bubble chamber', *Physics Letters*, **12** (1964), 281-285.
Blum et al. (1964)	W. Blum, 'Search for fractionally charged particles produced by 27.5 GeV/c protons', *Physical Review Letters*, **13** (1964), 353-355.
Boeck et al. (1964)	R.K. Boeck et al., 'Evidence for $(K\pi\pi)$ resonance with $T_z = \pm 3/2$ at 1270 MeV', *Physics Letters*, **12** (1964), 65-67.
Boeck et al. (1965)	R.K. Boeck et al., 'A search for heavy hyperon resonances produced by 5.7 GeV/c antiprotons in hydrogen', *Physics Letters*, **17** (1965), 166-168.
Boeckmann et al. (1965)	K. Boeckmann et al., 'Investigation of the reaction $p\bar{p} \to N_{33}^* \bar{N}_{33}^*$', *Physics Letters*, **15** (1965), 356.
Boeckmann et al. (1966)	K. Boeckmann et al., 'Elastic scattering, pion production and annihilation into pions in $\bar{p}p$ interactions at 5.7 GeV/c', *Nuovo Cimento*, **42A** (1966), 954.
Bondar et al. (1963a)	L. Bondar et al., 'Further evidence for the f^0-meson and a determination of its spin', *Physics Letters*, **5** (1963), 153-156.
Bondar et al. (1963b)	L. Bondar et al., 'Experimental results of the cross-section for $\pi^- \pi^0 \to \pi^- \omega^0$', *Physics Letters*, **5** (1963), 209-211.
Bondar et al. (1964a)	L. Bondar et al., '$\pi^- p$ interactions at 4 GeV/c. I. Four-prong events', *Nuovo Cimento*, **31** (1964), 485-519.
Bondar et al. (1964b)	L. Bondar et al., '$\pi^- p$ interactions at 4 GeV/c. II. Two-prong events', *Nuovo Cimento*, **31** (1964), 729-747.
Bott-Bodenhausen et al. (1966)	M. Bott-Bodenhausen et al., 'Time dependent interference effects in two-pion decays of neutral kaons', *Physics Letters*, **20**, (1966), 212-215.
Brandt et al. (1963)	S. Brandt et al., 'Negative pion-proton scattering at 10 GeV/c', *Physical Review Letters*, **10** (1963), 413-415.

Brown et al. (1962)	H.N. Brown et al., 'Observation of production of a $\Xi^- + \bar{\Xi}^+$ pair', *Physical Review Letters*, **8** (1962), 255-257.
Bruyant et al. (1964a)	F. Bruyant et al., 'Production of neutral particles in π^+d interactions at 6 GeV/c and I-spin of f^0', *Physics Letters*, **10** (1964), 232-234.
Bruyant et al. (1964b)	F. Bruyant et al., 'Cross-section for the charge exchange reaction $\pi^+ n \to p\pi^0$ at 6 GeV/c pion momentum', *Physics Letters*, **12** (1964), 278-280.
Buhler-Broglin et al. (1965)	A. Buhler-Broglin et al., 'An upper limit to the existence of charges (2/3)e in the cosmic radiation at 500 meters above sea level', *Nuovo Cimento*, **40A** (1965), 589.
Buhler-Broglin et al. (1966)	A. Buhler-Broglin et al., 'Search for charges 1/3e and 2/3e in the cosmic radiation', *Nuovo Cimento*, **45** (1966), 520-526.
Cabibbo & Gatto (1960)	N. Cabibbo and R. Gatto, 'Cross sections of reactions produced by high energy neutrino beams', *Nuovo Cimento*, **15** (1960), 304-310.
Cabibbo (1963)	N. Cabibbo, 'Unitary symmetry and leptonic decays', *Physical Review Letters*, **10** (1963), 531-533.
Caldwell et al. (1962)	D.O. Caldwell et al. 'Production of ϱ^0 mesons at 12 and 17 MeV', *Physics Letters*, **2** (1962), 253.
Caldwell et al. (1964)	D.O. Caldwell et al., '$\pi^\pm$-p and p-p elastic scattering at 8.5, 12.4 and 18.4 GeV/c', *Physics Letters*, **8**, (1964), 288-292.
CERN-Saclay (1962)	CERN and Saclay, 'Example of anticascade ($\bar{\Xi}^+$) particle production in $\bar{p}$-p interactions at 3.0 GeV/c', *Physical Review Letters*, **8** (1962), 257-260.
CERN History (1987)	A. Hermann et al., *History of CERN, I. Launching the European Organization for Nuclear Research* (Amsterdam: North-Holland, 1987).
Chadwick et al. (1963)	G.B. Chadwick et al., 'Study of the annihilation of stopped antiprotons in hydrogen: the reaction $\bar{p}p \to \pi^+\pi^-\pi^0$', *Physical Review Letters*, **10** (1963), 62.
Chew & Frautschi (1961)	G.F. Chew and S.C. Frautschi, 'Principle of equivalence for all strongly interacting particles within the S-matrix framework', *Physical Review Letters*, **7** (1961), 394-397.
Chew et al. (1964)	G.F. Chew, M. Gell-Mann and A.H. Rosenfeld, 'Strongly interacting particles', *Scientific American*, Feb. (1964), 74-93.
Chew & Low (1959)	G.F. Chew and F.E. Low, 'Unstable particles as targets in scattering experiments', *Physical Review*, **113** (1959), 1640-1649.
Chikovani et al. (1966)	G. Chikovani et al., 'Evidence for three new charged bosons of masses 1929, 2195 and 2382 MeV and narrow widths', *Physics Letters*, **22** (1966), 233-236.
Christenson et al. (1964)	J.H. Christenson et al., 'Evidence for the 2π decay of the K_2^0', *Physical Review Letters*, **13** (1964), 138-40.
Christofilos (1950)	N. Christofilos, 'Focusing System of Ions and Electrons', U.S. Patent No. 2, 736, 799 reprinted in *Livingston (1966)*.
Cnops et al. (1966)	A.M. Cnops et al., 'Test of C invariance in the three pion decay of the meson', *Physics Letters*, **22** (1966), 546-550.
Cocconi G. (1964)	G. Cocconi, 'Proton-proton elastic scattering at high energies', *Nuovo Cimento*, **33** (1964), 643-653.
Cocconi G. et al. (1961a)	G. Cocconi et al., 'Elastic and quasi-elastic collisions of protons with momenta between 9 and 25 GeV/c', *Physical Review Letters*, **6** (1961), 231-234.
Cocconi G. et al. (1961b)	G. Cocconi et al., 'Elastic and inelastic collisions of high-energy protons from hydrogen', *Physical Review Letters*, **7** (1961), 450-454.
Cocconi G. et al. (1963)	G. Cocconi et al., 'Differential cross-sections for the reaction $p+p \to \pi^+ + d$ between 2.1 and 8.9 GeV/c', *Physics Letters*, **7** (1963), 222-224.

Cocconi G. et al. (1964)	G. Cocconi et al., 'Inelastic proton-proton scattering and nucleon isobar production', *Physics Letters,* **8** (1964), 134–137.
Cocconi V.T. et al. (1960)	V.T. Cocconi et al., 'Mass analysis of the secondary particles produced by the 25 GeV proton beam of the CERN proton synchrotron', *Physical Review Letters,* **5** (1960), 19–21.
Colley et al. (1965)	D.C. Colley et al., 'An Ω^- particle produced by a 6 GeV/c K$^-$ meson in hydrogen', *Physics Letters,* **19** (1965), 152.
Cooper et al. (1964)	W.A. Cooper et al. 'Difference in mass between Y_1^{*-} and Y_1^{*+} isobars', *Physics Letters,* **8** (1964), 365–367.
Conversi et al. (1963)	M. Conversi et al., 'Search for time-like structure of the proton', *Physics Letters,* **5** (1963), 195–198.
Conversi et al. (1965)	M. Conversi et al., 'The leptonic annihilation modes of the proton-antiproton system at 6.8 (GeV/c)^2 timelike four-momentum transfer', *Nuovo Cimento,* **40A** (1965), 690–701.
Courant et al. (1952)	E.D. Courant, M.S. Livingston and H.S. Snyder, 'The Strong-Focussing Synchroton – A new High Energy Accelerator', *Physical Review,* **88** (1952), 1190–1196.
Courant et al. (1963)	H. Courant et al., 'Determination of the relative Σ-Λ parity', *Physical Review Letters,* **10** (1963), 409–412.
Cowan et al. (1956)	C.L. Cowan Jr. et al., *Science,* **124** (1956), 103.
Czyzewski et al. (1965)	O. Czyzewski et al., 'Antiproton-proton elastic scattering at 4 GeV/c and derivation of diffraction slope at infinite energy', *Physics Letters,* **15** (1965), 188.
Czyzewski et al. (1966)	O. Czyzewski et al., 'Charge exchange and the reaction p$\bar{\text{p}} \to \bar{\text{n}}n\pi^+\pi^-$ at 3.0, 3.6 and 4.0 GeV/c antiprotons', *Physics Letters,* **20** (1966), 554.
Dalitz (1963)	R.H. Dalitz, 'Strange-particle resonant states', *Annual Review of Nuclear Science,* **13** (1963), 339–430.
Danby et al. (1962)	G. Danby et al., 'Observation of high-energy neutrino reactions and the existence of two kinds of neutrinos', *Physical Review Letters,* **9** (1962), 36–44.
D'Andlau et al. (1963)	Ch. D'Andlau et al., 'Nouvelle détermination du rapport de branchement (modes neutres/ensemble des modes) de la désintégration du méson ω^0', *Compte rendu de l'Academie des Sciences,* **256** (1963), 1279–1281.
D'Andlau et al. (1965)	Ch. D'Andlau et al., 'Evidence for a non-strange meson of mass 1290 MeV', *Physics Letters,* **17** (1965), 347–352.
De Baere et al. (1966)	De Baere et al., 'K$^+$p elastic scattering at 3.5 and 5 GeV/c', *Nuovo Cimento,* **45** (1966), 885–894.
De Bouard et al. (1965)	X. de Bouard et al., 'Two-pion decay of K_2^0 at 10 GeV/c', *Physics Letters,* **15** (1965), 58–61.
Dehne et al. (1964a)	H.C. Dehne et al., 'Proton-antiproton interactions at 3.6 GeV/c and one pion exchange', *Physics Letters,* **9** (1964), 185–187.
Dehne et al. (1964b)	H.C. Dehne et al., 'Pion production without annihilation in antiproton-proton interactions at 3.6 GeV/c', *Physical Review,* **136** (1964), B843.
Dekkers et al. (1964)	D. Dekkers et al., 'Experimental study of the π^++d $\to$ p+p reaction between 0.65 and 1.95 GeV/c', *Physics Letters,* **11** (1964), 161–164.
Dekkers et al. (1965)	D. Dekkers et al., 'Experimental study of particle production at small angles in nucleon-nucleon collisions at 19 and 23 GeV/c', *Physical Review,* **137** (1965), B962–B978.

Deutschmann et al. (1964)	M. Deutschmann et al., 'Resonance production by 8 GeV/c positive pions on protons', *Physics Letters,* **12** (1964), 356–360.
Deutschmann et al. (1965a)	M. Deutschmann et al., 'Single π^0 production by 8 GeV/c positive pions and possible enhancements in the ($\pi^+\pi^0$) system', *Physics Letters,* **18** (1965), 351.
Deutschmann et al. (1965b)	M. Deutschmann et al., 'Slope of dσ/dt distributions in quasi-two-body interactions of 8 GeV/c positive pions', *Physics Letters,* **19** (1965), 608.
Deutschmann et al. (1966)	M. Deutschmann et al., 'Rare decay modes of ϱ, A_1 and A_2 mesons', *Physics Letters,* **20** (1966), 82.
Diddens et al. (1962a)	A.N. Diddens et al., 'High-energy nucleon-nucleon total cross-sections', *Physical Review Letters,* **9** (1962), 32–34.
Diddens et al. (1962b)	A.N. Diddens et al., 'High-energy proton-proton diffraction scattering', *Physical Review Letters,* **9** (1962), 108–111.
Diddens et al. (1962c)	A.N. Diddens et al., 'High-energy proton-proton scattering', *Physical Review Letters,* **9** (1962), 111–114.
Di Giugno et al. (1966a)	G. Di Giugno et al., 'Determination of the branching ratios among the neutral decay modes of the η particle', *Physical Review Letters,* **16** (1966), 767–771.
Di Giugno et al. (1966b)	G. Di Giugno et al., 'A measurement of the branching ratio $\omega \to$ neutrals/$\omega \to \pi^+\pi^-\pi^0$', *Nuovo Cimento,* **44** (1966), 1272–1275.
Dirac (1931)	P.A.M. Dirac, *Proceedings of the Royal Society,* **A133** (1931).
Dowell et al. (1962)	J.D. Dowell et al., 'Strange particle spectroscopy', *Physics Letters,* **1** (1962), 53–56.
Drell & Hiida (1961)	S.D. Drell and K. Hiida, 'Quasi-elastic peak in high-energy nucleon-nucleon scattering', *Physical Review Letters,* **7** (1961), 199.
Drevermann et al. (1967)	H. Drevermann et al., 'Study of six prong π^+p interactions at 5 GeV/c', *Physical Review,* **161** (1967), 1356.
Elementary Particle Conf. (1961)	*The Aix-en-Provence international conference on elementary particles,* 14–20 September 1961, Vol. I and II.
Elementary Particle Conf. (1963)	*Proceedings of the Sienna international conference on elementary particles,* Sienna, 30 September–5 October 1963, Vol. 1, Parallel Sessions.
Elementary Particle Conf. (1965)	*Oxford international conference on elementary particles,* 19–25/9/65 (The Rutherford High Energy Laboratory, 1966).
Elementary Particle Discussion (1963)	'A discussion on recent European contributions to the development of the physics of elementary particle', *Proceedings of the Royal Society A,* **278** (1964), 287–464.
EPHLC (1961)	'The BP3 heavy liquids chamber', *L'Onde Electrique,* **417** (1961), 976–1000.
Erwin et al. (1961)	A.R. Erwin et al., 'Evidence for a $\pi\pi$ resonance in the $I=1$, $J=1$ state', *Physical Review Letters,* **6** (1961), 628–630.
Fabiani et al. (1962)	F. Fabiani et al., 'Positron-electron annihilation in flight between 2 and 10 GeV', *Nuovo Cimento,* **25** (1962), 655–669.
Faissner (1964)	H. Faissner, 'Weak Interactions without Change of Strangeness (Experimental)'; talk given on the *III Internationale Hochschulwochen für Kernphysik: Weak Interactions and Higher Symmetries* held in Schladming from 24 February to 7 March 1964, published by Acta Physica Austriaca (Wien: Springer-Verlag, 1964).
Fermi (1934)	E. Fermi, 'Versuch einer Theorie der β-Strahlen', *Zeitschrift für Physik,* **88** (1934), 161–177.

Fermi School (1964a)	*Weak Interactions and High Energy Neutrino Physics,* Proceedings of the International School of Physics Enrico Fermi, Course XXXII, Varenna, 15–27 June 1964 (New York: Academic Press, 1966).
Fermi School (1964b)	*Strong Intractions,* Proceedings of the International School of Physics Enrico Fermi, Course XXXIII, Varenna, 6–18 July 1964 (New York: Academic Press, 1966).
Ferro-Luzzi et al. (1964)	M. Ferro-Luzzi et al., 'Evidence for $\varkappa(725)$ in K^+p interaction at 3 GeV/c', *Physics Letters,* **12** (1964), 255–259.
Ferro-Luzzi et al. (1965a)	M. Ferro-Luzzi et al., 'Evidence for a meson resonance with strangeness $+2$', *Physics Letters,* **17** (1965), 155–159.
Ferro-Luzzi et al. (1965b)	M. Ferro-Luzzi et al., 'The reaction $K^+p \to K^0 p\pi^+$ at 3 GeV/c', *Nuovo Cimento,* **36** (1965), 1101–1120.
Feynman & Gell-Mann (1958)	R.P. Feynman and M. Gell-Mann, 'Theory of the Fermi interaction', *Physical Review,* **109** (1958), 193–198.
Fidecaro G. et al. (1966)	G. Fidecaro et al., 'Determination of an upper limit for the partial width $\Gamma(\varrho\pi\gamma)$', *Physics Letters,* **23** (1966), 163–166.
Fidecaro M. et al. (1961)	M. Fidecaro et al., 'Experimental investigation of a neutral beam from the CERN 25 GeV proton synchrotron', *Nuovo Cimento,* **19** (1961), 382–392.
Fidecaro M. et al. (1962)	M. Fidecaro et al., 'Photon production in proton-proton collisions at 23.1 GeV', *Nuovo Cimento,* **24** (1962), 73–86.
Fidecaro M. et al. (1961)	M. Fidecaro, G. Finocchiaro and G. Giacomelli, 'Search for magnetic monopoles', *Nuovo Cimento,* **22** (1961), 657–660.
Fidecaro M. et al. (1962)	M. Fidecaro, G. Finocchiaro and G. Giacomelli, 'Absorption of 10 and 13.5 GeV photons', *Nuovo Cimento,* **23** (1962), 800–806.
Focacci et al. (1966)	M.N. Focacci et al., 'Mass spectrum of bosons from 500 to 2500 MeV in the reaction $\pi^-+p \to p+X^-$ observed by a missing-mass spectrometer', *Physical Review Letters,* **17** (1966), 890–893.
Focardi et al. (1965)	S. Focardi et al., 'Production and decay of $K^*(1400)$ in 3 GeV/c K^-p interactions', *Physics Letters,* **16** (1965), 351.
Forino et al. (1964)	A. Forino et al., 'Pion system in π^+d interactions at 4.5 GeV/c', *Physics Letters,* **11** (1964), 347–350.
Forino et al. (1965)	A. Forino et al., 'Double pion production in π^+d interactions at 4.5 GeV/c', *Physics Letters,* **19** (1965), 68.
Fowler & Samios (1964)	W.B. Fowler and N.P. Samios, 'The omega-minus experiment', *Scientific American,* Oct. (1964), 36–45.
Franzinetti (1963)	C. Franzinetti (ed.), *The 1963 NPA Seminars – The Neutrino Experiment* (Geneva: CERN 63-37, 1963).
Gatti et al. (1961)	G. Gatti et al., 'Total absorption lead glass Cerenkov counter', *Review of Scientific Instruments,* **32** (1961), 949–952.
Gell-Mann (1959)	M. Gell-Mann, 'Status of Weak Interactions', *Reviews of Modern Physics,* **31** (1959), 834–838.
Gell-Mann (1962)	M. Gell-Mann, 'Symmetries of baryons and mesons', *Physical Review,* **125** (1962), 1067–1084.
Gell-Mann (1964)	M. Gell-Mann, 'A schematic model of baryons and mesons', *Physical Letters,* **8** (1964), 214–215.
Gell-Mann & Ne'eman (1964)	M. Gell-Mann and Y. Ne'eman, 'The Eightfold Way', in *Frontiers of Physics,* (New York: W.A. Benjamin, 1964).
Giesch et al. (1963)	M. Giesch et al., 'Status of magnetic horn and neutrino beam', *Nuclear Instruments and Methods,* **20** (1963), 58–65.

Bibliography

Goldberg et al. (1965)	M. Goldberg et al., 'Evidence for a new $\pi^+\pi^-$ resonance at 1.67 GeV/c', *Physics Letters,* **17** (1965), 354-357.
Goldschmidt-Clermont et al. (1966)	Y. Goldschmidt-Clermont et al., 'Two-body channels in the interactions of 3, 3.5 and 5 GeV/c positive kaons on hydrogen: possibility of Regge-pole exchange', *Nuovo Cimento,* **46A** (1966), 539.
Goldsack et al. (1962)	S.J. Goldsack et al., 'Meson production by 16 GeV/c π^- mesons in hydrogen', *Nuovo Cimento,* **23** (1962), 941-953.
Gregory (1961)	B. Gregory, 'Les expériences au CERN', *L'Onde Electrique,* **417** (1961), 1034-1041.
Hagedorn (1960a)	R. Hagedorn, 'A new derivation of the statistical theory of particle production with numerical results for pp collisions at 25 GeV', *Nuovo Cimento,* **15** (1960), 434-461.
Hagedorn (1960b)	R. Hagedorn, 'Deuteron production in high-energy collision', *Physical Review Letters,* **5** (1960), 276-277.
Haque et al. (1965)	N. Haque et al., 'K^* resonances produced by 3.5 GeV/c K^- interactions in hydrogen', *Physics Letters,* **14** (1965), 338-342.
Harting et al. (1965)	D. Harting et al., '$\pi^\pm$-p and p-p elastic scattering at 8.5, 12.4 and 18.4 GeV/c', *Nuovo Cimento,* **38** (1965), 60- .
HBC 81 (1961)	'La chambre à bulle à hydrogène liquide CBH 81', *L'Onde Electrique,* **417** (1961), 1001-1041.
HEP Conf. (1956)	*Proceedings of the sixth annual Rochester conference on high energy nuclear physics,* 3-7 April 1956.
HEP Conf. (1959)	*Ninth international annual conference on high energy physics,* Kiev, 15-25 July 1959.
HEP Conf. (1960)	*Proceedings of the 1960 annual international conference on high energy physics,* Rochester, 25/8-1/9/60 (New York: Interscience Publishers, 1960).
HEP Conf. (1962)	*1962 International conference on high-energy physics at CERN,* CERN, 4-11/7/62.
HEP Conf. (1964)	*XII International conferrence on high energy physics,* Dubna, 5-15/8/64 (Moscow: Atomizdat, 1966), Vol. I and II.
Herz & Lock (1966)	A.J. Herz and W.O. Lock, 'The particle detectors, 1. Nuclear Emulsions', *CERN Courier, May* (1966), 83-87.
Instr. Conf. (1960)	*Proceedings of the international conference on instrumentation for high-energy physics,* LRL, 12-14 September 1960 (New York: Interscience Publishers, 1960).
Jones et al. (1966a)	L.W. Jones et al., 'The mass spectra of recoil nucleon states from inelastic ϱ^0 production by pions of 12 and 18 GeV/c', *Nuovo Cimento,* **44** (1966), 915-926.
Jones et al. (1966b)	L.W. Jones et al., 'The $\pi^+\pi^-$ interaction as studied in 12 and 18 GeV pion-nucleon reactions', *Physics Letters,* **21** (1966), 590-594.
Jones et al. (1968)	L.W. Jones et al., 'Peripheral di-pion production by pions of 12 and 18 GeV/c', *Physics Review,* **166** (1968), 1405-1430.
Jungk (1968)	R. Jungk, *The big machine* (New York: Sciber, 1968).
Kienzle et al. (1965)	W. Kienzle et al., 'Evidence for a singly charged boson of mass 962 MeV and narrow width', *Physics Letters,* **19** (1965), 438-441.
Krige & Pestre (1986)	J. Krige and D. Pestre, 'The choice of CERN's first large bubble chamber for the proton synchrotron (1957-1958)', *Historical Studies in Physical Sciences,* **16** (1986), 255-279.

Kuiper & Plass (1959)	B. Kuiper and G. Plass, *Fast extraction of particles from a 25 GeV proton synchrotron* (Geneva: CERN 59-30, 1959).
Lefebvres et al. (1965)	F. Lefebvres et al., 'Boson spectrum from 1.1 to 1.6 GeV and properties of A_2 meson observed by missing-mass spectrometer', *Physics Letters*, **19** (1965), 434–437.
Lederman (1963)	L.M. Lederman, *Neutrino Physics,* Brookhaven Lecture Series Number 23, 9 January 1963, BNL 787 (T-300).
Lederman (1967)	L.M. Lederman, 'Neutrino Physics' in E.M.S. Burhop (ed.), *High energy Physics II* (New York: Academic Press, 1967), 303–364.
Lederman (1970)	L.M. Lederman, 'Resource Letter Neu-1 History of the Neutrino', *American Journal of Physics,* **38** (1970), 129–136.
Lee & Yang (1956)	T.D. Lee and C.N. Yang, 'Question of Parity Conservation in Weak Interactions', *Physical Review,* **104** (1956), 254–258.
Lee & Yang (1960a)	T.D. Lee and C.N. Yang, 'Theoretical discussion on possible high-energy neutrino experiments', *Physical Review Letters,* **4** (1960), 307–311.
Lee & Yang (1960b)	T.D. Lee and C.N. Yang, 'Implications of the Intermediate Boson Basis of the Weak Interactions: Existence of a Quartet of Intermediate Bosons and Their Dual Isotopic Spin Transformation Properties', *Physical Review,* **119** (1960), 1410–1419.
Léveque et al. (1965)	A. Léveque et al., 'The parity of the $Y_1^*(1660)$', *Physics Letters,* **18** (1965), 69.
Levrat et al. (1966)	B. Levrat et al., 'Structure within the R(1675) boson and possible structure within the $A_2(1290)$', *Physics Letters,* **22** (1966), 714–718.
Lichtenberg (1965)	D.B. Lichtenberg, *Meson and Baryon Spectroscopy* (New York: Springer Verlag, 1965).
Livingston (1966)	M.S. Livingston, *The Development of High-Energy Accelerators,* Classics of Science, Volume III (New York: Dover Publications, Inc., 1966).
Lynch et al. (1964)	G.R. Lynch et al., 'The production mechanism for the reaction $K^+ p \to K^* p$ at 3 GeV/c', *Physics Letters,* **9** (1964), 359–362.
Maglic et al. (1961)	B.C. Maglic et al., 'Evidence for a $T=0$ three-pion resonance', *Physical Review Letters,* **7** (1961), 178–182.
Maglic & Gosta (1965)	B. Maglic and G. Gosta, 'A method for the search for unstable particles using Jacobian peaks in angular distribution', *Physics Letters,* **18** (1965), 185–189.
Maglic (1972)	B. Maglic, 'Discovery of two kinds of neutrinos', in *Adventures in experimental physics* (Princeton: World Science Communication, 1972).
Massam et al. (1965)	T. Massam et al., 'Experimental observation of antideuteron production', *Nuovo Cimento,* **39** (1965), 10–14.
Massam & Zichichi (1966)	T. Massam and A. Zichichi, 'Leptonic vs. hadronic quarks', *Nuovo Cimento,* **43** (1966), 227–229.
Meeting-Brookhaven (1962)	*Future Program for the CERN PS and the Brookhaven AGS,* Minutes of a Meeting held at the Brookhaven National Laborator, September 10-14, 1962.
Meson Conf. (1957)	*International conference on mesons and recently discovered particles,* Padua-Venice, 22–28 September 1957.
Morrison (1964)	D.R.O. Morrison, 'Experimental limit on the production of 24 GeV/c protons of long-lived particles with fractional charge', *Physics Letters,* **9** (1964), 199.

Morrison (1978)	D.R.O. Morrison, The Sociology of international scientific collaborations, in *Physics for Friends, Festschrift for Ch. Peyrou* (R. Armenteros, ed.) (Geneva: CERN, 1978).
Musgrave et al. (1964)	B. Musgrave et al., 'Study of the $Y\bar{Y}$ production in two-, three- and four-body final states by 3.0, 3.6 and 4.0 GeV/c antiprotons in hydrogen', *Nuovo Cimento*, **35** (1964), 735.
Ne'eman (1961)	Y. Ne'eman, 'Derivation of strong interactions from a gauge invariance', *Nuclear Physics*, **26** (1961), 222–229.
Neutrino Conf. (1965)	C. Franzinetti (ed.), *Informal Conference on Experimental Neutrino Physics*, CERN 20–22 January 1965 (Geneva: 65-32, 1965).
Neutrino-results (1965)	'Results of the CERN neutrino experiments, 1963 and 1964', *CERN Courier*, **1** (1965), 8–9.
Nimrod (1979)	J. Litt (ed.), *Nimrod. The 7 GeV Proton Synchrotron*, Proceedings of a Nimrod Commemoration Evening, Rutherford Laboratory, 27 June 1978 (Didcot: Science Research Council, 1979).
Nishijima (1957)	K. Nishijima, *Physical Review*, **108** (1957), 907.
Okun (1965)	L.B. Okun, *Weak interaction of elementary particles* (Oxford: Pergamon Press, 1965).
Paty (1965)	M. Paty, *Etudes d'interactions de neutrinos de grande énergie dans une chambre à bulles à liquide lourd* (Geneva: CERN 65-12, 1965).
Pevsner et al. (1961)	A. Pevsner et al., 'Evidence for a three-pion resonance near 550 MeV', *Physical Review Letters*, **7** (1961), 421–423.
Physics School (1962)	Ch. Frondsdale (ed.), *Weak interactions and topics in dispersion physics*, Lecture notes from the second Bergen international school of physics – 1962 (New York: W.A. Benjamin, Inc, 1963).
Pilkuhn (1964)	H. Pilkuhn, *On the production of antideuterons* (Geneva: CERN 64-40, 1964).
Plass (1965)	G. Plass, 'Informal Conference on experimental neutrino physics', *CERN Courier*, **3** (1965), 38–41.
Pomeranchuk (1958)	I. Pomeranchuk, 'Equality of the nucleon and antinucleon total cross-section at high energies', *Soviet Phys.-JETP*, **34(7)** (1958), 499–501.
Pontecorvo (1959)	B. Pontecorvo, 'Electron and Muon Neutrinos', *JETP*, **37** (1959), 1751–1757.
Puppi (1963)	G. Puppi, 'Pionic resonances', *Annual Review of Nuclear Science*, **13** (1963), 287–338.
Regge (1959)	T. Regge, 'Introduction to complex orbital momenta', *Nuovo Cimento*, **14** (1959), 951–976.
Regge (1960)	T. Regge, 'Bound states, shadow states and Mandelstam representation', *Nuovo Cimento*, **18** (1960), 947–956.
Regenstreif (1959)	R. Regenstreif, The CERN Proton Synchrotron (1st Part) (Geneva: CERN 59-29, 1959).
Regenstreif (1960)	R. Regenstreif, The CERN Proton Synchrotron (2nd Part) Ch.V. Injection (Geneva: CERN 60-26, 1960).
Regenstreif (1962)	R. Regenstreif, The CERN Proton Synchrotron (3rd Part) (Geneva: CERN 62-3, 1962).
Reich (1965)	K.H. Reich, 'External proton beams at the PS', *CERN Courier*, **10** (1965), 148–158.
Reines (1960)	F. Reines, 'Neutrino Interactions', *Annual Review of Nuclear Science*, **10** (1960), 1–26.

Sakurai (1958)	J.J. Sakurai, 'Mass reversal and weak interactions', *Nuovo Cimento,* **7** (1958), 649-660.
Schneider (1963)	H. Schneider, 'Mass and decay parameters of Ξ-particles', *Physics Letters,* **4** (1963), 360-361.
Schwartz (1960)	M. Schwartz, 'Feasibility of using high-energy neutrinos to study weak interactions', *Physical Review Letters,* **4** (1960), 306-307.
Scotter et al. (1968)	D. Scotter et al., 'Ω^- production by 6 GeV/c K^- mesons', *Physics Letters,* **26B** (1968), 474-476.
Seguinot et al. (1966)	J. Seguinot et al., 'Evidence for a singly charged boson of mass 1675 MeV and width Γ = 66 MeV', *Physics Letters,* **19** (1966), 712-714.
Selove et al. (1962)	W. Selove et al., 'Evidence for a $T=0$ π^+-π^- resonance at 1250 MeV', *Physical Review Letters,* **9** (1962), 272-275.
Seminar-CERN (1961)	*Particle Physics – CERN Seminars 1961* (Geneva: CERN 61-30, 1961).
Shutt (1967)	R.P. Shutt (ed.), *Bubble and spark chambers* (New York: Academic Press, 1967).
Smith et al. (1964)	G.A. Smith et al., 'Study of $S = -2$ baryon systems up to 2 BeV', *Physical Review Letters,* **13** (1964), 61-66.
Spark chamber Meeting (1964)	*Proceedings of the informal meeting on film-less spark chamber techniques and associated computer use* (Geneva: CERN 64-30, 1964).
Steel (1976)	E.W.D. Steel, *List of Publications covering bubble chamber experiments carried out at CERN during the period 1960-1974* Geneva: CERN 76-02, 1976).
Stirling et al. (1965)	A.V. Stirling et al., 'Small-angle charge exchange of π^- mesons between 6 and 18 GeV/c', *Physical Review Letters,* **14** (1965), 763-767.
Sudarshan & Marshak (1958)	E.C.G. Sudarshan and R.E. Marshak, 'Chirality Invariance and the Universal Fermi Interaction', *Physical Review,* **109** (1958), 1860-1862.
Tassie (1973)	L.J. Tassie, *The physics of elementary particles* (London: Longman Group Limited, 1973).
Theory Conf. (1961)	*International conference on theoretical aspects of very high energy phenomena,* CERN, 5-9/6/61 (Geneva: CERN 61-22, 1961).
Van der Meer (1961)	S. van der Meer, *A directive device for charged particles and its use in the enhanced neutrino beam* (Geneva: CERN 61-7, 1961).
Van Hove & Jacob (1980)	L. Van Hove and M. Jacob, 'Highlights of 25 years of physics at CERN', *Physics Reports,* **62** (1980), 1-86.
Veillet et al. (1963)	J.J. Veillet et al., 'Existence and spin of the proposed $f^0 \to \pi^+ + \pi^-$ resonance', *Physical Review Letters,* **10** (1963), 29-31.
von Behr & Hagedorn (1960)	J. von Behr and R. Hagedorn, *Graphs of Lab.-Spectra (at diferent angles (1°, 3°, 5°, 10°, 20°)) of particles produced in 25 GeV pp collisions according to a statistical theory* (Geneva: CERN 60-20, 1960).
von Dardel et al. (1960)	G. von Dardel et al., 'Total cross section or p, $\bar{p}$, $K^\pm$ and $\pi^\pm$ on hydrogen between 3 and 10 GeV/c', *Physical Review Letters,* **5** (1960), 333-336.
von Dardel et al. (1961)	G. von Dardel et al., 'Total cross sections for pions on protons in the momentum range 4.5 to 10 GeV', *Physical Review Letters,* **7** (1961), 127-129.
von Dardel et al. (1962)	G. von Dardel et al., 'Total cross section for pions on protons in the momentum range 10 to 20 GeV/c', *Physical Review Letters,* **8** (1962), 173-175.
von Dardel et al. (1963)	G. von Dardel et al., 'Mean life of the neutral pion', *Physics Letters,* **4** (1963), 51.

Bibliography

von Dardel (1964) — G. von Dardel, 'Neutrino Physics', *CERN Courier,* **1** (1964), 29–33.

Weak Interaction Conf. (1963) — *International conference on fundamental aspects of weak interactions,* Brookhaven, 9–11/9/1963.

Weak Interaction Conf. (1965) — *International conference of weak interactions,* Argonne National Laboratory, 25–27/10/65.

Wetherell (1961) — A.M. Wetherell, 'Recent work on strong interactions at CERN', *Review of Modern Physics,* **33** (1961), 382–389.

Willis et al. (1964) — W. Willis et al., 'Sigma leptonic decay and Cabbibo's theory of leptonic decay', *Physical Review Letters,* **13** (1964), 291–295.

Wu et al. (1957) — C.S. Wu et al., 'Experimental Test of Parity Conservation in Beta Decay', *Physical Review,* **105** (1957), 1413–1415.

Yamagucchi (1960) — Y. Yamagucchi, 'Interactions induced by high energy neutrinos', *Progress of Theoretical Physics,* **23** (1960), 1117–1137.

Zichichi et al. (1962) — A. Zichichi et al., 'Proton-antiproton annihilation into electrons, muons and vector bosons, *Nuovo Cimento,* **24** (1962), 170–180.

Zichichi (1964) — A. Zichichi, *CERN work on weak interactions* (Geneva: CERN 64-26, 1964).

CHAPTER 6

The construction of CERN's first hydrogen bubble chambers

Laura WEISS

Contents

6.1 Historical overview	271
6.2 The principle of the bubble chamber	275
6.2.1 The theory of bubble formation	276
6.2.2 The choice of a liquid	278
6.2.3 The technical aspects of a bubble chamber	279
6.3 CERN's bubble chambers	284
6.3.1 Preparatory steps	284
6.3.2 The hydrogen liquefier	286
6.3.3 The 10 cm bubble chamber	287
6.3.4 The 30 cm bubble chamber	289
6.3.5 The 2 m bubble chamber	292
6.4 Technical elements and construction	297
6.4.1 The buildings	297
6.4.2 The chamber body	299
6.4.3 The vacuum system and the cold shields	301
6.4.4 The cooling system	304
6.4.5 The expansion system	309
6.4.6 The optical system	313
6.4.7 The magnet	315
6.4.8 Safety	318
6.5 Conclusion	321
Notes	323
Bibliography	335

For my father

When Donald Glaser received the Nobel prize in 1960 for inventing the bubble chamber the device already had an established place in all high-energy physics laboratories. Not only had small experimental instruments been brought into service in most universities and American and European research centres, but very big chambers were also in use or under construction at Berkeley, Brookhaven, Chicago, Geneva, Dubna, Saclay, London,...[1] In May 1952 Glaser was looking for a new detection system for cosmic rays, but it was the development and widespread use of particle accelerators in the second half of the 1950s, and notably the Cosmotron and the Bevatron, which made of the bubble chamber the 'high-energy nuclear physics counterpart to the low-energy nuclear physics Wilson chamber'.[2] The bubble chamber was, in fact, perfectly adapted in its functioning to big accelerators. In addition, like them, it was a powerful and complex device, which required, first for its construction and then for its utilisation, teams of physicists and engineers who were specialized in different domains, from the physics of materials to optics, from cryogenics to the problems of safety.[3]

This chapter begins with a brief historical section intended to situate CERN's achievement in the international context of the day. Our choice is neither exhaustive, nor does it attempt to rank developments in an order of importance: we simply want to take note of several developments in the first half of the 1950s. Having done this, we go on to explain in broad outline the theoretical ideas underlying the functioning of the bubble chamber and its different components. Our aim here is to bring out the demands the construction of such a device entails, be that in terms of a knowledge of different fields of physics or of how to develop and organize a number of elements of considerable technical complexity. This is the focus in the body of the chapter, which deals with the construction of the first instruments of this type at CERN, and in which we hope to capture, to the extent that our documents allow, the specificity of the work that was done without continually referring to practices which were widespread at the time. Similarly, to avoid undue repetition in the final section, which describes more carefully the different technical components of CERN's chambers, we have concentrated on the most important of the first chambers built at CERN, and have only referred to the others when a special feature of their construction seemed worth mentioning. We cover then a period stretching from the mid-1950s to the start of the next decade, a period in which CERN brought three increasingly large hydrogen bubble chambers into service.

6.1 Historical overview

The bubble chamber was invented in 1952. A census probably taken in 1956[4] listed chambers built or to be built at universities in Birmingham, Liverpool, London (Imperial College and University College), Oxford and Cambridge, in Great Britain, by Bassi and Mannelli in Italy, by Steinberger at Columbia University's Nevis Laboratory, by Hildebrand in Chicago, and by Alvarez with his 4″ and 10″ instruments and his project for a 72″ at Berkeley in the USA. In May 1958, a report written at CERN to evaluate the construction of a big hydrogen chamber remarked that 'It is known that bubble chamber projects exist in about 20 separate European centres [...]'.[5] Of course the take-off of a new technology is not only to be measured by its general level of acceptance, by the fact that everybody, everywhere, wants to have it (which was doubtless the case with bubble chambers in the early 1950s, when many, even small and dispersed, groups were interested in the instrument, though not to innovate but either to copy what had been done elsewhere or to build small prototypes), but also by the magnitude of the investments big laboratories were prepared to make in it. That granted our aim is not simply to identify all the projects under way in the early 50s but, by describing the achievements of several leading groups, to show the enormous development in this field in less than ten years.

The history of the invention starts at the University of Michigan (and is associated with anecdotal details about bubbles in a beer glass![6]): in May 1952, Donald Glaser with his student David Rahm began to look at how liquids could be used to detect cosmic rays. By October they had shown that one could photograph the trail of bubbles in a small glass bulb 1 cm in diameter and 1 cm long filled with ether. After investigating various systems (sonic, counters, photoelectric cells...) for triggering the camera at the moment the particle arrived (when the chambers were later used with accelerators the sequence was reversed, and the beam was to arrive when the chamber was ready), Glaser presented his findings on 2 May 1953 at a meeting of the American Physical Society in Washington. The baton then passed to Luis Alvarez of the Lawrence Radiation Laboratory in Berkeley and Darragh Nagle at the University of Chicago.

While Glaser and Rahm went on with their research at Brookhaven,[7] concentrating on the analysis of photographs and so rather losing their leading position in the development of the device itself, the Berkeley group dedicated themselves forthwith to the new idea, and with such efficiency and enthusiasm that by the end of the 1950s one was already speaking of 'Alvarez's gamble'.[8] This was no exaggeration: 'To make a suitable particle detector, Alvarez had to transcend the intrinsic limitations of Glaser's technique, create several new technologies, walk confidently through unknown terrain, and raise a lot of money'.[9]

On 5 May 1953 two of Alvarez's collaborators, F. Crawford and Lynn Stevenson, began to build their first chamber, initially filled with ether, then with liquid nitrogen, and finally with liquid hydrogen,[10] being attracted by the theoretical

Notes: p. 323

simplicity of this wonderful 'sea of protons'—even though it was a long way from practical realization. John Wood,[11] an engineer—a point to be noted at a time when physics was only done by physicists, and symbolic perhaps of a change in attitudes and in methods of research—, soon showed that it was possible to photograph the tracks of bubbles in superheated liquid hydrogen. Wood also studied the possibility of using metallic chambers fitted with glass windows (called a 'dirty chamber'), since this was the only practicable way of building big chambers. With a chamber of 2.5″ internal diameter, Wood showed that the spontaneous ebullition which occurred where metal joined glass did not impede the observation of the tracks if the fall in pressure was very rapid. 'By late 1954, it was already clear that Berkeley, at least for some time, was going to dominate bubble chamber technology'.[12] D. Parmentier and A.J. Schwemin[13] continued in June 1954 with a chamber 4″ in diameter and 3″ deep, of simpler design, built quickly and fitted with a magnet which enabled them to determine the sign and the momentum of particles. In 1955 Alvarez could write provocatively that 'all that stood in the way of large, practical chambers, was a series of 'engineering details''.[14] Now he was ready to think big, and he set as his immediate goal a 30″ long chamber, whose dimensions he increased to 50″ × 20″ × 20″ after reflection. He also decided to build a 10″ chamber which, he felt, could be used to do some really interesting physics if it were placed in an 8 kGauss magnetic field. Its vacuum tank would be one quarter the size of that of the big chamber, and it could serve as a test bench for problems associated with the scale of the device, like the manipulation of large quantities of liquid hydrogen. For help with this in spring 1955 the group consulted engineers at the National Bureau of Standards' Cryogenics Engineering Laboratory at Boulder, Colorado (a division of the AEC created to prepare liquid deuterium and tritium for the hydrogen bomb tests in 1952). Now for the first time safety problems assumed considerable importance. An accident with the 10″ chamber confirmed that the precautions already taken were necessary and (almost) adequate, the ventilation system functioning sufficiently well to ensure that no one was hurt.[15] In consequence a checklist of the procedure was drawn up to protect the users and the chamber itself.

During these same years other American laboratories became interested in this new detector. Roger Hildebrand and Darragh Nagle at the Fermi Institute in Chicago followed a similar path and, what with Nagle being an expert on hydrogen, by August 1953 (so before the Berkeley group), and despite the major difficulties posed by the cryogenics and the dangers of working with this liquid, they had already shown that superheated liquid hydrogen boiled more rapidly in the presence of a γ-ray source, a necessary condition if one wanted to have a hydrogen bubble chamber.[16] Despite their success, however, they never managed to form tracks and they moved away from hydrogen for a while to concentrate on building big, heavy-liquid chambers giving better photographs.[17] The group came back to hydrogen chambers around 1955, and even though they got tracks of pion–proton collisions at the end of 1956, by now they were clearly behind the leaders as they were still working with glass-bodied instruments.[18]

At Columbia University Steinberger had also constructed a 6″ diameter liquid propane chamber in 1955, in glass and metal, which was used in a run at Brookhaven's Cosmotron in autumn 1955.[19] In July 1956 he went on to complete a bigger propane chamber in aluminium, with the aid, by the way, of a CERN physicist sent abroad to 'learn the art'.[20] Steinberger also used a 12″ diameter, 8″ deep liquid hydrogen chamber at the Cosmotron whose design ressembled that of his propane chamber but for the 'bellows' type of expansion system above it.

Shutt at Brookhaven was also working in this field, and early in 1957 he had made considerable progress with the construction of a large hydrogen chamber with piston expansion which was to be ready in 1960. Made of forged aluminium, of visible internal dimensions 20″ × 9″ × 10″, and containing 60 litres of liquid, it was suspended in a vacuum jacket and surrounded with cold shields to reduce heat exchange to the minimum.[21] At the time Bradner thought highly of Shutt's device: '[It] is an example of an instrument whose size and cost still permit design decisions to be made primarily on the basis of technical considerations'.[22]

All the while work advanced in big strides at Berkeley: the dimensions of the large chamber were increased further to 72″ × 20″ × 14″ and it was on the drawing boards in 1956,[23] when the young CERN began to think about its first bubble chambers. It was commissioned in March 1959, having cost the enormous sum of $ 2.1 million, of which $ 400,000 was for a special building for the chamber—a building which CERN took as its model when making its preliminary studies[24]—, and having absorbed 65 man-years of labour. As the 'competitors' at CERN put it: 'The chamber will have a visible volume of liquid hydrogen of 72″ × 23″ × 15″ and will contain about 500 litres of liquid hydrogen. To embark on such a project three years ago, when the largest operating bubble chamber was an affair containing only a few litres of liquid hydrogen may truly be classed among the most courageous of scientific decisions. Even today no comparable plan has reached a serious stage of design, far less of actual construction'.[25]

The problems the group had had with gaseous expansion for the 10″ chamber, in which only a thin 3 cm layer of liquid was sensitive, had led them to improve this aspect of the instrument so as to allow for rapid recompression which led to a spectacular decrease in the consumption of hydrogen.[26] All the same the rate of use of this liquid remained a weak point in the chosen system, though this did not unduly bother Alvarez, who wanted to use his chamber for doing physics at the Bevatron as soon as possible.[27] The 72″ thus used the same method of gaseous expansion, which had the advantage that it could be checked and repaired without emptying the chamber, and which used a 1550 W integrated liquefier to maintain a constant temperature. It had a single horizontal window 5″ thick, attached by inflatable joints which had been developed by the group, and capable of supporting a sealing force of 100 tons; through it the bubbles were illuminated and photographed at the same time. The use of the so-called "coathangers" allowed for a uniform dark field illumination. The magnet, consuming 5 MW, and weighing 137 tons, was completely

Notes: pp. 323 ff.

outside the vacuum tank, and was fitted with horizontal copper windings and a pole piece which increased the field in the chamber. It could run at 18 kG. The chamber was installed in a specially built building some 30m by 12m equipped with a 40 ton crane; as much of the piping as possible was buried underground.

European researchers were aware of the work the Americans were doing in this field, and in line with the tradition on the Old Continent small groups (two or three physicists often formed a team), almost everywhere,[28] began to build bubble chambers. The work seems to have been organized rather differently than in the USA, laboratories and universities which did not have frequent access to accelerators preferring to build small instruments, some of which had quite specific purposes and which drew on quite original techniques.

This was particularly true in Italy, where a kind of tradition was rapidly built of developing small chambers which could be transported from one accelerator to another. From the beginning, i.e. around 1956,[29] Italian researchers were very active with the prime aim of doing physics rather than developing the technology. They accordingly restricted themselves to small chambers, principally filled with propane, but also with helium, for example, which they took with them to accelerators, including those at CERN, to get as many photographs as possible to study. We find chambers built, often collaboratively, by groups in Rome, Bologna, Padua (0.5 litre pentane chamber), Genoa and Turin (1 litre chamber). Among the physicists involved we might mention Argan, Conte, Gigli, Gonella, Picasso, Tomasini... While CERN was building its 30 cm chamber in 1957-60, Bassi was building the 'Camera nazionale italiana' at Bologna, a device 25 cm in diameter with a bellows which was to see considerable service at CERN. Around 1962 a helium chamber built by Moneti's group in Rome was also brought to CERN. The Italians took back their chambers when their accelerator at Frascati was built.

However, it was Great Britain and France who were to be the European leaders, especially as far as equipment of some importance was concerned.

At Liverpool, where at one stage Kowarski thought to send a young CERN researcher in 1955 to study bubble chambers, Alston built a small propane chamber in 1957, and planned a small hydrogen chamber. Dodd and Henderson at University College, London[30] had a big freon chamber destined for Harwell on the drawing boards in 1959. In addition, in 1957, groups at Imperial College, London, at Birmingham and Liverpool Universities, and at the National Institute for Research in Nuclear Science combined to build the British National Hydrogen Bubble Chamber of 1.5 m. This project got under way early in 1959, the device was tested in England without a magnet, was then transported to CERN in pieces at the end of 1961, where it was brought into service in June 1964 before being taken back to Harwell for use at Nimrod.

In France two main laboratories worked in this field: that at the Ecole Polytechnique which successively built three heavy liquid chambers, BP1 of 5 litres, BP2, a medium-sized chamber of 40 litres with dimensions 35 cm × 25 cm × 20 cm,[31] and finally BP3 of 300 litres for use at the CERN PS. Grouped around Leprince-Ringuet, the decision to build BP3 was taken in July 1957. Lagarrigue directed the construction, the first orders were placed in summer 1958, assembly began at Saclay in September 1959, and the first photographs were taken as early as January 1960 at the 3 GeV Saturne accelerator at Saclay. The chamber was then sent to Meyrin, where it did its first experiments in July 1960, only one month later than originally planned.[32] BP3 was 50 cm × 100 cm × 50 cm, and was equipped with a 20 kG magnet consuming 4.5 MW of power which was attached to the body of the chamber in such a way that it could not be assembled without the magnet.[33]

In parallel, in March 1958, Francis Perrin, High Commissioner of the French CEA decided to build two hydrogen chambers at Saclay, deciding that the larger be ready in three years. Those responsible for the project—B. Gregory (scientific side), R. Florent (technical and industrial aspects) and R. Maillet—visited Shutt in Brookhaven, chose the chamber he was building as model, and were authorized to go ahead in October. The French had decided to copy the American 50 cm chamber, increasing its size to 81 cm, which meant doubling its volume. The first trials were held in January 1961, and the chamber was brought to CERN around the middle of the year.[34] It was anticipated to send it subsequently to Nimrod.[35]

6.2 The principle of the bubble chamber

Glaser described the bubble chamber as 'a new radiation detector in which ionizing events produce tracks consisting of strings of tiny bubbles in a superheated liquid'.[36] And although the principle of visual detection had been in use ever since the Wilson chamber was invented in 1911 and developed for work with cosmic rays,[37] as well as in Powell's nuclear emulsions,[38] it remained 'to find a particle-trapping medium which would be dense enough to afford the frequent collisions and precise tracks of the emulsion and flexible enough to be amenable to magnetic fields and to give the good-sized, single picture of the cloud chamber'.[39] The ideal medium for this would be a liquid.

A bubble chamber has three major advantages over the cloud chamber and its variants, on the one hand, and nuclear emulsions, on the other. In addition to the increased bulk of the liquid target, its sensitive depth is not restricted, it distorts the tracks only slightly and, above all, it is able to follow the rhythm of the accelerators. 'A liquid medium combines the high stopping-power and precise tracks of a photographic emulsion with the easy scanning of cloud chamber photographs, and moreover possesses advantages over both these methods of detection. The bubble chamber only records tracks made in a very short time interval, and so the background of unwanted tracks is reduced to a minimum. From curvature measurements on tracks produced in a magnetic field the sign of the charge and the

Notes: p. 324

momentum of the particle can be found'.[40] Thus the device rapidly replaced the cloud chamber and grew from a few cubic centimetres in 1952 to dimensions holding thousands of litres in the 1970s. Even allowing for his unbridled enthusiasm, there is no doubt that Henderson, writing in 1970,[41] was expressing the feelings of many physicists at the time: 'It is clear that bubble chambers will continue to have an important role to play in the elucidation of the problems of particle physics. No other detector gives such a complete, direct, unbiased view of complex high-energy interactions'.

6.2.1 THE THEORY OF BUBBLE FORMATION

Glaser's idea was to put a liquid into an unstable state so that the passage of a charged particle could trigger the formation of the microscopic bubbles which set off the boiling process. Theoretically,[42] the conditions which a liquid has to satisfy for this to be possible are relatively easy to calculate and to spell out: it should be non-conducting, so as to be ionizable, should have a low surface tension, so that the forces tending to collapse any cavity, including a bubble, are weak, and, by contrast, should have a high vapour pressure which would tend to maintain and increase the dimensions of the selfsame cavities. Finally, for practical reasons, the liquid has to be easily superheated at temperatures and pressures which can be readily attained and which are close to the critical point. Glaser's first choice, diethylene ether, superheated to 130°C and kept at a pressure of about one atmosphere, rapidly confirmed the accuracy of his predictions in a first test with glass tubes.[43]

The mechanism for recording gas bubble formation is based on the sensitivity of a liquid subjected to a sudden pressure drop. When the external pressure suddenly becomes substantially lower than the saturated vapour pressure, the liquid becomes superheated. Boiling does not, however, occur spontaneously without the presence of a nucleation source, i.e. without a microscopic cavity generated, in the case of the bubble chamber, by the passage of a charged particle (but which could just as well be an impurity in the liquid or an irregularity in the vessel).

Knowing the equilibrium state of a vapour bubble in its own liquid,[44] and making a few simplifications,[45] one can draw the curve separating the superheated region where the (uncharged) bubbles disappear from that where they grow, so arriving at a first set of acceptable parameters. In choosing them one must take care to avoid the 'foam' limit, i.e. a spontaneous boiling throughout the whole expanding liquid (as happens when, because of a too-powerful superheating, the statistical fluctuations of the molecular positions create momentary holes and can precipitate bubble growth).

To explain what happens when the passage of charged particles triggers the creation of vapour bubbles in a superheated liquid, Glaser began by imagining a situation completely analogous to that for drop creation in a Wilson chamber: a number of charges distributed equally over the surface of the bubbles causes a reduction in their surface tension by mutual repulsion, and the charged bubbles accordingly disappear less readily. This electrostatic theory was found to be false (Glaser himself subsequently proposed a more satisfactory alternative), but it had the considerable advantage of being very simple and of allowing very satisfactory predictions to be made of the pressure values to be used in bubble chambers. Roughly speaking the rule was to choose the pressure at 2/3 of the critical pressure and the temperature at 2/3 of the difference between the normal boiling point and the critical temperature of the liquid.

The so-called 'heat-spike' theory worked out by F. Seitz after using xenon as the sensitive liquid[46] is probably closer to the truth: one considers here the transfer of energy from a rapidly moving charged particle to the electrons of the atoms of the liquid. This causes a minute increase in energy in the form of heat precisely situated at a point in the liquid about one millionth of a centimetre big. Increase in heat means an increase in molecular motion and this creates a small rupture in the liquid's coherence which leads to bubble formation. For the bubble to grow it is also necessary that the dispersion of the heat by thermal conduction be less rapid than the local concentration of the heat. This theory explains why it is necessary to add a little ethylene or ether to certain liquids like xenon to make them sensitive: in effect the mono-atomicity of xenon does not allow it to transform the deposited energy into heat, and it tends to scintillate.[47]

The rate of growth of the bubbles is governed by the flux of heat in the liquid which provides the vaporization heat at the surface being formed. It depends on the thermal conductivity, on the latent heat, on the coefficient of thermal diffusion, and on the temperature of the superheated liquid, as well as on the density and the temperature of the vapour in the interior of the bubbles, assumed to be that which corresponds to the pressure after expansion.[48] In this process the time dependence of the radius of the bubble is similar to that of drops in a cloud chamber, but is much more rapid since a bubble that is to be photographed needs less material than does a drop. In addition one can take account of the fact that the bubbles climb and move towards the warmer parts of the liquid as they grow.

Detailed measurements made in hydrogen bubble chambers were in gross but satisfactory agreement with mathematical calculations. It was deduced that if one decreased the temperature of the bubble chamber so as to maintain the same sensitivity one had to increase the pressure drop, which in turn resulted in a decrease in the temperature of the vapour in the bubble. The saturated vapour density would then be lower, and all these effects combined would lead to an increase in the rate of growth of the bubbles, as was experimentally confirmed. A hydrogen bubble chamber working at 25 K with a bubble density of 10 cm^{-1} would have bubbles reaching a radius of 0.3 mm in a millisecond. At 27 K the rate of growth was an order of magnitude lower. The highest rates of growth were those obtained by Glaser in small glass chambers in his first experiment, where the liquid, a hydrocarbon, was very strongly superheated, i.e. very much expanded.

In big chambers this possibility of having a relatively slow growth after the

Notes: p. 324

explosive birth of the bubbles is particularly useful for getting good photographs. To have uniformly bright tracks it is necessary that all the bubbles on the same track be more or less the same size. The particles extracted from an accelerator arrived at the chamber in a burst of some 50 microseconds. If the growth time of the bubbles was comparable '[i]t would also have given Bevatron tracks with bubbles varying in size from a fraction of a millimetre to more than a centimetre in a single pulse'.[49] With times of the order of milliseconds at least this problem is resolved.

The density of the bubbles on a track is a measure of the specific ionization of the particles and so allows one, from a study of the track, to distinguish particles from each other. It is however strongly dependent on the temperature and the pressure of the liquid: a proton giving 24 bubbles/cm in pentane at 157°C will give 8 bubbles/cm at 155°C. A systematic analysis of tracks of known particles moving at known velocities[50] allows one to identify the particle and its velocity (whose square is inversely proportional to the bubble density) and adds further useful information to that obtained by studying the curvature of the particle trajectories in a magnetic field. Blinov and his colleagues[51] studied the problem closely and estimated a statistical error of 10% (inversely proportional to the number of bubbles counted) on a path of 5 cm, which gave a 5% error in the velocity. Other researchers were less positive: 'However, the method is very tedious, and so far its only application has been in making a qualitative distinction between pions and protons having the same momentum, which can usually be done at a glance, owing to the large difference in bubble densities'.[52]

6.2.2 THE CHOICE OF A LIQUID

Practically all liquids can be superheated to a metastable state in which they become sensitive to the passage of charged particles. Glaser has even said: 'No liquid that has been tested seriously has failed to work as a bubble chamber liquid'.[53] All the same it is interesting to combine high density and short radiation length[54] with temperature and pressure conditions which are practicable. Among the liquids studied one finds many that are corrosive or toxic, or which can be used at high temperatures or pressures. The bigger the chamber the more easy it is to satisfy the radiation length conditions, so that gradually a list of preferred liquids was built up.[55] 'The choice of liquids depends only on the physical objectives of the experiment and on engineering and economic considerations'.[56]

Among the pure liquids, and despite the necessity of advanced cryogenics and of the danger inherent in its use, there was above all liquid hydrogen. What better target in fact than pure hydrogen, this sea of protons which allowed the physicist to dispense with the tedious labour of identifying and eliminating parasitic tracks coming, for example, from protons dislodged from heavy nuclei by the incident particles! Its working conditions, about 27 K and 5 atm, increased the technical difficulties and the cost of the construction of hydrogen bubble chambers, especially 'dirty' chambers,[57] which were specially developed for this liquid (possibly in alternation with deuterium, which was valuable in all neutron studies). Helium was also a good candidate thanks to the simplicity of its nucleus (zero isotopic spin), and many groups chose it for its advantages over hydrogen of low operating pressure, less need for special safety precautions, and an expansion of only 1%.[58] Another rare gas, xenon,[59] with a radiation length of 3.1 cm, could detect neutral pions and gamma rays by the production of electron pairs and Compton electrons. It was used at $-19°C$ and 26 atm and had to be contaminated with 2% ethylene or ether to avoid scintillation. For the heavy liquids, among a very large range of compounds that were used,[60] propane[61] played a dominant role because of its easy operating conditions (60°C and 21 atm) and its high hydrogen density, and despite the inconvenience of having to take more than twice as many photographs as with hydrogen so as to separate out all the events associated with the carbon atoms. Freon was also widely employed.

Alongside these pure liquids a vast study was made of liquid mixtures[62] where one mixed, say, neon with hydrogen to facilitate the detection of gamma rays. Another path that was explored was the mixing of gases with liquids, where the bubbles were gas bubbles which grew much more slowly.[63]

To make a bubble chamber work one keeps the temperature of the liquid considerably above its normal boiling point, stopping the boiling by applying a pressure just above the saturated vapour pressure at this high temperature. On releasing the pressure the liquid is in a superheated state for a limited period of time. If a particle traverses it during this period it leaves a track of microscopic bubbles which rapidly grow to a visible size. At this moment a flash allows one to photograph the track, whereupon the chamber is returned to its previous condition by the application of pressure on the liquid. The temperature is kept constant throughout the operation.

6.2.3 THE TECHNICAL ASPECTS OF A BUBBLE CHAMBER

The theory outlined above described the production of a track of bubbles by the passage of charged particles as if it were a relatively straightforward matter. However, the construction of a system allowing for the appearance, then the photography, and finally the study of these tracks in a relatively controlled manner so as to draw conclusions of interest to high energy physicists, is technically very demanding and calls on a wide range of engineering knowledge and know-how. A brief classification of the different basic elements of a hydrogen bubble chamber will help us not to get lost in the innumerable details of a particular instrument.

The body of the chamber is a cylindrical or parallelepipedic vessel containing the sensitive liquid. It is furnished with one or two (and possibly three) windows for the illumination and photography of the bubbles. Glaser's first small chambers, called 'clean', were made entirely of glass, whereafter the technique was developed to

Notes: pp. 324 ff.

encompass 'dirty' chambers of special stainless steel alloys which ensured that the chamber was mechanically resistant, anti-corrosive, rigid, and non-magnetic. Copper was excluded despite its good mechanical and thermal properties: because of its excellent electrical conductivity any rapid change of the high magnetic fields needed to determine the charge and direction of particles crossing the chamber induced eddy currents in its body which were sufficiently great to cause a very disturbing increase in the local temperatures. The use of aluminium alloys, which had excellent thermal conductivity, had advantages as far as the spatial regularity of the temperature was concerned.[64]

One side of the body of the chamber is connected to the expansion system which causes the decompression and the compression of the liquid. The windows, of optical glass so as to get high-quality photographs, are difficult to make because of the high pressure inside the chamber. One gets around this problem by using very thick glass (up to 17 cm for the largest chambers), which complicates in turn the glass–metal joints which have to be absolutely leak tight. The problems associated with the very different coefficients of thermal expansion of these two materials have opened up a wide-ranging study on low-temperature leak-proof seals for which the construction of hydrogen bubble chambers was a key driving force. Beginning with a simple lead joint and ending with indium wires sealing a steel tube inflated by helium[65] under high pressure (inflatable gasket), numerous studies[66] were needed to develop systems which maintained their impermeability despite the stresses and the variations in temperature.

Of no interest for heavy liquid chambers which operate at or above the ambient temperature,[67] cryogenics was one technique which was crucial to the use of hydrogen chambers as here the temperature has to be kept to around 28 K to within about 0.1°C. There are two main ways of losing cold: static loss, decreased by improving the insulation, and dynamic loss, the heat coming from the expansion cycle of the chamber.

As for insulation, a vacuum vessel containing the chamber minimises conduction and convection of heat. To reduce heat radiation the chamber is surrounded with copper screens which are either cooled with liquid hydrogen or liquid nitrogen or, even better, one uses screens of both types in combination. Alternatively, or in addition, one can use a number of aluminated Mylar sheets to form a superisolating cover almost as efficient (and certainly simpler) than the hydrogen jacket. These screens were designed such that they play a role in the safety of the device, as they collect the hydrogen released when a window breaks.

The cooldown of the chamber is done gradually so as to avoid stress in the materials and especially in the glass which can lead to breakages if the temperature gradients are too large. For big windows the limit is about 4°C/hour, which corresponds to a cooling of the chamber which can take from two to four days, or even a week for very big chambers.[68] Before starting the chamber has to be 'purged' with hydrogen gas or helium to ensure that no oxygen impurities (air) or water remain in the circuit. Liquid helium chambers are even more difficult to cool down since they operate at 4.2 K and the heat of evaporation of helium is 10% of that of hydrogen. All the same in the early years one finds many small chambers of up to 100 cm[69] filled with this liquid,[70] as they offer the advantage of working at atmospheric pressure.

The temperature of the earliest chambers was controlled using a heat reservoir at a temperature lower than that of the chamber which was obtained by boiling liquid hydrogen at atmospheric pressure. A metallic conductor, a 'heat leak', led the heat from the chamber to the reservoir and the working temperature was adjusted by a small electric heater in thermal contact with the conductor. This system was costly in terms of liquid hydrogen, and despite the improvements possible by increasing the pressure of the reservoir (so forming a kind of pot of liquid hydrogen vapour), was no longer used with the bigger bubble chambers. More sophisticated systems were later employed, including at CERN.

Three expansion systems were tried to release suddenly the pressure on the liquid hydrogen and to put it in a superheated state: Alvarez's gaseous expansion, a piston in direct contact with the liquid (the method adopted by CERN), and the so-called bellows system, which was a modification of it.

The gaseous expansion system developed by Luis Alvarez's group at Berkeley[71] is the simplest but also the most expensive in terms of cold: a rapid valve at ambient temperature is placed at the end of a vertical tube connected to the chamber. Gaseous hydrogen flows into a reservoir when the valve is open. For recompression the hydrogen that has evaporated is replaced by hydrogen at a higher pressure coming from another container. One limitation of the system is that if the temperature gradient is too high only the top of the liquid is sensitive, though this can be got around by cooling the tube with a regenerator. The chamber is rapidly recompressed after a sensitive time of some 20 ms (5 to 10 ms to have a uniform sensitivity, 10 ms to allow for sufficient bubble growth).

Liquid expansion chosen by Charles Peyrou and his group at CERN is achieved using a piston of large cross-section which acts directly on the liquid in the chamber or in a neck extending from it. After adiabatic expansion[72] the piston is stopped and, unlike in the previous system, the volume is kept constant, which implies a small pressure increase; once the photographs are taken the system is recompressed. The movement of the piston is thus sinusoidal as a function of the time. The third system[73] is in fact a variant of liquid expansion in which a part of the chamber is mobile via flexible steel bellows.

On comparing the two main systems we see that the dynamic load is higher in the former. It is thus essential to eliminate all sources of vapour, the prime source of losses. To do this one has to minimize all roughness and discontinuities on the walls of the chamber which trigger the boiling of the superheated liquid. This in turn calls for a high-quality finish to the surfaces and a careful study of the joints. On the other hand, the piston must also be thoroughly cooled so that the heat produced by its

Notes: p. 325

friction in the cylinder can be rapidly evacuated, and its shape must reduce eddy currents to the minimum. There are two other difficulties with the piston: its inertia and above all the problem of the leak tightness of the seals.

A bubble chamber usually functions in parallel with an accelerator at its frequency of some 2 seconds. It is sometimes useful to have it work in a 'multipulse' mode, i.e. to have two or three beams of particles photographed by the chamber per machine pulse. This requires several expansions of the chamber per period, which is practicable. The method is very advantageous in that it reduces fixed costs (cooling, electric power, hours of work...) and enables one to get many more photographs per unit of machine time.

The optical system comprises illumination and photography. One method of the former is 'dark field illumination', lighting on a black background, where the light is focused on an absorbing surface situated between the camera lenses. The bubbles appear in this case as illuminated points on a black background. An inconvenience of a dark field system is the need for two big windows of optical glass, with all the sealing problems that that entails. If one wants to use only one window one has to apply the retrodirective system with a small light source placed at the centre of curvature of a reflecting spherical surface. The bubbles in the path of the light refract it towards the cameras. However one also has to 'trap' parasitic images of the same bubble: in Berkeley's 72″ chamber the mirror was replaced by 'coathangers' which absorbed all light rays which were not perpendicular to the surface. For focusing one uses optical condensors (lenses). Hydrogen and helium chambers raise special difficulties since the indexes of refraction of the liquids (1.09 and 1.025) are very close to that of the bubbles to be photographed. This gives very low light intensity for diffraction angles above 10°C. To reduce the angles in hydrogen chambers the shape of the light sources was carefully studied.[74] The flashes have to pulse several hundreds of thousands of times at a repetition rate of a few seconds.

'Bright field illumination' is an alternative method in which tracks appear as dark strings of bubbles against a bright background. A white screen diffusing the light can be used as the background for the chamber to improve the contrast. Other solutions, like using a layer of Scotchlite[75] on the internal walls of the chamber, which strongly reflects light having even large angles of incidence, permitted one to get very high quality pictures in very big bubble chambers.[76]

The lighting, which must be very bright during the brief moment that the photograph is taken, is supplied by one or more flashes, which are switched on very briefly (in 200 to 300 μs 80% of the light energy must be emitted) to minimise the effects of the movement of the bubbles.

To get a record of an event in the bubble chamber in its three dimensions on film one needs at least two cameras, though usually three or four are used to avoid stereo reconstruction problems (particularly tracks making a small angle with the stereographic axis). They are situated at the apexes of an equilateral triangle or of a square so as to facilitate the three-dimensional reconstruction of the particle tracks.

Since all the cameras take photographs from one side of the chamber only, the measurement of depth is always a problem. The bubbles, which are 100 to 200 micrometres large at the moment the picture is taken, are treated as if they are points. The size of the image is determined by the rules of diffraction and the quality of the lenses. The required conditions fix the maximum depth of field and the size of the film used.[77] The smallest film one can use without losing information is such that the diffraction image equals the film's resolution (number of lines/cm).

To rebuild the geometry of the system when analysing the tracks one traces fiducial marks on the optical windows of the bubble chamber. These marks, of which the number and position depend on the instrument for evaluating the photographs, are particularly carefully measured and serve as a basis for the programs which reconstruct the tracks.

Nonexistent in the early chambers, since other more fundamental technical problems had first to be resolved, barely interesting for liquids with a short radiation length, the addition of a magnetic field became increasingly important as the focus shifted from resolving technical difficulties to the wish to have photographs carrying the maximum physical information. The presence of the field permits one, first and foremost, to identify the sign of the charge of the particle, but also to measure its momentum[78] more easily than by using the density of bubbles on the track.[79] Usually generated by an electromagnet whose windings surround the chamber body, and directed perpendicular to the plane of the windows, the magnetic field was increased in strength along with the size of the chambers. At Berkeley the first bubble chamber equipped with a magnet (taken from a Wilson chamber!) was already planned in 1955 and measured 10″ in diameter. The main condition the field has to fulfil is to be as perfectly homogeneous as possible. To compensate for small variations precise maps of the magnet flux are established and used when calculating trajectories. The cooling of the coils is generally achieved by circulating water. At Berkeley cooling with liquid hydrogen was envisaged for the 72″ bubble chamber but was dropped for reasons of price and of safety. The possibility of pulsing the field, which is only needed for 1 ms per roughly one second, was also looked into with a view to decreasing the consumption of electricity, but it came up against problems of heating and of the considerable magnetic forces created by eddy currents. While for the small chambers the solution usually adopted is to place the magnet around the vacuum tank, so easing the removal of the chamber for checking and repairs, studies of how to build the magnet are needed for the bigger chambers to find the best way of gaining access to the chamber and to maximise the magnetic field. 'A compromise has always to be made between the intensity of the magnetic field through the chamber, its homogeneity, and the cost of the copper, the iron yoke, and the generator'.[80] This is all the more important since this is one of those items for which price increases proportionally to size,[81] and superconductors were welcome for this purpose once they were feasible.[82]

There are always a number of risks in a physics laboratory, and experimental

Notes: pp. 325 ff.

physicists rapidly become indifferent to those they are exposed to throughout their working day, whether they arise from the use of high electric currents, of elevated temperatures, of high pressures, of radioactive materials, etc. Unfortunately from time to time accidents remind one of the dangers which are trivialized just because they are always there. This also happened with bubble chambers, and it was only when large quantities of hydrogen began to be used that radical safety measures were taken. All the same, they were not always good enough to avoid accidents.[83]

The first systematic preventive measures were taken at Berkeley during the construction of the first big hydrogen bubble chambers in 1955. One of the great dangers with hydrogen is an explosion, which might occur immediately when a certain amount of the gas mixes with the oxygen in the air. Three kinds of preventive measures must be taken systematically: be sure that one accident does not cause another (a chain of accidents), avoid leaks of hydrogen into air, and avoid sparks or flames which might trigger an explosion of the hydrogen–air mixture. Liquid hydrogen boils very rapidly if the temperature is not kept very low. Thus if the temperature control system for the chamber does not work for any reason, the hydrogen pressure might increase quickly and could break a window, flow into the vacuum tank, or even come into contact with the air. A highly efficient ventilation system is needed which vents the hydrogen above the roof of the hall or into a safety sphere which can accommodate it all. The rate of cooling the windows to liquid hydrogen temperature must be scrupulously respected to avoid the ever-present risk of breaking the windows. The magnet, for its part, introduces its own set of dangers. It is crucial to protect it against a sudden power failure which would cause a sharp drop in magnetic flux and set up a potentially destructive electromotive force. This is done using auxiliary generators and inertial flywheels.

6.3 CERN's bubble chambers[84]

After building a small 10 cm experimental hydrogen chamber (HBC10), CERN constructed a 30 cm hydrogen chamber (HBC30) in 1958, and then a 2 m chamber (HBC200) which was commissioned in 1964. The last chamber BEBC, the Big European Bubble Chamber, which used very new technologies, especially for the magnet (superconductivity) was ready in 1973 and so falls outside the scope of this study. As for heavy liquid chambers, CERN produced one of 1 m in 1962, and then Gargamelle, mostly built in France. This device functioned in 1970 and its major achievement was the discovery of neutral currents.

6.3.1 PREPARATORY STEPS

If asked why CERN became interested in 1955 in the new technique of bubble chambers, any physicist or engineer who lived through the period will reply that it was what had to be done.[85] When the Scientific Policy Committee recommended in December 1954 that scientific work should be started well before the accelerators worked,[86] a programme for the track chamber section of the STS Division (Scientific and Technical Services) was submitted to the Director-General for his approval. This programme, based on the results of an experts meeting held in September 1954, made reference to an initial exploration of bubble chamber technique, which it recommended should be undertaken immediately,[87] along with a continuation of the effort on Wilson chambers which were to be used at the synchrocyclotron (SC).

In line with these directives, in summer 1955 the SC and STS Divisions at CERN joined forces to set up a group to build bubble chambers. Since the technique was similar in some respects to Wilson chambers, Charles Peyrou[88], who had been appointed by Gentner in 1957 returned to Kowarski's division in May and took over the leadership of the group. In parallel Lew Kowarski and Yves Goldschmidt-Clermont decided to send a young physicist in STS, Reinhard Budde, to Liverpool to learn bubble chamber technique by collaborating in the first British construction project. However, in autumn 1955 Edoardo Amaldi came to CERN inspired by a visit to Brookhaven where he saw Steinberger's first bubble chamber working. It was accordingly decided to send Budde to work with Steinberger, a professor at Columbia University, where he could simultaneously participate in experiments with the small 6″ chamber and in the construction of a 12″ propane chamber, which would work in July 1956.[89]

CERN's liquid hydrogen bubble chamber programme was adopted in mid-1956.[90] It involved the construction of a cylindrical chamber of 10 cm for trial purposes, built to learn the technique, to be followed by a chamber 30 cm in diameter to be used for doing physics, the most ambitious project yet undertaken in Europe. After Budde returned home with news of what the Americans were doing,[91] and sustained by the enthusiasm generated by the CERN symposium in 1956, the group got down to work.

Thus did CERN, without any experience, set off down this promising and difficult path more or less about a year later than the American groups who were working on this technology. We might ask whether it might not have been better to start by first building heavy liquid chambers, though on the other hand the area in which the group was weakest was cryogenics—here everything had to be learnt as soon as possible—, and the PS—which had to do physics at the research frontier very quickly—was not yet ready. In any event a certain realism pervaded CERN's efforts, stimulated by an awareness of what the Americans were doing.[92] 'Extrapolating the construction period of HBCs in the different laboratories of the United States for our situation at CERN is not easy. Because of the highly developed industry and because of the fact that all these laboratories are already fully working since many years, our initial steps may take somewhat longer'.[93]

Administratively the group was part of the STS Division in 1957, which was also responsible for cryogenic services, for the evaluation of track chamber photographs

Notes: p. 326

and for electronic calculation. A first change occurred in 1958 which saw the gradual incorporation of work on bubble chambers into the PS Division. On 1 January 1961 the new internal organization created the Track Chamber (TC) Division which was responsible for experiments with the 30 cm chamber and the construction of the 2 m chamber, as well as for the proper functioning and development of the hydrogen liquefier.

6.3.2 THE HYDROGEN LIQUEFIER

If Steinberger began by building a heavy liquid chamber, it was because it was difficult to see, in 1955, how to build a hydrogen chamber quickly even if the importance of the technique for experimental purposes had already become apparent.[94] In fact the knowledge of cryogenics in the newly constituted CERN group, as well as among most engineers and physicists at the time, was more or less nil. Only the AEC in the United States had really worked in this area for the development of the hydrogen bomb[95] and it was to them that Alvarez turned in spring 1955 when he set out to build his big hydrogen chamber. In Europe only a few laboratories seem to have been more or less competent: Cambridge, Grenoble, Leyden, Malvern, Oxford, Zurich.

The STS division, aware of the need to use very low temperatures in nuclear experimentation in general, and for hydrogen bubble chambers in particular, having studied the situation in Europe, decided to build a hydrogen liquefier of capacity 20 litres/hour and asked the University of Leyden, a major centre for low temperature work from the time of Kamerlingh Onnes, to take responsibility for the project. As a matter of fact only a research centre could successfully do this kind of thing at the time, as no industries in Europe or the United States built hydrogen liquefiers. Considerations of time and manpower thus justified this approach, all the more so since according to Professor Taconis, who was responsible for the project,[96] one or two years would be needed, after delivery of the machine, to train CERN's technicians in its use on account of the many safety regulations which had to be respected.

From mid-1957 the liquefier that was ordered worked satisfactorily despite considerable delays in its construction and delivery caused by an explosion of the whole machine at Leyden. It had already been necessary to modify it in anticipation of the demands of HBC30 and the 20 cm Bologna chamber.[97]

Early in 1958, and thinking ahead to the needs of the big chamber then under discussion, Pierre Amiot presented the results of his study on the production of liquid hydrogen.[98] An estimate of the needs was that much more difficult because CERN did not yet have a working hydrogen chamber. Basing himself on Berkeley's estimates for its 72″, he had to increase the loss of heat by radiation through the two windows (as opposed to one at Berkeley) in CERN's device. The length of the expansion mechanism of HBC30 (still on the drawing boards) gave an indication of the loss by conduction. Shutt's 6″ chamber gave an idea of the order of magnitude of the dynamic consumption. Three variants were examined and compared for costs and benefits: a size increase to 60 ℓ/hr, 100 ℓ/hr (for 90 cm and 1.1 m chambers respectively), and an entirely new installation (for a 2 m chamber). The last was, according to Amiot, 'on an industrial scale, and out of proportion to known European installations', while in addition 'European cryogenics experts were extremely rare'.[99]

Nevertheless it was this grandiose project for a low temperature facility which was described in the report in 1958 on CERN's big bubble chamber.[100] The overall cost was estimated at 1,090,000 SF. The same report conceded that one might prefer to build a smaller machine for the 2 m chamber (160 ℓ/hr, i.e.. still bigger than Amiot's two alternatives, which were inadequate for such a big device), for reasons of economy (cost: 665,000 SF). But it went on to cast doubt on the wisdom of such a policy: 'It must be realized that, although smaller liquefiers might well suffice, they allow no margin for safety, particularly for continuous operation, and that breakdown of the liquefier would then immediately stop the chamber. Furthermore, machine time is so valuable that it would be illogical to have to reduce the pulsing rate because the liquefying plant is too small to keep up with the consumption of the bubble chambers'.[101]

To conclude let us simply note that this project was never implemented on this grandiose scale, since the 2 m chamber was fitted with its own refrigerator/liquefier adapted to its considerable demands. As for the liquefier itself, which supplied liquid hydrogen to different users, its capacity increased more or less steadily—consumption doubled in 1960, then went from 13,000 ℓ to 83,500 ℓ in 1961 with the presence at CERN of various visitors!—, and in May 1963, with the installation of a new column, its output reached 100 ℓ/hr. In the interim Saclay had sent several loads of liquid hydrogen to help out between the end of 1962 and May 1963.

With automation achieved, and with lines installed for transporting hydrogen and nitrogen, the manipulation of the liquids was simplified and production was assured 'To the extent that, in so far as it is possible to see ahead, the present production of the liquefier ought to suffice for CERN's needs for several years'.[102] Accordingly in 1964 only the storage capacity of the system was improved and in 1965 facilities for the recuperation of gaseous hydrogen were installed.

6.3.3 THE 10 cm BUBBLE CHAMBER

The 10 cm bubble chamber[103] was above all an apparatus meant to enable the group to learn new techniques. That goal once given, it was decided to build a 'handy instrument', 'as perfect and as easily modifiable as possible'.[104] This is why the first chamber had no magnetic field. It was a cylinder of 10 cm internal diameter, 8 cm deep, volume 600 cm^3, closed by two disks of tempered optical glass of the borosilicate crown type, one of which allowed one to illuminate the liquid with a

Notes: p. 326

flash and the other to take the photographs. The chamber and the pressurized bath which controlled its temperature were of stainless steel, even though without a magnet field one could have used copper,[105] which was easier to work.

An attempt to evaluate the time needed to build the instrument with reference to the Americans' experience[106] indicated that with 10 active people in 1957 it should be operational between September 1957 and July 1959! The final result, obtained after almost two years of work from August 1956[107] to 1 May 1958[108] by a group whose number had been strengthened to about 26 people (who were also working on the plans for HBC30) was perhaps not as good as one might have hoped, as one can see from the 'many detailed modifications [which] had been made so as to improve the regularity of the chamber's performance'.[109] All the same, within the project's original terms of reference, the goal was apparently attained. In effect, without a big accelerator being available (only the SC was ready), and for such a small device, which was not intended to do physics, Peyrou's group preferred[110] to put the quality of the work ahead of the speed of construction, exactly the opposite of what Alvarez did, for example. They chose to dedicate themselves to planning a well-conceived bigger chamber which would run smoothly, and above all which would meet its design specifications.

The construction of the small chamber was based on a number of studies, many made by Nullens, but also by the nine other members of the group in 1957. Rather than starting from published work or the results obtained in other laboratories, the group tended to work ab nihilo as it were, basing itself on theory and on tests.[111] Indeed one often has the impression that the group took particular pride in getting results entirely on its own. Yet let us not forget that at the time the technique was relatively new everywhere, and that in all centres work progressed by mixing do-it-yourself techniques with advice from more advanced researchers.

From December 1956 to spring 1957 the orders for the principal components were placed, being the vacuum tank, the glass windows, and the four pumps, one a diffusion pump; at the same time the body of the chamber was constructed in the CERN workshop. In April 1957 the first assembly was made so that in the summer pressure and vacuum tests could begin, followed by tests of the cooling circuits and the vacuum, then of the control and measuring instruments, and finally of the low temperature pneumatic valves. On 1 May 1958 the first successful trial was made. After four further trials modifications were made, inter alia, to the piston and the cylinder.[112]

At the end of 1958 the chamber was placed in a pi-meson beam of 270 MeV at the SC and 6000 photographs were taken of which 60 contained events of interest. Since the main use of the chamber was not high-energy experimentation but the opportunity to test the functioning of a bubble chamber its entire construction was seen as a learning process: 'the smaller one shall serve for testing the above mentioned equipment and future developments of the HBC technique. It may also be used with the SC for experiments where the small size doesn't make the use unreasonable'.[113] This is why, from 1959, a systematic study of the extent of track distortion was made with this chamber along with the 30 cm chamber and in collaboration with the IEP (Instrument for the Evaluation of Photographs) group which was particularly keen to see the results. The chamber revealed no distortion (except very close to the expansion collar) i.e. in the useful region of the chamber the distortion was less than the curvature caused by the Coulomb scattering of particles in the hydrogen. In the same way systematic studies of ionization on the tracks obtained in HBC10 and HBC30 by measuring the mean gap length with a projection microscope suggested that the sensitivity of the device was homogeneous throughout its volume, and while the fluctuation in sensitivity from one photograph to another was not negligible it was relatively unimportant. If strict control was kept over the characteristics of the expansion, one could hope to measure the photographs without any reference tracks on the picture. What was otherwise certainly possible was to measure the ionization to identify the particles whose ionization approached the minimum, and this method proved effective to distinguish Λ^0 from K^0 particles when the kinematic relations were ambiguous, and to divide Σ disintegrations into those giving tracks of protons and of positive pions. Experiments were continued, particularly with a view to seeing whether it was possible to measure a rise in ionization in liquid hydrogen, as unlikely as that was. Unfortunately the results were negative.

From 1960 onwards the 10 cm hydrogen bubble chamber even disappeared from the CERN Annual Reports, and was no longer of anything other than historic interest. During its brief active life its most important aspect had been its construction, and the learning process that that involved. If the work was not always without its difficulties, if the construction took longer than originally foreseen, and if the path to development was strewn with problems, it is certain that this was an essential first step for CERN.

6.3.4 THE 30 cm BUBBLE CHAMBER

Planned simultaneously with the small experimental chamber, the drawings for 'another larger liquid hydrogen chamber of 30 cm diameter'[114] 'intended for use with the SC'[115] were already completed at the end of 1957. Its great novelty with respect to its predecessor was essentially the presence of a 15 kG horizontal magnetic field. However, the difficulties encountered with the first chamber led to certain changes, as we shall see shortly. An excellent device for measuring the path of the piston developed by Amato, and a small well-situated mirror which showed its movement, allowed one to analyze in detail what happened during the expansion cycle,[116] information which played a role in adopting the piston system for the big chamber.

The chamber was a cylinder having a horizontal axis 32 cm in external diameter and 15 cm deep, a total volume of 12.5 litres, highly polished walls, and was closed at both ends by two 3 cm thick windows of tempered glass.[117] On the top of the

Notes: pp. 326 ff.

chamber the expansion cylinder was a cylindrical tube with a vertical axis 11 cm in diameter, in which the piston moved, stopping 5 cm above the highest point of the chamber. At the upper extremity of the cylinder the body of the chamber was attached by a bridle to a supporting cylinder which was fixed to the vacuum tank.

Particular attention was paid to the ease of assembly, so as to ensure that it would be possible to carry out checks and repairs without removing the magnet. To this end the entire device—chamber, hydrogen reservoirs for controlling the temperature, valves and tubes—was attached at the top of the vacuum tank and could be removed therefrom without difficulty. The vacuum tank had a special shape, having two cylindrical arms which were placed in the coils of the magnet and which were closed with two glass windows. These arms rested on wheels which moved on rails placed below the magnet block, so enabling the system to be centred.

'A considerable effort was made to render the functioning as secure as possible and to increase to the maximum the ease and speed of the work of installing the chamber. To date it seems as though this effort was worth making'.[118] And while they could not eliminate minor accidents, of the kind only to be expected with this kind of complex machine, it is certain that the group seriously considered possible problems and took a good deal of trouble to develop a network of safety measures.

The assembly of the chamber proceeded along with the delivery of its different parts by the central workshop in 1956. As soon as possible, at the end of 1958/early 1959 tests were made[119] of the vacuum tank, the nitrogen jacket, the hydrogen reservoirs, and the pumping system.[120] In March/April 1959 it was tried with nitrogen[121] and with hydrogen on 7 May. It worked immediately and tracks of cosmic rays and Compton electrons were observed. Without further trials after the operational tests, the 30 cm chamber was placed at the SC at the end of August 1959—whereupon one of the pancakes in a magnet coil shortcircuited. Another run was made with a reduced magnetic field from 15-20 November in pi meson beams of 265 and 330 MeV from the SC, producing 100,000 photographs from more than 125,000 registered expansions.[122]

During these runs it was possible, as Morrison suggested,[123] to do rather less ambitious experiments calling for only a few thousand photographs, as opposed to the 50,000-80,000 needed by pion-proton elastic and inelastic scattering. As with the 10 cm chamber, the 30 cm chamber was used to make systematic measurements of distortion[124] and of ionization,[125] as well as for studies of bubble diameter and growth, of the stopping power of liquid hydrogen, and to test varieties of films. For the measurements of distortion preliminary data[126] indicated that the sagitta due to distortion was not much more than 0.05 mm, which was what one would expect in a cloud chamber of the same size. As for the work on ionization, the results were good:[127] by counting bubbles, by counting 'gaps', and above all by counting the mean gap length (with the method suggested in 1954[128]) it was possible to identify the particles producing the tracks. The dark field method gave practically no reflections, the contrast was good, and the bubbles were clearly visible. The measured error in the uniformity of the chamber (9.5%) was dominantly due to temperature and pressure gradients within it, while the reproducibility of the photographs was 8%. In experiments with 16 GeV/c negative pions it was possible, with the help of these figures, to identify the Λ^0 and the K^0 to a precision of ± 8% on tracks of 25 cm.

The 30 cm chamber was the only chamber available at CERN during the first six months after the PS first operated; it was exposed to a 16 GeV pi-meson beam from 21-24 March 1960,[129] after being tested and made ready for use.[130] Despite certain limitations,[131] like 'rather primitive beam transport', and the small dimensions ('medium size') of the chamber, 49,000 photographs were taken during the first run of some 50 hours. A few were distributed to a British group and the rest was used to study strange particle production by two CERN groups (IEP and HBC) and two Italian universities (Pisa and Trieste). The first preliminary results of this work were reported at the Rochester Conference and were published. In October of the same year the chamber continued its work and produced another series of photographs,[132] in another beam,[133] though in 1961, as foreseen, it was replaced by the 81cm Saclay chamber for the study of strange particle pairs which required a bigger chamber.[134] In mid-December it was taken to the North experimental hall for an experiment with 1.2-1.7 GeV negative kaons, but there was a delay in the arrival of new coils for the magnet.

When the 30 cm chamber was installed at the PS in March 1960 the valves gave some problems except when the chamber was used with a short pulse. At the end of August 1960, when the chamber was in a pion beam from the PS, further technical difficulties emerged[135] even though the 1959 Annual Report assured its readers that everything was 'almost perfect'.[136] The analysis of photographs was possible all the same. Problems occurred again in the October 1960 run, with the chamber working smoothly for shorter and shorter periods.[137]

By way of example, the cost of using HBC30 in 1960[138] was some 500,000 SF (100,000 SF being for the magnet coils) and 400,000 SF was put aside for it in the 1961 budget (still including 100,000 SF for the coils). These figures exclude the consumption of electricity, and of liquid hydrogen and nitrogen, but allow for possible use with deuterium.[139] Further research on the optical system was envisaged. Other important budget headings were the purchase and development of film, and improvements to the electronics.

The operation of HBC30 was stopped in spring 1962. During the part of the year in which it worked it took a further 150,000 photographs which led to publications, the device being 'very useful for experiments at medium or low energy in a purified beam'.[140] 150 km of film were consumed in the three years of its operation. The instrument was then shelved since 'that chamber is not being used and will not be used in the next future by any group at CERN, because of the existence of the larger chambers'.[141] In February 1965 a request to borrow it by the National Laboratory at Frascati for use on their new accelerator[142] was accepted without discussion,[143] whereupon HBC30 embarked on a new career[144] at a machine which was better adapted to its characteristics.[145]

Notes: pp. 327 ff.

6.3.5 THE 2 m BUBBLE CHAMBER

This time it was not just a question of building an instrument in the corner of a workshop, drawing on empirical knowledge, common sense, everyday experience, and trusting to luck. 'The construction of a large hydrogen bubble chamber is a project of considerable magnitude'.[146] The group realized the implications: 'To successfully complete a project of this importance one cannot simply improvise from day to day. From now on, from the start, and at every stage, we need an organization of resources such that each person knows what he has to do and nothing is overlooked'.[147] The big bubble chamber, whose characteristics were to be studied carefully, called for long-term planning.

Amiot and Nullens[148] proposed a work programme in four steps, estimating at each the man- (and brain-) power necessary: 'On the basis of a preliminary project, the head of the group, who in the CERN structure can only be an experimental physicist, helped by other physicists, draws up a list of responsibilities', identifying the requirements and the characteristics of the chamber (three months). Then follows 'the study of one or more projects conforming to the data above' by the technical service comprising four engineers (including one cryogenics expert), one or two engineers or specialists in optics, and two or three designers. This would take six months, followed by three months for the final drawings by five draughtsmen (including three designers). The chamber is built at the third stage, either by the main CERN workshop or by an outside contractor, but always 'under the supervision of the above engineers and designers'. Finally there is mounting and assembly, to be accomplished by ten technicians in the laboratory directed by four engineers. A special group comprising an engineer, a physicist, a designer, and two technicians would work on particular aspects, like the properties of metals at low temperature, and the manufacture of inflatable joints and of sealed cold valves. Even though the entire programme was actually drawn up for a 2 m bellows-type chamber with one or two horizontal windows—which was not, in fact, the final choice—, it foresaw that only one year would be needed to set up the project with twelve people (not counting the technicians), an estimate which was unrealistically optimistic. It shows, all the same, the need that certain researchers felt, in the light of the magnitude of the task, to tackle it step by step in order not to be swamped. Despite this, the group often seems to have been overwhelmed!

The first meetings were held early in 1958 and a study group was set up to define the project and to assess technical problems and construction costs. It was composed of those who had built the smaller chambers, so Peyrou, Vilain, Amiot, Bridge, Amato, Budde..., along with the members of the PS magnet group (e.g. Resegotti) as well as Ramm, Kowarski, and Adams. The decision that CERN should build a big chamber is studied in chapter 11 and elsewhere.[149] For the present purposes let us simply cite one or two important reports on the subject. The preliminary report discussed on 26 March[150] sketched the principal elements of the project. The final version of the most important study[151] that was made is dated 1 May 1958, and was laid before the ninth meeting of the Scientific Policy Committee on 23 May. It was a general study comparing, at the technical level, the advantages and disadvantages attached to the size of the new chamber chosen by CERN, and contained the germ of all the main choices that would be made on the construction of the chamber.

The key decision to be taken at this stage was the size of the chamber, as the construction principles were more or less known, and the report did not question the need for the device: 'Therefore the proposal for large chambers is reasonable and feasible'. Certainly, 'it must be realized that most of the [...] uncertainties arise from the relative novelty of the bubble chamber [...]. There are no generally accepted solutions to the design problems, i.e. there is no classical design', though the authors decided, somewhat regretfully, not to depart too far from orthodoxy: 'the chamber itself is quite conventional; one would like to contemplate a more brilliant design [...]'.[152] The choice of size offered was between 1.10 m and 2.00 m, and overall the report favoured the more ambitious venture: 'The conclusions of the study group are that the 1.10 m and the 2.00 m chambers would involve about the same number of staff and that the relatively small difference in the total cost of the two projects justifies the recommendation that, if any chamber is built by CERN, it should be a 2 m chamber'.[153]

Concerning fundamental issues of feasibility, the report raised five questions which it did not answer unambiguously. It was only on the functioning of the chamber that the answer was clear: yes, without doubt, it would show tracks. To evaluate its dynamic consumption, so to calculate the amount of hydrogen that had to be supplied, it was preferable to be generous so as not to have problems when the experimental work got under way: 'Also it is generally felt that it is better to start nuclear physics experiments, even with large consumption, than to spend too much time in trying to improve the performance of the chamber'. This phrase is indicative of the general tone of the study, in which the need to press ahead quickly is often stressed—to a certain extent in contradiction, it seems to us, with what actually happened: the construction finally took a very long time, was very carefully studied, and great care was given to every detail, almost as if in contempt of an altogether understandable haste to get some physics done at last. The concern to get uniform sensitivity defined the limits of temperature variation (1/10 K to 1/5 K was practicable), and fixed the maximum height of the chamber and the need for vertical windows (the upper wall allowed for good temperature control). As for the analysis of results, the question was: could one get distortion-free trajectories and measure directions precisely? The answer, which was studied systematically using the small chambers until 1960 (in 1958 results were still contradictory) was difficult, since while a uniform temperature avoided convection currents, there was movement of the liquid caused by the expansion. The envisaged solutions giving a uniform velocity

Notes: p. 328

field (which was difficult to obtain at the time[154]) were to move an entire wall or to use a system with several pistons sufficiently separated in the active volume. Finally, as for the cycling time of the chamber, it was still not known since the factors mentioned above affected it.

In the list of the work that remained to be done—build a chamber with a window that could withstand at least 13 atm at 27 K; isolate it thermally; design an expansion mechanism which releases the pressure in 5-10 ms with uniform velocity and negligible boiling; place it in a 10-12 kG magnet; design the chamber and the magnet so as to ensure an ease of assembly and of demounting; install constant-current generators and a cooling system for the magnet; provide sufficient liquid nitrogen and hydrogen for the cooling and functioning of the chamber; illuminate it and take photographs; take a particle beam to the chamber without having an intolerable background; provide for maximum safety; study the implications of the two previous points in choosing a suitable site for the chamber—the condition concerning the time of construction (to correspond to the date when the PS worked), and for which 'we must avoid designs based on extensive previous development and, although the solution may be less elegant, avoid unnecessary elements of uncertainty, in the interest of speed of construction',[155] was absolutely not satisfied, as the chamber was not ready before the end of 1964. Though it must be said that these requirements were that much less pressing with the arrival of the Saclay 81 cm chamber and the British national chamber of 1.5 m.

It is interesting to extract from this report the choices made in 1958 which were (or were not) implemented:
- height not to exceed 60 cm: implemented;
- vertical windows: implemented;
- liquid expansion: implemented;
- multipiston expansion: abandoned;
- dark field illumination with two windows: implemented;
- body of aluminium or stainless steel: the second was chosen;
- magnetic field of 10 kG minimum: implemented, largely exceeded;
- magnet installed on the ground of the PS experimental hall... at least if a special building is not built: a special building was erected;
- the liquid hydrogen for cooling and for temperature control will be supplied by different liquefiers: not implemented;
- the chamber will have cold safety tanks as protection if a window should break: implemented;
- 5 MW maximum electric power for the magnet: respected;
- deadlines (again!) which call for conventional solutions: not respected;
- an integrated transport system to handle the considerable weight: respected;
  all of which goes to confirm Peyrou's opinion: 'The bubble chamber which results from these considerations appears to be a reasonable instrument and, [...] any final chamber must bear a strong resemblance to the model considered'.[156]

The entire infrastructure needed for this project was not to be underestimated: 'Indeed the chamber will be such a large and complicated ensemble that it will require substantially more automation and monitoring than most ordinary physics instruments'.[157] Controls concern eleven systems[158] and special devices must still be tested before and during the final project. A number of subjects must still be studied: 'These problems (illumination [...] consistent with a practicable vacuum tank and magnet structure [...], handling of larger quantities of liquid hydrogen [...]) as well as others require further investigation but are not essential questions as far as the feasibility of building or operating the chamber is concerned'.[159] To diminish considerably the cost of later use it was necessary 'to understand much better than was the case [in 1958] the different phenomena arising in the course of gas or liquid expansion'.[160] Systematic experiments were thus made with the two small chambers to understand better 'the law of bubble growth, the light energy needed to get good photographs, the boiling phenomena on expansion and their consequences on the dynamic consumption, temperature conditions and the rate of expansion needed to get good tracks, etc.'.[161]

The overall estimate of the cost of the project, without the magnet, was 3.1–3.2 MSF, a cost which tried to accommodate the price of the major components which were to be constructed off-site.

In November 1959 almost two years had passed since the first studies were made but the construction had not yet started. The group was still working on the problems raised by HBC30 (an accident with the magnet) and above all on the design of the project. J. Vilain[162] drew up a three-page long 'enumeration of the points [to] check by calculation and experiment before setting out on any construction' which was to 'constitute a dossier on the construction of bubble chambers'. The list identified primarily mechanical calculations on the components (stresses, pressures, thermal variations, etc) and problems of manufacturing, cryogenics, and optics.

In fact during 1959 the team was essentially recruited, the new buildings necessary for the big hydrogen bubble chambers were planned, and preliminary contacts were made with firms. In 1960 the buildings were started and the first contracts placed. In 1961 the construction advanced steadily in outside industry. In 1962 different components were delivered to CERN, but one already had major delays in the fabrication of the vacuum tank and of the chamber body (the delivery of which was as late as 1964) and in the prototype window, which was ready at the factory in 1961 but which had not yet arrived at CERN in 1962. The CERN Annual Report for 1963 had to concede that 'To date CERN was behind American laboratories in the use of big bubble chambers',[163] but offered no justification for the new deadline of early 1965 which it gave for the functioning of HBC200.

Certainly when reading the accounts of the construction one cannot but be impressed by the number of important studies, tests, trials, and prototypes prepared by the group: 'several models were built'.[164] In addition, as we shall see below, a 1/10 scale magnet in monel was built in which key studies were made to measure in

Notes: pp. 328 ff.

advance the total field provided by the magnet and to check its uniformity, a model of the expansion system was tested, a model of the electromagnetic valve was developed, trials were carried out with inflatable gaskets, tests of the optics were made to eliminate reflections, the mechanical and magnetic properties of steels at low temperatures were analyzed, a prototype of the photographic apparatus was built... As the CERN Courier stressed, the bubble chamber created 'construction problems of an entirely different order' to its predecessors.[165]

HBC200 was a conventional chamber expanded by a piston placed above it, with vertical windows, direct illumination, and without a pole piece for the magnet, which gave a field of 1.7 tesla. The dimensions of the visible region were 200 cm × 60 cm (height) × 50 cm, giving a usable volume of 600 litres. In all 1150 litres of liquid hydrogen were contained in the chamber, of which 500 litres were in the converging cone and the expansion cylinder above the chamber. The windows were of unhardened borosilicate glass 17 cm thick, sealed to the chamber by two inflatable gaskets. The whole (chamber, reservoirs, magnet, etc...) weighed more than 700 tons. Later the chamber was capable of multiple expansion, and could make up to three cycles per PS pulse.

In December 1964, finally, the first trials were crowned with success and, after disassembly to make minor improvements (improved version of the stainless steel piston, later replaced by a titanium piston, addition of an auxiliary heater to accelerate reheating after the end of a run), the chamber finally entered into operation at the end of March 1965 in a 'very satisfactory' manner[166] (200,000 photographs were taken before a breakdown in an inflatable gasket). After the summer overhaul (pushed forward due to the breakdown which permitted new modifications), from mid-September onwards HBC200 provided many photographs for researchers to analyze (800,000 between September and December for different groups).

Naturally several improvements to the individual components did not give good enough results, and the replacement of faulty elements or further developments (use of deuterium, rapid cycling, perfection of the control system) were needed after the chamber was commissioned, and once a year anyway when the PS was shut down: '[the] ordinary maintenance programme [...] calls for scrupulous attention to detail if the chamber is to function properly'.[168] From 1973 onwards it seems as if the cases of 'fatigue' began to increase noticeably due to its age, and the chamber was finally withdrawn from service on 8 July 1977, seven days before the planned date. A break in the RCH piston joint caused the cancellation of the last experiment which was intended for teaching purposes.

In the twelve and a half years of its active life, until 1977, CERN's 2 m hydrogen bubble chamber produced 40,500,000 photographs (of which 6,700,000 were with deuterium), and consumed more than 20,000 km of film, 550 m in the cameras in less than an hour.[169] Around 1971 69 people worked at this photograph producing factory, which was operated around the clock, seven days per week, for runs which lasted many weeks without interruption. The number of operators was drastically decreased and its personnel progressively relocated in the two years before its withdrawal from service in 1977, leaving only 40 people working on it, which caused difficulties when one of the technicians fell ill. Teams of four people changed the films in the cameras, developed a sample to check the chamber's functioning, and ensured the proper operation of the entire device. HBC200 was useful for the study of the interaction mechanisms of high-energy particles and for experiments requiring high statistics on the properties of excited particle states. Even if it did not make any spectacular discoveries, it was a reliable and accurate instrument. Donated to the Deutsche Museum in Munich, which has not exhibited it to date, it was—in the words of its builder Charles Peyrou—'probably the best chamber in the world'.

6.4 Technical elements and construction

6.4.1 THE BUILDINGS

While the size and weight of the small 10 cm chamber, as well as that of the 30 cm, posed no special installation problems, so that the devices could be housed in a simple laboratory, from the time CERN began to build or to host bigger chambers they had to provide an adequate infrastructure, of which buildings were not the least important component. In fact more than half the cost of a bubble chamber programme is made up of 'facilities such as power, cooling water, liquid hydrogen, etc. which have to be available in CERN for visiting teams such as the big British bubble chamber'.[170] All of this meant a new experimental hall for the bubble chambers, needed above all for safety reasons and symbolic of a physics which gave pride of place to this detector.

The question of a suitable site was already raised in the first report on bubble chambers,[171] this being indissociable from the construction and use of big bubble chambers. One needed first of all a place to assemble the instrument which was close to the hydrogen compressor and all the equipment needed to manipulate the gas. The zone also had to be furnished with a good ventilation system for the hydrogen, and with access to a spherical gasometer for hydrogen, liquid nitrogen reservoirs, chamber control equipment, magnet generators and controls, water cooling facilities, cranes, a safe area for the personnel, and a source of high energy particles. If one insisted on maintaining the existing building (1958) it was important first to carry out suitable tests in it to study the risks of an explosion and to check the safety situation. Basing itself on the Berkeley approach, the report preferred to erect an ad hoc building for the CERN chambers and for other such detectors. This was not only needed if the big chamber was built, but for the British chamber, the Saclay chamber, and the Ecole Polytechnique propane chamber, whose arrival at CERN had already been agreed on. The report suggested that the most suitable and

Notes: p. 329

economic zone was close to the PS, which would be declared a bubble chamber site and used for all future developments. The buildings were to be supplied with rail systems for orientating the chambers, this being more suitable for moving them about over large distances than the 'feet' developed at Berkeley. Two separate working areas around the main beam were envisaged with a view to reducing costs. A common control area was foreseen for the bubble chamber areas, the gas compressor, and the liquefier; provision was also made for an office for the personnel in a safe area. The storage of the gas was, by contrast, to be relatively far away for safety reasons. The cost of the whole scheme was estimated at some 4 MSF.

The characteristics of the building depended on the magnet and its transport system and the beam paths, and these points were actively studied in 1958-59. In March 1959 the principle of having a special building was accepted: 'Granted the general agreement on the fact that the three big hydrogen bubble chambers cannot be used in the existing experimental hall [...]'.[172] It was to be a building which could accommodate two big hydrogen chambers, a beam deflector selecting the desired chamber. A long beam path of 600 m was also envisaged for beam analysis using time-of-flight counters. Other halls were foreseen for the generators supplying the beam transport magnets, the hydrogen compressors, the control rooms, the electric substations, and for the water cooling unit.

Construction began in 1960, after a first building for testing components of the big chambers was completed. It was in the east experimental area together with the buildings housing the compressors and the generators, 180 m from the accelerator, 'the largest [distance] possible compatible with the necessity of keeping a certain clearance between the buildings and the site main road [...], [and] long enough to operate, relatively cheaply, a good momentum selection of any beam coming from the machine as well as from any target located outside the PS ring [...]'.[173] As for 'the possibility of experiments with shortly living particles [these] cannot influence the choice of the bubble chamber location as these experiments are very likely to be performed by means of particles produced in interactions between some of the extracted beams and a target that will be located as near the bubble chamber as required [...] actually in no case is it advisable [...] for any so big bubble chamber to approach too much the Protonsynchrotron'.[174] The building was planned to be 57 m × 17 m, 11 m high (volume 17,000 m^3) and was equipped with two 30-ton overhead cranes. The external proton beam crossed the middle of the building between the two future chamber locations and three external targets permitting the extraction of the secondary beams which were immediately curved towards the chambers. Shielding walls were envisaged, one opposite the arrival of the beam, the other movable (with a crane) depending on needs. The building comprised a metal structure with walls of concrete elements which could easily be modified, two sliding doors 6 m × 6 m and two other doors 4 m × 5 m, and was equipped with a very efficient ventilation system. The floor could support loads of up to 85 tons/m^2.

The building was finished the next year and was fitted with explosion proof equipment, i.e. equipment which did not create sparks, to avoid all risk of an explosion in the presence of hydrogen. The lighting was pressurized, the two mobile cranes were driven by compressed-air motors, a special ventilation system renewed all the air in the hall every six minutes when the chambers were functioning normally (in emergencies all the air in the hall could be changed every 90 seconds). The compressor building which contained all the auxiliary heavy equipment had counting and control rooms, from which the experiments were watched. Finally the generator building, which contained the AC-DC current converters for the magnets was also put up. Due to difficulties in the course of the work the junction with the PS ring was only made in October 1962. This allowed experiments with bubble chambers to get under way, and was ready in good time for the British and the HBC200 chambers.

6.4.2 THE CHAMBER BODY

The heavy engineering side of the construction comprised the chamber body (the most complex component, with a highly-polished finish), and the different reservoirs (19.5 tons of stainless steel for HBC200), the jackets for insulation, and the optically flat windows. The body of the HBC200 was a parallelepiped 200 cm long, 60 cm high, and 50 cm wide, on top of which there was a pyramid-shaped trunk in which the expansion system was located. It was machined from a huge oval forged ring of stainless steel which could withstand a pressure of 13 atm. The beam entrance and exit windows were sealed to the chamber.[175] The total volume of liquid in the chamber was some 1050 litres kept at a pressure of 5-7 atm and a temperature of 25-27 K.

The main problem in choosing the material for the chamber derived from the extreme temperatures (the chamber contracted 6 mm during cool down to −248°C) and the strength of the magnetic field it would be subjected to. Already in HBC30[176] stainless steel which was chosen after studies on samples at the University of Grenoble,[177] had produced unexpectedly high precipitations of martensite[178] (2% change in the magnet charts) which fortunately did not affect its mechanical properties in a dangerous way. Much time and energy was thus devoted to this choice: in summer 1959, for example, two physicists in the group, G. Petrucci and J. Trembley went to the Fourier Institute in Grenoble to visit Professor Weil[179] to 'discuss the method permitting one to measure magnetic properties at low temperatures (20 K) of the stainless steel employed to build hydrogen bubble chambers'. There they learnt how difficult it was to obtain magnetically homogeneous stainless steel, particularly when it is cooled down suddenly, and how to construct an apparatus to measure the magnetization of samples with a precision of 0.1% (a technique applied successfully at CERN by G. Bonzano[180]). They took advantage of the trip to discuss with a representative of Ugine filters of sintered material to be used to block impurities in the liquid hydrogen which might otherwise enter the chamber while it was being filled.

Notes: p. 329

Things had moved on by 19 October 1959[181] when the group had a meeting with representatives from Uddeholm Steel who had prepared a first selection of austenitic steels[182] with a very low carbon content for the chamber body, after cooling trials to −189°C and −260°C (liquid nitrogen and liquid helium temperatures, respectively). A first choice of the steels was made, though further more extensive measurements — percentage martensite at low temperatures, influence of subsequent thermal treatment, degree of recarbonation,[183] magnetic qualities, breaking strength, resilience[184] — were still foreseen. These measurements were still underway the following summer[185] on steels and solders coming from Uddeholm, Zapp, Fischer, and Boehler, since the choice between forged and casted steel, alternatives which could affect the shape of the chamber, depended on the results of these tests, as well as on the capabilities of the various producers. 'In fact the metallurgical problem is intricate and leads to some development work prior to the actual manufacturing'.[186]

The means of disassembly and of suspension were also studied in detail[187] as these considerations also play an important role as far as the time taken to make necessary checks or to repair breakdowns is concerned. Because its total weight was 700 tons, and to assist in adjusting the position vis-à-vis the beamline and for disassembly, the chamber and its magnet rested on four trolleys each with three pairs of wheels. The complex could move on double rails on either side of the chamber by electric motors at a speed of 2 m/minute.[188] Chamber, vacuum tank, and upper platform were suspended or rested on a simple bridge which crossed the magnet at its highest point, perpendicular to the rails. The assembled chamber could be moved down the rails, or turned by lifting the trolleys and placing the magnet on a 'turntable'[189] some 3.5 m in diameter which could be rotated with a motor. When the desired position was reached four sheets of moulded steel were placed beneath the trolleys so transmitting the weight to the floor of the building. The system was never used, despite all the preliminary studies and construction work that were done. By the time the HBC200 was commissioned it was much easier to direct secondary beams into the chamber than to move the chamber around the hall.

Points of detail, like the number and position of the thermometers and pressure transducers (for which a further quick study seemed necessary) were still being discussed in 1961.[190] But perhaps because time was growing short it was agreed that 'The general feeling was that effort should not be devoted to developing new transducers unless there are very serious advantages to be gained'.[191] What is more the phrase 'if possible' or 'possibly' appears four times in one page of this text, signifying that the group wanted the construction to advance, even at the cost of the perfectionism which characterized the first texts.

It was probably this same anxiety which prevailed when T. Ball[192] visited the factory at St. Chamond on 13 September 1962 so as to check 'the state [...] of progress of the construction of various bubble chamber components': the body ought to be finally machined at the end of January 1963, a delay in thermal treatment having arisen due to the need to repair an oven; the dome of the chamber ought to be ready at the same time; the base of the chamber was scheduled for 15 November, etc. When he insists that 'this information is not to be taken as giving a new programme of deadlines for the construction of the chamber but as estimates of the time needed to complete each component under the best possible conditions', one detects the concern about deadlines, all the more so when he describes the work that has to be done for CERN at the Compagnie des Aciéries et des Forges de la Loire (C.A.F.L.), where 'tight and accurate planning is called for [...]. We have grave doubts on this point'. Were the difficulties over deadlines caused by the researchers, who were more familiar with small chambers than with a really industrial-sized instrument, and where an adequate schedule of orders ought to be drawn up at the start, or were they due to conditions in the factories, which were not used to having to meet the standards of precision demanded by CERN for a research instrument? In any event, at this time, September 1962,[193] a member of the group continued to survey the factories who could work for CERN, particularly in the area around Grenoble.

In 1963, after welding and tempering all the components, including the cold tanks, at 850°C,[194] machining finally commenced. The assembly of the whole system took place the following year, and was done extremely carefully, so that the chamber worked immediately.

6.4.3 THE VACUUM SYSTEM AND THE COLD SHIELDS

Being aware of the complexity of leak tightness problem, which should not be underestimated, from the start[195] the builders foresaw the need for supplementary developments in the project. These were all the more important since they had come up against difficulties with the experimental chamber caused by the very low temperature (which increased the density of the gas and called for an improvement in the seals) and because of the problems caused by the sometimes drastic changes in their properties which some materials underwent when cooled. Studies on how best to tackle these issues began in August 1957 and continued through to January 1958:[196] a first series of tests was made in September 1957 on HBC10 to confirm the tightness of the reservoirs and the channels and to check their mechanical resistance at high pressure. Measurements on the escape rates of helium, and the use of 'straingauges' and of stress coatings led to positive results. They confirmed that the weak component in all these systems, i.e. the gaskets, ought to be studied carefully. Hence various ways of welding were successively gone into — with tin, with silver at low temperatures, or by brazing with silver. These investigations, whose results were even presented to the 1st International Congress on Vacuum Technology in Namur (B) in June 1958,[197] showed that for steel–steel or steel–brass contacts, gaskets of pure indium (or possibly lead, less ductile but more easy to remove when one had fragile components) should be embedded between the two pieces to be joined, one having a V-shape and the other being flat. As for the glass windows to be sealed to the body of the chamber, problems were also raised by the hydrogen contained under

Notes: p. 329

pressure in the vessel, and by the different rates of expansion of glass and steel with temperature. Research suggested the possibility of using gaskets of indium wire, of copper tubes coated with indium and filled with asbestos, or of copper gaskets. Such findings were applied in the construction of HBC30, where the metal–glass join comprised two gaskets per window in copper coated with indium and placed in rectangular grooves. However, for HBC200 the solution chosen was that also used at Berkeley of two double gaskets of inflatable stainless steel tubes maintained at constant pressure by helium of 60 atm pressing against indium seals.[198] These gaskets continued to raise problems throughout the life of the chamber, and had to be modified more than once.

In a preliminary study on the vacuum system[199] R. Budde drew attention to five points concerning the pumping of the big chamber which, in his view, demanded special attention: (1) the cleanliness of the windows; (2) a pumping speed as high as possible; (3) maximum reliability; (4) minimal noise, notably by the forepumps; (5) the cost. A quick calculation showed that a surface of 1 m^2 at 20 K gave a rate of condensation for nitrogen of 100,000 litres/sec, a figure considerably in excess of the pumping speed of the fastest existing pump. The windows in particular might play this role of cryogenic pump, and became dirty as a result. 'Some people call this dirt just dirt, others call it "oil", and rather sophisticated ones [...] claim it is essentially water and N$_2$'. It made the taking of photographs that much more difficult if not impossible.

The system for the big chamber comprised a main vacuum tank which could withstand a pressure of 13 atm,[200] as well as two safety cold tanks for collecting the liquid hydrogen in the chamber if the windows broke. The chamber was surrounded by a jacket cooled by hydrogen which would ensure an even better insulation than that used on HBC30, the latter being equipped with a cold jacket of very pure copper connected to a copper reservoir filled with liquid nitrogen.[201]

The different vacuum tanks called for various separate pumping units: the main reservoir, for example, had three.[202] As for the different kinds of pumps, four models were considered, namely, mercury or oil diffusion, Getter, turbo-molecular Pfeiffer pumps, and cryogenic pumps.[203] The second and third types were rather novel at the time, and were not sufficiently reliable and certainly too expensive. While the last seemed rather tempting (though only as a back up, because their reliability depended on that of the compressor), the final choice of diffusion pumps was made because of their reliability, their longstanding presence on the European market, and their reasonable cost. This decision was not regretted afterwards as the pumps were practically trouble-free for 13 years, giving a vacuum of 10^{-5} or 10^{-6} torr, their only drawback being that one had to reheat the chamber partially two or three times a year to clean the windows.[204] The pumping units comprised the forepumps (two rotary pumps of two stages and two Roots pumps) and an oil diffusion pump topped with two 'baffles', one (watercooled) at ambiant temperature, the other at liquid nitrogen temperature, which reduced the pumping speed from 14,000 litres/sec to 4500 litres/sec at 10^{-4} torr. The mechanical pumps were placed as close as possible to the diffusion pumps, though certain connection lines were still 10 m long. Numerous pressure and vacuum gauges and control valves were foreseen, whose readings were transmitted simultaneously to the platform and to the control room, and a set of safety interlocks rendering any mistaken manipulation impossible. In case of an accident in the system the pumping could be rapidly switched so as to continue along another path.

The dimensions of the vacuum tank, discussed as from June 1959,[205] depended on the expansion system chosen (gas or with a piston) and above all on the mode of disassembly. For this it was hoped that a wooden model of the chamber would permit one to take a well-informed decision. The costs and benefits of two solutions proposed by Vilain were evaluated. From this year too there dates the study of a model under vacuum and under pressure which allowed one to pin down certain limits and to write with considerable optimism that 'the study of the vacuum tank has also advanced considerably'.[206] In April 1960 a few points remained to be settled,[207] and in August a first detailed description of the system was drawn up.[208] Finally, in January 1961, the final specifications were written.[209] The design foresaw a vacuum tank of stainless steel, suspended with the chamber from a bridge on the platform and supported by the magnet. It was provided with two extensions along the beam axis which served as high vacuum lines for three diffusion pumps, with openings for illumination and photography, with connections to the pumping units and the safety lines, and with paths for the tubing and the link to the expansion system.

The manufacture of the tank posed quite some problems, particularly when it came to welding the tubes to the plates covering the main frame, and this caused a considerable delay in delivery (October 1963). Finally the vacuum envelope was attached to the platform of the magnet and the next round of tests of leak tightness (the first were carried out at the factory) proved satisfactory. The forepumps and the Roots pumps had already been delivered in 1962, as well as the gauges and measuring instruments. The following year the stainless steel pump connections were delivered, tested, mounted on the main frame of the tank, and all the vacuum tests were repeated. The hydrogen jacket of four elements was made in the main workshop and was ready for installation in 1963, though in the first overhaul of the chamber made in 1965 it was to be abandoned (the eddy currents destroyed it when the magnet power failed) in favour of a supplementary system of cooling circuits which reduced the temperature gradients. This elimination was possible because the superinsulating system was very efficient and a poorer vacuum could be tolerated.

When the chamber was commissioned in December 1964, it was immediately necessary to add two more pumps for intergasket pumping of the tube gaskets, since the loss of hydrogen was 20 m^3/hour, a rate which did not permit one to get a vacuum in the envelope below about 1 torr.[210] This trouble was caused by the differential dilatation of the cold tanks, which were later connected by inflatable gaskets.

Notes: pp. 329 ff.

6.4.4 THE COOLING SYSTEM

This system comprises the cooldown of the chamber to the working temperature of 26 K and the control of the temperature, both highly demanding functions for a hydrogen bubble chamber. Heat arrives at the chamber by conduction and by radiation: the sum of the two is the static heat load. In addition, the expansion produces heat which gives the dynamic load.

In the three chambers the cooling is done by a pressurized bath of liquid hydrogen in thermal contact with the upper part of the chamber.[211] For HBC10 this bath was in contact with the top third, and to the extent that the distortion of the tracks due to convection currents was negligible,[212] the same system was to be applied to the following chambers. In HBC30 the pressurized bath was even a part of the chamber body itself; here the wall was reduced in thickness whenever possible to 3 mm to improve thermal contact, despite the low conductivity of the steel. With external walls also in stainless steel, and welded between them and the chamber, the bath surrounded the expansion cylinder and stretched along the vertical walls of the chamber.

The measurement of the consumption of hydrogen, of no importance in the small experimental chamber, was much more worthwhile with HBC30 where it was used to estimate the amount of liquid hydrogen needed for physics runs[213] and to evaluate the demands of HBC200, the figure here conditioning the choice of an incorporated refrigerator. To calculate the static load, conduction was assumed to arrive above all via the cylinder but also via the piston shaft, the tubes, the supports, etc...It was 45 watts, of which only a part arrived at the chamber as such, the heat coming from the cylinder being largely absorbed by the pressurized bath before it could reach the hydrogen in the chamber. In addition the heat from the piston shaft and the dynamic load primarily affected the upper and invisible layers of hydrogen in the chamber. Weak convection currents caused by heat inputs at the bottom of the chamber were largely induced by radiation through the windows (protected by the cold tanks) and by a little conduction coming from the tubes for filling and emptying the device.

The concern with dynamic consumption was so great that in May and June 1959, i.e. during the earliest tests of the chamber, two reports estimating it were written.[214] The first identified the presence of gas between the piston and the liquid by studying the expansion curve. The consumption of cold was considerable each time it was necessary to reliquefy some of the chamber liquid that had vaporized, and a calculation arrived at a figure of 41 watts per recompression, which, for the four test expansions per minute, corresponded to a consumption of liquid hydrogen of some 0.3 litres/hour. At a reasonable frequency of expansions this consumption rose to 1.55 litres/hour. The calculation was redone for HBC10 whose consumption was estimated to be 0.175 litres/hour at the same frequency.[215] With the help of the 'Amiot gadget'[216] another calculation was made and gave a similar figure. The second text written by J. Trembley calculated the difference in the period of the automatic control both with and without expansion, which was a measure of its dynamic consumption, and found that it was about 5% of the total consumption. Finally, the first experiments gave, for a photograph taken every two seconds, a dynamic consumption of 0.5 litres of liquid hydrogen per hour, and a static consumption of 4.5 litres.

For HBC200, with its 1050 litres of hydrogen at 26 K, the problem of the consumption was already raised in the 1958 report,[217] where the capacity of the refrigerator[218] was calculated by extrapolating from Berkeley data, and assuming that an improvement by a factor of two could be achieved, and that it would take two to three days to cool down the chamber. It was necessary to recalculate the static loss and to assess the power and space required for the compressors, which were to be housed in a separate building.[219] (A study on the production of nitrogen was to be made at the same time.) But it was the dynamic consumption which was of greatest concern as it affected the viability of the chamber and was needed to estimate the power of the refrigerating unit. In an amusing estimate of this magnitude made by Peyrou in November 1959,[220] of which the title, 'An attempt to rationalize the choice of a liquefier from the smallest to the largest', is an indication of the anxiety felt at the time, Peyrou used seven more or less serious hypotheses to arrive at seven ever larger values of the dynamic load of the future chamber (of which the highest had the same numerical value as the mass of the pi-meson!). This text, an instance of black humour, ended on an optimistic note. In 1965, after the chamber was commissioned, the load could finally be measured and with all the parasite effects was found to be 500 joules/cycle, which fell within the bracket originally foreseen.

In 1960 research on how to refrigerate the large quantities of hydrogen needed for the chamber was actively pursued inside the group, particularly with the arrival from Munich of a cryogenic specialist, F. Schmeissner, who was already being used as a consultant by CERN. Preliminary orders for adequate buildings for the machines were placed with the suppliers. Different kinds of dry compressors, rotating or with pistons, were studied,[221] several visits were made to hydrogen distillation plants,[222] the construction of a converter of hydrogen from the ortho- to the para-phase was planned[223] and that of a purifier pushed ahead,[224] and a set of vessels for storing 16,000 litres of very pure, very dry hydrogen gas at high pressure was put out to contract and then ordered from Loos Co.[225] The refrigerator was to be delivered by Sulzer Bros., though there was a price rise[226] to the tune of some 200,000 SF to accommodate CERN's safety specifications, which were more demanding than those prevailing in Swiss industry. For the construction and the trials the refrigerator was to be placed on a special platform also supplied by Sulzer, since the bridge for the bubble chamber was only to be delivered at a later date.

The installation was to have a low entry-pressure, which enabled one to use a dry compressor without oil, so eliminating high flow purifiers and low temperature dryers in the closed circuit of the refrigerator. Its anticipated cooling power was 4 kW at 23 K, which was more than enough for the chamber alone, but needed for the

Notes: p. 330

cooling of the hydrogen jacket. This system avoided there being any reflux of liquid hydrogen to the refrigerator. The chosen refrigerator incorporated a novel system of precooling of hydrogen, which was not done by evaporating liquid nitrogen but by the work executed by the hydrogen in cold turbines. The advantages of the system were that one could do without liquid nitrogen and use hydrogen at the relatively low pressure of 8 atm instead of 150 atm. The auxiliary elements of the refrigerator were included in the contract: dry gas reservoir, compressor for the circuit, piping between the compressor building and the bubble chamber building, the complete purification circuit for the hydrogen, two diaphragm compressors for the recompression of the purified hydrogen in the static cylinders, the safety circuits, and all the controls for the facility.

In 1961 the set of heat exchangers, liquid hydrogen reservoirs, and regulating valves for the refrigerator column were being mounted, and the safety and control systems were well advanced. In 1962 the leak tests under vacuum and under pressure were performed, and the control system was tested and optimized. The assembly of the main compressor got under way in February 1962.

The final specifications for the refrigerator, drawn up in January 1962,[227] were that it produce 4000 watts at 23 K and 7000 watts at a temperature between 150 K and 300 K. Refrigeration was to be obtained by admitting liquid hydrogen at 29.7 K and 8 atm into the cooling loops of the chamber, where it evaporated completely and was slightly superheated. As a physics run with the chamber could last several months, the refrigerator had to be guaranteed to work for at least six weeks without interruption. The installation was to run automatically, except during periods of reheating and recooling of course, which meant that it had itself to control the power and cooling needed, which could vary from 1000 to 4000 watts. In automatic operation the only checks needed were the verification of the instruments and the machines. The refrigeration equipment was not to be concentrated, but distributed between the compressor building (compressor, purification plant, equipment to recuperate the gas, high current, controls...), the bubble chamber building (refrigeration column and controls), and outside areas (gasometer, storage tanks), the piping being conveyed between the two buildings by a connecting gallery. Many on-site tests were envisioned and the circumstances of these, as executed according to the most recent requirements, would be specified in an agreement signed by the two contracting parties, CERN and Sulzer Bros.

A report by Schmeissner[228] evaluated this choice: the difficulties CERN had of getting hold of liquid nitrogen—no problem in laboratories like Berkeley or Brookhaven—, the cost in money and manpower of a supplementary facility for the liquefaction or refrigeration of nitrogen, the risks inherent in Joule–Thomson systems, the relatively low pressure of gaseous hydrogen, and the importance of having a high refrigeration efficiency (which counted heavily towards eliminating anything but small quantities of liquid nitrogen and choosing a turbine refrigerator instead, for which 'there existed already some experience about reliable performance and reasonable efficiency [...] in the cooling range desired'.[229] Even if the preliminary estimates of cold loss led to a usable power of 1300 watts, technical and economic reasons set the lower limit to the power decided on, giving a speed of cooling of 2 K per hour, so five to six days for the whole chamber.

In 1963 'a rigorous and detailed set of acceptance tests was carried out and the installation was accepted after various trials of its functioning which were crowned with success'.[230] The refrigerator performed even better than specified, going to 14 kW at 150 K and 7 kW at 22 K. It successively achieved the cooling of the chamber components (22 tons of steel, 2.2 tons of glass, and 2.5 tons of copper), liquefying its contents and carrying away the heat of radiation and of conduction as well as that produced by expansion during the chamber's work cycle. The 1963 CERN Annual Report gave it a good mark:[231] 'Thanks to a very simple and efficient regulation system proposed by CERN the installation adjusted itself without any difficulty, automatically and instantly, to all changes in the load in an interval between zero and the maximum possible value [...]. This facility represents a remarkable step forward granted the classical difficulties of precooling with nitrogen [...]'. One further detail: it could also be used as a liquefier producing 120 litres of liquid hydrogen per hour without using liquid nitrogen.

Temperature control is at once a difficult and essential problem in a hydrogen bubble chamber,[232] since it requires the removal of heat from the chamber (by evaporating the hydrogen in the pressurized bath) while keeping the temperature constant and homogeneous. As we have seen, homogeneity is achieved automatically by having the pressurized bath only at the top and sides of the chamber. In addition the heat exchange surface is sufficiently large for the temperature difference between the chamber and the bath to remain small. To keep the chamber's temperature constant one thus controls that of the pressurized bath with a sort of thermo-siphon invented by Amiot[233] (and nicknamed the Amiot gadget), made, in the case of the small experimental chamber, from several hundreds of meters of copper, monel, and stainless steel tubing. It was fundamentally a closed circuit of pressurized hydrogen, in this way avoiding both an overconsumption of hydrogen (heat-loss system) and considerable temperature gradients while the hydrogen was being transferred to the pressurized dewar (pressurized bath and reservoir system). The hydrogen gas which forms is led, as and when necessary, from the pressurized reservoir which cools the chamber across a valve in a heat exchanger filled with liquid hydrogen boiling at about 20 K, then sent back after liquefaction into the first reservoir. The valve opens in response to the pressure and allows for the controlled reinjection of the liquid hydrogen at the desired temperature.

After building and testing a prototype temperature control system at Taconis' laboratory in Leyden — the person who had built CERN's first liquefier — ,[234] and having applied it to HBC10, the system was used again on HBC30 and even on the Saclay 81 cm chamber while it was at CERN, the French system copied from Shutt's chamber proving inadequate.[235] For HBC30 two reservoirs were symmetrically located with respect to the supporting cylinder, and attached to its lower section. Made of welded copper, it contained three spirals of copper which served as heat exchangers: one to liquefy gaseous hydrogen, when the chamber was full, and the two others to reliquefy the hydrogen evaporating from the pressurized bath. The reservoirs and the pressurized baths were filled directly with liquid hydrogen using the same transfer line, while the chamber itself was filled with hydrogen gas at a pressure between 1 and 8 atm.

The temperature of HBC200 was controlled with four independent cooling loops. A first set of heat exchangers just above the visible zone controlled the temperature of the useful volume from a distance. Another, the true cooler, comprising four big copper wings, was situated in the expansion cone, adjusting the temperature and absorbing the heat flux in this region. A ring-shaped reservoir welded to the outside of the expansion cylinder, around and below the level of the piston controlled the temperature of the liquid immediately under the piston and absorbed the dynamic heat load. The hemispheric shape of the piston head allowed for a better control of the temperature in this critical layer of liquid and stopped the formation of vapour under the piston during the expansion–recompression cycle. Another ring-shaped copper reservoir, placed in the cylinder above the piston, ensured that the liquid-gas interface was located well above the piston and absorbed the main part of the static heat of the cylinder wall and of the piston head. The four cold loops were fed via a closed cooling circuit which first delivered liquid hydrogen at 8 atm to a supply reservoir in the column. From there the liquid was sent to the four loops where it was expanded, slightly superheated and its temperature controlled by a double-valve system. It then went to the loops of the safety tanks and the hydrogen jacket, returning to the refrigerator circuit. The system without precooling with nitrogen was more economical granted the situation at CERN (difficulties with transport, high cost of liquid hydrogen). On leaving the cooling loops its temperature was some 23–25 K, and its pressure from 1.5–2.5 atm. On returning it entered the refrigerator at 1.3 atm. During cooldown of the chamber the same amount of hydrogen gas passed through the loops, always entering at a pressure of some 8 atm. During the filling cycle the power was used partly to keep the chamber cold and partly to cool the filling gas already precooled by liquid nitrogen at 80 K and for the final condensation in the chamber.

The equipment for regulating the temperature control, delivered in 1962 and partially mounted, was tested at CERN in 1963 in a suitable device, the chamber having not yet been mounted: a brusque variation of the thermal load in the thermodynamic simulator only changed the temperature by 0.3 K, which was equivalent to a change in the chamber of less than 0.1 K, a result which was judged to be largely positive.

The cooling circuits contained removable cold joints and cold taps,[236] a solution taken up again after it had been abandoned for HBC30 because of technical difficulties, and because it had not given satisfactory results in practice with HBC10.

Studies, notably those made in collaboration with a German company, led to the development of a high-pressure valve which could be used above 7 atm. All the cold valves were located in a very large auxiliary vacuum tank, and the ensemble, ordered in 1961, was successfully tested in 1963.

Despite all this, after the chamber was commissioned, new cooling circuits had to be added in 1965 during the summer shutdown for safety reasons. They were intended to keep the chamber cold for a longer period in case of breakdown.[237]

6.4.5 THE EXPANSION SYSTEM

Of the three available expansion systems, that of Shutt was chosen from the start, a system in which a piston was in direct contact with the liquid (liquid expansion system) and in which a modification in the direction of Steinberger's bellows system was always possible. One of the reasons for the choice was the easier extension to the bigger chamber. This rapid expansion by piston, partially copied from cloud chambers, and rather cobbled together, was found to be very practicable, and was used again on HBC30 and HBC200. The distortions of tracks with this system were negligible, and convection currents in the liquid to regulate the temperature no longer affected the photographs.

In 1958 the choice of piston expansion does not appear to have been as readily accepted as it was for the two small chambers. Several other ideas were put forward: Amiot[238] foresaw 'combining expansion and thermal control, expanding by ejecting a controlled amount of liquid which will be compensated, on recompression by injection, by as equal amount of liquid as possible'. This solution was that much more miraculous in that it reduced considerably the liquid hydrogen needs and allowed one to keep the then available installation for liquefaction (simply adding another compressor). 'One sees immediately that a chamber functioning according to the above principle would need no liquid hydrogen other than that required for the precooling and initial filling'.[239] The only worry was that 'if it was impossible to achieve a thermal balance in this way, an auxiliary thermal control system would be needed'. If one could still get tracks by expanding the chamber from 6 atm (chamber pressure at 28 K) to about 4 atm—probable, according to an 'Berkeley Engineering Note'—one could use a safe and rather cheap rotary compressor for recycling. On the basis of a thermodynamic study and some preliminary calculations Amiot asked that a prototype be built functioning by hydrogen escape,[240] to be followed by a rectangular chamber 20 × 15 × 10 cm using the same principle, with a deadline of one year. 'The general study of a 2 m chamber [...] could be [thereby] considerably advanced'.[241] Even though, according to the author, this 'new concept in the field of liquid hydrogen chambers' was 'perfectly viable', and 'would lead to considerable savings in the budget estimate for cryogenics for a 2 m chamber' which 'would be then from 0.8 to 1 MSF', and even though 'until then no fundamental or radically destructive criticism [of it] had been made', it was never put into practice.

Notes: pp. 330 ff.

The report on big bubble chambers[242] which we have often quoted in this chapter looked into an alternative: the expansion system would comprise four pistons working in parallel, the closest approximation to an entirely mobile wall, an arrangement which could reduce distortion to a minimum.[243] Since the expansion cycle would not last for more than a few milliseconds, the piston could not move more than a few centimetres, which in turn meant four pistons (for the 2 m chamber) of a large diameter. Their independent pneumatic control system would be situated outside the magnet, and each piston would be connected by a long, non-conducting shaft to an external piston of the same dimensions, controlled by compressed air or helium. A synchronization of the two pistons was foreseen and a pneumatic shock absorption at the end of the expansion and of the recompression cycles. It was essentially the fear of not achieving perfect synchronization that led to this project being abandoned, as well as a subsequent system using two pistons.[244]

A year later, in May 1959, with the decision for the 2 m taken, Amato and Petrucci made a somewhat unconventional proposal, using electrodynamic forces:[245] inflatable elements filled with gaseous helium would be mounted, in alternation with narrow coils of electric wire, along the walls of the chamber. The passage of a current in the coils would draw them together, compressing the cushions of helium, and this would expand the chamber. An example calculated for a 500 litre chamber and about a hundred inflatable elements each 50 cm long seemed to yield a dynamic consumption less than that with the piston. It should again have been possible to combine expansion with temperature control since the helium cushions could also serve as heat exchangers. Once again, however, and despite the statement that 'In view of the fact that the present scheme seems to offer some advantages as far as costs and work are concerned, it would perhaps be interesting to study the possibility of its entering the practical stage', the solution was dropped for the more classical alternative using the piston.

To choose this the group was in the process of getting experience with the small chambers: HBC10 had a stainless steel piston, fixed to a stainless steel shaft, with a maximum displacement of 5 cm. Its dynamic tightness was assured with two rings of RCH,[246] a material which CERN was the first to use for this purpose and which proved to be excellently suited (100,000 trouble-free expansions with HBC30 whose piston was also coated with RCH).[247] After the first trials in May 1958, a new piston was designed and the cylinder modified to enable one to change it more easily.[248] For HBC30 piston and shaft were also of stainless steel, an aluminium disc attached to the shaft stopping the liquid being projected by expansion into the hot part of the cylinder. A sealed bearing guided the piston shaft and two pairs of joints pumped differentially ensured perfect leak-proofness around the piston. A second, warm piston was attached to the shaft and moved in a cylinder above the vacuum tank.

Additional calculations were needed for HBC200 particularly to evaluate the time constant of the chamber and an optimal shape for the upper part of the chamber (which is attached to the chamber by a cone).[249] This system needed valves working at pressures of 70-100 kg/cm^2 for very short periods. Trials[250] were thus made on Barksdale valves, 'the only reliable solenoid valves' (which had posed a number of problems for HBC30!), even though they were not commercially intended for such high pressures. The results were positive. Losses due to eddy currents during a complete expansion–recompression cycle of 20 ms were also calculated at the same time[251] so as to choose the most suitable material. While losses were about twice as large for a 3 mm stainless steel piston as for an anticorodal one of 5 mm, (and were below 210 W in both cases), the forces exerted on the piston were ten times larger for the anticorodal.

The cold chamber piston was driven by the movement of the warm piston in the upper expansion cylinder. For the experimental chamber this had an interior diameter of 60 mm, thick enough to simplify its manufacture and to assure its stability. It comprised two compartments, closed by two membranes, one filled with nitrogen at a constant pressure of 9 kg/cm^2, the second pushing the piston downwards with air at 12 kg/cm^2 supplied by a compressor. The evacuation of the gas contained in the second volume was achieved using two groups of two three-way 3/4″ Barksdale valves. A first pulse was sent for the expansion, followed by the other valve for the recompression. A delay of about 10 ms between the pulse and the drop in pressure was typical. A number of tests and discussions were held about the valves in general, which always gave trouble with the small chamber, and which unfortunately continued to do so for its successors.

For HBC30 the whole expansion mechanism was 80 cm high, with a stable temperature gradient from top to bottom (no convection currents), the surface of the liquid reaching above the cold piston, the remainder being hydrogen gas up to the top of the cylinder attached to the vacuum tank. In addition at the top the joints were at normal temperature and one could use rubber. If the piston moved rapidly, there was no pressure change in the cylinder, though there was one in the chamber. In the equilibrium position, in compression, the warm piston was at the bottom, pressing on the bottom of the cylinder through a rubber cushion.

Four Barksdale valves are activated to expand the chamber: nitrogen at 35 atm. coming from a reservoir arrives under the piston and pushes it upwards, so compressing the nitrogen above it. At a given moment the pressure under the piston decreases from 35 atm to 5 atm (with the opening of a valve). The pressure in the upper compartment (6 atm) thus pushes the piston downwards. Between the shaft and the warm piston there is a link such that the movement will be damped as it terminates. The piston stroke can be adjusted by changing the instant at which the valves function the second time. Four pairs of valves suffice for the commands. The total time taken for the cycle is 20 ms, and the chamber expands every two seconds. To see tracks the pressure in the chamber must fall to about 2 kg/cm^2 and the flash must be triggered when the pressure in the chamber is at its minimum, about 1–2 ms after the beam arrives. The movement of the piston is then 1–2 cm.[252]

Subsequently trials were undertaken[253] in July and then in October 1959[254] to find

Notes: p. 331

systems for absorbing the shock produced by the piston stopping, and another improvement was made with the construction of an electronic system using transistors, which avoided all dangers with hydrogen in the chamber, for the measurement of the levels, pressure, and position of the piston.

In 1961 the system for HBC200 conceived and designed by Scotoni was chosen, after some general trials. In 1962 tests continued with models to define more precisely the data for the final design, contracts were prepared and placed (including that for the gas compressors (nitrogen or helium) which operated the piston[255]) and construction began. In May the valves and the connection of the driving piston to the shaft of the main piston were completed. The rest of the system was to be ready in the summer.[256] In autumn 1963[257] the instructions for assembly and disassembly were drawn up after a trial assembly.

Expansion was obtained with a vertical piston situated in the upper part of the chamber. The piston's movement was controlled by nitrogen under pressure, acting on a small piston joined to the large one. At rest the same pressure of 10–20 kg/cm^2 was applied above and below the small piston on surfaces with different areas, giving a resultant downward force. When two valves admitted nitrogen under pressure below the piston, it moved upwards compressing the gas above it. The expansion of this gas then forced the piston downwards. When it arrived at its lowest point two other valves opened, clearing the volume beneath it until the original pressure was reached again. A careful timing of these two valves ensured a smooth arrival at the bottom position (damping). The piston then stopped in its initial position.

For a chamber volume of some 1000 litres and a dimension of the expansion cylinder of 2000 cm^2 the piston needed to move 5–10 cm for an expansion rate of 1–2%, doing so in 20–30 ms, which meant that it had to have a high initial acceleration. In addition, to empty and fill the chamber with the cylinder as required, the piston had to to be able to be lifted rather high, i.e. about 25 cm. It finally had a diameter of 500 mm, and consisted of a spherical dome mounted on a cylinder, followed by a shaft 5 mm thick and 200 mm in diameter close to the driving piston. The first piston that was used (which was changed once[258]) was of stainless steel and weighed 100 kg. It was then replaced by a piston of pure titanium, lighter (70 kg), and capable of making two expansions per PS pulse.[259] The small driving piston had an upper surface area of 900 cm^2 and a lower one of 600 cm^2. To limit and to control the volume above the driving piston there was a third regulating piston operated hydraulically which hardly moved during the expansion.

An emergency damping system was also foreseen below the driving piston, comprising pieces of a light alloy separated by cushions of nitrogen under pressure linked with small holes.[260]

The valves for filling the volume under the driving piston, as well as those which emptied the volume, were controlled by pilot valves, activated electromagnetically. The movement was measured by a cylindrical condensator in which the electrical capacity changed along with the movement.[261]

The chamber operated with an expansion of about 1% every two seconds, but could be run at three expansions per PS pulse. Its pressure fell from 5 atm to 2 atm and then in about 1.5 ms the beam passed through it and the photograph was taken, the bubble size being some 300 microns at that instant.

6.4.6 THE OPTICAL SYSTEM

As was the case for the expansion system, the choice of the optical system was not automatic for HBC200, even though the group was to end up adopting the solution used for the two small chambers: two vertical windows, one allowing illumination by the flash, the other one for photography of the particle trajectories by several cameras. In March 1958 several studies (which even included trials) indicated the feasibility of using a single window and/or horizontal panes,[262] and Amiot[263] even asked that HBC10 be modified to test this idea. However, the 1958 project[264] already put constraints on the illumination and the magnet: two vertical windows illuminated from behind simplified the optical problems but allowed no pole piece for the magnet. A modification to the illumination was envisageable, though the precision of the momentum measurement of the tracks was affected by using a single pole piece.

Illumination on a dark field, already used in HBC10 and HBC30, is best suited for bubble chambers, but implies that certain conditions have to be met: all the bubbles must refract the light equally, independently of their position; the intensity of the light must be adequate to allow the use of an ordinary film; the stereo angle must be sufficiently large to allow for a good reconstruction of the tracks and — because of hydrogen's low index of refraction — one needs a small angle between the illuminating beam and the camera. To see the whole of the chamber properly the stereo base must not exceed the height of the chamber, which means that for a height of 60 cm the cameras must be 2.5-5 m from the chamber.

In 1959 the analysis of the characteristics of the illumination of the chamber got under way,[265] using special flash lamps, by contrast to what was being done elsewhere[266] and on HBC30.[267] The requirements of the illumination — high luminous flux, narrow filter in the blue, point sources, short illumination (0.1 milliseconds at least), long lifetime — could not be met by standard tubes. This led the researchers to choose a system of three 2000 joule lamps, more costly, and produced by just one factory,[268] which had to flash once every three seconds.[269] Even better performance was achieved with those delivered to CERN in 1963.[270] Information was also gathered on anti-reflective coatings,[271] as multiple reflections posed serious problems for the uniformity of the illumination, and led in 1960 to the design[272] of a facility for coating the two big windows with antireflective layers, as well as 24 other smaller windows, the condensor lenses, nine other small lenses and various spare components. This facility was delivered to CERN in the summer of 1962 and first used for the 81 cm Saclay chamber and the 1.5 m British chamber.[273]

The light had to be very accurately focussed in one point in the plane of the camera

Notes: pp. 331 ff.

objectives with the help of three sets of two lenses (two lenses were used for HBC10 and HBC30, with a diaphragm to control the reflexions). This ensemble was already designed in summer 1960,[274] and was delivered, tested, and accepted in 1963. It was treated with antireflective coatings the following years in accordance with very careful instructions,[275] while a new and harder coating was applied to the windows during the summer shutdown in 1965 when the entire optical system was overhauled.

HBC10 and HBC30 had three cameras with Aposaphir objectives (for the experimental chamber: focal length 180 cm, diaphragm at f/22, placed 110 cm from the centre on a circle of radius 13.75 cm, giving an image on the scale of 1:5.1), while for HBC200 there were four cameras (objective focal lengths 180 cm) placed 2.70 m from the central plane of the chamber on a 60 cm diameter circle, giving photographs on the scale of 1:13.3.[276] These cameras, which had to operate in a high magnetic field, had the advantage over their predecessors of a more rapid film transport system (attainable frequency: 1 second). All the same, even if 'the quality of the images was always very good'[277] according to the CERN Annual Reports, the cameras were often improved as the years passed, notably to increase their speed to enable a more rapid cycling.

The film used on HBC30, ADOX KB17, was 35 mm in rolls of 30 m, non-inflammable and non-perforated. The chamber image on each photograph of 41 × 29 mm was slightly elliptical in shape with a diameter of 29 mm. A system of indicator lights[278] ensured that the cameras were loaded with film. With an expansion every 2.5 seconds it was necessary to change the film every half hour. A polaroid photograph was taken every 10-15 minutes to monitor the proper functioning of the chamber. A 50 mm film was chosen for the big chamber.

The two windows were a key element in the chamber and the group was aware of the problems which they posed: 'Their large size and the optical quality needed have considerably reduced the number of firms in a position to take an interest in the contract'.[279] Out of four firms, one decided not to submit an offer and two others could not deliver the hardened glass.[280] It was thus Schott at Mainz who was given the job[281] according to the specifications drawn up by CERN in May 1960[282] and after discussion on possible tests.[283] 'The amount of extra costs [were] unknown' since in addition to a 'lot of difficulties' given by the prototype window— which seems often to be the case for bubble chambers (the price of the windows for the 1.5 m British bubble chamber increased by a factor of 2.5, and that for the CERN propane chamber by 80%)—CERN made demands concerning polishing in particular which were not anticipated when the order was placed.[284] The first window finally arrived at CERN on 5 November, though having only average quality glass it was used on the illumination side of the chamber. The cooling of all the windows and lenses of the chamber was also studied[285] and even though in 1975 a window on the camera side which was damaged during disassembly had to be replaced, there was no fracture of the windows (as was the case with the propane chamber) in the life of HBC200.

Fiducial marks on the windows served as a reference system allowing the reconstruction of the particle tracks. For HBC30 the measurement of five marks before the windows were mounted, according to a report in February 1960,[286] was used along with a programme on the Mercury computer to give the coordinates after the films had been measured using an IEP. In 1962 a new programme which allowed one to make more complete measurements confirmed that there were few errors in the measurements with IEP and very little film distortion using these arrangements. However, at the end of 1961, the window was turned to suit the HPD. New measurements were then necessary,[287] and these showed some inconsistencies which were traced back to the symmetry of the marks in particular. For better precision it was then proposed to repeat the measurements of the marks before and after each experiment to avoid errors in track reconstruction.

These problems did not arise with HBC200. The marks were increased in size, cross-shaped, and similar to those used in the British chamber.[288] There were a dozen of them on the camera side, six for the IEP and six for the HPD, and four on the other window.[289]

6.4.7 THE MAGNET

While the small chamber did not have a magnet, HBC30 was furnished with an electromagnet which gave rise to considerable concern in the group. As there was no pole piece the vacuum tank was placed in the windings, a technique that was partially adopted again with HBC200. Problems of safety and of assembly were given particular attention during the construction. The magnet, intended to have a field of 15 kG, was designed by Morpurgo and tenders were expected in March 1957.[290] It was ordered from the Italian firm of Ansaldo-San Giorgio in Genoa in 1956. The measurements of the field of the chamber's magnet were made in July–August 1959[291] with an SC beam. New measurements were undertaken in December 1959[292] after an accident had made one of its 16 pancakes unusable, and then again in March 1960.[293] They gave values of the order 15 kG.

It is not easy to establish the exact date of the accident with the magnet.[294] After some disagreements with the manufacturers, and some concern about whether the magnet could be repaired, CERN decided to remake the coils completely. Technically speaking this was undoubtedly the best thing to do (and allowed an increase in the induction of the magnet from 16 kG with a consumption of 0.9 MW to about 19 kG with a consumption of 1.5 MW[295]), but it was obviously the more expensive solution (100,000 SF were added successively to the 1960 and then to the 1961 budgets for new magnet windings[296]). Unfortunately CERN benefitted very little from this improvement, as the chamber arrived at the end of its active life at about the time when these new coils were finally ready. The most striking thing in this whole affair is the endless beating about the bush by CERN which, though all the while saying how urgent the matter was—quite understandable since, at the time, this was the only

Notes: pp. 332 ff.

available chamber for the PS—, delayed more than two months before taking a decision, and then dilly-dallied for another eight months before calling again for tenders! Let us note though that the chamber went on working all the while though with a lower magnetic field.

The first project for HBC200[297] had already specified that the magnetic field should be as high as possible, and directed perpendicularly to the plane of the windows, of course. The very attractive idea of using a pulsed field, already proposed or tried elsewhere[298] was dropped at the start because of the difficulties inherent with a metal chamber body. The shape and the construction of the magnet could not be separated from that of the chamber even if they were independent in principle, for reasons of cost and of optimization of shapes: on the window sides one needed to have space for illumination and photography, on the beam side the minimum amount of solid material. For ease of access to the chamber, and bearing in mind that it would no longer be possible to extract it from the magnet along with the top of the vacuum tank—the customary solution with the small chambers—one had to envisage an ad hoc structure. The shape of the magnet yoke was of importance for economising current though this depended on the value of the field.[299] It also supported the windings, and held them against electromagnetic forces, it improved the uniformity of the field by compensating for asymmetry in the windings, and it limited the fringe fields. As for the windings, copper seems to have been better than aluminium because space was limited.[300]

Once again many studies and comparisons[301] preceded the final design of the magnet: 'The study of the chamber-magnet unit has been pursued so as to arrive at a suitable compromise between the necessarily contradictory requirements of magnetic performance and the ease of mounting and of operation of the chamber'.[302] Guido Petrucci, who with Morpurgo had been responsible for the HBC30 magnet, went first in August then in October 1959 to Ansaldo-San Giorgio to see to the repair of the damaged coils and to take the opportunity to start talks[303] on the design of the magnet for the big chamber. The type of steel, forged or moulded, was discussed, the second alternative being slightly less expensive and simpler, but not giving such a good guarantee on the homogeneity of the material, and failing to give reliable results in ultra-sound tests. In the end the most complicated parts were moulded, and the rest of the structure forged. The length and the type of copper, the insulation of the pancakes—bad experiences help!—, and the connections to the water and electricity supplies were also studied.

The factors influencing the design of the magnet were:[304] the maximum electric power of 5 MW (which was not respected), the height of the beam with respect to the floor of the experimental hall (1.25 m—a problem finally resolved by sinking the floor some 80 cm below ground level), the need for access to the chamber, the free space that had to be foreseen for the chamber, the optical system, the vacuum tank, the piping and the temperature controls, the maximum load the floor could support, and the maximum rate of flow of the cooling water. The flat shape of the vessels containing the cold gaseous hydrogen allowed one to come closer to the magnet windings and so to decrease the consumption of electricity.

The final designs, settled in June 1960,[305] comprised two parts, iron yoke and windings. The electromagnet 'consists essentially of a steel framework, [...] "magnet yoke" of a weight of some 350 tons and of two coils made of hollow conducting copper, cooled by water, and having a total weight of about 55 tons. The power to excite these windings will be about 6 MW (600 V and 10,000 A). Special boxes are foreseen for the windings. Two other much smaller coils are mounted on the iron yoke to compensate for the fringe field of the main windings where the particle beam enters the apparatus, thereby allowing even relatively low energy particles to reach the visible region of the bubble chamber. These [...] "compensating coils" are also of hollow copper cooled by water; their total weight is about 2.5 t and their excitation power is of the order of 300 KW'.[306]

The magnet and the windings could be separated into three pieces. The central yoke, to which the chamber components were attached, remained in place and allowed for ready access to the chamber. The two lateral hulls weighing about 100 tons were pulled away perpendicularly to the windows using auxiliary jacks on wheels running in the rails. To move the unit inside the building, or for a more complete disassembly, an entire transport system was foreseen, using jacks placed on trolleys and a rotating table. The magnet served to support the chamber itself along with its auxiliary apparatus like the vacuum system, the control platform,...

After lengthy discussions with different factories,[307] the contract for the magnet yoke was awarded to SFAC[308] (Société des Forges et Ateliers du Creusot, Schneider factory). It laid down very precise tests that would have to be passed and insisted that CERN would make the magnetic trials. The contract for the windings[309] was awarded to ACEC, who were also expected to go through a whole series of extremely strict tests of their materials. The construction was to be preceded by that of a prototype pancake whose mechanical properties were to be measured with an apparatus built by Bonzano,[310] and in which the insulation between the coils—which had caused the accident with the HBC30 magnet—was to be checked using a circuit conceived by Asner.[311]

Sometime before, in 1959, and so as to have a precise idea of the value of the magnetic field which one could get with such a magnet, the group had decided to construct a 1/13th scale model of Monel,[312] an alloy of nickel (70%) and copper (30%), which enabled them to reproduce 'at much lower values of magnetic field the same saturation state which is to be expected in the true magnet'.[313] The results of the measurements made on the model in May 1960 indicated that 'the absolute value of the field inside the visible region in the full scale magnet will amount at 6 MW excitation power, to 17250 gauss',[314] which proved to be in excellent agreement with the final measurements.

Despite the impression we may have that this time all was thoroughly foreseen and prepared, the construction of the magnet did not occur without numerous delays (as

Notes: pp. 333 ff.

for the chamber body). In August 1961, while the parts were undergoing final machining and the ultra-sonic tests were giving good results,[315] the visits made by Wittgenstein to ACEC,[316] to SFAC,[317] and then a meeting at CERN,[318] all brought to light a series of delays, due in part to the optimism of the manufacturers ('The factory expects to produce three pancakes a week [...] personally I think that a production of two pancakes a week would be remarkable'[319]), and also at CERN ('Our delay (about 4 weeks) in the delivery of results [..] only explains the increase in the delay in items a)-d)'[320]).

After having been satisfactorily tested, the different parts of the magnet arrived at CERN in spring 1962 and were assembled there to establish the magnetic field map. The results of the measurements done 'in vacuum' (without the chamber) in October 1962[321] were very positive: they differed by less than 1% from those obtained on the Monel model. The magnetic measurements were repeated in January 1965[322] with the chamber completely mounted and they gave an induction at the centre of the chamber of 17,385 G. This could be increased to 17,415 G with a uniformity of 2% when the compensating magnet was run at full available power. Other measurements gave 'mediocre' results,[323] and led to some quite 'minor' elements being changed, like the flowmeters in the coils' hydraulic circuits. To the best of our knowledge HBC200's magnet never gave any important problems subsequently.

6.4.8 SAFETY

Even though it is not really part of a bubble chamber, and even though in our technical account we have had to neglect more than one feature in an attempt to simplify and classify an instrument that is in no way simple, it seemed important to us to devote a separate section to safety since it placed several constraints on other parts of the apparatus and on the manner of working with the bubble chamber.

Of little importance for HBC10, because the small amounts of hydrogen and the absence of a magnet removed two main sources of danger, safety was already of some concern with HBC30. Here if large quantities of hydrogen flowed into the surrounding vacuum tank — the most feared accident — the gas was led via a 15 cm diameter hole in the vacuum tank along an evacuation line to a rupture disc which opened when the pressure in the tank exceeded the atmospheric pressure.

All the circuits in HBC30 (reservoirs, pressurized bath, chamber, cylinder) could be connected by a valve to a vent line ending above the roof of the building, as well as to a vacuum line and to a filling line. All the reservoirs thus had three independent circuits. The vent was used when large amounts of gaseous hydrogen were produced, as well as during the transfers of liquid hydrogen or in emergencies. The second line could be used for evacuation via a counterflow heat exchanger, whose other line was the chamber filling line. Finally the third line evacuated gases via a heat exchanger at the bottom of the cylinder, so carrying off some of the heat in the chamber produced by the cylinder.

Safety tanks were already envisioned in the initial HBC200 project.[324] They were to be kept at 25 K by the circulation of liquid hydrogen and sealed to the chamber in front of the windows, thus separating them from the main vacuum system. They were to collect the hydrogen from the chamber if a window broke, and so also protected the main vacuum tank and the magnet from an overpressure. In normal working conditions these safety tanks played the role of vacuum tanks, and helped in the delicate process of cooling the windows, during which too large a temperature gradient could lead to a breakage. Two vent lines were added to improve safety further, the smaller one being for the safety tanks themselves while the larger one was for the vacuum tank, serving to connect it to a big outside sphere in which gaseous hydrogen could be collected. Leak detectors were indispensable during assembly and testing of the chamber, and they were later built into the control racks.

This same report expressed considerable concern for the safety of visitors' chambers. As the CERN staff put it, 'It would be illogical for CERN to permit access to the site to any bubble chamber which was built to safety standards whose philosophy was more optimistic than the minimum safety codes to be laid down in CERN after a serious study'.[325]

For the control of HBC200 a set of data had to be continually recorded on paper or on an oscilloscope for the chamber and its associated equipment. Safety did not only increase the amount of work to be done but also the cost, since all the instrumentation had to be explosion-proof, electrical contacts could only be at ground potential, and an explosimeter with many channels had to constantly monitor the air and critical zones in the chamber, the liquefier, and the pumping system. A diesel generator was provided so as to ensure that the vacuum pumps and the chamber controls continued to function in case of power failure. This system was difficult to manage at the time (1958), and even in the early sixties, when the chamber was being built, as it was not possible to control it by central computer. Good automation, and a set of servo-mechanisms could not replace continual human surveillance. The collection of control data was done manually, a check being needed in the compressor building every two hours!

A complete 'warning and alarm system' was designed for the chamber in 1962, then circulated for information in November 1964 'since the exploitation of the chamber [...] needed ongoing supervision and permanent coordination of all the major components of the said chamber'.[326] Supervision and interaction between the different subsystems of the chamber (vacuum, magnet, chamber, expansion, cryogenics, and various supplies) was achieved by this alarm system, which comprised about a hundred different warning units, each controlling a function. In case of breakdown it sent a signal to one of three general alarm systems. The alarms were classified into three categories according to their importance: minor local faults with visual signal and pulsed sound, along with shutdown of the circuit in which the fault had arisen; in cases of danger for the chamber, there was a continuous visual and audible signal, and its functioning was stopped (i.e. expansion ceased) and the

Notes: p. 334

circuits concerned were shut down; a catastrophe and danger for the personnel were signalled by a hooter with a special sound, and the operators were called on to carry out emergency operations manually.

We have found a first 'Hydrogen Safety Code' drawn up in October 1959,[327] whose aim was 'to give some advice and safety rules concerning the use of hydrogen and deuterium', since '[the] past history of explosions and accidents involving hydrogen emphasises that this material is extremely unpredictable in its behaviour and should be treated with a great deal of respect'.[328] The text identified dangerous situations (hydrogen in contact with the air in a closed building, overpressure, proximity of an electrical system, of a motor, etc...), and gave the precautions to be taken when using hydrogen dewars. It was rewritten and extended several times, most notably in June 1964 when the chamber was about to start functioning.[329]

All systems for storing gas under pressure and even all systems which could contain gas had to have a carefully calculated pressure resistance. In January 1960 the group made a number of visits to various firms which specialized in the construction of such apparatus.[330] At the same time it began thinking about the huge safety sphere which was to receive the 1000 litres of liquid hydrogen which, escaping from the chamber after an accident like the dreaded breaking of a window, evaporated immediately, coming to occupy a volume 1000 times greater, i.e. 1000 m^3!

After discussion and calculation, the specifications[331] for the sphere were drawn up in February 1961. 'For safety of personnel and equipment a safety exhaust system consisting of rupture discs and relief valves, a safety vent line and a big safety dump tank built as a sphere, is required to handle and store the gaseous hydrogen in the event of such an emergency'.[332] This sphere, which would usually be kept under vacuum or filled with purified hydrogen at low pressure, had to be able to withstand an overpressure of 15 kg/cm^2 in the unfortunate event that it contained air at the time when the evaporating hydrogen entered it. Its specifications had to be approved by the 'Schweizerischer Verein von Dampfkessel Besitzern', who laid down minimum safety limits, at least when those of the country of manufacture were not stricter. 'This is perhaps more stringent than is found in normal industrial practice and only the highest quality workmanship and material, especially the welding, will meet this specification'.[333]

The 10 m diameter sphere was to have a geometric volume of 525 m^3 and had to be able to tolerate temperature fluctuations between -50°C and +40°C. It was to be placed on pillars fixed in concrete. A stairway allowed one access to the top of it where the safety and control instruments were located. The internal surface had to be extremely clean to avoid all contamination of the highly purified hydrogen which might be stored in it. After further discussions with the selected French manufacturers, 'Société de Chaudronnerie et de Montage, Tissot et Cie', intended to ensure that all of CERN's demands were met,[334] even if this meant spending more money, and with the 'Schweizerischer Verein von Dampfkessel Besitzern',[335] the big sphere was erected on the site during 1961, and underwent various pressure and vacuum tests in 1962.

Of course we could include in this section more detail on the valves and security piping chosen, describe the different leak detectors, look carefully at the 'special construction essential for security reasons' of the diffusion pumps[336] —but we risk going on indefinitely. Let us therefore simply conclude by saying that though the construction of HBC200 took longer than foreseen, at least safety concerns were not pushed aside.

6.5 Conclusion

This chapter has dealt with the construction of CERN's first three hydrogen bubble chambers: HBC10—a prototype used to gain experience—, HBC30—which worked very well for three years, and whose most important physics results were the first proof of predominant backward emission of hyperons in pion–proton collisions, and the observation that the transverse momentum of the particle produced was generally small, typically 0.3 GeV (HBC30 at the PS in 1960, 16 GeV/c, π^- interactions)[337]—, and HBC200—which lived through the period when the kind of physics favoured was the study of interaction mechanisms of high energy elementary particles and high-statistics experiments on the properties of rare excited states. Its shutdown in 1977, said the CERN Annual Report, represented 'the end of one era and the start of another', that being the exploitation of the SPS with BEBC, Gargamelle, Omega, and the two neutrino beams.

Although this will be treated more thoroughly in a third volume on CERN's history, we would like to give a quick overview of what happened after 1965, which seems to us to have been far more of a turning point for bubble chambers than 1977 (Gargamelle was to function only until October 1978, when a breakdown ended its life earlier than had been foreseen), so as to identify some parallels with the years we have dealt with in this chapter.

In this year 1965, when HBC200 started working, CERN and the CEA at Saclay, with the participation of the Ecole Polytechnique and the Faculté d'Orsay, signed an agreement to build Gargamelle. In this same year 1965 a mixed group of French, German, and CERN scientists began studies on the Big European Bubble Chamber (BEBC), the new CERN hydrogen chamber. The tripartite agreement between them was to be signed on 21 July 1967.

HBC200's success at CERN was undoubtedly one of the reasons why these detectors continued to be developed. However, as was the case for HBC200 at the end of the 1950s, these new big chambers were to pose completely novel technical construction problems to the physicists and engineers building them. BEBC in particular, with its size (3.7 m, 35 m^3) and the magnitude of its magnetic field, was such that no straightforward extension of conventional systems could meet these

Notes: pp. 334 ff.

requirements without involving prohibitive costs. As originally foreseen, it was felt that a complete rethinking was necessary, incorporating many then-recent developments in technology. And as when HBC200 was conceived, the scientists felt that the cryogenics and safety aspects of handling these large volumes of hydrogen no longer seemed as daunting as they had five years before. Similarly, they again pointed out that the proposed chamber would be an important step towards the realization of the enormous chambers that could be envisaged 'presuming that the bubble chamber technique still retain[ed] its usefulness'.

The construction of BEBC, like that of HBC200, was preceded by many studies and, in particular, the building of a 1 m^3 model. There were delays in its construction and assembly, partially due to budgetary constraints, and the first photographs were finally taken in 1974. But work was not over: new modifications were introduced, and during 1976 the chamber was prepared for its use around the SPS. Later, in response to the gradual loss of interest for 'pure' bubble chambers, BEBC was converted step-by-step into a hybrid detector by the addition of the External Muon Identifier, the Internal Picket Fence, the External Particle Identifier, the Track Sensitive Target... This went on until 1984, when work with the chamber stopped with little or no objection. It was dismantled in 1985, having worked around the SPS from the start of 1977.

These two chambers are certainly not comparable, BEBC being of another generation. Certainly HBC200 lived through the golden age of bubble chambers, and was among the most outstanding devices of its type, while the physicists working on BEBC had to stretch their imagination to the limits to keep it in the race by making the modifications we have just spoken of. All the same, we cannot but find some parallels between them. Both pushed technology to its limits when they were built. And even if, in the end, most experiments done with them were of a general kind, and more or less a repetition or a continuation of earlier experiments (though done with much better statistics), they were both among the best chambers in the world for upwards of 20 years. Considering that Europe had a certain amount of ground to make up in this area, that the resources allocated by CERN to the builders were not always adequate, and that the chambers were not just scaled-up copies of existing detectors, we have considerable respect for those who ensured that Europe was equipped with very good instruments. And though HBC200 did not make any major discoveries, we can always commend it for its remarkably regular and reliable performance as a piece of experimental equipment...

We would like to thank Dr. H.-P. Reinhard for his very attentive reading of an earlier version of this text, and Drs. C. Peyrou and D.R.O. Morrison for useful criticisms. We should also like to thank Dr. R. Budde for his help throughout the study.

Translated by John Krige

Notes

All of the primary material referred to below has been gathered together in chronological order and has been stored in the CERN archives, where it may be consulted.

1. See Glaser (1960) and Conference-CERN (1959).
2. See Siegbahn (1960), 526.
3. According to Galison (1985), this was not Glaser's intention. He always tried to continue working with his new detector as an independent and solitary researcher. The creation rapidly freed itself from its creator, however.
4. See *Bubble Chambers*, undated.
5. See CERN/SPC/64, 1/5/58, *Technical aspects of hydrogen bubble chambers for use with CERN Proton Synchrotron*, chap. VII, 4.
6. See Galison (1985), Glaser (1955), Glaser (1958), Glaser (1960), Alvarez (1955), Fretter (1955), Bradner (1960), Henderson (1970), etc.
7. See Brown et al. (1956), Glaser & Rahm (1955), and Galison (1985), 27.
8. See Krige & Pestre (1986), 258, and R. Budde (private communication). In 1964 Alvarez even bet Bernardini—the stakes were the price of a meal—that bubble chambers would take 70% of the machine time. In fact from the time when an accelerator produced several beams simultaneously it was difficult to know how to measure 70%: by adding the time of the various beams, possibly, though other factors were also important like the beam energy, the elapsed time, the number of researchers, etc. Alvarez settled the bet by inviting all those present to the Père-Bise in Talloires in France, as had been agreed!
9. See Heilbron et al. (1981), 86–87.
10. See Alvarez (1968).
11. See Wood (1954).
12. See Galison (1985), 36.
13. See Parmentier & Schwenim (1955).
14. See Alvarez (1955).
15. See Galison (1985), 39.
16. See Galison (1985), 23.
17. See Pless & Plano (1956).
18. See CERN–Conference (1956).
19. See Bugg (1959). It was a chamber that could be recycled twice per second, being expanded and compressed by the movement of a nylon-reinforced rubber diaphragm in the neck under the chamber. A second diaphragm, separated from the first by low-viscosity oil which ensured thermal isolation, and activated by the compressed air, produced a complete cycle of some 30 ms, of which 10 ms were used to establish equilibrium at the working pressure. The chamber was kept at 57 °C and at a pressure of 21 atm by heating elements wound around its body and temperature control was achieved with a thermocouple. It was placed in a horizontal magnetic field of 13.4 kG. Ilumination was assured by a single flash and three cameras placed at the apex of an equilateral triangle of side 25 cm took the pictures. The light was diffracted on the bubbles, and produced an image of the tracks on a dark background. The images on the 35 mm Linograph Ortho film were 1/10th of the actual dimension. Photographs were taken about 1 ms after the arrival of the particle beam so as to have visible tracks.
20. The person concerned was Reinhard Budde, a young CERN physicist.
21. Cooling was achieved using a conventional system: a hydrogen jacket followed by a nitrogen jacket. The chamber had two vertical windows of 3 cm thick tempered glass. The illumination was that used in Steinberger's 12 inch chamber: light focused by two lenses onto a point between four cameras placed at the corners of a 20 cm square about 1 m from the centre of the chamber. They took four photographs of 1/9th of the chamber. Direct illumination meant choosing a magnet without pole pieces which produced a 17 kG field uniform to 3% (consuming 1.2 MW). The copper windings weighed 3.5 t and the block of solid steel

20 t. Expansion was achieved using a piston driven by helium near the top of the chamber's neck. The cycle could be varied in pressure and in time to a minimum of 10 ms. The chamber temperature was controlled by a pressurized hydrogen reservoir in thermal contact with the aluminium body. The whole chamber and the temperature control system were later copied in the Saclay 81 cm chamber. It was modified when the chamber came to CERN because it was found to be too complicated. The hydrogen consumed was about 5 to 6 litres/hour to maintain the temperature, while dynamic consumption (due to the expansion) was 2 litres/hour at a frequency of 15 pulses/minute. The rapid expansion by piston allowed for a very low liquid pressure, and this gave a very rapid growth in bubble size. Good tracks were thus observed with a rate of expansion of 0.8% to be compared with rates of 2–4% with other hydrogen chambers, notably those at Berkeley. (Peyrou independently reached the same conclusion on the uselessness of large rates of expansion for the CERN chambers). The delays between the flashes were reduced to 25 μs, compared to 2–5 ms for most hydrogen chambers. There was little boiling near the piston, and little distortion of the tracks.

22. See Bradner (1960), 125.
23. See CERN–Conference (1956).
24. See note 5
25. See note 5, chap. VII, 1.
26. See Bugg (1959).
27. See Krige & Pestre (1986), 259.
28. See Galison (1985), 27.
29. See for example Argan & Gigli (1956a), (1956b), P. Bassi et al. (1956).
30. See Dodd (1956).
31. BP2 was of stainless steel, closed in a security tank which was furnished with a window allowing the photographs to be taken. The chamber was sensitive for 15–20 ms for 1 ms of beam time, and this very long sensitive time enabled one to choose optimum operating conditions. It was by passing the beam towards the end that the best tracks were obtained, as for hydrogen chambers. By contrast, the chamber was not equipped for control at a distance. For more information see CERN 59-24, 14-16.
32. See L'Onde Electrique (1961), 982.
33. The whole measured 3 m in length, 2.5 m in width, and 2 m in height, weighed 80 tons, and was supported on four jacks. 54 m^2 of space was occupied for a height of 3.2 m, and as screen one needed a 1 m thick concrete wall. For more information see CERN 59-24, 17-25.
34. See L'Onde Electrique (1961), 1001.
35. See Bloch et al. (1956).
36. See Glaser (1952), 665.
37. See for example Wilson (1962), Blackett (1948), Fretter (1955), Flügge (1958).
38. See for example Galison (1985).
39. See Glaser (1955), 47.
40. See Dodd (1956), 142.
41. See Henderson (1970), 5.
42. See for example Henderson (1970).
43. See Glaser (1960).
44. For the exact formula see for example Henderson (1970).
45. See, for example, Shutt (1967).
46. See for example Henderson (1970).
47. See Brown et al. (1956) and Galison (1985), 27.
48. See for example Henderson (1970).
49. See Alvarez (1956), 6.
50. See for example Glaser et al. (1956) or Dodd (1956).
51. See Blinov et al. (1957).

Notes

52. See Bugg (1959), 14.
53. See Glaser (1960), 539.
54. Length at which the energy of the electron is reduced by a factor $e = 2.71$.
55. See for example CERN 59-24, 34.
56. See Glaser (1960), 539.
57. See for example CERN 59-24, or Henderson (1970).
58. See for example CERN 59-24, 34.
59. See Brown et al. (1956).
60. As pure liquids one has WF_6, $SnCl_4$, CH_3I, trifluoromethan, C_5H_{12}, $C_6H_6O_2$, $SnBr_4$. See for example Alyea et al. (1957) or Pless & Plano (1956).
61. See for example Alston et al. (1957), Budde et al. (1956) or Blinov et al. (1956).
62. See Argan et Gigli (1956a), (1956b), (1958a), (1958b), Hahn (1956), Hahn & Fischer (1957), Hahn & Riepe (1958), CERN 59-24.
63. See Hahn (1956). Hahn's first choice was for carbon dioxide dissolved in hydrogen. Other trials were made with n-hexane always containing carbonic gas, which had the advantage that it dissolved easily. This idea seems to have been particularly carefully studied by researchers at the University of Genoa in Italy. Hahn's bubble chamber was a small brass cylinder 4 cm in diameter and 5 cm long closed at both ends with two resin (Araldite) windows, an original solution. The chamber was connected by a valve at high pressure to a mixing vessel, and the expansion was achieved with a piston moved by a compressed air in a separate expansion chamber. Before expansion the pressure was about 50 atm, and a camera photographed the bubbles.
64. As for the British national chamber, see Newport (1964). Peyrou remarked that this good thermal conductivity is not important for temperature homogeneity because heat is mostly transported by convection. According to him, CERN was the first to choose stainless steel and to cool only the upper part of the chamber, so taking this property into account.
65. See Derrick (1970), 247.
66. See for example Bridge (1960), Blinov et al. (1956), or Parmentier & Schwemin (1955).
67. See CERN 59-24, Bradner (1960) or Henderson (1970).
68. See Derrick (1970), 229.
69. See Glaser (1955), 47.
70. See note 5.
71. This system was adopted for the Berkeley 72 inch chamber.
72. That is to say, at constant heat since it was sufficiently rapid that there was no influx of heat.
73. This modification was developed by the 'Russians' according to Peyrou. However we do find it applied in the USA—see Nagle et al. (1956)—, and in the 25 inch Berkeley chamber, the 80 inch Brookhaven chamber and the H_2 chamber at Argonne National Laboratory. Subsequently it seems to have been used more and more.
74. Scotchlite is a trademark of the Minnesota Mining and Manufacturing Company. This material was also used for BEBC, which only had one window because of the enormous technical problems raised by its huge size.
75. Used, for example, on a chamber at the Rutherford High Energy Laboratory, Harwell, England, which was used with the NIMROD accelerator.
76. For example in the shape of a ring for the British national chamber or tubular with cylindrical lenses for the 80 inch Brookhaven chamber.
77. With a medium wave length of 5×10^{-5} cm and a film resolution of 300 lines/cm (taking the lens aberration into account), one can calculate the demagnification.
78. See above, section 6.2.1.
79. To obtain a 1% precision, one can calculate the trajectory in a 10 kG magnetic field, when the minimal measurable sagitta is about 20 microns. In reality the precision is lower because of Coulomb scattering.

80. See Bugg (1959), 43.
81. See Derrick (1970), 229.
82. See Derrick (1970), 247.
83. See Galison (1985), 40.
84. 9 bubble chambers worked at CERN (not counting those brought by visitors), but 4 of these were completely constructed outside CERN by national groups: HBC81 constructed at Saclay and brought to CERN in 1961, HLBC100 constructed at the Ecole Polytechnique de Paris and brought to CERN in 1962, BNBC150 constructed in Great Britain and brought to CERN at the end of 1961, and Gargamelle constructed by l'Ecole Polytechnique and the CNRS/Orsay laboratory (CERN/SIS-PU 76-07).
85. Interview Ch. Peyrou, June 1985.
86. Annual Report CERN 1955, 11.
87. Annual Report CERN 1955, 38.
88. Charles Peyrou was professor at the University of Bern (Switzerland) and was consulted by CERN's STS Division directed by Lew Kowarski for the construction of a Wilson chamber intended for cosmic rays experiments.
89. See above, section 6.1.
90. See Krige and Pestre (1986), 259.
91. R. Budde, *Some thoughts about bubble chambers*, undated.
92. See *Report on the group meeting held 22nd January 1957*, (SC/4666/Rapp/11), 14/3/57, 1, and see note 91, 2 and note 4.
93. See note 91, 3.
94. See Galison (1985), 36 and Alvarez (1955).
95. See Galison (1985).
96. See note 91.
97. Rapport Annuel 1958, 43.
98. P. Amiot, *Consommation des chambres à bulles et production d'hydrogène liquide*, 11/2/58 and see note 99.
99. P. Amiot, *Low temperature apparatus*, PA 22, 2/4/58.
100. See note 5.
101. See note 5, chap. V, 5.
102. Rapport Annuel CERN 1960, 52.
103. Rapport Annuel CERN 1957, 62.
104. Rapport Annuel CERN 1957, 62.
105. In the presence of a magnetic field the use of copper is hazardous because of the eddy currents which are immediately set up in a metal with such a high conductivity if the field changes rapidly. These produce strong local heating and can lead to the explosion of the chamber.
106. See note 91, 3.
107. *Notes on the meeting of the bubble chamber group, Tuesday, August 7, 1956, prepared by R. Budde*. Apparently there was only one. Meetings began again in November: *Internal Meeting of Bubble Chamber Group on 27th November 1956* (SC/3838/Bd/off/211). We also have the minutes of meetings on the 4/12/56 (No. 2), 11/12/56 (No. 3), 12/12/56 (No. 4), and from the 4/1/57 the number is included in the title: *Internal Meeting (No. 5) of the Bubble Chamber Group on 4th January 1957*. These meetings were held on 16-17/1/57 (No. 6), 29/1/57 (No. 7), 4/2/57 (No. 8), 11/4/57 (No. 9), 7/3/57 (No. 10), 22/3/57 (No. 11), 22/3/57 (No. 12). These documents were found in D.R.O. Morrison's personal papers and we would like to thank him here for allowing us to consult and to photocopy them.
108. Rapport Annuel 1958, 43.
109. Rapport Annuel 1958, 43.
110. See note 85.
111. P. Amiot, *Choix du cuivre pour récipients soudés*, 19/8/57, G. Nullens, *Chambre à Bulles de 100 mm*,

Notes

8/57, G. Nullens, *Chambre à Bulles de 10 cm. Résultats de tests*, 9/57, G. Nullens, *Résistance mécanique des soudures*, 1/58, and see also Bridge (1960).

112. See below for the details of the construction.
113. See note 106.
114. Annual Report CERN 1956, 40.
115. See note 91, 2.
116. See note 85.
117. Windows were manufactured by Schott Glaswerke in Mainz. According to Peyrou a good polish was sufficient; the high polish was not necessary.
118. Rapport Annuel CERN 1959, 53.
119. Rapport Annuel CERN 1958, 43.
120. A. Blin, *Essai du groupe de pompage C.H.B.30*, 28/1/59.
121. H. Filthuth & J. Trembley, *Chambre de 30 cm 1er test à l'azote liquide 19/3/59*, 29/4/59 and H. Filthuth & J. Trembley, *Chambre de 30 cm. 2ème test à l'azote liquide 24/3/59*, 29/4/59.
122. Ch. Peyrou, *Réunion sur les Barksdales*, PS (CBH), 1/11/60, and J. Bartke & H. Filthuth, *Technical information from the synchrocyclotron experiments with the CERN 32 cm hydrogen bubble chamber*, 7/1/60.
123. D.R.O. Morrison, *Additional measurements during PI + beam run*, 5/8/59.
124. D.R.O. Morrison, *Preliminary distortion measurements on 32 cm chamber*, 9/9/59.
125. D.R.O. Morrison, *Measurements of mean gap length and bubble density at different chamber conditions*, 7/11/59 (dated erroneously 7/11/69).
126. See note 124.
127. See note 125 and D.R.O. Morrison, *Particle identification by bubble measurements*, 7/9/60.
128. CERN Report BS 11, 1954.
129. Ch. Peyrou, *Bubble chamber experiment on PS*, 24/5/60.
130. H. Filthuth, *Request of time at SC for the 32 cm hydrogen bubble chamber*, 15/10/59 and CERN Hydrogen Bubble Chambers Group, *Preparation of the 30 cm hydrogen bubble chamber experiments at PS*, 10/2/60.
131. CERN Hydrogen Bubble Chamber Group, *Note about early experiments with the 30 cm hydrogen bubble chamber at the Proton Synchrotron* 31/8/59.
132. 16,000 photographs in August 1960 in a proton beam, 86,000 in October 1960 with protons and pi-mesons, 60,000 in 1962 in a pion beam...
133. H. Filthuth, *Note concerning the 32 cm HBC experiment in october 1960*, 22/7/60.
134. *Rapport d'activité de la division TC-1961*, 20/11/61.
135. Valve 4 which stopped working was replaced, but the new valve was no better. A study on the test bench only showed that the valves were working correctly though with differences in their delay times.
136. Though he added: 'In particular the joints and the valves specially studied at CERN gave complete satisfaction', which could be taken as critical of the others by an attentive reader—see Rapport Annuel CERN, 1959, 53.
137. Peyrou counted three successive breakdowns in Autumn 1960, though whether bad assembly or mechanical failure was responsible he could not say. In the end an experiment had to be done to choose the most suitable pulse length, long or short, and general rules were laid down in order to avoid further problems. They carefully distinguished new valves from valves in poor condition but worth keeping and from worthless valves.
138. *Budget 30 cm Bubble Chamber*, 11/8/59, *30 cm chamber*, 19/8/59.
139. If it had been possible to buy deuterium direct from the AEC the cost would have been considerably less than that allowed for in the forecasts.
140. Ch. Peyrou, *Chambres à bulles à hydrogène pour le PS*, 12/3/59.
141. CERN/TC/200 62-5.

142. Letter Mezzetti to Weisskopf, 2/2/65 (DIRADM20255). The Italians took back on this occasion the two chambers they had brought to CERN, the Bassi group's hydrogen chamber, and the Moneti group's helium chamber.
143. Answer by Burger to Mezzetti, 10/3/65 (DIRADM20255).
144. Letter from Mezzetti to Peyrou, 16/7/66 (DG20825). The chamber worked end June 1966.
145. See note 85.
146. See note 5, chap. VII, 1.
147. P. Amiot & G. Nullens, *Estimation des besoins financiers et en personnel pour réaliser une grande chambre à bulles du type 'accordéon' à glaces horizontales, dimensions 2000 × 600 × 500 mm*, 21/3/58.
148. See note 147.
149. See Krige & Pestre (1986).
150. CERN-PS/Nucl. Phys. 9, 1/4/58.
151. See note 5.
152. See note 5, 10.
153. This choice seems to have been determined by several elements: first and foremost by the physics to be done—'From these figures it is apparent that the chamber should be as large as possible', 'A chamber of 200 × 60 × 50 cm is probably a suitable instrument for the high energy physics at 25 GeV' (the PS)—but also because of international competition. The first paragraph is striking in this respect. It points out what is being built in other laboratoriers, i.e. the 1.8 m (72 inch) chamber under construction at Berkeley, the 2 m projects at Brookhaven (80 inch) and at Dubna, the British decision to build a 1.5 m chamber. Further into the report we read again: 'it should be noted that the Berkeley group is building a 1.5 m chamber for the 6 GeV Bevatron, the British group is building a 1.5 m chamber for the 7 GeV PS at the Rutherford Laboratory, and Saclay are contemplating a 1.0 m chamber for the 2 GeV Saturne. It could be argued that CERN should consider a bigger chamber than 2 m for the 25 GeV CERN PS, but the study group felt that any chamber greater than 2 m would be an unjustifiable risk for CERN in view of the lack of experience at present with any chamber greater than 40 cm. If the 2 m chamber proves as successful as people such as Alvarez expect, there may be good reason to build larger chambers in the future'. It seems then that, from the start, CERN was unable to reverse its decision and devote its efforts to a less ambitious project (even if Peyrou and his group perhaps would have preferred to build an intermediate model). One was apparently not sure what choice to make at this time, and under such circumstances it is less risky to do the same as everyone else. To 'excuse' CERN, in so far as it is necessary, it is worth pointing out that the same arguments were used elsewhere, a jump in chamber dimensions being justified by reference to what other groups were doing. Perhaps we have touched on a basic feature of scientific work here: to start an important project, which requires a lot of money, people, and work, one has to try to judge the probability of success, and one way of doing so is by comparing one's own project to those being undertaken by other groups.
154. This was done in the next generation of chambers planned in the mid-60s.
155. See note 5.
156. Ch. Peyrou, *Large Hydrogen Bubble Chambers of CERN. Some general considerations on the statement of the problem*, 21/3/58, 4.
157. See note 103, 10.
158. The eleven systems were: the chamber, its filling, and its pressure and temperature indicators; the pressurized hydrogen bath for cooling; the liquefier which maintained the temperature level in the bath; the nitrogen jacket; the precooling surface; the expansion mechanism and its pneumatic controls; the main vacuum system, the vacuum tanks and the pumping rings, the magnet, its generators and cooling circuits; the cameras and the illumination system; the security system: explosion-meters, aeration, and reverse reservoirs; the counter control and the electronics controlling the frequency.
159. See note 156, 10.
160. Rapport Annuel CERN 1959, 54.

161. Rapport Annuel CERN 1959, 54.
162. J. Vilain, without title, 6/11/59.
163. Rapport Annuel CERN 1963, 13.
164. Rapport Annuel CERN 1960, 52.
165. Courrier CERN 3/2, 1963, 24.
166. Courrier CERN 5/5, 1965, 67.
167. Rapport Annuel CERN 1965, 88.
168. Rapport Annuel CERN 1969, 57.
169. *Coup d'oeil sur le CERN No. 8*, July 1976 and Rapport Annuel CERN 1978, 77.
170. See note 150, 1.
171. See note 5.
172. See note 140, 1.
173. G. Petrucci, *About the positioning of the hydrogen bubble chamber building in the east experimental area*, 30/7/59, 3.
174. See note 140, 3.
175. See above, section 6.2.3.
176. Ch. Peyrou, *CERN 30 cm Liquid Hydrogen Bubble Chamber. Technical Description and Operation Performances*, CERN PS/CBH, 7/6/60 (Draft).
177. Rapport Annuel CERN 1957, 57.
178. Martensite is a solution of Carbon in Iron, a component of tempered steels.
179. G. Petrucci & J. Trembley, *Rapport de visite à l'Institut Fourier à Grenoble (Prof. Weil)*, 27/7/59.
180. G. Bonzano, *Description and operating instructions of the low temperature magnetic measurement apparatus*, 13/9/60.
181. *Résumé d'une discussion avec les représentants des Aciéries Uddeholm*, 19/10/59.
182. Austenite is a micrographic component of steels.
183. Transformation in carbonate.
184. Resilience is a number characteristic of the resistance to collisions.
185. J. Trembley, *Rapport préliminaire sur les mesures de perméabilité et de résilience d'aciers inoxydables*, 30/6/60, and *Mesures de perméabilité et de résilience à effectuer au cours du mois de juillet*, 30/6/60.
186. Hydrogen Bubble Chamber Group, *Description of the CERN 2 m hydrogen bubble chamber*, 2/3/60, 2.
187. Rapport Annuel CERN 1959, 54.
188. A.H. Ackermann, *Systèmes de transport pour CB2000*, undated.
189. *CERN 63-6*.
190. *Discussion of the chamber body CB*, 15/11/61.
191. See note 190.
192. Without title, 19/9/62, (TC/TB/jc). We have many papers on these control trips, for example F. Birchler, *Visite chez Forges et Chantiers de la Méditerranée, La Seyne (Var) France du 9-10 au 12/10/61, Rapport de visite No. 23*, 25/1/62 and *Rapport de visite No. 24*, 1962.
193. D. Jacobzone, *Rapport de voyage*, 5/9/62.
194. Rapport annuel, CERN, 1963.
195. See note 5.
196. See note 111.
197. See Bridge et al. (1960).
198. D. Jacobzone, *Avancement des travaux sur les joints gonflables de la chambre à bulle de 2 m*, 25/5/62, D. Jacobzone, *Joints gonflables et joints élastiques travaillant à froid (HBC2000)*, 26/2/63, and PS/INT/CBH200 60/14.
199. R. Budde, *Some remarks about the vacuum system of the 2 m HBC*, 24/7/59.
200. PS/INT/CBH200 60/14, 25/11/60, rewritten in 1961, CERN/TC/200 62/1, 5/1/62.
201. This nitrogen jacket reduced the heating due to radiation to the minimum, even though certain parts of the

jacket were as much as 20 °C warmer (due to the thinness of the copper). It had to be refilled every two or three hours and a simple copper plate, cooled by the hydrogen evaporating from the reservoir, stopped the heating from above.

202. H. Schultes, *The vacuum system of the 2m-HBC*, 17/8/62.
203. See note 188.
204. *Histogramme CBH2000*, 1965.
205. *Réunion du groupe de la chambre de 2m, jeudi 11 juin 1959. Projet*, 17/6/59.
206. Rapport Annuel CERN 1959, 54.
207. *Sujets à aborder pour finir l'étude du tank à vide*, 28/4/60.
208. G. Muratori, *Abstract of data and computations for the HBC vacuum system, (tentative layout)*, 2/8/60.
209. CERN/TC/200 62/1.
210. Rapport Annuel CERN 1964.
211. This was a CERN innovation. The contact with the upper part of the chamber alone allowed one to avoid having a colder zone at the bottom, convection currents cooling the whole of the liquid particularly quickly.
212. Rapport Annuel CERN 1959, 53.
213. See above, section 6.3.2.
214. *Estimation de la consommation dynamique de la chambre de 30 cm*, 29/5/59. J. Trembley, *Estimation de la consommation dynamique*, 23/6/59.
215. Dynamic consumption is not proportional to volume, HBC30 being 20 times as big as HBC10 by volume.
216. See above, section 6.2.3.
217. See note 5.
218. See note 202.
219. Rapport Annuel CERN 1958.
220. Ch. Peyrou, *On different ways to calculate the dynamic load of the 2 m chamber for piston expansion*, 12/11/59.
221. *Compressor-situation (Draft)*, 7/59.
222. F. Schmeissner, *Report on a trip to Ems on December 15, 1959*, 14/1/60.
223. *Hydrogen liquifier*, 27/1/60, (PS(CBH)JT).
224. H. Turner, *Hydrogen purifier-converter*, 16/5/60.
225. PS(CBH)/FSch/rmn (Draft), untitled, 15/9/60, *Protocol on meeting at CERN 29.11–30.11.1960*, and *The high pressure storage battery (Tender I-1882)*, 20/12/60.
226. H_2-*Refrigerator for 2m HBC to be delivered by Sulzer Broth*, 1/5/61.
227. CERN/TC/200 62/3, 30/1/62.
228. CERN/TC/200 62/5.
229. See note 225, 2
230. Rapport Annuel CERN 1963, 81.
231. Rapport Annuel CERN 1963, 81.
232. To have an idea of its order of magnitude let us remember that for the 80 cm chamber built by Saclay the specifications were to bring the temperature to 1/10th of a degree between 24 K and 28 K and to hold the temperature constant to 5/100th of a degree—despite the addition of static heat by convection and conduction from the surroundings, by radiation from the hot walls, and by conduction via the supports attached to the chamber, and despite the influx of dynamic heat due to the chamber expansion. See l'Onde Electrique (1961).
233. See Amiot (1958).
234. See above, section 6.3.2.
235. See note 85.
236. See also Bridge et al. (1960).
237. See above.

238. P. Amiot, *Sur un nouveau mécanisme de détente des chambres à bulles à hydrogène liquide*, 12/2/58, *Suite 1*, 10/3/58, *Suite 2*, 10/3/58.
239. P. Amiot, *Propositions de développements dans le domaine de chambres à bulles à hydrogène liquide*, 24/3/58.
240. See note 85.
241. See note 236.
242. See note 5.
243. Much later the 'Omega-bellows' would be a very good approximation of the mobile wall.
244. See note 85.
245. G. Amato & G. Petrucci, *Proposal of a new system for liquid expansion of hydrogen bubble chamber*, 11/5/59.
246. 'RCH Polytyäthylen Typ 1000' is a type of polyethylene with a very high molecular weight, higher than 1,000,000. It is an organic material having the rare advantage of resisting very low temperatures (20 K), see Bridge (1960).
247. See note 85.
248. Rapport Annuel CERN 1958, 43.
249. Ch. Peyrou, *Gradients de pression et constante de temps dans la détente des chambres à bulles*, 11/11/59.
250. J.V. Overhagen & J. Trembley, *Test on Barksdale valve for the expansion system of the 2 m chamber*, 17/6/60, and J. Trembley, *Mesure du coefficient de contraction d'une vanne*, 23/8/60.
251. A. Asner & G. Petrucci, *Eddy current losses in the expansion mechanism of the 2 m bubble chamber*, 26/8/60.
252. H. Filthuth, *Points à surveiller pendant l'expérience*, 10/11/59 and J. Bartke & H. Filthuth, *Technical information from the synchrocyclotron experiments with the CERN 32 cm hydrogen bubble chamber*, 7/1/60.
253. J. Trembley, *Essai des pistons balancés*, 6/7/59.
254. G. Amato, *Les mesures de niveaux, pression et position du piston dans la chambre à bulles à hydrogène de 30 cm*, 28/10/59.
255. CERN/TC/200 62/2 I-2298, 24/1/62.
256. P. Lazeyras, *Essai statique de la fixation de la tête du piston du système de détente*, 27/3/62 and *Système de détente de la CBH 2m*, 16/5/62.
257. CERN/TC/200 63/6.
258. See note 204.
259. See note 255.
260. See note 255.
261. See schema in CERN/TC/200 63/6.
262. P. Amiot & J. Trembley, *Résultats expérimentaux préliminaires sur un système d'optique à lumière polarisée pour CBH_2 liquide*, 13/3/58, and P. Amiot & G. Nullens, *Estimation des besoins financiers et en personnel pour réaliser une grande chambre à bulles du type 'accordéon' à glaces horizontales, dimensions 2000 × 600 × 500 mm*, 21/3/58.
263. See also Bridge et al. (1960).
264. See note 5.
265. D.R.O. Morrison, *Specifications*, 24/4/59.
266. D.R.O Morrison, *Flash-tubes for the two metre chamber*, 5/3/59.
267. HBC10 had a FT230 Xenon flash furnished by General Electric giving 100 joules. Half of this energy is dissipated in 28 microseconds. HBC30 had the same, of double power.
268. *Specifications for the power supply of the bubble chamber illumination*, 6/4/61, (TC/HPR/ag).
269. PS (CBH)/HPR/rmn.
270. Rapport Annuel CERN 1963, 84.
271. W.T. Welford, *Imperial College. Anti-reflexion coating of bubble chamber optics*, undated.

272. *Draft specifications for the anti-reflexion coating of the windows and lenses of the 2m bubble chamber*, 12/5/60.
273. Rapport Annuel CERN 1962, 106 and *Optics 2m Chamber*, 15/11/61.
274. W.T. Welford, *Notes on condensers for 200cm bubble chamber*, 27/4/60 and *Final contract for the condensor of the 2m hydrogen bubble chamber*, 10/2/61.
275. *Specifications for the coating of the lenses of the 2m Hydrogen Bubble Chamber*, 30/10/63.
276. *The optics of the CERN 2m Hydrogen Bubble Chamber*, 18/6/62.
277. Rapport Annuel CERN 1965, 81.
278. PS/INT/HBC30 59/1.
279. Rapport Annuel CERN 1960, 52.
280. *Tenders for the glass windows of the 2m chamber*, 12/7/60.
281. Jenaer Glaswerke Schott in Mainz was already in charge of the windows for HBC30 and HBC10.
282. *Draft specifications for the glass windows for the 2m hydrogen bubble chamber*, 6/5/60.
283. *Tests for the windows of the 2m Hydrogen Bubble Chamber*, 9/11/60.
284. *Costs of working the windows of the 2m Hydrogen Bubble Chamber*, 18/11/63.
285. H.P. Reinhard, *Cooling of the lenses of the 2m Bubble Chamber*, 20/7/60.
286. D.R.O. Morrison, *Optical data for the 30 cm hydrogen bubble chamber*, 2/2/60.
287. CERN/TC/Physics 62-3, June 1962.
288. H.P. Reinhard, *Fiducial Marks for the 2 m Hydrogen Bubble Chamber*, 26/11/63.
289. As early as 1960 Reinhard had calculated the *Errors produced by the front window of the 2m bubble chamber*, 27/6/60.
290. M. Morpurgo, *Considerations on magnets for the bubble chambers*, 20/3/58.
291. *Programme provisoire de la chambre de 30cm*, 9/7/59.
292. Brauner Greuther, *Magnetic field measurements in the 32 cm hydrogen bubble chamber*, 7/12/59.
293. J. Bartke, *Magnetic field in the 32 cm hydrogen bubble chamber, 16 GeV/c pi-run, March 1960*, 27/4/60.
294. On August 1959 Guido Petrucci went to Genoa to visit the Ansaldo-San Giorgio factory for the first time (G. Petrucci, *Visit to Ansaldo-San Giorgio, Genova, August 27, 1959*, 31/8/59) to discuss the repair of the magnet windings and — already — to prepare the project for the 2m chamber magnet. The disagreement between CERN and this manufacturer arose from the accident, whose cause has still not been precisely identified — and never will be. It is accordingly difficult to apportion responsibility, to know whether it was misuse by CERN which caused the short-circuit or a fault in the construction by the factory. On the sole pancake that was damaged the manufacturer proposed to replace the two turns of the windings with new copper, to overhaul the rest, and to mount the elements with new insulating rings. Tests were discussed, and it looked as though the work described would take three weeks. The manufacturer disassociated itself from the accident in two letters ('Granted that the magnet was not explicitly ordered so as to function in a particularly humid atmosphere and with a large difference between the ambient temperature and the temperature of the external windings, which can give rise to condensation' (letter Ansaldo-San Giorgio to CERN, 27/8/59)), and more thorough instructions to the users were subsequently given. Two days later, when disassembly was further advanced, the second letter from the factory informed CERN that it had discovered copper shavings in the windings probably coming from the work done by CERN ('Even if in this case the presence of the shavings had not yet created visible damage' — same letter). The cost of the repair without tests could vary from 7000FS to ten times that amount depending on whether one replaced a part or remade a new unit. The latter would take some seven months. Another visit was made to Genoa on 7 October (three weeks having passed without a decision being taken) to inspect the magnet coils and to have further discussions (on the magnet as well as the big chamber). As the cause of the accident had still not been found (a power surge was excluded with the installation used, and the arc which burnt four turns on the pancake could not be attributed to condensation), Ansaldo promised to have the repairs done by 15 October and accepted to absorb a (small) part of the costs (the transport by train and to keep to its initial quote), while CERN for its part asked that

the work be completed as rapidly as possible (they were counting the days). At CERN the way of protecting the magnet was quickly reviewed (R. Budde, *Protection de l'aimant de HBC30*, 14/10/59)—against lack of water, against condensation on the coils, against a too high temperature of the coils, against rapid decreases of the field, and by adding a safety button. The repairs to be made had still not been settled when, on 2 November, an engineer from the firm visited CERN to explain his findings on the accident. CERN was not held responsible but the cause of the accident remained a mystery unless, as CERN thought, it was attributed to an error in assembly at the factory. Since each assembly and disassembly could damage the insulation, the researchers preferred to use the magnet with 15 pancakes instead of 16, while waiting for the coils to be replaced by entirely new ones as soon as possible (five to six months delay). As for the manufacture of the new coils, Adams (Ch. Peyrou, *Résumé des discussions sur l'aimant de la 30 cm*, 2/11/59) thought that it should be regarded as an entirely new order which would not automatically go to Ansaldo unless the firm covered the costs, which it was obviously not willing to do. In the event an offer was made by Ansaldo before 15 November for the repair of the pancake with a guarantee that it would be done between 1 December and 1 January 1960, and 'CERN will accept or reject this offer weighting the advantages of having a complete magnet and the more or less serious risks consequent on the repair procedure proposed by Ansaldo'. CERN felt free and chose new coils, whose construction was once again confided to Ansaldo. In July 1960 (A. Asner, *Visit to Ansaldo-Elettrotecnico at Genova on July the 14th and 15th 1960*, 20/7/60)—the five or six months necessary for the construction had passed but nothing had been done yet—, there was another meeting in Genoa to discuss technical and commercial aspects of the new windings. Ansaldo guaranteed their proper functioning, increased the price slightly (giving a justification) to 85000 FS and promised the windings six months after 15 September 1960. Finally, on 5 August, the tender documents for the windings were drawn up. They carefully described not only the technical specifications of the two coils, each comprising 16 pancakes of 28 turns and their insulation, but above all the set of tests to be made on the pancakes, including the tests to be made by CERN on the assembled magnet. The writer added: 'The manufacturer must accept complete responsibility that the entire winding will satisfy all the requirements of its use, withstand all stresses arising during normal operation of the bubble chamber, and satisfy all the tests described, as explained in these specifications and covered by the guarantee clause'. (PS/INT/HBC30, I-1832, 60/11, 5/8/60, 2). 'One year after the 30 cm bubble chamber has been put into normal operation with the new windings, the mechanical and electrical properties of the different parts of this tender shall still meet all the requirements stated in these specifications'. And again the matter was urgent: 'CERN is interested in a short delivery time, such as 6 months from the date of order'. However, since the deadline for the submission of tenders was 30 September and since the firm to be chosen was again Ansaldo, it could have been quicker to accept their previous offer. Finally there was a last visit to Genoa on 13–21 December 1961 to check the situation (*Visit to Ansaldo-San Giorgio, Genova, from 13–21 December 1961*, 5/1/62). Here Bonzano confirmed that the tests were done and that all was in order as specified. The magnet windings were finally delivered in January 1962, only a few months before the chamber stopped working.

295. A. Asner, *Magnet of the 30 cm bubble chamber*, 29/3/60.
296. *Budget 30 cm Bubble Chamber*, 11/8/59 and *30 cm chamber*, 19/8/59.
297. See note 5.
298. See Oswald (1957), Colley et al. (1959) and *Report on the proposed high magnetic field bubble chamber of the MPI, München*, 17/2/60. The practical realisation of this system remains very problematic, even for small chambers. (Budde, private communication, August 1986).
299. See note 290.
300. See note 290.
301. G. Petrucci, *Magnet for the 2 m bubble chamber. Comparison between different types of magnets*, 10/58, G. Petrucci, *Memorandum about the magnet for the 2 m Hydrogen Bubble Chamber*, 16/4/59, *Requirements for the generators for the bubble chamber magnets*, 15/5/59, G. Petrucci, *Magnet for the 2 m Hydrogen Bubble Chamber*, 29/4/59, G. Petrucci, *2 m Hydrogen Chamber Magnet. Comparison*

between Berkeley and CERN data, 11/8/59, G. Petrucci, *2m Hydrogen Bubble Chamber magnet. Stresses on the winding due to the thermal expansion and electromagnetic forces*, 30/10/59, G. Petrucci, *2m Hydrogen Bubble Chamber magnet. Excitation winding data*, 17/11/59, and G. Petrucci, *2m Hydrogen Bubble Chamber magnet. Compensation coils data*, 16/12/59.

302. Rapport Annuel CERN 1959, 54.
303. Letter from Ansaldo San Giorgio to CERN, 29/8/59.
304. See note 290.
305. *PS/HBC 2m Hydrogen Bubble Chamber magnet*, 30/6/60.
306. CERN/TC/200 61/6.
307. G. Petrucci, *Visit to Ansaldo-San Giorgio, Genova, August 27, 1959*, 31/8/59, 2, G. Petrucci, *2m Hydrogen Bubble Chamber Magnet. Visit to OERLIKON, December 3, 1959*, 8/12/59, and G. Petrucci, *2m Hydrogen Bubble Chamber Magnet. Visit to ACEC (Charleroi), December 21, 1959, and visit of Mr. Dispaux to CERN, January 13, 1960*, 20/1/60.
308. See note 306.
309. CERN/TC/200 61/6 fin.
310. G. Bonzano, *Description of a device measuring the shearing stress on winding sections*, 13/9/60.
311. A. Asner, *2m Hydrogen Bubble Chamber Magnet. Circuit for testing the insulation between the turns of one excitation-winding-pancake*, 23/2/60 and A. Asner & G. Petrucci, *2m Hydrogen Bubble Chamber Magnet. Visit to OERLIKON, April 1, 1960*, 8/4/60.
312. G. Bonzano, *Magnetic characteristics of the MONEL*, 3/8/59 and G. Bonzano, *Visit to the Nickel-Contor in Zurich, August 10th, 1959*, 12/8/59.
313. G. Bonzano & G. Petrucci, *2m Hydrogen Bubble Chamber Magnet. Magnet model and preliminary magnetic measurements*, 13/11/59.
314. G. Bonzano & G. Petrucci, *2m Hydrogen Bubble Chamber Magnet. Measures on the magnet model*, 13/5/60, 2.
315. *Rapport d'activité de la division TC 1961*, 18/1/62.
316. *Visite de F. Wittgenstein à AEC, Charleroi, les 8 et 9 août 1961*, 14/8/61.
317. *Visite de F. Wittgenstein à SFAC, Le Creusot en date du 12/12/1961*, 19/12/61.
318. F. Wittgenstein, *Procès verbal de la réunion tenue les 5 et 6 février 1962 au CERN entre ACEC, SFAC, Maison Didier et le CERN*, 12/2/62.
319. See note 315, 2.
320. See note 316, 2.
321. G. Petrucci, *2m Hydrogen Bubble Chamber Magnet. Magnetic Measurements*, 8/11/62, G. Petrucci, *2m Hydrogen Bubble Chamber Magnet. Field Measurements*, 3/1/63, and G. Bonzano & E. Chesi, *2m HBC Magnet Circuitry for Magnetic Measurements*, 8/1/63.
322. CERN/TC/200 65/5, 10/3/65.
323. See note 322, 2.
324. See note 5.
325. See note 5, chap. VII, 4.
326. CERN/TC/200 64/2 (Edition provisoire) (KHR22160), 4/11/64.
327. T. Ball, *Hydrogen Safety Code*, 8/10/59.
328. See note 326, 1.
329. CERN/TC/GEN 64/5 (KHR22160), 16/6/64.
330. T. Ball, A. Burger & G. Traini, *Résumé of visits to MM. Reisholz, Mannesmann, F.A. Neumann and C. Spaeter, regarding High Pressure Gas Storage Vessels, Safety Sphere, and Dry Gazometer*, 11/1/60.
331. *I-1998*, TC/200 61/3, 9/2/61.
332. See note 331, first page.
333. See note 331, 1.
334. *Safety Sphere I-1998*, 20/3/61, *Safety Sphere (Tender I-1998), discussion with Tissot and Cie*, 7/4/61,

Safety Sphere, 6/4/61.
335. *I-1998, Safety Sphere (Tissot Tender), Meeting at 'Schweizerischer Verein von Dampfkessel-Besitzern', Zürich, on 7/4/1961*, 12/4/61.
336. Rapport Annuel CERN 1961, 97.
337. These results were not directly published (Morrison, private communication, February 1989).

Bibliography

Alston & Collinge (1957)	M.H. Alston, B. Collinge, 'A Propane Bubble Chamber', *Phil. Mag.* **2**, (1957), 820–829.
Alvarez (1955)	L.W. Alvarez, *The Bubble Chamber Program at UCRL*, 18 April 1955, doc. 00500370001–00500370021.
Alvarez (1956)	L.W. Alvarez, CERN Symposium **2**, Geneva 1956, 3–50.
Alvarez (1968)	L.W. Alvarez 'Recent developments in particle physics', Nobel Lecture, 11 December 1968, in *Nobel Lectures Physics 1963–1970*, (New York: Elsevier, 1972) 241–290.
Alyea Jr. et al. (1957)	E.D. Alyea Jr., L.R. Gallagher, J.H. Mullins, J.M. Teem, 'A WF_6 Bubble Chamber', *Nuovo Cimento* **6**, (1957), 1440–1488.
Amiot (1958)	P. Amiot, 'Device for thermal control of liquid hydrogen bubble chamber', *Nuclear Instruments* **3**, (1958), 275–277.
Argan & Gigli (1956a)	P.E. Argan, A. Gigli, 'A New Detector of Ionizing Radiation. The Gas Bubble Chamber. *Nuovo Cimento* **3**, (1956), 1171–1176.
Argan & Gigli (1956b)	P.E. Argan, A. Gigli, 'On the Bubbles Formation in Supersaturated Gas-Liquid Solutions'. *Nuovo Cimento* **4**, (1956), 953–957.
Argan et al. (1958a)	P.E. Argan, A. Gigli, E. Picasso, G. Tomasini, 'Remarks about gas bubble chamber sensitivity', *Nuovo Cimento* **10**, (1958), 177–181.
Argan et al. (1958b)	P.E. Argan, M. Conte, A. Gigli, E. Picasso, 'High Density and Large Atomic Number Systems for Gas Bubble Chambers', *Nuovo Cimento* **10**, (1958), 182–185.
Askar'ian (1955)	G.A. Askar'ian, 'Gas Bubble Chamber—A possible Recorder of the Elementary Act of Interaction of Ionizing Radiation with Matter', *Soviet Phys. JEPT 1*, (1955), 571.
Barrera et al. (1964)	F. Barrera, R.A. Byrns, G.J. Eckman, H.P. Hernandez, D.U. Norgren, A.J. Shand, R.D. Watt, 'The 25-inch liquid-hydrogen bubble chamber expansion system', *International Advances in Cryogenic Engineering* **10**, (1964), 251–258.
Bassi et al. (1956)	P. Bassi, A. Loria, J.A. Meyer, P. Mittner, J. Scotoni, 'On n-Pentane Bubble Chamber', *Nuovo Cimento* **4**, (1956), 491–500.
Bertanza et al. (1960)	L. Bertanza, P. Franzini, I. Mannelli, G.V. Silvestrini, 'A rapid Cycling Bubble Chamber', *Nucl. Instrum. and Methods* **9**, (1960), 354–356.
Blackett (1948)	P.M. Blackett, 'Cloud chamber researches in nuclear physics and cosmic radiation', Nobel Lecture, 13 December 1948, in *Nobel Lectures Physics 1942–1962* (Amsterdam, London, New York: Elsevier 1964), 97–119.
Blinov et al. (1957)	G.A. Blinov, I.S. Krestnikov, M.F. Lomanov, 'Measurement of the Ionizing Power of Particles in a Bubble Chamber', *Soviet Phys. JETP* **4**, (1957), 661–670.
Bloch et al. (1961)	M. Bloch, A. Lagarrigue, P. Rançon, A. Rousset, 'A 300-Liter Heavy

	Liquid Bubble Chamber', *Rev. Sci. Instrum.* **32**, (1961), 1302-1307.
Bolze et al. (1958)	E.M. Bolze, T.W. Morris, D.C. Rahm, R.R. Rau, R.P. Shutt, A.M. Thorndike, W.L. Whittemore, 'Liquid Hydrogen Bubble Chamber Expanded by a Piston in the Liquid', *Rev. Sci. Instrum.* **29**, (1958), 297-299.
Bradner (1960)	H. Bradner, 'Bubble Chambers', *Ann. Rev. Nucl. Sci.* **10**, (1960), 109-160.
Bridge et al. (1960)	H. Bridge, R. Budde, A. Burger, H. Filthuth, D.R.O. Morrison, C. Peyrou, J. Tremblay, 'Some Vacuum Problems at Low Temperaures', *Advances in Vacuum Science and Technology, Vol. II*, 1960 (London, Oxford, New York, Paris: Pergamon Press) 481-483. Presented at the 1st Intern. Congress on Vacuum Tech. Namur, Belgium, June 1958.
Brown et al. (1956)	J.L. Brown, D.A. Glaser, M.L. Perl, 'Liquid Xenon Bubble Chamber', *Phys. Rev.* **102**, (1956), 586-587.
Budde et al. (1956)	R. Budde, M. Chretien, J. Leitner, N.P. Samos, M. Schwartz, J. Steinberger, 'Properties of Heavy Unstable Particles Produced by 1.3 BeV Pi-Mesons, *Phys. Rev.* **103**, (1956), 1827-1836.
Bugg (1958)	D.V. Bugg, 'Use of freons in bubble chambers', *Rev. Sci. Intrum.* **29**, (1958), 587-589.
Campbell & Pless (1956)	N.P. Campbell, I.A. Pless, 'Optical system to simplify analysis of bubble chambers photographs', *Rev. Sci. Instrum.* **27**, (1956), 875.
Colley et al. (1959)	D.C. Colley, J.B. Kinson, L. Riddiford, '9-inch liquid hydrogen bubble chamber in a pulsed magnetic field', *Nucl. Instrum. and Methods* **4**, (1959), 26-29.
Conference-CERN (1956)	*The CERN Symposium 1956*, described by G.F. Dardel, 'The CERN Symposium 1956 (II)', *Nuclear Instruments*, **1** (1957), 10-16.
Conference-CERN (1959)	*International Conference on High Energy Accelerators and Instrumentation CERN 1959*, sponsored by the International Union of Pure and Applied Physics (IUPAP), Geneva 14th-19th September, 1959. Proceedings, L. Kowarski (ed.).
Von Dardel et al. (1958)	G. Von Dardel, Y. Goldschmidt-Clermont, F. Iselin, 'The present state of the Instrument for Evaluation of Photographs and future developments', *Nucl. Instrum.* **2**, (1958), 154-163.
Derrick (1969)	M. Derrick, 'The Bubble-Chamber Technique: Recent Developments and Some Possibilities for the Future', *Progress in Nuclear Physics* **11**, (1969), 223-269.
Dilworth et al. (1964)	C. Dilworth, D.R.O. Morrison, G. Mambriani, 'On the Measurement of Ionization in Hydrogen Bubble Chambers', *Il Nuovo Cimento,* serie X, **32** (1964), 1432-1444.
Dodd (1955)	C. Dodd, 'Radio-Carbon Content and Delayed Boiling of Liquids', *Proc. Phys. Soc.* **B68**, (1955), 686-689.
Dodd (1956)	C. Dodd, 'The Bubble Chamber', *Progress in Nuclear Physics* **5**, (1956), 142-156.
Dowell et al. (1960)	J.D. Dowell, W.R. Frisken, G. Martelli, B. Musgrave, 'The propane bubble chamber used with the Birmingham Proton Synchrotron', *Nucl. Instrum. and Methods* **7**, (1960), 184-188.
Flügge (1958)	S. Flügge, *Handbuch der Physik, XLV: Instrumentelle Hilfsmittel der Kernphysik II*, 1958, (Berlin, Göttingen, Heidelberg: Springer-Verlag).
Fretter (1955)	W.B. Fretter, 'Nuclear Particle Detection (Cloud chambers and Bubble chambers)', *Ann. Rev. Nucl. Sci.* **5**, (1955), 145-178.
Frisch & Oxley (1960)	O.R. Frisch, A.J. Oxley, 'A semi-automatic analyser for Bubble Chamber

	Photographs', *Nucl. Instrum. and Methods* **9**, (1960), 92–96.
Galison (1985)	P. Galison, 'Bubble Chambers and the Experimental Workplace', in P. Achinstein and O. Hannaway (eds.), *Observation, Experiment and Hypothesis in Modern Physical Science* (Cambridge: MIT Press, 1985), 309–373.
Glaser (1952)	D.A. Glaser, 'Some Effects of Ionizing Radiation on the Formation of Bubbles in Liquids', *Phys. Rev.* **87**, (1952), 665.
Glaser (1953)	D.A. Glaser, 'Bubble Chamber Tracks of Penetrating Cosmic-Ray Particles', *Phys. Rev.* **91**, (1953), 762–763.
Glaser (1954)	D.A. Glaser, 'Progress Report on the Development of Bubble Chambers', *Nuovo Cimento Suppl. Vol. XI, serie IX,* **2**, (1954), 361.
Glaser (1955)	D.A. Glaser, 'The bubble chamber', *Scientific American*, (February 1955), 46–50.
Glaser (1958)	D.A. Glaser, 'The Bubble Chamber', *Handbuch der Physik XLV* (Berlin, Göttingen, Heidelberg, Springer-Verlag, 1958), 314–341.
Glaser (1960)	D.A. Glaser, 'Elementary particles and bubble chambers', Nobel Lecture, 12 December 1960, in *Nobel Lectures Physics 1952-1962* (Amsterdam, London, New York: Elsevier, 1964), 529–551.
Glaser & Rahm (1955)	D.A. Glaser, D.C. Rahm, 'Characteristics of Bubble Chambers', *Phys. Rev.* **97**, (1955), 474–479.
Glaser et al. (1956)	D.A. Glaser, D.C. Rahm, C. Dodd, 'Bubble Counting for the Determination of the Velocities of Charged Particles in Bubble Chambers', *Phys. Rev.* **102**, (1956), 1653–1658.
Goldschmidt-Clermont et al. (1958)	Y. Goldschmidt-Clermont, G. Von Dardel, L. Kowarski, C. Peyrou, 'Instruments for the analysis of track chamber photographs by digital computer', *Nucl. Instrum.* **2**, (1958), 146–153.
Hahn (1956)	B. Hahn, 'A Carbon Dioxide-Hexane Gas Bubble Chamber', *Nuovo Cimento* **4**, (1956), 944–945.
Hahn & Fischer (1957)	B. Hahn, J. Fischer, 'Gas–liquid system for use in a gas bubble chamber', *Rev. Sci. Instrum.* **28**, (1957), 656–657.
Hahn & Riepe (1958)	B. Hahn, G. Riepe, 'Fluorocarbon Gas Bubble Chamber', *Rev. Sci. Intrum.* **29**, (1958), 184–185.
Heilbron et al. (1981)	J.L. Heilbron, R.W. Seidel, B.R. Wheaton, 'Lawrence and its Laboratory: Nuclear Science at Berkeley', *LbL News Magazine Vol. 6, No. 3*, Fall 1981.
Henderson (1970)	C. Henderson, *Cloud and Bubble Chambers*, 1970, (Methuens Monographs on Physical Subjects, London).
Hernandez et al. (1957)	H.P. Hernandez, J.W. Mark, R.D. Watt, 'Designing for safety in Hydrogen Bubble Chambers, *Rev. Sci. Instrum.* **28**, (1957), 528–535.
Jensen (1964)	J.E. Jensen, 'The Brookhaven National Laboratory. 80-inch liquid–hydrogen bubble chamber', *International Advances in Cryogenic Engineering*, **10**, (1964), 243–250.
Krige & Pestre (1986)	J. Krige and D. Pestre, 'The choice of CERN's first large chambers for the proton synchrotron (1957-1958), *HSPS 16:2*, (1986), 255–279.
Linlor et al. (1957)	W.I. Linlor, Q.A. Kerns, J.A. Mark, 'Bubble Chamber Pressure Gauge', *Rev. Sci. Instrum.* **28**, (1957), 535–541.
Nagle et al. (1956)	D.E. Nagle, R.H. Hildebrand, R.J. Plano, 'Hydrogen Bubble Chamber Used for Low-Energy Meson Scattering', *Rev. Sci. Instrum.* **27**, (1956), 203–207.
Newport (1964)	R.W. Newport, 'Cryogenic aspects of the British National Hydrogen Bubble

	Chamber', *International Advances in Cryogenic Engineering* **10**, (1964), 233–242.
L'onde Electrique (1961)	'Chambres à bulles françaises au CERN, *L'Onde Electrique XLI*, December 1961, 969–1046.
Oswald (1957)	L.O. Oswald, 'Propane Bubble Chamber in a high magnetic field, *Rev. Sci. Intrum.* **28**, (1957), 80–83.
Parmentier & Schwemin (1955a)	D. Parmentier, A.J. Schwemin, 'Liquid Hydrogen Bubble Chambers', *Rev. Sci. Instrum.* **26**, (1955), 954–958.
Parmentier & Schwemin (1955b)	D. Parmentier, A.J. Schwemin, 'Negative Pressure Pentane Bubble Chamber used in High Energy Experiments', *Phys. Rev.* **99**, (1955), 639.
Pless & Plano (1956)	I.A. Pless, R.J. Plano, 'Negative Pressure Isopentane Bubble Chamber', *Rev. Sci. Instrum.* **27**, (1956), 935–937.
Powell et al. (1958)	W.P. Powell, W.B. Fowler, L.O. Oswald, 'Thirty-inch propane bubble chamber', *Rev. Sci. Instrum.* **29**, (1958), 874–879.
Shutt (1967)	R.P. Shutt ed., *Bubble and Spark Chambers. Principles and Use*, (New York and London, Academic Press, 1967).
Siegbahn (1960)	K. Siegbahn, 'Physics 1960. Presentation Speech', 12 December 1960, in *Nobel Lectures Physics 1952–1962* (Amsterdam, London, New York: Elsevier, 1964), 525–528.
Slätis (1959)	H. Slätis, 'Survey article on bubble chambers', *Nuclear Instruments and Methods,* **5** (1959), 1–25.
Soop (1967)	K. Soop, 'Bubble Chamber Measurements and their evaluation', *Progress in Nuclear Techniques and Instrumentation, Vol. 2*, (1967), 217–328.
Wilson (1927)	C.T.R. Wilson, 'On the cloud method of making visible ions and the tracks of ionizing particles', 12 December 1927, in *Nobel Lectures Physics 1922–1941,* (Amsterdam, London, New York: Elsevier, 1965), 194–214.
Wood (1954)	J.C. Wood, 'Bubble Tracks in a Hydrogen-Filled Glaser Chamber', *Phys. Rev.* **94**, (1954), 731.

PART III

Planning and managing research at CERN

CHAPTER 7

The CERN system, its deliberative and executive arms and some global statistics on how it functioned[1]

Dominique PESTRE

Contents

7.1 The member states, the Council, and the executive: relationships strongly mediated by a stable group of people in the Council and in the Scientific Policy Committee	345
7.1.1 The Council and its committees from 1954 to 1966: who, where, when?	346
7.1.1.1 A flashback to the founding years	346
7.1.1.2 The control of key posts: who?	347
7.1.1.3 The Council: a group of men that remained stable for 15 years	348
7.1.1.4 The Finance Committee: a homogeneous technical team dependent on the Council	350
7.1.1.5 The Scientific Policy Committee: the voice of Science	351
7.1.2 Scientists and non-scientists in the Council and its committees: differences and points of friction	352
7.1.2.1 The scientists' attitudes to the Council and to the Finance Committee: between esteem and condescension	353
7.1.2.2 The non-scientists confronting the organization: the difficult task of control	357
7.1.2.3 The non-CERN scientists vis-à-vis the organization: from criticism to unavoidable collaboration	359
7.1.3 The strength of CERN vis-à-vis the member states: the existence of a convinced and powerful lobby	362
7.1.3.1 The defense of the Council's autonomy vis-à-vis the member states: the cornerstone of the CERN system	362
7.1.3.2 The functioning as lobby	367
7.1.4 A question in conclusion: Was CERN a political success?	369
7.2 The CERN executive: an ensemble strongly influenced by the scientific senior staff	370
7.2.1 The first organigramme, 1955-1957	370
7.2.2 Cornelis Bakker's two attempts at reorganization, 1957-1960	372
7.2.2.1 The first proposals for reform in 1957: a forced withdrawal	372

7.2.2.2 Bakker's second attempt, June 1959–April 1960: A solution that was never put into practice	377
7.2.3 John Adams' solution, May 1960–July 1961	380
7.2.3.1 The idea of having a Directorate	380
7.2.3.2 The choice of the twelve divisions and the nomination of their directors	381
7.2.4 Victor Weisskopf's way of managing CERN, 1961–1965	383
7.2.4.1 The nomination of Weisskopf and of the members of the Directorate	383
7.2.4.2 The functioning of the Directorate under Weisskopf's Chairmanship	386
7.2.4.3 Victor Weisskopf's philosophy of management	387
7.2.5 Concluding survey: a difficult history	390
7.3 CERN by numbers: some statistical data on the personnel and its activities	393
7.3.1 The structure by division	395
7.3.2 The structure by professional category	397
7.3.3 Visitors and research staff at CERN	399
7.3.4 CERN and its member states	401
Notes	410
Bibliography	809

CERN is an international laboratory financed by a dozen states. It is made up of two entities, a deliberative body, the Council, and an executive body, the Director-General and its staff. The deliberative body determines the policy of the Organization, it approves its plans for research and its budgets, it decides on any supplementary programme, and nominates the Director-General and the Senior Staff. The executive body, for its part, proposes to the Council the line of action it intends to follow and puts it into practice in the framework defined by the Council. Its functions are thus to anticipate, to prepare and to execute. This is the official image given by the Convention, an image which we are going to look at more closely in this chapter.[2]

The CERN Council is composed of two representatives per country who act on behalf of their governments. It does not meet very often — generally twice a year for 48 hours. It is assisted by three committees which, by shouldering many routine burdens, leave the Council that much freer and enable it to consider carefully the more important issues. The Committee of Council (CC), which is not mentioned in the Convention, was established in 1954. Initially conceived as a kind of secretariat of the Council comprising eight Council members, its role became more central in the following years. It identifies the major issues and difficulties before or between the Council sessions, it sounds out opinions and tries to formulate durable compromises. The Finance Committee (FC) takes care of all financial aspects linked with the functioning of CERN, particularly the preparation of the budget, and considers, when necessary, legal and administrative questions. After 1955, it was composed of one representative per member state. The Scientific Policy Committee (SPC) advises the Council on research programmes and options, and might be asked to consider priorities among the proposals made by the Director-General. It comprises scientific personalities nominated *ad personam* by the Council. On the executive side, the Director-General has an important initiating role to play and has the authority to organize the house as he thinks fit. He is responsible for the work done by the organization, he hires the staff — but for the Senior Staff —, and he prepares the budgets and manages them. At the start of the period we are dealing with here the laboratory was composed of divisions each led by a director. These directors, together with the Director-General, were collectively known as the leading board or the Group of Directors.[3]

One of the aims of this chapter is to go more deeply into these ideas, to pass them through the sieve of an historical study, to confront them with the reality of CERN in the years 1954/1965. More generally, we want to put the legal definition we have just given in a broader context, we want to draw into the picture all the actors connected with CERN in one way or another—state bureaucracies, national groups of high-energy physicists...—, we want to describe how this vast ensemble functioned, an ensemble which included, at one extreme, very high-level state officials, even Ministers, and at the other locally recruited ancillary personnel. We cannot of course study all aspects of this 'CERN system', and we had to overlook things like social policy, career structure, and the functioning of the administration, as well as the financial difficulties faced by states like Greece and Yugoslavia, the conditions for the access of new member states (Austria joined in 1959, Spain in 1961), or the relationships with non-member states like the United States and the countries of Eastern Europe. In fact we have restricted ourselves to two main aspects of CERN's functioning, two aspects which have a bearing on the most important relationships of power.

We begin by focusing on *the Council and its committees;* the analysis aims to grasp what is original in the way power is distributed between the four main components we have identified, the Council which deliberates and controls, the Director-General who proposes and realizes, the states who provide the finance, and the European scientific community which CERN is supposed to serve. During the twelve years we consider here, the distribution of powers around the Council did not change substantially, so that a chronological approach to this topic did not seem appropriate. Instead we have chosen to tackle it thematically, presenting three of its facets. Firstly, we have identified the people, the men who comprised the Council and its various committees. Then we have studied the nature of the relationships between different groups, notably between administrators appointed by the member states, the scientific members of the SPC or the Council, and the directors of CERN. Finally, we have considered the attitude of the Council to the states, introducing again the notion of the CERN lobby which we used in the first volume of our history.

The second main section of this chapter is devoted to the functioning of the executive. Very important changes were made in the organization of CERN —indeed the debates on this issue were very lively —and in this case we have adopted an approach by periods. We begin by discussing the reforms presented in 1957, and taken up again in 1959, by Director-General Cornelis J. Bakker who died in an accident in April 1960. We then go on to describe the solution advocated by the new Director-General, John Adams, the former director of the division which constructed the 28 GeV proton synchrotron. Finally we look at how the fourth Director-General, Victor Weisskopf, put these reforms into practice. Our attention is concentrated on two points in this section. On the one hand, on the balance between the Director-General, the divisional directors, and a very influential group known administratively as the Senior Staff; on the other, on the particular way in which the direction functioned, on the implicit and explicit rules of management which were adopted.

To round off these two studies, we thought it worth presenting statistical data which give a global view of CERN. Based essentially on an analysis of the scientific personnel, they enable one to have a better idea of what CERN was and of the investment which scientists from different European countries made in the laboratory and its facilities.

7.1 The member states, the Council, and the executive: relationships strongly mediated by a stable group of people in the Council and in the Scientific Policy Committee

Formally speaking, the Council and the laboratory interact in a simple manner. The direction of CERN presents its projects and its needs to the member states; these are studied by the administrations in the various countries who communicate their findings at the Council sessions; by a vote, usually requiring a simple majority, the representatives of the member states accept, refuse, or modify the proposals made by the direction; the laboratory runs within the political and financial framework thus defined. The real situation, however, is more complex than indicated by this schema, and for two main reasons.

The first is that the physicists have interests which are not only those of their home countries. Often convinced of the 'higher interests' of their discipline, they sometimes act independently or even against their domestic administrations. CERN is not therefore simply *a* laboratory confronting states who are strangers to it; it is itself part of a much greater complex—a European and international scientific community—extensive enough to include certain scientific counsellors who have the ear of governments.[4] The 'collective' interests of the group of scientists may however be in conflict with the interests of certain national laboratories when it comes down to distributing money. Scientific research involves competition between groups or individuals who seek to be the first in the race for 'discoveries'—and so for money. The behaviour of the European high-energy physics community in the years 1952–65 is thus one essential parameter if one wants to understand the relationship between CERN and the member states. Let us merely note for the moment that differences between countries are easily visible here— the French, German, or Italian scientific élites, for example, favoured CERN more systematically than their British equivalents during our period—, though an evolution of the whole over the passage of time is detectable—there was a movement towards an increasingly integrated system in which CERN was to play the role of a central laboratory, of a laboratory regarded by most physicists as being at the apex of the European pyramid. The debates between 1961 and 1965 on the second generation of accelerators for CERN were decisive in this evolution.

The second main factor making the relationship between CERN, the Council and the member states more complex than the Convention would lead one to think is that the Council is not simply an assembly representing states' interests. In particular, the 'political' delegates in it are at once national civil servants nominated to represent

Notes: p. 410

and defend their governments and men on whom the states rely to elaborate their national policies vis-à-vis CERN. Depending on how these men define their role, they can strongly influence the trajectory of the organization and can become, for example, its active ambassadors before their domestic administrations. The point to bear in mind in CERN's case is that the most important group in the Council, the group who in fact controlled it, behaved in this manner until the early 1970s. The central decisions taken thus continued to emanate from a mixed group consisting of the physicists–politicians and of this component of the Council, a whole whose presence dominated the founding years of the laboratory.[5]

7.1.1 THE COUNCIL AND ITS COMMITTEES FROM 1954 TO 1966: WHO, WHERE, WHEN?

To begin with we would like to present in a little more detail the *men* who were members of the Council, the Finance Committee, and the Scientific Policy Committee from 1954 to 1966.[6]

7.1.1.1 A flashback to the founding years

At the centre of the business that was to become CERN, and before the first intergovernmental conference called by UNESCO in December 1951, we found Edoardo *Amaldi,* an Italian nuclear physicist and cosmic ray specialist, Lew Kowarski, one of the French pioneers of nuclear energy, Cornelis J. Bakker, a Dutch accelerator specialist, and Peter Preiswerk, a nuclear physicist from Zurich. Alongside these scientists, supporting them politically and logistically we might say, we find Gustavo Colonnetti, the head of the Italian Consiglio Nazionale delle Ricerche, Jean *Willems,* his Belgian equivalent, François *de Rose,* a high-level French diplomat, and Francis *Perrin,* high commissioner in the French Commissariat à l'Energie Atomique. With them, but already out of the picture in 1954, there were Pierre Auger of UNESCO, Odd Dahl, who did not want to come to Geneva, and Raoul Dautry, who died in 1951.

The conferences of December 1951 and January 1952 added two names to this list of key actors, Jan H. *Bannier,* a Dutch science administrator, and Werner *Heisenberg,* the German Nobel prizewinner for physics. During the next 30 months this group of men became profoundly integrated, overcoming internal tensions — the problem of the site, of the Convention, etc. — and imposed collective solutions on the different governments. They progressively surrounded themselves with a 'layer' of others whom they were close to, who supported their initiatives, but who seem to have been less 'activist'. The most notable members of this group were Ivar Waller, a Swedish physicist, Alexander *Hocker,* a high official in the Deutsche Forschungsgemeinschaft, Paul Scherrer, one of the élite of Swiss physics lately converted to the project, and Albert Picot, a man of great ability in Swiss political life, member of the Conseil d'Etat of Geneva. The British also joined the venture; on the one hand, the young accelerator specialists, notably John Adams, on the other John Cockcroft, Nobel prizewinner and the Director of Harwell, Sir Ben *Lockspeiser,* the Secretary of the British DSIR, and Harold L. Verry, his right hand man. From the moment the British government supported the project it did all it could to ensure that it would succeed — but within the rules, with a view to avoiding any institutional or financial backsliding. This explains the role of the British in drafting major texts for the administration of the laboratory and in the creation of an Interim Finance Committee with Sir Ben in the chair.[7]

7.1.1.2 The control of key posts: who?

The first point to note about the years 1954-1966 is the ongoing presence of most of these names (and in particular those which we have italicized) in the key posts of the Council and its committees. By rotating their places some dozen people appear to have controlled all the decision-making nodes for more than fifteen years.

At the first Council session in October 1954 it was *Bannier* who proposed that *Sir Ben* be elected Council President. This was unanimously agreed to, as were (almost) all the elections we are going to describe here. For Bannier, Willems, or the French, it was a question of thanking the British—for having joined CERN, for contributing almost 25% of the budget, for having offered specialists for the construction of the PS, for having set up CERN administratively—, of honouring Sir Ben—who had put all his weight behind the scheme in the United Kingdom—, and, at the same time, of bonding Lockspeiser more subtly into the core of the organization. On the same occasion Jean *Willems* assumed the Presidentship of the Finance Committee, and Werner *Heisenberg* that of the SPC.[8]

These nominations being made for three years, the negotiations to replace these men began in 1957. The situation at CERN at the time was very difficult, as the organization was asking for a massive increase in its budget (of some 50%). In these circumstances the posts of President of the Council, and of Chairmen of the SPC and the FC, were exceptionally important. They were awarded unanimously to *de Rose* (Council), *Amaldi* (SPC), and *Bannier* (FC), three of the most active and most dedicated members of the old guard. The two Vice-Presidents of the Council, by the way, became *Willems* and *Heisenberg*.[9]

At its session in December 1960 President *de Rose* proposed to the Council that *Willems* take the helm, and this was agreed. The two Vice-Presidents were also appointed by acclamation—they were *Amaldi* and *Bannier*. *Hocker* was given the job of President of the Finance Committee; in the SPC it was *Powell,* British Nobel prizewinner in physics. The nomination of Hocker—the German delegate since the first conference in 1951/52—reflected the wish to have a representative from the Federal Republic of Germany, a country which had become both the third largest contributor to the budget and one of the most ardent supporters of the growth of

Notes: p. 410

CERN. Powell, for his part, had the triple advantage of being a long-standing scientific authority, of having initiated the system of multinational experimental collaboration around the PS, and of being British while not a member of his country's official delegation; close to the new Director-General, Victor Weisskopf, nominated *ad personam* to the SPC, he spoke for Science and was not constrained by the doubts which beset Her Majesty's Goverment. Through welcoming Hocker and Powell the initial core, with minimum risk, opened its doors to people formally external to it ... but already on its side.[10]

In December 1961 Hocker was asked by his government to take up other tasks at the nuclear research centre in Jülich. The Council named Gosta W. *Funke* as his successor, one of the most important Swedish administrators of science, Secretary-General of the Atomkommitten from 1945 to 1959, and Secretary-General of the National Science Research Council at the time. He had been Sweden's permanent delegate to the Council since 1955, and was one of the most active members of the FC. Certainly, he had often expressed the doubts of his government in the preceding years, but always so as not to impede the organization in any way. Personally convinced that CERN should grow, his nomination seems to have been motivated by the same rationale as was that of Sir Ben in 1954.[11]

At the end of 1963 the retiring President of Council, *Willems,* proposed that *Bannier* succeed him. This was accepted, and *de Rose* and Sir Harry *Melville* were nominated Vice-Presidents. Having replaced Lockspeiser as delegate in 1958, Melville was intended to reintroduce a British presence at the head of the Council. As for the SPC, it was *Leprince-Ringuet* who was unanimously accorded the Presidentship. He had been a member of the committee from its inception in 1954, was the director of the physics laboratory at the Ecole Polytechnique in Paris, and was grand artisan of the CERN bubble chamber programme. His main lieutenants, Lagarrigue and Gregory, were regularly working in Geneva with their chambers and, as far as he was concerned, CERN had to be regarded by the French authorities as top priority. On this point he fully agreed with Perrin and de Rose.[12]

Finally, in 1964, the Presidentship of the FC passed to the German delegate who had replaced Hocker, W. Schulte-Meermann.

About a dozen names have emerged in this rapid presentation of the reshuffling of key people in the Council and its committees, and of these four have recurred with remarkable frequency: Amaldi, Bannier, de Rose, and Willems. With Weisskopf, the Director-General of CERN from 1961 to 1965, we have to place these four people at the heart of all CERN business connected with the Council.

7.1.1.3 The Council: a group of men that remained stable for 15 years

About 60 people would usually attend a Council meeting during the years 1954–1966. Its decisive component was formed by the 24 or 26 official delegates designated by the governments of the member states. Each delegation was usually composed of a scientist and an administrator (or a diplomat), who were assisted by one or two advisors. The senior management of CERN—some twelve to fifteen persons led by the Director-General—were also present at Council sessions. Finally, the meetings were attended by the Chairmen of the Finance Committee and the Scientific Policy Committee.[13]

The most striking feature about the official delegates is their relative stability over a long period of time. Two things are particularly noteworthy, the continuing presence of the 'historic' delegates, often until their retirement, and the gradual, progressive integration of 'new blood'. This had important effects since it permitted the perpetuation and the reinforcement of the rather strong bonds—and even the affective links—which had been built up between the 'founding fathers' in the years 1951-54.

Let us be more precise about this relative stability. Two delegations were absolutely unchanged during our period, those of Denmark and Sweden. Bøggild and Obling represented the former until 1965, and beyond, while Funke and Waller represented the latter—and let us not forget that Waller was in contact with Auger as early as 1950! For France it was Perrin who did not budge from 1950 to the early 1970s; along with him there was de Rose who was posted to Madrid in 1953 by his Minister of Foreign Affairs, but who managed to return to the CERN Council three years later. Made responsible for atomic affairs at the Quai d'Orsay in 1957, he continued as French Council delegate until 1964 when, following a radical disagreement with General de Gaulle on defence policy, he was sent to Portugal as ambassador. His political 'exile' did not stop de Rose maintaining close personal contacts with CERN, particularly through Victor Weisskopf.[14]

Willems and Bannier were obviously the permanent delegates of Belgium and the Netherlands. Their scientific colleagues changed somewhat, on the other hand. The converse was true for Italy, where the unchanged and very active delegate was Amaldi, but where there was no such stability for the other delegate after Colonnetti's departure from the Council in 1956, at the age of 70 years.[15] We have some variation in the cases of the Federal Republic of Germany and of Switzerland, but nothing so serious as to damage our thesis of relative stability. Concerning the host state, Scherrer kept his post until he turned 71, and Picot until he was 78! Fierz and Chavanne succeeded them in 1962.[16] As for Germany, Hocker was representative from 1951 to 1961, when he was replaced by Schulte-Meerman; on the scientific side, a frequently ill Heisenberg was succeeded by Gentner, the former head of a CERN division, and by Jentschke, the head of DESY. Concerning Greece, Norway and Yugoslavia, three countries who were founder members, and Austria and Spain who joined later, the rotation of delegates appears to be frequent. These countries did not carry much weight in the Council, however, and a 'stabilization' of their delegates is perceptible from about 1961/62.[17]

This only leaves the United Kingdom. Her 'political' delegate only changed once during our period: Sir Ben Lockspeiser held the post from the beginning until 1957, when he was replaced by Sir Harry Melville. On the scientific side, Cockcroft was often absent, and his place was taken by two of his staff, Fry and Pickavance, though without real continuity until 1961. Then Adams took over. Without leaving the Council, he changed his Director-General's hat for that of official British delegate.

What is striking, then, is the permanent presence of a considerable fraction of the Council *over fifteen years,* and particularly of those people who had a decisive role either because of the weight their country carried or because of their previous functions in the organization. Indeed a statistical survey of the minutes of the some 40 Council meetings held from 1952 to 1965 shows that only the stable and 'historic' delegates speak at any length during the sessions. The others make much briefer declarations, usually serving to explain how they intend to vote.

7.1.1.4 The Finance Committee: a homogenous technical team dependent on the Council

The Finance Committee's function is to control the financial aspects of CERN's life on behalf of the Council and the member states. It discusses the budgets proposed by the senior management, it gives its opinion on certain expenditures before they are made, it looks at the distribution of industrial contracts between national industries, it considers the salary scales and the management of the personnel, and so on. For these reasons it is very closely associated with the political arm of the Council, it meets rather frequently —some six times a year—, and it is composed of at least one representative per country, often two.[18]

The first thing to note about the FC is that Council members who had high-level posts in their national administrations—and there were not a few of them at this time—did not take part in its proceedings. This was true of Lockspeiser and then of Melville for Britain, of de Rose for France, of Picot and Chavanne for Switzerland, and of Willems after 1960. On the other hand, if one of these men was the President of the Council he did tend to be present at the Finance Committee meeting held on the eve of the Council session. This was the case with Sir Ben, de Rose, and Bannier, though it was somewhat more unusual for Willems. As for the other countries, they tended to use their Council delegates—or at least their 'political' representatives—as their Finance Committee delegates as well. This happened with Hocker and Schulte-Meerman (FRG), with Bøggild and Obling (DK), and with Funke and Waller (S).

The second noteworthy feature of the Finance Committee is that, as in the case of Council membership, a certain stability is perceptible over fifteen years. All the same it is less marked than in the case of the parent body, particularly after 1960/61. Until this date the core of the FC is easily identifiable. Along with the Danish, German, and Swedish delegates we have just mentioned it included the two stalwarts who were its Chairmen, Bannier and Willems, and two faithful representatives from Britain and France, Verry and Neumann. Thereafter Willems was replaced by Levaux, Bannier was assisted by Hoogewegen, and small groups replaced Verry and Neumann: Hubbard, Walker, and Dunning for the British, Courtillet and Grondard for the French. The Funke—Schulte-Meerman backbone remained, however. As for the other countries, we see the same tendency to irregularity in FC membership which characterized their representation in the Council.

7.1.1.5 The Scientific Policy Committee: the voice of Science

In contrast to the FC and the Council, the members of the SPC did not represent their respective countries; they were nominated in their personal capacities and were required to judge issues strictly 'from the scientific point of view'.

In the early days, when the main task of the laboratory was to build accelerators, the SPC seldom met—twice a year between 1955 and 1957. In the following five or six years, during first the preparation and then the implementation of the experimental programme, the body had a tendency to meet rather more often (roughly three times a year). Thereafter the frequency of its gatherings increased even more (four times in 1964, five in 1965).

During the fifties about 15 people might be present at a meeting of the SPC. Eight or nine were CERN staff, the remainder being official delegates. Two people were outstandingly regular in their attendance from the start (1954) through to the early 70s, namely, Leprince-Ringuet and Amaldi. The latter was always invited between 1955 and 1957, and became a fully-fledged member in 1958 when he was nominated President. Along with these two, there were two Nobel prizewinners whose attendance could be counted on until 1958, Blackett and Heisenberg. Cockcroft and Scherrer, on the other hand, were less often present at meetings, and as for Alfvén and Bohr, who were members of the SPC until 1959 and 1961 respectively, they were often absent.

The SPC was enlarged a little in the 1960s, the old guard being progressively replaced at its own request. In 1963 the hard core was composed of Amaldi, Leprince-Ringuet, Møller, Perrin (who replaced Alfvén), Powell (who replaced Blackett), Adams and Gentner (who replaced Cockcroft and Heisenberg), and Gregory who was nominated in his capacity as President of the so-called Track Chamber Committee.[19] The committee grew again in size in 1964/65 to include Bernardini (I), Jentschke (FRG), and Wouthuysen (NL), as well as Butler, Cassels (then Puppi), and Ekspong in their capacities as chairmen of the three CERN experimental committees.

An interesting point about participation in the SPC meetings is that in 1958 the new President of the Council (de Rose) and the new Chairman of the Finance Committee (Bannier) decided to be present. This was a response to the budgetary crisis of the time, a crisis which had its origins in the new financial demands made on the organization's resources by the preparation of the PS experimental programme.

Notes: pp. 410 ff

De Rose and Bannier thought it best, under these circumstances, to discuss the problems in-house, as it were, and to speak directly to the scientists on the SPC. In 1961 the practice was continued by their successors, Hocker, Funke, and Schulte-Meerman from the Finance Committee and, after a break during Willems' reign, by Bannier in his capacity as Council President.

This presence of the 'political' leaders at the SPC meetings is important and it directs us along an avenue that is worth exploring further: the fact that, in contrast to the Finance Committee which had a lower status than the Council, the SPC appears as a body which shared its eminence. Gathering prestigious personalities—Nobel prizewinners and leading 'savants'—, speaking in the name of needs which others could not, or at least would not try to assess, the SPC became gradually recognized by the non-scientific members of the Council as a body with which a *balance* had to be found. In contrast to what the Convention might lead one to think, the SPC would not in fact be 'subject' to the Council but would form a kind of 'parallel power', a power derived from a technical expertise which no one denied.[20]

Our conclusion will be brief. In the mid-1960s we still find, in the Council and its committees, a barely unchanged core whose members were essentially those who had laid the foundations of CERN in the early-1950s. Playing a dominant role, they derived a certain legitimacy—and a certain power—from their historical role, a legitimacy and a power which was expressed through the structures and functioning of CERN's Council. Around them we find a number of people who were successively integrated into their ranks. By 1965 most of them had already been intimately associated with CERN for a considerable length of time.

7.1.2 SCIENTISTS AND NON-SCIENTISTS IN THE COUNCIL AND ITS COMMITTEES: DIFFERENCES AND POINTS OF FRICTION

Despite its stable dimension, the Council is not to be thought of as totally homogeneous in its behaviour vis-à-vis the organization. Quite clear differences appear even among the hard core, differences rooted in the kinds of links the various delegates had with their domestic authorities and with what they were doing in Geneva. While not the only one, the most notable of these differences was that between the scientists and the non-scientists, the divide between Amaldi, Perrin, Powell, (or Weisskopf as Director-General), for example, and Bannier, Hocker, de Rose, or Willems. No doubt both groups were collectively the hinge between the organization and the member states, no doubt they were placed at the foci of power in the CERN system, no doubt they shared a certain vision of their role which will allow us to speak in the following section of there being a *lobby* inside the Council. But this does not mean that certain splits did not arise, splits which can be traced back to the very different relationships which these two groups had with the high-energy physics community and with those who had appointed them. It is on these differences that we want to focus in this section.

7.1.2.1 The scientists' attitudes to the Council and to the Finance Committee: between esteem and condescension

The attitudes of the scientists, be they CERN staff or in the SPC, to the Council or the FC can, in our view, be organized around three themes: *a respect,* that much more noticeable when one observes the behaviour of scientists who have responsibility in the organization or in the Council and its committees; *various forms of incomprehension* or of denigration, ready to surface when there are tensions; *offensive attitudes of 'selling',* finally, which are brought into play when the scientific community regards a new project as decisive for European physics.

The esteem in which the scientists in the CERN direction or the SPC held their 'political' homologues seems undeniable, even if it is difficult to *measure* it since any source which is remotely official is *a priori* non-convincing: ceremonies associated with the handover of power or minutes of meetings always provide occasion for a dithyramb, both in political circles and in academia. We shall therefore only mention one of the many examples of reciprocal respect which we find in CERN documents of this kind, namely, a speech by Victor Weisskopf which he made in December 1963 in honour of Jean Willems who was relinquishing the Presidentship of the Council. The Director-General remarked on 'the tradition of mutual confidence and team spirit' which characterized the relationship between the laboratory and the Council, waxed lyrical about the paternal behaviour of Willems, 'who does not scold his children and never resorts to disciplinary methods, but is always ready to give advice and help', and spoke of Willems' goodwill when he, Weisskopf, put forward in private, in the rooms of the Hotel des Bergues, the main demands which he intended to lay before the Council.[21] As formal and codified as this text might be, it still shows traces of a particular state of mind, of a kind of personal relationship which might be considered somewhat surprising in an assembly as official as the Council. More private documents, and the correspondence in particular, confirm these first impressions: on the whole the tone is friendly, even warm, between certain CERN's Director-Generals and the main political personalities in the Council; and when there are moments of tension, it is usually frankness that dominates. Differences in personality were fundamental here. It was the relationships between Weisskopf, on the one hand, and Bannier and de Rose on the other, which seem to have been the most open while with Willems, for example, they were less close. Not that Willems did not act systematically in favour of CERN, on the contrary, he was one of those who felt the least need to 'weigh' issues. But his dealings with CERN reveal less personal dedication, less of an investment of time, particularly after 1960. Older, he emerges as an authority who frequently accepts what is asked of him while knowing

Notes: p. 411

that others will do what needs to be done. Finally, we should add that in some rather rare instances we have letters indicating a persistence of friendship beyond the demands of official obligations. For example there are the bonds built up in 1955/56 between the new Director-General Cornelis Bakker and the former French President of the provisional CERN Council, Robert Valeur, and those established between Victor Weisskopf and François de Rose: several personal letters written in 1965 to arrange meetings clearly indicate a real mutual esteem.[22]

This esteem which the scientists have is never blind, is never as total as it appears to be between the members of the physics clan themselves. It is limited precisely because the 'politicians' are not, after all, high-energy physicists; among those who practise *the* most fundamental Science—and so the most noble, the most prestigious of the sciences, the science that plumbs the depths of all the big questions—, there is a certain feeling of exclusiveness, a conviction that 'physicists are a better breed than the rest of mankind'.[23] Negatively, this is sometimes expressed in the behaviour of certain people by a measure of condescension towards the non-scientists—and notably the 'politicians', those who stop them doing their job properly, who always introduce unnecessary complications, who never grant even the budgetary minimum without looking sour, without bickering—and who want to control! These traditional attitudes of the scientists, for whom 'financial authorities always were a convenient bogey' (Gentner), sometimes manifest themselves in violent ways. Signs of them can be found, for example in a confidential memorandum sent by the DG to the divisional heads in 1957, in which he draws their attention to the ungracious way in which CERN staff were speaking of the members of the Finance Committee—'even those in lower grades'; or in an exchange of sharp letters in 1961 between one of the senior CERN physicists and the director of administration —in which the former's hostility towards the latter is all too evident.[24]

These sentiments, never totally absent though very variable in intensity between individuals, were, in Adams' words, 'almost traditional for scientists'. For that reason it is only worth mentioning them to show that the CERN staff were no exception to the rule.[25] More interesting perhaps was the 'spontaneous manipulation' of the Finance Committee or the Council by many physicists who were able to capitalize on CERN's role as the leading European laboratory. Since CERN always had to act *quickly*—CERN was to be the first in the world—and since CERN had to have the *best* men and the *best* equipment, it was always easy and often tempting to present the 'politicians' with a *fait accompli*. Without meaning to judge, it is certain that the scientists at CERN (or in the SPC) made abundant use of the argument that things were 'urgent and top priority'. To illustrate this thesis we offer two examples, one concerning *the way CERN thought the budget should be established,* the other dealing with the *argumentation* used by scientists, particularly when they wanted to convince the Council of the importance of a new piece of very expensive equipment.

In a symposium organized in 1956 at the National Physical Laboratory in England on the management of research establishments, D.R. Willson of Harwell pointed out that there were two ways in which to establish the budget of a scientific laboratory. It could either be laid down in advance by the financial authorities, and then broken down 'analytically' between the various services, or it could be built up 'synthetically' by the research centre in response to the demands made by the various group or divisional leaders. According to Willson, the first approach was typical of almost all research in peacetime and was based on the idea that it was up to governments to take the key decisions, and particularly to make the important financial choices. The second method, by contrast, was used almost exclusively in wartime, and assumed that costs were not a determinant factor and that the budget could only be properly established in relation to the needs of the field.[26] What is interesting about CERN is that, from the start, the scientists believed that budget estimates could only be made in the latter way—and it is this that dominated the long budgetary debates between 1957 and 1962 which John Krige describes in chapter 10. Unlike all other projects which had 'to compete for funds with other projects, [...]', wrote Elkington, 'CERN seemed to be exempt from this competition at least up to the mid-sixties'. When confronted with the arguments of certain Finance Committee delegates—and above all of the British who were being repeatedly told by *their* scientists that the spoilt child in Geneva could get almost anything it wanted while they had to live within restrictive budgetary allocations fixed by central government—, the staff always reacted by arguing that CERN's unique position demanded that its needs be fully met. Claiming that the laboratory had to be regarded as a special case—was it not, after all, the advance guard in the scientific war between Europe and the USA?—, always insisting that they only asked for the absolutely indispensable strict minimum—which laboratories competing for funds regularly disputed—, they acquired stock ways of arguing which bordered, in some cases, on blackmail.[27]

We will refine and conclude this point by looking at the way in which the scientists at CERN and in the SPC presented, in 1958, their building project for two big bubble chambers. From June 1957 to May 1958 the physicists and engineers at CERN discussed the equipment that they needed to exploit the PS from 1960 onwards. In May they concluded that the laboratory should construct a 2 m hydrogen chamber and a 1 m heavy liquid chamber, and that work on both should get under way as soon as possible. The FC and the Council still had to be convinced —and this was no easy task in May/June 1958. Firstly, because of the sheer scale of the project: preliminary estimates indicated that the whole programme (including supporting facilities, beam transport equipment, etc.) would cost over 20 MSF. Secondly, because no provision had been made for a project of this scale: the estimates made in September 1957 were of the order of 6 MSF for 1959 and 1960 for *all* the experimental equipment needed around the PS; a year later, the administration was asking for 9 MSF just for bubble chambers and for 1959 only! Finally, because, after

Notes: p. 411

a long and demanding debate, the Council had agreed in December 1957 to set a ceiling of 100 MSF on expenditure in 1958 and 1959, which left 44 MSF to provide for the entire organization's needs in 1959.[28]

To overcome these obstacles the CERN scientists deployed a number of what we shall call 'selling strategies'. In addition to a few classic and general arguments—like pointing to the American threat—, they used more refined and specific tactics, like demanding more than they really expected to get, appealing to scientific necessity, or claiming that the new demands were totally unforeseeable.

The first of these, demanding more than one expected to get, amounted to asking the SPC to choose between three different timetables for the construction of the two chambers, each with its own budget. One of them, the most ambitious, stretched the organization's resources to the limit. In its meeting on 8 May, the SPC 'naturally' recommended this crash programme to the FC and the Council. The direction of CERN then suggested that it might be able to manage with less, that it would be possible to delay somewhat the completion date of the 2 m chamber—a proposal which amounted to supporting the intermediate timetable. On 3 June 1958 the FC's Working Party recommended that this 'compromise' solution be adopted, a solution they said, which indicated the direction's spirit of conciliation.[29]

The 'appeal to scientific necessity' was typical of the SPC's rhetoric throughout this period. It amounted to changing a recommendation that was the result of a debate in which opinions were not necessarily unanimous into a categorical imperative, in transforming something that could be weighed against other budgetary choices, into an absolute and independent necessity which the 'politicians' could not but translate into a budgetary increase—and which only people who wanted to cut back CERN to nothing could think of refusing.[30]

'The appeal to the unexpected' was used when someone in the Finance Committee, for example, asked why these additional expenses had not been *foreseen* a few months earlier, in view, say, 'of the interest shown in that apparatus [the bubble chamber] for some time in the United States'. The standard reply to this kind of question was to *assert* that it '*was not* and *could not* have been foreseen before'—because of 'the many developments which had occurred during the last few months', or because 'the hazards connected with [their exploitation] had become apparent only recently'. In short, the 'tactic' amounted to holding events responsible for what was, in fact, a change in one's own appreciation of those events.[31]

Taken together these three arguments had rhetorical value at the interface between the scientists and the lay members of the Finance Committee. The willingness to 'retreat' from the full programme to the 'compromise', for example, led one FC member to remark that, 'far from being bold scientists unconcerned with financial problems', the members of the SPC had been acutely aware of the financial implications of their proposals. Admittedly it was unfortunate that those proposals put members of the Committee in a difficult position. But this could not be helped. The situation was *unexpected* and *exceptional*.[32]

7.1.2.2 The non-scientists confronting the organization: the difficult task of control

The behaviour of scientists thus seems to be dominated by their allegiance to physics. CERN, which is a key element for them, is *their* business, the business of their students and colleagues. This is true of the CERN physicists, as we have seen, but also of those in the SPC or the Council.

When these men are nominated to these managerial posts, it is generally because of their technical expertise, and their governments know that they will inevitably be both judge and party in their own cause. Indispensable—they are the only ones who have the knowledge—, and cossetted—because of the nuclear dream inherited from the war—, they simultaneously serve their government and their community; according to their perception of the situation at a given moment in time, they tend to favour one rather than the other, the essential determinant being how to develop physics *as such* and *their* country in one movement. In any case, they retain a certain autonomy towards those who employ them, towards the governments which nominate them; the main reason is that their legitimacy derives also from their equals, from the opinion which the international scientific community has of them—and not only from the states.[33]

The position held by non-scientists in the Council is quite different. Civil servants, members of the administration which has sent them to Geneva, they are expected to align themselves more closely with their government's policies than are the scientists. Necessarily more tied by their mandates, they do not derive their legitimacy from an external logic but from that internal logic which led them to be nominated as CERN Council delegates in the first place. In addition, not being technical experts themselves, they need to rely on the opinion of scientists to exercise the functions of control asked of them by their superiors. They are thus doubly 'vulnerable': situated at the interface between CERN and the states, between the scientists and their domestic authorities, their position is often very delicate.

In the case of the CERN Council, the situation of the non-scientists in the 'historic' core which we identified earlier has its own special features. Having built up affective links with their offspring, their dedication to this adventure meant that they wanted to make of CERN a real success, and would do all they could to ensure that it became the equal of the best American laboratories. In return they expected the organization to put its cards on the table, they demanded that it respect them and the rules of the game, that it tell them clearly and without deception what the real situation was. In short, they insisted that each talk straight with the other. The reason for this, sometimes explicit, 'arrangement', was that their credibility in their domestic state apparatuses was at stake; because they regularly stuck out their necks in favour of CERN, they needed a 'guarantee' that their authority at home was not going to be destroyed by hasty, thoughtless, or rash manœuvres behind their backs at CERN. One consequence was that, when matters became tense with the organization, these 'political' delegates to the Council were in an objectively weak

Notes: p. 411

position. If, for example, a CERN division flagrantly broke a rule with implications beyond the confines of 'internal' control, and by so doing put the delegates in a difficult position, there was very little they could do—in truth, they had no real means of applying pressure ... without putting CERN in danger. Unlike the executive which could *more or less* 'keep to the rules' with them, they were not able to make a *graduated response*. Placed in a reactive and conciliatory position, the only real means of action at their disposal was to protest verbally, to show their displeasure—and to accede.

An excellent observation point from which to study a situation of this kind is provided by John Krige in his chapter on CERN's contract policy with firms (chapter 11). Here he considers the attitudes of the FC delegates when judging how industrial contracts were to be allocated by CERN. Their official tasks were to ensure that this was done at the best possible price for the same quality, and that all European firms were treated on the same footing. John Krige's principal conclusions are that the CERN administration often bent the procedural rules laid down by the Council—notably between 1953 and 1959—, and that the Finance Committee found it extremely difficult to take any concrete action to stop such backsliding. Confronted by scientists for whom getting the job done quickly was a priority, and who made it clear that they would not let the formal rules impede their placing orders with the firms they had chosen, the delegates to the Finance Committee only rarely dared oppose them. Far from the site—they only came to Geneva six or eight times a year—, not always having the most reliable information—they depended on the CERN administration for that—, it was difficult for them to insist too much on what some considered mere procedural niceties. Thus it was only gradually that a better equilibrium was found. It took shape progressively, as the delegates to the FC came to understand a little better the way things worked in Geneva—and as the local physicists learnt how to collaborate with people who were 'extremely well disposed towards CERN', and who had 'nearly always approved the proposals which had been submitted' to them, as Bakker put it.[34]

It is possible to find other examples revealing difficult relationships outside the sphere of contract policy, but none of these indicate that what opposition there was was lasting, or that it soured dealings in several domains at the same time. They tended to occur in those sectors of CERN's life where the non-scientific members of the Council and of the Finance Committee had certain competences of their own and when they were in a position to discuss an issue on an equal footing with the Director-General and the administration. So as not to overburden the reader we will just quickly mention three situations of this type: when the salaries or allowances of CERN personnel were being discussed, the direction was often more generous than the Council would have liked—for example, on 9 December 1960, the latter refused to grant all of the allowance which the Director-General wanted to give to the members of the Directorate and to the Division Leaders; similarly, when the allocation of indefinite contracts to senior personnel was on the agenda, the Council stuck to the initial policy at CERN for a long time, *opposing* the award of indefinite contracts and *supporting* the rotation of senior posts; finally, the Council can be seen taking a hard line, if only temporarily, when the direction wanted to create additional posts which it thought unnecessary or too well-paid—as happened in 1964/65, for example, when the laboratory felt that it '[ought] to recruit a new CERN dignitary' (Willems), in grade 12, to handle the organization's public relations. This proposal gave rise to an exchange of sharp letters which Willems concluded by asking the Director-General to submit his decision for approval to the Council if he was unwilling to change his mind.[35]

7.1.2.3 *The non-CERN scientists vis-à-vis the organization: from criticism to unavoidable collaboration*

So far we have assumed that a kind of happy internationalism reigned among European physicists, that between 1953 and 1965 the majority of them were CERN supporters, that they routinely placed national or local interests second to the higher interests of the central laboratory that was CERN. This is a reasonable approximation to begin with, but should not blind one to the fact that in physics, as in many other research fields, each leader puts *his* activities and *his* teams first. If CERN was, for one of these 'bosses', part of his strategy, then what was good for CERN was good for him too. There was no reason, however, for that to happen systematically, particularly in the years which concern us here, when CERN had barely proven its worth and justified its claim to be the apex of the European high-energy accelerator pyramid.

Conflicts of interest were obviously strongest with that CERN member state which was most developed scientifically, the United Kingdom. Thus it is here that we shall begin our study of the tensions and the recriminations which existed between scientists in the member states and their colleagues at CERN.

In the mid-50s the United Kingdom always thought that 'as a matter of policy' it should be able 'to start work on a new machine every few years'. In 1957 the National Institute for Research in Nuclear Science was accordingly established, and the construction of Nimrod, a 7 GeV high-intensity synchrotron, was agreed to. In Sir Denys Wilkinson's view one of the most striking features of this decision was the complete absence of any reference to CERN, 'the astonishing insulation of thinking about the domestic programme from the the UK's participation in CERN'. This 'ignorance', this 'lack of interest' in CERN—and its implication in the form of repeated British complaints about the growing budget of the European laboratory—, was to last until about 1960, and was based on 'a strong undercurrent of feeling that [...] we must also make certain of being able to go on doing our own thing'.[36]

The debate at CERN in 1959 over the loan of the British and French bubble chambers helps us to refine these ideas. At the time CERN was perceived by the physicists in these two member states as a service centre, as a public facility whose task it was to furnish beams to *their* teams. It was not a laboratory which they thought to co-manage, or whose scientific programmes they saw as being partly their responsibility. They were outsiders, and CERN was simply something that they *used*. Correlatively, the CERN physicists tended to ignore totally these foreign intruders and to believe that CERN was, and ought increasingly to become, *their* own laboratory.[37]

In the early 1960s the notion that there should be a joint programme for the European high-energy physics community began to take shape, and with it the place that CERN should occupy in the scheme. The reasons for this slow and irregular movement were two. On the one hand there was the debate at CERN on future accelerators, a debate whose aim was to find a way of meeting the American challenge in the 1970s and which helped, from 1961 onwards, to tie the Europeans a little more to what would become their leading laboratory. On the other hand, there was the commissioning of the 25 GeV CERN proton synchrotron in 1960: it resulted in many well-known experimentalists — notably in Britain, France, and Italy —, coming to see CERN as the laboratory in which they had to work. They thus became increasingly involved with it and came to regard it as the common property of all European physicists.[38]

The results of this change of perception, which occurred rather quickly in France and Italy, only began to emerge in the United Kingdom from 1964/65 onwards. Of course some people were insisting in 1961 that the physicists should regard CERN as 'their intellectual centre', but they were in the minority and their influence on the country's investment programme between 1960 and 1964 was marginal.[39] NINA, the 4 GeV electron synchrotron, was agreed to in 1962 without any particular reference to CERN and/or a joint programmne. This at least is the view of *Geminus,* the well-informed columnist in the *New Scientist* at the time, and is also what one can infer from an article written by A.W. Merrison himself in the same journal in November 1962. To be more precise we would need to study archival sources; more specifically, we cannot say whether NINA was a deliberate choice made against, or in spite of the machines envisaged at the time at CERN — a local lobby carrying the day against a pro-Geneva faction —, whether the different projects were regarded as complementary — in which case unbridled optimism got the better of reasoned judgement —, or if the British physicists simply thought to seize the opportunity offered by a government that was ready to be generous at home if parsimonious in its attitude to CERN's budget.[40]

Contrariwise, the leading French physicists deliberately chose the alternative path: from 1962 onwards, and for six years, Perrin and Leprince-Ringuet postponed all decisions in France on a 45/60 GeV proton synchrotron so as not to impede the development of CERN. As for Italy, well-endowed with a large number of renowned physicists, but in a more difficult economic situation, it committed itself to CERN from the moment the accelerators were commissioned in Geneva. One of several proofs for this is the large experimental programme which the Italian physicists sent to Geneva in January 1960, to be honest the only programme of any coherence which was proposed for these first months of work on the new big machine.[41]

As for the other countries, they embedded CERN in the heart of their national programmes from the mid-1950s onwards. This is true of the small member states who quickly realized that CERN was their only way of gaining access to a world level; it is also true of the Federal Republic of Germany which was behind in this field of research at the time, and which declared through Hocker in November 1957 that 'it regarded CERN as an integral part of its national scientific pursuits and was accordingly prepared to treat it in the same manner as it treated them'.[42]

The general line of direction, the overall tendency, is thus clear. While physicists from national laboratories weighted the interest and the importance of CERN rather differently in the 1950s, the situation changed in 1960/61. And five years later CERN had become *the* European centre for high-energy physics. In consequence the sometimes sharp criticisms expressed in the 1950s decreased in number, and if they did not entirely disappear, they became more constructive in the following decade.

At a more everyday level—and so beyond the big questions of strategic choices—the physicists from the member states kept a close eye on the way in which CERN collaborated with its European visitors. Even if national interests did not play an important role in the life of the organization the balance of powers between the Director-General and the élite group of scientists in the SPC and the Council was always given *very close attention* by the principal actors.[43] On the whole, however, mutual recrimination was rather infrequent. Despite a systematic search, we have only found a few cases, of which we give two examples. In July 1963 Heisenberg complained, in an official letter to Weisskopf, that the Track Chamber Committee was discriminating against the less powerful countries—so Germany—in the distribution of bubble chamber photographs. The matter was investigated inside CERN and a detailed reply by Weisskopf seems to have settled the affair. In September of the same year, Francis Perrin raised the case of a French electronic experiment team which was not given access to the PS. Convinced that they were the victims of an injustice on the part of the physicists in the NP division, he demanded an explanation from Weisskopf—whose answer was passed back to him by the person responsible for research, the Frenchman Bernard Gregory. A solution was found three months later and the team was included in a collaboration with CERN.[44]

It only remains to add that it was more usually the 'scientific' and not the 'political' delegates who were involved in these incidents. For example, if we look at the behaviour of Perrin and de Rose in the well-documented debate held in 1959 on the loan of French bubble chambers to CERN, we notice that the former was less conciliatory than the latter, both in private (as seen from the archives of the French Ministry for Foreign Affairs) and in public (as emerges from British archives, for example). The reason for this is obvious and we have already given it: CERN was for

Notes: p. 411

physicists a *laboratory;* everything that happened there was of direct concern to them. The political delegates, by contrast, were less personally involved—and for them minor details could be overlooked in the interests of building a harmonious CERN, a CERN free of national tensions.[45]

7.1.3 THE STRENGTH OF CERN VIS-à-VIS THE MEMBER STATES: THE EXISTENCE OF A CONVINCED AND POWERFUL LOBBY

The most striking aspect of the relationship between CERN and the member states between 1951 and 1954 was the extremely reactive situation in which the latter were placed. From the very beginning the main actors effectively managed to act more or less autonomously, and to keep national bureaucracies at arm's length. In this process a lobby gelled, its cohesion lying in the novelty and importance of what was at stake—to be part of a collaborative scheme in *nuclear* science encompassing no less than a *dozen* states which aimed to beat the American monolith by building the biggest accelerator in the world. This out-of-the-ordinary scheme gave birth to feelings of adventure and of unity which suffused the organization and its Council for two decades. In this section—and leaving aside the differences which we have discussed previously—we want to concentrate on this, the mainspring of CERN.

7.1.3.1 The defense of the Council's autonomy vis-à-vis the member states: the cornerstone of the CERN system

To understand what we mean by the autonomy of the Council in its dealings with the states in the years 1950s and 1960s—and to show how this was defended by the hard core of the Council which we identified earlier in this chapter—the simplest thing to do is to describe the proceedings of a Council session. We have chosen that of December 1961 because it occurred when feelings on the level of the CERN budget were running high—and soon after the British government, wearying of the relentless growth in the organization's demands, decided to contact other governments directly in the hope that they might together take some joint action.[46]

During the Finance Committee meeting in October 1961 the Director-General obtained commitments to a budget level for 1962 of some 78-79 MSF from the representatives of all states bar Britain, Greece, and Spain. In an internal memo written on the 22nd, a British delegate, G. Hubbard, drew his superiors' attention to the growing isolation of the United Kingdom, who was almost alone in supporting a budget below 75 MSF. He attributed this to several factors, particularly 'the perennial British [economic] crisis'.[47] Three weeks later the Foreign Office sent its ambassadors in the member states a note asking them to approach the governments of the countries in which they were serving. Considering the well-nigh uncontrolled expansion of the budgets in previous years, and the risk that 'each successive year will bring further comparable increases', the British government suggested that the several governments get together and decide themselves, directly, 'what sum should be fixed as the firm upper limit' for the CERN budgets for the following three years. 'In the light of this', the note from the Foreign Office went on, 'a programme for the period and corresponding annual budgets should then be laid down by the CERN Council'. It added: '[...] it is essential that member Governments should make up their minds and that the Organization should realise that this total must be adhered to'. In short Her Majesty's Government was proposing to replace the 'synthetic' approach to establishing the budget by what Willson called the 'analytic' method in which the laboratory administration started from a fixed overall allocation. Four days before the Council met in December the Foreign Office informed its ambassadors that the United Kingdom would not vote for a budget for 1962 which exceeded 75 MSF.[48]

The Council dedicated all of the second part of its meeting on 19 December 1961 to the budgetary question. Just before getting to grips with the issue — and was this simply a question of chance timing? — Chairman Willems offered John Adams, the new British delegate, and Harold Leslie Verry, the United Kingdom's representative in the Finance Committee since 1953, the Council's warmest thanks for the work they had done for CERN. This gesture occasioned emotive replies from the two recipients, who had only learnt of the ceremony that same morning.[49] Willems then opened the debate on the budget asking 'whether the first great organization for scientific co-operation in Europe should now rest on its laurels', turning then to de Rose to elaborate on this question. In a long speech in which he referred to the historic meeting at UNESCO in December 1951, de Rose explained that 'the problem was more than a financial or even a scientific one — it was of a political nature'; that 'the Member States were at a cross-roads in the history of CERN and of scientific co-operation in Europe'; that the construction of CERN 'had taken place in an atmosphere of mutual confidence which he regarded as the main capital to be preserved', and that 'in order to preserve it, the delegations should display the greatest frankness'.

Willems then invited Bannier to say a few words (by the way, neither de Rose nor Bannier had official positions in the Council or the Finance Committee at the time!) Bannier began by saying how proud he was to have represented the Netherlands in the Council from the start, and by paying tribute to the memory of Director-General Cornelis Bakker. Following on de Rose who had sketched 'the broad lines of general policy', he proposed to treat 'more down to earth practical problems' — notably, the British aide-mémoire sent to governments in the member states via diplomatic channels. Recognizing that every government had the right to proceed in this fashion, Bannier explained that all the same the functioning of CERN was laid down in a Convention, a Convention that was translated into practice by a Council which represented the member states, and assisted by a group of outstanding scientific personalities who had the indispensable task of estimating CERN's scientific needs. For him, Bannier, and for his government, 'the CERN Council could not be

Notes: p. 412

by-passed for important decisions without demolishing the very foundations of CERN'. Describing the system of financing science used at the time in the Netherlands, Bannier went on to say that the idea of a fixed three-year ceiling decided in advance was a bad policy for an organization like CERN, since the unforeseen could not but play a crucial role in such a laboratory. Whereupon Willems thanked both speakers 'for their statements which dealt with principles essential to the life of CERN'—adding that 'the exact amount of the 1962 budget [was] a subsidiary question'.

The Danish delegate Obling then asked to speak. Personally aggrieved by the fact that his British colleagues had attempted to undermine his position by trying to sway the Danish government behind his back—and considering the British move as showing disregard for European ideals and as an attack on the model that CERN was for European collaboration[50]—, he strongly supported de Rose's and Bannier's proposals. 'CERN Council was the sovereign body for taking all decisions relating to the Organization', he repeated, and 'Diplomatic steps were likely to complicate problems and to make decisions even more difficult to take'. After further strong support, this time from the Italian delegate Ippolito, Chairman Willems proposed a solution: that a working group be established to study the question of CERN's long-term budgets, a group whose chairman, Willems suggested, should be...Bannier.

Sir Harry Melville took this opportunity to say a few words...and came out in favour of the Chairman's proposal—as did the other representatives in turn. As for Bannier, he suggested that 'his own participation [...] in the work of the Organization from the outset might arouse doubts as to his objectivity [as chairman]', a suggestion which Hocker quickly dismissed: 'there was no one better qualified than Mr Bannier', said the German delegate, to which Willems added, not without a touch of irony, that 'Mr Bannier loved CERN, but not blindly'. At this point in the proceedings the working group was established and the 1962 budget was voted at 78 MSF, Greece and Yugoslavia abstaining, Sweden and the United Kingdom voting against. Funke, who was the only delegate to explain his stance, made it clear that the main consideration informing his government's position was the wish not to outvote the largest contributor to the budget.

This account calls for little comment. It clearly reveals the jealously guarded autonomy of the Council and the functioning of the pro-CERN lobby at its heart, it shows that the Council regarded itself as the only possible decision-making nucleus, the nucleus which history itself had established and on which it had bestowed legitimacy, and it makes it clear that the Council members were apparently sufficiently powerful when dealing with their domestic authorities to neutralize any initiative aimed at bypassing them. In fact the British aide-mémoire seems to have elicited few official replies, and those that we have seen were negative, and perfectly in line with the speeches made by de Rose and Bannier. This debate also shows the weight of those men whom we identified earlier in this chapter and whom we have called the historical core of the Council or the pro-CERN lobby: launched by Bannier, de Rose, and Willems, the 'counter-attack' was carefully orchestrated so as not to involve formally those who, while sharing their sentiments, found it difficult to come out openly in defence of CERN. This was the case with Hocker, the Chairman of the Finance Committee—but of an FC which took no official stance because of the United Kingdom's position—, and with Funke who did no more than pay lip-service to an official position which differed significantly from his personal views.[51]

The determination of the Council to remain autonomous with respect to the states did not only manifest itself at times of major crisis. It was also present in the more routine defence of the 'flexibility' which the Convention bestowed on the Council. To illustrate this point we would like to stop a while on the *voting procedures,* notably those in force for the adoption of the budget.

According to the Convention of 1 July 1953 the Council had wide powers. It had every right, as we have seen, to expand the organization's programme of activities, or to suppress certain parts of it. In order to avoid institutional blockages and the possibilities of blackmail opened by a right to veto, the most common voting rule was simple majority, each member state having one vote. This, one should note, was also true for the budget vote.[52]

Major contributors to the budget tried to change this rule on two occasions during our period. In 1956 it was France who proposed an amendment to the Convention 'whereby the annual budget would have to be approved by a two-thirds majority [...], that should consist of States paying at least three-quarters of the total contributions to the CERN budget'. After a discussion in the Finance Committee and the Council, the French delegates however withdrew their note without a vote being taken. The reason they gave was that they were satisfied by the solemn undertakings made by President Lockspeiser who declared that 'though he could not commit his successor he could assure Council that, as long as he remained President, he would consider it most unwise to ask the Council to vote the budget, unless it secured the *quasi-unanimous approval* of the Member States' (our italics).[53]

Four years later, in December 1960, the budget was voted by eight states together representing 47,7% of the contributions, France, Greece, Sweden, the United Kingdom, and Yugoslavia voting against for different reasons. In a draft of a letter to Willems the British authorities thought of asking for an explanation and of getting the new Chairman of Council to undertake formally to respect the rule spelt out by Sir Ben four years earlier—a rule that had been rather liberally interpreted during de Rose's reign.[54] However, on 8 March, before this letter had been sent, the Italian delegate Toffolo submitted an amendment to the financial protocol. It asked that there be a two-thirds majority on the budget vote (but without financial conditions attached), and it was discussed preliminarily by the Council on 2 June 1961. Those who defended the existing system and who were, in fact,

Notes: p. 412

almost the only ones to speak—namely, Bannier, Willems, and Dakin, CERN's Director of Administration—, gave two kinds of argument against the Italian proposal. The first was of a judicial nature: they suggested that such a change would require a modification to the Convention, and not simply an amendment to the financial protocol, and so would have to be discussed in the various parliaments. This ghastly prospect (as it seemed) weighed heavily on the delegates who were not at all keen to have the issue brought up before 13 or 14 national assemblies. The second, more subtle, argument, showed that such a reform touched the mainsprings of the functioning of the Council itself. In essence it objected that the amendment entailed that any group of five states—and five meant *two or three* because of the chronic difficulties Greece and Yugoslavia had in meeting their financial obligations—would have a power of veto, so making 'the possibility of a compromise over the CERN budget still less likely than it had been in the past'. The logic of this argument was that a right to veto would make it possible for a few states who systematically wanted to reduce the CERN budget to be totally inflexible. (Bannier and Willems were thinking of Sweden and the United Kingdom without any shadow of a doubt). What is more, since a few states would now have the power to control the budget level, one could imagine their governments or Finance Ministers playing an increasingly important role, and even *imposing* non-negotiable positions on their delegates—so short-circuiting the Council and rendering it impotent since nothing important could be discussed in that forum any longer.

In the light of these considerations Toffolo explained 'that the Italian delegation had not made its proposal with a view to hampering the progress of CERN' and, at the following session, he withdrew the amendment. The Council contented itself with re-affirming, through President Willems, that it hoped to secure 'the greatest possible degree of agreement in any Council decision on the budget of the Organization'.[55]

The meaning of this notion of *greatest possible degree of agreement* (Willems) or *quasi-unanimity* (Lockspeiser) was not essentially arithmetic: it was not to be taken as saying that ten or eleven out of thirteen votes were needed. It simply meant that the core of the Council recognized that the majority was ready, *depending on the circumstances,* to consider the demands of minorities. By 'circumstances' one is to understand the desire never to put the 'good' atmosphere prevailing in the organization at risk, this climate being the *sine qua non* for getting the maximum from the states without pushing any of them so far that they seriously considered leaving. Depending on how the active core measured the risks involved in each vote, it behaved in a more or less conciliatory way vis-à-vis the minority. Hence the possibility that a dozen states would rally at a given moment to a lower budget proposed by only one of their number—as happened in December 1963, for example; or of a minority, calculated in terms of contributions, outvoting a majority—as was the case in December 1960. The interest of the system—pass the vote by a simple majority but aim for a consensus—lay then in the flexibility of functioning which it gave, in the possibilities for negotiation which it allowed in each case—it lay in the real power to take decisions which it bestowed on the historic core of the Council.[56]

7.1.3.2 The functioning as lobby

Certain moments are more crucial than others in the life of an organization like CERN. The debate in December 1961 on the British aide-mémoire was one perfect example of this, those in June 1963 and June 1964 on the accelerators to be built in the future were others. A notable feature is that, in these somewhat unusual situations, the dissent between members of the historic core of the Council which we described in section 1.2, tends to be completely watered down. In such moments the dozen or so people who constitute the pro-CERN lobby act as one man, in a coordinated fashion, and the other members of the Council find themselves in a situation in which it is very difficult to express substantial disagreement.

As the discussion in 1961 has already illustrated this point, we propose here only to describe, in a synthetic fashion, how major debates are conducted in the Council at these crucial moments.[57] Firstly, the main contributions to be made are always decided *before* the meeting. If the subject is of a political or institutional kind, as in December 1961, it is often Bannier, Willems, de Rose, Hocker or Funke who speak; if the subject is scientific or technical, it is more often Amaldi, Powell, Perrin, Møller or Leprince-Ringuet who have a say. In all cases the CERN Directorate, and especially Bakker and Weisskopf, come up in support. The second point to note is that the interventions made by the CERN lobby generally always go in threes—which ensures a good division of labour. The first speaker presents the problem which has to be debated or voted on. In June 1963, for example, during the first debate on the report presented by ECFA, it was Amaldi, its Chairman, who introduced the document. The second speaker usually speaks on matters of principle, so enlarging the scope of the discussion. De Rose did this in 1961, Powell in 1963. In the name of the SPC he gave a lyrical and enthusiastic speech which was nothing less than an ode to fundamental science and the growth of knowledge—concluding his panoramic overview of the recent evolution of high-energy physics by stressing all the benefits that human culture would derive by supporting CERN. The third speaker always has a more down-to-earth task, which is to reply to possible objections and to give detailed arguments. Bannier did this in 1961, Perrin in 1963. Speaking of money, the latter recognized the enormous financial effort called for, but invited the delegates to remember that they 'not only represent their government [at the Council] but also advise it at home about the decisions which have to be taken, on which CERN's future depends'.[58]

A third trait of these important debates is that other delegates very seldom contribute to the discussion. Generally they only express their views just before a vote is to be taken, and then those in favour usually speak first, followed by the others. This 'rule' is only dropped when the hard core of the Council thinks that it would not be opportune to outvote the minority. In this case, it tends to postpone the vote or, if that is impossible, to propose an acceptable solution which is then adopted unanimously, or almost.

There was one person whose behaviour did not regularly conform to this pattern after 1961. This was John Adams, former Director-General, and then serving as British delegate. An undoubted expert on CERN matters but, from 1961 onwards, owing his first allegiance to a government which was often unhappy about the level of the budget asked for by the organization, he was the only one who openly expressed some doubts about the 'imperative' needs of CERN or of high-energy physicists. Confident of his own position, knowing the ins and outs of the establishment, he was not hostile to CERN; but this did not stop him reminding the Council, often quite bluntly, that particle physics was not the only science in the world and that competition with the United States was not the only thing that mattered. Typically we find him in June 1963 remarking that 'From the scientific and technical point of view, there was no limit to the expenditure which could be carried out in the field of high-energy physics' and that 'the only limit which could be set was the share of the over-all expenditure on science to be devoted to high-energy physics as compared to that allocated to other scientific activities'.[59]

To conclude our quick survey of the methods and activities of the historic core of the Council, we must still speak of its role outside official meetings. Its members, who corresponded regularly with one another and with the Director-General, also took care of the organization's business in other contexts. This happened when some of them went to Vienna for meetings of the International Atomic Energy Agency, or when they participated in meetings of NATO (as did Bannier, de Rose, or Willems). They also served as contacts for finding possible money for use by CERN—e.g. with EURATOM in 1958—or by other high-energy physicists—e.g. a request made to NATO in 1963.[60]

Finally, in consultation with the organization, they also played a role in CERN's public relations. In their home countries they took it upon themselves to identify key decision-makers (Ministers, MPs), to feed them with information, and to arrange for them to visit the facility regularly. This role was particularly important when it was thought that senior administrators in a country had 'forgotten' CERN, or when a new political group took office. There are so many examples of this that it is hardly worth mentioning any in particular. The interested reader need merely page through the CERN Courier, the monthly journal first published by the organization in 1960, to get an idea of the members of royalty and other high state officials who visited CERN during our period.[61]

7.1.4 A QUESTION IN CONCLUSION: WAS CERN A POLITICAL SUCCESS?

A theme constantly found in talks on CERN is that the laboratory is a success, a success in that it is one of the rare cases in which European states have managed to pool their resources without some kind of impediment coming to suffocate the institution more or less immediately. Of course CERN has not been entirely free from pressure by the member states, but is has functioned without suffering any major setbacks, according to rules which have not been modified substantially, and to the overriding benefit of the scientists whom it serves.

There is no doubt that there is a political dimension to CERN's success thus characterized: it lies in the nature and composition of the Council, in the ways in which its members put into practice the institutional procedures which regulated the organism. Not constituted from above, by a juxtaposition of delegates nominated administratively by the states, we would say, the central core of the Council built itself up in the course of a battle which lasted several years. From these beginnings an unusual degree of cohesion was born, along with a determination to succeed, which bore fruit thanks to the favourable context in which it emerged, the context of a Europe looking for ways to unite, and fascinated by all that was nuclear.

This first consideration, as crucial as it may be, is perhaps not sufficient to account for CERN's 'political success' in the long term. For that, one has to accept that the great majority of states 'chose' to leave the organization to look after its own affairs for more than 15 years, without seriously intervening. And while there can be little doubt that this was so—the only notable exception being the United Kingdom—, this should not stop us trying to understand their attitude. Two things are worth mentioning in this connection. First, that states do not always have the same reasons, the same motives, for intervening and *controlling* more directly the activities of the bodies which they finance. A comparison with the European Atomic Energy Community (EURATOM) set up in 1957 will help to clarify what we mean.[62] In the first place, EURATOM was an integral part of the European Economic Community and was a child of political decisions at the highest level. CERN, by contrast, was less a response to political strategy than to the needs expressed and defended by European nuclear physics leaders. Another difference is that EURATOM, which worked in the field of nuclear energy was closely allied to industrial activity. CERN, by contrast, was not concerned with the direct application of its work, and its scientific results did not directly satisfy interests which states regarded as being of military or industrial importance. From whence centrifugal forces were reduced, each government not being primarily concerned to extract from the joint venture the maximum commercial spin-off possible. Thus the *kind of research* done at CERN (basic, but requiring heavy equipment) made for a certain autonomy vis-à-vis the states, for a more 'remote' control of its activities on their part, for the fact that CERN was rarely at the centre of concerns which called for 'top-down' decisions from the highest political strata.

Notes: p. 412

Our second remark is derived from the first: it is that this possibility of autonomy (jealousy guarded by the historic core of the Council) can only grow if CERN appears to be working properly, and not to be wracked by internal problems, if CERN seems to be efficient—in a word, if it is described as successful in political circles, notably by scientists. Once the leaders of European nuclear physics realized, early in the 1960s, that there was no alternative to building a central installation for the ever-heavier equipment their science demanded, and once they accepted that this installation would be CERN, they no longer submitted their national authorities to conflicting pressures, to pressures which, by working in opposite directions, gave the latter real opportunities for choice. Speaking with a single voice to the exterior, they ensured the continued good health and the success of the organization and left the member states watching developments from afar.

7.2 The CERN executive: an ensemble strongly influenced by the scientific senior staff

According to document CERN/100 dated 29 September 1954, which fixed the organigramme of the newly established organization, CERN was composed of six divisions. Two of them, the most important in the short to medium term, were to build the machines. These were the PS division, whose job it was to construct the 28 GeV proton synchrotron, and the SC division, responsible for the 600 MeV synchrocyclotron. The other divisions were Theory (TH), Administration (Adm), Site and Buildings (SB), which was to ensure the installation of a laboratory where before there had been little more than rolling farmland, and Scientific and Technical Services (STS), the division whose task it was to build experimental equipment. This structure was the outcome of two parallel movements: the gradual historical sedimentation of the central core of actors over the years 1951/53, on the one hand, and the more conscious attempts made at rationalization, in which the British played a notable role in 1953/54, on the other.[63]

7.2.1 THE FIRST ORGANIGRAMME, 1955-1957

The PS and SC divisions were the direct inheritors of two of the three working groups defined in 1951 by 'Auger's consultants'—namely, 'the planning group for a big machine', 'the planning group for an intermediate machine', and the 'group for the organization of the laboratory'. The first group, which became the PS division, was directed in 1954 by John Adams, a British accelerator specialist who succeeded Frank Goward when he died unexpectedly on 10 March 1953 aged 34 years. From its inception this division was very autonomous, being installed at the Institute of Physics of the University of Geneva, while the rest of CERN was gathered in barracks near Cointrin airport. Given the task of designing and building the more ambitious machine and, to be honest, the machine on which CERN's future rested, the PS division was well aware of its importance to the organization as a whole. It was already the largest group in terms of personnel, having 118 members in 1955. As for the SC division, responsible for the intermediary machine, it had 32 members at around this time. Early in 1955 it lost its original leader, Cornelis Bakker, who had managed the group since 1951; he was nominated to replace Felix Bloch as Director-General, who resigned after only a few months in office. On 10 June 1955 the Council appointed Wolfgang Gentner as head of SC, a German physicist from Fribourg.[64]

A Theory Division was envisaged for CERN rather early in the process of creation of the organization. It was set up to satisfy the wish expressed by European physicists that CERN should be, from the start, a *scientific* organization and not simply an engineering centre building two machines. Placed under Bohr in Copenhagen, it was resolved in 1953 that the division would have to move to Geneva after 3 years if the Council so decided. In September 1954, Møller succeeded Bohr and a small group was also formalized at CERN under the Director-General. Two years later, at its session in December 1956, the Council agreed to close down the division in Copenhagen and to transfer it to Geneva on 1 September 1957. This was done and Ferretti, who had led the Geneva group, was nominated director of the new division.

As for the three other divisions, they were built from the Laboratory Group which had been led by Lew Kowarski, a French nuclear physicist, ever since the meetings organized by Auger in 1951. Initially asked to plan and instal the various services which the new laboratory needed, Kowarski's group was cut in three. STS remained in his hands and, by 1955, it was wholly dedicated to preparing experimental equipment. The second 'slice' became the Administration division, and was put under Sam Dakin, a British civil servant from the Board of Trade. Its tasks covered general administration, staff policy and management, financial management, and contract policy. Finally there was the SB division, given to Peter Preiswerk, who had been Kowarski's right-hand man since 1952. Preiswerk was a nuclear physicist from Zurich and was, until this appointment, the only member of the initial nucleus assembled by Auger who had no directing post in the organization. The fact that this post was not given to a specialist in building reflected the conviction, widespread in the community, that scientists were capable of managing all of a laboratory's activities; it reveals, too, that the rights of those who had taken the first steps were clearly recognized; already symbiotically linked to the political representatives in the Council, they constituted a bloc from the time of the meetings held in December 1951 and February 1952, and were 'naturally', and without any real debate, appointed to key posts in CERN.[65]

The only newcomers to CERN in 1955 were Wolfgang Gentner and Jean Richemond, a senior French civil servant close to General de Gaulle, who replaced Dakin when he was called back to his home station. These appointments differed from the others we have mentioned in that they were made only after vigorous debate and on criteria of *political* equilibrium. The leadership of SC was in fact given to a German in response to complaints made by the FRG delegates that their country had never been represented in the top posts in the organization; similarly Dakin's post was given to a Frenchman partly in compensation for the loss of influence of France in CERN and in the Council—and in recognition of the debt the organization felt it owed France for the role the country had played in setting up the laboratory.[66]

This executive remained in place for two years, i.e. until the synchrocyclotron became operational.

7.2.2 CORNELIS BAKKER'S TWO ATTEMPTS AT REORGANIZATION, 1957–1960

In April 1957 Sir John Cockcroft asked the Council Chairman to make a critical survey of the functioning of CERN's divisions. The reason given for this was that, statutorily speaking, certain directors would soon have to be reappointed. In fact what Cockcroft wanted was a wide debate on the future of Kowarski's division, a division whose terms of reference had always created concern in the British camp.[67]

This proposal led the Group of Directors to initiate a far more wide-ranging debate on the future organizational structure of CERN, which 'should be in operation by the time of entry into service of the Proton Synchrotron' some two-and-a-half years later. At this time, in fact, the essential activity in Geneva would have to pass from a phase of *construction* to that of *exploitation* of the machines, to a phase characterized by the doing of research.[68]

7.2.2.1 The first proposals for reform in 1957: a forced withdrawal

In their first scheme dated October 1957, the Directors proposed that the divisions responsible for construction (PS and SC) and that concerned with preparing experimental equipment (STS) be replaced by two divisions. The first would bring together the research physicists; the second, called 'Applied Physics and Engineering', would be concerned with the completion and the daily exploitation of the two machines, with general technical development—studying new acceleration principles, designing new types of separators, etc.—and with the bringing into service of the sophisticated apparatus needed for experimentation. Alongside these two divisions those of Theory and of Administration were to be retained, and a division handling technical services was to be set up to deal with the site and its buildings, the transport network, security, and the main workshop. As far as Bakker was concerned this was only 'an outline of the desirable organizational structure', and he thought that the two big divisions, that of research and that of applied physics, could later be broken down into smaller units. On 29 October the SPC debated these ideas and recommended them as 'a basis for the future structure of CERN'. The committee asked, however, that the relations between, and the precise roles of, the two new divisions in the execution of experimental work be studied more deeply.[69]

One reason stands out among the Directors' motivations for the split between a research and an applied physics division. Most often mentioned by Adams, it was the growing concern felt by the Director of the all-powerful PS division to keep 'the staff of the PS Division together at least long enough to finish the machine', and it reflected the difficulty he was having 'to persuade them firstly that their type of work will continue in CERN after the machine is finished and secondly that CERN intends to employ them after 1960'. For him, the acceptance by the SPC and the Council of a reorganized CERN which included a *division of applied physics* 'should reassure them from the point of view of the continuity of the work'. This was not a new argument, and had already led Adams, some months earlier, to ask for a reinforcement of the subgroup working on accelerator research, to ask that some of his senior people be granted indefinite contracts, and to demand that CERN commit itself to additional building projects after 1960. It was also to lead him, a few months later, to support the construction of a very big hydrogen bubble chamber by CERN.[70]

In December 1957 and January 1958, the Directors and some members of the SPC met together to define more carefully the relationships between the two new divisions. In March 1958 their proposals were accepted by the entire SPC, only Leprince-Ringuet asking for more time to reflect on the scheme (he was disturbed to see that there was no place reserved in it for visitors). In the new version of the project it was clearly brought out that the work would remain organized by teams managed by group leaders belonging to the senior staff, with each team preparing an experiment or working on a particular technical development project. To clarify a controversial point the text added that '*smaller* research apparatus would usually be developed by the Experimental Physics group, but [that] some of the *bigger* items [...] would be taken over by the Applied Physics groups'. These latter would also remain in charge of the development of common heavy equipment (like separators) and for running specialized workshops.[71]

While accepted by the SPC, this arrangement was not to remain unchallenged, however. Unhappy at being left out of the picture, a part of the senior scientific personnel started grumbling in December 1957. In January 1958 the Director-General, in response to this turn of events, proposed that all people with grade 12 or higher—so some 30 staff members, the grades stretching up to 14—meet with him '[to further] direct contact between the General Management and the Senior Staff'. This was the first time that such a meeting was envisaged; it was held on 21 April 1958 with budgetary problems and the new organizational structure proposed by the Directors on its agenda.[72]

The meeting was noteworthy for the categorical refusal by the senior staff to accept any difference in principle between 'physicists', 'applied physicists', and 'engineers'. A week later six of the most influential people in this group wrote a long

Notes: p. 412

critical text developing this idea. They were Citron and Von Dardel, specialists in electronic detection and at the head of two ambitious projects, Goldschmidt-Clermont, Kowarski's deputy, Hine, Adams' right-hand man, Peyrou, leader in bubble chambers, and Schoch who was responsible for the new Accelerator Research Group. Their starting point was that the basic unit at CERN was 'the physics sub-group' and not the division. 'It is through the activity of the sub-group that actual work is done', they insisted, be that 'in the field of machine development, instrumentation, or nuclear physics'. Accordingly they felt that any organizational structure should give prime weight to the stability of these groups. They went on to say that it would be dangerous for CERN's future to limit these groups to one kind of activity for all time: research always called for shifts from phases dominated by the construction of equipment to phases in which it was used experimentally. They stressed finally that each accelerator imposed a specific type of practice, and that this limited the possibility of switching from one to another. In consequence, if two divisions had to be created, they ought not to separate 'physicists' from 'applied physicists' and 'engineers', but to separate people working around the PS from those working around the SC.[73]

Giving more detail, the authors added that two committees could be set up to bring the leaders of the groups together, a 'Nuclear Physics Committee', charged with thinking globally about research, and an 'Applied Physics Committee', whose task it would be to co-ordinate the equipment programmes. These committees, they said, need only step in to coordinate the programmes, the divisional directors of the 'PS area' and the 'SC area' being responsible for their execution, for financial management, for the career structure of their staff, and so on. In brief, and to summarize, the senior personnel wanted all the scientific and technical activities to remain focussed around each machine, and wanted decisions to be taken by themselves working together in two ad hoc committees. Depending on what they were doing at a given moment, the group leaders would participate in the committee planning experimental research or in that dedicated to the construction of equipment.[74]

In May the Group of Directors took note of this opposition, but refused to align themselves with it. Without explaining their attitude in writing, but apparently somewhat aggrieved, they decided purely and simply to withdraw the question of the future organizational structure from the agenda of the next Council session. For the time being, and until the PS came into service, nothing would be changed. In the short term the experimental physicists in Geneva would remain together in the SC division. On 23 May 1958 the Director-General informed the SPC and the Council of his new decision.[75]

Several questions are raised by this first debate, questions about the positions adopted and the motives at work, as well as about its development and its consequences. Let us begin by looking at the positions adopted.

The direction's project—in fact, and more precisely, the proposal made by Bakker, Gentner, Kowarski, and Bernardini, the Director of Research[76]—rested on a particular conception of what experimental work was and how functions should be distributed in a place like CERN. On the one hand, CERN had to construct machines and to get them to work for European physicists. This was the task of builders, engineers and applied physicists, the business of a pool of experts who had to be preserved for the future of Europe. On the other hand, it had to be a workplace for experimentalists exploiting this equipment. Drawing implicitly on the typical features of the currently most widespread form of experimental work, electronic experiments, and on the way of dealing with the smallest machines then operating, the idea here was that of groups of physicists working separately, and more or less autonomously. Each group would use its habitual mode of detection, would construct its equipment itself—which would still be very light—and would do its experiment.

The counter-project articulated by the senior staff—and dominated in fact by Adams' men and a few physicists who shared their views for various reasons[77]—, was based on the idea that work around the PS involved a *change in scale*. Granted the complexity of the machine, physicists could not simply turn up with their equipment and start working: a narrow link with the masters of the PS would be needed both for the installation of beams or protecting walls—hundreds of tons of material—and for the building of experimental equipment whose size would increase proportionately. Hence their idea that the unity of the PS area should be preserved in a division which brought together classical experimentalists and people who were more technically oriented.

The difference between these two projects, however, did not merely lie in a difference of opinion about what experimenting with the PS was, or would be. It was also rooted in the balance of forces inside CERN. Firstly, there were personal ambitions: several scientists in Adams' division thought that they would like to do physics with the big new machine themselves. Taking Alvarez at Berkeley as their model, they felt that as builders they were entitled to certain privileges when it came to using the equipment they had helped install.[78] Then there were the claims to power: the senior staff saw themselves as the real driving force of CERN, and believed that any organizational structure should recognize the centrality of this fact. Hence their insistence that the groups which they were directing in Adams' division or around the SC were *the basic units* at CERN; hence their rejection of divisional separations which were 'too clean'—the new PS 'division' which they envisaged was to collect together some four-fifths of the senior staff, which amounted to saying that it was barely a division at all; hence, again, their demand for a very broad ensemble, managed by the two committees they had in mind, committees which gathered the senior staff and which decided collectively on the distribution of labour.[79]

What are we to make of the evolution of the debate, and in particular of the directorate's stopping the process in its tracks in May 1958? One might imagine that

Notes: pp. 412 ff.

they did this simply because they did not think that it was urgent to take a decision since the PS would only become operational in 1960. The senior staff had put their finger on a real problem, and the management preferred to preserve for the time being a structure which permitted them to complete the PS and to experiment with the SC. However, while not false, this hypothesis does not go far enough: obviously the management did not want—nor was it able—to act against the wishes of the majority of the senior staff ranged against it. In this sense, the non-decision of 1958 is above all to be interpreted as the recognition by the directorate of a banal fact in every organization—that, *de facto,* the senior staff has a kind of veto right over all decisions concerning how the house is to be managed.

It only remains to discuss the effects of this non-decision. Since no division was clearly put in charge of preparing the PS experimental programme—this would have meant combining potential users and builders, at least initially—, an *ad hoc* Committee cutting across divisions was set up by the directorate in summer 1958. It was baptized the *Executive Committee for the PS Experimental Programme* and was presided over by the Director-General. Consisting essentially of senior scientific staff drawn from the PS, SC, and STS divisions, this committee met very regularly until February 1960, so some months after the PS was commissioned. It considered various problems concerning the future use of the big machine, it circulated various texts, and it studied a number of projects—yet all the same in 1960 a general lack of preparation for using the PS was evident. This lack was both in regard to experimental equipment (of which there was virtually none ready) and to infrastructural material needed in the experimental zones (magnets, quadrupole lenses, power supplies, often ordered but not always delivered, and any way too few in number).

To understand this paradox one must first remember the intrinsic complexity of the task, and the radical unknown that was work with such a powerful synchrotron in the absence of any tradition. One must also realize, however, how heavy organizational structures are. Confronted with long-established divisions, each having its own budget, its system of values and above all its priorities, a committee like the Executive Committee operated more like a social club than like a homogenous body working towards a definite goal. Without independent financial means, without real power over people, the priorities it defined had little chance of being taken seriously when they were put against the distinct goals each division set itself. Adams and the PS staff, for example, probably would have liked to prepare the best PS experimental equipment possible, but their first job, the thing they had to do before all else, was not that: it was to build, as quickly as possible, the best accelerator in the world. Physics, after all, was physicists' business. Similarly, the physicists were primarily thinking about the experiments they wanted to do with the new synchro-cyclotron. This work was fascinating and was proving more exciting than originally anticipated—we would even say that this good fortune was a disaster for the preparation of the PS experimental programme—, and they were not keen to turn aside from it and to think about what an as yet unfinished PS would need. All the more so since the machine was not under their control and was located in a quite different sector on the site. Without a division endowed with the necessary resources and having this as its primary task, it is hardly surprising that the Executive Committee remained largely ineffectual.

7.2.2.2 Bakker's second attempt, June 1959 - April 1960:
A solution that was never put into practice

In June 1959, so some six months before it was expected to bring the PS into operation, the CERN direction again opened the debate on the future organizational structure. Two texts dated 10 (Adams) and 24 (Dakin) drew together their first ideas.[80]

Instead of the six existing divisions and the five foreseen in 1957/58, a *dozen* were now envisaged. The Theory, Administration, and Site and Buildings divisions remained intact, as before; this time Kowarski's STS division was also left untouched. By contrast the SC and PS divisions were split up. The former was broken down into one division whose task it was to operate the SC, and into two research divisions. The latter, the existing PS division, was fragmented into four, one for PS machine operation and improvement, one 'to undertake new engineering work in the PS area', one whose task was the construction and use of bubble chambers, and a fourth 'to investigate new ways of accelerating particles and to form a pool of knowledge of accelerator theory'. New names were given to these divisions. Apart from the specific divisions (PS and SC operation, PS engineering, SB and Administration), all the others were given names which carefully avoided making reference to engineers or applied physicists: they were called Physics I, Physics II, etc.[81]

The logic of these new proposals seems rather simple: it was to avoid making a choice by shifting the formulation of the problem as it had been posed the year before, it was to put forward a solution which would not rekindle the splits of 1958, and which smoothed over personal susceptibilities concerning the allocation of senior posts. Distributed between the six 'physics divisions', the research teams were formally placed on an equal footing in the heart of the organization. And, as the senior staff had suggested in 1958, their leaders would come together in a committee to plan research and in two committees to organize the proper functioning and development of the PS and the SC.

On 24 June Dakin raised a question which he thought crucial—and which he believed would also occur to the experienced members of the Council: 'if the new [division] leaders are going to have substantially the same duties [...] as the present Division directors', he wrote, 'the Director-General will be unable to handle all the business which will result [...]'. He went on: 'There is a principle, pretty well established and proved by practice, that the largest number of responsible lieutenants

Notes: p. 413

which any one man can handle is about 5'. Dakin also remarked that the three interdivisional committees or their chairmen would not solve the problem because they could not 'remove from the DG any of the burden of decision, since they [would] have no [hierarchical] authority'. One ought then to think of building into the organigramme one or two posts of deputy to the DG, whose job it would be to take care of certain functions or certain divisions, or of transforming the directors of the most important divisions into super-directors who would be responsible for a group of divisions.[82]

Not wishing to let the delicate issues which had been carefully sidestepped come in again through the back door (by recreating major splits of the kind 'research' and 'applied physics' or 'PS area' and 'SC area') the Group of Directors chose to ignore Dakin's proposals: no formal structure would be created between the DG and the divisional heads. They did, however, confirm the setting-up of a 'group leaders committee' to coordinate the research programme of CERN and 'two housekeeping committees', one dealing with the PS and the other with the smaller machine. The chairmen of these two committees, chosen from among the divisional directors, would regulate 'in the first instance' issues which might arise *between* divisions, each director being nevertheless, and solely, responsible to the Director-General.[83]

During the summer this scheme became the official proposal of the Director-General and was submitted to the SPC in October 1959. The SPC accepted the new layout, as it had done with the previous one at the beginning of 1958; it did however raise one question which it thought important—exactly that of the intermediary hierarchical level which it thought indispensable if one had ten or twelve divisions. In its view the housekeeping committees would not be sufficient for this purpose.[84]

It was only at the end of November 1959 and in January 1960 that the idea was taken up seriously again. As in 1958, it was once more the senior staff who took the initiative. This time, however, it did not speak as one man. The majority defended a position similar to that which it had articulated 18 months before. Deriving new inspiration from the poor preparation of the experimental work around the PS, it insisted that a central *authority* was essential in the *PS zone,* an authority able to impose its will on people who would be working there, and who would be competing fiercely for money, for machine time, for workshop facilities, etc. Those who held this view believed that a super-director was indispensable in the PS area; he would prepare the budgets with the divisional directors, and he would be responsible for the technical divisions as well as for those doing research.

The second tendency, which emerged around mid-December 1959, seems to have been supported by fewer people than was the first. It held that if an intermediary was to be nominated he ought to be a director put in charge of *research,* 'with the standing of a deputy DG'. To give an idea of the kind of person they had in mind, Dakin wrote to his friend Verry that they were thinking of a theoretician, someone like Salam, Levy, or Ferretti. During a meeting arranged by Dakin at the end

December/early January 1960, it appeared that the scientists coming from Adams' division preferred to have a PS area director, while the physicists attached more importance to a research director who was deputy to the DG—which is hardly surprising. What is perhaps a little more surprising, on the other hand, is that almost no-one suggested combining the two ideas.[85]

In response to this situation, and under the impulse of the physicists in the Leading Board—Bernardini, Gentner, Kowarski, Preiswerk—, the direction put forward a compromise: that of creating *two assistants* to the Director-General, initially called 'coordinators', one for coordinating 'heavy' activity around the PS, the other for organizing research and dealing with visitors. These two posts were characterized both in terms of their *functions* and in reference to specific *divisions,* each coordinator 'taking charge' of a clearly identified subset of the latter. This arrangement raised doubts about the exact power relations between the Director-General, the two coordinators, and the divisional directors. It was one of the points which the SPC clearly put its finger on when it met on 19 January 1960.[86]

During this crucial meeting the SPC tried to go further than the direction and to force it to draft texts which were more explicit. Firstly, and against the proposals made by some German physicists, it decided that the coordinators could not be *elected* by the divisional directors or the senior staff (as would be the dean of a university department). Of course the Director-General 'would ascertain' that they 'would be acceptable to the Divisional Directors', but the DG would *propose* and the Council would *appoint*. This was obviously the only possible procedure if one wanted the coordinators to have real power. Similarly, the SPC agreed 'that the line of authority should go clearly *through them* to the Divisions for which they were responsible', and that Bakker's text ought to be rewritten 'to remove any ambiguity'. The irony of all this is that, after a long detour, the debate arrived at a position very close to that which the Group of Directors had proposed 20 months earlier: beneath the Director-General there would be two large subgroups, each directed by a kind of deputy DG, the one for applied physics (now limited to the PS area, it is true), the other for research (and responsible for the two machines). The main difference was that the budgets, and their distribution, were now in the hands of the dozen divisional directors, and not in those of the two 'deputy-directors'.[87]

On 6 February 1960 a special Council session studied the proposals made by the directorate and reviewed by the SPC. After discussing the respective areas of competence of the divisional heads and the coordinators, as well as the exact grade of the latter, the Council approved in principle the proposed reorganization, and invited the Director-General to call for applications for the posts of coordinators. It expressed one condition—that the Finance Committee consider the financial implications of the new system which multiplied the number of independent budgets inside the laboratory.[88]

On 10 March the Finance Committee stressed the budgetary inflexibility introduced by having so many divisions. Without opposing the project, it suggested

Notes: p. 413

several minor modifications which the Group of Directors tried to take into account in the weeks that followed. The latter also went through the different candidates for the posts of coordinators and named John Adams and Giuseppe Cocconi as interim coordinators for applied physics around the PS and for research, respectively. The Group also suppressed the mechanism for coordination which had been in operation for two years and which was now redundant, the Executive Committee for the PS Experimental Programme.[89]

One month later, on 23 April 1960, the Director-General was killed in an aircraft accident during a trip to the United States. The new interim Director-General immediately put forward his ideas for reorganization.

7.2.3 JOHN ADAMS' SOLUTION, MAY 1960–JULY 1961

Within a few days of Bakker's death Council Chairman de Rose proposed that John Adams be CERN's interim Director-General. One of the first things Adams did was to reconsider the organigramme elaborated during Bakker's term of office. On 20 May he submitted his own proposals to the SPC and to the Committee of Council.[90]

7.2.3.1 *The idea of having a Directorate*

The partition of CERN into twelve divisions was kept by Adams, even if he adapted the boundaries between them so as to disturb the existing structures in the PS and SC divisions as little as possible. He also recognized the need for an intermediary component between the DG and his twelve divisional directors, but refused to create a new hierarchical level. Adams thus insisted that the directors would only be responsible to the Director-General, though the latter would be assisted by a *Directorate* of four or five people. As in the case of the staff surrounding the Supreme Commander in the Army Headquarters, each member of the Directorate would not be a division leader acting directly on the battle field; his tasks would be to reflect, to anticipate, to be sure that everything went well in the sector under his supervision—research, applied physics, SB or Administration—but without having any hierarchical authority over the directors of the relevant divisions. Not taking responsibility for any division, said Adams on 27 May, the members of the Directorate 'would thus have a greater freedom of mind to examine questions of general interest to CERN'; and if they were to coordinate the work of several divisions, that task would always be specific and would be given to them on an *ad hoc* basis by the DG.[91]

On 27 May the SPC and the CC, meeting jointly, approved these new arrangements notwithstanding the risk of fragmentation of authority which, in their view, they embodied. With this in mind they proposed that the size of the Directorate be kept as small as possible, and that it did not use the vote to come to a decision. In the event of non-unanimity, they repeated that the DG himself had to decide—he was the sole formal authority, after all—, but that the divergent opinions should be reported to the Council for information.[92]

In September 1960 a precise and more comprehensive text settled the last details. In parallel to the twelve divisions and the Directorate, it established a system of four committees to define the experimental programme of the two machines. This arrangement was the outcome of complex negotiations with the visitors, the organization of work with bubble chambers playing the decisive role in the process. At the same time Adams put forward a list of names for the various posts in the Directorate. This precipitated intense consultation—not to say haggling—with the senior staff which lasted several months.[93]

7.2.3.2 The choice of the twelve divisions and the nomination of their directors

The *precise* definition of the responsibilities of each division followed a somewhat tortuous path which was as much the result of the desire to set up a 'rationally' functional system as of the history of the different groups—the inertia of structures—and of personal ambitions. There were a number of constants in this process, in particular the arrangements for three of the four 'peripheral' divisions, Theory, Site and Buildings, and STS (renamed DD, or Données et Documents). Their status was never seriously in dispute, even if Bakker had thought, in the reform he proposed in 1957/58, to integrate Kowarski's STS division in a larger research division. In a structure of twelve, by contrast, there was less reason for removing this division: its work was clearly defined—it had been cut back in 1959 to the the design of instruments for measuring bubble chamber photographs and to computing—, and by keeping it one 'solved' the Kowarski problem. A CERN pioneer, Kowarski had steadily been stripped of his prerogatives to the benefit of the SC—electronic experiments equipment—and the PS division—construction of heavy equipment like bubble chambers. In the context of the bickering which went along with the definition of the twelve divisions the direction thought it wise to bring his loss of influence to a halt at the boundaries of DD—at least for the moment.

The fourth 'peripheral' division, that of Administration, was also left unchanged for a long time. However, under pressure from the Committee of Council and the Finance Committee who felt that the scientific directors of the organization underestimated the tasks of management, Adams accepted, at the last minute, to hive off finance from the rest of the division, though refusing a request by the CC for a split into three—finance, general administration, and personnel. In September 1960 a Finance Division was set up under Tièche, and the rest of the staff were grouped under Sam Dakin. It was only in 1964, after Dakin had left, that this division of administration and personnel passed into the hands of a non-member of the Directorate, G. Ullmann.[94]

Notes: p. 413

How were the two big divisions of SC and PS broken up? In 1959/60 the SC division consisted of two distinct subgroups. The smaller was responsible for the machine, the other was composed of physicists doing research. In June 1959 it had been proposed to formalize this difference by creating a division of SC machine operation—a proposal which was never challenged.

The first idea on what to do with the *physicists* was to divide them into two divisions, one gathering together specialists in electronic detection, the other a few specialists in nuclear chemistry, and those people working with emulsions or Wilson chambers. The main reason for this choice was that it would help disperse a very large group comprising strong personalities who were felt to behave far too individualistically. This proposal held sway until June 1960, the reorganization proposed by Adams included. However, when it finally came down to *appointing* the two directors, the trouble began.[95]

To get around the problem of personal sensitivities some people like Bernardini proposed that the problem be tackled differently, and be looked at from the point of view of CERN *physicists as a whole*. This approach meant including the bubble chamber specialists then in Adams' division in the arrangement. More precisely, Bernardini's idea was to have an electronic experiments division, and one dealing with track detectors (emulsions, Wilson chambers, bubble chambers), the latter research division being given to Charles Peyrou, the leader of the hydrogen bubble chambers in the PS division. Peyrou refused immediately: while he had nothing against directing a *research* division dealing with track detectors, he was not going to abandon the *construction* of the 2 m hydrogen chamber. Bernardini's plan thus ran aground, and everybody was back to the original problem: how to distribute the physicists of the former SC division.[96]

In July Adams suggested putting them in one division working at the PS and in one working at the SC. Lundby or von Dardel were proposed as director of the division of 'electronic' physics around the PS, and Citron or Lundby for the corresponding SC grouping. Preiswerk was thought suitable for the post of director of the division of SC machine operation; as in the case of Kowarski a place had to be found for this founder physicist of CERN. A month later, with negotiations deadlocked, Adams went back to the idea of having a single division of physics derived from the ex-SC division which he called, somewhat strangely, Nuclear Physics, and which he handed to the most longstanding CERN man of the four, Preiswerk. Two deputies were named, Lundby for the SC zone and von Dardel for the PS zone. As for Citron, Adams thought to give him the task of division leader of SC machine operation. Citron refused, and a more 'technical' person, Lapostolle, replaced him. Later, at Weisskopf's suggestion, Citron became secretary of the committee charged with defining the PS experimental programme, the NPRC.[97]

The history of the groups coming from the PS is a little more simple. Of the four units which Adams considered in his memorandum of 8 June 1959, two had a solid tradition and they developed unproblematically into the two divisions foreseen at the time, Accelerator Research, which remained under Schoch, and Engineering which was always Grütter's domain. A PS machine operation division was also set up. Launched by Zilverschoon in 1959/60—who chose to remain at the leading edge of accelerator research—, Pierre Germain, one of the old PS hands, was asked to direct it. The fourth and last group from the former PS division which Adams thought could serve as the base for a new division was the 'bubble chamber group'. It was, in fact, a recently formed conglomerate comprising Ramm and his troops—construction of a propane chamber, electrostatic separators, etc.—and Peyrou's group, which dealt with hydrogen chambers, and which had been transferred from Kowarski's division (STS) to Adams' a short while before.[98]

Until the end of 1959 it was always thought to have just one bubble chamber division. In January 1960 two were mentioned in the documents, one predominantly concerned with engineering, and responsible for building the 2 m chamber, the other mostly concerned with physics, and responsible for doing research with the available chambers and for building the less complex propane chamber. While relatively 'coherent', this scheme could not survive the personal ambitions of the main protagonists. Adams accordingly yielded to institutional realities and, in September 1960, proposed that there be two bubble chamber divisions. One was to deal with hydrogen chambers and was given to Peyrou and the other, called Nuclear Physics Apparatus, was to be responsible for beam transport and for heavy liquid chambers. It remained in the hands of Ramm, who had been dealing with this kind of equipment since 1958. One interesting point is that both divisions were responsible for the construction *and* the organization of the experimental work with these detectors. Dissatisfied with the splits imposed by personal rivalries and with having two divisions handling bubble chamber research, the Directorate specified that this structure was only temporary. Within a year, one would once again have a research division and an engineering division. A pious hope, for the necessary modification was not made: until 1965 two parallel divisions were concerned with constructing and using track chambers at CERN.[99]

7.2.4 VICTOR WEISSKOPF'S WAY OF MANAGING CERN, 1961–1965

The long organizational debate we have just described came to a definite end at the end of 1960 with the official nomination of the divisional directors. On the 1 January 1961 the new structure came into force. In the meanwhile since John Adams could no longer stay at the head of CERN, a new Director-General was appointed by the Council at its session in December 1960.

7.2.4.1 The nomination of Weisskopf and of the members of the Directorate[100]

When Adams was nominated DG in May/June 1960, he had already accepted a post of importance in the United Kingdom. Having accomplished his magnum opus—to deliver a top quality PS within the deadline set at the start of the project—, he had preferred to become the director of the controlled fusion laboratory at Culham than to stay in a second-level post, behind Bakker, at CERN.

After Bakker's unexpected death de Rose, Amaldi, and Perrin went to London to see whether the Atomic Energy Research Authority would consider releasing Adams from his obligations in Culham. Even though the answer was no, in June the SPC and the CC unanimously agreed to repeat the request officially. As research activities were coming to take on a greater importance than in the past, the SPC also contacted Victor Weisskopf asking him to accept to be Directorate Member for Research. He agreed in June, and was given half-time leave by MIT.[101]

The Council at its session on 14 June unanimously supported the proposal of the SPC and the CC, and de Rose and Amaldi returned to London to meet Lord Hailsham, the UK Minister for Science. He remained unmoved over the question of Adams, but did accept to leave him at CERN's disposal on a part-time basis from the end of September 1960 to July 1961. De Rose, Amaldi, and Cockcroft thus decided to contact Casimir, the Research Director at Philips. Unofficially reticent in May, Casimir definitely declined their request that he stand for the post of Director-General before the extraordinary Council session in July.[102]

At this meeting Adams was nominated Director-General for a year, and Bernardini and Weisskopf were appointed as Directors of Research. In this way a minimum of continuity was maintained and new blood was injected into the organization. At the same time, on 20 July, de Rose sent a second letter to the member states, asking them to submit names for the post of Director-General. These were to arrive in Geneva before 15 September, and the final decision was to be taken by the Council in December.[103]

On 28 September 1960 Adams told the states of the replies received. France proposed Pierre Auger, Italy put forward the name of Gilberto Bernardini, Greece suggested Victor Weisskopf, and Denys Wilkinson was nominated by the United Kingdom. As for the scientists, Cockcroft proposed Eklund (Sweden), Heisenberg proposed Bjørn Trumpy (Norway), and Amaldi put forward Segré's name, Nobel prizewinner, and fellow countryman who had become an American citizen. During the subsequent meetings of the SPC and the CC no unanimous decision was possible. Even though he seemed to be the most highly qualified candidate in high-energy experimental physics, Segré's name was dropped from the list; wanting Weisskopf to stay at CERN as Research Director, the SPC felt that it would be unwise to have two Americans nominated. At its session in December 1960, and after having considered various combinations, the Council rallied to the view held by a two-thirds majority in the SPC and the CC—to nominate Victor Weisskopf Director-General. He was to take office on 1 August 1961.[104]

As most of the debates were held in closed session, and since the related correspondence we have found is also somewhat sparse, it is not easy to build a picture of the shifts and attitudes of each participant. Three things are certain, though. Firstly, that political considerations in the broadest sense played a very important role in the process. As in the case of the appointment of Felix Bloch six years before, and of Bernard Gregory in 1965, national considerations were crucial in shaping allegiance to one candidate or the other. These were by no means the only determinant of the result, however, and the 'scientific' quality of the applicants was also taken very seriously. This explains the importance of the Americans who were thought to be more in touch with the leading edge of physics. The only exception was perhaps Casimir—even if he was not a nuclear or high-energy physics specialist. In 1965, by contrast, when the time came to replace Weisskopf, the two candidates in the running, Bernard Gregory and Gianpietro Puppi, were not up against anyone from the United States. In five years, the situation had changed considerably.

It remains to mention the importance of personal considerations. Since this was the start of a period in which the laboratory had equipment which put it on equal footing with its American rivals, the post of Director-General was one of prestige and one which, potentially, could leave its mark on history. Negotiations thus touched sharply on personal ambitions and sensitivities, one common attitude being that no one wanted to play second fiddle or be defeated in a direct vote. In fact before the Council session in December 1960 several candidates let it be known that they wanted their names withdrawn.

After Weisskopf's election in December the Council put the final touches on the process of reorganization. In June 1960, Dakin had been nominated Directorate Member for Administration; six months later, the Council ratified another nomination, that of Mervyn Hine as Director of Applied Physics and confirmed the appointment of Bernardini, who had resigned for a while, in the post of Research Director. This 'come back' of Bernardini allowed Italy to be represented in the senior posts, as Amaldi noted when asking that an effort be made to keep a measure of geographic equilibrium between the various leading positions.[105] The Directorate, in fact comprised Weisskopf and Bernardini, and three British citizens—Adams, Dakin, and Hine—, while among the twelve divisional directors four were French—Lapostolle for the SC machine, Kowarski for Données et Documents, Mallet for Site and Buildings, and Peyrou for hydrogen bubble chambers—, three were Swiss—Grütter for Electrical Engineering, Preiswerk for Nuclear Physics, the division dealing with 'electronic' physics, and Tièche for Finance—, two were Belgian—Germain for the PS machine, Van Hove for Theory—, two were British—Ramm for the propane chamber and beam transport, Dakin for the Administration—and one was German—Schoch, the director of the division of Accelerator Research. For this last it was quickly agreed that the post should rotate annually, with Johnsen (Norway) and Zilverschoon (The Netherlands) interchanging with Schoch.

Notes: p. 414

7.2.4.2 The functioning of the Directorate under Weisskopf's Chairmanship

In contrast to his predecessors, Weisskopf did not formally modify the structure of CERN's direction. He only made minor changes to the organigramme between 1961 and 1965, and these were essentially conjunctural and linked to the movement of people.

In June 1962 the Council acceded to Bernardini's request to be relieved of his functions in the Directorate. It nominated Gianpietro Puppi, an Italian specialist in electronic detection to replace him. Puppi in turn left this position at his own request in December 1963 and, on Weisskopf's recommendation, the Council allocated the post to a French bubble chamber specialist, Bernard Gregory (the change took place in September 1964). The Council also accepted two other requests made by the Director-General in 1963/64, namely, to replace the Director of Administration whom he could not get along with—another Britain, George Hampton, took Dakin's place—, and to nominate Pierre Germain to the Directorate to deal with technical coordination.

The same year three individual cases concerning divisional directors were dealt with. Firstly the division of Electrical Engineering was dissolved, and, on Weisskopf's advice, Grütter took a post in the United States. Then the leadership of DD changed hands. Kowarski also accepted a position across the Atlantic at the University of Purdue, and the top post passed to his deputy, Ross Macleod, a British experimental physicist with experience in the application of digital measuring and computing techniques to the analysis of experiments. Finally, since Weisskopf judged that Preiswerk was not fully suited to his post, two people were put at the head of the turbulent NP division. From September 1964 onwards, Wolfgang Paul, professor at the University of Bonn, along with Preiswerk, shared the responsibility of managing work with electronic detectors at CERN.

Finally, George Ullmann took over the division of Administration-Personnel in 1964 and Peter Stanley that of PS Machine Operation in 1965. Thus, Germain and Hampton became full-time members of the Directorate.[106]

On the face of it, then, Victor Weisskopf kept the essentials of the system elaborated by Adams. *De jure,* he remained the sole hierarchical superior of the divisional directors and was surrounded by a sort of brain trust of three or four people, the Directorate. *De facto,* his 'style' of management led him to behave differently to what his predecessor had had in mind. To describe this, we might say that Weisskopf's system was a *hybrid* of the direct hierarchical system which Bakker had tried to put into practice (and which Gregory was to adopt later) with Adams' system. We might even say that this hybridization varied with the members of the Directorate.

The crucial element to note in this hybridization is the importance of the role played by Mervyn Hine in the new CERN system. One can in no way capture the scope of his activities by reducing them to those implied by his official title: Directorate Member for Applied Physics. Hine's functions were far broader and were spread over two main areas. For one thing, he served as Weisskopf's right hand man, being responsible for overall planning in the long term, and for the elaboration of the main budgetary needs in particular. In this role he was in direct contact with the Finance Committee, he prepared the papers which Weisskopf defended before the Council—he was, in fact, the one person without whom Weissskopf could not have managed the laboratory. Hine's second role was to serve as a super-director of the technical divisions until 1963, whereupon he shared this part of the work with Pierre Germain. In this capacity he certainly had to think about and define major choices—for example, on computer policy—but he also 'controlled' the divisional directors on behalf of the Director-General. Less interested in this aspect of CERN, quicker to delegate his power, Weisskopf thus authorized a 'bypass' of the official rules which Adams did not—and would not have—tolerated.[107]

A similar phenomenon is detectable in the administration, where first Dakin and then Hampton were in positions of superiority over the directors of the Finance and the Personnel-Administration divisions. By contrast, in the research divisions, the hierarchical situation was seldom found, and Bernardini and Puppi, and Gregory after them, were mainly concerned to coordinate and plan the experimental work. This is partly because this domain was the one in which Weisskopf himself intervened all the time when he was not dealing with the Council, the public relations of the organization, or the big issues affecting CERN. In addition they were also confronted with directors who jealously defended their autonomy—people like Peyrou, for example—, or with a division like NP which was itself subdivided into quasi-independent territories each controlled by a member of the senior staff. These men took it upon themselves to identify important research topics and to organize, with those in the division of PS machine operation, the work on the shop-floor.

7.2.4.3 Victor Weisskopf's philosophy of management

Weisskopf's period of office is regarded at CERN as a sort of peak in the laboratory's lifetime, as a moment of renewal. Unlike what one finds with Bakker, and to a lesser extent with Adams, CERN staff speak spontaneously of Weisskopf's personality, of his way of behaving, of his style of managing people. Contrary to what we have done in the previous sections, we are thus going to pause on this phenomenon of *collective* memory and try to grasp the reasons for it.[108]

When Weisskopf arrived at CERN in 1960 he had several advantages over his predecessors. The first was that he was a new man arriving from MIT, a scientist of international authority who brought with him the prestige of American science. Renowned for his pre-war work in Europe, he afterwards made his name in the United States. A theoretician, he was also familiar with a major scientifico-technical project, that of Los Alamos. With him at the helm CERN people came to think that

European weaknesses would be overcome, that they were going to be guided through the stormy competition that was to come with the PS. Weisskopf's second big advantage emerged with the passage of time: it was his character, at once charismatic and seductive. A cultivated Austrian, as Van Hove has said, he had an 'open and warm' approach to people deriving from a 'profound humanism'. Widely read, and a lover of music, Weisskopf was as capable of charming diplomats in the Council or visiting Ministers, as he was of strolling through a workshop and chatting to technicians. Speaking German, French, and English equally well, he was able to please the Latins as readily as he could the Nordics and the Americans.[109]

These particular aspects of Weisskopf's personality made the enormous impact they did because of the very difficult period CERN passed through between mid-1960 and the end of 1961. There were the tensions with the representatives of the visitors in the SPC during summer 1960—a momentary crisis of confidence between CERN and the key leaders of European physics; there was the interpersonal rivalry for posts as divisional heads during the autumn—a kind of crisis of government and of succession; there was the persistent difficulty of organizing electronic experiments from winter 1960/61 to the following autumn—which revealed a kind of 'individualism' quite opposed to the *esprit de corps* (already mythologized) of the former PS division; there was the violent opposition between 'physicists' and 'builders' regarding the ISR from spring 1961 to the end of the autumn—and, finally, the brutal shock caused by the failure of the neutrino experiment, the symbol of a Europe on the way to a Nobel prize. The euphoric atmosphere created by the commissioning of the PS had given way to doubt and hesitation, to the feeling that no one quite knew what to do or how to do it, to the fear that the organization was crumbling for want of clearly-defined goals. This was particularly true where the physics and the research programme were concerned, the domains most prone to uncertainty; two years after the PS first worked (auto-)criticism was rampant.[110]

The arrival of Victor Weisskopf has to be seen against this background, a new man, certainly, but also a new period. At the start of the 1960s, when the organization, without perhaps knowing it, was looking for a unifying force to guide its actions and for the reassurance that it was capable of holding its own against the people at Berkeley and Brookhaven, Victor Weisskopf arrived from America with the armoury needed to impose his will on most of the CERN groups. Commanding respect, he found himself able to regulate more easily internal conflicts (or at least to de-escalate them), to restore partly physicists' selfconfidence —to become, in fact, the father figure that part of the CERN personnel had been 'waiting for'.

Weisskopf's way of managing CERN emerges clearly from his first speech to the personnel on 7 August 1961. Here he revealed a determination to communicate his faith in the potentialities of CERN and his concern for the particular problems of the different groups. On the first point, for example, he tried to bring out that CERN was 'a place in which most fantastic scientific experiments are carried out' to which each, no matter what his level in the hierarchy may be, made a contribution. He also insisted that CERN was 'a place in which true international collaboration has become a reality', in fact, 'the only true scientific laboratory because it is international'. On the second point, coming across as both lucid and confident, he began by speaking of the disarray which he felt had beset a large part of the organization after the PS worked and after the 'neutrino' setback. 'As always after great human achievement', said Weisskopf, 'there came a feeling of let-down after success'. He spoke first to the *builders,* explaining that their job had only just begun because they still had to think up 'those apparatus whose nature I don't know because it will be invented soon'. Stressing the quality of their past achievements, he exhorted them to think like physicists—'look at the great men in this field, Panofsky, Fermi, Lawrence, and you will see what I mean'—, and to display imagination and flexibility. 'Too much string and sealing wax is out of place in a high energy lab', he insisted, but he added: 'too much engineering perfection will give us wonderful instruments but after the Russians and the Americans have made all the interesting discoveries'.

To the *physicists* Weisskopf said that the work would be difficult, that they would 'be frustrated not only by nature but also by [their] own inexperience'. Expanding here again on what he thought to be the particular weakness of the Europeans, he wrote: 'we definitely do not devote enough time for studying and understanding fundamental problems'. '*This is wrong*', he added, promising to do everything he could to remedy the situation. Finally, addressing himself to the entire *subordinate staff* he asked for their indulgence: 'it is very difficult', he said, 'for the Director-General to get in contact with the staff'. To overcome this he declared that his office would always be open to everyone. 'Don't worry in the least about disturbing me', wrote Weisskopf, 'I like to talk to people and I have a large staff of women to protect me if it really becomes necessary'.[111]

In practice Weisskopf tried to put physics at the centre of everyone's concerns. He did this by giving a weekly talk to all the staff, for one thing, and by reintroducing seminars and by trying to impose particular lines of research on the experimentalists, for another.[112] As for management, he practised an 'open door' policy. Faithful to Oppenheimer's precepts at Los Alamos, he adopted a 'paternalist' attitude, regularly visiting the divisions 'whenever an important discussion or seminar took place, whenever an experiment reached the decisive moment'.[113] Finally, he encouraged all staff members to send him their ideas in writing. These letters, of which we have found about a dozen, are often hard on the organization but indicate the faith which their authors placed in their Director-General. Criticizing the pettiness of someone or other, or various forms of waste, they not only reveal the limits of this kind of 'paternalism' as a way of managing an organization but also the capacity which people attributed to Weisskopf to smooth over difficulties between managers and their subordinates, or between users and the people responsible for the daily running of CERN.[114]

Do we then say that Weisskopf's period in office was more or less ideal? No, for notwithstanding the Director-General's personality, many tensions persisted between those whom he called the 'barons of CERN'. On the other hand, there is no doubt that their respect for Victor Weisskopf's scientific excellence enabled the latter to overcome some of the conflicts which repeatedly surfaced between the members of the Senior Staff. One of Weisskopf's favourite anecdotes is that all he often had to do was to turn up for a short while at a meeting of the committee whose task it was to organize concretely the experimental work around the PS to smooth over the niggling between various groups. There where Stanley, for example, had failed to impose what he thought was the best compromise, the Director-General succeeded in less than half-an-hour.[115]

In other cases, by contrast, Weisskopf had to give way to the divisional directors and to the Senior Staff—as had Bakker and Adams over the reorganization. Sometimes he accepted their demands to the extent that they seemed to offer a good way of improving the practical efficiency of the experimental work. A good example of this is the dropping, at the end of 1961, of the 'privilege' Adams had accorded to the visitors in the Experiments Committees. Directed since the end of 1960 by someone who was not a staff member so as to give some weight to the non-CERN physicists, Weisskopf agreed to give back the chairmanship of the Electronic Committee to Preiswerk, as the Senior Staff wanted. In other cases Weisskopf made the insistent demands of the Senior Staff his own, even against his will, so as not to come into direct conflict with them. An example of this is the staff contract policy. Despite his personal desire to keep a maximum of flexibility inside CERN and to have policies facilitating the rotation of senior personnel in Europe, Weisskopf asked a Council even more reticent than he to approve the reforms asked for by the Senior Staff. Having dedicated a *very* large number of meetings to this topic so vital to them, the Senior Staff concluded that the policy of granting employment contracts of limited duration (three years, renewable) ought to be replaced, for the most part, by a system of indefinite contracts. And this was a battle they won.[116]

7.2.5 CONCLUDING SURVEY: A DIFFICULT HISTORY

To round off this section we would like to stress how difficult and painful an apprenticeship the history of the executive from 1954 to 1965 was. An apprenticeship, because all had to be created and invented, without a history to draw on; difficult, because the launching of CERN was a response to an abstract and voluntarist step, that of putting European physics immediately back among the best; painful, because the choice of men did not always allow for continuity and because death too played its part.

The first Director-General, Felix Bloch, was persuaded to take up a post in which, in fact, he had little interest. Sought after by the European Nobel prizewinners, who wanted one of their own to direct what was to be the beacon of high-energy physics on the Old Continent, Felix Bloch arrived from California in September 1954. He found a building site dominated by engineers and construction companies, and he was burdened with more adminstrative work than he had expected. Within less than a month he realized that it was impossible for him to do the kind of physics that interested him, and he handed in his resignation. Alerted by this experience, the Council and the SPC then settled for a less ambitious choice, nominating Cornelis Bakker as Director-General of the organization. One of the founder members, more a specialist in machines than in physics, Bakker was better suited than Bloch to the needs of the time — even if he lacked the latter's renown.

Bakker's task was not an easy one. On the one hand, John Adams built his empire as he thought fit, without reference to the Director-General. A strong leader of 'his' men, very capable both politically and as an engineer, Adams quickly united his division and rendered it at once autonomous and indispensable to the realization of all important projects. A professional, both technically and organizationally, he created a division that grew steadily in size and whose own interests — to build the PS — he often put before the 'collective' interests of CERN — e.g. to make the Executive Committee for the PS Experimental Programme function properly. On the other hand, Bakker had to deal with a cohort of physicists who knew that CERN had been created for them. A kingdom divided against itself by strong personalities, it nevertheless tried to manage its own affairs independently of the 'engineers', rather like a group of professors in a university faculty.

When Bakker died in April 1960 the field was left open to the physicists but above all to Adams and 'his praetorian guard', as Amaldi once put it. A third force made its presence felt, however, the SPC, concerned to protect the interests of the physicists in the member states. It had already made its voice heard the previous year in the negotiations over the loan of the bubble chambers.[117] During summer 1960 it insisted that visitors be treated in the same way as CERN physicists as far as experimental work with the PS was concerned. In reply — and aware of the importance of maintaining good relations with the Council and the SPC —, the new Director-General Adams organized a system of experimental committees which was very open to the exterior. Convinced also of the impossibility of governing without a minimum of support from the Senior Staff, he modelled the organization of CERN on the forces actually present. Finally, wanting to control the divisions directly himself, he established a Directorate to assist him, but having no stated executive role.

Impressed by Adams' achievements, favouring a continuity in the period of transition ensuing on the commissioning of the PS, but at the same time keen to have a famous physicist at the head of CERN, the SPC and the Council would have liked a permanent solution with Adams as DG and Weisskopf as Research Director. Confronted, however, by Britain's refusal to leave Adams at CERN's disposal — and by a number of personal hesitations and ambitions — the Council resigned itself to an alternative which did not include the former PS division leader. In December 1960 it appointed Victor Weisskopf Director-General.

Notes: p. 414

Preceded by an outstanding reputation, Weisskopf dedicated his energies to making CERN a centre of excellence in physics. Relying on Mervyn Hine to do the important tasks which he found less interesting, he created a style of management more supple than that of his predecessors. Quicker to delegate power, adept at finding compromises, he managed to overcome, to a certain extent, the balkanization of the organization inherited from Bakker's term of office. He did not, however, succeed in abolishing the principal baronies altogether nor in fundamentally undermining the decisive weight of the Senior Staff.

If a last word is to be said on the relationships of power inside CERN, it must then be on this intermediate layer of personnel known administratively as the Senior Staff. This group which, as we have seen, aborted Bakker's first attempt at reform, which largely managed to impose its views during the nominations to the posts of divisional directors under Adams, and which obtained indefinite contracts for the scientific staff during the Weisskopf era, was set up very early on. Clearly defined from a hierarchical point of view, it enjoyed the privileges and immunities attached to the status of the Diplomatic Corps—obtained from the sometimes reluctant Swiss authorities soon after the creation of CERN[118]—, and a list of its members was included in the annual reports of the organization immediately after the organigrammes of the Council and the executive. Most (about 95%) of its members were scientists, comprising senior applied physicists and engineers, mostly from Adams' division, and high-level research physicists, particularly those in the divisions SC and STS. The former were those who built the laboratory and got it to run, those who were concerned with medium-term planning. The latter, the 'users', were more preoccupied with the present and the state of 'international physics'. With the passage of time (i.e. during the years 1963–65) this difference tended to disappear—on the one hand because all the staff were granted the same status in 1963, on the other because their tasks grew to resemble one another with the arrival of more and more visitors: having to deal with these 'outsiders', the service and planning dimensions became increasingly important for everyone.

Confronted with the Senior Staff, the successive CERN directors appear rather weak. More precisely, they seem to have avoided any head-on opposition, preferring systematically to abandon projects than to resort to authoritative measures. In some cases, one might argue, the top management should have imposed its choices against the particular interests of the Senior Staff or a part thereof. The usual answer at CERN to claims of this kind is of the dialectic of master and slave type, arguing that no one can rule against those whose task it is to carry out the orders, that the authority of a leader has to be based, first and foremost, on its voluntary acceptance by his men. This basic response is not false, but cannot alone account for our case. It has to be supplemented in two ways.

Firstly, by noting that CERN's directorate was composed of scientists managing their equals and not of professional managers who dealt with 'others' who were their 'inferiors'. The corollary of this situation, at least in Europe, was the difficulty of saying no, of making clearly stated choices, for fear of being accused of favouritism or of victimization. Paradoxically, while this was 'understood' by the members of the Senior Staff, they still complained that the Director-Generals or the Research Directors often used roundabout arguments to mask their negative choices. For example, they mentioned the ritual appeals to a lack of machine-time or to the shortage of beams as 'excuses' for refusing an experiment, or the sudden creation of a general rule which enabled the 'authorities' to deny something to someone without doing so directly.[119]

Secondly, there is the classical appeal to the originality of scientific work or, more exactly, to *the image* which each person has of it and to the consequences drawn from that by the laboratory's directors. In brief, the argument here is that the most important thing to do in a research laboratory is to preserve the creativity of the scientists, which means creating an environment with as few restraints as possible. Scientists being, by and large, hardworking and dedicated to their job, administrative controls are to be kept to the minimum and the organization of their work is to be left in the scientists' own hands. To do good research people have to be left free to follow their own paths, at their own rhythm, in a voluntarily accepted framework, the task of the direction being only to build a climate of mutual confidence. Hence the importance for a Director-General not to oppose his leading scientific staff. That matters are perhaps not that simple was recognized very early on by those responsible for administering science; all the same, these attitudes were an essential component in the management philosophy of CERN's Director-Generals—and that because the Senior Staff *and* the Director-Generals shared them.[120]

7.3 CERN by numbers: some statistical data on the personnel and its activities

This introductory chapter to part III is intended to give a first picture of the organization, to present, in particular, its modes of direction and management. To conclude it we want to do something a little different, something which will give us a rather more *global* view of CERN. To this end we have built up a series of statistical data concerning primarily the CERN staff and the visitors between 1954 and 1965. Our concerns here were two. On the one hand we wanted to follow the evolution of the whole, to identify the main groups, the main tendencies; on the other hand, we wanted to see who had worked at CERN, to know how the different European states and the various national physics communities 'invested in' or 'used' CERN.

Table 7.1[121] presents the most coarse-grained data. Covering an unusually long period—1955–1985—it shows the evolution of three basic parameters, the number of staff, the number of visitors, and the budget. We notice that the growth in personnel is quasi-exponential from the start until about 1975, whereupon it more or less levelled off, CERN losing a few percent of its staff during the last ten years. The

Notes: p. 414

budgetary curve is obviously of roughly the same shape. Until 1961-62, it is the growth of staff that is the driving force, pulling a somewhat chaotically shaped expenditure curve upwards. The two then develop in parallel until about 1973/74. The visitors curve, by contrast, has a different form: while they were slow to start working at CERN—there were no more than about a hundred visitors in 1960/61—their number has steadily grown, and at an increased rate (1300 in 1975,

Table 7.1
The CERN budget in constant Swiss Francs,
and the numbers of staff and of visitors

Year	Budget in 1985 prices	Number of Staff	Number of Scientific Visitors
1952/54	23.33	144	11
1955	63.58	270	21
1956	99.34	405	30
1957	152.81	576	49
1958	136.85	697	60
1959	131.40	908	54
1960	156.97	1092	71
1961	164.55	1241	134
1962	186.67	1660	172
1963	209.38	1779	223
1964	242.83	1996	247
1965	274.87	2251	315
1966	340.19	2489	387
1967	470.40	2764	423
1968	522.61	3019	451
1969	568.71	3201	520
1970	565.15	3311	656
1971	631.30	3372	853
1972	737.01	3547	970
1973	849.74	3658	1021
1974	865.06	3784	1118
1975	843.37	3788	1289
1976	831.01	3692	1421
1977	771.74	3639	1593
1978	747.94	3648	1826
1979	703.46	3588	2148
1980	700.44	3604	2400
1981	703.15	3573	2604
1982	710.59	3531	2890
1983	717.89	3452	3321
1984	722.10	3518	3633
1985	724.52	3531	4077

Source: Communication to the author by the CERN administration (1986)

more than 4000 ten years later). The years 1953/54–1965 thus emerge, when seen in a long-term perspective, as those in which continual growth was *set in motion,* a finding confirmed by John Krige in his study of finance policy and my own work on the structures of research.[122]

7.3.1 THE STRUCTURE BY DIVISION

Before continuing our commentary we need to clarify one point. The data which we have used to build our tables sometimes vary significantly from one source to another. The main reason for this is that, for the staff as such, there was a regular recruitment of people who were called temporary or supernumerary — often so as not to come into conflict with certain Council directives — but who were, in fact, often permanently employed. A lot depends, therefore, on how and when these people are included in the statistics.[123] The problem is more complicated with the visitors. Belonging to a laboratory other than CERN, they came to work in Geneva either alone or in a team, they seldom received a salary from the organization, and they appeared that much less systematically in the official statistics. If they followed the house rules their presence was registered with the administration; this, however, was not always done with enough precision for us: visitors frequently came and went and it is difficult to estimate their actual presence on the site or to give estimates in men/year. In short, it is essential to regard our figures as giving orders of magnitude, not to be surprised by possible variations between one table and another — and to take careful note of the source used. This is particularly so for visitors: table 7.6 shows that variations of 20 or even 30% are possible.

The staff structure by division before the reforms of 1960 is shown in tables 7.2 and 7.3. The division building the proton synchrotron is obviously the biggest

Table 7.2
Distribution of CERN staff and fellows
by division on 31 December each year

Division	1954[1]	1955	1956	1957	1958	1959	1960
PS	50	118	143	168	221	296	350
SC	12	32	61	109	147	184	223
STS	10	41	61	58	46	60	72
SB	–	17	54	163	202	215	253
Adm	28	48	60	73	84	95	109
Total[2]	114	260	396	597	733	886	1052

Source: CERN Annual Reports
Notes: (1) As on 1/10/54
(2) Includes the DG's services and Theory

Table 7.3
Distribution of directing staff, scientists, engineers, and fellows
by division on 31 December each year

Division	1955	1956	1957	1958[1]	1959[1]	1960[1]
DG	1	1	1	1	1	2
PS	41	45	48	59	72	78
SC	13	21	26	32	48	49
STS	15	20	26	17	20	22
TH	6	10	6	6	7	6
SB	2	3	4	6	7	7
Adm	4	4	4	–	–	–
Total	82	104	115	121	155	164

Source: CERN Annual Reports
Note: (1) Scientists, engineers, and fellows only

throughout this period. It employed about one-third of the personnel in all categories, and half the 'scientific staff' (i.e. the physicists and engineers). The second largest division was that of the synchro-cyclotron. It comprised people building the machine and the bulk of the research physicists. In 1959/60 it accounted for one-fifth of the CERN staff and almost one-third of the scientific staff. Kowarski's division STS is striking for its slow and difficult growth and for its high proportion of scientists. In 1957 it 'lost' part of its prerogatives (and of its members!) to the SC division—the preparation of electronic experimentation—and the next year the preparation of bubble chambers was transferred to Adams' division.

After the 1960 reforms (table 7.4), four divisions catch the eye in terms of numbers

Table 7.4
Distibution of staff and fellows (and of staff, fellows and visitors)
in the four main divisions on 31 December each year

Division	1961	1962	1963	1964	1965
TC	66	122(176)	201(308)	248(323)	299(377)
NP	164	196(294)	212(291)	235(326)	218(365)
MPS	167	199(202)	220(224)	274(278)	297(303)
SB	330	386(386)	391(391)	505(505)	560(560)
Total CERN staff	1236	1664	1775	1996	2191
Staff and visitors	1437	1902	2049	2246	2530

Source: CERN Annual Reports
Remark: 122(176) means 122 CERN staff and fellows, and 176 staff, fellows and visitors.

of staff—indeed they account for two-thirds of the total. The division of Site and Buildings had become the biggest after the fragmentation of Adams' division, even if it had a high population of 'junior' staff: in 1960, there were only seven scientists and engineers in it out of a total of 164 at CERN. Next we have the MPS division, whose task it was to run the PS and to prepare the experiments around the machine, and the two big research divisions, NP, responsible for electronic experiments, and TC, which handled hydrogen bubble chambers. We note that the visitors are largely concentrated in these two divisions—186 out of 274 of them in 1963—, and that the Track Chamber division was the biggest scientific division in 1965. One reason for this was undoubtedly its role in research; another was the fact that it was building heavy equipment. In 1965 the 2m hydrogen chamber was brought into service, and the CERN-Germany-France collaboration to build the giant bubble chamber which became BEBC, the Big European Bubble Chamber, got under way.

Along with these four divisions the division of Données et Documents— ex-STS— is also clearly a division of some importance. It had 120 members in 1961, and almost 200 in 1965 of which about 40 were visitors. The divisions SC Machine and NPA— separators and propane chambers—had each between 80 and 100 staff, while the Accelerator Research division doubled its tally between 1961 and 1965 (from 50 to about 100). This last shared with the Theory division the privilege of having the largest proportion of scientific personnel and of Senior Staff (there were 10 members of the Senior Staff in the NP division in 1962, 7 in MPS and 6 in the divisions AR,TC, and TH).[124]

7.3.2 THE STRUCTURE BY PROFESSIONAL CATEGORY

We have already alluded to this categorization in speaking of the scientific personnel and of the Senior Staff (who, we will remember, were personnel of grade 12 or higher on a scale of 14). In table 7.5 the people working at CERN have been grouped in six 'professional' categories borrowed from the CERN Annual Reports, namely, engineers and scientists, technicians, administrative staff, ancillary staff, CERN and Ford fellows (Ford fellowships were distributed by CERN), and visitors. The criteria for this distribution are not explained in the reports, and there is considerable ambiguity about the counts for 'technicians' and 'ancillary staff'. To see what we mean consider the data for 1964, the year in which CERN changed its classification system: from 1008 the number of technicians 'changed' to 536 (without any explanation being given) and that of ancillaries increased from 442 to 871. It is therefore impossible to give meaningful figures on the distribution of staff between senior technicians, technicians, qualified manual workers, and various ancillary personnel.

One of the most apparent features of the years 1956-1966 is the relative stability of the three units for which we have reasonably reliable data: scientific personnel and engineers, administrative staff, and other kinds of personnel. The first and the second, whose percentage decreased slightly between 1956/57 and 1965/66, moved respectively between 20 and 16% and between 15 and 11%. If one includes fellows and scientific visitors the fraction of high-level scientific staff increases. Between 1960 and 1965 it comprised from 26 to 27% of the total personnel present on the site, and often close to half of the personnel in certain scientific divisions (Theory is obviously an exception, for all its personnel bar a few secretaries were scientists). As for the third group, the technicians, manual workers, and ancillaries, they represented three-fifths to three-quarters of the staff between 1956 and 1965.

These data illustrate the character of CERN. In the first place, it is a laboratory

Table 7.5
Distribution of personnel by category on 31 December each year

Category	1955	1956	1957	1958	1959	1960
Scientists & Engineers[1]	83	103	115	121	157	170
Technicians	102	156	241	307	420	527
Ancillary Staff	35	82	128	157	226	241
Administrative Staff	49	62	77	88	105	127
Fellows	18	27	50	61	58	71
Visitors						90
Total Staff[2]	269	403	561	673	908	1065
Total on the site						1226

Category	1961	1962	1963	1964	1964	1965	1966
Sc & Eng	190	264	284	321	305	349	413
Tech	601	780	856	1008	536	604	717
Anc Staff	287	431	422	442	871	922	987
Admin	158	189	213	225	298	316	373
Fellows	75	81	81	79	79	73	
Visitors	126	157	193	171	171	266	
Total	1236	1664	1775	1996	2010	2191	2490
On Site	1437	1902	2049	2246	2260	2530	

Source: CERN Annual Reports, among which Annual Report (1962), 149
Remark: CERN changed its classification Technicians/Ancillary Staff/Administrative Staff in 1964 (without explaining either the old or the new criteria for classification)
Notes: (1) The differences with Table 7.3 are notably due to the inclusion of 'supernumeraries' paid by CERN
(2) The differences with Table 7.2 are due to the use of different sources and to the inclusion of 'supernumeraries'

doing *big science,* where one needs a large number of technicians and qualified workers to construct equipment (accelerators, bubble chambers, etc.). Then it is a laboratory *in full expansion:* in 1965, it welcomes more and more visitors and has a very big Site and Buildings division, 95% of whose staff are 'technicians' and 'ancillaries'. Finally, it is a *research facility* at the leading edge of its field: highly qualified people — notably engineers and people with doctorates in physics — comprise a quarter of those working on the site.

7.3.3 VISITORS AND RESEARCH STAFF AT CERN

Table 7.6 gives three series of data from different sources on visitors and fellows. We discuss each in turn. The figures given in the *Annual Reports* only begin in 1960, even if we know that dozens of visitors worked on the site before this date. There also seems to be an anomaly in the figure for 1964, as the number of visitors decreases to 171 compared to 193 the year before, and 266 in 1965. The two other series, however, show a steady growth and overall they evolve in parallel. The data provided by the CERN administration in 1986 (at our request) are on average higher than those given by Mervyn Hine in a report written in August 1965 and those in the

Table 7.6
Visitors, Fellows and Staff according to sources

Source	Category	1955	1956	1957	1958	1959	1960	1961	1962	1963	1964	1965
Annual Report	Fellows (CERN and Ford)	18	27	50	61	58	71	75	81	81	79	73
	Visitors						90	126	157	193	171	266
Administration in 1986	Visitors	21	30	49	60	54	71	134	172	223	247	315
CERN/SPC/ 213,1965	Visitors				12	39	54	72	118	152	186	214	211[1]
	Visitors from non-member states				11	25	28	28	39	44	54	55	51[1]
CERN/SPC/ 213,1965	Scientific Staff	83	103	115	121	157	170	190	264	284	321	349
	Research Staff and Fellows			25	37	60	85	90	107	138	142	162	178
Pestre,1988 estimate	Visitors			?	?	50	54	72	125	160	200	230	315

Sources: Anual Reports, CERN/SPC/213, data given by the Administration in 1986.
Note: (1) From January to August only.

Annual Reports. Extrapolating from these three series one arrives at the following 'reasonable' figures: 50 visitors in 1958, 54 in 1959, 72 in 1960, 125 in 1961, 160 in 1962, 200 in 1963, 230 in 1964, and 315 in 1965 (Hine's figure of 211 for this year was only based on 8 not 12 months). We could possibly add the fellows who were usually at CERN for one or two years to these data: their numbers changed from 60 in 1958 to 80 in 1962/64, before dropping back when the Ford fellowship scheme was stopped.

There are two jumps in the number of visitors. The first occurred when the PS came into service, i.e. in 1960 and 1961; the second in 1965 and 1966 when there was a massive influx of visitors specializing in electronic detection methods and an expansion of external and mixed teams (CERN-visitors). Another way of studying this presence of visitors is by comparing their number with that of CERN's *research* staff and fellows. (Let us remember that most of the visitors (at least 90%) came to CERN to do research; between 1960 and 1965, in fact, 85% joined the divisions TH, NP and TC.) Until 1960 the research staff and fellows at CERN outnumbered the visiting researchers: there were roughly 3 or even 4 of them to 2 visitors. In 1961/62 the numbers were about balanced and, from 1963 onwards, the difference increased steadily in favour of visitors. In this year there were 4 visiting researchers for about 3 paid by CERN, while by 1965/66 the ratio was something like 3 to 2.

Table 7.7 gives a snapshot of the situation on 31 August 1965. It confirms that the total CERN researchers (76) and fellows (66) was some two-thirds of the number of visitors doing research (90% of 244). It also shows that the entire scientific staff of CERN, including the engineers (so 334 people) was about the same size as the number of people of the same level working temporarily at CERN (244 visitors and

Table 7.7
Distribution by country of CERN scientific staff, fellows, and visitors as on 31 August 1965

Countries	CERN Research Staff	Fellows	Visitors	Other CERN Scientific Staff	Total CERN Scientific Staff
France	9	6	56	37	46
Germany	10	9	30	51	61
Italy	16	12	26	25	41
United Kingdom	20	9	37	59	79
Total of the 4	55	36	149	172	227
Total of the other 9 member states	19	30	44	79	98
Non-member states	2	–	51	7	9
Overall total	76	66	244	258	334

Source: CERN/SPC/213, 9

CERN by numbers   401

66 fellows). Finally it shows that non-Europeans, and Americans in particular, were well represented among the visitors (51 out of the 244). In short one might conclude that, by 1965, CERN had become an open laboratory offering its facilities to scientists not only in Europe but throughout the world.

7.3.4 CERN AND ITS MEMBER STATES

Our analysis here is based on a set of twelve tables, six in this chapter (tables 7.7 to 7.12) and six in chapter 8. Briefly, what we now want to do is to describe how each national group contributed to CERN's development and benefitted from it. Table 7.8 allows for some general remarks before we visit each country in turn. It gives the percentage of different national groups in the scientific staff in 1965 (line 1), in the group: scientific staff plus fellows plus visitors (line 2), in the total member state population (line 3) and in relation to the CERN budget (line 4). The data discriminate between the four big countries (France, Germany, Italy, the United Kingdom), the nine smaller states which are grouped together, and Switzerland, as there are certain distinct features to be noted about the host state. The next four lines in table 7.8 give the ratios of lines 1 to 4. Thus the figure 4.9 in the line 'Ratio of 1 to 3' means that there are 4.9 more Swiss in the CERN scientific staff than in the combined population of the 13 member states.

Table 7.8
Comparison by country of the scientific personnel and the visitors working at CERN with that country's contribution to the CERN budget and its population on 31 August 1965

Percentage relative to the 13 member states	9 'small'	4 'big'	France	Germany	Italy	UK	Switzerland
CERN scientific staff (1) (Phys & Eng)	30%	70%	14.1%	18.8%	12.6%	24.3%	9.2%
Staff + Fellows + Visitors (2)	29%	71%	18.5%	17.1%	13.5%	21.4%	7.9%
Country's population (3)	30%	70%	16.1%	19.5%	17.1%	17.4%	1.9%
Contribution to CERN budget (4)	23%	77%	18.6%	22.7%	10.8%	24.5%	3.2%
(1)/(3)	1.	1.	0.88	0.96	0.74	1.4	4.9
(2)/(3)	~ 1.	~ 1.	1.15	0.88	0.79	1.23	4.23
(1)/(4)	1.3	0.9	0.76	0.83	1.17	1.	2.9
(2)/(4)	1.3	0.9	1.	0.75	1.25	0.87	2.5

Source: CERN/SPC/213, 32.

Remarks: (1) The figures in the first four lines are as a percentage of the total for the member states
(2) The 'big four' are France, Germany, Italy, and the United Kingdom

This table shows clearly that certain major equilibria were respected in CERN's recruitment policy even if there was no formal policy of 'just return' for each country ('just return' means a strict proportion between persons recruited and contribution to the budget). In contrast to what happened with industrial contracts, more attention was given to the national breakdown of appointments to ensure that the differences were, on average, never 'too' large. More precisely, they were controlled and tended to favour the smaller countries. As a result people recruited in the nine 'small states' occupy about 30% of the posts though their collective contribution to the budget amounts to 23%. If one adds Italy to this group—remembering that it is a country with a large population and a relatively small budgetary contribution—we find that 10 states paying one-third of the budget received 43% of the posts.

Using now the total population of each country as our basis for comparison, the two extreme cases which emerge are Switzerland (over-represented in the personnel) and Spain (under-represented), with Norway, the Netherlands, and Belgium successively occupying the best positions after Switzerland. Among the 'big four' Italy is worse off than the other three: Italians are distinctly less numerous than the Germans and the British in the *personnel*, and are also less numerous than the French among the visitors. The country does however have a very good ratio of presence at CERN to contribution to the budget. For this reason, and because its personnel were very much in evidence in the *research* work, we shall begin our tour of the member states with her.

The *Italians* were the most numerous, in the long term, in the Theory division. From 1956 to 1965 they occupied 43 posts counted in man-years against 36 for the French and 26 for the Germans (see table 7.10) and they were the most numerous among the staff of grade 10 or higher in this division (see tables 7.11 and 7.12). In electronic experimentation, their presence was first strongly felt around the SC where

Table 7.9
Distribution of research staff and fellows at CERN by scientific activity and year

Number of Physicists	1956	1957	1958	1959	1960	1961	1962	1963	1964	1965	Growth 1961–1965
Bubble chambers	8	11	12	17	20	25	46	52	59	65	+ 40
Electronic experiments	14	22	33	45	52	54	64	60	68	78	+ 24
Theory	3	4	15	23	18	28	28	30	35	35	+ 7
Total	25	37	60	85	90	107	138	142	162	178	+ 71

Source: CERN/SPC/213, 11 and 13
Note: 'Electronic experiments' also includes a few CERN emulsion specialists and nuclear chemists

Table 7.10
Distribution of research staff and fellows at CERN by scientific activity and country

Man-years 1956–1965	Bubble chambers	Electronic experiments	Theory	Total
France	48	48	36	132
Germany	92	76	26	194
Italy	18	111	43	172
United Kingdom	66	94	20	180
The other 9 member states	91	161	94	346
Total	315	490	219	1024

Source: CERN/SPC/213, 11 and 13
Note: 'Electronic experiments' also includes a few CERN emulsion specialists and nuclear chemists

Table 7.11
Staff in grade 10 or higher in 1962 by division and country

Country	PS + Engin.	SC	Acc. Res.	Data Hand.	NPA	TC	NP	TH	Fin. + Adm.	SB	Directorate	Total
Austria	–	–	–	–	–	–	–	–	–	–	–	0
Belgium	D	–	–	1	–	–	1(+2)	D	3	–	–	9
Denmark	–	–	–	–	–	1	–	–	–	–	–	1
France	3	D+2	1	D+3	1	D+4	2(+1)	2	2	D+2	–	27
Germany	5	2	4	3	1	5	7(+2)	1	–	–	–	30
Greece	–	–	–	–	–	–	–	–	–	–	–	0
Italy	2	2	1	–	2	1	7(+1)	4	–	1	1	23
The Netherlands	–	–	2	–	2	–	5	–	1	1	–	11
Norway	–	–	D+1	–	–	1	2	–	1	–	–	6
Spain	–	–	–	–	–	1	–	–	–	1	–	2
Sweden	–	1	–	–	–	–	1	1	–	1	–	4
Switzerland	D+5	1	3	1	–	1	D+2	1	D+4	2	–	23
United Kingdom	4	2	6	6	D+1	4	5(+2)	1	D+4	3	2	42
Other countries	3	–	–	–	–	–	1	–	–	–	DG	5

Source: *List of Staff Members*, 1/10/62 (part-time posts excluded)
Remarks: (1) D+5 means that the divisional director plus 5 staff members are from the country concerned
(2) In NP, 7(+2) means 7 members doing electronic experiments and 2 in emulsions, nuclear chemistry, or Wilson chambers

Table 7.12

Staff in grade 10 or higher in 1965 by division and country

Country	PS	SC	Acc. Res.	Data Hand.	NPA	TC	NP	TH	Fin. + SB Pers.	Direc-torate	Total
Austria	–	–	–	–	–	–	1	–	–	–	1
Belgium	3	–	1	–	–	1	–	D	2	1	1	10
Denmark	–	–	–	–	–	1	–	–	–	–	1
France	3	2	3	1	1	D+7	3(+1)	2	2	D+5	1	33
Germany	3	2	D+5	4	3	10	D+6(+1)	1	D+2	–	–	39
Greece	–	–	–	–	–	–	–	–	–	–	0
Italy	2	D+1	2	1	2	2	13	3	–	1	–	28
The Netherlands	1	–	4	1	2	1	4	1	1	1	–	16
Norway	–	1	1	1	1	–	2(+2)	–	1	–	–	9
Spain	–	–	–	–	1	1	–	–	–	–	–	2
Sweden	1	1	–	–	–	–	1	1	–	–	–	4
Switzerland	5	–	–	–	3	3	D+4	1	D+6	3	–	27
United Kingdom	D+8	2	6	D+9	D+2	6	8(+2)	2	2	3	2	55
Other countries	5	1	–	–	–	1	2	1	–	–	DG	11

Source: *List of Staff Members*, 1/8/65 (part-time posts excluded)

Remarks: (1) D+5 means that the divisional director plus 5 staff members are from the country concerned

(2) In NP, 3(+1) means 3 members doing electronic experiments and 1 in emulsions or studying nuclear structure

they were almost as powerful as the British. In the long term, indeed, they became even more numerous: in man-years there were 111 Italians from 1956 to 1965 (against 94 British and 76 Germans), and 13 out of the 28 Italian *members* of CERN of grade 10 or higher were in the NP division in 1965. By comparison, there were 8 British and 7 Germans of this grade concerned with electronic experimentation at this time. If we add that Bernardini and Puppi were two of the four leaders of research at CERN from 1960 to 1965, that Cocconi and then Salvini played an important role in the organization of electronic work from 1960 to 1962, and that Bernardini was the pivot of the neutrino experiment, we get a rather impressive picture.

Work with bubble chambers followed a different pattern, apparently. Table 7.10 gives 18 man-years for Italy against 92 for the Germans and 66 for the British among CERN's research staff. All the same, it would be hazardous to conclude from this that the Italians were absent since, in contrast to electronic experimentation, it was the *visitors* who dominated bubble chamber work at CERN. Indeed, with their 14 groups analysing bubble chamber photographs, the Italians were, along with the French, one of the two main users of chambers at CERN. Without having important chambers of their own, they easily established privileged links with the French teams and collaborated with builders from their fellow 'Latin' country (Peyrou at CERN, Gregory and Lagarrigue in France). The results were striking: from 1960 to 1965 the

Italians handled 22.4% of the photographs taken at CERN against 23.3% for the French and 17.7% for the British (tables in chapter 8).[125]

To understand this remarkable situation there are three things to be borne in mind. The fact that Italian physics was the child of the great Fermi—one of the few if not the only contemporary physicist who was at once a great theoretician and an outstanding experimentalist—and that, in contrast with the situation in Germany, Italian physics was not destroyed by fascism and by the war; then the fact that it was very closely allied to American physics in the 1940s and 1950s and that high-energy physicists (men like Bernardini, Cocconi, and Puppi) worked across the Atlantic for a long while before returning to Europe; finally the fact that Italy did not launch a major domestic accelerator construction programme—primarily for financial reasons—and that this 'pushed' her physicists to work in centres like Berkeley, Brookhaven, or Saclay in 1959/60. Top quality men, long tradition in high-energy physics, familiarity with working in foreign laboratories like those in the United States—these seem to have been the three keys to Italy's 'success' at CERN.

The Italians were, by contrast, absent from technical divisions not centred on research—in 1962 there were 3 of them out of a total of 57 above grade 10 in Engineering, PS, AR, and DD, while there were 5 out of 73 in 1965; nor were there Italians in the Administration and Site and Buildings divisions.

The *British* case was largely symmetrical with this: Britain was particularly well represented in the technical divisions and her 'main effort [was] concentrated on the two National Institute Machines' (Nimrod at Rutherford and Nina at Daresbury) and not at CERN. This relative 'dispersion' of effort, the most extensive in Europe in the mid-1960s, sometimes led the British to be rather mean about their contribution to CERN and to use the PS at Geneva rather less than the other countries from 1964/65 onwards. All the same, the United Kingdom was the leading European financial and scientific power in Europe between 1956 and 1965, and its presence was undoubtedly felt in the higher grades at CERN. In 1965 the British occupied 24.3% of the scientific staff posts against 12.6% for the Italians and 14.1% for the French (table 7.8), and they had 55 people of grade 10 or higher against 39 Germans and 33 French (table 7.12).[126]

As for applied physics, one must remember that the construction of the PS was largely thanks to them, that this domain was the largely unchallenged territory of Mervyn Hine after 1960, and that the British were the major power at CERN in terms of machine construction. In 1962 they occupied 6 out of the 14 posts of grade 10 or higher in the Accelerator Research division, and in 1965 there were 15 British as against 9 Germans out of a total of 55 in the PS and AR divisions together. In addition, after Kowarski's departure, they were in a substantial majority in the new DD division led by Macleod: if one considers the personnel above grade 8 in 1965 in the field of *computing,* we find 23 British... and 17 representatives from other states—which reflects also how ahead the United Kingdom was in this field vis à vis

Notes: p. 414

other European countries.[127]

The situation is more nuanced in research, above all if we bear in mind the enormous weight Britain had in European high-energy physics in the decade after the war. From 1960 to 1965 British physicists' participation in bubble chamber work was important without being exceptional—the French and the Italians handled more photographs—, and their contribution to electronic work was rather substantial—they occupied 94 man-years in this domain from 1956 to 1965 and 13 posts above grade 8 out of a total of 60 in NP in 1965. However, a British document dated February 1966 recognized, with regret, that 'it is believed that British physicists are less numerous among the Visitors engaged in electronic experiments [than their number in the CERN personnel]'.[128]

In the case of the *French* physicists we find the greatest asymmetry in the use of detection techniques. Leaders of the bubble chamber 'mafia' at Geneva, the French were largely absent from electronic experimentation.[129] Until 1959 they do not appear on the lists of scientific personnel at CERN specialized in this technique and there are only 3 of them above grade 10 in this domain in 1965 (plus 6 fellows, it is true) out of a total of 46. One might argue that these figures are hardly significant since the French were always under-represented on the scientific staff—including in the area of bubble chambers, see table 7.10—and that they preferred to come and work at CERN as visitors (in 1965 there were 56 of them working at Geneva as visitors against 37 British and 30 Germans). One might also point out that three groups of about a dozen people each (coming from the CEA and the Institut de Physique Nucléaire) were working at CERN under the auspices of the NP division in 1964–1965, and that it was a French electronic team which was the first visiting group to work on its own at the PS in 1964.[130] This is not false but it does not alter our first remark significantly, a remark which is to be explained by the strategic choice made in France in favour of bubble chambers at the end of the 1950s, particularly in Louis Leprince-Ringuet's laboratory. French physicists, on the other hand, were very numerous in the Theory division, in line with a tradition going back to the early days of CERN and to the presence of Abragam in Geneva in 1953.

As for Italy, France was, by contrast, rather poorly represented in the divisions dominated by engineers constructing big equipment, even if P. Lapostolle was the director of the SC Machine division for several years and if G. Charpak won renown for his contributions to detector technology. On the other hand, the French dominated the very large Site and Buildings division directed by C. Mallet. To conclude, let us not forget that the 'absence' of a clear decision in France in favour of a successor to Saturne freed considerable resources which were directly available for CERN, and meant that this country was one of the main proponents of a policy of strong growth both as far as accelerators and big detectors were concerned. In 1964, for example, France decided to construct Gargamelle for CERN (a heavy liquid chamber for neutrino physics) and, along with Germany, was one of the driving forces for that which became BEBC, the biggest hydrogen bubble chamber ever built in Europe.[131]

The *Germans* were the second biggest group, after the British, in grade 10 or higher. Their presence is noticeable in machine construction, where they shared the leading posts with scientists from northern Europe, like Kjell Johnsen, and with the Dutch and the British. In 1965 they were as numerous as the British in the AR division, whose historic head was Arnold Schoch. Like the Italians, on the other hand, they were totally absent from the financial management of CERN — which was jealously guarded by the Swiss — and, until the nomination of Ullman in 1964, barely present in the administration — which the British traditionally occupied. This situation, which was not due to chance, was the historic consequence of Germany's political position in Europe immediately after the war, at the time when CERN was born.

As for research, German physicists were the most numerous group among the *scientific staff* and the *fellows*. From 1956 to 1965 they occupied 194 man-years, against 180 for the British, 172 for the Italians, and 132 for the French (see table 7.10). In this same category of CERN personnel, they were in the majority in TC, they were more numerous than the Italians in NP in 1962, and Wolfgang Paul was appointed director of this division in 1964 alongside Peter Preisswerk, the Swiss physicist from Zurich. This strong presence among CERN staff contrasts sharply with a very weak representation among the *visitors* and *visiting teams*. In electronic detection — and as opposed to what we find with the other three 'big' countries — there was no German visiting team on record before 1964, and bubble chamber groups in the FRG only asked for 8% of the photographs taken in Geneva — against 23.3% by those in France and 22.4% in Italy (tables in chapter 8). This only serves to throw the transformation in the later years into sharper relief. In 1964 DESY, the 6.5 GeV electron synchrotron, entered into service. At the start of 1966 there were 650 people working around it, of whom 150 were scientists. In the same period the number of groups in the domain of electronics multiplied rapidly (30 to 35 physicists in Aachen, Bonn, Heidelberg, Karlsruhe and Munich) and Germany decided to collaborate with CERN and France in building the 60 m^3 BEBC. In a word, Germany had emerged from the more or less number two position in which it had remained after the prohibitions imposed on it by the postwar nuclear world.[132]

Switzerland emerges as something of a special case. In 1965 she contributed 3.2% to the CERN budget and had 1.9% of the member state population. All the same in 1965 there are only a few less Swiss than French and Italians in grade 10 or higher: 27, for 28 Italians, 33 French, and 16 Dutch. Completely dominating the Finance division which was placed under C. Tièche, and the transient Engineering division under F. Grütter, they were also rather numerous in the higher grades in NP, where Preiswerk was the historic director. The presence of the groups from Geneva and

Zurich in this division and of those from Berne and Fribourg in the field of bubble chambers completes the picture.

To understand this strong representation at CERN of Swiss scientific personnel one needs to bear two things in mind. Firstly that Switzerland was one of the group of '9 small' countries, but among them was one of the richest and the most advanced scientifically— like Norway and the Netherlands, for example. Secondly that it was in a privileged position by virtue of being the host state; this was particularly clear in the award of industrial contracts, but it also had implications here. Concerning the first point, one must remember that the '4 big' countries sought to be generous with the '9 small' ones in terms of scientific personnel; table 7.7 shows that about one-third of the posts were awarded to them. The paradox of this situation, in which the aim was to be generous to the 'small', is that it was the most powerful of them—the Swiss— who reaped the greatest benefits. The relative gap between 'big' and 'small' thus widened, particularly as far as the 'weaker small' countries were concerned.

This second remark also applies to the nationals of the *Netherlands,* who form the second group from the '9 small' countries. They played a very important role in the divisions responsible for new accelerators and for the construction of sophisticated equipment (to name a few, C. Zilverschoon, S. van der Meer, B. de Raad, and W. Middelkoop), as well as in the NP division where there were 5 above grade 10 in 1962 in a total of 34, and 4 in 1965. The *Belgians* were a little less numerous than the Dutch in the top grades and more evenly distributed between the divisions. Two names in particular spring to mind, P. Germain, Adams' successor at the head of the PS division, and L. Van Hove, Director of the Theory division from 1960 onwards. By contrast, we find rather few Belgians in the experimental groups. This is not surprising. According to a Belgian document in 1966, *in toto,* there were only seven Belgian senior physicists working with bubble chambers and 14 on theory—while only two had been engaged in electronic experimentation at Saclay since 1963. In the case of the Dutch the respective figures were 20, 27, and 3.[133]

The *Norwegians* were just behind the Swiss as far as their presence in the senior grades relative to their population was concerned (4% in the Senior Staff as against only 1.2% of the population in the member states). The best known of them was Kjell Johnsen, who was appointed to build the ISR in 1965. On the other hand there were *very few* Norwegians among the visitors even if two Norwegian groups collaborated on bubble chamber work at CERN, and worked around Saturne with the Ecole Polytechnique chamber in 1965 and 1966. There was no national group doing electronic experiments. The majority of the five *Swedish* groups, by contrast, worked with electronic equipment; they were at CERN in 1965, particularly around the SC where they built ISOLDE and specialized in nuclear structure. As a matter of fact, the Institute of Physics at Lund commissioned a 1.2 GeV electron synchrotron in 1963. On the other hand, unlike the Norwegians, the Swedes were under-represented on the scientific staff having only four against Norway's nine in 1965.[134]

The last four countries—*Austria, Denmark, Greece,* and *Spain*—were hardly represented at Geneva in the mid-1960s (see table 7.12). In January 1966 the Academy of Sciences in Vienna created an institute of high energy physics which incorporated the bubble chamber group associated with Hans Thirring and which planned to create a second group to work in Geneva and at DESY in 1967. A Danish group of five physicists collaborated in bubble chamber work while another somewhat larger team was set up in 1965 to collaborate in the ISOLDE project, and in the study of nuclear structure with other Scandinavian countries, France, and Germany. As for Spain, four groups from Valencia, Madrid, Seville, and Saragossa collaborated at CERN in the emulsion and bubble chamber groups.[135]

To finish we want to make two general remarks. Firstly, that there is a tendency to 'national homogenization' in the divisions, a tendency which decreases markedly as one passes from the peripheral divisions—where it is very clear—to the research divisions via the construction divisions. The most obvious examples are the Finance division and Grütter's Engineering division (both directed and dominated by Swiss personnel), the administration before Ullman's appointment (which stand out for having an important British presence), and finally, the SB division, which was administered by the French. The divisions where national 'imbalances' were least marked were those of NP and Theory (see tables 7.11 and 7.12).

It is hardly surprising to find this happening in the 'service' divisions and in divisions having very few high-grade personnel, like in Administration, Finance, SB, and Engineering. In the interest of efficiency it is often important for a director to surround himself with people he knows and to whom he can relate easily. The phenomenon is only more noticeable in our case because the national dimension serves to highlight it. It is also related to everyday modes of behaviour: one spontaneously 'chooses' to work with those who share one's culture and one's way of life. An analysis of bubble chamber collaborations shows this very clearly. From 1960 to 1965 the French and the Italians worked together in eleven collaborations, the British and the Germans in ten. By comparison, the Germans and the Italians were never in the same collaboration, while the British only collaborated twice with the Italians, and five times with the French. As for the French and the Germans, they worked together in three collaborations. In short, people from northern and southern Europe obviously preferred to work with colleagues from the same 'geographical' zones.

A quick second remark: the distribution of ancillary staff and technicians by country is *very* unbalanced, and differs profoundly from the far more equilibrated situation we find with the scientific staff. As is only to be expected, the recruitment was often local (the Franco-Swiss region around Geneva), and that much more localized when one sought relatively unqualified personnel. Staff in these lower

Notes: p. 415

grades were thus dominated by the French and the Swiss. This is yet another factor explaining the preponderance of nationals from these two countries in certain peripheral divisions. In SB, for example, there was good reason to build an essentially French (or Swiss) framework to deal with predominantly French (or Swiss) personnel.

Translated by John Krige

Notes

1. The main sources for this chapter are the official minutes of the Council and its various Committees (Committee of Council, Finance Committee, Scientific Policy Committee). They may be complemented by listening to the magnetic tapes of the meetings. For the executive aspect, the DG files, notably the correspondence files, the Group of Directors and Directorate series, and some Director of Administration files have to be added. As for the bibliography, we would only refer to two collective works, Management (1957) and Hill (1963). Composed of texts written by professional managers of scientific institutions, they throw critical light on the idealistic way the functioning of big science units is generally perceived.
2. History of CERN (1987), 219–221, 246–252.
3. This description of the Executive is valid for the period following Felix Bloch's short directorship; for the transition, see John Krige, 'The nomination of the first Director-General and its aftermath', History of CERN (1987), 261–273.
4. Let us mention Werner Heisenberg in Germany, Francis Perrin in France, etc.
5. See Krige and Pestre, 'The how and the why of the birth of CERN', History of CERN (1987), 523–544.
6. The sources here are the minutes of meetings. In general, we do not give specific references; the lists of participants are always given at the start of each proceedings.
7. History of CERN (1987).
8. Council Minutes, 1954-1955, 14, 17–18.
9. Council Minutes, 1957, 50–52.
10. Council Minutes, 1960, 74–77; See also FC Minutes to appreciate the role of Hocker.
11. On Funke's role, see FC minutes and Council Minutes; for his nomination, Council Minutes, 1961, 45–46.
12. Council Minutes, 1963, 57–59; for Leprince-Ringuet, see *infra* and chapters 10 and 14.
13. The provisional Council held 9 sessions between 1952 and 1954, and the CERN Council 31 between 1954 and 1965. This section is built from a study of these 40 meetings.
14. On de Rose, the information is from the Ministry of Foreign Affairs, and an interview with the author in October 1983 in Paris.
15. Amaldi became Council member in 1959 only. He nevertheless took part in all provisional Council sessions as Secretary-General, and in several CERN Council sessions between 1954 and 1958 as Assistant Director-General in 1954/55 and as Chairman of the SPC in 1958.
16. More precisely, Picot was replaced by Sernaclens between 1959 and 1961; in 1962, Fierz succeeded Scherrer and Chavanne succeeded Sernaclens.
17. Supek, from Yugoslavia, was rather permanent in the Council; Yugoslavia left CERN in 1961.
18. Certain delegations tended to have several people in FC meetings, others only one as envisaged by the rules. What follows is built from FC minutes.

19. Salvini and Cassels, who were the first two chairmen of the Electronic Experiment Committee, attended the SPC very rarely before 1964.
20. This growing autonomy of the SPC vis-à-vis the Council goes hand in hand with a close link with the direction of the laboratory.
21. Council Minutes, 1963, 60.
22. Letters Valeur to Bakker, 9/12/55, Valeur to Bakker, 7/6/56, Weisskopf to de Rose, 1/2/65, de Rose to Weisskopf, 15/2/65, Weisskopf to de Rose, 22/7/65 (DG20821).
23. Quotation from, amongst many, letter Weisskopf to Powell, 19/1/63 (DG20783).
24. See CERN/DGO/Memo/202, 26/6/57, and the discussion which followed during a Group of Directors meeting (CERN/GD/132,4); the quotation from Gentner comes from this document. Concerning the sharp exchange of letters, see Memorandum to Dakin, 31/10/61 and the answer DIR/ADM/PERS/09(1), 1/11/61 (DIRADM20416).
25. Quotation from CERN/GD/132,4.
26. Willson (1957).
27. See personal and confidential note by Verry to Willems, Bannier, Neumann, 5/3/57 (*Policy on Budgetary Control*, DIRADM20382), *Remarks on CERN's financial policy*, CERN/GD/126, 20/5/57, etc. About the British, DSIR, Research Grant Committee, *Report from the CERN Panel...*, NP37, 12/2/58, letters Skinner to Verry, 29/10/58, Cockroft to Verry, 2/11/58 (SERC–NP31A). R.G. Elkington's remarks are in a report *Notes on the first meeting of the Bannier Working Party*, 28/2/62 (SERC–B125). For a strong attitude by Bakker during a joint CC-SPC meeting on 29 October 1957, see CERN/SPC/53/Rev. 2, 7/1/58, 6-9.
28. This point has been studied with care in an article by Krige and Pestre (1986).
29. CERN/SPC/67, 8/5/58; CERN/SPC/70/Draft, 1/8/58; CERN/FC/291, 4/6/58; CERN/FC/297/Draft, 7/7/58; CERN/WP/FC/298, 14/7/58.
30. Krige and Pestre (1986), 277-78.
31. Quotations respectively from CERN/FC/297/Draft, 7/7/58, 5; CERN/FC/287, 27/5/58, 1 (recommendation from the SPC to the FC); Council Minutes, 1958, 37 (remark by the Director-General); CERN/FC/297/Draft, 7/7/58, 6 (remark by John Adams).
32. CERN/FC/287, 27/5/58; CERN/FC/291, 4/6/58; CERN/WP/FC/297/Draft, 7/7/58, CERN/FC/303, 19/7/58.
33. The power of the governments remains essential in the last instance; if there is strong political disagreement (with Joliot in France or Blackett in the United Kingdom), they can veto any delegation.
34. See chapter 10 and 11; quotation from CERN/GD/132, 4.
35. On point one, see CERN/382, 18/11/60 and Council Minutes, 1960, 68. On point 2, see CERN/449, 8/6/62, Council Minutes, 1962 21, CERN/498, 30/5/63, Council Minutes, 1963, 27, *Working Group on appointment policy, Report*, 31/8/72. On point 3, private communication of the letters.
36. Sir Denys Wilkinson, in Nimrod (1979), 7-20, quotations on pages 9 and 20.
37. See chapter 8, section 8.2.
38. See chapter 12, section 12.6.
39. Wilkinson, *op. cit.* note 36, 20.
40. Geminus, *New Scientist, 300* (1962), 365; Merrison, 'A powerful electron accelerator for British physicists', *New Scientist, 312* (1962), 312-313.
41. On France, see chapter 12, sections 12.4, 12.5 and 12.6; on Italy, see chapter 8, section 8.4.
42. CERN/FC/252 (Draft), 22/11/57, 8. The only case which is a little more specific is Sweden.
43. This is readily seen by reading the letters they exchanged, or by going through the regular reports sent by the Director-General to the Council delegates, and by listing the nationals who are CERN fellows, research associates, etc.; for France for example, see DG20821.
44. Letters Heisenberg to Weisskopf, 25/7/63; Weisskopf to Heisenberg, 7/8/63; Perrin to Weisskopf, 6/9/63; Weisskopf to Perrin, 11/9/63.

45. See chapter 8, section 8.2.
46. Council Minutes, 1961, 24–49, session of December.
47. *Note* G. Hubbard to the Secretary of the DSIR (through Mr. Elkington), 22/10/61 (SERC–B125). For the importance of this point for the UK Foreign Office and Treasury, see various documents dated March–April 1961 in SERC–B124.
48. *Note* addressed to *Chanceries in Europe*, Foreign Office, S.W.1, 10/11/61; *Telegram* from Foreign Office to Embassies in Member-States, 15/12/61 (SERC–B125).
49. Council Minutes, 1961, 34–36.
50. This is reported in an internal British document, Hubbard to A. Stoddart (Foreign Office), 8/2/62 (SERC–B125).
51. Unless otherwise stated, the information and the quotations in this section come from Council Minutes, 1961, 37–43. About cool reactions in Member Sates to the UK Aide memoire, see the official answer of the Dutch government, 18/12/61 (SERC–B125).
52. History of CERN (1987), 220–221.
53. Council Minutes, 1956, 23–24.
54. Council Minutes, 1960, 71; *Draft letter to Monsieur Willems*, 15/2/61 (SERC–B124).
55. Council Minutes, 1961, 17–18, 30, 47.
56. Council Minutes, 1960, 71; 1963, 57–58.
57. What follows is mainly written from Council sessions in December 1961, June 63 and June 64.
58. Council Minutes, 1963, 15-23; the talks by Powell and Perrin are given *verbatim* on pages 16-18 and 19-21.
59. Council Minutes, 1963, 22. It was largely the debate about the new accelerators which led Adams increasingly to adopt such a stance; see chapter 12, sections 12.4 and 12.5.
60. About contacts with EURATOM, see Annex to chapter 9; on contacts to get money from NATO for a summer school in high energy physics, see letter Weisskopf to W. Allis, Scientific Advisor of NATO, 19/2/63; on contacts for an International Institute of Science and Technology, letter Bannier to Willems, 9/12/63 (reproduced in CC/528, 10/12/63).
61. About well co-ordinated initiatives to mobilize energies in a particular country, we would mention the campaign launched in Germany in 1964 (see DG20822). For an official recognition of the importance of 'diplomatic' visits, see *Concil Minutes, 1962*, 34 (on Denmark) and 35 (on Sweden).
62. About EURATOM, see Polach (1964) and Cadres Juridiques (1970).
63. CERN/100, 24/9/54.
64. History of CERN (1987), 273–282 and 261–273; CERN, Rapport Annuel 1955, 50.
65. History of CERN (1987), 209–230.
66. See correspondence in files DG20821 and 20822. The political nature of his nomination was clearly repeated to the author by Jean Richemond; about Gentner's nomination, see also History of CERN (1987), 416–421. Both were officially nominated during the Council session of 10/6/55.
67. Letter Cockcroft to Sir Ben, 16/4/57 (copy in CERN/GD/124, 3/5/57); GD is the file of the Group of Directors (of Divisions).
68. Quotation from *Future Organizational Structure of CERN*, CERN/SPC/46, 3/10/57,2.
69. Proposition made by Bakker in CERN/SPC/46, 3/10/57; minutes of the SPC held on 29/10/57, CERN/SPC/54, 25/3/58.
70. See the various interventions by Adams in CERN/GD/142, 17/10/57; GD/149, 6/11/57; GD/155, 11/12/57; CERN/SPC/54, 25/3/58, 9-11; Conseil, minutes, 19/12/57, 43. Quotations from letter Adams to Bakker, 15/10/57 (DG20551).
71. CERN/GD/157, 20/12/57; GD/162, 3/2/58; CERN/SPC/61, 24/2/58; SPC/63/Draft, 2/4/58 (meeting of 4/3/58), quotation p. 12.
72. CERN/GD/162, 15/1/58 (GD20963); CERN/SSM/6 (Senior Staff Meeting), 14/3/58; CERN/SSM/8/Draft, 25/4/58 (meeting of Senior Staff held on 21/4/58), quotation p. 2.
73. CERN/SSM/8/Draft, 25/4/58; SSM/6/Add.1, 29/4/58 (*Comments on the proposal for a new*

organizational structure (SSM/6) and an alternate proposal for a structure).

74. SSM/6/Add.1, 29/4/58, 6-9.
75. See the table given by Bakker to the SPC meeting held on 23/5/58 (CERN/SPC//70, 1/8/58, 9).
76. See letter J.M. Cassels to Sir John Cockcroft after a visit at CERN on the 11 and 12 October 1957, 12/10/57 (SERC-NP21).
77. We have in mind A. Citron, who arrived with Gentner but worked regularly with the PS division people; or C. Peyrou, who was responsible for bubble chambers.
78. See for example Ramm and Zylverschon's report after a visit in California, Memorandum to Bakker, 23/10/57, NTD1 to NTD5 (DG20551).
79. See the series CERN/Ex.C/1 to CERN/Ex.C/76. CERN/Ex.C/1 are the proceedings of the first meeting of the Executive Committee held on 23/9/58. For a longer analysis, see chapter 8, section 8.4.
80. *Major Groupings of the PS Division*, by J. Adams, 8/6/59, PS/Int. DIR. 59-6 (reproduced in CERN/GD/248, 9/6/59); *A proposal internal organization of CERN*, by S. Dakin, CERN/GD/251, 24/6/59.
81. Quotations from Adam's note given note 80, 1-2.
82. Quotations from Dakin's note given note 80, 2
83. CERN/GD/255, 30/6/59 (meeting of the Group of Directors); CERN/GD/251/Rev., 24/7/59 (new version of Dakin's note from which the quotations are taken); CERN/GD/258, 30/7/59 (meeting of G.D.).
84. See CERN/GD/251/Rev.2, 18/9/59; CERN/SPC/98, 28/9/59; CERN/SPC/101 (meeting held on 16/10/59), 19/11/59.
85. This condenses a far more complex set of events; see CERN/SSM/39, 14/12/59 (SS meeting, 27/11/59); note by Merrison, undated (DIRADM20004); CERN/GD/278, 15/12/59 (*Comments of the Senior Staff* [...]); CERN/GD/281, 21/12/59; letter Dakin to Verry, 5/1/60 (DIRADM20005). At the beginning, the Senior Staff tried to define only one position; this proved more and more difficult and two major groupings appeared at the end of 1959 when Dakin organized a survey among the Senior Staff (see Annex 1 of CERN/GD/281).
86. *Suggestions for the future structure of CERN*, signed by Bernardini, Gentner, Kowarski, Preiswerk, 7/1/60 (DIRADM20005); CERN/SPC/98/Rev., 12/1/60; CERN/SPC/105 (meeting held on 19/1/60), 16/3/60.
87. CERN/SPC/105, 16/3/60, 3-13.
88. Council minutes, 1960, 9-15.
89. CERN/FC/405, 25/1/60; FC/413/Draft (meeting held on 10/3/60), 28/3/60, 3-9; FC/418, 2/5/60 (*Financial Arrangements in connection with the future internal structure of CERN*). About the first coordinators and the Executive Committee, see chapter 8, section 8.4.
90. CERN/CC/354 (extraordinary meeting of Committee of Council held on 3/5/60), 24/5/60; letter de Rose to Member States, CERN/6600, 3/5/60; the new proposal by Adams is CERN/344/Rev., appended to CERN/CC/352, 20/5/60.
91. CERN/CC/352, 20/5/60, 2-5; quotation from Adams during a joint CC and SPC meeting held on 27/5/60 (CERN/CC/363, 13/6/60, 4).
92. CERN/CC/363, 13/6/60, 3-9.
93. CERN/344/Rev.3, 29/9/60, 19 pages, text for a joint CC and SPC meeting to be held on 3/10/60. For detail on the organization of the experimental work, see chapter 8, section 8.4.
94. CERN/CC/363, 13/6/60, 9; CERN/344/Rev.2, 3/6/60, 1; CERN/GD/311, 23/8/60, 1; CERN/344/Rev.3, 29/9/61, 3-4; CERN/344/Rev.4, Annex, 21/11/60, 4.
95. See the various versions of CERN/344 mentioned in the previous notes and in note 96.
96. CERN/GD/307, 8/7/60 (*Proposals for filling CERN Senior Posts temporarily*); CERN/GD/311, 23/8/60 (GD meeting); letter Bernardini to Adams, 12/9/60 (DIRADM20005); CERN/344/Rev.3, Annex, 29/9/60, 1-5; letter Citron, Michaelis to Adams, 18/10/60; CERN/344/Rev.4, Annex, 17/11/60, 1-5;

CERN/344/Rev.4, Annex/Rev., 14/12/60; Council minutes, 1960, 66–67.
97. All documents cited note 96, plus NPRC, meeting held on 21/11/60 (CERN/NPR/2, 5/12/60, DG20877), letter Citron to Weisskopf, 23/11/60 (DG20880).
98. Note by Adams in CERN/GD/248, 9/6/59.
99. See particularly CERN/344/Rev.4, 17/11/60, 4–5 about expectations among the group of directors for 1962 onwards.
100. For the whole section 2.4.1, see the very precise chronology of events drawn up by Monique Senft for a previous history project (CHIP10037).
101. SPC and CC met on 27/5/60 to discuss possibilities other than Adams for the post of DG. See Senft, *op. cit.* note 100, 19–20.
102. Senft, *op. cit.* note 100, 20–31.
103. Senft, *op. cit.* note 100, 32–37.
104. Senft, *op. cit.* note 100, 38–45.
105. Bernardini withdrew from the Directorate in connection with the negotiations for the post of DG; see for example letters Bernardini to de Rose, 23/11/60 and 25/11/60 (SC22823); about Amaldi, Council minutes, 1960, 67.
106. This information is derived from Council minutes and Annual Reports.
107. About Hine's role in financial matters, see chapter 12. It was also confirmed during various conversations with Pierre Germain, Kjell Johnsen, Mervyn Hine, Léon Van Hove and Victor Weisskopf.
108. This phenomenom emerged in more or less every interview with CERN staff, and surfaced again clearly during a seminar held at CERN by our history group with the CERN Staff in November 1987.
109. Partly derived from a long talk with Léon Van Hove on 27 November 1987.
110. See chapter 8, section 8.4, and chapter 11, section 11.3; see also the conclusion of this volume.
111. *Speech delivered by Professor Weisskopf to Staff Members*, on 8 August 1961 (DG20517); all quotations are from this 5 pages long text.
112. See chapter 8, section 8.5; DG correspondence files (DG20940 and 41); Hine correspondence (MGNH22133) and file DG20779.
113. Quotation from a letter Weisskopf to the authors, 14/9/87.
114. A perfect example of these letters is letter J. Augsburger (SB division) to Weisskopf, 1/5/63 (DG20601) and the answer, 14/6/63 (DG20939).
115. As reported by V. Weisskopf in an interview with the authors at CERN, 11/8/87.
116. On Experiment Committees, see chapter 8, sections 8.5 and 8.7. In addition to references already given in note 35, point 2, see DG20870, DIRADM20416. (Senior Staff meetings documents), WOL23124 (staff association), and interview G. Ullmann with D. Pestre, 24/9/87.
117. See chapter 8, sections 8.2 and 8.3.
118. On the active role played by CERN Senior Staff to get the privileges and immunities attached to Diplomatic Status, see DIRADM20041.
119. This is often invoked by active scientists during interviews.
120. On early doubts about this image, see Management (1957) and Hill (1963).
121. Sources are given with each table.
122. Chapters 10 and 12.
123. *De facto*, we tried to include them as often as possible.
124. Drawn from Annual Reports.
125. This point is detailed further and substantiated in chapter 8, section 8.6.
126. Quotations from *Preliminary Report of the United Kingdom,* CERN/ECFA/66/2, 2/5/66, ANNEX II, pages 29–31. For more precision, see chapter 12, sections 12.4 and 12.5.
127. References as in tables 7.11 and 7.12.
128. *Op. cit.* note 126, 31.
129. 'Mafia' is used here in a 'friendly' way—as most CERN scientists do when they use it.

130. For details on the electronic experiments, see chapter 8, section 8.7.
131. On French overall policy vis-à-vis CERN, see chapter 12, sections 12.4 and 12.5. BEBC became a tri-partite enterprise (CERN–France–Germany).
132. A Citron, *Preliminary Report of the Federal Republic of Germany*, CERN/ECFA/66/2, ANNEX II, 7-9.
133. These last figures concern the whole of the community bar Ph.D students. See *Preliminary Report of Belgium*, and *Preliminary Report of the Netherlands*, CERN/ECFA/66/2, ANNEX II, 3-4 and 16-17.
134. *Preliminary Report of Norway*, and *Preliminary Report of Sweden*, CERN/ECFA/66/2, 2/5/66, ANNEX II, 18-20 and 24-26.
135. CERN/ECFA/66/2, 2/5/66, ANNEX II, 1-2, 5-6, 21-23.

APPENDIX TO CHAPTER 7

Another aspect of CERN's European dimension: the 'European Study Group on Fusion', 1958-1964

Dominique PESTRE

7A1.	The origin of the proposal: stakes and motives	418
7A2.	The June 1958 Council session, and the political refusal to collaborate with Euratom	421
7A3.	The activities of the study group, 1958–1964	422
7A4.	Conclusions to be drawn from this episode on the nature and place of CERN in Europe at the end of the 1950s	424
Notes		426
Bibliography		809

The history of nuclear fusion for civil applications or, more precisely, of fusion by magnetic confinement, began around 1950/51.[1] During the early years it remained exclusively in the hands of some American, British and Soviet scientists. With the nomination of Lewis Strauss as the head of the AEC in 1953, money was suddenly abundant in the United States—something which also seems to have happened in Britain and the Soviet Union. However, due to the strict secrecy regulations prevailing at the time, the three communities remained completely isolated from one another. This was the period when knowledge on fusion was still very empirical, when one was more or less groping in the dark, where no theory was accepted by more than one group at a time and for more than a few months. This situation, typical of the years 1953–1957/8, arose for several reasons: from the enthusiasm and confidence of nuclear physicists ensuing on their achievements during the war years—everyone hoped that fusion could be controlled in less than 10 years; from the conviction that 'brute force technology' would succeed—the model obviously being the tactic adopted by General Groves in the Manhattan project; from the effects of the policy of secrecy which stopped anyone having a grasp of the field as a whole, and left each thinking that their rivals were poised for success.

In this context 1958 was first the year of great hopes, then that of a traumatic disappointment, finally that of a radical change in the understanding of the whole field. In brief, 1958 was a watershed. The trauma, perhaps more important in the West than in the East, was precipitated by the realization in April 1958 that the fundamental discoveries made eight months earlier by the British with Harwell's Zeta machine were false, discoveries which had raised hopes that the end of the tunnel was in sight. This was followed in the United Kingdom and the United States—the first exchange of researchers between the two countries was authorized in April—by grave doubts about the suitability of the method followed in all countries since the early 1950s. In September, during the Geneva conference, the truth dawned on an incredulous community: on this occasion the veil of secrecy was lifted, and scientists from the West and the East discovered that they were all a long way from confining plasmas, that they had each worked empirically by trial and error because they were convinced that magnetic confinement was just about to be achieved by the others—in short that each had thought that, since a technical solution was imminent, there was no need to check its theoretical foundations. The decade of the 1960s was thus marked by a number of characteristics each the opposite of what had happened during the first ten years. Basic research into plasma physics and magneto-hydrodynamics—rather than the technological attempt to build a 'fusion reactor'—was made top priority; university courses were set up, graduates were trained, specialist journals were established—and the policy of secrecy was abandoned; in consequence, in a few years, according to Joan Lisa Bromberg, there was a rapid improvement in the quality of the work done—and the growing conviction that controlled fusion had every chance of not being realizable before the year 2000.[2]

This is the background against which, at CERN early in 1958, the European Study Group on Fusion was born.

7A1. The origin of the proposal: stakes and motives

On 23 April 1958, François de Rose, recently appointed Director of Atomic Affairs at the Quai d'Orsay (Foreign Affairs) in Paris and then President of the CERN Council, asked to speak at the meeting of the Committee of Council. Under the item 'Other Business' he reported that he had been in contact during the last four or five months with Louis Armand, a top-level civil administrator in France and the President of Euratom. (Let us remember that, set up in 1957, the European Atomic Energy Community was an institution created by the Six (Belgium, Federal Republic of Germany, France, Italy, Luxembourg, the Netherlands) having as main goal research and production of industrial importance.[3])

In the course of their discussions, Armand and de Rose considered the possibility of forging links between Euratom and CERN 'in the sphere of fundamental research'. On the advice of Director-General Bakker, de Rose 'had made a new approach to Mr Guéron, Chief Scientific Advisor to Euratom, and told him that CERN might carry out some fundamental research on behalf of Euratom, particularly in plasma physics'.[4]

In the subsequent discussion Bakker spelt out his ideas in more detail. Believing that it could be interesting for CERN to work under contract with Euratom, he proposed that a study group be set up in Geneva 'to prepare a research programme for the next three to five years, to make budgetary estimates [...] as well as a forecast of the necessary staff and of the various facilities required'. No one raised any objections, and de Rose proposed that the views of the SPC be heard before he again got in touch with Guéron.[5]

A few weeks later John Adams summarized the origin, as he saw it, of what was to become the European study group on fusion. At the start of the year, he said, European specialists on fusion had asked CERN to organize a meeting, a meeting which took place at the laboratory in March 1958. There they discussed for several days 'the results of plasma physics experiments aimed at controlled thermonuclear reactions' and, in particular, 'the most notable results', those recently obtained at the AERE (Harwell) with Zeta. Granted the significance of the British findings, those present expressed the wish that CERN should organize further meetings of this type.[6]

In the wake of these two initiatives—and whatever the relative importance of each or their possible connections—a meeting between Guéron and the directorate of CERN (in fact Adams and Bakker) was arranged for 30 May. The proposal to create a 'Euratom-CERN Joint Study Group for Fusion Research' emerged from it. Expected to meet every month as from July—and so to start its work before the Atoms for Peace Conference in Geneva in September—the group was to submit its final report before the end of the year. Its prime aim was 'to note and evaluate plasma physics research programmes' then existing in the world and 'to consider and make suggestions for coordinated European fusion programmes' and 'of the means by which [they] could be carried out'.[7] The Adams-Bakker-Guéron agreement was submitted to the Committee of Council on 19 June 1958, approved, and passed on to the Council the next day.[8]

It is interesting to identify the motives of the different parties in this affair: after all, the proposal was of some significance in the euphoria of the early months of 1958. As the two preceding parallel stories have indicated, one must distinguish at least two groups of people and of motives. As for Adams, the high-energy physicists, and the European specialists in plasma physics, the arguments they put forward to explain their move were very similar to those used between 1950-1952 for the creation of CERN itself: Europe was behind in this new and crucial field of research; a significant financial investment was required and specialized people were in short supply; hence, if there were 'one research facility vital to the work and yet large enough to need the joint financial resources of European governments', the question arose as to whether 'following the pattern of CERN, it would be advantageous to set up one European laboratory for fusion research'.[9]

The emergence of these concerns at this moment can be traced to two changes in the international political situation. On the one hand the shock in the West at the success of the Soviet Sputnik (and the feelings of urgency that went with it), on the other 'the partial lifting of the security restrictions at the beginning of 1958' which was, according to Adams in March 1959, 'the main reason for these first discussions at CERN'. He added: 'it was hoped that very much more information would be released at the Atoms for Peace Conference', so justifying the management's decision to act very quickly.[10]

Two other more specific considerations have to be added if one wants to understand the rapid mobilization at CERN itself. The first is associated with a feature of the organization which we have spoken of on other occasions[11]: around 1957-1958 the CERN management was concerned about the future of the engineers who were building the PS. For one thing they wanted to keep this pool of top-quality expertise at the ready and to stop it breaking up; they also had to find them sufficiently interesting and challenging things to do. It is here that the proposal for a CERN-Euratom collaboration on controlled fusion became important. As Bakker stated in April—and this was his first argument for supporting de Rose—'research on plasma physics would be [...] likely to keep at CERN the staff which it would be unfortunate to lose'. Adams, for his part, said during the SPC meeting in May that

Notes: pp. 426 ff.

'the staff engaged in building the proton synchroton would [...] be very well suited for that kind of activity'.[12]

The second more particular kind of motive that we want to mention concerns the attitude of the politicians. Most European states, aware that they were considerably behind the United Kingdom and the United States in a field thought to be industrially and militarily very rich, hoped to take advantage of the imminent relaxation of secrecy to draw up a reliable balance sheet of the situation—and if possible get the relatively backward countries on the Continent to cooperate. It is obviously this need to define French activities in a favourable international context that inspired the network of administrators behind the scheme, notably de Rose, Armand, and Guéron. Their major concern was to extend and to homogenize the scientific policy of their country: to have the research arms of CERN and of Euratom collaborate could not but be good for Europe—and so for France—, and for science—and so for national industry.[13]

In conclusion one cannot but be struck by the close similarity with the circumstances surrounding the birth of CERN. In both cases it was a break in the international situation which stimulated the resurgence of ideas for European collaboration. In 1949 it was the explosion of the Soviet bomb and the end of the American monopoly in nuclear research; in 1958 it was Sputnik, and the declassification that went along with the preparations for the Atoms for Peace conference. This similarity suggests that Europe's position in international scientific competition had not changed fundamentally in 1958, that Europe was still largely dependent. Another similarity: in both cases there were two parallel initiatives (that of Adams and the scientists and that of the network of French administrators), initiatives which were combined into a single and rather ambitious proposal. One should note though that this time, thanks to CERN, the fusion of the two was far more rapid; it took only two or three months, while eight years earlier it had taken one and a half years; it was only in December 1951 that the Amaldi-Auger and the de Rose-Willems-Colonnetti networks really got together.

There was another difference: this time people thought that there would very likely be direct industrial applications—which was not the case in 1951-52. As Blackett remarked the first time the SPC discussed the question, 'work on fusion and plasma physics must of necessity have practical aims and would lead to problems of ownership and patents'.[14] And what of the military aspect? There is no reference to this in the sources we have found at CERN, though it is hard to believe that no one was aware of such implications—again in contrast to what happened in 1950/52. The list of laboratories that participated in the Atoms for Peace conference, and the national groups which took part in the activities of the CERN study group on fusion from 1958 to 1964 (see below) serve to make the point: military laboratories were very numerous among the former, and not absent from the latter.

7A2. The June 1958 Council session, and the political refusal to collaborate with Euratom

The notable feature of the joint proposition made to the Council by the CERN management, the SPC, and the CC is its very open character: collaboration with other European organisms (on 19 June de Rose mentioned, besides Euratom, the possible participation of the European Nuclear Energy Agency, ENEA), an ambitious programme (coordination of European programmes, even perhaps the creation of a new laboratory), a possible expansion of the activities of CERN itself (remember Bakker's words to the Committee of Council in April, or Scherrer's opinion, 'in favour of the pursuit of research into fusion questions at CERN'[15]). In one week the Council was to cut all these back to far more modest objectives.

In fact the Council was split down the middle on 20 June. The DG's proposal was supported by the delegates from the member states of Euratom, while all the others refused to vote in favour of the text. The Scandinavians and Yugoslavia did not want to consider the question—their governments had not yet had time to study it closely—, the United Kingdom asked that the aims of the study group be pared down—it was not to make proposals for coordinating programmes or building new laboratories—, and Switzerland and Greece proposed an amendment which affirmed that the non-member states of Euratom would in no way be committed by the conclusions reached by the group. Faced with this deadlock, it was decided to convene an extraordinary session of the Council one week later.[16]

Adams opened the session on the 27th, followed by Waller (S) who asked that a study group be established by CERN alone. Fry and Verry (UK) picked up the idea, adding a tight definition of its objectives. The minimalist British line carried the day in the face of the impossibility of getting any kind of agreement on the initial proposal. The text adopted affirmed that the Council:
- agreed that it would be useful to undertake a study 'to evaluate the plasma physics research programmes aimed at fusion',
- agreed that a study group should be set up to do this evaluation,
- considered that this study should be conducted by CERN alone,
- accepted that Euratom, ENEA or the OEEC could send observers,
- and thought it premature 'to consider if research of the kind under discussion [was] within the provisions of the Convention, or in the interest of CERN'.[17]

In general we do not like to 'reduce' explanations too quickly. We prefer first to describe the arguments exchanged in their variety and their complexity without distinguishing between 'pretexts', 'rationalizations', and 'real' reasons. Here, however, it is possible to go more directly to the point. On the one hand, there was the position of the states which were members of Euratom; they argued that there should be no delay since the study group was only being asked to put forward technical proposals, since it cost nothing (metaphorically and financially) to set it up,

Notes: p. 427

and since Europe was lagging behind: something had to be done quickly if one wanted to put the right questions to the Americans and the Soviets during the Geneva conference in September. The others replied that it was a matter of political choice since Euratom was involved and would be the main source of funds, that matters were not that urgent (did it really matter if one started two months before or two months after the conference?), and that CERN had to 'protect its complete freedom vis-à-vis other organizations'.[18] Obviously these arguments do not lay bare the two real issues at stake: to link or not to link with Euratom seen as the Trojan Horse of the Europe of the Six, in full expansion and wanting to rationalize its programmes; to have or not to have a coordinated European programme, this point being of particular concern to the British. As in 1950–1953, in fact, their technological advance put them in a position where they had far more to give than to receive.[19]

One last point deserves attention. These two debates in the Council on 20 and 27 June unfolded in a well-known and familiar way. Those who were backing the most ambitious programme (de Rose, Bakker, Adams) tried to trigger a process, to get others involved in a dynamic which they hoped would pick up steam—all the while insisting that this was just a technical stage, completely reversible, and involving no long-term obligations. (In passing let us remember that Auger behaved in the same way in 1950–52, both vis-à-vis his scientific opponents and the governments.) The others refused to accept the need to act quickly, fearing to be caught up in a process which would be manipulated by their opponents. The pressure being rather strong (a similar situation arose at the meeting in Paris in December 1951), a minimalist solution was accepted by all, a solution which juxtaposed the alternatives and from which each party hoped to gain the maximum benefit. A study group was set up, for example, which satisfied the pro-Euratom camp, but its nature and its objectives were reduced, something the other party imposed.[20]

7A3. The activities of the study group, 1958–1964

From summer 1958 to March 1959, the date when it submitted its report to the Council, the CERN Study Group on Fusion met three times: on 25 and 26 September, just after the 'Atoms for Peace' conference, so as to assimilate the new information; on 11 and 12 December to discuss European work in the field and to assess the possible interest of a joint laboratory on fusion; on 5 and 6 March to discuss the report to be submitted to CERN.[21]

Its first conclusion evidently took up the result reached by the participants in the Geneva conference, a result that was entirely unexpected only six or eight months before: namely, that the hopes 'of achieving a quick break-through in the problem of producing a fusion reactor' were unrealistic; and that the main task now—if not the only one—was 'to accelerate the understanding of the [fundamental] physics of plasma'. Its second, more precise, conclusion was that the European effort differed from that in the United States in that 'the operating costs for fusion research [were] but a small fraction of those of the USA'. This was apparently because the European groups, often recently set up, had 'concentrate[d] their efforts on smaller experimental devices'. Important differences existed all the same between the United Kingdom, on the one hand, and Germany, France, and Sweden on the other. The report also remarked that, as in all European research in physics, there was a severe shortage of qualified manpower. 'The first need, therefore, [was] to train more physicists', the report insisted, the second being to multiply the exchanges between European laboratories. Finally, the report raised the question which lay behind its creation, that of the establishment of a joint laboratory. Asking whether there was a very large piece of equipment which would be indispensable for research in plasma physics and 'which might be considered as the raison d'être of a European laboratory', it concluded that, in contrast to what one might have thought when Zeta had its finest hour, this was not so. From this the report concluded that, at least in the near future, no supranational laboratory seemed necessary. We should add, though, that the British were the most 'convinced' on this point.[22]

Though having completed the work for which it was set up, the group decided not to dissolve. It wanted to continue reviewing 'the European programmes for fusion research', thinking about a possible central laboratory, and organizing 'regular scientific discussion meetings'. To do this, it asked to retain CERN's patronage— failing which it would create 'a European Society for Controlled Thermonuclear Research', an independent scientific society which would sponsor the future work of the group.[23]

The group met seven times in the following years. As far as we know, the last meeting was held at Aachen, in Germany, on 12, 13, and 14 May 1964.[24] On the whole the discussions were technical, and dedicated to exchanging information and reports on the various European national programmes—in fact 'the original discussions on planning the research work in Europe' ceased from the fourth meeting. All the same Adams concluded in June 1962 that 'the Study Group meetings [did] provide a useful forum for the European groups by which some duplication and waste of effort in this research [were] avoided and a great deal of enthusiasm [was] spread'. 'Failing a more formal relationship', he added regretfully, 'the Study Group meetings [formed] an essential scientific link for Europe'.[25]

Before closing this episode in European nuclear fusion, there are three remarks we would like to make about the meetings held between 1959 and 1964. Firstly, on the precise composition of the group. Representatives from most western European study groups on fusion participated in all or part of the ten meetings. Thus we find two Belgian groups, one Danish group, four French groups, seven German groups (including one from the Siemens company), an Italian group, five Dutch, a Norwegian, two Swedish, one Swiss, seven groups from the United Kingdom

Notes: p. 427

(including staff from the Office of Naval Research in the American Embassy in London), and numerous CERN physicists.[26] At the end of the period more than one hundred participants were present at meetings held alternatively in the Federal Republic of Germany, in France, in Italy, in the Netherlands, and in the United Kingdom. We would also like to draw attention to the military and industrial connections which, though not absent, were weak and largely disliked by Adams, the Group's secretary; to the ongoing links with Euratom which was always represented by observers; to the looser ties to the EEC; and finally to the very strong connections with the nuclear establishment in the United States.[27] Initially, between 1958 and 1960, Amasa S. Bishop represented the US Atomic Energy Commission and took part in the meetings. In January 1961, after Bishop had left his European posting, Arthur E. Ruark, the head of the 'Controlled Thermonuclear Branch, Division of Research' of the US AEC, asked Adams if Edward A. Frieman of Princeton could replace him in the meetings of the CERN Study Group. This was arranged by Adams, and the reports were sent directly to the United States.[28]

Finally we would like to touch on CERN's precise attitude to the request in March 1959 that it be the Group's 'godfather'. In April and June 1959 Bakker suggested that the Group be independent. Since no specific joint action was recommended in its report CERN's role as a catalyst seemed to the Director-General to be over. The delegates of the member states in the Council, on the other hand, while appreciating his attitude, did not go along with it. Assuming fusion research to be crucial for their countries, they proposed that the question be reconsidered at the end of the year, and that CERN continue to play its role as European mentor.[29] In October and November 1959 Bakker repeated his position: CERN needed to devote all its efforts to the PS experimental programme, and since no CERN group was working systematically on plasmas, he would prefer it if the European fusion specialists founded their own scientific society and became autonomous. The latter, by contrast, put pressure on the Council to maintain CERN's patronage even if this was merely 'nominal' and 'more or less analogous to the role which UNESCO [had] played in the creation of CERN'.[30] On 2 December 1959 the Council acceded to the Group's wishes, and accorded it its patronage for another year. This was renewed in December 1960 and December 1961. The last mention of the European Study Group on Fusion in the CERN documents was on 13 June 1962. On that day the Council took note of a report written by Adams.[31]

7A4. Conclusions to be drawn from this episode on the nature and place of CERN in Europe at the end of the 1950s

The role played by CERN in this episode can firstly be described as that of a *home base,* the natural centre towards which partisans of European scientific solutions turned, and as *a model* which could possibly be repeated because it was judged to be effective and sufficiently autonomous of changes in the politico-diplomatic microcosm. This perception of the situation was obviously that held by the scientists, as well as by people like de Rose. As soon as the opportunity presented itself —the lifting of secrecy—and in the light of their conviction that the European states were behind, these men repeated the steps they had taken seven or eight years before and which had 'succeeded' so well. This time though, these same actions did not lead to the creation of another central laboratory. Certainly for 18 months—from early 1958 to summer 1959—the matter was seriously considered. At CERN the idea of directing some of the laboratory's efforts to this new field of research was even looked on very favourably. Matters were reappraised in mid-1959, however, and though the Group prolonged its own life while formally keeping more or less the same objectives, its real purpose was transformed: it was no longer thinking of a central laboratory, nor of coordinating the activities of the European groups, but of a simple venue for exchanging ideas and information, a neutral point of contact for European laboratories.

In Adams' last report to the Council he paused to reflect on this 'setback'.[32] He began by considering at some length the question of 'whether a European laboratory for fusion research would have helped the work or not'. His response was ambiguous but sufficiently nostalgic for one to get an idea of what he really felt. Of course he accepted the fact that fusion research did not yet call for equipment which 'unduly strain[ed] the national resources of the Member States', but he maintained that 'a greater impetus could have been given to this work had the existing effort been concentrated in one laboratory'. He was also convinced that, if CERN had 'decided to go ahead with plasma physics as well as nuclear physics', the encounter of both specialities in the same organization would have produced outstanding results. If this did not happen, if the opportunity had been lost, the cause was pusillanimity: 'only a determined and unanimous action with clear advantages in mind could have overcome [it]', explained Adams. Unfortunately, he concluded, the scientists in the European Study Group on Fusion did not dare propose such an initiative in their March 1959 report—and most politicians accepted this, relieved.

This explanation, which is not false, has its limits. Three other elements have to be taken into consideration if one wants to understand the 'non-creation' of a European fusion laboratory in 1959. The first, as in 1950–52, was the unequal development of Europe. Scientific and economic life being competitive, if one 'partner' is way ahead of the others all attempts to integrate efforts in a united whole tend to collapse—unless a very marked technical advance seems to be on the cards (e.g. a 30 GeV PS of the 'strong-focusing' type, which even interested the British). In 1959 nothing like this seemed in sight. The second element was that many thought that economic, and indeed military, benefits were just around the corner. The centrifugal tendencies for strict national control were then reinforced—and capitalized on by certain states. Finally, there was CERN itself: anything that threatened its existence was, of course, unacceptable to the directorate. More precisely, any move that seemed to be to the organization's advantage was warmly welcomed by them; but if the same initiative turned to be a complication (because of a change in the context, for example), it was pushed into the background as much as possible. Thus in 1958 there were high hopes for plasma accelerators. If one adds the fear of losing senior staff, one can easily understand Bakker's positive reaction to the proposals made by de Rose and Adams. In spring 1959, by contrast, the Accelerator Research Group completely reoriented its activities, and it did not look as though anyone at CERN would be short of work in the future: high-energy physics was in full expansion. Nothing was then to be allowed to impede CERN's taking its rightful place in this development—and interest in fusion research waned. To paraphrase Marx, one might say that initially, in 1958, European physicists had nothing to lose but their delay. At the end of 1959, by contrast, they had something much more important to lose: CERN itself. And in their view fusion, even European fusion was not worth the sacrifice.

The only people willing to take the 'risk' were those, like de Rose, who were not directly responsible for managing CERN, or a man like John Adams who was already interested in fusion and who was ready to play a role in this new adventure (in fact he became director of the British laboratory at Culham in 1960). For them, as for Armand *vis-à-vis* Euratom, the real stakes transcended those of CERN as laboratory. They treated it rather as a perfect prototype, as an excellent medium for installing their new organism—in brief they reasoned as statesmen favouring ambitious European solutions, and not simply as high-energy physicists. Wanting to repeat what they had achieved in 1950/52—even if the research had military and industrial connections in the medium term—, they envisaged a new laboratory shared by European states, and even considered stretching CERN's activities to have it enter this new field. Most states refused to support them—and fusion remained above all a national affair.

Translated by John Krige

Notes

1. This introduction is based on Bromberg (1982).
2. Bromberg (1982), 89–105.
3. CERN/265, 8/5/58 (CC, meeting, 23/4/58), 10–12. About Euratom, see Polach (1964).
4. CERN/265, 8/5/58, 11. The contact with Guéron is explained by the fact that 'Mr. Armand had become seriously ill shortly afterwards' (p. 10).
5. CERN/265, 8/5/58, 11–12.
6. Council, minutes, 27/6/58, 25–26. Same version in CERN/SPC/88, 31/3/59, 1. The quotations are from this last document.

7. CERN/269, 5/6/58 (*Minutes of a meeting held at CERN on 30 May 1958,* Guéron, Bakker, Adams being present).
8. CERN/280, 7/7/58.
9. The quotations are from CERN/452, 5/6/62. This is the final report drafted by Adams for the Council. Similar formulations can be found in the joint document EURATOM-CERN (CERN/269, 5/6/58, 1, which mentions 'the creation of a European centre'), or in Council, minutes, 27/6/58, 25-28. About investment and manpower, see CERN/SPC/88, 31/3/59, 18-20.
10. CERN/SPC/88, which reproduces CERN/FSG/7, 24/3/59, *Report of the CERN Study Group on Fusion Problems,* 1.
11. See Krige & Pestre (1986), 266. See also (CERN/CC/354, 23/5/60, 5).
12. CERN/265/Draft, 8/5/58, 11, for Bakker; CERN/SPC/70/Draft, 1/8/58 (SPC meeting, 23/5/58), 8, for Adams.
13. See for example CERN/SPC/80, 31/3/59.
14. CERN/SPC/70/Draft, 1/8/58, 8. On industrial connections, see also CERN/452, 5/6/62, 2, and the request by Siemens to participate in 1961 (letter to Adams, 23/2/61, DG 20625).
15. For EANE, see CERN/280/Projet, 7/7/58 (CC, meeting, 19/6/58), 8-9. Scherrer's opinion is given by Picot (Council minutes, 27/6/58, 27). Finally see Bakker during CC meeting held on 1/12/59 (CERN/CC/342/Projet, 19/1/60, 9).
16. Council, minutes, 20/6/58, 19-23.
17. On the fact that the negotiation was very intricate, see CERN/269/Add. 1, 20/6/58; Add. 2, 25/6/58; Add. 3, 27/6/58; Add. 4, 27/6/58; CERN/278, 27/6/58; CERN/278/Rev. 1 and Rev. 2, 10/7/58; Council, minutes, 27/6/58, 25-29.
18. Council, minutes, 20/6/58, 19-23 and minutes, 27/6/58, 25-29. The quotation is from Picot, 27/6/58, 27.
19. The detailed analysis for 1950-53 is in *History of CERN, Volume 1,* chapters 12 and 13 by J. Krige.
20. About Auger, *History of CERN, Volume 1,* chapter 6 by D. Pestre.
21. See CERN/FSG/3, 13/10/58 (MGNH 22077); CERN/FSG/9, 12/3/59 (DG 20625); CERN/FSG/7, 28/1/59 (DG 20625), *Report on the CERN Study Group on Fusion Problems, Draft,* signed J.B. Adams, 29 pp.; CERN/FSG/7, 24/3/59, 2nd Draft, attached to CERN/SPC/88, 31/3/59, 30 pp.
22. CERN/SPC/88, 31/3/59, quotations from pp. 17, 17, 20, 20, 21, 24.
23. CERN/SPC/88, 31/3/59, quotations from pp. 27, 27, 29, 29.
24. The last report we found is *European Study Group on Fusion, 10th Meeting,* Aachen-Jülich, 12, 13 and 14, May 1964 (DG 22793). A list of the first eight meetings is given by Adams in CERN/452, 5/6/62, Appendix I.
25. CERN/452, 5/6/62, quotations from pp. 2, 3 and 3.
26. See CERN/FSG/9, 12/3/59 (DG 20625); CERN/FSG/11, 17/8/59 (MGNH 22077); CERN/FSG/12, 8/2/60; CERN/FSG/13, 11/10/60 (DG 20625); CERN/FSG/15, 18/5/62 (DG 22793) and above all CERN/452, 5/6/62, Appendix II.
27. Letters H. Maecker, Siemens, to Adams, 23/2/61; Adams to Maecker, 3/3/61 (DG 20625).
28. Correspondence between Ruark, Morton Jr. and Adams, first semester 1961 (DG 20625). See also the lists of participants in documents cited note 26.
29. CERN/304/Add. 1, 27/4/59; CERN/SPC/92, 8/5/59, 6; CERN/CC/318 (Projet), 22/7/59, 9-10; Council, minutes, 26/5/59, 28-29.
30. Memo Dakin to Bakker, DIR/ADM/PS/04, 12/10/59; Memo Bakker to Dakin, CERN/DGO/Memo/721, 13/10/59; CERN/328, 29/10/59 (Adam's report); CERN/328*, 20/11/59; CERN/CC/342 (Projet), 19/1/60 (CC, meeting, 1/12/59, 9 from which the last quotation comes). The first is from Adams, letter Adams to Dakin, 13/11/61 (DG 20625).
31. Council, minutes, 1-2/12/59, 62-63; minutes, 8-9/12/60, 75; minutes, 19/12/61, 34; minutes, 13/6/62, 23; CERN/452, 5/6/62, Adams report.
32. All quotations are from Adams report, CERN/452, 5/6/62.

CHAPTER 8

The organization of the experimental work around the Proton Synchrotron, 1960-1965: the learning phase[1]

Dominique PESTRE

Contents

8.1 Setting up research around the Synchro-cyclotron, 1957–1960	432
8.1.1 Formulating a policy, 1956–1957	433
8.1.2 The implementation of the policy for the SC	434
8.2 The arrangements concerning the loan of big bubble chambers to CERN; 1957–1959	436
8.2.1 The agreement for the British National Hydrogen Bubble Chamber	437
8.2.2 The agreements for using the two French chambers	439
8.2.3 The issues at stake and the importance of these debates	442
8.3 The debate in 1959 around 'National Participation in Research at CERN': towards the disappearance of truck teams	444
8.3.1 First proposals: the idea of mixed teams	444
8.3.2 The reactions in the member states: thinking more globally about the organization of research	446
8.3.3 The revision of document SPC/93: a shaky compromise	448
8.3.4 Some reflections on this debate	449
8.4 The choice of experiments: the laborious setting up of the first committee system, 1959–1960	450
8.4.1 Remarks on some differences between electronic experimentation and experimentation with bubble chambers	451
8.4.2 The preparation of the PS experimental programme before the accelerator was commissioned, 1958–1959	454
8.4.3 The coordination of the emulsion groups, 1958–1960	456
8.4.4 The setting up of bubble chamber collaborations and of the Track Chamber Committee, the TCC, 1960	458
8.4.5 The first system for taking overall decisions on the experiments: Adams' project in autumn 1960	461

8.5	The changes made to Adams' decision-making system for experiments, 1961–1962	465
	8.5.1 The lingering problem of the Electronic Experiments Committee, the EEC	465
	8.5.2 The question of a science policy: the NPRC's relations with the specialized committees (EmC, TCC, EEC)	468
	8.5.3 Some concluding remarks on the process which led to the committee system finally adopted by CERN	471
8.6	Experimenting with bubble chambers: the strength of the system of collaborations, 1960–1965	473
	8.6.1 The functioning of the bubble chamber collaborations	473
	8.6.2 Some statistics on bubble chamber work at CERN from 1960 to 1965	476
8.7	Electronic experimentation: the struggle for access to the PS, 1960–1965	480
	8.7.1 The NP division and the CERN teams	480
	8.7.2 House-physicists and visitors in electronic experimentation: a sometimes difficult relation	484
8.8	Conclusion	489
Notes		494
Bibliography		809

The organization of experimental work in the more important high-energy physics centres assumed new forms at the beginning of the 1960s. Between 1959 and 1962 at CERN, and around the same time at American institutes like Brookhaven, Berkeley, or Stanford,[2] a start was made on codifying the rights and obligations of the various parties competing for time around the machine (laboratory management, in-house physicists, individual visitors, outside teams authorized to experiment, ...) The process of learning to co-exist was slow and regularly conflict ridden, the changes in habits being imposed by two factors. There was first a question of *size*—so of unitary cost of the basic equipment, be that accelerators or bubble chambers. From the 1950s onwards the former, for example, became so big that one no longer thought of constructing many machines of the same kind, even in a single continent. Then there was the nature of the *fundgivers*. From being local (groups of universities or states) they became federal in the United States and intergovernmental in Europe—and this imposed 'certain forms' of cooperation. Two extremes, two models, surfaced repeatedly during discussions: that of the facility as a *service centre* at the disposal of all the physicists in the region concerned, and that of a *laboratory in the conventional sense of the term,* sole master of its destiny, and so treating 'visitors' as little more than privileged guests.

To date very few studies have been made on *the concrete organization of experimental work* in large research units. Usually it is existential questions which have grasped the imagination—that of the relationship between individual and collective work, for example, or of the restrictions on creativity imposed by big science. We have in mind the nostalgic remarks of Alvin M. Weinberg for whom 'parts of basic science have already acquired [in 1972] some of the facelessness that characterizes applied science', Robert R. Wilson's 'my fight against team research', or Kowarski's words ('by the combined process of team work and of boss selection, Big science began [in the 60s] to influence 'pure' research in favour of playing safe and putting less insistence on markedly original innovation').[3] Other studies have sought to describe the daily operation of big laboratories. In many cases it is the directors themselves who did this, and the tendency is for the accounts to be somewhat acritical: tensions are glossed over, recipes for good management proliferate, and the real problems raised by teamwork and decision-making mechanisms are seldom if ever gone into.[4] Finally, there are the works of practising physicists who recount various aspects of their experience, sociologists who are interested in the functioning of the scientific community in its transition 'from the workshop to the factory', and journalists fascinated by the power struggles which are a feature of 'big science'.[5] Two titles from this variety of material seem to us to merit special mention. The first for its originality and for the precision of the enquiry— and because it is directly relevant to our topic: a report by Kowarski on 'user relations in the U.S. accelerator laboratories' which appeared in 1967. The second since it includes different and not always docile accounts in a genre where this is not usually the case, that of laboratory anniversaries: here the 'symposium on the History of the ZGS' held at Argonne on 13 and 14 September 1979.[6]

That granted, no detailed study has yet been made of *the concrete setting up of experimental work* around big accelerators in the 1960s and 1970s. This is the kind of study we shall undertake here, the case being that of CERN between 1959 (the year in which the 28 GeV proton synchrotron first worked) and 1965. The case is particularly interesting because it was historically the first, CERN having to deal with all these questions from 1958–59 onwards. Its 'precociousness' arose from the multinational nature of the organization and from the fact that, for the first time since the war, European high-energy physicists could enter into competition with their American colleagues on an equal footing.

We have been particularly interested in two aspects of the organization of experimental work: that of the *formation of experimental teams* (notably of collaborations which were usually composed of groups coming from laboratories in different countries) and that of the *decision-making mechanism* which controlled the access to the machine, a complex mechanism that was progressively put into operation at CERN between 1958 and 1965. Our overall approach to these questions is chronological. By way of introduction we spell out the system which was adopted between 1956 and 1959 around the 600 MeV Synchro-cyclotron, CERN's first accelerator which came into service in 1957. In sections 8.2 and 8.3 we go on to describe the long and stormy discussions held in Europe between 1958 and 1959 concerning the organization of future work around the PS; here we pay particular attention to the problems raised by the British and French requests to bring their bubble chambers to CERN. The problem was not simply that of several groups wanting to use a single accelerator but also that of using detection equipment which had become enormous and difficult to move. In sections 8.4 and 8.5 we study the decision-making mechanism for experimentation which was adopted at Geneva between 1960 and 1962. Several solutions were tried, the central issue being to maintain the equilibrium between the interests of CERN physicists and those of the member states, and between those of the bubble chamber physicists and of specialists in other experimental techniques. In section 8.6 we detail the solutions arrived at in practice at CERN for organizing bubble chamber teams while in 8.7 we look at some features of experimentation with so-called 'electronic' detectors.

8.1 Setting up research around the Synchro-cyclotron, 1957–1960

CERN had a double character for those who built it. It was to be a laboratory in the full sense of the term, that is to say, having its own scientific life and its own research teams, but it was also expected to be 'open', at the disposal of the entire European high-energy physics community. One year before the commissioning of the 600 MeV synchro-cyclotron (SC) the Scientific Policy Committee (SPC) set up a working party to look into the implications which this conception of CERN's role had for experimental work in Geneva.

8.1.1 FORMULATING A POLICY, 1956–1957

In its report on the organization of experimental work at CERN, dated 28 February 1957, the working party evoked two general principles each accompanied by a 'nuance'.[7] The first: the Director-General of the organization, and he alone, was to have the final power to decide about any experiment that was to be done at Geneva— though he was only to exercise this right after extensive consultation with the outside teams, the SPC, and the Council. The second principle: scientific excellence was obviously to be the sole criterion of choice for an experiment—it was only the intrinsic value of an experiment that counted and there was no question of there being quotas for different countries. The working party did concede, however, that a certain fraction of the time at the SC (estimated at 30% for the first year) was to be set aside for the best visiting teams. And it proposed that the choice of their experiments should pass through three stages: first 'through appropriate national bodies designated by the Member States', then at CERN where the best would be considered by an Advisory Committee nominated by the Council, and finally by the DG who, in the last resort, would decide on the proposals made by the Advisory Committee 'on such grounds as disputed value, undue burden on CERN facilities, etc.'.

This set of proposals was accepted without much difficulty by the Council in the ensuing months. Only one point was debated at any length. Raised by the Swedish delegates, it essentially concerned the high transport costs which the countries which were remote from Geneva would have to bear. Appealing to fairness, the Swedes asked that CERN itself pay these costs. The issue provoked a polemic which expanded to include the question of possible daily allowances, a polemic which the Council finally put a stop to by agreeing on 28 June 1957 that 'no allowance of any kind would be paid [by CERN] to members of visiting teams'.[8]

CERN's and the Council's philosophy on how to organize research seemed then to be rather simple, at least in theory, in 1956/57. CERN was a *laboratory* in its own right, and not only a service centre for member states physicists. It thus had *its own research equipment* and it was managed by an *independent executive* whose powers were clearly recognized by all. At the same time, being a collaborative facility constructed by and for the member states, it offered their scientists privileged access

Notes: p. 495

to its equipment. And their right to have a certain amount of machine time set aside exclusively for their use was recognized.

The Advisory Committee was established in June 1957. It met for the first time in August in the presence of seven visitors' representatives and ten CERN staff, when it discussed the proposals for SC experiments.[9] Their number being rather low, all were accepted. The committee met six times between 1957 and 1960, and put the agreed policy into practice.[10]

8.1.2 THE IMPLEMENTATION OF THE POLICY FOR THE SC

Two main points have emerged from the study of the implementation of this policy during the years 1957-1960. Firstly, the decision-making process for the experiments was not managed by a single body but by two, acting in parallel. The CERN management decided on its own what experiments would be done by CERN teams (whether or not they included the odd visitor), while the Advisory Committee, for its part, only studied the proposals coming in from the visiting teams. But it was the CERN teams who installed the basic infrastructure needed for experimenting, the best example being the μ-meson channel built by A. Citron's group.[11]

The second important finding is that the visitors who came to work at CERN did so almost exclusively within the framework of what was usually called *truck teams*. This means that they came as a group rather than as individuals to CERN, that they came from one, or at most two, national institutions, and that they brought their experimental equipment with them. Thus, by way of example, we find the Italian or French teams coming to CERN with their small bubble chambers, and preparing the beams in collaboration with CERN staff. The corollary was that there were very few mixed teams (CERN staff plus one or two national laboratories), even though this possibility had been envisaged by the working party set up by the SPC in June 1956.[12]

This conception of a sharp *divide* between CERN groups and national groups of visitors, both in the formation of teams and in the decision-making structure, was universally accepted until about 1959 as the system best suited to experimental work at the laboratory. It had the double advantage of satisfying the immediate interests of the teams (which were still rather small and were still as strangers one to the other) and of being simple to put into practice: it did not require that people of different origins, and who barely knew one another, were forced to work together. This is why, when the Advisory Committee began, at the end of 1958, to discuss how to organize research around the proton synchrotron it accepted that the same system be extended to the new machine. 'As a general policy', said an internal paper, 'it seems best for CERN to ask all visiting teams to be completely self-contained as regards staff, apparatus and tools'.[13]

In the light of what we have said, one may wonder whether there were not conflicts over the 30% time quota reserved for outsiders. Indeed there were, but they were of little import.

Two logics of argument, two kinds of attitudes were, in effect, discernible, that adopted by the CERN staff, and that of the representatives of the member states and of the visiting teams. For the former, the only criterion to be used in choosing any experiment was scientific excellence (interest of the proposal, quality of the team members, knowledge of the machine,...). Thus they only accepted that there be a quota of time for the visitors with some hesitation. Certainly, being aware of the importance attached to it by the member states' representatives, they did not ask that this quota be suppressed. On the other hand, they opposed any attempt to increase it above 30%, appealing to scientific quality. For them — but no one apparently disagreed with them on this point — 'research by the CERN groups was more efficient than that of visiting teams that often are sent out by institutes having as yet little experience in high energy physics'.[14]

Matters looked a little different to the representatives of the member states in the Council and the Advisory Committee. While they appreciated the importance of the criterion of scientific quality, this did not stop them regularly asking that the quota be reconsidered. To justify this they insisted that the gap between CERN teams — permanently on site, and increasingly masters of the machine and its equipment — and the visiting teams — necessarily less well prepared to use CERN's accelerators — could only widen if one was not careful. If the quota was never increased, some said, CERN would not fulfil one of its purposes, which was to provide physicists from the member states with the opportunity of doing top-level high-energy physics research.[15]

Despite its frequency, and the importance of the principle at stake, this debate had few *concrete* effects. The main reason was that visitors put forward relatively few proposals for experiments. The 30% quota was thus sufficient during the first two years and the Director-General could conclude at the Council in May 1959 that he saw no real opposition between the two points of view. 'Under the circumstances', he said, 'the two methods of approach led to the same result'.[16]

To conclude let us ask whether there were conflicts, or tensions, between national groups. Basing ourselves on the documents that we have, it seems that they were few and far between — which is only to be expected since there was little competition for the use of the machine. However, there are two episodes which lead us to suspect that, in a different context, things might have turned out differently. During the fifth meeting of the Advisory Committee (which was not attended by the Italian delegate Puppi) some of the proposals put forward by Italian groups were refused. This led the (Italian) Director of Research around the SC, G. Bernardini, to conclude that on this occasion the time granted to Italian teams was 'rather small'. At the following meeting a new Italian delegate, M.

Notes: p. 495

Conversi, was present. He immediately took up Bernardini's remark, stating that 'the last sentence of the Minutes [contained] an understatement, as the cyclotron time allocation to the Italian teams was 'very small' and not 'rather small''. Which leads one to think that national suspicions (which were if anything encouraged by the 'truck team' system) were never far from being mobilized.[17] The second example concerns the same country, from which two teams had been authorized to do their experiments simultaneously in 1959. In April Puppi said that he would prefer one of the experiments to be postponed, adding that he wanted the time set aside for it transferred to the other Italian teams — which the Advisory Committee accepted. Here one might well imagine that, if other teams had been waiting in line to take their turn, the reply would not necessarily have been the same. According to CERN rules, machine time did not belong to one country but to the community of visiting truck teams.[18]

8.2 The arrangements concerning the loan of big bubble chambers to CERN, 1957–1959

The way of practising research which we have just described began to be put in question at the end 1958/early 1959, at least by some people, and only as far as research around the proton synchrotron was concerned. The doubts were raised by the British proposal to bring a 1.5 m hydrogen bubble chamber to CERN.

To understand the new situation created by this offer, one must keep three things in mind. Firstly, if CERN wanted to host this chamber it had to prepare itself. Significant human and financial resources were needed for the purpose, investments which had to be counted in terms of years of work and of millions of Swiss francs. It was thus not simply a question of the British arriving with their equipment, experimenting, and then going back home. Secondly, once installed, material of this kind called for specific security arrangements and could not easily be disassembled or moved. Its bulk, its complexity, the time necessary to bring it into operation, demanded that it be installed and left in one place for rather a long time. This meant that a considerable amount of space in the experimental halls was immobilized for CERN and the other member states. Finally, the costs of operation were considerable, and here too one had to think in terms of millions of francs annually. Which meant, of course, that it was difficult to stick to a strict truck team system.[19]

Before launching into the detail of the negotiations, let us simply say that these questions were of general validity. They were at the heart of all the problems surrounding the organization of experimental work around giant accelerators from 1960 onwards.

8.2.1 THE AGREEMENT FOR THE BRITISH NATIONAL HYDROGEN BUBBLE CHAMBER

On 30 April 1957 a subcommittee of the British Department of Scientific and Industrial Research (DSIR) discussed for the first time the construction of a very large national bubble chamber. Two months later it concluded that a 'large hydrogen chamber' ought to be built in the United Kingdom by a consortium of five universities plus Harwell, and that its construction should start 'as soon as possible' so that the British could use the device in Geneva immediately the PS came into operation. The chamber could afterwards be brought back and installed at Nimrod, the British proton synchrotron then under construction, or left at CERN if a second chamber was built in the United Kingdom.[20]

In August 1957 CERN was told of this proposal, and arrangements were quickly made to discuss technical points regarding the characteristics of the chamber itself and the infrastructure needed to host it at Geneva: building, system for moving the chamber, ventilation and security system, power supplies, liquid hydrogen plant, etc. In December 1958, with the technical studies well advanced, the first systematic attempts were made to spell out the financial and scientific terms of the chamber's use at CERN.[21]

The first suggestion was that made by the British Treasury, who proposed, in December 1958, that the chamber be *rented* to CERN for about 18 months, i.e. for the period separating the presumed operation dates of the chamber and of Nimrod. This idea—which Cockcroft immediately dismissed as being absurd—was rejected by scientists in both the United Kingdom and in Geneva. In fact it overlooked one principle dear to physicists' hearts: that those who construct a new piece of experimental equipment (or, to be more precise in this case, have it constructed for them by engineers) do not do so for the pleasure of building something but for it to be used to do physics. The construction period is thus for physicists an investment and a necessity, never an end in itself, and the idea of renting out the device to colleagues had of course no appeal to them.[22]

This is why CERN proposed, in January 1959, another way of tackling the problem based on sharing the costs/sharing the benefits between the laboratory, the United Kingdom, and the other member states. This meant resolving two questions: the composition of the group which would decide on the first experiments, and that of the team that would do them. The proposal worked out by the CERN management on 22 January 1959 was submitted to Butler and to Moore, the two people responsible for constructing the British chamber. This offer suggested that the chamber be serviced by a group of British technicians, that the experimental programme be decided by CERN and that the experimental teams be 'partly CERN staff and partly from Member Countries'. Put differently, CERN was proposing that the chamber be used as much as possible on a European basis, i.e. without giving the British any clear advantages.[23]

Notes: p. 495

The British physicists found this arrangement rather unjust. '[...] To win the acceptance of the scientists and subsequently of the Treasury', argued Butler and Moore, 'it would be necessary to be more definite about the research benefits which the United Kingdom would get out of bringing the bubble chamber to Geneva'. Accordingly they proposed that a part of the time on the chamber be set aside for the exclusive use of a British truck team.[24] (It is worth noting that, in the very first version of the document that was subsequently sent to Butler and Moore, Dakin also envisaged this possibility. It was cut out on Director-General Bakker's insistence.[25])

At the request of the British, the CERN management modified the first draft of the agreement. On 25 February Dakin proposed the following changes:
- the chamber would be 'accompanied by a team of technicians *and scientists* to operate and *to do research with it*' (Dakin's own emphases),
- the programme 'of *beams* to be *put through it*' would be proposed by a group of scientists 'of CERN and Member Countries, certainly including the United Kingdom' (which, after further negotiation, became 'of a few scientists of CERN and UK and possibly including a few from Member States'),
- in principle, 'the UK would have a special claim to the use of a proportion of the film [i.e. the photographs taken with the chamber], say up to 25%'.[26]

Doubts persisted in the United Kingdom nevertheless. Since a part of the time was not specifically allocated to the British, they insisted that the *composition* of the team which would use the chamber be spelt out explicitly. For Butler and for the DSIR, the group 'of technicians and scientists to operate and to do research with [the chamber]' had to be mentioned as *constituting this team* 'together with some members of CERN and other staff of Member States'. Without a stipulation of this kind the agreement would be 'so inattractive to the Treasury so as to produce a request that CERN should pay for the use of the Bubble Chamber'.[27]

At this stage in the negotiations Dakin—in whom Butler seems to have placed a good deal of faith—put forward an argument which convinced all parties. He insisted on three points. Firstly, he referred to the conditions being laid down by the French concerning their own bubble chambers. 'I repeated', he wrote on 13 May 1959 in a note for Bakker, 'that we could not possibly alter this wording now since the French were wishing to send bubble chambers, and we thought would welcome any chance of getting their operation more closely under national control'. He thus first played on national rivalries to justify his refusal to concede what the British wanted. Then he pointed out that a careful reading of the text implied in fact 'a very substantial British contingent in the whole [experimental] team', as well as a not unimportant role in the choice of the beams. Put differently, Dakin was suggesting that the new British demands were somewhat formal and that they could be regarded as already covered by the existing agreement. Finally, he suggested that the text put to the Council '[was] not an agreement or even a draft of an agreement, but a statement of certain broad principles' which would have to be 'adapted' in direct discussions between CERN and the chamber builders, i.e. without other member states getting involved in the affair. The details, said Dakin, could be thrashed out at this time.[28]

Reassured by these interpretations — Butler felt able to say to his British colleagues on 13 May that, 'although [he] was not entirely happy with the present draft basis of agreement [...], he considered that in the light of discussions he had 'behind the scene' with Bakker and Dakin, it was the best bargain the UK group could get from CERN — at any rate on paper' —, the British delegation accepted the text which was unanimously voted by the Council on 26 May 1959 as document CERN/311.[29]

The British Treasury were not satisfied, however. The agreement foresaw that the United Kingdom would finance the construction of the chamber, its transport and the personnel responsible for its functioning, while CERN would take care of the infrastructure needed to host it, the auxiliary equipment, and the operating costs. On this the Treasury 'took the view that the scientific arrangements proposed in effect would place the bubble chamber entirely under the control of CERN and that the 25% of the results which the UK would get [...] was hardly more than the UK could have expected [...] since they contributed 25% to the budget' of the laboratory. Hence in August they proposed 'that CERN should pay for the transport of the bubble chamber [...], for its erection and dismantlement'. And discussions were reopened between Dakin and S.M. Smith, the DSIR's Finance Officer.[30]

After some months, and to cut a long story short, Smith accepted that the supplementary costs borne by the British — the transport of the chamber — were not as onerous as those borne by CERN who had to keep the chamber running for the benefit of, amongst others, the British physicists. Smith thus withdrew his demand, and CERN/311, 'Arrangements with the British Chamber Group', remained unchanged.[31]

8.2.2 THE AGREEMENTS FOR USING THE TWO FRENCH CHAMBERS

Negotiations concerning the French chambers got under way in March 1959. Two groups in Paris were involved at the time in building bubble chambers. The first was that at the physics laboratory of the Ecole Polytechnique, directed by Louis Leprince-Ringuet, and specialized in propane chambers. Its biggest chamber, 1 m in length, was decided on in 1957 and Leprince-Ringuet thought to use it first at Saturne, the 3 GeV French proton synchrotron, and then in Geneva early in 1960. On 28 April 1958 he officially informed Bakker of his wishes.[32]

The second group was that of the Service de physique corpusculaire à haute énergie at Saclay, which was specialized in hydrogen chambers. In 1958 it was building four detectors of this type one of which, the 80 cm, was scheduled to be ready at the end of 1960.

On 21 March 1959 Francis Perrin, 'High Commissioner' of the French CEA, told Bakker of the 'projects for the utilization of the CERN proton synchrotron by

French teams specialized in bubble chamber technique'. Reminding the DG of the offer made by Leprince-Ringuet a year earlier concerning the 1 m propane chamber, Perrin added that he hoped that CERN would also welcome French hydrogen chambers.[33]

Initially the French envisaged using their chambers at Geneva in the truck team framework. For them it was consistent with 'CERN's fundamental principles' that visiting teams were allowed to do experiments at the laboratory 'with instruments and techniques which [were] their own'. The chambers which they proposed to bring were not yet particularly complex—Leprince-Ringuet said, for example, that the 1 m propane chamber was 'relatively easy to handle' and that its security posed no major problems—, they would first be tested at Saturne before they were dispatched to CERN, and their operating costs were not likely to be onerous.[34]

In the light of these considerations the French proposed an agreement which differed from that then being negotiated for the British chamber. It had two main components. The first: the French teams would take financial and scientific responsibility for the transport and installation of their chambers, and for their operation throughout their stay at Geneva, the duration of this stay to be decided in consultation with the CERN management. The second: the use of the chambers would vary over time. For six months they would be reserved for the groups who had built the equipment, namely the French teams and the people at CERN whose task it was to prepare the beams. For a second period (at least equal to the first) the detectors would be treated as the common property of all European physicists. In this way one would both protect the cooperative dimension which CERN embodied and the privileges which were justly due to those who had invested time and money in building big detectors.[35]

By and large CERN reacted negatively to these offers, its idea being that 'the equipment [should be] regarded as CERN equipment for the duration of its stay in CERN even if it was supplied by a Member State'. For the laboratory, as in its negotiations with the British, the cooperative aspect was to take precedence over any priority that was due to the builders. Its argumentation was thus straightforward. Was one of CERN's tasks not to ensure that 'the interests of other Member States [are] taken care of' (Bakker) and that the CERN accelerators, 'the expensive facilities built and paid for by all the Member States of CERN' (Adams), were to be used in a balanced way by all countries? French demands to have a degree of preferential treatment were thus unacceptable. For the CERN management the impossibility of installing at Geneva all the big equipment built in the member countries left them no other option.[36]

A solution was found for the propane chamber in June 1959. Actually CERN had decided a year earlier to build a 1 m propane chamber of its own. Two chambers which were roughly identical in overall features were thus being built in parallel in spring 1959. The *builders* (Ramm at CERN and Lagarrigue in Paris) thus put forward the idea 'that close contacts be kept', 'that a decision as to which chamber was to be used first should be made when the operating date was more certain than now', and that, whatever that decision was, 'the first experiments would be carried out by a mixed team from the French and CERN propane chamber groups'. Announced at CERN by Ramm, this proposal was approved by the management, and Bakker could write to Leprince-Ringuet on 15 June 1959 that he was in agreement with 'the most important principle in [his] proposal', namely, 'that the first series of experiments on whichever propane chamber was the first to operate at CERN would be carried out by a group from the Ecole Polytechnique and a scientific group from CERN'. Bakker added: 'it goes without saying that the CERN group could include some visiting physicists'.[37]

An agreement of this kind between two parties had to be submitted to the SPC. As it stood the proposal largely ignored physicists from the 'other' member states—and it was violently attacked at the SPC meeting of 16 October, notably by Amaldi, Bernardini, and Heisenberg. The first, for example, repeated CERN's official policy: in principle heavy equipment brought to CERN 'should be available for use by scientists from all the other Member States'; one was therefore in no position to decide in advance who would do the first set of experiments. Scientific interest could be the only criterion. Each stuck to his line—Leprince-Ringuet, for his part, insisted that the fact that one had built such equipment 'must carry with it some form of reward'—and the meeting was long and tempestuous. Seeking a compromise, the SPC agreed on a set of general principles which could serve as a framework for redrafting the agreement on the propane chamber. These principles affirmed that:[38]

'1) proposed experiments with large equipment [...] would be discussed by an Advisory Committee [as for the SC];

2) in reaching its decisions, the Committee should take into account the effort made by the national team concerned in building the equipment;

3) the Committee and the national team would be responsible for ensuring that full use was made of the scientific resources available in Europe [...]'.

Negotiations restarted, then, after the SPC meeting between Leprince-Ringuet and the CERN management. A new and somewhat convoluted text was produced which included the following two paragraphs:[39]

> 'b) The programme of experiments for the bubble chamber will be worked out, for presentation to the Director-General, by French, CERN and any other scientists interested in experiments with the propane chamber. The particular experiments will be done by teams, always including the French representatives necessary for the operation of the chamber, and, in any case in which they wish, by the French physicists who naturally have a preponderant interest at least during the first period. c) During the second period, experiments will be done with a multinational team comprised of French, CERN and other Member State staff'.

On 2 December 1959, after another spirited discussion, the Council approved 'the proposed arrangement for the French propane B.C.'. All the same it was to be

Notes: p. 496

'interpreted' as Professor Amaldi had specified during the meeting; for him, 'it seemed dangerous to admit French physicists [...] simply because they had expressed the wish to take part in the work'[40]...

The debate over the Saclay 80 cm hydrogen chamber was no less intense. To avoid repetition, let us simply say that in this case the CERN management refused to envisage the time being divided into two parts (as for the propane chamber) and that, supported by the majority of the delegations, it insisted that the conditions adopted for the British chamber should apply. On 6 January 1960 Perrin accepted these arguments and agreed to this solution.[41]

8.2.3 THE ISSUES AT STAKE AND THE IMPORTANCE OF THESE DEBATES

On reading the preceding pages one cannot but conclude that the debates around the British and French chambers essentially came down to hard bargaining between rival groups, each wanting to be at the centre of the action that lay ahead—namely, research with the first European accelerator capable, since the war, of competing with American machines. More generally, any negotiation around a *single* piece of equipment to be used by *many* whose demands largely exceed the supply have these features—and we find them in 'national' laboratories in the United States.[42] Be that as it may, in Europe in 1959, three factors complicated the negotiations.

1. Firstly, they took place between the representatives of countries which were longstanding rivals, and who still harboured strong suspicions about the motives of the other national groups. An element of *a priori* mistrust was thus added to the competition between scientists, a mistrust rooted in a sense of otherness far more radical than if the competition had been confined to the national level.

One might object that this was not the first time that Europeans had to work together, and that, for example, they succeeded quite well when building the PS. This is not false, but misses the point. The PS was constructed by one team, collected together in one place, whose task it was to build a unique instrument—and who were in a race to be the best against the specialists at Brookhaven who were building the 30 GeV AGS. This determination to be first generated a solidarity in a team whose efforts were further integrated by the incontestable skill of its leader, John Adams.

The situation was quite different in the case of the debate in 1959 about the French and British bubble chambers. Here the competition was between European groups, and it was between teams and laboratories which as yet had no experience of working together. Adams' entire division was in Geneva, but Butler was in London, Leprince-Ringuet was in Paris, and Amaldi was in Rome.

2. The second factor complicating the debate was that, at the time, each party was reasoning with a particular model in mind, the tried and trusted truck team system

adopted for the SC. This is evident in the case of Leprince-Ringuet or Perrin, as we have seen, but it was also true for the British. Speaking, for example, about the Treasury's idea of renting the detector, Cockcroft said in December 1958 that this was unacceptable because it 'strikes at the root of the truck team principle'. Similarly, when Smith drafted the minutes of the British Bubble Chamber Management Committee meeting held on 20 February 1959, he wrote that 'Butler would get written into the agreement an assurance that 25% of running time would be allocated to UK Truck Teams'.[43]

Even more important, though, is the fact that it was this conception of experimental work which motived the construction of the different chambers. It was because each country thought that it would come and do the kinds of experiments that it liked at CERN, and with its equipment, that it launched an independent national construction programme.[44] CERN's policy was informed by a similar logic. Between January and May 1958, while it was defining its own bubble chamber construction programme, it knew what the British and the French were building. The management also knew that they intended to bring their chambers to CERN.[45] All the same, it did not take these facts into consideration, its interest being to equip itself as if it were an autonomous laboratory. Certainly, in March 1958, the SPC discussed the offer made by Blackett for a CERN–UK collaboration so as to speed up the construction of the British chamber, but the CERN management was not keen. It wanted to build its own hydrogen chamber. Similarly, desiring to have a detector ready when the PS first worked, it chose to build a 1m propane chamber, without even referring to the fact that a chamber of the same type was being built at the Ecole Polytechnique. The possibility of coordinating construction programmes was far from everyone's minds, then, in 1957 and 1958.[46]

Hence the misunderstanding and the asperity which characterized the debate 12 or 18 months later. And it was only at this time that a minimal notion of planning began to emerge, and Bakker, Amaldi, or Leprince-Ringuet could say that it would be 'highly desirable [for the future] that two countries should not construct the same sort of large experimental apparatus for research work at CERN'. But in a sense this was too late for this first generation of bubble chambers.[47]

3. The third factor complicating the negotiations was that the parties had to draft texts which laid out principles rather than trying to solve immediate and concrete questions which had emerged in practice. This problem was linked with the fact that CERN was a new intergovernmental body in which each state wanted to protect itself from the others by laying down precisely the rights and obligations of all. When the interests of visitors were at stake the states wanted to codify beforehand the rules which they expected the management to apply. In this context the 'legislative' aspect — with the risks of dogmatism that it carried — was given far more weight than was the case in ordinary national laboratories.

The rather abstract nature of the arguments used derived from this constraint, as did the difficulty of reaching a 'fair' compromise between the diverse positions. The arguments of those who wanted to bring heavy equipment to Geneva were, after all, 'justified': they asked for the right to favourable treatment in view of the time and money they had invested. But so too were the arguments on the other side: the space was limited, the PS had been financed by all, and it was difficult not to use any equipment installed at CERN on a communal basis. The proof that these two attitudes were 'well-founded'—at least when formulated in this abstract and antagonistic way—is that they were easily reversed. The French, for example, could remark, during the debate on the arrangements for the British bubble chamber, that it seemed normal to them that the UK physicists do a set of their own experiments with their detector. All the same they accepted the arrangement which made no allowance for this. The British, for their part, were dissatisfied with this same arrangement, but we never saw them intervening in the debate on the French chambers.[48]

We find the same ambiguity in the attitude adopted by the CERN management. When appealing to principles, it always said that no privilege should be granted to teams bringing equipment, and that it was its responsibility to ensure that 'the interests of other Member States were taken care of'. It 'forgot' all about this on one occasion, however: in June, as we have said, it acceded to the wishes of those of its staff who were building a propane chamber, and asked—justice required it—that they be among the first to use the device.[49]

8.3 The debate in 1959 around 'National participation in research at CERN': towards the disappearance of truck teams

We have just seen that the discusssions on how to use the French and British bubble chambers at Geneva led to the idea that teams which were not homogenous should be set up. The debate, however, did not stop at this point. Beginning in April 1959, it was extended to cover all kinds of experimental groups—and not simply those using bubble chambers—which were envisaged around the PS.

8.3.1 FIRST PROPOSALS: THE IDEA OF MIXED TEAMS

The British were the first to raise the more general question. In a letter of 11 February 1959 Sir John Cockcroft asked that 'the question of the arrangements for visiting scientists to work with the proton synchrotron' be included on the agenda of the SPC meeting in April. He proposed 'that it should be possible to form mixed teams of CERN staff and visiting scientists to work for periods of 3 months with a mixed team and then return to their own department and later on to come back again'.[50]

In anticipation of this discussion the CERN Senior Staff met in March. An internal document dated 2 April summed up their position. They stressed two points in particular. Firstly, that since 'there are special habits of mind and working which are necessary and which take some time to acquire before anyone can be useful round one of the big machines', visitors should not stay at CERN for short periods of time. 'Two years', wrote Dakin, 'was the minimum stay for work round [...] the proton synchrotron', failing which 'the duller work of preparing experiments is likely to fall on CERN staff'. Secondly, that 'if continuity [in the work] was to be ensured', the maximum number of visitors should not exceed two per team of five or six. For the CERN physicists this meant that very little space would remain for the truck teams.[51] All the same the document remained cautious on this last point. It accepted that a large range of teams could be envisaged 'from normal truck teams (if these are practicable round the PS)' to teams in which CERN physicists were preponderant.[52]

In the subsequent meeting of the SPC the visitors approved of the idea of mixed teams. There was, nevertheless, a difference of opinion over the minimum period visitors should stay at the laboratory. For Massey or Amaldi, for example, it was possible to rotate people in a team, the idea of sending someone to Geneva for two years seeming to be too difficult to implement in the framework of the European university system.[53]

All of this appearing to be somewhat vague, the CERN management was instructed to draw up a more complete document to be submitted to the member states. The first version (GD/152) was drafted by Dakin on 24 June. It was discussed by the Group of Directors the next week, after which a definitive version dated 14 July (SPC/93) was distributed to the member states.[54]

These documents began by defining the framework for discussion. Firstly, it took into account nothing but research 'with apparatus and instruments provided by CERN' since 'it is assumed that, with the very high cost and complication of the most advanced apparatus now in use, there will be a strong tendency for the truck teams [...] to disappear'. Secondly, it was said that 'teams composed simply of visitors from Member States would in nearly all cases be themselves quite impracticable' since the help of CERN scientists was needed for the setting up of the experimental apparatus.[55]

Three organizational principles flowed from these basic considerations. The first was that an average quota of posts ought to be reserved for scientists from the member states, this quota being estimated at about 50%. Similarly, it was suggested that 'a due proportion of the visitors must be given duties as team leaders or deputies'. The second principle was that 'the visitors should remain members of the staff of their home institutes [...] but [that] they will have to come to CERN for certain minimum necessary times', estimated at 'two to three years for the PS'. It was only in exceptional cases that a shorter stay would be acceptable. The corollary of this was that the category of research associates — scientists from the members states who were paid by CERN to work at the laboratory for two or three years — would be abolished. The third and last principle was that 'there should be ample and frequent

Notes: p. 496

discussion at the working level between physicists from Member States and CERN during the planning period'. In this way the choice of experiments and the formation of the teams could be arranged with the minimum of conflict.[56]

Two remarks have to be made at this point. The first is that some historical residues of the truck team system persisted among the CERN Senior Staff. On the one hand there were the CERN physicists, the only people thought to be really capable from a scientific point of view, and on the other the scientists from the member states, equally capable in abstract, but necessarily less efficient because of their limited knowledge of the PS and of the specificities of the physics that could be done with it. '[The] policy on visiting scientists', commented a CERN document, ought therefore to be 'something of a *compromise between the political need* to give Member States what they will feel as an adequate direct participation in the research going on in CERN, *and the fully efficient conduct of that research*' (introductory statement to GD/152, our emphasis). Hence the insistence on the notion of a quota, already suggested in the first discussions in April, an arithmetic solution to the problems raised by the dual nature of CERN.

The second point to note is how much everything remained centred on CERN itself. Of course the laboratory would welcome visitors from other centres, but at all times it was to keep control of the situation. For CERN Senior Staff, only they would be capable of building sophisticated equipment, and only they would be present in all teams.

8.3.2 THE REACTIONS IN THE MEMBER STATES: THINKING MORE GLOBALLY ABOUT THE ORGANIZATION OF RESEARCH

On 15 July 1959 document SPC/93 was sent for comment to the member states. Two months later CERN had received nine replies. These were gathered together and distributed for information at the end of September under the labels SPC/93/Add and Add2.[57]

All responses had three points in common. Firstly, they recognized that the question raised by the document was a crucial one, as was the principle that 'there should be ample and frequent discussion at the working level [...]'. This was indicative of a latent fear in the member states that they would not be fully associated with CERN's scientific life. The two other remarks, by contrast, were openly critical. The demand that people should stay at the laboratory for two or three years was dismissed as simply unrealistic. As the British put it, 'Such a proposal would cut him [the visiting scientist] off from his parent organization [...]. It would make him virtually a member of the CERN staff in all but pay'. The person who drafted the reply, ironically taking up a phrase in SPC/93, added 'The panel does not consider that a British university would regard it as sufficient recompense for a long period of secondment on full pay that it would be able 'to count on the return of the visitor bearing his laurels in the shape of experience and publication''. The second criticism we find in the bulk of the replies concerned the idea of abolishing research associates. As many of those who wrote felt that it was inconceivable to release university staff for long periods, they demanded that this category be retained. It was the only scheme which allowed for the secondment of people for the long periods that were deemed to be indispensable.[58]

A rather more general point was raised by Bannier and de Boer (The Netherlands), Berthelot and de Rose (France), Heisenberg (Germany) and the British DSIR. All asked why 'the only scheme on which a visiting scientist could be accepted for work with one of the large machines should be as *individual* for a period of 2-3 years', why the document considered it 'appropriate to exclude national *teams*[59] altogether for the future' (our emphases). On the contrary, they insisted, for a particular experiment 'a University (or other research organization) might make an agreement to maintain a group of a defined number of physicists at CERN for a specified period of several years', the internal organization of this group being its business.[60]

It was, however, Berthelot who made the most comprehensive critique of the paper. Identifying two imperatives — 'do good physics at very high energy', and 'associate actively the greatest number of member states' physicists with this physics' — he suggested that two directions could be followed. The first was that spelt out in SPC/93: 'CERN, builder and manager of the huge European accelerator, also become the only constructor of the equipment needed to exploit it'. Hence it welcomed individuals into its teams. According to Berthelot, this direction would cripple national laboratories since their best personnel would leave them for 2-3 years, and because national centres would no longer participate in the development of new experimental techniques. In Berthelot's view this would damage both Europe as a whole and CERN in particular: after all, he pointed out, 'we often see original ideas concerning experimental equipment emerging outside the big centres'. His conclusion was clear: even if this 'philosophy' had some advantages, it was certainly not the best solution.

The other approach, and the one favoured by Berthelot, was less detached from the actual situation, more pragmatic. He suggested that 'CERN renounce the idea of doing everything itself and seek an equilibrium in the distribution of activities between those which it would undertake itself and those which would fall to laboratories in the member countries'. Considering the problems in the sequence in which they would arise in practice, he distinguished several phases. Firstly, one would have to establish regularly *the list of experiments*. Here, the ultimate decision was to lie with the Director-General assisted by a special committee. It was his task too to designate the people responsible for doing the experiments by choosing 'between the different candidates which [had] emerged during the consultations'. It was only then that the *definite composition of the teams* would be settled. They would consist of CERN groups as well as individuals and national laboratories. 'No *a priori* codification of their composition' was to be made: that was the responsibility of the leader of the experiment. Similarly, the construction of the necessary

Notes: p. 496

equipment was to be dealt with by the *team*. CERN would help, of course, but certain laboratories had highly competent people, and it was 'reasonable to take advantage of these men's experience and to let them build the apparatus they were qualified to build'. Contracts 'between CERN and the interested laboratories — with a financial contribution by CERN and a possible distribution of tasks' could well be imagined. In other words, 'one ought to keep things fluid', CERN's task being to coordinate the efforts regarding the provision of equipment in Europe as a whole, and not to try to do everything itself.

As for the *phase of experimenting* at Geneva, here too Berthelot thought that there was no need to be dogmatic. Each team obviously had to be as efficient as possible, but there was no point in repeating over and again that CERN physicists had to participate in every experiment or that visitors had to come for at least two years. The first comment identified something 'which would be obvious to the experiment leader', while the second 'was a matter which was purely internal [to the team] and was not something which had to be subject to a rule'; it was up to the team to keep 'a sufficient permanence in number and in qualification to ensure that it made the best use of machine time'.[61]

In the light of these sometimes radical criticisms, a revision of SPC/93 was obviously called for.

8.3.3 THE REVISION OF DOCUMENT SPC/93: A SHAKY COMPROMISE

This document was revised twice by the CERN management in September 1959. On 16 October a new version, SPC/93/Rev, was laid before the SPC. Several amendments were proposed, and a new paper (CERN/322) was accepted without changes by the Council on 2 December.[62]

The most striking difference between SPC/93 and SPC/93/Rev. was the recognition that the 'universities and institutes in the Member States wish now, and will doubtless continue to wish, to bring to CERN apparatus which they have constructed themselves [...]'. A measure of suppleness was thus introduced into the document. Its backbone remained untouched, all the same, and the radical proposals made by Berthelot were overlooked. SPC/93/Rev thus reaffirmed, in its preamble, that 'quite apart from CERN staff who will be operating the machines, [and] those providing peripheral facilities such as beam transport [etc...], any experimental team should contain a proportion of CERN experimental staff'; it maintained that '[even if] there is no wish to be dogmatic about the length of stay', the question should not be ignored: ultimately one should envisage adapting the principles of functioning of the universities to the new situation created by CERN rather than the contrary; finally, the document still suggested that the category of research associates be suppressed.[63]

The SPC met on 16 October 1959. Its first unanimous decision was to keep the research associate system, this being ratified by the Council in December. Thereafter the discussion was stormy, and interwoven with the debate concerning the use of the French bubble chambers. As a result some of the more interesting questions raised in the responses to SPC/93 were only touched on. Nevertheless a final version of SPC/93/Rev. was submitted on 10 November. In comparison with the preceding version it emphasized the importance attached to the fact that all the teams be international, and that they would use 'apparatus constructed by CERN or brought by the member states'. It also specified that the arrangements for using this material could vary (those for the French and for the British chambers, for example), and expressed the hope that, henceforth, 'any proposal having in mind the construction of an item of equipment to be sent to CERN would be announced and discussed in detail from the outset'. Finally it asked that, as requested by the SPC, an Advisory Committee for the PS be created. As for the SC, its task would be to watch over the interests of the visitors.[64]

8.3.4 SOME REFLECTIONS ON THIS DEBATE

The most important point to bring out — to the extent that it is not again necessary to insist on the 'legislative' and 'principled' character of this debate — concerns the two major axes underlying it. In contrast to what was happening at the same time in the discussions over the French and British chambers, this conflict did not pit visiting groups against another — on the contrary they were united —, but the CERN Senior Staff against the representatives of the member states. The former persisted in declaring that they alone had the human and material resources needed to do physics at the leading edge of research, while the latter reacted violently against this opinion and refused to be treated as second class physicists, or to be reduced to the role of an auxiliary force. To understand this entrenched division is to understand above all the attitudes of CERN's senior physicists, that of the visitors always being defined in reaction to them.

The Genevan physicists were first driven by their experience around the 600 MeV SC. They thought, as we have seen, that research done by CERN groups was always better than that done by outsiders, the main reason being that the latter had insufficient contact with the machine. Convinced that this difference could only increase in the case of the PS, they were emphatic that visitors should only come to CERN as individuals, whereupon they should be integrated fully into the CERN teams (hence their demand that they stay in Geneva for a long time). One might interpret their attitude as arrogant — as the outside physicists did — but it also seems to be symptomatic of a certain anxiety. To experiment at 28 GeV with the most powerful machine in the world — and without having any intermediate experience at the time — was in fact a leap into the unknown, a challenge sufficiently daunting in itself that it was deemed essential not to complicate matters by the presence of visitors. The effects of the decision, taken in 1952, to compete with the Americans by beginning at once to build the biggest machine in the world were now being felt.

Notes: pp. 496 ff.

Hence a measure of uncertainty and of anxiety about what lay in store. Hence the reticent and cautious attitudes of CERN people when speaking of the organization of work around the PS.[65]

One can also put forward another, more prosaic, reason for their attitude. No matter how great the challenge presented by working with the PS, the rewards would be proportionate—the possibilities of new discoveries in an energy range which was as yet unexplored. Seen from this angle, it was preferable to retain as much latitude as possible concerning experimental work, to have as few limitations as one could on the use of the accelerator—and of course not to multiply the number of competitors who also wanted time and space in the experimental halls. Hence a lack of enthusiasm for strangers, for a demanding and yet poorly prepared additional contingent.

A word by way of conclusion—deriving perhaps from the fact that we know how the story ended. All discussants spoke of 'experimental teams' without being more precise. But there was not 'one' kind of experimental team and it is quite likely that different people were speaking of different things. The teams referred to in CERN documents—and by G. Bernardini in particular—were small (5-6 people), and typical of so-called 'electronic' detectors. When Berthelot was speaking, by contrast, he seems to have had larger teams in mind using, for example, big equipment like hydrogen bubble chambers. Possibly then we have here a dialogue of the deaf facilitated by the abstract character of the debate.

8.4 The choice of experiments: the laborious setting up of the first committee system, 1959-1960

Now that we have studied these debates it would be worth going on to describe what actually happened in the following years. However, to do this properly one first needs to have some idea of the committee system that was set up to decide the experimental programme. We shall look at this now, postponing our remarks on the actual way in which teams were organized to the last two sections of the chapter.

Before getting under way one important comment must be made. No general discussion was held in 1959 on how, and by whom, the experiments at the PS would be decided on. There were several ideas in the air—notably that of a procedure based on the system then in use at the SC[66]—, but no overall view had been expressed in November 1959 when the PS first worked. As a result partial and pragmatic solutions were put forward until summer 1960, solutions which were successively or simultaneously tried without there being any overall plan. The account that we will now give will accordingly be rather descriptive and close to the actual course of events—here, more than elsewhere, to understand the final result arrived at largely comes down to describing its genesis[67]—, and it might seem rather fragmented. To make it more cohesive would, however, mean imposing *a posteriori* a pattern on the events which did not exist at the time. For a while the reader will therefore have to contend with a number of parallel 'stories' stitched together to form a patchwork rather than a coherent whole.

One other point: to grasp the problems dealt with in setting up a committee system to decide on the experimental programme one must have some idea of what it means to do an experiment in high energy physics. We shall therefore begin with a rapid presentation of this dimension of the problem, particularly for readers who are not specialized in such matters.

8.4.1 REMARKS ON SOME DIFFERENCES BETWEEN ELECTRONIC EXPERIMENTATION AND EXPERIMENTATION WITH BUBBLE CHAMBERS

Two main experimental traditions existed in high-energy physics in the 1950s and 1960s. These were totally different from one another, made use of quite distinct kinds of know-how, and were carried out by different people (and even by different laboratories).[68]

The first was that using electronic forms of detection, i.e. based on the measurement of electronic impulses to identify the particles, their position, and their momentum. The Geiger counter, the central tool for nuclear physics before the war, was the father of this mode of detection. Poorly adapted to the high flux of particles produced by accelerators, it was superceded in the years 1945–55 by scintillators (connected to photomultipliers) and Cerenkov counters, and by the development of complex electronic circuitry which allowed for an extremely precise time resolution (some nano-seconds in 1958–1960). Around this time another type of electronic detector came onto the scene, the spark chamber; it allowed for a better spatial resolution than the previous systems. Used during the sixties with scintillators (which served to trigger the chamber at the right moment) this device became an important tool for electronic detection.[69]

The second tradition was that of visual detectors which used changes in the thermodynamic equilibrium in a gas or a liquid (changes occasioned by the passage of a charged particle) to make that particle visible. This group comprises essentially the Wilson or cloud chamber, the first in the tradition, and the bubble chamber, which played a key role in high-energy physics experimentation in the 60s. Photographic emulsions, which were the stock-in-trade of cosmic ray specialists, but which were also used around accelerators, may also be considered as part of this tradition, even if they were somewhat marginal to it.[70]

We do not intend to go into the technical aspects of these two traditions. Our aim is simply to study the effects on experimental practice. And though of course there was no absolute determinism, it is certain that the technical aspects of the work generated different organizational forms. Let us be more precise.

Of prime importance in the organization of a bubble chamber experiment is the chamber itself: it is a heavy and expensive piece of equipment, as we have said, difficult to move and not easily modified (at least not without a hold-up of several months), and it demands exceptional security measures if it is a hydrogen chamber. It is these devices that therefore impose the greatest number of constraints in the experimental halls. They are also instruments which serve simultaneously as target — the nuclei of the liquid with which they are filled and with which the incoming particle beam interacts — and as detector — the tracks of charged particles produced in the collisions inside the chambers are photographed. As detector they are 'universal' in the sense that they record all the interactions which occur inside them, without any preselection. Each photograph thus contains many very different events, and each can be investigated in all its aspects. It is up to the physicist to choose once the photographs have been taken the process or processes he would like to study. As a result, in terms of the organization of the experimental work, the phase of data collection around the accelerator (what is called a run) is entirely distinct from the data analysis phase; the latter — studying the photographs — can be done anywhere in the world, and remote from the accelerator and the chamber in particular. The same film can also be studied successively by different groups, either to reinvestigate a particular type of event already looked at, or to consider interactions of a type different to those for which it was used before.[71]

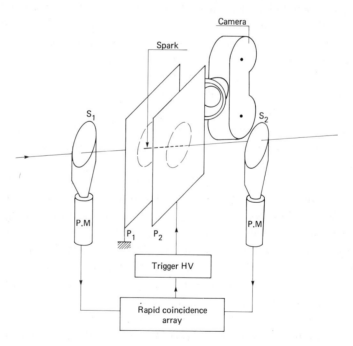

Fig. 8.1 Functioning of a spark chamber. S_1 and S_2 are two scintillators which trigger the high voltage applied between plates P_1 and P_2. The spark initiated by the ions formed as the particle crosses the chamber is photographed.

The technical part of a bubble chamber physicist's work is thus so reduced that 'very few of them were involved in the functioning of the chamber, the supervision of the beams or the development of the photographs'.[72] This is the job of technicians, specialist engineers, and a handful of physicists designated by their colleagues. And since it is not physicists either who construct these highly complex bits of equipment that are the bubble chambers, they are rather like 'white-collar workers' who practise their physics far from the 'shop floor'.

Without wanting to anticipate too much, let us merely note that this technique easily allows for the collaboration of many teams interested in the same chamber and a given type of incident particle. Since the time around an accelerator is limited, and since a chamber can take thousands of photographs a day, people can get together and impose the construction of a beam (increase in power in the committees choosing experiments) and can organize the run collectively (economy of time and money)—the ultimate goal being to share the photographs and the subjects under investigation.

In contrast to experimentation with bubble chambers where the detector is a given and where the real research only begins after the data is taken, an electronic experiment gets under way well before the detector is exposed to the beam of particles and calls on the physicists who are doing the experiment to (re)construct a particular detector each time. During our period no 'hermetic' electronic detection system was as yet in use (i.e. a system which permitted the researcher to recompose all the interactions in the detector in all their complexity). Thus scientists had to imagine a particular arrangement for each experiment which permitted them to answer a specific question—and to build the set-up themselves with the help of technicians or engineers. Of course they often reused the same basic components but the manual (or 'blue-collar') dimension was still very much there.

Before taking any data it was then necessary for an electronic team to test and to calibrate the different components in the set-up, to assemble it on the spot, to test the whole around the accelerator—and to make the modifications necessary to ensure an optimal functioning. And while data was being taken it was essential that at least a part of the team be present to check that everything was going as planned and to adapt the detector by feedback. An electronic team was thus a highly-integrated whole united by one or two strong personalities, being those who proposed the experiment and visualized the design of the detector. The groups tended to remain autonomous and rather small, to follow a type of questioning in which they were past masters, and to prefer certain experimental layouts to others. They also tended to perpetuate themselves, as each data-taking phase (which was never completely distinct from the analysis phase, and was sometimes concomitant with it) raised new questions and new ideas for changes to the detector system. In this sense—and even if there was a beginning and an end to each experiment chosen by the selection committees—electronic experiments, unlike those with bubble chambers, had something of the character of a process which was open to the future.

Notes: p. 497

8.4.2 THE PREPARATION OF THE PS EXPERIMENTAL PROGRAMME BEFORE THE ACCELERATOR WAS COMMISSIONED, 1958-1959

The question of what programme of research to follow around the proton synchrotron began to be looked into in summer 1958. At the end of August the Director of Research around the SC, Gilberto Bernardini, and the directors of the SC and PS divisions proposed to the Group of Directors that an 'Executive Committee for the PS Experimental Programme' be created. It was to be composed of some 15 people, half of them physicists and half of them engineers. Accepted by the Senior Staff and the SPC, this committee first met on 25 September 1958. Its precise goal was to define 'the responsibilities of the various teams developing the numerous instruments and facilities needed for the experimental PS programme' and to ensure that the equipment needed for experimenting would be ready in time. This committee was not to concern itself with the experimental programme as such—'another Committee, with a more scientific bent, could be formed in the future to discuss the scientific aspects of the [...] programme', said the SPC in September—the idea being that these aspects could only be studied later, just before the commissioning of the machine, and by physicists. [73]

The Executive Committee for the PS Experimental Programme existed between September 1958 and March 1960. It met 21 times (every three to four weeks) and, as foreseen, devoted itself essentially to the material preparation of the programme. It also debated the construction and furnishing of the three experimental halls, the supply and distribution of electric power inside them, the preparation of internal targets and the beam ejection programme, the system for transporting beams from the accelerator to the detectors, the installation of bubble chambers and the construction of those being built by CERN, etc.[74] However, since no other committee was established before 1960, it also considered one or two experiments envisaged by electronic groups at CERN and the proposals of the European emulsion groups. Without really deciding on the final programme, in June 1959 it 'approved', for example, the plans for two preliminary experiments proposed by von Dardel and by Lundby (intended primarily to study the beams emitted by internal targets); in September it looked into a third experiment proposed by Peters (from Argonne) and asked Fidecaro 'to take over the work on beam surveying'; in October it approved 'the first experimental lay-out' (the arrangement of the beams in the experimental halls) and discussed the idea of holding a big meeting of European physicists; and in November it settled the date, 20 January 1960.[75]

The aim of this meeting was to allow external teams 'to submit their plans and their needs for facilities' and to avoid giving them the impression 'that they were being kept in the dark'. On 6 November an official letter was sent to the member states asking them 'to let me [the Director-General] know, by say the end of November, what proposals your country will be making [...]'. During the next two months CERN received replies from Sweden, Germany, the United Kingdom and, above all, from Italy. These were discussed at the 20th meeting of the Executive Committee on 13 January 1960.[76]

The striking feature of this meeting, which was largely devoted to accepting or refusing experiments, is that most of the elements of the 1959 debate surfaced again in rapid succession. Cocconi, the first coordinator of the PS experimental programme declared, for example, that 'for 1960 precedence should be given to CERN teams [...even if] CERN experiments should be placed on the same footing as outside experiments and judged purely on scientific merit'. Preiswerk, the first leader of the NP (Nuclear Physics) Division affirmed 'that many of the proposed experiments [by outside groups] could better be done by the existing teams [at CERN], but with participation from outside laboratories to make them mixed teams'. To which Adams replied 'that one could first set up the teams and then present them with a list of proposed experiments'. Which Bernardini in turn contested, at least 'for counter experiments when one had to know the specific problems'.[77] Nothing much came of this meeting, though rapporteurs at CERN were nominated: they were to meet the non-CERN group leaders and report on the feasibility of the experiment which they proposed.[78] Despite the confused nature of this debate and the absence of a consensus on how to organize the experimental work around the PS, one thing was clear: everyone realized that before long the rules for choosing between the teams which would be allowed to do experiments would have to be laid down. If nothing had yet been done it was because the committee 'with a more scientific bent' (and which was to take care of physics questions) had never been created. And while no one quibbled with the principle established in 1956 that the Director-General was ultimately responsible for making the choice, this alone was obviously not enough: his decisions would have to be prepared, as in the case of the SC.

There is an important conclusion to be drawn from this first engagement with the problem: despite the presence of the Executive Committee, hardly anyone seems to have been really ready to experiment early in 1960. The CERN groups were on the spot, around the PS, and were making a number of preliminary studies; but only one of them, that led by von Dardel, apparently had reasonably sophisticated equipment. 'No one this year', wrote Hine in August 1960, 'will be doing anything which Bernardini would call an experiment, as opposed to making apparatus tests or survey measurements'.[79] The situation was hardly better outside CERN and among the proposals received only those coming from Italy were reasonably detailed. As for the remainder, it suffices to quote the words used by the British delegate in his response to the Director-General in December 1959: the visiting groups, he said, 'have found it almost impossible to formulate anything very precise at this stage, and the following are only general indications'.[80] As a result, after a meeting held at the end of March for 'the allocation of PS machine time [...] up to the middle of July [1960]', Dakin could write that 'it appeared, rather to everyone's surprise, that with the experiments which are, or will be ready to start in that time, and with a certain amount of parasiting, there should be about 20% machine time to spare'.[81]

Notes: p. 497

We are not going to consider the reasons for this situation in detail in this chapter. Let us simply note that they probably lie in the absence of a European tradition in experimenting at very high energies—the machine which started working in 1957 was only a 600 MeV synchro-cyclotron, while the Americans had had 3 and 6 GeV synchrotrons at their disposal since 1952-1954—, in an almost total lack of knowledge of what the secondary beams produced by the PS would be—some six months were needed after the machine worked to draw up the list—, and finally in the fact that the machine had not yet been 'run in' at the beginning of 1960.[82]

A machine like the PS is actually never available on the day that the production of a beam is officially announced, on the day that the first particles have been accelerated as foreseen. The machine remains 'capricious' (to borrow a word used by H. Collins[83]) for several months, and those in charge of it do not yet understand its behaviour properly and are not yet able to 'reproduce' its performance routinely. (Adams made the point nicely when he told the SPC on 19 January 1960 that 'For the time being only Dr. Hine was able to set up the machine properly'.[84]) Thus a considerable fraction of the machine time had to be set aside in this inital period, not for the physicists, but for the engineers. To study the phenomenon of instability, for example, (which was important with the injector, in the case of the PS) and to carry out immediately the required modifications to certain components of the machine; to standardize the mode of procedure and to educate technicians 'for routine operation'; to test targets and to be able to produce bursts of particles of variable length as required by different types of experiment; to set up, finally, the secondary beams and to study the radioactivity in the machine area.[85] This is why only 5.5 hours per week were available for physics in February 1960 although the machine was operating for 13.5 hours per day. In May, the figure had climbed to 14 hours per week, plus seven hours for target tests; in July it was up to 19 hours and tests continued. By November 1960, and one year after the PS first worked, 75% of its operating time was set aside for physicists, comprising 60 hours contained in 5 days of 16 hours of running time. In other words by then, and in contrast to the situation which had prevailed early in the year, only 20 hours of running time were kept for technical development.[86]

8.4.3 THE COORDINATION OF THE EMULSION GROUPS, 1958-1960

European emulsion specialists began to get organized in parallel with the debates in 1959 and the preparations and plans being made in the Executive Committee. On 20 and 21 November 1958 they were the first to hold a meeting on the work to be done at 25 GeV. This initiative was triggered by physicists outside CERN and, in particular, by the most renowned of them all, Cecil F. Powell at Bristol. Their aim was to 'secur[e] the best use of nuclear emulsions', and to give 'to small laboratories the chance to make and use exposures [...]'. More precisely, they wanted to ensure that 'well-designed experiments were made in good time and that the best methods of recording and analyzing data were available'. To do this emulsion specialists recognized that collaboration 'was the best method'. The meeting set up three working parties and called a general meeting of all European groups for 15 and 16 January 1959.[87]

The January conference was attended by 51 physicists from outside CERN and was held at Geneva with Powell in the chair. The first point it studied was '[the] processing and properties of Emulsions'. Its conclusion was that it was not necessary for the time being 'to set up any new organization for the production of special emulsions'. It would be better to capitalize on the experience of plate manufacturers (like Ilford) and to work with them on improving the product. The two other points tackled were the analysis of plates with microscopes and the possibilities of mechanizing this. So as not to put a stop to a collaboration which all deemed valuable, an Emulsion Group was established at CERN. Its role was to serve as secretariat and executive arm for the coordination of European groups which were working with the technique. The first experiments that were to be done were defined four months later.[88]

During the following eighteen months regular working meetings were held and numerous 'collaborations' were set up (20 with groups in the member states, 12 with outside groups), while regular contact was maintained with 52 institutes. The Emulsion group at CERN played a quite specific role in this highly integrated system which did not subsequently change. As secretariat it centralized the experiment proposals, suggested how they (and the teams) could be grouped, and proposed adaptations. As executive arm, and from the very beginning of 1960, it supervised the exposures themselves. In some cases it did this with the support of outside physicists or technicians; in others it simply offered to help outside groups who came to do their experiment themselves at the PS. A 'link man' was supplied in such cases 'to assist in resolving the technical problems which arise'. Finally it made 'many surveys of radiation intensity and composition' (from the time the PS first worked) and contributed to the installation of equipment of use to all groups at CERN (it set up a scattered-out proton beam through the shielding walls, it installed a pulsed magnet, etc.).[89]

One point needs to be stressed about these European groups specialized in emulsion technique: the original and 'pioneering' nature of the coordination which they established. It was original in the sense that it involved integrating the maximum number of experimental groups in a collective task, in a deliberate and determined way, and against the spontaneous tendency of the groups to compete; it was pioneering in the sense that the groundwork was done in 1959 in parallel with the stormy debate on 'mixed teams', and in that the coordination of emulsion physicists was to serve as a model for the bubble chamber experiment committee and, by ricochet, for the electronic committee. In our view there are three rather different reasons why this happened. Firstly, because collaborating was a well-established

Notes: pp. 497 ff.

tradition among emulsion specialists; since 1953 CERN itself had in fact been the chosen site for the creation of large European ventures using balloons and emulsions for the study of high-altitude cosmic rays. For this reason the most active people in the milieu simply needed to transpose the habits they had learned in a different context first to the SC and then to the PS.[90] The nature of the technique also seems to have been decisive, in that an emulsion experiment calls for far more limited resources than the other two techniques. In contrast to a bubble chamber experiment, for example, far less space is immobilized and security is of no concern; in contrast to an experiment with counters, emulsion experiments are prepared far more rapidly and are 'based on steadiness and routine', to use Bernardini's somewhat harsh phrase. Perceived by the specialists using the other techniques as, in general, less well adapted to complex experimentation around accelerators, emulsion physics tended to be somewhat marginalized in an advanced high-energy physics laboratory and to be done 'parasitically'. Grouping together was a way of ensuring that one was not 'forgotten'. Finally, there was the fact that this technique was practised at the time by a large number of groups in Europe, particularly in the 'small' countries, and that these groups were themselves small. There was considerable demand, then, a demand which feared that it would not be satisfied in a world dominated by big organizations.[91]

8.4.4 THE SETTING UP OF BUBBLE CHAMBER COLLABORATIONS AND OF THE TRACK CHAMBER COMMITTEE, THE TCC, 1960

The European bubble chamber specialists followed a path more or less parallel to this even if delayed in time. They first met on 21 January 1960, two months after the first particles had been accelerated to design energy in the PS. The day before they had taken part in the meeting organized by the Executive Committee to discuss the experimental programme around the 28 GeV machine. Six reports were presented on this occasion, two presenting the PS and giving '[a] preliminary investigation of the PS beams', one describing the proposals for the coordination of emulsion work, and three spelling out the characteristics of the French and British chambers. Being in Geneva anyway, the European bubble chamber specialists decided to meet and to set up a permanent committee 'for organizing CERN's PS Bubble Chamber experiments'. On 8 March 1960 this committee met for the second time in the presence of 34 people, 23 from CERN, four British, four French, and three Italians.[92]

The meeting began with a presentation of the programmes foreseen for the two chambers then ready to work at Geneva. It was Peyrou, its builder, who spoke of CERN's 30 cm chamber. At the end of March, he said, 'a beam of 15 GeV/c negative particles which contains mainly π^-' would be sent through it. Three objectives were defined: to study '(a) any new type of event, (b) hyperons, anti-hyperons and K^0, (c) systematic study of multiple production'. Next the programme envisaged for the 1 m

Ecole Polytechnique chamber was outlined by Lagarrigue, who was in charge of it. The device was to be installed as from September in π^- and p beams of about 20 GeV, then in 6 GeV antiprotons. After a quick discussion on these proposals, the meeting considered how to distribute the photographs from the first run, which was going to be done a few days later with the 30 cm chamber (proposal by O. Frisch of Cambridge). C. Franzinetti, from Pisa, expressed his interest in the three subjects advocated by Peyrou; D. Roaf, for Oxford and Imperial College, was only interested in topic (c). As no one else wanted to work on these photographs, Franzinetti, Peyrou, and Roaf agreed that CERN and Franzinetti's groups would study topics (a) and (b), while the British groups would concentrate on question (c).[93]

Two weeks later a letter signed by Bakker but drafted by Peyrou and Kowarski, was sent to all groups that were thought to have an interest in bubble chamber experiments. Foreseeing an expansion of the committee set up in January to include all the member states, it called for a meeting to be held on 17 June. A preliminary questionnaire was appended. It asked of each group if it was interested 'in the taking of pictures' or only in their analysis, if it envisaged collaborating and on which programme, if it would describe the scanning tables and other such equipment that it had, etc.[94]

The June conference was opened by John Adams, the new Director-General. Referring to the example set by emulsion collaborations, he quickly stated that 'Mixed teams should be set up as well for the bubble chamber experiments'. They ought to be established before the experiment so that 'the analysing group will be the ones who designed the run and took part in it'. The meeting then turned its attention to the beams to be built in autumn 1960 and spring 1961 for the 81 cm Saclay chamber. Finally, at Adams' suggestion an official Bubble Chamber Committee was set up 'as an advisory body to the Director-General'.[95]

This new committee met on 22 July. C.C. Butler (Imperial College), C. Franzinetti (Pisa) and H. Filtuth (CERN) described the work done since March by collaborations around the 30 cm chamber. They then explained what this collaboration would like in the next run (protons of variable energy, possibly πs of 18 GeV), and expressed the wish to continue working together on the same terms as before. M. Deutschmann, of Aachen, asked to what extent the Munich group, and the Aachen–Bonn–Hamburg collaboration could be linked to the British groups to study jets in hydrogen (point (c) in the March proposal). Lagarrigue then detailed his preferences for the Polytechnique's chamber (Paris): during the exploratory run in August 10,000 photographs would be taken with 25 GeV protons and 10,000 with 18 GeV π^-s. He remarked that the analyses would be carried out by an Ecole Polytechnique–CERN (IEP and PC groups) collaboration.[96] Franzinetti said that Italian groups were very interested, and would like to be associated with this first set of experiments.[97]

At the end of the meeting all the requests to collaborate (and notably those of Deutschmann and Franzinetti) were accepted, apparently without any difficulty. Three 'collaborations' were formed. One, composed of CERN (IEP and PC groups), of the Ecole Polytechnique, and of groups from Milan, Turin, and Padua, was to work on high energy p and π^- in the propane chamber. At the meeting on 3 October this 'mixed' team (in the 1959 sense) stated that it was functioning smoothly. The analysis of photographs taken with the 30 cm chamber was allocated to two 'sub'-collaborations. One was the collaboration CERN (HC group), Pisa, and Trieste, the other, known as the 'jet collaboration', combined the British with the Germans. At the same time other groups asked, from October onwards, if they might have copies of the film so as to familiarize themselves with the work of analysis (H. Winzeler of Berne, J. Meyer of Saclay, G. Ekspong of Stockholm).[98]

In October the chambers' programme for 1961 was spelt out and a 'catalogue of experimental proposals' was started. These had to be sent to the committee secretariat which would circulate them before each meeting. B. Gregory, who was in charge of the 80 cm chamber on behalf of the French, specified that he was open to all suggestions for collaboration—so going beyond the letter of the agreement on the hydrogen chamber. He added: 'the groups [...] should realize, however, that their help in carrying out the experiment will be needed'.[99]

In the next few days the CERN management formalized what had been empirically put in place. The committee—rebaptized the Track Chamber Committee (TCC)—was the only body having the power to propose to the Director-General 'the combination of laboratories to work on specific experiments and to take part in bubble chamber runs'. All requests or proposals had therefore to be sent directly to the chairman of the TCC, Bernard Gregory. For this reason, on 12 October 1960, Adams asked the Italian groups in Milan, Padua, and Turin, to make official their collaboration with Lagarrigue and the CERN groups on the Polytechnique chamber 'so that everybody knows what is happening'.[100]

The system was subsequently generalized. Thus on 16 November a new collaboration was built up around the 80 cm chamber. Gregory's proposals for the beams that would be required having been accepted (slow antiprotons) he was nominated 'convener of the $\bar{p}$ collaboration'. Three months later—and when the propane chambers had been set aside for the neutrino experiment—two big 'collaborations' remained. The '$\bar{p}$ collaboration' was composed of three units: Rome and Trieste using the photographs to study interactions in flight, Cambridge, Oxford and Padua to study annihilations when $\bar{p}$s were stopped without visible production of Ks, and the Ecole Polytechnique and CERN to study the same process when it was accompanied by K production. The 'high energy' collaboration, for its part, the descendant of the historic run in March 1960, was now formed of four analysis groups. As for the second half of 1961, the study of some thirty experiment proposals which had been received led to the definition of two new projects: the building of a fast $\bar{p}$ beam (2 or 3 GeV/c, to be used with the 80 cm chamber filled first with hydrogen and then with deuterium) and, in the north hall, K$^-$ beams (stopping studies in the 30 cm chamber...).[101]

The choice of experiments

There is no point in going on much longer with this description. One thing is sure, though: at the end of 1960 the decision-making system and that for setting up collaborations were clearly defined. To illustrate this we propose to anticipate a little, and to present a document dated 21 February 1963. Essentially banal—there is nothing special about it compared to dozens of others that we have seen—its form is explicit and easily allows one to see that the system empirically put in place during 1960 was now a permanent feature of the CERN landscape. Entitled 'Combined proposals for high energy negative pion and proton beams in the British National Bubble Chamber', it was put forward by eight groups and began with a paragraph which we shall quote in full:

> A number of proposals have been made for experiments with high energy proton and negative pion beams in the British National Hydrogen Bubble Chamber. At the suggestion of the Track Chamber Committee, discussions have taken place between groups to combine their proposals with the aim of finding common beam energies and of reducing the numbers of photographs required by arranging for exchange of the films. All groups expressed a willingness to cooperate and to exchange films quickly. Since all groups wished to separate 'very peripheral' or 'elastic type' inelastic events from the others, the common desire was for the highest possible beam energy obtainable to make this rather small separation as large as possible.

The document went on to list the groups interested and the events which they wanted to study, the number of photographs thought necessary, the way in which these would be distributed, the characteristics of the beam required (cross-section, intensity, design), etc. To conclude let us simply note that this was a new collaboration; in another field it might already have been constituted. In this case it would have had to explain why it had been led to ask for another run, for another beam (or the same one), for another chamber (or the same one), etc.[102]

8.4.5 THE FIRST SYSTEM FOR TAKING OVERALL DECISIONS ON THE EXPERIMENTS: ADAMS' PROJECT IN AUTUMN 1960

The measures taken in 1958 and 1959 by the Executive Committee and the emulsion specialists had no implications for the definition of a global policy for the exploitation of the new synchrotron. Aware of this fact, the CERN management took a number of preliminary steps in December 1959. Its first idea was to set up a committee 'of users' representatives' comprised of seven CERN staff. Five were engineers and represented 'the machine group' (i.e. those responsible for the daily functioning of the PS) and the 'PS running-in group' (charged with running in the machine, improving it, rendering it reliable). The two others were physicists (G. Cocconi and C. Peyrou). The task of the committee was to establish regularly 'the total number of runs and their subdivisions between the various classes of users'.

For the first two months the committee allocated 96 hours to the procedures for the daily setting up of the machine, 124 h for the machine group, 132 h for the study of the machine by the running-in group, and 115 h for physics. These 115 hours were divided into weekly sections of 5.5 h (Tuesday evening), essentially reserved for electronic experiments, and into one exceptional run with the 30 cm chamber which was to last for three continuous days at the end of March.[103]

The system was modified three months later, in March 1960. Two 'coordinators' were then nominated, Cocconi for research and Adams for the machine; together they replaced the users' committee and distributed the time between the groups. For the research programme Cocconi had to discuss matters regularly with the physicists using the PS (essentially the CERN Senior Staff), and these meetings *de facto* replaced those of the Executive Committee set up in 1958 and dissolved on 1 March 1960.[104] For the 'visitors' the CERN management still envisaged convening the Advisory Committee for the PS, the strict equivalent of that which had functioned for the SC and whose definitive composition had been agreed to by the SPC on 19 January. The first meeting was planned for 2 June. On 16 May Adams took a major decision and postponed the meeting. He proposed instead that several very open working sessions be held with the 'visitors', as open in fact 'as those which have already enabled us to get very useful results in the field of emulsions and of bubble chambers'.[105]

The reasons for this postponement of the June Advisory Committee meeting—which were never explicitly spelt out—seem to lie in the incompatibility between the prevailing decision-making procedures. The system 'Cocconi and CERN senior physicists on the one hand, Advisory Committee on the other' was taken over from the SC. It distinguished the visitors, between whom the Advisory Committee distributed a certain amount of machine time, from the CERN physicists, organized as they thought fit, and who distributed their portion of the time under Cocconi's direction. But a lot had happened between January and early May: the emulsion coordination and the committe for organizing bubble chamber work around the PS were both functioning independently, 'transversally', one might say, to the official structure. Created simultaneously by CERN and non-CERN physicists, they initiated a new form of work (the collaborations) which undercut the foundations of the Advisory Committee (separate visiting teams). Hence the difficulty.

In June John Adams drew several conclusions from the existence—and practical efficiency—of the Emulsion and Bubble Chamber committees. On 17 June, opening the meeting of bubble chamber specialists, he accepted their necessity: only they could study in detail the experimental proposals sent to CERN and 'advise [the DG] on the carrying out of experiments [...]'. He went on to propose that their composition be formalized (they would consist of equal numbers of CERN physicists and outside representatives)— and he killed the Advisory Committee in the same breath.[106]

The SPC met a few days later, and expressed its dissatisfaction with the ways things had worked during the preceding six months. Certainly, when questioned by Dakin, Leprince-Ringuet could say that he was 'very happy' with the organization of bubble chamber work, but there were other things that bothered the SPC—namely, CERN's excessive weight in the definition of the overall programme.[107] The theme was first taken up with respect to the SC, where the complaints reached a peak early in 1960—even if visiting truck teams had recently been granted 50% of the time on the machine. Then it was the turn of the CERN electronic teams which were accused of being closed in on themselves, and 'still organized and staffed on the assumption that they should be self-supporting and independent units'. 'One may incidentally remember', remarked Bernardini by way of generalization, 'that the Rome He-chamber did not [initially] find a very well paved road to CERN and the Padua people were for a while under the impression that at Saclay they would find a more friendly atmosphere'. Finally there was the official decision-making mechanism itself, which was entirely under CERN's control. The users' committee established in February and Cocconi's physicists' committee were CERN bodies—as were the two coordinators. And since the Advisory Committee had never met... From whence the anxieties in the SPC and the proposal of Amaldi and Leprince-Ringuet to make provision for a visitors' representative at the highest level in the new organigramme, the CERN Directorate.[108]

Adams took the bull by the horns on getting back down to business in September, and reorganized the whole system. Basing himself on the bubble chamber committee, he proposed 'to extend the idea to the counter experiments and set up three Committees', one per technique used (emulsions, bubble chambers, electronic). All propositions, whether they came from within or without CERN, were to be submitted to one of these committees for their opinion. 'The Chairmen of the three Committees, together with certain of the senior research staff of CERN will then together form a Nuclear Physics Research Committee'—the NPRC. Its aim would be to define the entire programmes for the PS and the SC, the practical details for each machine being the responsibility of a 'Scheduling Committee'. 'In order to make the laboratories outside CERN feel that they are getting a fair share of the facilities at CERN', Adams proposed that the chairmen of the three specialized committees be visitors. The chairman of the NPRC, on the other hand, would be one of CERN's research directors.[109]

These ideas were first put forward in a document concerning 'the future internal structure of CERN'. The paper was discussed by the SPC on 3 October. Adams emphasized that his new proposal 'would [...] give the physicists of the member countries the chance of playing a much greater part in CERN's work than had so far been the case'. Everyone warmly approved, and 'Professor Leprince-Ringuet stated that he was very favourably impressed by the Director-General's proposals and that he admired their clarity and simplicity'. On 12 October the names of Powell, Gregory, and Salvini, an Italian physicist, were put forward by Adams as

Notes: pp. 498 ff.

chairmen of the Emulsion Committee (EmC), the Track Chamber Committee (TCC) and the Electronic Experiments Committee (EEC). No one formally objected, and early in November the first meeting of the Nuclear Physics Research Committee (NPRC) was fixed for the 21st. The whole system was to come into operation in 1961.[110]

To conclude this section we want to draw attention to the two sources from which this decision-making structure had descended. The first was the positive experience of the Emulsion coordination and of the Bubble Chamber Committee. Impressed by their efficiency and by the fact that they provided a simple means of integrating CERN and non-CERN teams, Adams decided to replicate the model. One question that arose immediately was whether it was wise to have three committees distinguished by technique rather than simply one which was independent of the technique used, as at Brookhaven. The remark was pertinent, and underscored the ambiguity of the NPRC and its terms of reference. Was it to 'harmonize' the requests coming from the three committees simply by dividing the time between them, or was it to play a more active role in the organization of research? We shall see that this question was discussed for the best part of 1961. One thing is sure though. In autumn 1960 no Director-General could reasonably bypass the Emulsion and Bubble Chamber committees. These committees had filled the vacuum left in November 1959, when the machine was commissioned, by the absence of a clear decision-making structure. By October 1960 their presence could no longer be ignored. This alone does not account for Adams' choice, however. The second reason is to be found in the attraction which the Emulsion and Bubble Chamber committees had for the visitors and for the SPC. The fears of non-CERN physicists that they would be confronted with an opaque system which overlooked their needs and was dominated by the Senior Staff in the organization had evaporated.[111] To extend the Emulsion and Bubble Chamber models seemed to them the simplest and most satisfactory solution. Considering also the now rapidly increasing demands to do experiments and the increase in the machine time available for them, one has all the elements for the play. Sixty hours were available for physics in the autumn, as we have seen, and people were even speaking of running the accelerator 24 hours a day instead of 16. There was no place left for amateurism, and the formalization of strict rules for choosing between experiments was an urgent necessity.

One last word. To generalize the systems enshrined in the Emulsion and Track Chamber Committees was to suppose that the concrete organization of experimental work with electronic detectors was amenable to collaborations of a similar kind. Adams' proposal had the advantage of being simple and the beauty of a symmetrical arrangement. But perhaps it did not make sufficient allowance for the nature of electronic experimentation itself. At least that is what certain people were quick to point out.

8.5 The changes made to Adams' decision-making system for experiments, 1961–1962

8.5.1 THE LINGERING PROBLEM OF THE ELECTRONIC EXPERIMENTS COMMITTEE, THE EEC

The critical remarks we have just touched on intensified between November 1960 and January 1961 when the time came for the Electronic Experiments Committee to meet. Early in November Adams asked Salvini to convene 'his' committee. On 13 December the Italian physicist arrived in Geneva and discussed the matter with the Directorate and with the leader of the Nuclear Physics (NP) Division — which was responsible for electronic experiments at CERN. A month later, having fixed the first meeting of the EEC for the end of January, Salvini again discussed the scheme with three key people in the division, Lundby, Preiswerk, and von Dardel, and then with all the group leaders in the division (18 people in all). Four days later, on 21 January, he cancelled the meeting of the EEC and asked Adams in a long letter to reconsider his experiments committee system. What did he say?[112]

Firstly, that there was no real need for an electronic committee modelled on the Track Chamber Committee (i.e. 'a rather large body with members from CERN and members appointed from outside'). 'As we know', wrote Salvini, '[the TCC and EmC] have to distribute films and plates among the European physicists, which is at least one basic reason of their existence'. Unfortunately, this did not apply to electronic experiments. He also doubted whether it was a wise policy to ask of a necessarily partisan EEC 'to decide on the value of the priority of different experiments'. Salvini argued that this was a task for an independent group (the NPRC for example) and not for those few group leaders who would be nominated to sit along with the visitors on the EEC. Finally he raised the problem of external physicists, Adams' justification for setting up the EEC in the first place. Salvini's view here was that individual contacts with physicists in the NP Division would replace the role envisaged for the committee. Less bureaucratic, this solution would be facilitated by the group leaders who, according to Salvini, were all willing to form mixed teams. The alternative proposal which he thus made, in agreement with the NP Division, was to centralize experiment proposals in Preiswerk's hands (whether they came from group leaders or from outside CERN) and to have their interest and feasibility studied by a small group which could meet easily and in response to concrete questions. As members he suggested Preiswerk, the two coordinators, and himself. Its work being thus prepared, the NPRC could then choose (as was its purpose), and submit its decisions to CERN's research directors.[113]

The underlying reasons for these proposals emerge from the minutes of the meeting of group leaders with Salvini on 17 January. On the one hand, there were the technical chracteristics of electronic experimentation. Taking up the arguments which Bernardini had already developed in a long memorandum to the Directorate,

Notes: p. 499

all the physicists present emphasized the need to be immersed in research around the PS if one was to be able to judge the interest and the feasibility of the proposals. In their view the know-how gained in this way was decisive here, and even at CERN meaningful decisions could only be taken 'by people who know the problems and the technical details'. In addition, in contrast to what was required for experimenting with heavy equipment like bubble chambers, 'flexibility is essential to give the groups the opportunity to try out new ideas'. Long-term planning was easy with polyvalent devices like bubble chambers (the best beam in the best chamber), but this was not so for counters. It was therefore not possible to have 'a large and complicated committee' which would be inflexible and which could not 'work fast enough to deal with proposals of experiments which are actually going to be performed'.[114]

It seems however that it was not only the specific characteristics of the technique that were at issue in this debate. That dimension was mixed with a certain way of working which the senior physicists in the NP Division wanted to preserve. Until that time this group had actually functioned rather like an aristocracy—in the original sense of the word—and had had direct access to the machine. Never really having to make difficult choices, these scientists had been free to perform most of their preferred experiments at the PS, the selection being made inside the house. In addition, they had usually only experimented with their own group, though some did work with individual visitors who came to CERN for reasonably long periods of time as invited professors or research associates (von Dardel with D.H. Frisch of MIT, for example, or Farley with R.L. Garwin, another American).[115] All the same now, around 1960–61, they began to feel that they were under attack. Firstly, by the bubble chamber specialists who were asking for more and more machine time (one group leader defending the idea of an EEC said: '[it] is an organization to protect counter groups against bubble chamber-, emulsion- etc. groups'); then by the visitors, who wanted to come to Geneva and to have some machine time for themselves; and finally by the directorate, who wanted to instill some 'self-discipline' into the senior physicists at CERN and to force them to make sometimes painful choices themselves. Hence their demand to keep things as they were, to leave the NP Division in charge of the experimental work—with the promise that they would integrate the visitors. A comment made by one of the group leaders during the meeting on 17 January nicely illustrates this feeling. One reads in the minutes: '[X] thinks that visitors are the main problem. He wants to deal with it inside CERN—internal arrangements with CERN groups. He is against national representatives. Experiments should not be allocated on nationality basis'.[116]

The responses of the Directorate and of Adams to these reactions were, by and large, negative. Insisting again that 'we have not had a blameless record with respect to visitors and [that] we want to make sure that they are integrated into the mixed teams in the future', Adams maintained that CERN needed an electronic committee 'similar to the other two'. He simply agreed to reduce its size. Four of its members would be those proposed by Salvini, but three specialists from outside CERN would also be nominated, and Hine and the Research Directors (Bernardini and Weisskopf) would take part in meetings.[117] This committee met for the first time on 1 March 1961, and as foreseen studied the 20 odd experiment proposals that had been submitted, including that for the neutrino experiment. It met again in April, May, and June. On 25 September, during its fifth meeting, the problem discussed in January surfaced again and all the group leaders were invited to attend—'because there was a feeling of too little contact between the EEC and the group leaders'.[118]

The criticisms the latter made in September were globally those they had levelled at the beginning of the year: in brief, the system introduced inflexibility into the programmation and the running of the experiments. They remarked, in particular, 'that it was unrealistic to ask the counter groups to present their programmes 6 or 9 months ahead as the development of this programme depends on the results obtained at each stage'. The only person to oppose them was Hine, Director for Applied Physics, who remarked that 'At the PS 6 or 9 months advance planning is however sometimes necessary for technical reasons'—which the PS coordinator, the physicist von Dardel, thought to be exaggerated.[119] The debate was thus reopened inside the organization and at the October meeting of the SPC. Reminding those present on this occasion of the considerations which had led him to propose the creation of the EEC, Adams noted that the most important of them had not, in fact, been met: only a few isolated non-CERN specialists of electronic detection had come to work at the PS since 1960. Weisskopf, the new Director-General who had taken up office three months earlier, did not disagree, and promised to give his ideas on how to reorganize the experiment committees in the near future.[120]

He presented his project in January 1962.[121] Concerning the Electronic Experiments Committee (the Emulsion and Track Chamber Committees were to be left untouched) he proposed that 'The functions of the EEC [be] largely taken over by the NP Division'. The decisions would be prepared by 'Preiswerk and a few of his staff', these being 'assisted by an external member who reviews the situation periodically together with them'. The tasks of this supple structure (still called the EEC) would also be reduced. While it would always be expected to pass an independent judgement on the feasibility and interest of the proposals it received, it no longer had to make the choice. This would become the responsibility of a strengthened NPRC.[122] On 11 April 1962 all was in place and Preiswerk, the new chairman of the EEC, made public 'a new system for submitting proposals to the EEC'. A 'letter of intent' was first to be sent to the leader of the NP Division who would study it. Very quickly, since he was working alone, 'he may encourage the group to work out a preliminary proposal'. Once this was ready the proposal would be studied in the weekly seminars which the division had inaugurated at the end of 1961. The definitive project would emerge from this, on the basis of which machine time would be allocated to it. In other words, and notwithstanding some minor differences in emphasis, the proposals made at the end of 1960 by the NP Division

Notes: p. 499

group leaders had been adopted. The insistence on the autonomy of the division—buttressed by the appeal to the specificity of the technique—had taken precedence over Adams' determined efforts to set up a system which opened outwards to welcome non-CERN physicists.[123]

8.5.2 THE QUESTION OF A SCIENCE POLICY: THE NPRC'S RELATIONS WITH THE SPECIALIZED COMMITTEES (EmC, TCC, EEC)

The reorganization undertaken at the end of 1961 did not primarily concern the EEC but the NPRC. We must therefore spend some time studying this committee.

In Adams' arrangements of September and October 1960, the aims of the NPRC were only spelt out in rather general terms, namely: to gather together the proposals coming up from the three committees, and to decide, on this basis, on the global programme. During the first meeting of the NPRC on 21 November 1960 its chairman, Victor Weisskopf, proposed that two kinds of meeting be held. One variant, to take place bi-annually, was to assemble a very large number of people; its purpose would be to elaborate a research policy for CERN in the medium term, 'deciding for example such general points as whether the PS effort should be concentrated on work at high energies or at lower energy [...]'. The other, smaller type would meet every two or three months, and would concern itself with the immediate programme 'by which he [Weisskopf] meant receiving, and assessing, at regular intervals, the detailed proposals of the committees for emulsion work, track chambers, and electronic experiments, and any proposals which may come from other sources'.[124]

These ideas were widely debated between November 1960 and February 1961.[125] Three 'solutions' emerged from this first exchange of opinions. The first was that advanced by A. Citron, one of the group leaders who was influential in the NP Division, and who had been proposed by Weisskopf to be the secretary of the NPRC. Citron suggested keeping the three specialized committees, giving them a more or less fixed proportion of machine time and abolishing the then-redundant NPRC altogether. Thus it would be possible, if the three committees met often enough, to take decisions quickly, and to avoid the need for anyone to have recourse to parallel 'unsocial' practices.[126]

Weisskopf proposed a different solution. In the face of Citron's objections—which doubtless represented the views of most of the NP group leaders—Weisskopf was inclined to think that one of their premises was correct, namely, that the existing system was too cumbersome and slow. 'On the whole', he wrote to Hine on 5 January, 'I think we have too many committees'. Nevertheless, and against the future secretary of the NPRC, he criticized above all the first layer in the structure: the three specialist committees made it impossible to elaborate a *coherent* scientific policy for CERN. He therefore preferred to keep just one rather small committee which would meet frequently, a committee 'that really goes into the details of every experiment proposed, and discusses its value and its priority' — the NPRC.[127]

John Adams, the Director-General, took up a third position. Satisfied with the functioning of the existing committees (TCC and EmC), and seemingly less responsive to the particular problems created by electronic experimentation, he continued to defend the established system. Accepting the risk of overburdening the daily running of the laboratory, he proposed that the *four* committees be kept and that they meet each month. On 6 February the Directorate rallied to his position and settled the final composition of the NPRC. It was to comprise the four scientific members of the Directorate (Adams, Hine, Van Hove, Weisskopf), the chairmen of the three specialist committees (Gregory, Powell, Salvini) and seven CERN Senior Staff, essentially the divisional leaders, 'those with responsibility to carry out the programmes'.[128]

The NPRC thus constituted met for the first time on 8 and 9 March 1961. It discussed short-term (choice of experiments) and medium-term (beams to be built in the second semester, tests of detectors,..) problems. The striking point is, however, that it never got around to 'comparing' the various proposals or to drawing up a list of priorities which transcended those advocated by the specialized committees. The very long minutes of this and the three following meetings of the NPRC offer numerous instances of this point — examples brought out clearly by the fiercely ironic style of the committee's highly sceptical secretary, A. Citron.[129] Thus he was to write in March:

> On the question of relative priority of counter and bubble chamber experiments in June, PEYROU commented on the bubble chamber programme. Two π^- runs at different energies will be devoted to the problem of peripheral collisions, which is very exciting. Many pictures are required to single out the interesting events. The pencil beam is supposed to give the first Ξ production events. BERNARDINI proposed highest priority to Lundby and Cocconi [NP division], next to the bubble chamber, lower priority to von Dardel and Fidecaro [NP division]. GREGORY was not satisfied with this. Arguments drifted away from physics and no conclusion was reached'.

Or again:

> On the problem of priorities GREGORY was now prepared to say that all bubble chamber experiments were first priority. Their experiments were preselected from a large number of proposals. In order to make this more plausible, GREGORY discussed a larger list of experiments including some which had not been proposed [...].[130]

The results of these meetings in which the chairmen of the specialist committees played a dominant role were many. Firstly, a strong tendency emerged to divide the machine time by committee, en bloc — even if no formal decision was taken on this question before January 1962; then a propensity to draw up a complex and very compressed programme so as to get as many experiments done as possible — a refusal

Notes: pp. 499 ff.

to choose, in a way; then too an ad hoc approach to the installation and transformation of the beams, this following more than ever the sometimes contradictory choices of the specialist committees; finally, the participants became really dissatisfied with the number of meetings which often treated very similar problems and where the debate was frequently reopened on experiments already approved in principle.[131] At the fourth meeting, on 25 September 1961, the functioning of the NPRC and of the three specialist committees was thus again on the agenda. R. Cool, of BNL, was invited to present the Brookhaven system. More centralized and autocratic (it comprised just one committee with eight members), more 'independent' of pressure groups (by virtue of its composition), it met every week, and was seen by many Europeans to have considerable advantages—which in no way meant that they were personally willing to abandon the benefits of the previous system. And at the next SPC meeting Weisskopf proposed another reform.[132]

Taking up again the idea that he had had at the beginning of the year, he suggested giving the NPRC the main decision-making power. His argument was that it alone could have a global view and could think, for example, on the overall policy for beams. In this schema, the role of the specialist committees was simply that of gathering together proposals, organizing teams, and advising the NPRC as to which experiments they thought interesting and feasible. Downstream from the NPRC, he proposed strengthening the role of the Scheduling Committee. Reduced in number, and enjoying considerable autonomy, its aim would be to put into practice the directions and the choices defined by the NPRC at the 'strategic' level. In February 1962 it was rebaptized the Experiments Planning Group and was inserted into the framework of the MPS division, the division responsible for running the PS. As for the NPRC itself, Weisskopf suggested that it meet twice a month and rather briefly, and that its size be reduced. On this last point he was forced to backpedal repeatedly, having to accommodate his view to the pressures on him from the various divisional leaders.[133]

On 20 February 1962, and after four months of consultation, Weisskopf presented his final project to the SPC. The Committee discussed it briefly on 3 March but sidestepped the problem. It simply declared through Amaldi that 'the new system should be tried out before the Committee attempted to formulate any comments on it'.[134]

We should like to make two remarks on the practice followed in 1962. The first is that The Emulsion and Track Chamber Committees were not affected by this redistribution of powers. Notwithstanding certain additional constraints imposed by an NPRC which sometimes played a more future-oriented role, they continued to be the real organizers of the work, both as concerns the setting up of the teams and choosing the experiments. Nothing was going to change them fundamentally after two years of success. This was also more or less 'recognized' by the NPRC on 19

January 1962, when the Committee accepted to distribute the machine time equally between bubble chambers and counters, keeping 10-15% aside for emulsions. The only condition was that 'If very urgent and important experiments turn up, the period under discussion could be extended in order to accommodate such experiments, regardless of the technique, by which the experiment is proposed'.[135]

The second comment is that at first the NPRC tried to introduce important changes into its way of working. Downstream, it really tried to have as little as possible to do with scheduling, something that it had not managed in the first year of its existence. Upstream it tried to impose 'research lines' on the EEC (in 1962 the 'research line' defined by the EEC and the NPRC was the systematic study of p-p, π-p etc. scattering).[136] This idea was a response to two criticisms that had been made in summer 1961, namely that 'there were too many parallel users at the PS for maximum efficiency', and that 'in several cases, problems had been raised by experiments, but then had been left open' because no machine time had been kept in reserve. The idea therefore was to step back from day-to-day experimentation, to organize a closer supervision of experimental work, to prefer 'a smaller number of complete experiments to a great number of incomplete ones'.[137] At the end of 1962, the majority view was that up to 50% of the machine time should be reserved for series of experiments which dealt with specific questions. As for the remainder, the policy was 'to look for the promising, isolated, experiments which do not fall within a research line'.[138]

8.5.3 SOME CONCLUDING REMARKS ON THE PROCESS WHICH LED TO THE COMMITTEE SYSTEM FINALLY ADOPTED BY CERN

To conclude these two sections devoted to the establishment of the decision-making mechanism for experiments at CERN we would like to *summarize* the reshuffling of the committees so as to have the main lines of force emerge. To a first approximation we might formalize what happened by seeing it as a series of attempts to get people to work together in a context of very hot competition with the United States. Four major poles—though not only they—structured the forces at work: two modes of experimentation on the one hand, two attitudes *vis-à-vis* the institution (from within or from without CERN), on the other. Each of these forces in turn played a major role, the 'referees'—by which we mean those who fixed the rules of the game and who had enough power to impose compromises—being the two Director-Generals.

In the first months of 1960 the dominant model in the minds of the counter specialists at CERN was that of separate teams. More precisely, they thought that CERN teams could receive individual visitors who would stay at Geneva for periods of two to three years—this was the idea behind CERN's proposals in 1959—, and they were very hesitant about hosting external teams, particularly in the the early years.

Notes: p. 500

As for bubble chamber and emulsion specialists, the attitude 'imposed' by the techique was that resources and know-how were to be shared between CERN teams and others; thus the idea quickly spread that a profound mixing of groups was preferable, a mixing which was encouraged by the 'heavy' nature of the equipment and which was strongly advocated by the new Director-General John Adams.

All the same there was no sense of urgency — the PS was still in the hands of the engineers — and no overall system was put in place.

In autumn 1960 Adams put forward a homogenous solution. Less receptive than Bernardini or Weisskopf to the claims to specificity being made by the counter specialists (Adams was not a physicist in the strict sense of the word), but more attentive to the demands of an SPC concerned that CERN was not sufficiently open to outsiders, in January 1961 he chose to generalize the bubble chamber committee system, and to reject Salvini's proposals (he saw them as being, above all, an attempt by the group leaders in NP Division to maintain their privileges). The visitors and the SPC were satisfied.

Having no other choice, the physicists in NP Division went along with this. Sceptical nevertheless, they remained convinced that the arrangement would impede their work efficiency.

In autumn 1961 Victor Weisskopf became CERN's new Director-General. His arrival, and the difficulties which the EEC and the NPRC had had since the beginning of the year, changed the situation. More concerned about practical efficiency and flexibility in the organization of the experimental work ('It is no good in this field to be excellent and always late' Weisskopf was to say a short while later), more responsive to the arguments of the electronic specialists with whom he had collaborated as a theoretician, the new DG agreed to break the symmetry introduced by his predecessor. He thus gave responsibility for electronic experimentation to the NP Division itself, but tried to strengthen the NPRC so as to contain the dispersion and the individualism which reigned in the NP Division — and to simplify somewhat the whole decision-making process.

All the same he did not succeed in reducing the size of the NPRC as he would have liked — because of rivalry between CERN's division leaders — nor could he, or did he wish to, change the system and the power of the Emulsion and Track Chamber Committees — with which both visitors and CERN staff were quite satisfied.

As a result it was accepted to divide the machine time equally between the two dominant modes of detection — which can be seen as something of a setback for the NPRC.

This balance of forces did not undergo any major changes until the nomination of the new Director-General in December 1965; the system set up in 1962 was thus that which controlled the access to the machine in the following years.

8.6 Experimenting with bubble chambers: the strength of the system of collaborations, 1960–1965

Now that we know how decisions on experiments were taken at CERN we want to come back to the question of the *nature* of the experimental teams which worked around the PS. In this section we are going to focus on the bubble chamber teams, our main aim being to confront the organization which was concretely set up with the big questions which had been raised in the debates in 1959. Concerned essentially with whether or not the builders of the detectors should have particular advantages when it came to using them, the discussions on the French and British chambers never arrived at an unambiguous conclusion.

8.6.1 THE FUNCTIONING OF THE BUBBLE CHAMBER COLLABORATIONS

This key question of 1959 was 'resolved' in practice: *de facto,* some priority was always given to those who had constructed big equipment. This was never formalized or defined as a right but no one ever openly disputed it.

This can be seen from the fact that the backbone of the *first* collaboration, as well as its leader, were always supplied by the team that had built the equipment. It happened in 1960 with Peyrou (for the 30 cm chamber), Lagarrigue (for the 1 m propane chamber) and Gregory (for the 80 cm hydrogen chamber). The same thing occurred later with the British bubble chamber (Butler)[139] and with CERN's propane chamber. In this last case we have a letter from the Director-General to the builder, Colin Ramm, which illustrates how banal this 'rule' was. Weisskopf had received 'the draft of [Ramm's first] planned experiments with the heavy liquid bubble chamber', and replied on 10 August 1961 that he had discussed 'the physics details with some members of the Track Chamber Division', that he was 'extremely glad that [Ramm's] bubble chamber group is now going into research work', and that in his view it would be 'excellent for this group to perform an experiment 'from run to publication''. Obviously, then, it was one of Ramm's privileges to be the first to experiment with his chamber (in a collaboration, to be sure) and to choose the first experiment (even if this had to be discussed by others).[140]

The possibility of choosing more or less freely the conditions for the first experiment(s) was, in practice, the second privilege the builder had. This question always came down to the choice of the beam. *De jure* the matter was laid before the TCC which studied all proposals; *de facto* we have not come across *a single case* in which a decision was taken against the wishes of the group that built the device. The proposals made by Peyrou and Lagarrigue in 1960 were accepted despite a number of different suggestions made by G. Puppi in March and by Y. Goldschmidt-Clermont in October. Similarly, Gregory proposed that the first beams to be put through the 80 cm chamber be beams of anti-protons just when the British were insisting that 'the production of separated K-mesons is of the greatest importance at CERN'. In

Notes: p. 500

November 1960, the Committee settled the issue 'after a detailed discussion considering the interests of the different groups as well as the feasibility'. Its conclusion was that 'the first run should be done with slow and stopping antiprotons'.[141] The same can be said for the British chamber whose preliminary programme was drawn up by physicists in the United Kingdom and first presented by Butler in November and December 1961. For the next round of experiments, by contrast, other interests were always taken into consideration and different collaborations using other beams were always accepted. It was in this way that the 'high energy' collaboration initially formed around the 30 cm chamber was able to extend its programme and work with the 80 cm chamber.[142]

Clearly, then, it was accepted that the builders had the right to certain initial advantages, as did those who had brought material which, while less important, was just as indispensable (the Padua separators in 1961, for example), and those who installed a beam—Bernardini suggested in July 1961 that 'there should be less of a feeling that a beam, set up by a group, was the property of that group alone'.[143]

Official confirmation of the fact lay in the way in which the post of chairman of the Track Chamber Committee was awarded. The first man nominated, at Adams' suggestion, was Gregory—which is not a coincidence when one remembers that the French chambers were the only important ones which were operational early in 1961. Eighteen months later the question arose of who was to replace Gregory. This time it was *explicitly* linked to the imminent arrival of the British chamber, and Weisskopf offered the chairmanship of the TCC to Butler, the head of the British contingent. Butler accepted but added that since 'the peak use of the British bubble chamber at CERN' would only occur during the second half of 1963, he thought it 'appropriate if you [Weisskopf] should decide to ask me if I am willing to be Chairman for that period'. This was done, and Gregory remained in his post up to that time.[144]

All the same there was no question of the CERN management leaving the field open to unbridled competition between the groups. Starting in the first months of 1960, it insisted that 'mixed teams' be created systematically, and that this be done as openly as possible. It was for this reason that the management sent its detailed questionnaire to all the European groups in April, that it organized the conference in June, that it 'rebuilt' the Track Chamber Committee in September (to make it 'transparent'), and that it took the trouble to make public all the existing collaborations.[145] It also installed a 'catalogue of ideas', which was published regularly and, in support of those physicists who were insisting that external groups participate in the preparatory stages of an experiment 'including preparation, set-up of beams, taking of the pictures, etc', it opposed the model for using the chambers which Gregory advocated in spring 1960. Referring to the example of BNL, Gregory's idea was to have the chambers function as in a service centre, where each experimental group did its own runs as it chose, once its proposal had been accepted by the TCC.[146]

Does this mean then that to be accepted in a collaboration all one had to do was to ask to join it? No, since the competitive aspect intrinsic to scientific research never disappears. The rule adopted in practice at CERN was that there was a double right of precedence, a double priority right. The collaboration responsible for the run had the privilege of 'owning' the film. Thus, if an external group asked to analyze a portion of the photographs 'these requests are a matter to be first considered by the collaboration'. The particular groups in the collaboration who were studying the type of events which the outside group also wanted to look at had then to be consulted. Their agreement to 'share the topic' with the newcomer was also necessary.[147] Let us give just one example to illustrate this procedure, that of a meeting of the TCC held in October 1963. On this occasion a group from East Berlin had asked to analyze some of the interactions produced in the high-energy collaboration. With the prior agreement of CERN and of Hamburg who were already studying this kind of event, and after an internal discussion in the collaboration, the request was accepted and the photographs sent — 'with a warning however that from now on further requests will require close attention' since the collaboration was already very large.[148]

Are we to conclude from this that the organization of the teams always proceeded smoothly? That would be too much. Clearly there were tensions, threats, resentments at times. We have come across two clear examples. The first was that of Butler, not too pleased to see Adams prefer Gregory as the first chairman of the TCC at the end of 1960, and fearing that the Frenchman would abuse his position and impose his model of organization based on what was being done at BNL.[149] The second was that of the German physicists whose spokesman was Heisenberg. In July 1963 he complained to Weisskopf that his colleagues were being discriminated against in comparison to the CERN physicists in the distribution of films. CERN replied that this was not so. Budde accepted that it was 'probably true that it helps in receiving bubble chamber film, if one has a chamber and an operating crew containing a few physicists on the CERN site', but he insisted that this was in no way a necessary condition. 'Many groups from countries having no chamber of their own have sent people to CERN for building beams (e.g Italy, U.K., Netherlands, etc.)', he went on, 'and have received as 'creamy' film as the groups who own chambers'. For Budde, the relative marginalization of Germany was not the result of a systematic bias introduced by the way collaborations were formed but the result of the lower level of investment of German physicists in the work (participation in collaborations, in the TCC...). Matters rested there.[150]

These examples excepted, it is the 'open' aspect of the bubble chamber collaborations, their friendliness, which has struck us — and that in contrast to the relations between the electronic groups. In our view the main reason for this lies in the nature of the technique itself, notably in the fact that a huge number of films were produced in one run which could be used by many to study a variety of topics. Obviously the nature of the technique did not *impose* the particular form of organization which we have just described, but it did make it more probable once a

Notes: p. 500

catalyst was there which stimulated or called for an organization of this kind. In CERN's case there were two such catalysts. Firstly, the debates in 1959, the anxiety in the SPC in summer 1960—and the conclusions drawn from them by Director-General Adams. Secondly, the self-interest of the various national groups: in the long term to collaborate was the best solution since they needed one another. This was the case for the British vis-à-vis the French in the early years ... and for the French vis-à-vis the British when the United Kingdom's chamber arrived. As a result we find rather large, planned collaborations, which were set up fairly easily once the rules had become internalized, and which seldom if ever left someone out of the picture: in practice it was always possible to be integrated into a collaboration and to receive photographs.

8.6.2 SOME STATISTICS ON BUBBLE CHAMBER WORK AT CERN FROM 1960 TO 1965

To enrich our analysis and to allow the reader to get an overall picture of bubble chamber work we have collected together a certain amount of statistical data on the collaborations which existed at CERN between 1960 and 1965.[151]

Granted the way the work was organized (the collaborating groups were rarely present on the site), it is of little interest to count the people who came to work in Geneva. The result would simply confirm the omnipresence of the groups who had built the chambers and who kept them running for the different collaborations. We have thus chosen three other indices. Firstly, the number of photographs taken by each chamber during the six years in question—so as to establish those that were decisive. Then the number of photographs studied in each country—as this will give us an idea of the investment which different national groups made in bubble chamber work. Finally, the size of the collaborations, as measured by the number of institutions taking part in each run.

France brought two bubble chambers to CERN, as we know. The Ecole Polytechnique heavy liquid chamber functioned at Geneva from August 1960 to October 1964 and took 2 million photographs—so 17% of the total number of photographs taken at CERN between 1960 and 1965. The 80 cm Saclay hydrogen chamber, for its part, had an outstanding career, not only because it worked satisfactorily but also because the other hydrogen chambers, the 1.5 m British chamber and the 2 m CERN chamber, were ready so late. Installed in February 1961, the Saclay chamber took 6 million photographs out of an overall total of 11.5 million. It was still in Geneva at the end of 1965.[152]

The British hydrogen chamber arrived at CERN in 1963. It was used for 18 months, from January 1964 to June 1965, and took 1.3 million photographs. It was then moved to Nimrod. The two CERN chambers respectively took 1 million photographs—Ramm's heavy liquid chamber[153] —and 700,000 photographs—the 2 m hydrogen chamber. (Let us not forget that the latter only started working in

March 1965 and that our statistics only go up to November of that year; the 2 m subsequently took some 40 million pictures.[154]) The Italians brought a 20 cm helium chamber from Rome, and they took 120,000 photographs with it between December 1961 and March 1962.[155]

We must also mention the beam equipment brought to CERN in 1961 and which was invaluable in the light of the lack of material in the laboratory at the time. Crucial were the separators brought by Cresti from Padua, the first to be used at CERN, and a number of quadrupole lenses loaned by Saclay. Subsequently, most of the material needed for building beams was provided by CERN's NPA Division.[156]

The one striking point to emerge from these data is the importance of the French contribution, without which the situation at CERN would have been rather poor during the first four years after the PS functioned. The role of this country's teams in bubble chamber work emerges even more clearly if we mention their decision, taken in 1964, to construct along with CERN a giant heavy liquid chamber (Gargamelle) and their participation in the tripartite France-Germany-CERN 'Big European Hydrogen Chamber' (BEBC) accepted one year later.[157]

A study of the distribution of photographs largely confirms the role of the French and the weight of Italian and British visitors in bubble chamber work at CERN. Our method of estimation calls for one comment, however. To establish just who used the photographs is a more or less impossible task since neither the details of the distributions nor the rotation of the photographs between the groups were recorded. We have thus followed the practice used by Mervyn Hine in a report he wrote in 1965, and distributed the number of pictures taken in each run equally between the participating groups (this number being known). Our data are thus approximate and proportional to the number of institutions present in the collaborations. This probably leads us to underestimate the number of photographs actually handled by CERN (the laboratory often being considered in the data as a single institution whereas several of its groups often participated in experiments) and by the French teams (less numerous than in the other 'big' countries, but often having more powerful data-handling equipment). An example: 104,000 photographs were taken in the runs in August and October 1960 with the Polytechnique chamber. Six institutions took part in these runs (CERN, Collège de France, Ecole Polytechnique, Milan, Padua, Turin). Accordingly in our tables half the pictures are attributed to the Italian teams. In reality it seems that CERN and the Ecole Polytechnique alone handled more than half of the photographs.[158]

These words of warning having been made, let us now turn to the results which are to be found in table 8.1 (the distribution of the photographs by country and by year). The features which emerge most clearly are the importance of France and Italy in the long term, the weight of the CERN and the British teams, and the relatively low participation of visitors from Germany (only 8.6% of the photographs) and from the nine 'small' countries (9%). In fact the German teams were present in almost twice

Table 8.1
Distribution by country of bubble chamber photographs taken at CERN (in thousands)

Country \ Year	1960	1961	1962	1963	1964	1965	Total 1960–65
France	27	67	535	517	346	1046	2538
Germany	7	68	93	187	252	258	865
Italy	81	104	418	693	253	895	2444
United Kingdom	36	129	366	285	701	412	1929
CERN	31	82	292	353	250	495	1503
9 'small' countries	29	60	179	288	146	246	948
Non-member states	22	39	85	159	167	202	674
Total	233	549	1968	2482	2115	3554	10901

Source: CERN/SPC/213, 28/11/65, 18.

fewer collaborations and runs than the French and Italian teams, the British occupying an intermediate position. French and Italians visitors organized ten runs each either alone or with CERN, the British three and the Germans only one (with the Saclay chamber in the case of the last two countries). The number of scanning tables per country in 1965 (tables which allow one to reconstruct the paths of the particles from the photographs before going on to analyze them), confirms this point: there were 45 in France and in Italy, 34 in the United Kingdom, 14 in Germany, and 13 in the 'small' countries.[159] Most of our indices therefore converge to give a picture of three countries who extensively used the possibilities offered by CERN in this domain, the teams from the smaller countries and the Federal Republic of Germany being rather slower to use the organization. Signs of change were, however, perceptible in 1965.

There is another consideration which is worth noting. It is whether one has statistical evidence showing a preference for their chambers by the groups who built them or who brought them to Geneva. The reply calls for a distinction to be made between the 80 cm Saclay chamber and the two CERN hydrogen chambers (30 cm and 200 cm), on the one hand, and the British chamber and the two propane chambers (CERN, Ecole Polytechnique), on the other. The use of the Saclay hydrogen chamber does not seem to have been too heavily biassed by its origins; it was more or less treated on a communal basis. This is probably to be understood by the fact that it was the only 'basic' chamber in the CERN programme for three years. From the first months of its functioning, for example, it was used more by collaborations without French teams than by those including the French, and this remained statistically the case until October 1965. The Italians, for example, worked a little less than did the French with this chamber. Similar remarks are to be made about CERN's two hydrogen chambers, the 2 m chamber in particular being used far more by outside than by CERN teams.

As for the British chamber whose operation was so long delayed, it was always used by very large collaborations. All the same British teams were present in all the collaborations in 1964, and in six runs out of the nine organized with the detector in 1965 (the three other runs were managed by a Genoa-Hamburg-Cracow-Milan-Saclay-Warsaw collaboration for the study of 11 GeV πs). In 1964 the British also received more photographs than in any other year (700,000 against 285,000 in 1963 and 412,000 in 1965), this being indicative of their above-average presence in the collaborations.[160] This situation seems to be explained by two factors. Firstly—and even if this is more marked here than for the other hydrogen chambers—by the advantages always granted to the builders in the initial period (remember that this chamber only stayed at CERN for 18 months). Secondly, by British physicists' impatience to use at last a chamber which had come on stream so late.

The information we have on the size of the collaborations in the period 1960–65 is summarized in table 8.2. It gives the number of runs carried out each year according to the number of teams who participated in them. It confirms, firstly, that work was slow to get under way in 1960 and 1961, notably because of the delay in the provision of separators. Then that the model of very large collaborations was the solution to this problem of 'scarcity': on average six groups worked together on each run in 1960, and eight in 1961. Finally, that the number of runs increased in 1962 and almost doubled again by 1965—and that this went along with a reduction in the number of teams per bubble chamber collaboration. The increased choice reduced this from 6–8 in 1960 to 3–4 in 1965.

Table 8.2
Number of bubble chamber runs per year;
distribution by number of groups involved in the run

Number of teams per run	1	2	3	4	5	6	7	8 and more	Total number of runs	Mean number of teams per run
1960					1	1	1		3	6
1961				1		1	1	2	5	8
1962	4^2	1	1	2	5	1	2		16	3,9
1963	2	2	7	5	2				18	3,2
1964	3	2	4	4	2	2	1		18	3,6
1965[1]	2	7	9	2	8	4			32	3,6

Source: CERN/SPC/213, 37–44.

Notes: 1) Up to 31/10/1965.
2) In particular, runs with the Rome He chamber.

8.7 Electronic experimentation: the struggle for access to the PS, 1960–1965

Before describing the specificity of electronic experimentation at CERN, it is perhaps necessary to remind the reader of what an experiment using counters is. In contrast to the operation of a 'global' data collector like a bubble chamber, an electronic experiment is selective in its aims and, at least in the early sixties, remained close to the image of an experiment which lay people usually have. Generally set up to test a particular theory, using an apparatus to be rebuilt for each particular experiment, and taking data itself at the machine, a counter experiment at the time was still performed by a small group of people, united by a specific project and a certain mode of practice, and centred on one or two strong personalities.

8.7.1 THE NP DIVISION AND THE CERN TEAMS

By way of introduction we thought of determining the size of the electronic teams working in Geneva. To do this we used the only source which was stable in the long term, the publications—recognizing that these limit the information to the number of physicists.[161] For 1960 we found an average of six names per team, with a weak variation around this average (six teams were given permission to experiment at CERN in 1960). The following year the average lay between six and seven,[162] the minimum being three physicists (in the case of one team) and the maximum being seven (for five teams). At the other end of our period, for 1964/65, the average oscillated between eight and nine researchers per team, the span having also increased (from four to eleven). There is no comparison then with bubble chamber collaborations which gathered together some four to eight teams, each of which comprised several physicists and a technical team whose job it was to study the photographs. We do see, though, a slight increase in the average size of electronic teams from 1960 to 1965, a tendency confirmed by a document written by the EEC chairman in 1969. Considering institutional sources (and not publications) he gave for 1968 an average of nine to eleven physicists per group depending on the type of group (CERN, mixed, visiting).[163] There is an exception to be noted, the neutrino experiment, which was left out of previous calculations. The number of participating scientists here was far greater, and reflected the importance and the weight of these experiments for CERN. In 1964-65 40 physicists were involved, 20 electronic specialists, and 20 bubble chamber scientists.[164]

Each electronic team was then an autonomous entity. Its problem was to gain access to the machine, access for which it was in direct competition with other electronic groups—and here the idea of a 'collaboration' in the bubble chamber sense of the word had little meaning (at best one shared a beam). The consequence of this was a sort of 'individualism' which expressed itself at each stage of the process of experimentation.

It arose, firstly, when the material was being prepared, when each group wanted to have access to everything that was technically possible. In 1961, for example, Adams 'had some time [...] to look around CERN a bit'. What struck him was that 'each experimental group [in the NP Division] feels the need to possess its own hydrogen target (in addition to its own electronics and even its own spark chamber)'. A good director (and probably an excellent engineer), he went on to explain to the NP Division leader: 'I wonder whether all this is economic or really necessary. It may be that all the targets are different but it seems that several parts are at least common. For instance, the Track Chamber Division is continually supplying transfer lines, valves and other parts to your groups'. In his reply, Preiswerk agreed that many components could probably be shared and that a minimum of rationalization was possible; all the same he stressed that it was of the essence of this type of experimentation 'to get familiar with the technique and to get the feelings and ideas in which way spark chambers [for example] could be incorporated in an experimental set-up'.[165] Three years later the same wish for regrouping and for standardization was to resurface in some parts of the organization—but the group leaders once again successfully repulsed any moves towards 'centralization'.[166]

Next, this 'individualism'—and the tensions which it sometimes created—was visible around the machine whether it was a matter of installing equipment and testing it or of taking data. Different people behaved here in different ways, and one must not take the examples we are going to give as necessarily typical. The strongest tensions were generally created by the same groups or individuals, some of them using this in a deliberate fashion as a way of getting what they wanted. Two examples will suffice. In October 1961 Preiswerk sent a memorandum to 'All physicists of the NP Division'. He remarked that 'Most groups are complaining about the lack of discipline displayed by other groups', adding that 'Equipment is taken away or altered without notifying the group to which it belongs. This often leads to severe losses of running time when this equipment is complicated'.[167] During tests and runs the conflicts were centred most frequently on the distribution of the beams. Either because one group asked for additional time to 'finish' its data taking, or because it had had to share a beam with another group (as often happened in parasite runs).[168] In November 1964 the experimental team 'PAPLEP'—a team which seems to have complained rather regularly—sent a protest note to the Director of Research objecting to the behaviour of the 'Ξ-parity' team which 'has obtained 4.1 times more parasite time than us and 1.5 times more user time than us' in a beam which they were supposed to be sharing. PAPLEP insisted that Gregory persuade the coordinator (who did not accept PAPLEP's figures) 'to be more helpful regarding our needs of PS time'.[169]

Tensions between group leaders also appeared during the decision-making process for experiments. We find here classical 'lobbying' activities, practices undertaken to short-circuit the normal ways of taking decisions. At the end of 1960, just when he was trying to convince Weisskopf to simplify the CERN committee system, Citron

Notes: p. 501

identified four kinds of such practices. It is worth repeating them even if his letter is not to be interpreted literally.[170]

> a) *Fait accompli*
> While the discussions in the various committees are going on, the physicist prepares his apparatus and tests it in parasitic runs. At a certain moment he then declares that everything is ready and if he is only given such and such a number of shifts he will have his result. Nobody will dare to refuse him machine time under such conditions.
>
> b) *Smoke screen*
> The physicist puts up a plausible proposal to the committee in order to secure himself an allocation of machine time. Once he has got the machine time, he does whatever experiments he considers most interesting at the given moment. [...].
>
> c) *Wholesale booking*
> This is a variant of the smoke screen tactics, practised by people who have much spare time or many students. They work out proposals for almost all possible experiments. Then they are pretty sure that by the time the proposals are agreed to, one of them will still be interesting and they will do that one.
>
> d) *Brute force*
> The physicist addresses himself to an influential member of the Directorate, say, and simply jumps the queue.

In this highly competitive atmosphere, where the hope of a Nobel prize drove more than one American or European physicist, three particular elements at CERN reinforced the centrifugal tendencies inherent in the detection technique—tendencies also perceptible in laboratories other than CERN. The first was *the personality of the leader of NP Division,* Peter Preiswerk, who was initially not chosen to guide the research but, according to Adams in 1964, 'merely [to] keep the whole complicated Division running smoothly'.[171] Rather quickly—and other documents confirm this point—Preiswerk lost 'the confidence of the group leaders and the other physicists in the Division regarding scientific judgement and also technical questions'. Whether or not this was justified does not concern us here, but it certainly had decisive effects. 'There was nobody in the Division', summarized the Director-General in 1965, 'who could talk to the people with enough authority in order to bring about a certain collaboration and common study. Preiswerk had not enough personal authority to prevent an unsound increase of small proposals for new experiments, and was not able to foster the necessary objective discussions among group leaders in order to plan a reasonably feasible programme which does not overlap too much'.[172] This is the reason Weisskopf gave when he called on Wolfgang Paul in May 1964 to share the management of the division, 'to lead and influence the scientific research here at CERN in a better way than has been the case to date'. Paul accepted the proposal in the Autumn and his nomination was ratified by the Council in December 1964. As a result, and for two years, the division had two leaders.[173]

To attribute everything to Preiswerk's 'failings' would be, however, to go too far; another kind of explanation for the particularly clear difficulties in electronic work at

CERN emerges frequently in minutes of meetings and in the private correspondence. This was *the size of the NP Division,* and the large number of group leaders and of 'professor[s] [piled] upon professor[s], all recruited above the working level and all trying to influence the Division in a direct or indirect way'.[174] This set of people (around 20 in 1964/65) formed the all-powerful core of electronic experimentalists, they constituted (to use a phrase we have already introduced) an aristocracy which collectively controlled this type of work and which was united by one thing: the determination to reject any kind of outside intervention. Without the accumulated experience of their American colleagues, they were sometimes described as 'performing' badly in comparison to them in the experimental work around the PS, as being less quick to seize technical opportunities—indeed 'reluctant to take up suggestions of any kind'. 'CERN was lacking in seminars where the newest information on the progress of physics is presented in a digestible form' said Roberts of Argonne in 1961 (such seminars were inaugurated in 1962), while Bernardini declared that it was very difficult 'to convince people [in NP] to give up experiments which turn out to be really silly, or to establish a cooperation between two groups [from whom] we had the same proposition'. He concluded: '[they] did not want to resign to the high priority they had established because they had sent me a piece of paper of two lines saying I want to do this experiment'.[175]

These criticisms persisted in 1964 and 1965, and various people put forward different solutions: split the NP Division in three smaller ones ('therefore a better coherence and more collective responsibility inside a division') and increase the scientific authority of the new division leaders (Citron's proposal in 1964); reduce the number of teams allowed to work at the PS ('by amalgamating those which are understaffed') and strictly limit the number of experiments which each team can do simultaneously (Geibel, 1965); (re)open the EEC towards the outside (so as to break the pattern of local functioning) and adopt a model closer to that of the TCC (Van Hove, 1966), etc.[176]

There is yet another factor which one needs perhaps to bear in mind if one wants to understand properly the problem of electronic experimentation at CERN: *the complex and diluted nature of the decision-making mechanism for experiments* which many physicists regarded to be remote from the real problems and the real actors, and so tending to make poor judgements and to be unduly swayed by lobbying. Let us remember that after 1962 a proposal had always to go via Preiswerk (and Paul) to the group leaders who met regularly to discuss them, then to the EEC which drew up the list of experiments which it deemed 'scientifically interesting and technically practicable', and finally to the NPRC—and to the DG if the proposal was refused and if someone appealed against the decision. The intermediaries were thus multiple, as were the possible allegiances. For some the effect of this was the practice of 'research politics rather than research policy'—to quote a very critical Geibel—, a tendency to divide machine time as much as possible—what Geibel called a 'policy of cheap protons'—, at the limit a fear to accept ambitious proposals

Notes: p. 501

which would absorb too much machine time and which accordingly trod on too many people's toes. Against this way of understanding the problem, we would like to suggest that the situation faced by the experiment committees was more complex, that the question was perhaps not primarily that of a weak system of decision — that the 'policy of cheap protons' reflected the state of European physics. If the machine time was distributed between so many electronic groups, it was also because they remained small and without outstanding proposals. An indirect proof can be found in the neutrino experiments; there clear priorities were always given by the EEC and the NPRC.[177]

That granted, we do not want to suggest that the quality of the work was uniformly poor — as one can see from the survey presented in chapter 5.[178] What we have tried to do here is to show how electronic work functioned concretely at CERN, and to do so in all its complexity, without trying to gloss over the human dimension. Of course those who had to judge between competing proposals generally did what they thought to be best. All the same, everyone wanted to be the first to do the decisive experiment and from that perspective (almost) any means of achieving one's ends was 'acceptable'. This was at least the case for some groups who accepted that getting machine time to do an experiment was, of course, a question of scientific merit, but also and plainly one of 'politics', and as such to be treated with total cynicism.[179]

8.7.2 HOUSE-PHYSICISTS AND VISITORS IN ELECTRONIC EXPERIMENTATION: A DIFFICULT RELATION

To conclude this section we want to look at the relationships between physicists at CERN and the visitors who came or wanted to come and work at Geneva. This was a very delicate question in 1960/61 — and remained so at least until 1965/66 — and so we first want to give some statistical data, leaving aside as much as possible the conflicting judgements which they gave rise to (by the visitors on the non-openness of CERN — and vice versa).

Table 8.3 presents some information on the ratio visitors/CERN physicists as derived from official sources. It shows that the ratio was not negligible in the years 1960-65 and that it seems to have increased in later years. The data is nevertheless not all that reliable, and we have repeated the exercise using publications. Taking institutional affiliation as given by authors as our key criterion,[180] we get a ratio of about one visitor to two CERN staff in 1960, of one to one in 1961 and 1962, of a little more than one to one in 1964, and rather closer to three to two in 1965.[181]

Table 8.3

CERN research staff, fellows and visitors in the NP Division

Physicists in NP Division (but for 1960)	RA 1960 Exp. Phys.	Th. Phys.	RA 1961	RA 1962	Preiswerk April 1964	Doc. NP Oct. 1964	RA 1964	RA 1965 PS	SC	Hine 1965
CERN research staff	60	9	?	40	36	43	61	34	9	29
CERN fellows	23	25	20	} 60	} 69	} 86	} 91	} 100	} 38	20 } 79
Ford fellows	10	5	8							
Unpaid visitors	50	0	101							

Sources: *Rapport Annuel*, 1960 (p. 14), 1961 (p. 39), 1962 (p. 53), 1964 (p. 42), 1965 (p. 48); P. Preiswerk, *Development of Electronic Experiments*, 9/4/64 (NP 22914); *Comparison of CERN Research Staff and Visitors as at 10th October 1964*, doc PE/F/VS, ss. date (DG 20741); CERN/SPC/213, 48 (Hine).

Remark: Granted the possible variations in definition from one source to the other, only the orders of magnitude need be remembered; there are no data for 1963 in the Annual Reports.

Table 8.4

Classification of the electronic experiments teams at the PS by their sources of finance

Electronic detection at the PS	1961	1962	1963	1964	1965	1966	1967	1968
CERN team	5	5	7	6	3	3	4	3
Mixed team	1	3	3	4	8	10	11	10
Visiting team	0	1	1	1	1	4	4	6
Total	6	9	11	11	12	17	19	19

Source: CERN/SPC/272, 25/2/69 (Memorandum of the EEC Chairman to the SPC).

The essentials do not however lie there but in the *changes in the nature of the teams* over time. Let us be more precise. In 1960 and 1961, official CERN documents and publications show that all groups — with perhaps one exception — were so-called CERN groups, i.e. led by one or two CERN physicists and including only a few individual visitors. In 1962/63 a new type of team decribed as 'mixed' made its appearance (i.e. comprising a CERN group and at least one visiting team). As examples here we may cite the CERN-Harwell collaboration or the CERN-Bologna-Liverpool-Michigan (Ann Arbor) collaboration which studied the elastic diffusion of π-p and p-p at 8, 12, and 18 GeV/c with a liquid hydrogen target and nine spark chambers. At this time only one team was clearly a visiting team (i.e. without the participation of a CERN group), and that was Roberts' team from Argonne. Two years later, in 1964, CERN teams were still in the majority around the PS but there were some four or five mixed teams (out of 15) and a Saclay-Orsay collaboration was in Geneva studying $\pi^- p \to \pi^0 n$ charge exchange at energies between 2 and 18 GeV/c. This was then the first visiting team from a member state. The next year two collaborations were 'visiting', the remaining teams being distributed more or less equally between CERN teams and mixed teams. In 1966 the changeover took place: mixed teams were clearly in the majority, and the ratio of mixed and visiting teams to in-house teams was now 2:1.[182]

The paper drafted by the chairman of the EEC which we mentioned a few moments ago provides us with an independent confirmation of this reversal of matters between 1964 and 1966. Drafted in 1969 it contains a table which distributes the electronic teams working at the PS according to their source of finance (see our table 8.4). The differences with the preceding results are negligible. Let us simply draw attention to the increasing presence of so-called mixed teams as from 1962 — the model: six CERN teams, three mixed teams, one external team remained stable until 1964 — and note that the balance seems to have begun to shift in 1965, when only three teams out of twelve were purely CERN-financed.[183]

The overall picture is thus reasonably clear. In 1960 and 1961 the leadership was in CERN's hands. All teams were formed around a CERN nucleus but accepted individual visitors. During these years American visitors were numerous (19 out of 70 in 1961). Between 1962 and 1964 teams based on CERN staff were always in the majority (at least three out of five) though some mixed teams hosted external groups who came to work at CERN each year. In 1965/66 a new pattern, a new way of working around the PS emerged: the rule now was the mixed team, and visitors were perhaps two, three or even four times more numerous on the CERN site than the local physicists. And we now also find European teams who were capable of experimenting on their own and who had built a large part of their material at home — an arrangement that was to become more widespread later.

This picture calls for a comment: it was very similar to the kind of development outlined by Research Director Bernardini in a memorandum to the Directorate dated *August 1960*. Reminding them how much electronic experimentation 'depended essentially on individual characteristics and orginalities', he concluded that it was urgent for CERN to recruit top-quality people, to have working around the PS inventive physicists who had a thorough knowledge of the available techniques. He agreed with Hine that 'most ideas come to the few who have them after they have worked on the spot for some time and have really got a feeling for the problems involved'. Unfortunately, he added, 'the contribution from European scientists is somewhat crippled by a lack of experience and lack of confidence'. Bernardini's conclusion was simple: CERN should not encourage visiting teams for the moment. Instead initially it should invite the most brilliant individuals and have them inspire the CERN teams (here he was thinking predominantly of American physicists). In a second phase, really mixed teams could begin to emerge centred around the nuclei already present in Geneva. Thereafter, but only in a third phase, external teams capable of working alone around the PS could be invited to come to the laboratory. The correspondence between the envisioned model and reality is clear — though this of course does not mean that what we have here is simply the successful implementation of a policy defined in advance. On the contrary, Adams' attitude in 1960/61 seems to suggest that the Directorate was not particularly sympathetic to Bernardini's philosophy.[184]

The correspondence was not perfect either: Bernardini thought that the evolution he had described would be spread over about three years — it took six or seven instead. And this leads us to one of the criticisms most frequently expressed by the visitors: the dislike of the physicists in NP Division to share the machine, and their preference for CERN teams hosting individual visitors. According to them, if it took so long to install the pattern which emerged in 1965/66 this was not so much due to the weakness of the visiting teams but to the difficulty they had of breaking down the NP Division's tendency to remain closed in on itself. And it is true that in 1966 the new CERN management thought that 'the EEC and the procedure to be followed by outside counter groups for being accepted at CERN' was one of the key problems to be solved — and to be solved urgently.[185]

To conclude this study we want to look at the participation by country in electronic work at CERN. The number of collaborating institutions is given in table 8.5. It confirms the emergence of the new pattern in 1965 (13 institutions who collaborated as opposed to an average of 7 or 8 from 1960 to 1964), it shows that the British were the first to make full use of the CERN machines even if we note a certain decrease in 1964 and 1965 (which might be connected to the start-up of Nimrod, their 7 GeV synchrotron), it indicates the important part played by Italian and then French

Notes: p. 502

Table 8.5
Number of institutions that collaborated in
electronic experiments at CERN

Institutions	1958	1959	1960	1961	1962	1963	1964	1965
France	1	1	1	2	1	2	3	4
Germany	–	–	–	–	–	–	–	3
Italy	1	–	2	3	3	2	2	4
United Kingdom	3	2	2	2	2	2	1	1
9 'small'	–	–	1	1	1	1	1	1
United States	–	–	–	–	2	1	–	–
Total	5	3	6	8	9	8	7	13

Source: CERN/FC/213, 25.

Remark: Applies to SC and PS activities but for nuclear structure studies.

groups, but it confirms the late arrival of German groups (very few prior to 1965) and the very low level of participation by the nine 'small' countries ('the' laboratory in table 8.5 is frequently the Ecole Polytechnique in Zurich).

We have also studied the weekly records drawn up by the CERN administration to monitor personnel movements. They consist of lists of names of arrivals (visitors included), departures, and changes of status. We have selected from this data the arrival of visitors who came to CERN to do an electronic experiment with the PS or the SC, for all countries in 1961, 1963, and 1965, and for the four major ones in 1962 and 1964. Part of the data is presented in table 8.6.[186] They strongly confirm the importance of British visitors (though less so the 'reduction' in 1964-65), and the decisive role played by the French and the Italians (let us remember in this connection that the latter were by far the most numerous among the CERN staff who were specialized in counter physics).[187] The significant number of non-member states' physicists, and Americans in particular in 1961 (19 arrivals out of 70) is also evident, as is the sometimes considerable importance of Sweden and Switzerland and the low participation of scientists from the other seven 'small' countries (one arrival in 1961, four out of 94 in 1963, 13 out of 157 in 1965), and finally the relatively small number of German visitors before 1965 — even if the gap does not appear to be as great as when one counts institutions (33 arrivals between 1961 and 1964 against 82 for France, 57 for Italy, and 85 for the United Kingdom).

Going into details, we should point out that the high figures — notably by month or by quarter — correspond to the arrival of a team. The high American figure for 1961, for example, is related to the arrival of Roberts' group (14 arrivals out of 19); those for the British are related to the presence of teams from Harwell (1962), Imperial College (1963), and the Rutherford Laboratory (1965), those for the French to their participation in the Argonne group (1962) and the presence of Falk-Vairant's group from Saclay (1964), those for the Germans to the arrival of teams from DESY, Heidelberg, and Karlsruhe (1965), etc. For the Swedes, the 1965 figure is to be correlated with their groups working at the SC, notably on the construction of the on-line isotope separator (ISOLDE) for nuclear chemistry work.

8.8 Conclusion

To conclude we would like to give a panoramic overview of the organization of experimental work at CERN. To do this we propose to summarize the information by period.

Until 1958 the way of organizing experimental work at Geneva hardly differed from what one found in a university laboratory. It was the obvious way: the CERN physicists arranged access to common equipment between one another, they felt no need to have an 'experiments committee', the bosses (Gentner and Bernardini) were those who arranged the necessary compromises... The number of in-house physicists was more or less proportional to the experimental possibilities—in line with the

Table 8.6
Number of visitors entering CERN and joining the NP Division

	1961	1962					1963	1964	1965				
		J-M	A-J	J-S	O-D	Total			J-M	A-J	J-S	O-D	Total
France	6	3	6	15	8	32	14	30	0	7	9	3	19
Germany	4	1	2	0	5	8	8	13	0	2	23	3	28
Italy	14	5	1	2	8	16	10	17	3	6	13	1	23
United Kingdom	16	8	2	9	19	38	19	12	1	8	14	4	27
Sweden	1						2		0	6	5	6	17
Switzerland	3						14		0	3	3	1	7
9 'small'	5						20		0	17	13	7	37
United States	19						11		4	1	2	1	8
Non m-states	25						23		7	5	7	4	23
Total	70						94						157

Source: Weekly reports *Mouvements du Personnel CERN*, ADM/PERS/Contrôle, then PE/PERS/Contrôle, Personnel division.

Remark: We only have the number of *arrivals*. If the same person arrived several times in the same year he was counted each time he came. And if he remained more than a year without leaving CERN, he was not counted.

common practice of a university only to recruit as many researchers as its equipment could support—so no major tensions arose.

This picture is 70% true—since 30% of the machine time was set aside for visiting teams. Tolerated by the CERN staff, these 'foreigners' represented for them the price that had to be paid to the states financing the laboratory. To have access to the 600 MeV SC these outside teams had to convince two committees, one which vetted their proposals in their home countries, and the other, nominated by the representatives of the member states decided in the last instance. Once admitted, a team arrived with its equipment, did its experiment, and left; in short, it used CERN as a service centre.

There was then some 'sharing' at CERN from the start. This 'sharing' was related to the scarcity and the size of the equipment: unusual, it was attractive and demanded by many. This necessary 'sharing' was also a consequence of the mode of financing. Through their representatives in the Council and the SPC the states insisted that their nationals should be able to come and experiment in Geneva. Not that they rejected the idea that CERN was an 'independent' laboratory, but they also demanded a part of the machine time for their physicists. The fact that we also have many states involved here must be stressed: compulsion is far less effective if only one country is involved. The example of parallel American laboratories which were financed largely from federal funds is interesting here: they were not subjected to the same kind of pressure as CERN experienced; that only happened later, when the equipment became really unique, with no possibility of its being copied.

The year 1959 saw the first serious perturbation of this system. This was for no abstract reason nor because a matter of principle was at stake. The reason was on the contrary specific, almost concrete: part of the experimental work around the PS was to be done with huge hydrogen bubble chambers. These detectors were heavy, dangerous, and almost as immobile as the accelerators themselves. Complex and costly, one could never install many of them in the space available for doing experiments; as a consequence visitors could no longer come as truck teams, as they had been doing up until then.

The simple solution, and the one preferred by the CERN staff, was that CERN should build this kind of equipment itself: as a laboratory, this was one of its tasks, just like building accelerators. Generously, the CERN physicists would also keep a quota of time for visitors. This was the way Alvarez did things at Berkeley, and the scheme he repeatedly recommended to his visitors from Geneva, it was the common sense solution, the solution which was consistent with the practice of equipping one's own house oneself. However, CERN staff had to cope with two 'difficulties'. Firstly, CERN was not ready: it had started too late, was short of resources, and so could not supply its own bubble chambers. The British and the French, by contrast, were already en route. Not wanting to be behind in a technique which had been known to be crucial for the past one or two years, groups in these countries were building very big chambers to be used, in part, at Geneva. Wanting to come, and to be 'compensated' for their investments, they reactivated the idea of CERN as a 'central' laboratory, available to all. This was the theme most coherently expressed by Berthelot, one of the most important leaders in France who was based at Saclay, when asked his opinion in summer 1959.

Three poles crystallized in the debate about the use of the British and French chambers. The CERN staff wanted the present situation to be seen as only temporary; in time CERN would build all important equipment. In consequence CERN physicists were to remain the privileged backbone of any experimental collaboration. The other physicists refused to accept this scenario. The British and the French wanted the builders to be privileged users — even if because of the heavy nature of the equipment they did not ask for the sole rights. Physicists from other countries, by contrast, while also recognizing the right for all to participate fully in the construction of experimental material, demanded that there be no privileged users: scientific merit was to be the sole criterion.

To ensure CERN's survival — and none of the protagonists would have wanted otherwise — provisional agreements were drawn up. They were drafted in ambiguous terms and subject to diverse interpretations. In the short term a few little privileges would be granted to those bringing the three chambers; by and large, however, their use was to be as communal as possible. In the long term, it was agreed to try to have at least a minimum of coordination; decisions on the heavy equipment to be built in the future were to be planned, and those who wished to take initiatives were not to confront others with *a fait accompli*. Nevertheless nothing was settled regarding the mechanisms for such decisions, nor about who would equip CERN.

We arrive then in 1960, the first year in which the PS worked. It was the year of running in, of tensions, of first practical solutions, and of misfires too.

In November 1959 the PS produced its first beam. This in no way meant that the machine was ready for the physicists. Up until summer 1960 it was the engineers who had built it who worked most of the time on the accelerator, and it was only in November 1960 that three-quarters of its running time was available to experimenters. For this reason there was no great sense of urgency at the start of the year, even if the first ways of organizing work were important: they could easily turn into habits.

What is striking about spring 1960 is that very few physicists were really prepared. The equipment available to do experiments was rudimentary, and the resources were lacking to build good quality secondary beams. Attention was primarily focussed on surveying the particles produced at internal targets. Two philosophies clearly emerged concerning the concrete organization of the work, however. One was that of the physicists who were specialized in electronic detection and of the engineers who studied, prepared, and improved the machine. These two groups together occupied about 80-90% of the accelerator. Comprising only CERN staff or long-stay visitors, they amicably shared the available time. A family affair, the management of the PS was organized as in any 'ordinary' laboratory under the supervision of Adams, the leader of the builders, and of Cocconi, the physicists' representative. Applicants were still in short supply—because of the lack of preparation—and each group had the time it wanted to become familiar with the new order of magnitude which the PS represented.

At this time the emulsion and bubble chamber specialists were using very little machine time. CERN's 30 cm hydrogen chamber and the Ecole Polytechnique's 1 m propane chamber were only fully operational by the summer, or even the autumn of 1960. For these physicists, however, management was not to be the sole prerogative of the CERN family. 'Outsiders' were there, from January onwards, preparing for the arrival of their chambers and wanting to take part in runs. An organizational scheme which differed from that in force at the SC was thus elaborated, a new practical arrangement which allowed for a real integration of CERN staff and visitors. In contrast to what had happened in 1959, there was now no debate on principles; the solution emerged gradually, by the systematization of empirically tried and tested micro-solutions; in brief, one learnt to walk by walking.

A mixed committee of bubble chamber physicists from CERN and the member states emerged from informal working meetings. Its aim was to satisfy people as much as possible by installing chambers and choosing beams; it was then to pull together the wishes of each and to bring into being collaborations which collectively managed the runs and the analysis of the film. Under the influence of two previous students of Leprince-Ringuet, Peyrou at CERN and Gregory for the visitors—the British chamber was not yet ready—, a policy for systematically integrating all the groups who wanted to participate was implemented. The procedure was in place at the beginning of the Autumn.

How are we to account for this novel practice? By the nature of the technique, to begin with: each chamber/beam pair enabled one to produce films at will. If one agreed to share the pictures liberally—which was unusual since one gave away possible 'discoveries' to competitors who often contributed little if anything to setting up the experiment—such big collaborations could function. The nature of the technique was not enough then: it had to be coupled with a 'political' will. This existed at CERN, facilitated by the shared nationality of Peyrou and Gregory and by the role of buffer which Adams played *vis-à-vis* his British colleagues and friends.

As for the counter specialists, their tendency to 'keep things in the family' persisted, and the contrast with the organization set up by the bubble chamber people generated considerable dissatisfaction in the member states. A prestigious branch of physics, electronic experimentation was managed by obscure methods, opaque to outsiders. In July 1960, the SPC told Adams of their disquiet. Feeling that the anxieties were justified, and the prevailing practice dangerous for the atmosphere at CERN and in the Council, the new Director-General made a clear choice: this type of work would be managed by an Electronic Experiments Committee of CERN staff and visitors modelled on the Emulsion and Bubble Chamber Committees. Clearly then Adams, with the Council's support, abandoned the idea of CERN as an independent laboratory, an 'ordinary' laboratory, a laboratory serving *its* physicists before all others. These were still there, of course, in numbers; they were however no longer allowed to be alone, to be privileged, to be 'at home'.

The years 1961-1962 were those during which the new rules were learnt—with the exception of one sector, electronic experiments. The rejection of Adams' system got under way in the CERN division containing the counter specialists in January 1961. They explained that the nature of their work, which called for an ongoing presence and constant modifications, could not be subjected to the cumbersome decision-making system Adams had proposed; being the only people permanently on the site, only they were really competent. They insisted therefore that they be left to organize the work themselves and promised to integrate visitors—a proposal which the new Director-General Victor Weisskopf accepted in the interest of efficiency. Arrangements were thus kept in the family, at least for the time being. Meanwhile, the visitors were coming to know the machine and would soon be in a position to replace the Genevans, or at least to hold their own against them. At the end of 1962, however, they were still somewhat marginal.

It was during these years, by contrast, that the bubble chamber specialists' manner of working stabilized, became normal. Around the PS, on the shop floor, the relations with those who were running the accelerator were also sorted out. To experiment also meant, from CERN's point of view, to use as best one could the space and the basic equipment. This was dealt with by the accelerator division at the heart of which, and under the influence of its leader Pierre Germain, there was a growing tendency to play a service role. Being aware of the difference between themselves and the physicists, the engineers in this division took it upon themselves to draw up medium term plans. In contrast to physicists, who were always preoccupied by the immediate needs of their experiment—and its top priority!—they tried to anticipate what was needed to accommodate the greatest number of experiments possible. Operational as from 1962, Munday's group became the indispensable interface between the physicists and the engineers, between the experimental groups and the NPRC—the committee chaired by the DG which had to settle the experimental programme—and, on the ground as it were, between the experimental groups themselves, who were often ready to use all manner of means to get their own way at the expense of others.

Towards the mid-1960s, finally, an equilibrium was established. In electronic experimentation non-CERN groups began to work in Geneva. Having acquired by this stage the know-how, the novel equipment, the good ideas, and the money, they came to work at CERN, either in collaboration with CERN groups or on their own. In the wake of this development which was not without its tensions—the number of requests grew increasingly—the Electronic Experiments Committee was more officially reoriented towards the outside in 1966.

The NPRC, for its part, saw its role increase as the number of choices it had to make multiplied. Initially in an uncomfortable position between the three specialized committees it became more and more able to identify priorities. The system remained rather cumbersome though, symptomatic of a lack of leadership; in contrast to its equivalent at BNL, it did not have an almost absolute authority. This did not stop one essential development taking place under its guidance: by the mid-1960s CERN had become an atypical research facility in that its research programme had become largely *European* and as such was no longer primarily conceived and realized inside the organization itself. The house-physicists in fact represented little more than 20-25% of the experimenters working at Geneva.

In 1965-66 this picture of what CERN was had become internalized. It was not yet on the scale which we find ten or fifteen years later, when the CERN physicists were reduced to 4-5% of the total, and were simply an auxiliary force. All the same the changeover occurred around the mid-1960s, when the schema set up in 1958-59 disappeared. Indeed the CERN management inscribed this new openness in the staff rules. Having to decide on the distribution of indefinite contracts to scientific personnel, it chose to give them more or less automatically to those (essentially the engineers) who accepted to keep the laboratory going and to be selective in awarding them to CERN physicists. From now on those who *used* CERN to do physics had to 'go back home' regularly, the majority to the European universities from which they had come.

Translated by John Krige

Notes

1. There is no precise bibliography on this topic. However, see Cockcroft (1965), Crozon (1987), Galison (1987a), Gaston (1973), Hagstrom (1964), Kowarski (1967, 1977), Krige & Pestre (1986), Lemaine et al. (1982), Swatez (1970). In terms of archives, we mainly used the CERN DG and DIRADM files for British and French Bubble Chambers, and for the experimental committees (NPRC notably); NP (Nuclear Physics) files for the experimental committees and for the NP division; TC and EEC documents for the corresponding experimental committes.
2. Argonne was the first to be confronted with similar problems in the USA; see History of ZGS (1980) and Kowarski (1967). Add Galison, Hevly & Lowen (1988) and Westfall (1988).
3. The themes have often been dealt with in scientific milieux; see Kowarski (1977), 397; Weinberg (1972), 425; Wilson (1972), 468.
4. Adams (1965), Cockcroft (1965), Pickavance (1965, 1968).
5. For example Crozon (1987) and Morrison (1978) for physicists; Gaston (1973), Lemaine et al. (1982), Swatez (1970) for sociologists; and Taubes (1987) for the last.

6. Kowarski (1967), History of ZGS (1980). Recent studies should now be added, notably Westfall (1988) and Galison, Hevly & Lowen (1988).
7. See *Preparation for experimentation with the SC machine*, CERN/SPC/12, 11/11/55; SPC meeting held on 21/11/55 (CERN/SCP/14, 7/12/55, 4); SPC meeting held on 23/6/56 (CERN/SCP/26, 22/8/56) during which the working group was set-up (Bernardini, Heisenberg, Leprince-Ringuet, Skinner); *Visiting Scientists,* CERN/215/Draft 1, 28/2/57 (group report); Draft 1/Add. 1, 10/4/57; Draft 1/Rev., 7/6/57 (submitted to CC and Council on 28/6/57); Conseil, minutes, 28/6/57, 17-19. Documents CERN/215/... are more important; quotations are taken from them.
8. Conseil, minutes, 28/6/57, 17-19.
9. Conseil, minutes, 28/6/57, 17; the Advisory Committee comprised Skinner (UK), Puppi (I), Berthelot (F), Paul (FRG), Wouthuysen (NL), Fierz (CH).
10. *Report on the first meeting of the Advisory Committee on 'truck teams',* CERN/SPC/48, 10/10/57 (meeting held on 23 August; Bøggild (DK) took part in this meeting). For the meetings of the Advisory Committee, see CERN/AC series.
11. Written from documents CERN/AC/1, 26/9/57 to CERN/AC/27, 22/7/60.
12. CERN/AC/5, 21/7/58, 3-4; CERN/AC/8, 4/11/58, 2-4; CERN/AC/18, 29/6/59, 3-4, etc.
13. *General Services Planned for the PS Experimental Programme,* CERN/AC/7, 4/11/58, 2.
14. CERN/AC/18, 29/6/59, 4. The quotation is given as the answer made by CERN directors to a question put by Skinner.
15. For example, CERN/SPC/93/Add., 22/9/59; in this document, de Rose used the notions of 'invités' and 'maitres de maison'. See also the Council meeting held on 26 May 1959 (minutes, 26/5/59, 15).
16. Quotation, Conseil, minutes, 26/5/59, 15.
17. CERN/AC/27, 22/7/60, 3.
18. CERN/AC/18, 29/6/59, 3.
19. For a general presentation of the question, see Crozon (1987), 166-178, Galison (1985a), Galison (1987a), Krige & Pestre (1986). For this particular case, documents CERN/SPC/64, 1/5/58, CERN/SPC/67, 8/5/58, CERN/PS Int. DIR 58-10, 22/10/58 (MGNH 22069).
20. *Proposed Cooperative Project for the Construction of a Large Bubble Chamber,* DSIR, NP-22, 26/6/57 (SERC-B22).
21. Letter Butler to Bakker, 12/8/57 (SERC-NP31), Memo DG to Adams, etc., DGO/Memo/371, 27/5/58, Letter Skinner to Adams, 15/5/58 (DG 20570), CERN/PS Int. DIR 58-10, 22/10/58 (MGNH 22069), Letter Butler to Bakker, 29/10/58 (DG 20570).
22. Letter Cockcroft to Melville, 15/12/58 (SERC-NP29), Letter Skinner to Adams, 17/12/58 (DG 20570), on which Skinner wrote by hand: 'I was sorry to hear that Verry raised the question of rent [...]. I would like to say that none of the scientific people involved was consulted about this and that it is certainly contrary to our ideas [...]'.
23. *British Bubble Chamber,* CERN 5940, 22/1/59 (DG 20570). Add Memorandum Dakin to Bakker, DIR/ADM/PS/03, 20/1/59 (DG 20570). The interest of this version is that it is annotated by Bakker himself. For preparatory documents, see Dakin files in DIRADM 20253.
24. Memorandum Dakin to Bakker, DIR/ADM/PS/03, 25/2/59, written after a second visit by Butler and Moore on 24 February. For British demands see also letter Smith to Verry, 23/2/59 (SERC-NP 29).
25. This simplifies a complex set of events. See DIRADM 20253 for the period 12/1/59 - 2/2/59. The quotation is taken from the memo cited note 23 which has to be compared with CERN 5940 cited note 23.
26. Memo cited in note 24; the other version is in CERN/311, 27/4/59.
27. *Note on a telephone conversation with Prof. Butler, 13 May 1959,* signed by Dakin (DIRADM 20253).
28. *Note* cited in note 27 for the first quotations; Memorandum Dakin to Bakker, DIR/ADM/PS/03, 28/5/59 for the last one (DIRADM 20253). The idea is explicitly taken up in a letter Dakin to Butler, 29/5/59 (DG 20570).

29. *CERN, Loan of U.K. Bubble Chamber,* internal memo written by H.L. Verry, 13/5/59 (SERC-NP 29). For the role played by the Council, see Minutes, 26/5/59, 26-28.
30. Memorandum Dakin to Baker, DIR/ADM/PS/03, 12/8/59, which reports Smith's words (DIRADM 20253).
31. See DG 20570 and DIRADM 20253 for details.
32. Letter Leprince-Ringuet to Bakker, 24/4/58 (DG 20565).
33. Details in letter Perrin to Bakker, 21/3/59 (DG 20565).
34. Letter Leprince-Ringuet to Bakker, 25/5/59 (DG 20565); Conseil, Minutes, 26/5/59, 26-28; Letter Perrin to Bakker, 26/6/59 (DG 20565).
35. Letter Leprince-Ringuet to Bakker, 25/5/59 (DG 20565); See also *Note of Conversations au CEA,* 3/10/59, CERN/DGO/Memo 716 (DG 20565).
36. First quotation from Bakker during an Executive Committee meeting, 3/6/59 (CERN/Ex. C/42, 17/6/59, 4); for the rest, see CERN/Ex. C/54, 22/10/59, 6.
37. Memorandum Ramm to Adams, 29/5/59 (DG 20562); letter Bakker to Leprince-Ringuet, 15/6/59 (DG 20934); declaration by Ramm during an Executive Committee meeting, 3/6/59 (CERN/Ex. C/42, 17/6/59, 3).
38. SPC, meeting held on 16/10/59, CERN/SPC/101, 19/11/59, notably pages 4 and 5.
39. *Dispositions envisagées pour la chambre à bulles à propane française,* CERN/324, 2/11/59, 1.
40. Conseil, minutes, 1-2/12/59, 56-60.
41. Letter Perrin to Bakker, 6/1/60. For details, see DG 20565 file.
42. Kowarski (1967) and the post-scriptum in a letter Citron to Weisskopf, 3/1/64 (DG 20630) which describes 'an (angry) users Meeting' at Brookhaven.
43. Letter Cockcroft to Melville, 15/12/58 (SERC-NP 29); report Smith to Verry, 23/2/59 (SERC-NP 29).
44. See the letters mentioned in notes 23, 32 and 33; Krige and Pestre (1986), 265-274.
45. The bubble chamber construction programme was drawn up at CERN between January and May 1958.
46. This is described in Krige and Pestre (1986).
47. SPC, meeting held on 16/10/59, CERN/SPC/101, 19/11/59, 3.
48. See, for example, de Rose's speech at CC meeting on 25 May (CERN/CC/318, 22/7/59, 8), or that of Perrin at the Council held on 26 May (Conseil, minutes, 26/5/59, 26-27); about the British, they seem to have *never* spoken during the debate on French chambers.
49. Memorandum Ramm and letter Bakker to Leprince-Ringuet given in note 37.
50. Letter Cockcroft to Bakker, 11/2/59, reproduced in CERN/GD/238, 2/4/59 (DG 20965).
51. The reason is that everybody thinks in terms of *quotas*. If one admits that 30% of users are to be visitors, but that the latter will be 2 in each mixed team of 6, it is no longer possible to accept truck teams.
52. Paragraph built from *Visiting Scientists (Paper by the Director of Administration),* CERN/GD/238, 2/4/59 (GD 20965).
53. SPC meeting held on 17/4/59, CERN/SPC/92, 8/5/59, 6-7.
54. *Mixed Teams,* CERN/GD/252, 24/6/59; *Leading Board Meeting,* 30/6/59 (CERN/GD/255, 6/7/59); *National Participation in Research - CERN,* CERN/SPC/93, 14/7/59; circular letter CERN/6008 signed by Bakker, 15/7/59 (DG 20965).
55. CERN/SPC/93, 14/7/59, 1.
56. CERN/GD/252, 24/6/59, 2 and 3; CERN/SPC/93, 14/7/59, 2, for the first principle; CERN/SPC/93, 3-4 for the second; CERN/GD/252, 3 and CERN/SPC/93, 4-5 for the third.
57. Answers from Belgium, France, Norway, The Netherlands, Germany, Sweden, United Kingdom (CERN/SPC/93/Add., 22/9/59); note by W. Heisenberg (CERN/SPC/93/Add. 2, 9/10/59).
58. Quotations from CERN/SPC/93/Add., 22/9/59, 20 and 22.
59. The meaning is truck teams building their own equipment.
60. Quotations from the British text, CERN/SPC/93/Add., 22/9/59, 20-23.
61. CERN/SPC/93/Add., 6-14; this is the longer text received by CERN during the summer.

62. CERN/SPC/93/Rev., 18/9/59; CERN/SPC/101, 19/11/59 (meeting of the SPC held on 16 October), 2-7; CERN/322, 10/11/59; Conseil, minutes, 1-2/12/59, 55.
63. CERN/SPC/93/Rev., 28/9/59, 1-6.
64. CERN/SPC/101, 19/11/59, 2-7; CERN/322, 3/11/59; Council minutes, 1959, 53-57.
65. We should also mention the advice not to trust others which Alvarez from Berkeley gave his CERN visitors in 1957 and 1958. For him, the experimental programme was to be defined only by the physicists in the laboratory, not by the visitors. See for example the handwritten manuscript marked confidential and written by C.A. Ramm and C.-J. Zilverschoon, without title, part entitled *Non Technical Discussions,* 5 pages numbered NTD1 to NTD5, without date (MGNH 22044).
66. The Council of December 1959 approved 'the proposals of the Director-General concerning the creation of a Consultative Committee for the PS experimental programme' (Conseil, minutes, 1-2/12/59, 66). This system, modelled on that of the SC, was adopted the same day the Council approved of the document on national participation of the research at CERN (CERN/322) which envisaged the setting-up of 'mixed teams'.
67. On this dimension of *process,* see Pestre (1988a); see also Allison (1971), Crozier and Friedberg (1977), Schilling (1961).
68. For two noteworthy studies on these questions, see Crozon (1987) and Galison (1987a). Add the remarks made by J. Krige in chapter 9.
69. For more detailed analyses, see Crozon (1987), 122-128, and Galison (1987a).
70. Crozon (1987), 165-178, and Galison (1985a).
71. On this aspect, see chapter 9.
72. Crozon (1987), 176.
73. *Proposals on PS Experimental Programme,* CERN/SPC/71, 1/9/58; CERN/SPC/75, 4/9/58; CERN/SPC/76, 25/9/58 (*Draft minutes,* SPC, 4/9/58). The quotations come from CERN/SPC/76, 3. For details on the organizational structure at CERN, see chapter 7.
74. See the collection of documents CERN/Ex. C/1 to 76. CERN/Ex. C/1 is the minutes of the first meeting of the Executive Committee held on 23/9/58, CERN/Ex. C/76, those of the last one held on 1/3/60.
75. CERN/Ex. C/45, 7/7/59, 5-7; CERN/Ex. C/51, 5/10/59, 7-8; CERN/Ex. C/54, 22/10/59, 3-7; CERN/Ex. C/57, 18/11/59, 4-6. Experiments with emulsions had been considered in the previous months (CERN/Ex. C/20, 28/11/58).
76. CERN/Ex. C/54, 22/10/59, 7; *draft* of a letter sent by the D.G. to Member States, written by Dakin, 30/10/59 (DIRADM 20262). The answers by Member States are in CERN/Ex. C/62, 63, 65, 65/Add., 65/Add. 2.
77. CERN/Ex. C/70, 26/2/60 (meeting of the Executive Committee held on 13/1/60), 5-6. Preiswerk was nominated Director of NP division when it was created on 1/1/61.
78. Their reports are in CERN/Ex. C/68 (10/2/60), 68/Add. 1 (19/2/60) and 68/Add. 2 (24/2/60).
79. Memorandum Hine to Dakin, 12/8/60 (DIRADM 20227).
80. Letter Verry to Bakker, 18/12/59, reproduced in CERN/Ex. C/65, 8/1/60, 1-3. The answer of W.K. Jentschke (Hamburg), M. Deutschmann (Aachen) and W. Paul (Bonn) is as vague (CERN/Ex. C/65, 4).
81. *Note for the File PS Research Programme,* 29/3/60, signed Dakin (DIRADM 20262).
82. For a longer discussion on that point, see chapter 13, *Concluding Remarks.*
83. Collins (1985).
84. CERN/SPC/105, 16/3/60, 17 (meeting held on 19/1/60).
85. On these tasks, see amongst others CERN/Ex. C/76, 24/5/60.
86. *Schedule of PS until Easter,* CERN/Ex. C/71, 26/2/60. In October 1961, 92 hours out of 112 were for physics; in October 1962, 113 hours out of 128 (reports made in June and December to the Council).
87. *Notes on a discussion on the use of nuclear emulsions during the testing period for the PS,* 21/8/58 (JBA 22794); *Emulsion work with the CERN 25 GeV Proton Synchrotron, minutes of a meeting held at CERN on November 20th and 21st, 1958,* CERN/Ex. C/20, 28/11/58; *id., Report of a meeting held at CERN,*

Geneva, on 15th and 16th January, 1959, CERN 59-13, March 1959, 1-3. Quotations come from this last report.

88. Report CERN 59-13 cited note 87, quotation page 2.
89. Report CERN 59-13 cited note 87; *Facilities for PS Emulsion Experiments by Visiting Groups [...],* CERN/SPC/119, 15/9/60; *The relationship between outside emulsion groups and the CERN group for the prosecution of experiments at CERN,* EmC 61/16, 9/10/61; *Nuclear Emulsion work using the CERN accelerators, December 1959–April 1963, by J.C. Combe and W.O. Lock,* CERN 63-22, 12/6/63.
90. History of CERN (1987), 218-219.
91. On the great number of those groups in Europe see *Preliminary Reports on High-Energy Physics in Member States,* CERN/ECFA 66/2, Annex II, 1-31.
92. CERN/PS/EP/12, 26/11/59 (DIRADM 20262) for the programme of the meeting held on 20/1/60; Note by J.B. Adams *(2nd Bubble Chamber Meeting),* 26/2/60 (DG 20562); *Committee for Organizing CERN PS Bubble Chamber Experiments, March 8th, 1960,* PS/INT/HBC/COM 60/1.
93. *Committee ...,* PS/INT/HBC/COM 60/1, cited note 92; this is the minutes of the meeting held on 8/3/60.
94. Kowarski was Director of STS division, Peyrou leader of the bubble chamber group. He became Director of the new Track Chamber division on 1/1/61. For this paragraph, see Memorandum Dakin to Bakker, 19/4/60 (DG 20562), circular letter CERN/6581 (DG 20562), *Questionnaire* (Annex of letter 6581), in DG 20873. The answers are in DG 20562.
95. *Meeting on Bubble Chamber Experiments at C.P.S.,* 17/6/60, PS/INT/HBC/COM 60/2, 14/7/60.
96. The IEP group partly worked on techniques for the analysis of track-chamber pictures, the PC group on the construction of the propane chamber.
97. *Working Party for Bubble Chamber Experiments at CPS, Meeting of the 22nd July 1960,* PS/INT/HBC/COM 60/3, 7/9/60.
98. *Working Party ...* cited note 97; *Working Party ..., Meeting of the 3rd October 1960,* PS/INT/HBC/COM 60/5, 3/11/60.
99. *Working Party ... 60/5,* cited note 98, 4.
100. *The Internal Organization of CERN [...],* CERN/344/Rev. 3, 16/9/60 in which the Track Chamber Committee appeared for the first time; letter A.J. Hertz (Istituto Nazionale di Fisica Nucleare, Sezione di Milano) to Adams, 7/10/60, asking for his approval for the arrangement 'which a number of Italian laboratories have made informally with the Lagarrigue-Ramm bubble-chamber groups'; answer Adams to Hertz, 12/10/60, giving the official approval of CERN; letter Adams to Butler, 12/10/60.
101. *Working Party ..., Meeting of the 16th November 1960,* PS/INT/HBC/COM 60/14; *Meeting on Track Chamber Experiments at CERN, 21st December 1960,* TC/COM 61/4, 17/1/61. The membership of the TCC was fixed on that occasion (12 CERN staff, 1 German, 3 British, 3 French, 3 Italians, 3 from small countries); *CERN Track Chamber Committee, Meeting of the 25th January 1961,* TC/COM 61/5, 9/2/61. For an overview of the work done in the first semester of 1961, see CERN/TC/COM 61-27, 18/8/61.
102. CERN/TC/COM 63-16, 21/2/63. At the beginning, the proposals were kept separate and are now in various files. After 1961, lists are kept in TC/COM files.
103. *Schedule of PS until Easter,* CERN/Ex. C/71, 26/2/60. The machine group was represented by G. Brianti, P. Germain and P.H. Stanley, the 'running-in group' by Adams, F. Grütter being there for the 'engineering group'. Document Ex. C/71 says of the committee 'of users representatives': 'It is felt that this committee represents correctly the different classes of users for the next six months'.
104. CERN/Ex. C/76, 24/5/60 (meeting held on 1/3/60), 1-2; *Leading Board Meeting,* CERN/GD/299, 30/3/60, 4.
105. On 22/4/60, the *Advisory Committee* for the PS was still planned (CERN/GD/302, 21/4/60, 2 and circular letter by Bakker, CERN/6581); on 16 May 1960, Adams sent a letter to inform each member that the meeting planned for 2/6/60 was postponed (letter 6620 in DG 20935).

106. PS/INT/HBC/COM 60/2, 14/7/60 cited note 95, 1. The bubble chamber committee was formed of Berthelot (F), Butler (UK), Deutschmann (FRG), Ekspong (S), Houtermans (CH), Kluyver (NL), plus one physicist to be nominated by Italy, plus Gregory and Lagarrigue for the chambers on the site; on the CERN side, 9 people plus Adams, the DG.
107. *Memorandum 'Note of a Conversation'*, Dakin to Adams, 10/8/60; *Draft Memorandum 'Research Programme, Organization, and Visiting Teams'*, Dakin to Adams, 11/8/60 (DIRADM 20227).
108. This was discussed at length by Bernardini, Dakin and Hine; for Dakin, see note 107, 11/8/60, 5; *Memorandum* Hine to Dakin, 12/8/60; *Memorandum* Bernardini to Dakin, 15/8/60 (DIRADM 20227). This group of 3 notes is 20 pages long.
109. CERN/344/Rev. 3 cited note 100; letter Adams to Butler, 12/10/60 (DG 20873).
110. Joint Meeting of the Scientific Policy Committee and the Committee of Council, 3/10/60, CERN/CC/377 Draft, 4/11/60—from which Leprince-Ringuet's remark is taken, page 6; letter Adams to Butler, 12/10/60 (DG 20873); CERN/DIR/7, 10/11/60; NPRC, meeting held on 21/11/60, CERN/NPR/2, 5/12/60.
111. Among these numerous anxieties, see the British ones expressed in *Comments by the UK Advisory Panel, Annex A*, 26/11/60 (SERC-NP 38); or Adams himself some years later when he said that 'in the early part of 1960 [...] many physicists in Europe felt that CERN was not welcoming visiting scientists as much as it should' (Adams (1965), 251).
112. CERN/DRI/7, 10/11/60, 2; letter Salvini to 'the Members of the Directorate', 14/12/60 (NP 22914); CERN/DIR/19, 4/1/61; *Preliminary Meeting of the Electronic Experiment Committee (EEC), Draft Minutes*, 16/1/60 (NP 22914); *NP Division, Draft Minutes, Group leaders' meeting held on January 17th, 1961*, 23/1/61 (NP 22914); letter Salvini to Adams, 21/1/61 (NP 22914).
113. Documents cited in note 112; the quotations are from letter Salvini to Adams, 21/1/61 (NP 22914).
114. Speeches by Coccini, Farley, Middlekopp respectively, *NP Division ..., op. cit.* note 112, 2-3; Bernardini memorandum is that mentioned in note 108.
115. See for example memorandum by Bernardini cited note 108, 3.
116. Speeches by von Dardel and Fidecaro in *NP Division ..., op. cit.* note 112, 1-2.
117. CERN/DIR/26, 1/2/61; letter Adams to Salvini, 1/2/61 (MGNH 22063).
118. *First Meeting of the Electronic Experiments Committee, 1/3/61, Draft Minutes* (71p!)(NP 22914). The minutes of the other meetings are also in NP 22914. Quotation by Weisskopf, fifth meeting.
119. Speeches by Farley and Hine, *Draft Minutes of the Fifth Meeting of ther Electronic Experiments Committee, 25/9/61*, 1 (NP 22914).
120. SPC, Meeting, 27/10/61, Draft Minutes, CERN/SPC/149/Draft, 27/11/61, 7-8. The debate has been cut in the official minutes. It is thus better to listen to the tape.
121. We simplify here more complex events; see *Decisions taken at the Seventh Meeting of the NPRC*, 8/11/61 (EDWS 22313); add *NP Division, Minutes of the Group Leaders' Meeting*, 6/11/61 (NP 22891).
122. Quotations from *Decisions taken at the Ninth Meeting of the NPRC on January 10th, 1962*, 16/1/62 (EDWS 22313). Add *Reorganization of the Committee System (by the Director General)*, CERN/SPC/155, 20/2/62.
123. *Decisions taken at the Twelfth Meeting of the NPRC*, 26/4/62 (EDWS 22313); on the seminars held in NP division, see *NP Division ..., op. cit.*, note 121, 1-2.
124. *Nuclear Physics Research Committee Meeting, 21 November, 1960*, CERN/NPR/2, 5/12/60 (DG 20877). The quality of these minutes are poor. We used those written by W.M. Gibson for Powell, Combe and Lock, *NPRC, Notes on first meeting*, held 21st November 1960 (EDWS 22313). The quotations are from Gibson.
125. Several dozens of notes, memoranda and letters.
126. Letter Citron to Weisskopf, 23/11/60 (DG 20880). Add letters Hine to Weisskopf, 5/12/60 (DG 20490), Adams to Weisskopf, 3/1/61 (MGNH 22063), Memorandum Dakin to Adams, 5/1/61 (DIRADM 20227).
127. Letter Weisskopf to Hine, 5/1/61 (MGNH 22063).

128. CERN/DIR/26 *(Directorate Meeting, 30/1/61)*, 1/2/61, 1-2; *Notes on Nuclear Physics Research Committee (From Director General)*, CERN/DGO/Memo/1008, 8/2/61 (DG20951); letter Fidecaro to Adams, 15/2/61 (DIRADM 20227).
129. The fact that Citron was skeptical is excellent for us; even if there is no reason to take what he says for granted (no more than for anyone else), his position makes his description informed but not indulgent, something perfect for any historian. This way of writing the minutes was criticized by Salvini (Draft Minutes of the Third Meeting of the EEC, 21/6/61 (NP 22914).)
130. *Draft Minutes of the First Meeting of the NPRC, held on March 8 and 9, 1961 at CERN,* 10 and 12 (DG 20877).
131. The minutes of NPRC meetings are in EDWS 22313.
132. *Draft Minutes of the Sixth Meeting of the NPRC, held on 27th and 28th of September at CERN,* 18/10/61 (EDWS 22313), 7-8; CERN/SPC/149/Draft, 27/11/61 (meeting held on 27/10/61), 7-8.
133. The EEC reform is parallel to the NPRC one and archival references are identical for both (see notes 121 and 122). The evolution of NPRC membership can be followed through these documents.
134. CERN/SPC/155, 20/2/62 and CERN/SPC/157/Draft, 5/4/62 (meeting held on 3/3/62), 14.
135. *Decisions taken at the Tenth Meeting of the NPRC on February 14th, 1962,* 1-2, section: 'Points of procedure' (EDWS 22313).
136. See for example *Report on a discussion meeting of the NPRC on November 12th, 1962,* NPRC 17, 12/11/62, pages 1 and 2. Here one could remark that the notion of 'direction of research' is limited to electronic experiments – 'since the basic task [for bubble chamber] is to put a specific bubble chamber in a particular beam' (page 2).
137. For a long discussion on these points, see minutes, Fourth EEC meeting, 20/7/61, 6-7 (NP 22914) and minutes, Fifth NPRC minutes, 26/7/61, 7.
138. *Op. cit.,* note 136.
139. To check this, just follow the way people describe each chamber's programme at TCC meetings from November 1961 to January 1964. The *first* proposal for the British chamber is *A preliminary Study of the Physics Programme of the 1.5 m Hydrogen Bubble Chamber at CERN,* 23/10/61, signed C.C. Butler (DG 20886). On 8 November, he said that the chamber 'should be ready by about September 1962. The main interest lies in experiments with high energy negative beams [...]', etc. See CERN/TC/COM 61/50, 14/11/61.
140. Memorandum Weisskopf to Ramm, CERN/DGO/Memo/1107, 10/8/61 (DG 20562).
141. PS/INT/HBC/COM 60/1, 8/3/60, 3 for Puppi's criticisms; PS/INT/HBC/COM 60/5, 3/11/60, 5 for Goldschmidt-Clermont. For the British preferences, *Bubble Chamber Experiments at CERN during 1961,* 3 pages, written after a meeting at Imperial College on 31/10/60, without date; for decision and quotations, PS/INT/HBC/COM 60/14, 1/12/60, 2.
142. References note 139.
143. Draft Minutes of the Fourth Meeting of the EEC, 20/7/61, 6 (NP 22914).
144. On Adams' choice and Butler's surprise, see letters Adams to Butler, 12/10/60, Butler to Adams, 14/10/60, Adams to Butler, 28/10/60, Butler to Adams, 2/11/60 (DG 20873); on the situation in 1962, letter Weisskopf to Butler, 18/5/62, Butler to Weisskopf, 22/5/62.
145. See section 4.4.
146. This conception was described by Gregory in March 1960; it worried Butler considerably. Adams insisted on having a more 'collaborative' system. See bubble chamber committee meetings cited note 92.
147. This policy was spelt out by Gregory during the TCC, meeting held on 6/9/61 (CERN/TC/COM 61/37, 26/9/61, 5).
148. CERN/TC/COM 63-61, 7/11/63, 4.
149. Letters given note 144.
150. Letter Heisenberg to Weisskopf, 25/7/63; Memorandum Van Hove to Weisskopf, 5/8/63; Note Budde to Weisskopf, 7/8/63; letter Weisskopf to Heisenberg, 7/8/63.

151. The basic source here is the document compiled by Mervyn Hine in 1965 for the SPC: *Statistics on Research Collaboration between CERN and other countries,* CERN/SPC/213, 28/11/65.
152. CERN/SPC/213, 33. Same source for the following paragraph.
153. Amongst other things, this chamber was lengthened to 1.2 m.
154. Figures given by Crozon (1987), 173.
155. The Italians also brought a 18 cm hydrogen chamber to the SC; it took 260 000 photographs and was still at CERN in 1965.
156. See CERN/SPC/213, 28/11/65.
157. On these two chambers, see DG 20566 to 568 and DIRADM 20257 (on Gargamelle) and DG 20561 to 63 and DIRADM 20558/59 (on BEBC).
158. See section 4.4. *supra.*
159. Written from CERN/SPC/213, 33–36 and 37–44.
160. Same references as in note 159.
161. As *lists* of publication, we used those given in CERN Annual Reports. For 1960/61, we found around 80% of the original publication, for 1964/65 around 70%. The number of physicists in each team is calculated from the number of signatures.
162. 9 teams 'recomposed' by us out of 13 authorized to experiment.
163. *Prospects for the investigation of high-energy phenomena using electronic techniques at CERN,* CERN/SPC/272, 25/2/69, 2.
164. This calculus is made as were the previous ones. The articles used for it are Bernardini et al. (1964), Bernardini et al. (1965), Bienlein et al. (1964), Block et al. (1964).
165. *Hydrogen Targets,* CERN/DGO/Memo/997 Adams to Preiswerk, 11/1/61; *Spark Chambers,* CERN/DGO/Memo/994 Adams to Preiswerk, 11/1/61; *Memorandum,* Preiswerk to Adams, 8/2/61 (DIRADM 20274).
166. Note I. Pizer to Preiswerk, 20/12/63; *NP Division, Minutes of the Group Leaders' Meeting held on February 4th, 1964,* 24/2/64, 1 (NP 22891).
167. *Memorandum* by Preiswerk, 5/10/61, 2 (DIRADM 20274).
168. For one example, see *Request for machine time on π^0 lifetime experiment,* 5/3/62, *Memorandum* von Dardel Group to Weisskopf, 9/3/62 (DG 20888).
169. Note NP/1498/860, PAPLEP to Gregory, 10/11/64 (DG 20588).
170. Letter Citron to Weisskopf, 23/11/60 (DG 20880).
171. Adams explained his attitude of 1960 in a letter to Weisskopf, 31/12/64 (DG 20588). He added: 'Preiswerk's job as Division Leader was mainly managerial – certainly that was my idea in first proposing his appointment to the Council'. For more details on Preiswerk's nomination in 1960, see chapter 7.
172. Letter Weisskopf to Adams, 26/1/65 which explains the nomination of Paul (DG 20588).
173. Letter Weisskopf to Paul, 20/5/64 (DG 20940). Add letters Melville to Weisskopf, 10/12/64, Adams to Weisskopf, 28/1/65 (DG 20588), Annual Report (1965).
174. Amongst others, see meetings of EEC, 20/7/61, 6–7 (NP 22914), NPRC, 26/7/61, 7–11 (interventions by Van Hove and Salvini – EDWS 22313), EEC, 25/9/61, 1–2 (interventions by Van Hove, Roberts and Bernardini – NP 22914), *Memorandum* Preiswerk, 5/10/61 (DIRADM 20274), SPC meeting held on 27/10/61 (tapes and CERN/SPC/149/Draft, 27/11/61), letter Gregory to Weisskopf, 11/11/63 (DG 20486), *Report to the Director General* by Abdul Ghani, without date (DG 20588), *A proposal to reform the NP Division* by A. Citron, 12/1/64 (MGNH 22006), letter J. Geibel to Weisskopf, 13/4/65 (DG 20868), letter Gregory to Paul, 24/11/65 (DG 20588), *Memorandum* Van Hove to Gregory, 12/4/66 (DG 20875). The quotation comes from Citron's note, 12/1/64.
175. The quotations are respectively from EEC meeting, 25/9/61 (attributed to Van Hove, Bernardini, Roberts), SPC meeting, 27/10/61 (tape, speech of Bernardini). References are in note 174.
176. Notes or letters by Citron, 12/1/64, Geibel, 13/4/65, Van Hove, 12/4/66, cited note 17.
177. References given note 174.

178. See also Irvine and Martin (1984) and chapter 13, *Concluding Remarks*.
179. Box DG 20588 gives several examples. Sometimes, the pressures are more directly 'political'. This is the case with Heisenberg's letter in 1963 (see section 6.1 supra), and Perrin's the same year in favour of a French 'electronic' experiment. See letters Perrin to Weisskopf, 6/9/63, Weisskopf to Perrin, 11/9/63, CERN/DGO/Memo/1397 from Weisskopf to NP Division, 'to inform you that the Directorate has decided to allow the CERN–Ivry experiment to be carried out at the PS'. Those pressures seem very rare, however.
180. The problems here are the changes in the composition of the teams and the delay due to publication.
181. We had more difficulties to 'recompose' the teams for 1965; therefore the figure is less precise for that year.
182. Study made from Annual Reports and publication.
183. CERN/SPC/272, *op. cit.* note 163, 2.
184. Memorandum Bernardini to Dakin, 15/8/60; Hine to Dakin, 12/8/60 (DIRADM 20227). For Adams in 1960/61, see section 5.1, supra.
185. Memorandum Van Hove to Gregory, 12/4/66 (DG 20875).
186. Processing of weekly reports *Mouvement du Personnel CERN,* ADM/PERS/Contrôle, archives of CERN Personnel Division.
187. Hine gives 111 man-years for the Italians in electronic experiments from 1956 to 1965 (CERN staff and fellows), 94 for the British, 76 for the Germans and 48 for the French (CERN/SPC/213, 28/11/65, 13).

Collection of photographs and documents
of historic interest

The Synchro-cyclotron magnet coil on its way to Geneva, December 1955 (Photo CERN).

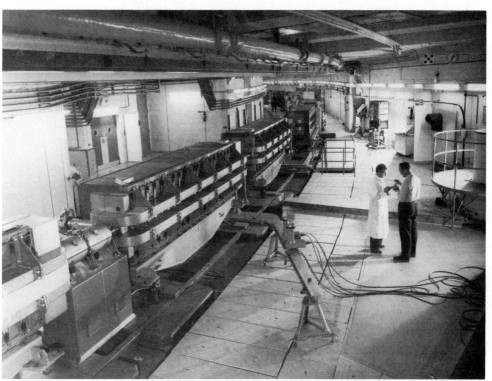
The completed PS ring, 1959 (Photo CERN)

General view of the PS linac, 1958 (Photo CERN).

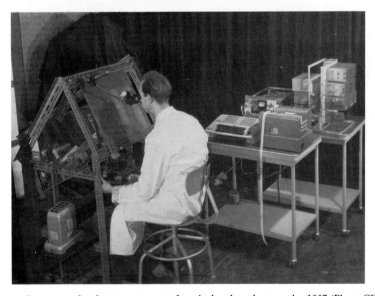
Prototype instrument for the measurement of track chamber photographs, 1957 (Photo CERN).

A separated K-meson beam being installed in the North hall, 1961 (Photo CERN).

Some of the motor-generator sets supplying power to beam transport magnets in 1960 (Photo CERN).

Charles Peyrou with his first 10 cm hydrogen bubble chamber, 1957 (Photo CERN).

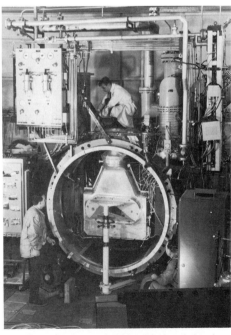

The Saclay/Ecole Polytechnique 81 cm hydrogen bubble chamber being installed at the PS, 1961 (Photo CERN).

The CERN 2 m hydrogen bubble chamber, commissioned in December 1964 (Photo CERN).

The spark chamber used for neutrino experiments at CERN in 1964 (Photo CERN).

Bending and focusing magnets packed into the South hall, 1962 (Photo CERN).

The 10 m long electrostatic separator, one of which was first used in a beam in December 1961 (Photo CERN).

An aerial view of CERN with the PS ring in the right foreground, March 1959 (Photo CERN).

The CERN site at the end of 1965 (Photo CERN).

Oppenheimer and Bakker (Photo Jean Mohr).

De Rose and Adams (Photo CERN).

From left to right, Victor Weisskopf, Leon Van Hove, and Bernard Gregory (Photo CERN).

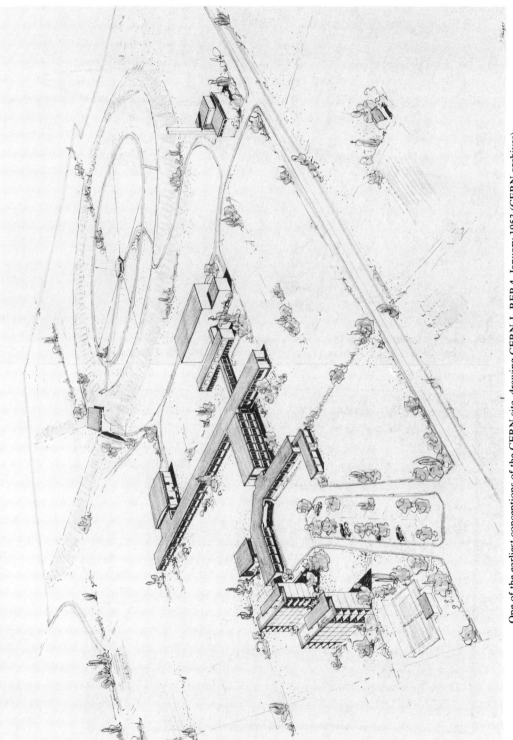

One of the earliest conceptions of the CERN site, drawing CERN-L-REP 4, January 1953 (CERN archives).

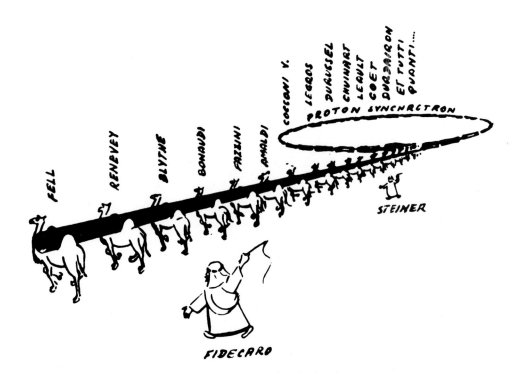

A cartoon found posted on a board at CERN after a party of about 80 people held to celebrate the successful construction of the first separated anti-proton beam at the PS, 1961 (CERN archives).

> Société Genévoise de Physique Nucléaire
> S.A. au capital de 100.000 F. entièrement versés par le CERN
>
> Physiciens loués à l'heure et au mois
> — Prix Nobel reversés à la trésorerie du CERN
> Grand choix de spécialistes en tous domaines y compris la spallation
> — Chambres à bulles reparées "while you wait"
> Pensions de famille à Rochester et à Kiev
> — Pas de comptes à rendre au Comité des Finances
> Frais généraux strictement limités à 350%
> — Qualité | Discretion | Économie

A facetious "advertisement" devised by Lew Kowarski during a meeting of the CERN Management Committee in January 1961 (CERN archives).

CERN/GD/246
4 June, 1959

ORGANISATION EUROPÉENNE POUR LA RECHERCHE NUCLÉAIRE
CERN EUROPEAN ORGANIZATION FOR NUCLEAR RESEARCH

LEADING BOARD
10 June, 1959

"The Stay and the Staff"
Isaiah iii 1

REPORT OF THE SENIOR STAFF STUDY GROUP
ON FUTURE APPOINTMENT POLICY

"Hypocrite, first take the beam out of thine eye" Matthew

"Stand not upon the order of your going,
But go at once!"
Macbeth iv 119

"I will stay with thee,
And never from this palace of dim night
depart again."
Romeo and Juliet iii 106

"One shall be taken, and another left."
Matthew xxiv 40

"Turn over a flat petrification, and what beetles will crawl out."

Lew Kowarski's irreverent annotations added to the citations on the cover page of a paper prepared for the Leading Board, June 1959 (CERN archives).

UNIVERSITA' DEGLI STUDI - ROMA
ISTITUTO DI FISICA "GUGLIELMO MARCONI"

ROMA, January 21, 1961
Piazzale delle Scienze, 5

Dr. J.B. ADAMS
General Director
C E R N
Geneva 23
SVIZZERA

c.c. Prof. G. Bernardini - Geneva
Mr. S.A. ff Dakin - Geneva
Dr. M.G.N. Hine - Geneva
Prof. W. Weisskopf - Geneva

Dear Doctor Adams:

I did my best to have the Electronic Experiments Committee efficiently working, but still there are problems, which I am trying to list in the present letter.

I shall divide this letter into three parts. In part I, I summarize the meeting that Preiswerk, von Dardel, Lundby, Harting and myself had on Jan. 16th.; in part II the solution preferred by the groups working with counters techniques is summarized; in part III I take the freedom to make a few general comments.

I - According to what I said in my letter of December 14th, we (Preiswerk, von Dardel, Lundby, Harting acting as secretary, Salvini) had a meeting yesterday, January 16th, on the programme and the aims of the Electronic Experiments Committee.

In this meeting we discussed the SC and the PS program.

For the SC there was a request of one external group (Rome) and we could allocate machine time to it. In fact, it does not seem at present, that the SC is going to be too overcrowded (which is not, in itself, the best thing).

./.

Extract of a letter from Salvini to Adams explaining the difficulties he had had in setting up the Electronic Experiments Committee, January 1961 (Reproduced with the permission of Prof. Salvini and the CERN archives).

UNIVERSITA' DEGLI STUDI - ROMA
ISTITUTO DI FISICA "GUGLIELMO MARCONI"

ROMA, January 21, 1961
Piazzale delle Scienze, 5

For the PS there were no new requests, and everything has already been taken care of by the co-ordinator.

We considered the question of the members which should be added to the Counter Committee, to start with Professor A. Merrison, who would be very welcome. Once we appoint him as being proposed by the U.K. panel, we have to appoint other members, which would make the final committee rather large, more or less like the others two. At this point, we all began to have doubts, and they were the following:

- Has the Counter Committee the same reason of existence as the track chamber and the emulsion committee? As we know these last two have to distribute film and plates among the european physicists, which is at least one basic reason of their existence. I have no difficulties in understanding why the track chamber committee may be a solution of the scientific problems connected with that technique, while the same is not true with the counter committee.

- Must we really ask the counter committee to decide on the value of the priorities of the different experiments? This is doubtful: the division of physics into three committees according to the different techniques makes it difficult to prepare a real priority list in each committee, and one feels that it shall be necessary any way to go over it again at the Nuclear Physics Research Committee.

- Is there some easier way for having physicists from outside bound and involved in the scientific programme at CERN, and for avoiding difficulties with external groups?

I asked therefore Prof. Preiswerk to let me take part in the group leaders meeting of the Nuclear Physics Division, in order to discuss the problem with the group

./..

cc. Members of the Directorate

CERN/DGO/Memo/997   11 January, 1961

To   : Professor P. Preiswerk

From : Director-General

Hydrogen Targets

    I am sorry to worry you about many problems all at once but I am going to England again tomorrow for two weeks.

    I have noticed that we are constructing many hydrogen targets at CERN and we are rapidly reaching the stage that our hydrogen liquifier output is inadequate to supply all the targets together with the H.B.C. It seems to me, as an outsider, that each experimental group feels the need to possess its own hydrogen target (in addition to its own electronics and even its own spark chamber). I wonder whether all this is economic or really necessary. It may be that all the targets are different but it seems that several parts are at least common. For instance, the Track Chamber Division is continually supplying transfer lines, valves and other parts to your groups.

    I think that it would be useful to have a programme for hydrogen targets for CERN so that a policy can be established for them. Perhaps you and Gilberto could discuss this with your groups and let the Directorate know the result.

    I should add that my remarks are more in the nature of an enquiry and should not be taken as a criticism. I have had some time over Christmas to look around CERN a bit and these points have just occurred to me.

J.B. Adams

Memorandum from the Director-General remarking on the multiplicity of hydrogen targets being built in Preiswerk's NP Division, January 1961 (CERN archives).

Now comes the period of exploitation of the machine; physicists will "take over"; the builders of the machine have no other duty but to run it for the physicists. A highly creative job now becomes a service job; the engineers are demoted. So we run into the problem of the position of the engineers in our activities. I f we comment on him.

My idea is completely different. First; high-energy physics cannot be divided into physicists and engineers. In this field the engineer must be a physicist and the physicist an engineer. Look at the great men in this field, Panofsky, Fermi, Lawrence, and you will see what I mean.

Second; the exploitation of a machine such as the PS is as challenging a task of engineering as the construction. We need bubble chambers, separators, beam transport equipment, higher intensity and many other things; all tasks of creative construction. Even more we need those apparatus whose name I don't know because it will be invented soon. The trouble only is that the final aim of all these activities cannot be so easily defined as in the first period, where everybody was working for one thing: 25 GeV protons.

But we will have nevertheless conflicts between two points of view. ~~This leads easily to a different emphasis among Divisions.~~ The socalled physicists would like to make use of existing facilities as fast as possible, and the socalled engineers would like to improve them and to make new ones. However these differences are only secondary, not fundamental. They might and should even occur within one individual. We must have short range planning and long range planning, but we should not overdo either of them. Too much string and sealingwax is out of place in a high energy lab; too much engineering perfection will give us wonderful instruments but after the Russ. + Am. have made all interesting discoveries.

Extract from Victor Weisskopf's notes for his first talk as Director-General to the CERN staff, August 1961 (CERN archives).

PERSONAL

23 December 1961

I am sorry that I could not attend your meeting about CERN the other day. Unfortunately I did not receive an invitation from your office until 4.30 p.m., when the meeting was already in progress.

I understand that the conclusion of the meeting was to set up a study group to advise you on the magnitude and distribution of the support required for university, national and international laboratories working on high energy nuclear physics research. It would therefore be premature for me to put forward my views to you at this stage.

There is, however, another matter that arose during the recent session of the CERN Council that has implications for this country other than the obvious embarrassment of being the only major Member State of CERN voting against the 1962 budget. The matter is the following.

CERN is not simply a co-operative effort in high energy nuclear physics; it is the first experiment in European co-operation in scientific research. So far it has been well supported and it has been a successful experiment that has led European governments to plan similar ventures in other research fields, of which space research is the most notable.

If, for reasons of national policies or financial limitations, it is found that CERN cannot be supported any longer by its Member States, and particularly by the larger States, then the new co-operative ventures will be re-examined. You may remember that initially the French government voted against joining with the U.K. in the European rocket launching plans involving the Blue Streak vehicle. Largely due to the efforts of M. Francois de Rose and his successful conversion of General de Gaulle to these ideas, the French government reversed its original decision, and negotiations are now under way for the setting up of the organisation called ELDO. However, M. de Rose is one of the two French delegates to CERN and a past President of the CERN Council, and he made it clear to me that unless CERN continues to receive adequate support from its Member States, the French delegates would seriously consider advising their government to withdraw from ELDO, and from other similar ventures.

/The

The Rt. Hon. Viscount Hailsham, P.C.,
2 Richmond Terrace,
Whitehall,
London, S.W.1.

Letter from John Adams to Lord Hailsham concerning British policy on the CERN budget, December 1961 (Reproduced with the permission of the UK Science and Engineering Research Council and Lord Hailsham).

The views of the majority of the Member States were that CERN must be developed at a rate that maintains the laboratory in the forefront of its subject, and provides facilities for European physics at least as good as those existing and being planned in the U.S.A. This was the original purpose of setting up CERN. Certainly the French view is that unless CERN can be developed in this way, they will go ahead with extending their national laboratories, even to the extent of building a national accelerating machine twice the size of the CERN proton synchrotron. I am very much afraid that this will mean the death of CERN and the stillbirth of other European efforts towards co-operative scientific research. Since no single country in Europe can afford to support these expensive fields of research on a comparable scale to the U.S.A. and U.S.S.R., we, in Europe, must inevitably fall behind these two continents in these subjects.

I think that the gloom that pervaded the private and public sessions of the CERN Council was perhaps premature, but it indicates the very close relationship between what happens at CERN and what will happen to the other European organisations. The point is that, for better or for worse, physics research, because it is becoming international, is now a matter of foreign policy. The French government recognised this in setting up a department at the Quai d'Orsay under M. de Rose to deal with these matters. It is not an exaggeration to say that the fact that the U.K. has been twice outvoted on the CERN budget has raised doubts about the U.K. in the minds of the Common Market countries that will affect the terms on which we may eventually join with them. It was evident at the CERN Council meeting that, not only did all but the Swedish government disagree with the implications of the note recently sent by our own Foreign Office to the Member States of CERN, but that it produced a strong reaction unfavourable to the U.K., amounting to a resentment in some States as to the manner in which it was presented.

I report these matters to you at length because so many of the CERN Council delegates were clearly very worried about them. M. de Rose particularly asked me to make the French views known in this country.

Yours sincerely,

J. B. ADAMS

others to think more hopefully about an accelerator higher than 100 GeV. Apparently, very few people in France would rather be in favour of a national machine of 60 GeV rather than a larger European machine, and, in any case, their views seemed unlikely to prevail.

Perrin claimed that Europe should not lose the opportunity of being a pioneer in the 200-300 GeV range, and he was pretty confident that necessary money could be found in Europe. This would, of course, be a CERN enterprise, at least as far as the over-all management was concerned.

The Director-General said he had even tried to damp Perrin's enthusiasm by stressing the many difficulties such an ambitious scheme would encounter, but the French were full of dynamism and extremely hopeful. At the same time, they were convinced that the storage ring project should go ahead at CERN.

As regards the big accelerator, they would very much like to have it located in France and would be quite willing to offer the site and even to build or to help in constructing the necessary buildings. As a matter of fact, they were so generous that the Director-General had to warn them that if they went too far the accelerator would appear to be a French machine, and this would create an undesirable situation. They understood these reasons but were nevertheless quite ready to do their utmost.

The prospects of a very large accelerator built in Europe should be thoroughly investigated and it would require some time before one would know for sure whether its construction could be launched. Meanwhile, the French had to face another problem: that of the morale of the group actually working on the 60 GeV plans. If the group knew that their future was rather uncertain, they would lose their enthusiasm. That is why Perrin would very much like to arrange a very close collaboration between the French and the AR Group, so as to give his people the assurance that should the 60 GeV project not be taken up they would be used on the big-scale project. What he proposed to do was to have separate units in Paris and at CERN; the former would work on the 60 GeV project, the latter on the big-scale machine and a third unit would be engaged on the storage ring project. In principle, the Director-General thought that this was not a bad proposition, but he had been careful to remain rather vague and not to commit himself, as he wanted first to know what the AR people would say.

Dr. Johnsen remarked that that kind of collaboration could create some difficulties if the people in the Paris group were not of the same level, in so far as seniority and experience were concerned, as those in the whole group in Geneva, and this could lead to quite a lot of trouble later on, if an amalgamation was made.

Extract from the Director-General's report on his talks with Professor Perrin in Paris, November 1962 (CERN archives).

CERN/TC/COM 63-16
21.2.1963

COMBINED PROPOSALS FOR HIGH ENERGY NEGATIVE PION AND PROTON BEAMS IN THE BRITISH NATIONAL BUBBLE CHAMBER

Introduction

A number of proposals have been made for experiments with high energy proton and negative pion beams in the British National Hydrogen Bubble Chamber. At the suggestion of the Track Chamber Committee, discussions have taken place between the groups to combine their proposals with the aim of finding common beam energies and of reducing the numbers of photographs required by arranging for exchange of the films. All groups expressed a willingness to cooperate and to exchange films quickly. Since all groups wished to separate "very peripheral" or "elastic type" inelastic events from the others, the common desire was for the highest possible beam energy obtainable to make this rather small separation as large as possible.

There has been a request, by Cambridge, for photographs with 15 GeV/c protons to allow the energy dependence of certain processes to be studied. While this is very desirable, they are agreeable to this exposure being considered as of second priority.

It is expected that a number of groups which have proposals for other experiments in the BNHBC, will wish to have some film for preliminary trials. As 1000 photographs may be expected to be yield of the order of 1000 events, it is proposed that an extra 10.000 photographs in the π^- and proton beams be taken to meet such special requests.

The numbers of photographs required per group will be classified as:

"First Request" - a small basic number of photographs which will be exhaustively studied.

"Final Number" - the number of photographs that it is hoped to measure after films have been exchanged.

It is suggested that if a film is being measured it is not available for exchange, but if it has been scanned only it is, in general, available.

PS/3687/jc

An example of a proposal put to the Track Chamber Committee, February 1963 (CERN archives).

Genève, le 1 mai 1963

Monsieur le Professeur,

      Plus de dix fois j'ai recommencé le rapport dont je vous ai parlé, mais toutes ces choses, que je connais fort bien, sont bien difficiles à exprimer, délicates mêmes, et je ne sais comment le faire.

      Je pense cependant que le malaise qui règne sur les petits qui forment mon milieu est dû à plusieurs raisons.

      Tout d'abord ces membres du personnel, qui forment une part importante de l'effectif, ignorent trop la destination, l'utilité de leur travail et ne voient plus bien à quoi servent les efforts qu'ils fournissent. Une espèce d'indifférence naît de cette ignorance.

      D'autre part, ayant dès le début travaillé sans réserve pour aider, même dans une modeste mesure, au succès final, ils avaient espéré qu'une organisation tenant compte des réalités serait mise sur pied, dotant ainsi l'Institut de moyens à l'échelle des nécessités.

      Dès l'inauguration du PS, un peu avant même, une sorte d'angoisse s'est abattue sur le personnel. L'augmentation de travail, le départ de certaines personnes, la lutte pour "le pouvoir", la désagrégation de l'équipe qui en quelques années avait créé le PS ont jeté le désarroi chez beaucoup.

      Il a manqué à ce moment-là, un programme <u>connu de tous</u> qui aurait montré à chacun le nouveau but à atteindre et les moyens pour le faire.

      Pour nous par exemple, ex atelier PS, le fait de quitter la Division pour laquelle nous avions fait tant d'efforts et d'être rattachés à une Division anonyme en quelque sorte, n'a pas été heureux. Je n'en prendrai comme preuve que la situation actuelle où, partageant notre travail à peu près également pour les deux Divisions NP et NPA, deux groupes se forment avec des tendances très nettes; il faudrait maintenant peu de chose pour que l'esprit renaisse, qui permet aux gens de travailler sans compter.

                                                                       ./.

Letter from J. Augsburger, ex PS Workshop, to DG Victor Weisskopf remarking on the change in spirit at CERN, May 1963 (Reproduced with the permission of Prof. Weisskopf and the CERN archives).

D'autre part encore, l'atmosphère générale a considérablement changé; alors qu'au début il était connu de tous que (je cite) "nous avions au CERN cherché à faire la meilleure des Institutions, afin que le personnel, libéré de tous soucis, puisse se consacrer entièrement à son travail", actuellement ces paroles ne sont plus de mise. Mille mesquineries viennent émailler la vie quotidienne.

Est-ce vrai qu'il y a eu déjà deux suicides?; et, si c'est vrai, ne peut-on penser qu'une activité satisfaisante les eût empêchés? A ce sujet, le fait que la mort d'un collègue de la première heure ait échappé à l'Administration a été cruellement ressenti par beaucoup.

Détail de valeur: la grande liberté qui règne sur l'emploi du matériel gêne beaucoup de personnes et ne les incite pas à l'économie, c'est-à-dire au dévouement. Un contrôle terriblement sévère, non pas à la sortie des magasins, ce qui n'épargne rien, mais lors de l'emploi et après l'emploi, persuaderait les gens de la qualité de l'Organisation.

Sur un autre plan, je crains également que les cours de formation n'aient pas été appréciés par tous de la façon qu'il faudrait.

Certains n'ont pas vu la générosité du geste et ont pensé, sans voir plus loin, qu'il y aurait là des heures de travail qui allaient disparaître, alors que nous en avions tant besoin.

A ce sujet, je me permets de proposer, comme je l'ai fait du reste, que ces cours soient donnés en dehors des horaires normaux et payés - pourquoi pas ? - comme heures supplémentaires au vu des résultats d'examens.

Ce sont-là quelques unes des choses que je connais et qui sont de mon petit cercle. Je serais très heureux si ce très modeste exposé peut vous aider à redonner à notre Institut l'esprit qui doit y régner.

Veuillez je vous prie croire, Monsieur le Professeur, à l'assurance de mon entier dévouement.

(signé) J. Augsburger

NP/1498/860   10.11.64

To   Prof. B. Gregory - Director of Research

from PAPLEP Group

Re   Correct figures of the time sharing in the m_4-beam

        We have pointed out many times to the co-ordinator that
the time sharing between the g-parity and PAPLEP was not correct.
In your memo dated 26th October you conclude that this sharing is
indeed correct (within 10%). For this reason we present in the
attached table the details of our computations, from which it follows
that the g-parity has obtained 4.1 times more parasite time than
us and 1.5 times more user time than us in the period which goes
from week 29* to 41, inclusive.

        Parasite time is vital for us because the new high-intensity
beam obliged us to change our 100 nsec electronics into a 10 nsec one,
and also to add extra μ-triggering counters; furthermore the availability
of new heavy-plate chambers implied a new γ-μ calibration of the set-up.
Moreover, the new two "electron-sandwiches" are of vital importance
for the "e-e" channel and need careful checks and calibrations.

        We have been ready with our apparatus and we could not get
parasite-time, in spite of the fact that the incorrect sharing was
already pointed out to the co-ordinator at the very beginning of October
when we could have obtained our legitimate week; instead this was
given to the g-parity. Consequently, because of the very small amount
of parasite time given to us, we are not in a position to use any
"user time" in December.

        We would be extremely grateful to you if you could convince
the co-ordinator to be more helpful regarding our needs of PS time.

c.c   Prof. V.F. Weisskopf - D.G.
      Prof. W. Paul - Research Leader of NP
      Prof. G. Puppi - Chairman of EEC
      Prof. P. Preiswerk - NP Division Leader
      Dr. W.O. Lock - PS Co-ordinator

---

* We do not include week 28 because this week was used for
   "beam-studies".

Memo from the Paplep collaboration to the Director of Research complaining about a lack of beam time, November 1964 (CERN archives).

CHAPTER 9

Planning the infrastructure for the PS experimental programme[1]

John KRIGE

Contents

9.1 A few technical essentials	506
9.1.1 Bubble chambers	506
9.1.2 Beam transport systems	507
9.2 Some reflections on the SC bubble chamber programme (1955-1956)	508
9.3 Defining big equipment for the PS experimental programme	511
9.3.1 June-December 1957: the growing conviction that CERN should build its own large hydrogen bubble chamber	512
9.3.2 January-May 1958: fixing the size of the bubble chamber	518
9.3.3 The propane chamber	522
9.3.4 The beam transport system	525
9.3.5 'Selling' the programme	526
9.3.5.1 Demanding more than one expects to get	527
9.3.5.2 The appeal to scientific necessity	528
9.3.5.3 The appeal to the unexpected	529
9.4 Acquiring the material for the PS experimental programme	530
9.4.1 Organizational aspects	530
9.4.2 The first order of beam transport magnets	532
9.5 The beam transport crisis in 1961 and its causes	535
9.5.1 A brief chronology	536
9.5.2 The roots of the beam transport problem	540
9.5.3 A few remarks on the CERN bubble chambers	543
9.6 Data-handling facilities for track-chamber pictures	545
9.6.1 Bubble-chamber picture analysis: a few basic concepts	546
9.6.2 Measuring instruments	547
9.6.2.1 The Iep	547
9.6.2.2 The HPD	548
9.6.2.3 The European dimension	552
9.6.3 Computers	554
Notes	561
Bibliography	809

The CERN proton synchrotron first reached its design energy of some 25 GeV on the 24 November 1959. This achievement was the culmination of seven years of research and development, and it heralded the transition to a new phase in the laboratory's life, the transition from a period in which the machine was constructed to one in which it was to be exploited as a useful tool for high-energy physics. An accelerator is only a means to an end. To do physics with a machine like the PS it had to be surrounded with auxiliary equipment, equipment to generate secondary beams, to 'transport' them away from the main accelerator, to study their reaction products, and to analyze the resulting data. In consequence, as John Adams pointed out at an important symposium at CERN in 1956, an 'intricate organization of the experimental programme [was] necessary for full exploitation' of the potential of a big machine.[2]

Two factors made these organizational problems particularly challenging ones for the young CERN to handle. Firstly, there was the scale of the proton synchrotron itself, far greater than anything that had been built on the Continent before. In fact most European physicists in the early 1960s found themselves in a world where primary beam energies were much higher, experimental set-ups much more complicated, and equipment much bigger and more expensive, than anything that they had ever confronted before. And, unlike their more fortunate American colleagues, they had to deal with this situation without having worked on 'intermediate' energy machines like the 3 GeV Cosmotron and the 6 GeV Bevatron. They were catapulted from working at a few hundred MeV to doing research at tens of GeV.

The second challenge facing the laboratory was the competitive one. Brookhaven's AGS, based on the same technology and a little more powerful than the CERN machine, was due to be commissioned soon after the big accelerator in Geneva. For a short while then European physicists would find themselves transported from a state of relative backwardness as far as accelerators were concerned to having the most powerful machine in the world at their disposal. Here was an opportunity to skim the cream in a new energy range, here was an opportunity to make up, in a leap, the ground lost to the United States since the second world war—if one could capitalize properly on the early lead.

The debates on what big equipment CERN should have to exploit its proton synchrotron got under way inside the organization in summer 1957. It was decided to launch a large hydrogen bubble chamber project and to construct a medium-sized propane chamber in summer 1958. In spring 1959 the first order for beam transport magnets was placed, and work on building a number of separators to purify the beams got under way. In parallel with these developments equipment was acquired to facilitate the automatic measuring and analysis of bubble chamber tracks. In summer 1958 CERN took delivery of its first computer. By summer 1960 (and as the AGS reached its design energy of some 30 GeV) the CERN PS primary beam intensity had been tuned to 2×10^{11} circulating protons, the properties of the secondary beams coming off its internal targets had been surveyed, about one third of the machine time was available for high-energy physics—and the first grumblings about the lack of adequate experimental equipment began to be heard. By 1961 dissatisfaction and disillusionment was widespread, and openly expressed by outside representatives at the Scientific Policy Committee. In April it was the lack of adequate beam transport equipment that preoccupied them. In October they questioned the efficiency of CERN's data-handling facilities, and argued that its computing power should be improved. Computing capacity was again criticized as inadequate in 1963. And if no one complained about the need for a big hydrogen bubble chamber, it was only because physicists working at CERN could use an 81 cm device loaned to the laboratory by Saclay. In fact it was not until June 1964 that CERN had a hydrogen bubble chamber comparable in size to those available at Berkeley in March 1959 and at Brookhaven in June 1963.

There is no doubt that CERN did not fully exploit the potential of its big machine, and its lead over American laboratories, in these early years, and that the lack of certain items of experimental equipment was one important reason for this. One 'big' discovery after the other eluded its grasp—the ω^0 at Berkeley reported in August 1961, the two neutrinos at Brookhaven in June 1962, the Ω^- at Brookhaven in February 1964. The world standing of CERN suffered accordingly. Using citation analysis to illustrate 'the full effects of [CERN's] not being ready to mount a comprehensive experimental programme...', Irvine and Martin have shown that between 1961 and 1964 publications from the Brookhaven AGS were cited about four times more frequently than those produced at the CERN PS, and that Brookhaven scientists published about three times as many highly-cited papers as did those in Geneva.[3] As Weisskopf was to put it so trenchantly to the Council in June 1962, 'it is no good in this field to be excellent and always late'.[4]

It is against this background that we propose to study, in this chapter, the preparations for the PS experimental programme at CERN which got under way in the 1950s, and their implications in the early 1960s. Our focus, we should say at once, is not on the physics content of that programme—part II of this book was set aside for that purpose—but on its infrastructural needs. More specifically, we want to study in depth the causes for the so-called 'equipment gap' at CERN by looking at the decision-making processes in the late 50s and early 60s about the acquisition of material to exploit the PS, and their consequences in the years that followed. The reader will of course find scientific, technical, and institutional factors interwoven in these processes. After all, they concerned decisions *about what equipment the laboratory needed to do good physics*—and we have never lost sight of that cardinal fact in our account.

Notes: pp. 561 ff.

The chapter contains six main sections. We begin with brief descriptions of what a bubble chamber and a beam transport system are. We then study the preparations made to exploit CERN's smaller accelerator, the 600 MeV synchro-cyclotron, which preceded those made for the PS and which influenced them in certain important ways. This is followed by a lengthy analysis of the decision to launch a big hydrogen bubble chamber project at CERN (9.3), and a study of the considerations leading up to the placing of its first order for beam equipment (9.4). In section 9.5 we explore the beam transport crisis that arose in the Geneva laboratory in 1961, while section 9.6 is devoted to the steps taken to provide CERN with automated measuring instruments and computers adequate to its share of the pictures taken by bubble chambers at the PS. Our main finding is that, while there were undoubtedly periodic mismatches between equipment supply and demand (e.g. shortages of beam magnets, lack of computing power), the early exploitation of the CERN PS suffered most from the fact that really good separators to produce purified secondary beams were only available at the very end of 1961, two years after the machine first worked.

9.1 A few technical essentials

9.1.1 BUBBLE CHAMBERS

The bubble chamber is one of a number of 'visual detectors' used for recording the tracks of particles produced initially by high-energy accelerators. It is based on the principle that under suitable conditions such particles leave a visible trace of microscopic bubbles on passing through a superheated liquid. A wide range of liquids can be used for this purpose, the most usual being liquid hydrogen, low molecular weight hydrocarbons, and some organic halogen compounds.

The principle underlying the bubble chamber was first devised by Donald Glaser at the University of Michigan in 1952.[5] Its embodiment in a scientifically useful device proved to be difficult, and sometimes hazardous. It required that a liquified gas contained in a chamber with transparent windows be repeatedly subjected to sudden changes in pressure. These were synchronized with a camera which filmed the processes occurring inside the chamber. This technology was complex and the development of bubble chambers thus passed through several stages, the earlier versions being rather small (i.e. around 10 cm) and filled with the relatively safer heavy liquids. These were the forerunners of the huge liquid-hydrogen chambers subsequently built.

Some of the technical demands made by these big chambers can be gauged from a consideration of the parameters of the CERN 2 m chamber that was first used in 1965. The heart of the chamber was a stainless steel vessel weighing 22 tons and containing 1650 litres of liquid hydrogen at a temperature of -247 °C and under a pressure of some 6 atmospheres. The front and back walls of this vessel were optically clear glass windows 200 cm wide, 60 cm high, and 17 cm thick. The usable volume of liquid hydrogen, some 600 litres, was illuminated by a flash tube and lens system through one of the windows, and the particle tracks were photographed through the other by an array of four cameras. The equipment was enclosed in an electromagnet of overall dimensions 3.6 m × 6.5 m × 4.16 m high, and weighing over 400 tons.[6]

Several groups in Europe, the United States, and the Soviet Union were building bubble chambers in the mid-1950s.[7] The most outstanding and advanced of them all was that led by Luis Alvarez at the Lawrence Radiation Laboratory in Berkeley. Alvarez (who won the Nobel prize for physics in 1968 for his 'decisive contribution to particle physics [...] made possible through his development of the technique of using hydrogen bubble chambers and data analysis') launched a major hydrogen bubble chamber construction programme in 1954. By 1957 he was doing experiments with a 10 inch (25 cm) chamber, a 15 inch (about 38 cm) chamber was undergoing preliminary tests, and a giant 72 inch (1.8 m) chamber was nearing completion. The same laboratory also had a 30 inch (75 cm) propane chamber, technically so sophisticated as to be regarded as 'the last word' in big propane chambers by one admiring CERN visitor.[8]

9.1.2 BEAM TRANSPORT SYSTEMS

Most of the nuclear physics experiments around a (fixed-target) high-energy accelerator are performed, not with the primary accelerated beam, but with secondary beams of quite different particles. The simplest way of creating such particles was to intercept the primary beam with a target — usually Be or Al foil or wire in the case of the CERN PS — which was flipped up into its path. Apart from the elastically scattered protons, the main particles emitted from an accelerator like the PS in this way were positive and negative pions (i.e. π^+, π^-) and kaons (i.e. K^+, K^-) and antiprotons. To give some idea of their relative abundances we may quote the values reported by Cocconi at the Rochester conference in 1960, just after the first survey had been made at the CERN PS. With the targets and energies then in use the kaons were from 10% to 30% of the pions produced, and the antiprotons some 1% of all particles emitted up to about 10 GeV. All these particles were generated in useful quantities over a wide range of momenta and of angles to the position of the target.[9]

The aim of a beam transport system is to bring the desired particles emitted at the targets in the primary beam up to the detectors arranged around the accelerator. Until the mid-1950s this was achieved by suitably combining two basic elements: quadrupole magnets or 'lenses' to focus the beam and bending or analyzing magnets to select particles of a particular momentum from it. The 'output' of a beam transport system was thus a mix of particles with the same momentum. This did not

Notes: p. 562

matter very much if the particle of interest was the most abundant component, and/or one's detectors could discriminate between particles on the basis of velocity, as could electronic detectors (e.g. counters). With the advent of bubble chambers in the latter half of the fifties neither of these conditions was satisfied: interest turned to studying the interactions of the far less abundant kaons and antiprotons with hydrogen in the chamber, and the cleaner the incident beam the easier would be the subsequent data analysis.[10] As a result essentially two ways of producing purified beams were developed. In the simpler, the so-called electrostatic separator, a momentum-analyzed beam passed at right angles through crossed magnetic and electric fields, the latter being established between two horizontal parallel plates. The resulting separation of the particles was proportional to the field strength and length of the electrodes, and inversely proportional to the momentum of the particles. In the technically more complex radio-frequency separators, which were still very much in the design stage in the early 1960s, separation was achieved by having 'unwanted particles travel with the velocity of a suitable wave in the structure [so being] deflected out of the way, whereas the wanted ones slip[ped] in phase by 2π in their passage and so receive[d] no net angular deflection'.[11]

To study relatively unstable particles one either had to build short beams, or arrange to eject the primary proton beam from the accelerator with a suitable magnet, and to transport it up to another target near the detector. Typically this might be done with 'slow' kaon beams. The decay length of K-mesons was 8 m per GeV/c of beam momentum, which meant that if one wanted to study interactions with, say, 3 GeV/c kaons the detector had to be less than 24 m away from the point of production of the particles. In the mid-1950s, when most of the work around accelerators was done with counters, the beams built were relatively simple and relatively few analyzing magnets and quadrupole lenses were required. In 1956 Lofgren reported that seven analyzing magnets and six sets of quadrupoles (each set a triplet) were available at the Bevatron; the Cosmotron was equipped to roughly the same level.[12] With the increase in the complexity and in the number of experiments run in parallel, and with the demands made by bubble chambers for well-focussed separated beams, beam design became increasingly sophisticated, and the amount and cost of 'standard' beam equipment in a major high-energy physics laboratory rose sharply.

9.2 Some reflections on the SC bubble chamber programme (1955-1956)

The preparations for the SC experimental programme got under way in 1955. During the course of that year a number of groups were set up at CERN to look into the material equipment which would be needed. Their tasks included 'the design and construction of the deflecting systems, magnetic channels and analyzing equipment', a study of the 'kind of chambers (expansion, diffusion, or bubble) [which] will be required for the experiments', and the consideration of 'scintillation counters, Cerenkov counters, and γ-ray spectroscopy'. On 22 November 1955 an ad-hoc committee met in Geneva 'in order to decide on the kinds of experiments which can and ought to be prepared' for the SC. Aside from the CERN staff involved, invitations to attend were sent to Amaldi, Bernardini, Cassels, and Siegbahn.[13]

By June 1956 the groups were ready to lay their ideas for the SC 'Research Programme for 1956 and 1957' before the SPC. They made a number of proposals for machine development (beam extraction, improvement to intensity), for experiments to be performed (with π- and μ-mesons), and, of importance for us, for the development of track chambers. 'We intend', said the report, 'to build a bubble chamber which will have all the specific advantages of this instrument (high density of hydrogen and fast recycling) and will be able to take over the type of work hitherto performed by the diffusion chambers'.[14]

This was not the first venture by CERN staff into the construction of a track chamber. In September 1954 a group of experts recommended that the laboratory build a cloud chamber to study the lifetimes of K-mesons.[15] The so-called 'Track Chambers' section in Lew Kowarski's Scientific and Technical Services Division devoted themselves to this task, and by June 1956 they had completed the construction of a sophisticated double cloud chamber which they then handed over to the 'Cosmic Ray Group'—unfortunately just when 'a great deal [had] already been done [on K-meson lifetimes] by the workers with accelerators'.[16] In parallel, and in line with another recommendation of the 1954 expert panel, arrangements were made at the end of 1955 to train Reinhart Budde, a new CERN recruit, in bubble chamber techniques in the United States.[17] Around August 1956, with the work on the cloud chamber over, and stimulated by reports at the CERN symposium in June on progress made not only in the USA and the USSR but also in France and in Italy, work on CERN's first liquid hydrogen bubble chamber got under way.[18]

We do not intend to describe this early programme in any detail. However it is interesting for the light it throws on CERN's approach to the building of bubble chambers. Two points in particular merit attention.[19] Firstly, it was stressed that bubble chambers were novel and complicated devices and that the technique 'was quite different from the cloud chamber technique'. Hence it was concluded that the construction of the (hydrogen) chambers at CERN should proceed step by step. A prototype would first be built. Of diameter 10 cm, its main purpose would be, to quote Budde, 'to allow all necessary tests and experiments of different expansion-, cooling- and other equipment to be made' and to study 'future developments of the HBC technique'. After three months of 'running time to get the most important information [from the 10 cm chamber]', 'construction of the larger chamber [30 cm diameter]', 'intended to be used with the SC', 'could be started [...]'. It was felt then that the technology of bubble chambers had to be cumulatively learnt by building and using successively larger devices—in contrast to what happened with the cloud chamber, whose technology was more or less standard, and where CERN built what

Notes: p. 562

was essentially a sophisticated copy of the multiplate chamber used at the Pic du Midi.[20]

Secondly, it was assumed without question at CERN in this period that the construction of 'instruments' in general, and of bubble chambers in particular, was the responsibility of nuclear physicists. It was a physicist, Charles Peyrou, who was in charge of the 10 cm and 30 cm construction projects. It was a physicist, Reinhart Budde, who spent part of 1956 gathering 'background information on the various developments in that field during a stay in the United States, where he studied cryogenics, optics and the control of bubble chambers at Columbia University as well as in various other American laboratories interested in this novel technique'. And it was physicists who played a leading role in the first group building the 10 cm chamber: according to Budde, around November 1956 it comprised four physicists, one 'engineer-physicist', and six technicians, and was due to be strengthened by the addition of two more physicists in February 1957.[21]

This conception of the role of the physicist was simply one dimension of a more general division of labour between them and engineers prevailing at CERN at this time. The task of the engineer was to build a machine; the task of the physicist was to use it. The engineers developed the technology needed to provide the physicists with an intense and stable beam; the physicists developed the 'instruments' adapted to the kind of experiments they wanted to carry out with that beam. This division of labour was mirrored in CERN's first organizational scheme. There were two machine divisions—the PS and SC Divisions—whose job was essentially to build the accelerators, and there was the STS (Scientific and Technical Services) Division which was responsible for building the instruments needed to exploit them. And whereas the leading staff in the PS group were all engineers (men like Adams, Hereward, Hine, Johnsen, and Schmelzer), those in the STS group were all primarily nuclear physicists (e.g Budde, Goldschmidt-Clermont, Kowarski, and Peyrou—the last moving from the SC to the STS division when he took on the bubble chamber construction projects).

Note, however, that the engineer's task was considerably expanded, at least in John Adams' mind, after he had heard a session on beam ejection and the transport of 'secondary' beams at the 1956 symposium. Summing up after the meeting he was struck by the fact that the anti-proton yield at the Bevatron had 'been increased by a factor of 15 mainly by improvements in the beam transport system'. 'By considering that a machine project is not finished until the largest fraction of ejected particles not only leaves the machine but actually passes through the counter telescopes', Adams wrote, 'large increases in the efficiency of utilization of the machine can be achieved, equivalent to increasing the beam current by orders of magnitude'. In short in 1956 the boundaries of the engineer's task were becoming rather blurred, at least in Adams' mind, and were beginning to go beyond the 'mere' production of a good circulating primary beam.[22]

To conclude this quick survey there are two points we should like to make. Firstly, *any* laboratory which launched a bubble chamber programme in the mid-1950 had to accumulate the know-how for itself from scratch, gradually getting a better grasp of how the device functioned. As Alvarez explained at the 1956 symposium, it was only after the operation of his first 6.3 cm device was 'understood' that he went on to construct a larger 10 cm chamber. This was technically more sophisticated than its predecessor, so that its operation 'became more of a science and less of the art it had formerly been'. It was followed in turn by a 25 cm chamber, and by the giant 72 inch chamber commissioned in 1959. Shutt's group at Brookhaven did the same thing. They began by building a small 5 inch prototype hydrogen bubble chamber in 1955, moving on to a 20 inch chamber, and finally on to an 80 inch chamber for use around the AGS.[23] In short, granted the novelty and complexity of the technique, a laboratory which, in the early 1950s, seriously considered using big hydrogen bubble chambers had to proceed stepwise: it was not yet possible to 'copy' what others had done.

A second point to remember is that, at this stage at least, CERN's physicists felt that it did not matter if their small bubble chambers were 'late' for the SC—while for the PS, by contrast, a bigger chamber would need to be ready in time. Budde, for example, tackled this question indirectly in his report. He estimated that the 10 cm chamber would be ready around September 1957, and felt that the construction of the 30 cm chamber should start three months after that, calculating that with the available staff it could be first filled with liquid hydrogen around July 1959. There was then no doubt in Budde's mind, nor apparently any concern, that by staggering the construction of the two small chambers in time, CERN's 30 cm chamber would only be available long after the SC first operated. His thinking concerning experimentation around the PS was different. The bubble chamber programme for this machine, he said, 'should be started not later than 1958, since construction periods of 200 man.months are involved, [and] it is necessary to enlarge the staff by a considerable number of people'. For Budde then it was apparently desirable to have a bigger chamber ready for the PS when it worked in 1960, and he assumed that, with the knowledge the staff had gained by 1958, and with sufficient human and material resources at their disposal, it would be possible to build one in time to meet this deadline.[24]

9.3 Defining big equipment for the PS experimental programme[25]

The first serious efforts to plan the PS experimental programme got under way in mid-1957. And although all aspects of the programme were tackled from the outset—the experiments to be done, the beam transport equipment needed, the counter and bubble chamber material to be built or acquired—, the early discussions

Notes: p. 562

were in fact overshadowed by one major concern: whether or not CERN should launch itself into a major hydrogen bubble chamber construction programme and, if it should, how big that chamber should be.

9.3.1 JUNE-DECEMBER 1957: THE GROWING CONVICTION THAT CERN SHOULD BUILD ITS OWN LARGE HYDROGEN BUBBLE CHAMBER

Around the summer of 1957 a number of informal meetings were held to discuss the PS experimental programme. One of them took place on 4 June with Bruno Ferretti, the first leader of the Theory Division in Geneva, in the chair. Seven of the dozen people present at the meeting were theoreticians, the remainder were experimental physicists.[26]

The aim of the meeting was to draw up a preliminary list of possible experiments to be done with the PS taking into account scientific interest and practical feasibility. This preliminary list would be sent to selected specialists in Europe, and would be discussed further at the international conference on mesons and recently-discovered particles to be held in Venice-Padua from 22-28 September 1957. It was hoped that a 'final programme' would emerge from these deliberations that would 'form a basis of further work by the Theoretical Division and by the physicists planning the lay-out of the experiments'. However, under the influence of Charles Peyrou and Oreste Piccioni, an experimental physicist who had designed an important beam ejector at the Cosmostron, most of the meeting was devoted less to a consideration of possible experiments than to a discussion of the kind of equipment that would be needed to exploit the PS. The desirability of having certain very large and expensive beam transport magnets was looked into, as were the methods of enriching some beams using separators. Peyrou stressed the advantages of using large bubble chambers for the study of secondaries at the energies attainable with the PS. He 'propose[d] something like a liquid H chamber of 100 cm × 50 cm × 30 cm', and also suggested that a propane chamber of similar dimensions could be built using the same magnet. 'Agreement [was] not complete' at the end of the day, and those present 'intended to resume the discussion after the list of interesting experiments will have been drawn up, so that one will be able to decide what is the actual need for various devices'.

About a month later John Adams drew up a 'programme of work, covering the years 1958, 1959, 1960, aimed at bringing the CERN Proton Synchrotron into operation in 1960 as a complete high energy nuclear physics tool [...]'.[27] In it the PS division leader distinguished carefully between the machine programme—where 'the need for the several machine components [was] self-evident'—, and the experimental programme—where 'the need for any particular piece of experimental apparatus [was] very much a matter of opinion'. Not content to leave matters at that, he went on to draw up a possible 'maximum programme' of work. It foresaw the formulation of 'a list of likely experiments for the C.P.S. by the end of 1957'. This was to be followed in 1958 with work on beam transport systems and various other items of experimental equipment, including a large bubble chamber (it was 'by no means certain', wrote Adams, 'that a large hydrogen bubble chamber [was] the best track chamber for the C.P.S., and [...] other bubble chambers using different liquids, for example propane or tungsten hexafluoride, should be considered seriously').

In mid-September Paul Amiot, a member of Kowarski's group, was sent to the USA for a month to study the state of bubble chamber technique. He was specifically instructed to look into cryogenic problems raised by liquid hydrogen bubble chambers and to report on the requirements in terms of personnel and money for the successful construction of big chambers.[28]

Ferretti's list of experiments was duly presented at the Venice-Padua meeting towards the end of September, followed by a talk by Mervyn Hine on the PS experimental facilities. The former apparently raised little interest: in the subsequent discussion on it 'there were two remarks' 'about the list of experiments'. The discussion on Hine's paper was apparently more lively. It was opened by Owen Chamberlain, who explained the nature and cost of the beam programme at the Bevatron. This was followed by a talk by Peyrou in which he gave, in Hine's words, 'the physical reasons [...] which [led] to the conclusion that the hydrogen bubble chamber [was] the best track chamber for these energies, and that with a length of 1.5 m it [was] very powerful for work on hyperons, which [would] all decay inside such a chamber'. 'Peyrou's next favoured chamber', according to Hine, 'would be a large propane chamber [...]'. The session was terminated by a talk by von Dardel on counters and instrumentation, though 'because of lack of time there was not much discussion [...]'.[29]

Back in Geneva, Ferretti summarized the situation as he saw it in mid-October. Adams' maximum programme required '60 people, plus mechanical design effort' for three years if all the equipment was to be ready by 1960. 'This may be too large an effort and perhaps not the most reasonable [...]', Ferretti went on, proposing that, to clarify matters, the list of experiments be discussed further 'in order to pick up those which [were] really likely to be performed in the first two years'. At the same time he thought it would be wise to 'design some of the likely experiments in some detail [...]' so as to get an idea of the merits of different apparatus 'for overcoming the experimental difficulties' that might arise.[30]

This step was soon taken. Using Ferretti's list as a basis, a group of CERN physicists drawn from the PS, SC, and STS Divisions selected several kinds of particle interactions that would help them to identify the beam transport facilities that they needed and to compare 'the relative merits of counters and bubble chambers'.[31] The selection was laid before a large group of senior staff who met to discuss 'P.S. Nuclear Physics' on 28 November 1957.[32] The meeting picked out two experiments using pion beams for further investigation, and set up three groups to look into their requirements. Two of them were to work on the design of an experiment on the total and differential scattering cross sections of pions against protons— one concentrating on 'the target, intensities, beam collection and related problems', the other on 'the counters used for defining the incoming pions and [to] distinguish them from other particles'. The third group were to do the same for inelastic π-p interactions (total and differential cross sections, production of hyperon pairs, decay modes of anti hyperons) 'from the point of view of beam requirements, bubble chamber and data handling'.

The senior staff, including the Director-General, met again on 20 December 1957.[33] A theoretician began by reporting on 'the progress of the calculation on the pion yield, using Fermi statistical theory'. Von Dardel then 'sketched a preliminary layout of the analysing and focusing magnets needed' to study π-p processes with counters, while de Raad and Resegotti, two engineers in Adams' group, described 'a first version of the focusing lens and bending magnet providing the momentum analysis' in von Dardel's beam. Thereafter Peyrou talked to a long report in which he discussed the size, filling liquids, and general beam requirements of a bubble chamber and Goldschmidt-Clermont described what was needed for the analysis of bubble chamber photographs. After commenting on these various reports, the gathering agreed that experiments on proton-antiproton interactions using counters and bubble chambers should be studied next. Thereupon it turned all its attention to bubble chambers. The results of its deliberations were summarized in a memorandum by Kowarski and Peyrou dated 13 January 1958.[34] In it they proposed that a large bubble chamber project 'should be declared urgent and started at once' at CERN. Two weeks later another large staff meeting was was held in Geneva.[35] Now only bubble chambers were on the agenda, and for the next few months the formulation of the CERN bubble chamber construction programme was the dominant concern of Peyrou, Adams, and a number of engineers in the PS group.

The discussion at the meeting on 20 December and the proposal emanating from it early in January were symptomatic of a break. Whereas before there had been considerable uncertainty about whether or not CERN should embark on a major hydrogen bubble chamber construction programme, now those doubts were set aside. What was the cause of the anxiety and how had it been dispelled?

The root of the uncertainty was the risks involved, the fear of making an over-hasty commitment with damaging consequences for the experimental programme as a whole. To begin with, the technology of liquid hydrogen bubble chambers was proving extremely difficult to master. The year before Budde had indicated that the small and relatively simple 10 cm prototype would be ready for filling by September 1957; in fact it was not ready until May 1958. Amiot confirmed that US laboratories were having similar difficulties: Shutt's group at Brookhaven had underestimated the date of completion of their 20 inch chamber by 'about 50%' which, said Amiot, 'seems to be a quite general phenomenon'.[36]

Then there were the needs in terms of staff and resources. Alvarez had used about 50 people for three years on his 72 inch chamber. 'CERN could hardly expect to use less than 30 people for 3 years, even assuming that much of the technological work is already done', wrote Adams. This was not only a big effort in absolute terms. It also meant that half of the staff envisaged in Adams' 'maximum programme' would be devoted to this one sector—proportionately far too much according to prevailing opinion at the Venice-Padua conference.[37]

Finally, there was the fear that, after having invested all this time, staff, and money—Amiot estimated that Alvarez's device would cost $2.5 m—, a very big bubble chamber would be outdated when it was finally built. The danger of this happening with any piece of equipment that took a long time to build was raised by Adams in July and repeated by Ferretti in September. It was confirmed by Glaser when Amiot visited him during his trip in September/October: 'On the subject of constructing very big liquid hydrogen chambers, one hesitates to embark on the construction of apparatus which are very difficult to build and which require a high budget and a large construction staff, all the more so when there is a risk that new detection methods will be discovered which render such apparatus obsolete'. Hence the attraction of a propane chamber which, as Peyrou said in Venice-Padua, 'would be much easier to make and handle, and had almost all the advantages of hydrogen except for the presence of the carbon nuclei'—though even here Peyrou felt that the drawbacks were 'much reduced' in a large chamber, 'and consequently a major argument against propane drops out' (Peyrou).[38]

Two ways of dealing with this uncertainty, two ways of clarifying the equipment needs of the PS experimental programme are discernible in the debates we have described. The one saw the question as subsidiary, as depending on the answer to the more 'fundamental' question 'What experiments should be done?'. In practice this meant that one first decided on the experiments that were 'likely' to be done at the PS, and then defined the equipment needs from that perspective. This was the approach favoured above all by the theoreticians. Recruited specifically to guide the experimentalists—a major preoccupation of the SPC throughout 1955 was to attract a 'leading theoretician' to Geneva who could 'make a real contribution to the direction and development of ideas useful to the general research programme of CERN'—, and enjoying the prestige, the power, and the responsibility, of drawing up the experimental programme, they 'spontaneously' sought to 'derive' particular equipment from specific experiments.[39] It was also an approach which had some appeal for John Adams: as he put it in July 'the *proper* start of the experimental apparatus construction programme [was] first to formulate a list of likely experiments and from an analysis of this list to decide upon the most generally useful apparatus' (our emphasis).[40]

Notes: p. 563

Alongside this approach there was another, an attitude which was shaped by more general considerations like the possible evolution of high-energy physics in the next decade, and the kind of equipment which any competitive experimental programme would call for. We find this mode of thought among experimental physicists like Piccioni or Peyrou. It is also present in Adams' thinking, particularly regarding bubble chambers. Listing in his July 1957 *Note* the various elements constituting his provisional 'maximum programme', he carefully distinguished the need for items which, he said, should be 'based on likely experiments' (beam transport systems, fast electronics, and electronic detectors) from bubble chambers which, he implied, were essential to *any* experimental programme. Similarly, in November/December, when asked to study bubble chamber needs for a particular experiment, Peyrou rejected the question as irrelevant: his report, he said, 'was not concerned with the bubble chamber only with the pion-proton experiments in view but on a more general basis'.[41] The break in December which we have identified was symptomatic of the dominance of this way of thinking over that favoured by the theoreticians, and a rejection of the intellectual and social hierarchy which it implied as far as settling equipment needs were concerned.

In our view the key factor securing this dominance, the main consideration dispelling the uncertainty, the main argument convincing the CERN senior staff that a big hydrogen bubble chamber was needed come what may, was what we might call Alvarez's 'bubble chamber philosophy'. This was developed in an open letter dated 17 September 1957 to Edwin McMillan, who was soon to succeed Lawrence as the director of the Berkeley Laboratory.[42] In it Alvarez argued that the days of emulsions and counters as detectors at high energies were numbered, that the bubble chamber was 'going to be the source of most of the high energy physics information in the future [...]'. His reasoning was that at high energies most of the more common and scientifically interesting processes were inelastic processes, and that a physicist 'could hardly get started in a serious investigation' of such processes 'if he [did] not have a visual detector'. Alvarez realized that his views were not shared by everyone, that emulsion physicists were only just beginning to come round to them (or so he said), and that counter physicists were even more sceptical. But he remained convinced that they too would come to appreciate that 'in the high energy field the bubble chamber [was] the universal detector'. If Alvarez was right there was no need to fear that the bubble chamber would quickly become obsolete. If Alvarez was right there was no doubt that to remain competitive in the sixties any serious high-energy physics laboratory would have to be equipped with hydrogen bubble chambers. The weight of his arguments was felt at CERN. In October Patrick Blackett told an SPC meeting that 'he essentially agreed with Alvarez on the importance of bubble chambers in high energy research'. In November Gilberto Bernardini, the Director of Research on the Synchro-cyclotron, told a meeting of the senior staff that 'he concurred with Alvarez that the weight should be put on bubble chambers, unless new techniques render counter experiments more attractive'.[43] And by 20 December any doubts about whether the risk of building a big hydrogen chamber were worth taking had been overcome, or at least set aside.

The attraction of Alvarez's argument was reinforced by a growing sense of urgency, the feeling that, with the PS due to start in some two years time, it was essential to start planning a big bubble chamber project as soon as possible. This sense of urgency was heightened by the recognition that the time was coming when CERN would be called upon to set priorities, and that if mistakes were made it would be difficult to come to the Finance Committee with a last minute request for additional resources. Indeed the meeting of the senior staff on 20 December took place the day after the Council had voted for the first time to impose a financial ceiling on CERN's expenditure for 1958 and 1959, a policy which was intended (by the British anyway) to signal to the CERN administration that funds for the organization were limited, and that they were to adjust their scientific programme to fit in with a fixed budget laid down two years in advance.[44] In this pressured climate one could not go on talking indefinitely in the hope that all doubts would be removed. One had to make choices, and draw up preliminary cost estimates. Alvarez's argument had a crucial role to play here — it reassured the CERN staff that priority should be given to a big hydrogen bubble chamber project.

To summarize. The decision taken at the end of 1957 to launch a big hydrogen bubble chamber project at CERN involved the crossing of a psychological barrier, the overcoming of the fear that the resources needed would be 'wasted'. When CERN had launched its hydrogen bubble chamber programme in 1956 it had been more or less taken for granted that the laboratory would construct a big chamber for the PS, and those involved set out systematically to build up their knowledge of the technique. But when the time came, in summer 1957, to be more specific about what was needed, they at first drew back, intimidated by the scale and complexity of a large hydrogen chamber project, and were strongly inclined to build a propane chamber instead. To clarify their thoughts some of the senior staff, and the theoreticians in particular, tried to dispel the uncertainty by 'deriving' the need for a chamber from a list of likely experiments. In parallel there were experimentalists and engineers who, while perhaps not rejecting this line of attack out of hand, felt that a big bubble chamber was needed anyway. Faced, in December 1957, with the need to decide quickly on what should be done, and finding that theory was unable to play the directing role which some had hoped of it, they drew sustenance from Alvarez's 'bubble chamber philosophy', they were reassured that hydrogen bubble chambers would be a kind of 'universal' detector in the high-energy physics of the future — and they crossed the anxiety threshold which until then had stood in the way of their taking a definite decision.

Have we not placed too much weight on Alvarez's thinking, have we not exaggerated the importance of his prognosis at the expense of more 'objective' considerations, like the actual scientific and technical progress being made with

Notes: p. 563

hydrogen chambers? We do not think so.⁴⁵ Certainly Alvarez's 10 inch chamber first worked during this period, and this might have reassured those at CERN who were struggling to get their little chamber working, but this did not entail that it was worth scaling up to 72 inches, as Alvarez planned to do. It is also true that in October 1957 the Berkeley group reconfirmed the non-conservation of parity with this chamber, but no other scientific result of note appeared that might have generated enthusiasm for the construction of very big chambers. In this respect our period seems to differ from that starting around the end of 1959, when bubble chambers truly came into their own with the increasing interest and importance of the study of resonances. In short scientific and technical advances as such were at best of marginal importance in getting the CERN senior staff to decide, in December 1957, to launch a big hydrogen bubble construction programme.

One last comment. Did the 'support' for Alvarez's arguments mean that there was a loss of interest for experiments with counters at CERN? Certainly not. Alvarez's argument that the bubble chamber was a 'universal' detector pivotted on the claim that the only interesting physics to do with high energy accelerators was the study of high-energy inelastic interactions; he admitted that at low momenta counters were better than bubble chambers. One could easily accept then that the bubble chamber was an essential, even the universal, detector at high secondary beam momenta, all the while insisting on the interest of working with counters at low energies. Nor was this to 'squander' the potential of a high-energy accelerator as Alvarez seemed to think—at least not for a counter physicist like von Dardel. As he emphasized, 'in many cases experiments in the low momentum range will only be possible with a higher energy machine either for intensity reasons or because of a high threshold energy'.⁴⁶

9.3.2 JANUARY–MAY 1958: FIXING THE SIZE OF THE BUBBLE CHAMBER

Having accepted in December 1957 that CERN ought to build its own large bubble chamber, one of the first tasks was to settle its size. The most striking feature about the debate on this parameter is that its value grew steadily between January 1958, when the subject was first seriously discussed, and May when an official paper was laid before the SPC.

In the first proposal made in the Kowarski/Peyrou memorandum dated 13 January 1958 it was concluded that 'A project of about 90 cm should be declared urgent and started at once'.⁴⁷ This recommendation was discussed at a meeting of the CERN senior staff on 30 January, at which it was quickly agreed not to rule out the possibility of building bigger devices: 'it was decided to set up a study group which would investigate in more detail the alternatives of a '90 cm' (60-100 cm) chamber, a '180 cm' (150-200 cm) chamber, or both chambers in succession'.⁴⁸ The study group presented a preliminary report of its findings to a meeting of the SPC in March.⁴⁹ It was now considering four variants—75 cm, 100-110 cm, 150 cm, and 150-200 cm.

Defining big equipment

The SPC could not reach a decision on the best size for the chamber and resolved to have a special meeting to discuss the matter further before the next Council session in June. In anticipation of this meeting the bubble chamber study group prepared a bulky report for the SPC dated 1 May 1958. Appended to it were three brief papers originally written by Adams.[50] The main report discussed the design for two possible sizes of bubble chamber, 1.1 m and 2 m, as well as the magnet systems, the optical devices, the low-temperature installations, the beam requirements, and the buildings involved in the project. It also attempted to estimate the cost, the staff, and the time of construction for both. Adams' appendices highlighted different aspects of the whole programme—the network of ancilliary facilities, the beam transport system, and the detector itself. It was recommended that 'if any chamber be built by CERN, it should be a 2 m chamber'. Between January and May, then, the size of CERN's bubble chamber had doubled, from '90 cm' to 2 m.

How had this happened? Part of the reason was scientific. It was generally agreed that, from the point of view of the science that one could do with it, there was no upper limit to the desirable size of a chamber. There were two specific reasons for this. Firstly, because bubble chambers were primarily intended to study inelastic processes, and the probability of secondary interactions (which were both interesting in themselves and as a means of identifying the incoming particle) increased with chamber size. Secondly, because the accuracy of particle momentum determinations increased as the square of the (linear) chamber dimensions (in a constant magnetic field). In brief, from a scientific point of view, the bigger the chamber was the better.

From this it follows, of course, that the size of the bubble chamber could not be derived from scientific considerations alone. At best they could suggest a minimum below which the scientific potential of a particular accelerator would not be adequately exploited. But apart from that there was, to quote Alvarez, 'no logical way to decide on the overall length'. Put differently the length was also fixed in the light of other kinds of reasons, it was set, as Peyrou put it 'by purely practical considerations, not necessarily of an economic nature'.[51]

The two 'practical' considerations which dominated the *January* proposal were the need to build a large hydrogen chamber that would be ready in time for the start-up of the PS (i.e. in 1960), and, secondly, to do so without recruiting additional staff from outside the organization. It was these factors that led the Kowarski/Peyrou memorandum to conclude that 'No big B. Ch. project should be started inside CERN', and to recommend instead that a 90 cm project be started at once. A much bigger chamber would not have been ready in time. Alvarez had taken between three and four years with a staff of 50 to build his 1.8 m device. CERN could not hope to do better: a bubble chamber considerably larger than 1 m could take as long as five years to build unless one recruited new personnel immediately. There were two arguments against doing so. First, no provision had been made for this in the 1958 budget. Second, and more importantly, to recruit more staff while the PS was still being built would cause a 'temporary swelling' of applied physicists and engineers at

Notes: p. 563

CERN and would exacerbate the problem of what to do with the staff coming free when the PS construction programme was completed.[52]

This consideration was of paramount importance. Throughout 1957 concern grew about the long-term future of the staff, particularly those building the PS. Discussing in July 1957 the schedule for building CERN's experimental equipment John Adams 'assumed that any solution involving recruiting new staff [was] not desirable', adding that it was 'logical' 'to use the C. [CERN] P.S. staff as they become available from machine work to design and build the experimental apparatus' in collaboration with CERN nuclear physicists. He explained why in the same note. 'From the point of view of keeping together in CERN a strong and experienced applied physics and engineering effort this solution is attractive, and it should be noted that both the Cosmotron and the Bevatron started in this fashion'. The argument turned into a threat at the SPC meeting in October: unless his staff were offered further suitable work at CERN, the PS Division leader warned, they may not stay on long enough to complete the new accelerator, let alone maintain it and improve it once it had worked.[53]

The '90 cm' hydrogen bubble chamber was one way of reconciling these various factors. It was large enough to do scientifically useful work, its design could be based on that of CERN's smaller chambers, it could be started in 1958 within the limits of the budget, it could be pushed ahead rapidly in 1959 with staff released from the 30 cm project, and with people coming free inside in the PS division, and it would be ready in 1960, just as the PS became ready for doing physics. Why then was it jettisoned in favour of a bigger 2 m chamber?

The change of policy was indicative of a change of priorities, more specifically of a willingness to relax the time constraint on the availability of CERN's big hydrogen chamber. Put differently, if in January the need to have the chamber ready in time for the PS had been the dominant concern in the Kowarski/Peyrou memorandum, by May the experimental physicists accepted to build a bigger chamber — which would be late. Three main considerations apparently informed this reshuffling of priorities.

Firstly, let us remember, there was the scientific argument, the recognition that the bigger the chamber the better it would be from the point of view of the scientific results that one could hope to obtain with it.

Secondly, there was the desire to ensure that CERN remained competitive in the medium term.[54] It was stressed that Berkeley was building a 1.8 m chamber, that 2 m chambers were envisaged at Brookhaven and Dubna, and that the British were designing a 1.5 m chamber. Clearly if CERN wanted to keep up with the world leaders in nuclear physics it needed to construct a hydrogen chamber at least as big as theirs — and this would be impossible if they chose instead to devote their available resources to building a smaller chamber to be ready by 1960.

Finally, there was the hope that the needs of physicists in the short term would be met by having the 1.5 m British chamber available at the CERN PS soon after the accelerator was commissioned. During the summer of 1957 serious thinking began in the UK about building a big hydrogen bubble chamber as a collaborative project between Imperial College, Birmingham and Liverpool Universities, and Harwell. In October 1957 Patrick Blackett officially informed the SPC of this development and suggested that, while the chamber was ultimately destined for use at Harwell around its new 8 GeV accelerator which was then under construction, it could first be brought to CERN, and then shipped back to Britain when the national machine (Nimrod) was ready.[55] It was planned to have the 1.5 m chamber built around 1960 or 1961 — so it could fill the gap between the start of physics at the PS and the availability of CERN's own big hydrogen chamber. With their fears about being late thus assuaged, CERN physicists were that much more inclined to build a chamber at least as big as any on the market.

We cannot leave matters at that, though. If experimental physicists like Gilberto Bernardini, Alec Merrison, and Guy von Dardel came to 'accept' that CERN build a 2 m chamber which would be late it was also because they were more or less 'forced' to do so by those who were actually going to build the device, John Adams as spokesman for the PS engineers and applied physicists, and Charles Peyrou. In other words the decision to build a 2 m chamber was also indicative of the power relations inside the organization, of the preponderant influence over policy wielded by those who built big apparatus as opposed to those who were to use it.

Adams was always ready to relax the time constraint on construction. It was more important to him as division leader to protect the future of his staff and to ensure that, as engineers who were building one of the most technologically advanced accelerators in the world, they were given suitably challenging work to do. Hence his insistence in his July 1957 *Note* that it was 'logical' to 'use the C.P.S. staff as they [became] available from machine work to design and build the experimental apparatus' — even if that meant that 'the machine [would] not be fully exploited in 1961 and 1962'. Hence too his opposition to von Dardel's claim at the March 1958 meeting of the SPC that it was important to have a large hydrogen chamber available when the CERN PS started up — CERN, said Adams, should push to the 'technological limit'.[56]

Similarly Charles Peyrou was less concerned than other experimental physicists that CERN's big chamber be late, was more inclined to follow Adams' line of thinking. Indeed at the meeting on 30 January at which the Kowarski/'Peyrou' memorandum was discussed, the memorandum recommending that no chamber greater than 1 m be built at CERN, it was Peyrou who 'emphasized' instead that 'several alternatives must be investigated, for example a 90 cm chamber in 1960 or a 150 cm chamber in 1961, *taking the risk that this may not be ready in time for the initial operation of the synchrotron*' (our italics).[57]

Why was Peyrou relatively unconcerned about being late? Apart from the arguments we have mentioned (remaining competitive in the medium term, arrival of the British chamber in the short term) there was also his reluctance to support any move that might delay the development of 'his' 30 cm chamber. As a physicist he had taken on this project because he hoped to have a useful little device for work at the SC, his group was gradually accumulating the technical know-how required to construct it, and he was not keen to see the few people that he had put on the job diverted from it now. Indeed it is striking that, throughout the debates among the senior staff in December, Peyrou refused to consider releasing staff from the 30 cm project so as to accelerate the design and construction of a bigger device. There were arguments for this of course (his staff, which consisted essentially of physicists and technicians, was not professionally suited to take on a large engineering project, one had to finish the 30 cm before extrapolating to larger sizes, orders had already been placed for the main components and there was little to be gained by cancellation now, work at the SC with the 30 cm chamber would prepare Europe's inexperienced physicists to use bigger chambers later at the PS, and so on) but the fact remains that underlying them there was a determination not to construct a big CERN chamber at the expense of the small one. And if those responsible for building CERN's chambers — Peyrou and the PS engineers — were willing to countenance such delays, what leverage did the experimental physicists, the users, have to persuade them to do otherwise?

9.3.3 THE PROPANE CHAMBER

At the end of April 1958, in addition to drafting the 'appendices' to the May report, Adams prepared another paper for the SPC.[58] In it he set out a possible programme for building the experimental apparatus for the PS. And he suggested that, in addition to building a 2 m hydrogen chamber, CERN should construct a 1 m propane bubble chamber.

The argument was a simple one. The hydrogen bubble chamber would only be ready about one or two years after the machine had provided 'some sort of beams of particles'. Accordingly some thought had been given to having a 'more modest piece of equipment' available when the PS worked. A 1 m propane chamber seemed the ideal 'intermediate tool'. It was considerably simpler in design than a hydrogen chamber, it was safer, and it was almost certain to work. What is more it could be built with relatively few staff and it was not particularly expensive — its total cost was estimated at 2 MSF.

The most striking feature of this proposal is the apparent haste with which it was drawn up: indeed it seems almost to have been 'tacked on' to the hydrogen chamber proposal. The only quasi-scientific argument given for the propane chamber was hardly persuasive — that it could be used along with other existing detectors (like cloud chambers) 'for the initial beam measurements'. And although there had been considerable support for building a propane chamber in autumn 1957, the study-group set up at the end of January 1958 considered only liquid-hydrogen chambers. There is no evidence that any thorough technical study had yet been made of the project. Indeed the first technical report on the propane chamber made by two of Adams' staff, Ramm and Resegotti, was only circulated in November 1958. Around this time too the cost of the chamber was substantially increased, from 2 MSF to 6.67 MSF.[59] In short the inclusion of the propane chamber in the May 1958 plans for the PS experimental programme seems to have been something of an afterthought.

We suggest three reasons for its re-emergence at this point. First, the project would go some way towards satisfying physicists who felt that CERN should have its own bubble chamber ready in time for the PS, and whose wishes had been overruled in the 2 m hydrogen proposal. Propane chambers had clear limitations compared with hydrogen chambers—the liquid was denser, which increased the amount of scattering, and it was sometimes difficult to distinguish interactions with protons against a background of events with carbon nuclei. Nevertheless, as Peyrou frequently stressed, these drawbacks were less important in large propane chambers, so that 'useful scientific work' could be done with a 1 m chamber at 28 GeV.

Safety considerations may also have played an important role in Adams' thinking. Early in April there was an informal conference on hydrogen bubble chambers held at Imperial College in London at which various groups explained their plans.[60] CERN was represented by Peyrou, Ramm, Vilain, and Bridge (the last a visitor to the laboratory from MIT). According to one of the Americans present—probably Gow from Berkeley—'there [appeared] to be serious worry regarding the safety aspects of hydrogen chambers in the minds of responsible CERN personnel. It was suggested, for example', Gow went on, 'that large chambers (probably including propane chambers) should be banned from the experimental hall'. This idea, he added, was roundly condemned by those present, notably the British and Soviet physicists. The inclusion of the propane chamber in the May report written a few weeks later may have been intended to meet this objection without backpedalling completely on the question of security.[61] It was still planned to put up a separate building for big hydrogen bubble chambers (see below), but it was suggested that small hydrogen chambers 'and bubble chambers of a less explosive nature, such as propane chambers' could be used in the existing experimental halls. The large propane chamber, in other words, may have been intended to reconcile the builders' concern for safety with the experimentalist's concern to install a 'useful' chamber near the machine.

The third possible reason for the sudden re-introduction of the propane chamber concerns the question of staff. We have already mentioned the anxieties expressed by Adams about the long-term future of the members of his division. While he was arguing forcefully for a 2 m hydrogen bubble chamber others were warning that it was difficult to see 'who in CERN [was] willing or able to undertake the complete 2 m chamber project'.[62] Soon after the SPC had agreed to the propane project at its meeting on 23 May 1958, Ramm officially informed Adams that 'he, and most of the PS staff, [were] not interested in building the hydrogen chamber'. It was a five-year

Notes: p. 564

project, of limited attraction from a technical point of view, and the possibility of doing physics with the device was not sufficient compensation for the time spent building it. 'On the other hand', Ramm went on, 'the propane chamber [was] a sufficiently short project to be attractive, and he would be quite prepared to build it'.[63] If Adams knew this when he wrote his paper in May—and he probably did—[64] we can reasonably infer that his need, not only to keep his staff, but to keep them happy, was another reason why he proposed that CERN build a 1 m propane bubble chamber.

The senior staff ended with a comprehensive CERN bubble chamber programme: a 1 m propane chamber to be ready in time for the PS, a 2 m hydrogen chamber which would take some four or five years to build (and a 30 cm hydrogen chamber which would first be tested at the SC). What is striking about this programme is the more or less unquestioned assumption that CERN should build bubble chambers, the conviction that it should have its 'own' big chambers regardless of what groups in the member states were doing. For example, in April 1958 Louis Leprince-Ringuet, director of the Ecole Polytechnique physics laboratory in Paris, officially informed Bakker that he would like to bring a 1 m propane chamber to Geneva—a proposal that the Director-General simply ignored while CERN planned to build its own essentially identical device.[65] Similarly, no account was taken of French plans, formulated early in 1958, to build an 81 cm hydrogen chamber at Saclay.[66] And a suggestion that CERN staff collaborate with the British group to help their 1.5 m project out its staffing difficulties, and so speed it up, was also strongly resisted on the grounds that such participation 'would tie up staff that could otherwise be used to build CERN's own bubble chamber', that CERN was not prepared to cooperate if the interesting work was done by the British group while CERN staff did the heavy engineering, and so on.[67] There was no question then: CERN had to and would construct its own bubble chambers alongside and independently of those being built in the member states—even at the cost of duplication or delays for everybody.

In taking this position regarding the construction of big, complex, and dangerous equipment CERN was in fact behaving as if it were one national institute among others. It would build its 'own' big bubble chambers independently of what other European groups were doing. These would be available for the use of all at CERN—where else could they profitably be exploited?—though the builders would expect to have special privileges over their use (choice of the beams, selection of the photographs, etc.). CERN, its staff were saying, was not to be seen simply as 'service station' for Europe; it was to be regarded as a laboratory in its own right, with its own equipment, and its own research programme, just like any national laboratory. Indeed we find the same desire to construct one's 'own' big equipment (to be shared with other CERN users, certainly, but on terms laid down by the builders) in the groups building big hydrogen chambers in France (81 cm) and in Britain (1.5 m). Each group, no matter how willing to 'collaborate' with the others, no matter how willing to 'share' their equipment, were building that equipment on the assumption

Defining big equipment

that they would have a certain priority over its use, that they would be the first to explore the riches of the new energy range opened up by the PS with 'their' detector. Though the competition between scientists and laboratories is 'friendly' it is still competition in the full sense of the word.[68]

9.3.4 THE BEAM TRANSPORT SYSTEM

The only other technical aspect of the May report of relevance to us here concerns the provision made for beam transport equipment. The scheme proposed was dominated by the conviction that, for safety reasons, big hydrogen bubble chambers could not be installed in the existing experimental halls. As Adams explained, when these were designed it was assumed that bubble chambers and targets containing some 60 litres of liquid hydrogen would be used around the machine. To cope with any accidental escape of this gas, and the subsequent risk of an explosion, the halls were designed with sloping roofs having hydrogen gas outlets at their highest points. These measures were quite inadequate to handle the kind of emergency which might arise if not 60, but 600 litres of liquid hydrogen were accidentally released into the atmosphere from one of the large hydrogen bubble chambers then being envisaged. As a result CERN, wrote Adams, had been obliged 'to review the whole question of whether such large and dangerous devices should be mixed in with the other experimental apparatus in the halls, or whether a separate installation should be made, remote from the existing halls'.[69] The May report proposed that the second alternative be implemented, and that an area on the north side of the machine (what subsequently became known as the East area) be set aside for the testing and operation of big hydrogen bubble chambers. It entailed the provision of transport facilities for beams with a minimum length of about 100-120 m, the distance between the envisaged targets in the PS ring and the new bubble chamber building.

Five main beams were foreseen.[70] Apart from three high energy beams (of scattered out protons, of mesons, and of neutrons), the layout made provision for an ejected proton beam (needed to produce relatively low-energy K-mesons at a target near the bubble chambers) and a separated antiproton beam of up to 3.5 GeV/c. In the May report it was estimated that the capital costs of the five beams would be some 2.25 MSF, and that 14 staff would be engaged for three years in designing, constructing, and testing the various channels. The overall cost of the beam transport system was calculated to be just under 4 MSF.

Three points are worth noting about this proposal. Firstly, it was clearly intended to undertake a major development effort to build the electrostatic separator for the purified beam. The device envisaged was enormous—the crossed electric (600 kV) and magnetic fields were to be applied over 45 metres between electrodes 10 cm apart. The report stressed its complexity and the problems its construction would pose. To achieve electric and magnetic fields homogeneous 'to at least 0.1%' there had to be 'very tight tolerances' on the dimensions of the device. The electrodes had

Notes: p. 564

to be 'carefully shaped and polished'. The vacuum tank surrounding them had to be of 'very rigid construction' so as to keep the spacing between the plates accurate and constant 'to better than 0.1 mm'. And it was necessary 'to construct a short full-scale model' to study the combination of field strength and high vacuum at which electrical breakdown and sparking across the electrodes occurred.[71] Here was the major technical challenge in the entire beam transport system. There was no question of trying to copy something fairly standard; a major development project was envisaged.

The second point to note is that, as far as magnets were concerned, the beams designed were simple and the elements used were few: the high-energy meson beam, for example, was built up from one momentum analyzing magnet and three focussing triplets. Their total length was about 18 m and the particles simply drifted in a tube under low vacuum for much of the remaining 80-100 m. This was in stark contrast to the beams designed for the South hall just six months later, for example, which were far more complex. Whatever the reasons for these rudimentary beam layouts, their effect was to underestimate hopelessly the costs of an adequate beam transport system for a PS experimental programme.

Finally it should be noted that the proposal made in the May report was, to quote Adams, 'the first instalment of the general beam transport and analysing apparatus that will be needed for nuclear physics experiments around the PS'.[72] In fact, apart from the little beam designed by von Dardel and de Raad in 1957 as part of the exercise based on Ferretti's list of likely experiments, no other sustained attempt had been made to determine CERN's beam transport needs. In particular, nothing had yet been done about providing equipment for the South and North experimental halls, in which the first experiments with the PS would take place. There was perhaps no cause for alarm just yet: Hine had attended the Venice-Padua conference in September 1957 full of ideas and queries about possible beams, only to find that 'No one, not even Brookhaven (Collins) has gone any great way into these questions as affecting a 25 GeV machine, and only Berkeley has any definite views even at the 6 GeV level'.[73] But now in summer 1958, with the PS due to work in some 15 months time, and with delivery dates of up to 18 months or more on custom-made magnets, it was becoming urgent to place a first order for beam elements if one wanted to have adequate equipment available to exploit its potential.

9.3.5 'SELLING' THE PROGRAMME

Drawing up a bubble chamber programme was one thing; providing convincing arguments for the Council and its committees that the money should be found for the scheme was quite another. And although these bodies were generally keen to assist the scientists in developing the laboratory, circumstances required that particularly good arguments be prepared on this occasion.

First, there was the sheer scale of the project—two bubble chambers, beam transport equipment, and supporting facilities (buildings, generating plant, liquid-hydrogen plant, etc.). Preliminary estimates indicated that this programme would take up to five years to complete, that almost 80 staff would be involved in it, and that capital and staff costs would be over 20 MSF, considerably higher than the construction costs of the SC itself.[74]

Then no provision had been made in earlier budget estimates for a project on this scale. The long-term budget drafted by the CERN staff in September 1957 and covering the period up to 1960 talked of 'expanding' and 'speeding up' the PS experimental programme. The total equipment cost envisaged for this 'expansion' during 1959 and 1960 was some 6 MSF, which included a provision for bubble chambers and counters. Six months later 7-8 MSF were being requested for 1959 alone, and by September 1958 this figure had climbed to over 9 MSF just for bubble chambers.[75]

Finally, as we mentioned before, there was the growing pressure in the Council itself to limit expenditure. After a long and demanding debate the supreme body had agreed in December 1957 to couple the budgets for 1958 and 1959, and to set a ceiling on expenditure of 100 MSF for the two years. Of this 56 MSF were committed for 1958, leaving the balance of 44 MSF for 1959. Now, no sooner had the ceiling been set than Council members were being asked to go back to their domestic financial authorities for more money. This, as the British representative to the Finance Committee remarked, was 'very bad policy'.[76]

To overcome these obstacles the CERN senior staff deployed a number of strategies or persuasive devices to secure what they wanted. Such strategies are commonplace at the interface between the scientific community and its funding sources, even if they change according to the context. In the United States one may '[wave] the red flag to stimulate the appropriation of funds for research';[77] similarly in Europe one may speak of the American threat. However, apart from these general appeals to the need to remain competitive, more refined strategies may also be used. We select three from among those invoked to raise funds for the PS experimental programme in summer 1958: demanding more than one expects to get, appealing to scientific necessity, and claiming that the new project could not have been anticipated (the appeal to the unexpected).

9.3.5.1 Demanding more than one expects to get

At its ninth meeting on 23 May 1958 the CERN management presented the SPC with three alternative programmes for building experimental apparatus for the PS. To choose between them was to choose between different completion dates of three items of equipment—the 1 m propane chamber and beam transport material, facilities for large hydrogen bubble chambers, whether built by CERN or brought by visitors, and the 2 m hydrogen chamber.[78] Programme A, the maximum possible within the existing 1959 ceiling, involved completing the propane chamber in

1960-61, more or less in time for the PS, completing facilities for hydrogen chambers by 1961-62, and having the CERN 2 m chamber ready by 1963-64. Programme B involved advancing the completion date of the second item by one year (to 1960-61), and overspending the 1959 budget by 3.5-4.0 MSF. Programme C, the crash programme, foresaw completing all items as fast as possible. It meant having a propane chamber and hydrogen bubble chamber facilities available for the first work around the machine in 1960-61, and the CERN 2 m chamber ready soon thereafter in 1961-62. To finance it the 1959 ceiling would have to be broken to the tune of 7-8 MSF. The SPC was asked to accept programme C. It duly did so, recommending that the Finance Committee agree to the additional expenditure needed in 1959.

Whatever its merits, did anyone really expect the crash programme to be accepted? We think not, and that it was primarily an opening gambit in the negotiating process. We say this because, as we have seen, the decision to build a 2 m chamber at CERN was deliberately a decision to put time considerations second to the wish to build the biggest chamber that was technically feasible without recruiting new engineers and applied physicists. Since building the propane chamber and the supporting facilities for the hydrogen chambers was likely to absorb all the available staff coming free inside the PS division in 1960, it is hard to see how anyone among the CERN senior staff seriously expected the 2 m chamber to be ready by '1961-62'.[79] Indeed, no sooner had the SPC agreed to recommend the crash programme than Adams and Bernardini, among others, suggested that 'a delay of the CERN [hydrogen] bubble chamber could be accepted as a compromise if the British chamber came to CERN in 1961-62 and remained there for two years'—precisely the plan that had been agreed to among the senior staff a few months before.[80]

The SPC's recommendations were duly discussed at a meeting of the Finance Committee's Working Party on 3 June 1958. Caught between its respect for the SPC's views and the 100 MSF ceiling imposed by the Council, the Working Party agreed to the 'compromise', recommending something close to programme B on the clear understanding that facilities for using the British chamber would be ready on the site late in 1961, and that funds would be made available so that CERN's 2 m chamber could replace it around 1963-64.[81]

9.3.5.2 The appeal to scientific necessity

When presented with the proposed bubble chamber programme on 23 May 1958, the SPC concluded that 'CERN, from a scientific point of view, must have a 2 m bubble chamber at the earliest possible date'. There was also general agreement that *'from the scientific point of view* the committee must strongly recommend the full programme [...]' (our emphasis). It is of course the task of the SPC to evaluate the scientific interest of the proposals laid before it: there is nothing to surprise us in that. What is of interest is the rhetorical power of the imperatives, the implication that conclusions reached and recommendations made may be logically derived, as it

Defining big equipment

were, from scientific considerations, and have all the force of objective, incontrovertible truths.

There are two comments to be made on this strategy. Firstly, that scientifically speaking there was no one, necessary, course of action to follow in deciding the structure of CERN's bubble chamber programme. Different groups had different ideas about what was scientifically 'necessary' and put forward programmes which embodied correspondingly different sets of priorities. For several experimental physicists, scientific imperative called for a bubble chamber that was ready in time to do experimental work on the PS as soon as it produced its first low-intensity beam—and they proposed building a chamber of about 1 m. For some of the applied physicists and engineers it was 'necessary' to push to the technological limit—and they wanted to build a 2 m chamber. The SPC's proposal simply amounted to an endorsement of this latter way of thinking, and to the adoption of the priorities it represented. To imply that it was 'necessary' was to imbue it with a transcendence and absoluteness which it did not, in fact, have—and to shut out the disagreements over timing and chamber size which had so preoccupied the senior staff earlier in the year.

Secondly, our account has shown that both of these recommendations were informed by a number of non-scientific concerns—the need to keep staff and to sustain their morale, tactical considerations, and so on. More deeply, they reflected the determination among the senior staff to build up the organization into a quasi-autonomous laboratory in its own right, with its own complex of facilities, career structure, and research programme. It was as much CERN's scientists' needs as some abstract 'scientific' need that lay behind the SPC's recommendations.

9.3.5.3 The appeal to the unexpected

Granted that a financial limit had just been imposed for expenditure in 1958 and 1959, the request for funds to support a programme that would break the ceiling inevitably frustrated some Finance Committee members. As a Swedish delegate put it at one meeting, 'He was somewhat surprised... that large bubble chambers should not have been provided for in the estimates for 1958-59, in view of the interest shown in that apparatus for some time in the United States'.[82] The standard reply to this kind of question by the CERN senior staff appealed to the unexpected. It was 'the many developments which had occurred during the last few months in the field of experimental technique' that explained the need for the new programme. Additional halls would be needed to house hydrogen bubble chambers 'because the hazards connected with [their exploitation] had become apparent only fairly recently'. In short a big hydrogen bubble chamber project '*was not* and *could not have* been foreseen before' (our emphases).[83]

In a sense all of this is true. In 1957 the CERN senior staff had not yet decided to launch a big hydrogen bubble chamber project and had not yet made a technical

Notes: p. 565

study of its needs. Nor could they as long as they, like many others, were unsure as to whether the time and resources needed for such a scheme would be well spent. There were 'new developments' then, but primarily in the sense that, under the influence of Alvarez's 'bubble chamber philosophy', the pressure of time, and a new mood in the Council, it was decided in December 1957/January 1958 that a large hydrogen bubble chamber would be an essential piece of equipment. Until that decision had been taken the resource needs of the project 'were not' and perhaps even 'could not' have been foreseen by the CERN senior staff. The rhetorical value of the formulations used by the CERN management lies in their implying that *no one* could have anticipated the situation which was now causing difficulties for the Finance Committee — so exonerating the senior staff from any blame for it.

These three lines of argument — which will surprise no one — seem to have been effective, both at the interface between the scientists and the lay members of the Finance Committee and in the dealings of official CERN delegates with their governments. The willingness to 'retreat' from the crash programme to the 'compromise' led one FC member to remark that, 'far from being bold scientists unconcerned with financial problems', the members of the SPC had been acutely aware of the financial implications of their proposals. There should be 'no discussion' in the FC about their 'validity' — the 'experts' in the SPC were the best judges of what was scientifically necessary. Admittedly delegates had been put in a difficult position vis-à-vis their domestic financial authorities. But this could not be helped. The situation was unexpected and unexceptional.[84]

That granted one should be careful not to exaggerate the importance of these rhetorical 'tricks', to infer that the CERN management used them to *manipulate* the Finance Committee. By and large the members of the FC and the Council were as determined as the CERN management to develop the laboratory, and they essentially agreed that CERN launch a major hydrogen bubble project with its supporting facilities. They did not need to be persuaded of this. What they wanted rather were supporting arguments which they could use in their home countries, if need be, to convince their financial authorities to break a ceiling imposed only six months before. This they managed to do. At its twelfth session held on 3 December 1958 the Council agreed that CERN spend 55 MSF in 1959 — 11 MSF more than the figure laid down the year before. Over 9 MSF of this was earmarked for the PS experimental programme, the bulk of it for the bubble chamber project.[85]

9.4 Acquiring the material for the PS experimental programme

9.4.1 ORGANIZATIONAL ASPECTS[86]

By summer 1958 some of the most influential physicists and engineers at CERN had spent upwards of a year discussing the equipment, and particularly the detectors, that would be needed to exploit the PS. Now, with a rough idea of *what* was required, they turned their attention more closely to *who* would be responsible for providing it. Various discussions along these lines were held during August 1958 between SC Division leader Gentner, PS Division leader Adams, and the Director of Research (at the SC) Bernardini, as well as at two meetings of the Group of Directors. Early in September a paper was prepared for the SPC explaining a provisional set of 'organizational measures' which '[had] to be taken in order to proceed with the preparation of [the PS] experimental programme'.

The main organizational unit proposed was the detector. Five kinds of teams were envisaged: several counter teams, a propane bubble chamber team, a hydrogen bubble chamber team, a photographic emulsion team and a cloud chamber team. These teams were to be 'led by competent physicists, each with its own budget and a staff suitable for constructing the apparatus and carrying out the experiments'. The nuclei of the staff for these teams were essentially to be drawn from the PS and SC Divisions. There were already counter, emulsion, and cloud chamber groups working around the SC, and it was suggested that these remain the responsibility of that Division. Bubble chamber teams would be situated in the PS Division, where they could draw on in-house experience 'in carrying out big-scale projects'. To coordinate the work between these various teams and the two divisions it was proposed that an Executive Committee for the PS Experimental Programme be set up as soon as possible (it first met a few weeks later on 25 September 1958). The Committee was to be chaired by the Director-General and to be composed of almost all members of the senior staff: the PS and SC Division leaders and their closest assistants, leaders or representatives of the counter teams and the propane and hydrogen bubble chamber teams, and engineers and physicists involved in providing the infrastructural support for the programme. The role of the Executive Committee was primarily to co-ordinate the material needs of the experimental programme; it was foreseen that 'broader policies' on its scientific aspects 'be discussed in another interdivisional body set up by the Director-General'. In fact no committee of this kind was established until 1960.[87]

Whose names were put forward as team leaders? Von Dardel for one of the counter teams (it was envisaged to form others), Gibson for the emulsion team and Ballario for the cloud chamber team—all of them already responsible for teams of this kind at the SC. Ramm, the leader, assisted by Resegotti, de Raad and C. Germain (all members of the PS Division) were to form the nucleus of the propane chamber team. No leader of the 2 m hydrogen chamber team was explicitly identified. Peyrou's group, it was noted, was still engaged in building the 30 cm chamber, and the paper simply suggested that 'The PS Division [would] see with Peyrou how best to make an immediate start with the 2 m. chamber and provide such assistance with the 30 cm. chamber as may become advisable'.

Notes: p. 565

What were to be the precise responsibilities of the teams as far as equipment was concerned? A distinction was drawn here between apparatus specific to the team, notably its detector, and apparatus common to all teams, like targets, beam transport systems, shielding blocks, electrical supplies, cooling water, and other services. Each team was responsible for the construction (or assembly) of its own detector. As for the common equipment, here its task was 'to work out the requirements of its experiment' and to submit its findings to the Executive Committee. 'The design and ordering of the actual apparatus', the paper suggested, 'could probably best be done by the PS engineers'. The implementation of this arrangement, particularly as concerns the acquisition of beam transport material, is the topic of our next section.

9.4.2 THE FIRST ORDER OF BEAM TRANSPORT MAGNETS

By mid-November 1958 two groups, the CERN propane chamber group led by Colin Ramm, and Guy von Dardel's counter group, had submitted worked out proposals for beams in the South hall. The needs of nuclear emulsion physicists were explored 'in an informal and preliminary manner' at an unofficial meeting held on 20 and 21 November 1958, and attended by CERN staff and 15 outside users. 50 emulsionists met again at CERN in mid-January 1959 to specify their needs more precisely.[88]

Ramm's group proposed five beams which effectively filled the South hall. The design of three of these was more or less frozen: an elastically-scattered high-energy proton beam, which could easily be converted into 'a rather pure beam of π^--mesons' (beam 1), a neutral beam containing neutrons and γ-rays up to 25 GeV/c (beam 2), and a separated beam about 100 m long—the maximum possible within the confines of the South hall—producing antiprotons of top momentum 2.5 GeV/c. The purification of the beam was to be achieved in three successive elecrostatic separator tanks whose details were as yet unspecified. The design of beams 4 and 5 was more tentative and was said to be subject to modification in the light of further discussions. The first was a separated K-meson beam. The beam was to be purified by passing a mixture of kaons and π^--mesons through a 'degrader' (essentially a material absorber) in which they lost different amounts of momenta. The pions were then eliminated using a bending magnet, leaving charged kaons of 500 MeV. The intensity of this beam was rather unsatisfactory—about 90% of the kaons decayed in the channel—and steps were under way to improve its design. Beam 5 was a mixture of the main charged particles produced at targets in the machine, with a maximum momentum of 6 GeV/c. Its detailed layout was only to be settled after 'further discussions with counter groups'.

Von Dardel's counter group designed one beam for the South hall: a 'well-collimated and analyzed' beam of antiprotons produced either directly at the target or from antihyperon decays in the vicinity of the target. The particles varied in momentum from about 1-4 GeV/c; they were detected by plastic scintillator and gas Cerenkov counters. Von Dardel 'justified' having a second antiproton beam by pointing to its higher energy and suggesting that it could be quite useful for bubble chamber experiments, particularly if counters were used to localize the regions of the bubble chamber where the antiprotons entered it.

The emulsion physicists, who tended to work 'parasitically' on existing beams, took Ramm's proposals as a point of reference, and tried to visualize the kinds of experiments they would like to do with them. Their only quibble was with the kaon beam. They complained that the energy of the planned beam (500 MeV/c) was too low and, pointing out that Berkeley was trying to build a 1.1 GeV/c beam, remarked that they would want 'to use a momentum-analyzed 2-6 GeV/c beam containing K-mesons as soon as possible'.[89]

These proposals were considered by the Executive Committee for the PS Experimental Programme at its meetings on 25 November and 9 December 1958. On 13 January 1959 John Adams mentioned that a list of magnets based on these deliberations 'had been prepared and was being discussed internally within PS from the point of view of generators and design'. The proposed list was placed before the Executive Committee at its meeting on 10 February. It recommended that CERN order a total of 50 magnets: 37 for Ramm's beams as they stood, seven for von Dardel's beam, and six spares. The Committee, however, decided to leave aside the six spare magnets for the time being, and at its meeting on 25 May 1959 the Finance Committee authorized CERN to place an order for 18 H-shaped bending magnets and 26 circular quadrupole lenses with the Swiss firm Oerlikon.[90]

The most notable feature of this first order for magnets is the decision to standardize on just two lengths for both the magnets and the lenses (they were all to be either 1 m or 2 m long) and on one aperture size (14 cm gap for the bending magnets and 20 cm diameter for the quadrupoles). CERN's first order for magnets differed substantially in this respect from that placed at Brookhaven for the AGS (see table 9.1). There, as Hine reported after a visit to the USA in April 1959, 'the magnets proposed as standard [were] more varied than the first CERN standard set, and [tended] to be rather shorter, higher in field levels, and [...] [to] include some wide aperture quadrupoles', the last because the beam designers were concerned that the intensity of the primary beam would be far lower than that of the Cosmotron and they wanted 'to push for the highest collection and transport efficiencies'.[91]

The decision at CERN to standardize and to simplify the first order for beam transport magnets seems to have been shaped by three main considerations. Firstly, there was an element of cautiousness, a reluctance to overcommit resources on costly items of equipment when future needs were difficult to predict. As Ramm explained in introducing his beam layout, 'the transport channels are inherently expensive and time consuming in construction and

Table 9.1
The first order for magnets placed by CERN and BNL, being a comparison of the equipment available in each laboratory around September 1961 for use with their strong-focusing machines.

	CERN	Brookhaven
Bending magnets	5 of 14 cm gap × 1 m	4 of 15 × 45 cm × 90 cm
	13 of 14 cm gap × 2 m	5 of 15 × 45 cm × 181 cm
		3 of 15 × 75 cm × 181 cm
Quadrupole lenses	18 of 20 cm × 1 m	5 of 20 cm × 120 cm
	8 of 20 cm × 2 m	4 of 30 cm × 75 cm
		6 of 30 cm × 150 cm
		6 of 15 × 60 cm × 90 cm
Totals	18BMs, 26QLs	12BMs, 21QLs

Notes

1. We have assumed 1 inch = 2.5 cm;
2. Data for CERN from CERN/FC/357, 15/5/59; data for BNL from F. Grütter's report, (see note 92). The fourth type of lens at BNL was a rectangular quadrupole designed by Panofsky; in his report (see note 91) Schnell gave their length as 90 cm and specified that they were for use with beam separators.

adjustment, and it is therefore of fundamental importance that a system designed at the present time must be of considerable experimental use some years later'.[93] One way of dealing with this uncertainty was to place a bulk order for a number of relatively inexpensive and easy-to-build magnets of 'general' use, adding special magnets as and when the need arose. Hence Ramm's suggestion that 'only a minimum order for magnets should be placed now', and the Executive Committee's decision to reduce the list of 50 magnets to the 44 already foreseen: the remaining six 'spare' magnets, its minutes tell us, 'would probably be replaced by special magnets needed by the experimental groups'.[94]

This cautiousness was probably inspired, at least partly, by the prevailing financial climate at CERN. For it was around this time, spring 1959, that the British were insisting, not simply that there be a ceiling on CERN's expenditure, but that the laboratory operate with a level budget for the next three years. It was the first time that the UK delegation had interpreted its ceiling policy in this way, and it doubtless had important repercussions on the spending habits of the CERN staff.[95]

The composition of the first order for magnets cannot simply be put down to cautiousness in the face of financial constraint, however. The decision also reflected — and this is our third point — the power of the PS group, and of Ramm (supported by Adams) in particular, to impose their view of what was needed on the experimental physicists in the other groups. Von Dardel's beam as originally designed had one of Ramm's standard bending magnets, three special magnets, and one C-shaped and one E-shaped magnet. As for quadrupoles, it had two 'standard' lenses 1 m long and two lenses 1.5 m long.[96] This diversity was suppressed. After sustained pressure from Adams to make lists of magnets (and their power supplies) 'of the same type' as those proposed by Ramm's group, von Dardel changed his request to '3 standard type BB bending magnets [1 m] and four standard type QB quadrupoles [1 m]'.[97] It was these that were included in the original order for 44 magnets. The emulsion physicists were even less fortunate. After their meeting on 20 and 21 November 1958 Gibson asked the Executive Committee that, over and above the standard magnets for Ramm's and von Dardel's (modified) beams, 'a pool should be available in addition'. He also recommended that the magnets and lenses for the 'high-energy' kaon beam that the emulsionists had asked for be added to the original list.[98] His requests were ignored. So too were suggestions for special magnets made by visitors, or after visits to see what was being done abroad, e.g. suggestions that CERN acquire spare C-shaped magnets (Yuan), very high field magnets (Lofgren), or rectangular aperture 'Panofsky-type' lenses (Hine, after his visit to Brookhaven in April 1959).[99] In short Adams and Ramm effectively decided that the first order for magnets would contain the minimum number of standard items, as defined by them, and stubbornly refused to add more or different magnets to what they felt to be adequate.

9.5 The beam transport crisis in 1961 and its causes

We now move forward in time. We have studied the decisions taken at CERN to launch a big hydrogen bubble chamber project (in spring 1958) and to buy a first set of standard beam bending magnets and quadrupole lenses (in spring 1959). Our attention is now directed to the implications of the timing and nature of those decisions on the implementation of the PS experimental programme, and to the claim that, because essential equipment was not available when it was needed, European physicists could not properly exploit the potential of the most powerful accelerator in the world, nor capitalize on the competitive edge which they initially had over their rivals at Brookhaven. As the lack of beam transport material caused the most concern we shall devote the bulk of this section to that aspect of the problem, and say relatively little about the commissioning of the CERN bubble chambers.

Notes: p. 565

9.5.1 A BRIEF CHRONOLOGY

When the CERN proton synchrotron first reached its design energy in November 1959, there was very little equipment available to do experiments with it: as Cocconi put it, 'only three small deflecting magnets, no quadrupoles and no separators'.[100] On the other hand, at this stage of its development no one was yet in a position to do serious physics. For one thing the machine first had to be 'run in', 'debugged' (see chapter 2, section 2.1), and the technical staff who were going to be responsible for its routine maintenance had to be trained. As a result the bulk of the machine time during the initial six or nine months after its birth was taken up in tuning the accelerator and in becoming familiar with its idiosyncracies. For another, when physicists were able to get their hands on the PS their main task was beam surveying. This was the first time that anyone had bombarded targets with protons of as much as 25 GeV, and the energy spectra and angular distributions of the particles produced in these collisions had to be established before laying out beams and starting research.

By August 1960 the situation had changed. There was more machine time available for the experimenters (about 20 days out of 60 during July and August). The intensity of the primary beam had been increased from 10^{10} circulating protons in November 1959 to 2×10^{11}. It had become possible to operate two targets simultaneously so allowing bubble chamber and counter experiments to be run in parallel. The bubble chamber physicists had been meeting regularly for over six months to plan their experimental programme about the PS. The CERN 30 cm hydrogen bubble chamber had taken its first set of photographs in a high-energy pion beam and the 1 m Ecole Polytechnique propane chamber was at CERN and ready to do its first runs. As for beam equipment, deflecting magnets, quadrupoles, and their generators were just beginning to arrive and to be installed. And the first hints that the lack of beam equipment was imposing restrictions on the experimental programme began to surface in public. Cocconi, who had unsuccessfully asked the Executive Committee to increase the first order for beam transport equipment in October 1959, implied as much in his 'Progress Report on Work With the 25 GeV Proton Synchrotron' which he presented to the 1960 Rochester conference held from 25 August–1 September.[101] A fortnight later the problem was explicitly acknowledged by Mervyn Hine at a conference on instrumentation for high-energy physics held at Berkeley. 'Until now' said Hine, 'experimental work with the CPS [CERN PS] has been limited by lack of equipment, such as bending magnets and separators, and by the amount of time it has been possible to run the machine'. Indeed, to judge from a paper presented by Ramm at the same conference, the runs with the propane chamber were apparently being made with a beam which was deflected away from the main ring by two bending magnets placed inside the shielding around the PS, and which then drifted clear across the South experimental hall into the detector without passing through any additional elements.[102]

Even as Hine was commenting on the lack of separators at CERN, a temporary solution to the problem was being provided by Marcello Cresti of Padua University. Writing to John Adams from Berkeley on 1 September 1960, Cresti explained that his institute was building an electrostatic separator based on the Berkeley design. It comprised two tanks each 3.5 m long and was expected to give 'good separation [...] for K's and antiprotons up to momenta of the order of 1 GeV/c'. Cresti offered to bring his separator to CERN as soon as it was ready, and to leave it there for 'as long as it [was] useful'. Adams' reply was quick and positive: 'We have no doubt', he wrote to Cresti on 20 September, 'that it would be a great advantage to the experimental programme of the CPS if your separator came to CERN as early as possible'. Separators were essential beam equipment for bubble chamber physics, and the 10 m tanks being built under Ramm's supervision were far from ready.[103]

In October an order for nine more bending magnets and twelve quadrupole lenses was placed with Oerlikon, the supplier of the first set. The lengths remained fixed at the 'standard' agreed on the year before: the order comprised five 2 m and four 1 m bending magnets, and eight 2 m and four 1 m quadrupoles. The elements were for use in the East area, which was due to be opened up in summer 1962, and which was intended to house first the big British hydrogen bubble chamber and later CERN's 2 m hydrogen chamber.[104]

By February 1961 all of the bending magnets and quadrupoles from the first order had been delivered. Cresti's separator had also arrived in Geneva, somewhat later than he had hoped, and tests on the two tanks were 'just about to begin'.[105] As for CERN's own electrostatic separators, it was reported at the end of 1960 that the detailed design of their elements had continued, that assembly of the first tank would commence in February 1961, and that it was 'hoped to install the three tanks and associated components for a separated beam in June 1961'.[106] In spring 1961 then, and about 18 months after the PS first worked, it was possible to build four or five fairly standard secondary beams at CERN. On the other hand no separated beams could be built.

On 29 April 1961 the beam situation in the laboratory was the subject of a long and anguished debate at the nineteenth meeting of the SPC. The discussion was triggered by a paper prepared for the occasion by Gilberto Bernardini, the Directorate Member for Research. It had come out again and again in meetings of the Directorate and in committees, said Bernardini, that the beam situation at the PS was 'very unsatisfactory'. 'The quality and intensity of all external beams [had] been severely limited for a long time by the scarcity of magnets, and in particular, by the lack of particle separators', he pointed out. In consequence, Bernardini went on to say, 'a large proportion of possible experiments could not be done owing to the lack of beams'. This, he added, was the main cause 'for the lack of a good programme on strange particle physics'.[107]

All the physicists who spoke in the ensuing debate echoed these sentiments. Amaldi, while enthusiastic about the low-energy antiproton beam being built with

Notes: pp. 565 ff.

Cresti's separator, was distressed by the fact that CERN could not build two 'clean nice beams at the moment' and that no work on strange particles had yet been possible. Gregory remarked that 'it had become clear from discussions in the Track Chamber Committee that the beam problem was of paramount importance'. The only pure beams available to date were of high-energy protons and pions. The shortage of equipment had forced the TCC to restrict proposals for experiments to be run at end 1961/beginning 1962 to the use of a fast antiproton beam in the South hall and of a relatively low energy (0.6 to 2 GeV/c) kaon beam in the North hall, which was due to be opened up shortly. Even then, Gregory added, the TCC realized that 'its requests were reaching the limit of possibilities since they would involve using half the bending magnets and quadrupoles available at CERN for bubble chamber work and greatly reduce the number of magnets and quadrupoles used by counter groups up to date'. Salvini drew attention to the urgent need for an extracted proton beam (as he colourfully explained, rather than getting milk by going to the cow, the milk could be distributed to various consumers). Perrin emphasized that the shortage of beam transport was a 'very crucial problem for CERN'.[108]

The CERN management, notably Directorate Member Mervyn Hine and Director-General John Adams, while stressing that CERN actually had quite a lot of beam transport magnets, and that more were on order, did not deny that the problem was a real one. Hine remarked that CERN did not have enough conventional beam equipment or enough small separators like Cresti's 'to enable us to have any feeling of freedom and elbow room in constructing an experimental programme'. It was clear, he added, that 'for the next year every time anyone thinks of an experiment the first question he will have to ask is can I find the necessary magnets for it'. The debate ended with the DG being asked to prepare a report on the status of beams for the next meeting.

The first bubble chamber results using Cresti's separators were obtained in a low-energy (below 1 GeV/c) antiproton beam in June 1961. In mid-July Mervyn Hine prepared a paper for the SPC describing the beams and beam transport situation for the PS for 1961-1963. And although by now it was clear that Ramm's 10 m tanks were not going to be ready before the end of the year, and far later than had been thought, it looked as though the magnet 'shortage' was not that severe. In his paper Hine explained that, towards the end of 1961, the low-energy antiproton beam in the South hall would be improved using the 10 m separator tanks, and that the number of simple beams in the hall would be reduced. The magnets thus released would then be used to set up two K-meson beams in the North hall, which would by then be ready for physics. One of the K-beams would use a Ramm separator, the other would be a medium-intensity unseparated beam for several planned counter experiments. 'The existing magnets and generators and the three separator tanks now being assembled and tested', Hine added, 'seem to be just sufficient for the beginning of the programme, even after the lenses borrowed from Saclay [eight quadrupoles-JK] have been returned [...]', though 'the layout and efficiency of some beams would be improved if we had more lenses less than 1 m long, and also separators intermediate between the 3 m Padua tanks and the new 10 m tanks'. As for 1962, Hine noted that the East area would be opening up, and that the second order of 21 magnets was due to be delivered during the first six months of the year.[109]

In line with these findings Hine instructed the Engineering Division to look into the possibility of having 'classical' electrostatic separators of about 5 m in length entirely manufactured by outside firms. By August a set of parameters for such devices had been drawn up and two firms had been asked to submit offers. Contact was also made with a third firm which had developed a 564 cm long separator called 'Anatole' for Saclay.[110]

By the end of the year the first three 2 m long bending magnets ordered for the East area in October 1960 had been received; the rest of the magnets were to be delivered by May 1962. A third order for 55 beam transport magnets was also placed in December 1961. 39 of them were the CERN standard 1 m and 2 m length bending magnets (9) and quadrupole lenses (30). The remaining 16 were 0.5 m long quadrupole lenses. In December 1961 too the first of the three 10 m long separator tanks was installed in a 3 GeV/c antiproton beam in the South hall.[111]

In March 1962 Hine again surveyed the separator situation at CERN. The 10 m tanks, he said, were 'proving rather satisfactory in operation', and were quite suitable for high-energy beams. Again he repeated that a case could be made for having shorter units as well, both for use in lower energy beams, especially those calling for two stages of separation, and in order to increase the ease of construction of beams. Hine indicated that something like 5-6 m long separators were 'reasonable in many cases, except for low-energy K beams where about 3 m would be better'. Discussing the initiatives taken by the Engineering Division to purchase 'rather conservatively rated separators from outside firms', Hine noted that C. Germain, who had been responsible for the 10 m tanks, had also recently proposed that CERN build a 3 m long module using many of the components developed for the big separators. These would certainly perform better than 'Anatole', and would cost less, but would require more CERN input and a slightly longer delay — 9-10 months quoted for delivery of the commercially made tanks, 12 months estimated for building a CERN tank 'if adequate drawing office and assembly effort [were] available'. In the event, 'after some confusion as to what types really were needed', it emerged that there was really no longer an urgent need for 'short' separators. In March 1962 it was decided to build two more 10 m units of the existing kind for the East area, and to 'start to make' additional 3 m and 6 m long units in-house using the same components. The former were to be ready early in 1963, and the latter were to follow after that.[112]

In June 1962 Director-General Weisskopf once more addressed the Council on the question of the 'equipment gap', this time emphasizing that it was closing. The new 10 m separators, he said, had turned out to be 'extremely good in quality', and the 3-4 GeV/c separated antiproton beam that CERN had built with them had the

Notes: p. 566

540   *PS experimental programme*

'highest energy at present of any antiproton beam in the world'. Another 'novelty' for CERN, Weisskopf added, was the opening up of two new beams of K-mesons, one a separated beam which also 'made it possible to exploit CERN in a better way'.[113] The 'beam transport crisis' was over.

9.5.2 THE ROOTS OF THE BEAM TRANSPORT PROBLEM

To analyze the roots of the beam transport crisis we need to distinguish carefully between the difficulties created by shortages of magnets and those arising from the fact that the 10 m electrostatic separator tanks were only available from December 1961 onwards. In our view the first problem was far less important than the second. In other words, we believe that, insofar as there was a beam transport 'crisis' at CERN in the early 1960s, it was due to fact that a good, separated high-energy beam could not be built until the very end of 1961.

In the previous section we identified several periods when experimental physicists felt that CERN could have done with more magnets. And while their complaints were doubtless justified, their importance should not be exaggerated. There certainly was very little beam equipment available during the first six months after the PS first worked. But one must not forget that serious experimenting was not yet possible, and that the set-ups needed for beam surveying were rather rudimentary—a few magnets feeding particles to a simple counter set-up may suffice.[114] And although CERN was still short of magnets in September 1960, the problem was soon resolved—by November, when physicists could use the PS for 75% of the time, the bulk of the equipment had been delivered and there was little free space left in the South hall. Indeed in a report prepared for the Management Board in February 1961, Hine remarked that he had found that 'on the average, 4 (nearly) independent beams exist[ed] at any one time, using some 25 magnets' in the South area—far less than the 44 which had been delivered—, adding that 'Parkinson's law [...] ensured that all magnets available [were] nominally in use'.[115]

As for the vociferous complaints at the SPC meeting in April 1961, the main problem here as far as magnets were concerned seems to have been, not the situation prevailing at the time, but the one that would arise when the North hall was opened up for experiments later in the year. This was precisely the problem voiced by Gregory and reiterated by Hine. But again one must be careful. As we have seen, by suitably redistributing elements (and remember that the beam for the top-priority neutrino experiment required no magnets) one could build one outstanding separated $\bar{p}$ beam in the South hall and a reasonable separated K-beam in the North hall with the magnets one had—but only after the 10 m separators were available. Which is why in July 1961 Hine could consider returning the lenses loaned to CERN by Saclay before taking delivery of the second order of magnets.

We have touched the core of the problem here. The shortages of magnets that arose at CERN in 1960 and 1961 were undoubtedly a nuisance, and could have been averted if the first order had been more varied, particularly by including shorter lenses, and if a second order had been placed with the needs of the North hall particularly in mind. On the other hand we must not forget that these 'mismatches' between supply and demand were, firstly, only of a temporary nature—they lasted perhaps three months at most. Nor should we ignore that to some extent they were only to be expected—Alvarez's results with his 72 inch hydrogen bubble chamber

Table 9.2
Beams and detectors used to discover some of the most important resonances with accelerators at the end of the 50s and early 60s[a].

Date[b]	Particle	Lab.[c]	Detector[d]	Particle	Beam Momentum (GeV/c)	No. of Seps.
09/02/59	Ξ^0	LBL	15 in HBC	K^-	1.15	2
31/10/60	Y^*	LBL	15 in HBC	K^-	1.15	2
16/02/61	K^*	LBL	15 in HBC	K^-	1.15	2
12/05/61	ϱ meson	BNL[e]	20 in HBC	π^+	0.9, 1.09, 1.26	1
11/05/61	ϱ meson	BNL[e]	14 in HBC	π^-	1.89	?
12/05/61	Y_0^*	LBL	15 in HBC	K^-	1.15	2
31/05/61	Y_0^*	LBL	15 in HBC	K^-	1.15	2
14/08/61	ω^0 meson	LBL	72 in HBC	$\bar{p}$	1.61	3
15/09/61	ω^0 meson	LBL	72 in HBC	$\bar{p}$	1.61	3
10/11/61	η^0 meson	LBL	72 in HBC	π^+	1.23	?
02/01/62	η^0 meson	LBL	15 in HBC	K^-	0.76, 0.85	2(?)
19/02/62	$\bar{\Xi}$	CERN	81 cm HBC	$\bar{p}$	3.0	1
19/02/62	Ξ	BNL	20 in HBC	$\bar{p}$	3.3	2
27/06/62	Ξ^*	LBL	72 in HBC	K^-	1.8	2
02/07/62	Ξ^*	BNL	20 in HBC	K^-	2.24, 2.5	2
22/08/62	f^0 meson	BNL	20 in HBC	π^-	3.0	2
02/11/62	f^0 meson	CERN	1 m HLBC	π^-	6.1	0(?)

[a] This list has been compiled from the chronology given by Six and Artru[116], which contains '*all* resonances found in the years 1960–1961, but only some typical ones in the following years'.

[b] The date is the date on which the paper in which the finding was published was received by the journal—see references given in Six and Artru.

[c] LBL: Bevatron, Lawrence Berkeley Laboratory;
BNL: AGS, Brookhaven National Laboratory (but see [e]);
CERN: CPS, CERN.

[d] HBC: Hydrogen Bubble Chamber;
HLBC: Heavy Liquid Bubble Chamber.

[e] These two results were obtained using the external proton beam of the Cosmotron.

Notes: p. 566

from 1959 onwards completely transformed experimental practice, calling for far more sophisticated beams than anyone had built before. Finally, these magnet shortages were of marginal significance — marginal in the sense that, no matter how many magnets they had, if physicists did not also have good separators they could not build the kinds of beams that they were crying out for in 1961/62, the kinds of beams that first Berkeley and then Brookhaven were using to discover one resonance after the other in the early 1960s (see table 9.2).

Why did CERN not have separators of its own in 1960 and 1961, and have to fall back on Cresti's tanks which, though invaluable, were too short to enable physicists to exploit the full potential of the PS at 25 GeV? The roots of the answer are to be found in the context in which the decision to build separators was taken, the framework in which CERN launched a big hydrogen bubble chamber programme in 1958 (see section 9.3).

This framework was marked by three features. Firstly, and consistent with Alvarez's arguments, it was assumed that bubble chambers at the high-energy machine would be used with high-energy secondary beams.[117] To achieve 'good' separation at high beam momenta the electrostatic separators had to be correspondingly long. Secondly, it was assumed at the time that there was no great urgency to build them. In summer 1958 it was thought that the separators would be used in what became known as the East area, an area set aside for using the big British and CERN chambers. Since it was known that the British chamber was only due to arrive during 1961 at best, it looked as though CERN had three years in which to get its separators built and working. This was reinforced by the prevailing attitudes among those promoting CERN's hydrogen bubble chamber programme, notably John Adams. For Adams, as we have seen, there were distinct advantages in pushing technology to its limits, even if that meant being 'late'. Finally, it was assumed that, since separators were pieces of heavy equipment, engineers would build them. In particular it was understood that the engineers 'coming free' from Adams' PS group would find here a challenge sufficiently rewarding to keep them at CERN once the big machine worked.

It was in this context that a group at CERN led by C. Germain embarked on what amounted to a research and development project. Setting aside the need to have something that 'worked' ready 'on time', they chose instead to build a separator which was at once sophisticated and flexible, and that would ensure CERN's competitiveness in the medium term. Just how complex the resulting device was can be seen from a comparison of the separators used at BNL and CERN early in 1962. The vacuum chamber in the separators in the Brookhaven 3.3 GeV/c antiproton beam which was used to discover the antihyperon in December 1961/January 1962 (see table 9.2, item Ξ) was a little under 1 m in diameter and 5 m in length. The gap between the separating plates was 5 cm and the maximum operating potential across them (i.e. before sparking) was about 400 kV.[118] The separator used in the parallel experiment at CERN (one of the three built by Ramm's group) was far bigger and more intricate. Its vacuum tank was 1.4 m in diameter and 10 m long, the gap between the electrodes was variable between 6 cm and 26 cm, and it could be 'made to fit a diverging or converging beam, or a curved mean trajectory in the case of low energy particles'. Provision was made to place one megavolt across its electrodes at large spacings.[119] Around the end of 1960 it was hoped to start installing these tanks in separated beams in June 1961. In the event the first was only commissioned in December 1961 after months of 'difficult and painstaking' work – and by June 1962 Weisskopf could announce that CERN had built the highest energy antiproton beam in the world. Its separators had turned out to be 'extremely good in quality', but at a cost: they were ready 'late'.

9.5.3 A FEW REMARKS ON THE CERN BUBBLE CHAMBERS

To round off our study of the acquisition of big equipment by CERN for the PS experimental programme we would like to squeeze in some observations on the timing of the arrival of its two big bubble chambers, the 1 m propane chamber and the 2 m hydrogen chamber.

The heavy liquid chamber was built more or less on schedule. Of standard design, it underwent its first operational trials in December 1960, about five months after a similar chamber built by the Ecole Polytechnique had made its first exploratory run at the CERN PS. It was placed in the neutrino beam in 1961, and took 150,000 pictures in preliminary runs between May and July that year.[120] On the other hand, the CERN 2 m hydrogen chamber was 'late'. It was first cooled down on 8 December 1964, and it gave its first tracks in the early hours of 13 December 1964. By contrast Brookhaven's slightly bigger 80 inch chamber operated successfully for the first time on 2 June 1963. And what with the British 1.5 m chamber only taking its first photographs in June 1964, it can safely be said that, but for the 81 cm hydrogen chamber loaned by Saclay, and installed at the PS early in 1961, CERN's, and Europe's, entry into the high-energy accelerator field would have got off to a very shaky start.[121]

How are we to explain CERN's 'lateness' compared to Brookhaven? The obvious answer is that it took much longer to build the big hydrogen chamber. A set of main parameters for a CERN 2 m chamber was formulated in spring 1958, and the project was accepted by the Finance Committee in June that year. The decision to build a large chamber at BNL was taken early in 1959.[122] Thus the time between conception and completion was some six to seven years in the case of the CERN chamber and about four to five years at Brookhaven. BNL seems to have started later and finished earlier.

There were two main reasons for this difference. Firstly, it took quite some time to actually get started on the CERN chamber: by March 1959 no project leader had yet been found to take charge of the 2 m, and work on it was only going ahead

Notes: p. 566

'slowly'.[123] This was partly because in the early stages of the project Peyrou and his group were more interested in completing the 30 cm chamber and doing some physics with it than in launching out into a new venture. It was also because the 10 cm and 30 cm were 'university-scale' instruments, which had been constructed by a few physicists and technicians on site; the 2 m was a piece of heavy equipment requiring the collaboration of engineers and of industry. Peyrou and his colleagues had to learn to deal with both.

The second reason for the relatively slow rate of construction of the CERN 2 m chamber was lack of staff, and of resources in general, at least until 1962. With the reorganization of CERN in 1961 the 2 m chamber simply became one of the activities of Peyrou's Track Chamber (TC) Division, which had global responsibility for the hydrogen bubble chamber programme at CERN (scientific research with CERN's 30 cm chamber, construction of the 2 m chamber, liason with French and British visiting groups, etc.). TC had a total staff of 54 people at the end of 1961. By contrast, early in 1962, there were 68 people 'actively working' just on BNL's 80 inch chamber.[124]

This relative lack of staff was the consequence of a deliberate policy decision to give the 2 m project a relatively low priority: CERN, it was felt, could manage first with Saclay's 81 cm chamber, and later with the British 1.5 m chamber. This go-slow policy was seriously reconsidered in 1962. In May that year Weisskopf pointed out that it was essential that CERN's 2 m chamber be completed at the end of 1964, adding that 'it seems today that this date can be met only if the 2 m project gets all necessary support from now on. It is no longer possible', the DG went on, 'to slow down this project on the grounds that the British chamber is replacing it in the near future'. Weisskopf reinforced this shift in priorities in a key speech to the Council the next month: speaking of the 'de-emphasis of the 2 m bubble chamber in the last two years', he added that 'we intend now to push it in order to catch up again'.[125] With this change in policy the accelerated progress of the 2 m was assured, and its completion by the end of 1964 was made possible.

We want to make one last point before closing this major section of the chapter. It will doubtless have occurred to some readers that one main reason why the 2 m chamber was slowed down was lack of money, that it was a victim of the 'restrictive' budgetary climate prevailing in the early 1960s. In fact this was the argument used by Weisskopf to the Council in 1962: 'The construction of our own very large bubble chamber is going ahead slowly', he said, 'slowly again because it was pushed back many times for budgetary reasons'. The same argument was used to 'explain' the difficulties CERN was having in building good beams in 1961: 'the shortage of beam transport', Perrin told the SPC in April that year, 'was clearly due to the budget limitations imposed by the Member States'.[126] We are extremely sceptical about this argument, and regard it as being primarily a bargaining counter, quite legitimate of course as a tactic for raising money, but generally not worth pursuing in depth. And we say this because the argument is in fact tautologous. Any laboratory, or at least any active and lively laboratory, which is not given as much money as it wants will have to set priorities and in doing so will inevitably decide to promote some projects at the expense of others. As Weisskopf himself put it in May 1962, 'Whatever the budget the Council will decide for CERN, there will be more projects planned than the budget can bear'.[127] That granted it is always open to the management to say that something or other has not been done 'for budgetary reasons': the statement is necessarily true, and simply means 'we have made this a relatively low priority'. The important analytical question then becomes: Given that the management was expected to make choices within a certain budgetary envelope, why did it make the choices that it did? We believe that the reader will find above the main elements needed to answer this question as far as the provision of beam transport material and of the 2 m hydrogen bubble chamber are concerned.

9.6 Data-handling facilities for track-chamber pictures[128]

We now turn to study the provision of equipment needed to handle CERN's share of the photographs produced by track chambers. From the start it was recognized by people like Lew Kowarski (the head of the STS division in the 1950s) and Yves Goldschmidt-Clermont (who worked closely with him) that computers would be an essential part of this process. The main data-handling facilities thus came down to instruments for converting selected information from track-chamber photographs into machine-readable form, and computers of sufficient capacity to deal with the increasing number of events of interest which had to be analyzed. The organization of this section of our chapter mirrors this double aspect. After the usual brief 'technical' introduction, we study first the instruments developed at CERN for measuring track-chamber film and then the steps taken in the laboratory to acquire sufficiently powerful computers. We shall see that, by and large, there were no major delays in this part of the PS experimental programme. This was partly due to the foresight of those who launched and maintained it, partly due to the relatively small size and cost of the measuring instruments developed for it, but perhaps above all due to the basic policy assumption which underpinned it: that CERN would only analyze a fraction of the track-chamber photographs produced on the site, the remainder being distributed to collaborating institutions. In consequence, and despite some hiccups, the supply of data-handling facilities at CERN managed to keep up with the demand throughout our period.

Notes: p. 567

9.6.1 BUBBLE-CHAMBER PICTURE ANALYSIS: A FEW BASIC CONCEPTS

A typical bubble chamber photograph comprises a large number and variety of tracks. Each track is a somewhat discontinuous string of bubbles generated by charged nuclear particles as they traverse the superheated liquid in the chamber. If the chamber is not immersed in a magnetic field most of the tracks are essentially straight, their path modified by scattering on the nuclei in the chamber. In the presence of a magnetic field the tracks of charged particles are curved into the familiar helices, again disturbed by scattering. A minority of tracks display distinctive features – a sudden change of direction or a number of 'rays' diverging from a vertex, for example. These indicate that the incoming particle has interacted in the chamber, creating products which can go on to react further. Such processes are characteristic of an event. For present purposes, let us define an event as being made up of one or more vertices from which individual particle tracks emanate.

In the late fifties – early sixties a reasonably large and sophisticated bubble chamber like the Alvarez group's 15 inch (38 cm) liquid hydrogen device produced about a million photographs a year when used on the Bevatron. Of these from 5-10% contained strange particle events, the kind of event for which bubble chambers were predominantly used at the time. Manual procedures for measuring and analyzing such events had been developed by cloud-chamber physicists but, as Hugh Bradner (LBL) pointed out in 1958, 'the analysis of only the [200] strange particles produced in 24 hours of bubble chamber operation could keep one physicist using the old analysis techniques, busy for almost a year'.[129] To process the 50-100,000 events of interest being churned out annually by one bubble chamber at the end of the 50s, and the mega-event experiments which were typical of the early 60s, more rapid, automated procedures were obviously needed. As a result, as Kowarski put it in September 1960, the evolution of data-handling technique in this area was 'towards the elimination of humans, function by function'.[130]

Six main processes were involved in the handling of track-chamber pictures. They were scanning (selecting the pictures containing the events one wanted from all those produced in a run), measuring (measuring the coordinates of the points on selected tracks of interest), geometrical reconstruction (reconstruction of the 'actual' tracks in three-dimensional space), kinematical analysis (interpreting the measured event using momentum and energy conservation), experimental analysis (e.g. performing statistical analyses on the results), and book-keeping (maintaining an up-to-date record of the photographs handled and the results obtained).

Two main instruments were developed at CERN during our period for automatically measuring bubble-chamber film and outputting the information in machine-readable form. The first, the so-called Iep (Instrument for the Evaluation of Photographs), was similar, though less automated, than the very advanced devices being built at the end of the 50s by scientists at the Lawrence Radiation Laboratory in Berkeley. The initial research and development for the second, the so-called HPD

(Hough-Powell Device), named after its inventors, was done at CERN itself, from where the idea rapidly spread across the Atlantic in the early 60s.

9.6.2 MEASURING INSTRUMENTS

9.6.2.1 *The Iep*

Towards the end of 1956 Yves Goldschmidt-Clermont, a Belgian physicist who had joined Kowarski's group in 1953, along with Guy von Dardel and Charles Peyrou, began to think seriously about the kind of device CERN might need for dealing with the output of track chambers. Early in 1957 Goldschmidt visited a number of major American installations to refine his ideas. Later that year he and the other physicists, along with F. Iselin, took the opportunity provided by a small European meeting on instrumentation to describe 'the present state of the instrument for the evaluation of photographs and future developments' at CERN.[131]

The heart of their measuring system was a standard, movable microscope platform (or stage) carrying a lens which was situated above the rolls of bubble-chamber film, and which projected an enlarged image of the tracks onto a screen. The stage was displaced along the track of interest. When the image of a point on the track fell on a cross-hair engraved on the screen its coordinates were digitized and automatically punched on computer input tape.

In the simplest version of an Iep the operator followed the track to be measured by moving the stage manually, and used a magnifying glass to help centre it on the cross-hair. In more sophisticated instruments both of these steps – track centring and track following – were automated. The first machine to do this was Berkeley's 'Franckenstein', named after its designer J.V. Franck.[132] At CERN a device called the sambatron was developed in the late 50s for automatic track centring. The sambatron converted the distance between the image of the centre of the track and the cross-hair into an error-signal. This was fed to an electronic circuit which automatically displaced the stage until the track was aligned (zero error signal). Automatic track following was but a short step from this: by pressing a suitable pedal the operator could move the stage down the track, keeping it centred with a device like the sambatron, and measuring coordinates as she (and women were generally recruited to do these tasks) went. Rough guidance was provided by a manually operated joystick, which the operator used to steer the automatic system across weak regions of the track, to ensure that it did not set off along a crossing track, and so on. The sambatron was never a great success.

From 1958 onwards the STS division gradually built up its stock of Ieps and scanning tables, the technical staff to operate and to maintain them, and the general-purpose geometric and kinematic programmes to analyze their output.[133] By summer 1962 this 'measuring factory' was equipped with eight Ieps of various degrees of sophistication and about six scanning tables. It was staffed by 66 people (including visitors), half of whom were scanning and measuring technicians. There were nine programmers and eleven physicists assigned to the complex.[134] By 1964 the factory's Ieps had apparently reached their maximum measuring speed and its output was quoted as being 50,000 bubble chamber events per year.[135]

How efficient were the Ieps? Could CERN's scanning and measuring factory effort keep up with demand? The answer seems to be yes. In fact we have found only one complaint about its performance, and that complaint was hotly contested by no less than the Chairman of the Track Chamber Committee himself. At the meeting of the SPC in October 1961 Amaldi 'observed that several Italian scientists who had recently been working at Berkeley considered that Laboratory's scanning effort to be one or two orders of magnitude greater than the European scanning effort', including CERN.[136] As a result, he claimed, CERN had missed the discovery of the ω^0, recently announced by Alvarez's group using their 72 inch hydrogen bubble chamber, which was also present in some 'wonderful pictures' taken in Geneva but not yet fully analyzed. This interpretation of the events was disputed by Gregory. 'The delay in producing results from the bubble chamber pictures', he said, 'was basically due to the fact that these pictures had been taken later at CERN than at Berkeley'. In April 1963 he again expressed his confidence in the factory: 'only Berkeley', Gregory claimed this time, 'had better facilities than CERN for the analysis of bubble chamber pictures, the East Coast laboratories being about similar to CERN'.[137] Of course this is not to say that the measuring factory was trouble-free. Indeed in a wide-ranging survey written in 1962 Macleod and Montanet identified a number of technical and organizational improvements that were called for.[138] But by and large the Ieps apparently did what the physicists asked of them – which, as matters turned out, was to analyze just about all bubble chamber film measured at CERN until 1965.

9.6.2.2 The HPD

Around September 1959 Paul Hough, a TV specialist based at the University of Michigan, visited CERN for a year. While there he teamed up with Brian Powell, a physicist from Imperial College, London. Within a few months they had worked out the idea and some of the hardware for a new kind of measuring machine which, they said, could also be transformed into an automatic scanner.

The HPD's key innovation was the method of coordinate measurement. It was achieved by sweeping a spot of light, generated by a perforated spinning disc, back and forth in the y-direction across a bubble chamber film. The film was attached to a precision table which was slowly advancing beneath it in the x-direction. The position of the spot and of the table were digitized when the light was absorbed by the image of the track, and the coordinates were fed to a computer.[139]

The HPD as thus described recorded the coordinates of all the bubbles on the negative. Some means had to be found of selecting only those of interest to the physicist. Several ways of doing so were tried at CERN in 1960 before the idea of rough digitizing was hit on. This involved manually measuring a few key coordinates of the track of interest at a digitized scanning table. These were fed to a computer which fitted a parabolic zone some 200-400 μm wide through the points. Points within the zone, or 'road', recognized as belonging to the track were averaged, and these 'filtered' points were then passed on to the geometric and kinematic stages of analysis.[140]

Hough was quick to encourage the development of the HPD in applied areas of research (for example, he pointed out to a colleague at IBM that it could read a page of a book in ten seconds, that it could be used to analyze photographs in missile work, and so on). In anticipation of his return home in autumn 1960 he began to look 'for a way to take advantage in the United States of our development work at CERN'. It occurred to him that a joint venture between CERN and an American laboratory would be one way to speed up the development of the measuring instrument, to the mutual benefit of all concerned. In July he proposed the idea to Ralph Shutt at Brookhaven; in September he raised it with Luis Alvarez's hydrogen- and Howard White's propane-chamber groups at Berkeley.[141] Reactions were positive, and Kowarski was asked to come up with an official proposal from CERN as soon as possible.

Around end September/early October 1960 Kowarski spelt out to the CERN management the advantages of this arrangement as he saw them. At the moment, he said, CERN held a leading position in the development of the HPD: 'no other, even remotely similar gadget [existed] in any other laboratory'. This lead would count for nothing, he went on, if CERN tried to compete 'against a resolute American effort which [would] be deployed very soon'. A cooperative venture, on the other hand, would safeguard the lead 'for some time ahead', and could also solve some of CERN's financial and technical difficulties connected with the development of the HPD. In particular the laboratory could tap Berkeley's and Brookhaven's 'electronic and programming thinking as applied to bubble data' in which, said Kowarski, they were considerably ahead of Europe.[142]

The management were not enthusiastic about the idea. While encouraging Kowarski to maintain the 'necessary contact', they felt that it was not yet clear 'what kind and what size of [data reduction] effort should be undertaken at CERN, taking into account its budgetary latitude and the foreseeable processing requirements at CERN and in Europe'.[143] At the same time, across the Atlantic, LBL and BNL were having doubts of their own. The Alvarez group decided to put their energies into other devices. And at Brookhaven it was 'far from clear' what was to be done.[144] By December 1960 matters had been clarified. Edwin McMillan, Berkeley's Director, encouraged White's propane-chamber group to go ahead with the HPD, while Ralph Shutt confirmed Brookhaven's interest in interlaboratory collaboration at the technical level.

In April/May 1961 a group of scientists from the three laboratories, plus some representatives from Britain's Rutherford laboratory, gathered in Geneva to test a

Notes: p. 567

prototype HPD coupled to an IBM 709. By the end of May the use of the instrument as 'an input device for measurement and numerical storage of photographic data in the memory of an on-line computer' had been demonstrated to the satisfaction of all those present. At a meeting of the interested parties on 16 and 17 May it was recognized that 'future developments [would] have to diverge to a certain extent [...]'. Each laboratory would produce its own working prototype tailored somewhat to local needs, and an exchange of personnel would be encouraged particularly in the area of programming.[145]

In practice this signalled the onset of considerable cooperation between the two American laboratories, and the slackening of links with CERN. LBL offered to build the optical and mechanical system for BNL (a task that Kowarski had originally thought could be undertaken in Geneva). At the same time, early in 1961, both US laboratories decided to acquire an IBM 7090, a computer six times faster than CERN's IBM 709. This difference in speed considerably simplified the electronics interfacing the flying-spot device with the on-line computer. It also made for a much more efficient use of the computer by the HPD, which was reflected in the structure of the programs being written under White's leadership to handle digitized output. By the end of September 1961 it was clear to Kowarski and Powell that CERN could no longer contribute much to the now all-American collaboration, though 'we sure could use some help'.[146]

At the end of 1961 it was generally believed that it was only a matter of time before the HPD would go into production, at least at BNL and LBL – Howard White estimated that it would be ready to do physics there by about mid-May 1962. He was still rather optimistic six months later: 'volume measurement of data was expected to begin in the autumn' at Berkeley, he said in July 1962. A month later Macleod and Montanet hoped that CERN's HPD would be 'operating sufficiently well for small scale production to begin at the cadence of about 1 hour per day at the beginning of 1963'. These estimates proved to be unrealistic. In January 1963 Lew Kowarski reported that over the past year it had become clear that there had been a 'major miscalculation' at all three laboratories, none of which yet had the HPD ready for daily use in physics. The basic hardware, Kowarski pointed out, was essentially ready; it was the software, the programming, that was causing the holdup.[147]

The main source of the problem was the FILTER programme. Its task was to select those coordinates belonging to the track of interest from the 2000 or so which fell into the zone or 'road' in which the track lay. In practice it was proving extremely difficult to devise a program which could reliably distinguish coordinates belonging to the event from those associated with crossing tracks, parallel tracks, background noise, and so on, all of which were also fed to FILTER ('gated') if they fell in the road.[148]

Each of the three laboratories took a different approach to this problem. BNL installed a cathode-ray tube to monitor visually the performance of FILTER as it travelled down the road selecting event coordinates. This helped the operator 'to force good filter output', and it helped the programmers debug the system. By the summer of 1964 the laboratory had begun production on this basis. Berkeley also adopted a pragmatic policy. Despite the limitations of the HPD, in July 1963 they decided to go ahead and use it for the large-scale analysis of pictures from the 72 inch chamber 'even at the cost of allowing occasional invalid points to be output from regions of confusion'. CERN, for its part, took what D. Tycho called 'a much more difficult or at least a more extreme approach' than either of the two American laboratories: a group around Gerry Moorhead in the DD division decided to rewrite FILTER completely. They were taking this step, Moorhead explained in August 1963, because they (and others) felt that FILTER could 'never work properly unless it [was] modified so as to converge towards a track following system', i.e. a system in which the previous history of a track was used to predict its subsequent trajectory. The program was essentially ready by end 1964, but due to delays caused by the changeover to the new CDC 6600 computer (see section 9.6.3), the HPD was not used for the bulk measurement of bubble chamber film at CERN until summer 1965.[149]

The difficulties created by programming for bubble chamber film can be gauged from the relative speed with which the HPD was commissioned for measuring spark-chamber pictures. The information in these much simpler photographs was 'far less dilute' and so that much easier to process. They contained almost no background, desired events were often 'readily and obviously distinguishable' from undesired ones, and their relative abundance could be controlled by using triggers to discriminate against unwanted output.[150] This simplified programming enormously, and in fact by spring 1964 the HPD/computer system had 'completely and automatically scanned, measured and analysed' 200,000 spark chamber pictures 'at the rate of 1000 per hour'. A year later the device, which some judged to be 'relatively too powerful' for sparks, was complemented by a somewhat less accurate variant called Luciole, in which the spot was generated by a commercially produced cathode-ray tube. By April 1965 Luciole had 'processed some 100,000 events successfully with a maximum speed of over 2,000 events per hour'; a count taken on a set of seven experiments analyzed with it by the end of 1968 arrived at a figure of over 2.6 million spark-chamber pictures successfully handled.[151]

The most striking feature about the development of the new technology that was the HPD is the relative lack of urgency about it felt at CERN. This is evidenced by the reluctance of the CERN management in 1961 to invest substantial resources in it. At Berkeley up to 25 people were attending regular work-in-progress meetings on the HPD in September 1961. At Brookhaven, and despite a 15% cut to the laboratory's budget that year, Hough reported that qualified people had been assigned to every problem and that $170,000 had been allocated to the development of the HPD. At CERN, by contrast, between five and ten staff were put on the project in 1961, and

Notes: pp. 567 ff.

about $100,000 were allocated in the 1962 budget for all post-Iep developments – HPD, cathode-ray tubes, spark data processing etc. The HPD just ticked along, 'without having much to suffer from budget difficulties [...], but without any acceleration either'.[152] Similarly, in 1963, the CERN programmers felt free to eschew 'makeshift' solutions to the problems posed by FILTER, so delaying the onset of the use of HPD for the bulk analysis of bubble chamber photographs. At BNL and LBL, by contrast, it was decided to find ways of working within its limitations in the interest of getting physics results with the HPD as soon as possible.

Why was CERN willing to delay putting this new tool into service? Presumably because the Ieps were up to the task of measuring the bubble chamber photographs handled at the laboratory during our period. That in turn was due to the fact that CERN never measured more than a fraction of the pictures produced there, the rest being distributed widely among collaborating institutions. This European dimension was a prime consideration shaping the management's policy in this area, and profoundly affected the priority it gave to the development of data-handling techniques.

9.6.2.3 The European dimension

The recognition that bubble-chamber film taken at CERN would be shared with national laboratories was seen to imply that, in addition to developing its own measuring equipment, it should take a lead in helping member states build up theirs. To this end a conference 'to catalyse an interchange of information between CERN and various groups working on measuring devices', and 'to enquire what steps could be taken towards initiating a co-operative effort in this field [...]' was held at the laboratory on 15 and 16 September 1958. It was attended by some 60 visitors, about 50 from western Europe, and seven from the United States. Alvarez's group, who had displayed their 'Franckenstein' at the United Nations Atoms-for-Peace conference just before, was particularly well represented.[153]

The meeting confirmed that Kowarski's group was by far the most advanced in Europe. Three institutions building big bubble chambers, the Ecole Polytechnique, Birmingham University, and Imperial College, were developing instruments similar to CERN's Ieps, but did not envisage anything more than assisted track centring. Two others, Saclay and the Cavendish, had developed instruments which had no aids like track centring or track following: automation here amounted to punching the coordinates on tape. At the Max Planck Institute in Munich and at Bologna University the entire process was still essentially manual.

At a discussion held at the end of the meeting there was a 'general feeling' that, as a first step towards remedying this situation, 'some sort of standardization of instrumentation' was called for. At a second, smaller meeting held at CERN on 23 February 1959 it emerged that at least France, Germany and Italy, though probably not Britain, were interested in purchasing a mass-produced device which was not too expensive. A 'standardization and market exploration committee' was set up with Goldschmidt-Clermont as secretary, and it presented its final proposals in December 1959. The committee recommended an instrument offered by a Paris-based firm, La Société d'Optique et de Mécanique de Haute Précision (SOM). The cheapest version was to cost 134,000 SF (about \$35,000), and was like one of the simpler Ieps. The stage was moved manually, and an enlarged oscilloscope display of the track region was used to assist (manual) track centring.[154]

In mid-1960 Goldschmidt made a survey of the equipment available in Europe for the measurement of bubble-chamber photographs. He found that, by the end of the year, there would be 37 groups working on such photographs, using 57 scanning projectors and 28 measuring machines. He repeated the exercise early in 1962. By now there were some 200 physicists in Europe spread across 32 groups and CERN working on the analysis of bubble chamber pictures. They were 'well supplied' with scanning tables, scanners and machine operators, and measuring instruments, and were, he estimated, able to measure about 100,000 events per year, corresponding to '10^6 bubble chamber photographs for the type of events considered in the present [CERN] experimental programme'. This capacity, he pointed out, would increase substantially in the future. In fact Goldschmidt gave data showing that the number of physicists in Europe (including CERN) working on the analysis of track chamber photographs was expected to increase by some 50% by 1963, that the number of scanners and scanning tables would roughly double, and that the number of digitizing projectors would roughly treble, allowing for the measurement of well over 300,000 events by 1963, 'equivalent' to over 3 million bubble chamber pictures.[155]

Europe's capacity for dealing with bubble-chamber film was again estimated at the end of 1965, this time by Mervyn Hine. His figures reveal clearly that CERN had now become simply one institution among many in Europe where bubble chamber film was being processed, and not a particularly big one at that. For example, Hine reported that there were 109 manual film measuring devices in CERN and the member states, and 56 digitized scanning tables: only nine of the former and three of the latter were at the Geneva laboratory. Hine also indicated the important role which CERN had played in facilitating the dissemination of the scanning and measuring equipment first developed in the laboratory. He pointed out that many outside institutes had adopted a measuring instrument built by SOM (now called SOPELEM) based on the specifications drawn up at CERN by Goldschmidt's committee in 1959, and that the company had produced about 50 such devices. He also remarked that a Norwegian firm was building 20 scanning tables under CERN's supervision for handling the film from the 2 m hydrogen chamber, based on similar tables designed and built in the laboratory.[156]

The European effort in the handling of track-chamber pictures was not restricted to the use of Ieps. There were about ten HPD's built in Europe and, as with the Iep, the mechanical part of the 'flying spot device' was commercialized (by Sogenique, a subsidiary of a Geneva-based instrument company). Computer programs were also shared, particularly between CERN, Rutherford, Munich, and Bologna, and two important meetings on programming for the HPD were held during our period, one at the Collège de France, the other in Bologna.[157]

It was generally assumed that CERN would only process about 30% of the bubble chamber film produced at 'its' accelerators. Consistent with this policy some people in the laboratory actively encouraged and assisted European universities and institutes to acquire data-handling equipment of their own. By the end of 1965 over 10 million bubble chamber pictures had been produced in Geneva. No less than 85% of them had been analyzed by institutes in the member states.[158]

9.6.3 COMPUTERS

The first steps to acquire a computer for CERN were taken by Lew Kowarski in 1955. At the time he was apparently not sure himself that such a machine was necessary in a laboratory like CERN: as H.L. Anderson put it in a letter to him which he circulated widely, 'For a physicist, a certain potential barrier has to be overcome before he is willing to take the trouble and the time to learn how to use a machine' like a computer. Accordingly in the first paper on the topic which he laid before the SPC in November 1955 Kowarski asked the committee's opinion 'as to the advisability of purchasing the electronic computer', and as to the 'validity of the reasoning' which led him to 'favour' ordering a Ferranti Mercury. The SPC set up a subcommittee to look into the matter, it confirmed Kowarski's choice, and the Finance Committee ratified the purchase of the British machine at a special meeting held on 20 December 1955.[159] Originally promised for shipment in February 1957, the Mercury arrived about 15 months later, and passed its acceptance tests in October 1958.[160]

Why a Ferranti Mercury? Firstly, Kowarski thought it best that CERN purchase rather than hire a computer. This effectively excluded IBM. The American firm offered CERN an IBM 701 which, in Kowarski's view 'would be at least comparable to the Ferranti Mercury (definitely less modern, but faster and much bigger)'. All the same he advised against it. It would cost 55,000 SF per month to hire, so that after 18 months 'the outlay would be equal to the purchase price of a Mercury [about 1 MSF], without giving CERN any ulterior rights'. A second point in favour of the Mercury was that it was being bought by other European laboratories. CERN, said Kowarski, was 'a relatively small organization which [could] not hire staff in order to make a contribution to the rapid development of scientific electronic computation'. Hence the need to choose a computer which could easily share programs with other centres whose work lay 'in a field similar to CERN's'. As Harwell and Saclay were installing Mercury machines, it seemed wise that CERN do the same, so facilitating an eventual 'exchange of services' between the three sites.[161]

When the Mercury was bought it was thought that its main use would be for performing physical calculations. And indeed the machine was well-suited 'for those problems in which the amount of calculation [exceeded] by a large factor the amount of data to be transferred'. On the other hand, as this remark implies, its input/output facilities were so slow that they could effectively squander its computing speed. This made the machine particularly unsuitable as a companion to the Ieps: when running Iep programs it could spend as much as a quarter of its time punching out results. To get around these difficulties the Iep group rented time on an IBM 704 in Paris and, at the end of 1959, ordered three magnetic tape decks to improve the overall performance of the Mercury.[162] But these were obviously only temporary palliatives to a problem which could only grow more important with time.

In May 1959 representatives from about ten European computing laboratories which were equipped with the Mercury met at CERN to discuss future computing needs. Asked to recommend a 'solution' to 'CERN's computer problem', the majority of them suggested that the laboratory buy a second Mercury at once, ordering another bigger machine in about summer 1960. A few months later, in September 1959, Kowarski drafted a paper for the SPC outlining the options. He made only passing reference to the possibility of buying another Mercury, showing instead a distinct preference for hiring an IBM 704 machine. This suggestion was quickly confirmed by the SPC the following month, and accepted more or less without dissent by the Finance Committee shortly thereafter.[163] In the event the contract with IBM was subsequently altered to cover a slightly newer machine, and an IBM 709 was delivered to CERN in November 1960, and put into operation the same month. It had arrived just in time: in May 1960 Kowarski reported to the SPC that 'the Mercury was already saturated which was sooner than expected'.[164]

Why the change in policy from buying a Ferranti to renting an IBM, the latter being an option specifically rejected a few years earlier? Firstly, there was the desire to 'join the club' of users with similar needs to CERN's – though this time both the club members (major American laboratories) and the main need (handling bubble-chamber output) were very different. Just before Kowarski drafted his paper Art Rosenfeld (LBL) had explained the 'complex specialized programmes for picture evaluation' being developed in the USA and whose use was 'spreading in all leading high energy centres all of which [had] an access to a big IBM machine (704 or better)'. By acquiring a similar machine, said Kowarski, CERN could hope to 'make full use of the work already done in the U.S.' (12 man/years on program development in Berkeley alone during the previous academic year), and to 'participate in the future common development'. What is more the IBMs were much faster than the Mercury: the latter's cycle time was 60 microseconds compared to 12 microseconds for the 709, while both the 704 and

Notes: p. 568

the 709 had a 'fast memory' capacity of '32,000 words' compared to the Mercury's '1000 words'. The American machines were also equipped with input and output devices which made them 'particularly well suited to handle a large volume of data'.[165]

Why rent rather than purchase? The main argument given was financial. By the end of 1959 CERN had already absorbed all the reserves in its budget for the next three years, and the savings made by renting a machine rather than buying one, as originally planned, could be used to replenish them. And the terms of the deal struck with IBM were particularly attractive. In September 1959 it was prepared to offer CERN a 40% 'educational' discount on the rental of an IBM 704. Later it offered even better terms for an IBM 709 'in view of the scientific importance of CERN'.[166]

Kowarski made the opening moves to secure funding for CERN's next computer in September 1962. Data-handling for track-chamber pictures again provided the main rationale for the acquisition. Kowarski identified 'two major new demands', in particular, which would be made on computing time in 1963: 'the processing of spark chamber photographs, and the entry into service of the Hough-Powell digitizer which [would] open up new possibilities for the processing of bubble chamber pictures'. To satisfy these demands he proposed that CERN rent an IBM 7090, a transistorized machine which was considerably faster than the valve-based 709 (cycle time 2 microseconds compared to 12 for the 709). Kowarski recognized that there were many other fast computers coming onto the market, some of them European. However, he felt that the time was not yet ripe for installing them. CERN needed something that was proven and needed it quickly. Figures from Berkeley indicated that expenditure on computing was now doubling annually. This suggested that the 7090 would in turn have to be replaced in two or three years time; it may then be interesting to acquire one of the new, relatively untried machines just becoming available.[167]

The Finance Committee were extremely reluctant to endorse this proposal. By now it was becoming clear that computers would have a major role to play in a nation's industrial and military development, and everyone was concerned to ensure that European states were not left on the sidelines. The British delegation, in particular, felt that the decision was being rushed and that not enough consideration had been given to European alternatives. The committee decided to set up an expert panel to advise them. It reported back in December, and recommended that CERN should acquire an (American) CDC 3600 built by the Control Data Corporation if it had to have a new computer by December 1963, adding that a (British) KDF 9 built by English Electric might prove to be an equally good choice if the machine could be delivered later. In response the management agreed that the CDC 3600 'was the best machine now being put on the market' – but insisted that an IBM 7090 was the most suitable 'stop-gap to cover the needs of CERN for a period of about two years'.[168] They got their way. In September 1963 the IBM 709 was replaced by an IBM 7090, so increasing the total computing capacity available at CERN by a factor of about four.[169]

What arguments were advanced for this choice? Firstly, the need to have additional capacity quickly, not only to handle track-chamber film with the new automated instruments but also well in advance of the Rochester conference to be held in July 1964. This effectively eliminated the KDF 9, which could not be delivered before spring 1964 and for which certain essential peripheral equipment still had to be developed. With the British computer out of the running, two arguments counted heavily against having a CDC rather than an IBM. The first was the wish to retain membership of the American user's club. Both Brookhaven and Berkeley were developing programmes to handle HPD output for use on an IBM 7090. And the case was reinforced by the fact that, the United Kingdom apart, IBMs were widely used throughout Europe. The 7090 was installed at the Ecole Polytechnique and at Saclay in France – the 'homes' of two of the main bubble chambers then in use at the PS – , groups in Italy could use Ispra's 7090 and Bologna's 704, soon to be replaced by a 7094, Munich had a 7090, and so on.[170] By acquiring the IBM CERN could both profit from the programming developments on the HPD being made in the USA, and collaborate better with its member states.

The second argument for having an IBM was linked to this: that if a CDC was acquired CERN would have to rewrite its programs. The expert panel felt that this would be worth the effort, and that the money saved on rental could be used to pay the additional staff needed for the purpose. The CERN management insisted that the panel had underestimated the difficulties involved, and that it would be very bad policy to make such a change-over just when 'we are in the middle of programming for the Hough-Powell Device, on whose early success so much of our future progress of European data handling depends, and when we are beginning to tackle the new and most serious problem of data analysis from spark chambers'.[171]

1963 was a turning point as far as the provision of data-handling equipment for CERN was concerned in that two far-reaching policy decisions were taken. Firstly, a number of major organizational changes were made in accord with the ideas discussed by a new Data Handling Policy Group set up early in the year by Director General Weisskopf.[172] Secondly, after an extensive review of CERN's needs and the computers available on the market, a specially constituted committee, including member states' representatives, recommended what machine the laboratory should acquire next. Their proposal was ratified by the Finance Committee in December the same year.

The main organizational change stimulated by the Policy Group was to split off from the DD division physicists who still wanted to remain active experimentalists, along with the responsibility for established measuring instruments like the Ieps and their basic computer programs. This activity was an 'historical residue' from the previous decade when data handling had been the responsibility of Kowarski's STS

Notes: pp. 568 ff.

division, renamed DD (Données et Documents) in 1961, and it was felt that it would be more 'rational' for those who primarily wanted to do physics to be located in one of the experiment-oriented divisions. As for the refashioned DD itself, it was proposed that it concentrate on the development of new techniques for data-handling, like the HPD and Luciole, on the development of the programs needed to exploit them, and on the acquisition of the computers for the laboratory.[173]

In line with these proposals Kowarski's deputy, and the head of the DD/Exp section, Yves Goldschmidt-Clermont, along with others who wished to continue doing physics, were transferred to Peyrou's Track Chambers division on 1 June 1963. At the same time, in May 1963 Hine informed the staff that Ross Macleod, who had played a leading role on the computing side, would be appointed Deputy Division Leader of DD, and that he would manage the division during Kowarski's imminent one-year absence at Purdue.[174] Kowarski was never again to have a position of authority at the laboratory. Macleod was later confirmed as Acting Division Leader until June 1965, whereupon the Council converted the appointment into divisional head, with responsibility for the 'operation and development of the CERN central computing and data processing facilities'.[175]

The need to explore CERN's future computer requirements was explicitly recognized by the expert panel who was asked, at the end of 1962, to look into the advisability of replacing the IBM 704 with a 7090. The panel insisted, along with the management, that this course of action could never amount to more than an interim solution, and at a meeting held in Amsterdam in February 1963, considered that 'very early attention should be given to a careful study, in consultation with co-operating institutes in the Member States, of what should replace' the 7090 in its turn.[176]

In response to this initiative a so-called European Committee on the Future Computing Needs of CERN met for the first time on 1 May 1963, and immediately established a Continuing Committee to study the problem in depth. The Continuing Committee was chaired first by Kowarski and then, when he left for Purdue in September, by Hine. It had six other members, four of them users 'representing' each of the major member states, and Macleod and Lipps from CERN. The committee met six times between May and October 1963, considered ten internal working papers, heard reports from visits to laboratories and computer manufacturers in Europe and the USA, and analyzed replies to its request for provisional offers for computing systems from manufacturers on both sides of the Atlantic. It presented its findings in October 1963 in a 30-page report backed by eight appendices totalling over 100 pages. The Committee recommended that CERN purchase a CDC 6600 as soon as possible.[177]

The dominant consideration informing this choice was the conviction that CERN's next machine had to 'incorporate the latest advances in computer design', notably 'the time-sharing concept'. The October report explained what this meant: that several programs could be stored simultaneously in the machine's memory, programs which were 'obeyed at different times in such a way that no part of the computer was unnecessarily idle'. This meant a substantial increase in 'accessibility' – the computer could be used by several people simultaneously – , but also in 'capacity' – from the 32 K of the IBM 7090 to at least a 64 K and preferably a 128 K memory store. With a time-sharing computer having a capacity of 128 K CERN could hope to satisfy the growth in demand for the analysis of bubble and spark chamber experiments, which was expected to double each year, and which would lead to the saturation of the 7090 by the end of 1964, and to a total capacity equivalent to several 7090s by the end of 1966. It could also envisage connecting the main machine to smaller computers hooked up on-line to a counter experiment or, say, being used to control film-measuring instruments like the HPD. Finally, with a sufficiently large time-sharing computer CERN users could have access to 'several programme testing stations, each consisting typically of an input output typewriter and paper tape reader and punch [...]. A user at each station [could] test his programme, communicating with the computer via the typewriter, quite independently of the other users and without being aware that he [was] not the sole user of the computer'. The Continuing Committee's report noted that four such 'consoles' were in use on the Ferranti Atlas computer at Manchester, while Livermore thought that up to a hundred could be provided with a CDC 6600.[178]

It fell to Mervyn Hine to steer the recommendation that CERN buy a CDC 6600 through the Finance Committee. Again, it was the British delegation that had to be convinced, and understandably. The only other possibly suitable time-sharing computer on the market was Ferranti's Atlas II, and while the British admitted that the CDC 6600 'was a much bigger and faster machine', they were not convinced that their alternative was 'really inadequate for the needs of CERN' from a technical point of view.[179] On top of which it was considerably cheaper and did not require purchasing a piece of strategically important scientific equipment outside CERN's member states. The United Kingdom's objections were apparently handled to their delegate's satisfaction in a private discussion he had with Hine at the 56th meeting of the Finance Committee on 16 December 1963.[180]

In January 1965 a CDC 6600 computer was delivered to CERN. It was not exploited fully that year due to delays in the delivery of the time-sharing operating system, 'combined with technical troubles in the machine itself'. By mid-October 1965 the situation had deteriorated so much that the machine was handed back to the company for overhaul, and CDC decided to install a 3400 computer on site until the time-sharing system was working properly. It was only at the end of 1966 that the CDC 6600 was in operation 20 hours a day. By then it had been realized that computing had become 'an integral part of the scientific work of CERN' and that

Notes: p. 569

another independent machine was needed to provide backup capacity in the laboratory.[181] The CDC 6600 cost 24 MSF, and was the biggest single piece of equipment that CERN had ever bought.

Were the measures taken at CERN to build up the laboratory's data-handling capacity first in 1955, then in 1959, then in 1962, and then again in 1963, adequate to keep up with the demand for computer power coming from the experimentalists? It would seem not, to judge from a series of remarks made by management and users alike in 1963. In February and in June that year Weisskopf told the Finance Committee and the Council that the evaluation of pictures in an important experiment to compare the parity of sigma and lambda particles 'had not proved easy' because the IBM 709 had broken down during the data handling process, and CERN had had to borrow Munich university's 7090. This, said the DG, was 'a striking illustration of CERN's lack of data evaluation facilities which should be remedied as soon as possible'. As 'another proof of the inadequacy of CERN's data handling facilities' he went on to mention the lack of a suitable computer to handle half a million spark-chamber pictures taken in a counter experiment relating to the diffraction peak in pion-proton scattering. CERN had not evaluated its film, whereas Brookhaven had by using a spare Merlin computer. Gregory backed up this claim at a meeting of the SPC in April the same year. While CERN's capacity to handle bubble chamber pictures was on a par with that in US East Coast laboratories, he said, 'most American laboratories were very much better equipped on the computer side, in particular for the analysis of spark chamber data'.[182]

We do not doubt that CERN could have used more computing power than that provided by the Mercury and the IBM 709 early in 1963. Yet we feel that this so-called 'lack' must be put into perspective. In particular it must be stressed again that nothing short of a revolution was taking place as far as the introduction of computers into high-energy physics laboratories was concerned. With the demand for computer time doubling annually (see figure 9.1), and with new uses for computers constantly being found, it was inevitable that predictions of what was needed would be overtaken by events, and that temporary 'bottlenecks' would arise as a result. This happened at CERN in the first part of 1963 – until the IBM 7090 was installed in September. It happened at Berkeley and Brookhaven in 1964 – both of which, wrote Hine in a confidential trip report in September that year, had 'run into a severe lack of computer capacity [...]', adding that 'Argonne will be facing it soon'.[183] Gregory and Weiskopf were right to point to CERN's 'inadequate' computing capacity early in 1963. But this did not indicate any deep-seated failing in the provision of data-handling equipment, nor did it call for any more drastic steps than those already being taken in the laboratory to provide for the future. And we should not forget that their statements were made at a time when the member states were soon to be asked to invest an unprecedented sum of money in a new generation of computers for CERN.

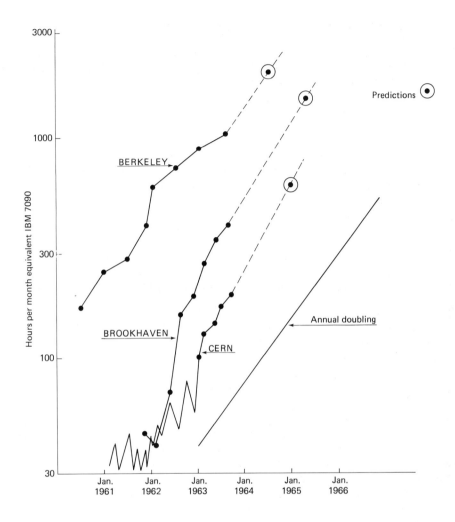

Fig. 9.1 Growth of Computer Use at High Energy Physics Laboratories. Source: Appendix IV to the Report of the European Committee for the Future Computing Needs of CERN (CERN/516, 25/10/63).

Notes

1. The documents used for this chapter can all be found in the CERN archives. Files DG20551 and DIRADM20408 were particularly useful for the study of the bubble chamber decision making process. On the purchasing of beam transport magnets we made considerable use of the minutes of the Executive Committee for the PS Experimental Programme in file DIRADM20417. Three boxes of papers in Lew Kowarski's collection at CERN, files LK22482, LK22483, and LK22484 were invaluable for developments of the Ieps and the HPD. On computing we drew heavily on several papers by Mervyn Hine in box

DG20598. Other primary and secondary literature which we have used is included in the bibliography for parts III and IV of this book.

The chapter is a revised and improved version of three published texts: Krige and Pestre (1986) on bubble chambers, Krige (1989) on beam transport, and John Krige, *The Development of Techniques for the Analysis of Track-Chamber Pictures at CERN* (Geneva:CERN Report CHS-20, 1967).

2. Adams (1956), 6.
3. Irvine and Martin (1984). The material cited is from the second of their two papers.
4. Council, minutes, 13/6/62, 17.
5. For a non-technical description of the discovery see Glaser (1955) and (1964). For the application of the chambers at Berkeley and Brookhaven see Alvarez (1970), L.W. Alvarez, *The Bubble Chamber Program at UCRL*, 18/4/55, stencilled typescript, unpublished but widely circulated, and Shutt (1966). For studies of the topic see Galison (1985a), Gaston (1973), and Swatez (1970).
6. The figures are taken from the CERN Courier, January 1965, 3.
7. Laura Weiss surveys the developments in chapter 6.1.
8. P. Amiot, 'Rapport de mission aux EUA', 1/12/57 (MGNH22044).
9. Cocconi (1960), 802.
10. Crozon (1987), chapter 6, and Galison (1987b) give good descriptions of bubble chambers and electronic detectors and the different practices they demand.
11. The choice of words is Hine's. See M.G.N. Hine, *Report on a Visit to the U.S., April 1959,* Report PS/Int.TH 59-7, 12/5/59, 42.
12. Lofgren (1956), 501.
13. CERN/SPC/12, 11/11/55.
14. CERN/SPC/16, 5/6/56.
15. CERN Annual Report, 1955, 38.
16. CERN Annual Report, 1956, 13, 39.
17. CERN Annual Report, 1955, 38. See also chapter 6.3.1.
18. See chapter 6.3.1.
19. To glimpse these attitudes we shall rely heavily on a couple of key documents, notably Budde's report written towards the end of 1956 after his US visit and the contribution by John Adams to the CERN symposium in June 1956. See Adams (1956) and R. Budde, *Some Thoughts About Bubble Chambers,* undated, but probably around November 1956. The document is in the collection of papers deposited by Laura Weiss in the CERN archives.
20. The information on the cloud chamber is from CERN Annual Report 1956, 39, and 1955, 38. For Budde, see previous note.
21. The description of Budde's task in the US is from CERN Annual Report, 1956, 39-40. See also note 19.
22. Adams (1956), 3.
23. The quotations are from L. Alvarez, 'Liquid Hydrogen Bubble Chambers', Accelerator Conf. (1956), volume II, 13-15. On Brookhaven's chambers see Louttit, Instrumentation Conf. (1960), 117-120, and Shutt, Instrumentation Conf. (1962), 71-94.
24. See note 19.
25. All of this section, but for 9.3.4, is based on Krige and Pestre (1986). The main differences lie in the stress placed on decision making in conditions of *uncertainty* (and in seeing Ferretti's 'rational' approach to the choice of equipment, and the approach swayed by Alvarez's bubble chamber philosophy, as two ways of dealing with uncertainty), and in a more explicit articulation of the priorities of various groups when it came to choosing the size of the bubble chamber.
26. For the minutes see *Informal meeting on the experimental programme for the proton synchrotron,* 4/6/57 (DG20551).
27. J.B. Adams, *The Programme for the CERN Proton Synchrotron,* report CERN-PS/DIR, 1/7/57(JBA22798).

28. For the aims of Amiot's trip and his trip report, see note 8.
29. The quotations are all from Hine's report on the discussion at the Venice-Padua conference, PS/Theory/MGNH-Note 18, 1/10/57 (MGNH220880).
30. B. Ferretti, *Preliminary Report on the Experimental Programme of the CERN PS*, CERN/SPC/51,18/10/57. See also Adams' report, note 27.
31. The suggestions were contained in a list appended to a page written by G. von Dardel, *The PS Research Problem. I. 24 October 1957*, document PS Nuclear Physics Group/1 November 1957 (DG20551). Von Dardel was the secretary of the group referred to here and its only full time member.
32. The minutes, from which the subsequent quotations in this paragraph are taken, are PS Nucl. Phys./3, 29/11/57 (DG20551).
33. The minutes are PS Nucl. Phys./6, 20/1/58 (DG20551). The report of the bubble chamber study group is CERN-PS/Nucl.Phys./5, 18/12/57 (DG20551). The other two study groups presented a combined report, PS Nucl. Phys./4, 12/12/57 (DG20551).
34. The memorandum from Kowarski and Peyrou to the 'PS Research Group' is entitled *Bubble Chamber Programme 1958-1959*, and is dated 13/1/58. It is Appendix I to PS Nucl. Phys./6, 20/1/58 (DG20551).
35. See *Minutes of Meeting on Bubble Chambers*, CERN-PS/Nucl. Phys. 8, 3/2/58 (DG20551).
36. See his report, note 8.
37. See Adams' report, note 27 for the quotation, and Hine's report, note 29, for feelings expressed at the Venice-Padua conference.
38. For Glaser's fears that the bubble chamber would become obsolete, see Amiot report, note 8; for Peyrou's arguments for a propane chamber, see Hine's report, note 29.
39. The SPC's idea that experimental work should be guided by theoreticians is found in CERN/SPC/5, 25/7/55.
40. See Adams' July 1957 report, note 27.
41. For Adams, see note 27; for Peyrou, see the minutes of the meeting cited in note 33.
42. Letter, Alvarez to McMillan, 17/9/57 (DG20551).
43. For Blackett, see the minutes of the seventh meeting of the SPC held on 29/10/57, CERN/SPC/54, 25/3/58, 8. Bernardini's remarks are from the minutes of the meeting cited in note 32.
44. For the important changes in the member states' attitudes to the financing of CERN which occurred during the second half of 1957, see chapter 10.1.2 and 10.1.3.
45. To check this point we scanned the following journals for the period 1957-1960: *Nuclear Instruments and Methods, Review of Scientific Instruments, Review of Modern Physics, Reports on Progress in Physics, Progress in Nuclear Physics*. It was also confirmed by Charles Peyrou in a conversation with us.
46. See note 31.
47. See note 34.
48. See note 35.
49. The (draft) minutes of the SPC meeting held on 4/3/58 are CERN/SPC/63/Draft, 2/4/58.
50. The main report is CERN/SPC/64, 1/5/58. The appendices are CERN/SPC/64(A),(B),(C), undated. Preliminary drafts of the latter, written by Adams, were discussed at a Group of Directors meeting on 2/5/58. Its minutes are CERN/GD/181, 29/4/58 (DIRADM20408).
51. For the remark by Peyrou see note 35. The comment by Alvarez is from his internal paper *The Bubble Chamber Program at UCRL*, 18/4/55. In September 1955 Alvarez added some corrections by hand to his text, and it is this version that we have used. It is in file (LK223867). It is fascinating to see Alvarez putting the claim we have quoted into practice in this paper. After a long description of the kinds of processes one may study in a bubble chamber he arrived at a figure of 37″ as the minimum length for it. In April 1955 he then wrote: 'Since there is no logical way to decide on the overall length, we have arbitrarily increased it from 37″ to 50″'. Six months later he crossed out 'increased' and '50″' by hand, and replaced them with 'doubled' and '72″'. The same views were expressed at an international conference held at CERN in 1959 (see Accelerator Conf. (1959)). For example, W.H. Evans, who was involved in building the 1.5 m British

chamber remarked that the final decision about its size was shaped by many factors, such as the physics desired, engineering feasibility, economic resources, available personnel, and so on (p. 445). D.C. Rahm from BNL was more provocative: 'We gazed into a crystal ball to fix the size of our chamber', he said (p. 441).

52. See the Kowarski/Peyrou memorandum cited in note 34. It is important to appreciate that the authors did not challenge the idea that there should be no 'temporary swelling' of CERN's staff to build a big hydrogen chamber.
53. For Adams' report see note 27. For his threat see CERN/SPC/54, 25/3/58, 10-11.
54. Within the limits of technical feasibility, of course. One technical factor limiting the chamber to 2 m was the maximum size of the glass windows – see Kowarski's report on the work of the study group presented on 26/3/58 in CERN/PS/Nucl. Phys. 9, 1/4/58 (MGNH22069).
55. For Blackett's offer, see the minutes of the meeting, CERN/SPC/54, 25/3/58.
56. Adams' July 1957 report is cited in note 27. His interest in pushing to the technological limit is found in the minutes of the eighth meeting of the SPC, CERN/SPC/63(Draft), 2/4/58.
57. The quotation is from the minutes of the meeting cited in note 35.
58. The original version of this paper, prepared for the meeting of the Group of Directors on 2 May 1958, is numbered CERN-PS/30/JBA, 28/4/58, and was issued as CERN/GD/182, 29/4/58 (DIRADM20408).
59. The technical report written by Ramm and Resegotti is PS/Int.EA-58-3, 12/11/58 (MGNH22070). For the figures (2 MSF in May, 6.67 MSF six months later), see CERN/SPC/67, 8/5/58, 3, and budget figures appended to CERN/289, 16/10/58.
60. See unsigned *Report on International Conference on Hydrogen Bubble Chambers Held at Imperial College April 1-4, 1958*. Dominique Pestre and I found this paper in the DC files in the archives at LBL.There is a copy of it in the CERN archives, box (USA/KP/H/4). We infer that the author was Gow since he was the only UCRL representative present.
61. For the May report, see note 50.
62. Comment written, probably by Kowarski, on the back of a copy of CERN/GD/182, 29/4/58 (DIRADM20408). The paper was prepared for the Group of Directors meeting on 2/5/58.
63. As reported in the minutes of a meeting on big bubble chambers held on 6 June 1958, PS/Int. NP-13, 10/6/58 (MGNH22069). At the same meeting Peyrou indicated that he might be willing to lead the 2 m project. He never did, and in fact the chamber was completed without ever having a formal 'project leader' (see section 9.5.3).
64. Note that Kowarski's comment (note 62) was made at a meeting that took place *before* Adams drafted his paper.
65. Letter, Leprince-Ringuet to Bakker, 24/4/58 (DG20565).
66. 'Numero consacré aux chambres à bulles francaises au CERN', *L'Onde electrique,* **41** (decembre 1961), 417.
67. See the minutes of the ninth meeting of the SPC on 23/5/58, CERN/SPC/70/Draft, 1/8/58.
68. For the arguments used by the British and French hydrogen bubble chamber builders, see chapter 8.3.
69. Adams' fears, and the quotation, are from CERN/SPC/64(A), 1/5/58.
70. See CERN/SPC/64(B), 1/5/58.
71. The details are in CERN/SPC/64, 1/5/58, chapter 6.
72. CERN/SPC/64(B),1.
73. The remark is made in his report cited in note 29.
74. CERN/SPC/67, 6/5/58, 3.
75. The information in this paragraph is drawn from the draft long-term budget, CERN/FC/235, 25/9/57, 16, CERN/SPC/67, 8/5/58, 3, and CERN/FC/303, 19/9/58, 15.
76. H.L. Verry at a meeting of the Finance Committee Working Party on 3/6/58, minutes CERN/WP/FC/298, 14/7/58, 5.
77. See Greenberg (1971), 215. Galison (1985a), 339-40, has quoted George Kolstad (chief of the Physics and

Mathematics Branch of the AEC Division of Research) as appealing to Russian developments to raise money for bubble chamber equipment for the Alvarez group.

78. The alternative programmes were explained in CERN/SPC/67, 8/5/58. the minutes of the SPC meeting are CERN/SPC/70/Draft, 1/8/58. A preliminary draft of the building programme came before the Group of Directors on 2/5/58 – see CERN/GD/182, 29/4/58 (DIRADM20408).
79. Adams pointed out that it would require a major recruiting effort if the 2 m chamber was started in 1959 in a paper prepared for the Group of Directors' meeting on 2/5/58 – see CERN/GD/182, 29/4/58, 5.
80. See minutes of the ninth meeting of the SPC, CERN/SPC/70/Draft, 1/8/58, 5.
81. The minutes of this meeting, the eighth of the Working Party, are CERN/WP/FC/298, 14/7/58. Its recommendation, CERN/FC/291, 4/6/58, was accepted by the Finance Committee at its twenty-first meeting on 19 June 1958, minutes CERN/FC/297/Draft, 7/7/58.
82. Minutes of the FC meeting held on 19/6/58, CERN/FC/297/Draft, 7/7/58, 5.
83. These remarks are, successively, from the meeting cited in the previous note, from a recommendation from the SPC to the FC (CERN/FC/287, 27/5/58, 1), from a statement made by the DG to the Council (Council, minutes, 9/10/58, 37), and from remarks made by John Adams (CERN/FC/297/Draft, 7/7/58, 6).
84. This paragraph is composed of elements from CERN/FC/287, 27/5/58, CERN/FC/291, 4/6/58, CERN/WP/FC/297/Draft, 7/7/58, and CERN/FC/303, 19/7/58.
85. For details see the draft budget for 1959, CERN/FC/303, 19/9/58, 2, 15. Programme B required some 3.5-4 MSF in 1958. The increase came from a doubling in the cost of the propane chamber for 1959, and from an agreement to go a little beyond Programme B by making some 1.7 MSF immediately available to start work on the big hydrogen chamber. Dominque Pestre has described what he calls the CERN lobby, and its willingness to help CERN, in chapter 7.1.3.
86. This section is based on the paper entitled *Proposals on PS Experimental Programme,* CERN/SPC/71,1/9/58.
87. A thorough discussion of the committees set up to manage the PS experimental programme has been given by Dominique Pestre in chapter 8. Chapter 8.4.2 is especially pertinent here.
88. For von Dardel's and Ramm's proposals, see respectively CERN/Ex.C/13, 20/11/58 and CERN/Ex.C/14, 21/11/58. (DIRADM20417). For the emulsionists see CERN/Ex.C/20, 28/11/58 in the same file, and Emulsion Meeting (1959), 2.
89. See Emulsion Meeting (1959), 8.
90. The minutes of the three meetings mentioned here are, respectively, CERN/Ex.C/19, 27/11/58, CERN/Ex.C/25, 29/12/58, and CERN/Ex.C/33, 18/2/59 (DIRADM20417). Adams' made his remark at the seventh meeting, minutes CERN/Ex.C/27, 29/1/59. The minutes of the FC meeting which authorized the order are CERN/FC/365/Draft, 23/7/59.
91. See M.G.N. Hine, *Report on a Visit to the U.S., April 1959,* Report PS/Int. TH 59-7, 12/5/59.
92. The two trip reports used in constructing this table were F. Grütter, *Beam Transport Magnets and Power Supplies at Brookhaven, September 1961,* Report ENG/Int.DL 61-15, 13/10/61 (LK22457), and W.Schnell, *Notes on a Visit to Brookhaven, February 2 to 7, 1961* (KHR22138).
93. See CERN/Ex.C/14, 21/11/58 (DIRADM20417).
94. See CERN/Ex.C/29, 4/2/59 and CERN/Ex.C./33, 18/2/55 (DIRADM20417).
95. These contextual elements are described fully in chapter 10.2.
96. See von Dardel's proposal in CERN/Ex.C/13, 20/11/58 (DIRADM20417).
97. This sentence is from CERN/Ex.C/19, 27/11/58 (DIRADM20417).
98. For Gibson's requests see CERN/Ex.C/25, 29/12/58, and CERN/Ex.C/27, 29/1/59 (DIRADM20417).
99. Hine's trip report is cited at note 91. The rebuffals were made at various meetings of the Executive Committee: CERN/Ex.C./33, 18/2/59, CERN/Ex.C/42, 17/6/59, CERN/Ex.C/48, 11/8/59, CERN/Ex.C/54, 22/10/59 (DIRADM20417).
100. See Cocconi (1960).

101. For Cocconi's complaint to the Executive Committee see CERN/Ex.C/54, 22/10/59. See also note 100. For other pertinent information see M.G.N. Hine, 'Features of the CERN Proton Synchrotron of Interest to Experimenters', Instrumentation Conf. (1960), 214-222.
102. Hine (previous note), 222, Ramm (1960).
103. Letter Cresti to Adams, 1/9/60, and reply 20/9/60 (DG20573).
104. Memorandum Peyrou to Adams, *Beam Transport Equipment for the East Side,* 3/10/60 (DG20562). See also CERN/FC/524, 12/12/61.
105. As reported in a survey on *Beam Transport for the PS 1961-1962* made by Hine for the Management Board, CERN/MAN/17, 23/2/61 (DG20866).
106. CERN Annual Report, 1960, 46-7.
107. The quotations are taken from Bernardini's paper, CERN/SPC/128, 20/4/61, and his opening address to it at the SPC meeting, minutes CERN/SPC/133/Draft, 7/7/61.
108. The quotations in this paragraph and the next are from the official minutes of the SPC meeting (see previous note) supplemented by a tape recording of the proceedings themselves.
109. See Hine's *Notes on Beams and Beam Transport for the CERN PS 1961-63,* CERN/SPC/137, 19/7/61.
110. See Memorandum from A. Asner to the DG, Directorate Members, and 11 CERN Senior Staff on *4-5 m Electrostatic Separators and H-V Supplies [...],* 14/8/61 (DG20556).
111. This information is from the CERN Annual Report, 1961, 72, 103.
112. Memorandum, M.G.N. Hine to DG and others on *Programme for Electrostatic Separators,* 2/3/62(DG20490). The decisions on what should be done are reported in letter Hine to Butler 14/3/62 (MGNH22133). See also Hine's Note on the *Choice Between the Two Electrostatic Separators 'Anatole' and 'Bernard',* 19/1/62 (DG20490) and G.L. Munday's Memorandum to Germain and others, *Electrostatic Separators [...],* MPS/EP/5, 28/2/62 (KHR22155).
113. The quotations are from the verbal record of Weisskopf's address to the Council, minutes, 13/6/62, 14 et seq.
114. It is true that Weisskopf did say that BNL's 'investigations of the composition of the secondary beams coming from internal targets were done much more systematically than at CERN' (CERN/422, 24/11/61, 6), though we believe that this was due more to the inexperience of the CERN staff than to a lack of equipment.
115. See the report for the Management Board cited in note 105.
116. Six and Artru (1982).
117. This was implicit in Hine's address to the SPC in April 1961 (see note 108), and he repeated the point more than 20 years later in a discussion with the author. The problem, he told the SPC, had arisen because 'the beam transport orders that were placed [...] were based on an assumption that it was to be mainly high-energy beams, they were designed [...] with the idea of doing high-energy physics rather than low-energy physics'. This 'desire' to do low energy physics was, of course, determined by what others were doing at the time. Physicists at CERN would have been quite happy to build high-energy beams in 1961 – if they had had long enough separators.
118. The information on the BNL separators is from Baltay et al. (1961) and (1963). See also Leitner et al. (1963).
119. For information on the CERN separators see Ramm (1960) and Germain and Tinguely (1963).
120. On the CERN propane chamber see Ramm and Resegotti, 'The Principles of the Design of the CERN Propane Bubble Chamber', Instrumentation Conf. (1960), 127-32, and CERN Annual Report, 1960, 44-45; 1961, 101-102.
121. The data on the British and CERN chambers is from CERN Annual Report, 1964, 76, 81. The data on the BNL chamber is from Brookhaven Annual Report, 1963, xvi.
122. For the CERN chamber see section 9.3.2. For Brookhaven's chamber see Androulakis et al., 'Brookhaven National Laboratory 80″ Hydrogen Bubble Chamber Status and Plans', Instrumentation Conf. (1962), 100-9.

123. For the lack of a project leader see CERN/FC/337, 12/2/59 and CERN/FC/344, 31/3/59.
124. The data for CERN is from the 1961 Annual Report, for Brookhaven from the paper cited in note 122.
125. Memorandum, *Priority Scale of CERN Activities,* unsigned, but apparently written by Weisskopf on 10/5/62 (MGNH2064).
126. Council, minutes, 13/6/62, 17, for Weisskopf, SPC meeting minutes, note 107, for Perrin.
127. See the memo cited in note 125.
128. This section is based on my far more extensive *The Development of Techniques for the Analysis of Track-Chamber Pictures at CERN* (Geneva: CERN Report CHS-20, 1987). I have also benefitted considerably from Galison (1985a).
129. The quotation is from Bradner & Glaser (1958), 421. See also Bradner (1960a) and Rosenfeld (1959).
130. In his Introduction to the session 'Reduction of Data from Bubble Chamber Film', Instrumentation Conf. (1960), 223.
131. For his trip report see Goldschmidt (1957). The outline of CERN's initial Iep programme is in Goldschmidt et al. (1957), and von Dardel et al. (1957).
132. For information on the equipment available at Berkeley see Bradner and Solmitz (1958) and Iep Meet. (1958).
133. The evolution of CERN's Iep building programme is described in CERN/SPC/54, 25/3/58, CERN/SPC/55, 14/2/58, CERN/SPC/78, 17/11/58, and CERN/SPC/90, 31/3/59. See also Memorandum, Goldschmidt-Clermont to Peyrou and Bassi, 30/4/59 (LK22483).
134. See memo Macleod and Montanet to Goldscmidt-Clermont entitled *DD (Exp) Estimates for 1963,* 24/8/62 (LK22483).
135. The performance in October 1964 is mentioned by Hine in his memo to the members of the Data Handling Policy Group (see later), reference no. DHPG12, 6/10/64 (DIRADM20324). He suggested that it was a maximum in CERN/SPC/182/Add., 20/4/64.
136. The minutes of the meeting are CERN/SPC/149/Draft, 27/11/61.
137. Gregory made these remarks at the meeting just cited and at the 26th meeting of the SPC – minutes CERN/SPC/170, 16/5/63.
138. See the memo cited in note 134.
139. The classic papers on the HPD are Hough & Powell (1960a) and (1960b).
140. Roads were first established by coating the interesting parts of the film with red dye (letter Hough to Shutt, 16/2/60 (LK22482)). Later it was decided to draw the roads on a transparent strip beside the picture but rigidly attached to it, and to measure their coordinates with an additional flying spot (letters Hough to Shutt and to Gosselin, 12/5/60, both on (LK22482)). In August the idea was to mask out the unwanted areas of the bubble-chamber image by placing on top of it another film on which the roads appeared as transparent zones in an opaque field – Hough & Powell (1960b). Rough digitizing is described in Hough & Powell (1960a), a paper given in mid-September 1960.
141. See letters Hough to Shutt, 16/2/60 (from which the quotation is taken), 4/60, and 18/7/60, letter Hough to Gosselin (IBM), 12/5/60, and the proposal to develop 709 Matched IEP-Y to Kowarski and others at CERN, and to Rosenfeld and others at Berkeley, 12/9/60 – all on (LK22482).
142. The framework for the collaborative project was spelt out by Kowarski in two memos, entitled *New Developments in Data Processing for Track Chambers* and *Note on the Proposed Collaboration CERN-U.S.A. on Data Processing,* both on (LK22482).
143. For discussions at the level of the CERN management see CERN/DIR/2, 14/10/60, CERN/DIR/3, 20/10/60, CERN/DIR/4, 27/10/60, and CERN/DIR/6, 5/11/60, all in (DG20862). See also letter Kowarski to Thorndike, 28/10/60 (LK22482).
144. Letter Thorndike to Kowarski, 25/10/60 (LK22482).
145. For a survey of these developments see CERN/SPC/132, 26/6/61, and Kowarski's presentation of it to the SPC in CERN/SPC/133/Draft, 7/7/61. The official report of the meeting held in May is Benot et al. (1961).

146. The quotations are from letters Kowarski to Russell, 22/9/61, and to Hough, 4/10/61 (LK22482).
147. Letter White to Kowarski, 14/11/61 (LK22484), White et al. (1962), memo Macleod and Montanet (note 144), and Kowarski's presentation to the ECFA meeting on 17-18/1/63, FA/EC/2, 65-7 (DG20783). See also Franck et al. (1962).
148. For a general idea of the problems being encountered in programming for the HPD see the papers presented at FSD Conf. (1963) and FSD Conf. (1964).
149. For quotations on the situation at BNL see Abrams et al. (1964) and Hough (1963). For that at LBL see letter Russell to Kowarski, 26/2/63 (LK2384), and White (1965). For the quotation about CERN by Tycho see FSD Conf. (1963), 131. Moorhead described their approach in Moorhead & Krischer (1963)
150. See Roberts (1963), 31 for the quotations. There is a far more extensive description of the development of spark chamber processing at CERN in my Report CHS-20 (see note 128).
151. The performance figures of the HPD in 1964 are from *CERN Programme 1964-1968: Data Handling and Computing Equipment*, DIR/AP/136, 2/4/64 (MGNH22062), for Luciole in 1965 from report by B.Stumpe, CERN/DD/DA/65/5, 2/4/65 (MGNH22002), and for Luciole in 1968 from a memo by P. Zanella written in May 1969 and entitled *Summary of Spark Chamber Photographs Measured so Far* (in the personal collection of B. Powell).
152. For CERN see letter Kowarski to Hough, 4/11/60 (LK22482). For LBL, see typically the informal minutes of FSD Meeting No.2, 19/9/61 (LK22484) chaired by White and attended by 19 people. For BNL see letter Hough to Kowarski, 5/9/61 (LK22482).
153. CERN's general policy vis-à-vis Europe is outlined in discussions held at the eighth and eleventh meetings of the SPC – CERN/SPC/63/Draft, 2/4/58, and CERN/SPC/84, 2/12/58. For the conference see Iep Meet. (1958).
154. The proceedings of the smaller meeting were published as Iep Conf. (1959). The *Final Report of Standardization Committee* recommending the design proposed by SOM is dated 10/12/59 (LK2483).
155. The figures are from Goldschmidt (1960) and from his memo CERN/TC/COM 62/5/Rev, alternatively DD/IEP/61/45, dated 20/2/62 (LK22483).
156. Hine's data and the quotations are from his *Statistics on Research Collaboration Between CERN and Other Countries*, CERN/SPC/213, 28/11/65. See also CERN/TC/COM/65/32 for information on the scanning tables.
157. The information in this paragraph was supplied by B. Powell and G. Moorhead in a conversation with the author. See also FSD Conf. (1963), (1964).
158. See Table X in Hine's report cited in note 156. The table was constructed 'by assuming that the film for any particular run was divided equally between the participating groups' (p. 27).
159. Kowarski's paper proposing that CERN purchase a computer was CERN/SPC/13, 11/11/55, to which he attached an extract from Anderson's letter. The minutes of the SPC meeting at which it was first discussed are CERN/SPC/14, 7/12/55. The subcommittee which it set up met on 22/11/55; for their report see Annex of letter Richemond to Finance Committee, 8/12/55 (SC22840). The minutes of the FC meeting on 20/12/55 ratifying the purchase are CERN/FC/108, undated.
160. For the delivery dates of the Mercury see CERN Annual Report, 1958, 56.
161. All of the quotations in this paragraph are from the sources quoted in note 159.
162. Difficulties with the input/output system are described by Kowarski in CERN/SPC/92, 8/5/59, and in CERN Annual Report, 1958, 61.
163. The *Report on the Meeting to Discuss the Possible Extension of CERN Computing Facilities held on 21 and 22 May 1959* was shown to me by Gerry Moorhead. For the rest of the material see Kowarski's paper for the SPC, CERN/SPC/99, 25/9/59, the minutes of the SPC meeting, CERN/SPC/101, 19/11/59, and the minutes of the FC meeting, CERN/FC/399/Draft, 20/11/59.
164. See minutes of meeting, CERN/SPC/112, 10/6/60.
165. Rosenfeld's ideas are in Rosenfeld (1959). For other quotations see CERN/SPC/99, 25/9/59, CERN/SPC/101, 19/11/59, and CERN Annual Report, 1960, 85-7.

166. For the budget difficulties at the end of 1959 see chapter 10.2.3. The terms offered by IBM are spelt out in CERN/FC/390, 10/11/59. See also CERN/326, 29/10/59, and its two addenda. Changes to them are discussed in CERN/FC/437, 16/9/60, and its addenda. The FC meeting agreed to these changes, and the rental of the 709, on 4/10/60 – see minutes CERN/FC/447, 24/10/60. See also CERN Annual Report, 1960, 85.
167. The background to this initiative was the attempt to place CERN's medium-term plan on a more systematic basis, in accordance with the needs of the Bannier procedure which we describe in detail in the next chapter. A survey of the laboratory's requirements for the next four years led Hine, in May 1962, to write that the IBM 709 would soon need replacing by a 7090 (CERN/443, 31/5/62). The official proposal by Kowarski to the Finance Committee is CERN/FC/552, 27/8/62. The debate in the meeting was in fact conducted in terms of acquiring the very similar IBM 7094 (cf. CERN/FC/561, 5/10/62), an idea which was later dropped.
168. The British objections, and the decision to set up an expert panel are in CERN/FC/561, 5/10/62. For the expert's reports see CERN/FC/579, 7/11/62 and, more definitively, CERN/FC/592, 17/12/62. It was Weisskopf who said that the CDC was the best machine on the market and that the 7090 was merely stop-gap – see minutes of FC meeting on 18/12/62, CERN/FC/598, 8/1/62. The management's case was spelt out by Kowarski in CERN/FC/586, 26/11/62, and by Weisskopf in CERN/FC/594, 17/12/62. It should be pointed out that the expert panel partly based its case for the CDC 3600 on (mistaken) rumours that both BNL and LBL were acquiring this computer specifically for data-handling.
169. See CERN Annual report, 1963, 90.
170. This last information was in an additional document to those just cited, CERN/FC/597, 10/1/63.
171. The quotation is from Weisskopf's CERN/FC/594, 17/12/62.
172. The founder members of the Data Handling Policy Group were Hine, Kowarski, Peyrou, Preiswerk, and Puppi. Hine was its first chairman – see his report on *Development in Facilities and Organization of Data Handling and Computing in CERN*, DIR/AP/120, 27/5/63 (DG20598).
173. The term 'rationalization' was used in the CERN Annual Report, 1963, 90. For the reorganization proposed see the report cited in the previous note and Hine's memorandum to the Policy Group on data handling and computing dated 18/2/63 (DG20598). For what was actually done, see CERN Annual Report, 1963, 90.
174. Memo from Hine to all staff members, fellows and visitors, DIR/AP/121, 29/5/63 (DIRADM20324).
175. Council, minutes, 16-7/6/65, 30.
176. See the report on the meeting, CERN/FC/600, 28/1/63.
177. The Report of the committee, from which the organizational details are taken, is CERN/516, 25/10/63. Its Appendices are CERN/516/Appendices, 25/10/63.
178. This entire paragraph is based on the report quoted in the previous note.
179. Hine's paper is CERN/FC/653, 1/11/65. The opening discussion in the FC can be found in CERN/FC/657/Draft, 6/12/63. Some of the UK's objections to CERN's argumentation are in CERN/FC/665, 16/12/63.
180. The minutes of this meeting are CERN/FC/668/Draft.
181. CERN Annual Reports, 1965, 90, 1966, 107-8.
182. For the statements by Weisskopf see CERN/FC/611, 25/3/63, and CERN Council, minutes, 20/6/63, 10. For Gregory see CERN/SPC/170, 16/5/63.
183. See Hine's report on his visit to the USA circulated widely among the CERN senior staff and dated 9/9/64 (DG20598), at p. 27.

CHAPTER 10

Finance policy: the debates in the Finance Committee and the Council over the level of the CERN budget[1]

John KRIGE

Contents

10.1 The emergence of the idea that there be a ceiling imposed on CERN's expenditure	573
10.1.1 1955 and 1956: the attempts to formulate long-term plans	573
10.1.2 Summer 1957: the problems posed by the rising costs of research	578
10.1.3 December 1957: the first ceiling is imposed	581
10.1.3.1 Early divisions	581
10.1.3.2 The 1958 budget is voted	584
10.2 1958 to 1960: the continuation of the ceiling policy and its use as a means to stabilize the CERN budget	586
10.2.1 May and June 1958: financing a bubble chamber programme	587
10.2.2 Autumn 1958–May 1959: breaking the 1959 ceiling to finance the 'New PS experimental programme'	589
10.2.3 1959 and 1960: accommodating the rising costs of research to a three-year level ceiling	592
10.2.4 CERN's inability to plan ahead	594
10.3 1961: the three-year ceiling is broken and the foundations are laid for a new policy	596
10.3.1 The confrontation between the CERN 'lobby' and the British	596
10.3.2 Some remarks on the factors at work in this conflict	602
10.4 The Bannier Report and the budget debates from 1962 to 1965	604
10.4.1 The Working Party and its Report	604
10.4.2 The practical application of the Bannier Report's recommendations from 1962 to 1965	606
10.4.2.1 The budget debates in 1962 and 1963: the 'equipment gap'	607
10.4.2.2 The budget debates on the 'improvement programme' in 1964 and 1965	613
10.5 Concluding remarks: the significance of the Bannier Report	620
Notes	623
Bibliography	809

There are several ways in which one can study the financing of an international scientific laboratory. For example, one might concentrate mainly on the procedure *inside the laboratory* itself whereby the estimates for one or more subsequent years are compiled by the divisions, and combined into a global budget for the organization by senior management. The grounds on which such estimates are made, the accounting methods used to present various items of expenditure, the tactical discussions about the level at which the Director-General should pitch his budget claim to the Finance Committee and the Council, the internal conflicts within and between divisions over who should absorb cuts and how—all of this is part of what the financial management of a laboratory like CERN entails. In the same way one may choose to study the corresponding negotiating process *inside the member states,* the process whereby the national delegates to CERN secure the necessary funds for the CERN budget from their domestic financial authorities. Of interest here would be the kinds of arguments they used to justify an increase in expenditure, the instructions they were given and how they interpreted them, the internal debates over national science policy and CERN's place in it, and interdepartmental conflicts over the importance of CERN and the level of funding it merited. Again, although we shall frequently catch glimpses of these discussions, they are not our main concern here. Instead we have put at the centre of our analysis the *interface* between the CERN administration and the member states, we have chosen to focus on the various methods of managing the overall financing of CERN in the short to medium term which were devised by national delegates in consultation with the CERN senior staff. This means that we shall follow year by year the negotiations between the delegates and the administration over the precise level of the CERN budget, as well as look at the sometimes profound differences of opinion between various national groups on policy towards CERN, the compromises that they reached, and the conflict that, in 1961, threatened to damage the core of the organization itself.

The United Kingdom delegation played a key role in these debates. This was primarily because, for much of our period, Britain was both a major contributor to the CERN budget and something of a reluctant partner in the CERN adventure, always doubting whether the laboratory really gave value for money, and constantly seeking ways of imposing limits on its expenditure. For the British the main role of financial policy was to impose some form of external financial control over the laboratory's expenditure. This set them (and their supporters, generally the Swedes) clearly apart from the majority. For the latter, financial considerations, while obviously important, were essentially a secondary concern. They were determined to build CERN into one of the best laboratories of its kind in the world, and by and large they were willing to give the administration what it said it needed to achieve that objective. For them policy was something supple, a set of broad guidelines on appropriate levels of expenditure to be interpreted in the light of particular needs and conjunctural circumstances. This chapter studies how and why these two groups, starting from essentially different premises, finally came to agree on a set of principles which provided a suitable framework for the financial management of CERN.

The debates on the budget studied here fall into two main parts: that stretching from 1957 to 1961, and that covering the period from 1962 onwards. During the first phase the Council, on the insistence of the British, attempted to contain CERN's expenditure by imposing a two-, and then a three-year ceiling on its expenditure (section 10.2). This policy proved extremely difficult to apply in practice at the time, mainly because of major uncertainties surrounding the costs of research in the new energy range opened up by the PS, and when the British again tried in 1961 to stabilize CERN's budget for three years, they were attacked head-on by some of the most influential members of the Council (section 10.3). The conflict was resolved by setting up a Working Party on the CERN Programme and Budget chaired by J.H. Bannier, and between 1962 and 1965 its recommendations provided the backdrop against which the budget debates took place (section 10.4). In the concluding section (10.5) we pause to reflect on the significance of the Bannier Report, which undoubtedly coincided with a turning point in the attitudes of the British, in particular, towards the financing of CERN. To begin with though a few words on the early, failed, attempts by the Finance Committee and the management to base CERN's expenditure on long-term plans and the emergence of the idea of imposing ceilings (section 10.1).

10.1 The emergence of the idea that there be a ceiling imposed on CERN's expenditure

The first serious initiative to impose some form of 'external' financial control on the CERN budget was taken by the British delegation in summer 1957. Before describing it it is useful to have some idea of the context against which it emerged, a context characterized by the more or less explicit assumption that CERN could and should have as much money as the administration thought was needed for the laboratory to fulfil its mission.

10.1.1 1955 AND 1956: THE ATTEMPTS TO FORMULATE LONG-TERM PLANS

Officially at least, the dozen member states who signed the CERN Convention on 1 July 1953 did so with a particular picture of the laboratory's spending pattern in mind. A preliminary version of its needs had been circulated in October 1952; the data were updated by the Executive Group of the 'provisional' CERN in the document CERN/GEN/5, the *Second Report to Member States on The Organizational and Financial Implications of Future European Co-operation in Nuclear Research,* which was issued some two months before the Convention was signed.[2] This official picture had two main features. Firstly, it assumed that during the construction period of the laboratory, lasting about seven years (i.e. until mid-1960), annual expenditure on CERN would be constant at about 17 MSF per annum. Thereafter CERN/GEN/5 indicated that annual expenditure would decrease considerably to a new level once both machines began to operate. The image that governments had of CERN's expenditure pattern was then one of two plateaux separated by a sharp drop around 1960, as the laboratory moved from the phase of building the machines to exploiting them. Thus when the British Chancellor of the Exchequer, for example, agreed in December 1952 that the United Kingdom join CERN he saw the country as 'incurring the potential financial commitment estimated at £348,000 a year for seven years [the first plateau] and thereafter operating expenses of the order of £134,000 per year [the second]'.[3]

Alongside this official picture there were the expectations of the government representatives in the 'provisional' CERN Council itself. By October 1953 most of them did not doubt that the Proton Synchrotron would cost far more than originally estimated and that the total construction cost of the laboratory (120 MSF = 7 × 17 MSF) would be exceeded as a result. In October 1952 Odd Dahl, the first PS Group leader, told the Council that a 30 GeV strong-focusing machine would cost the same as a conventional machine of 10 GeV, i.e. some 55 MSF. Shortly thereafter the PS Group estimated that the machine would in fact cost about 70 MSF including its buildings. In the light of this development the Council insisted that the design energy be reduced to 25 GeV so as to cut costs. At the same time it accepted that the final price could only be known when the contracts for the major items of equipment had been awarded.[4]

Within months of the laboratory formally coming into being (September 1954) the Finance Committee decided that the sooner the figures in CERN/GEN/5 were replaced by something more realistic the better. Accordingly it asked the administration to draw up a new estimate of the laboratory's long-term expenditure.[5] The first such document, entitled 'Capital Investment Programme 1952–1960', was released by the administration on 4 November 1955; a slightly modified version was issued about a month later. The exercise was repeated towards the end of the following year when a revised Capital Investment Programme for the period up to 1960 was drawn up by CERN.[6]

There are two important points to remember about this early period. Firstly, that none of the delegates to the CERN Council or the Finance Committee were particularly perturbed by possible increases in the costs of laboratory. In October 1953 the Council President, Sir Ben Lockspeiser, was confident that governments 'would certainly understand' that it was difficult to give reliable estimates of expenditure for the PS in particular. Belgian delegate Jean Willems agreed.[7] Correlatively, the main purpose of having long-term plans drawn up in 1955 and 1956 was *managerial*. On the one hand they were meant to help the Finance Committee appreciate the long-term implications of the budgetary decisions they were taking in any given year so as 'to give to Member States a chance of estimating to what extent they need increase their national contributions [...]' to CERN.[8] On the other they were intended to obviate the 'procedural problems' which arose when domestic financial authorities did not have enough time to accommodate estimates of CERN's needs in their national budgets. The financial year of most national treasuries was six months out of phase with that of CERN, and in any given year they liked to know by July what their contributions to the CERN budget for the following calendar year would be.

The second noteworthy feature about these early attempts at long-term planning is their total failure. This is evident from Figure 10.1, which graphically illustrates the wide divergence between the administration's estimates made at the end of 1955 and of 1956, and CERN's actual expenditure. Indeed after this initial setback all serious

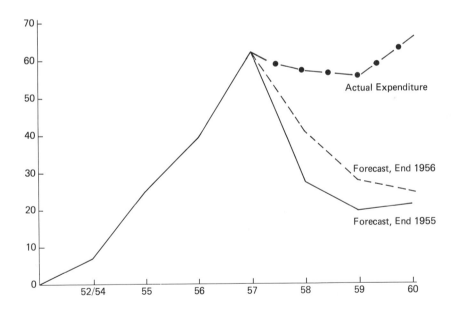

Fig. 10.1. The long-term forecasts made by the administration at the end of 1955 (from CERN/FC/89/Rev., 9/12/55), at the end of 1956 (from CERN/FC/150/Rev., 15/10/56), and the actual expenditure curve (from the CERN Annual Reports).

Notes: pp. 623 ff.

efforts at long-term forecasting of the laboratory's needs were abandoned until the early 1960s.

Why the divergence? The major factor upsetting, not to say completely demolishing the CERN estimates was the costs of research, and of those arising from the use of bubble chambers in particular. Pioneered by Luis Alvarez at Berkeley in the mid-1950s, these detectors had to be embedded into a sophisticated and expensive infrastructure which included well-focussed and purified secondary particle beams, properly ventilated and protected experimental areas, and complex data-analysis equipment. Figure 10.2 gives one an idea of the corresponding costs. It shows the money spent on so-called current research (detectors and associated apparatus, track chamber film and its analysis, computing, theoretical studies, visitors, fellows, etc. and buildings) and on beam transport (magnets, separators, power supplies, cooling, shielding, and buildings). We see these items growing from about one-sixth of the CERN budget in 1957 (some 10 MSF) to almost three-fifths in 1960 (some 38 MSF).

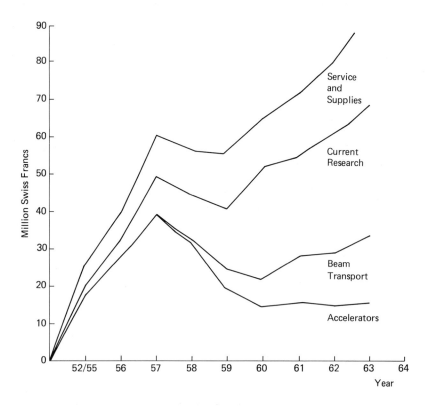

Fig. 10.2. The distribution of CERN's expenditure between various activities (Source: CERN Annual Report, 1963, 149).

The recognition that they needed to revise drastically their ideas about what was needed for future research erupted into the consciousness of the CERN staff at the 'CERN symposium on high energy accelerators and pion physics' held from 11 to 23 June 1956.[9] They came away from the meeting convinced that CERN was in danger of lagging behind its competitors in diverse aspects of high-energy physics, notably the conception of new machines and the implementation of new techniques for exploiting existing machines. The development with the most immediate impact on the budget was the 'unexpected rapidity' of the evolution of 'research techniques in the π-meson field', 'particularly in the United States and the U.S.S.R. [...]'. Scientists in these countries, said SC division leader Gentner, 'had successfully carried out most of the experiments CERN had planned to do with its own machine. It was therefore necessary', he went on, 'to carry out far more complicated experiments which would call for very expensive apparatus'.[10] A number of emergency measures were immediately taken to deal with the situation. 650,000 SF of the current year's resources were allocated for capital expenditure on experimental equipment for the SC group. At the same time the Group drastically revised its long-term plans (see table 10.1). It increased its overall estimates for each following year, including 1957, by between 75 and 100%, to cover expenses like 'the so-called "liquid hydrogen bubble chamber" technique' and 'ultra high-frequency detection apparatus', to expand its staff, and to associate 25 research fellows annually with the group.[11] Even then the Director-General made it clear that this was only the tip of the iceberg. The discussions on the SC experimental programme, said Bakker, were 'merely ushering in the problem of experimental research as a whole'. No one had any idea yet what the costs of research on the PS would be—though it was 'likely that discussions would begin in the near future' on this subject—, and the PS group had as yet made no specific allocations for it in the new long-term plan.[12]

Table 10.1

The SC Group's estimates of its overall expenditure and staff needs for 1957–1960 made at the end of 1955 (the 'probable' figure given in CERN/FC/89/Rev., 9/12/55) and again at the end of 1956 after the CERN symposium (from CERN/FC/150/Rev., 15/10/56).

Financial year	Estimate made in	Total exp. (MSF)	Staff & fellows
1957	1955	3.1	53
	56	5.4	109
1958	1955	2.5	57
	56	4.4	110
1959	1955	2.0	61
	56	3.9	105
1960	1955	2.0	65
	56	3.8	105

Notes: p. 624

10.1.2 SUMMER 1957: THE PROBLEMS POSED BY THE RISING COSTS OF RESEARCH

The Finance Committee delegates were unanimous in their concern about the rising costs of research. In response the French delegation called for a number of technical changes in accounting methods. These were intended to stop the CERN administration smuggling in what would, in fact, amount to an unauthorized expansion of the programme. They also suggested that the voting procedure on the budget — which could be passed by a simple majority — be changed so as to give more weight to the votes of the major contributors, an idea which they subsequently withdrew when Council President Sir Ben Lockspeiser stated publicly in December 1956 that as long as he was at the helm 'he would consider it most unwise to ask the Council to vote the budget, unless it secured the quasi-unanimous approval of the Member States'.[13] As for the United Kingdom, their delegation decided to tackle the problem at its roots, to try to impose some form of 'external' financial control over CERN's expenditure.

Britain's aim, as explained by their senior delegate to the Finance Committee H.L. Verry, was to try to formulate a budget policy for CERN which was 'intelligible and statesmanlike', and which would 'reassure the financial authorities of the Member States without unduly hampering the healthy existence of the laboratory'. He circulated his first considered thoughts on the matter privately to Finance Committee members Bannier (NL), Neumann (F), and Willems (B) on 5 March 1957. He then drafted a more formal paper which he laid before the thirteenth meeting of the Finance Commitee on 1 and 2 May, where he spoke at length on the issue.[14]

'At the moment', said Verry, going straight to the heart of the matter as he saw it, 'if I were asked to explain CERN's financial policy, I would say it hadn't one'. In Verry's view this situation had arisen because the original framework for limiting expenditure was now obsolete. When the laboratory had been planned, said Verry, the member states, and particularly the United Kingdom, had thought to limit their liability by restricting its main activity to the execution of a so-called 'basic programme'. 'The programme', he went on, '[was] purely a programme of buildings, and building two accelerators; it [said] nothing about the extent of the research [...]'. This was now seen to be a serious omission for a laboratory like CERN, which was working in a field where there were 'limitless possibilities in the way of research opening out almost every day [...]', and in which there was 'very, very hot competition with two countries who [could] spend almost unlimited amounts, the United States and Russia'. In short, with the shift from construction to research, one 'could quite ingeniously build up a tremendous budget completely within the basic programme', which was rapidly 'becoming a complete fallacy for financial safeguard'.[15]

Verry's intention at this stage of the proceedings was not to propose solutions but to raise an issue. In the paper he laid before the Finance Committee he recommended that the matter first be examined by the Committee's Working Party in consultation with the Director-General and a representative of the Scientific Policy Committee. He also expressed the hope that the FC would be 'able to put recommendations to the Council in time for the latter to take a decision at its June Session'.[16]

During the following weeks CERN's Director of Administration, Jean Richemond, drafted a lengthy reply to the British paper, which was discussed and modified by the Group of Directors on 22 May 1957. Bakker laid their agreed version before a meeting of the FC Working Party on 28 May. It was briefly discussed on this occasion, and again on 26 June 1957, when the entire meeting of the Working Party was dedicated to the CERN paper and to Verry's comments on it which he had circulated beforehand.[17] It was mutually agreed that there was no point in trying to put a recommendation to the Council just yet, and it was decided instead that the Working Party and the CERN administration should draft an 'agreed note' outlining the situation as they now saw it. This 'Joint Note by the Director-General and the Finance Committee' entitled 'CERN—Financial Policy' was issued on 29 August 1957. In it the Director-General and the Finance Committee (in fact Verry) 'both recognize[d] that since ultimately the scale of research depends on the capacity of the Member States to find the money, it may well be found that the only practical course will be to set a financial ceiling, below which research projects would have to be accommodated by some scheme of priorities'. Thus was the foundation laid for a policy of 'external' financial control over CERN's expenditure.[18]

The wording of this 'agreement' embodied a compromise between the attitudes of the British delegation, who felt that the time had come to impose limits on CERN's expenditure, and the administration and some influential members of the Finance Committee, who felt that it had not. There was a 'sharp difference of outlook between Bakker and us', wrote Smith to Verry inside the DSIR earlier in the year. 'Bakker seems to assume', Smith went on, 'that provided he has given the SPC an estimate of the cost of any new proposals and provided the SPC has taken the financial implications into account (whatever that may mean in practice) he has discharged his obligations to the FC—all that remains is for the FC to raise the money for those things that the SPC has looked on and seen to be good'. As a result, as Verry explained to the FC in May, the committee risked degenerating 'into little more than a rubber-stamping body'.[19] A change of attitude inside the organization was needed if this situation was to be rectified. In particular, CERN had to understand, said Verry, 'what should have been obvious', that it 'depend[ed] on the contributions of Member States for its finance, and that there must be a limit to what can reasonably be expected'. Concretely this meant being prepared to set priorities, that the 'FC has got to look at the broad financial picture first and it [was] for the SPC to work out its policy in that framework'.[20] In short the British were asking that the 'standing policy', whereby the administration, in consultation with the SPC, laid down a scientific programme, which the member states through the FC were expected

Notes: pp. 624 ff.

to finance 'according to requirements' (Willems) be replaced by one in which CERN be made 'to fit its programme within the bounds of well defined appropriations' (Willems).[21]

In reply the administration insisted that the British were making unrealistic and counterproductive demands. Firstly, they claimed that they could not give any idea of what the research programme might cost. 'It was well-nigh impossible', Bakker told the FC Working Party, 'to estimate the cost of research expenditure, even for the Synchro-cyclotron, since new ideas in that field were still in full development'. The same held true, even more so, for the other machine: it was 'extremely difficult to estimate the cost of the research equipment which should be needed for the Proton Synchrotron'—in fact, Bakker went on, the equipment needed 'was still largely unknown'. Secondly, the administration insisted that neither they nor, in their view, the Scientific Policy Committee should bother about financial considerations in drawing up the research programme. 'For the next year or two at least', Bakker said, 'it would be necessary to concentrate on planning the research equipment without considering the question of cost. The whole question was far too fluid for the cost element to be worked out and any financial considerations were likely to act as a brake on the development of new ideas'.[22] People like Bannier and Willems were inclined to agree. As far as they were concerned it was premature to try to use financial policy to shape scientific policy—in fact, Bannier mused, one might have to admit that 'scientific policy, in this field, [was] inherently unstable', and that what Verry was asking for was 'unanswerable'.[23]

There are several reasons why it was the United Kingdom delegation, in particular, that took this initiative at this time. Firstly, there was the government's ongoing lukewarmness *vis-à-vis* international organizations. In 1956, Verry said, the UK had contributed 'some six million pounds sterling, very largely in hard currency [to such bodies], and CERN was among the half dozen largest contributions'. It must not be forgotten, he pointed out, 'that some Governments now instinctively associate the thought of International Organizations with ever-increasing budgets', and that if CERN's needs were 'to continue to command the sympathy of, at any rate the U.K. Treasury, the Organization must show that the size of the annual budget is under effective control, and based on an intelligent and realistic policy'.[24]

Then there were the criticisms being voiced by the domestic physics community: that CERN was 'in fact l'enfant gâté' of their government. Among 'national workers', said Verry, CERN 'has the reputation of being able to get anything it asks for, as compared with the difficulties encountered by the national institutes in getting State grants'. As a result scientists in Britain, 'animated perhaps by quite understandable envy [...]' were suggesting that CERN was 'lavishly equipped'—the people in universities, Verry said, had to deal 'with bits of string and nails' and had 'to improvise apparatus'. Matters were not helped by the attitudes of the CERN staff who tended, said Verry, 'to look down somewhat condescendingly on the less well-equipped and perhaps make-shift apparatus of national institutes'. Nor by the 'boasts, or rumours of boasts, that in a few year's time the staff will have reached the thousand level'.[25]

Then—though this was not mentioned aloud by Verry at the FC meeting—, there was the financial squeeze engendered by Britain's determination to develop its national programme in parallel with CERN. On 14 February 1957 the foundation of NIRNS—the National Institute for Research in Nuclear Science—was announced in the House of Commons. According to one of the key planning documents, the centre was to cost some £10m (about 120 MSF), £8m of which was for 'two large machines of different types', one being the 7 GeV high-intensity proton synchrotron, Nimrod.[26]

These arguments point to a fundamental difference between the financing of CERN in Britain and that in at least some of the other member states, a difference which persists to this day and which continues to differentiate the United Kingdom from some of its partners: it is the source of money for CERN in the national budget. In countries like France, Italy, and (at least at first) Germany, membership of CERN was seen by high state officials primarily as a political gesture, and resources for it were made available by the departments of Foreign Affairs. Money for CERN was thus 'independent' of money voted for national science expenditure. In Britain by contrast, there was less enthusiasm for international scientific collaboration, and the main motive for joining CERN was said to be scientific and economic: it was the cheapest way of doing physics at 25 GeV. As a result the DSIR played a dominant role in presenting the case for membership to Ministers in 1952, and the funds for the laboratory were part of its overall (civilian) science budget. In practice this meant that universities and national research institutes competed with CERN for funds from the same source—and generally lost out. As Verry explained, 'Whether it realizes it or not [CERN] stands in a very, very privileged position when one considers that in everyone's own country scientific societies or organizations come to the Governments for money, and that they are usually either turned down or given half of what they ask for. CERN is in the happy position of nearly always getting what it asks for'.[27]

10.1.3 DECEMBER 1957: THE FIRST CEILING IS IMPOSED

When the Finance Committee Working Party and the Finance Committee met respectively on 15 and 16 October 1957 they had before them the draft budget for 1958 prepared by the CERN administration. It amounted to 59 MSF—the same as the expenditure anticipated for 1957 and almost 50% more than the amount predicted in the long-term forecast prepared the year before (40 MSF—see Figure 10.1). In the words of Jean Willems the estimate 'put CERN in the most serious situation it had ever had to face since it had been created', and in fact it led President Lockspeiser to call for an extraordinary Council session in November.[28] How that 'situation' was handled and, more specifically, the struggles led by the British to 'resolve' it by imposing a ceiling on CERN's expenditure form the subjects of this section.

10.1.3.1 Early divisions

In their initial responses to the 1958 estimates, the Finance Committee delegates split roughly into three groups. Firstly, there were the British delegates Verry and Smith, who made much of the running at the Working Party meeting: in their view, the administration had to make substantial cuts in the proposed 1958 budget. They were furious that the laboratory had *already* committed more funds for 1958 than the 40 MSF foreseen in the long-term plan ('the placing of orders depended on the funds available', said Verry, 'a fact that CERN should bear in mind'), they were unconcerned about the effects of slowing down the programme ('CERN's position differed little from that of a Government department carrying out a programme on the assumption that funds would be available', said Smith. 'If those proved insufficient, certain parts of the programme had to be interrupted'), and they were totally unmoved by the administration's plea that a major reduction in the 1958 budget would be 'catastrophic' for CERN ([Verry] 'wanted to be given a precise answer [as to the effects of cuts] and he could not be content with dramatic or melodramatic considerations that might be put forward by the Director of Administration'). In short, and consistent with the policy that they had 'agreed' on with the Director-General in August, the British wanted CERN to draw up a list of priorities—and insisted that the administration study the effects of reducing the 1958 estimate from 59 MSF to 50 MSF.[29]

Secondly, there were a number of delegates whom, while more or less inclined to support the British line, were unwilling to commit themselves at this stage. This was true, for example, of Obling (DK), who 'felt that the figure was high and [that] it would probably be necessary to reduce it', but 'found it difficult to suggest any cuts', of Neumann (F) who, while remarking that the administration had already pruned its estimate by about 3MSF, 'wished to know whether additional cuts might be possible', and of Campiche (CH), who asserted that 'the Swiss financial authorities would welcome certain budgetary cuts, provided they did not affect the programme'.[30]

Finally, there were the attitudes of Willems (notably at the Working Party) and of Bannier (when the full committee met the next day). For them it was 'pointless to scrutinize the budget in order to effect a few cuts' (Willems). In their eyes the task at hand was not to force the CERN administration to reduce its 1958 budget—and Willems appealed to the glowing statements about CERN's achievements made by members of an expert Staff Advisory Committee to justify this view. It was to find a way around the *procedural difficulties* caused by the increase in the 1958 estimates several months after most domestic financial authorities had fixed their allocations for that year. As Bannier put it, 'If the delegations had been able to anticipate such a development the 1958 Budget would not raise any serious issue'.[31]

What was the solution they put forward as an alternative to cuts? As sketched by Willems and developed by Bannier it was that CERN be allowed to spend roughly what it needed in 1958, and that the deficit be absorbed in the allocations for CERN the following year. The administration quickly prepared an estimate of their needs for the next two years (about 110 MSF), and in the light of this figure Bannier asked that the Director-General prepare more careful estimates for 1958 and 1959 'for which it might be possible to obtain from the Member States a total appropriation of 100 Million Swiss Francs'.[32]

Was this intended to be a ceiling which was not to be broken? Neither Bannier nor Willems were thinking in those terms at this stage. Their main aim, to repeat the point, was not to find a way of imposing limits on CERN's expenditure but to devise a technical 'trick' to cover the 1958 deficit.

This approach to the problems posed by CERN's rising budgets is indicative of the gulf which separated British policy, as voiced by Verry, from the attitudes of the members of the CERN lobby, as expressed by Bannier and Willems (see section 7.1.3). For the latter financial considerations were secondary, and were to be subjected to the 'higher' goal of building the organization into a laboratory which could compete on an equal footing with the best in the world. Within that perspective, for a man like Willems, it was premature to think of imposing ceilings—such a policy 'anticipated the time when the Member States which had pledged themselves to build up CERN would request the Organization to adapt its activities to available national financial resources', a time which, in his view, had plainly not yet come. 'No one', he said rhetorically, and with one eye on the recent commissioning of the Synchrocyclotron, 'could countenance the possibility of halting the first joint European scientific enterprise at the moment when it was entering on the phase of practical achievements'. In short, unlike the British, Bannier and Willems were not ready to reverse the prevailing policy of treating CERN as a 'privileged' beneficiary of funds.[33]

Correlatively, they differed too on the role of the Finance Committee. As far as Bannier was concerned, its delegates' prime role was not to challenge such privileges but to defend them before their national domestic authorities. Its members, said Bannier in October, and in a thinly-disguised reference to the British contingent, 'should not come to CERN Committee meetings with preconceived ideas' about the amount of money they were prepared to spend on the organization in any given year. It was only part of their task to criticize the budgets and programmes laid before them by the administration; they also 'had to secure the financial resources CERN needed to fulfil its task'. That granted, delegates were to bear in mind, said the Dutch member, that they were not simply 'representatives of the Member States [...] they were at the same time CERN's representatives with their respective governments'. And their task was 'to present CERN's point of view to their own governments'.[34]

Notes: pp. 625 ff.

The mode of thought evinced by Bannier and Willems was apparently not unique to some CERN delegates, and was indicative of a profound conviction that one's own country's interests were best served by European collaborative ventures. We find similar behaviour much later by some member states' representatives to the European Space Agency, 'who consistently interpreted their instructions in as favourable a light for ESA as was humanly possible' and who, in the words of an ESA Director-General, 'voted whenever possible for the European solution' 'without betraying their national masters'.[35] It was precisely their scepticism about the merits of 'European' solutions that set the British at CERN apart from some other delegations and from men like Bannier and Willems in particular.

10.1.3.2 The 1958 budget is voted

The vote on the 1958 budget was the culmination of several rounds of negotiations which began on 29 October with a joint meeting between three FC delegates (Neumann, Obling, and Verry), and four representatives of the Scientific Policy Committee (Blackett, Heisenberg, Leprince-Ringuet, and Scherrer). It was followed by meetings of the Finance Committee, the Committee of Council, and the Council between the 13 and 15 November, the aim at this stage being to have 'a discussion on the needs of CERN, so that all delegations should be fully conversant with the situation when they sought final instructions from their governments'. A similar sequence terminated with a Council meeting on 19 December 1957, when the 1958 budget figure was voted, along with a global estimate for 1958 and 1959.[36]

The most important development during this process was the adoption by the Council of the ceiling policy, and the insistence by the British and French delegations in particular, that if CERN was to be given 100 MSF for 1958 and 1959 together this figure was not to be exceeded. Support for the British policy put forward in the summer was first manifest at the joint meeting with the SPC at the end of the September (Neumann 'felt that a ceiling must be accepted for the CERN budget and that such a ceiling could hardly be much higher than 100 million francs for 1958 and 1959', while Obling stated that 'his personal view was that a budget of 100 million for 1958 and 1959 would be the maximum Denmark and the other two Scandinavian countries would be able to agree to').[37] It remained an essential component of the French position throughout the debate (in December Morat confirmed that a vote for 100 MSF was to be regarded 'as more than a declaration of intention. It must be tantamount to the adoption of a ceiling that was not to be exceeded').[38] In the face of this determination the Council provisionally agreed in November that CERN should be given roughly what it wanted for 1958 (56 MSF), and that its expenditure for 1958 and 1959 'could and should be held at a figure not exceeding 100 million Swiss francs', i.e. it pegged expenditure for 1959 at a maximum of 44 MSF. A resolution to this effect was passed at the December session with only Sweden voting against.[39]

How widespread was the support for this policy? Bannier and Willems were apparently not happy about it: what they had proposed as a conjunctural solution to a technical problem (as they saw it) had assumed the weight of Council policy. But the most consistently outspoken opponent was the German delegate Hocker, engaged at the time in domestic discussions about the Federal Republic's national science policy. For him a ceiling was antipathetic to the very nature of scientific research itself. 'It was extremely difficult', he said 'to prepare a scientific Budget more than a year in advance and [...] practically impossible to work out accurate long-term estimates'. Far from being compatible with ceilings, 'scientific research demanded great flexibility in financing so that new opportunities to carry out important research could be seized as soon as they arose', he added. And he recommended that the CERN administration be given what they said they needed for 1958 (58.9 MSF) and that no figure be laid down for expenditure in 1959.[40] In short, some fundamental doubts about the wisdom of imposing ceilings remained, even if the idea had the support of the two most powerful countries inside the organization (as well, it seems, as Italy and the Scandinavians). These doubts were to surface again repeatedly during the next few years.

To conclude, a few words about the apparently rapid shift in the British position, particularly during the round of discussions in mid-November. At the Finance Committee meeting on 13 November Verry said that 'his instructions did not enable him to accept a Budget above 45 million Swiss francs for 1958 or 90 million Swiss francs for 1958 and 1959 [...]'. The majority of the other delegations, by contrast, were prepared to support a figure of 100 MSF. These positions were apparently so firm that the Chairman of the FC abandoned the search for a recommendation that would have unanimous support.[41] However, two days later, at the extraordinary Council meeting on 15 November, the British came in line with the majority, and supported a proposal allocating 56 MSF to CERN for 1958 and a ceiling of 100 MSF for 1958 and 1959. Why had the British changed so quickly, and a month before it was 'necessary' for the UK delegation to show a willingness to compromise?

One reason for this seems to have been a determination by Lockspeiser, in particular, not to be bound by his official 'instructions'. When Verry stuck to the 90 MSF at the FC meeting on 13 November he was apparently adopting the rigid line laid down by the Treasury; when Lockspeiser voted for the 100 MSF two days later he was apparently expressing a personal determination to support the majority figure. In fact Sir Ben even went so far as to hint that he would allow his delegation to be outvoted in December if his domestic financial authorities were unwilling to revise their figure upwards.[42]

What considerations informed Lockspeiser's attitude? For one thing, it was probably a question of priorities: having got the Council to accept the principle of the ceiling, it was diplomatically fitting to 'compromise' on the less important question of the actual limit for 1958 and 1959. More fundamentally, though there was the desire to avoid confrontation, the wish, as Verry put it, not to be seen as

Notes: p. 626

acting unilaterally in a matter of concern to all the member states. This, in Verry's view, would not only damage the United Kingdom's prestige and her standing in European affairs. It would also damage the organization itself.[43]

Granted this concern for the wellbeing of CERN, we need to be more subtle than heretofore about the differences between 'the British' (and Verry and Lockspeiser) on the one hand, and 'the CERN lobby' (and Willems and Bannier) on the other. We need to remember that neither 'the British', nor any other national group for that matter, were necessarily a united bloc, and to appreciate the opposing pulls which informed the attitudes of their representatives in the Council and its committees. More particularly, the task of Verry and Lockspeiser was that of trying to reconcile the demands of a hostile Treasury and the scepticism of a resentful physics community— which they fully understood—, with their personal and professional dedication to CERN—which they shared with other members of the lobby. This meant that they constantly had to steer a course between the pressures at home to have CERN's costs reduced and their conviction that Britain's interests were best served by it playing a leading role in making the laboratory one of the best in the world.

10.2 1958 to 1960: the continuation of the ceiling policy and its use as a means to stabilize the CERN budget

As an aid to following the discussion let us begin with a quick survey of the main budget decisions taken by the Council between 1957 and 1960:

(1) As we saw in the previous section, towards the end of 1957 the CERN administration put forward a budget for 1958 of about 60 MSF, 50% higher than the figure foreseen in the long-term plan drawn up about a year before. In December the Council agreed to grant the laboratory 56 MSF for 1958, but limited the expenditure for 1959 to 44 MSF.

(2) Towards the end of 1958 the CERN administration put forward a budget for 1959 of about 55 MSF, 11 MSF higher than the figure stipulated in the ceiling set the year before. It added that it would need about 69 MSF for 1960. In December the Council agreed to grant CERN the 55 MSF it said it needed for 1959, on condition that there be a ceiling of between 100 and 120 MSF on expenditure for 1959 and 1960;

(3) In May 1959 the Council agreed to fix the ceiling for 1959 and 1960 at 120 MSF, i.e. it set the ceiling for 1960 at 65 MSF, just 4 MSF below the administration's request. At the same time it imposed a ceiling of 65 MSF on CERN's expenditure for 1961 and for 1962, i.e. it attempted to stabilize the CERN budget for three years;

(4) CERN stayed (roughly) within its budgetary ceilings in 1960 and 1961. However, towards the end of 1961 the CERN administration put forward a budget of about 80 MSF for 1962, 15 MSF above the ceiling laid down in 1959. After an explosive debate in the Council in December, reflecting fundamental differences of opinion between the member states, it was agreed to give CERN 78 MSF for 1962. At the same time a

working party was set up to study and make recommendations about CERN's programme and budget.

The period we are now discussing was thus marked by three features. Firstly, a regular increase in the budget which the CERN administration said it needed (primarily due to increases in the estimated costs of research). Secondly, the attempt by the 'Council' to restrict this growth by the imposition of ceilings. Thirdly, despite this policy, CERN got more or less what it said it wanted between 1958 and 1960, and when the policy began to 'bite' in 1961 it was jettisoned after a head-on confrontation between the British and the French. It is these developments that constitute the leitmotif of the next sections of this chapter.

10.2.1 MAY AND JUNE 1958: FINANCING A BUBBLE CHAMBER PROGRAMME

On 23 May 1958 the CERN Executive laid a paper before the SPC outlining 'a programme for building experimental apparatus for the CERN proton synchrotron' covering the next three to four years. Its cost was estimated at some 22 MSF. The bulk of this—some 20.5 MSF—was set aside for a comprehensive bubble chamber programme, comprising the construction of a 1 m propane chamber and services (2 MSF), a 2 m liquid hydrogen bubble chamber (7 MSF), a liquefier, a special building, and other supporting facilities for hydrogen chambers (7.5 MSF), and beam transport equipment (4 MSF). Three ways of spreading the costs corresponding to three rates of implementation of the programme were identified. At one extreme the 'minimum programme' kept within the 1959 ceiling, but at the cost of having only the propane chamber and its facilities ready by 1960/61, the first year of the PS machine's operation, and no CERN hydrogen bubble chamber until 1963/64. Then there was an intermediate programme, which involved exceeding the 1959 budget by about 4 MSF. It had both the propane chamber and facilities for hydrogen chambers available in 1960/61—but left the 2 m chamber to 1963/64. Finally there was the crash programme by which CERN could have its own propane chamber by 1960/61, could also host 'outside' hydrogen chambers by 1960/61, and could have its own 2 m chamber in operation by 1961/62—but which required exceeding the 44 MSF ceiling for 1959 by 7-8 MSF.[44]

The SPC endorsed the CERN Executive's request that the crash programme be implemented despite its financial implications. At the same time, recognizing that this may be financially impossible because of the constraints imposed by the ceiling, a more modest alternative was not ruled out. A British team led by C. Butler was hoping to build a 1.5 m hydrogen chamber of its own to be ready by the end of 1961, and was willing to loan it to CERN for 18-24 months, before taking it back to England and using it at Nimrod. In the light of this possibility several delegates, including Adams and Leprince-Ringuet, 'considered that a delay of the CERN bubble chamber could be accepted as a compromise [...]' if the Council was unwilling to vote the funds for a crash programme. In other words, the SPC was willing to accept the intermediate programme.[45]

These proposals were first discussed in a preliminary way on 3 June by the Finance Committee's Working Party. Various ways out of the 'impasse' created by trying to reconcile the SPC's recommendation for the crash programme with the 1959 ceiling were considered — breaking the ceiling, paying 1960 contributions in advance, raising a loan—, only to be rejected. Thereupon the Working Party concluded that the 'compromise' discussed in the SPC was the best possible 'solution' to the difficulties it faced: the scheme was financially attractive in that it enabled CERN to spread costs, and it was scientifically attractive in that it provided for a big hydrogen chamber around the PS around the end of 1961. No decision on how to finance what was 'in fact a new programme of 18 to 20 million' was reached, though it was stressed that governments would need to have some idea of the 1960 estimates before they could even consider it.[46]

The Council met on 20 June, and having heard the SPC's recommendations and the 'compromise' which the FC proposed, some leading scientific delegates (Perrin (F), Scherrer (CH)) as well as the new French Council President Francois de Rose insisted that the crash programme not be ruled out *a priori*. 'Whether only supporting facilities would be begun in 1959 or whether, on the contrary, a start should be made in the construction of the [2 m] bubble chamber itself', said de Rose, would be determined by the money national financial authorities were willing to make available for CERN. The item concluded with de Rose asking the Council delegates to attend an extraordinary Council session in October armed with instructions from their governments.[47]

There are two points we want to mention about this opening debate on financing the PS experimental programme. Firstly, no one questioned that CERN had to be properly equipped with bubble chambers, and big hydrogen bubble chambers in particular, as quickly as was financially and technically possible. For the SPC and many of the member states, it was felt that without such a device CERN could not hope to remain competitive in the early 1960s: Alvarez was about to commission a 1.8 m hydrogen chamber in Berkeley, and 2 m chambers were being contemplated at Brookhaven and Dubna.[48] The British had an added motive for supporting the initiative. They hoped that Butler's 1.5 m chamber would be the first big hydrogen bubble chamber used around the CERN PS, and that it would give British physicists—who may well have expected to have privileged access to the device which they had built—an opportunity to skim the cream in the new energy range.[49]

The second point to note is the determined attitude of the French delegation to see the PS properly exploited. Perrin and de Rose were plainly not bothered about breaking the ceiling, feeling that it was more important that the CERN 2 m chamber be constructed as quickly as possible. And although, just as in 1957, their domestic financial authorities were initially inclined to be cautious, their view did apparently move closer to that of the

French Council delegates during this debate. On 3 June 1958 Neumann told the Finance Committee that 'It was hardly possible to face the financial authorities of the Member States with a further request for funds so soon after the budget ceiling had been agreed to'. Two weeks later, while still worried that costs might spiral out of control (if the 1960 budget went above 50 MSF on account of the bubble chamber programme there would be 'serious objections' from France, said Morat), the French delegation to the FC was now prepared to see the 1959 ceiling broken.[50]

This dedication of the French Council delegates coincided with the return to power of de Gaulle in May 1958, which was followed by a period of heavily increased funding for science by the French government. Indeed during the years that were to come the French delegation led by Francois de Rose (until he left CERN in 1964) and Francis Perrin was to emerge as one of the staunchest supporters of CERN, and as the undisputed leader of a majority group in the Council who constantly backed relatively high budget figures for the Organization. It is a point we shall come back to again.

10.2.2 AUTUMN 1958–MAY 1959: BREAKING THE 1959 CEILING TO FINANCE THE 'NEW PS EXPERIMENTAL PROGRAMME'

During the summer of 1958 the CERN administration drew up its estimates for what was now called the 'New PS experimental programme' and the corresponding CERN budgets for 1959 and 1960. These were laid before the Finance Committee Working Party on 7 October 1958, whereupon the administration was asked to extend its estimates until 1961. On the next day it laid some hastily assembled data of the cost of the programme over the years 1959–1961 before the full committee. It looked as though the PS experimental programme over these three years would involve the recruitment of some 200 extra staff, and would cost about 80 MSF (56 MSF for capital expenditure, 24 MSF for staff and general expenses). The corresponding estimate of the CERN budget for 1959 was some 55 MSF—11 MSF above the ceiling laid down the previous year; that for 1960 was 69 MSF, making a total of 124 MSF for the two years. It also seemed as though 'CERN's normal rate of expenditure' would be around 60 MSF annually.[51]

Within four months then the cost of the PS experimental programme had quadrupled—from about 22 MSF in May to 80 MSF in September. The increase had come about partly because the programme had been somewhat extended beyond the original items foreseen in May, but far more importantly because the estimated costs of individual items in the original programme had escalated sharply. For example in May the total cost of the propane chamber had been put at 2 MSF over 2 years; now it was estimated to cost 3.3 MSF in 1959 alone. Similarly, in May the total cost of beam transport equipment had been put at 3.9 MSF over 2–3 years; now it was estimated that 3.1 MSF be spent in 1959, and it was stressed that this was 'by no means the total cost of the beam transport system for the P.S.'.[52]

Notes: pp. 626 ff.

The various member states separated out into two main blocs as they worked out their positions on these developments over the next two months. The one comprised the British and the Scandinavians. The Finance Committee delegates from these countries were emphatic that cuts could and should be made to the budget without damaging the basic programme, and they appealed to the opinion of their national physics communities to back up their claim. Thus Obling 'remarked that scientists in Denmark thought it might not be essential to have a propane bubble chamber' — and indeed everyone knew that a similar instrument was being built in parallel at the Ecole Polytechnique, and would be 'loaned' to CERN at about the same time as its own 1 m propane chamber would be ready. The 'scientific advisers to the United Kingdom delegation', for their part, suggested that staff could be reduced by 10%, and that an envisaged cloud chamber as well as the proposed increase from two- to three-shift working on the SC were not essential. Informed by such arguments, the delegates from Britain, Norway and Denmark were prepared to see the 1959 ceiling broken, but in return insisted that another ceiling be imposed on expenditure for 1959 and 1960. The Swedish delegate went further. Objecting that 'the decision taken by the Council in December 1957, to set a ceiling of 100 million Swiss francs for 1958 and 1959, could not be discarded after less than one year [...]', he warned the FC in December that if the Council voted in favour of 55 MSF for 1959 his government 'might have to reconsider Sweden's situation in the Organization'.[53]

There was considerable opposition to these attitudes among the other member states, both as concerns the amount of money that CERN should be given and the more substantive question of the ceiling policy. Concerning the first, several delegations made it clear that they were against cuts to the programme, and that they favoured a budget for 1959 and 1960 together of 120 MSF — just below the 124 MSF that the Director-General had asked for. In support of their line some of them referred to a paper prepared by Edoardo Amaldi, then Chairman of the SPC, who argued that any reduction of the budget below 115 MSF could only be achieved by delaying or abandoning the 2 m hydrogen bubble chamber.[54] As a result Hocker (FRG), for example, felt that the ceiling had to be set at at least 115 or 120 MSF, and 'since the difference between those two figures was only 5 million', he added, 'the higher figure should be agreed'. The French and the Dutch were similarly inclined. In November Neumann reported that his government could approve 55.5 MSF for 1959 and would probably be willing to pay up to 120 MSF for 1959 and 60 together, while Bannier announced that his delegation could also approve a budget of 120 MSF for the two years.[55] As for the idea of again imposing a ceiling, both de Rose and Bannier expressed their grave misgivings about the move. 'Keeping too strictly to a ceiling', said the Council President in October, was bound to lead CERN into difficulties: 'the future in high-energy physics could not be foreseen with enough certainty to prepare a precise budget two years in advance', de Rose explained, and if CERN committed itself with the member states on an uncertain basis it 'would run the risk of forfeiting its good name when it was essential for the Organization to keep the reputation for honesty which it deserved'. Bannier echoed these sentiments just before the Council voted the 1959 budget in December. He had already had occasion, he said, to warn all delegations of the dangers involved in setting a definite ceiling for the years 1959 and 60. 'In point of fact', Bannier went on, 'the setting of a two-year ceiling had not so far proved to be a happy experience, since it was the very notion of a ceiling which was now causing difficulties to the Finance Committee and the Council'.[56]

These differences were not easily resolved. On 3 December 1958 the Council adopted the budget for 1959 at 55 MSF by 9 votes to 1 (Sweden). At the same time it accepted that a ceiling be imposed on expenditure for 1959 and 1960, though the delegates could not agree on a precise figure. Instead they set it at between 110 and 120 MSF, the final level to be decided after a committee of experts had examined CERN's staff proposals and the administration had given more information on the 'total financial implications of the new experimental programme'.[57]

The expert panel concluded in February that CERN's staff estimates were 'adequate and economic' in the light of the envisaged programme, and that CERN salaries were 'not extravagant'.[58] The executive's overall figures for the cost of running CERN from 1959 to 1962 were laid before the FC early in March. After spending some 55 MSF in 1959, said Bakker in an accompanying statement, CERN would need about 66.5 MSF annually for the next three years.[59] The programme he proposed was ratified at an enlarged meeting of the SPC on 16 and 17 April 1959 which was attended by about 20 senior physicists from the CERN member states, as well as by Bannier and by de Rose, the latter concluding that 'if the Member States decided on a budgetary reduction, it would be tantamount to amputating CERN, not of a limb, but of something like a lobe of the brain, so that the scientific life of the Organization would be very deeply affected'.[60] A month later the Council agreed without dissent that the ceiling for 1959/60 be set at the higher figure of 120 MSF, i.e. it awarded CERN 55 MSF for 1959 and 65 MSF for 1960. At the same time, taking its cue from Bakker's figures, it 'bore in mind' that the implementation of the PS experimental programme in its entirety and its completion by the end of 1962 would 'involve annual budgets of approximately 65 million Swiss francs (as an average) in each of the three years 1960, 1961, and 1962'. Despite the caveats surrounding this debate, and the careful wording of the final resolution which studiously avoided any reference to a ceiling, everyone knew that a new three-year ceiling was being imposed on CERN.[61]

There are two points to remember from this somewhat rapid survey of these debates. Firstly, that the near-unanimous votes on the budget in the Council in December 1958 and May 1959 did not signify that there was a consensus on basic points of policy. They were the consequences of an 'unbalanced' deal in which a purely financial concession was granted in return for an 'agreement' on a matter of

Notes: p. 627

principle: the funds immediately needed by the laboratory were conceded by the British in return for the imposition of a long term ceiling. It was an expedient intended to avoid confrontation, and which in fact did little more than paper over fundamental differences between those who wanted to set limits on CERN's expenditure and those who felt that, in the light of past experience, the time was not yet ripe for doing so. The second point to note is that in 1959 the United Kingdom delegation's aim was to try to stabilize CERN's expenditure for the next three years. The 195 MSF they were prepared to accept were to be divided equally between the budgets for 1960 to 1962. This meant that CERN was expected to live on 65 MSF per year just at the time when serious experimental work around the PS was getting under way. This attempt to impose a regime of level funding at this stage of the laboratory's development was to bedevil the CERN administration and the Finance Committee from the start.

10.2.3 1959 AND 1960: ACCOMMODATING THE RISING COSTS OF RESEARCH TO A THREE-YEAR LEVEL CEILING

Within months of the Council agreeing in May 1959 that CERN could spend 65 MSF in 1960, the administration was forced to take drastic steps not to exceed the ceiling. The problem was again due to increases in the costs of research equipment. As Bakker explained, 'as work [had] continued on the more complete design and estimation of the various items' in the PS experimental programme, it had 'become clear that the provision for certain major items of construction and operation was seriously insufficient'—to the tune of ~6 MSF for beam transport and power, ~4 MSF for the 2 m hydrogen chamber, ~2 MSF for building and operating the propane chamber, and ~1 MSF for SC groups researching on the PS. To deal with this situation it was decided that, rather than change the planned PS experimental programme, the administration would commit in 1960 all the reserves set aside in the 195 MSF which had been allocated for 1960, 1961, and 1962. This meant that, while keeping within the ceiling for 1960, 'CERN was [...] no longer in a position to meet any contingencies in the next three years'.[62]

A somewhat similar situation arose in 1960 concerning the 1961 estimate, though this time the ceiling was exceeded. Again it was an underestimate in the costs of research that was the culprit or, more precisely, in the number of staff needed to operate the big machine as a research tool. The original figures, said the administration in June 1960, had been based solely on experience gained in constructing accelerators. Now, after six months of exploiting the PS for research, 'We know much more [...] about how many staff are required to operate and maintain the machine [...] on three shifts. We know much more about the real load of picture evaluation and computing [...]. We have learnt a good deal about the load of development and construction work for the necessary heavy ancillary equipment, beam separators and so on [...]'. To satisfy these newly-appreciated needs the administration asked that it be allowed to employ 34 additional personnel in the current year (1960) — the FC authorized them to recruit 23 —, and 100 new staff in 1961, bringing the total complement at the end of that year to 1115. And in the 1961 budget vote taken at the end of 1960 the Council agreed to a level of contributions of 67.7 MSF (which corresponded to a budget of 69.7 MSF), Bannier reporting with not a little irony that, 'After long discussions with the Director-General and the CERN Administration, the [Finance] Committee had come to the conclusion that this increase [2.7 MSF] was due to the rise in prices and the cost of living', and so 'justified'.[63]

The vote on the 1961 budget was unique in that, for the first time in the laboratory's history, it was not only passed by a slender majority, but against two of the major contributors to the CERN budget. On 3 November 1960 the delegates to the Finance Committee met to consider the 1961 estimates in anticipation of seeking final instructions from their governments. When Chairman Bannier made his usual survey of their positions towards the end of the meeting, eight states (Austria, Belgium, Denmark, Federal Republic of Germany, Italy, the Netherlands, Norway, and Switzerland) said that their governments were prepared to accept a budget entailing contributions of 67.7 MSF. The delegates from France, Britain, and Sweden, on the other hand all indicated that their domestic financial authorities would probably find it difficult to reach this figure. This was confirmed when the Committee met again on the morning of 7 December 1960, the eve of the Council session. Neumann and Verry repeated that they had been instructed not to vote for a budget involving contributions in excess of 67 MSF. And though Greece rallied to their position, and the delegates from Sweden and Yugoslavia said that they were not prepared to outvote the two largest contributors (24.40% (UK), 20.57% (F)), the majority felt less inhibited. The next day the Council adopted — by 8 votes to 5 — a level of contributions for 1961 of 67.7 MSF.[64]

How was this possible? Essentially because the problems raised by the increase were apparently essentially technical, by virtue of which the delegates from France and Britain did not object to being outvoted: as Neumann put it, the CERN allocation for 1961 had already been made by his national authorities and the delegation now felt 'bound' by them. The British delegation, for their part, apparently also hoped to score a point against their domestic financial authorities. To quote Sir Harry Melville (from the tape recording of the Committee of Council meeting of 7 December 1960): 'For technical reasons [my delegation] would prefer to be outvoted. I don't know what France would say [general laughter]. There are shades [?] of dealing with Treasuries which [sic] sometimes its better to demonstrate your unwillingness and then be overruled [laughter]'. Hence President de Rose felt free to advise the Committee of Council that the 'gentlemen's agreement' that the budget should not be voted against the wishes of major contributors should be used 'with discretion'. 'I would be the last to appeal to this consideration as a motive for

Notes: pp. 627 ff.

steering the Council away from something that is important for the organization', he added. In the light of this situation the Chairman of the Finance Committee (Bannier), who had begun by suggesting that the contributions for 1961 should not exceed 67 MSF, closed the debate by saying that 'he would be willing to propose that the Council should adopt a budget with total contributions of 67.7 millions'. Which they did.[65]

10.2.4 CERN'S INABILITY TO PLAN AHEAD

Before going on to study the turning point in finance policy at the end of 1961, we want to reflect for a moment on why the CERN administration found it so difficult to make accurate forward estimates of its costs at the end of the 1950s. We have already indicated the extent of the problem as far as the SC Group was concerned (see table 10.1). Table 10.2 gives some truly startling figures for the Site and Buildings division: they show its estimates for 1958 and 1959 roughly tripling between 1955 and 1956, and then almost doubling again between 1956 and 1957, i.e. we have as much as a six-fold increase in the estimates put forward by this division over a span of only two years. Finally, to give just one more example, there are the costs of the research programme as a whole which, as we have just seen, increased by a factor of four between May and September 1958, followed by further increases in 1959 within months of Bakker putting forward the 195 MSF figure for 1960–62, and by further increases again, this time for staff, in 1961. This was not planning, this was *ad hocery*.

Table 10.2
The S&B Division's estimates of its overall expenditures for 1958 and 1959
as made at end of 1955 (the 'probable' figure given in CERN/FC/89/Rev., 9/12/55),
at the end of 1956 (from CERN/FC/150/Rev., 15/10/56), and at the end of 1957
(from CERN/FC/234/Rev., undated). The actual expenditure is taken from the CERN Annual Reports.

Financial year	1958			1959		
Estimate made in	'55	'56	'57	'55	'56	'57
Estimated exp. (MSF)	4.8	14.8	28.2	3.3	8.8	14.2
Actual exp. (MSF)			26.3			18.5

One key underlying reason for this was the dramatic changes taking place in the state-of-the-art, notably the relatively sudden and unexpected transformation in the scale (and so cost) of the equipment needed to exploit big accelerators at the end of the 1950s. This was precipitated by the emergence of the hydrogen bubble chamber as an essential tool for the study of strange particles, and changes in technique on this scale were apparently as unanticipated at Brookhaven as they were at CERN. For example, it was as late as 1959 that BNL planned to increase its available experimental area around the AGS by 58,000 ft^2 (about 6,000 m^2)—roughly a factor of four larger than originally envisaged—to accommodate the kind of equipment physicists were going to need.[66]

The uncertainty consequent on these changes was compounded at CERN by internal organizational difficulties. As we saw in chapter 9, until the summer of 1957 there was no serious consideration given to the PS experimental programme. What is more, when the bubble-chamber programme was agreed to by the Council a year later, it proved extremely difficult to find anyone to take charge of building its main component, the 2 m hydrogen chamber. In addition the study of beams and how best to assemble them was not given the prominence it might have had, and the development of data-handling techniques followed its own more or less independent path. The confusion reigning at the time is nicely captured by Bakker's reply to Neumann at the end of 1958 when he was asked to justify an allocation for 'additional expenses' in the 1959 budget estimates. Bakker did not know just what the money would be used for, but he assured the French delegate that anyway 'if the money was not spent on one type of equipment it would have to be spent on another'.[67] Without organization, without co-ordination, without individuals willing and able to take overall responsibility for its major components, there was no way in which CERN could hope to produce a coherent and realistically costed equipment programme for exploiting the PS.

Finally one cannot ignore the fact that the CERN staff apparently did not really *bother* to make reliable estimates of their needs. The absence in the 1958 estimates drawn up at the end of 1957 of any reasonable provision for equipment to exploit the PS even as the machine's experimental programme was being discussed, the leap in the 1959 estimates from 33 MSF to 46 MSF in just a month (between September and October 1957), the near quadrupling of the costs of the overall programme between May and September 1958—all of these suggest a startling degree of *indifference* regarding research costs among the CERN staff concerned. Whence this attitude? It was the correlate in the minds of scientists and engineers of the lavish and uncritical expenditure by governments on nuclear science in the post-war period. Scientists had simply come to expect that money was no object, to assume that they could have whatever they wanted, to behave as if funds were unlimited.[68] In the particular case of CERN such indifference to expense was given considerable rein by the Finance Committee, and notably by those of its members who formed the core of the CERN lobby. Indeed it is striking that throughout our period the Finance Committee *never* appears to have refused a request by the administration to exceed an earlier estimate. Instead meetings were arranged with a view to finding 'weighty scientific and technical arguments to convince the financial authorities in the various Member States that the amounts required by the Director-General were vital for the progress of the Organization', and mechanisms were devised (like coupling the budgets of successive years) to solve the procedural problems raised by an unexpected shortfall in resources.[69] In a climate like this the administration was free to 'manage' the laboratory in a piecemeal way, dealing with its material

Notes: p. 628

requirements as and when they arose rather than trying to anticipate and budget for them in advance. In short another reason why the CERN staff did not plan properly was simply that they had become used to living in a world of plenty, and that there was little or no external pressure on them to change their ways.

The administration's ability to get what it wanted when it liked from the FC is nicely illustrated by the following little example. At issue was a relatively small amount of money—some 350,000 SF—for relatively marginal items—additional parking space for cars (250,000 SF) and the levelling of a sports field (100,000 SF). The significance of these requests was that they were made just after the administration had increased the cost of the PS programme from 22 MSF to 80 MSF, was asking that the 1959 budget ceiling be broken to the tune of 11 MSF, and had specifically been asked by the FC to make savings. Instead it increased its 1959 estimates even more. Yet, rather than being reticent, Director-General Bakker unashamedly defended the allocations for the sportsground ('It was [...] thought that the Council of CERN might [...] like to make an effort and provide the staff with certain recreational facilities [...]') and for the parking places ('[...] on account of CERN's remoteness from the dwelling areas in Geneva, a very large proportion of staff members had to have their own cars, albeit second-hand ones, and parking space had to be made available for them').[70] And rather than being unyielding, the Finance Committee merely rapped the administration over the knuckles for its tactlessness, asked for further arguments to justify the expenditures—and authorized CERN to go ahead and provide the recreational and parking facilities that it wanted. At no stage in the debate did it, or apparently could it, say no.

10.3 1961: the three-year ceiling is broken and the foundations are laid for a new policy

10.3.1 THE CONFRONTATION BETWEEN THE CERN 'LOBBY' AND THE BRITISH

Immediately after the vote on the 1961 budget was taken by the Council in December 1960 CERN's new Acting Director-General, John Adams observed that, with the level of contributions set at 67.7 MSF, CERN would not be able to recruit sufficient staff in 1961 to exploit the PS properly in 1962. If this situation was not to persist into 1963, he said, 'provision for the necessary increase in funds and staff would have to be made in the 1962 budget'. Adams went on to suggest that the administration present its provisional budget estimates for 1962 in March, rather than the usual September, of the following year. In doing so the DG was effectively warning the delegates that, if the big accelerator was to be exploited 'as efficiently as possible', that if CERN was to hold its own in the 'friendly but nevertheless real rivalry' (Amaldi) with Brookhaven, which had just commissioned its own big PS, the 1962 ceiling would have to be broken.[71]

The administration laid its first detailed estimates of what it would need for 1962 before the Finance Committee at its meeting on 27 April 1961. Rather than present a single estimate, the executive proffered three alternatives. The so-called alternative A (69.1 MSF) took the 65 MSF ceiling as its point of reference, adding the increases in staff and for inflation which the Council had already accepted. If adopted, this alternative, said the administration, meant slowing down 'the rate of carrying out some of the scientific programmes of the Laboratory agreed by the Council [...]'. Alternative B (82.9 MSF) involved recruiting 200 more staff in 1962, and the implementation of the programme 'at a rate which [would] completely justify the capital invested in the Laboratory to date and which [would] keep CERN in the forefront of high energy physics [...]'. Finally, there was alternative C (74.7 MSF), the 'compromise' solution. This involved recruiting 100 more staff in 1962 and exploiting the machines in a way that was 'far from ideal [...] but nevertheless self consistent from the physics point of view'. The directorate, as one would imagine, favoured alternative B.[72] At a meeting on 29 April the Scientific Policy Committee, for its part, came to the conclusion that 'alternative A would not be disastrous for CERN but would seriously curtail the experimental work with the CERN PS', that alternative B would 'obviously be the best', and that alternative C, 'though acceptable' would leave out some essential items of beam transport equipment. Hence a fourth alternative was born, alternative D (78.2 MSF), which (ultimately) added 3 MSF for the development of experimental facilities and 0.5 MSF for preparatory work for a new accelerator project to alternative C.[73]

These figures were discussed in the Finance Committee on 27 April and 1 June.[74] Hocker, the new Chairman of the FC, reported the results of their deliberations to the Council on 2 June 1961: a majority of the delegations 'seemed' to favour compromise C (~75 MSF), and some were even 'willing to consider the possibility' of the SPC's more expensive alternative D, whose importance was immediately stressed by Heisenberg, by Perrin, and by Powell, the new Chairman of the SPC.[75] At the same time, since 1962 was the last year covered by the previous three-year estimate, the Director-General was asked to draw up forecasts of expenditure for 1963 and 1964. To help him Perrin made a suggestion which signalled a fundamental change in thinking about the financial management of a big science facility: he introduced the idea that *laboratory expenditure, rather than being stabilized, should grow from one year to the next.* Experience in managing scientific budgets in France, said Perrin, had shown that if a budget was kept constant from one year to the next it had the effect of reducing activities. Correlatively, 'apart from any new project, it was generally necessary to allow for a certain annual percentage of increase in order to keep activities at a given level'. The Director-General immediately came out in support of 'the idea of a changing but finite increase of expenditure each year', and agreed to prepare rough over-all estimates for 1963 and 1964 along these lines.[76]

It fell to the new Director-General, Victor Weisskopf, to present the administration's estimates for 1962 and its forecasts for 1963 and 1964 to the Finance Committee on 19 October 1961 and to the Committee of Council a week later. Weisskopf essentially argued that CERN's budgets should be seen as lying on a straight line with a positive slope extending from 1960 to 1964. Taking two points on this line as fixed—x/y coordinates 1960/65 MSF (the amount allocated by the Council for expenditure in that year) and 1962/79.5 MSF (the SPC's alternative D plus 1.5 MSF for a salary increase)—Weisskopf argued that CERN should be given 86 MSF for 1963 and 93 MSF for 1964. Put differently, he was asking the Council to support a steady rate of growth for CERN from 1960-64 of about 8.5% per annum on average, and to allocate the laboratory 260 MSF overall for the three years 1962-64. To back up this claim he, and particularly Mervyn Hine, who had entered the Directorate as member for Applied Physics in January 1961, developed Perrin's argument that stabilization meant stagnation in a number of imposing documents and oral presentations at the meetings of the various committees of the Council.[77]

The administration's proposals immediately polarized the FC delegates. One group—the majority—accepted the policy of growth: to quote Obling (DK)—though we could as well have chosen Campiche (CH), Hoogewegen (NL), or Schmid (A)—'it was unrealistic to fix a stable budget for a research institute'. This group was also prepared to give CERN the 78.2 MSF requested by the SPC (alternative D) for 1962, though not the additional 1.5 MSF for a 5% salary increase. The other, much smaller, group—of which the United Kingdom was the most prominent member—was at this stage unwilling to see CERN's budget increase steadily: British policy, said the UK delegation, 'would be to stabilize expenditure at around 75 million Swiss francs' over the three-year period 1962-64. The British authorities, then, were prepared to break the 1962 ceiling but only on condition that expenditure remained constant at the new level for three years.[78]

Britain's refusal to accept that CERN's budgets grow from 1962 to 1964 was interpreted by the administration and by influential members of the CERN lobby as a fundamental threat to the well-being of the laboratory. Do we want 'to use this huge laboratory to the extent which is rational, which is reasonable for the purpose that we all had in mind when we created it' or do we not, asked de Rose rhetorically, is our aim 'to compete or not to compete' asked Hine. And with a view to changing the attitudes of those deemed hostile to CERN it was decided to gather what Powell called the 'absolutely essential ammunition' needed to launch an assault, not only in Britain, where the greatest menace lay, but also in the other member states which may have been tempted to follow her lead.[79] To this end the FC and the CC commissioned two documents during their meetings at the end of October. One, to be written by the Director-General in language which was easily intelligible to non-scientists, was to compare the present state of CERN physics with that being done in similar laboratories like Berkeley and Brookhaven (de Rose). The other, to be written on behalf of the SPC, was to 'explain the stage reached by CERN at the moment and the likely development of high energy physics in Europe' (Hocker), and to show that 'from a scientific point of view, an overwhelming case [could] be made' for a 'forward-looking policy' involving rates of expansion of at least 12% per annum (Powell).[80]

How did the British Council delegates react to these developments? Sir Harry Melville welcomed the fact that matters were at last coming to a head. We had made a 'fatal' mistake, he said, when, two years ago, we set an 'absolutely flat ceiling' for three years; what was now needed was an (upwardly) 'sloping ceiling'. That granted, as far as Sir Harry was concerned, the question now was 'Are our people going to support CERN as it ought to be supported or are they not?' We cannot come to Council meetings year after year with lower figures than the others, said Melville: 'this is not a satisfactory way of collaborating in international organizations', and 'the sooner we face up to this the better'.[81]

By the time the Finance Committee met again on 14 November the British had modified their position somewhat. They were sticking to their figure of 75 MSF for 1962 (against Weisskopf's 79.5 MSF) but were now prepared to give the laboratory some 240 MSF for 1962-64 (against Weisskopf's 260 MSF). Although, in effect, this allowed for an annual rate of growth at CERN this was not the framework within which the British Treasury was thinking: their concern was to fix a three-year expenditure envelope. And to put teeth into their proposals the UK government let it be known that they were 'contemplating' negotiating a figure for 1963 and 1964 with other governments (i.e. bypassing the Council) and inserting it in 'a kind of financial protocol annexed to the Convention'.[82] And as British resolve firmed, so did that of some smaller states —Austria, Denmark, Norway,...— begin to waver. 'And then came the bombshell', to quote Weisskopf. The Italian delegation 'read a formal declaration of the Italian Government' stating that, primarily for procedural reasons, they could not vote for a budget for 1962 above 75 MSF. With Sweden and Spain already backing Britain, the chances of the Council accepting a budget of 78 MSF in December seemed slim indeed.[83]

Weisskopf immediately intensified the campaign launched the month before. He asked Powell to convene an emergency meeting of the SPC at which he would ask those present to give him 'the power to write letters on their behalf and on my own to the Ministries of Foreign Affairs of all Member States', and he personally impressed upon Bohr (who seldom came) and Perrin how important their presence would be. He suggested that Willems see Paul-Henri Spaak and ask him 'to intervene in England and Italy on behalf of CERN', and that Bannier approach Prince Bernhard 'for an intervention of a similar kind'. And he sent long letters to Adams, Blackett, Butler (the leader of the British hydrogen bubble chamber project and the chairman of the panel responsible for advising the DSIR on CERN), Cockcroft, and Melville, in which he explained that a 'very critical situation' had arisen, and that, in his view, 'a sum below 260 million [for 1962-1964] would mean the end of a vigorous high-energy research laboratory in Europe'.[84]

The British government did not sway from the path it had chosen. On 20

Notes: pp. 628 ff.

November the UK Foreign Office, acting on its earlier threat, took the unprecedented step of circulating an Aide Mémoire to the other member states in which it suggested that the participating governments 'should determine [...] what sum should be fixed as the firm upper limit, at present price levels, of the resources that can be made available to C.E.R.N. over the next three years', leaving it to the Council to lay down the annual budgets within this limit. They proposed 'a ceiling figure of not more than 240 million Swiss francs', and insisted 'that member Governments should make up their minds and that the Organisation should realise that this total must be adhered to'. Possible procedures for achieving that aim were spelt out in a *Note by the United Kingdom Delegation* which was formally submitted to the Council in mid-December. The *Note* reiterated the UK's conviction that CERN should operate with three-year ceilings which 'once fixed [...] should be binding', and that such limits 'should have the formal sanction of at least the Financial Protocol and possibly of the Convention'.[85] In addition, around the last week of November, the DSIR provisionally stopped the shipment of parts of the 1.5 m British hydrogen bubble chamber to CERN.[86]

The last round of meetings to settle the 1962 budget got under way on the morning of 18 December 1961. During these two days the deep hostility and resentment felt by the members of the CERN lobby to the line adopted by the British government burst into the open. Choking with anger at the meeting of the Committee of Council, Bannier roundly condemned the Foreign Office's suggestion that the upper limit to CERN's expenditure for the next three years be decided by national governments in consultation with one another. This policy, said Bannier, would 'diminish the authority of the international body which is the CERN Council', which 'should always make the final decision [on the budget] according to the voting procedure laid down in the CERN Convention'. To do otherwise, to 'by-pass' the Council on important decisions of this kind would, in Bannier's view 'demolish [...] the very foundations of CERN'.[87] He was backed up by de Rose, who went so far as to threaten political retaliation if the British persisted with their announced policies.

De Rose's main target of attack was the rates of growth implied in the British suggestion that CERN's budgetary ceiling for 1962–1964 be set at 240 MSF. Assuming that the 1962 budget was voted at 78 MSF, this implied that CERN should have 80 MSF for 1963 and 82 MSF for 1964: a growth rate of 2.5% per annum. This figure, said de Rose, was to be compared with annual rates of expansion in big American laboratories of between 15% and 18%, and of something like 12% in France. If CERN 'were to fulfil in Europe the duties which it had been given with regard to national laboratories', and to 'keep up in the competition with big laboratories in the United States and the Soviet Union', he went on, 'the only suitable rate of expansion would be one comparable to national laboratories'.[88]

De Rose backed up his arguments by threatening to advise his government to withdraw from other collaborative ventures if the French delegation did not get its way on CERN. Firstly, he drew attention to the fact that negotiations were then under way to set up two new European organizations for scientific cooperation, viz. ESRO (space research) and ELDO (launcher development). If CERN's budgets were to be kept down to make money available for these other organizations, as the British had implied in their Aide Mémoire, if CERN was to be 'sacrificed in favour of new organizations' (de Rose), he would propose that France not join them. 'I do not know if my advice will be followed or not', said de Rose, 'but I know that it was followed in the case of France joining the "Blue Streak"'—here referring to de Gaulle's change of attitude from opposition to support for the use of a British rocket as launcher for ELDO.[89] De Rose's message to the UK government was clear then: if you do not follow my policy on CERN's development, I will tell my government not to support you in the formation of ELDO—and I am likely to be heeded.

De Rose's second threat was intended to put the British in the unpleasant position of being seen to wreck the expansion of CERN. In an off-the-record remark to the Committee of Council he pointed out that the French government was then negotiating with the Swiss authorities for an extension of the CERN site into France. Despite the complexity of having an organization straddle the border between two sovereign states, negotiations were proceeding favourably. However, these arrangements were naturally 'based on the hypothesis that CERN would develop': if it did not, if the Council voted a three-year ceiling of 240 MSF the next day, 'I could not go back to my government and say I continue this proposition', said de Rose.[90]

Both de Rose's and Bannier's contributions were warmly commended by many delegates to the Committee of Council on 18 December. By special request, both were repeated at the full Council meeting the following day, where they followed immediately after the presentation of farewell gifts by the Council to Verry and to Adams. Feted for their contributions to the organization in the past, condemned for their policy concerning its future development, threatened with grievous sanctions if they did not fall in line with the majority, the British delegates were at once embarrassed, outmanoeuvred, and isolated. When the vote on the 1962 budget was taken it was 10-2 for a 78 MSF level of contributions—only the Swedes supporting the British, and even they implied that if they had realized earlier the size of the majority in favour of this figure they might have joined its ranks.[91] As for the level of expenditure in the subsequent years, the British *Note* suggesting that a three-year ceiling for 1962-64 be formally enshrined, perhaps even in the Convention, was ignored, and the question of the budgetary contributions to be made for 1963 and 1964 was left unresolved.[92] Instead it was agreed to set up a small group whose task it would be to 'draw up proposals for exploiting CERN's research potential, to evaluate the financial implications of those proposals, especially in connection with the rate of growth of the budgets, and finally to consider suitable means of fixing and controlling expenditure'.[93]

Thus was the Bannier Working Party on the CERN Programme and Budget born.

Notes: pp. 629 ff.

10.3.2 SOME REMARKS ON THE FACTORS AT WORK IN THIS CONFLICT

To conclude this section we want to try to explain why it was that, at the end of 1961 in particular, the debate over the CERN budget became so acrimonious, threatening to tear apart the very fabric of the organization. To do this we shall focus on the British position and, by comparing the situation in the United Kingdom with that in some of the other member states, and in France in particular, we shall try to explain why the differences between Britain and her partners at CERN escalated into a head-on conflict.

Let us begin by asking why the Foriegn Office tried to bypass the CERN Council, suggesting that CERN's budget levels be agreed between governments. One reason is fairly obvious: the move was symptomatic of the growing impatience inside the government, and notably the Treasury, with the DSIR and its representatives at CERN, whom they saw as too closely associated with the CERN lobby. The Treasury, for its part, was increasingly frustrated by the apparent inability, or unwillingness, of the UK delegation to impose its budget line in the Council, and was determined to find a way of enforcing ceilings. The vote against the British at the end of 1960 confirmed it in its view that informal arrangements made in the Council, like the 'agreement' not to outvote a major contributor, were not guarantee enough, and early in 1961 it favoured changing the budget voting rules, either by changing the Convention or the Financial Protocol (as the Italians suggested in March 1961), to avoid the same thing happening again.[94] The DSIR and 'its' CERN representatives, for their part, preferred to arrange matters internally, between themselves and inside the Council. Hostile to Treasury attitudes—it was 'obsessed by [Britain's] economic stagnation' and it 'resent[ed] the cost of [CERN's] success' wrote one DSIR official angrily—, they opposed the procedural reforms backed by the Treasury, suggesting instead that what was needed was appropriate 'machinery for planning' the laboratory's finances.[95] And while the Treasury apparently accepted the DSIR's view in summer 1961, its patience seems to have snapped in October, when the CERN administration asked for a budget for 1962 that was 20% above the ceiling agreed two years before. Taking matters into its own hands, the Treasury decided to try to strip the CERN Council— and with it the CERN UK delegation—of some of their powers.

A number of other more fundamental factors tended to reinforce this interdepartmental conflict. Firstly, there was the relative lack of pressure on the British government from the national physics community, the lack of a scientific statesman in the UK to plead the laboratory's case directly at the highest levels of government. In France there was Perrin (and Leprince-Ringuet), in Germany there was Heisenberg, in Italy there was Amaldi.[96] In Britain there was no scientist of comparable status and influence in political circles who was as dedicated as these men were to the development of the laboratory in Geneva. Without such people to promote CERN at this time the process of policy formation inside Britain was highly bureaucratized: the mechanisms used by the physicists to transmit their views on CERN to the government were predominantly formal, and so inevitably lacked 'punch' and a sense of urgency. Where a scientific statesman would have gone straight to the top and personally discussed and fought for his concerns, the UK physics community made recommendations in committee which were passed up a bureaucratic chain of command. This institutionalization of scientific advice to the British government on CERN is clearly evident in our case. In France at the end of November 1961 'arduous discussions were in progress between the scientific and financial authorities on the question of CERN's long-term budget prospects', the aim being to secure the agreement of the latter 'to one essential basic principle, namely that CERN was far from having reached its optimum state of development and that its activities could not therefore be regarded as stabilized'. By contrast, in Britain at the end of November 1961 'the panel of scientists [...] responsible for advising the Department of Scientific and Industrial Research had unanimously agreed that it was essential to give proper support to CERN [...]', but 'was awaiting the results of the [emergency] meeting of the Scientific Policy Committee [held on 25 November] before submitting its views officially'.[97]

The general economic situation in Britain at the time also played a part. As one commentator has put it, 'The postwar performance of Britain's economy has entered the annals of history as a tale of woe, of constant balance of payments crises [of which there was one in 1961 - JK], of overspending and living on credit'.[98] Britain's investment in Research and Development was squeezed in parallel: already higher than any other European country (as percentage of GNP) it grew very slowly around this period (from 2.2 to about 2.4%, between 1962 and 1967). In this context it is not surprising that the Treasury baulked at the demands coming from a range of European joint enterprises, some of them entirely new (ESRO and ELDO), some of them well established, but exceeding all previous expectations (CERN). The situation in France was quite different. On the one hand, after the general recession in Europe in 1957-8, the economy grew extremely rapidly—at between 4 and 7% per annum between 1959 and 1963. At the same time the Gaullist government increased its R&D investment enormously—from about 1.5% of GNP in 1962 to over 2.1% in 1967, the steepest rise in this period of any country in the world.[99]

Finally one cannot overlook the fact that the policies adopted by the British and French delegations at CERN—particularly concerning the relevance of competition with American laboratories—reflected more general perceptions of their place in the world. For the British, there was still that 'special relationship' with the trans-Atlantic superpower, that form of dependence which had been established during the second world war. For the French, and the Gaullists in particular, there was a general dislike of 'Anglo-Saxons', and a determination to be independent of American influence, a firm resolve not to be an American satellite. In this climate it was only 'natural' to find a UK Treasury official wondering whether or not British

Notes: pp. 630 ff.

physicists could get the same benefits they would have at CERN for less cost from the programme at Berkeley—and to see the French diplomat de Rose determined to build CERN into a laboratory which could 'keep up in the competition with big laboratories in the United States and the Soviet Union'.[100]

10.4 The Bannier Report and the budget debates from 1962 to 1965

10.4.1 THE WORKING PARTY AND ITS REPORT

The Bannier Working Party—or the 'Working Party on the CERN Programme and Budget' to give it its official title—met four times between mid-February and end-April. All but one of its members—Seippel, a Director of Brown, Boveri & Co. (CH)—, were high-level science administrators: Bannier, Director of the Dutch ZWO, Durán, Vice-President of the Spanish Junta de Energia Nuclear, Elkington, Director of Establishment and Finance in the DSIR (UK), Gentner, Director of the Max-Planck-Institut für Kernphysik (FRG), Ippolito, Secretary-General of the Italian CNRN, Maréchal, Principal Scientific Advisor to the French Government, and Obling, Chief of Section in the Danish Ministry of Education. The Secretary (Goldschmidt-Clermont) was provided by CERN, and Hine and Dakin (Head of CERN Administration) attended all meetings. A representative from the British Treasury was invited to the Working Party's second meeting.[101]

The group's terms of reference instructed it to 'make appropriate proposals for the purpose of securing a reasonable exploitation of CERN's research potential', evaluate their financial implications, 'preferably by setting a rate of increase for the budgets', and 'examine the case for having fixed three-year budget estimates or put forward alternative proposals [...]'. It presented its findings and made its recommendations in a 40-page long report, to which was appended three annexes (a report by the SPC, a report by the DG, and two reports dealing with the development of high-energy physics in the United States). The Bannier Report was laid before the June 1962 Council session where it was very positively received.[102] It had three key components.

Firstly, the signatories to the Report accepted without any apparent difficulty that CERN's budget should grow. What is more the Report recognized that that growth could occur along two axes which it carefully distinguished from each other. The existing facilities were to be extended—and here the Working Party recommended rates of growth for the next four years. There might also be 'major new developments'— and as far as the Working Party was concerned these were to be regarded as an additional cost and subject to 'separate policy decisions of the Council'.[103] There was no longer any question then of expenditure on CERN stabilizing; on the contrary the Working Party foresaw the possibility of an expansion both in degree and in kind of the laboratory's facilities.

Secondly, there were the rates of growth themselves. The Working Party concluded that for the next two years it would be 'reasonable' if CERN's budget grew 'within the range of 10 to 14%' per annum. More specifically, it pointed out that in 1963 and 1964 a growth rate of 13% seemed to be 'the most reasonable rate of increase' 'if cuts in the programme [were] to be avoided'. Thereafter the rate of growth for 1965 and 1966 '[might] fall off but not to less than 10%'. In short the Report recommended that CERN grow at an average of 11.5% over the next few years. All of these figures were exclusive of any major new developments, as we mentioned a moment ago, and of inflation, which the Working Party thought might add another 2.5% to the figures.[104]

Finally, there was the so-called Bannier procedure for forward planning. It was based on the idea that there be a four-year rolling programme and budget, with a rigid two-year ceiling imposed for the first two years. More technically, in any year X, the Council was to make a 'firm determination' of the Organization's budget in real terms for years $X+1$ and $X+2$, and a 'provisional determination of the figures for $X+3$ and $X+4$'. The figures for the first two years were firm in the sense that the 'real' expenditure for year $X+2$ agreed on in year X could only be revised later 'without dissent in the Council'. By contrast, the 'provisional determination' of the figures for the next two years was to be regarded as 'non-binding', following Director-General Weisskopf's insistence that it was 'impossible to define even a small percentage of the work which [would] be done three or four years hence'.[105]

How were the rates of growth recommended in the Bannier Report arrived at? Apparently two considerations were dominant. Firstly, they were established with reference to the arguments developed by Mervyn Hine in a number of papers prepared for the Finance Committee and the Working Party in which he spelt out the forward needs of the laboratory as judged by the management. Secondly, they were based on a comparison with rates of growth foreseen for comparable national and American facilities. In fact, while it is difficult to be certain, in our view it is probably the latter consideration that carried most weight. 'National laboratories', de Rose told the December 1961 Council session, 'were developing at a rate of between 10 and 15% per year according to the country', the figure being 12% for France. As for the United States, in the Bannier Report Hine presented a graph depicting the anticipated rates of expansion of a number of American high-energy physics institutes, and in particular the laboratories at Argonne, Berkeley, and Brookhaven. Summarizing the data, the Bannier Report concluded that an average annual increase of 13% per annum over the next four years for CERN was 'not far out of line with—although on the whole slightly lower than—the American laboratories'. As a matter of fact, this figure of 13% annually from 1963 to 1966 was just the forecast rate of growth for the operating costs of the accelerators at BNL over the same period.[106]

If we are inclined to put more weight on these figures than on those proposed by the management, it is because Hine's estimates of CERN's needs escalated steadily as the Working Party got into its stride. In *October 1961,* so a few months before the Working Party was set up, Hine indicated to the Finance Committee that to bring the CERN PS into 'full operation' by 1964, and to develop the CERN SC and general scientific work at CERN 'at a healthy rate', CERN staff and visitors would have to grow by some *35%, or 370 people,* over the end-1961 figure. With contributions in 1962 voted at 78 MSF, this was correlated with budgetary increases of about 10% per annum in 1963 and in 1964. Hine added that he thought that the budget might then 'level off or at least expand more slowly' from 1965 or 1966 onwards.[107] By *March 1962* Hine had revised these 'rather preliminary evaluations' upwards after collating the findings of a 'series of small working parties covering the whole range of CERN activities'. In a paper prepared for the Bannier Working Party he indicated that to reach its 'grown-up state', i.e. the state 'at which the potentialities of the major existing facilities [were] fully exploited' the number of staff and visitors would have to increase by some *60%, or 630–650 people,* over the 1961 figures, after which 'it would seem to be justifiable to say 'Level off [...]''. Hine did not commit himself to a time scale or annual rates of increase in this paper but, according to an internal British document, at this time CERN was proposing 'a rate of growth of 13% for the next two years and 10% thereafter'.[108] Finally, in *May 1962* Hine prepared another 'Draft Four-Year Programme for CERN' to accompany the Bannier Report. Here it was argued that personnel at CERN had to increase by about *80%, or 900 people,* over the 1961 figures to reach a state of full exploitation. From this Hine concluded that 'an expansion rate of 14–16% per annum for the next few years would apparently be the optimum [...]', and that a 'minimum programme' would require a rate of expansion of staff 'of at least 13% per annum over four years, and adequate financial support'—which, Hine implied, meant more than a 13% increase in the budget in 1963.[109] Faced with these figures, around April Bannier came down in favour of rates of growth for CERN of 13% per annum for the next four years—and was not at all happy to see the administration turn this into a 'minimum programme' shortly thereafter. The British, for their part, became increasingly sceptical—as one DSIR official put it afterwards, CERN's arguments for the rates of growth seemed to him to be 'a rationalization of their increasing programmes rather than the real explanation for the rises'—, and though not inclined to neglect Hine's calculations entirely, they apparently plumped for the figures they said CERN had put forward in March.[110]

10.4.2 THE PRACTICAL APPLICATION OF THE BANNIER REPORT'S RECOMMENDATIONS FROM 1962 TO 1965

The first test of the strength of the member states' commitment to the rates of growth recommended by the Bannier Working Party (BWP) was not long in coming. Indeed indications that the administration would be looking to push the Report's provisions to the limit, and even to exceed them, were implicit in Hine's May 1962 figures and in the Director-General's Progress Report to the June 1962 Council session. Weisskopf's speech took place against the background of the announcement that scientists at Brookhaven had beaten CERN to the discovery that the neutrino exists in two different forms. Asking 'Why did we not succeed when they succeeded so impressively?', he suggested that part of the problem was budgetary, that due to the very low increase in the allocation from 1960 to 1961 an 'equipment gap' had opened up at CERN.[111] This issue was to dominate the debate on the allocations for CERN's 'basic programme'—i.e. the programme covering the construction and operation of the PS and the SC—during 1962 and 1963, whereupon it was displaced by another. In April 1964 the Director-General reported to the Finance Committee that the discovery of the omega particle—which confirmed experimentally the theoretically predicted existence of the SU_3 symmetry law, and 'which represented a great step forward in this field'—'had not been made at CERN, but at Brookhaven'. This provided a suitable introduction to the subsequent debate in the Committee over a range of new developments foreseen for CERN, including an 'improvement programme' for the PS, which was discussed in 1964 and 1965.[112] Alongside these developments, and increasingly overshadowing them there was the discussion of another project whose financial implications began to emerge in 1963: a 'supplementary programme' for the next generation of accelerators in Europe.

In the following two sections we are going to study how the delegations in the Finance Committee and the Council dealt with the financing of CERN's basic programme— the supplementary programme as such is handled independently by Dominique Pestre in chapter 12, and enters our story primarily as an important background element. More precisely, then, our main aim at this stage of our exposition is to explore the *context* in which the recommendations of the BWP were first put into practice, and to study how and why they served to limit the amount of money the member states were prepared to allocate to the development of CERN.

10.4.2.1 The budget debates in 1962 and 1963: the 'equipment gap'

The main argument used by Hine and Weisskopf in 1962 for treating CERN generously was that the rate of growth in the number of personnel (i.e. staff, fellows, and visitors) between 1959 and 1962 had exceeded the increases in the allocations for capital equipment in the same period, so leading to a relative shortage in the facilities required for doing physics. And the cause of the problem, the reason why CERN had been 'living beyond its means', to quote Weisskopf, was not (as one may perhaps have expected), that the laboratory had recruited staff too quickly, but that the rates of increase in the budgets had been too low. To quote the numbers used in the presentation to the FC in September 1962, the number of CERN staff (i.e. excluding 'scanning girls' and auxiliaries) had grown at an average rate of 12% per annum, while that of fellows and visitors had grown at an average of some 30% p.a.

Notes: pp. 631 ff.

Combining these figures, one found that whereas the total number of personnel at CERN (i.e. including fellows and visitors) had increasesd on average by some 15% p.a. between 1959 and 1962, the laboratory budgets over the corresponding period had only increased by 8% p.a. in real terms. As a result the cost per head had 'drifted down to what we find to be an unhealthy value' (Hine) (57,600 SF instead of 61,500 SF), and CERN was faced with an 'equipment gap' (Weisskopf) of some 8 MSF—to keep within its past budgets it had postponed the construction of buildings and it had not ordered enough ancillary material, particularly for beam transport (magnets, separators) to keep pace with the growth in the number of personnel, and of visitors in particular.[113]

How was this situation to be overcome? The ideal scenario favoured by Hine and Weisskopf, and underpinning what they called the CERN Four-Year Programme, was to increase personnel within the limits proposed by the BWP, so reaching 'full exploitation' of the existing facilities (some 2,200 people) by late 1965/early 1966, all the while increasing the budget more rapidly. In figures the Programme foresaw the number of personnel working at CERN growing sequentially by 14/13/12/9% from 1963 to 1966, the corresponding rates of growth of the budget being 23% and 17% for the first two years, followed by 10% and then 9% thereafter.[114]

The delegates to the Finance Committe, while expressing sympathy for the problem CERN was facing, made it clear from the start that budgetary increases above the limits proposed by the BWP were out of the question, and the administration and the SPC had no option but to toe the line. It was a pity, said Bannier when the extent of the problem was first explained to the Committee in September, 'that at the time the Working Party was preparing the report, the figure of the backlog, about 7 million, had not been stressed more clearly, as the Working Party might have suggested some special measure. *Now, it was too late [...]*' (our emphasis). In fact the best that the Dutch, French, and German delegations could offer at this stage (and remember that they were usually among the most generous contributors to CERN) was to make a 'substantial effort for 1963'— meaning a budgetary increase of 13% in real terms over the 1962 figure.[115]

An extraordinary session of the Council was held in October to discuss the problems raised by the equipment backlog. The Chairman of the SPC opened the proceedings by detailing the Committee's reactions to the Report of the BWP. While stressing that 'A budget increase within the limits set by the Bannier Report' would 'force' 'certain painful cuts [...] upon the Organization', the Committee did not ask that those limits be broken. Preferring not to squander CERN's 'principal asset', 'the goodwill of governments', the SPC recommended 'that the greatest possible provision should be made for the support of CERN, within the limits set by the Bannier Report [...]'.[116] The Chairman of the Finance Committee then reported on the positions so far taken by the delegates on the rate of increase for 1963. The UK was considering 12%, France, Belgium and the Netherlands, and Spain were thinking in terms of 13%, Italy and Switzerland were a little above that, and Austria indicated that it could accept 14%. The meeting ended after a large number of delegations had expressed their support for the proposal's made by the Bannier Working Party, and the Director-General had once again impressed upon them 'that CERN should be given the means of clearing up its backlog as soon as possible'.[117]

The most important development over the next two months was the evolution of the British position. A week before the decisive December Council session their FC representative, R. St. J. Walker, wrote to Weisskopf to tell him that his government was now prepared to support a rate of growth of 13% for 1963. At the same time Walker stressed that the United Kingdom delegation would vote for this figure 'provided—and only provided—' that the firm estimate for 1964 was approved at +11% on the 1963 level (in real terms). Walker also indicated that his domestic financial authorities were keen to see the provisional estimates for 1965 and 1966 settled at the forthcoming Council session, and was 'prepared to support increases of up to 10% for 1965 and 9% for 1966 provided that other members are agreeable to increases of this size'. In sum, and against the recommendations of the BWP, the British were proposing that CERN's rate of growth fall steadily from a 'high' of 13% in 1963, rather than remaining fixed at 13% for the first two years, and that it drop to a value in 1966 which was below the lower limit specified in the Report.[118]

The British proposals for the rates of growth for 1963 and 1964 (13% and 11%) were accepted with little debate and no formal opposition in the Finance Committee on 18 December 1962.[119] And although staunch members of the CERN lobby, like de Rose and Bannier, were plainly disappointed by the line being taken by the United Kingdom, at the Council session the following day they did little more than plead with the UK delegation to reconsider its views. De Rose made 'a few rather gloomy comments' about the situation, stressing that the funds which the member states now thought to provide for the laboratory would not enable CERN 'to fulfil the mission for which it had been created', and indicated a change in policy which would have 'very serious consequences'. Bannier remarked that the Dutch delegation 'was extremely concerned to see that it had not been possible to accept fully the practical proposals put forward by the Working Party for the next few years', at the same time 'begging' countries like the UK 'to do their utmost during the next few years to give CERN the support necessary for full exploitation'. When it came to the vote the Council unanimously adopted rates of growth for 1963 and 1964 of 13% and 11% respectively. As for the provisional forecasts for 1965 and 1966, it was clear to Walker that too few delegations had instructions for the question to be settled just yet, and the Council agreed that the matter should be discussed at its session in June the following year. The British figures of 10% and 9% remained on the table.[120]

The negotiations around these figures in the months before the June 1963 Council session were dominated by a new factor: the possibility of a supplementary programme to advance design studies on the next generation of accelerators for Europe. Ever since 1957 a small but growing number of CERN scientists had been

Notes: p. 632

studying various possible new accelerators. Towards the end of 1962 it was felt that activities in this area should move beyond mere technical studies to the definition of a project and the raising of financial support for it. In January 1963 a group of European physicists met, constituted themselves as the ECFA—the European Committee on Future Accelerators—, and set up a Working Party under Edoardo Amaldi whose task it was to propose short- and long-term accelerator programmes for Europe. As reported by the Director-General to the Finance Committee in May 1963, ECFA planned to submit its recommendations to the Council the next month, and in December 1963 a request would be made to spend some 3.8 MSF in 1964 'to cover a supplementary programme for design studies'. If that money was granted work would continue on two plans: 'one for the construction of storage rings on the newly acquired French part of the [CERN] Site, and another for a larger machine in the 150–300 GeV range, to be built elsewhere'. At the same May 1963 meeting of the Finance Committee Hine presented his budget forecasts for 1965 and 1966: he called for increases of '*at least* 12% per year (in real terms), assuming that design studies for new accelerator projects [were] supported for 1964 and afterwards by a supplementary programme'.[121]

The British immediately made it clear that they were not prepared to increase their 'bid' beyond the figures they had put forward the previous December. In justification they insisted that the rate of growth should be settled bearing in mind total expenditure on the basic and the supplementary programmes. It looked as though the latter would cost 3.8 MSF in 1964—which meant an 'actual increase of 16–17% over 1963'. What is more, their FC delegate added, if design studies were continued on a big new machine and the storage rings were started, the global budget by 1966 'could be expected to be at least 160–170 million'. Backed by these arguments, the British coupled their bid with a threat: if the increases for 1965 and 1966 exceeded 10% and 9% respectively, (equivalent to budgets of 115 MSF and 125 MSF in 1963 prices), they would not participate in the supplementary programme.[122]

Several delegations differed from the UK at the May 1963 meeting—after all the Bannier Report had explicitly tried to disassociate expenditure on new items from the rate of growth of the underlying, 'basic' programme. With this in mind men like Bannier and de Rose insisted that the administration's estimates were in line with the Report's recommendations, and were willing to back higher figures (e.g. 12% then 11%), even if they were not prepared to give Hine the 12%/12% he had asked for. However, when it came to the vote at the June Council session, it was the British who carried the day by seven votes (Italy, Belgium, Spain, and the Scandinavian countries voted with them), to four (France and Germany, Austria and Switzerland). Greece and the Netherlands abstained, Bannier's argument being that 'the figures provisionally determined were not of vital importance, because they could be modified when they were converted into firm estimates', i.e. at the December Council session.[123]

The Directorate immediately set about trying to get the Council to reconsider its decision. On 7 August 1963 the Director-General circulated a letter to all delegates pointing to the discrepancy between the sequence of the rates of growth for CERN proposed in the Bannier Report (13%/13%, then at least 10%/10%) and those now adopted (13/11/10/9%).[124] This was followed by a substantial paper by Hine which was laid before the SPC and the FC on 10 October, and the Council the following day. In it Hine argued that 'the Council's figures will not support CERN even if the physics growth is cut to about half its present rate', and that, 'if confirmed, will involve a complete change in the style and scale of work in the Laboratory in a rather short time'. CERN, said Hine, needed, in 1963 prices, 119 MSF (+13.8%) in 1965 and 134 MSF in 1966 (+12.6%), which, if coupled with the sums actually agreed on for 1963 and 1964 were equivalent to rates of increase of 13/13/12/11% over the four years. Summing up at the October Council session, Weisskopf said that the Administration was simply 'asking for consistency with the Bannier framework, so that the necessary minimum could be done to maintain the momentum which European physics had gained'.[125]

The British remained convinced that it was not worth risking funds for the supplementary programme by increasing the rates of growth they had suggested for the basic programme. Focussing attention on the net rates of increase of the basic programme, they brought two new arguments in defence of their line. Firstly, that if a supplementary programme was adopted for 1964 it would effectively add about 1.5-2% to the rate of increase agreed on for that year (11%). This was because certain items budgetted for under the existing basic programme, notably the development of the French site, would be charged to the supplementary programme if it was adopted, and this would release about 3 MSF for other purposes. Secondly, Walker insisted that it looked as though the 'equipment gap' might be closing. When the 1963 budget had been adopted, he pointed out, it had been said that only 80 new staff were to be recruited during the year so as to redress the imbalance between staff and equipment, and to push up the cost per head. In fact, he said, over 100 new staff had already been taken on by October 1963.[126] Reinforced by such arguments, and notwithstanding the pressure put on the UK delegation at the Council meeting in October — Leprince-Ringuet spoke of it being 'a great mistake to spoil the ship of CERN for a ha'p'orth of tar', de Rose compared CERN's position to 'that of a farm in which large investments had been made to get the soil ready, but [that now that] it was time to sow the seed, the farmer said: "I must economize on grain because I am short of money"' —, Sir Harry Melville was emphatic. 'Everyone was keen to see a supplementary programme launched', he said, and 'he feared that, if the Administration were to press for very much higher budgets than had been contemplated, it might be most difficult for the Member States to get the money for the supplementary programme, which was equally necessary'.[127]

De Rose's pleas at the October Council meeting were followed by a forthright statement by the French delegation at the Finance Committee meeting in November. Praising the work done by the Bannier Working Party, M. Courtillet added that its findings 'deserved to be followed without question. The French Delegation', he went on, 'had always held this view and had no intention of modifying it [...]'. To pursue a more restrictive policy, said Courtillet, 'would be equivalent to ignoring the fruits of the financial sacrifices made in the past to create one of the most remarkable instruments of research and international co-operation in the world'. France, he concluded, would support sequential rates of growth of 12/11/10% for 1965/66/67 corresponding to budgets of 117 MSF, 130 MSF, and 143 MSF—and giving a total for 1963-66 inclusive which was 'very close to the minimum programme set forth in the Bannier Report [...]'. The French statement was immediately welcomed by the Dutch delegation, who announced that they intended to support rates of growth of 12/10/10% for 1965/66/67. Switzerland also announced that she could support 12% for 1965. In fact, by the time of the December meeting of the Finance Committee, the majority of delegations announced that they could support the French sequence of 117/130/143 MSF.[128]

In the light of these developments the United Kingdom decided to increase its offer of 115/125/135 MSF (10/8.7/8%) for 1965/66/67 by 1 MSF all round to 116/126/136 MSF (11/8.6/8%). The British emphasized again that their participation in a supplementary programme would be jeopardized if these figures were not accepted. They also repeated that, in their view, these figures were 'within the terms of the Bannier recommendations, when allowance had been made for the fact that there would be a supplementary programme in 1964 [...]'.[129]

Britain's willingness to seek a compromise was matched by a similar gesture by the French at the Committee of Council meeting on 16 December 1963. There was no point, said de Rose, in risking 'the future of physics for the next fifteen years' for the sake of the next two or three by 'trying to force the adoption' of the higher budget figures, even though a majority of governments considered them to be appropriate. 'It was desirable to make the task of the United Kingdom Delegation easier and enable it to adhere to the new programme', he said. Thus de Rose proposed that the firm estimate for 1965 be agreed at Britain's figure of 116 MSF on condition that it be revised if the supplementary programme for 1965 was not adopted. This recommendation was approved by the Council the following day with one abstention for technical reasons (Denmark). The Council also agreed to postpone settling the provisional estimates for 1966 and 1967 until June the next year.[130]

We want to draw attention to two aspects of the positions adopted by the member states' delegates in these negotiations. Firstly, there was the United Kingdom's determination to force down the rate of growth of CERN's budgets from year to year, and its willingness not to be bound by the bottom limit of 10% for 1965 and 1966 recommended in the Bannier Report. This policy was adopted despite the fact that in summer 1962 the UK 'revised its whole nuclear physics programme, including the CERN programme, and the upshot had been to give CERN first priority for financial support'. It was indicative, firstly, of the low growth rates in the country's

R&D expenditure in this period which we have commented on before—indeed if the growth rates below 10% that the British were putting on the table seemed unduly parsimonious to the majority of other delegations, they were considerably higher than the 2% rate of increase which UK universities were granted after the mid-1962 policy review.[131] Britain's position was also based on the advice being given at home by John Adams, former CERN Director-General and member of the UK's Council delegation since December 1961. In July 1962 Adams told DSIR officials that he thought that Hine's staff figures for the SC and for technical services were inflated and that, in general, CERN was 'taking too self-centred a view and was becoming increasingly unwilling to build collaborative relations with the national programmes'. As for the overall rate of growth of the budget, Adams felt that 12% would be reasonable for 1963, and told the DSIR 'that the rate of increase should then fall off to about 8% in 1966 [...]'. The British, then, were committed to seeing CERN grow—but not as fast as most other member states were prepared to countenance.[132]

The second point to bear in mind is that, however disappointed the members of the CERN lobby were by Britain's attitudes, they were not prepared to force the issue as at the end of 1961, or to vote against the British in any major budget decision. One obvious reason for avoiding confrontation, at least in 1963, was the fear of alienating a potentially major contributor to the supplementary programme (Britain's share would be the maximum 25%). Another consideration was possibly the determination not to do anything that might upset the ratification of the Conventions establishing ELDO and, more especially, ESRO, CERN's 'sister' European organization. These Conventions had been signed in London (29 March 1962) and in Paris (14 June 1962) respectively, soon after the explosive budget debate in the CERN Council in December 1961. The process of ratification continued throughout the period we have discussed—ELDO's Convention entered into force in February 1964, that of ESRO three weeks later—, passing through a particularly delicate period in mid-1963, when a quick ratification of ESRO's Convention by Britain was deemed highly desirable.[133] In these circumstances it would have been damaging to say the least if the United Kingdom's major partners at CERN, who were also to be her major partners in ESRO, again isolated and embarrassed the British delegation in a budget debate.

10.4.2.2 The budget debates on the 'improvement programme' in 1964 and 1965

The so-called PS improvement programme was essentially an assortment of projects collectively intended to upgrade and extend the laboratory's facilities so as to enable it to compete with American and Soviet machines in the late 1960s, and to service an increasing flow of visitors from the member states. It was first seriously discussed by the SPC and the FC in spring 1964. Here Hine argued that part of the programme could be covered by the administration's existing estimates of what it

Notes: p. 633

would need to reach full exploitation of the laboratory by 1967. Additional money was needed, however, for three '"extra-Bannier items"', as Weisskopf called them: a new linac and other modifications to improve the PS intensity—'more or less forced on CERN by Brookhaven's plans' (Hine) for similar improvements to the AGS—, a new experimental area with its beams and ancillary equipment—'essential' 'if many countries were going to rely on CERN to supply the accelerator part of their national programmes' (Hine)—, and, finally, a new large bubble chamber—'(U.S. projects for very large H_2 chambers - 40 m^3 at Brookhaven - must not be ignored)' (Hine).[134] As discussed in the Finance Committee in April 1964 the cost of this programme amounted to about 100 MSF to be spread over three to five years. Two-thirds of the money was for the first two projects; CERN did not intend to construct the big bubble chamber itself, hoping instead 'to join in building and operating a 4 or 6 m heavy liquid bubble chamber' (Weisskopf) with French scientists. Translated into budgetary terms for the immediate future, Hine proposed that in 1966 CERN would need about 130 MSF for its existing 'slowed-down "Bannier" programme'—the French figure already on the table—*plus* 18–23 MSF for new equipment, and about 140 MSF plus 29–34 MSF in 1967 (the French were proposing 143 MSF). As an alternative he suggested that the PS linac be omitted or considered as part of the storage ring project, in which case CERN would need 8 MSF in 1966 and 15 MSF in 1967 for 'extra-Bannier' items.[135]

These proposals made no immediate impact on the provisional estimates for 1966 and 1967 which the various FC delegations had proposed in December the previous year, and which were supposed to be (belatedly) agreed on by the Council at its June session. At a meeting of the Committee on 3 June 1964 France repeated its offers of 130 and 143 MSF for 1966 and 1967, and Britain again indicated that it preferred 126 and 136 MSF. A survey of those present suggested that all countries bar Spain supported the higher figures, and at the end of the meeting it was generally felt that they would in fact be agreed on unanimously when the FC made its final recommendations to the Council two weeks later.[136] This was not to be, however. On the eve of the Council session the Italian delegation unexpectedly announced that for technical reasons they would have to support the lower British figures, though they hoped to be able to vote for the higher French figures when the provisional determinations were turned into firm estimates. And although seven member states (F, CH, NL, N, S, A, FRG) together representing over 56% of the contributions supported the 130/143 MSF sequence, the Council agreed 'not to come to a vote which would have unpleasant results for several Member States' and to leave the provisional determinations for 1966 and 1967 'undecided'.[137]

The main reason for this postponement was the uncertainty surrounding the costs of the supplementary programme. Indeed CERN's future programme was *the* overriding issue at the June 1964 Council meeting. After almost two days of discussion the delegates concluded, in the words of Council President Bannier, that although a 300 GeV machine 'was the key to the future of high-energy physics in

Europe', it was still not clear 'whether Europe should just *have access* to a big machine or should actually *build* it' (our emphases). 'In the meantime', Bannier went on, 'high-energy physics would still have to be developed and this could best be done by improving the present facilities at CERN and building the storage rings. The decision on these two things', the President suggested, 'should be taken at the December Council session'.[138] Until it was, there was no point in sticking to the letter of the Bannier procedure if it meant outvoting Britain and Italy—all the more so as the estimates were only provisional anyway.

During summer 1964 the CERN management broke the improvement programme down into seven topics, each of which was studied in depth inside the organization. By October Hine was in a position to draft a new set of estimates of its costs for the Finance Committee. The overall estimate had now doubled to 200 MSF, though it was stressed that 'reasonable programmes with different scientific emphases could also be made with fewer items and a smaller total, particularly when the 25 GeV facilities offered by the ISR [were] taken into consideration'. Hine then proposed two alternatives which were essentially similar in content to the schemes advanced earlier in the year, one costing about 130 MSF, the other about 100 MSF. The costs of these 'new capital projects' were then superimposed on a budget curve for the 'running expenses' ('the logical continuation of the budgets considered in the Bannier Report') of the laboratory up to 1969, with the following results:[139]

Year	1965	1966	1967	1968	1969
Running expenses (MSF)	120.5	132	139	146	153
Rate of growth (%)		9.5	5	5	5
New capital projects (MSF)	1	5-7	18-20	29-37	?
Total (MSF 1964 prices)	121.5	137-9	157-9	175-83	?
Total (MSF 1963 prices)	116	131	150	167	
Rate of growth (%)		~12.8	~14.6	~11.5	

Notes:
- The figure for 1965 had already been fixed by the Council
- The costs for new capital projects in 1969 depended strongly on the programme chosen, those for the years before less so, according to Hine
- The figures for the overall rates of growth and the total in 1963 prices are based on the lower of Hine's estimates

Hine's new figures call for comment. Firstly, it is striking how he has slowed down the rate of implementation of the new programme: if in April he was suggesting spending something like 50 MSF in 1966 and 1967, now he has reduced the allocations for those two years by half. What is more he has brought his estimates more in line with the French (and majority) figures of 130 and 143 MSF (1963 prices), i.e. 136 and 150 MSF in 1964 prices which had been on the table in June. This does not simply reflect a willingness to 'compromise': it was also symptomatic of the pressures in 1964 in the United States and Europe to cut back expenditure on high-energy physics, pressures which also partly explain the Council's decision in June to postpone taking a decision on the 300 GeV machine for the time being.[140]

The Finance Committee continued to react cautiously to these figures when they were first presented to them in October 1964: no decision had yet been taken on the supplementary programme and no one was willing to increase their proposed allocations until they knew just what other expenses they would have to bear from 1966 onwards. A new element did however enter the picture in November. The three Scandinavian countries insisted that more than just money was involved in the choice between the set of figures supported by the majority (136 and 150 MSF for 1966 and 1967) and those supported by Britain (132 and 143 MSF): a question of 'principle' (Norway), of 'policy' (Sweden) was at stake. Their point was that, if one accepted the data produced by Hine, the higher set of figures put forward by France the year before made provision for a start to the improvement programme; the lower set being defended by Britain did not. And the Swedes at least made it clear that they wanted 'enough intensity and enough beams and bubble chamber equipment to take advantage of CERN', and that they would support the higher figures if enough other countries did.[141]

The British delegation's initial reaction to this development was to affirm that they were 'not convinced that more than 132 MSF was needed for 1966 to enable a start to be made on the improvement programme' — to which the Director-General replied that if CERN had to work within such a figure he would 'instruct the PS Machine Group not to go ahead'.[142] By December the United Kingdom had decided to climb down. At the Finance Committee meeting on the eve of the Council session the British delegation increased its offer for 1966 to 135 MSF; their French counterparts, welcoming the UK's 'attempt to reach a compromise' invited all their erstwhile supporters to join them in voting for this new figure.[143] It was approved by the Council on 16 December 1964 with one abstention (Italy — for internal financial reasons). The Council also agreed that 6.8 MSF be spent in 1965 to continue preparatory work on a supplementary programme, pending a definite decision by a sufficient number of member states on whether or not to proceed. The provisional determinations for 1967 and 1968 were left in abeyance.[144]

The administration put forward its new figures for the costs of the basic programme, including improvements, in April 1965. Here Hine indicated that CERN would need 166 MSF for 1967 (made up of 147 MSF of 'running costs' and 19 MSF of improvements) and 186 MSF for 1968 (154 MSF plus 32 MSF for the improvement programme) — essentially the figures that he had put forward the year before, adjusted to 1965 prices.[145] They corresponded to rates of growth for the basic programme of first 17% then 12% — to be compared to the majority offer of 158 MSF (+11%) for 1967 which was still on the table.

How did the administration justify this short-term jump well beyond the upper limit laid down by the BWP? They argued that, while 11% per annum was a suitable average rate of growth over five years for the basic programme, more than this would be needed in 1967 and 1968 if the improvement programme was not to be seriously delayed. 'Our proposals', Weisskopf explained, 'take account of the unavoidable fact that at the start of a programme of capital projects there is bound to be a higher than average rise in budgets for one or two years, followed by a "plateau" period of several years when the growth rate [was] less than average'. The kind of policy the member states seemed to want, a 'strict 11% policy', while leading to the same overall expenditure on the basic programme between 1966 and 1971, would give the Director-General 20 MSF less than he had requested for 1967 and 1968.[146]

When these figures were first debated in the Finance Committee in May 1965 the member states were unwilling to come anywhere near Hine's 1967 figure (166 MSF). Instead all the major contributors to the budget, including the United Kingdom, rallied to the French figure of 158 MSF that had been suggested the year before.[147] Their position softened somewhat in response to strong opposition by the administration and the Scientific Policy Committee. By June several delegates, including the British, announced that they were willing to reconsider, and the French proposed increasing the figure for 1967 by 1.5 MSF to 159.5 MSF.[148] After pleas from Weisskopf and Leprince-Ringuet (for the SPC), the Council decided in June 1965 to postpone again the provisional determination for 1967. Instead it hoped to settle the firm estimate for that year and the provisional determinations for 1968 and 1969 when it met in December.[149]

If the Council were convinced that they could settle the allocations for CERN three years ahead in December — a goal that had consistently eluded them until then —, it was because at this June 1965 session they approved 'in principle [...] a supplementary programme for the construction of intersecting storage rings'.[150] Hine also supplied them with 'firm' estimates of the overall costs of the basic and supplementary programmes. The improvement programme, he said, could be financed with the budgets for the basic programme he had put forward in April increased by some 5% to 7% in 1969, 1970, and 1971 to cover unforeseen items and costs. As for the ISR, he committed CERN to building it by 1971 for 332 MSF (in 1965 prices), plus a provision of 15% overall for unforeseen items and costs.[151] Armed with these data, Council delegates were asked to return in December with instructions as to whether they wished to participate in the supplementary programme or not.

The ensuing debates in the Finance Committee were concerned less with the total amount to be allocated to CERN for 1967 to 1969 inclusive, which differed little from what Weisskopf wanted, than with how to distribute it over the three years. In November 1965 the Director-General asked that money for the basic programme be allocated in the sequence 163 MSF, 181 MSF and 200 MSF from 1967 onwards. The

Notes: p. 634

French delegation proposed that he be given 162 MSF for 1967, and 'of the order of' 180 MSF for 1968 and 197 MSF for 1969. The British, for their part, preferred a sequence of 160, 180, and 198 MSF, rates of growth of some 12.5% for the first two years, and 11% thereafter. The Council settled the matter a month later.[152] At its December meeting the 'firm estimate' for 1966 became a budget of 142 MSF at 1965 prices. The British sequence of allocations for the basic programme was ratified. The Council also approved the draft budget for the ISR. It adopted a figure of 20.8 MSF for 1966 a firm estimate of 68.6 MSF for 1967, and provisional determinations for 1968 and 1969 of 73.3 MSF and 74.6 MSF respectively (in 1965 prices). At the same time, accepting that it was necessary to continue preparatory studies on the 300 GeV machine although 'the date for actual construction was receding' (Melville), the Council authorized the DG to spend 4 MSF over the next three months for this purpose, and agreed to reconsider the status of the project around March the following year.[153]

The most striking feature of the budget debates in 1964, and particularly in 1965, is the willingness of the member states to invest heavily in CERN. Certainly the precise extent of that commitment remained in doubt until the supplementary programme had been settled. But once that was done there seemed to be no shortage of money for the laboratory. Apart from accepting that the existing facilities in Geneva should expand at about 5% annually in real terms, the Council supported an improvement programme expected to cost about 130 MSF in 1965 prices, and the construction of the Intersecting Storage Rings which was estimated to cost about three times as much again.[154] In concrete terms in December 1965 the Council voted to spend about 750 MSF in 1965 prices for 1967 to 1969 inclusive—an average annual expenditure which was roughly double that spent on CERN that year—, while at least two of the member states, France and Germany, were willing to participate in the construction of new big bubble chambers and 'were prepared to give them considerable financial and technical support'.[155]

At the most general level this 'generosity' towards CERN was inspired by two factors—the conviction among national high-energy physics communities that CERN had to be supported as *the* centre in Europe for doing research at a level comparable to that being pursued in the United States, and the willingness of domestic financial authorities to support that objective in a period of national economic growth, and of increased expenditure on Research and Development. Within that general framework it is the change in attitude towards CERN in 1964 and 1965 of the Scandinavian countries and of Britain, in particular, that deserves further comment.

The main reason for strong Scandinavian support for the laboratory's expansion was the recognition in 1963 that it was only by participating in CERN that they could hope to keep in touch with leading research in the field. They reached this conclusion after making an assessment of their capabilities in the light of the recommendations of the ECFA Working Party. The Working Party, which published its report in June 1963, recommended that Europe embark on a two-pronged programme of accelerator construction: a so-called *summit* programme 'in the region of highest energies', of which the ISR to be built at CERN was one component, and a new very high energy PS the other, and a so-called *base of the pyramid* programme. This was supposed to supplement the former and to comprise 'suitable national or regional accelerator projects in the Member States'.[156] Acting on this recommendation the Scandinavian (and Benelux) countries sat down to discuss the possibilities of regional machines during the latter half of 1963. They concluded that they lacked the experience, the manpower, and the resources to launch such a venture. As Funke explained to the FC in November 1963, 'the Scandinavian countries had considered building an accelerator of about the same size as Nimrod, and had discovered that this would be an extremely expensive enterprise which would cost Sweden more than her combined contribution to the present CERN and any future super-CERN. It therefore seemed to be a better proposition to contribute more to CERN and develop the experimental possibilities there', he went on.[157] Consequently, whereas before the Scandinavians had been relatively withdrawn in budget debates and, in the case of Sweden, had tended to side with the United Kingdom in supporting relatively low budget figures, even to the point of being outvoted in the Council, these countries were now often the first to come out in favour of the relatively high rates of growth put forward by the French. They were also all strong backers of the improvement programme: Danish physicists were 'in favour of improving the PS to make it as competitive as possible with other similar machines' (Bøggild, April 1964); 'The Swedish authorities attached great importance to developing the present CERN and making the best use of the facilities available and keeping them up to date' (Funke, June 1964); 'Norway was not in a position to do independent research work, and it relied on CERN as a high standard research institution which could feed its universities with work and to which it could send visitors' (Ibsen, April 1964).[158] Finally two Scandinavian countries, Norway and Sweden, were among the first six member states to express support for a supplementary programme in 1964.[159]

We do not have direct documentary evidence explaining the distinctly more positive attitude towards the financing of CERN taken by the British in 1964/65. One might surmise though that three factors played a role. Firstly, the growing recognition of the importance of the laboratory in the eyes of British physicists, stimulated in particular by the use they made of it in 1964/65 to do bubble chamber physics. In January 1964 the 1.5 m British National Hydrogen Bubble Chamber did its first runs in Geneva, where it stayed for 18 months before being transported back to the UK to work around Nimrod. This coincided with a marked increase in the British presence at CERN. Butler, the leader of the British bubble chamber project, replaced Gregory as the Chairman of the Track Chamber Committee during the course of 1964, while on the experimental floor British groups were to be found in all

Notes: pp. 634 ff.

collaborations with 'their' chamber in 1964 and in six of the nine runs done in 1965. The number of bubble-chamber photographs distributed in the UK rose accordingly, from 285 million in 1963 to just over 700 million in 1964 and 412 million in 1965.[160]

The other two reasons why the British attitude may have changed during these two years are of a politico-diplomatic nature. On the one hand, there was their increased isolation in budget debates. Ever since 1957 Britain had played a significant, critical role in the Finance Committee as leader of a minority group which insisted on the need to limit expenditure on CERN, and which consistently defended a lower set of budget figures. With the change in policy in the Scandinavian countries, in particular, in 1963 the balance of power in the Committee shifted markedly in favour of the French, and the British delegates found themselves alone and speaking for no one but themselves. Hence the need to compromise, and the interest of siding at least temporarily with the majority so as to be better able to regain a leading role later. Another 'political' factor may have been the changes in British politics occurring at the time. In October 1964 Harold Wilson's Labour Party was voted into power on a platform which made science a high priority and which stood for the 'white heat of the technological revolution', a programme which set them distinctly apart from their Conservative Party opponents at that time.[161] If this change of tone permeated quickly through the state bureaucracy it might well have encouraged a reappraisal of attitudes in the Treasury towards the financing of CERN.

10.5 Concluding remarks: the significance of the Bannier Report

The Bannier report has a special place in the history, one might even say in the mythology, of CERN. Hailed in the Council in 1962 'as a kind of Bible for the consideration of budgetary methods in large centres', the idea of having a four year rolling programme and budget is still standard practice in the laboratory today, almost 30 years later.[162] This persistence can easily lead one to *reduce* the historical significance of the Bannier report to the formulation of a new *procedure* for long-term forward planning, to concentrate on its technical achievements at the expense of its political importance. Indeed if one then goes on to contrast the stormy debate in 1961 with the relatively mild disagreements over the budget in the Finance Committee and the Council between 1962 and 1965, it is tempting to attribute this 'changed tone' to the introduction of the Bannier procedure as such, to see the importance of the Bannier Report as establishing a 'rational' system of planning which depoliticized the budget debate.

This interpretation, in our view, inverts cause and effect. The importance of the Bannier Report is not that it devised a technical solution to what had become a political problem. On the contrary, *the Report was essentially a political agreement between the delegates from the member states*—in the sense of an agreement on certain key questions of policy—, and it was only to the extent that such an agreement was possible that the Bannier procedure could be implemented. There is one simple index of this: the fact that it was only possible to instal the four year forward planning system in its entirety at the end of 1965, when the uncertainty surrounding the nature of the supplementary programme had been settled. In short *the Bannier procedure could only 'work' as long as there was an underlying consensus on important matters of policy.*

In the context of the time, to speak of consensus is to speak of an agreement on basic principles between a majority group led by the French and a minority led by the British. From 1957 to 1961 the British delegates to the Finance Committee had insisted that there be limits imposed on CERN's expenditure, and in 1959 they had successfully persuaded the Council to impose a three year regime of level funding on the administration. In 1961 the Treasury tried once again to stabilize the budget for a three-year period, this time backing it with an aide mémoire suggesting that the ceiling be agreed 'between governments' rather than by the Council. If the Bannier Report was a breakthrough it was because, by 'accepting' its recommendations, the British rejected these policies. Put differently the key to its success lay in the fact that *the United Kingdom delegation agreed that CERN's budgets should grow annually.*

This 'local' consensus was, in its turn, an expression of more general economic and political developments in Europe. During the late 1950s and early 1960s most European economies improved markedly, and there was a new willingness in many countries, Gaullist France being the outstanding example, to increase substantially the proportion of the GNP devoted to scientific research and development. The success of the Bannier Report cannot be separated from this broader politico-economic context. It was in fact just one of a number of expansionary measures which included the construction of important electron accelerators in Britain, Germany, and Italy, and the decision to set up ESRO at a cost of 306 million 'accounting units' (about £110 million) over eight years.[163]

The stress we have placed on the underlying agreement in the Finance Committee in the immediate post-BWP period must not be exaggerated. In particular it would be wrong to infer that, if before there was conflict and confrontation, now there was consensus and cooperation. To accept the Bannier Report was to accept that CERN's existing facilities be quantitatively expanded at a rate which would fall over the next four years from a high of some 13% per annum to a low of some 10% per annum. There was still considerable scope for disagreement, both within that framework—over the rates of growth and how rapidly they should decline—, and outside of it—when policy decisions calling for a *qualitative* change in facilities (the improvement programme, the supplementary programme) were at stake. In fact the confrontational tactics that were used in 1961 over the issue of growth vs stability were also brought into play in 1962 and 1963 over the rate of growth—we find again the threats (e.g. by the British that they would not support the supplementary programme) and the counterthreats (e.g. by the 'lobby' that if the UK did not support CERN sufficiently it would be reduced to a second-rate institution).

Does this mean that there was no 'change in the tone' of the budget debates

Notes: p. 635

between the pre- and post-Bannier epochs, as is so often said? That would be going too far, at least as far as our brief period is concerned. For while there may have been threat and counterthreat, no one was prepared to force a confrontation with the British in 1963, for example, when the estimates for 1965 to 1967 were being negotiated. At one level this was for technical reasons: there was very little difference between the figures on the table for 1965, and all parties readily agreed to postpone a decision on the more contentious estimates for 1966 and 1967. More fundamentally, though, it indicated a *political* choice by France and the majority not to isolate the British when their support was dearly needed for a supplementary programme at CERN (and when the situation in ESRO was particularly delicate). To generalize, the tone in the budget debates at CERN both before and after 1961 was determined essentially by the actors' assessment of the policy/political issues at stake (if any), and by the tactics they chose to adopt in a specific situation—and not primarily by the availability or otherwise of a technical mechanism for planning ahead.

To emphasize the political significance of the Bannier Report is not to deny the importance of the Bannier *procedure:* it is simply to put it in perspective. And while it may be true that the tone of the budget debates rose and fell depending on whether or not major matters of policy were felt to be involved, it is equally certain that the long-term planning required by the procedure considerably eased the process of negotiation within an agreed policy framework. The key factor here was the 'willingness' of all parties, the administration and the member states' delegates, to adhere to a two-year ceiling, to accept that the estimate of expenditure for year $X+2$ agreed on by the Council in December of year X, was indeed 'firm'. This not only meant that there were no last minute appeals in year $X+1$ for an increase in the budget in year $X+2$ (as had happened in 1957 and 1961), with the 'procedural difficulties' that that entailed. It also meant that the debate on the budget during any year dealt with the appropriations required two, three, and four years hence. The delegates therefore had enough time to work out their positions in consultation with the CERN administration, with one another, and with their domestic financial authorities: technically derived crises were now avoidable.

The workability of the Bannier procedure from a technical point of view, as from a political point of view, was dependent on a number of contextual factors. In particular if CERN was able to plan two, three and four years ahead, and even reasonably commit itself to an overall construction budget for the ISR in 1965, it was essentially because the laboratory now had a *history*. One could draw on past experience both in Geneva and abroad in making forecasts, estimating future needs by extrapolating from past trends. To this must be added the key role played by Mervyn Hine. Appointed to the Directorate in 1961 as the member for Applied Physics, from the summer of that year onwards Hine devoted himself to planning CERN's expenditure with diligence—the papers on CERN's future needs which he submitted to the SPC, the FC, and the Council fairly streamed from his pen—, and with a growing confidence in his ability to make reliable and durable estimates—as captured in his slogan that he was able 'to do it right and tell it straight'. Which is not to say that he, the Director-General, and the administration did not seize every possible opportunity to extract the maximum possible resources from the member states, nor find one argument after the other why the delegates to the Finance Committee should accept rates of growth above the 14% limit 'laid down' in the Bannier Report...

What of the recommendations of the Bannier Report as serving as a tool for 'financial control'? The question is interesting if only because it misses the point: it persists in seeing the achievements of the Working Party as primarily technical. Certainly the limits to growth which the BWP recommended were an essential point of reference for the FC delegates in their negotiations, and certainly no major contributor to the budget during our period was prepared to countenance an increase from one year to the next of more than some 13% — for the 'basic programme'. And therein lies the rub. For by specifically *excluding* 'major new developments' from its recommended rates of growth the Bannier Report deliberately avoided trying to impose *a priori* checks on the overall expansion of CERN, deliberately left the 'control' over much expenditure *'to separate policy decisions of the Council'* (our emphasis). And its publication coincided with the start of an exponential growth in the CERN budgets which was to last for 15 years, and which only began to level off (in real terms) around 1975.[164]

Notes

1. The main sources used for this chapter were the official papers laid before the CERN Council and its committees, notably the Finance Committee, files on the Bannier Working Party and its Report in the DIRADM and MGNH series in the CERN archives, and a set of letters, memoranda, and reports dealing with the financing of CERN in the SERC archives, Hayes, Middlesex, notably boxes NP31A, B124, and B125.
2. The *First Report to Member States on the Organizational and Financial Implications of Future European Co-Operation in Nuclear Research,* document CERN/GEN/3, was dated 25/10/52, and was followed six months later by the far more substantial *Second Report...,* document CERN/GEN/5, 5/5/53. They are both in file (CHIP10014).
3. Memorandum, R.A. Butler to The Most Hon. The Marquess of Salisbury, 29/12/52, copied to Eden and Cherwell (PRO-DSIR17/551).
4. Dahl's first figures for the cost of a 30 GeV strong-focusing machine (i.e. some 55 MSF) were reported to the 'provisional' CERN Council at its Third Session from 4-7 October 1952, in which he said that it 'may be constructed faster and cheaper' than a 10 GeV machine of conventional design. The minutes of this meeting are document CERN/GEN/4, 15/2/53 (CHIP10014). Soon thereafter a group of PS engineers along with John and Hildred Blewett from Brookhaven costed the machine at 69.2 MSF — reports PS-MHB1 & 2 dated 23/10/52 in the CERN archives. This figure was discussed in the Council at its seventh session from 29-31 October 1953 (minutes CERN/GEN/12, dated 20/3/54, in CHIP10014), where Dahl reported that a 30 GeV machine would probably cost at least 70 MSF, and the PS Group was instructed to reduce the design energy to contain costs.
5. This was one of several recommendations to the Council emanating from the Finance Committee and a working party that it set up immediately after the Convention came into force, and which were

summarized in CERN/FC/25 Rev., 25/1/55.
6. The Capital Investment Programmes for 1952-1960 drawn up in 1955 were CERN/FC/89, 4/11/55 and CERN/FC/89/Rev., 9/12/55, while the final version produced the following year was CERN/FC/150/Rev., 15/10/56.
7. See minutes of the 7th Session of the Provisional CERN Council, CERN/GEN/12, 20/3/54 (CHIP10014).
8. The quotation is from CERN/FC/25/Rev., 25/1/55.
9. For the full reference see Accelerator Conf. (1956). Summarizing the first week of the symposium in the introduction to Volume 1 of the proceedings John Adams had this to say on papers by L.W. Smith on the operation of the Cosmotron, by E.J. Lofgren on the running of the Bevatron, and by a Soviet group on the use of their 680 MeV synchro-cyclotron. They demonstrated, said Adams, 'the intricate organization of the experimental programme necessary for full exploitation. Nuclear physicists not acquainted with machine physics and more used to a University research time scale must have been horrified at the rigors and discipline to which high energy nuclear physicists are submitted these days' — Accelerator Conf. (1956), 6. We said more about the unpreparedness of CERN physicists to exploit their machines in the previous chapter.
10. The first quotation is from the SC division's chapter in the draft budget for 1957 — CERN/FC/143, undated, but around September 1956. The quotation from Gentner was made at a meeting of the FC Working Party held on 2-3 October 1956 — minutes CERN/FC/162, 18/10/56, 25, in which he explicitly connected the need to extend the SC research programme with the 'disclosures' made at the CERN Symposium and during a visit he and Bakker had paid to Moscow.
11. The quotations are from the chapter on the SC in CERN/FC/143 (see previous note), the number of fellows from CERN/FC/150/Rev., 15/10/56, 16.
12. Bakker to the FC Working Party meeting on 2-3/10/56, minutes CERN/FC/162, 18/10/56, 9. As we saw in the previous chapter Adams and his group began to take an interest in the equipment to be used around the PS in summer 1957. As a result their estimates for 1959 more than doubled from 7.5 MSF at the end of 1956 (CERN/FC/150/Rev.,15/10/56), to 17 MSF (including 4 MSF for experimental equipment) at the end of 1957 (CERN/FC/234/Rev., undated) and increased by another 40% to 23.3 MSF (including 7.6 MSF for experimental equipment) at the end of 1958 (CERN/FC/303, 19/9/58).
13. The French delegation was particularly concerned about transfers occurring between main budget headings, notably transfers from 'Staff' and 'General' expenses to 'Capital' expenses — see CERN/FC/162, 18/10/56, and CERN/FC/173/Add.1, 10/12/56 for a corresponding proposed amendment to the Financial Rules drafted by the UK delegation. The French proposal that the Convention and the Financial Rules be changed so that the budget could only be passed by a two-thirds majority consisting of states who contributed three-quarters of the total contributions was made in CERN/FC/176, 6/12/56. Sir Ben gave his word at the December Council session (minutes, 14/12/56, 24). Officials in his Department, the DSIR, felt that the proposal had 'more theoretical than practical value' as there was 'no danger' that 'budgets would be forced through by a bloc of smaller states against the wishes of larger contributors [...]', letter Smith to Verry, 15/11/56 (SERC-NP31). For more detail on the UK attitude see a long memorandum on the topic which was 'copied' on 20/11/56 (SERC- NP31).
14. Verry's first thoughts circulated on 5 March 1957 are in a paper entitled *CERN Policy on Budgetary Control* (DIRADM20382), while his more formal paper is CERN/FC/188, 12/4/57. His intervention at the FC meeting on 1-2/5/57 is recorded in the minutes, CERN/FC/198, 10/5/57. Eliane Bertrand also sent the British delegate the verbatim record of his intervention after the meeting; it is attached to letter CERN/3748, 16/5/57 (DG20804).
15. All of the quotations in this paragraph are verbatim from Verry's intervention — see note 14.
16. See CERN/FC/188, 12/4/57. As its load increased the Finance Committee had its meetings prepared by a working party which generally met the day before. These meetings were attended at this time by the chairman of the full Committee, an FC delegate from Britain, France and Germany, and two or three

other full committee members 'representing' various national groupings (the Scandinavian countries, Belgium and the Netherlands, etc.). The meetings were also attended by senior members of CERN's management.

17. Richemond's reply drafted for the meeting of the Group of Directors on 22/5/57 is paper CERN/GD/126, 20/5/57, the minutes of the meeting at which they discussed it is document CERN/GD/127, 16/7/57, both in (DG20963), and the (only slightly modified) version agreed on by the Directors is document CERN/FC/203, 22/5/57. Verry's reply, dated 13/6/57, is to be found on (DIRADM20382). The minutes of the FC meeting on 28 May are CERN/FC/212, 18/6/57, those of its Working Party on 26 June are CERN/FC/224.
18. The *Joint Note* is CERN/FC/227, 29/8/57.
19. Letter Smith to Verry, early 1957, written after reading the minutes of the December 1956 Council session (SERC-NP35). Unfortunately I neglected to record the precise date of this letter when taking my notes; there may have been other minor errors in transcription. The quotation from Verry is from his text dated 5/3/57—see note 14.
20. The quotations are respectively from Verry's text of 5/3/57 (see note 14) and Smith's letter to Verry (see note 19).
21. Willems identified these 'two extremes', as he called them, at the meeting of the FC Working Party on 28/5/57—minutes CERN/FC/218, 18/6/57. The distinction is very similar to that explained by Willson (Willson (1957)) between synthetic and analytic methods of drawing up budgets. The first, in which the administration of a laboratory summed its various needs, and expected the financing authority to provide the necessary resources (the practice in operation at CERN at the time) was typical of cases in which a Government thought that research in a particular field was so important that it had to be pushed ahead as rapidly as possible; it would, said Willson, 'be rare in peacetime'.
22. The quotations are all Bakker's remarks made to the FC Working Party meeting on 26/6/57, minutes CERN/FC/224.
23. See Willems' remarks to the FC on 16/10/57 (minutes CERN/FC/242, 30/10/57), and letter Bannier to Verry, 19/6/57 (SERC-NP35).
24. The quotations in this paragraph are all from Verry's paper of 5/3/57 and the verbatim record of his intervention at the FC meeting on 1-2/5/57—see note 14.
25. See previous note.
26. This information is from D. Wilkinson, 'Events Surrounding the Construction of Nimrod', Nimrod (1979), 7-20.
27. The quotation is from Bertrand's record of what Verry said at the FC meeting early in May 1957—see note 14. For more on Britain's reasons for joining CERN see my chapters 12 and 13 in History of CERN (1987), particularly section 13.5.
28. The first version of the draft budget for 1958 was document CERN/FC/234, 25/9/57. It was accompanied by a *Draft Long Term Budget 1952-1960*—CERN/FC/235—which no one took seriously. Its forward estimates were totally unrealistic (e.g. 33.2 MSF for 1959, which was immediately replaced by 45.8 MSF in CERN/FC/234/Rev., without date, when the FC instructed the administration to make a 'fairly exact' estimate of its expenditure for that year). The first set of figures was discussed by the FC Working Party on 15/10/57, minutes CERN/FC/244, 1/11/57, and by the full committee the following day, minutes CERN/FC/242, 30/10/57. Willems' assessment of the situation was made at an FC meeting in November, minutes CERN/FC/252, 22/11/57.
29. These quotations are from the minutes of the FC Working Party meeting on 15 October 1957, CERN/FC/244, 1/11/57. In fact the administration had already made it clear that they were strongly opposed to budgetary cuts. At the request of Lockspeiser they had prepared a paper for this meeting looking into the consequences of cuts (CERN/FC/234/Add.3): they concluded that all that could be cut would be items 'for which no expenditure is yet committed', i.e. CERN's 'most progressive works'.
30. All quotations from the FC Working Party meeting cited in the previous note.

31. See the meeting of the Working Party and of the Finance Committee cited in note 28. The Staff Advisory Committee—G. Lehmann (F), E. Telschow (FRG), and D.R. Willson (UK)—submitted its official report as CERN/243, 11/11/57.
32. This idea was proposed by Willems at the Working Party meeting, and extended by Bannier when the full committee met the next day—see note 28.
33. The two quotations are from the minutes of the Working Party meeting, CERN/FC/244, 1/11/57.
34. The last quotation is from the minutes of the FC meeting of 13/11/57, CERN/FC/252, 22/11/57. The others are from that of the meeting held on 16/10/57, minutes CERN/FC/242, 30/10/57.
35. The quotations are from an article by ESA's Director-General from 1975 to 1980, Roy Gibson, entitled 'By the way...', and included in Space (1984), 51-53.
36. We give the references for the minutes of the successive meetings of the Council's committees. The joint meeting of the SPC and FC representatives held on 29/10/57—CERN/SPC/53/Rev.2, 7/1/58; FC meeting held on 13/11/57—CERN/FC/252, 22/11/57; Committee of Council meeting held on 14/11/57—CERN/250, 27/1/57; FC meeting held on 18/12/57— CERN/FC/264, 11/3/58. The quotation is from the President's introductory remarks to the the Council session on 15/11/57—Council, minutes, 15/11/57, 25.
37. The previous note gives the detailed reference for the minutes of this meeting.
38. Morat said this at the meeting of the FC on 18/12/57—see CERN/FC/264, 11/3/58.
39. For the resolution passed in November, see Council, minutes, 15/11/57, 32. The resolution passed a month later differed in that the 100 MSF was described as 'the *estimated* total of contributions for the years 1958 and 1959' (our emphasis)— Council, minutes, 19/12/57, 38. Although this wording apparently implied that the two-year figure was not a ceiling, it was interpreted as such by everyone during the budget debates in 1958.
40. Hocker made his remarks at the 17th meeting of the FC on 13/11/57, minutes CERN/FC/252, 22/11/57. Similar sentiments were expressed by Telschow, the Director of the Max-Planck Society for the Advancement of Science, at the FC Working Party meeting on 15/10/57—minutes CERN/FC/244, 1/11/57.
41. For the remark by Verry, see the minutes of the meeting, CERN/FC/252, 22/11/57. The FC Chairman's report for the Council in which he explains that the UK and, following them, Sweden, did not have instructions which allowed them to fall in line with the majority is CERN/244, 14/11/57.
42. At the meeting of the Committee of Council on 14/11/57, Lockspeiser stated that, despite his earlier commitment not to have the budget voted against the wishes of a major contributor, 'He was bound to apply the provisions of the Convention [...]'. 'As President of the Council, his first duty was to be impartial [...]', Sir Ben added, going on to say that 'if it did not prove possible to reach a unanimous decision concerning the budget [i.e. because of UK opposition] he regarded it as his duty to call for a vote and ascertain what the exact position was'. It was Sir Ben who then put forward a draft resolution to the effect that there be a ceiling of 100 MSF on CERN's expenditure for 1958 and 1959.
43. Memorandum, Verry to Lockspeiser.
44. The essential document outlining the programme, and from which all the information in this paragraph is taken, is CERN/SPC/67, 8/5/58. For more information the reader is referred to the previous chapter.
45. The minutes of the SPC meeting on 23/5/58 at which these matters were discussed are CERN/SPC/70, 1/8/58. Its recommendations were summarized in CERN/FC/287, 27/5/58.
46. The minutes of the Working Party's meeting are CERN/WP/FC /298, 14/7/58. Its recommendations to the full committee are document CERN/FC/291, 4/6/58. The full committee met on 19/6/58—its minutes are CERN/FC/297, 7/7/58.
47. The quoted words of de Rose are in Council, minutes, 20/6/58, 17.
48. As Amaldi put it, 'work ought to start on CERN's bubble chamber, as well as on supporting facilities, not later than 1959, if CERN was not to leave the fruits of early exploration in the 25 GeV range to United States or Soviet scientists'—CERN/WP/FC/298, 14/7/58. De Rose expressed similar sentiments at the

Council meeting later in June: 'large bubble chambers should unquestionably be brought into service at CERN', said the President, where 'The twelve Member States had pooled their resources in order to build the best installations in the world for nuclear physics [...]', Council, minutes, 20/6/58, 17.

49. The notion of having privileged access to equipment one had built derived partly from the truck team system being implemented at the time at the CERN SC, and was to surface explicitly during negotiations on the use of the British chamber at CERN in 1959—see chapter 8.2.1.

50. For the quotation from Neumann see CERN/WP/FC/298, 14/7/58. For Morat's views see CERN/FC/287, 27/5/58.

51. The draft budgets for 1959 and 1960, which included the new estimates for the PS experimental programme, is CERN/FC/303, 19/9/58. These figures were discussed by the Working Party and the full FC committee on 7/10 and 8/10/58—minutes CERN/WP/FC/333, 22/1/59, CERN/FC/312, 20/10/58. We have not been able to find the revised figures assembled overnight by the CERN administration for the FC meeting on 8 October; the data in the paragraph is based on what delegates to the meeting said.

52. The quotation is in CERN/FC/303, 19/9/58, 12.

53. The views of their scientific advisers were reported by Obling and Verry at the FC meeting on 5/11/58—minutes CERN/FC/323, 14/11/58. The Swedish delegate's threat was made at the FC meeting on 2/12/58—minutes CERN/FC/331, 19/12/58.

54. Amaldi's paper is CERN/289, 16/10/58. The paper was not well received by the FC delegates who had to plead CERN's case with their domestic authorities, and they were not slow to express their disappointment with it—see CERN/FC/323, 14/11/58, 8. Perhaps they were so disappointed because the paper was so blatantly tactical: for example, it did not even seriously discuss the cut proposed by Obling and the Danish physicists, i.e. stopping work on CERN's 1 m propane chamber, which according to Amaldi's own figures would have saved 6.67 MSF in 1959 and 1960.

55. See CERN/FC/323, 14/11/58 for the lines adopted by the Dutch, French, and German delegates.

56. For de Rose, see Council, minutes, 9/10/58, 39; for Bannier, Council, minutes, 3/12/58, 59. De Rose's view may have been shared by the French financial authorities at the time: Neumann expressed similar sentiments at the meeting of the FC on 2/12/58, minutes CERN/FC/331, 19/12/58.

57. The resolution is reproduced as Annex 1 to the minutes of the Council meeting held on 3/12/58 (at page 67).

58. The *Report by the Staff Panel to the Chairman of the Finance Committee* is document CERN/FC/337, 12/2/59. Its members were all drawn from major industries in the member states. They were H.B.G. Casimir (Philips, NL), O.W. Humphreys (General Electric, UK), and K. Steimel (AEG, FRG).

59. The first version, discussed on this occasion, was CERN/FC/335, 19/2/59.

60. The report by Amaldi, the Chairman of the SPC, on this meeting, and de Rose's assessment of its significance, are in the minutes of the FC meeting held on 18/4/59, CERN/FC/352, 28/4/59. The minutes of the SPC meeting itself are CERN/SPC/92, 8/5/59.

61. The resolution voted by the Council on 26/5/59 is Annex 1 to the bound version of the Council minutes for 1959, at p. 31.

62. The information in this paragraph is based essentially on the DG's report to the FC meeting in October, document CERN/FC/381/Rev., 19/10/59. The final quotation is from Bannier's report to the December Council session—Council, minutes, 2/12/59, 50. The only saving to the scientific programme that was mentioned concerned the cloud chamber. Bakker reported that there was 'a prospect that arrangements may be made for a Member State which would share in the construction, operation and cost of the chamber'. No FC or Council delegate seems to have objected strongly to this. This was doubtless because it raised no immediate financial problems—though as Bannier pointed out to the Council in December 1959, the lack of contingencies created a 'far from healthy' situation, which was aggravated by the fact that CERN was now supposed to be staying within a three-year ceiling. It must also be said that with the successful commissioning of the PS in November no delegate was probably prepared to say that the PS experimental programme should be cut back—or at least not just yet.

63. The paper presented by the administration in June laying out its staff estimates for 1960 was CERN/FC/432, 3/6/60. It was followed by a new set of estimates for 1960-61 in September—CERN/FC/444, 21/9/60—in which 100 new staff were requested for 1961, the breakdown between divisions being explained in CERN/FC/444/Add., 24/10/60. Bannier's remark to the Council is in minutes, 8-9/12/60, 69.
64. The minutes of the FC meeting on 3/1/60 are CERN/FC/456, 28/11/60, those of the meeting on 7 December CERN/FC/465, 1/2/61.
65. The official record of the proceedings of the Committee of Council meeting is CERN/CC/390, 24/4/61. The far more comprehensive tape of the meeting, from which the remarks by Melville and de Rose are taken, is available in the CERN archive. Why, as in 1959, was the ceiling exceeded without causing a dispute over policy? The main reason seems to have been the fact of Spain's accession, which took place officially on 1 January 1961. Her contribution to the budget for 1961 was about 2.7 MSF—apparently just the difference between the 65 MSF ceiling and the 67.7 MSF level of contributions voted by the Council (see CERN/FC/441, 26/9/60, and CERN/FC/456, 28/11/60). As Bannier explained to the Council, this made it 'possible to share the financial burden of the Organization among a greater number of Member States and consequently to raise the budget total without affecting the contributions of the Member States. Despite this fortunate event', he went on to say, 'some delegations were unable to approve a budget corresponding to total contributions of over 67 million'—Council, minutes, 8-9/12/60, 69.
66. For the information on the additional buildings put up at BNL, see BNL Annual Report, 1960, 28 (Report BNL 632 (AS-14)).
67. See the minutes of the joint meeting of SPC and FC representatives held on 29/10/57, document CERN/SPC/53, 11/11/57.
68. Ernest Lawrence, the head of the Radiation Laboratory in Berkeley, provides a stunning example of this changed perception of what scientists could expect in terms of government funding: he increased his estimate of the laboratory's budget for the first postwar year of operation by a factor of 10 between 1944 and 1945—see Heilbron et al. (1981), 46-47. See also Krige (1988) for a discussion of a similar development immediately after the war in the UK.
69. The quotation is from a remark made by Verry at the joint meeting referred to in note 67.
70. Bakker made these remarks at the FC meeting on 5/11/58—minutes CERN/FC/323, 14/11/58.
71. The cited comments by Adams and Amaldi are in Council, minutes, 8-9/12/60, 71.
72. The document spelling out these alternatives is CERN/FC/478, 13/4/61.
73. The minutes of the SPC meeting at which these alternatives were discussed are document CERN/SPC/131, 7/7/61, and the committee's recommendations to the Council are CERN/393, 5/5/61.
74. These were the 38th and 39th meetings of the FC and their minutes are CERN/FC/478, 13/4/61 and CERN/FC/492, 21/7/61, respectively.
75. For Hocker's report, see Council, minutes, 2/6/61, 9.
76. For Perrin's intervention, see Council, minutes, 2/6/61, 12. Perrin was contrasting this scheme of setting global rates of growth with the prevailing practice at CERN of trying to make detailed estimates for future years. He suggested that the rate of growth could fall off progressively in successive years, proposing that the budget for 1962 be 10% higher than that for 1961, that for 1963 7% above that for 1962, and that for 1964 4% higher than that for 1963.
77. Weisskopf and Hine argued their case to the Finance Committee on 19/10/61 (minutes CERN/FC/511, 8/11/61) and to the Committee of Council a week later (CERN/CC/415, 1/12/61), as well as in the introduction to the 1962 draft budget (CERN/FC/497, undated) and in Hine's *CERN Programme and Budget Forecasts for 1962-1964* (CERN/FC/506, 6/10/61)—undoubtedly the most sophisticated document of its kind yet produced inside CERN. Hine's estimates were also discussed by the SPC at its meeting on 27/10/61 (minutes CERN/SPC/149, 27/11/61). In these various forums they argued that a laboratory's budget had to grow for three reasons: to raise the level of activity around its machines to the point of 'full exploitation', to keep up with inflation, and because of the sophistication factor, i.e. the fact

that there was a tendency for the equipment in use around the accelerator to become increasingly more complicated and expensive. They accepted, with Perrin, that growth rates could fall off progressively, but only after a laboratory had reached a 'full grown' stage, i.e. when the staff, visitors, and equipment had increased to such a level that the potential of the accelerators was 'fully exploited'.

78. The quotations in this paragraph are from the meeting of the Finance Committee referred to in the previous note.
79. The quotations are all from the tape recording of the Committee of Council meeting on 26/10/61 (CERN archives).
80. For the quotation from Powell and de Rose see previous note; similar statements are to be found in the official minutes, CERN/CC/415, 1/12/61. Hocker made his request to the SPC at its meeting on 27/10, minutes CERN/SPC/149, 27/1/61. The DG's report to the Council, *The Place of CERN in High-Energy Physics*, is Annex II to CERN/422, 24/11/1961. The SPC's report which emerged from its meeting on 25/11/61 is CERN/420/Rev., 5/12/61.
81. These quotations are taken from the tape—see note 79.
82. The British position is that spelt out by Smith to the FC meeting on 14/11/61, minutes CERN/FC/520, 13/12/61, supplemented by Weisskopf's report of the meeting to Melville, letter Weisskopf to Melville, 16/11/61 (DG20937).
83. The Italian position has been extracted from the sources quoted in the previous note. The announcement came as a 'bombshell' because, as Weisskopf explained, at the Committee of Council meeting some three weeks before Amaldi had said 'that Italy would fully support the 78 million budget. It seemed to us therefore at that time', he went on, 'that we could count on a 78 million budget since the opposition of Britain and Spain only would probably crumble in view of an overwhelming majority', letter Weisskopf to Melville, 16/11/61 (DG20937). We do not know the reasons for the change in Italy's position, though it apparently came as a surprise to Amaldi, who immediately set about reversing it (letter Weisskopf to Melville, 16/11/61 (DG20937)). It should be noted though that the Italian government, like the British government, was becoming increasingly concerned about the rising costs of CERN, and in March 1961 proposed that the Financial Protocol be amended so as to ensure that the budget be acceptable to a two-thirds majority of the member states (see CERN/CC/408, 4/10/61). They discussed this with their British colleagues in the Foreign Office in July, and though British officials welcomed the initiative — all the more so after the vote against them in December 1960—they felt that the matter were best handled by a change to the Convention (see, for example, letter Hubbard (DSIR) to Annand (Treasury), 4/11/61, and letter Hainsworth (Foreign Office) to Count Roberto Riccardi (Italian Embassy, London), 28/8/61, both in (SERC-B124)). The Italians subsequently withdrew their proposal (CERN/FC/514, 2/11/61) in favour of a recommendation which was adopted by the Council at its December 1961 session. A resolution was passed asking the FC to ensure that any draft budget it submitted to the Council was 'supported by a majority of the Member States so composed that the sum of the contributions of these Member States exceeds the sum of the contributions of the Member States who would oppose the budget' (Council, minutes, 19/12/61, 47).
84. The information in this paragraph is extracted from letters Weisskopf to Blackett, Bohr, Melville, Perrin, and Willems, all dated 16/11/51, and all in (DG20937).
85. We have not found a copy of the Aide Mémoire itself. We do have a confidential memorandum from the Foreign Office to the British Chanceries in all the CERN member states, dated 10/11/61, explaining the official position. The quotations are taken from this. We also have the response from the Dutch Ministry of Foreign Affairs, dated 18/12/61, which provides supplementary information. Both are in (SERC-B125).
86. It is not clear just why the British stopped the shipment of parts of their bubble chamber, though Weisskopf himself 'suspect[ed]' that certain 'unhappy wordings' he had used in his letter to Melville on 16/11 may have been 'one of the causes' (letter Weisskopf to Cockcroft, 27/11/61 (DG20937)). In that letter Weisskopf implied that unless CERN was given at least 260 MSF for 1962-64, against the UK's 240

MSF, they would consider cutting the British bubble chamber out of their research programme in order to achieve the necessary economies (letters Weisskopf to Melville, 16/11/61, 27/11/61, both on (DG20937)). In fact two days before Dakin, himself British, had told the FC that 'Although he had always felt reluctant to mention this, a saving of some 2 million could be effected if preparations to receive the British bubble chamber were stopped in the East area', CERN/FC/520, 13/12/61.

87. The minutes of the FC meeting are CERN/FC/527, 14/2/62, that of the Committee of Council which was also held on 18 December, CERN/CC/435, 23/1/62. The remarks made by Bannier are taken from the minutes of the CC meeting (my translation from the original French), 19/12/61, 38-39. See also the tape recording of the CC meeting which is in the CERN archives. Bannier's attitude was shared by his government which, in an Aide Mémoire from the Ministry of Foreign Affairs in The Hague, stated that 'where procedural matters are concerned it should be borne in mind that for consultations between member governments on the C.E.R.N. programme and financial requirements, a Council was set up assisted by a Scientific Policy Committee and a Finance Committee', adding that 'the Netherlands Government do not feel the need for consultations between governments in another form [...]', Aide Mémoire, 18/12/61 (SERC-B125).
88. The quotations in this paragraph are taken from (my translation of) the minutes of the CC meeting, CERN/CC/435, 23/1/62, and from Council, minutes, 19/12/61, 37-38.
89. See minutes of CC meeting, CERN/CC/435, 23/1/62. De Rose's threat about French withdrawal from ELDO and similar ventures was reported by Adams to Viscount Hailsham, his 'science minister', in a letter, 28/12/61 (SERC-B125).
90. See the tape recording of the Committee of Council meeting held on 18/12/61 in the CERN archives.
91. The minutes are Council, minutes, 19/12/61, 27-49. Dominique Pestre has discussed the significance of this historic Council session as indicative of the influence and workings of the CERN lobby in chapter 7.1.3.1.
92. The British note is CERN/430, 15/12/61. It was 'backed' by a telegram from the Foreign Office to the Chanceries in the member states sent on 15/12/61, asking the recipients to inform the governments to which they were accredited that at the forthcoming Council session the UK could not vote for more than 75 MSF for 1962, and that it wanted budgetary maxima to be fixed for CERN at least three years in advance (SERC-B125).
93. This is the form of words used in the Council, minutes, 19/12/61, 40. The version laid before the Council was less formal but did not differ in substance — see CERN/433, 18/12/61.
94. For the attitudes in the Treasury see the exchange of letters between Hubbard (DSIR) and Annand (Treasury) on 11/4/61 and 14/4/61 (SERC-B124). See also note 83.
95. The DSIR's preference for planning machinery over procedural changes is explained in letter Smith to Crawford, 4/8/61 (SERC- B124); its recommendations in this regard were spelt out in some detail early in 1961 in a note for the FC entitled *Long-Term Financial Control* (CERN/FC/476, 18/4/61). The anger over the Treasury's attitudes was expressed by Hubbard in a memo to the DSIR Secretary on 22/10/61 (SERC-B125). Hubbard was a UK Finance Committee delegate and one of the advisers to the British CERN Council delegation and had to deal personally with the diplomatic and emotional consequences of Treasury and Foreign Office policies. He explained what that meant: Britain he said, has tended 'to declare far too conservative a policy to command support, to die in the last ditch in defence of our declared policy, and to see the moderates waver and deprived of our leadership, go over to support the expansionists'. 'What we considered the ultimate in generosity', added the DSIR official, was seen by our co-partners as simply indicating that 'we were being mean as usual' — memo Hubbard to Elkington, 3/11/61 (SERC- B125).
96. To get some idea of the political weight these men had remember that it was Amaldi who apparently got his government to reverse its decision to support the UK at the end of November — see note 83.
97. The quotations are from interventions by Courtillet at the FC meeting on 18/12/61 (CERN/FC/527, 14/2/62), and by Powell at the emergency meeting of the SPC (CERN/SPC/151, 10/1/62).

98. Laqueur (1982), 224.
99. The figures for 1962 are from Freeman & Young (1965), 71, while those for later years are from OECD (1984), 27. France's growth rate is taken from Laqueur (1982), 214.
100. The statement by de Rose was made at the Committee of Council meeting on 18 December—minutes CERN/CC/435, 23/1/62. Doubts about whether it was necessary to compete with the USA were a constant theme in the Treasury's dealings with the DSIR. Early in 1962 John Adams made a concerted effort to bring the point home to officials in the department, stressing that CERN was expressly set up to have facilities comparable to those across the Atlantic, and that 'If we do less than they do [...] [and if] due to inferior facilities, we take second place in research, [...] it is questionable whether it is worthwhile investing in it at all [...]', letters Adams to Griffiths, 5/1/62, 19/1/62 (SERC-B125).
101. For this information see Bannier Report (next note), 2-3.
102. The *Report of the Working Party on the CERN Programme and Budget* is document CERN/442, 29/5/62. For convenience we shall simply refer to it as the Bannier Report. There is essential background material in two files in the DIRADM series, DIRADM20354 and 20355, and in the collection of papers deposited by Mervyn Hine in the CERN archives, in particular file MGNH22104. These papers are particularly interesting for they contain the information on costs and levels of investment in physics research for a number of European countries and some major American laboratories which were used in the Bannier Report. For the reception of the Report in the Council, see Council, minutes, 13/6/62, 8-14.
103. Bannier Report, note 102, 39. The idea that important new developments be considered separately was Bannier's—see his *Comments on the Draft Report* (CERN/WP/PB/26) which was apparently circulated to members of the Working Party some time in April—see letter Bannier to the members of the Working Party, 25/4/62 (DIRADM20355). The *Comments* are also in this file.
104. Bannier Report, note 102, 38-39.
105. Bannier Report, note 102, 39. In fact there was some dispute over the weight that should be given to the estimates for years $X+3$ and $X+4$. The British felt that these figures, while not firm, should at least be 'considered as goals to work to, with reasonable assurances that the initial indications received [would] not be too different from the actual figures'—CERN/WP/PB/24, 10/4/62 (DIRADM20355). Weisskopf objected strongly. In his view, two years was the longest period for which one could make 'binding' predictions for the work to be done in a laboratory like CERN, and the figures for years $X+3$ and $X+4$ should 'only be regarded as advance information to the Council of our intentions and of the possibilities for development of its Laboratory' (letter Weisskopf to the Bannier Working Party, CERN/WP/PB/16, 30/3/62 (DIRADM20354), from which the quotation in the text is also taken). The Report did not resolve this dispute: it followed the British proposal that spoke of the estimates for years $X+3$ and $X+4$ as 'provisional determinations' but in the preamble to the recommendation said that the figures were non-binding. The matter was settled in practice: national financial authorities tended to put the UK interpretation on the 'provisional determinations', and as a result the administration, the FC members and the Council delegates became increasingly wary about backing such figures unless they were reasonably sure that they could be sustained.
106. Bannier Report, note 102, 19-22. Hine's data are referred to in the next three notes. For the quotation from de Rose, see Council, minutes, 19/12/61, 37.
107. CERN/FC/506, 6/10/61. The staff figures for end 1961 (1050) were taken from CERN/WP/PB/17 (next note). They excluded the staff of the Site and Buildings division, which were omitted from his calculations, said Hine, 'since so much of it [was] in the form of auxiliaries and other types of labour, whose proportions change with time'. In FC/506 Hine gave two sets of figures for the 1963 and 1964 contribution levels: 86 MSF and 93 MSF (as calculated from divisional forecasts, and excluding inflation) and 89 MSF and 100 MSF (as calculated by extrapolating the overall past expenditure, including inflation).
108. CERN/WP/PB/17, 30/3/62 (DIRADM20354). This file also contains the findings of the internal working parties who studied CERN's needs. The British document is memo R. St. J. Walker to Atherton, 24/7/62

(SERC-B125).

109. CERN/443, 31/5/62. To retain comparability with the two other sets of figures for staff in this paragraph we have excluded Site and Buildings from our calculations, which are derived from Table I in this report. If this division is included staff was to grow from 1332 at end 1961 to 2400 to reach a stage of full exploitation—so still 80%.
110. See Bannier's *Comments,* note 103, for the fixed rate of expansion of 13%. The British opposed this in the Working Party, proposing instead that the rate of expansion for 1963 and 1964 be 'between X and 13%', adding that 'for the following two years it would seem reasonable to plan for a falling rate of growth', though of not less than 'Y%'—see *Mr Elkington's Comments on the Draft Report,* undated, but around end April 1962 (DIRADM20355). Bannier spoke of his 'fears' on reading, in CERN/443, that the 'rate of development considered to be optimum by the Working Party appeared to be reduced to the status of minimum rate' at the FC meeting on 12/6/62—minutes CERN/FC/547, 19/7/62. The critical remark about Hine's figures was made by R. St. J. Walker in the memorandum cited in note 108.
111. Council, minutes, 13/6/62, 14-18.
112. The minutes of the FC meeting held on 22/4/84 are CERN/FC/687, 14/5/64.
113. This paragraph is essentially based on Weisskopf's exposé to the FC at its meeting on 11/9/62. The minutes of the meeting are CERN/FC/561, 5/10/62, and the DG's talk was issued separately as an Annex to them and as CERN/460, 2/10/62. Hine's figures for the drop in the cost per man are from the first page of the 1963 draft budget, CERN/FC/571/Rev., dated 1962.
114. To simplify the presentation of this debate we give in this paragraph the Four-Year Programme as explained by Weisskopf to the CERN Council in December. His statement was released as CERN/469, 3/12/62. See also Council, minutes, 19/12/62, 54-57. For the calculations underlying Weisskopf's figures see Hine's *Proposed CERN Programme and Budgets for 1963-1964,* CERN/FC/565, 1/10/62, and his *Provisional Programme and Budget Forecasts for 1964, 1965 and 1966,* CERN/FC/676, 4/12/62.
115. See minutes of the FC meeting cited in note 113.
116. Powell's remarks are verbatim in Council, minutes, 12/10/62, 27-30. He was speaking to a *Report of the Scientific Policy Committee to the Council,* CERN/461, 8/10/62.
117. The report made by FC Chairman Funke to the Council is in minutes, 12/10/62, 30-35. The quotation from Weisskopf is on p. 35.
118. Letter, Walker to Weisskopf, 12/12/62 (DIRADM20342).
119. The minutes of this meeting are CERN/FC/598, 8/1/62.
120. The December Council debate on the budget, from which these quotations are taken, is Council, minutes, 19/12/62, 43-48.
121. The minutes of the FC meeting on 14/5/63 are CERN/FC/620, 7/6/63. Hine's estimates were justified in *Budget for 1964 and Forecasts for 1965 and 1966,* CERN/FC/612, 1/5/63. There is a detailed discussion of the birth of ECFA, of 'its' report, and of early reactions to it, in sections 12.4.3, 12.5. The report itself is entitled *Report of the Working Party on the European High Energy Accelerator Programme,* FA/WP/23/Rev.3, 12/6/63.
122. For the quotations see the minutes cited in the previous note.
123. See Council, minutes, 20/6/63, 23-25. Note that the vote on the estimates made in accordance with the Bannier procedure followed the rule laid down at the end of 1961, i.e. they had to be passed by a majority of the member states so composed that their contributions to the budget exceeded the contributions of those who opposed them—see Council, minutes, 19/12/61, 47. See also note 83, and CERN/FC/556, 4/9/62.
124. Letter Weisskopf to Council delegates (?), CERN/8709, 7/8/63 (DG20939).
125. The paper was entitled *Considerations on CERN Budgets for 1964-1967,* CERN/SPC/174, 1/10/63. It was supported by *CERN Cost Statistics 1960-1963,* CERN/FC/640, 1/10/63. The remark by the DG is from Council, minutes, 11/10/63, 41.
126. The minutes of the FC meeting of 10/10/63 on which this section is based are CERN/FC/649, 28/10/63.

The administration's estimate of the cost of the supplementary programme in 1964 is CERN/FC/630, 3/9/63. As for the staff figures there was a lively exchange between Walker and Hine at the FC meeting. Hine insisted that 'the number of staff in a position of any responsibility was being held down to the figures contemplated at the beginning of 1963' — to which Walker replied that 'the number of staff in grades 12 and above had gone up from 60 at the end of 1962 to 75 now, and that the number of staff in grades 5 to 11 had risen from 1036 to 1148 [...]'. Ullmann then apparently attributed the increases in grade 12 and higher primarily to 'internal promotions'. While the evidence we have does not enable us to 'take sides' in this dispute it is valuable as yet another example of the research that the British delegation put into their arguments on the budget.

127. The quotations are from Council, minutes, 11/10/63, 42-44. The debate on the budget at this meeting was once again intended to embarrass the UK: an introduction by Hine was followed by a characteristically florid appeal by Powell, soon followed by the interventions of Leprince-Ringuet and de Rose. The difference with a similar situation in December 1961 was that this time no one was apparently prepared to use political threats to try to force a change in the UK's attitude. We suggest why in a few moments.
128. The minutes of the FC meeting on 13/11/63 are CERN/FC/657, 6/12/63, those of the meeting on 16/12/63 are CERN/FC/668, undated draft version.
129. The offer was made at the FC meeting on 16/12/63, from whose minutes (see previous note) the quotation is also taken.
130. The minutes of the Committee of Council meeting are CERN/CC/532, 20/4/64. The Council debate on the budget is Council, minutes, 17/12/63, 51-52.
131. The quotation and the figure of 2% are taken from Walker's minuted contribution to the October 1962 Council session, minutes, 12/10/62, 32.
132. Adams's view that CERN was becoming self-centred was reported in a memo Hubbard to Walker, 12/7/62 (SERC-B125), in which we are also told that Adams was extremely critical of Hine's role in the organization. The growth rates are quoted in memo Walker to Atherton, 24/7/62 (SERC-B125). Both of these memos are reports of what Adams reputedly said. For his own, later, views, on suitable rates of increase for CERN, see his submission *To all [UK] CERN Panel Members,* 7/8/62 (JBA22776). Here Adams's main concern is not to challenge his Minister's maximum growth rates for CERN of 12/12/10/10% successively from 1963 to 1966, but to insist, against the CERN administration, that staff should not rise at the same rate, but 3% more slowly.
133. After a 'contentious meeting' of potential member states of ESRO in May 1963, it emerged 'that a great deal depended on speeding up the ratification procedure' of the ESRO Convention. 'Already there had been some suspicion about the UK's attitude to ESRO as it had already ratified ELDO some months earlier (March 1963)', and in this climate, perhaps, some CERN Council delegates were being particularly cautious not to give the UK offence. The quotations are from Massey and Robins (1986), 137.
134. All the quotations bar the last are from the minutes of the FC meeting held on 22/4/64, CERN/FC/687, 14/5/64. The remark about the 40 m^3 bubble chamber is in supporting document CERN/SPC/182, 9/4/64. See also CERN/SPC/182/Add., 20/4/64.
135. The data given here is based on the SPC documents cited in the previous note. The administration stressed throughout this debate that their plans were much less ambitious and costly than the sought-after developments at Brookhaven.
136. The minutes of the FC meeting held on 3/6/64 are CERN/FC/703, 10/8/64. The statement that it looked as though the FC would reach unanamity at its meeting on 17/6 (minutes CERN/FC/707, 1/9/64) is not evident from the official minutes of either: it was made by FC Chairman Funke to the Council, minutes, 18-19/6/64, 32.
137. The Council debate from which the quotations are taken is Council, minutes, 18-19/6/64, 31-32.
138. The quotations are from Council President Bannier's concluding reflections on the lengthy debate on CERN's future programme, Council, minutes, 18-19/6/64, 12-31, at p. 29. When Bannier spoke of European physicists having *access* to a big machine he was alluding to discussions then under way to build

an 'international' accelerator co-operatively by Europe, the USA, and the USSR. The scheme was stillborn—see section 12.6 for more detail.
139. These figures, as well as the earlier quotations in this paragraph, are from CERN/FC/710, 6/10/64. The work of the seven study groups is explained in CERN/SPC/189, 13/10/64.
140. This interpretation is developed in section 12.6, where Dominique Pestre highlights the growing resentment in other areas of science about the increasing costs of high-energy physics, which went along with some questioning within the ranks of physicist/administrators themselves (notably Alvin Weinberg, the Director of Oak Ridge National Laboratory) (see especially section 12.6.1.1). The adjustment for inflation, or 'Cost Variation Index' used to convert 1963 figures into 1964 prices was 3.7% (CERN/522, 25/11/63).
141. The minutes of the FC meeting on 20/11/64 are CERN/FC/741, 8/12/64.
142. This exchange occurred at the FC meeting cited in the previous note.
143. The minutes of the FC meeting on 14/12/64 are CERN/FC/747, 1/2/65.
144. The Council discussions and votes on these items are to be found in Council, minutes, 15-16/12/64, 45-49, 51-52. The budget for the supplementary programme in 1965 is CERN/FC/722, 12/10/64, and CERN/569/Rev., 1/12/64.
145. The documents on which Hine based his demand were CERN/FC/764, 15/4/65 and CERN/SPC/198, 20/4/65. Earlier in the year Hine had given provisional estimates of the costs of the ISR and 300 GeV projects as well—CERN/SPC/196, 19/2/65. The Cost Variation Index for 1965 vis-à-vis 1964 was 5.24% (CERN/568, 1/12/64. This was later reduced to 5.13% (CERN/624, 30/11/65).
146. The argument is spelt out in detail in paper CERN/595, 26/5/65, which Weisskopf prepared for the June Council session. We have again slightly violated a strict chronological sequence here in the interests of simplicity of exposition.
147. The minutes of the FC meeting early in May are CERN/FC/777, 3/6/65.
148. When the FC met in June it had Weisskopf's Council paper CERN/595, 26/5/65 before it and its delegates were told that the SPC had 'expressed great disappointment' concerning its 1967 and 1968 figures—see minutes, CERN/FC/785, 7/7/65.
149. The Council proceedings are Council, minutes, 16-17/6/65, 28-30, 33.
150. Council, minutes, 16-17/6/65, 24.
151. Hine's data were in CERN/599, 2/6/65. See also CERN/FC/788, 12/7/65.
152. The minutes of the November meeting of the FC are CERN/FC/816, 24/11/65.
153. These decisions are to be found in Council, minutes, 15-16/12/65, 52-57. The minutes give the figure of 21.7 MSF for the ISR budget in 1966, in 1966 prices. As the Cost Variation Index for 1966 vis-à-vis 1965 was 4.2% (CERN/624, 30/11/65), this is equivalent to 20.8 MSF in 1965 prices.
154. An overall figure of some 130 MSF for the Improvement Programme was given by Hine in CERN/SPC/196, 19/2/65. The data for the ISR is from the sources cited in note 151.
155. The possibility of member states contributing to the cost of building large bubble chambers was discussed throughout 1964 and 1965. In the event two were built: the heavy liquid chamber Gargamelle (essentially constructed in France) and the Big European (Hydrogen) Bubble Chamber, BEBC (based on a tripartite arrangement between CERN, France, and Germany). These decisions and the arrangements surrounding them will be analyzed in a separate chapter in Volume III of the History of CERN. In the interim the interested reader is referred to the CERN Council minutes. The quotation is from Council, minutes, 25/3/65, 15.
156. The quotations are from the *Report of the Working Party on the European High Energy Accelerator Programme,* FA/WP/23/Rev.3, 12/6/63, 3,
157. The quotation is from the minutes of the FC meeting held on 13 November 1963, document CERN/FC/657, 6/12/63. In similar vein in April 1964 Bannier reported that discussions on building regional accelerators in the Scandinavian and Benelux countries had failed—CERN/FC/687, 14/5/64. He explained why at the next FC meeting: when the Netherlands and Belgium had studied the possibility of

building an 8-10 GeV machine they had found that it would cost more than their joint contributions to the improvement of the existing CERN facilities, plus the storage rings, plus a 300 GeV accelerator, and perhaps even a 1000 GeV machine—see CERN/FC/703, 10/8/64.
158. The remarks by the Danish and Norwegian delegates are in CERN/FC/687, 14/5/64, those by the Swedish delegate in CERN/FC/707, 1/9/64.
159. See Council, minutes, 11/10/63, 36, where Belgium, France, the Netherlands, Norway, Sweden, and Switzerland—five small countries plus only one big one—agreed to contribute financially to a supplementary programme costing 3.8 MSF in 1964.
160. This information is extracted from section 8.6.2.
161. See Sked and Cook (1980), 215.
162. Council, minutes, 11/6/62, 11.
163. The electron accelerators were built at Daresbury (Nina), Hamburg (DESY), and Frascati (Adone). The cost of ESRO is taken from Massey and Robins (1986), 126.
164. The period immediately following the publication of the Bannier Report also coincided with a marked improvement in the salaries and working conditions of the CERN staff. The principle that CERN higher-grade salaries be comparable to those in a 'similar' international organization, and ESRO in particular, was entrenched (see a report submitted to the Director-General by a group of experts on *The Pay and Allowance Structure and Conditions of the European Organization for Nuclear Research,* August 1964, attached to CERN/FC/715, 9/9/64). CERN salaries increased by between 5 and 10% per annum each year during our period, according to CERN's statistics—well above the annual rise in the consumer price index in Switzerland in the same period (3.2%, on average, from 1960-1965—OECD (1970), 25). Finally, the policy of granting indefinite contracts to CERN staff, previously restricted to the minimum, was extended particularly so as to enable 'experienced and highly-qualified supporting staff at all levels from technicians to senior engineers and physicists helping with technical problems' to make a career at CERN (CERN/SPC/153, 22/2/62. See also CERN/449, 13/6/62, another 'early' document on this point).

CHAPTER 11

The contract policy with industry[1]

John KRIGE

Contents

11.1 Laying the foundations of CERN's contract policy: the gap between principle and practice	640
11.2 The competitive award of technical contracts: a critique of the 'rational model'	643
11.2.1 The award of the contract for the PS magnet blocks	644
11.2.2 The point of view of the firm	649
11.3 The award of building contracts: the pitfalls of *a posteriori* control	650
11.3.1 Some remarks on the organization of building work	650
11.3.2 The contracts for the two main scientific buildings	653
11.3.2.1 The SC building	653
11.3.2.2 The PS building	655
11.3.3 The evolution of the building programme up to 1959: the advantage of being installed on the site	658
11.3.4 The CERN Administration and the Finance Committee	661
11.4 Some global statistics on the distribution of contracts awarded by CERN between 1955 and 1965: the advantages of being the host state	664
11.4.1 Competitive and noncompetitive contracts	665
11.4.2 The distribution by country	668
11.5 Final remarks: the tenacity of CERN's contract policy	671
Notes	671
Bibliography	809

Having studied the various policies adopted by the Finance Committee to regulate the overall level of the CERN budget, we turn in this chapter to a somewhat different aspect of the financial management of the laboratory: its policy for awarding contracts for the purchase of material, equipment, and supplies from industry. This topic is more 'mundane' than that explored in chapter 10 in the sense that it did not raise fundamental questions of national policy vis-à-vis CERN, nor was it the cause of major disputes between the member states, or between the governments financing the 'house' and those responsible for running it. It is important all the same, for two main reasons. Firstly, because of its topicality. CERN's relationship with industry, of which its contract policy is one dimension, is of ongoing interest to its member states who are increasingly concerned to see what, if any, benefits their industries derive from a centre which prides itself on pushing technology to the limits—and which works in one of the most expensive fields of basic scientific research. Indeed, as recently as the mid-1980s the 'Abragam Committee' set up at the request of the Council 'to conduct an in-depth comprehensive review of CERN' singled out this aspect of the laboratory's management, including its contract policy, for special consideration.[2] Correlatively, ever since the early 1970s CERN itself has actively promoted a debate on the issue. In 1974 it organized a meeting attended by over 200 people, most of them businessmen, at which some of the more advanced technologies arising from high-energy physics were displayed. A year later the organization published the research findings of H. Schmied, who tried to quantify 'the technical and economic benefit to the manufacturing industries involved in CERN contracts [...]', with particular reference to high-technology areas. And in 1984 a follow-up study, confirming and extending the results of its predecessor, was also published under CERN's auspices.[3]

The second important reason for studying CERN's contract policy is its novelty. More precisely, in contrast to other European organizations like ESRO (space research) which were set up later, CERN's contract policy does not make provision for the member states to receive a 'just return' on their 'investment' in the laboratory.[4] On this principle, the percentage (by value) of the contracts awarded by the organization in any of its member states has to correspond (more or less) to the percentage contribution which that country makes to the organization's budget. No such restriction applies in the case of CERN: here the official policy is that 'normally' a contract must be awarded competitively to the lowest bidder, regardless of the country in which the firm is situated.

The main focus of our study is not the policy itself—that has remained essentially unchanged for over 30 years,—but its implementation in practice. What we want to do, in other words, is to study how (i.e. by what procedures) and where (i.e. to which firms and in which countries) CERN awarded its contracts in the period under review. This task is complicated in two ways. Firstly, there is very little published literature that we can draw on for guidance. Schmied and his associates concentrated on the benefits accruing to industry from CERN contracts, not on the procedures whereby the laboratory awarded them. And although considerable attention has been given to the procurement of advanced technology in the United States, these have tended to focus on cost-overruns and malpractices in the defence-related sectors of the economy. These have contract policies (e.g. the cost-plus principle) which are very different from those used at CERN (except in a few special cases), and which raise entirely different questions.[5] The second complicating factor is the sheer number of the contracts which CERN has placed—almost 25,000 in 1965 alone—and, more fundamentally, the confidentiality with which the organization surrounds the awards that it makes. To overcome this difficulty we have restricted our documentation primarily to the papers laid before the Finance Committee by the executive in our period. These papers contain useful information on the awards of the most important contracts as well as statistical data produced by the purchasing office which throw considerable light on certain global features of CERN's contract policy. What is more they can hardly be regarded as confidential any longer: the CERN/FC/... series are among the most 'public' of the organization's official documents, are reproduced in the hundreds, and are distributed all over Europe and indeed beyond.

To keep the study within reasonable limits we have chosen to concentrate our attention on two areas of contract activity. Firstly, we study the allocation of a major technical contract, that for the PS magnet blocks, which was arguably the most important award made in our period, both from a financial and a scientific point of view: at over 8 MSF it was far and away the biggest single item CERN purchased (until it bought the CDC computer in 1963), and at the time it was felt that the very success of the laboratory itself hinged on the PS magnets meeting extremely strict performance specifications. Secondly, we study the award of building contracts, with particular emphasis on the first stage of the building programme (i.e. up to 1959). The main reason for this is its financial importance: 55% of CERN's capital expenditure between 1955 and 1959, and one-third of it in the subsequent six years, occurred in the area of site and buildings. Finally we present some statistics, derived from official sources, which show how CERN's contracts were distributed (by value) between competitive and non-competitive awards and between the various member states.

Our main result is that a gulf existed between principle and practice, between policy and procedure. Indeed from the start the Finance Committee, even while laying down the competitive 'rule', left the executive considerable freedom to interpret it pragmatically (section 11.1). As a result a wide variety of ways and means of purchasing equipment evolved, depending on the nature of the goods, their cost, specific local considerations, personal preferences and prejudices, and so on (sections

Notes: pp. 671 ff.

11.2 and 11.3). And we find that, globally speaking, in our period a *formal* competitive policy was honoured far more in the breach than in the observance, and that the emphasis on pragmatism coupled with the rejection of the policy of just return benefitted some states—and the host state in particular—far more than others (section 11.4).

One final word before we get under way. While our results indicate that there was something of a gap between what is usually said to be CERN's contract policy and what actually happened in practice, this is not to say that the contract procedure at CERN was shot through with malpractices of various kinds or that the laboratory squandered money. Mistakes certainly were made, particularly in the building programme in the early 1950s as we shall see, but this was a rather particular and unusual situation. Insofar as our analysis is critical then, it is not primarily aimed at the day-to-day execution of contract policy, but at a widely held and unreflective cluster of impressions of what that execution entails.

11.1 Laying the foundations of CERN's contract policy: the gap between principle and practice

The foundations of CERN's contract policy were laid in the very early days of the organization's life, even before the Convention establishing the laboratory was ratified. It was the British who took the initiative early in 1953, within weeks of their obtaining the status of official observers to the provisional CERN Council. Their delegates proposed a set of financial regulations for the new organization, including a contract policy modelled on that in force in the Department of Scientific and Industrial Research. These proposals were discussed during 1953, first in the so-called Administrative and Financial Working Group, and then in the Interim Finance Committee (IFC) which superseded it. A provisional set of rules was agreed on by the interim Council at its seventh session in October 1953.[6]

The only important point of disagreement in these deliberations was precisely the one we mentioned in our introduction, namely whether there should be some formal mechanism for ensuring that industries in the member states got some return on their government's investment in CERN. Several delegates supported the idea whereby no country was to receive an 'unduly high proportion' of the more valuable orders. This proposal was vigorously opposed by the British at the seventh session of the interim Council. They insisted that if CERN did not give preference to the lowest tender the organization would become enmeshed in 'inextricable political difficulties'. Their argument carried the day and the proposal was dropped.

Two points of principle were then established in this period. Firstly, that as a rule CERN's practice would be to award contracts competitively. And secondly, that invitations to tender would normally be limited to firms located within the member states. A procedure for calling for tenders on a wide geographic basis was also provisionally agreed on. Tenders were to be invited from two to three firms in each country where there were suitable firms. These firms were to be selected from lists built up by the CERN management in consultation with the staff and with the Finance Committee, whose members were expected to add the names of capable contractors in their own countries. This was the only measure taken to ensure that the interests of the member states were protected within the framework of the competitive policy.

In parallel with these debates on policy various members of the Executive Group—the small core of people responsible for planning CERN at the time—were taking a number of steps to ensure that there were no delays in acquiring equipment once the setting up of the laboratory had been officially sanctioned. By July 1953 Bakker, who was responsible for the SC Group, had already called for tenders for certain components of his machine and Kowarski and Preiswerk, who were jointly in charge of the so-called Laboratory Group, had discussed preliminary plans for the layout of the site and buildings with a consulting engineer and an architect.[7] At the end of the year the PS group, at that time led by Odd Dahl, moved to Geneva and was authorized by the Interim Finance Committee to spend 200,000 SF immediately on a magnet model (70,000 SF 'supported by bid from Sècheron, Geneva'), test equipment and essential stores.[8] Six months later—and three months before the Convention had been formally ratified— excavation work on the site was under way, and Bakker was given permission to call for tenders for the SC magnet frame and coils.[9]

Once the Convention was ratified at the end of September 1954, and the 'permanent' organization came into being, the Finance Committee set up a working party to put the finishing touches to various formal aspects of the laboratory's management, including contract policy. The directorate's ideas on contract procedure were laid before the working party on 17 December 1954. They were elaborated on the eve of the meeting by division leaders John Adams (Proton Synchrotron) and Peter Preiswerk (Site and Buildings) in discussion with Sam Dakin, CERN's Chief Administrative Officer.[10]

The CERN personnel wanted to see changes in two main areas: the competitive principle, which they thought unduly restrictive, and the tendering procedure, which they found unnecessarily cumbersome. The former, they said, was 'completely uneconomic' for relatively small purchases, was far less effective than 'informal bargaining' for some materials, and amounted to little more than a 'costly ritual' for specialized equipment (since the researcher needed to discuss the detailed design and specifications with a particular supplier). As for the latter, they insisted that the experience of the past year had shown that the practice of sending the documentation to too many firms had 'produced the most expensive and undesirable results' and that, for building contracts in particular, there was no point in calling for tenders from firms other than those in Switzerland and its neighbours. In short, those responsible for implementing CERN's contract policy argued that awards should be

Notes: p. 672

made competitively 'where practicable', and not 'normally' as proposed in the financial rules, and that usually only about six contractors should receive 'the full tender forms and specification', the bulk of them Swiss where civil engineering was concerned.

The administration's proposals were discussed at length at the meeting of the Finance Committee's working group in mid-December. In the light of the discussion a revised paper was drafted and submitted to the Finance Committee who approved it in January. Three important points were settled in these deliberations.[11]

Firstly, the Finance Committee 'strongly emphasized that competitive tendering must [...] be regarded as the method to be used wherever possible [...]'. This principle was formally enshrined in the first clause of Rule 10(b) of CERN's Financial Rules. The version agreed on in February 1955 stated that at least three competitive tenders should normally be obtained for the purchase of material by CERN, and that the contract be allocated to the firm whose offer was the lowest which complied with technical and delivery requirements.

Secondly, taking account of the objections raised by the CERN staff to a too-rigorous application of the competitive policy, the Finance Committee agreed that in practice 'certain modified procedures could be used where technical and administrative requirements made them necessary'. More specifically, they accepted that the word 'normally' in the Financial Rules was not to be taken to mean 'that full and formal competitive tendering [would] always be used': it was to be 'interpreted rather in the light of [...] the principles and methods of contract procedure which [were] being developed in practice' at CERN. The Committee also accepted that there 'be a certain streamlining of the procedure for drawing up tender lists'.[12]

Finally, measures were taken to enable the Finance Committee to monitor the allocation of contracts. Any contract worth more than 500,000 SF had to be approved by the Committee before the order was placed. Any contract worth more than 100,000 SF also had to be justified to the Committee beforehand, but only if the administration wanted to dispense with competition or did not want to accept the lowest bid made. Finally, contracts worth between 5000 SF and 100,000 SF and allocated non-competitively had to be reported afterwards to the Committee and reasons given. This was to be done every quarter by the Purchasing Office, who was also asked to differentiate the number and total value of contracts placed after 'full competitive tendering' from those placed 'by other procedures' (e.g. comparing catalogue prices). In practice the administration also tabulated the value of the contracts awarded in each country in these reports.[13]

The general attitude of the FC, then, was that CERN should be left free to purchase equipment and supplies with the 'maximum administrative economy and speediest procedure'. This meant two things. Firstly, the Committee was willing to accept that in practice it was not necessary for the administration always to resort to 'full and formal competition' (i.e. calling for tenders from a large number of firms widely distributed throughout the member states) particularly for relatively low value contracts.[14] From the start, then, it was tacitly understood that contract procedure could deviate markedly from contract policy—that the day-to-day practice of purchasing need not necessarily have to conform to the principle enshrined in the financial rules. Secondly, it meant that, barring special cases, the Committee monitored CERN's contract awarding procedures *a posteriori*. Except where relatively large amounts of money were at stake, it did not get involved in the purchasing procedure itself, leaving that to the discretion of the CERN staff. All that the Committee required was that the Purchasing Office report back to them on what it had done, giving reasons—and even that last stipulation was effectively dropped around 1960 when it was felt to be both cumbersome and unnecessary.

11.2 The competitive award of technical contracts: a critique of the 'rational model'

There is a simple, step-wise process for awarding contracts which is regularly associated with the idea that they are allocated competitively: for convenience we shall call it the rational model. Schematically, it runs something like this. First, the client—the scientist, the engineer, the purchasing office—, draws up specifications that are as precise as possible for the item that is desired. Then, these are incorporated along with a deadline for submission into a tender document which is sent to enough firms of repute to ensure that there is sufficient competition for each tender. The deadline once passed, the tenders are opened at a special meeting called for that purpose. They are compared with one another, and the contract is awarded to the lowest bidder who satisfactorily meets the technical and delivery requirements of the client. On this schema the procedure for the award of a CERN contract has thus three main features: it is formal, in that it follows a number of clearly differentiated steps, it is faceless, in that it is essentially bureaucratic and does not involve ongoing contact between the client and the potential suppliers once the specifications have been drawn up, and it is 'objective' in that the final decision is based solely on a comparison of quantifiable entities (cost, time, technical specifications, etc.). Most fundamentally of all, perhaps, it assumes that the clients *know what they want* from the start of the process. In this section we shall compare this 'rational model' with a short description of the process leading to the award of the single most expensive, time consuming, and scientifically crucial contract placed by CERN's engineers in our period: the contract for the manufacture of the PS magnet blocks.[15]

Notes: p. 672

11.2.1 THE AWARD OF THE CONTRACT FOR THE PS MAGNET BLOCKS

The heart of the PS magnet system comprised 100 magnet units (and one extra reference unit) arranged on the circumference of a circle of radius 100 metres. Each unit was about 430 cm long, and consisted of ten C-shaped blocks of laminated steel. The field was produced by two exciting coils horizontally encircling the upper and the lower pole faces of each unit. They were connected in series to the magnet power supply which had a peak rating of 3000 A at 10 kV.

By the summer of 1954 the PS group was able to give CERN delegates a rough idea of what the manufacture of the blocks would involve: about 3500 tons of low carbon sheet steel between 1 mm and 10 mm thick, machined to tolerances of 0.05 mm on the pole faces. These specifications had been firmed up by the end of the summer, when two main techniques of magnet manufacture were under serious consideration. The first involved using 1 mm thick steel sheets insulated from one another by paper and bonded together with araldite (a synthetic resin). No bolts were necessary. In the other, 10 mm thick laminations insulated with 0.5 mm of fibre or 'prespan' were bonded together with tie bolts.[16]

In October 1954 a letter was sent to 27 large, well-equipped firms of established reputation in eight member states inquiring whether they were interested in fabricating the blocks (13 were), and whether they had a preference for 1 mm or 10 mm thick laminations. To compare the relative merits of each two full-scale models were ordered. The Compagnie Electro-Mécanique (which was developing the magnet for a new synchrotron at Saclay) was asked to build a thin-lamination model, and the Dutch firm of Werkspoor was invited to build a model with 10 mm laminations. By December measurements on the first of these models were under way at CERN; the thick laminations model was expected later.[17]

In addition to these formal contacts PS engineers established informal links with the interested contractors, visiting factories and meeting representatives in Geneva. During these exchanges the firms explained in more detail how they proposed to manufacture the blocks—type of steel, thickness of laminations, nature of insulation, technique of bonding, characteristics of endplates—and gave a rough idea of their delivery times and costs. In the light of these discussions in April 1955 the PS engineers drew up a shortlist of nine firms which were anxious to do the job and which, in their opinion, were technically capable of carrying it out. By July the list had been reduced even further. With the agreement of the Finance Committee, the PS engineers decided to continue technical negotiations with only a restricted number of firms, and hoped to call for tenders in September 1955.[18]

It soon became clear that this deadline could not be met. The model studies had shown that it was preferable to use laminations of about 1 mm thick. However it had emerged that more precise information was needed on the properties of the magnets (like permeability and coercive force) and their dependence on the characteristics of the steel used (like chemical composition, physical state, and 'age'). By mid-October 'intensive work' was in progress on improving measuring techniques, and wide-ranging discussions were under way with steelmakers and their experts. A programme for studying a variety of steels was launched and additional models using different types of steel and laminations of varying thickness up to about 2 mm were ordered.[19]

These investigations revealed that no commercially available steel was sufficiently pure to meet CERN's specifications. The best one could do was to use either a low carbon content 'car body' steel or a low silicon steel, and to even out the effects of their impurities by systematically mixing the manufactured sheets before block production began. It was hoped that in this way the fluctuation of the average magnetic characteristics from block to block could be reduced by as much as a factor of ten.[20]

On 9 March 1956 invitations to tender were sent to eight European blockmakers: Compagnie Electro-Mécanique, Forges et Ateliers de Constructions Electriques de Jeumont, and Schneider Westinghouse (F), AEG (FRG), Ansaldo San Giorgio (I), Werkspoor (NL), and English Electric and Richard Thomas and Baldwins (UK). The firms were called on to quote for sheets varying between 1 mm and 2 mm in thickness, to propose one or more steels/steelmakers with which they would work, and to comment on various ways of mixing the steel sheets to obtain the necessary uniformity in the blocks. CERN made it clear that, although the tenderer had to accept responsibility for the mechanical properties of the steel delivery, its engineers would accept responsibility for the steel's magnetic properties.

Four blockmakers submitted offers—Ansaldo San Giorgio, CEM, Jeumont, and Werkspoor. Between them they quoted for one or more of four different steel thicknesses (1 mm, 1.5 mm, 1.8 mm and 2 mm) and used a total of nine different steel suppliers. After studying their replies the four were invited to submit further tenders by mid-July. Those that had not originally quoted for steels and thicknesses which particularly interested the CERN engineers were invited to do so. At the same time all contractors were asked 'for a more elaborate price breakdown so that detailed comparisons of the costs of production could be made which would reveal any unnecessarily expensive procedures'.[21]

The results of this second round of tendering were discussed extensively by CERN staff with the firms during the latter half of July and in August 1956. The procedures to be used for manufacturing the blocks were gone over carefully and various ways of reducing expenses were explored. Ongoing research at CERN now indicated that it would be essential to fabricate a full-scale model of a magnet unit for further tests, and at the end of August all of the contractors were asked whether they would be in a position to produce such a model by the end of the year. After having given the firms another opportunity to submit final comments, the CERN staff drafted a major paper summarizing their findings and putting forward their proposals (CERN/FC/144 of 14 September 1956). It was laid before the Finance Committee at its 10th meeting on 4 October 1956.

Notes: p. 672

The introduction to CERN/FC/144 was written by John Adams. In it he proposed that the contract for manufacturing the magnet blocks be awarded to Ansaldo San Giorgio of Genoa. No steelmaker was recommended. Instead Adams proposed that Ansaldo choose their steelmakers in consulation with CERN. He gave six reasons for preferring the Genoese blockmaker above its competitors: it had 'adequate press and die facilities' which were 'immediately available', it had proved that it was 'capable of making precision dies in its own factory', it had made '11 successful magnet blocks for the P. S. Division', it would reduce costs by 'up to 4% of the quoted price' by not charging CERN for rejected blocks, it could 'immediately make a prototype magnet block' on receipt of the order, and, finally, 'using as a basis of comparison the price of fabrication of the blocks Ansaldo offer[ed] consistently lower prices than the other firms'. These arguments were backed up by a 30-page analysis by Colin Ramm and Cornelis Zilverschoon of the history of the tender. Here they exposed the strengths and weaknesses of each blockmaker who had submitted an offer and described their experiences with nine steelmakers. They also gave tabulated data which showed the steps whereby an original 44 firms were whittled down to five, which detailed the results of their studies on the magnetic properties of a wide number and variety of steel samples, and which specified the first and 'final' offers received, the former being the results of the original call for tenders while the latter was the cost arrived at 'after technical modifications requested by CERN'.

The French delegate to the Finance Committee raised substantial objections to the procedure advocated in CERN/FC/144 in an official memo circulated a few days before the October meeting. The CERN staff, noted Mr Neumann, was asking the Committee to agree to the award of a contract whose precise value was not known because a steelmaker had not yet been chosen. This seemed unreasonable, all the more so since the cost of the steel comprised about 50% of the value of the contract. A more satisfactory procedure would be for CERN first to decide on what steel it wanted. That done it should ask each blockmaker to fabricate a final model using this steel and then to submit its tender. This 'normal procedure' was the only one which would put all the competitors on an equal footing, and which would enable CERN to obtain the best conditions of price and quality. For this, and a number of other reasons, Mr Neumann regretted that he could not 'go along with the proposition of the PS division to award the contract to the firm ANSALDO as long as the problem of the steel remained unresolved'.[22]

At its meeting on the 4 October 1956 the Finance Committee conceded the procedural point to the French delegate—it could not authorize the allocation of the award until the steelmaker had been chosen; it conceded the substantial point, however, to the PS division leader—the blocks would be manufactured by Ansaldo. To choose the steel, the blockmaker, in consultation with CERN personnel, would ask for quotations from all steelmakers whose product was technically suitable. These figures would be incorporated in Ansaldo's final tender and would be used by the Finance Committee in authorizing a contract. Mr Neumann accepted this proposal in the light of technical and financial data 'which had just been made available' by Adams.[23]

The results of the contacts made with five steelmakers were laid before the Finance Committee at its eleventh meeting on 29-30 October 1956. Ansaldo's lowest quotation for the manufacture of the blocks — 8.65 MSF — used 1.5 mm thick plates supplied by the Societa Italiana Acciaierie Cornigliano. The Italian government granted an additional rebate of 0.64 MSF to the blockmaker for using domestic raw materials, which brought the price down to 8.01 MSF, almost 10% below the next tender. The firm also had the advantage of being in Genoa, which made it easier for Ansaldo to collaborate with it, and simplified the life of the PS staff who would have to be resident at both the steel- and blockmaking works for a considerable period. The only remaining doubt concerned the quality of Cornigliano's steel. The company proposed to use a special steel first produced the month before, and the PS group were not sure that its magnetic properties would meet their specifications. To get around this problem it was suggested that Cornigliano provide CERN with 200 tons of its new steel starting 15 November. If the sample met specifications Ansaldo would contract the steelmaking to Cornigliano; if it did not, CERN and the blockmaker were free to approach other steelmakers. As it turned out the magnetic properties of the new 'low carbon rimming steel' were quite satisfactory for CERN's purposes, and the contract for the PS magnet blocks was awarded to the Ansaldo/Cornigliano combination.[24]

The most striking feature of this account is the extent to which it differs from the simple 'rational model' we sketched at the start of this section. The process was not *formal:* one step did not terminate one stage and initiate another. The first official call for offers was not the only one. It was followed by a second, after which part of the contract was awarded, and then a third, after which the steelmaker was chosen — and the cost was negotiated between successive tenders. Nor was the process *faceless:* suppliers were in contact with the CERN engineers at every stage of the process, discussing both the technical and the financial aspects of the contract. The stringent tolerances imposed on the properties of the steel by the strong-focusing principle raised problems which could 'only be solved in close cooperation with the manufacturer', and this meant that over a dozen magnet models were fabricated, sophisticated measuring equipment was developed, and more than 400 steel samples from 12 firms were studied by CERN and the steelmakers.[25] At the same time CERN staff were constantly looking for ways of reducing costs in consultation with the suppliers, going so far as to arrange for Ansaldo to acquire the araldite from the Swiss firm CIBA on very favourable terms, so reducing the contract by 200,000 SF.[26] Finally, the process was not simply *'objective':* the decision to award the contract did not flow 'logically' from a set of unambiguous, quantifiable parameters. Compromises had to be struck: the steel with the best magnetic properties was excluded because it was far too expensive, and elaborate measures were taken to ensure homogeneity by mixing the laminated sheets. Early research showed that sheets should be no thicker than 2 mm, and preferably 1 mm, but since cost increased as thickness diminished 'The final compromise [was] a thickness of 1.5 mm where the experience gained in two years of study of 1 mm block-plates [would] still be profitable'.[27] Correlatively there was a considerable amount of room left for personal judgement and 'feel'. The amount of enthusiasm the firm showed for the job, its willingness to cooperate with CERN on research into the steels, its reactions to the suggestions made by the CERN engineers, and so on, all played an important, if unquantifiable role in narrowing down the field of enterprises with which the PS group were prepared to continue discussions and negotiations.[28]

The inadequacy of the simple 'rational model' of how CERN awards contracts lies primarily in the fact that it assumes that the customers have a more or less clear idea of their technical requirements at the start and that the suppliers can more or less unambiguously cost the product sought after. This is sometimes possible— notably when one is buying a 'standard' item of equipment 'off the shelf'. And the model will 'fit' to the extent that that kind of situation can be reproduced. Put differently, the more expensive, unusual, and complex the device, and the less experience and in-house knowledge CERN has of its technical features, the more the process of awarding contracts will probably resemble that which we have described for the PS magnet blocks. In short, while the prevailing image of the process whereby CERN awards its contracts is not false, it is only partial, and probably typical for only a very specific kind of material.

Are we not exaggerating the point? Have we not generalized too quickly from a single example and, considering the scientific and financial stakes involved, an exceptional one at that? We do not think so for two reasons. Firstly, and fundamentally, because the 'rational model' shares a number of key features of an abstract picture of decision-making (full awareness of what one wants, optimization of a limited number of unambiguous variables) which completely obscures the complexity of the actual process whereby decisions are reached. In particular — and this is clear in our case — the model fails to grasp the change in the situation as perceived by the actors over time, and the reorganization of their knowledge and their preferences that that entails. We shall say no more about this here; the point is richly illustrated and elaborated by Dominique Pestre in chapter 12.[29]

Along with this theoretical argument there is a more practical reason for thinking that our criticisms of the 'rational model' are justified: the fact that an organization like NASA, which shares with CERN an interest in purchasing highly sophisticated technological equipment, specifically made allowance for a very different kind of contract involving a quite different kind of procedure: negotiation. To quote from an administrative history of the first five years of the Space Agency's life,[30] 'If the items to be procured can be clearly and completely defined in specifications and drawings, formal advertising for competitive bids is possible', and an 'Invitation for Bid' is sent to each potentially interested supplier. The contract is awarded to the lowest bidder who is 'responsible' and whose bid 'meets all requirements'. This procedure is clearly homologous with that described in the 'rational model'. On the other hand, in NASA's case, 'if the items cannot be well defined' the much more elaborate 'negotiation route' is followed. Here a restricted number of qualified suppliers are asked to submit a proposal—which 'is infinitely more complicated and expensive to prepare than a bid'. These proposals are evaluated in terms of quality (design, cost, schedules, etc), and of the technical and the managerial competence of the proposer, whereupon 'a decision is made [...] on the supplier to do the work. After selection, negotiations are begun to iron out the details of the contract'. The 'negotiation route', doubtless somewhat 'idealized' here, has much in common with the process actually followed by the PS engineers in acquiring their magnets. The very fact that another 'similar' organization found it necessary to make allowance for this kind of approach—and that CERN's failure to do so has recently been criticized[31]—must cast doubt on the generalizability of the 'rational model'. So too does the relative infrequency of the Invitation-for-Bid procedure inside the American space agency. According to our source, 'In NASA, 90 percent of the procurement dollar is spent via the negotiation route'.

11.2.2 THE POINT OF VIEW OF THE FIRMS

So far we have studied the process of awarding a technical contract essentially from the point of view of the CERN staff. We now want to make a few remarks about the position of the tendering firms. Our documents, which are mostly internal to the laboratory, are not very helpful here; however, there is no doubt that some firms have felt, and still feel, that they have been badly treated by CERN. Although we have very little evidence of hostility from firms in the particular case of the award of the contract for the PS magnet blocks, the example does throw some light on a topical and important issue.[32]

One reason for suppliers to be dissatisfied was that, in anticipation of a possible deal, the firms had to spend time and money on research for their clients without any compensation if they failed to win the contract. The investment could be considerable. In the case of the PS magnet blocks some firms made special melts of steel and rolled plates in quantities up to 300 tons. One spent nearly a year on metallurgical researches alone. In Adams's view such expenditures were 'a normal industrial risk', 'part of the overhead of large firms competing for tenders', and he refused to pay anything to one of the unsuccessful steel contractors who asked him to do so.[33]

Another commercial risk run by firms in a contract like this, which called for a considerable amount of research and development, was that of having their technical innovations disclosed to their competitors. CERN staff visited the factories of all major contenders for the award, talked to their experts, studied their samples, and were given access to the results of their research. No tenderer could be sure that its client, inadvertently or otherwise, would not divulge technical information of commercial importance to one of its rivals.[34]

Finally, there was the problem of 'unfair competition', of firms feeling not just disappointed, but cheated when the results of the tender were known. This would have been particularly understandable in our case since, once Ansaldo was chosen to make the blocks, the dice were heavily loaded in favour of Cornigliano, a steelmaker not simply in the same country but also in the same town. To increase the resentment of the unsuccessful suppliers, the steelmaker which was finally chosen was among those who had made no particular effort to address itself to CERN's problems. For the first half of 1956, while some of its competitors were going to great lengths to develop special melts and processes, Cornigliano 'expressed an unwillingness to offer other than their standard production steel [...]', which was quite unsuitable for CERN's needs. In fact the firm did nothing to enhance its chances until it was clear that its Genoese neighbour was going to be awarded the contract for the blocks—whereupon it promptly announced that it could produce a superior quality product. In effect then Cornigliano only began to study ways of improving its steel once it was fairly sure that it would get a big order: all of its rivals were pipped at the post by a competitor who had only run the last lap of the marathon![35]

11.3 The award of building contracts: the pitfalls of *a posteriori* control

No study of the kind we are making here would be complete without a thorough analysis of the award of building contracts. The value of the contracts alone—over 50% of the capital expenditure at CERN up to 1959 and about 30% for the next six years—makes this imperative. The analysis is also important for the light it throws on a problem we discussed in the previous chapter, namely, the difficulty CERN had, at least until the early 1960s, of making reasonable future estimates of its expenditures. As we can see from Table 10.2, Site and Buildings found forward planning extremely difficult—its estimates of expenditure for 1958, for example, tripled between the end of 1955 and of 1956, and doubled again between 1956 and 1957. Finally, our findings are significant because they reveal that, at least in the area of buildings, there were systematic 'geographical' biasses in favour of certain enterprises and certain countries in the allocation of contracts—a 'perverse effect', one might say, of the rejection of a policy of just return.

11.3.1 SOME REMARKS ON THE ORGANIZATION OF BUILDING WORK

CERN's building programme passed through three stages during our period. During the first and most important, which lasted until 1959, the buildings for the two accelerators and associated laboratories, the main administrative block, the main workshop, the power station, and a variety of other infrastructural buildings, were put up. The main development during the second stage was the east (or bubble chamber) experimental area, and during the third phase (i.e. starting around 1961) a number of relatively minor additions and extensions to the existing buildings were made. The organizational framework of the first stage differed markedly from that of the next two, and was somewhat atypical of CERN practice at the time, in the sense that the laboratory effectively handed over responsibility for the programme to outside experts.

At its first meeting in May 1952 the provisional CERN Council decided to appoint Lew Kowarski as head of the Laboratory Group which was 'in charge of studying the general structure' of the proposed new facility. At the next meeting of the Council in June Kowarski, who was based in Paris, announced that a second office would be set up in Zurich under his deputy, Peter Preiswerk, with responsibility for the 'general lay-out of the site, [and the] main features of the buildings'. Both of these appointments were made, at least partially, in the light of these men's past interest and dedication to the European laboratory project—as nuclear physicists neither had any particular expertise in the area of site and buildings. Accordingly, they immediately appointed two highly reputable Zurich-based consultants, who began work on 1 July 1952. The architect, Dr. R. Steiger, had installed a big betatron at Zurich hospital, and was assisted by a civil engineer, Mr. Hubacher.[36] In October 1952 the Council adopted Geneva as the site for the laboratory, and by mid-January 1953 Steiger and Hubacher had drawn up a preliminary site plan. It is of particular historical interest, and is reproduced in the set of photographs in the centre of the book.[37]

As soon as the Convention establishing the 'permanent' laboratory was signed on 1 July 1953 the Interim Finance Committee decided, 'in the interests of economy, and in order to avoid costly delays', to authorize Steiger to produce a 'completely detailed set of proposals with sufficient information on which tenders for the building work could be invited'.[38] The interim Council stressed that this engagement entailed no obligation on their part to award the Zurich architect the final contract for the building programme itself. However, by March 1954, with the Convention not yet ratified, and with increasing pressure from the machine groups to get the building work under way (see later), the Executive Group, with the support of a number of senior Council delegates, proposed that the construction of the laboratory be definitively entrusted to architect Steiger. No one objected to this proposal at the ninth meeting of the provisional Council, and on 2 June 1954 a formal contract was signed between CERN and Steiger.[39]

There are two interesting points about this contract. Firstly, there is the extent of the powers it granted Steiger. It stipulated that he take charge of the architectural and civil engineering works related to the 'preliminary project, definitive project, estimates, final specifications, tenders, general management of the work, verification, [and] supervision of the building site' for all the buildings in the first phase of the programme, as well as their fixed furnishings, and the roads, canalization, and gardens. It was assumed, though not stipulated in the contract, that

Notes: pp. 673 ff.

this would be done within the general framework of CERN's contract policy, though Steiger was authorized to award 'directly' contracts for small works up to a value of 8,000 SF. In short, CERN effectively placed total responsibility for the construction of the laboratory in architect Steiger's hands, and only 'controlled' it indirectly at the financial level.

In the light of this it is surprising — and this is our second point — that little if any debate surrounded the appointment of Steiger. The Executive Group recommended that 'the post of Chief Architect [..] not be put out to competition', and the interim Council agreed, doubtless because they saw no good reason to do otherwise. The architect had to be Swiss — or at least acceptable to the Swiss government —, there was enormous pressure on the Council to get ahead with the buildings — and Steiger had already drawn up its detailed plans —, and his competence was not in doubt — the Italian Council delegate Pennatta was 'highly impressed' by him, and 'after consultation with other experts' recommended that Steiger be granted the work.[40]

In October 1954 the permanent Council of CERN agreed to split Kowarski's Laboratory Group into three divisions, and nominated Preiswerk to head one of them, called Site and Buildings (SB).[41] During the first phase of the building programme SB was little more than a customer. The architect and his engineers designed the buildings in consultation with their clients and supervised their construction; SB was only responsible for the maintenance and security of the finished product. However, as the next stage of construction got under way around 1959, the architect's role diminished and the division's activities expanded. Steiger was used as a consultant for the East Experimental Area, and drew up some of the plans; it was SB, however, who designed the technical installations and basic fittings, and who supervised their execution. By the time the third phase of the building programme was launched (around 1961) the architect's role was reduced even further to advising on 'aesthetic questions and on general site planning'; detailed building design and civil engineering had also become part of SB's portfolio of responsibilities.[42]

Along with these changes — and indeed making them possible — went the introduction of properly qualified personnel into key positions in the division. In 1958 Preiswerk moved back into experimental physics and SB was placed under the direct responsibility of the Director General, with F.A.R. Webb — the site engineer who had joined CERN in October 1956 on release from the British Ministry of Works — as its nominal head. A year later Charles Mallet — a French civil engineer with two decades of experience in North Africa — was appointed division leader, and Webb was made his deputy.[43]

To conclude this quick survey it is worth noting that this shift of responsibility from Steiger to SB apparently did not occur smoothly, and there seems to have been considerable tension between him (and Preiswerk) and the staff in SB. Apart from disputes of an essentially technical kind — e.g. not everyone shared Steiger's ideas and standards — , some people resented the scope of his influence — one SB member referred to him publicly as the 'All Powerful Architect' — , and the closeness of his relationship with Preiswerk — the same man on the same occasion spoke of group meetings as 'a kind of small, intimate gathering, conducted almost entirely in Swiss-German'.[44] Whether or not these feelings were 'justified' there is nothing really to surprise us here. Steiger was undoubtedly powerful — with a nuclear physicist in charge of SB essentially for 'historic' reasons, things could hardly have been otherwise. And his relationship with Preiswerk was doubtless very close — they came from the same city, the physicist relied heavily on the expertise of the architect and, after all, they worked together for years to bring CERN 'concretely' into being.

11.3.2 THE CONTRACTS FOR THE TWO MAIN SCIENTIFIC BUILDINGS

These were valuable contracts from two points of view. Not only were they the most expensive building projects launched at CERN at the time. They also gave two or three consortia an early foothold on the site, which was to be of considerable importance to them later.

The decision to proceed with the construction work on the CERN site was taken by the interim Council at its ninth session on 8-9 April 1954. They did so with some misgivings. The Convention establishing the organization had not yet been formally ratified, and it was not clear whether the interim body had the authority to take a step of this magnitude. On the other hand both Odd Dahl, who was still leading the PS Group, and Peter Preiswerk insisted that it would be fatal to delay the launch of the construction programme any longer. 'It was true', said Dahl to the Council, 'that the starting of foundation work was contingent on the question of the ratifications; nevertheless, it was of the utmost importance that construction work should start as soon as possible [...]'. If it did not get under way in the spring, bad weather, which made work on the site difficult during the winter months, could set it back by as much as a year. The Council accepted the argument and found a way around the formalities. To press ahead as rapidly as possible it was agreed that tenders for the building of the roads — the first work to be done — could be restricted to local firms, and five were invited to submit offers late in April. The contract was awarded to the lowest tender early in May, and on the morning of 17 May 1954 the Geneva firm Spinedi started levelling the earth at Meyrin.[45]

11.3.2.1 The SC building

The first call for tenders for the SC building was made by architect Steiger in consultation with Preiswerk soon after work on the roads began. Despite the fact that it was official policy to call for tenders from a wide diversity of member states, only Swiss and French firms were asked to submit offers.[46] This restriction was

Notes: p. 674

defended by Preiswerk at a meeting of the Interim Finance Committee (IFC) shortly thereafter, where he successfully persuaded those present that 'in practice, the Swiss and French firms within easy radius of the site were likely to be best placed for dealing with the present work', notably the excavation of the SC building. This mainly involved earth moving, and it seemed reasonable to assume that the cost of bringing the necessary heavy equipment to Geneva effectively excluded geographically remote firms. While seeing the merits of this argument, the IFC insisted that this case was to be treated as 'exceptional', and demanded that for the *construction* of the SC building 'there [had to] be full opportunity for competitive tendering'. Secretary-General Edoardo Amaldi was thereupon 'asked to communicate with all Governments so that lists of suitable firms could be sent to him without delay'.[47]

Throughout this debate the IFC was not aware that Steiger and Preiswerk had *already* called for geographically restricted tenders for both the excavation *and* the construction of the SC building. This only dawned on them on 12 July, when the Chairman (Lockspeiser) and Secretary (Verry) met with Amaldi in London to survey the offers already received. Steiger was contacted by telephone via the Swiss delegate to the IFC, and asked why he had done this. He implied that there was no point in tendering only for the excavation, since before digging could get under way 'a commitment for building work [had to] be undertaken'.[48]

Steiger's argument placed those at the London meeting in something of a dilemma. On the one hand, they were constrained by the competitive policy agreed on by the IFC a few weeks before, which required that tenders be called in a variety of member states. On the other hand, they now realized that if they enforced this policy they would probably have to wait several months before a start could be made on excavations for the SC building. In the event they decided not to exceed their powers. On 13 July Amaldi informed Steiger that he was not to place an order for the SC hall on the basis of the geographically restricted tenders received, but was to await the results of an approach to a 'wider circle of firms from other Member States'. At the same time, Amaldi asked the architect to 'place an order confined to the excavation work of the S.C. building, [...] in order to save delay', if he felt that this could, in fact, be done.[49]

Ten days later Amaldi met with Kowarski, Preiswerk, and Steiger. At this meeting it emerged that, technically speaking, it was actually only necessary to choose a firm *to lay the foundations* of the SC building before starting the excavations. 'It was not necessary to know who [would] build the part above ground'. Here was a way to get started on the building—though possibly at some cost, for it was 'reasonable to expect' that splitting the construction between two different firms would mean additional expense. Preiswerk and Steiger were authorized to choose a firm to lay the foundations of the SC building from among their limited selection of French and Swiss firms. They were also asked to send out invitations to tender for the rest of the building 'to as many firms and countries' as the IFC wanted.[50]

By 25 July Amaldi had received lists of building contractors from five member states—Belgium (8 firms), Italy (9), the Netherlands (18), Switzerland (28), and the United Kingdom (4). Shortly thereafter France (7) and Germany (7) also submitted names. In a letter dated 5 August he asked all non-Swiss firms specialized in construction work if they were interested to tender for a number of buildings on the site, including the upper-structure of the SC building. The deadline for replies was 6 p.m. on 11 August. Around the same time the Chairman and Secretary of the IFC agreed that the firms Locher (Zurich), Cuénod and Induni (Geneva) could proceed with the foundations of the SC building up to ground level at a cost of just under SF 700,000. In Steiger's original call for tenders this group had together made 'the lowest offer, as well for the foundations as for the whole building and [had] an excellent reputation'.[51]

The plans and details of the SC building were officially displayed in Rome, Milan, Paris, Brussels, and The Hague from 9 to 11 August; they were also sent directly to two of the four British firms who expressed an interest to tender. By the deadline on 11 August 16 (non-Swiss) firms out of 54 had said that they were interested to tender for the upper-structure of the SC building. They were asked to submit their offers by 25 September.

The Interim Finance Committee met for the last time on 6 October 1954—the day before the first meeting of the Council-proper of the new organization. The results of the tenders for the section of the SC building above ground were laid before it. The lowest tender was that of the firm who had received the contract for the foundations, and the consortium of Locher, Cuénod, and Induni was thus invited to construct the entire SC building. The contract was worth some 1.3 MSF.[52]

11.3.2.2 The PS building

The PS reinforced concrete works were the most valuable building contract awarded in the mid-1950s. The scope of the contract was briefly described in document FC/39 prepared by the SB division early in March 1955. It covered the concrete and masonry works and sewer system for five buildings: the 200m-diameter ring to house the magnet, the linear accelerator building, the experimental halls on the 'North' and 'South' sides of the ring, a laboratory wing, and a generator room. Sand, gravel, and cement were to be supplied from the neighbourhood of the site. The overall cost of the works was estimated to be about 11 MSF.[53]

Document FC/39 included a brief introduction drawing the Finance Committee's delegates' attention to a 'big problem': the PS engineers needed their experimental hall built by June 1956 to receive the first magnet elements, and wanted the ring building ready about four months later. This schedule called for 'an extremely rapid execution of work'. It could only be met if a number of conditions were satisfied, including the implementation of 'a quick procedure for the selection of firms invited to submit tenders'. The SB division proposed three ways of doing this: by reducing the number of non-Swiss firms invited to tender, by giving contractors a 'relatively short period' to send in their tenders, and by streamlining the process whereby the Finance Committee exercised its powers regarding adjudications.

What were the implications of these suggestions for the contract for the PS reinforced concrete works? 33 non-Swiss firms had expressed an interest in tendering for this job. The administration suggested inviting just six—one each from Belgium, France, Germany, Italy, the Netherlands, and the United Kingdom—along with five Swiss firms. As for the time schedule, SB proposed that 'only 4 weeks [...] be granted to the firms to submit their offer'. Adjudications would 'have to be decided at the latest in the beginning of June' 1955, and the work should start some two to three weeks later.[54]

The Finance Committee considered these proposals at its third meeting on 23 and 24 March 1955, in the presence of several members of the CERN senior staff and the architect. The Committee accepted that tenders for building contracts could be geographically restricted, though several members did feel that, for the particular case of the contract for the PS building, more than six firms from outside the host country should be invited to submit offers. In the face of considerable opposition from the CERN staff, it was finally agreed that 'up to two firms might be invited to tender' in each of the six non-Swiss countries. The Dutch representative, J.H. Bannier, also questioned the time schedules. He thought that there was too little time given to prepare and submit tenders, as well as to start the work once the adjudication had been made. Against this the administration insisted on the need to go ahead with the maximum possible speed if the building was to be ready in time, and Bannier did not press the point.[55]

Tender forms for the reinforced concrete works for the PS building were sent out on 27 April 1955. The work was broken down into two components: the ring and linac buildings (section A), and the experimental hall, laboratory wing, and generator and power house building (section B). 16 firms were invited to tender: five Swiss, one British (the British delegate to the Finance Committee did not propose another), and two from each of the five other non-Swiss countries close enough to Geneva to be considered seriously. The deadline for the submission of offers was 28 May.

Nine firms, all located in France, Italy, or Switzerland, submitted offers in time, more often than not combining themselves into consortia. The lowest offer received was that from the Italian consortium of Gandini & Vandoni, and Guffanti: they wanted 3.53 MSF for the PS ring and 3.84 MSF for the hall, making a total of 7.37 MSF in all. Next in line came the Swiss group of Zschokke, Spinedi, and Losinger with 4.60 MSF (ring) and 5.41 MSF (hall), a total of 10.01 MSF. Nevertheless, and although the Italian consortium was considerably lower than its Swiss competitor, the Finance Committee agreed—without opposition from the Italian delegation—to split the work. On the recommendation of Steiger and the SB division, the rings were awarded to Zschokke et al. and the halls to Gandini et al. for a total cost of 8.44 MSF—over 1 MSF more than if the whole contract had been awarded to the Italian group.[56]

Why was the contract for the PS ring not awarded to the lowest tender? The main argument put to the Finance Committee for favouring the Swiss consortium for the ring was that this contractor offered better value in terms of price and quality. In terms of price, because the Italian group 'did not appear to have made sufficient allowance for the various elements of increased cost' which this demanding structure entailed. In terms of quality, because a visit by engineers from Steiger's office and from SB to some of the buildings put up by the Italian firms 'showed a quality of construction which [was] not equivalent to that desired expressly for the difficult task of ring construction'. Indeed it was supposedly not good enough for the construction of the hall either, but the consortium had ensured the CERN visitors that their standards for this more conventional building could be satisfied. Hence the recommendation to split the work between the two groups.[57]

It is difficult to regard these as serious arguments rather than rationalizations. For example, is it not strange to refuse a lower bid on the grounds that the enterprise had underestimated costs? Surely that was all to the best for CERN, and if the organization had some reason to protect Gandini et al. from their own 'mistakes' all it had to do was to point them out to the Italian firm and ask for a revised tender. Indeed, we have reason to believe that even before the tenders were called the PS engineers had effectively decided what firm they wanted to erect the PS buildings for them.

Early in 1955, and long before tenders were invited for the PS buildings, SB awarded a contract for a 16 m long life-sized model of the PS ring. It was to be used to study various aspects of the ring's construction and to make a detailed analysis of its costs. The firm that built this model would obviously be in a privileged position when it came to constructing the real thing. The job was awarded noncompetitively to Conrad Zschokke of Geneva.[58]

Why the preference for this enterprise? Its competence apart, primarily because it was a Swiss firm. This simplified dealings with the architect, who obviously preferred to work with a consortium that was familiar with local building regulations. It also avoided the bureaucratic delays encumbent on the use of a non-Swiss builder. Gandini et al. ran into them at once: the Italian authorities were unhappy about their exporting equipment and materials into Switzerland, and the Swiss authorities were unhappy about their using Italian labour on a building site in Geneva. 'We knew all that at the beginning', wrote Adams to Verry on 13 October 1955. 'In fact', he went on, 'that was our main reason for wanting to give the contract to a Swiss firm'.[59] Finally, there seems to have been a distinct feeling inside the PS group, reinforced by national stereotypes, that only a Swiss firm—or at least nothing as 'disordered' as an Italian firm—could be trusted to build something as delicate and sophisticated as the PS ring. Indeed at one point Adams' lack of confidence in Gandini et al. was so strong that he actually proposed withdrawing the contract for the halls from them several months after it had been awarded.[60]

Notes: pp. 674 ff.

An important point to retain from this section is the advantage which Swiss firms had in winning these building contracts. A variety of factors worked together to this end. There were local and national bureaucratic constraints—the architect had to be acceptable to the Swiss government, and foreign firms had to navigate their way through a complex and time-consuming bureaucratic procedure before being allowed to work in the country. There were the preferences of the architect—a Swiss of great experience who obviously preferred to entrust such an important and exacting project to firms in his own country which he already knew. There were the demands made by the scientists themselves—they wanted to press ahead as rapidly as possible, and it was much quicker to discuss matters face-to-face with 'local' firms than to pass through the laborious formal tendering procedure, or to overcome the bureaucratic obstacles faced by foreign enterprises. They also felt that Swiss construction firms were better than most—though the popular image of the Swiss as outstandingly reliable and competent was perhaps as important here as informed professional judgement. Finally, there was the question of the cost, of the fact that, being closer, Swiss building firms were necessarily advantaged since they did not have to transport heavy construction equipment over long distances.

This last factor deserves further comment. While it is obviously true in a general sense—and was the basic reason why the calls for building tenders were geographically restricted— one should not exaggerate its importance. For one thing, Italian tenders for the PS were substantially below those of the Swiss consortium which was awarded the ring. For another, even when Swiss firms did submit the lowest bids, as was the case for the SC, the fact remains that the odds were heavily biassed in their favour by the interpretation placed by Preiswerk on the principle of geographically restricted tendering. We shall come back to this point later. For the present we simply want to stress— without any intention to judge—the important role played by official regulations, as well as personal factors like past experience, trust, and perhaps even prejudice, in the award of two of the first three major building contracts at CERN to Swiss firms. As in the case of the PS magnet blocks, though the contracts were awarded competitively, this did not mean that the 'laws of the market' were the only factor determining the outcome.

11.3.3 THE EVOLUTION OF THE BUILDING PROGRAMME UP TO 1959: THE ADVANTAGE OF BEING INSTALLED ON THE SITE

As the first phase of the building programme gathered momentum so were the major consortia on the site joined by a myriad of relatively smaller contractors putting up roofs, fitting doors and windows, laying cables and piping, doing carpentry and painting, erecting flagpoles and marble stairways.... And while we cannot study this rich tapestry in detail, there is one aspect of it that we should like to look at here—the benefits reaped by firms already installed on the site, as measured by the value of the additional work they were called on to do.

One way in which this happened was simply by enterprises having the value of their existing contracts go up. Table 11.1 lists some of the more important examples. It shows increases in the worth of the same contract by as much as a factor of two (or even three in the case of the French firm Tunzini) — these increases, of course, never being subject to a further round of competitive tendering.

Table 11.1
Increases in the value of some major CERN building contracts in the mid-1950s (in MSF)[61]

Contract	Firm	Original price	Awarded	Revised price	Agreed
PS ring	ZSL	4.60	Jun 55	5.78	Jun 57
Power house, etc.	ZSL	1.07	Sep 55	1.29	Oct 56
PS halls	GVG	4.12	Sep 55	5.54	May 57
Main workshop, etc.	GVG	1.15	Sep 55	2.17	May 57
Air conditioning	Tunzini	1.24	Oct 56	3.46	May 57
Main admin. bdg.	GVG	1.05	Dec 56	2.25	May 57

Key: GVG = Gandini et al.   ZSL = Zschokke et al.

An important reason for these escalations was that the original tenders were let on the basis of somewhat tentative and sketchy plans (scale 1:100 or even 1:200), and that the cost of the job increased during the course of construction.[62] The design of the foundations of the magnets in the PS ring is a case in point: it underwent continuous development for more than a year until a suitable structure was found.[63] Another reason for the price rises was simply the mistakes arising from inexperience. The lay-out of the PS air-conditioning system, for example, was drastically changed after being studied by Tunzini's experts — and its cost trebled.[64] Then again it seems that some prices were initially kept low by the management in order to facilitate their passage through the Finance Committee, only to leap ahead once the contract had been agreed to. Preiswerk, for example, persuaded the Committee to give Gandini et al. the contract for the reinforced concrete work on the main building noncompetitively on the grounds that, at 1.1 MSF, it could reasonably be regarded as an extension of their existing contracts. Within months a detailed costing of this standard building led him to come back to the Committee for 1.2 MSF more.[65]

The second way in which contractors already installed on the site were advantaged over potential rivals was through their being given additional work, much of it for relatively small jobs awarded noncompetitively. All of the firms we have already mentioned benefitted in this manner. Between 1955 and 1959 Zschokke et al., for example, were awarded 58 contracts noncompetitively together worth about 0.5 MSF, as well as being given a lot of new work 'competitively'. By way of a change, and to bring out how wide the scope of this phenomenon was, let us follow in a little more detail the fortunes of another firm — Ateliers de Constructions Mecaniques S.A. based at Vevey near Lausanne on Lake Geneva. Early in 1955 this

Notes: p. 675

firm was awarded a contract worth some 100,000 SF for the structural steel on the SC building in the face of stiff and comprehensive international competition. On the basis of that success it was given a contract worth 58,500 SF for the metal construction at 'Adams Hall', a provisional building for the PS division, without having to compete for it. It was also one of three Swiss firms invited when geographically restricted tenders were called for the roof over the PS experimental hall—which was allocated to it in mid-1955 at a total cost of 450,000 SF. This was the first of a number of large contracts awarded competitively to the firm for work on the PS building: late in 1955 it was asked to do structural steel work worth 347,000 SF, and around mid-1957 it was awarded two further contracts together worth just under 334,000 SF for an installation comprising gangways, rail tracks for travelling cranes, etc. inside the building. Along with the latter came five smaller noncompetitive contracts for cranes and other lifting devices worth about 120,000 SF. The firm also did structural work elsewhere on the site (the stores, the main auditorium) as well as building one or two rather specialized pieces of equipment (the base for the SC magnet, the vacuum tank for the PS linac). To summarize, between 1955 and 1959, Ateliers de Constructions Mecaniques were awarded 27 contracts on the CERN site worth almost 1,7 MSF. 17 of these, together amounting to some 350,000 SF, were allocated noncompetitively.[66]

Several practical considerations account, at least partly, for this concentration of work in the hands of an enterprise once installed on the site. It was inadvisable to have different firms working side by side in the same area: the zone became congested, it was difficult to control their work, and there was inevitably friction between them. There was the question of the time that would be saved by placing orders with a firm already on the spot. And there was the advantage of knowing beforehand that the job would be done properly by people with whom one had worked before and who had come to understand one's needs.[67]

And again, what of the question of cost, of the competitive advantage accruing to a firm which had already installed its heavy equipment at Meyrin, and did not have to bear the burden of transport costs for heavy civil engineering work? On the face of it this consideration was important, even determining, for the large contracts that were awarded competitively and which, after all, accounted for about 95% by value of those to Zschokke et al. and Gandini et al., and some 80% of the work allocated to ACM (Vevey), for example. But one must be careful, for two reasons. Firstly, because the administration often classified as competitive a job allocated non-competitively to a firm, arguing that its past competitive record indicated that no one could do better for the job in question. It is therefore difficult to know how many jobs were simply 'repeat orders' and so difficult to have an idea of the real cost benefits accruing from not having to set up a building site again. More importantly, though, where we do have clear evidence that a new job was tendered for, it seems that there was no systematic correlation between being installed at Meyrin and submitting a lower offer than a potential newcomer for a major building contract. This is a point we shall return to shortly.[68]

If we combine the findings of this section with our results in the one before we see immediately that major Swiss consortia played an important, even dominant, role in the first phase of CERN's construction. And what is true for them is all the more valid for the many enterprises called on to do smaller jobs which were generally not awarded 'competitively' (in the sense that formal tenders were called for). Here local firms (i.e. those close to Geneva including, of course, firms in neighbouring France) were obviously the best placed (cost, administrative efficiency), and were often able to win further contracts 'automatically' (the need to standardize on certain items of equipment like electrical installations, doors, windows, taps, etc). And among the local firms it was the Swiss firms that the architect preferred. Indeed using — or, according to some British officials in 1958, abusing — the powers granted to him by his contract, Steiger awarded about 12 MSF worth of small (i.e. less than 8000 SF) noncompetitive contracts, all of them essentially to Swiss firms.[69] In fact, and although we do not have reliable data for the national distribution of non-competetive contracts, we can reasonably estimate that almost 70% of CERN's building work between 1955 and 1959 was performed by Swiss firms.

11.3.4 THE CERN ADMINISTRATION AND THE FINANCE COMMITTEE

Apart from the factors already identified, there were two arguments put forward, and practices followed, by the administration which encouraged monopolization and biassed the award of building contracts in favour of Swiss firms. One was the principle of geographically restricted tendering, the claim that there was no point in calling for bids for major construction work from other than local firms since those far away from Geneva could not compete.[70] The other was the request that a new valuable building contract be awarded noncompetitively to a firm already on the site, on the grounds that the contractor had made the lowest offer for a previous piece of work, and was willing to do new work of a similar kind at the same unit price. Sometimes this argument was backed up with a threat: that to call for tenders would actually be counterproductive, since the contractor on the site knew that he could safely increase unit prices and still be the cheapest.[71] Neither of these arguments was seriously challenged by the Finance Committee during the first phase of the building programme. However, when the control over the programme shifted to the Site and Buildings division in 1958/59, its new head Charles Mallet made a deliberate effort to throw building contracts open to a wider constituency. Firstly, the value of the work to be done and its duration were increased by grouping buildings together.[72] Secondly, the system of geographically restricted tendering was abolished for large contracts. In 1960 25 firms in 11 member states were invited to tender for the main contract in the East Experimental Area, and the practice was subsequently repeated: 48 firms in 13 member states for the cover on the east apron

Notes: pp. 675 ff.

in 1961, 43 firms for laboratories and a workshop in 1963, 49 firms in all member states but Greece for the new AR laboratories and experimental halls in 1964, and over 50 firms for the ISR in 1965.[73] By virtue of these initiatives we have data which we can use to assess the validity of the administration's claims—and of the Finance Committee's wisdom in not challenging them.

Table 11.2 presents the results of the major building contracts awarded competitively during our period, as defined by those contracts for which the administration drafted a paper for the Finance Committee. To complete the picture, the competitive position of the two firms most active on the site in the first phase of the building programme is also given.

Table 11.2

The firms who successfully competed for building work in the first half of the 1960s, and a comparison with the performance of those consortia who were most active in the PS area in the late 1950s (5/6 means fifth out of six firms that submitted tenders).[74]

Date	Building	Successful Contractor	Country	Value (MSF)	Positions	
					GVG	ZSL
3/60	PS 3rd Wing & Workshop	Eiffel	F	0.63	5/6	4/6
6/60	PS East Expt. Area	ZSL and Chemin	CH + F	3.07	6/6	1/6
10/60	Accel. Res. Bdg.	Gandini et al.	I	0.69	1/6	3/6
12/60	East Junct. PS ring	Zschokke et al.	CH	1.10	tenders not called	
12/61	Cover, East Apron	Stahlbwk. Muller	FRG	1.46	---	2/6
1962	Labo. 12	Constrn. Moderne Fr.	F	0.13	4/4	3/4
1962	Technol. Wkshop	Constrn. Moderne Fr.	F	0.33	3/4	4/4
1962	Labo. 4	Constrn. Moderne Fr.	F	1.04	2/5	4/5
5/63	NPA Extn., Tech. Wkshop.	Société Aixoise	F	1.61	2/5	4/5
5/63	Substatn.	Gandini et al.	I	0.37	1/5	4/5
4/64	Roads, Parkings, etc.	Zschokke et al.	CH	1.19	--	5/6*
8/64	Accel. Res. Hall etc.	Société Aixoise	F	4.09	2/6	3/6

* ZSL was only awarded this contract after a heated debate in the FC. Its offer was about 10% above the lowest bid—see note 74.

Key. GVG = Gandini & Vandoni and Guffanti
      ZSL = Zschokke, Spinedi, and Losinger.

The first point we want to look at is the geographical one, the countries of origin of the successful contractors. And we see immediately that, although a concerted effort had been made to interest building firms in all member states in work at CERN, in fact only those from Switzerland, or countries bordering on the host state (France, Germany, and Italy) were able to compete.[75] Thus—and this is hardly surprising—there is apparently little to object to in the principle of geographically restricted tendering, at least if one's sole concern is to get the best value for money with the greatest administrative efficiency.

We cannot leave matters at that, however. For it is not the principle that is dubious, but the interpretation that was put on it in the 1950s. For in practice the administration used it to legitimate inviting many more Swiss firms to tender than those in any other member state, including those surrounding Switzerland. This biassed the whole process in favour of the host state, if only because it seems that the spread in offers for a contract from firms *within* a country was far greater than the spread in the lowest offers *between* countries.[76] By 'sampling' six firms in Switzerland—and only two in each of France, Germany, and Italy —the odds that a Swiss firm would submit the lowest offer were substantially increased.

Table 11.2 also indicates how dubious was the administration's claim that firms already on the site were best placed to win further contracts—and the conclusion that they drew from that: that there was no need to call for tenders for a new project. In fact, newcomers clearly could compete, a point that did not escape the attention of the Finance Committee.[77] Zschokke et al. and Gandini et al. were regularly 10-30% more expensive than the firm submitting the lowest bid. And this did not only happen in their cases. Eiffel, for example, which was awarded the third wing of the PS building in March 1960, was fifth out of six when tenders were called for the Accelerator Research Building about six months later. In short, there does not seem to have been any systematic correlation between presence on the site and submitting the lowest offer for a new, large, building contract.

Granted the controversial nature of some of the policies adopted by the administration during the first phase of the building programme, one may well wonder why the Finance Committee essentially went along with them. One reason may have been linked to the organizational framework in which the building programme was situated. CERN effectively contracted out the work to Steiger and, having given him the responsibility for the job, the Finance Committee's authority over how he did it was accordingly restricted. Another reason for the Finance Committee's relative indifference may have been that delegates were not particularly concerned to protect their national building industries. Construction was neither a particularly glamorous sector of the economy—one commentator has claimed that in the early 1950s in Europe 'the building industry was on the whole inefficient and conservative in outlook, consisting mostly of small firms'—nor, and perhaps more to the point, was it short of work.[78] Faced by a critical postwar situation, governments allocated on average 3-5% of their GNP to housing construction in the

Notes: p. 676

late 1940s, while during the 1950s the number of houses in western Europe roughly doubled. Finally, and perhaps most important of all, the attitude of the FC was consistent with their policy of minimal interference in the internal management of the laboratory. Like the CERN senior staff its delegates wanted above all to see a laboratory built to the highest standards with the best equipment and in the shortest possible delay, a laboratory of which Europe could justly be proud, and which would enable it once again to compete on an equal footing with similar institutions in the United States—and they trusted them to get on with the job, restricting the FC's role as much as possible to one of *a posteriori* control. As this section has shown, for all its strengths, this policy was not without pitfalls of its own.

11.4 Some global statistics on the distribution of contracts awarded by CERN between 1955 and 1965: the advantages of being the host state

The data that we are going to present below are based on the quarterly reports of the CERN purchasing office. Bearing in mind the two main dimensions of CERN's contract policy, these reports distinguish systematically between contracts awarded competitively and noncompetitively (the latter being further subdivided into contracts below 5,000 SF, from 5-100,000 SF, and architect's contracts) as well as breaking down the data according to the value of the contracts awarded in each country. Three limitations of these comprehensive, official statistics should be mentioned at once.

Firstly—and the point must be stressed again here—they used a rather restrictive definition of the concept of a competitive award: it was an award made 'after full competitive tendering'. This means that a purchase made after comparing catalogue prices of various suppliers, for example, was treated in the same way as a purchase placed without doing any market research whatsoever. For statistical purposes they were both counted as noncompetitive awards.

Side by side with this restriction on the meaning of competitiveness there was an anomaly we have already alluded to: the quarterly reports classified as competitive some purchases which were certainly noncompetitive. For example, so-called development contracts—like that for the PS linac—in which only one firm was approached to build a more or less unique and sophisticated piece of technical equipment, were categorized as competitive. So too were repeat orders for equipment like beam transport magnets originally acquired after a call for tenders, even though the new orders were 'automatically' placed with the same firm.[79]

Finally, there are a number of sources of ambiguity in the data given for the distribution by country. For one thing, it is not clear whether a piece of American equipment bought, say, from a Swiss agent was classified as a purchase made in the United States or in Switzerland. For another, it seems that when a contract was awarded to a consortium of firms from more than one country, it was counted as a

Distribution of contracts

payment to the country in which the main partner was based. We have no way of judging the extent of any of these practices nor, correlatively, of compensating for 'distortions' that they might have introduced into the data. What we are presenting here, in other words, is not the 'truth' but, more importantly perhaps, the official picture of how CERN's contracts were awarded, the picture painted for the Finance Committee by the administration whose task it was to apply the policy on a day-to-day basis.

11.4.1 COMPETITIVE AND NONCOMPETITIVE CONTRACTS

Figure 11.1 is a histogram depicting the distribution by value between these two categories during the first decade of CERN's life. The most obvious features of the

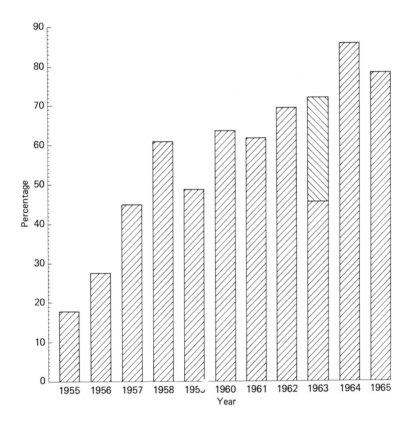

Fig. 11.1 The percentage of contracts awarded noncompetitively by CERN between 1955 and 1965. The higher value in 1963 is that obtained when the cost of the CDC computer is removed.
Source: Quarterly Reports of the CERN Purchasing Office.

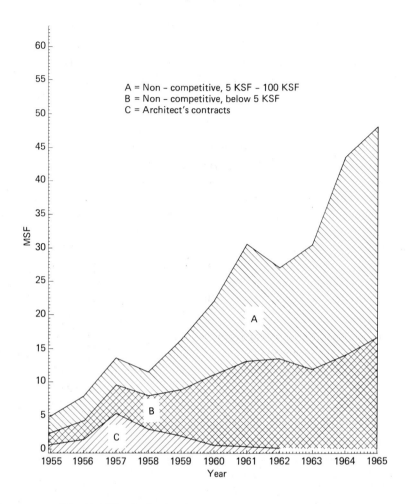

Fig. 11.2 The distributions by value of noncompetitive awards.
Source: Quarterly Reports of the CERN Purchasing Office.

figure are the high proportion of noncompetitive awards, and the general upward trend over time in this category of purchase—and the contradiction that these entail between official policy and the day-to-day management of the laboratory. Financial Rule 10(b)(i) lays down that 'not less than three competitive tenders shall normally be obtained for the purchase of plant, equipment, supplies and other services [...]'. Figure 11.1 shows that the majority of the contracts (by value) were only allocated competitively during the first three years of CERN's life. Thereafter at first about 60%, and then upwards of 80%, of the value of contracts placed by CERN was allocated without calling for tenders.[80]

As a first step towards understanding why this should be so, let us next consider Figure 11.2, which breaks down the noncompetitive awards by value between the various categories used by the Purchasing Office. It shows that, until about 1960, the bulk of the noncompetitive awards were low value awards (i.e. below 5,000 SF each by the Purchasing Office, and below 8,000 SF each by the architect), and that in fact from 1957 to 1965 the annual value of this kind of contract remained more or less in a band lying in the range 10-15 MSF. By contrast, from 1960 onwards the annual value of relatively high value awards (i.e. from 5000 SF to 100,000 SF) allocated noncompetitively grew enormously, from about 10 MSF in 1960 to over 30 MSF worth of business in 1965. By this stage a full two-thirds (by value) of CERN's noncompetitive awards were being allocated in this higher bracket.

These findings are confirmed if, instead of dealing with the value, we look at the number of the contracts awarded. In 1956 there were 271 contracts worth between 5,000 and 100,000 SF awarded by CERN; 220 of these (i.e. over 80%) were awarded noncompetitively. In 1960 the number of awards in this category had grown to 795, of which 732 (over 90%) were noncompetitive awards. By 1965 'all' contracts worth less than 100,000 SF were being awarded without 'full and formal tendering': 1,537 were let in the bracket lying from 5,000 to 100,000 SF and only one of these was reported as being allocated competitively.

By disaggregrating the area under the bars in our histogram we see then that the upwards trend in the 1960s was caused predominantly by an increase in the number and value of noncompetitive contracts worth from 5 to 100,000 SF. Two factors account for this development: the changing needs of an expanding staff, coupled with the administration's contract-awarding behaviour. As the laboratory moved from construction to operation, and the number of staff increased—it roughly doubled between 1956 and 1960, and doubled again between 1960 and 1965, when it reached some 2,000 people—, there was a decrease in the absolute value of purchases of heavy and expensive technical equipment, as well as of major building works. More and more people spent more and more money on relatively less expensive items. At the same time the administration and the scientific staff, never particularly enthusiastic about the bureaucratic complexities and time delays imposed by calling for tenders, were increasingly reluctant to award competitive contracts worth less than 100,000 SF as the volume of orders expanded. Figure 11.1 is the outcome of these two tendencies.

One last word on the significance of the 100,000 SF figure. This was the value of an award below which the CERN staff could allocate the contract noncompetitively without first getting the Finance Committee's approval—all they had to do was to add a few words afterwards explaining their actions in the quarterly reports of the Purchasing Office, and even that practice was abandoned in 1960. Faced with an FC that seldom if ever seriously challenged their purchasing practice, an increasingly confident administration exploited this concession to pragmatism to the full.

11.4.2 THE DISTRIBUTION BY COUNTRY

Table 11.3 presents the data we need for this section, as derived from the Purchasing Office's quarterly reports. A notable feature of the data is the rough correlation between distance from the site and the percentage value of the contracts awarded. Thus we see the three Scandinavian countries receiving together 1.9% by value of the contracts awarded, the Low Countries 5.5%, the United Kingdom 6.0%, Germany a little over 15%, France almost 23%, and Switzerland an astounding 40%. A kind of exception to the rule might be Italy, with 7%, and Austria, though the data we give for this country are not strictly comparable as it only joined the organization in July 1959. We notice too that the host state reaped more than a ten-fold return on the government's 'investment' in CERN. The next best performance in this respect was that of France, where the ratio return: expenditure was about 1:1.

Table 11.3

The distribution by country of the value of contracts awarded by CERN in the years 1952-1965 inclusive. The second column gives the absolute values, and the third the percentage by value of the orders placed *in Europe*. The last two columns give the percentage contribution each state made to the CERN budget in 1955 and in 1965.[81]

Country	Total, 1952/65 (Swiss francs)	% Europe	% of CERN budget 1955	% of CERN budget 1965
Switzerland	157,007,323	40.2	3.71	3.19
France	89,398,727	22.9	23.84	18.57
U.S.A.	60,641,784		----	----
F.R. Germany	59,946,190	15.3	17.70	22.74
Italy	27,862,985	7.1	10.20	10.78
United Kingdom	23,414,724	6.0	23.84	24.47
The Netherlands	13,570,734	3.5	3.68	3.92
Belgium	7,716,114	2.0	4.88	3.83
Sweden	5,612,287	1.4	4.98	4.23
Yugoslavia	3,020,921	0.8	1.93	----
Norway	1,542,277	0.4	1.79	1.47
Austria	1,192,014	0.3	----	1.95
Denmark	467,569	0.1	2.48	2.07
Other non-m.s.	428,458		----	----
Canada	381,778		----	----
Spain	25,545	0.0	----	2.18
	452,229,430			

Two factors account for the relative magnitude of the purchases made in Switzerland. Firstly, and particularly in the early phases of the laboratory's life, there were the advantages firms in Switzerland had during the *building programme*. Here the benefits of proximity on keeping prices down were amplified by the Swiss regulations, the preferences of the architect and the first head of the Site and Buildings division (both Swiss), and the interpretation placed on the principle of geographically restricted tendering. The other reason why Swiss enterprises did so well at CERN during our period was because they, like firms in neighbouring France, were well placed to win *noncompetitive contracts*. If instead of the formal mechanism of calling for tenders one relies on informal practices (telephone calls, on-site discussions with representatives, personal contacts, etc.) for clarifying specifications and prices, local firms obviously have a considerable advantage — as has been recently stressed by firms themselves in distant countries.[82] Figure 11.3 confirms the point. It shows that these two member states together won almost 70% of this very lucrative dimension of CERN's business between 1955 and 1965, the distribution shifting from 6:1 in the host states' favour to something like 2:1 as the

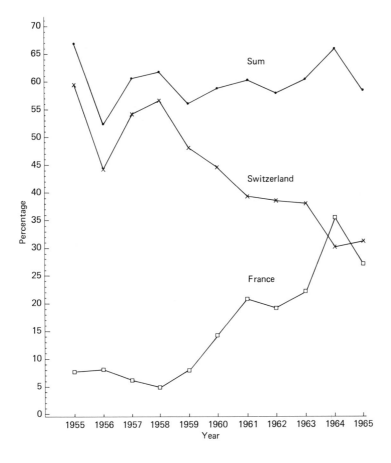

Fig. 11.3 Percentage of noncompetitive contracts awarded by CERN to firms in France and Switzerland between 1955 and 1965.
Source: Quarterly Reports of the CERN Purchasing Office.

Notes: p. 677

laboratory's focus of interest changed from construction to operation; by the end of our period France and Switzerland were each winning about one third of the contracts awarded noncompetitively.

Having spent some time discussing the main beneficiaries, let us now look at some of those countries which won relatively few CERN contracts. The Scandinavian countries and the United Kingdom stand out here: the ratio of percentage by value of contracts awarded to percentage contribution to the budget was about 0.2/0.3 for Norway, Sweden, and Britain, and even lower for Denmark. This was particularly galling for the United Kingdom because it paid more into the budget than anyone else at this time, while its performance was markedly inferior to that of the other 'big three' (the corresponding ratio for France was about 1.0, and those for Germany and Italy were about 0.6/0.7).

Their rather poor performance at CERN was a frequent source of concern to officials in the UK and in Scandinavia. Representatives from the DSIR and, later, the Ministry of Technology (which was set up by a new Labour government in 1964) were frequent visitors to the site, and produced a number of domestic internal reports of what they found.[83] And in the early 1970s the Scandinavian countries made a concerted effort to be compensated for their distance from Geneva, and were in fact successful in forcing a minor revision of the rules for awarding contracts.[84]

The British investigators identified a number of reasons why their industry was not performing well at CERN, factors which can be seen as supplementing the geographical dimension, though of course to what extent we cannot say. For example, they reported that British firms were not keen to tender—either because many of them had full order books and were unable to take on additional work, or because 'they preferred to work to British codes of practices with U.K. labour and without the language problem to cope with', or because they apparently had 'a preference for straightforward jobs especially susceptible to mass production and likely to lead to a repeat order'.[85] It must also be said that the disadvantages of distance were probably amplified by the negative attitude towards their national industry by some senior British members of the CERN staff, who felt that it was unprofessional and incompetent. Adams, for example, has allegedly said that 'When I go around the Continent and want to talk about things that I want in these specifications, I see the managing director who knows all about it. I can talk to him. When I come to England, it's "Oh, see George down the corridor there"'.[86]

As for the Scandinavian countries, in October 1973 their three delegates to the Finance Committee addressed a letter to its chairman suggesting that the costs of transport could not be excluded as one reason for their poor performance at CERN, which had not improved since our period. Although the CERN administration argued, with supporting data, that there was 'no evidence' to suggest that asking for tender prices including delivery costs 'systematically place[d] any country at a disadvantage', the issue continued to be discussed in the Finance Committee for another five years. Finally, in 1979, it was agreed that under certain conditions and within certain limits the contract could be awarded to the firm that submitted 'the lowest FOB tender', i.e. the lowest tender excluding the costs of delivery to CERN.[87]

11.5 Final remarks: the tenacity of CERN's contract policy

To conclude we should like to make a few comments about the tenacity of CERN's contract policy over time, about the fact that, although contract policy has been scrutinized by the Finance Committee on several occasions in the 35 years of the organization's life, no fundamental changes to it have been made.

The main reason for maintaining the competitive policy is not financial — although doubtless it has considerably helped to reduce costs on many items that CERN has bought. Nor is it administrative — as we have seen, at least in our period, very few contracts were actually placed via the formal procedure that it entails. It is kept, we believe, for essentially *political* reasons: it represents an agreement between the member states that no consistent effort be made to distribute contracts between them in relation to their contributions to the budget. The *practical implications* of the 'competitive' policy are thus best understood *negatively,* as meaning that there is no policy of just return in the award of contracts for plant, equipment, and supplies.

If CERN, as an international organization, is able to uphold the principle of no just return so vigorously it is only because domestic authorities and industries in the member states do not, as things stand, see the technology they supply to the laboratory as being of particular importance for national commercial or military interests — or if they do, they ensure that the components in which it is installed (e.g. parts of a detector) are built in national institutions. In fact the recent Review Committee reported that the overall value of CERN contracts placed in the member states and deemed to be 'technically interesting' did not exceed, on average, 60 MSF per annum — so about 7-8% of CERN's present annual budget.[88]

Notes

1. The main source for this chapter was the papers presented by the CERN staff to the Finance Committee, CERN/FC/..., which are held in the CERN archives. Two kinds of material in this collection are particularly pertinent: quarterly reports of the Purchasing Office, which give data on how, why, and where contracts were awarded, and documents laid before the Committee justifying the more valuable awards.
2. The Abragam Committee, so named after its Chairman, was officially known as the CERN Review Committee. It held 14 plenary meetings between September 1986 and November 1987, and its final report was released as CERN/1675 — see Abragam et al. (1987).
3. For these various initiatives by CERN see High Tech Meet. (1974), Schmied (1975) and Bianchi-Streit et al.

(1984). See also Hine & Taylor (1976), Schopper (1985).

4. Some information on the principle of 'juste retour' as it applied in ESRO is provided in Schwarz (1979).
5. For a brief bibliography see the essay by Roland (1985). Marshall & Meckling (1962) give some data on cost-overruns in the US Army based on a survey they did on behalf of the RAND Corporation.
6. We have described the life of the provisional CERN in detail in chapters 7 and 8 of History of CERN (1987). The original set of Financial Regulations proposed by the British is document CERN(WG)4, undated; the DSIR contract procedure is explained in CERN(WG) No.8, undated. These papers were discussed at the first meeting of the Administrative and Financial Working Group held on 30/4-1/5/53 (Minutes CERN (W.G.) (1953) 1st Meeting) where the basic outline of the rules defining CERN's contract policy were laid down. All these documents are in (DG 20534). The policy was then discussed by the Interim Finance Committee on 27-28/8/53 (Minutes CERN (I.F.C.) (1953) 2nd Meeting) (on file CHIP 10025), and referred to the interim Council, who proposed a policy for adoption by the 'permanent' Council at its 7th session held 29-31/10/53 (For Council minutes see CHIP10014).
7. These developments were reported to the first meeting of the Interim Finance Committee on 1-2/7/53. Its minutes are CERN (I.F.C.) (1953) 1st Meeting (CHIP10025).
8. For the needs of the PS Group see CERN/IFC/13, Annex B, 4/12/53. The IFC recommended that they be granted 200,000 SF in January and February 1954 in document CERN/IFC/17, 19/12/53 (CHIP10025).
9. See History of CERN (1987), chapter 7.9. For the arrangement with Bakker see the minutes of the 8th meeting of the IFC, CERN/IFC/46, 30/6/54 (CHIP10025).
10. The document laid before the Finance Committee Working Group which met on 17/12/54 (Minutes are CERN/FC/23, 20/12/54) was CERN/FC/15, undated. For Dakin's discussions see *Note of a Discussion with Prof. P. Preiswerk,* 16/12/54, and *Note on Contract Policy,* 15/12/54 (for Adams' views). Both are in (DIRADM20060). Adams's ideas are also expressed in his *Note on the Purchasing of Equipment,* CERN-PS/Admin.15/JBA, 24/9/54 (DIRADM20393).
11. The Finance Committee working group was guided by a British paper objecting to CERN/FC/15—see *U.K. Comments on Draft Memorandum on Contracts Procedure,* 4/12/54 (DIRADM20060). The paper which emerged from the deliberations on 17 December (for minutes see note 10) was CERN/FC/15 Rev., 12/1/55. It is summarized in CERN/FC/25, 12/1/55. See also CERN/FC/15 Rev. Add., 4/2/55. The minor objections raised by the Finance Committee at its meeting on 25/1/55 are detailed in the minutes, CERN/FC/33, 10/2/55.
12. The delegates to CERN had always accepted that there would be many cases in which competition would be limited or impracticable—see the *UK Guiding Principles for the Selection of Firms and Treatment of Tenders,* CERN/IFC/24, 18/2/54 (CHIP10025), for example.
13. CERN/FC/25, 12/1/55, and CERN/FC/15 Rev. Add., 4/2/55, give more detail on the kind of information the quarterly reports should carry.
14. The quotations are from the material cited in note 11.
15. There is one key source which has been used throughout this section: it is *Adjudication of the Contract for the Supply of the Magnet Blocks for the CERN Proton Synchrotron,* CERN/FC/144, 14/9/56, which had an introduction by John Adams to a major report by Colin Ramm and Cornelis Zilverschoon. We shall only refer to it below in exceptional cases.
16. Preliminary information on what was required for the PS magnet blocks was provided in CERN/94, 3/6/54. For the technical alternatives described here see Zilverschoon's report to the 50th PS staff meeting, 14/12/54 (informal minutes in PS Lib 4).
17. For this paragraph see Zilverschoon, *ibid.*
18. The evolution of the list can be followed in CERN/FC/43, 13/4/55, CERN/FC/44, 13/4/55, and CERN/FC/57, 4/7/55. 44 firms in all were asked whether they were interested in submitting offers.
19. This paragraph is based on Ramm's report to the 67th PS Staff Meeting, 18/10/55 (PS Lib 5).
20. See particularly CERN Annual Report, 1955, 16.
21. The quotation is to be found in CERN/FC/144, 14/9/56, 11.

22. Neumann's comments are in a *Note de la delegation francaise au sujet de l'adjudication de fourniture des blocs de l'electro-aimant du synchrotron a protons* dated 1/10/56 (DIRADM20251).
23. The (provisional) minutes of the 10th meeting of the Finance Committee are document CERN/FC/158, undated. The meeting passed a resolution proposed by Mr Neumann authorizing the procedure described in our text; it is CERN/FC/155, 4/10/56.
24. The document proposing the award of the contract to the Ansaldo San Giorgio/Cornigliano combination is CERN/FC/144/Rev., 29/10/56, which was discussed at the 11th meeting of the Finance Committee, (provisional) minutes CERN/FC/170, undated.
25. The quotation is from the minutes of the 76th PS Staff Meeting, held on 20/3/56 (PS Lib 5).
26. CERN bought the araldite from CIBA on behalf of Ansaldo, and arranged for it to be imported into Italy free of customs duties—see CERN/FC/144/Rev., 29/10/56, and memo Schou Olsen to Tièche, 17/4/59 (DIRADM20251). In similar vein, in August 1956, when it looked as though Ansaldo might have to import its steel from outside Italy, Adams asked Italian delegate Pennetta whether it would be possible to arrange for the blockmaker not to pay customs duties on the purchase—see letter Adams to Pennetta, 1/8/56 (JBA22594).
27. See the minutes of the 83rd PS Staff Meeting, 6/11/56 (PS Lib 5).
28. The importance of the impression a firm made on its CERN clients can be seen from Adams' attitudes towards British contractors, about whom he was particularly scathing. One of them effectively ruled out its chances because when asked for proposals on how they would go about building the magnet they did 'not put any work into the inquiry and, therefore, had no ideas at all on how they would go about doing the job'. He condemned them roundly for their 'lackadaisical attitude'— see letter Adams to Verry, 27/7/55 (DIRADM20249). See also the interview of Sir Ben Lockspeiser by Margaret Gowing, 5/6/73, at page 12 (CERN).
29. For a useful bibliography of material on this see Pestre (1988) at note 58.
30. Rosholt (1966), 61-5, from which all the quotations in this paragraph are taken.
31. In Abragam et al. (1987) we read that '*various improvements* [to contract procedure] *were suggested by a large number of firms. The most frequent was that CERN should conduct thorough-going technical negotiations with some of the firms while the tenders were being evaluated* so as to take better account of their technical competence, reputation, credibility and experience, and not merely the amount of their bids' (emphasis in the original).
32. Some examples of firms feeling badly treated by CERN which we have come across are the steelmaker who felt that he should be paid for the samples he had made (see letter Adams to its Director General, 28/2/57 (DIRADM20251)), the manufacturer of electronic racks who felt that CERN was using his patented designs (see memo of a discussion by Verry with the manager, 16/6/59 (SERC-NP13A)), and the manufacturer of pulse height analysers who insisted that CERN had ignored his device in preference for one produced in the United States (see letter Hine to de Rose, 23/1/64, and other documents in DIRADM20393).
33. He made this remark in the letter quoted in the previous note. This is still a bone of contention with CERN's suppliers. Abragam et al. (1987) recommend that 'firms agreeing to bear the costs inherent in any technological development should be given a minimum guarantee of being awarded the subsequent series production contract' (at p. 37).
34. This concern was clearly expressed by firms in their submissions to the Review Committee—see Abragam et al. (1987).
35. The information on Cornigliano's attitudes is from CERN/FC/144, 14/9/56, 15.
36. For information on the two consultants used by Kowarski and Preiswerk, see report CERN/L. Rep. 5, 11/1/53 (LK22448).
37. The plan dated 12 Jan. 1953 is document CERN/L. Rep. 4 in (LK22448).
38. The quotation is from the minutes of the first meeting of the IFC, CERN (I.F.C.) (1953) 1st Meeting, 9/7/53 (CHIP10025).

39. The Executive Group's recommendations concerning the appointment of Steiger are spelt out in paper CERN/IFC/34, 19/3/54; a copy of Steiger's contract was circulated as CERN/IFC/43, 17/6/54 (CHIP10025).
40. For Pennatta's opinion on Steiger see the minutes of the first IFC meeting (cited in note 38) and those of the 6th meeting, CERN/IFC/33, 18/3/54 (CHIP10025).
41. The activities of these various groups are described in more detail in chapter 9.2.1.
42. For the changes in the role of SB see CERN Annual Report, 1961, 116; see also Mallet's intervention at item 6 of the 31st Finance Committee meeting (minutes CERN/FC/413/Draft, 28/3/60). CERN/FC/714, 15/9/64 and CERN/539 and CERN/539/Corr., 14/5/64 set out proposals for the role of the division in building the ISR. The idea of limiting Steiger to a consultant on aesthetics is mentioned in CERN/DIR/91 Annex, 20/11/61 (DG20601).
43. For biographical details on Mallet see CERN/317, 24/6/59, and for Webb see the after-dinner speech made on 17/11/61 on the occasion of his departure from CERN, *Discours de M. Rouel...* on (DG20601).
44. The critical comments on Steiger were made by Rouel on the occasion referred to in the previous note. That he should have done so at a farewell dinner indicates that he was confident that his views were widely shared by his CERN colleagues—and that Steiger and, more interestingly, Preiswerk were not invited.
45. The steps taken by the Executive Group in the interim CERN to persuade the delegates to the IFC and the Council to start construction can be followed in letter Amaldi to Lockspeiser, 7/4/54 (DG20927), minutes of the 7th meeting of the IFC held on the same day, document CERN/IFC/40, 21/4/54 (CHIP10025), and the minutes of the 9th Council session, CERN/GEN/14, 30/6/55 (CHIP10014). The information about the construction of the roads is from a *Report on the Beginning of Construction Work at the Site at Meyrin* which was presented to the Executive Group on 17/5/54, document CERN/LZ/, in (DG20600).
46. See *Report on the Construction Work at the Site at Meyrin* from Preiswerk to the Executive Group, CERN/L/17.6.54-2, 17/6/54 (DG 20600).
47. These quotations are from the minutes of the 8th meeting of the IFC, document CERN/IFC/46, 30/6/54 (CHIP10025).
48. Letters Amaldi to Preiswerk, 12/7/54, and to Steiger, 13/7/54 (DG20548).
49. *Ibid.*
50. See a memorandum annexed to Amaldi's *Weekly Report No.78,* 30/7/54 (CHIP10029).
51. The detailed information in this paragraph and the next is provided in Preiswerk's Report on the Construction Work at the Site to the Executive Group, CERN/L, 24/9/54 (DG20600).
52. The minutes of the 9th and last meeting of the IFC are document CERN/IFC/51, 20/10/54 (CHIP10025), from which the quotation is taken. By 25 September one Belgian and four Italian firms had submitted offers for the SC upper-structure. Unfortunately we do not know what the situation was when the IFC considered the tenders received ten days later.
53. The document is entitled *Building Contracts Programme for 1955 and Reports on the Procedure Adopted for Tenders and Adjudications,* CERN/FC/39, 2/3/55. This is the date on the cover sheet. Some documents in the file are dated 9/3 or 10/3/55.
54. The above material is contained in the document cited in the previous note. The number of firms that wished to be invited to tender for the PS building is given in 'Adjudication No 12', *Report on the Procedure and the Results of Tenders for the Excavation of the First Lot of the P.S. Building,* which is part of CERN/FC/39.
55. The minutes of this meeting are CERN/FC/42, 25/3/55.
56. See Adjudication No 26, 'Adjudication of the Reinforced Concrete Work for the Proton Synchrotron Building', in *Reports on the Procedure Adopted for Tenders and Adjudications for the Construction of the Buildings at Meyrin,* CERN/FC/48, 8/6/55.
57. The minutes of the Finance Committee at which this award was discussed (the fourth, on 8/6/55) are CERN/FC/55, 4/7/55.
58. The contract is reported as Adjudication No 16, *Rapport sur la Procedure Adoptee pour la Commande de*

la Construction d'un Modele d'Essai d'une section du batiment Annulaire du Synchrotron a Protons, dated 2/3/55, in CERN/FC/39, 2/3/55.

59. Letter Adams to Verry, 13/10/55 (JBA22594). The argument that the use of the Italian consortium would involve delays actually cost them another contract, that for the reinforced concrete work for the power house. Gandini et al.'s offer was some 15% below Zschokke et al.'s (CERN/FC/75, 26/9/55). At the 5th meeting of the Finance Committee Preiswerk remarked that 'no valid technical reason existed for favouring one or other of the two firms', and gave the 'additional delay' of the Italian consortium as the reason for wanting to award the work to their more expensive competitor (CERN/FC/92, 24/10/55).

60. These attitudes run throughout Adams' letters to Verry on 13/10/55 and 14/10/55, in the second of which he proposed withdrawing the contract to Gandini et al. (JBA22594).

61. The data presented in the rows of this table are, respectively, from CERN/FC/48, 8/6/55 and CERN/FC/214, 17/6/57 (PS ring), CERN/FC/75, 26/9/55 and CERN/FC/149, 18/9/56 (power house), CERN/FC/48, 8/6/55 and CERN/FC/190, 27/5/57 (PS halls), CERN/FC/75, 26/9/55 and CERN/FC/194, 27/5/57 (main workshop), CERN/FC/148, 18/9/56 and CERN/FC/189, 18/4/57 (air conditioning), CERN/FC/182, 13/12/56 and CERN/FC/191, 23/4/57 (administrative building).

62. The practice of placing contracts on the basis of sketchy plans was mentioned by Steiger at the 14th meeting of the Finance Committee (CERN/FC/212, 18/6/57). The plans for the PS ring were drawn to 1:100 (CERN/FC/214, 17/6/57), those for the halls at 1:200 (CERN/FC/190, 27/5/57).

63. The concern about the foundations of the PS magnet and the evolution of its design can be found in CERN Annual Report, 1955, 18, and 1956, 46, and in CERN/FC/214, 17/6/57. At the 90th PS staff meeting Mervyn Hine remarked on the need to design the machine 'to work with reasonable certainty', meaning that 'at that stage of knowledge playing safe about apertures, foundations, etc. and trying to make the machine as perfect as possible, i.e. not relying on correcting devices as an essential part of the operation'; Minutes of CERN-PS Staff Meeting, 5/3/57 (PS Lib 5).

64. The details of the changes are described in CERN/FC/189, 18/4/57. Steiger described the role of Tunzini's engineers to the Finance Committee at its 13th meeting (CERN/FC/198, 10/5/57).

65. Preiswerk presented his case to the 12th meeting of the Finance Committee (CERN/FC/182 — provisional minutes, undated). If we judge Preiswerk harshly, it is because within a matter of months the cost of the contract doubled after 'detailed specifications were drawn up'. This suggests that, deliberately or otherwise, he gave unrealistic information to the Finance Committee who thereby treated the work as an 'extention' of the Italian consortium's existing contracts.

66. The SC and PS awards for the Vevey firm are Adjudication 14 in CERN/FC/39, 2/3/55 and Adjudication 21 in CERN/FC/48, 8/6/55, respectively. Other data on this firm is from the quarterly reports, from which one can also learn of the additional work given to Gandini et al., Locher et al., and Zschokke et al.

67. Much was made of some of these practical advantages at the 14th and 15th meetings of the Finance Committee (CERN/FC/212, 18/6/57 and CERN/FC/223 Draft, for the meeting on 27/6/57). See also CERN/FC/459, 24/11/60.

68. See Table 11.2 hereafter.

69. For the architect's preferences see particularly the minutes of the 6th meeting of the Adjudication Board, 29/2/60 (DG20548). An official in the British DSIR who studied the quarterly reports carefully complained to Verry that he detected a 'consistent policy of splitting orders to evade financial limits' and claimed that the architect's awards contained 'particularly cynical examples of it' (letter Smith to Verry, 19/8/58 (SERC-NP39A), and several other letters by Smith in the same box). Preiswerk's spirited rejection of this allegation is reported in letter Verry to Smith, 23/10/58 (SERC-NP39A). For a possible example of the use of this tactic by the administration see CERN/FC/279, 5/5/58 and its Add.1, 16/6/58, which show that a single order for cables worth about 545,000 SF was split into two; this particular case was raised by Verry at the 20th meeting of the Finance Committee (minutes CERN/FC/290, 30/5/58).

70. The idea that a measure of 'restricted competition' was acceptable for construction work on the site gained ground at the 8th meeting of the IFC, 23/6/54 (CERN/IFC/46, (CHIP10025)). Soon thereafter the

676   *The contract policy with industry*

71. The argument that it was not worth tendering was used by Jean Richemond, the Director of Administration, at the 14th and again at the 15th meetings of the Finance Committee, minutes CERN/FC/212, 18/6/57, and CERN/FC/223, undated. Preiswerk made a similar claim at the 14th meeting.
72. The first attempt to group a major complex is reported in CERN/FC/425 and its two addenda, dated from 7-13/6/60. A procedure extending the value of tenders to 4-6 MSF and the duration of the work to 12-18 months is outlined in CERN/FC/608, 27/2/63; a proposal to push the value to 10-12 MSF and the duration to about two years is made in CERN/FC/711, 28/8/64. The work to be done for the ISR is set out in CERN/FC/778, 31/5/65.
73. The references for this point may easily be identified among the sources given in the following note.
74. The data in this table is extracted, row by row, from the following sources: CERN/FC/409, 7/3/60 for the PS third wing, CERN/FC/425/Add.1, 10/6/60, for the East Experimental Area, CERN/FC/453, 27/10/60, for the Accelerator Research building, CERN/FC/459, 24/11/60, for the east junction, CERN/FC/523, 8/12/61, for the east apron cover, CERN/FC/608, 27/2/63, for all the contracts awarded in 1962, CERN/FC/616, 6/5/63 for the NPA extention and the substation, CERN/FC/677, 3/4/64, for the roads and parking areas, and CERN/FC/711, 28/11/64, for the accelerator research hall.
   For the arguments for awarding the east junction without tendering to Zschokke et al., see CERN/FC/459, 24/11/60. The debate over the roads contract was held at the 58th meeting of the Finance Committee (minutes CERN/FC/687/Draft, 14/5/64) with supporting document CERN/FC/677, 3/4/64.
75. Austria also touches Switzerland, of course; that it apparently did not benefit as much as the other neighbours is, at least partly, an artifact of our data, which covers 13 years for France, Italy, and Germany, and only some 5 years for Austria. That consideration apart, joining in 1959, when the first building phase and many major engineering contracts had already been placed, also disadvantaged Austrian firms, which had to break into an already well-established framework.
76. It is difficult to get hard and fast evidence for this point precisely because the geographical range of tenders was frequently restricted, and even when it was not Swiss firms were more inclined to submit tenders than those elsewhere. However, for the excavation of the PS area, for example, 43 firms were invited to tender, the bulk being Swiss and French. 14 offers were received, 6 from Swiss firms. The Swiss offers varied by a factor of two, from some 160,000 SF (the lowest offer) to 331,000 SF (10th out of 14); within that range we have offers from firms in Germany and Italy, the best being 240,000 SF from Guffanti & Co. in Milano.
77. The point that 'newcomers had quoted lower prices than certain firms which had already worked for CERN or were still on the site' was made by both Neumann and Verry at the 31st meeting (minutes CERN/FC/413/Draft, 28/3/60). The suggestion that a new set of contracts be awarded noncompetitively to CMF was made by Dakin at the 50th meeting (minutes CERN/FC/611/Draft, 25/3/63), and refused. The results of the subsequent tender opening are reported in CERN/FC/616, 6/5/63.
78. The remarks about conditions in the European building industry after the war are taken from Laqueur (1982), 253 et seq.
79. The award for the PS linac is reported as competitive in the report for the second quarter of 1955, CERN/FC/70, undated. CERN/FC/3, 8/11/54, explains that this is to be a development contract placed with Metropolitan-Vickers 'without seeking competitive tenders'. For an example of when a repeat order is classified as competitive, see the contract for three additional power supplies for feeding the beam transport magnets. CERN/FC/723, 8/10/64, argues that it should be given to Siemen's, who had already supplied five of these devices, without calling for tenders; CERN/FC/756, 17/3/65, covering the last quarter of 1964, reports this contract as competitive. In the late fifties such repeat orders were included among the 'supplements' to contracts originally awarded competitively, and were counted as competitive — see, for example, CERN/FC/420, 3/6/60, for the last quarter of 1959.

At the 70th meeting of the Finance Committee (CERN/FC/799, 27/10/65), the British delegate registered his concern at the fact that 'a large majority of the purchases reported were non- competitive'. He was however comforted by the thought that many of those falling into categories B and C (see Figure 11.2) were 'running orders, which had been competitive in the first instance, and also unavoidable repeat orders for equipment that had originally been chosen competitively'. He asked that henceforth these categories be distinguished from 'the fully competitive category' and from 'genuinely noncompetitive' orders. The results can be seen in the report for the last quarter of 1965, CERN/FC/831, 13/12/65.

80. For Financial Rule 10(b) dealing with contract policy, see CERN/80/Rev.3, 1/2/55.
81. The data in the table are taken from CERN/FC/869, 10/6/66 (for column 2), and from the CERN Annual Reports, 1955, 1965, for columns 5 and 6.
82. Thus in Abragam et al. (1987), p. 138, we read that 'Firms in distant countries emphasized the advantage Host State firms enjoy in being close to CERN. Physical distance complicates and renders more costly the direct contacts which many firms considered essential to a proper grasp of the specifications'.
83. These accounts are to be found in various boxes at the SERC archives. Particularly useful are a letter from Melville to Turnbull in the Office of the Minister of Science, 15/2/60 (SERC-NP9), and a *Report of a visit to CERN made by Mr P.M. Lavelle, Mintech, and Mr. J. Hutchinson, S.R.C. on 20th and 21st April, 1967* (SERC-NP91).
84. For the Scandinavian letter see Annex 1 to CERN/FC/1701, 10/4/74.
85. For the unwillingness to tender, see letter Copisarow to Viscount Hailsham, 6/2/60 (SERC-NP9), report on a visit to Schou Olsen by Muston, 20/1/61 (SERC-NP13A), and Melville's letter to Turnbull (note 83). Dakin, in letter to St. J. Walker, 22/5/63 (DIRADM20393), produced data showing that British firms refused 25 out of 33 invitations to tender for a total of eleven major engineering projects.
86. Adams' views were attributed to him by Lockspeiser in his interview with Gowing—see note 28. The general point has been confirmed by J. Freeman (private communication).
87. The history of the discussion in the FC on the effects of transport costs on adjudications is summarized in CERN/FC/2302, 15/10/79, while the procedure agreed on can be found in CERN/2221/Rev., 1/2/79.
88. The figure is given in Abragam et al. (1987), 34.

CHAPTER 12

The second generation of accelerators for CERN, 1956—1965: The decision-making process[1]

Dominique PESTRE

Contents

12.1 New acceleration principles and future accelerators: a chronological account of the evolution of the field, 1956-65	683
12.1.1 The 1956 International Conference on accelerators	684
12.1.2 The situation in autumn 1959	689
12.1.3 The American swing to very high energies, 1959-61	692
12.1.4 Other approaches considered between 1959 and 1961	696
12.1.5 The big choices, 1961-65	697
12.2 The time of new ideas: the CERN Research Group on Accelerators, 1957-1960	699
12.2.1 The activities of the Accelerator Research Group from 1957 to 1959	700
12.2.2 The Accelerator Research Group in 1960: towards Storage Rings for the CERN PS	703
12.3 The tempestuous period: the first important year of the debate between scientists, 1961	706
12.3.1 The proposal put forward by the builders, December 1960-April 1961	706
12.3.2 The pendulum swings back, April-July 1961	709
12.3.2.1 The SPC meeting on 29 April	709
12.3.2.2 Arguments for and against the proposal to build the ISR	712
12.3.2.3 The SPC meeting in July 1961	715
12.3.3 Why the conflict between 'Builders' and 'Physicists'?	716
12.3.4 The consequence of the American summer, September-November 1961	719
12.3.4.1 The American 'shock'	719
12.3.4.2 The concrete decisions taken at CERN at the end of 1961	722
12.3.5 A global schema for understanding the decision-making process in 1961	726

12.4 The calm after the storm: the emergence of CERN's
'New Study Group on Accelerators', December 1961–January 1963 — 729
   12.4.1 Setting the scene — 729
   12.4.2 The activities of the CERN Group: long-term projects,
the electron model, the 'defense' of the ISR — 730
   12.4.3 Towards the creation of the European Committee for
Future Accelerators, ECFA — 733
12.5 The European high-energy project: ECFA's first report,
January–June 1963 — 736
   12.5.1 Mobilizing the community — 736
   12.5.2 The significance of the report and the reasons for the choice — 739
   12.5.3 The way scientists argue — 742
12.6 The 'selling' of the ISR to the Member States and the postponement
of the 300 GeV project by the Council, June 1963–December 1965 — 745
   12.6.1 The problem of priorities, summer 1963–spring 1964 — 745
      12.6.1.1 The new framework of the debate — 746
      12.6.1.2 Some of the positions in evidence at the time — 747
   12.6.2 A decisive moment: the first steps by the SPC and the Council
towards having 'only' the ISR, June 1964 — 750
      12.6.2.1 The SPC debate on 2 June — 750
      12.6.2.2 The first steps taken by the Council towards having 'only' the ISR — 752
   12.6.3 The Council makes its final choice, June 1964–June 1965 — 753
      12.6.3.1 The Director-General defines respective roles and powers — 754
      12.6.3.2 The vote on the supplementary programme,
the Intersecting Storage Rings, by the Council — 755
   12.6.4 The physicists' last stand, June–December 1965 — 756
12.7 Conclusions — 759
Annex 1 — 763
Annex 2 — 764
Notes — 765
Bibliography — 809

In 1952 the provisional organization decided to equip CERN with a 25 GeV strong-focusing proton synchrotron, the CERN PS. Seven years later, towards the end of 1959, the machine produced its first proton beam meeting the design specifications. By then the question of what the next generation of accelerators should be was already on the agenda: indeed it had been of interest since 1956. What we want to do here is trace the history of the first ten years of this debate, up to the official decision of the CERN Council to construct the Intersecting Storage Rings (ISR) in December 1965.

There are several ways of tackling this kind of history. One of the most prevalent is to take as reference point *one* of the machines that was actually built (the ISR, for example, or the 400 GeV synchrotron) and to work backwards, extracting from the flux of events those which seem to lead to the final outcome.[2] The ensuing account is often grossly anachronistic — most of the arguments and the problems are perceived, not in their *context*, but in the light of the result — and tends 'spontaneously' to hagiography — more often than not the search for antecedents and precursors goes hand in glove with the need to allocate credit. This is a kind of history which is popular in scientific institutions, it is usually a history in awe of Science, it is often the story of a great technical and human adventure whose successive results are treated as if they were objective facts being steadily accumulated in a linear fashion.[3]

Another current way of writing this kind of history is to treat the process above all as a politico-financial saga, as a conflict between powerful and unscrupulous personalities and groups. Here the objective light of reason is dimmed by the cold and piercing cynicism of interests, and scientists and politicians share roles in a game of mutual deceit. Here there is no longer any question of taking beliefs and arguments seriously, all is attributed to deliberate manoeuvering, and the reader is left pitying the naive who are always the losers at the expense of the more cunning. Here then we do not have an edifying account of how Scientific Reason progresses, but an account — edifying in its own way, one must admit — which teaches that in science, like everywhere else, one generally finds the same motives at work, those of a perpetually villanous human nature.[4]

There are more fruitful ways of proceeding than either of these. To date they are to be found in studies of political decision-making processes — those of Allison on the Cuban missile crisis or the pioneering work of Lindblom, for example —, in analyses of decisions in the heart of complex organizations — those of Crozier, March and Olsen, Gernoux, Jamous, etc. — or of decisions on military technology — amongst which we could mention Armacost, Morris, Schilling, or MacKenzie and Spinardi. These approaches have however been very rarely used in studies of decisions concerning big scientific equipment though we could cite Darmon and Lemaine, Edge and Mulkay, Jurdant and Olff-Nathan.[5] What these studies show is the importance of knowing thoroughly the contexts in which the actors are situated; the importance of the time dimension in any decision; the need not to define in advance what will count in the end, what will finally be decisive; the interwoven nature of explanatory factors which one is tempted to treat as if they were quite different from one another — the intellectual and the material, the scientific and the social, the technical and the professional, etc. As Rudwick has said of scientific knowledge — and we can generalize this for our purposes — one must treat it as 'both artifactual and natural, as a thoroughly social construction that may nonetheless be a reliable representation of the natural world'. Our analysis must thus focus on the network of actors, on what they were not aware of as well as on the strategies which they developed in the light of their 'intellectual' or 'material' interests, on the succession of local microdecisions which together made up the process in its entirety, on the adjustments and the reinterpretations of 'what was at stake' which the main protagonists constantly made.[6]

A last brief comment before getting under way. In what follows we present an historical monograph, a case study. The narrative mode, which demands chronological precision, will thus be dominant, though we shall interrupt it regularly to present some remarks and thoughts of a more general kind. As the aim of the study is to follow the events in all their richness, we have chosen the widest possible variety of documents. We have taken into consideration the published literature — scientific journals, manuals... —, the 'grey' literature which circulates between groups of specialists, the organizational literature — annual reports, minutes, proposals and even tape recordings of various meetings — and, finally, the most private types of documents, like letters. Our resulting documentation was formidable and it was obviously impossible to present it in its entirety. To have done so would not only have been wearisome for the reader; most likely he would not have been able to see the wood for the trees. We have therefore adopted a mode of exposition which combines different levels of precision, different degrees of enlargement, different time-scales.[7]

In most of the text we have kept our 'reading' of events at a constant level; this permitted us to follow the interactions between all the main actors but not to study them individually; in some places on the contrary, when the situation demanded it, we went deeper into our documentation to consider the process of formation of scientific 'micro'-judgements, of 'micro'-decisions, of 'micro'-consensus. One of those situations arose between early 1961 and the autumn of the same year; we have studied that year in depth because not only were these months crucial to an understanding of our topic but also because the documentation was extremely good: CERN's research director was in the United States at the time, receiving treatment for an automobile accident, and a copious private correspondence complements the public documents.

The chapter is structured as follows. In the first part we present in broad outline the debate on future accelerators which took place among the international physics community. Our account, which is based essentially on published material, is rather brief and is intelligible to non-specialists; some additional information is presented in the notes and can be skipped by the lay reader. The main reason for this section is that until about 1961 Europe and America were out of phase; Europe was to some extent *dependent* in high-energy physics and as regards new acceleration principles. Certain aspects of the American situation and of the technical solutions put forward across the Atlantic were thus needed; they more or less shaped the perception of the field, and partly determined the way of thinking and of reacting in Europe.

In the second section we study the beginning of the debate at CERN, from 1956 to the coming into operation of the CERN PS and the Brookhaven AGS (autumn 1960, roughly). What we stress here is the variety of technical solutions investigated and the formation of a nucleus of specialists at CERN — the Accelerator Research Group.

In the third section we turn our attention to the eight to ten month period we mentioned a moment ago (March - November 1961): it is the period in which the first project was submitted for approval, a project elaborated in the group — now division — charged with accelerator research; it is also the period in which the first confrontation occurred between scientists, a confrontation of which the terms rapidly shifted in September 1961 in the light of results achieved in the United States during the summer. This period is interesting because it enables us to study in detail a discussion which took place only between scientists about a project which, at this time, they were not primarily concerned to 'sell' to the politicians.[8]

We continue our account in the fourth and fifth sections of this chapter at a somewhat coarser-grained level. Here we present the second stage of the debate between scientists. It is dominated by the creation of ECFA, a committee of physicists which brought together the élite of the European high-energy physics community. ECFA's aim was to spell out *the* overall policy for Europe. This period closed in the summer 1963 with the presentation, 'from a strictly scientific point of view', of the wishes of the Europeans.[9]

The last section of the chapter deals with the period mid-1963 to the end of 1965, which was as decisive as 1961. The representatives of the European member states — held until then at arm's length — now entered the picture. This was the time when the scientific projects were 'sold' to the Council members and to the politicians in the CERN member states.

12.1 New acceleration principles and future accelerators: a chronological account of the evolution of the field, 1956-65[10]

This study will not be exhaustive. It will simply supply the minimum information needed for a proper understanding of the European decision-making process. Some

Notes: p. 765

important parts of the research conducted between 1956 and 1965 are thus not presented (electron machines, for example) or are only touched on (like plasma accelerators). The choice has been made in the light of what is needed in the following sections of the chapter. We should add that the American situation is dealt with more thoroughly than that in the Soviet Union, not simply because of the dominant role played by the United States but, above all, because European scientists had more contact with their colleagues across the Atlantic than with those in the USSR.[11]

12.1.1 THE 1956 INTERNATIONAL CONFERENCE ON ACCELERATORS

An opportune point at which to begin our chronological survey is the symposium on 'high energy accelerators and pion physics' held at CERN from 11 to 23 June 1956.[12] It was during this conference that the CERN management gained the impression that the organization 'had fallen behind in machine design'. True, as far as the conception and realization of classical synchrotrons and strong-focusing synchrotrons were concerned, 'Adams and his team had made highly pertinent contributions to the debate'. However when new principles were involved 'the discussion was restricted to the two big ones [USA and USSR], plus a few questions raised by the English'.[13] Hence the deep concern which gripped CERN once the conference was over — 'we not only had done little work', remarked a document laid before the SPC in October, 'but it was even difficult to follow the work that other groups had carried out' — and the action subsequently taken — the setting up of a special research group to keep in touch with new ideas.[14]

This is not the only reason for beginning our account at this point. Another is that this symposium was the first in a series of international conferences on high energy accelerators, conferences which, from 1959 onwards, alternated with the Rochester conferences. Finally, the years 1955-57 were particularly rich in new ideas: in fact, they initiated a period of intense theoretical and practical activity dedicated to the design of new accelerators.[15]

In trying to imagine what a new accelerator would be like, physicists and accelerator builders have generally to choose, roughly speaking, between two main options. They can continue along the lines of the existing technologies — for example, by envisaging the construction of a proton synchrotron whose energy or intensity would be stepped up tenfold —, or they can try new principles — like putting into practice the idea of phase stability and transforming a cyclotron into a synchro-cyclotron. In the first case they develop 'an existing species' — to use John Adams' metaphor —, in the second they create a new one, competition between them being resolved by the 'survival of the cheapest'.[16] That being said, it is striking that the dominant view at the 1956 conference was that the 'strong-focusing proton synchrotron' species, which was soon to emerge at CERN and at Brookhaven, was not the species most likely to survive in the future; strong-focusing machines of 300 or 1000 GeV would, technologically and financially speaking, be monsters. Put differently, at this time there was an implicit conviction that one or more new principles were going to be developed, and that what had happened in the 30s with the cyclotron, at the end of the 40s with the synchro-cyclotron and the weak-focusing synchrotron, and in 1952/53 with strong-focusing synchrotrons, would be repeated in the near future.[17]

The three most interesting new accelerator concepts were put forward in the session entitled 'New ideas for accelerating machines'. One of these — that for the so-called FFAG synchrotron — had already been discussed in the literature for two or three years. The two others, by contrast, were entirely novel.

The first group of suggestions was made by the Soviets (notably Budker) who proposed that an annular plasma be used to guide the protons one wanted to accelerate. In this system the magnetic field inside a narrow ring of high-intensity electrons is used to steer the particles which can then be accelerated to high energies. The advantage lies in the powerful guiding field which one obtains, and which enables one to reduce the accelerator's radius of curvature proportionately. According to Adams, 'an accelerator of a mere 3 metre radius could be made to hold 100 GeV protons in the orbit'. Certainly both the stability of the toroidal plasma itself, and the means of accelerating the protons within the plasma, were challenging problems and 'no one was expecting rapid confirmation of the principle in all its aspects'. But the enthusiasm of the theoreticians for the idea was such that people hoped for 'notable advances in the next year'.[18]

The second important idea debated at length in Geneva was that put forward by the MURA (Midwestern Universities Research Association) accelerator research group. The essential feature of their device — the Fixed Field Alternating Gradient accelerator (FFAG) — was the use of a magnetic field which did not change with time. This was achieved, *inter alia*, by a special configuration of the magnet poles. Compared to a classical synchrotron, Judd explained,

> [it is] no longer necessary to pulse the magnet current in synchronism with the increasing energy of a batch of particles being accelerated, so as to keep the batch always near to the radius corresponding to a circular orbit; rather, the magnetic field could be fixed in time while particles could be accelerated at a rate depending only on how rapidly energy could be supplied to them.[19]

The main interest of this system was the accumulation process involved in the FFAG principle, and the associated possibility of reaching *intensities* as much as 1000 times greater than those expected at the time with accelerators of the CERN PS type. The disadvantage lay in the cost, which was also several times greater than that of a CERN-type PS of the same energy (a few years later a 10 GeV FFAG accelerator was estimated to cost $100M, against about $30M for the Brookhaven proton synchrotron, the 30 GeV AGS).[20] The only papers on this subject at the conference were those presented by the MURA physicists, the leading light among them being

Notes: pp. 765 ff.

K.R. Simon.[21] Their ultimate goal was to construct a proton synchrotron of 20-50 GeV, i.e. in the energy range of the machines then being built. This was a particularly attractive proposition at a time when the main concern was to improve *intensity* and very few people were interested in going to energies above 50 GeV. Scientists had now realized that, with the intensities available at the Cosmotron and the Bevatron, one needed long experimental runs and the statistical precision of the results was often limited. Physicists thus began to ask for machines with 'denser' beams, machines which could provide much more refined and detailed data, particularly on the interactions of known strange particles.[22]

The third approach to the construction of new accelerators which interested the conference was the practical possibility of colliding particle beams. In an accelerator like a synchrotron (the CERN PS for example), the accelerated particles interact with a target fixed in the laboratory frame of reference. The energy available during the collision is that measured in the centre of mass of the system target nucleon/incident particle. It thus only increases with the square root of the accelerator energy, the target being at rest. On the other hand, if two particles of the same energy moving in opposite directions collide head-on, the available energy is proportional to the sum of the two energies measured with respect to the laboratory system. Put simply, and to paraphrase Kerst and his colleagues, two accelerators of some 22 GeV arranged tangentially in such a way that the proton beams collide head-on at the point of contact, are 'equivalent' to one machine of 1000 GeV in which the protons hit a fixed target![23] That said, one must not forget that the 'equivalent' *intensity* of a two-beam collider is very much lower than that of an ordinary proton synchrotron. The number of interactions between the beams is in effect considerably reduced, since their density is in no way comparable to that of a solid or liquid target. Until early 1956, then, no one seriously regarded colliders as being of any practical interest.

One of the first ways of getting around this problem was once again devised by the MURA group — which was only logical, since an FFAG machine was expected to deliver high-intensity beams. On 23 January 1956 ten members of MURA sent a note to *Physical Review* suggesting that two 'fixed-field alternating-gradient accelerators could be arranged so that their high-energy beams circulate in opposite directions over a common path in a straight section which is common to the two accelerators' (see Figure 12.1). Believing that 5×10^{14} particles per machine would be necessary to obtain a physically interesting rate of interaction during the collision, they concluded that this type of arrangement seemed, at first sight, to be feasible.[24] At the same time Crawford and Stubbins showed experimentally that it was 'possible to store the charge of from five to seven beam pulses' in the 184-inch Berkeley synchro-cyclotron. They thus also confirmed that one could probably construct 'a ring of protons by successively bringing up several groups of particles to the same final energy by frequency modulation', so augmenting the beam intensity. The notion of beam accumulation — with its corollary: the possibility of achieving collisions between beams — was taking shape.[25]

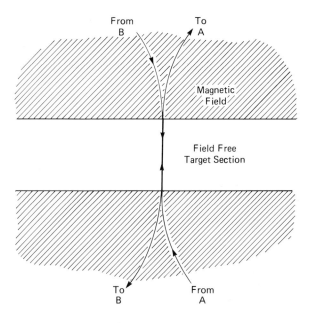

Fig. 12.1. Taken from Kerst et al. (1956). Submitted to *Physical Review* on 23/1/56.

A few weeks later G.K. O'Neill of Princeton proposed separating the functions of accumulation (to increase the intensity) and acceleration, the advantages being that it was cheaper and had a greater flexibility. (Remember that, with a machine like the FFAG, the two processes took place in the same machine, the same vacuum chamber). Taking a normal synchrotron ('any accelerator having a strong, well-focused external beam'), O'Neill suggested injecting alternatively the particles into two 'storage rings' (formed of simple 'focusing magnets','of solid iron and simple shape', 'containing straight sections one of which is common to both rings'), and to 'store' the particles until one had two beams sufficiently intense for an interaction (see Figure 12.2). There were two major unknowns in this scheme: the transfer of the beams from the main accelerator to the storage rings, and the means to be used to store successive bunches of particles in the rings.[26] In 1956 O'Neill had no particular ideas about how to tackle the first problem, though he thought it was soluble. He did, however, put forward a possible solution for the second.[27]

For the rest, the Geneva symposium discussed less revolutionary, less radical, devices, notably linear accelerators, injection techniques, and electron synchrotrons. Among the 80 papers read, two others merit particular attention: that of W.K.H. Panofsky who offered his 'speculations concerning the multi-BeV applications of electron linear accelerators', and that of W.M. Brobeck of the Berkeley Radiation Laboratory on the 'design of high energy accelerators'. What makes these two papers particularly interesting *in 1956* is their insistence, unique at the time, on very big

Notes: p. 766

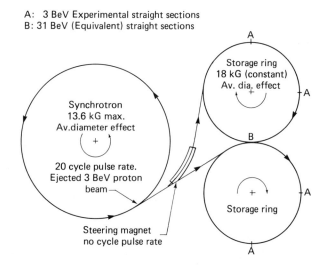

Fig. 12.2. Taken from O'Neill (1956a). Submitted to *Physical Review* on 23/4/56.

machines of the order of several kilometres with energies far greater than those of existing accelerators, and built on the basis of known principles. In short, these articles are noteworthy because their authors were willing to consider the possibility of building 'monsters'.[28]

In his contribution Brobeck identified three possible directions which research into the design of accelerators could take. These were 'towards higher energies, higher currents, and higher ratios of observable information to background'.[29] As the latter pair were the province of MURA and of O'Neill, Brobeck restricted himself to the first, and reported that for the past year Berkeley had made 'a rather detailed study of a machine for the next step in energy above that of the CERN and Brookhaven synchrotrons'. Although Brobeck's conclusions were still very general, one of them was seen as of crucial importance. Since the magnets constituted a considerable part of the cost, it was advisable to reduce their size. That in turn depended on the cross-section of the vacuum chamber in which the protons circulated. Accordingly, if one could reduce the size of the chamber, one could make substantial savings. The idea put forward by Brobeck was that, by using two machines 'in tandem', 'it may be possible to take more advantage of the contraction of the beam with increasing energy than has been done in the past'. Protons would first be accelerated to a given energy in a synchrotron with a large opening, and then transferred into a second synchrotron of much larger radius but with a reduced vacuum-chamber cross-section — 'which would be adequate after the contraction that would occur in the first stage [of acceleration]'. In this way the cost of the magnets in the big ring could be considerably lower than that of a simple scaled-up version of the Brookhaven AGS or the CERN PS. All the same, there was one problem with this system of two successive rings: the transfer of the beam 'from one machine to the other on a single turn' — a problem of the kind O'Neill faced with injection into his intersecting rings.[30] Obstacles remained, then, on a path whose outlines were only slowly emerging.

12.1.2 THE SITUATION IN AUTUMN 1959

The 1956 symposium created a great stir: numerous ideas were raised, some of them with great enthusiasm. Definite solutions were still a long way off though, and the years which followed were to see European, American, and Soviet physicists pursue all the suggestions made in June 1956. Plasma accelerators, FFAG machines, and colliders, in particular, were widely studied.

We shall not discuss developments surrounding the first type in any depth because, even if considerable work was done on very dense electron rings and on annular currents in plasmas between 1956 and 1959, it quickly led to the conclusion that this approach 'did not offer particularly encouraging prospects' in the short to medium term.[31] Essential questions of a fundamental physics kind remained unresolved. Indeed the hopes held in 1956 that one could construct plasma-based accelerators receded along with the enthusiasm for research into controlled fusion. By 1959/60 everyone agreed that this approach was of no immediate interest.[32]

The situation concerning FFAG accelerators was quite different. Though many practical problems still had to be reckoned with, and though the difficulties we mentioned earlier remained (notably the cost), two converging elements gave the partisans of these machines grounds for optimism. Firstly, both the theoretical and experimental work of the MURA group seemed to be bearing fruit — at least in the eyes of some. In August 1957 their second model — a 'spiral sector electron accelerator' — came into operation, and they began the construction of a third, more complex, 50 MeV model whose aim was the 'production and [experimental] study of large accelerated currents and beam handling methods with these currents'. Eight months later, in March 1958, they began to design a 'two-way[33] proton FFAG accelerator' of 15 GeV. Positive results and new solutions were also coming both from CERN and the Soviet Union (where this type of machine was called a 'phasotron'). The initiative was thus being supported and was apparently rich with promise.

A second reason for being confident was that high intensity was recognized as *the* priority by the bulk of physicists. The report of the 'Special Panel of the [US] President's Science Advisory Committee' published on 16 May 1959 recalled that 'there does not now appear to be a clear need for extension of the energy parameter for protons beyond [30 GeV]' while there was a need 'for a high intensity proton accelerator at an energy of 8 GeV or above'. Similarly, in his opening speech at the 1959 conference D.L. Judd insisted on the urgency of high-intensity machines. For

Notes: pp. 766 ff.

further confirmation one need simply cast an eye over the accelerators then envisaged or under construction in the world. There was Nimrod, the British 7 GeV PS designed explicitly for high intensity, whose construction began in 1957; there was the 12 GeV ZGS at Argonne, whose construction got under way in 1959, and which had a design intensity of 10^{12} protons per pulse (while one was expecting 10^{10} at the CERN PS and the Brookhaven AGS); at the Geneva conference there was also talk of a Princeton-Pennsylvania 3 GeV PS whose pulse frequency would be 19 per second (which would permit one to reach more than 10^{12} protons per second), of a 'very high pulse rate' 'Southern Regional Accelerator', as well as of several Soviet accelerators.[34] Clearly, then, intensity remained the dominant preoccupation.

It only remains to speak of the colliders, which attracted considerable attention at the 1959 conference — half of the opening session was devoted to them for example. By contrast, we may note that *no* paper was specifically devoted to a high-energy PS of some hundreds of GeV, even if the issue was partially raised by Judd on the first day when drawing a comparison with intersecting storage rings.[35] Two lines of research were covered by the generic term collider: one was directly connected to FFAG machines — hence its attraction — and the other was that pursued by O'Neill in which, as we have said, the functions of storage and acceleration were kept distinct.

The first approach was backed primarily by the MURA group in the United States and by A.A. Kolomenskij's group at the Lebedev Institute in Moscow. Certainly both of these groups were ready to consider that an FFAG machine could serve as an *injector* into storage rings as envisaged by O'Neill — and by virtue of the density of its beam such a machine constituted a good injector; certainly it was possible to set up *two* FFAG accelerators tangentially with the beams colliding head-on at the point of contact — which was the essence of Kerst's proposal of January 1956; however, both groups were attracted by a third solution, namely, that of having a *single* FFAG accelerator of the so-called 'radial sector' type. In such a machine one could, thanks to the symmetry between positive and negative magnet field sectors, circulate two beams of identical particles in opposite directions in the same chamber, the beams crossing one another at regular intervals (see Figure 12.3). Proposed by Kolomenskij, on the one hand, and by Okhawa working for MURA on the other,[36] the idea stimulated several theoretical studies (for example, on instabilities generated by the presence of two beams circulating in the same chamber), the building of experimental models (MURA's 50 MeV electron model, a 100 MeV electron model at CERN, etc.), and numerous comparisons with colliders of simpler conception (this, for example, was the aim of Judd's introduction to the 1959 international conference).[37]

The possibility of separating acceleration and accumulation continued to fascinate an indefatigable O'Neill. From 1957 onwards the practical realization of storage rings was one of his main preoccupations. To achieve this he, along with Panofsky, proposed to use Stanford's linear electron accelerator, the Mark 3 of 500 MeV, and to join two tangential storage rings to it (see Figure 12.4). Construction was already

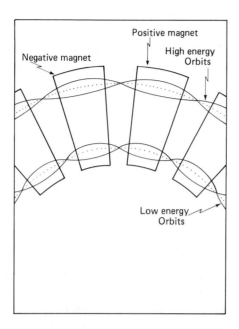

Fig. 12.3. Taken from Ohkawa (1958). Submitted to *Review of Scientific Instruments* on 19/8/57.

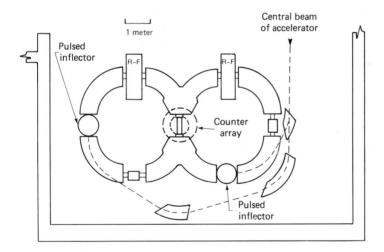

Fig. 12.4. Taken from O'Neill (1959b). Communicated to the international conference in 1959.

well under way in 1959.[38] In parallel O'Neill continued to envisage the possibility of 'Proton Storage Rings'. Here the problem was a little more complex. To begin with one had to develop a good ejector/injector from the accelerator into the rings — a crucial point, as we have said, and a recurring leitmotiv in this period. What is more, success depended on the possibility '[of] the trapping of many pulses of protons into a d.c. magnet', and the ability to store new beams without perturbing those already in the ring. When one had electrons dissipative effects, like orbit radiation, 'can cause the phase space occupied by a bunch of particles to shrink indefinitely', so playing an essential role in the storage process. With protons, by contrast, if one could not obtain 'a similar damping' (as one sought to do during a period) one would be compelled to use phase space as much as possible — 'to stack in phase space' as O'Neill put it in September 1959.[39]

Some approaches had thus been more or less set aside by 1959, though the landscape was still blurred.

12.1.3. THE AMERICAN SWING TO VERY HIGH ENERGIES, 1959-61

Many of the participants found the 1959 conference less interesting and stimulating than that of 1956. In comparing the two it was clear to Christofilos, for example, that 'in the recent meeting practically no new idea on accelerators was disclosed' — which is further proof, if that were needed, that research into new species of accelerators to be built in the immediate future was still everyone's major concern. Christofilos also repeated that 'the general attitude seems to be that nothing is expected from very high energy machines (above 50 BeV)' — a view which was shared by almost all.[40] This situation, so characteristic of the years 1956-59, was to change dramatically in the next two years, not so much in the sense that novel ideas were advanced, but rather in that the very high energy range (200-1000 GeV) obtained using established technologies was to become the centre of interest.

Until the summer of 1959 no proton accelerator above 35 GeV was seriously considered anywhere in the world — except for the 'synchro–phasotron of 50 GeV' at the research institute for electro-physical equipment in Leningrad[41] — and only a few ideas which were not taken up and some summary reports touched on the technical problems raised by fixed-target machines of more than 100 GeV. We have mentioned the work done at the Radiation Laboratory in 1955-57 and some general observations made by Brobeck in 1956; to this should be added the report presented by Matthew Sands of the California Institute of Technology (Caltech) to the accelerator conference organized by the MURA group at Madison (Wisconsin) in June 1959.[42] For the first time someone had made 'some specific calculations and [had tried] to find some optimization' for a strong-focusing proton synchrotron of 300 GeV in which two synchrotrons were arranged in tandem — Sands used the term 'in cascade'. Starting from certain reasonable assumptions — for example, that the shape of the magnets at the CERN PS and the Brookhaven AGS was sound — and invoking an optimization criterion — namely, 'to minimize the weight and cost of the magnet' —, Sands calculated the 'ratio of the apertures and the ratio of the diameters of these two synchrotrons', and from these the main parameters of the magnets. Result: 10 GeV and an aperture of 13 cm for the first machine, 300 GeV, an aperture of $4 \times 1 \text{cm}^2$, and a radius of about 1 km for the big ring.[43]

However, the time was not yet ripe for 'monsters' of this size in the summer of 1959. Before that could happen, before they could be seen as good solutions by a substantial proportion of physicists and engineers, two things were needed. Firstly, the operation of the CERN PS and the American AGS. In November 1959 and June 1960, respectively, these machines demonstrated that the strong-focusing principle was perfectly reliable and, above all, that the beams one obtained were *of remarkable intensity*: having anticipated about 10^{10} protons per pulse — which was immediately achieved — the builders increased this number by a factor of 10 in some months (3×10^{11} in the summer of 1960 at CERN).[44] Of course, in theory all were *expecting* to see these accelerators work as planned. However, the psychological impact of *seeing* the machines operate and of seeing the principle of strong focusing mastered was considerable. Whereas before there was a discernible element of fear, of doubt, now there was a feeling of relief, of joy. And even if the accelerator was still not well-understood —'for the time being', said Adams on 19 January 1960, 'only Dr. Hine was able to set up the machine properly' — confidence prevailed, and permitted all manner of boldness, all possible extrapolations.[45] Hence a renewal of interest in synchrotrons, *notably in California*, which was without a machine of this type at the time and was 'falling behind' the East Coast and Europe in the energy race.[46]

Contact between Americans and Soviets was the second factor facilitating the development of accelerators of some hundreds of GeV. Several informal discussions were held at Kiev from 18 to 24 July 1959 and at CERN during the international conference in September; at one such gathering, on 15 September in Geneva, Europeans were also present.[47] The decisive step, however, was taken at the meeting in November 1959 between McCone, President of the American AEC, and Emilianov, his Soviet counterpart. The agreement they drew up on that occasion mentioned 'the design and construction [in common] of an accelerator of large and novel type'.[48]

Research on very big machines, on synchrotrons of the order of 100 to 300 GeV, thus got under way again at the Radiation Laboratory and at Caltech. In August 1960 R.R. Wilson of Cornell University organized a meeting during the Rochester conference to take stock of the situation. Three key questions were on the agenda: the (scientific) desirability of machines of more than 100 GeV, their cost, and the possibility of doing experimental work at these energies. According to Amaldi, who brought back news of the discussions to Europe, Wilson 'concluded with a few remarks which were rather optimistic with respect to both the feasibility and the cost'.[49] In addition, in September, Matthew Sands published a report of some 30 pages entitled 'A Proton Synchrotron for 300 GeV'.[50] In September, too, a working meeting of American and Soviet physicists held at the American Institute of Physics in New York decided 'to explore further the feasibility and desirability of

Notes: pp. 767 ff.

accelerators for energies > 300 GeV'. The two groups agreed to compare results a year later during the third international accelerator conference.[51] In the light of these accords, a minimal division of labour was established for the American laboratories. The Californian physicists were asked to continue with their work on machines in the energy range 100-300 GeV, while Brookhaven was to undertake the study of the 'intercontinental' accelerator (300-1000 GeV). During the next twelve months this activity was to dominate the work on future accelerators in the United States, and probably also in the USSR. The unfolding story of proton synchrotrons of several hundreds of GeV was picking up steam.[52]

At Caltech events were largely dominated by Sands and his group. He began his September 1960 report on 300 GeV synchrotrons by justifying his project: scientifically, physics at 300 GeV would be possible — and easier than with intersecting rings;[53] economically, the cost would be roughly the same as that of an FFAG device of some 15 GeV, i.e. about $100M; technically, the machine was 'very likely feasible'. To show this Sands took up again the arguments he had presented a year earlier and arrived at similar conclusions. Concerning the apertures, for example, the more precise calculations made in 1960 gave a figure of 17cm for the synchrotron injector, and 3 cm for the big ring —though Sands felt that the first should be reduced to 12cm and the second increased by some 50% for a number of more intuitive and subjective reasons.[54] Concerning the intensities which could be expected with such machines, Sands simply reminded his readers that his design of two synchrotrons in series 'provides both a higher injection field and a higher maximum theoretical intensity', and that recent proposals for increasing the real intensity had just been suggested. He did not give, however, any precise figure.[55]

A similar feasibilty study was also made at Brookhaven National Laboratory, this one for a machine of 300-1000 GeV. The work began towards the end of 1960 and in May 1961 a preliminary report appeared under the name of J.P. Blewett.[56] Its main conclusions were twofold and deserve quoting in full: 'First, no single factor has appeared to bring the feasibility of a 300-1000 BeV accelerator into question. Second, a great deal of detailed design study remains to be done' — particularly 'with remote control', 'safety', and 'beam extraction'. The conclusion also remarked that 'Intensities for all three machines [400, 700, 1000 GeV] should be of the order of 10^{13} protons per pulse, or 4×10^{12} protons per second'.[57]

During the summer the level of activity increased sharply. In June a conference 'on theoretical aspects of very high energy phenomena' was held at CERN.[58] Thereafter a group of American and European physicists worked together at Berkeley from 15 June to 15 August. About 30 seminars and a dozen reports were presented to a gathering of some 40 specialists, the rest of the time being 'spent in work on particular problems by individuals and small groups, with interruptions for frequent spontaneous discussion periods involving larger numbers of people'.[59] In August the centre of gravity moved to Brookhaven where 16 seminars were held and the May report was subjected to the scrutiny of 25 'accelerator experts'. The result was the publication of a second report of some 150 pages entitled 'Design Study for a 300–1000 BeV accelerator' — after which the third international conference on accelerators opened at the Plaza Hotel in New York.[60]

The summer of 1961 would undoubtedly deserve a chapter of its own in a history of accelerators.[61] Though this cannot be done here, we can give a few of the key elements. First point: along with the studies we have just described there were probably as many reports, seminars, and other conferences tackling the question of *how to do physics* at 300–1000 GeV, in which the scientific interest of this energy range, the kinematics of the primary and secondary beams, the conception of the experimental areas, etc. were looked into. These topics were the dominant concern of the Berkeley seminars, and at the end of August a report of some 200 pages was published by BNL ('Experimental Program Requirements for a 300–1000 BeV Accelerator').[62]

Second point: these studies precipitated a renewal of interest in proton linear accelerators. Up until this time no linac had operated above 70 MeV.[63] However, if one wanted to use linacs as injectors for 300–1000 GeV synchrotrons they had to be big enough to produce 2 to 10 GeV protons, an increase in size which called for an entirely new design. This technique was favoured on the East Coast — particularly at Brookhaven and Yale — while Matthew Sands deemed a synchroton injector to be a safer bet.[64] Accordingly there was an extension of research in this domain in 1961. A two-week conference on linear accelerators was held at BNL in April, and was repeated at the same venue in August 1962.[65]

Third point, this one concerned with acceleration technique: these studies resulted in 'important contributions to the synchrotron art'. For example, there was the idea (and the technique) of introducing straight sections along the accelerating ring as a means of facilitating beam extraction and experimental work, there were Robinson's propositions 'for constant frequency acceleration', there were Symon and Tollestrup's suggestions for accelerating antiprotons, etc.[66]

It only remains to give the conclusion to which all this work done in 1961 led. It was simple, and was clearly stated by Blewett in August: 'we believe', he said, 'that there is no question of the feasibility of accelerators for energies above 300 BeV'. The technical solutions were many and the only problem was 'to choose the most effective and least costly [one]'.[67] In a certain sense this is a disturbing result. It merely seems to repeat the statements made by Sands in 1959 and 1960 or Blewett's own words before the summer: in both cases the speakers *believe* that synchrotrons of 300–1000 GeV are *feasible*. It would be superficial to leave it at that, however, because behind very similar words[68] there can lie quite different images, degrees of precision and of 'conviction'. More specifically, the same words conceal meanings which evolve. With time, for example, participants in the debate came to think that certain problems were solved (e.g. for Sands the system of two synchrotrons in cascade), or that the values of some parameters were fixed or their limits specified

Notes: pp. 768 ff.

(e.g. the size of the vacuum chamber). This enabled them to concentrate on new questions whose solutions, be they theoretical or practical, reinforced the credibility of the whole and/or led them to abandon certain posibilities — in any event enabled them to define more precisely what they thought soluble even if still 'unknown'.[69]

The clearest result of this process, a process in which the participants become increasingly involved, is that beliefs become more and more unshakeable, that the number of the 'convinced' increases as the number of studies multiplies, and that finally the majority of the scientific community recognizes the soundness of certain lines of argument initially only adhered to by a few. As each new stage of development is reached, as each particular question is studied in depth, one's picture of the machine improves in quality, its grain becomes finer. And with that the number of those willing to share in the work grows too.

A last consequence of this step-wise development is the gradual lifting — at various stages in the process depending on the individual — of what might be called 'psychological obstacles' which, at certain moments, make it difficult for one to think about certain questions. We have pointed out, for example, that there was no paper on proton synchrotrons above 100 GeV at the international conference in 1959. The main reason for this was that no one — or almost no one — dared consider such machines, that no one — or almost no one — dared envisage constructing 'monsters' which would simply be copies of the CERN PS or the Brookhaven AGS scaled up tenfold. To do so was unimaginative, it revealed a penchant for the use of brute force, and anyway the cost was prohibitive. To be freed from this 'mental prohibition', a leap, a reorganization was needed. It was Panofsky who played this role for Sands: 'it was a little frightening' he said by way of excuse in September 1959, 'at first to think of a 1 mile diameter machine but Panofsky has removed most of those psychological disadvantages'. (Two years later, now surer of the viability of his approach he said simply: 'Once the mental adjustment is made to the fact that absolute size is no impediment, the A.G. synchrotron is an obviously open path to very high energies'.) Subsequently Sands himself — and he was not the only one — played this role for many others, so that one can say that, by the beginning of 1961, all 'mental blocks' of this type had disappeared in the United States. Things were different in Europe though, as we shall see.[70]

12.1.4 OTHER APPROACHES CONSIDERED BETWEEN 1959 AND 1961

In the months following the international conference in 1959[71] MURA and CERN continued work on FFAG-type machines. There was a notable change at CERN some nine or ten months later. Towards the summer of 1960 the construction of a 'bidirectional' electron accelerator ('two-way model of 100 MeV') was stopped and the emphasis in the accelerator research division shifted: a model to study the simple storage of electrons in a ring was preferred along with the theoretical investigation of colliding storage rings for 25 GeV protons coming from the PS. In December 1960 a first project of this kind was submitted; it was widely debated in 1961.[72] A roughly parallel development is detectable at MURA. The 50 MeV 'two-way FFAG' electron model was completed in December 1959. In subsequent months it worked 'up to energies of approximately 27 MeV, at which point both beams were completely lost'. In April 1960, thinking that corrections were needed for the magnetic field produced, the MURA physicists dismantled the model. Sixteen months later, in August 1961, '[it] successfully operated [but] *in the one-way mode* [...] accelerating particles to 51 MeV'.[73] In parallel, the group continued to work on the feasibility of a bigger project, something like a high-intensity FFAG proton accelerator of about 10–15 BeV.[74] The March 1958 idea of having an accelerator–collider for protons of 15 BeV of the '2 way FFAG' type was jettisoned (as too expensive and complex) and, after summer 1959, efforts were redirected towards a 'one-way FFAG' accelerator of 10 BeV with an intensity of 2×10^{14} protons/second. MURA submitted a project along these lines to the conference in 1961.[75]

As far as simple storage rings in which collisions could occur are concerned, three groups should be mentioned: that at CERN (of which we shall speak shortly), that at Frascati in Italy, and that led by O'Neill and his collaborators.[76] From 1959 to September 1961 the construction of the 'Princeton–Stanford Colliding-Beam Experiment' went ahead. In September O'Neill, reporting on the state which the work had reached, said that the experiment would take place in two months.[77] The Italians, for their part, put Ada, the Anello di Accumulazione, into operation in 1961. This was a ring 'intended to obtain colliding beams of e^- and e^+ of 250 MeV', the e^-e^+ pairs being produced at a target inside Ada by 'a 1 GeV γ-ray brehmsstrahlung beam' coming from a synchrotron. The particles were stored and their paths followed, though there was no opportunity to study collisions, the intensity being too low. The Italians also investigated the possibility of having a ring which was bigger than Ada, a kind of 'big Ada', an 'Adone'.[78]

A final point to bear in mind. The building of these colliding ring(s) was accompanied by numerous theoretical studies on the physics one could do with such machines. We encountered a similar situation with the 300–1000 GeV synchrotrons — it is present here again, even more so in a certain sense: for detection was a far more delicate operation (the problem of background noise); for it required the conception of entirely new kinds of equipment (the problem of spatial arrangement); for many physicists were not particularly keen on these colliders, and they had to be convinced. We shall come back to these points at length when we study the situation at CERN a little later.[79]

12.1.5 THE BIG CHOICES, 1961–65

We can be relatively brief here as the 'gap' between Europe and the United States was now reduced. We simply draw the reader's attention to four points.

Firstly, physicists had virtually abandoned studies on possible plasma accelerators.

Notes: pp. 769 ff.

Around 1959/60, as we have said, it was widely believed that these machines were not practicable in the short to medium term. In 1963, at the Dubna conference, one still finds one or two papers on the topic, mostly by the Soviets. Two years later, at Frascati, they had disappeared — even if F.E. Mills (MURA) presented a few comments on the 'coherent acceleration of protons'.[80]

Secondly, a similar fate befell FFAG-type proton synchrotrons. Traces remained at the meetings at Dubna and Frascati but by then such projects were in their death throes. How does one explain this? Following Greenberg,[81] we could have recourse to an essentially 'socio-political' explanation: MURA's setback was that of a group which failed to survive in the American political and scientific jungle. Confronted with competitors who were used to 'lobbying', who were more powerful, and who had better connections in the corridors of power, MURA was never able to sell its idea. We find this explanation inadequate, marginal in fact — at least if one is speaking simply of the choice between *proton* accelerators. Certainly, one can hardly doubt that the Californian and East Coast physicist 'lobbies' were formidable and had better contacts in Washington than the MURA group. On the other hand, Greenberg's argument does not explain why the idea for FFAG machines was dropped in Europe and the USSR at the same time as it was on the wane in the United States — a fact which cannot easily be attributed to the influence of Ramsey or of Rabi. Indeed in our view *scientific judgement* was dominant: briefly, we would say that MURA did not manage to convince the majority of physicists at the time of the *scientific validity* of its ideas. From 1956 the interest in the FFAG machine lay in the intensities one could hope to achieve with it. However, from 1961/62 onwards, intensity no longer occupied the dominant place in the scale of expectations of theoretical and experimental physicists, a place it still had in 1959: energy had joined it, had in fact supplanted it. Thus one reads in the Ramsey report published in 1963 — and it would not be difficult to multiply the examples — that 'Proton energy is the single most important parameter to be extended', or that 'The highest priority in new accelerator construction should be assigned to a succession of large steps toward the highest attainable energy'.[82] Finally, and this is perhaps even more important if we want to understand why physicists' expectations changed, other more 'economical' means of achieving 'good' intensities gradually made their appearance. Let us not forget that, from the end of 1961, everyone knew that the CERN PS and the Brookhaven AGS could furnish almost 10^{12} protons/pulse and probably more in the next three to five years, and that the majority of the specialists in the world were convinced that one could have 10^{13} protons/pulse with 300 GeV machines. Put differently, strong-focusing proton synchrotrons were seen to have *both* advantages: in essence they were 'multipurpose'.[83] Hence their irresistable attraction over and against MURA's proposals.

The latter were buried in 1963 by the Ramsey Panel in the United States. In the most diplomatic of terms they *authorized* 'the construction, by MURA, of a super-current accelerator' — adding all the same the following condition: 'without permitting this to delay the steps toward higher energy'. Granted this 'qualification', and granted the prevailing financial constraints, the 'recommendation to build' was strictly equivalent to a death sentence: the money would never be found.[84]

Does this mean that we are back where we started, to the situation prevailing in 1956? That Adams' hope of seeing 'new species' rapidly replace the old had been disappointed? Partly yes, in the sense that it was techniques which were already proven in 1956 which were on the way to victory. And partly no because, as we have already seen, to go from 30 to 300 GeV did not mean simply scaling the drawings by a factor ten. And partly no again, because intersecting rings of the O'Neill type had entered the picture.

Which brings us to our third point: the idea of using intersecting rings had been accepted though, to be honest, for electrons rather than for protons. Without dwelling too long on this point, which is beyond the scope of this chapter, it should be said that the interest aroused by electron and electron/positron colliders between 1961 and 1965 is quite striking. To give some examples. We have the Italian project Adone reconsidered at the end of 1961 and in 1962, after which its practical realization got under way.[85] We find Ada moved to the linear accelerator at Orsay near Paris for the purposes of experimentation. We have the Orsay physicists setting up their own electron/positron collider of some 500 MeV (ACO).[86] And we see the Princeton–Stanford colliding rings come into operation at the end of 1962. (Their performance was mediocre, however; 6×10^9 electrons were stored with a lifetime of 3 min, new and disturbing effects appeared immediately (in particular, a rise in pressure in the vacuum chamber caused by synchrotron radiation), the rings were dismantled and reassembled in 1963, and results were obtained in 1965, though with no more than average energies and intensities).[87]

As for protons, only CERN was seriously interested. In 1965 it was the only laboratory presenting papers on this topic — and no less than five at that —, Brookhaven having sullenly refused the Ramsey Panel's recommendation to 'construct' intersecting rings 'at the earliest possible date'.

One final point to conclude, and doubtless also the most important, a point which, like the former, will be considered at length in the following pages: strong-focusing proton synchrotrons were the main preoccupation for a majority in Europe and in the United States. It was thought to build them in the energy range 100–300 GeV and, to the protagonists in this story, they increasingly came to be seen as the cornerstones of any future accelerator programme.[88]

12.2 The time of new ideas: the CERN Research Group on Accelerators, 1957–1960

As we have said earlier, the conviction that CERN — in its capacity as the 'main' European laboratory — ought to study the machines of the future dates from the

Notes: p. 770

1956 symposium. Adams gave the alarm in a three-page report submitted to the SPC in October. For him CERN had to undertake 'research on machine design' if it wanted to regard itself 'as a complete centre such as Brookhaven and Berkeley'. To this end he proposed the creation, within his division, of a group of some twenty people, three-quarters of whom would be visitors from Europe and America, i.e. scientists who were not CERN staff members.[89] The SPC approved Adams' proposal at its meeting on 12 and 13 October 1956, and the December Council session established the Accelerator Research Group under A. Schoch, who had been responsible for the PS Theory Group until then.[90]

12.2.1 THE ACTIVITIES OF THE ACCELERATOR RESEARCH GROUP FROM 1957 TO 1959

The group was rather small until 1959 (at least in comparison with the rest of the PS division) and somewhat marginal in the organization. This was because its activities were of very low priority, the construction of the PS being task number one. This situation was to change dramatically at the end of 1959 when the PS entered into operation, when it was seen to perform exceptionally well, and when a section of the personnel responsible for its construction found themselves freed from their earlier obligations.

Two independent documents enable us to gain a rough idea of the composition of the group in the years 1957-1959.[91] The first point to note is its regular growth from early 1957 (five physicists or engineers in June) to spring 1958, when it tended to stabilize at 12-15 members. The second point to note is that, as suggested by Adams in 1956, the number of personnel who were permanent CERN staff remained small. Four names appear in 1957 and 1958 (A. Schoch, the Group Leader, J.G. Linhart, N. Vogt-Nielsen, M.J. Pentz),[92] and three more were added to the list at the end of 1958/early 1959 (E. Fischer, P.T. Kirstein, F.A. Ferger). They were supplemented by a larger group of visitors whose composition changed regularly. About a dozen collaborated during the first thirty months, each for an average period of nine to ten months.[93]

This situation was not without its difficulties. In May 1958 the SPC's attention was drawn to them: 'The studies undertaken on new acceleration principles deal with a hitherto little known field; they constitute a long-term problem and call for uninterrupted efforts. Experience has shown that fellows or research associates working for short periods in such a group did not derive full benefits from the work and were not in a position to contribute fully to this work'. The conclusion was simple: it was proposed 'to use more permanent staff' and to transform the fellows and research associates of the group into CERN staff.[94] Stressing the exceptional nature of the step, the Finance Committee (FC), the Scientific Policy Committee (SPC) and the Committee of Council (CC) approved this proposal — and the Council somewhat hesitantly accepted it at its meeting on 26 June 1958.[95]

(A remark of a more general kind is appropriate here. This was not the first time that CERN had been confronted with the problems raised by having people work part-time on long-term projects. A similar situation had arisen in 1952 in Odd Dahl's group whose task it was to design the CERN PS, when J.P. Blewett disputed the usefulness of this kind of collaboration. More surprising perhaps, the 'New CERN Study Group on Accelerators' set up by the Council in *1962* did the same thing again, recruiting some temporary personnel.[96] How is one to explain this attitude? By noting simply that having recourse to part-time workers is not always a *choice*. It is often the only way of getting going on very long-term projects for which there is no guarantee that they will ever be implemented).

The scientific activities of the group were laid down in February 1957. 'Clearly', wrote Schoch, 'two of the most interesting lines of work are the plasma accelerators and intersecting beam machines and these will be studied first'. In fact for the next nine months all efforts were directed to the plasma accelerators.[97]

In the same report of February 1957 Schoch remarked that '[since] the work on plasma accelerators involves a branch of physics that is new' those engaged in it should be taught its basic principles. To this end A. Schlüter of Göttingen came to Geneva to give a series of lectures introducing the physics of plasmas. From the summer onwards experimental studies were made using a betatron comprised of air coils. The toroidal chamber of the betatron (having an orbit radius of 26cm) was filled with a hydrogen plasma whose properties were 'measured', in particular the number of electrons circulating per centimetre of the chamber length and the intensity thus obtained.[98]

At the end of 1957 work on the 'beam-stacking proposal of the MURA group' got under way: the first report was issued in February 1958.[99] Its announced goal was to study another means of producing 'high intensity beams of relativistic electrons' and, possibly, to come to grips with 'intersecting beam machines'.[100] To look into this aspect more thoroughly the group turned its attention, during the course of the year, to a model whose first specifications were established in November 1958. It was to be a synchrotron of the '2-way FFAG' type, with a 'radial sector' type magnet structure.[101] By March 1959 the final design study was ready and the group was poised to construct it. Financially, the project needed some 2.5MSF per year from 1960 to 1962 — so about 3% of CERN's total budget.[102]

The arguments put forward to convince the SPC and the Council of the usefulness of the project were of essentially two kinds. Firstly the achievements and the projects of the competitors were spelt out — 'a 2 MeV one-way FFAG model' already built at Dubna, 'a two-way 40-MeV electron machine' by MURA, whose stated aim was to construct 'a 15-GeV 800 amp intersecting beam proton machine', 'a storage ring device for electrons up to 500 MeV to go with the Stanford linac', i.e. O'Neill's project, etc.[103] Secondly the envisaged long-term goals were described, namely, to produce circulating currents of hundreds or thousands of amps. Because of limitations imposed by space-charge forces 'such currents may be obtained only', the report went on, 'by either an accumulative process ("beam stacking") or by neutralization of the space charge, i.e. by using a plasma'. However, enormous experimental problems had emerged with this latter procedure and it seemed easier to stabilize 'non-neutralized high current beams, and to build them up by stacking the contents of successively accelerated "buckets"'. Hence the proposal put to the SPC 'for concentrating the main effort of the group' on 'a FFAG two-way machine for 100 MeV maximum electron energy and the order of 100 amps stored current'.[104]

In the following months this 'concentration' of effort became increasingly marked. In his report of November 1959 Schoch only devoted one paragraph to plasma betatrons and the associated theoretical studies. He gloomily remarked that the beam intensity 'has not been increased substantially' and that 'it is believed that the electrons are lost quite early in the accelerating period'.[105] A year later the CERN Annual Report only had this to say about these activities: 'Devoting more substantial effort to the problems of plasma accelerators does not seem justified' at least 'if one is aiming at devices to be useful in the not too distant future'.[106]

To conclude our presentation of the activities of this group from 1956 to 1959, we

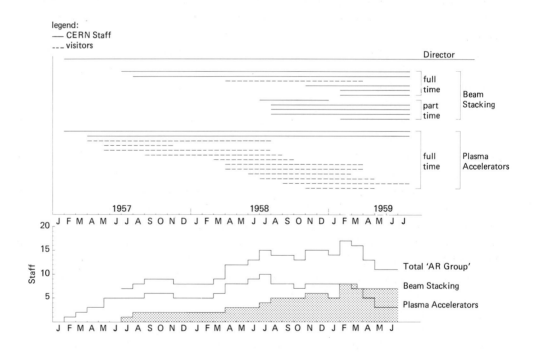

Fig. 12.5 Accelerator Research Group, CERN, Staff

present two graphs (Figure 12.5). The first reveals the rather rapid rotation of the personnel in the group as well as the growing importance of beam stacking studies vis-à-vis work on plasma accelerators. Each line represents a physicist or an engineer (technicians are not included) working in the group during the given period; CERN staff members are depicted with a solid line, others with a broken line. Beam stacking activities clearly made two spurts, first in the summer 1958, then at the start of 1959. The second graph is connected to the first and shows, at each point in time, the approximate number of physicists and engineers in the group. Here the key date is obviously March 1959, the month in which the group submitted its report to the SPC recommending that efforts now be concentrated on electron models of the 2-way FFAG type. Early in March there were still eight physicists working on plasma accelerators. By 1 May there were only three, and the short-term contracts were not extended.

12.2.2 THE ACCELERATOR RESEARCH GROUP IN 1960: TOWARDS STORAGE RINGS FOR THE CERN PS

Between 15 and 31 December 1959, with the PS now operating, the AR group was reinforced by a dozen of Adams' best specialists. Among them there were Kjell Johnsen, one of the original CERN staff recruited by Odd Dahl in 1952, H.G. Hereward, P.M. Lapostolle, and C.J. Zilverschoon. Once these men were permanently installed in Schoch's group, its activities were dramatically redirected within a few months.[107]

In his annual report for 1959 Schoch again dealt primarily with the progress of the work on the two-directional electron machine of FFAG design and of energy 2 × 100 MeV. The first models of the magnets had already been made ('azimuthally profiled pole pieces have been completed', he said in November, 'and are nearly finished being machined'), 'a full scale model of an RF acceleration cavity' was built, and the studies and tests on the injector and the vacuum system were well advanced.[108] Nevertheless, a few months later, the construction of this accelerator was halted, and the group decided to build a different model consisting of two separate elements, an electron accelerator and a simple storage ring.[109] For this ring the vacuum chamber could be 'of relatively small aperture' (so permitting 'the use of small bending magnets and focusing lenses'), and the electron energy would only be some 2 MeV.[110] Technically speaking one thus went from a gigantic vacuum chamber with cross-section of some 2.5m to one of about 10cm by 5, from a sophisticated RF system not yet tested to one that was relatively banal, and to magnets whose complexity was correspondingly reduced. Obviously the new model had but one goal: to serve as an experimental model for the study of beam storage.[111] By way of comparison let us remember that the decision taken in 1958 to build a model of a bidirectional FFAG accelerator was a response to many more interests and objectives: it was intended as a response to *all* the important challenges then

Notes: pp. 770 ff.

prevailing in the field of acceleration *technique*. In fact at the conference in the summer 1959 the CERN group still felt that simple storage rings of the O'Neill type 'would provide less scope for *accelerator research* than would the FFAG accelerator'. It added: 'the fact that [a two-way machine] also introduces a *complication* [in comparison with an FFAG machine with one beam] is, for an experimental accelerator, not entirely a disadvantage' (our emphasis).[112]

The choice of an energy as low as 2 MeV was not without significance either: it enabled one to reproduce rather faithfully the situation which one would encounter if one tried to accumulate *protons*. At 2 MeV the storage of electrons took place 'under the restriction of Liouville's theorem' (as was the case for protons), i.e. without the advantages of dissipative effects produced by more energetic electrons.[113] It is no surprise, then, to see this reorientation in terms of *model* followed, during the course of the year, by a much more ambitious project: the accumulating rings for the PS, which took its final shape in December 1960. On 22 December Hereward, Johnsen, Schoch, and Zilverschoon published a report entitled 'Present Ideas on 25-GeV Proton Storage Rings' in which they suggested the construction at CERN of two tangential or near-tangential rings, an arrangement which would allow the study of head-on collisions of protons of 25 GeV (see Figure 12.6).[114]

In less than a year we have thus moved from exploratory work along sophisticated avenues of research with a medium term outlook, to a project which would lie at the heart of the second generation of accelerators at CERN. At the end of 1959 one was still trying to take on board the widest possible range of techniques, not wanting to dismiss any approach which might possibly bear fruit in the next five to ten years. A year later one was in the process of putting together a model which would enable a quick decision to be taken on a specific project on the scale of the PS itself.

Two main reasons seem to us to explain this 'precipitation' (in the chemical sense) of the '25 GeV Proton Storage Rings' between the beginning and the end of 1960. The first is connected with the evolution of the context in which the European laboratory was immersed and with a particular perception of that context by its members: beginning in autumn 1959, and even more so from spring 1960, there was the impression in CERN that Europe was once again in danger of lagging behind the USA and the USSR in the medium term. We have mentioned that it was this kind of feeling which led to the creation of the research group on accelerators in September 1956; at the end of 1959 it was the American-Soviet negotiations concerning a joint machine which played this role, and provoked the fears. To get some idea of the extent of the concern we need merely look at the meetings of the CERN direction: from summer 1959 the question of 'international cooperation in the field of high-energy physics accelerators' was on the agenda of *all* Scientific Policy Committee meetings and of *all* meetings of the Committee of Council and of the Council itself.[115] Similarly, from summer 1960, the urgent need for CERN to decide on its next accelerators was regularly mentioned by the most active members of the

The time of new ideas

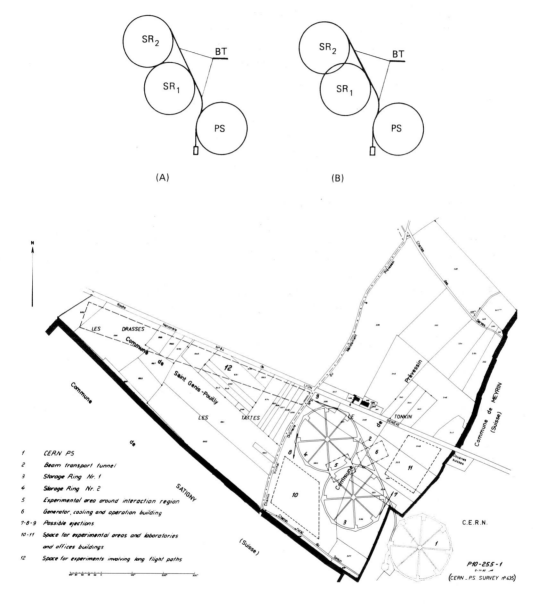

Fig. 12.6 Taken from Hereward, Johnsen, Schoch, Zilverschoon, *Present Ideas on 25-GeV Proton Storage Rings,* CERN, PS-Int. AR/60-35, 22/12/1960.

'CERN lobby'. Drawing attention to all that was happening in California and in Dubna, they insisted that it was time to think of the near future, time to think of the machines that would replace the PS, time to make the first contacts in the political

Notes: p. 771

sphere. In short, they stressed that the international situation demanded a reaction, and a quick one at that.[116]

The second reason why the prototypes were abandoned and the successors to the PS were immediately envisaged was the arrival of new blood in Schoch's group. The group was still relatively marginal in summer 1959. At the start of 1960, however, its nature changed: a group of *builders* was added to its existing core, a group which had just finished building the PS and was swathed in the aura of success, a group whose concern was not to 'potter about' but to get going as soon as possible on a major, concrete project. This group comprised some of the closest collaborators of John Adams, the director of the all-powerful PS division, of the division in charge of all the important projects. On their arrival, these builders looked again at what had been done in the light of the remarkable behaviour of the PS (let us remember that its intensity, in particular, was 10^{11} protons/second in summer 1960) and immediately agreed that it would be interesting to 'link' this achievement in some way with one of the new ideas studied for several years by Schoch's group, beam accumulation. From the point of view of physics, linking a good synchrotron injector with accumulating rings gave one the remarkable possibility of proton–proton interactions at 50 GeV in the centre-of-mass; from the technical point of view, it offered a number of interesting challenges. It enabled one to integrate further two great traditions at CERN, the pragmatism of the PS engineers — a tradition lacking amongst MURA's theoreticians[117] — and the innovativeness which characterized the machine theoretician who was Schoch. The result was the abandonment of the complications of the FFAG system and the orientation towards a project which was realizable in the short term, a project which was feasible and also original, refined, and not simply calling for brute force — and a project which was not too costly either.[118]

12.3 The tempestuous period: the first important year of the debate between scientists, 1961

12.3.1 THE PROPOSAL PUT FORWARD BY THE BUILDERS, DECEMBER 1960 - APRIL 1961

On 22 December 1960, the first report suggesting what the new generation of accelerators for CERN could be was released. Signed by Schoch and the three new arrivals in his group, this text of some 15 pages outlined a project for building accumulating rings for 25 GeV protons. The authors reached two conclusions: that the scheme was achievable in practice — 'it is already now felt that a storage ring system is a feasible project' —, and that it was worth making an in-depth study with a view to eventual construction — 'one should rather soon take some practical steps necessary to make it possible to realize such a project'.[119]

This first feasibility study was communicated to the directorate and submitted to the CERN scientists for their consideration.[120] Three months later, on 30 March 1961, Mervyn Hine wrote a second report dealing with CERN's long-term programme. Its aim was 'to accompany the AR Division report on storage rings'. Both documents were submitted to the SPC for its opinion. [121] In this document Mervyn Hine argued for the growth of CERN, an altogether new thing in the organization.

Here, for the first time, someone put forward the idea that CERN was confronted with a stark choice: grow — which implied the construction of new machines to replace or to complement the PS — or become second rate. Describing the projects being undertaken or being decided on in the United States — notably SLAC —, and explaining the hopes of the American and Soviet physicists — 100-300 GeV PS in California, the high intensity accelerator of MURA, the Soviet-American machine of more than 500 GeV, etc. —, Hine concluded that, unless the Europeans did something in their turn, the centre of gravity of fundamental physics could not but shift 'back again to the USA, after rather few years in mid-Atlantic'. This was a situation which the author refused to countenance.

Buttressed by this postulate — growth as a necessity — Hine tried to demonstrate that the Intersecting Storage Rings (ISR) were the only possible solution for CERN. The first possibility he considered was that CERN waited on developments in the intercontinental project 'which Europe could then join'. Hine rejected this solution arguing that negotiations could go on for a long time, that they might not lead to an agreement, and that if they did and the venture was launched the most gifted Europeans would leave. In fact, he added, already 'all the senior accelerator staff in CERN have been receiving pressing invitations from several USA groups to join them'. In these circumstances it would be as ill-advised to mark time as to decide unambiguously to halt the growth of CERN.

'We are left with [two] alternatives', Hine went on, 'to start on a new accelerator project [fixed target machines, synchrotrons or linear accelerators] or to find ways of developing the existing CERN installations so effectively that the status of CERN can be maintained without a new accelerator'. What was one to make of the first option? Hine felt that it was a poor solution since the costs would be considerable — between 500 and 1000 million Swiss Francs for a synchrotron of any interest[122] —, since one would inevitably have to find a site other than Geneva, which was already overcrowded, since the people needed to build it would have to be drawn from CERN and, finally, since the project would probably be 'in direct competition with a similar American project'. In short, for a number of politico-financial reasons the only solution remaining on the list was the one described as 'Intensive development of existing CERN facilities'.

What did this scheme involve? The accumulating rings proposed in December, which alone would permit 'a relatively cheap entry into a field of very high energies (1 TeV)' and which '[would] greatly increase the flexibility and usefulness of the PS at its present energy'. A remark should perhaps be made here. It is not a coincidence

Notes: p. 771

that Hine did not call the ISR a new accelerator, preferring to speak of the 'Intensive development of existing CERN facilities'. By using this formulation he implied that, as far as CERN's Convention was concerned, the ISR was not a 'new programme' (with all that that entailed in terms of negotiations between its member states). What is more, since the idea of improving the PS was on everyone's lips, it was convenient to associate the new proposal with it.

The many reactions of the CERN experimentalists and theoreticians to this document were predominantly hostile. For them, it implied that only one possible alternative was being seriously considered and that the decision would be taken without a real study of the question by the future users, the physicists. Consequently, in the days following the drafting of Hine's text, an ad-hoc meeting of the CERN senior staff was called on the insistence of the physicists; it was held on 14 April.[123] Two points emerged clearly from the discussion: many thought that it was unwise to force the Council to take a quick decision, and most of the scientists present expressed their preference for 'a high energy accelerator' rather than intersecting rings.[124] Forced to beat a retreat as precipitate as it was unforeseen, John Adams, then Director-General, drew up a new text for the SPC. After having summarized the earlier official position, he pointed out that 'whether it is better to extend the experimental facilities of the CERN PS in the manner proposed [by Hine] or to build a new accelerator is clearly a matter of physics judgement and, inevitably, of opinion'. That granted he asked the SPC 'to approve that CERN should engage in a new accelerator programme', 'to consider whether the new programme should be [...] storage rings or whether a new very high energy accelerator is preferred' and, finally, to approve the suggested timetable, i.e. submission of the same questions to Council at the end of 1961, creation of a study group for the chosen project in 1962, final adoption of the project by the Council at the end of 1962.[125]

Two features of the situation as it appeared in April 1961, on the eve of the SPC meeting, must therefore be borne in mind. Firstly, Adams' new text allowed for the possibility of a choice. Whereas Hine had argued that the ISR was the *only* serious solution, Adams considered two possibilities, the ISR and a more classical fixed-target machine. However, it must be said that he did not hide his preference for storage rings, nor did he hide the fact that, to date, this was the only project into which CERN had put any effort. What is more, Adams did not change the time scale: for him, as in Hine's text written a few weeks before, the SPC was asked to take a stand quickly. In other words, even if Adams' text allowed for a choice, it was heavily biassed in favour of the ISR.

Not surprisingly, then, the physicists continued their 'agitation'. At the initiative of Léon Van Hove, the head of the Theory division, a conference was scheduled for June with a view to understanding better what was required 'from the physics point of view' — i.e. what the physicists' wishes were. This development stirred Adams into sending another note to the divisional directors on 28 April. In it he asked them all to trust the directorate 'to present the problems fairly to the Scientific Policy

Committee' and he insisted that the aim of the SPC meeting the next day was 'simply to start them thinking seriously about the subject and no decisions between the alternatives [were] expected until at least the end of this year'.[126]

The second point calling for comment is the sense of urgency expressed by Schoch and his colleagues in December, by Hine in March, and by Adams on 14 April. It is easy to understand why, as machine builders, they adopted this stance — the desire to get their teeth into a new project, the impression that the Americans were once again getting ahead, the feeling that, since the ISR project was well on its way, it was unnecessary to wait, etc. On the other hand one cannot but be surprised by the fact that they did not prepare the ground among the physicists, the members of the SPC, and the delegates from the member states. None of these had even begun to think about the question at this time. The last, for example, were convinced that major expenses for equipment — like the PS — were a thing of the past, and had not even accepted that the annual budget for operating the laboratory would not be constant but would have to grow gradually. To ask them suddenly to double or triple expenditure on CERN — and to expect them to decide this in a matter of months — was to ask the impossible.

12.3.2 THE PENDULUM SWINGS BACK, APRIL–JULY 1961

The spring was punctuated by three important events. On 29 April the SPC debated at length Adams' report and the documents issued under the name of Hine and the four members of the AR division. From 5 to 9 June Van Hove's international conference on very high energy phenomena took place, and arrived at a number of rather clear conclusions. On 21 July the SPC met again.

12.3.2.1 The SPC meeting on 29 April

Adams opened the discussion on point 5 on the agenda with a long and very subtle intervention. Taking the disruption which had occurred during the previous four weeks into account, he stressed that his aim was simply to stimulate discussion, not to terminate it. He explained that if the ISR alternative was presented in some detail while that for a new accelerator was not, this was only due to a shortage of staff. 'However', he went on to say, 'this has been done in the United States' and to remedy the deficiency he suggested sending some members of the AR division across the Atlantic 'to work with the American design teams engaged in these projects, this summer'.[127]

The first thing that the members of the SPC forcibly stressed in reply was that it was crucial not to lose sight of CERN's priorities. For Amaldi, Cockcroft, Leprince-Ringuet, Perrin, and Powell, to mention only the principal contributors to the debate, in the context of CERN's budgetary difficulties the most important thing was to exploit the PS to the full. This was to be understood in a restrictive sense: for

Notes: pp. 771 ff.

them the ISR was a new machine, not an 'improvement of the PS'. Thus objective number one, the task to which all available resources had to be dedicated, was to do the best possible physics with the existing machine — which meant, for example, finishing off the East experimental hall, improving the beam system, running the PS 24 hours a day, etc.[128]

The second thing that the SPC members did was to refuse not only to choose what kind of accelerator CERN should build, but even to decide *there and then* if the construction of such a machine should get under way at the end of 1962. This was partly because of the prevailing mood in politico-financial circles, and partly because many European physicists were not yet convinced that an increase in accelerator energy was essential, or at least urgent 'from a physics point of view'. As a result the SPC proposed that CERN set up a study group whose aim would be to assess the scientific merits of any future development and, that done, to propose, if necessary, the best way, or ways, forward. To this end the Committee authorized the Director-General to recruit ten additional staff in 1962 and resolved to request the Council to make an additional 500,000SF available in the budget. Finally, to show that it was interested in this venture — the SPC understood the argument that CERN should not fall behind the United States — the body recommended that this sum be kept aside for this purpose no matter what other adjustments there may be within the budget.[129]

This being settled, the members of the SPC spent considerable time discussing the suggestions made in the various documents laid before them. With one or two notable exceptions — and this is the third characteristic feature of this meeting — they rejected in advance the idea of building a *monster*. By this time the word was commonplace and was used constantly in meetings and in correspondence. It meant, in Dakin's words, a big machine 'based on a not very imaginative application of known principles, and which by the time it is built may be found to be out of date because someone has been a bit more clever a year or two after the plans were settled'.[130] The example *par excellence* — and the case at the back of everyone's minds — was Dubna's 10 GeV PS which was started just before the strong-focusing principle was formulated, a fate which CERN escaped by the skin of its teeth in 1952. In fact behind this notion we find a mixed bag of fears and uncertainties, an assortment of impressions never clearly formulated, but very influential all the same. For example, there was the conviction that each step forward in the past was always accompanied by the implementation of a new principle, by the development of a new 'species' as Adams called it in 1956. Of course there was nothing 'absolute' about this, nothing obligatory for the future, but the idea was widely held in Europe early in 1961. Similarly, the notion of monster conjured up an image of the use of brute force, which was roundly condemned. Perrin and Amaldi, for example, contrasted it with the ISR which seemed to them to be more alluring, less unreasonable, more intelligent in a way. The 'monster' thus connoted at once a feeling that it was not urgent to increase available energies, a hope for a radically new idea (like the superconducting magnets of which people were speaking), a rejection of needless excess (be it in terms of ambition or of cost), a criticism of a lack of imagination or of the urge to go ahead just for the sake of going ahead. Bernardini, in fact, spoke for the large majority of the SPC when he said that 'nobody would like to do it' in Europe, 'it' being 'the American type monster' then being studied at Brookhaven, a 300–1000 GeV PS.[131]

One last remark. Since the term served rather to draw together an assortment of anxieties, its use varied considerably from one person to the next. Among the older generation, the usage was rather broad and it applied as easily to very high energy synchrotrons as to high intensity synchrotrons (50 GeV, 10^{14} p/s). For a Scherrer or a Perrin, for example, monsters also encompassed 'the inconvenience caused by residual activity which would entail remote control operation on a very large and expensive scale'.[132] At the other extreme, for young people like Gregory in the SPC or Ramm inside CERN itself, the term was not used in such a negative sense. Ramm was able to envisage, in the medium term, a machine of 300 or 1000 GeV and Gregory announced himself ready to consider very high intensity synchrotrons — even if he admitted that he found it difficult to see 'good scientific reasons' for a 300 GeV.[133] Put differently, what we find here is a situation not dissimilar to that we met when the proposal to build the CERN PS was being debated in 1951-53. Then the 'opponents' to the biggest accelerator in the world were drawn from the European 'establishment' of the day, were relatively older, and were called Bohr, Kramers, Chadwick; the defenders were Amaldi, Auger, Perrin, even Cockcroft. Ten years later the 'opponents' to the new monsters were the Amaldis, the Cockcrofts, and the Perrins — and the less negative were drawn from the younger generation ... and the machine builders. If the latter were not particularly keen on a 300 GeV, it was not because they thought of it as a monster, but because they preferred the ISR.

Does this mean that the members of the SPC had no positive views on the next generation of machines? No, that would be false. Some of them — like Amaldi, Perrin, and partly Cockcroft —took the ISR seriously because it was an antimonster, because it was a relatively cheap solution whose immediate interests (sample the physics possible at 50 GeV in the centre of mass) and wider significance (learn the technique of beam intersection, the only solution in the long term) were not negligible, and because it would allow CERN calmly to await the next generation of fixed target machines.[134] Others — like Salvini, Gregory, and partly Bernardini — doubted the interest of the ISR, spoke of high intensity machines (which were easily justified both technically and scientifically), and did not necessarily reject the monsters.[135] Others again — like Leprince-Ringuet, Powell, or Scherrer — expressed no particular opinion and adopted the stance agreed on by the entire SPC: it was not the construction of the new machine that was urgent, but the full exploitation of the PS.

Notes: p. 772

12.3.2.2 Arguments for and against the proposal to build the ISR

Before taking up our narrative, we shall pause to consider some of the more important arguments deployed by protagonists and opponents of the suggestion that CERN build the ISR. The debate on this proposal got under way in January 1961, flared up during April, continued vigorously in May, and subsided temporarily in the weeks after the conference organized by Léon Van Hove in June. It is noteworthy for its acerbity, for the very 'mixed' nature of its arguments, for the fact that it took place only between scientists — and for the unambiguity of its outcome: the CERN physicists deemed the ISR project to be of little interest.

The arguments advanced by the ISR's partisans were relatively straightforward.[136] Broadly speaking they were of four or five kinds. They said, firstly, that the ISR would allow for an improvement in the PS's flexibility in that each storage ring could be used to set up new experimental areas — a need strongly felt at the time —, that each made it possible to have 'very long pulses [...] and/or very intense short pulses (10^{14} protons per pulse)', and that the rings could be used to accumulate particles other than protons, like antiprotons, deuterons, ions (He^4, Li^6, C^{12}), and even electrons. It must be added, however, that, as regards this last consideration, the defenders of the ISR became increasingly prudent with the passage of time.[137]

The second type of argument for building the ISR dealt with its aspect as a collider. It was described as 'a window open on the future', as '[one] possibility of probing into the very high energy region', that of 50 GeV in the centre of mass, and so equivalent in energy 'to experiments on stationary targets bombarded with 1300 GeV protons'. The same was true of d-d, d-p, ... collisions.[138]

The third kind of argument was used in the debate when those we have just mentioned were contested on the grounds that the machine imposed limitations on experimental techniques. It ran thus: the ISR is not perfect and does not allow one to do everything. But it does constitute an excellent *intermediary tool* to be deployed while one is in no position to build a monster and has to wait. It could be installed *quickly* — unlike other projects which, for technical or political reasons, would take much longer to mature; it was *cheap* and its construction did not preclude building costly fixed-target machines in five to eight years time; finally, it was an irreplaceable stopgap which would allow one to take a properly-informed decision about a giant machine. In short — because one could do a lot of good physics with them — the Intersecting Storage Rings were good value for money and had a high ratio of quality to timing.[139]

A fourth category of arguments put forward was that the ISR could be installed *at CERN itself* without calling for any marked increase in staff. Big projects, on the other hand, required the recruitment of new personnel who were difficult to find on the labour market — at least without destabilizing and competing with other major projects like the space programme — and the selection of a site other than Geneva. If that were done CERN would be digging its own grave since the new venture would attract the best people away from the parent laboratory. [140]

The most violent attacks against the ISR — and actually the first which appeared — were launched against the very principle of the device, the intersecting beams. The experimentalists in particular pointed out that the ISR would not provide beams of secondary particles (π, K, $\bar{p}$ ...) — just the beams they needed to do most if not all their physics (the search for resonances, for example); it was said that research into 'rare events', which was also at the heart of the new physics, would be very difficult, even impossible in the intersection zone; it was repeated that one could no longer use a variety of experimental techniques (because of the configuration imposed by the machine) and that most of the techniques then in use would have to be jettisoned. Yes, in 1959 O'Neill had proposed building bubble chambers which completely surrounded the intersection zone, but most physicists did not take this seriously: from the point of view of *doing physics*, the ISR was not an accelerator but just *one experiment*, an experiment on proton–proton scattering, and this was an intolerable restriction.[141]

Those against the ISR also answered other arguments in its favour. They disputed some of its alleged 'advantages' — the fact that it could be used with ion beams, for example, which the physicists deemed of secondary importance, not to say devoid of any interest whatsoever; they questioned the 'impossibility' of certain measures — for example, the construction of a synchrotron of several hundreds of GeV near CERN; and they insisted that certain 'gambles' were just not worth taking — for many physicists the ISR was not a good intermediary tool: on the contrary, 'if CERN is to build storage rings', one said, 'it will probably have to wait for *two* accelerator generations (i.e. something like 15 years) before new sources of secondary particles become available' — and this was totally unacceptable.[142]

Of course the protagonists of the ISR replied in their turn to these objections, and so on. We shall only mention one reply to a reply here, namely, that which took on the claim that it would be difficult to do experiments with the ISR. For Hine, this fear was simply symptomatic of a tendency for the physicists not to see beyond their noses. 'I say short-sighted', he wrote to Weisskopf in May, 'because it is all in terms of existing detector techniques: nobody remembers that the PS was decided upon before hydrogen bubble chambers existed at all, not to speak of separators'. It was up to physicists to *imagine* appropriate detectors, then, rather than to reject the ISR in advance in the name of existing techniques.[143]

To conclude, we want to draw attention to the intimate interconnection of the arguments employed and to the fact that there was not *one* logic of debate shared by all, but various large constellations of themes opposed one to the other. Two such dominated.[144] That of the 'builders', which was the most sophisticated and which took the widest range of parameters into account. Schematically it may be

Notes: p. 772

characterized thus: in a certain — relatively low — price range, and within a certain — relatively short — time scale the ISR offered the greatest number of advantages. The second, that of the future users, rested on *one* key argument: the need for secondary beams, without which no serious physics was possible, and which the ISR would never furnish. In this situation one cannot really talk of there being a discussion. For one group the attitude of the other was narrow, conservative, uninformed, while their own reflected a willingness to innovate responsibly, to take a calculated risk. For the second group, if the approach of the first revealed a taste for adventure (and had the advantage of elegance, some would add), it also suffered from a fatal flaw: it did not allow one to study the most interesting topics in particle physics.[145]

At Van Hove's conference the position of those hostile to the ISR was reinforced: the future users, the physicists, confirmed their first reactions. Summarizing the conclusions drawn from the meeting on 13 June, Van Hove wrote that 'Most of the high energy problems which can now be predicted to be of central interest in the coming years require good, intense beams, not only of protons, but of the unstable particles, with an energy as high as possible but usually not an order of magnitude higher than now available'. 'More than for a transition to much higher energies', he added, 'there is a great need for methods of detection and data analysis which would be highly selective and suited for very rare events', i.e. calling for high intensity. Of course he mentioned the uncertainty inherent in predicting needs ten years hence — particularly when some crucial data at 25 GeV were still awaited. All the same he concluded that while the ISR undoubtedly had merits, they were 'not based on physics considerations'.[146]

In a longer and so more nuanced 'Summary Lecture' written shortly thereafter, Van Hove took the opportunity to survey the important theoretical issues more carefully. Regarding weak interactions, for example, then of central importance to CERN because of the neutrino experiment, he wrote that 'a wealth of valuable information could be obtained from high energy ν reactions, if at least the neutrino flux can be made sufficiently intense' and, more generally, 'many important developments [...] would be greatly helped by a moderate increase of energy and of intensity for π and K beams'. Concerning strong interactions, he insisted again on the need for 'numerous accurate data', stressing the possibility 'of higher symmetries among the strongly interacting particles' and that 'some or most of the so-called elementary particles would in fact be composite'. All this was nevertheless rather speculative and Van Hove ended on a note of caution: a large *flexibility* in use — many secondary beams, a variety of experimental arrangements — and a good *intensity* — needed for high-quality data — ought to be the objectives. Hence his conclusion: that though the importance of having very energetic primary protons was proven, an 'energy increase should not go at the cost of the flexibility and intensity requirements mentioned above'. Which quite simply amounted to an outright dismissal of the Intersecting Storage Rings.[147]

12.3.2.3 The SPC meeting in July 1961

Everyone interpreted the results of the June conference in this way. In a document prepared for the SPC by the CERN directorate, for example, the formulation was: 'To gamble the future of the Laboratory on say p-p interactions and a few others even at a c.m.s. energy of 50 GeV was less attractive than say the safer course of building a 100 GeV proton synchrotron with an intensity of 10^{13}-10^{14} particles per second'.[148] Even more explicitly, Adams opened the SPC meeting by remarking on 'a general lack of interest for the idea of storage rings' during the conference. Whereas in April, he went on to say, one was hesitating between the ISR and 'monsters', the stress now seemed to be on high-intensity machines.[149]

Half of the SPC meeting in July was devoted to this question of the 'long-term programme of CERN'. The negotiations between the USA and the USSR dominated the proceedings, even though the members were sceptical about its chances of success. The first question they considered at length was that of the type of intercontinental machine they wanted — if such a machine was agreed on, and if the Europeans participated in its construction. At the end of June Adams had said that 'In private discussions with physicists from America and the Soviet Union it appeared that a 300 GeV PS of normal intensities [i.e. 10^{11-12} protons/pulse] might now be considered'. On 21 July, having received J.P Blewett's report of 15 May which described the studies made at Brookhaven, Adams spoke of a proton synchrotron of 500-1000 GeV of the highest intensity possible, so at least 10^{13} protons/second.[150]

The second question discussed by the SPC was what to do at CERN if the intercontinental machine was not built. The dominant feeling was that, in this case, one should follow the conclusions reached at the June conference and envisage an average-energy PS (50 or even 100 GeV) of very high intensity (10^{13}-10^{14} p/s). One of the alternative projects mentioned during the meeting was a 600 MeV superconducting proton linac, defended by Salvini and vigorously attacked by Amaldi and Leprince-Ringuet. While they accepted that this machine would supply very intense pions and muons, they emphasized that all the physics of K-mesons, antiprotons, and neutrinos would be impossible. This, as Amaldi pointed out, would go against 'one of the conclusions of the Conference [...]: not to limit the choice of beams'.[151] To this a further question was attached: When are we to decide that an intercontinental machine is not going to be built or, put differently, what are we to do if the negotiations continue indefinitely? The consensus on this point was that the technical study of the European machine was to be undertaken immediately by the AR division, though a decision about whether or not to construct it need not be taken for 18 or even 24 months — after all the April conclusion was still valid: there was no hurry, and the choice would be that much easier after one had some firm results from the PS at 25 GeV.[152]

Third question: what was to be done if the intercontinental machine was agreed to and the Europeans were partners in the scheme? Here opinions varied considerably.

Notes: pp. 772 ff.

One tendency insisted that it would be difficult to construct two machines in parallel for financial reasons, because of the lack of manpower in Europe, and because the world machine could not but be situated in Europe for political reasons.[153] The consequence of this line of reasoning was not to envisage a new European accelerator if a positive decision was taken on an intercontinental synchrotron. Hine, on the other hand, inspired a different way of looking at the matter. He emphasized that without their own machine the Europeans would be reduced to a secondary role in the tripartite collaboration. Accordingly he proposed that the order of priorities be reversed: the Europeans should go ahead and construct their own machine anyway, participating in the other according to their possibilities. This position prevailed at the end of the meeting and, at the suggestion of SPC Chairman Powell, it was resolved that CERN should devote all the necessary attention to 'a 50–100 GeV proton synchrotron with an intensity of 10^{13}–10^{14} p/sec'. In the meanwhile the effort needed to keep in touch with the project for an intercontinental synchrotron of '300–500 GeV with a relatively high intensity of, say, 10^{13} p/sec' would be sustained.[154]

This debate leads us to ask what has become of the word 'monster' used so frequently a mere three months before. The answer is simple: it has virtually disappeared, which indicates that people had become accustomed to the scale of equipment it signified. All the same it is true that machines of 300–500 GeV were still generally thought of as being beyond the range of Europe alone, that they remained associated with a tripartite project — and a project which was very much at the hypothetical stage. In this sense the new habit of thinking was only partial: the shadow cast by the monsters was still at work. And as a result the envisaged increase in energy for a European machine was relatively moderate: the SPC agreed on 50 to 100 GeV.

12.3.3 WHY THE CONFLICT BETWEEN 'BUILDERS' AND 'PHYSICISTS'

The debate we have described was a debate between individuals who defended very different points of view. All the same, two or three major attitudes can be identified, attitudes which map onto particular 'socio-professional locations' vis-à-vis knowledge and the power relations inside the organization. This is hardly surprising if one does not insist on treating scientists as disembodied spirits inhabiting an abstract realm of ideas and arguments. Very schematically, what scientists do is to develop, to refine, and to test models which take account of the regularities or differences revealed in the experimental set-ups they deploy. The models they elaborate in this work flow from formal rules, heuristic demands, and particular traditions which differ between specialities (mathematical physics, pure or more phenomenological theoretical physics, experimental physics using 'electronics' or 'bubble chambers', accelerator construction, conception of new detectors...), between schools, between generations, between professional formations etc. In consequence their ways of approaching new questions, of formulating and of resolving them have every likelihood of bearing the imprint of these 'locations' and of the histories which led to them.

In the debate which concerns us here one rift emerges with particular clarity, the rift already alluded to between those whom we have called 'the builders' and those whom we have called 'the physicists', be they theoreticians or experimentalists. This division is not only easily detected by reading the minutes of meetings or correspondence, it was also recognized by the actors themselves.[155] We do not therefore need to justify it further. More interesting is to establish the logic underlying the dichotomy, the reasons why it took the specific form which it had. To do this, we could first ask why, until March 1961, the physicists showed so little interest in the work being undertaken in the Accelerator Research division. The answer is simple. It is because it was of no relevance whatsoever to their concerns at the time. All their energies were absorbed in 'doing physics', and in defining the PS experimental programme in particular. Was this just shortsighted, as some builders would have us believe? We do not think so. It was in the logic of things, that of physicists impatient to get to grips with *their* 'frontier'. Do not forget that until 1960 they had worked in the energy range 3–6 GeV; that with the coming into service of the PS the horizon expanded to 25–30 GeV. How can one imagine that, in this situation, they could already be fascinated by energies of 300 GeV and willing to spend their precious time considering machines of the generation $X+2$ — just as generation $X+1$ was coming on-line, and the physics they knew was that of generation X! 'Logically', one might say, they first wanted to *use* the new PS, and not study what may be needed in 1970–80. As they stressed, they knew very little about what happened at 30 GeV — and to extrapolate directly from 3 to 300 was too arbitrary, it was a leap which was too great and could not be justified 'from the physics point of view', to borrow the hallowed phrase.[156]

As for the builders, once the 30 GeV machine had been commissioned, they began to look in the opposite direction: what the physicists saw as a goal (using 30 GeV machines) they saw as an achievement (since the machine worked), something over and done with. Now one needed to look ahead, there was no time to be lost and the next machine must be discussed *immediately*: after all it could take eight or ten years to build. The different attitudes of the physicists and the builders can thus be partly understood by relating them to their specific positions vis-à-vis knowledge: when a new accelerator comes into service — and when the excitement and the impatience peak — it is as if there was a gap of 'two machine generations' between them. From whence their different attitudes as to what is 'urgent'; from whence the initial lack of interest of the physicists; from whence the reciprocal charges of ' narrowness'.

This is not, however, the only source of the differences between the two groups. Even when they look at the same problem — here the machine of generation $X+2$ — their interests lead, pull them in directions which are not necessarily identical. That

Notes: p. 773

which motivates, that which preoccupies, that which drives each group to burn the midnight oil, is not the same. For the one, it is above all the machine, for the other, the use of the machine; for the one, it is above all the technology of the machine, its conception, for the other, the experiments permitted or forbidden by the technology; for the one it is the object as object which fascinates them, for the other the possibility of playing with that object. Of course each is aware of the logic of the other: the subtlety of the ISR project seduced physicists like Perrin and Amaldi, and the demands of flexibility and intensity convinced Adams in July 1961. But there are reflexes, fascinations, interests, — this time in the *double* sense of being interested in something and having a vested interest in it — and the two groups tend to move apart.

A second question: was there an *a priori* hierarchy in the ways of approaching problems, a hierarchy which was accepted by all at the time, at least in principle? Yes, it was that the physicists had prior rights in decisions of this kind since the machine was being built *for them*. It was in the name of the physics that they expected to do — the 'ends' as defined by the physicists — that the machines had to be chosen — the 'means' as proposed by the constructors — and not the contrary. What we find affirmed here — and notably in June — is that there was a ranking of scientific knowledges, a ranking which implied a hierarchy of competences. Certainly those who constructed the PS played a crucial role in the daily functioning of the organization and no one would deny the importance of Adams or of Hine. In many domains — finance policy, the preparation of budgets, the estimation of long-term needs, the construction of big detectors (bubble chambers), the preparation of beam separators etc. — it was they who managed the organization and got it to function as an institution. In a choice of this kind, on the other hand, it was the physicists, the future users, those for whom CERN was established, who were accorded rights of priority.

A final question: what refinements ought we to add to this analysis in terms of two groups? How far can we reasonably extend it? As always, one must first remember that explanatory models have limits. Each person has in fact an original intellectual and emotional configuration, a specific 'internal heuristic programme'. Hence one finds spectra of attitudes which are never completely discontinuous. There are also polarities other than those between 'builders' and 'physicists'. We have remarked on the generational differences, but there are others. Mutual distrust and nuances of opinion are perceptible, for example, between theoretical and experimental physicists. Take Melville writing to Weisskopf about the fears of British experimentalists: 'They stressed the danger', he wrote, 'that long term plans might be formulated by theoreticians on the one hand, and engineers on the other, without sufficient consultation with experimentalists'. The most marked rift — that between constructors and physicists in our case — is thus perturbed, blurred, by a range of sub-rifts.[157]

One last point: one must not *freeze* the image. We are not dealing with 'programmed' automatons whose sole purpose is simply to vanquish those who oppose them. These men discuss issues within precise mental, social, and professional frameworks it is true, but they also seek to make the 'right' choice — and this makes them open to technical suggestions and arguments complying with the theories and paradigms current in the field. Hence the changes of opinion, the evolutions in the appreciation of alternatives, the refinements in the images, the models, the ideas which these men create or accept. The following episode in our story nicely illustrates this last point. It shows how technical studies made in the United States led to a radical modification in the *European* list of available technical 'means', in the divisions between European groups...and ultimately in the perception of what was at stake.

12.3.4 THE CONSEQUENCES OF THE AMERICAN SUMMER, SEPTEMBER–NOVEMBER 1961

In the first part of this chapter we explained at length the nature of the work done at Berkeley and Brookhaven from July to September 1961. What we want to do now is to explore European perceptions of the American studies and to investigate their effects.

12.3.4.1 The American 'shock'

What the Europeans 'learnt' from their American colleagues was above all a way of posing problems, a way of reacting to certain questions. Three things struck them particularly.

Firstly, the conviction, widespread in the United States according to them, that the most important thing to do with any new accelerator was to increase the available *energy*. Secondly, the way of *arguing* for this increase. As Hine put it when summarizing the talk made by the theoretician R. Serber, the thing to do was 'to look at the past and say that every time we've increased centre of mass energy by a certain amount, it's paid off much more than we ever imagined at the time, and then ask if there is any reason why this won't be so in the future'.[158] In a report to the SPC Hine added that Serber 'warned that asking theoreticians to justify it [an increase in energy] in detail might do more harm than good'. Anyway, he pointed out, American theoretical work had produced no more than that undertaken in Europe in June.[159]

The third thing which struck the Europeans was that their American colleagues were unanimous in thinking that the machine of the next generation should be a PS (of the type then at CERN) but with an energy of at least 200 or 300 GeV. For them, 'it was considered to be a general purpose accelerator of vast capabilities, and justifiable as such', 'a practical engineering proposition' which was better than all other machines in terms of energy *and* intensity for both secondary *and* primary

Notes: p. 773

beams.[160] As Adams explained to his British colleagues, what the Europeans learnt during the summer was that 'the AG machine principle really seemed to imply [...] that one can't talk about high energy and high intensity machines as alternatives these days; one has to think of them as the same thing, and in fact the higher the energy the more intensity one is likely to get. This is a complete reversal of where we were before [during the Van Hove conference], when people said "Well let's not go for energy, let's go for intensity", so they dropped the energy in an attempt to make a high intensity machine'.[161]

These three elements being given — and to speak of a 'shock' for most Europeans is not to exaggerate — certain conclusions were easily drawn. Before presenting them, however, it is worth asking three questions about the situation in the United States.

First question: did this American way of justifying an increase in energy imply a lack of interest in physics questions which were so dominant at Van Hove's conference organized at CERN in June? Paradoxically, perhaps, the answer is no. To take the case of Serber himself, it is worth noting that, having encouraged people in his introductory speech to the international accelerator conference in 1961 to rally to the argument of 'the generosity of Nature' — 'and not to ask theorists for further justification' —, *the heart* of his talk was precisely dedicated to grasping the interest of a 300 or 1000 GeV machine as a means of 'resolving' certain fundamental problems in physics.[162] And indeed it is commonplace to find the two types of discourse going on in parallel. Put differently, even if the Serber argument was seen as adequate *justification* for outside consumption, it did not stop the physicists immediately thinking of the physics they could do, of how they would do it, or of what they could hope to find. And of course these arguments in their turn could be used to support or to undermine the argument of the 'generosity of Nature' itself.

Second question: when the Americans studied the points already raised at CERN in June did they arrive at the same results as Van Hove or did they rather show that it was crucial to increase the available energy, that this was the most important variable? The answer here is somewhat delicate. In one sense the results were very close in that the same problems — with some slight differences in emphasis — were identified as being those for which it was essential to increase the energy considerably, and in that all agreed that it was important to have multiple and intense beams of different particles.[163] The conclusions differed all the same in that the Americans were more 'optimistic' about the results that one could get at high energies and in that they more often tended to see encouraging signs in the doubt and uncertainty surrounding certain problems. In other words, confronted with a group of theories and interpretations about which there was not an overall consensus, and with incomplete experimental results at 30 GeV, the Americans tended to believe that high energies would settle the question while their European colleagues, stressing different aspects, were far more sceptical, and loath to commit a large amount of money to a venture whose results, in their eyes, might not merit it.[164]

Third question: Was there not, one might ask, a more conjunctural reason why the Americans at this time so frequently repeated that Nature would be generous, and so 'flippantly' (in European eyes) affirmed loud and clear that one did not need precise reasons for building a high-energy machine? To be blunt, was it not because a PS of 300–1000 GeV appeared, during the summer, to be THE ideal machine satisfying different demands (notably that of intensity) that one was so easily convinced of the 'scientific interest' of this energy range?[165] To this we would reply yes, in all likelihood, for the distinction between the ends (as defined by the physicists) and the means (proposed by the builders) no longer had much meaning: after all there was only one way of satisfying everyone's needs. That granted, how was one to fix the only independent variable, the energy? By the amount of money one could obtain since, scientifically, 'the bigger the machine the better, and people who talk about a smaller one are merely faint-hearted!' Seen from this angle, the Serber argument served to ratify the new situation.[166]

What then do we find in Europe after the summer — since obviously the account above has shown that the logic was somewhat different before.

During the European debate in the Autumn, a marked attitude of prudence and scepticism was perceptible, a resistance to 'knock down' arguments (like that of Serber) or to those which encouraged reckless expenditure. In this connection it is worth noting Hine's efforts to convince his British colleagues that physics at very high energies was, in fact, cheap, that Space cost society much more, and that physicists should therefore not hesitate to ask for all the money they needed.[167] Hence the feeling of many Europeans that they were being unduly pressured in the energy race by the Americans; hence their suggestions for a 'moratorium' — why not wait for better results at 25 GeV? why not decide later, when more is known?; hence the proposals for a joint programme with the United States, which would tackle the problem as a whole, and in a coherent way; hence the frustration which many felt at the insatiable urge of the Americans to build bigger.[168]

The second thing to keep in mind is that the conviction that there existed ONE obvious multipurpose machine for the future was not yet shared by all. Such a machine existed — but only after the terms of the debate had been suitably simplified. The partisans of the ISR stuck to their claims that their machine had a better quality/price ratio and that it was more 'refined'. What they suggested was that there could be a different *frame of reference* for judging future projects, another financial-technical paradigm. They were in the minority, it is true, but they did not climb down, and insisted that the rules of the debate could be reconsidered, particularly because a 300 GeV, 10^{13} p/s PS would be *awfully expensive*.[169]

The last point we would like to stress is that, despite the attraction exerted by a PS of several hundreds of GeV, not everyone had the same reactions, convictions, impressions. During a debate in the United Kingdom, for example, Cassels refused to believe in the need to compete in the energy race, and refused, at least for the time being, to believe that Nature was necessarily bounteous. 'We were told this morning', he said, 'that every step in energy has led to new and exciting physics. Well this isn't quite true in the step from 6 to 25 GeV'. At the opposite extreme Salam, already famous for his contributions to theory, insisted that THE question was that of substructures, of hidden symmetries, and that projectiles of hundreds of GeV were necessary: 'the trouble with CERN', he said in reply to Cassels, 'is that the energy is too low. I am sorry, but this is the truth'.[170] Finally Hine or Adams defended a PS of 300–400 GeV without in any way losing their interest in the ISR. One can see then — though we lack sufficient examples to make a typology — that no opinion was globally held in Europe, even after 'the American summer'.

12.3.4.2 The concrete decisions taken at CERN at the end of 1961

On 18 October Mervyn Hine summarized the work done during the summer in a report submitted to the SPC on behalf of the directorate and entitled 'The CERN Long-Term Programme'. The Committee discussed it ten days later and instructed the Director- General to inform the Council of its conclusions. On 30 and 31 October more than 30 European accelerator specialists gathered together as the European Accelerator Group (EAG) to consider the overall situation. Finally, on 27 November a report by the AR division entitled 'Proposed Programme for a CERN Study Group on High Energy Projects' was published under K. Johnsen's name. After being informally studied by several members of the SPC, this collection of documents was submitted to the Council in December 1961 where its main proposals were ratified. They fell into two groups: those dealing with a possible intercontinental machine and those discussing the programme of the CERN working group for the next 18 to 24 months.

In the SPC meeting in October it became clear that the question of whether Europe should participate in technical studies for an intercontinental machine ought to be held distinct from that regarding CERN's future accelerators, the former being subject to too many political uncertainties. In September, as we said earlier, the Soviets did not go to BNL. Nevertheless a short while later, at a meeting at the IAEA in Vienna and at a Pugwash conference in Stowe, Vermont, they confirmed their wish to continue working on a joint machine. For practical reasons, and at the suggestion of Amaldi and Perrin, the SPC proposed that the DG should represent Europe for all technical matters concerning the intercontinental project, and should choose his own advisors in consultation with the member states. The Council accepted this proposal on 19 December 1961.[171]

As for the CERN long-term programme, in October the SPC confirmed that it was obviously too early 'to make any valid recommendations about whether CERN should undertake the construction of a new accelerator', and it expressed the wish that the working group which it had proposed in July be established forthwith with a budget of 500,000 SF.[172] Organized by the AR division, this group would gather CERN physicists, visitors, and physicists and builders from other European laboratories. The SPC paid scant attention to its exact programme. If Bernardini raised the 'very naive question', to wit, 'what is it supposed to study?' no one risked a reply of any precision. Instead the SPC adopted Hine's idea that one should first debate the project during the EAG meeting on 30 and 31 October.[173]

Two questions arise at this point. Firstly, why did the SPC not really discuss the working group's programme? Probably it was due to the technical complexity of a debate in constant evolution, to the multiple parameters involved, to the impression that all the necessary information could not be had. Let us not forget that the SPC, as influential as it may have been on paper, was not a group of accelerator experts. It comprised the elite of the European nuclear and particle physics community, men who were physicists rather than builders, and already somewhat advanced in age. What is more they had been told in April that the ISR was probably the best idea, in July that an average energy (50-100 GeV), very high intensity (10^{13}-10^{14} p/s) PS would be preferable, and in October that a '300 GeV' offered more advantages despite its high cost. Confronted with this information, confronted with the difficulty of forming a stable picture of the prevailing alternatives, confronted also with persistent uncertainty in the member states regarding the possibility of increasing the CERN budget, the SPC opted for caution. Anyway since a *decision* was not urgent it was wiser, this time, to wait for the results of other expert discussions, first in each country, then at the European level. It would take time to digest the results of the summer.

Second question: what was the European Accelerator Group? It was set up the year before at a meeting organized on 16 and 17 December 1960 by the Frascati laboratory, which was anxious at the time to debate its experimental programme and that of CERN. In the days following this mini-conference, Winter, 'who certainly represent[ed] the accelerator interests, at least in the [French] CEA', Quercia, his equivalent at Frascati, and Adams, decided that European accelerator builders should meet regularly. 'Clearly', Adams wrote to Winter on 3 January 1961, 'the primary aim is to discuss new accelerators'.[174] The first meeting was held at Saclay on 30 and 31 May, the second at Abingdon (Harwell) at the end of October.[175] At this meeting Adams presented an introductory report on 'the European High-Energy Physics facilities', and J.P. Blewett one on 'the design studies in the USA for a 300-1000 GeV machine', mentioning in passing the need to free oneself of the 'psychological difficulties associated with such large amounts of money [...]'. Thereafter Mills gave a report on MURA, Hine, Merrison, and Jentschke discussed the lack of qualified personnel in particle physics, Johnsen spoke on 'Intensity Problems in High Energy Machines', and there were a number of subsequent papers on related issues (space-charge questions, induced radioactivity, etc.).[176] We see then — and this is why it was worth giving the list — an overall shift of interest from May to October. In May one was discussing storage and intersecting rings at length (Quercia, Johnsen, Hereward — 50% by volume of the minutes) and synchrotrons of average energy and high intensity (Bruck, Blaser, and Laslett). In October, apart

Notes: p. 774

from MURA's report, these issues were no longer dealt with, and 'monsters' were centre stage.[177]

Did a programme for the working group at CERN emerge from this meeting of the EAG ? We do not think so; the conclusions reached were of a very general kind, insisting in particular on the need to coordinate the various levels of decision-making, international, European (CERN), and national. On the other hand, on 27 November 1961 the AR division, with the support of the Director-General, defined the exact tasks of the group. It wrote under Johnsen's signature:

'The task of the study group is to work on the design and use of the following two possible projects:
  a) A set of storage rings for the CPS
  b) A very large proton synchrotron, typically one of 300 GeV'.

Concerning point (a) — and justifying this choice — Johnsen added: 'Storage rings have been studied during the past few years and an electron model is under construction. This work should continue'. And on (b): 'A 300 GeV synchrotron is agreed by physicists in Europe and USA to be a very good general purpose successor to the present 25-30 GeV synchrotrons'.[178] Thereafter Johnsen listed the particular problems that had to be tackled. For the ISR, 'Inflection Techniques', 'RF Problems', 'The Terwilliger scheme [...] to reduce the size of the stack in the interaction region', the arrangement of the interaction region, vacuum techniques, site problems, experimental possibilities, and so on.[179]

The reasons for studying a 300 GeV PS — and no longer one of 50-100 GeV, 10^{13}-10^{14} p/s — are easily understood. But why the comeback of the ISR? It was almost forgotten in Europe in July, quickly discussed but not studied in any depth in the USA during the summer,[180] and ignored in the European Accelerator group in October. So why is it back on the list?

A part of the answer could be that there is no 'comeback' because the project never disappeared: considering *other* possibilities from May to October did not imply abandoning a serious investigation of the ISR for the CERN PS.[181] One could temporarily look into other alternatives without forgetting about the ISR option. This argument seems acceptable as long as one does not try to stretch it too far: the ISR never disappeared *for a certain number of people* (notably those in the AR division, and Kjell Johnsen in particular, the product champion of the storage rings), even if the majority felt that there was little to be gained by continuing such studies. The proponents of storage rings, we should add, did not remain inactive, organizing matters so as to be sure that the ISR at least remained on the agenda. In July, for example, O'Neill presented a very enthusiastic report on their behalf to the Berkeley seminar; a few days later he impressed upon the Director-General of CERN 'that the question was not to be dismissed lightly after such a superficial review [that done in June at CERN]', and he asked if he might come to Geneva '[in order] to interest experimentalists'.[182] But it was in September, in the one session dedicated to intersecting rings at Brookhaven, that the discussion was the liveliest, to the extent that on returning to Geneva Hine got in touch with O'Neill 'to see if we can arrange for a visit to CERN'. In short, during this period, the proponents of the ISR were not dormant and they set about trying to reverse the dominant trend. 'I felt you could give us a good deal of technical help', Hine wrote to O'Neill on 20 October, adding 'more important still, [I felt you could talk] as an experimental physicist who was enthusiastically *for* colliding beams rather than against them as CERN people here seem to have been'.[183]

This active policy of trying to keep alive a project buried by many is not surprising. It had, as always, two main objectives. On the one hand, to try to modify the rules of the game, to try to lay down the criteria in terms of which the choice between different projects would be made. As Hine said at Brookhaven in September, if one did not ask '"Would you rather have storage rings or a big accelerator" but rather "In terms of physics value for money are the storage rings a reasonable investment"', the answer changed completely. The other main objective was to convince other scientists that it was worthwhile for their understanding of scientific questions and for their future careers to take the time to study the physics that one could do with the ISR. In this way the picture which the community had of this machine could be refined — and the protagonists could hope to enlarge the circle of its partisans. This process, this 'manoeuvre' is thus both an interest-seeking strategy *and* leads to the generation of new knowledge which is accepted as such by the main protagonists.[184]

There is still one other reason for the 'reappearance' of the ISR in the picture. Twenty years ago, in a different context, and considering a completely different question, Allison showed beautifully that the information available to political decision-makers is always elaborated and structured by specific organisms, that it frequently reflects schemata already embedded in the long-term projects held dear by each section of an organization (in his case, the air force, the navy, the Pentagon, etc.). He also argued that, since the implementation of a decision has consequences for such bodies, the constraints, the traditions, indeed the simple habits of these various sectors carries real weight in the elaboration of options.[185] Bearing this in mind, we can imagine that a similar phenomenon was at work in our case. A genuine expertise in storage rings had been built up at CERN, particularly in the AR division. A notable fraction of the senior staff was convinced that the ISR was technically 'sweet' and no one said that it was of no interest — 'I don't think we are against it in principle', said Van Hove in September, specifying that the problem was that it would be impossible to construct only the ISR.[186] In short, because of what the AR division was, with its past and its research tradition, because 'big' projects spanning the medium or long term had always been the concern of the PS people, the *builders*, because we were still in a preliminary phase, at a stage when it was wise not to burn one's boats too soon, we can more easily understand why Johnsen proposed a double

Notes: pp. 774 ff.

726   *The second generation of accelerators*

programme on 27 November — and why the various members of CERN's directorate accepted it.

12.3.5 A GLOBAL SCHEMA FOR UNDERSTANDING THE DECISION-MAKING PROCESS IN 1961

To conclude this long and detailed section, we would like to survey the main lines of the account. At each stage of the decision process we ask the same kind of questions i.e. who were the actors — their projects, their strengths, their weaknesses —, what was at stake, and how issues were settled.

– December 1960

The only actors were the builders, members of Adams' old PS division, and those regrouped in the new Accelerator Research division in particular. They had the idea that a new accelerator was to be decided on quickly. Why?

1. Because they had just finished the PS (a great technical success), because they were ambitious (for CERN!), because it took seven to ten years to design and build an accelerator;

2. Because they knew that the Americans and the Soviets were speaking of a giant machine in common, because they saw Europe as being in a weak position again, because they were afraid of missing the bus;

3. Because the ISR was a 'technically sweet' machine based on a new idea (a new European species), because it satisfied the criteria deemed to be crucial (high energy for a modest price), because all the alternatives were either too costly or performed less well;

4. Because the Americans were not speaking of the ISR so that the two continents would construct different machines (global rationalization);

5. Because, unlike the physicists, the 'builders' thought in terms of making a career at CERN itself, and then favoured a quick decision.

Their strength lay in the fact that they were the only ones to have defined a project (though this could become a weaknesses if the tide turned against them). They also had the ear of the new Director-General, Adams (same group — old PS division; same tradition — predominantly engineers, heavy and sophisticated equipment; same preoccupations — CERN as a centre equipped with the best machines). In brief they saw themselves as the serious side of CERN, those who thought deeply of its future.

There was no real issue at stake yet, no debate. The text drafted in December 1960 was only an invitation to discussion, not a call to decision.

– April 1961

With the publication of Hine's report on 30 March (obviously with the support of

Adams who would not have been against CERN deciding on its new accelerator under his directorship) what had formerly been on the margins of most people's consciousness became central.

Other actors then defined their positions. The physicists, in particular, reacted violently: there was no urgency for a decision, all the more so since they had no project of their own (to ask for a decision now was thus unfair).

Their strength lay in:

1. The right which they demanded, and which all conceded, to speak first, and the power to reject any project which they would judge inappropriate to their goals (the demands of experimental practice);

2. Their weight in the Senior Staff;

3. The important network which they represented in Europe (the power of the high-energy physics community);

4. Their sharing the tradition of being physicists with one of the top policy-making organs at CERN, the SPC.

The outcome of this was the calling of the conference organized by Van Hove and the retreat of the CERN directorate (Hine-Adams). The latter realized that the question 'How is one to do physics with the ISR' was to be reconsidered (they realized that they had gone too far without proper consultation); what is more (as befits their role as members of the directorate), they did not want to risk dividing the organization deeply.

A new actor appeared, the SPC, a body attached to the Council and the governments which financed CERN, and which had to be convinced of the soundness of any important project. Composed of the European nuclear 'establishment', it often arrived at the same conclusions as the CERN physicists. It declared that there was no need for haste and that the operation of the PS was top priority. Less homogeneous than the other groups, it rejected 'monsters' and deemed the ISR intelligent; finally, it retained its trust in the management which had successfully built the PS.

At the end of April the physicists' values predominated: the means (the machines to be built) were to be subordinate to the ends as defined by the physicists (what physics to study what phenomena at high energy?). In the mean time the priority (as defined by the SPC) was to make the best use of the PS.

– July 1961

The physicists were now rather sure of themselves. They ended Van Hove's June conference categorically affirming the importance of flexibility of use, and the need for varied and intense secondary beams: very high energies, for their part, were judged of secondary interest. They arrived here:

1. Because to date they only had a little experience at 25–30 GeV, and to extrapolate from 3 GeV to 300 GeV or more was somewhat delicate (the trends are

confused, difficult to identify);

2. Because they were suffering from a lack of secondary beams at CERN at the time, and the neutrino experiment had shown dramatically the imperative need for good intensity;

3. Because they had not studied the ISR machine, nor could they imagine how to experiment with it;

4. Because physics on a fixed target machine of 300–1000 GeV would be very expensive (by European standards).

The conviction that the ISR had much to recommend it persisted among the builders; faced with the conclusions reached by the physicists, they 'accepted' that CERN consider a PS of 50–100 GeV, 10^{13}–10^{14} p/s; they did so without conviction, all the same, without taking this conclusion too seriously. Since, in principle, they accepted the dominant role of the physicists, and since there was no urgency, the stakes were not particularly high.

As for the 'decision-makers', the SPC accepted (shared?) the physicists' conclusions; it reaffirmed the need to proceed stepwise (incrementalism), without eliminating any possibility, nor any group (it, too, had a directing function). The new DG, Weisskopf, was not really in touch with the problems yet.

By the end of July the physicists had gained satisfaction and had perhaps even convinced some of their opponents. The key element in their 'victory' was the recognition that the means ought to be subject to the ends. It was understood that the matter would only be settled later (a decision in 1962), and this made it easier to rally the support of the SPC and of the undecided.

– October 1961

A new situation emerged after the American studies made in the summer.

1. The balance of forces: the Europeans were convinced that the Americans would construct a PS of some 300 GeV (problem: how to remain competitive?); the 'monsters' triumphed politically;

2. The list of technical means (the accelerators one could envisage): the alternative put forward in June seemed out of fashion; the 'monsters' triumphed technically;

3. The scale of costs: that of these machines was enormous; the partisans of the ISR could once again mention their idea (good quality/price ratio);

For Johnsen or Hine, the ISR remained their pet project: they organized its defence; all the same the '300 GeV' seemed inevitable to them: it was the conservative aspect which attracted the physicists; they recognized that this machine was multipurpose.

For the physicists the matter was clarified, even if it was not their chief concern: it was inconceivable not to have a machine which performed as well as that of the Americans. The '300 GeV' was, for them, THE solution (if there was some money left over, why not the ISR?);

The SPC, not feeling any urgency, simply ratified the proposal made by the AR division with the support of the DG. It could do so all the more easily since it was the solution which left as many options open as possible ('How to decide without really choosing', Schilling): the CERN group, attached to the AR division, would study the two projects in parallel for the time being.

Apparently there was no conflict at this stage since the '300 GeV' 'resolved' the earlier dilemma which arose from seeing energy or intensity as alternatives (eliminated as a false problem); the study of the ISR seemed to be a concession to an influential minority in the accelerator field; they would have to show, however — in anticipation of a final decision — that one could do 'good physics' with the ISR — and convince the physicists of it; whence the importance of the invitation to O'Neill.

12.4 The calm after the storm: the emergence of CERN's 'new study group on accelerators', December 1961 – January 1963

12.4.1 SETTING THE SCENE

To understand the turn of events in 1962 and 1963 we first need some additional contextual elements. About the intercontinental machine, to begin with. At the SPC meeting in March 1962 Hine reported that, although the possibility of a world machine was still under discussion, the Americans were becoming increasingly uninterested. This was particularly so at BNL (with whom the Europeans had the closest ties), where the prospect of collaboration with the Soviets no longer raised much enthusiasm. In response to a suggestion that one might have a western Europe – USA machine, Hine replied that BNL seemed to be turning towards Berkeley, and that the business was increasingly becoming purely American. In short, the thing to do now was to equip oneself to compete.[187]

Europe, however, was not a cohesive bloc. Its different national components were either constructing their own machines or expecting to do so. Certainly, unlike the United States, a single centre (CERN) was recognized, and other laboratories were not vying for absolute supremacy (as was the case with Brookhaven or Berkeley). On the other hand, with the summit taken care of at the European level, the competition at national level remained . We spoke of the Italian installations at Frascati in section 12.1. While on electron machines, we should also mention DESY in Germany. Approved in the second half of the 50s, the 6 GeV electron synchrotron was expected to enter into operation in 1963. A similar situation arose in Britain with its proton synchrotron Nimrod. Agreed to in 1957, it was to start operation in 1964.[188] But it is two decisions being taken in Britain and France *at around this time* (i.e.: 1962) that particularly interest us. The United Kingdom was still the major European scientific and technical power. Less integrated into the construction of Europe, closer to the United States (an always-available alternative if there was trouble with the

Notes: p. 775

continentals), she remained determined to develop her national programme in a balanced way. In high-energy physics this meant that her own programme was not to be 'subject' to the common programme, that of CERN. Thus in addition to Nimrod initially conceived as a high intensity machine to complement the high energy device in Geneva, the British scientists recommended, in November 1960, the construction of an electron synchrotron. The project was detailed in 1961 and the final decision was taken in 1962: in July 'the Minister for science announced financial approval' for a 4 GeV synchrotron.[189] Because of this recently agreed on large and expensive accelerator construction programme, British physicists placed themselves in a delicate situation between 1962 and 1965, when CERN made its demands.[190]

The second case, that of France is roughly symmetrical. For reasons no less obvious — and notably the desire to favour whatever enabled Europe, and so France, to remain at the research frontier at a level at least equal to that of the United States — it was the pro-European tendencies, those which sought a common solution, which were dominant. Having a lot of money for research during these years of Gaullist grandeur, the French physicists also had to decide on a successor to their 3 GeV proton synchrotron Saturne, which was completed in 1958. Two projects were on the table, the one, already mentioned, was linked to the linear electron accelerator at Orsay, the other, which was supported by the more 'nationalistic' physicists, saw France independently equipped with a new giant machine — like Germany or Britain. Their idea was to build a 60 GeV proton synchrotron with high-energy injection (1.5 GeV) from a rapid-cycling synchrotron, which would allow one to achieve a minimum intensity of 3×10^{12} p/p. The initial project was published by the CEA in February 1962, and the government's decision was expected in 1963. Nevertheless, in contrast to the United Kingdom, influential personalities like Francis Perrin were not ready to give their support to this scheme if it impeded, for financial reasons, a CERN '300 GeV' project. The business accordingly remained very complex and the tensions it induced were to emerge during the debate on CERN's future machines.[191]

12.4.2 THE ACTIVITIES OF THE CERN GROUP: LONG-TERM PROJECTS, THE ELECTRON MODEL, THE 'DEFENSE' OF THE ISR

During its session in December 1961 the Council accepted that the new study group responsible for projects on future accelerators be officially set up at CERN: its work got under way early in 1962. Some half-dozen high-level scientists formed its backbone, namely, Johnsen, the driving force, Schoch and Zilverschoon, two members of the senior staff, and Resegotti, de Raad and Schnell. Along with them there were the visitors: Cocconi, who was invited for the summer, Burhop, from University College London, Simon from MURA who came for a year's sabbatical leave, Parrain, Bronca, and Neyret from Saclay, who were 'loaned' by Winter, the head of accelerators at the CEA. They were assisted by other CERN staff on specific technical problems: Grütter (the director of the electro-mechanical division) on power supplies, Bonaudi (MPS division) on questions of lay-out, Munday (also MPS) on vacuum, etc.[192]

As we said above, the official task of the group was to prepare precise *technical specifications* for the ISR and for a 'large proton synchrotron'.[193] In practice this turned into defining the fundamental parameters, studying the various alternatives, and dealing with technical problems (RF, site, etc.). As for the ISR, already studied carefully in 1960 and 1961, the notable feature was the suggestion of a different set of parameters: the rings were to be concentric with eight intersecting zones 'at an angle of ~15° in long straight sections' (see Figure 12.7).[194] The advantage of this arrangement was 'that one gets more experimental areas with the possibility of preparing and running more experiments independent of each other', the disadvantage being the increase in size.[195] As for the 'large proton synchrotron', no precise proposal emerged during 1962: the work remained at the level of a feasibility study comparing basic parameters with one another. In November 1962, for example, the injection system was still undecided ('3 GeV linac' or '200 MeV linac +

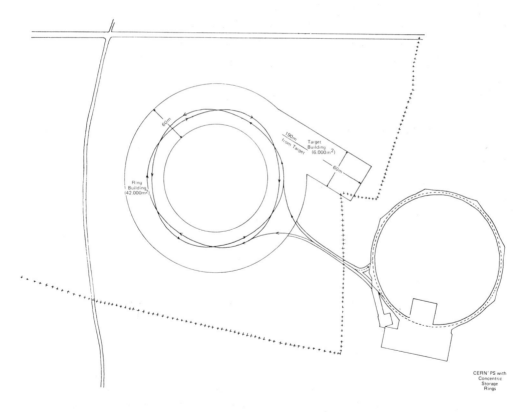

Fig. 12.7. The ISR Project as reported in AR/Int.SG/62-11, 3/11/1962.

Notes: p. 775

6 GeV booster synchrotron'), as was the final energy: 300 GeV was the figure studied most closely, but 150 was also considered. One important point, contentious the year before, was now 'accepted': beams could be ejected at will, so that it would be possible to use only external targets, which in turn meant that the dangers posed by intense radioactivity around the machine could be avoided without having to install costly systems for manipulation at a distance.[196]

In parallel with these studies on the second generation of accelerators, part of the AR division worked on the 2 MeV electron storage ring model agreed on in 1960. The point to note here is the relatively slow rate of progress of the work. The parameters were fixed at the end of 1960, and the orders given to industry in 1961. The magnets and the quadrupoles had arrived at CERN early in 1962, but the tests were only under way in June '[because of] an unexpected delay in delivery of the stabilized power supply for the magnets'. Problems arose in preparing the vacuum chamber and there were constant delays in the delivery of the 2 MeV Van de Graaff which only arrived in January 1963.[197] It was mounted in April, but a leak appeared during the bake-out which stopped one getting better than 10^{-8} torr, while $10^{-9}/10^{-10}$ torr were needed (the vacuum in the PS was about 10^{-6}). Shut down, reassemble, test again.[198] A first beam circulated in the ring at the end of December 1963. In January 1964, the beam was regularly lost during the acceleration of pulses coming from the Van de Graaff. In the following months the complex, including the Van de Graaff, was improved. Finally, in summer 1964, more promising results were obtained. In particular, it was possible to stack up to 20 pulses (the theoretical maximum was 50) and to study their behaviour and the stability of the whole.[199]

A third activity of the group was to organize a two-week seminar from 26 March to 6 April 1962 on the 'experimental use of proton storage rings'.[200] Of course this was not exactly innocent and we know that the 'colliding beam enthusiasts' (to use Johnsen's phrase at the SPC in March 1962) had been trying to arrange a meeting of this kind since summer 1961. Officially the aim was 'to see whether it was worth continuing on this line [ISR]'. 'If it was obvious that the idea ought to be killed because of the small physics prospects', wrote Johnsen, 'then it should be killed as soon as possible'.[201] Unofficially, however, the aim was *to make* the CERN physicists confront this question, think about it — if not to do the work for them. In other words, the objective was to convince them of the soundness of the ISR, to prove to them — because none of the organizers had any real doubts about the outcome of the seminar — that experimentation was technically possible with intersecting rings, that the quality of the data would be adequate and reliable, that the results could be used by the theoreticians, and that their interpretation would not be an order of magnitude more complex than those presently being dealt with.[202]

Between 25 and 30 people took part in these discussions, very few of the active participants being CERN *physicists* (Van Hove was the notable exception). The formal list of those present effectively comprises the names of physicists from

Frascati and Orsay (the electron ring 'enthusiasts'), O'Neill of course, Mills and Jones from MURA, a few other Americans and Europeans and, above all, the core of the CERN group in charge of accelerators. As one would expect the questions raised were things like possible experimental arrangements (Jones, de Raad, Resegotti), data-handling (O'Neill, Taylor, Rodnick), ways of going beyond the limits of p-p (Johnsen, Drell). O'Neill's conclusions in his 'Summary of the Discussions' were even more 'convincing' (and 'convinced') than those he drew at Berkeley in July 1961 - even if the physicists did not seem to have shared either his beliefs or his interest. As we shall see the same insistent doubts continued in 1963 and 1964.[203]

12.4.3 TOWARDS THE CREATION OF THE EUROPEAN COMMITTEE FOR FUTURE ACCELERATORS, ECFA

Towards the end of 1962, a number of reasons, differing considerably from person to person, led to a reconsideration of the role of Johnsen's study group at CERN, the idea being that it should have a somewhat larger structure which would carry more *political weight*. Ever since 1960 the successive groups at CERN in charge of future accelerators were essentially technical and composed of scientists. During the last part of 1962 it began to be felt that a new step ought to be taken: *the time had come to 'commit oneself' to a project and to look at how to 'sell' it to outsiders.*

A catalyst was needed before this sentiment took the shape of an offensive proposal from the Director-General.[204] The requisite shock was provided by Perrin when he explained to Weisskopf, at the end of October, that if Europe did not *rapidly* decide in favour of an accelerator 'up to 300 GeV', the French '[would] go ahead fast with the construction of a 60 GeV machine'. As far as he was concerned, CERN ought to be in charge of constructing the giant machine, and it should be built near the existing site, if possible. Perrin pointed out that to implement the plan a decision was needed in 1963, and the final agreement had to be reached at the end of 1964 at the latest. To maintain the 'morale' of the Saclay group working on the 60 GeV PS — it is always difficult to keep one's momentum if there is a risk that the machine will not be built — Perrin also asked that the links between the groups at CERN and at Saclay be reinforced.[205]

We have said that French demands for rapid and important decisions can be traced back to the pursuit of a very active science policy by de Gaulle's government, and to the fact that the last accelerator built in France was becoming obsolete. Perrin thus found himself in a difficult position. If he continued to postpone a decision on a national accelerator in the name of having a European machine, he risked coming in for increasingly violent criticism.[206] For one thing, the British did not have such scruples and were putting the communal scheme in danger by building their electron synchrotron NINA. For another the French physicists had managed to obtain the money needed to make an in-depth study of their 60 GeV machine (1.5 MFF for

Notes: pp. 775 ff.

1963) and the negotiating process with the government was already well under way. To many (in France) it was too great a gamble to turn one's back on these positive developments for a hypothetical European project.

At first the CERN directorate seemed somewhat peeved by this 'eruption' of French activity, and only Van Hove 'considered it a tremendous stroke of luck that one of the big Member States was so eager to go ahead with the project of a very large machine'.[207] Johnsen, for his part, remarked that while the communal nature of CERN excluded any favouritism vis-à-vis a particular state, he was not against more effective forms of collaboration with physicists 'from those member states that [wanted] to participate' in all important projects. Since it seemed to him that, in any event, one had to move, some day or other, from the 'study stage to the design and possibly construction stage', he put forward a solution: that his group should have 'a more independent status', that it should become fully European.[208] In this way one could collaborate more easily with Saclay —and anyone else — and stimulate a more global debate on the future. Early in November, 'after some more discussion in the Directorate', it was decided '[to] start probing the reactions of the scientists occupying a position equivalent to that of Professor Perrin in other countries, with a view to arousing their sympathy and interest'.[209]

Another consequence of this French initiative was the organization by Burhop of a meeting early in November 'to discuss physics' and 'to try and forecast possible developments in experimental techniques, and how those will affect the design and choice of accelerators' (ISR *and* big PS). This time, however, 'a more restricted [meeting], with invited physicists from outside CERN' was also envisaged, with a view to comparing the projects, be they national or communal. As Adams put it when the proposal was put to him, 'this meeting would be very useful in providing background information on plans and ambitions in Europe'.[210]

On 19 November the CERN Directorate met again to discuss the 'delicate' problem posed by 'the French initiative'.[211] This time it decided to adopt a firmer line. On the one hand, it envisaged calling together the European high-energy physics elite in January, instructing it to prepare a report on 'a comprehensive programme for European physics within the next 10-15 years'. On the other, Weisskopf suggested mobilizing the most faithful members of the CERN lobby — '[people] interested in this subject and [who] have governmental connections' to quote Adams' definition — the aim being to find 'the best way in which to prepare a sympathetic response from the various European countries'.[212] All necessary steps were quickly taken. A 'sub-committee for the January meeting on new accelerators' (composed of Weisskopf, Burhop, Hine, Puppi, Schoch and Van Hove) met several times to prepare the 'background documents' and to establish a list of those to be invited, while Weisskopf, for his part, wrote to those whom he called 'the Founding Fathers' 'to have some informal exchange of view among people who are beyond the pure scientific level'. The list was impressive, and Adams, Melville, Powell, Perrin, de Rose, Amaldi, Heisenberg, Møller, Bannier and Willems were invited to a preliminary dinner on 19 December and to a more formal meeting the following day.[213]

On 30 November the project assumed the shape it was to have in 1963. A 'meeting of senior experimental and theoretical [European] physicists' was called for 17 and 18 January 1963. Since they had to survey the 'national projects' — and could not do this in two days — they were expected to establish 'a small Continuing Committee of physicists from Member States and from CERN'. It was this committee in fact which compiled the report submitted to the SPC and the Council in 1963, with the prior approbation of the European high-energy aristocracy. The preliminary list of its members included Adams, Butler, Merrison, Pickavance, Salam (from the United Kingdom), Perrin, Leprince-Ringuet, Berthelot, Gregory, Lagarrigue (France), Heisenberg, Gentner, Jentschke (FRG), Amaldi, Conversi, Salvini (Italy), CERN staff like Bernardini, Hine, Johnsen, Puppi, Van Hove, etc.[214]

On the evening of 19 December, after the Council meeting, the 'Founding Fathers' gathered at the *Béarn* in Geneva. While savouring a Consommé au Xérès, a Soufflé au Jambon, a Coquelet du Pays, and a Bombe Glacée, they had 'preliminary discussions' which, we hope, did not interfere 'with the social as well as gastronomic pleasures'. The following day they discussed various possible ways of contacting governments, and on 18 January the terms of reference of the Continuing Committee were laid down.[215]

Four weeks later, as foreseen, 50 physicists from among the best of Europe's high energy specialists met at CERN. Those present decided to transform themselves into a European Committee for Future Accelerators (ECFA), to meet twice more before the summer, and to submit a report on the whole of Europe's high-energy needs to the Council in June. To this end they established a working group of nine people under the chairmanship of Edoardo Amaldi. That evening 'the Working Party of ECFA' —otherwise called 'the Continuing Committee' — met. The phase of taking the decision and selling it had begun.[216]

These three months of activity reveal — the specificities of the French case apart — an evolution in the way the situation was perceived. Since 1960 one had essentially been talking of machines, of physics, of doing experiments. In autumn 1962 the centres of interest began to shift. To quote Weisskopf speaking to the Council, the feeling was gaining ground that 'the work [at CERN] had reached the point where the situation should be reviewed [...] in order to submit various possibilities to the governments and to begin to consider the financial implications of such projects'.[217] With the American 'Panels' in mind, CERN's Director envisaged that a sort of European equivalent — ECFA — would recommend a common policy.

The second thing that these developments reveal is the growing awareness that CERN's projects would only be supported if they were inserted into an overall scheme. Unless there was some kind of balance between national and European accelerators there was always a danger that tension, rancour, and misunderstanding would arise. The European Accelerator Group played an important role in

Notes: p. 776

heightening this awareness. It met twice in 1961, in May and in October, the purpose of the latter being to discuss the American work done during the summer as we said earlier. It also met twice in 1962, first in May in Amsterdam when Wilson (Cornell) took the chair, then on the 3, 4, and 5 December at Frascati. This session was quite different from the others in that technical matters were given only scant attention and questions about European science policy in the medium term dominated. Thus the state of the work of CERN's study group was considered, Zilverschoon's 'Ground Investigations related to Possible Geographical Location of a Future European Machine' were presented, and, most important of all, there were three reports by Amaldi, Gugelot (NL), and Pickavance on 'Manpower and Funds for High Energy Research in Europe'. The meeting ended with a closed session (which lasted four hours) of the directors of the major European laboratories. Hine having reported on the 'tactics' adopted by the CERN directorate since November, the participants tried to imagine the different national machines that it would be good to have in Europe in the future. It is largely here, we believe, in these meetings of the European Accelerator Group, that the notion of the *interdependence* of the national and the CERN programmes was born.[218]

The third and last element we want to bring out is how much diffidence there still was at the end of 1962 towards a big synchrotron (something between 100/150 and 300 GeV) for Europe. Certainly one had been speaking about it since autumn 1961, and CERN's study group had published five preliminary reports on the topic in 1962. Certainly the *physicists* had not lost interest in the machine. All the same the French proposal that a rapid decision be taken caused some to draw back. This happened at CERN in October, as we have said, but also with Bannier, who felt the costs were prohibitive, or with Adams, who scoffed at 'the present French enthusiasm for accelerators'.[219] Obviously the desirability of building a machine of some kilometres (a monster?) was still far from evident at the end of 1962. And if some, like Perrin, Van Hove, or the the CERN Directorate in November, wanted to reverse this tendency, they needed to generate a powerful *dynamic*. This was the task given to ECFA.

12.5 The European high-energy project: ECFA's first report, January–June 1963

12.5.1 MOBILIZING THE COMMUNITY

The meeting of the 'Founding Fathers' on 19 and 20 December 1962 considered two notes distributed by Adams and Amaldi. The former described the stage reached in the international negotiations: Adams' conclusion was pessimistic, if only because the question of the *site* was so delicate. Amaldi's note took up again the ideas he had presented a couple of weeks before at Frascati on 'Manpower and Funds' in Europe for high-energy phyics.[220]

ECFA's first report

A consensus on three major points was reached at this meeting of the CERN lobby. Firstly, while the possibility of a minimal division of labour with the United States was kept in mind, all agreed that Europe must define *its* programme and put it into practice. In any event it was not to be held back in the hope of participating in a wider collaboration. Then, 'in order to secure a balanced effort in high-energy physics in Europe', the group accepted that there be a *pyramid* of European accelerators. CERN's machines would form the summit, the others (national, or 'regional' for the smaller countries) being seen as a collective unit. This notion of a pyramid was a response to the idea that Europe needed 'rational' and 'harmonious' development, that it needed a 'plan', particularly if it was to catch up with the Americans. Everyone agreed, nevertheless, that there was to be no coercion regarding the proposals for the 'base'.[221] Finally, the group started discussing 'the means of obtaining the funds'. Separating the 'real' arguments (for which read the scientific arguments) from the official ones (the reasons needed for 'those holding the purse strings'), the participants considered different scenarios. Amaldi proposed 'to proceed in the manner that had been used to start CERN', and de Rose emphasized the value of the competitive argument. 'In presenting the case to governments', he said, 'scientists should not forget to recall that the USSR was building a 70 GeV machine'. Bernardini refined this by stressing that one had to know how to play on both competition *and* collaboration. 'From the point of view of policy', he said, 'the matter should be presented in terms of co-operation or of competition since both approaches had general appeal'. Other ways of how 'to get the best deal' were subsequently suggested. For the present let us only take note that the participants were well aware that there were several levels of argumentation.[222]

Differences of opinion on the type of machine to recommend emerged among the physicists during the 'historic meeting' of ECFA on 17 and 18 January 1963.[223] Opening the meeting, Weisskopf reminded those present of its nature and its purposes. It was not, he said, a simple gathering of CERN people but the first official meeting of European high-energy physicists and 'CERN was anxious to act mainly as consultant and helper in this connection'. The goal was thus not merely to discuss science, as in the past, but *to put forward a programme* for the medium and long term which took account of 'scientific aspects [...], man-power, finance and time scales'. He concluded by adding that June 1963 could be thought of as 'a rough, if optimistic, time-table' for the submission of the first report.[224]

The main disagreements during this first ECFA meeting concerned the choice of accelerators for the base of the pyramid, the rate of growth of expenditure on the discipline, and above all — as on 20 December — 'the energy and intensity of the big accelerator'.[225] Opinions on this were ranged about two poles. One line, strongly supported by the British (Powell and Adams in December, Adams and Pickavance in January) advocated a PS of some 90–120 GeV. Their most important argument was one of cost, though Adams did include 'the [way] physics arguments stood at the moment'.[226] The other line was more diffuse. Its proponents suggested building 'the largest possible machine' (which meant 200–300 GeV), and at least some of them did not mind if this entailed a delay of several years. Many CERN scientists adopted this position — Weisskopf, Van Hove, Burhop, Ramm, etc. — as did most of the French. Their reasons ranged from technical advantages (the intensity of secondary beams, for example) to the need for Europe to remain competitive with the USA. Van Hove was one of those who particularly stressed the scientific side. He favoured an increase in energy in the centre of mass of at least two or three, and felt that more than a tenfold increase in the laboratory system would make experimental work difficult. Seen in this way 300 GeV seemed just about right.[227]

This was not, however, the only alternative (the ISR, for example, was always mentioned for its originality), and in the light of such uncertainty ECFA decided that it was impossible to take a decision there and then and that 'it would be worth asking the Continuing Committee to look seriously into the question'. From the outset its terms of reference included the study of the optimal energy for a PS ecumenically defined as lying between '200 ± 80 GeV'. The Committee's conclusions on this point were expected at the end of March.[228]

It was also agreed that the Committee 'should consist mostly of accelerator users as opposed to accelerator builders', the latter only being the secular arm of the body, as it were. The list of members drawn up after negotiation was Amaldi (I) — the Chairman — and Cassels (UK), Gregory (F), Paul (FRG), Wouthuysen (Benelux), Ekspong (Scandinavia), Puppi, Johnsen, and Burhop (CERN).[229]

This 'Working Party' met four times between 18 January and 11 March. At its last meeting it agreed on a preliminary report which it submitted to ECFA on 1 and 2 April. This report only dealt with the 'summit' and affirmed the need for *two* machines (as 'Neither in itself would be complete without the other', it said). The first project was an ISR with concentric rings, the second a PS with high-energy injection. A synchrotron of 90–120 GeV with injection like that used on the CERN PS was thus put aside, though the report refused to settle a design energy. That, it said, was for ECFA to decide.[230]

On 1 and 2 April ECFA discussed the different parameters of the synchrotron. According to Chairman Amaldi's resumé of the discussion, 'a substantial majority favoured mentioning only 300 GeV in the recommendations'. Elaborating, ECFA declared that 'the choice of [that] energy [...] involve[d] a compromise between considerations of the richness of the physics programme, technical considerations relating to the separation and identification of high energy particles, and the cost, scale of effort and time required for construction'. It added that anyway 'some flexibility should be maintained' as 'definite decisions have not yet been taken' in the United States.[231]

With the summit of the pyramid now established, the Working Party was asked 'to discuss the remainder of it' between then and the end of May.[232] The Party's (final) report was ready by the deadline set, viz. 7 June, when ECFA met again for the third and last time. The SPC, in its turn, met to discuss the paper in the afternoon.

According to the terms of the recommendation which it submitted to the Council — and notwithstanding a lively debate on the energy of the PS and the time scales for decision and construction — 'the meeting unanimously accepted the conclusions of the Working Party of the European Committee', specifying that 'if the large machine, in the range around 300 GeV, is to be built, this should definitely be decided some time in 1965', while the authorization for the ISR should be obtained at the end of 1964.[233]

The Working Party's report approved by the SPC was some 15 pages long, and dealt with the programmes for 'the summit' and 'the base', with the question of data-handling, with the problem of communication between European laboratories, and with the costs of such a programme and the personnel needed to execute it. Sixty pages of annexes contained the information on which the report was based. For the 'base of the pyramid' the group identified the 'best' machines that it would be wise to construct in Europe, namely, two 'pion factories' (supplying intense beams of π mesons), a 'kaon factory' (a high-intensity proton accelerator of some 10 GeV producing K mesons, anti-protons, and neutrinos), and an electron synchrotron of more than 10 GeV. The 'total' European programme (described in annex VI) would call for an annual expenditure of about 1600 MSF in 1977, while that of the 'summit' (CERN) was estimated at 500 MSF.[234]

12.5.2 THE SIGNIFICANCE OF THE REPORT AND THE REASONS FOR THE CHOICE

One striking feature emerges on reading this material: the fundamental optimism of the report which the physicists submitted in the name of ECFA. We have seen that the sums of money asked for were considerable. Let us remember that CERN's budget in 1963 was 95 MSF (to be compared with the 500 MSF envisaged for 1977), and that the transition from the one to the other required an annual growth rate of 13.6% *for twelve consecutive years*. Or again, that the Bannier report tabled the previous year *recommended* an annual growth rate of 10–14% for the next two years, followed by a minimum of 10% for the two subsequent years — figures which, in this same year 1963, the Council refused to consider as practicable because they were too high.[235] It is also noteworthy that, in terms of machines, the report did not make a real choice and simply juxtaposed all the proposals. It recommended the ISR *and* the 300 GeV PS for the summit, and a list of all desirable machines was drawn up for the base. In fact the only 'choice' made was the PS energy — 300 GeV. The assumption underpinning the report was thus that resources were quasi-infinite: high-energy physics was too important, the obligation to stay in the field against the Americans too crucial, for everything not to be done here as it seemed be done over there, across the Atlantic.

European scientists were not that naive, one might say. They knew quite well that they would not get all they asked for and, being used to bargaining with state

Notes: pp. 776 ff.

officials, they had learnt to pitch their initial claims high. We have no doubt that this is partly true. As a matter of fact we have already pointed out that this was one of the three major 'selling' strategies adopted at CERN — 'demanding more than one expects to get' — in the decision taken on bubble chambers in 1958.[236] The debate in the SPC would also confirm this point: one issue there was to know which machine to 'sell' first, the ISR or the 300 GeV, which goes to show that some delegates thought that they could not have both simultaneously.[237] That granted, we remain convinced that hopes were running high in the community and that physicists really believed that they could get the best part of what the ECFA report asked for.

The 'choice' of 300 GeV as the energy for the PS remains to be considered. The shift of opinions from 90-150 GeV to 300 GeV between December 1962 and April 1963 can be understood in two ways. Firstly, there was the discovery of the recommendations made by the Ramsey Panel in the United States. On 29 January Weisskopf informed Amaldi that 'the committee strongly supports the construction of storage rings [...]. They [also] gave up the plan of a simple 100 GeV machine for Berkeley, considering it too simple a step forward. Then [sic] wanted to recommend a 200 GeV Machine for Berkeley'. Granted the importance of the information for the Europeans' choice Weisskopf suggested that 'three representatives (Ramsey, Gell-Mann and Panofsky) should come to Geneva on April 2', i.e. during the second ECFA meeting.[238] This was done and at the meeting Ramsey confirmed the earlier reports (he spoke of an ISR for Brookhaven and of a '150-300 GeV proton AGS' for Berkeley), and outlined the 'base' of the American accelerator pyramid.[239] A kind of tacit 'agreement' was reached in which the spreading out in time of the 'proposed completion dates of [world] accelerators' was encouraged. In this way a PS of 150 GeV could be built at Berkeley by 1972, followed by a 300-350 GeV for Europe in 1977, followed in turn by a 600 GeV for Brookhaven in 1982, etc.[240] One result of these deliberations was then to increase the design energy of a future European PS.

This, however, is not reason enough for understanding the shift in opinion. Another consideration, linked to cost estimates, also played a role. In execution of the tasks given him by the Working Party, Zilverschoon published two 'draft cost estimate[s]' for proton synchrotrons of 150 and 300 GeV.[241] Taking the cost of the 28 GeV PS as starting point, and assuming injection at 3 and at 6 GeV by 'booster synchrotrons', he arrived at global figures of 942 and 1460 MSF for the two machines. More significantly, though, he worked out the distribution of these sums over six and eight years respectively, using the temporal pattern of expenditure of the 28 GeV PS as a guide, and showed that the cost of both machines would be of the *same order* during the first six years. It was then tempting — and, judging by the frequency with which the argument was subsequently used, many succumbed — to choose 300 'rather than' 150 GeV. The 300 GeV was 'proportionally' cheaper in terms of cost/GeV, and anyway the additional burden was only felt in the seventh and eighth years. As for the fact that the 300 GeV would only be available two years later, this was generally regarded as being acceptable.

Given the importance which this argument assumed — and its somewhat paradoxical appearance — it is worth trying to test its reliability. We know that expert information is never neutral and that, in this domain, a well-chosen (or badly chosen!) hypothesis can strongly change an estimate. In the Working Party Gregory had already expressed doubts about these numbers, doubts which he repeated during ECFA's second meeting. It was true, he said, that 'he found no flaws in the present reasoning', but he thought that 'Another working party might make a different estimate of cost'.[242]

It would take too long to repeat Zilverschoon's work and to confront it with what actually happened with the SPS. However, in comparing the two estimates made in February 1963 it is at least possible to make two remarks. For *the machine itself and its buildings* there seems little to quibble with in Zilverschoon's estimates. They look like this (in MSF):[243]

Year	1965	1966	1967	1968	1969	1970	1971	1972	total
150 GeV	15	62	140	156	140	82			595
300 GeV	15	85	169	199	183	158	104	65	978
Difference	0	23	29	43	43	76	104	65	383

Where Zilverschoon's figures are perhaps more dubious — and this Gregory specifically brought out — is in the estimates given of peripheral costs and, more particularly, in the cost of material for doing experiments (beams, calculating capacity, data-handling). Zilverschoon gave the following figures for these headings:[244]

Year	1965	1966	1967	1968	1969	1970	1971	1972	total
50 GeV	0	1	8	36	62	98			205
300 GeV	0	1	7	8	14	35	87	128	280
Difference	0	0	−1	−28	−48	−63	+87	+128	+75

We should like to make two remarks on this table. Firstly, regarding the total expenditure envisaged, the figures seem far too low. As a matter of fact it was conventional at this time to estimate the annual cost of 'using a machine' as at least equal to its (average) annual construction cost. A quick glance at the previous table shows that this 'rule of thumb' has not been respected here. One might object that the figures do not cover *all* the equipment, and notably not the detectors — but that is simply another point which could be held against the author. What is more this

Notes: p. 777

does not explain why the ratio between the figures (205:280) is only 1:1.4 while that for the machines is 1:1.7 (597:980 MSF). The second point to note is that, since Zilverschoon concentrated the expenditure in the last two to three years (and all the more so for the 300 GeV), he could counterbalance the differences in the first table. The 'reason' for choosing this temporal pattern was probably that the costs for equipment on the PS were very low up to 1957. That said, there was little justification for using the PS data as a point of reference here. By 1963, everyone agreed that the preparation of experiments with that machine *had been started much too late* and that this explained the poor use that was made of the PS in 1960, 1961, and even 1962. CERN's 2m hydrogen chamber, for example, was still not completed in 1963 — and it was small compared to what would be needed for a PS of 150 GeV or 300 GeV.

We do not want to be emphatic about this. But one thing is certain: a strictly equivalent situation had arisen in 1958 during the decision on the bubble chambers. Having to choose between two chambers of 1.1m and 2m the study concluded that there was little difference in cost... and that the time to build them was roughly the same. And on that occasion this conclusion also counted heavily in favour of the choice for the 2m that was ultimately made.[245]

12.5.3 THE WAY SCIENTISTS ARGUE

To conclude, and before considering the reaction in the Council and in government circles to the ECFA report, it seems useful to make a few general analytical remarks on the kind of arguments we have met during the long debate between scientists. Remember that this began in December 1960 with a report by the accelerator research group proposing the ISR — and that it was temporarily closed with the report presented by ECFA's Working Party in June 1963.

The *arguments* debated most extensively, statistically speaking, are those which we may describe as 'arguments of a scientific kind'. They are not homogeneous, and we can distinguish at least two or three different ways of tackling issues. The first tries to determine the machine that is to be built in the light of physics problems, in relation to the question 'what experimental programme does one have in mind?'[246] It is this type of enquiry which, in 1961, led the physicists to reject the ISR, and it is this that enables us to grasp why the first ECFA meeting in January 1963 was opened by a talk by Van Hove on 'The physics that could be done with Future High Energy Accelerators'.[247] The results arrived at in this way always have one chronic weakness however — that of trying to extrapolate ten years ahead from theories which one knows will be largely outdated by that time. As a result another way of reasoning tends to exist alongside this. It consists in trying to imagine, not a specific experiment, not the theories one would like to test, but the most flexible machine, the machine which would cover the greatest possible number of developments in the long term. This is why the physicists so liked the 'big synchrotron' once the builders had shown them, during summer 1961, that such machines were essentially multipurpose (high energy and high intensity, excellent secondary beams, ease of experimentation, etc.). And once *one* kind of machine is widely recognized as being *the* multipurpose device, the substance of the decision is reduced to fixing *one* parameter, the size. This was true for the cyclotrons and synchrotrons built since 1930, and it was true for the bubble chambers in the 50s and 60s.[248] In such cases it is regularly argued that the bigger the piece of equipment, the better it is 'from a scientific point of view'. The upper limit is then fixed in the light of economic considerations — and 'psychological' ones, to repeat Sands: it is fixed at the point where the physicists (or builders) begin to feel uneasy about being able to manage such a big leap into the unknown, where the fear arises that one cannot conveniently carry out experiments, or interpret the 'results'. This was held against the ISR in 1961, against a 300 GeV (Puppi, Bernardini), and *vis-à-vis* a machine bigger than 300 GeV (Van Hove) in 1963.[249]

We should add that, in addition to the scientists' desire to make an 'optimal' choice in the most 'rational' and 'scientific' way possible, these continual debates are also motivated by a twofold practical necessity: to facilitate the design of the machine (experimental zones, lay-out, ...) and to trigger reflection on the experimental equipment that will be needed (one freezes its characteristics as late as possible but, ideally, it ought to be ready simultaneously with the accelerator).

Other considerations are as decisive as these in the judgements made by scientists when they discuss big equipment. These weigh particularly heavily when one is dealing with machines which the majority feel to be multipurpose. They concern the technical limitations surrounding the construction of their components,[250] the availability (always negotiable, of course, and never definitely given) of time and manpower, and finally the cost and the time required for construction. We shall briefly elaborate on the last consideration, as we have already touched on the question of cost in discussing Zilverschoon's report.

The interest of having a big enough piece of equipment (a 300 GeV instead of a 100 GeV for example) lies in the superior experimental possibilities it offers to the physicists. The main disadvantage is that, the bigger it is, the longer it takes to build, and the later experimentation will begin . This notion of 'timing' (when will the machine be ready?) is often decisive because it is intimately associated with the competitiveness between scientists, between laboratories, between continents. It is thus important that the future users weigh up short and medium term interests (an average sized machine ready rather quickly) against long term ones (a more sophisticated machine, but not ready for many years).[251] Since at this time Europe was somewhat behind the United States (as measured, for example, by the number of accelerators functioning or on the drawing boards) it tended to position itself with reference to what there was across the Atlantic. In the straightforward case of the bubble chambers in 1958, the Europeans tried to do *as well as* their American rivals

Notes: p. 777

(a 2m hydrogen chamber). In the more complex case of the accelerator programme for the 1970s, no one took up a fixed position in the debate: as one party made a move, the other modified its position.

Finally it must be said that it is not just anyone who says anything at any time. We have already remarked on the rift between builders and physicists in 1961 — and on its roots. In 1963, when the energy of the big synchrotron was the issue, this 'professional' difference was dormant. The physicists were more involved than the builders on that occasion, and the latter, believing that synchrotrons of some hundreds of GeV were technically feasible, left the future users to settle the exact energy. All the same the ISR remained a potential source of conflict. If it became essential to make a choice, to make a financial sacrifice, it was likely that phycisis would 'offer' the ISR in expiation.

Another kind of difference is that between countries, particularly between British and French scientists. During the SPC meeting on 7 June, for example, Perrin insisted on the urgency of a decision (an urgency reflected in the SPC's recommendation), while Adams denied it ('the Ramsey Panel had come up against a good deal of opposition', he said in justification of his line). In fact Adams proposed that one *rediscuss* the energy of the PS and asked if there was really 'enough man-power to build the storage rings and the big accelerator at the same time'.[252]

We could draw a simple conclusion from this and say that here, as everywhere, the social world obviously 'determines' attitudes and judgements. This is not false, but what is interesting is the limits of this determination (Puppi and Van Hove, for example, had very different opinions in January 1963, as did the two British scientists Adams and Cassels in April.) The relative autonomy of individual 'judgement' is also illustrated by people's changes of mind. Gregory, for example, said on 25 February, during a Working Party meeting, that 'he had first been in favour of a 120 GeV machine based on a scaling-up of the CERN PS but had been impressed by the arguments in favour of high energy injection', so that he now preferred a machine of 150 or 200 GeV; later, in April, at the SPC, he said that he was now in favour of a 300 GeV machine 'because of the considerable gain in the intensity of secondary beams'.[253] As we see — and this is hardly surprising — personal judgements and socio-professional positions never strictly overlapped.

To conclude, we would like to stress that there is no one way of arguing, no one type of argumentation which would always be dominant in such a debate, that whatever is decisive is so only in a precise context, a context which is itself defined by the accumulation of preceding layers of micro-decisions. We have suggested, for example, that two key new arguments led to the choice of 300 GeV between January and April 1963 — the competition with the United States and Zilverschoon's cost estimates. While this cannot be doubted, one must also realize that the range of possibilities was very narrow at the time. The choices 'ISR plus PS' and 'kind of PS' had already been made beforehand, and for quite different reasons. If we were to remove the preceding framework, if we were to treat March and April in isolation, we could certainly arrive at a simple explanation (for example that no really scientific or technical consideration influenced the decision for a PS of 300 GeV). This conclusion unfortunately would ignore the debate in 1961 and the majority opinion that *the* 'multipurpose' machine of the future should be a high energy PS, that, in theory at least, 'the bigger the device, the better', so that, as a result, the *last* choice, that of the energy of the machine, 'becomes an economic rather than a physical [question]'.[254] The simple explanation thus would fail by transcending the limits of its applicability, forgetting that there is not ONE explanation but networks dependent on the scale chosen by the person making the study.

12.6 The 'selling' of the ISR to the Member States and the postponement of the 300 GeV project by the Council, June 1963 – December 1965

The events that took place during the succeeding 24 to 30 months were particularly complex because of the multiplicity of the actors (the range now covers European Ministers to CERN physicists), because of the variety of their centres of interest and the different alliances which that entailed, and because of the inertia of the decision process itself. So as not to lose the reader — and even though the subject really merited it — we do not give an analysis as detailed as that we made for 1961. We prefer to limit ourselves to the conclusions of our analyses and to concentrate our attention on the evolution of the *process* of negotiation–decision. In other words the study is focused on the actors and their respective positions, on the issues at stake and the attitudes adopted, and on the actions and their consequences for the whole of the decision-making system. In each period we give only the minimum information necessary, either for understanding the context and its changes, or for reporting on the other activities of CERN.

12.6.1 THE PROBLEM OF PRIORITIES, SUMMER 1963–SPRING 1964

On 20 June 1963 the Council met and heard Edoardo Amaldi's report on the work done by ECFA. He took the opportunity to thank the personnel of CERN, and particularly Burhop, Hine, Johnsen, Puppi, and the study group in the Accelerator Research division. Powell then put forward the SPC's point of view. After enthusiastically explaining the importance of the high-energy field for the history of western thought, no less, Powell concluded by stressing that it was time steps were taken inside the member states to discuss ECFA's report, and to envisage the launch of a *supplementary programme* in 1964.[255]

Let us remember that CERN functions by programmes. Only one existed in 1963, that laid down in the Convention of 1953 and called the *basic programme*: the SC and the PS already built were its two main components. The Convention, however, also made provision for *supplementary programmes*. To be set up these needed to be voted for by two-thirds of the member states; however, to stop an uninterested minority blocking the launch of any new programme, a vote in favour of a supplementary programme 'did not imply that one had to participate financially [in it]'. Only the states that wished to need join the new project, contributing to its costs and deciding on its means and its ends. In short, there was a legal possibility of having 'à la carte programmes'.[256]

The supplementary programme mentioned by Powell on 20 June 1963 was intended 'to go further in determining the design parameters of the storage rings and of a 300 GeV proton synchrotron'. Presented in an annex, its budget was 3.8 MSF for 1964, so about 3.5% of the CERN budget.[257] Called together again to debate this point in October, the Council approved, without any opposition, the principle of the new programme. In its December session all the member states, excepting Spain and Greece and notwithstanding some reservations expressed by the United Kingdom, declared that they would participate in the supplementary programme.[258] Thanks to this money the CERN study group on future accelerators grew notably in size, increasing from 16 members during summer 1963 to 55 in May 1964 — when it submitted its first detailed technical report on the ISR (a report of some 300 pages which it submitted to the Council). An equivalent report on the 300 GeV (500 pages of text plus a mass of figures) appeared early in December.[259]

12.6.1.1 The new framework of the debate

The context in which the debate about CERN's future programme was conducted remained more or less homogenous from the summer of 1963 to the spring of 1964. Three features distinguished this period from the preceding six months. Firstly, the discussion no longer took place primarily amongst the aristocracy of the high-energy physics community, as represented by ECFA, but in the SPC and the CERN Directorate. This meant that the majority of those involved in the debate were still scientists, but scientists who had direct and significant institutional weight for the 'summit' programme, that of CERN. What is more, these scientists included the most faithful nonscientific members of the CERN 'lobby' in their deliberations, people like Bannier, the Council President, Willems, one of his predecessors in this role, and Funke, a key figure in the Finance Committee. In fact until it met on 18 and 19 June 1964 the Council as such was still not directly in the picture. The second point to note is that the physicists' 'ideal programme' was no longer the obvious programme to be considered. This meant that the state of mind of European political bodies had to be taken into careful consideration, with a view to finding ways of having the accepted programmes voted for. The third and last important contextual characteristic of this period is that the high-energy physics community began to feel that the political situation was deteriorating and that the objectives spelt out in the

Amaldi report of June 1963 could not be satisfied in their entirety.

One source of anxiety about the future was financial. The first blows were struck in the United States in summer 1963 — budgetary restrictions were imposed on BNL, for example. This created some alarm in Europe. The most disturbing signals were those emanating from the United Kingdom: her opposition to the other member states at the end of 1963 to the CERN budgets for the period 1965-1967, the declaration by the Advisory Council on Scientific Policy in November 'that the United Kingdom should not increase still further its commitment to high-energy physics', etc.[260] Summarizing the situation in the *New Scientist* in May 1964, Howard Simon wrote: 'Here [at CERN], hopes are not as high as they might be and the reason seems to boil down to Britain' — 'though no CERN official would say so in as many words', he added.[261]

There was another reason why 'high-energy physicists [seemed] to be on the defensive everywhere' early in 1964: the growing resentment of scientists in other disciplines. Since the period was one in which belts were being tightened, and since physics was becoming more and more expensive, they tended to think that 'high-energy physics has had enough cake for a while'. This demand for a greater 'fairness' in the distribution of funds found expression in a number of published articles, including the very famous piece by Alvin Weinberg, himself a physicist at Oak Ridge, which was published in *Physics Today* in March 1964 — and the little polemic in the CERN Courier which was sparked by Weisskopf's complaint about the unfavourable way in which Weinberg 'had applied his criteria to high-energy physics'.[262] It was also manifest in the calls for moderation: 'to obtain the necessary funds', said the Director-General in June 1964, 'high-energy physicists should refrain from competing with representatives from other disciplines [...]. The scientific community as a whole', he went on, 'should agree on the steps to be taken for the general development of Science'.[263]

In short, a new situation was emerging. The optimism of the early months of 1963, as symbolized in the ECFA report, was giving way to a growing pessimism about what could actually be obtained. New rifts thus began to appear — and the old were reawakened.

12.6.1.2 Some of the positions in evidence at the time

The first divide that again came to the fore was as ever that between the representatives of the United Kingdom and the majority of the other delegations and the CERN Directorate. Not that Adams, Powell, or Sir Harry Melville shared the opinions of the Advisory Council on Scientific Policy (which some British physicists called the Association of Chemists for the Suppression of Physics). But the situation in Britain, they felt, demanded prudence.[264] Thus during the Council session in June 1963 Adams recommended that further thought should be given to the programme advocated by ECFA before presenting it to the member states. Powell similarly spoke

Notes: pp. 777 ff.

of the need for in-depth studies so that, 'in the course of *the next two or three years*, the Council [could] be *informed* of all the issues which [would have] a bearing on the decision that it [would be] called on to take' (our emphasis).[265] By contrast, at this same Council session, Amaldi confirmed December 1964 and December 1965 as the ideal dates for giving definite approval to the ISR and 300 GeV projects — a timetable which Hine repeated in his report on 'Long-term High-Energy Physics Programmes in Europe' which he submitted to the SPC on behalf of the Directorate in February 1964. As for Perrin, increasingly torn between his wish not to do anything which might impede a European decision for a 300 GeV and the pressure exerted by those French physicists who wanted to avoid their country being the only major European state to sacrifice its national programme, he continued to insist on the urgency of a decision on the 300 GeV.[266]

This point of friction was significant granted the political weight of the United Kingdom and the reticence — or doubts — of some of the small states. All the same it was not the most important. A second, cutting across the first and independent of national groups, appeared in the early months of 1964. Its appearance (its reappearance, in fact) was provoked by a shift in attitude by the CERN directorate.

Confronted with the situation which we have just described, the CERN Directorate — Weisskopf, Director-General, Gregory, member for Research, Hine, member for Applied Physics, Germain, member for Technical Coordination, and Hampton, member for Administration — put forward the idea that there ought to be a choice (or at least *priorities* defined) between the two machines in the 'summit programme'.[267] Of course the Direcorate remained convinced that, in principle, *two* machines were needed: they formed a unique programme together. However, granted the danger that the Council would repeatedly postpone the 300 GeV project — notably because of its cost — ,Weisskopf and the Directorate suggested that it may be wise to get the Council to accept the ISR first, and to no longer associate it organically with the decision on the 300 GeV. The correspondence of the two most important people in the Directorate, Victor Weisskopf and Mervyn Hine, helps to refine their argument. The latter seems to have been more in favour of the idea than the former, and that for two main reasons. On the one hand, he wanted to save CERN, to ensure that it had the means to remain a laboratory of international status. From this point of view the 300 GeV seemed a poor bet since everyone thought that it ought not to be built in Switzerland. The ISR, by contrast, being inseparable from the PS, could only be built in Geneva. Hine's second reason, linked to the first, was that he wanted to keep Kjell Johnsen's group united. This outstanding professional team, he thought, ought rapidly to be given the wherewithal to exercise its talents. Hence his wish to have a quick decision in favour of the ISR, the cheaper, and so more acceptable, machine.[268] Weisskopf, by contrast, seems to have been more hesitant in summer 1964. While appreciating Hine's arguments, he was also very much aware that the 300 GeV was the more important machine, granted the greater range of facilities that it offered. 'On the one hand', he wrote to Bannier, 'I understand that there is a danger of an indefinite postponement of the larger machine'. 'On the other', he added, 'I do not see how we can say no to a possible offer to give us the storage rings, even if the offer is not accompanied by a promise of the larger machine'. He concluded by telling Bannier that 'this is a political problem, where we need very much your wisdom and advice'.[269]

Nevertheless on 6 May 1964 Weisskopf officially recommended that the SPC adopt Hine's 'tactic', and approve that a decision be taken at once in favour of the ISR.[270] He gave three reasons for this in his memorandum. Firstly, that the ISR was innovative, that it opened the way to a new type of physics, and that its quality/price ratio was excellent — the classical arguments then. Weisskopf even went as far as to say that it would make no difference to annual expenditure if one delayed the construction of the 300 GeV by a year and a half, and began instead to build the ISR. 'This is not a reason to recommend that postponement', he was quick to add, 'but it helps to demonstrate the relative smallness of the amount necessary for ISR [...]'.[271] Weisskopf's second argument was, as the saying goes, that half a loaf is better than no bread. In the event that the 300 GeV was refused by the Council,— 'an event, we hope, that will not occur', wrote the Director-General —, one should have at least one new machine. And then there was a third reason that had often been used by the key members of the SPC themselves (Adams, Amaldi, Perrin) before Hine did so: 'it will give the CERN–Meyrin Laboratory a much longer period of vigorous research and will keep better people at Meyrin'. In short, to 'kill off' Meyrin was impossible.[272]

This shift by the Directorate in favour of a *de facto* priority for the ISR revived — as one would expect — the rift between the active physicists (both at CERN and in the member states) and the 'builders'. The latter raised no objection to the Directorate's proposal: the situation demanded that a choice be made and they *believed* in the ISR and in the future of CERN–Meyrin. The physicists, by contrast, rallied strongly to the opposite position. The heart of the future programme, they said, should be the 300 GeV, and everyone in Europe, not least of all Hine and Weisskopf, knew that. If therefore there was a risk that the whole programme could not be accepted, 'the ISR project should not stand in the way of a speedy approval of the 300 GeV machine or of PS improvements'. If money had to be saved, this was the way to do so. 'It is proposed by them', said Weisskopf in summary, 'to give up the ISR, in order to obtain the other objectives in a shorter time and with more certainty'.[273] In support of their line, the physicists appealed to the conclusions reached in 1961 and to what they felt were the unambiguous results reached by ECFA.[274]

On the eve of the SPC and of the Council meetings in June 1964 — and without pretending to be exhaustive, because each protagonist was weighing in his own way the risks associated with different requirements — four or five major positions were identifiable. The CERN Directorate first. They wanted a rapid decision and accepted the risk (for the 300 GeV) of an independent vote for the ISR. Strongly against this line were the European physicists: for them only the 300 GeV was scientifically decisive — the rest was mere verbiage. The 'establishment' in the United Kingdom — Powell, Adams — wanted to slow down all projects, and the 300 GeV above all: it was too expensive. Backstage the British *physicists* protested: it was just because one could not have both machines, that one should *only* ask for the 300 GeV. Perrin, for his part, was pushing for a combined vote: there should be a *single* programme with the 300 GeV at its core, and if it was decided to have the ISR first it was because the design was ready, because it was cheap, and because it was vital for the survival of the Geneva laboratory. Finally Amaldi, another central figure in the decision, continued to think that both machines could be obtained but 'accepted' — political realism left him little choice — an independent decision on the intersecting storage rings.

12.6.2 A DECISIVE MOMENT: THE FIRST STEPS BY THE SPC AND THE COUNCIL TOWARDS HAVING 'ONLY' THE ISR, JUNE 1964

12.6.2.1 The SPC debate on 2 June

In summer 1964 there was no time left to go on talking — or at least that is what the CERN Directorate felt. At the start of May, confronted by the persistent and tenacious refusal of the physicists to come round to their view of the situation and of what the priorities should be, Weisskopf made an explicit appeal to the SPC to accept the Directorate's line. Beyond the SPC it was of course the Council's approval which was ultimately sought after. 'It is important for us [the CERN Directorate]', wrote Weisskopf in a memorandum for the SPC dated 6 May, 'that we have the full support of the Scientific Policy Committee in this matter'. He added: 'We would like to *share the responsibility* with the Scientific Policy Committee for the programme [...] including the proposal to start the construction of ISR at the earliest possible date' (our emphasis).[275]

The discussion at the SPC lasted almost all day on 2 June. The physicists point of view was defended by Cassels, Ekspong, and Van Hove, and to a somewhat lesser extent by Bernardini and Gregory, all of them active physicists. This did not stop the SPC as a whole — Leprince-Ringuet (Chairman), Amaldi, Bernardini, Jentschke, Perrin, Powell, Wouthuysen (members), Butler, Cassels, Ekspong, Funke (ex officio members) — from 'accepting' the Director-General's request. It was Amaldi who found the appropriate formulation when he said 'that [...] there seemed to be a clear majority in favour of a programme consisting of the 300 GeV accelerator and the storage rings [note the order] with a decision on the storage rings to be taken in December'.[276]

Three remarks are worth making before we study the attitude the Council adopted to the resolution submitted to it by the Scientific Policy Committee. Firstly, one must appreciate the very heterogeneous nature of the 'majority' Amaldi spoke of. Those who comprised (or supported) it did so for different motives and with contradictory hopes, since one finds among it the CERN Directorate and its plea for realism — a strong argument in the eyes of the permanent members of the SPC —, the partisans of the ISR, the 'minimalists' like Powell, who believed that it would be difficult even to get the ISR, and the 'maximalists' like Perrin for whom the ISR was only the hors d'oeuvre.

The second point we would like to make is that one must not be misled by the resemblances between this debate and that of 1961. Certainly one has the supporters of the ISR on one side, those of the 300 GeV on the other — and even a conciliatory 'decision' reaffirming, as in 1961, the need for both. The resemblance ends there, however. For one must not overlook the time factor, which makes of a decision a *process* marked by a certain irreversibility. In 1964, for example, the discussion was not as 'free', as 'open' as it had been three years before: the field of possibilities was reduced progressively, and even the 'juxtaposition' of the two machines no longer had the same meaning as in 1961 when the physicists had accepted that the CERN study group *also* investigate the ISR alongside the 300 GeV. What is more the criteria of judgement themselves had changed. In the context prevailing in 1964 it was not simply what was needed 'from the physics point of view' that mattered, it was no longer enough to say that the 300 GeV was THE multipurpose machine. What was cardinal now was to be able to 'sell' the project to the member states, and to be able to do so *immediately*. Seen from this angle the 300 GeV had a number of serious disadvantages. Technically, in the first instance, because the CERN study group had completed the design study of the ISR (wholly by chance?) but had not yet finished its report on the 300 GeV. Politically too, because the ISR was relatively cheap, and one knew where to build it. The French government had offered the land (adjacent to that of CERN) and no one could overlook the fact that the site of the future 300 GeV (which would not be Geneva) was a source of tension between member states.[277] Mentally, one should also say, the defenders of the 300 GeV had invested less in the nitty-gritty of the negotiations: it was not they, it was the AR division, it was Johnsen, it was Hine, who were daily involved in these questions. And in contrast to the physicists, they were perhaps not that convinced of the priority of the 300 GeV. Institutionally, finally, the protagonists of the 300 GeV were peripheral to the nucleus who would actually take the decision — at least if a consensus could not be restored and if there was open conflict. The Directorate included Gregory, but not Van Hove, for example, and the SPC consisted mostly of physicists of a different generation who could not ignore the need for a rapid decision (the demand for realism) and the feelings of the 'politicians' in the Council. In this somewhat defensive atmosphere it is therefore not surprising that the physicists' position became somewhat uncomfortable. In a way, insisting in June 1964 on the principle which everyone accepted one year earlier — that the ISR should be sacrificed if it posed a threat to the 300 GeV — could only appear to be *extremist* since it looked like an all-or-nothing attitude. Hence the tendency to resign oneself to the course of

Notes: pp. 778 ff.

events without losing hope for the medium term — which was quite clearly Gregory's position.

Our third remark: an additional element enabled the Directorate's proposition to gain hold more easily — and enabled the French maximalists to risk seeing the 300 GeV pushed back another two to three years. This was the fact that Berkeley's 200 GeV machine recommended by the Ramsey Panel had got caught in the crossfire between scientists and politicians in the Midwest, and was suffering the effects of a bad press for high-energy physics. Europe therefore did not have to act *quite as quickly* as one had believed, and an additional delay for the 300 GeV was admissible. Weisskopf's argument that building the ISR would only mean setting back the 300 GeV by a year and a half was particularly apt here. While the moment at which a piece of equipment must be ready is always crucial for physicists (and much more so than for the builders, as we have said), it is a little less pressing if there is no competitor.

12.6.2.2 The first steps taken by the Council towards having 'only' the ISR

On 18 and 19 June 1964 the Council dedicated more than two-thirds of its two working days to 'CERN's future programme' and 'for the first time, governmental reactions to the Amaldi Report proposals were expressed'.[278] The meeting was carefully prepared by the SPC on 2 June. Bannier, who had taken part in this meeting in his capacity as Council President, began by giving a solemn tone to the proceedings. The various committees of the Council, he said, had progressively become too important, and it was time that the Council was given back its full authority. That said, he used his opening speech to expand at length on the medium and long-term policy of CERN.[279] Powell, the only remaining Nobel prizewinner in the SPC after the departure of Bohr, Cockcroft, and Heisenberg, then presented a report on the 'Importance of High-Energy Physics, its place in European Science and Culture'. Its aim was to 'establish whether high-energy physics was the basis of science in general and the balance to be struck between different disciplines in a civilization like that of Europe'.[280] Leprince-Ringuet then introduced the concrete programme envisaged for the next decade. Presented 'as an indivisible whole, even if its three components could be distributed through time', he mentioned the 300 GeV —'to abandon this project would be to render vain all those efforts made to date for CERN' — the ISR, and the PS improvement programme which was directly stimulated by a similar project then being adopted at Brookhaven (new linac injector, etc.).[281] Hoping that the delegates had grasped the point Bannier summed up: the 300 GeV was the heart of the programme, even though the SPC 'deemed it possible to delay a decision on this point for a year or two'. Adams then concluded, explaining the improvements foreseen for the PS in more detail and the financial implications of the whole programme.[282]

The subsequent discussion caught the CERN Directorate and the SPC somewhat off-guard. Their hopes for a rapid decision on the ISR and a more or less tacit agreement in principle on the 300 GeV were both disappointed. While everyone seemed interested in the ISR, particularly the non-scientific Council members, no decision was possible within the next six months. Only France and the Netherlands thought to be able to give a reply by the end of the year. As for the 300 GeV, it rapidly emerged as most improbable, even in the medium term.[283]

The lengthy minutes of this meeting enable us to catch a glimpse of the motives of the non-scientific members of Council. The grounds for their support for the ISR — and their scepticism about the 300 GeV — were somewhat different from those of the SPC, the Directorate, and the members of the Accelerator Research division. Three things counted particularly for them. Firstly the cost: the ISR was a financially acceptable scheme, while the alternative was simply unrealistic; then the fact that the politico-institutional parameters were known: they could see quite clearly what the process involved and that all necessary steps would be taken at CERN-Meyrin; finally — and this was crucial — there was the picture they formed of the machines: by comparison to the 300 GeV the ISR seemed extraordinary. It was technically sophisticated (they had been told why for the past three and a half years), it was novel (because the Americans were not planning to build it anymore), and it enabled one to attain exceptionally high energies ('equivalent' to 1700 GeV!). If this implied that there were some constraints (experimental, for example), that was only to be expected. If the ISR dominated the Council meeting — with the sure and effective help of Adams — it was because it seemed to the non-specialists to be the 'ideal' machine, the machine which covered most possibilities — rather like the 300 GeV for the physicists. In other words, it was for their own reasons — or, at least because they weighted a number of reasons in their own way —, that the new arrivals in the discussion supported the ISR and tended to ignore the 300 GeV. Seeing the matter partly from their governments' point of view, using their own criteria of judgement and of rationality — namely, to remain competitive for as low a cost as possible — they defined their priority. As for the 'package' that the SPC proposed (the indivisible whole which Leprince-Ringuet tried to foist on them) — they never really took it seriously.

12.6.3 THE COUNCIL MAKES ITS FINAL CHOICE, JUNE 1964 - JUNE 1965

In the aftermath of the Council meeting Weisskopf and Amaldi called a new meeting of ECFA for October 1964. The conclusions reached in June had caused considerable dismay and Weisskopf was putting it mildly when he said that there had been 'some opposition to the storage rings project' among 'physicists both at CERN and in the Member States'. To tell the truth the rumblings of discontent were such that it was unthinkable not to give the physicists another opportunity to air their views.[284]

Notes: p. 779

12.6.3.1 The Director-General defines respective roles and powers

Two questions were raised by the fact that the decision on the 300 GeV had been postponed without a date being fixed. According to a preparatory document for ECFA prepared by Weisskopf and Amaldi the first was 'what delay in the 300 GeV machine will in itself seriously reduce the scientific value of the whole proposal?', while the second was 'will the financial decisions on the future of CERN-Meyrin [PS improvement *and* the ISR] seriously increase the unavoidable delays [...] in starting the big machine, or even jeopardize the whole project in its present form?' To the European physicists, Weisskopf and Amaldi explained that while the answer to the first question hinged on scientific judgement — and so on the opinion of the physicists and of ECFA —, that to the second demanded a 'political evaluation of the decision-making habits of governments' — and so depended on the opinion of the Committee of Council and of the SPC.[285] They then drew attention to a basic institutional rule: in any case, and particularly if the reply to the second question was negative, it was up to the *Director-General* to propose to the Council 'where a cut should be made', and for the latter to accept or to refuse. The scope of ECFA's deliberations was thereby clearly limited from the outset. It was not up to it to 'sell' the project, nor to evaluate the risk of certain tactics. Calling ECFA was thus a response to the physicists' wish to discuss the issue again — also providing an opportunity for Amaldi and Weisskopf to describe the complexity of the situation and to explain how the margin for manoeuvre had been reduced during the past months; but it was also a means to remind people of the rules of the game for any decision: the Director-General would consult ECFA, the Committee of Council and the SPC, he would then make his proposals — and the Council would choose. The need to decide now on key issues — and the impossibility of reaching a consensus — had forced relationships to be formalized and *relations of power to be explicitly spelt out.*[286]

It is difficult to know just what happened at this ECFA meeting as we only have one brief report on it written by Amaldi.[287] Two points emerge from his summary. Firstly, and the words are Amaldi's, 'all participants had been in favour of the 300 GeV accelerator', while opinions on the ISR remained divided. All the same, and to quote Amaldi again, ECFA decided to maintain its June 1963 position (*two* machines at the summit) and 'not to allot an order of priority to the various projects'.[288] Reading between the lines of this cautious formulation, two things could be perceived. On the one hand, that the European physicists were still worried about the effects that 'selling' the ISR would have on the 300 GeV programme. Indeed, as Van Hove pointed out, granted the reticence of the politicians to finance a *European* 300 GeV, 'there was actually a danger that the United States might revive the idea of *intercontinental* co-operation once their 200 GeV machine had been approved', so leaving the Europeans having just the ISR.[289] The other consideration informing Amaldi's summary was that it had become almost impossible to reaffirm too clearly the priority for the 300 GeV for this would put at risk the measures already taken, beginning with the Council meeting in June, to sell the ISR to the governments. Hence the formula 'not to allot an order of priority', although everyone *knew* that the reality was much more complex. To come out *now* clearly in favour of a 300 GeV would be totally inacceptable to the Council members: a change in policy at this stage, in mid-stream, would seriously undermine their credibility in governmental circles and the prevailing general confidence in CERN and in its seriousness. In some ways the process of selling the ISR had already gone too far, had already picked up too much momentum to be stopped without doing considerable damage. Did this mean that the physicists ceased protesting about what was happening before their very eyes, despite themselves, one might say? Not at all, but they now had to accept that *some* things could no longer be changed, that the direction in which things were going was now too settled for them to hope to change it radically.

An excellent illustration of the somewhat irreversible character of the process now under way is provided by the reactions to a resolution proposed by the British physicists in February 1965 (so a few months later). Continuing to be relatively autonomous from the rest of Europe, and doing things in their own good time, they gathered together early in 1965 to specify their recommendations on the CERN programme to their government. To the astonishment of all the continentals (Weisskopf said: 'it would have been desirable for the UK physicists to inform their colleagues [...] of the tactics they intended to adopt'), they announced that 'they wished to make quite sure that their Government approved first of all the big European accelerator'. The members of the SPC who were present told Powell and Cassels that this position was indefensible, and the pressure was so strong that the British delegates had no other choice but to align themselves with the tactics adopted by the majority. Of course physicists like Gregory 'stated that [they] sympathized with the anxieties of the United Kingdom physicists', but everyone had to accept that there was no going back — and the demands of the British physicists had to be quietly shelved.[290]

12.6.3.2 The vote on the supplementary programme, the Intersecting Storage Rings, by the Council

Nothing unexpected happened after the ECFA meeting. The following day the SPC confirmed the attitude that it had taken earlier, and in December 1964 Leprince-Ringuet presented, in the name of the SPC, a 'single programme' comprising 'three parts which could be modulated in time' comparable, he said, to 'a great symphony in three movements'. The prelude was the PS improvement programme, the andante the ISR, and the finale the 300 GeV. For the time being the SPC asked only for a decision on the prelude and the andante.[291] Four countries (Austria, the Federal Republic of Germany, France, and the Netherlands) declared themselves in favour of the ISR, but no one made any commitment on the 300 GeV.

Notes: p. 779

The Council decided to hold an extraordinary meeting in March 1965 to decide on setting up a supplementary programme. The session was not held: despite the undertakings of the Belgians and the Norwegians, no definite decision could yet be made. In May, during the SPC meeting, the Director-General declared that — with the exception of Greece, Spain, and the United Kingdom — 'it appeared that all the other Member States would vote in favour of [the ISR]', some of the smaller countries making their support conditional on a positive response from the United Kingdom. Finally, in June, Sir Harry Melville declared that he was 'pleased to announce that Her Majesty's Government had decided in principle that the United Kingdom would participate in the ISR project, provided that the project received general approbation'. A resolution was then voted in which the Council approved the principle of a supplementary programme for the ISR, a programme which was submitted to the Member States for examination. Those who wished to participate financially would take the final decisions in December 1965.[292]

12.6.4 THE PHYSICISTS' LAST STAND, JUNE–DECEMBER 1965

After the June Council session the rhythm and the nature of the last decisions to be taken seemed to be clearly established. After almost five years of discussion, the end of the road seemed to be in view. But this was not to take account of the physicists (mostly experimentalists) who were suddenly faced with a brute *reality*: the ISR was no longer simply an eventual possibility to be conceded around the table, it was now the *only certainty* in the future of European high-energy physics. They wanted therefore to look again at the ISR project, to cast a new eye over it: if *300MSF were now available* (or almost) it was worth considering seriously the use to which they could be put.

These discussions were held in the summer, and provoked such a fury of despair in Kjell Johnsen that he threatened to stop everything there and then. He clearly did not understand the way in which physicists worked, he did not understand why they had waited so long to think seriously about the ISR. Summarizing the state of the discussions on 4 October, Weisskopf declared that 'a certain number of physicists thought it preferable to construct another type of machine [than the ISR], provided the same amount of money was available'. And if the ISR had to be built after all, their view was that 'it would be necessary to consider a certain number of improvements [...] to make experimental work easier'.[293] Among the alternatives proposed, Gregory kept two for the SPC in October 1965. The first, generally felt to be less interesting than the second, was to construct a 50 GeV PS without 'stacking' and to collide its beam with that of the 25 GeV PS 'at one point and at a small angle'. The second solution involved using two rings of 25 GeV and 50 GeV fed by the PS (See Figure 12.8).

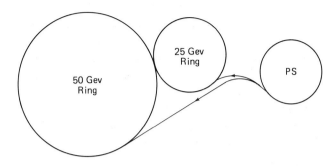

Fig. 12.8. Project for (50 + 25) GeV Proton Rings as presented in CERN/SPC/209/Draft/Add., 13/12/1965

It is out of the question to discuss the whole of the debate and the arguments over these new propositions. In essence it involved reorganizing the elements we are already familiar with. Let us simply note the following:

1. That from the point of view which the actors regarded as being 'strictly scientific'— i.e. to evaluate for each of the solutions put forward '[the] magnitude of interaction rate', '[the] flexibility for doing experiments', etc. —, more or less everyone came to similar conclusions.[294] Opinions diverged, however, when it came to 'comparing' the relative advantages and disadvantages of the different solutions. Here the members of the Directorate (Germain, Gregory, Hine) along with Johnsen strongly backed the ISR, while Schoch, Puppi, and a large number of physicists, particularly those in NP division, preferred the new projects. In addition certain members of the SPC (Jentschke, Powell) regretted the absence of detailed studies '[on] experiments [which] could be performed with the storage rings' — an argument which Butler and Gregory forcefully disputed.[295]

2. That as for the detailed technical aspects and cost estimates, Adams, Perrin, and a clear majority of the SPC felt that no meaningful comparison could be made with the ISR for two years. The ISR project could be launched — but not the others. To take them into account *now* would amount to postponing any decision for 18 to 24 months, a step which carried high political risks.[296]

3. That confronted with the reply that the new projects (and particularly the solution shown in Figure 12.9) need actually only be studied 'for about six months', the Director-General remarked that '[it] would be difficult to carry [them] out because of staff problems and problems of a psychological nature [...]'. 'It was very difficult', he explained, 'to find capable staff to carry out studies which seemed likely to reach a dead end'. After a number of other exchanges the SPC accordingly concluded that it was 'in favour of continuing the ISR project, without excluding the possibility of studying as far as possible any new idea which might arise'. As a result

Notes: p. 779

'[it] formally support[ed] the Director-General in continuing work on the storage rings project [...]'.²⁹⁷

Two months later, on the eve of the Council session of December 1965, the old and the new Director-Generals (Weisskopf and Gregory), supported by Amaldi, Gentner, and Leprince-Ringuet (and in the absence of Adams, Perrin, and Powell), managed to get the SPC to recommend to the Council that the ISR be constructed immediately. It is quite clear that for all those entrusted with the future of the organization, be they in the SPC or the Directorate, the discussion had gone on long enough. 'As Niels Bohr had put it', remarked Weisskopf, 'there was complementarity between action and thought'. 'In [my] opinion', he added, 'the time to act [is] now'.²⁹⁸ The next day, 15 December 1965, and having heard the recommendations of the Finance Committee, all the CERN Member States bar Greece declared themselves willing to participate in the ISR programme. The day after another far more modest supplementary programme (5 MSF for 1966) was agreed on, of which 1 MSF were allocated for the first three months. Its goal was to continue the study of the 300 GeV project.²⁹⁹

ECFA met again in 1966 for the third time.³⁰⁰ It confirmed the imperative need for a proton synchrotron of 300 GeV, the only dissenting voice being that of Werner Heisenberg. In 1967 Austria, Belgium, and France announced that they were willing to contribute financially towards its construction. For its part, the delicate question of the site was raised in all its complexity in committee after committee. At the Council in June 1968 the British delegates informed their colleagues that Her Majesty's Government would not participate in a 300 GeV project. By contrast, before the end of the year, Germany, Italy, and Switzerland had all joined Austria, Belgium, and France: a project with six participants — open to others — was thus taking shape.

Early in 1969 John Adams agreed to be the Director of the 300 GeV programme. He advocated a set of precise procedures to resolve the various outstanding questions. The scheme ran aground at the Council session in December 1969, the problem again being that of the site. In summer 1970 the situation seemed to be deadlocked.

Adams then came up with a new, and entirely different idea: install the 300 GeV at Geneva — a simple solution to the site problem —, and use the CERN PS as an injector — a way of making important economies. The result of this was the agreement of the United Kingdom — with the aid of the Secretary of State for Education and Science, Margaret Thatcher. With the four major contributors on board, it was possible to 'convince' the smaller states which had not yet agreed to the scheme (the Netherlands, Sweden) to join in. On 19 February 1971, ten member states accepted to finance the new accelerator.

Four years earlier the American authorities had agreed to finance an accelerator of the same type with an energy of 400 GeV. Its construction was begun at once. On 1 March 1972 it came into operation on the plains of Illinois, about 50 kilometres from Chicago. The equivalent European machine (which was also finally upgraded to 400 GeV) only worked in 1976: it was four years behind.

12.7 Conclusions

To conclude we would like to draw attention to two issues of a somewhat different kind. The first concerns *the way* in which the decision was taken, and the dimension of a process which it manifested in particular. The second, drawing on a comparison with similar processes in the United States, stresses what seems to us to be *specifically European* in this decision.

Someone wanting to understand the 'essence' of the decision which we have presented above might be tempted to summarize or to visualize the description using the model of the parallelogram of forces familiar in statics. From this perspective, the way to explain the decision would be first to compile a list of the different actors who played a part, of the different forces at work; then to evaluate their respective importance or their relative weights; finally to show the goal pursued or the objective sought by each — the final choice being the resultant of these various vector-forces. Unfortunately such an approach has a radical weakness in that it gives a picture, a reconstruction, of the events which is 'too' rational and which ignores the *time* dimension, the dimension of process which a decision of this type always involves. In fact, it is never a matter of taking an all-embracing decision, of tackling a problem in its entirety at a given moment in time. If one wants to speak in terms of a resultant one must also think of it as the outcome of a succession of intermediate decisions each of which had only limited consequences, as the last link in a chain of local choices distributed through time and often mainly a response to particular situations. In other words, the system of negotiation in which the actors are involved must also be seen as having its own dynamic (each element, for example, always tends to place constraints on the next, to reduce the freedom of choice in the future), a dynamic which the implicit model of the balance of forces overlooks.

The reason why this dimension of process is often ignored is that the spontaneous image one has of a rational decision — and surely a decision on the equipment they need by scientists is a prototype? — rests on the idea of *freedom of choice* so dear to western thought since at least the 17th century. Proceeding in a linear, logical, fashion a Rational Man first defines his objectives; in the light of his ends he surveys the means available and compares them; he then implements the optimum choice. At each of the steps, separate in principle even if they necessarily follow one after the other, he is supposed to be able to inform himself completely (presupposition of total information), to be aware of the effects that his actions will have (presupposition of transparency of action) and to be logical and coherent in his final decision. In brief,

Notes: pp. 779 ff.

he makes an optimo-rational choice from an exhaustive list of well-defined possibilities.[301]

The study we have made nicely illustrates the limits of this kind of hypothesis. In effect:

1. Successive choices are made by actors on the basis of alternatives which are often implicit, limited in number, and uneven in quality. Let us remember that the proposal for the ISR in December 1960 was made by a group of scientists who did not separate means from ends (a new accelerator was needed/it had to be the ISR). The physicists refused it because they felt that there was no urgency and that another possibility had not been considered, a high-intensity synchrotron. After the summer, all the same, this alternative was outmoded since another kind of machine had appeared on the scene, a more polyvalent machine, a high-energy synchrotron. We see that the information the actors have is never complete, nor is it stable, and that the list of 'means' is never closed — on the contrary, it is almost always specific proposals or provisional decisions which stimulate and give birth to new political and technical solutions. An eloquent confirmation of this fact lies in the attitude of the physicists in summer 1965. It was only when the final decision to construct the ISR had been taken by the member states that they seriously looked at the machine and proposed... to replace it with something different which had never been considered before.

2. These attitudes are partly related to the fact that different actors make very different 'investments' in the debate. Time is in short supply, as we all know, and the actors do not devote the same attention to the same questions at the same moment. Builders, for example, are far more preoccupied with the next generation of accelerators than are physicists, and we know why. That this latter group often has much less information than the former, that they follow the decision-making process more episodically and from a greater distance, that they generally move in *reaction* to the proposals made by the builders or to intermediate decisions — none of this is surprising. And if as a result they intervene 'too late' (as in 1965), that is largely in the logic of things.

3. Our account has also shown that there is not *one* way of posing a problem, but many, that the aims, the stakes, and the criteria considered as decisive vary across individuals and with time. There is no question, then, of there being *one* rationality of decision-making which is at work from start to finish, but of rationalities, each as 'respectable' as the other, rationalities which are now opposed to one other, now succeed one another. One result of this is that those in a position of having to 'judge' are inclined to be cautious, to proceed piecemeal, and to search for partial consensuses which avoid as far as possible a confrontation between the conflicting opinions of rival groups. To *juxtapose* projects and to postpone for as long as possible the moment when one 'really has to choose' is thus common practice. As Schilling has remarked, one is often trying to find ways of 'deciding without ever really choosing'.

4. It is never the same criteria of choice, the same determinations, the same groups, which are decisive. It is the *moment* at which one finds oneself in the process which brings the real stakes to the surface, and so defines the actors and the arguments which will carry the greatest weight. This can, of course, lead to somewhat uncertain results or to what we customarily call 'perverse effects'. Take the turn of events in 1964. At the start of the spring the fear that one would get nothing at all led the CERN Directorate to propose that the ISR and 300 GeV projects should be staggered in time. The Director-General somewhat hesitantly submitted a proposal to this effect in May. In June the SPC made the idea its own and forwarded it to the Council who approved it a few days later. During the summer the European state machineries ground into motion to set aside the money for the ISR. In the autumn the physicists reacted strongly — only to discover that they were powerless to reverse the course of events: each state was thinking in terms of financing the ISR, and was against all projects of the 300 GeV type! An interlocking sequence of microdecisions had thus produced a macro-result which was contrary to the repeatedly expressed wish of those who, from the start, had been designated as having the right to take the final decision — namely, the physicists. With respect to the logic which the actors said they wanted to follow (the final choice would lie with the users) the result is paradoxical. It is not so if one accepts a dynamic notion of what a complex decision-making process entails.

Apart from these very general remarks, we would like to come back to a conclusion more closely related to the particular case which concerns us here. In the majority of studies devoted to decisions on big equipment (in Europe as in the United States), we generally find several groups or laboratories which are in direct competition with one another vis-à-vis governments, and which are fighting to have as much as they can of a more or less finite budget. In these battles each group tends to define *its* project, a project on which it places its bets and which is based on equipment different to that of its rivals (an electron linac, an FFAG proton accelerator, etc.). In the inital stages each group tries to drum up maximum support in and outside the scientific community, to win over the greatest number of allies possible (ministers, high civil servants, MPs, etc.). The philosophy underlying these attitudes is that anything goes for those who want to succeed and who want to lead a major and costly project. We find this kind of thing in the case of French radioastronomy studied by Darmon and Lemaine, of British radioastronomy studied by Mulkay and Edge, of the Franco-German projects in the 60s which we have looked at, and in the works that have been done on American high-energy physics in the 50s and 60s (on the 200 GeV project, on Panofsky's ...).[302]

Ways of doing things at CERN appear rather differently. The originality lay in the fact that the dominant figures of European high-energy physics 'accepted' that there was a kind of *hierarchy* of laboratories, understood that a certain amount of *centralization* was unavoidable — and as a result a ruthless competition with no

Notes: p. 780

holds barred was ruled out from the start. This hierarchy, this centralization, plainly underpinned the notion that there be a pyramid of accelerators, a European pyramid which ECFA officially accepted in 1963 and which was to be crowned by the European machines which were to be chosen collectively, democratically, we would dare to say. In this sense, not all was allowed for the apex of the pyramid; in particular, one was not free to play a national machine off against a European machine. Of course everyone had the right to do what he wanted at home, and one should not idealize the European 'way of doing things' for the top of the high-energy field. Here as so often there was virtue in necessity. But this does not alter the fact that, for the 'summit' accelerators that were to lead the European assault, scientists of different countries 'accepted' to speak as one to their governments.

This helps us to understand a striking feature of the process we have investigated, the fact that it was divided into *two* chronological phases in which *two* successive and different groups controlled the project. The first phase lasted from 1961 to mid-1963 and was under the sole and direct control of the scientists. During this period, one talked physics and technology, political leaders in the Council did not intervene and nobody was trying to get money for the projects. This 'selling' was to be organized by another group, though only in a second phase, once the scientific choice was made by the physicists and engineers. Here the Council had to play the key role. Composed of an equal number of high-level scientists and of diplomats, and animated by a very strong *esprit de corps* and an unflagging dedication to CERN, it was its function to deal with national bureaucracies, it was its *raison d'être*.

This rather unusual characteristic of European high-energy physics induced unexpected effects during the decision-making process, as we have said. It also helps to understand that, contrary to what all studies have shown regarding cost estimates for new and ambitious projects, those produced by CERN were generally reliable, at least for *construction* costs. The main reason for this is that there was no serious competition, and so no vital need to underestimate the real costs of the machine so as to 'corner the market'. It was then in the organization's interests to 'tell things straight'; to show that one's estimates were exact, and were properly made, was to set oneself apart from almost all the others who fudged their figures in the knowledge that it was almost impossible for an agency to cease financing the construction of big equipment once it got under way. By contrast, there were no unpleasant surprises in store when one dealt with European high-energy physicists — these were people on which governments could count. If the CERN Council felt able to get everyone to respect this discipline — and this it managed to do — then that was the most intelligent solution for the community as a whole. In the long term it was the policy that fostered the very best of relations with the governments of the member states.

Translated by John Krige

Annex 1

Accelerators Conferences

International Conferences

CERN, Geneva, 11-23/6/56, *CERN symposium on high energy accelerators and pion physics* (Geneva: CERN, 1956).

CERN, Geneva, 14-19/9/59, *International Conference on high-energy accelerators and instrumentation* (Geneva: CERN, 1959).

BNL, Long Island, 6-12/9/61, *International Conference on high energy accelerators* (US AEC-BNL (?), 1961 (?)).

JINR, Dubna, 21-27/8/63, *International Conference on high energy accelerators,* American version CONF-114, TID-4500 (US AEC, ss.d.)

Frascati, 9-16/9/65, *V International Conference on high energy accelerators* (Roma: CNEN, 1966).

European Conferences

Remark: since no proceedings were published, we give in brackets the archival reference of the proceedings we used. We add brief comments.

Frascati, 16-17/12/60 (DG 20780). Initially Italian, this mini-conference was later enlarged a little. Its goals: the experimental programme at CERN; future accelerators, CERN and Frascati.

Saclay, 30-31/5/61 (DG 20780). Report entitled: *European Accelerator Group, First meeting,* CERN/EAG/1, 11/6/61. Very interesting for intersecting rings and high intensity proton synchrotrons.

Harwell, 30-31/10/61 (DG 20780). Report entitled: *European Accelerator Study Group, Second Meeting,* CERN/EAG/2, 1/12/61. Most important aspect is high energy synchrotrons (above 100 GeV). Appearance of the idea of a European Programme.

Amsterdam, 14-15/5/62 (DG 20780). Report entitled: *European Accelerator Study Group, Third meeting,* CERN/EAG/3, 5/7/62. All kinds of machines considered: linac, 'π factories', ... Studies on the physicists (problem of manpower) in various European countries.

Frascati, 3-5/12/62 (DG 20780; DIRADM 20102). No minutes; apparently the last conference of this type. Very few scientific papers. Rather a survey of the stage reached by studies in CERN and in the USA. Several communications on the 'availability of funds and people' in Europe.

Apart from these, see:

CERN, 5-9/6/61, *International Conference on Theoretical Aspects of Very High-Energy Phenomena* (Geneva: CERN 61-22, 1961).

Working Group/Seminars/Conferences, USA, 1959-1962

Madison (Wisconsin), June 1959, MURA Conference on Accelerators. We do not have the minutes; see Judd (1959), 7-14; Sands (1960), 3; J.P. Blewett (1962), 8.

Rochester[*], August 1960, working meeting organized by Wilson. No minutes; see CERN/SPC/123; J.P. Blewett (1962), 9.

BNL, April 1961, Linear Accelerator Conference; see Linac Conf. (1961).

LRL, 15/6-15/8/61, The Berkeley high-energy physics summer study; see Berkeley Study (1961), Smith (1961).

BNL, August 1961, 1961 Summer Study on 300-1000 BeV Accelerators; see 300-1000 GeV (1961) and the series of reports BNL, Accelerator Department, 1961 Summer Study, IA-1 à IA-19.

BNL, 20-24/8/62, Conference on Linear Accelerators for High Energies; see Linac Conf. (1962).

---

[*] Within the framework of the conference 'on High energy physics', 25/8-1/9/60.

Annex 2

List of official reports on American policy in high-energy physics

1. REPORT OF THE ADVISORY PANEL ON HIGH-ENERGY ACCELERATORS TO THE NATIONAL SCIENCE FOUNDATION — OCTOBER 25, 1956, known as the Haworth Panel Report.

2. SUPPLEMENT TO THE REPORT OF THE ADVISORY PANEL ON HIGH-ENERGY ACCELERATORS TO THE NATIONAL SCIENCE FOUNDATION — AUGUST 7-8, 1958.

3. PIORE PANEL REPORT — 1958, U.S. POLICY AND ACTIONS IN HIGH-ENERGY ACCELERATOR PHYSICS, REPORT BY A SPECIAL PANEL OF THE PRESIDENT'S SCIENCE ADVISORY COMMITTEE AND THE GENERAL ADVISORY COMMITTEE TO THE ATOMIC ENERGY COMMISSION, published on 16 May 1959.

4. PIORE PANEL REPORT — 1960, submitted in December 1960.

5. REPORT OF THE PANEL ON HIGH-ENERGY ACCELERATOR PHYSICS OF THE GENERAL ADVISORY COMMITTEE TO THE ATOMIC ENERGY COMMISSION AND THE PRESIDENT'S SCIENCE ADVISORY COMMITTEE, published on 20 May 1963, known as the Ramsey Panel Report.

6. AEC'S REPORT ON POLICY FOR NATIONAL ACTION IN THE FIELD OF HIGH ENERGY PHYSICS, submitted on 25 January 1965, known as the Seaborg Panel Report.

Remark: all of these can be found in *US Panels (1965)*.

Notes

1. The archives used for this study were primarily those at CERN, to which we have had free and essentially unconditional access. Additional material was obtained during two visits to BNL (Brookhaven National Laboratory, Long-Island, NY) and LBL (Lawrence Berkeley Laboratory, Berkeley, CA) by John Krige and myself in 1984, and by Uli Mersits in 1985. Three kinds of studies are pertinent for this text. To begin with, those dealing with decisions on big scientific equipment. There are very few of these. To be mentioned are those of Darmon (1981), Darmon & Lemaine (1982), Edge & Mulkay (1976), Galison, Hevly & Lowen (1988), Gibbons (1970), Mulkay & Edge (1973), Westfall (1988, 1989). Then there are the books and articles written by accelerator specialists on the evolution of their field. These describe the machines envisaged and built throughout the world, rarely the decision making process. Note 10 below gives a first list of them. Finally, there are studies of decisions taken in other domains. They are of uneven quality. Those which we found useful are given in note 5 below. The first version of this text, circulated widely in 1987, was carefully read by the principal actors, notably Kjell Johnsen. We should like to thank them all.
2. In the case of CERN, see for example ISR (1974), 400 GeV (1977a), 400 GeV (1977b), ISR (1984).
3. More specifically, we do not want to write two successive histories, one dealing with the ISR, the other with the 300 GeV project. If one goes about the study in this way it is impossible to understand the logic of the debate and of the decisions taken between 1956 and 1965.
4. An interesting case here is Greenberg (1971), chapters X and XI, mainly about MURA.
5. In addition to the references given in note 1, add Armacost (1969), Allison (1971), Allison & Morris (1975), Bernoux (1985), Crozier & Friedberg (1977), Coulam (1977), Jamous (1969), Klein (1962), Lindblom (1959), Mackenzie & Spinardi (1988), March & Olsen (1975), Marshall & Meckling (1962), Morris (1985), Schilling (1961), Sfez (1981).
6. Rudwick (1985), 455.
7. See the founder article by Braudel (1958). For a slightly different formulation, see Rudwick (1985), 3-16. See also History of CERN (1987), chapter 14.
8. The case is thus 'pure', hence its interest. One can compare it later to the arguments used vis-à-vis the non-scientific world. Writers often do not distinguish the various levels of discourse, so drawing rather strange conclusions.
9. To use the community's hallowed phrase.
10. To write this section we have had recourse above all to the proceedings of conferences on accelerators held during this period. A list is given in annex 1. We have also worked from review articles written between the end of the 50s and the end of the 60s, consulting *Advances in Physics, Annual Review of Nuclear Science, Contemporary Physics, Il Nuovo Cimento, Physics Today, Plasma Physics–Accelerators–Thermonuclear Research, Reports on Progress in Physics, Review of Modern Physics, Review of Scientific Instruments*. H. Blewett (1961), H. Blewett (1967), J.P. Blewett (1956), Hubbard (1961), Johnsen (1964), Judd (1958), MURA 50 MeV (1964) were used as well. See also the talk given at BNL by J.P. Blewett (1962) and Livingston & Blewett (1962). Finally, we have turned to articles, conferences, and original reports for all important documents.
11. Moreover, we had ready access (at CERN) to the American reports. The relative importance of the USA and the USSR is difficult to measure since the Soviets did not participate in the important international conference in 1961. For an attempt at a comparative 'measure' see Irvine & Martin (1985).
12. Accelerator Conf. (1956), title.
13. CERN/SPC/33, 12/10/56 (first quotation); unsigned document (probably written by the Director of Administration, Richemond), without title, in DIRADM 20478, for the last two quotations.
14. CERN/SPC/33, 12/10/56.
15. From 1958 onwards the Rochester conferences on high-energy physics were held in even-numbered years. Accelerator conferences thus took place in 1959 (at CERN), in 1961 (at Brookhaven), in 1963 (at Dubna), in 1965 (at Frascati), etc. For more on these conferences, see annex 1.

16. Adams (1956), 1.
17. For example, see J.P. Blewett (1956).
18. See Accelerator Conf. (1956), 68-100. The quotations are from Adams (1956), 1-2.
19. Judd (1958), 189. See also J.P. Blewett (1962), 3.
20. J.P. Blewett (1962), 3-4.
21. We are not particularly interested in searching for the 'fathers' of ideas. The situation is complex here. K.R. Symon is on the list, though in 1956 A.A. Kolomenskij drew attention to his work in 1953 with V.A. Petukhov and M.S. Rabinovich (Accelerator Conf. (1956), 66; reference to the report in Accelerator Conf. (1959), 113); J.P. Blewett, for his part, emphasized the parental rights of the former director of BNL, along with those of several other people: 'It is not generally known that this machine [FFAG accelerator] was first *invented* [our emphases] by our former Director Dr. Haworth, in 1952. It was *reinvented* [id.] in Japan and in Soviet Union and finally in an improved version by Keith Symon' (J.P. Blewett (1962), 3). Judd (1958), 189, made a more subtle remark: 'It is interesting that one aspect of their thinking [MURA] led them through a full circle back to the original work of Thomas mentioned above, which had meanwhile been given an experimental test at the University of California Radiation Laboratory in Berkeley on two models of constant-frequency cyclotrons which accelerated electrons to semirelativistic energies.'
22. See Adams (1956), 2: 'If the cosmic radiation can be likened to an apéritif, the first wave of machines now appears to have been the entrée. The main course consisting of the high intensity machines is yet to come. There seems at the moment less justification coming from the Cosmic Ray experiments for machines exceeding 50 GeV.'
23. Kerst et al. (1956), 590. This is true for protons. For electrons the idea is, in theory at least, even more interesting: two beams of 21.6 GeV 'would be equivalent' to a machine of 1.87 million GeV. See also Kerst (1956).
24. Kerst et al. (1956). Quite a few people think that 'the possibility of doing high-energy experiments with colliding beams was first suggested by the MURA group in the USA in connexion with their proposal for a two-way FFAG accelerator'. This is not so (see below part 1.2). The quotation is from Johnsen (1964), 446.
25. The work of Crawford and Stubbins is reported in Kerst et al. (1956), 590. The quotations are from this article. For confirmation see Brobeck (1956), 63.
26. O'Neill (1956a). See also Judd (1958), 207. For readers interested in searching for the 'ancestors' of an idea, see O'Neill (1956b). He mentions the names of Brobeck and Lichtenberg there. More extensive information is given in Judd (1958), 207 and 215 and in Jones (1959), 22, note 4.
27. O'Neill described his solution thus (O'Neill (1956a), 1418-1419): 'In order to prevent the beams striking the inflectors on subsequent turns, each ring contains a set of foils, thick at the outer radius but thinning to zero about one inch inside the inflector radius. The injected beam particles lose a few MeV in ionization in the foils; so their equilibrium orbit radii shrink enough to clear the inflectors after the first turn. After several turns, the beam particles have equilibrium orbits at radii at or less than the inside edge of the foils.' In the months that followed Symon and his coworkers showed that this system could not work — which Kolomenskij confirmed. On this episode see the *Discussion* following O'Neill (1956c), 134-135.
28. Neal & Panofsky (1956); Brobeck (1956). We will not describe the former paper here; it deals with electron machines.
29. The third direction pertained to research on colliding beams, the first to fixed target machines. Both aimed at very high energies.
30. The quotations are from Brobeck (1956). The question of 'ancestors' was not raised at this meeting. It was, however, brought up during the discussion following Judd's communication in 1959 (Judd (1959), 12). Kolomenskij mentioned Salvini and Robert Wilson. Sands, in the same discussion and in Sands (1960), 4, confirmed Wilson and attributed to M.L. Oliphant and T.A. Walton 'a cascade system of a

cyclotron and a synchrotron'. For Salvini, see his communication in 1956 (Accelerator Conf. (1956), 40-41).

31. Report by Schoch, the leader of the accelerator research group, CERN/268(A), 5/6/58, 8. For a similar, though somewhat more nuanced conclusion, see Judd (1958), 209.
32. Report by Schoch, CERN/378(B), 10/11/60, 9. All the same, seven papers on this topic were read at the conference in 1959.
33. See below for this new type of FFAG accelerator. The information comes essentially from MURA 50 MeV (1964), a special number of *The Review of Scientific Instruments.* The first model built by MURA was for 400 keV electrons and was of 'radial sector' design. It came into operation in March 1956. See also the communication of the 'MURA staff' in 1959, on the 50 MeV model (Accelerator Conf. (1959), 71-74), with photographs.
34. Piore Panel, US Panels (1965), 137; Judd (1959), 7-8; on the machines projected or under construction see Accelerator Conf. (1956), 343-346 (on Nimrod), 359-361 (on ZGS), 362-365 (on Princeton-Penn.), 366-371 (on the Southern Regional Accelerator). See also Accelerator Conf. (1961) and (1963) where the state of progress is described. See also History ZGS (1979).
35. It was only in Judd (1959) and the ensuing discussion that very high energy proton machines were mentioned. A point about vocabulary: around 1959 'high energy' usually meant 10-30 GeV. An accelerator of more than 50 GeV would be called an 'ultra-high energy machine'. This terminology was to change in 1960/61.
36. Ohkawa (1958); Kolomenskij's article is referred to in Accelerator Conf. (1959), 98, for example. We have not looked at it.
37. On the problems raised and their possible solutions, see Barbier et al. (1959), 100, or Judd (1958), 206. On the models, MURA 50 MeV (1964) and Barbier et al. (1959). See also Judd's introduction to the conference, Judd (1959). The models in question were never built in their entirety.
38. For a description see O'Neill (1959b), who gives the reference of the original report published at Stanford, HEPL 170, June 1959. See also Judd (1958), 207.
39. O'Neill's description is in O'Neill (1959b), 128-134. O'Neill (1959b), Judd (1958), Schoch, *A Discussion of Colliding Beam Techniques,* undated, probably the second half of 1960 (DG 20576) have been used to write this paragraph. The quotations are from Schoch, but for the last (O'Neill (1959b), 135, during the discussion). The attempt to achieve a 'damping' for protons 'similar' to that for electrons refers to O'Neill's solution presented in note 27.
40. N.C. Christofilos, *Trip Report,* 1/12/59, UCRL-5806T (LBL-DC file, Travel Report). We have said that Christofilos's remark was symptomatic of the mood of the times. All the same *some* scientists were fascinated by very high energies (and storage rings) prior to 1959. Christofilos mentioned Panofsky in particular.
41. Accelerator Conf. (1959), 373-374.
42. Sand's report was communicated to us by C. Westfall. It is also mentioned by Judd (1959), 6-7, by Sands (1960), 3, (Internal MURA Report N° 465), and by J.P. Blewett (1962), 8.
43. Based on Sands' intervention in the *Discussion* following Judd's paper, Judd (1959), 12-13.
44. See for example CERN, Rapport Annuel, 1960, 12.
45. The importance attached to the successful commissioning of the CERN PS and the Brookhaven AGS, and their 'good' intensity, strikes one when reading the Piore Panel report of December 1960 (cf. US Panels (1965), 126-127). This was also confirmed during our interviews by K. Johnsen and M. Hine. J. Krige has remarked on its significance in Great Britain in 1952: the commissioning of the Cosmotron had a major impact, and substantially changed the terms of the debate 'for and against' the United Kingdom joining CERN. See History of CERN (1987), chapter 13. Subsequently (ISR, ...) the same phenomena can be observed. The quotation from Adams comes from document CERN/SPC/105, 16/3/60, 17.
46. This is obviously relative, Panofsky being about to get the money for his 2 mile long electron linear accelerator (estimated to cost $105 M in Accelerator Conf. (1959), 349-359). For the Radiation

Laboratory and for Caltech, by contrast, the problem was posed in these terms.
47. Document CERN/SPC/100, 23/9/59 is a source which gives a considerable amount of detail.
48. The Committee of Council document CERN/CC/342/Projet, 19/1/60, point 3 is rather definite on this point. It is corroborated in many American publications. See, for example, 300–1000 BeV (1961), preface, or J.P. Blewett (1962), 8–9. The terms of the agreement are reported by J.P. Blewett (1962), 9. Note that we have mentioned two *external* reasons for this renewed interest in very high energy proton synchrotrons. They are not the only ones: one should also take into account the internal logic of research on accelerators (difficulties with plasma machines, with the FFAG scheme...), the competition between groups (California/East Coast, e.g.), etc.
49. See CERN/SPC/123/Draft, 17/2/61, 14–15, Amaldi's report to the SPC meeting of 4 and 5 November 1960. For a confirmation of R.R. Wilson's meeting, J.P. Blewett (1962), 9. There is no trace in the proceedings of the conference itself. Blewett's account in 1962 was more categorical than that of Amaldi some weeks after the conference. According to Blewett in 1962, 'it was agreed at Rochester that accelerators for energies > 300 BeV were both desirable and feasible'.
50. Sands (1960).
51. 300–100 BeV (1961), preface, confirmed by J.P. Blewett (1962), 9, from which the quotation is taken. One finds a corroboration by the Europeans in CERN, Conseil, 8–9/12/60.
52. See 300–1000 GeV (1961), preface; Blewett (1962), 9. The Soviets did not participate in the 1961 conference. Accordingly it is difficult to know how much effort they dedicated to very high energy machines. All the same: 'We continue to hear intriguing and tantalizing news reports from the Soviet Union. Apparently a 1000 BeV design study was reported at a meeting in February [1962] of the Russian Academy of Sciences' (J.P. Blewett (1962), 9).
53. This whole paragraph is based on Sands (1960). Note that a comparison with competitive projects is always found in American reports. For a particularly good example of charming words concealing cutting comments see J.P. Blewett (1962).
54. 'The results obtained in the above treatment give a startlingly small aperture for the Main Ring. It also appears somewhat unreasonable that the stored energy in the Main Ring should be only one-half that of the Booster Ring, and that the cost of the Main Ring magnet should be much smaller than the cost of the tunnel which houses it. Such intuitive and subjective feelings have led to the tentative proposal of an aperture for the Main Ring about 1.5 times larger and for the Booster Ring about 2/3 as large as the values derived above.' Sands (1960), 22.
55. Sands (1960), 26–28.
56. J.P. Blewett (1961). This report of 60 pages, written by J.P. Blewett, was preceded by a series of 'Internal Reports' (IA) in the BNL Accelerator Department. A complete list is given as an annex to 300–1000 BeV (1961), 124–127. The report dealt with successively '[the] particle dynamics in AG synchrotrons', '[the] choice of basic parameters', '[the] injection', '[the] magnet', '[the] RF acceleration', '[the] vacuum system', then the buildings, the site, and the costs.
57. J.P. Blewett (1961), 48.
58. Organized by Léon Van Hove, this conference was a response to a slightly different logic to that prevalent in the USA. See below, part 3. Published proceedings: Theory Conf. (1961).
59. Berkeley Study (1961); citation page iii. Four Europeans (Daniele Amati, Kjell Johnsen, Donald H. Perkins, William Walkinshaw) took part in this collective work. Cocconi too appears in the proceedings.
60. The seminars held at BNL are numbered IA-1 to IA-19 (*1961 Summer Study on 300–1000 BeV Accelerators*), dated from 31/7/61 to 30/8/61. Independent reports were also published (there is a list in 300–1000 BeV (1961), 125–127). The second report dated 28 August 1961 and submitted to the US Atomic Energy Commission, is 300–1000 BeV (1961). The proceedings of the *International Conference on High Energy Accelerators* held at BNL from 6 to 12 September 1961 are Accelerator Conf. (1961). See also H. Blewett (1961).
61. Though only if such a history was not restricted to a list of the 'major discoveries'. The importance of this

summer of 1961 will appear more clearly later in this text (see part 3 below).
62. Same references as in notes 58, 59 and 60. Add Experimental Program (1961).
63. Introduction to Linac Conf. (1961), 3.
64. See Linac Conf. (1961), 3-4; Linac Conf. (1962), i-ii, v; J.P. Blewett (1962), 10-11. Both the Rutherford and the Radiation Laboratories were extremely interested in these projects. BNL and Yale University organized the 1962 conference on linacs.
65. The solutions finally adopted were borrowed from the 'Sands model'. See Sanford (1976) on Fermilab; CERN/563, 11/12/64 for CERN.
66. Extract from 300-1000 BeV (1961), 119-120. See Collins (1961), Robinson (1961), Symon & Tollestrup (1961). For additional examples, see J.P. Blewett (1962), 12-13.
67. 300-1000 BeV (1961), 49.
68. One can argue that different motivations lie behind the same words: for example, the wish to 'sell' his idea in a 'competitive' market, the wish to raise funds for an in-depth 'feasibility study'. In this case, it is in one's 'interests' to conclude that the project is 'likely feasible'. This interpretation is not without merit and it probably partly explains some of the initial formulations (1956-1959). All the same it is not only this that is at stake. The exchange of ideas, for example, between Kolomenskij, Judd and Sands at the CERN conference in 1959, an exchange reproduced in the proceedings, proves this clearly: scientists were trying, between themselves, to arrive at the 'best' solution by successive approximations.
69. The various solutions to the injection problem illustrate that systems studied by different groups each have advantages and disadvantages, and raise doubts, certainties, and hopes of different kinds. For example, an FFAG injector brought with it the problem of an as yet unproven technique, as did a possible linac of some GeV. A classical synchrotron thus had distinct advantages as an injector. On the other hand, this device shared with the FFAG synchrotron the ejection/reinjection problem, which had not succeeded experimentally in 1961 (even through judged to be soluble in the very short term). The length of the 'pulses' worked to the advantage of a linac, though partisans of synchrotron injectors thought the problem could be quickly solved (multi-turn injection, ...), etc. For an analysis of this biassed in favour of linacs, see 300-1000 BeV (1961), 32-46.
70. Sands made his first statement in the *Discussion* following Judd's paper at the 1959 CERN conference (Judd (1959), 13); the second is in Sands (1961), 145. Compare the debate in the SPC at CERN at the end of April 1961, which was largely organized around the notion of 'high energy monster' (CERN/SPC/133, 10-15, particularly 14-15). See below, part 3, for a more detailed analysis.
71. Since we have not studied the Soviet scientific literature, and since the Soviets were not at the 1961 conference, we do not go into their work in this section.
72. For the detail, see below, part 2.
73. See MURA staff (1961b), from which the first quotation is taken; MURA 50 MeV (1964), second quotation, our emphasis.
74. The 'spontaneous' explanation one has for this 'choice' of MURA has two components: firstly, and in short, the idea of the FFAG is MURA; then MURA does not have a high energy proton synchrotron: it therefore cannot orientate itself towards simple storage rings.
75. See the paper by F.T. Cole in 1959 (Accelerator Conf. (1959), 82-86); MURA staff (1961a), who proposed 'a 10 BeV FFAG accelerator'; MURA 50 MeV (1964). The date at which the 'two-way' project was abandoned is not clear from our sources (around summer 1959). The reason given in MURA 50 MeV (1964), 1394—'the difficulties of doing experiments with colliding beams'—does not, at first sight, seem to us to be particularly convincing.
76. Excluding the Soviet Union.
77. O'Neill (1961b).
78. Accelerator Conf. (1961), 256-264, quotations on page 256. See also European Accelerator Group, First meeting, 30-31/5/61, CERN/EAG/1 (DG 20780), pages 23-53. In this seminar at Berkeley in summer 1961 (Berkeley Study (1961), 74-80), O'Neill said of the experiment with Ada: 'A countable number of

electrons (about 14) has been stored, with mean lifetimes of 2 to 4 min at pressure of about $0.5 - 10^{-6}$ mm'.

79. For the time being, see O'Neill (1961a).
80. Accelerator Conf. (1963); Accelerator Conf. (1965), 444–447.
81. Greenberg (1971), 207–265.
82. Ramsey Panel, 1963, US Panels (1965), 94 et 103.
83. On the importance of 'multi-purpose' machines, see Krige & Pestre (1986). As for the evolution of the idea of what counts as 'high intensity', see J.P. Blewett (1963), 89.
84. Ramsey Panel, 1963, US Panels (1965), 103–104. See also Greenberg (1971), 243–265.
85. Accelerator Conf. (1963), 309–328.
86. Accelerator Conf. (1963), 365–371; Accelerator Conf. (1965), 271–275.
87. Accelerator Conf. (1963), 378; Accelerator Conf. (1965), 266–271. On the difficulties with the Stanford rings, see the report of Mervyn Hine, *Note on Instabilities in Storage Rings* (CERN/SPC/190, 4/12/64), his report on the same topic written a year later, *Progress Report on Studies on Performance Limitations in the ISR* (CERN/SPC/212, 1/12/65) and the important ensuing discussion at CERN.
88. To conclude let us remember that we have only considered high energy machines, i.e. of several GeV at least, from 1956 onwards. All the, sometimes dramatic, improvements made to lower energy machines have been overlooked.
89. Report by Adams entitled *Research Work on Machine Design and Allied Problems*, 12/10/56, CERN/SPC/33. Proceedings of the SPC meeting held 12–13/10/56, CERN/SPC/35, 6–7.
90. Report for the SPC entitled *Research Group on Accelerators*, 18/2/57, CERN/SPC/36. See also Pentz (1967), 1.
91. Document CERN/SPC/87, 31/3/59 and Pentz (1967). Areas of obscurity emerge when we confront these two reports with each other and with the group's annual reports. It is difficult to know the criteria used (are they referring to physicists working with the group?, paid by the group? etc.) and the information itself is sometimes contradictory. The figures we give are thus only approximate.
92. F. Schneider is also given as a member of the group from 1/4/57 in CERN/SPC/87, 31/3/59, 11 but does not appear in Pentz (1967), 20–21, before the 15/12/59. In the same way, M. Barbier and A. Susini seem to have worked with the group—if we are to believe the signatures on scientific articles (see note 112 for example).
93. Calculations made from the 11 cases mentioned with sufficient detail in CERN/SPC/87, 31/3/59, 11.
94. Report to SPC, CERN/SPC/69, 16/5/58.
95. CERN/SPC/70/Draft, 1/8/58, 7–9 (meeting of the SPC, 23/5/58); CERN/272, 12/6/58 (about FC); CERN/272*, 13/6/58 and CERN/280/Projet, 7/7/58 (about CC); CERN/279 (about the Council meeting held on 20/6/58).
96. See History of CERN (1987), 280–281, for 1952; infra, part 4 for 1962.
97. CERN/SPC/36, 18/2/57, 2.
98. CERN/SPC/36, 18/2/57, 2; CERN/SPC/43/Draft, 27/3/57, 4; CERN/225(A), no date, 13–14; CERN/249(A), no date, 8–10; CERN/SPC/58, 17/2/58, 1–2, for the description of the betatron.
99. CERN/SPC/63/Draft, 2/4/58, 10–11; The first report on the *Study Programme on Beam Stacking Processes* is CERN-PS/MJP 1, February 1958 (M.J. Pentz, *Introduction*). See for example CERN/SPC/58, 17/2/58, 3.
100. Defined in CERN/SPC/63/Draft, 2/4/58, 10 and CERN/SPC/81, 11/11/58, 1.
101. CERN/SPC/81, 11/11/58, 1–2.
102. CERN/SPC/87, 31/3/59, 1. See also the proceedings of the SPC meeting held on 17/4/59 (CERN/SPC/92, 8/5/59, 5–6).
103. CERN/SPC/87, 31/3/59, 9–10.
104. CERN/SPC/87, 31/3/59, 3–8.
105. CERN/319, 18/11/59, 10.

Notes

106. CERN/378(B), 10/11/60, 9 and 10.
107. Pentz (1967), 20-22 which gives the list of personnel. Four members of the Senior Staff joined Schoch in 1960: Hereward, Johnsen, Lapostolle and Zilverschoon. At the time there were 16 in the PS division and about 40 in CERN. See CERN Annual Report (1960), 25.
108. CERN/319, 18/11/59, 10-11.
109. The first mention of this change is to be found in the Council, minutes, 14/6/60, 31. At this stage one envisaged building *two* storage rings, and the aspect of collisions was still stressed. A few months later in a text by Schoch entitled *A discussion of colliding beam techniques* (DG 20576), only one ring was mentioned. This would become the model later built under the name of CESAR. Schoch's text was written towards the end of 1960, between September and November; there is no date given on it.
110. Schoch, *op, cit*. note 109, 10-11.
111. Schoch, *op. cit.* note 109. During an interview with U. Mersits and D. Pestre on 11/2/86, K. Johnsen claimed that the model 2×100 MeV seemed to everyone to be much too complicated, perhaps even impossible to get working, at the end of 1959.
112. Barbier *and al.* (1959), 100 et 101.
113. See for example Schoch, *op. cit.* note 109, 10-11; CERN/378(B), 10/11/60, 9-10; CERN Annual Report, 1960, 43-44; Hereward et al. (1960). Note that at 2×100 MeV, with electrons, one could hope to do physics. The model could thus also be a small research apparatus. This was obviously not the case with a 2 MeV ring.
114. Hereward et al. (1960).
115. See for example the documents CERN/327, 5/11/59, CERN/327*, 18/11/59, CERN/327*/Add., 1/12/59, CERN/CC/342/Projet, 19/1/60, Council, minutes, 1-2/12/59, 61, CERN/CC/354/Projet, 23/5/60, CERN/CC/362/Draft, 12/6/60, Council, minutes, 14/6/60, 27-28, CERN/SPC/123, 17/2/61, Council, minutes, 8-9/12/60, 76. See also Section 1 of this chapter.
116. The best example is to be found in the minutes of the SPC meeting of 4 and 5 November 1960 (CERN/SPC/123, 17/2/61).
117. This view was expressed by K. Johnsen during a recent conversation with the author.
118. See CERN/378(B), 10/11/1960 and CERN Annual Report, 1960, 43-44.
119. H.G. Hereward, K. Johnsen, A. Schoch, C.J. Zilverschoon, *Present Ideas on 25 GeV Proton Storage Rings,* document PS/Int. AR/60-35 (DG20576); quotations p. 13. The term ISR (Intersecting Storage Rings) would only appear later. All the same the project was exactly that described here as '25 GeV Proton Storage Rings'. We thus use both formulations interchangeably.
120. Introduction to the report cited note 119; memorandum Schoch to Adams, Bernardini, etc., 22/12/60; memorandum Bernardini to Schoch, 17/1/61 (DG20576).
121. Memorandum Hine to Director-General, etc., 30/3/61 (DIRADM20102); *Possible Long Term Programmes for CERN, Draft,* Paper by M.G.N. Hine, 30/3/61, CERN/DIR/39. For a slightly different version with the same title see CERN/SPC/125, 14/4/61. All the quotations in the following paragraphs are extracted from these two reports.
122. To be more precise, Hine envisaged 'A.G. proton synchrotron in the 100-300 GeV range', 'high intensity proton synchrotron in the 5-25 GeV range', and then 'large electron linear accelerator, say 50 GeV', 'large proton linear accelerator', 'large electron synchrotron'.
123. This meeting is mentioned in C.A. Ramm, *Some further thoughts on the future programme of CERN,* 20/4/61 (DG20576). Some unsigned manuscript notes taken on the occasion are in (DG20576). Box DG20576 contains a large numbers of notes and letters exchanged in April on this issue. Do not forget that the role of the SPC is to advise the Council on technical matters.
124. Conclusions drawn by John Adams himself in a memorandum sent to divisional heads on 28 April 1961 (CERN/DGO/Memo/1056). See also *Long-term Accelerator Programmes for CERN,* paper by the Director-General, 14/4/61, CERN/SPC/125/Add.
125. CERN/SPC/125/Add., 14/4/61, 1 and 2.

772   *The second generation of accelerators*

126. CERN/DGO/Memo/1056, 28/4/61, 1 and 2.
127. SPC, 29/4/61, Draft minutes (CERN/SPC/133/Draft, 7/7/61), 11.
128. CERN/SPC/133/Draft, 7/7/61, 12-14. These minutes are not particularly informative. As a result we listened to the tape recordings of the meetings, which are available in the CERN archives. According to Perrin the full use of the PS would require about a 30% increase in the budget.
129. CERN/SPC/133/Draft, 7/7/61, 13-15, plus the tapes.
130. Letter Dakin to Hubbard, 16/5/51 (DIRADM20102).
131. This analysis of the implicit usage of the word 'monster' is derived from the tape of the SPC meeting of 29/4/61 (from which our remarks on Amaldi and Perrin came, as well as the Bernardini quotation) and from a large number of letters, for example, Ramm to Adams, 13/4/61, Ramm to Weisskopf, 21/4/61, Hine to Weisskopf, 8/5/61 (DG20576), Bernardini to Weisskopf, 12/5/61 (DG20889), Weisskopf to Hine, 15/5/61 (DG20486), Ramm to Weisskopf, 16/5/61, Hine to Weisskopf, 19/5/61 (DG20576), etc.
132. CERN/SPC/133/Draft, 7/7/61, 15.
133. Tape recording of the SPC of 29/4/61.
134. CERN/SPC/133/Draft, 7/7/61, 13-15. 'In 5-6 years' time, but not before, Europe might contemplate building a new large accelerator on an up-to-date technological basis', said Perrin, so avoiding a repetition of the mistake made by Dubna.
135. The tape recording is more explicit on these points than the official minutes. See also letter Bernardini to Weisskopf, 12/5/61, 3 (DG20889).
136. One finds them in the reports cited in notes 119 and 121, as well as in letters, some of which are cited in note 131.
137. Quotations from letter Adams to Lofgren, 12/1/61 (DG20861); report by Hine, CERN/DIR/39 cited note 121. On electron storage see letter Hine to Weisskopf, 8/5/61 (DG20576).
138. Report cited note 119, page 4. Taken up, for example, in Hine's report CERN/DIR/39 cited note 121.
139. The differences are most notable where cost estimates are concerned — even if everyone recognized that the ISR was the cheapest solution. For the extreme positions, see Accelerator Conf (1961), 287-290.
140. See letter Hine to Weisskopf, 8/5/61 (DG20576) and CERN/399, 23/5/61, 1 and 2. One finds the most radical arguments of this kind in Hine's report of 30/3/61 cited note 121, page 4.
141. See the correspondence mentioned note 131 and the tape of the SPC meeting of 29 April 1961, where all these arguments are repeated. For a résumé of them, letter Van Hove to Director-General, etc., 13/6/61 (DG20576).
142. The quotation is from Van Hove's letter quoted in the previous note.
143. Letter Hine to Weisskopf, 8/5/61 (DG20576).
144. There were others, like that spelt out by Ramm in his letters cited in note 131.
145. CERN/SPC/133/Draft, 7/7/61, 12; letter Van Hove to Director General, etc., 13/6/61 (DG20576).
146. Letter Van Hove cited note 145. In a note also dated 13/6/61 and entitled *Future plans for CERN,* Van Hove defined what was 'urgently needed': 'differential cross-section for proton–nucleon elastic scattering' up to 25 GeV, and 'total cross-sections for π^-p, $\pi\bar{p}$, [...], K^+p, $K\bar{p}$, $\bar{p}p$'.
147. Van Hove (1961a), 409-10, 410, 420, 421, 422.
148. We have three successive versions of the document entitled *Recent Developments affecting the Long-term Programme of CERN* which were drafted after the June conference. The first, numbered SPC/130, is in box DIRADM20102; the second, also numbered SPC/130 is dated 26/6/61 (DG20576); it became CERN/SPC/131, 30/6/61, submitted to the SPC. The differences between them are stylistic and in what each stresses. The quotation is from SPC/131.
149. The quotation is from the tape recorded minutes of the SPC meeting of the 21/7/61. The official version (CERN/SPC/140/Draft, 27/9/61, 3) gives a little different and abridged picture.
150. CERN/SPC/131, 30/6/61, 4; tape of the meeting of the SPC held on 21/7/61.
151. CERN/SPC/140/Draft, 27/9/61, 5-7.
152. CERN/SPC/140/Draft, 27/9/61, 8-9; CERN/SPC/139, undated.

153. It should be noted that CERN as an organization has a kind of automatic defence mechanism, which implies that it has difficulty thinking of an 'intergovernmental' machine as being anywhere in Europe but in Geneva.
154. CERN/SPC/139 and SPC/140/Draft, 9-12. The most reliable source because most detailed and contemporaneous with events still remains the tape of the meeting of the SPC held on 21/7/61. Documents SPC/140 (and probably SPC/139) were only written in September and, in our view, have a tone a little different to that we picked up by listening to the tape.
155. To give some examples, Ramm wrote to Weisskopf on 16/5/61 that 'only those directly interested in the intimacies of machine design support the storage rings', Hine for his part spoke 'of the nuclear physicists who are not impressed by storage rings'. See letters Ramm to Weisskopf, 16/5/61 (DG20576), Hine to Weisskopf, 8/5/61 (DG20576). For a severe critique of 'physicists' by a 'builder', see Hine (1962), 293-294.
156. For this aspect of necessary prudence the reader is referred to Van Hove (1961a), for example page 422: 'We should stress, however, that any such considerations are bound to be extremely conservative, since they are based on questions and problems suggested in the most immediate way by our present state of knowledge and since they do not take into account the well-known and drastic limitation of our power of imagination'.
157. Letter Melville to Weisskopf, 14/8/61 (DIRADM20102); see also DSIR and NIRNS, Joint Consultative Panel for Nuclear Research, minutes of the second meeting, 24/10/61, 50-51 (DG20831) and CERN/SPC/149/Draft. The tape recording is far more explicit than the minutes. In their work on LEP which deals with a different period and scientific context (the end of the 70s), Baudouin Jurdant and Josiane Olff-Nathan speak of a rapprochement of the builders and theoreticians vis-à-vis the experimentalists; see Jurdant & Olff-Nathan (1981).
158. DSIR and NIRNS, Joint ..., *op. cit.* note 157, 14. This is a constant theme in Europe after the summer. See also European Accelerator Study Group, Second meeting, CERN/EAG/2, 1/12/61, and proceedings of the meeting held at Harwell (UK) on 30 and 31/10/61, notably Adam's report, 3-7.
159. CERN/SPC/144, 4. Serber's original text, to which Hine is faithful, is in Accelerator Conf (1961), 3-5.
160. The phrase is Hine's in CERN/SPC/144, 6. To understand better this 'unanimous' opinion of the Americans, the reader is referred to section 1.3 of this chapter, *the American swing to very high energies, 1959-61.*
161. DSIR and NIRNS, Joint ..., *op. cit.* note 157, 25. In report CERN/SPC/144, 7, Hine wrote: 'A more important result of recent work is a better understanding of what are likely to be the *limitations on the accelerated current,* and secondary beam intensities, of alternating gradient machines [...]. In general, it is clear that machines designed for a high energy, or rather with a large radius, are intrinsically high-intensity machines, both because of the greater circumference to fill with protons and, more importantly, because one is forced to a higher injection energy.'
162. Accelerator Conf (1961), 3-5, quotation p. 5. The study of the meetings of the SPC of April, July, and October, for example, would allow us to generalize. One should also refer to section 1.3 of this text for information on the contents of the Berkeley seminar (July-August) and on the *two* reports published by Brookhaven at the end of August. If one still feels the need for a third proof, see the contribution made by Powell in DSIR and NIRNS, Joint ..., *op. cit.* note 157, 34.
163. Compare for example Serber (1961) with Van Hove (1961a) or Van Hove (1961b).
164. One of the best examples of these differences in 'psychological' or 'mental' attitudes to the same question is provided by the way people tackled the question of the cross-sections of strongly interacting particles as one approaches very high energies. (Remember that the dominant theory at the time was derived from Pomeranchuk's hypotheses, and predicted an asymptotic curve — 'total cross-section' of particles A and B 'became' constant and equal to that of the antiparticle A with the particle B, etc. In 1961, though, it seemed that certain experimental results obtained with the PS and the AGS indicated 'that the asymptotic regime [...], if it exists at all, is not quite reached at the energies now available' — Van Hove (1961a), 418).

With the protagonists confronted with an *open* question of this type one can follow the differences in emphasis by comparing the articles published in 1961 in Europe and the USA. For example, Hove (1961a), 418, Serber, 'The Future of High-Energy Physics', Accelerator Conf. (1961), 5, etc.

165. Hine wrote in October: 'The theoretical work [in the USA, ...] did not seem to arrive at any different conclusions. Not many reasons could be given for high energies as such [...]. A much stronger practical reason for a high-energy accelerator to do physics of the kind which can now be discussed seriously by theoreticians is the great gain in intensity of secondary beams in the 5–50 GeV region which a high-energy machine provides' (CERN/SPC/144, 4). See also CERN/SPC/144, 6.

166. The quotation is from Hine in DSIR and NIRNS, Joint ..., *op. cit.* note 157, 14. See also Serber's reply to H. Blewett during the international conference at Brookhaven in September. On the question of the choice of energy ('100 BeV, 500 BeV, 1,000 BeV or what'), Serber's reaction was: 'I think the question becomes an economic rather than a physical one. We'd like to go up as far as is technically feasible and still not ask for something too unreasonable from the government' (Accelerator Conf (1961), 11). Note that the argument 'the bigger the machine, the better' is often present once the *type* of machine has been settled. For a similar development regarding the size of bubble chambers, see Krige & Pestre (1986).

167. DSIR and NIRNS, Joint ..., *op. cit.* note 157, 14–15.

168. See for example the discussion among British scientists in DSIR and NIRNS, Joint ..., *op. cit.* note 157, 31–55, particularly pp. 51–53. The term 'moratorium' is Powell's. We mentioned earlier that the willingness to coordinate the European and American programmes was habitual in Europe.

169. See the rather lively debate which followed Schoch's paper at BNL in September (Accelerator Conf (1961), 287–290).

170. DSIR and NIRNS, Joint ..., *op. cit.* note 157, 51.

171. The SPC meeting's minutes are CERN/SPC/149 Draft. There is also a tape recording available. See also Hine's report to the SPC, CERN/SPC/144, 18/10/61, which gives information on the IAEA and the Pugwash conference, as well as that of Adams in DSIR and NIRNS, Joint ..., *op. cit.* note 157, 24–31.

172. CERN/SPC/149/Draft; the quotation comes from *Draft Recommendation by the SPC to the Council, Long-term Programme for CERN,* CERN/416.

173. CERN/SPC/149/Draft and the tape recording. The quotation is not given in SPC/149.

174. See file CERN-DG20780, and particularly Riunione da tenersi a Frascati il 16–17 dicembre 1960, *Programma Preliminare,* letters Winter to Adams, 23/12/60, Adams to Winter, 3/1/61, Adam's circular letter, 2/2/61. The remark about Winter was made by Adams during the SPC meeting in October 1961 (see tape).

175. *European Accelerator Group, First meeting,* Saclay, 30–31/5/61, *European Accelerator Study Group, Second meeting,* Harwell, 30–31/10/61 (DG20780).

176. The quotation from Blewett is in *European ...,* Harwell, 30–31/10/61, *op. cit.* note 175, 7.

177. As J.P. Blewett, in that bitter-sweet tone which he frequently affected, could not help saying at a public meeting held a short while later (J.P. Blewett (1962), 8): 'The arguments are still in progress. At a meeting of a European co-ordination group on accelerators held last May, much enthusiasm was expressed for the CERN colliding beam proposals. At the next meeting, held in October of last year, colliding beams were not mentioned, and all the emphasis was on super-high-energy accelerators. But I understand that a study group now in operation at CERN still has considerable enthusiasm for the colliding beams. [...] Whether or not their ideas will gain more acceptance remains to be seen.'

178. Report of the AR division signed Johnsen, *Proposed Programme for a CERN Study Group on High Energy Projects,* 27/11/61, CERN/AR/Int. SR/61-29. The quotations are from p. 1 of this text. The DG's approval is in the text which he submitted to the December Council session: *Introduction to the CERN Long-term Programme (by the Director-General),* CERN/428 (no date).

179. CERN/AR/Int. SR/61-29.

180. See section 1. Concerning the USA, J.P. Blewett said on 30 October: 'Since the MURA debate in 1958 [1959 in fact], colliding beams have received very little consideration in the USA. O'Neill has been their

sole proponent and, although he has not suffered in silence, the studies for a 1000 GeV machine have been carried out on the tacit assumption that extension of the present methods to higher energies is the appropriate next step' (*European ...,* Harwell, 30-31/10/61, *op. cit.* note 175, 12).

181. This, amongst others, was Weisskopf's argument. See letter Weisskopf to O'Neill, 4/8/61 (DG20630).
182. See section 1.3; letters O'Neill to Weisskopf, 19/7/61 (DG20630), Cocconi to Weisskopf, 7/8/61 (DG20630). The quotations are from the first letter.
183. Letters Hine to O'Neill, 18/9/61, O'Neill to Hine, 18/10/61, Hine to O'Neill, 20/10/61 (DG20630).
184. Hine's remark was made at the Brookhaven conference in September (Accelerator Conf (1961), 287). Van Hove's reply, which was also focused on the *scope* of the question, was: 'The question that was discussed [at CERN] was really the following: assuming, as is likely, that our governments would, in the years to come, offer us at most one new machine, and not more, what would the physicists at CERN like this machine to be. It was in answer to that particular question that we felt that the storage rings were not the most attractive proposal' (p. 288).
185. Allison (1971).
186. Accelerator Conf (1961), 288.
187. CERN/SPC/157/Draft, 5/4/62 (SPC meeting, 3/3/62), 9; here again the tape recording is much more reliable than SPC/157/Draft. Note the rather 'subjective' nature of this information, the Americans feeling that it was the Europeans who were only marginally interested.
188. Nimrod (1979), 7-20
189. *New Scientist,* **278** (1962), 608; *Nature,* **195** (1962), 436; *New Scientist,* **300** (1962), 365; Merrison (1962), from which the quote is taken.
190. 'Geminus' in *New Scientist,* **300** (1962), 365, wrote of the electron synchrotron: 'Even where the interests of high-energy physics are in question, there is no assurance that the best results are to be had by the means now chosen. It may easily be, for example, that still more could have been achieved (for the same people who will now benefit) by using the money as a catalyst for a further expansion of the European Centre for Nuclear Research (CERN)'.
191. See Memorandum Hine to D-G, Johnsen, 5/2/62 (4 pages); *Notes on visit to Saclay, 2 May 1962,* signed Hine, 7/5/62 (4 pages) (DG 20783); letter Hine to Perrin, 1/6/62 (MGNH 22133); *Projet d'accélérateur à Gradient alterné 60 GeV-3×10^{12} protons par cycle,* 14/2/62, an internal report of the French CEA, Service de Physique appliquée (DG 20783). Hine's source was usually Winter, which gives a bias in favour of the builders. We have only mentioned Perrin, but the attitudes of Leprince-Ringuet, Gregory, etc., were also mentioned.
192. See a long report by Johnsen in CERN/SPC/157/Draft, 5/4/62. See also the reports written in 1962 of which a list is given in CERN/527, 6/12/63, 45-51.
193. CERN/444, 1/6/62, *Progress Report,* 49-51.
194. *European Accelerator Study Group, Third meeting,* 14-15/5/62, CERN/EAG/3, 5/7/62 (DG 20780), 65-70; see also A. Schoch, *Characteristics of CERN-PS Storage Rings,* March 1962, AR/Int. SG/62-11, 3/9/62.
195. CERN/466, 4/12/62, *Progress Report,* 51-54, quotations on p. 53.
196. CERN/466, 4/12/62, *Progress Report,* 54. Beam extraction progressed considerably in 1962 and 1963. For a quick overview, see *Courrier CERN,* **3** (1963), 63-64.
197. *Courrier CERN,* **2** (1962), 3; CERN/444, 1/6/62, 49; CERN/466, 4/12/62, 51.
198. *Courrier CERN,* **3** (1963), 18, 46-47; **4** (1964), 34.
199. *Courrier CERN,* **4** (1964), 34, 124; CERN Annual Report, 1963, 124-125; 1964, 114-115.
200. Title of a 'report of a working party meeting from March 26 to April 6, 1962' at CERN; 17 texts gathered under the title (CERN) AR/Int. SG/62-11, 3/9/62.
201. Johnsen's report to the SPC of March 1962 (CERN/SPC/157/Draft, 5/4/62, 12).
202. See Schoch, *Introduction;* Schoch, *Questions examined,* AR/Int. SG/62-11, 3/9/62; Jones, *Experimental Utilization of Colliding Beams,* AR/Int. SG/62-2, 12/3/62.

203. AR/Int. SG/62-11, 3/9/62, notably O'Neill, *Summary of the Discussions [...]'*, 17.1-17.9. For Berkeley, see O'Neill (1961a).
204. See for example letter Powell to Weisskopf, 22/10/62 and Weisskopf to Powell, 25/10/62 (DG 20783), Weisskopf originally suggested to raise the matter again at the June 1963 Council session.
205. CERN/DIR/151, 8/11/62, *Report of the Director-General on his talks with Professor Perrin in Paris*, 3 pages; Memo Johnsen to Weisskopf, 25/10/62 (DG 20783); letters Weisskopf to Perrin, 22/11/62 (DG 20783), Hine to Perrin, 23/11/62 (MGNH 22133).
206. More detailed reports by Perrin and de Rose on 20 December 1962 during the *Accelerator Meeting* (Minutes, 7/1/63, DG 20783).
207. CERN/DIR/151, 8/11/62, 3.
208. Memo Johnsen to Weisskopf, 25/10/62 (DG 20783).
209. CERN/DIR/151, 8/11/62, 3
210. Memorandum Puppi to Division Leaders, 29/10/62 (DIRADM 20102); letter Adams to Weisskopf, 21/11/62 (DG 20783).
211. CERN/DIR/155, 23/11/62 (Meeting held on 19/11/62), 3-4. In a later version CERN/DIR/155*, 'the French initiative' became 'the various national initiatives'.
212. CERN/DIR/155, 23/11/62, 3-4; letter Adams to Weisskopf, 21/11/62 (DG 20783).
213. See boxes DG 20783 and 20577 for these meetings and the exchange of letters. One could for example read letter CERN/8187 sent on 30/11/62 by Weisskopf to the 'Founding Fathers'.
214. See for example the circular letter mentioned in note 213 and *Notes of Meeting of Sub-Committee for January Meeting on New Accelerator Programme, November 30, 1962,* 10/12/62 (DG 20577), from which the quotations and the phrase 'Founding Fathers' come. The list of physicists is obviously much larger in the note mentioned above.
215. One can find the menu in DG 20783; the quotation is from letter Bannier to Weisskopf, 4/12/62 (DG 20783) where he agrees to the meal with the 'Founding Fathers' and adds: '[...] we can have preliminary discussions (not interfering, I hope, with the social as well as gastromic pleasures)'.
216. *ECFA, 17-18 January, 1963, First Session, minutes,* FA/EC/1 (DG 20783).
217. Council, minutes, 19/12/62, 53.
218. *European Accelerator Study Group, Third Meeting,* 14-15/5/62, CERN/EAG/3, 5/7/62 (DG 20780). There are no proceedings of the Frascati session. See however *Meeting of the European Accelerator Study Group, Frascati, 3-4-5 December, 1962, Detailed Programme,* November 1962 (DG 20780); E. Amaldi, *Manpower and Funds for High Energy Research in Europe,* undated (DIRADM 20102); on Hine's role, see *Notes ....., op. cit.* note 214.
219. *Accelerator Meeting,* Geneva, 20 December, 1962, *Minutes,* 7/1/63 (DG 20783), 6, concerning Bannier; letter Adams to Weisskopf, 21/11/62 (DG 20783).
220. J.B. Adams, *Notes on Future High Energy Physics Research Facilities,* 11/12/62 (DG 20783); for Amaldi, see note 218.
221. *Accelerator Meeting,* Geneva, 20 December, 1962, *Minutes,* 7/1/63 (DG 20783), 2. On 17 January 1963, introducing his paper, Johnsen said: 'Recently it has become rather fashionable to talk about the pyramid of accelerators.' (*ECFA, First session, Proceedings,* FA/EC/2, undated, 29).
222. Quotations taken from *Accelerator Meeting, op. cit.* note 221, 2, 5, 6.
223. The remark is Weisskopf's (*ECFA, First Session, Minutes,* FA/EC/1, 29/1/63, 3).
224. Ibid.
225. *Op. cit.* note 223, 5; compare with *Accelerator Meeting, op. cit.* note 221, 2.
226. In December, Powell mentioned 70-100 GeV; in January, Adams mentioned figures between 100 and 150 (FA/EC/1, 8, 10), insisting on 120 (FA/EC/1, 14).
227. *Accelerator Meeting, op. cit.* note 221, 2 for the quotation; for Van Hove, FA/EC/1, 14.
228. The first quotation is of Hine speaking of the PS's energy (FA/EC/1, 14); the figures are in the conclusion to FA/EC/1, 16, a conclusion written by Amaldi.

229. FA/EC/1, 15.
230. For more details see the box DG 20577. The submitted report was *ECFA, First Report of Working Party,* FA/WP/8. The quotation is from the version FA/WP/8-Rev 2, 10/4/63, 20. It is worth looking at the *Draft Minutes of the Third Meeting of the Working Party,* FA/WP/7, 28/2/63 (DG 20577).
231. For the debate inside ECFA and for Amaldi's conclusion, see *ECFA, Second Session,* Geneva, 1 and 2 April, 1963, *Minutes,* FA/EC/4, 19/4/63, 7. As for the ECFA report, see *Interim Report of European Committee on Future Accelerators,* FA/WP/13-Rev. 1, 10/4/63 (DG 20577); quotations pp. 1, 4.
232. FA/EC/4, 19/4/63, 10.
233. The final ECFA report is FA/WP/23. The version FA/WP/23/Rev. 2 was studied by the SPC, and the version published subsequently was FA/WP/23/Rev. 3. There are *minor* differences between these texts. For the SPC of 7 June 1963, see CERN/SPC/172/Draft, 29/7/63. For the quotations, see *Recommendations of the SPC to the Council on Future High-Energy Accelerators,* CERN/503, 14/6/63, 1 and 2.
234. FA/WP/23/Rev. 3, 12/6/63.
235. CERN/442, 4/6/62. In December 1962, the 1963 budget was fixed at 13% above that of 1962 while the 1964 budget was to exceed that for 1963 by +11%. Six months later, the estimates for 1965 and 1966 were fixed at +10% and +9% respectively.
236. Krige and Pestre (1986), 274-275. The two other 'selling strategies' were what we have called 'The appeal to scientific necessity' and 'The appeal to the unexpected' (pp. 275-279).
237. CERN/SPC/170, 16/5/63 (meeting of the SPC of 2/4/63), 6.
238. Letter Weisskopf to Amaldi, 29/1/63 (DIRADM 20102).
239. In February, Weisskopf proposed including Soviet physicists in the second meeting of ECFA (*Draft minutes ....,* op. cit. note 230). Panofsky, Ramsey, Kolomenskij and Yablokov took part in the second closed session of the second session of ECFA held on 2 April 1963 (FA/EC/4, 19/4/63, 11-19).
240. See document FA/WP/14, 3/4/63, *Suggestions for collaboration among appropriate bodies from the USA, the USSR and the member states of CERN on future accelerator programmes.*
241. *Draft Cost Estimate for a 300 GeV Proton Synchrotron,* AR/Int. SG/63-4, 6/2/63 (respectively 150 GeV, AR/Int. SG/63-5, 6/2/63).
242. *Draft minutes ....,* op. cit. note 230, 3 and FA/EC/4, 19/4/63, 6.
243. *Draft cost ....,* op. cit. note 241, 'table II' for the two reports.
244. *Draft cost ....,* op. cit. note 241, 'table III' for the two reports.
245. Krige and Pestre (1986), 266-271.
246. Adams (1961), 1.
247. FA/EC/2, without date, 11-29
248. Krige and Pestre (1986), 260-265. See in particular what we have called 'Alvarez's way of arguing' or 'Alvarez's "bubble chamber philosophy"' (page 264).
249. For a very clear case, see Puppi during *Accelerator Meeting ....,* op. cit. note 221, 7/1/63, 7. For Van Hove, FA/EC/1, 29/1/63, 14.
250. Krige and Pestre (1986), note 33, page 269.
251. The more important the equipment, the less crucial is this notion of 'timing'. It counts enormously for bubble chambers, but far less for accelerators of hundreds of GeV. Compare Krige and Pestre (1986), 268-271 and Hine (1962), 293-294. See also Pestre (1988).
252. CERN/SPC/172/Draft, 29/7/63, 4, 6.
253. *Draft minutes ....,* op. cit. note 230, 3; CERN/SPC/170, 16/5/63, 6.
254. The quotation is from R. Serber, Accelerator Conf. (1961), 11. He concluded: 'We would like to go up as far as is technically feasible and still not ask for something too unreasonable from the government'.
255. Conseil, minutes, 20/6/63, 15-24; for Amaldi, 15-17; for Powell, 17-19.
256. The quotation is from the Conseil, minutes, 20/6/63, 20. For a more general introduction to the idea of 'CERN programmes', Pestre (1984b). On the limits of this system, and particularly on the wish to avoid

CERN becoming too 'à la carte', see also Dufour (1981). On the historical origin of these clauses, see Krige, chapter 8 of History of CERN (1987).
257. Conseil, minutes, 20/6/63, 19-20; CERN/SPC/178/Draft, 17/1/64 (meeting of the SPC, 10/10/63), 11. The remark is by Johnsen, and is in the latter document.
258. Conseil, minutes, 11/10/63, 37-38, 48; Conseil, minutes, 17/12/63, 57.
259. Compare CERN/527, 6/12/63 (*Progress Report by the Study Group on New Accelerators to December 1963,* by K. Johnsen, formerly AR/Int. SG/63-44, 29/11/63) with CERN/542, 12/5/64 (*Rapport sur le projet d'Anneaux de Stockage à Intersections pour le Synchrotron à protons du CERN,* formerly AR/Int. SG/64-9, 12/5/64). The technical report on the 300 GeV is CERN/563, 2 volumes, 11/12/64 (*Rapport sur le projet d'un Synchrotron à protons de 300 GeV par le Groupe d'Etudes du CERN sur les nouveaux accélérateurs,* formerly AR/Int. SG/64-15).
260. Letter Goldhaber to Seaborg, 31/7/63 (CHS 10085); Conseil, minutes, 17/12/63, 57-58; Geminus, 'It seems to me', *New Scientist,* **373** (1964), 93, from which the quotation is taken; *id., New Scientist,* **389** (1964), 299; H. Simons, 'Britain holds the key to expansion', *New Scientist,* **393** (1964), 542.
261. Simons, *op. cit.* note 260, 542.
262. The first two quotations are from H. Simons, *op. cit.* note 260, 542; A.M. Weinberg, 'Criteria for Scientific Choice', *Physics Today,* **17** (3) (1964), 42-48; 'Deux lettres ouvertes échangées entre MM. Weisskopf et Weinberg', *Courrier CERN,* **7** (4) (1964), 88-89, 95; quotation from Weisskopf page 88.
263. CERN/SPC/188/Draft, 19/10/64, 3 (meeting of SPC, 2/6/64).
264. This name for the Advisory Council is reported by Simons, *op. cit.* note 260, 542.
265. Council, minutes, 20/6/63, 18, for Powell; on the same occasion he said that 'we need not be particular to a year or two', adding that nevertheless 'there should be no unnecessary delay in the serious consideration of the problem' in each member state.
266. Conseil, minutes, 20/6/63, 16, for Amaldi; 20-22 for Perrin; CERN/SPC/177, 17/2/64, 3, for Hine. For Perrin in the first months of 1964, the reader is referred to his statements (and those of Gregory) during the SPC of 2/6/64 (CERN, SPC/188/Draft, 19/10/64, 3).
267. We have simplified the evolution of the debate in that a new preoccupation entered it, namely, *to improve the existing machines,* and the PS above all. In 1964 a third component was thus added to ECFA's 'summit' programme (ISR and 300 GeV), the improvement of the installations. Note that P. Germain was also the Director of the Proton Synchrotron Machine division.
268. Letter Hine to Weisskopf, 28/5/64; notes for the Director General for the Council session, Draft, undated, signed by Hine; letter Hine to Bannier, 15/6/64 (MGNH 22133).
269. Letters Weisskopf to Bannier 21/5/64; to Adams, 21/5/64 (DG 20940).
270. *Memorandum on CERN Accelerator Policy,* by V. Weisskopf, CERN/SPC/183, 6/5/64.
271. Memorandum quoted note 270, 5-6, quotation page 6.
272. Memorandum quoted note 270, 5 and 6 for the two quotations; Weisskopf again took up the last argument during the Council session in June 1964 (Conseil, minutes, 18-19/6/64, 13); before him, see Adams (Conseil, minutes, 20/6/63, 22), Amaldi (CERN/SPC/188/Draft, 9), Perrin (*id.,* 9) and again Adams (Conseil, minutes, 18-19/6/64, 25).
273. See CERN/SPC/180/Draft, 8/4/64 (meeting SPC, 25/2/64), 4-5; CERN/SPC/185/Draft, 1/6/64 (meeting SPC, 20/4/64), 8-9; CERN/SPC/183, 6/5/64 (Memorandum by Weisskopf), 5, from which the quotations are taken.
274. In an ECFA document dated October 1964, we read that 'The Amaldi Report makes it clear that the ISR project is no substitute for the 300 GeV machine and should be sacrificed if it were clearly causing an unacceptable delay to the latter' (*Notes for the Meeting,* 9/10/64, FA/EC/6, 5, DIRADM 20097). See also letter Weisskopf to Adams, 23/7/64 (DG 20940).
275. Memorandum by Weisskopf cited note 270, 5.
276. For the SPC of 2 June 1964, see CERN/SPC/188/Draft, 19/10/64, 3-12; for Amaldi and the conclusion,

see pages 11 and 12.
277. See letter CERN/8120, 12/11/62 informing the member states of the gift. Add Conseil, minutes, 19/12/62, 50-51; Conseil, minutes, 20/6/63, 29; 11/10/63, 46-47; 17/12/63, 62. For what follows, see Conseil, minutes, 15-16/12/64, 57-58; 16-17/6/65, 38.
278. Conseil, minutes, 18-19/6/64, 12-31; the quotation is from the summary in FA/EC/6 (*op. cit.* note 274), 2.
279. Conseil, minutes, 18-19/6/64, 8.
280. Conseil, minutes, 18-19/6/64, 14-16; the quotation is from Leprince-Ringuet's introduction to Powell's speech, *id.,* 14.
281. Conseil, minutes, 18-19/6/64, 17-18, quotations page 17. Concerning the fact that the PS improvement programme (already mentioned note 267) corresponded closely to that for Brookhaven's AGS, see CERN/SPC/178/Draft, 17/1/64 (meeting of SPC, 10/10/63), 11, where Amaldi 'asked whether a study could be made of the respective advantages of building storage rings at CERN and of building a new 1 GeV linac injector, as proposed by Brookhaven [...]'; CERN/SPC/180/Draft, 8/4/64, 5-6; CERN/SPC/183, 1; Conseil, minutes, 18-19/6/64, 21 (Perrin), 22 (Weisskopf), 23 (Amaldi), etc. From the summer of 1963 the Europeans knew very well of Brookhaven's interest in improving the AGS and of its firm refusal to build the ISR; in this connection see letter Goldhaber to J.S. Toll, 15/10/63, reproducing extracts from the letter Goldhaber to McDaniel, 10/10/63 (CHS 10085).
282. Conseil, minutes, 18-19/6/64, 18, for Bannier; 19-21, for Adams.
283. Conseil, minutes, 18-19/6/64, 21-31; see also the interpretation in FA/EC/6, 9/10/64, *op. cit.* note 274, 2 et 5-6. The authors wrote of the 300 GeV: 'the timetable was amended to talk of a decision at the end of 1966. Even that date appeared very early for many governments and some held strongly the view that all possibilities of wider collaboration be explored before considering seriously a new European Laboratory' (page 2).
284. Explanation given by Weisskopf himself, CERN/SPC/191/Draft, 5/2/65 (meeting of the SPC, 20/10/64), 3.
285. FA/EC/6, *op. cit.* note 274, 5.
286. FA/EC/6, *op. cit.* note 274, 5.
287. CERN/SPC/191/Draft, 5/2/65, 4-5.
288. CERN/SPC/191/Draft, 5/2/65, 5, for the two quotations.
289. This view of Van Hove's was reported by Wouthuysen to the SPC (CERN/SPC/191/Draft, 5/2/65, 7).
290. CERN/SPC/199/Draft, 26/4/65 (meeting of SPC, 16/2/65), quotations on pages 12, 13 and 14. On the relative 'autonomy' of the British in the mid-1960s, see Goldsmith and Shaw (1977), 47. They speak of the United Kingdom around 1965 as 'suffering from the traditional suspicion that anything done on the Continent could only be regarded as entertainment'.
291. CERN/SPC/191/Draft, 5/2/65, for the SPC of October 1964; Conseil, minutes, 15-16/12/64, 45-46 for Leprince-Ringuet.
292. Conseil, minutes, 15-16/12/64, 46-52; 25/3/65, 8-18; 16-17/6/65, 23-25 (the quotation from Melville is from page 24); CERN/SPC/203/Draft, 28/5/65 (meeting of the SPC, 4-5/5/65), 8 (from which the remark by Weisskopf is taken).
293. See the introduction by Weisskopf to the SPC of 4 and 5 October 1965 (CERN/SPC/209/Draft, 23/11/65, 7); about Johnsen, letter Weisskopf to Johnsen, 9/7/65, 3 pages (DG 20941).
294. CERN/SPC/209/Draft, 23/11/65, 8-9.
295. Constructed from the debate in the SPC in October, CERN/SPC/209/Draft, 23/11/65, 9-14; quotations pages 9 and 11.
296. For Adams, see page 11; for Perrin, pages 11-12 (*op. cit.* note 295).
297. *Op. cit.* note 295, 12 (declaration by Weisskopf) and 14 (résumé by the chairman of the SPC, Leprince-Ringuet).
298. CERN/SPC/217/Draft, 1/3/66 (meeting of SPC, 14/12/65), quotations page 5.

299. Conseil, minutes, 15–16/12/65, 55–57 (ISR); 57–60 (300 GeV).
300. This résumé of the 'end of the story' is not intended to be precise; its aim is simply to identify certain dates. We have followed Goldsmith & Shaw (1977), 43–62.
301. Criticisms of the myth of 'the rational decision' are now well known. The most important are given in note 5.
302. References given note 1; add Greenberg (1971) and Pestre (1988).

PART IV

Concluding remarks

CHAPTER 13

Some characteristic features of CERN in the 1950s and 1960s

Dominique PESTRE

Contents

13.1	CERN in the mid-60s: cruising steadily ahead	784
13.2	The Council and the member states: an unusual relationship	786
13.3	The 'corporate spirit' of CERN: an important element making for cohesion	789
13.4	Learning to work together: a problem more or less solved around 1965	792
13.5	Learning to do 'big physics': a process still at work in the mid-1960s	794
13.6	American and European physicists: an historical difference	799
13.7	CERN and the history project: some lessons	803
Notes		806
Bibliography		809

13.1 CERN in the mid-60s: cruising steadily ahead

The dominant impression one comes away with when looking at CERN in the mid-60s is that of an organization which is cruising steadily ahead. After fifteen tumultuous years CERN seems to be on its way, safe in the conviction that the future could be managed using a framework which had been tried and tested in the past. The election of the new Director-General Bernard Gregory during 1965 is a good index of this new trend. In contrast to what had happened in previous cases, the atmosphere seems to have been less tense, the choice easier to make. Of course there was the usual battle over two men — which is only normal — but the debate was always kept within rather narrow limits: one was choosing between people of the same kind, in the light of considerations which were accepted by all, between two European physicists who had already made their mark and who already had some experience of the intricacies of CERN management. The debate thus seems to have been less conflictual than in 1954 when the 'Nobel prizewinners' opposed the 'CERN pioneers' and had Felix Bloch nominated, and even than in 1960 when to too many the appeal to the United States still seemed to be essential.[1]

Several parameters can be used to measure the fact that CERN had achieved a more 'stable' rhythm of development, was less prone to the jolts consequent on repeated improvisation. Two are worth mentioning to set the ball rolling, namely, the relationship between CERN and the member states during these years, and that developed between the physicists and the scientific visitors to the Genevan laboratory. As far as the states financing the organization were concerned, a policy of growth was accepted by all in 1962, after a period in which CERN had to fight daily for the budgets it wanted against national administrations who were most unhappy to see costs continue to rise after the 28 GeV PS had been commissioned.[2] This turning point, which was only obtained after a tough battle in the core of the Council, amounted to a political defeat for the United Kingdom and meant two things for the future. Firstly, that if CERN was to remain the leading European high-energy physics laboratory, its budgets could not be stabilized. Confronted by developments in America, the member states recognized that they could not but follow the growth that was taking place across the Atlantic. This policy adopted in 1962 was also associated with a new budgetary procedure. In contrast to what had happened for the previous ten years, it no longer seemed possible to draw up the budget 'synthetically' by summing the needs as identified each year by the division leaders. This was because it was impossible for the Finance Committee to control the overall outcome, and above all because of the enormous difficulty of foreseeing one's needs in a field where the leading edge of research evolved extremely rapidly. What the group chaired by J.H. Bannier proposed, by contrast, was to fix the annual rate of growth as calculated from national expenditures in related fields. Of course the actual determination of the budget for the following year required that one take the requests of each division into account, but this only served to refine an order of magnitude already agreed on by the states.[3]

As far as the visiting physicists were concerned, the 'normalization' of CERN's functioning is clearly indicated by the fact that electronic experiments too became a communal activity. This remained firmly in the hands of the CERN physicists until around 1964, when a marked change was perceptible: now conceived on a European scale, almost all the teams became 'mixed' teams most of whose members — and not simply the CERN groups as before — were fully responsible for the conception and the realization of a part of the detectors and of the experiments. At the same time the number of visitors grew steadily, and the Electronic Experiments Committee itself was opened to physicists from national laboratories who were increasingly mastering experimentation around the 28 GeV PS. In short, in the midst of the 1960s, CERN had become part of the patrimony of its member states.[4]

The fact that CERN, in the mid-1960s, was cruising along and apparently growing steadily was not only thanks to the organization itself, and the experience that it had acquired. Three contextual elements made this possible, three things facilitated the development. The first was that the field of high-energy physics (as scientific field) was itself riding the crest of a wave — and this stimulated a new optimism in the physics community. Consider, by way of illustration, the success of the SU(3) classification and the idea of quarks which followed on it — as well as the conceptual reorganization which these concepts brought with them into the rich and active intellectual life of the discipline in 1962/65.[5] Then one must remember the universally euphoric state of the European economies at the time, and the expansion in the proportion of the budget which the states were prepared to devote to research. In 1963 France spent a little more than 1.5% of its GNP on Research and Development; four years later the figure had reached 2.2%. During these same years the rate increased from from 2 to 2.2% for the Netherlands and from 1.4 to 1.8% for Germany, while Switzerland continued to oscillate between 2.2 and 2.5%. The main country suffering from economic difficulties was the United Kingdom. All the same one must remember that it was still the richest European nation by far (it was still paying 25% of CERN's budget in 1965) and the one who always had spent, and still was spending, the highest percentage of its GNP on Research and Development.[6] The third and last element to bear in mind is that, for the political class in general, the 1960s were still the nuclear age and the time of big programmes — which was to the benefit of the big science that was high-energy physics —, and that they were the years in which fundamental research was perceived as an almost direct motor of industrial and economic development. Uncritically and forcibly advocated by the

Notes: p. 806

physicists, this way of seeing things posited an almost automatic connection between investment in basic research (as defined by academics) and national technological achievement. Implicitly modelled on the history of the bomb which was seen as being an outcome of the disinterested and non-directed research done by the Joliots and by Hahn and Meitner, this notion remained very active in the organs of state. Of course a number of pragmatically oriented studies began to throw serious doubt on the idea that there was some automatic spin-off from basic research, but their influence was limited in a period of economic prosperity, at a time when one did not yet have to make painful choices, and when stress was placed on modernity, on a victorious science, and on the American model. What is more, since these themes had only recently been concretized in ministries and state bureaucracies—whose job it was to manage science and its spin-offs—, the framework in which CERN could see its budgets grow regularly was established. The situations in Germany and France are very good examples of how ministerial or interministerial structures for science were built up between the end of the 1950s and the start of the 1960s. These cases also show how the very existence of these new ministerial departments engendered its own dynamism.[7]

13.2 The Council and the member states: an unusual relationship

Four aspects seem to us to need careful study if one wants to enrich the analysis and to define more accurately the original features of CERN in the years 1953-1965. The first, which we will go into now, concerns the nature of the relationship between CERN and the member states or, more precisely, between the national bureaucracies, the CERN Council, and the managers of the laboratory. The interest of considering the specificities of the Genevan case is that we do not find here a simple relation between a state apparatus and a laboratory, between a political network and a scientific network, but between several of these on account of the multinationality. It is also that all the actors, be they politicians or scientists, were always quick to say that CERN functioned extremely well, whether one compared it with national laboratories or whether one treated it as a European organization and compared it to the European Economic Community, to Euratom, or to the European Launcher Development Organization.

What characterized the 'CERN system'[8] between 1950 and 1965 was undoubtedly the existence of a central group, half scientist, half political, which was extremely powerful and which was blessed with a large degree of autonomy vis-à-vis the state apparatuses, namely, the CERN Council. Formally located between national state bureaucracies which paid for CERN and which gave its members their 'directives' and the CERN Director-General whom it appointed but who was the real master of the laboratory, the Council knew how to make itself the central pivot of CERN's policy. While legally comprising delegates appointed by the national governments, it appears in fact to have been a body which was not administratively constituted from above. At its core lay a group of virtually immovable men who rotated the powerful posts between themselves. Consisting essentially of personalities who played a leading role in CERN's birth in the years 1951/52, this group enjoyed a kind of historical legitimacy which the states never challenged—except for the United Kingdom which tried once, in 1961, and failed. Welded together through a struggle that had lasted for years, determined to see their child prodigy succeed completely, they became known as the 'founding fathers'.

Aware of the balance of forces between countries and within each country, careful not to give offence to anyone, this group always tried to achieve unanimity in the Council, thereby aiming to give governments as little opportunity as possible to intervene directly, or to complain. However they always carefully avoided having this search for consensus become a formal institutional procedure; the rules for taking decisions and for voting in the Council never required unanimity, even for the adoption of the budget. In this way the historical core managed to maintain a real feeling of unity and adventure inside the body of the Council—and ensured that no state alone could decide to block the functioning of the organization (one or two 'recalcitrant' governments could always be outvoted). Insisting in addition on the double nature of the Council—the organ representing the states and responsible for controlling CERN, but also the body expected to advise the same states on matters concerning CERN and high-energy physics—they saw their meetings as providing an opportunity for collective reflection and elaboration, particularly on how best to plead for the development of the organization before their domestic authorities.

Underpinned by this wish for cohesion, and by the desire to see CERN grow as best it may, this group thus kept on with the original adventure into which it itself had been launched at the very beginning of the 1950s, keeping the states at their distance— but for the 'greater good' of each government. As it operated in a field considered as prestigious, high-energy physics, and offered an example of stable collaboration and technical efficacy, it raised little opposition in the member states and was able to rely on the support of the more determined governments to drag along the more hesitant ones, the situation varying according to the hazards of international politics, the changing economic circumstances, and the evolution of domestic policies for science.

This very brief summary calls for one important refinement. United as it was around its shared roots and the determination to see 'its' laboratory succeed, the Council always worked very closely with the European high-energy physics establishment. Around 1962/63 this mixed group of diplomats and influential scientists conceived the project of integrating CERN more obviously into a European 'pyramid' of institutions and laboratories, whose development could only be achieved collectively, in planned harmony of one with the other. There was a tactical dimension to this wish to associate all European laboratories with CERN's work; it amounted to having everyone accept CERN's place at the apex of the European

Notes: p. 806

accelerator pyramid and to avoid making enemies of those who had paid for it. Given a place apart, outside any direct competition for money, CERN had a unique and specific task all the same: to be as good as the best American institutions. In 1963 this way of seeing things was ratified during the first meetings held by the European Committee for Future Accelerators (ECFA).[9]

The apparently 'ideal' CERN system experienced a serious difficulty around 1964/65 when it was called on to settle a crucial question, that of the accelerators to replace the 28 GeV CERN proton synchrotron. To grasp the unexpected effects which the system engendered we have to recall the way the Council thought it should 'sell' big equipment to the governments. Convinced of its power, its members tended to believe that the task of the scientists was to decide what was needed from a 'strictly scientific point of view', leaving it to them to raise the money. Explicitly articulated as the best way of doing things, this approach had as a consequence a split between two kinds of negotiations carried out successively by two different groups.[10]

In contrast to what we often find in similar negotiations, then, it was not the same people who followed the decision process from start to finish.[11] In an initial phase the discussion on the accelerators to be built for the 70s was held only between scientists in terms of their 'ideal' programme. It led them to propose the construction of two European machines, the Intersecting Storage Rings (ISR) to be linked to the 28 GeV PS, and a 300 GeV proton synchrotron. After a long debate the ISR were considered by European physicists to be of lower priority than the 300 GeV; they were kept in the programme all the same as they had been defended as offering some very unusual advantages at a 'reasonable' cost (particularly by the engineers/machine builders), but above all because it was assumed that both machines would be built. In a second phase, which got under way in June 1964, and during which the Council delegates dealt with their national administrations, it emerged that everything would not be financially possible. In consequence the ISR, which were much cheaper than the 300 GeV, were soon on the way to being accepted, while the decision on the big machine was pushed further and further into an indefinite future. Deliberating in a temporarily more difficult situation, and in terms of their own assessment of the two projects, the Council delegates and the domestic financial authorities de facto made a 'choice' between the ISR and the 300 GeV — leaving the physicists standing on the sidelines. Placed during this phase of the negotiations outside the decision-making circles, the latter could do little more than restate their preference and try to deflect a process which had now acquired its own dynamic, and which was unfolding predominantly at ministerial level. The policy, 'positive' in itself, that CERN ought to be considered as a special case to be treated with particular benevolence, generated then a 'negative' or 'perverse' result in the eyes of the actors who had always been regarded as having the right to the last word — the physicists: in 1965 they found themselves with only one machine guaranteed for the future, the ISR, just the machine which they deemed to be of lesser importance.[12]

As regards the future, this 'non-decision' for a 300 GeV machine in Europe was the only major disappointment for CERN's physicists in the middle of the 60s.

13.3 The 'corporate spirit' of CERN: an important element making for cohesion

The second important thing to remember about CERN in the years 1952-1965 is the presence, inside the personnel, of a strong 'corporate spirit'. By this we mean a set of widely-shared values which gave to the people at CERN the feeling of belonging to a quite unique and unusual whole. These values and feelings do not exist to the same degree in all organizations. In some cases they are weak and barely active, and can even carry the risk of fragmentation or give the impression that some groups are excluded. At the other extreme they can have dynamic consequences, engender solidarity, and tend to mobilize energies. In such cases the personnel have a very positive attitude to 'their' organization, a stimulating picture which they reconstitute in their everyday speech or root in a more or less mythical history of the organization's origins. The point to remember about CERN is that an internal culture of this kind existed in the 1950s, that it was partly lost in the 60s — and that we find traces of it again in the 1980s in the guise of a golden age often spoken of, never precisely dated, but now past.[13]

We have already spoken of the strong and positive image outsiders very quickly had of CERN. Linked to the nuclear aspect and to the modernity of all big science, endowed with the aura of the European organization which had escaped from the burdens of nationalism, this 'outside' image gave CERN the status of a model which people sought to repeat—for example, in space research at the start of the 1960s — or with which other organizations sought to collaborate—as in the case of Euratom in 1958. The internal culture of CERN which we are going to try and capture now was obviously connected to this public image so often evoked in Europe in the 1950s and 1960s. On the other hand it was not the exact replica of it, and three traits seem to characterize it.

The first is a consequence of the kind of work actually done at Geneva— high-energy physics—and of the way in which this field of research was generally perceived or interpreted. Embedded in that long tradition in the physics community which saw this search for ever smaller structures as the most important, the most decisive, the most fundamental of all, high-energy physics was regarded as the foundation of all knowledge; always probing deeper, always building ever more powerful 'microscopes'—the accelerators—, it was the most exciting of adventures, the new frontier of knowledge. Characterized within a classical reductionist perspective, that of the nesting of the sciences, high-energy physics was thus placed at the base of the hierarchy of knowledge, in a position which could not but command respect, attract the best people, and justify all requests for funding.

Notes: p. 806

Because it had this privileged position high-energy physics was also thought to enrich humanity in ways transcending its simple technical framework. Since it touched on the most basic of principles and contributed to the ongoing destruction of common sense which had been crumbling ever since the emergence of relativity and of quantum mechanics, it could not but feed the imagination of philosophers and epistemologists and lead to the reshaping of our world picture. From this perception of the field of research that was that of CERN, from this strong image which one finds at the core of the thought of people like Cecil Powell and Victor Weisskopf, a feeling of excellence was born, even of exclusivity, a feeling of adventure which was shared by a large part of the personnel.[14]

The second feature of the CERN spirit which one finds to be very active in Geneva in the 50s and 60s is connected to the international aspect of the organization, to the fact that CERN was supranational. The conviction here was that the organization had bypassed the blockages and the immobility which bedevilled other European organisms. Thanks to the quality of its 'founding fathers' and of the rules that they had established, and notwithstanding certain epidermic reactions of defiance towards the 'controllers' in the Finance Committee, the organization knew that it had won the esteem of the member states without being at their mercy. This situation generated feelings of cohesion among the staff, feelings which were reinforced by the sentiment that the national origins of the staff members had only minor centrifugal effects on the daily functioning of the organization. CERN of course always carefully respected a certain geographic and national equilibrium in the recruitment of scientific personnel, but it never had to apply any such rule strictly. For example, it always tended to favour the smaller states and accepted wide margins in its recruitment quotas.[15]

This self-image captured a real phenomenon. Two little things have to be added, however. Firstly, that it would be wrong to conclude from this that the national dimension disappeared totally at CERN. For one thing appointments at the level of upper management remained a largely political affair; for another 'spontaneous preferences' were evident, be that in the recruitment of their colleagues by certain divisional directors or in the composition of experimental collaborations— statistically speaking, the 'Latins' worked more with the 'Latins' than with the 'Nordics', for example. The second thing to say about the insistence, in Geneva, on the absence of conflict between national groups is that it was a corollary of the claim that CERN was a 'success' in its dealings with the member states. Since the staff were 'insensitive' to differences in geographical origin, states had very reduced means of exerting pressure; since the states had few ways of objecting to the wrong being done to one of their nationals, and so to enter the life of the organization, CERN was distanced from the confrontations which continued to affect European inter-state relations. CERN was thus able to embody the Essence of Science — universalism. In this respect Victor Weisskopf ideally symbolized the supranational unity of the laboratory, be that vis-à-vis the personnel, the high officials in the Council, or the distinguished visitors. An Austrian emigré to the United States who came back to lead a European organization, he represented the unique nature of CERN, the fact that the organization itself was a kind of new state perceived by its members as a universal entity situated above and beyond national rivalries.[16]

A third feature of this internal culture which unified CERN between 1953 and 1965, and imbued it with a determination to succeed, was the pioneer spirit, the feeling that one was participating in an ambitious adventure, that conceived by the 'founding fathers' in 1950: to reinstate Europe at the pinnacle of Science, to make of CERN the spearhead of a continent's efforts to meet the American challenge. Thus the 50s were dominated by the race against Brookhaven for the construction of the 28 GeV PS. Setting off at roughly the same time as the competitor in New York State, CERN swelled with pride in 1960 when the European machine was ready ahead of its American rival, and appeared to perform better. This was followed in 1961 by a mood of disappointment consequent on the 'mistakes' made in the preparation of the experimental programme, and on the revolution in work-habits demanded by the new organizational structure. The previous members of the PS group in particular, now dispersed through several divisions, lost some of that sense of unity which had been theirs for almost a decade. Confidence came back in the later years, however, once the physicists had learned to manage the new scale of equipment associated with the new accelerator.[17]

These three elements constitutive of what we have called CERN's corporate spirit — to be at the cutting edge of Knowledge because doing high-energy physics, to be a model which had overcome national rivalry, and to be Europe's spearhead in its bid to regain scientific and technological supremacy — were quickly crystallized in writing by the principal actors. As one of the first pioneers, Kowarski played a major role in the codification of this history. In 1955 he published a text identifying what he took to be the key features of CERN's development in the Bulletin of the Atomic Scientists, a text which he filled out six years later at the directorate's request in a brochure published by the laboratory. This self-image was spread in parallel through commemorative ceremonies, inaugurations, and annual reports which ensured that it was dispersed widely.[18] Pivoting on the idea that CERN was exceptional, the picture it gave was that of a destiny that was ineluctable and free of major setbacks, of a journey that was watched over by the tutelary heroes of European physics. Shielding out the divergences and the refusals, the image was that of a controlled forward movement initiated by Louis de Broglie and Niels Bohr. This historical account rapidly acted as a cement, as a powerful support to internal solidarity — it helped the staff to situate themselves in History, to give a positive connotation to them and to what they did, to give them an identity which was expressed through a collective dynamism.

Notes: p. 806

13.4 Learning to work together: a problem more or less solved around 1965

The third main feature of CERN's life that it is worth stopping to consider is that the Europeans had to learn to work together—but without any tradition to guide them. Coming to work in Geneva meant in effect, for the scientific and managerial staff, a radical uprooting on the one hand, and the insertion into a group without history on the other. More precisely, it meant leaving a familiar world for one composed of individuals who were strangers to one another—and to make out of this heterogeneous mass one group. In contrast to most other laboratories, CERN did not have an organic and continuous development starting from a central personality who gradually accumulated students and colleagues around him, as was the case at the Radiation Laboratory in Berkeley, for example. This was built up around the incontestable personality that was Ernest Lawrence, scientific leader and laboratory head who ran the place and raised money for it. At CERN, on the contrary, the scientists and engineers who came to build the laboratory were nominated to do so— Adams and Hine by Cockcroft, Johnsen by Odd Dahl, etc.—and found an empty place where they met people from other countries, people who had themselves been designated to go there.[19]

Against this one might object that CERN was neither the first nor the only laboratory to be in this situation. BNL, for example, was also created ex nihilo by a group of 12 American universities. This is true, but overlooks two things; that the scientists on the East coast were not strangers to one another, that they shared the same political and cultural environment, that their leaders had a common recent past in wartime projects; and that the authority responsible for controlling them and assuring their means of existence was not internally split and was derived from one and the same administration. The members of the SPC or of the Council, by contrast, while obviously keen to succeed, came from a dozen communities who were initially suspicious of one another. A comparative study of the archives of the British Foreign Office and those of the French Ministry of Foreign Affairs, for example, would confirm the point: the search for the hidden motives of the traditional enemy was for a while an essential activity, both in the various ministerial departments and in the upper echelons of the scientific communities in the two countries.[20]

During its early years, then, the multinational character of CERN engendered a rather formal atmosphere of control in the organs responsible for the European laboratory in the different member states. Having to deal with an altogether new entity, these authorities tended to insist that CERN draw up very precise texts setting out its future programmes and how it planned to organize them (for a clear contrast one might look at Herbert York's account of the freedom the leader of Livermore laboratory had during these same years[21]). Not always trusting one another, the various national groups did all they could to ensure that the interests of their country and their physicists were not overlooked. And on the whole it was the scientists—rather than the 'politicians'—who were the more suspicious. Directly involved in the daily activities of the laboratory, and affected by the rules of access to the machines, it is they who emerged as the most demanding. A good example of this initial period of mutual mistrust between national groups—and one of the last—would be provided by the very long and very important debate in 1959 over the use of the bubble chambers loaned to CERN by France and by the United Kingdom. Here one sees the scientists of different countries opposing one another on the drafting of a text intended to spell out very precisely the use of these detectors. The futility of these debates—and of the texts which were finally 'adopted'—was apparent almost immediately. Once these same physicists had to work concretely together around the PS they found new forms of collaboration which led them to forget the resolutions which they had drawn up so laboriously. Within a few years, in a few months even (in the case of the bubble chamber specialists), a priori suspicions evaporated and one community of European physicists came to replace a sum of national communities.[22]

There is another aspect that needs mentioning. The learning process did not only concern the relations between different national groups, but also those between visiting physicists from the member states who came to experiment for a while at Geneva and the house-physicists, the CERN physicists—and this irrespective of nationality. Imposed by the size and cost of the equipment built at CERN, this relationship raised problems which were systematically tackled at the European laboratory before anyone else had to deal with them.

From its inception CERN had been characterized in two ways: as a laboratory having its own research teams—and so supposed to function as any other laboratory— and as a service centre which constructed equipment for the use of physicists from national laboratories—and so reserving a place for them under the control of the SPC and the Council. When the 600 MeV SC came into operation in 1957, the Council and the CERN directorate chose the simplest and in some sense the most 'natural' way of dividing the installations between internal and external groups, the one that did not call for any real integration of the experimental groups: machine time was simply split between visitors and CERN physicists.[23] With the commissioning of the 28 GeV PS in November 1959—the most powerful machine in the world—and the beginning of real big science with the arrival of bubble chambers—equipment that had to be used collectively if only because it was dangerous, bulky and difficult to move around—the competition for the use of the facilities in Geneva intensified. Convinced of their superiority—because of their presence on the site and of a better knowledge of the demands imposed by the PS—the CERN physicists were initially inclined to monopolize the accelerator. Confronted by the protests made through the SPC by physicists who were not staff members, CERN's Director-General imposed a better equilibrium at the end of 1960, using the ideas of 'mixed' teams and of formal experiments committees whose functioning had to be transparent to visitors.[24]

The example drawn on was that of experimentation with bubble chambers, the nature of the technique having much to do with this. The European physicists using this device invented the system of collaborations composed of all teams interested in studying the same type of phenomenon, and divided the photographs and the topics up between themselves. Put into operation in 1961, this system was subsequently adopted as standard practice in the international community. By contrast it was rather more difficult to strike a balance in electronic experimentation. This was still basically an activity of very small groups, and as much of its know-how could only be acquired on the laboratory floor, the CERN physicists were more easily able to keep the 'outsiders' at bay—at least until the mid-1960s. Appealing to efficiency, they managed more or less to control the electronic experiments committee. Nonetheless, pursuant on a degree of homogenization of the quality of CERN's best physicists and of visitors, and after years of working with the same equipment and the same techniques, a more realistic form of coexistence was established between 1964 and 1966. At this time the mixed team became the norm—i.e. one composed of visiting groups and of CERN groups—replacing the practice whereby CERN teams 'generously' hosted one or two visitors.[25]

There are thus two important things to remember about the organization of experimental work. Firstly, that CERN was the first laboratory to have to confront directly the 'users' problem. An issue at Geneva from the very outset, this question of how to establish links between the physicists at the laboratory and outside physicists who had an unquestioned right to experiment was resolved for bubble chambers in 1960-61, and between 1964 and 1966 for electronic experiments. It only really surfaced on the American scene around the mid-1960s—via several user's revolts like that at BNL in 1964—although the principle had been accepted for the ZGS at Argonne and above all for the enormous accelerator which Panofsky built at Stanford: the machines, after all, were built with federal dollars, and the laboratory had to give some time and space to physicists from other centres in the country.[26] The second thing to remember is that the most difficult thing to arrange at Geneva was not to get physicists from different countries working together—as one might have thought—but to have the visitors working with the 'indigenous population'. Thus while the archives are replete with complaints or latent conflicts between 'guests' and 'hosts', we find few traces of tension between national groups of physicists after 1960. The latter was something of a delicate and sensitive issue until dispersed by the actual practice of working together; it was just there, by contrast, on the laboratory floor, that the former was nourished.

13.5 Learning to do 'big physics': a process still at work in the mid-1960s

The need to work together in an organization created out of nothing demanded another kind of apprenticeship, even more complex and even more time-consuming—namely, learning how to experiment effectively with very heavy equipment on the scale of that in the best American laboratories, but without having the benefit of a tradition, without having transited through medium-sized devices or intermediary energies. Expert at working at several hundreds of MeV or with cosmic rays, the Europeans suddenly found themselves face to face with 28 GeV protons and, in consequence, with experimental constraints two to three orders of magnitude greater than any they had previously had to contend with.

We shall have to spend rather more time on this fourth and last feature of CERN since it was the one which was debated most at Geneva between 1961 and 1965, and one of those most often treated in subsequent writings about the laboratory. The most usual form which these discussions took was to try to explain CERN's failure to adapt rapidly to this new scale of experimentation, to try to understand the difficulty it had in competing with the Americans in the race to the 'big discoveries'—in short they tried to grasp why the organization was always late in doing crucial experiments. The most celebrated and oft-repeated examples are those of the experimental confirmation of the two neutrinos by physicists working at BNL in 1962, the identification the following year at the same laboratory of the Ω^- predicted by the new SU(3) classification, and finally the acquisition of experimental evidence in 1964 for the violation of CP parity, again at Brookhaven. The clear initial superiority of BNL in the production of important scientific results, followed by a more comparable performance after 1964, was confirmed by two British researchers a few years ago by using the Science Citation Index and counting the articles cited more than 30 or 100 times in the four years after their publication.[27] More interesting for our purposes, however, is that Irvine and Martin took the opportunity of their study to ask about 200 American and European physicists the reason for this initial 'gap' between CERN and BNL. In their view four kinds of factors were involved.

To begin with there were errors of judgement made by the CERN management in planning the equipment needed to exploit the accelerator. Already spoken of at CERN in 1961, this 'unpreparedness' came down to a lack of magnets, quadrupole lenses, and separators to build secondary beams, and a delay in the building of big detectors, primarily, but not only, bubble chambers. Another oft-mentioned reason was that there were far fewer qualified experimentalists in Europe than in the United States, that the former 'lacked experience', and had difficulty elaborating a research programme focussed on the most important physics questions. In 1962 CERN's Research Director, Gilberto Bernardini, gave this reason as the most important one during a Council meeting.[28] Then there were the effects of CERN being multinational, effects particularly noticeable in the 'cheap proton' policy—have the maximum possible number of experimental groups working around the machine—and the structure of the committees responsible for settling the experimental programmes. Cumbersome and slow to take decisions, CERN's procedures were compared to those at Brookhaven, seen as much more supple and quick to react

Notes: pp. 806 ff.

because concentrated in a few hands, and in those of Maurice Goldhaber, the Director, in particular. More 'cautious' and 'democratic', the CERN-system was less efficient than the more 'autocratic' procedure favoured at BNL. Finally 'cultural' differences between Americans and Europeans were mentioned, the former being described as more bold and speculative in their approach, the latter as more conservative, more likely to proceed gradually. This was supposedly revealed in the tendency at Geneva to 'overdesign' equipment, that is equipment which though more reliable in the medium to long term, was always available to experimentalists that much later. (The converse of this difference would be that the Europeans always produced more systematic and refined results—as Irvine and Martin confirm.) In the accounts given by American actors, these last two themes are seen to reflect differences in 'style'. In contrast to the Europeans, the Americans describe themselves as knowing how to get around organizational restraints, as more quick to adapt and to turn a mistake to their advantage, as more capable of grasping the essentials of what has to be done and to ignore 'junky' research—in short to be more alert and clever than their European colleagues in what is, above all, a high-pressure race to be the first to make big discoveries.

If we confine ourselves to impressionistic evidence, and if we admit of course that all that really matters is to succeed in the 'Nobel' race, these accounts can seem convincing. In essence they seek to explain the difference in the production of a few results which the community deems to be decisive by identifying a number of 'gaps' between the United States and Europe. The difficulty, however, is that such very global arguments tend to explain 'too much', that when confronted closely with specific questions they appear as sometimes 'true', sometimes 'false'—and as sometimes 'quite irrelevant'. Accordingly—and to show what we mean—we want to consider in turn, and in some detail, the exact problems which the experimentalists at CERN had to confront in the early 60s.

Let us begin by looking at the question of the standard equipment necessary for the installation of secondary beams. Since there is no ideal stance from which one can judge the preparations that CERN 'ought' to have made, the best alternative is to compare CERN and BNL. The first remark to make is that CERN does not seem to have been later than BNL in placing its first order for magnets and quadrupole lenses, that this order was not quantitatively smaller, but that it was far less varied (magnets of the same length, relatively fewer quadrupoles, etc). The second noteworthy point is that CERN apparently did not place its second order as quickly as did Brookhaven, nor in the light of the limitations of the first. In seeking to explain these developments we have been led to conclude that no one of sufficient authority at CERN placed a high priority on the design of beams, no one dedicated himself to keeping up with developments in this field, that no one thought 'standard' beams sufficiently important for him to prefer them to more noble tasks like building bubble chambers or more sophisticated equipment. At BNL, on the contrary, the secondary beam problem was studied in detail from 1959 onwards. If there is an equipment gap to be found here, it is then to be associated with an underestimation of the importance of the problem, with the fact that no one at CERN saw the implantation of 'everyday' equipment of this type as being a particularly important job, that it was rather 'forgotten' in the distribution of key tasks in the organization.[29]

Turning now to less standard beam material, like electrostatic particle separators, it is interesting to note that both CERN and BNL initially planned to build similar devices (10 m tanks), and did so at about the same time. However CERN's separators were ready much later than Brookhaven's, so seriously impeding the bubble chamber programme until early in 1962. The problem here was twofold. On the one hand, starting from nothing, a major Research and Development effort had to be undertaken in Europe; BNL, for its part, relied on its experience, changed its plans and decided to build smaller (about 5 m long) devices which it modelled on those in use at the Cosmotron (the Cosmotron was a 3 GeV PS commissioned in 1952). This does not tell the whole story however, since the Europeans chose (be it consciously or not) to construct sophisticated, multipurpose separators which took far longer to build than the conventional ones ordered at Brookhaven.[30] This brings to mind the general argument that the 'Europeans' tended to 'overdesign' their apparatus. The trouble with formulating the problem like this—speaking of the 'Europeans' and of the 'Americans'—is that it obscures differences which were more fundamental, more at the root of the specific problem we are trying to solve here, namely, that a greater gap existed in Europe between physicists and engineers, a gap which was inscribed in the structure of the laboratory and its power relations. This structural given made it possible for engineer/builders at CERN to act with considerable autonomy once a task had been given to them, it allowed them to indulge a tendency to seek technological perfection—and to be relatively insensitive to the demands of the physicists for whom big discoveries often meant acting quickly, for whom it was often more important to have an 'imperfect' piece of equipment ready at the right moment than a 'perfect' one ready when the dust of the battle had settled. This clearly happened in the case of electrostatic separators for secondary beams at CERN: the engineering division which was building them did so at its own pace, without really worrying about physics.[31]

Before coming back to this difference between physicists and engineers in Europe, we would like to continue with our analysis of the precise problems facing European scientists in 1960-62. Now we want to focus on big detectors, and to try to understand why there was a notable time-lag between the operation of big bubble and spark chambers at CERN and BNL—not to speak of Berkeley. Our impression is that we need to introduce another element here—the fact that European physicists, understood in the narrow sense of the term, were not leaders in the field of experimental high-energy physics in the 50s and 60s, and that, as a consequence, they were not the brains behind new instruments directly connected to the art of

Notes: p. 807

experiment. At that time at CERN the development of new kinds of detectors did not generally spring from ideas generated by local practice but was more usually based on the importation of concepts born and tried out elsewhere. This dependence had two kinds of effects. Unlike a group which innovates, betting on its idea and doing a lot of R and D before being able to convince others that its equipment works and is qualitatively superior, groups which are not at the heart of the initial research tend to wait and see if the idea is worth taking up. This happened in Europe (*and in many American groups* vis-à-vis Alvarez) in the case of bubble chambers in the mid-50s, and at CERN (*and other American centres*) vis-à-vis the groups of Roberts, Cork, and Cronin with spark chambers some time later.[32] In both cases the first reaction was to keep an eye on developments, and the decision to 'take the plunge' was only made when the advantage of the devices were there for all to see. This is not an indication of a hesitant 'nature' or of a systematic propensity to be conservative, but merely a banal attitude for any group which is not at the heart of the action — as were almost all European physicists, limited until then to working at low energies. Moreover, when the decision is taken to enter the field, the followers tend to position themselves with regard to the leader(s) and their latest choices, and to skip the intermediate stage: this often seems to be the only way to avoid always having obsolete equipment. On the other hand, it accentuates the 'backwardness' in the short to medium term since the most advanced equipment takes longer to build. These considerations were clearly at work in CERN when it was decided to build directly a 2 meter hydrogen bubble chamber, the key argument being that Europe's central laboratory could not afford to have equipment that was inferior to what the Americans were building. As a result CERN was unfortunately without its own big hydrogen bubble chamber in the first years after the PS worked.[33]

Finally let us consider big and sophisticated equipment which is a response to needs which can be formulated well in advance and which are less directly linked to the development of experimentation. We are thinking here of things like RF separators whose principles were known but which required several years of R and D, or of the neutrino horn invented by Simon van der Meer. Here the Europeans were often leaders and revealed a considerable capacity to imagine and to innovate. We find the same thing in the design and construction of accelerators, the 28 GeV PS, for example, being a novel machine based on the newly-discovered strong-focusing principle which the Europeans unhesitatingly chose to build before the Americans had launched their programme. The conception of the Intersecting Storage Rings, advocated at CERN from 1960 to 1965, was a response to an innovative drive of a similar kind.[34] In such cases the image of 'Europeans' as more conservative and less innovative than 'Americans' does not fit. Employed indiscriminately, and without specifying how and to whom it is supposed to apply, the distinction stops one seeing the more important difference which we have already identified. In fact, one must distinguish between instances in which physicists strictly speaking were decisive, in which instrumental developments were associated directly with the art of detection and rooted in physics questions, in which the conception of new equipment emerged organically from the demands of experimental practice, — and here 'the Europeans' appear to have been somewhat behind though this does not necessarily mean that they were 'lacking' something by 'nature'; and between cases in which, by contrast, the aim was rather to solve R and D problems, to develop and to improve radically devices whose features were already glimpsed, even if vaguely — and here 'the Europeans' do not seem to have been 'backward' at all. On the contrary, they might be considered too bold, too innovative — indeed unwilling to devote themselves to projects which did not seem to pose a sufficient technical challenge, like standard beam transport equipment.

13.6 American and European physicists: an historical difference

We have now arrived at the heart of the matter. What happened in the United States between the 30s and the 60s — a phenomenon from which the Europeans were largely excluded — was the emergence of a profound symbiosis previously unknown in basic science, a fusion of 'pure' science, technology, and engineering, it was the emergence of a new practice, a new way of doing physics, the emergence of a new kind of researcher who can be described at once as physicist, i.e. in touch with the evolution of the discipline and its key theoretical and experimental issues, as conceiver of apparatus and engineer, i.e. knowledgeable and innovative in the most advanced techniques (like electronics at that time) and able to put them to good use, and entrepreneur, i.e. capable of raising large sums of money, of getting people with different expertise together, of mobilizing several kinds of human, financial, and technical resources. The most successful examples of such men were to be found around Lawrence, who was one of the first to orientate his group in this direction. It was men like Alvarez, Lofgren, McMillan, Panofsky, and Wilson, for example, who became the masters of the new physics and who imposed their rhythm on world science.[35]

What characterized them was a pragmatic and utilitarian approach notable for its clear stress on 'getting numbers out', an approach which preferred results and practical efficacy to means and aesthetic harmony, an approach which was rooted in 'an educational philosophy that emphasize[d] the empirical, the experimental practice', and which was kept alive by the institutional arrangements in American universities where theoreticians, experimentalists, and apparatus builders were encouraged to work together.[36] Then there was the experience of the war, which meant lavish financial and technical means, which meant multidisciplinarity and the linking of people with different educational backgrounds, which meant the imperative to succeed at whatever cost using all the technical and industrial resources available. This reinforced the 'full-blooded empiricists', the 'radical pragmatists', the people for whom all was permissible methodologically, who preferred a heuristic

Notes: p. 807

emphasizing improvisation and risk. In 'physics' this stimulated phenomenological approaches and discouraged 'a sustained focus and effort on fundamental theory'; in 'practice' it brought the 'engineering' side of laboratory work to the fore, notably the ability to use industrially available material in new and interesting ways.[37] Subsequently, once these methods had proved their indisputable efficacy, there was the added bonus of the cold war and the growing importance of applied research. The American system for supporting science—notably by the Department of Defence and its famous summer schools—as well as the plethora of unexpected and exciting experimental results generated with the new means at hand, consecrated this technical approach to treating problems, far and away the most efficient means for imposing structure and order on a field dragged forward by experimental and technological practices. As a result, fifteen years after the war, the gulf between the United States and Europe was impressively wide.[38]

By contrast, European physicists in the years 1945-1960 appear above all as the heirs to a tradition which continued to attach great importance to 'pure' science and to keeping 'applied' science separate, which kept fundamental theory—something refined—apart from experimental phenomenology, which still tended to regard technique as of lesser importance in the elaboration of knowledge. Without the 'stimulus' of a war effort European physicists did not become apparatus builders before all else, and—even in Britain[39]—remained people for whom the conception of big and sophisticated equipment did not derive directly from their expertise. Being experimentalists in the classical sense of the word, they did not become managers who were immediately able to handle the new scale of activity demanded by big science. An experiment—and even if nuances are needed depending on whether we are talking of electronic detectors or track chambers—remained something one did in the short term, on a human scale, something which was not in itself a permanent race to use equipment constantly having to be changed. In short experimenting remained primarily the practice of an art, secondarily the mastery of techniques.[40]

It is for this reason that the engineers who worked around them enjoyed so much autonomy. Indispensable by virtue of the size of certain undertakings, they were put in charge of 'all big equipment'—and this often led them to become the 'real bosses' of the laboratory. De facto the physicists had to go through them, and only they were ready to manage centres employing one or two thousand people and in which investments were made on a 5 or 10 year basis. Kept, by contrast, on the periphery of physics proper—because after all that was not their main preoccupation—they remained detached from the urgency of research and the needs which grew from it. Excellent at designing equipment whose goals were clearly defined, they could not imagine, starting from an experimental practice they did not have, the new detectors which the Americans invented. Capable of being the first to instal a 28 GeV PS embodying a new focusing principle, they were not in a position to 'invent', to 'think of', to 'have the idea for' a bubble or spark chamber. Standing back from day-to-day experimental practice—and constrained by certain cultural and educational traditions peculiar to Europe—they tended to prefer technology per se, to be 'pure' technicians, to refuse boring and unimaginative tasks, to demand the licence to explore new avenues, to work on challenging projects. For want of an interface, since no Alvarez or Panofsky existed in Europe, a hiatus was always possible. And more: since this intellectual and professional difference was inscribed in the organization's structure, the phenomenon was amplified, perpetuated, rigidified.[41]

Now that we have identified the crux of the problem, the core of the difference between CERN and the equivalent American laboratories, we are in a position to take up again two or three of the traditional arguments used to account for the 'delays' or the 'setbacks' at CERN in the years 1960-65. We shall first consider the multinational character of the body, which was said to impede the functioning of its experimental programme (because of the heavy experimental committee structure it demanded), and to have allowed physicists to perform many experiments of little interest at the expense of giving priority to decisive experiments of lasting significance (the so-called 'cheap proton' policy).[42]

In discussions of this kind the notion of 'multinationality' is used ambiguously. It essentially mixes two rather different ideas with one another, two realities which must be kept distinct, namely, the fact that CERN brings together several sovereign states, and the fact that it is a unique and central laboratory for a polycentric physics community. In the early years, as we have already shown, the first aspect was dominant; CERN was a new business, the wishes of its various Member States did not yet converge, and the relationships between CERN and the national physics communities were somewhat formal. Once concrete experimental work got under way around the PS, however, and once it was officially accepted (in 1961/62) that budgets should grow as a matter of policy, there was a radical shift, and the second aspect became prominent. What CERN then had to deal with was less rivalries between national groups—rivalries which would have to be managed by a complex system of experimental committees—than the rapid emergence of a new problem which all big science laboratories had to handle a few years later, that of accommodating numerous visitors in a place already occupied by a local scientific staff which preferred to remain alone.

One might object that, even if it was not CERN's multinationality that was responsible for its experiments committee system, this system was very cumbersome all the same and introduced rigidities into the laboratory's functioning. Highly decentralized since comprising three committees specialized in different detection techniques—Track Chamber, Emulsion, Electronic—, it suffocated rapid adaptations to changing circumstances and impeded the implementation of a central policy concerned above all to do crucial experiments.[43] This conclusion, too, would be too hasty however, and actually would invert cause and consequence. Our impression is that the committee system reflected the conditions prevailing in the European physics community, that it took the form it did because at the time there were no physicists in Europe having the aura of a Lawrence or a Goldhaber, because

Notes: p. 807

it allowed a community not always that sure of itself to reduce the risks inherent in any autocratic system of management. In this sense it was not so much the source of a 'too' prudent experimental programme as the structural counterpart of the situation in which European physicists were at the time, be they from CERN or from national laboratories, namely, of not yet quite knowing how best to use equipment of the scale of that in Geneva. If the CERN of the years 1960-65 appeared to be a 'big photocopier' reproducing and improving novel work done elsewhere,[44] it was because the European physicists were learning how to experiment, were learning new ways of doing things which their American colleagues had been familiar with for almost two decades. Their 'conservatism' and their 'prudence' on the one hand, 'the heaviness of their decision-making system' on the other, were thus merely two manifestations of the gap between the two practices, two manifestations of the apprenticeship which the Europeans were serving—in the absence of a master, and by a method of trial and error.

To conclude let us speak of money, insofar as this was also a reason frequently given for CERN's being 'late'; the organization's Annual Reports constantly refer to the low budgets granted CERN by its member states between 1957/58 and 1961/62.[45] We shall be brief here because, while it is true that CERN was less well-endowed than its competitors in these years, we do not think that this was a decisive reason for the 'delays'. For even if higher budgets had, for example, allowed a more rapid recruitment of personnel, and so permitted the acceleration of certain projects, the situation we described above would not have changed dramatically. For the present purposes it suffices to mention, by way of confirmation, the acquisition of standard beam transport equipment and the drawing up of the experimental programme: it is difficult to see how anything here would have changed in an important way if budgets had been considerably higher than they were. The situation was a little different for the big hydrogen chamber. All the same, if only because there were other chambers at Geneva, there is no reason to believe that much would have changed even if the 2m had been commissioned one or even two years earlier. Finally, one should bear in mind that resources are never infinite (except perhaps in wartime!), and that the real question is that of what choices to make. In this connection remember that no CERN division was solely responsible for preparing the PS experimental programme before 1961. The construction of bubble chambers fell to John Adams' division whose job it was to finish the PS, and the choice he made was to favour the completion of the accelerator above the construction of the detector. We thus come back to the characteristics of the European context which we have already discussed, notably the rift between engineers and physicists which was crystallized in CERN's organizational structure.[46]

13.7 CERN and the history project: some lessons

This second volume is the last of the CERN history project as proposed by Armin Hermann in 1982, and it is perhaps not without interest to end it with a few words on the work itself and on the relationship that we have had with the actors about whom we have written. This project being the first of its kind in Europe, though others are beginning to take shape on the horizon, the point possibly deserves some attention.[47]

The first thing we would like to recall is that the history we have written was not an official history of the organization—if one understands by that a project 'ordered' by an institution, a project for which it puts up the money, and over which it has certain controlling rights. While CERN undoubtedly showed considerable interest very early on in having its history recorded, the fact remains that our project was not a direct product of those sentiments. By the same token our group was financed independently of CERN, the organization's contribution, for which we are grateful, being to provide us with office space, to absorb some of our current expenses (e.g. telephone), and to have opened completely the CERN archives. The value of this gesture cannot be denied: we have had access to all the collections gathered by CERN, without any previous selection being made by laboratory staff. Finally, the institution has never tried to control our work, nor has it tried to restrict our freedom of enquiry or of judgement. The point is worth stressing because we are aware of many other cases where things work rather differently.[48]

Some readers of our first volume have called our History of CERN an official history, all the same. The main reason is that it seemed as if the organization had recruited and paid us since we were living and working on the spot, at CERN. If we did this, if we chose to instal ourselves at Geneva it was because of the presence there of a large number of the main actors in our story and because of the availability of the archives. The importance of this will be evident to anyone who knows the vast amount of paper work that a big organization can generate, particularly a multinational organization which has to be extremely careful and conscientious about keeping records of its decisions. To have immediate access to dozens of tons of documents stored just beneath our feet was an irreplaceable advantage and one which led us to choose Geneva as our working base.

Admittedly working in the heart of the organization which he studies can be a serious handicap for any historian. Even if there are no direct pressures, the immersion in a milieu which reacts to his presence and his results creates a climate which can subvert his intellectual autonomy. His choice of topics, for example, his definition of the relevant questions, the simple fact of 'seeing' some things and 'not seeing' others— all this can arise all too easily just because he is there, at the laboratory itself. Nevertheless—because there is no neutral location, and because we felt that it was possible to overcome these 'objective' difficulties—we took the risk of working at Geneva, hoping to cope with the situation as we found it. It is now up to our readers to say whether we have succeeded or not.

Notes: p. 807

Our relationships with the principal actors in CERN's history have passed through several phases. Initially there was a kind of mutual incomprehension springing from two main sources. CERN scientists tended to think that they were the only ones who really knew the history of the organization since it was they who had built it. Having already established a mental picture of what that history was like, they expected us to begin by interviewing them with a view to getting their versions on record. Convinced, in addition, that written documents only gave a very poor and very partial picture of what actually happened, and that it was difficult to get at the real events which lay concealed behind carefully packaged (by them!) formulations, CERN scientists believed that their memories were a more fundamental source and, above all, one whose contents were unavailable anywhere else. For our part, we began with a certain scepticism about what is generally called oral history (i.e. a history mainly written from interviews with the chief protagonists) and wanted to base our work on archival documents. This feeling was quickly reinforced by the variety of sources we found at CERN and in European archives, and by their quality. We were sceptical of oral history because we feared that we would become ensnared in the mythical accounts of their own birth and development which all organizations secrete, that we would become trapped in the network of institutional memories and land up simply repeating 'standard histories'. This seemed to be particularly likely at CERN, since the context was highly structured and since its rather short life has been marked with countless commemorative events which have effectively produced an 'official' version of its past. Any coherence we might find in confronting interviews was thus as likely to be the consequence of 'artificial' memory—the spontaneous repetition of old stories told over and again—as of 'real' recollections. For these reasons we made very few interviews when we first arrived, relying almost exclusively on those made some ten years earlier by Margaret Gowing and her co-workers. Which, we should say, numbered some 50.[49]

These early 'misunderstandings' sharpened somewhat when our first preliminary reports for volume I appeared. Published by CERN's Scientific Information Service, and largely distributed by the organization to scientists, historians, sociologists, etc., our aim was to receive comments and criticisms before settling on the final versions of our texts. The resulting tension seems to have been caused primarily by the fact that our archival work had led us to cast considerable doubt on some of CERN's standard accounts of its origins. As most of those we spoke to in Geneva had not lived through the period of creation, their first reaction was to distrust the new version. Not being able to 'remember' where the 'truth' lay by themselves, they tended to see us as troublemakers. Difficulties also arose because the physicists at CERN and in the Member States had 'spontaneously' thought in terms of having a scientific and technical history before all else, rather than a political history of the origins and the foundation of the laboratory. The latter were of secondary importance compared to what was the 'essence' of CERN, to wit, the production of scientific knowledge. Finally, our direct contacts with the principal CERN personalities had remained limited—because we had not embarked on a systematic program of interviews—, and a feeling of being left out of something that concerned them deeply also surfaced. As a consequence our early results encountered some vigorous criticisms.

Relationships became closer with the holding of two one-day seminars at CERN, one with the people who were involved in the creation of CERN between 1949 and 1954 (Amaldi, Auger, Rabi, de Rougemont..), and the other with CERN scientists who had lived through the period covered by volume II (1954-1965).[50] Things improved because, on the one hand, our results appeared to be substantiated and 'informed' to those who were personally familiar with events—and hence less easily dismissed with a wave of the hand; on the other hand, because, for our part, we better grasped the merits of a more systematic oral history, at least at this stage of our work. More self-assured, and encouraged by the interest of many discussions held with the actors on our texts, we undertook a new series of interviews. This time the aim was less to glean factual information, which we could derive from other (more reliable) sources anyway, but to confront our interpretation of events with the impressions held by those who had lived them, and to enrich our understanding of the community's view of its own history. From this a more open and much less guarded relationship began to develop.

It seems to us that three lessons can be drawn from this experience. Firstly, that it would have been possible and preferable to begin with a number of in-depth interviews which, apart from anything else, would have improved our relationships with the actors and, above all, given us access to new documents. As 'one-sided' as their accounts may be (and they are not more so than any other account) the chief protagonists supply points of reference and grist to the intellectual mill which can only be extracted with difficulty from the written sources alone. Margaret Gowing undoubtedly left us a large number of these, but direct contact would probably have helped us avoid several unhelpful detours. The second lesson to be drawn is that archival information is irreplaceable and that oral information is often imprecise, dubious, or merely wrong. Memory does not simply select; it reorders events, simplifies them, 'rationalizes' them. This is obvious where 'facts' are concerned; it is also true when it comes to characterizing the issues at stake or to reconstructing a dynamical process. Contrary to what many may think—but the point is banal—, a close reading ('between the lines', if you like) and a comparison of documents from several sources often throw into relief what the authors thought to hide. It has the further advantage that one sees the same thing from several points of view at once, so enabling one to detect blind spots in the vision of particular individuals. Finally, we have learnt that regular discussions with the actors, based on previous work and focusing on specific points, are invaluable. Particularly when they start from a

Notes: p. 807

written text, such contacts can suddenly cast light on an aspect only dimly glimpsed till then, and can allow one to detect inevitable distortions which arise in the intellectual reconstruction of the past.

Translated by John Krige

Notes

1. About Bloch's nomination, see History of CERN (1987), 261-73; about Weisskopf's, see chapter 7, section 7.2.4.
2. On these aspects, see chapter 10, sections 10.1 to 10.3.
3. See chapter 7, section 7.1.3.
4. See chapter 8, notably 8.5 and 8.7.
5. As example, see Crozon (1987) or Pickering (1984).
6. See OCDE (1968), OCDE (1984).
7. For a general presentation of the question, Salomon (1970) and Salomon (1986). About the oversimplified model which depicts science and technology as an 'assembly line', see Glantz (1978) and Wise (1985). About the criticisms which were made in the 1950s see for example Management (1957). On the new ministerial structures set up in the 1950s and early 1960s in Continental Europe, Pestre (1989) and Prost (1988).
8. On this notion of 'CERN system', see chapter 7, notably 7.1.3.
9. Details in chapter 12, notably 12.4 and 12.5.
10. On this conviction, chapter 12 and sections 12.5, 12.6. See also Pestre (1988), 125-30.
11. For a contrast remember the ways in which the Radiation Laboratory at Berkeley and SLAC at Stanford were run under Lawrence and Panofsky. See Heilbron, Seidel & Wheaton (1981), Galison, Hevly & Lowen (1988), Leslie (1987).
12. See chapter 12.6.5 and Pestre (1988).
13. On the notion of corporate spirit, see Hamon & Torres (1987). After writing this text we read the excellent book by Sharon Traweek (Traweek (1988)) which considers questions of this kind.
14. To appreciate this, have a look at Perrin's, Powell's, or Weisskopf's speeches at CERN Council sessions (for example, CERN Council minutes, 1963, 15-23). See also chapter 7, section 7.1.2 for a deeper analysis.
15. See chapter 7, section 7.3.
16. On these aspects, see chapter 7, notably 7.1.3 and 7.2.4.
17. See sections 8.4 and 8.5; see also letter J. Augsburger to Weisskopf, 1/5/63 (DG20601).
18. Kowarski (1955) and Kowarski (1961). See also the several booklets edited by CERN for the anniversaries of the organization and the commissioning of its main machines; CERN Courier; the booklets in honour of the main personalities in CERN's history, etc.
19. See History of CERN (1987), 223-25 and 282-85.
20. For an example, see History of CERN (1987), 331-35 (on the French side), 246-252 (on the British side). See also chapter 8, sections 8.2 and 8.3.
21. York (1987), 62-84.
22. Compare sections 8.2 and 8.6 in chapter 8, sections 10.2 and 10.4 in chapter 10; see also 7.1.2 for more general thoughts.
23. See chapter 8, section 8.1.
24. See chapter 8, sections 8.4 and 8.5.
25. See chapter 8, section 8.7.

26. For an introduction to the American situation regarding 'users', see Kowarski (1967). See also History of ZGS (1980), Galison, Hevly & Lowen (1988) on SLAC, Westfall (1988) on Fermilab.
27. Irvine and Martin (1984).
28. For a perfect example of such debates at the time, see Council minutes, 13/6/62, 14-19.
29. See chapter 9, sections 9.4 and 9.5.
30. See chapter 9, section 9.5.2.
31. On the institutional reality of these differences, see chapter 7, section 7.2; on the autonomy of the 'engineers', see the debate about the ISR and the 300 GeV PS in chapter 12, sections 12.3 and 12.6.
32. See chapter 9, sections 9.3.1 and 9.3.2; see also Krige & Pestre (1986).
33. Krige & Pestre (1986); chapter 8, section 8.6.
34. See History of CERN (1987), 273-82 on the PS; chapter 12, section 12.3 on the ISR.
35. For this section, see Holton (1985), Schweber (1985), Schweber (1986), the whole special issue of Historical Studies in the Physical and Biological Sciences, 18:1 (1987), and autobiographical works like Alvarez (1967) or York (1987). Remarks of the same kind should apply to entrepreneur-engineers like Vannevar Bush or Frederick Terman at Stanford.
36. This is directly inspired by Schweber's remarks on American theoretical physics (op. cit., note 35).
37. This is also inspired by Holton (1985).
38. Amongst hundreds of works see Godement (1979), Kevles (1988), and the whole issue of HSPS cited in note 5.
39. See for example Hoch (1988).
40. Weisskopf was well aware of the situation when he gave his first speech at CERN as Director-General (after having spent a year there already). See chapter 7, section 7.2.4.3, where his speech is commented on.
41. The story of the decision to build the ISR conveys much of this feeling. See for example Pestre (1988).
42. See Irvine & Martin (1984) reporting the opinions of many physicists.
43. Chapter 9, section 9.4.
44. The phrase is Pierre Germain's in an interview with the author held on 14.11.87 (along with G. Munday and P. Standley).
45. A perfect example is the complaint by the laboratory that the reduced budgets led to an equipment gap in 1962 (Council minutes, 19/12/62, 54-7).
46. See chapter 7, section 7.2.2 and chapter 8, section 8.4.2.
47. Without being exhaustive, we think here of the German projects to write the history of the national nuclear institutions, and the French project for a history of the CNRS.
48. We have in mind the histories of military programmes and institutions.
49. For more thoughts on oral history, see Histoire Orale (1980), Histoire Orale (1986), and Pestre (1989).
50. They were held at CERN on 20/9/84 and 17/11/87.

Sources and Bibliography for Parts I, III and IV

John KRIGE and Dominique PESTRE

Archival Sources

The CERN archive was the main source used for parts III and IV of this book. Of particular importance were the DG and DIRADM files, papers deposited by Mervyn G.N. Hine, and official documents prepared for the Council and its Committees. They were supplemented by material deposited in the UK Science and Engineering Research Council archives (Hayes, Middlesex), and at the French Ministry of Foreign Affairs, Paris.

Printed Primary Sources

300-1000 BeV (1961)	Accelerator Design Group, BNL, report, *Design Study for a 300-1000 BeV Accelerator* (Upton: BNL, 28/8/1961).
Abrams et al. (1964)	K. Abrams, 'The Status of HPD Programs at Brookhaven National Laboratory', *FSD Conf. (1964)*, 97-112.
Accelerator Conf. (1956)	*CERN Symposium on high energy accelerators and pion physics, Proceedings* (Geneva: CERN, 1956).
Accelerator Conf. (1959)	*International Conference on high energy accelerators and instrumentation, Proceedings* (Geneva: CERN, 1959).
Accelerator Conf. (1961)	*International Conference on high energy accelerators, Proceedings* (US AEC-BNL (?), 1961(?)).
Accelerator Conf. (1963)	*International Conference on high energy accelerators, Proceedings of the* American version, CONF-114, TID-4500 (US AEC, undated).
Accelerator Conf. (1965)	*V International Conference on high energy accelerators, Proceedings* (Roma: CNEN, 1966).
Adams (1956)	J.B. Adams, 'The Design of Accelerating Machines', *Accelerator Conf. (1956)*, 1-6.
Adams (1961)	J.B. Adams, 'Introduction', *Theory Conf. (1961)*, 1-3.
Adams (1965)	J.B. Adams, 'CERN: The European Organization for Nuclear Research', *Cockcroft (1965)*, 236-261.

Amato et al. (1963)	G. Amato et al., 'One Stage Separated K-meson Beam of 1.5 GeV/c Momentum at the CPS', *Nuclear Instruments and Methods,* **20** (1963), 47–50.
Anders et al. (1962)	H. Anders et al. 'A Preliminary Study of a Cathode Ray Tube Device for Scanning Spark Chamber Photographs', *Instrumentation Conf. (1962),* 414–418.
Anders et al. (1963)	H. Anders et al., '"Luciole". A Cathode Ray Tube Flying Spot Digitiser for Measurement of Spark Chamber Pictures', *Nuclear Electronics Symp. (1963),* 349–356.
Atomic Energy Conf. (1958)	*Second United Nations International Conference on the Peaceful Uses of Atomic Energy,* Geneva, 1–13/9/58 (Geneva: United Nations, 1958).
Aubert et al. (1963)	B. Aubert et al., 'Low Energy Separated Beam at the CERN PS', *Nuclear Instruments and Methods,* **20** (1963), 51–54.
Baltay et al. (1961)	C. Baltay et al., 'Design and Performance of a 3.3 BeV/c Separated Antiproton Beam', *Accelerator Conf. (1961),* 452–458.
Baltay et al. (1963)	C. Baltay et al., 'The Separated Beam at the AGS – Performance with Antiprotons and π^+ Mesons', *Nuclear Instruments and Methods,* **20** (1963), 37–42.
Barbier et al. (1959)	M. Barbier et al., 'Studies of an Experimental Beam-Stacking Electron Accelerator', *Accelerator Conf. (1959),* 100–114.
Benot et al. (1961)	M. Benot et al., *An Analysis of Results from the Prototype H.P.D.* (Geneva: CERN 61-31, 1961).
Berkeley Study (1961)	*The Berkeley High-Energy Physics Study, 15/6/61–15/8/61* (Berkeley: LRL, UCRL-10022, 1961).
Bernardini et al. (1964)	G. Bernardini et al., 'Search for intermediate boson production in high-energy neutrino interactions', *Physics Letters,* **13** (1964), 86.
Bernardini et al. (1965)	G. Bernardini et al., 'Lower Limit for the Mass of the Intermediate Boson', *Il Nuovo Cimento,* **38** (1965), 608.
Bienlein et al. (1964)	J.K. Bienlein et al., 'Spark Chamber Study of High-Energy Neutrino Interactions', *Physics Letters,* **13** (1964), 80.
H. Blewett (1961)	H. Blewett, 'The Brookhaven International Conference on High-Energy Accelerators', *Physics Today,* **14**:12 (1961), 31–37.
H. Blewett (1967)	H. Blewett, 'Characteristics of Typical Accelerators' *Annual Review of Nuclear Science,* **17** (1967), 427–468.
J.P. Blewett (1956)	J.P. Blewett, 'The Proton Synchrotron', *Reports on Progress in Physics,* **19** (1956), 37–79.
J.P. Blewett (1961)	J.P. Blewett, *Design Study for a 300–1000 BeV Accelerator, Preliminary Report, 22/5/61* (Upton: BNL, Accelerator Department, IA-JPB-15, 1961).
J.P. Blewett (1962)	J.P. Blewett, *Accelerators of the Future* (Upton: BNL Lecture Series, number 18, 13/6/62)
J.P. Blewett (1963)	J.P. Blewett, 'Accelerator Design Studies for Production of High Center-of-Mass Energies', *Accelerator Conf. (1963),* 89–93.
Block et al. (1964)	M.M. Block et al., 'Progress Report on Experimental Study of Neutrino Interactions in the CERN Heavy Liquid Bubble Chamber', *12th International Conference on High-Energy Physics, Dubna, 1964,* Vol. II (Jerusalem: Israel Program for Scientific Translation, 1969), 7.
Bradner (1960a)	H. Bradner, *Problems and Techniques in the Analysis of Bubble Chamber Photographs,* talk given to the American Physical Society, 29/1/60 (Berkeley: UCLR-9104, 1960).

Bradner (1960b)	H. Bradner, 'Capabilities and Limitations of Present Data-Reduction Systems', *Instrumentation Conf. (1960)*, 225–228.
Bradner & Glaser (1958)	H. Bradner and D.A. Glaser, 'Methods of Particle Detection for High-Energy Physics Experiments', *Atomic Energy Conf. (1958)*, vol. 14, 412–422.
Bradner & Solmitz (1958)	H. Bradner and F. Solmitz, 'On the Analysis of Bubble Chamber Tracks', *Atomic Energy Conf. (1958)*, Vol. 14, 423–426.
Brobeck (1956)	W.M. Brobeck, 'The Design of High Energy Accelerators', *Accelerator Conf. (1956)*, 60–63.
Chambre à bulles (1961)	'Numéro consacré aux chambres à bulles françaises au CERN', *L'onde électrique*, **41** (Dec. 1961).
Chew (1961)	G.F. Chew, 'Summary of Theoretical Conclusions from 1961 Berkeley Summer Study', *Berkeley Study (1961)*, xi–xiv
Chick (1960)	D.R. Chick, 'Plasma Research', *Contemporary Physics*, **1** (1960), 169–190.
Cocconi (1960)	G. Cocconi, 'Progress Report on Work with the 25 GeV Proton Synchrotron', *HEP Conf. (1960)*, 802.
Cockcroft (1965)	J. Cockcroft (ed.), *The Organization of Research Establishments* (Cambridge: Cambridge University Press, 1965).
Collins (1961)	T.L. Collins, *Long Straight Section for AG Synchrotrons* (Cambridge Electron Accelerator Report, CEA-86, July 1961).
Courant & Cool (1959)	E.D. Courant and R. Cool, 'Transport and Separation of Beams from AG Synchrotron', *Accelerator Conf. 1959*, 403–412.
de Shong (1962)	J.A. de Shong Jr., 'Spark Chamber Track Measuring System', *The Review of Scientific Instruments*, **33** (1962), 859–865.
Deutsch (1962)	M. Deutsch, 'A Spark Chamber Automatic Scanning System', *Multiparameter Analyzers Conf. (1962)*, 167–170.
Deutsch (1965)	M. Deutsch, 'Guided Flying Spot Systems for Spark Chambers', *Instrumentation Conf. (1965)*, 69–72.
Emulsion Meeting (1959)	*Emulsion Work with the CERN 25 GeV Proton Synchrotron* (Geneva: CERN 59-13, 1959).
Experimental Program (1961)	High Energy Study Group, BNL, report, *Experimental Program Requirements for a 300–1000 BeV Accelerator* (Upton: BNL, 28/8/1961).
Faissner et al. (1962)	H. Faissner et al., 'The CERN neutrino spark chamber', *Instrumentation Conf. (1962)*, 213–219.
Farley (1965)	F.J.M. Farley (ed.), *Progress in Nuclear Techniques and Instrumentation*, Vol. 1, (Amsterdam: North-Holland, 1965).
Farley (1967)	F.J.M. Farley (ed.), *Progress in Nuclear Techniques and Instrumentation*, Vol. 2 (Amsterdam: North-Holland, 1967).
Franck et al. (1962)	J.V. Franck, P.V.C. Hough, and B.W. Powell, 'Realisation of HPD Systems at Three Laboratories, *Instrumentation Conf. (1962)*, 387–392.
Frautschi (1961)	S. Frautschi, 'Report on the CERN Conference: "Theoretical Aspects of Very-High-Energy Phenomena", 23/6/61', *Berkeley Study (1961)*, 3–7.
FSD Conf. (1963)	*Programming for HPD and Other Flying Spot Devices*, Collège de France, Paris, 21–23/8/63 (Geneva: CERN 63-34, 1963).
FSD Conf. (1964)	*Programming for Flying Spot Devices*, Bologna, 7–9/10/64 (Geneva: CERN 65-11, 1965).
Gelernter (1961)	H. Gelernter, 'The Automatic Collection and Reduction of Data for Nuclear Spark Chambers', *Il Nuovo Cimento*, **22** (1961), 631–642.

Germain & Tinguely (1963)	C. Germain and R. Tinguely, 'Electrostatic Separator Technique at CERN', *Nuclear Instruments and Methods,* **20** (1963), 21–25.
Glaser (1955)	D.A. Glaser, 'The Bubble Chamber', *Scientific American* (Feb. 1955), 46–50.
Glaser (1964)	D.A. Glaser, 'Elementary Particles and Bubble Chambers', *Nobel Lectures in Physics* (Amsterdam, 1964), 529–551.
Goldschmidt (1957)	Y. Goldschmidt-Clermont, *The Analysis of Nuclear Particle Tracks by Digital Computer* (Geneva: CERN 57-29, 1957).
Goldschmidt (1958)	Y. Goldschmidt-Clermont, 'The Analysis of Nuclear Particle Tracks by Digital Computer', *Discovery,* April (1958), 148–152.
Goldschmidt (1959)	Y. Goldschmidt-Clermont, 'Instruments for Picture Evaluation: A Survey', *Accelerator Conf. (1959),* 523–532.
Goldschmidt (1960)	Y. Goldschmidt-Clermont, 'Recent Developments in Europe in Bubble Chamber Data Production', *Instrumentation Conf. (1960),* 233–238.
Goldschmidt (1964)	Y. Goldschmidt-Clermont, 'Progresses in Data Handling for High Energy Physics', *HEP Conf. (1964),* Vol. 2, 439–462.
Goldschmidt et al. (1957)	Y. Goldschmidt-Clermont, G. von Dardel, L. Kowarski, and C. Peyrou, 'Instruments for the Analysis of Track Chamber Photographs by Digital Computer', *Instrumentation Meet. (1957),* 146–153.
Gould (1961)	C.L. Gould, *Vacuum System for a 1000 BeV Accelerator, 14/2/61* (Upton: BNL, IA-CLG-14, 1961).
HEP Conf. (1960)	*1960 Annual International Conference on High Energy Physics at Rochester* (Rochester: University of Rochester, 1960).
HEP Conf. (1964)	*XII International Conference on High Energy Physics,* Dubna, 5–15/8/64 (Moscow: Atomizdat, 1966).
Hereward (1962)	H.G. Hereward, 'Current Development of High-Energy Proton Synchrotrons', *Instrumentation Conf. (1962),* 9–11.
Hereward et al. (1960)	H.G. Hereward, K. Johnsen, A. Schoch and C.J. Zilverschoon, *Present Ideas on 25-GeV Proton Storage Rings* (Geneva: CERN PS/Int. AR/60-35, 22/12/60).
Hine (1959)	M.G.N. Hine, 'CERN 25 GeV Proton Synchrotron', *Accelerator Conf. (1959),* 383–384.
Hine (1962)	M.G.N. Hine, 'The CERN Proton Synchrotron, 1954–1962: Forecasts and Reality Compared. An informal talk', *Rendiconti della Scuola Internazionale di Fisica «E. Fermi»* (New York: Academic Press, 1962), 287–294.
Hine (1964)	M.G.N. Hine, 'Financing High-Energy Physics', *CERN Courier* **4** (1964), 105–110.
Hine & Germain (1961)	M.G.N. Hine and P. Germain, 'Performance of the CERN Proton Synchrotron', *Accelerator Conf. (1961),* 25–38.
Hough (1963)	P.V.C. Hough, 'Attempts to Obtain High Rate, High Quality Measurement Events Despite FILTER II Imperfections', *FSD Conf. (1963),* 133–135.
Hough & Powell (1960a)	P.V.C. Hough and B.W. Powell, 'A Method for Faster Analysis of Bubble Chamber Photographs', *Instrumentation Conf. (1960),* 242–245.
Hough & Powell (1960b)	P.V.C. Hough and B.W. Powell, 'A Method for Faster Analysis of Bubble Chamber Photographs', *Il Nuovo Cimento,* **18** (1960), 1184–1191.
Hough & Powell (1965)	P.V.C. Hough and B.W. Powell, 'On Getting Results from an HPD System', *Instrumentation Conf. (1965),* 253–269.

Hubbard (1961)	E.L. Hubbard, 'Heavy-Ion Accelerators', *Annual Review of Nuclear Science,* **11** (1961), 419–438.
Humphrey & Ross (1964)	W.E. Humphrey and R.R. Ross, 'Operation of the SMP Data-Analysis System', *HEP Conf. (1964),* 403–408.
Iep Conf. (1959)	*International Conference on Instruments for the Evaluation of Photographs,* CERN, 23/2/59 (Geneva: CERN 59-19, 1959).
Iep Meet. (1958)	*International Meeting on Instruments for the Evaluation of Photographs, Summary of Proceedings,* 15–16/9/58 (Geneva: CERN 58-24, 1958).
Instrumentation Conf. (1960)	*International Conference on Instrumentation for High-Energy Physics,* Berkeley, 12–14/9/60 (New York: Interscience Publishers, 1961).
Instrumentation Conf. (1962)	*1962 Conference on Instrumentation for High-Energy Physics,* CERN Geneva, 16–18/7/62, *Nuclear Instruments and Methods,* **20** (1963), 1–520.
Instrumentation Conf. (1965)	*Purdue Conference on Instrumentation for High-Energy Physics,* Purdue University, 12–14/5/65, *IEEE Transactions on Nuclear Science,* NS-12 (1965), Number 4.
Instrumentation Meet. (1957)	*CERN Meeting on Nucleonic Instrumentation for High-Energy Physics,* CERN, 30/9–2/10/57, *Nuclear Instruments,* **2** (1958), 69–270.
Johnsen (1961)	K. Johnsen, 'Some Thoughts on Beam Intensity in a 300 GeV Synchrotron, 7/8/61', *Berkeley Study (1961),* 204–213.
Johnsen (1964)	K. Johnsen, 'Features of the next generation of proton accelerators', *Proceedings of the Royal Society A,* **278** (1964), 439–451.
Jones (1959)	L.W. Jones, 'Experimental Utilization of Colliding Beams', *Accelerator Conf. (1959),* 15–22.
Judd (1958)	D.L. Judd, 'Conceptual Advances in Accelerators', *Annual Review of Nuclear Science,* **8** (1958), 181–216.
Judd (1959)	D.L. Judd, 'Studies at Berkeley and MURA on Future High Energy Proton Accelerators', *Accelerator Conf. (1959),* 6–11.
Kerst (1956)	D.W. Kerst, 'Properties of an Intersecting-Beam Accelerating System', *Accelerator Conf. (1956),* 36–39.
Kerst et al. (1956)	D.W. Kerst, F.T. Cole, H.R. Crane, L.W. Jones, L.J. Laslett, T. Ohkawa, A.M. Sessler, K.R. Symon, K.M. Terwilliger, N. Vogt–Nielsen, 'Attainment of Very High Energy by Means of Intersecting Beams of Particles', *Physical Review,* **102** (1956), 590–591.
King (1961)	N.M. King, 'Theoretical Beam-Handling Studies at the Rutherford Laboratory', *Accelerator Conf. (1961),* 422–432.
Koester (1961)	L.J. Koester, 'Automatic Scanning of Spark Chamber Photographs', *Spark Chamber Symp. (1961),* 529–530.
Kowarski (1960)	L. Kowarski, 'Introduction', *Instrumentation Conf. (1960),* 223–224.
Leitner et al. (1963)	J. Leitner et al., 'Performance of the AGS Separated Beam with High Energy Kaons', *Nuclear Instruments and Methods,* **20** (1963), 42–46.
Linac Conf. (1961)	*Linear Accelerator Conference, minutes* (Upton: BNL-IA AvS-1, 1961).
Linac Conf. (1962)	*Conference on Linear Accelerators for High Energies* (Upton: BNL, 1962).
Livingston & Blewett (1962)	M.S. Livingston and J.P. Blewett, *Particle Accelerators* (New York: McGraw-Hill, 1962).
Lofgren (1956)	E.J. Lofgren, 'Bevatron Operational Experiences', *Accelerator Conf. (1956),* Vol. I, 496–509.
McCormick & Innes (1960)	B.H. McCormick and D. Innes, 'The Spiral Reader Measuring Projector and Associated Filter Program', *Instrumentation Conf. (1960),* 246–248.

Macleod (1960)	G.R. Macleod, *Digitized Protractors for the Measurement of Track Chamber Photographs* (Geneva: CERN 60-14, 1960).
Macleod (1962)	G.R. Macleod, 'The Development of Data Analysis Systems for Bubble Chambers, for Spark Chambers and for Counter Experiments', *Instrumentation Conf. (1962)*, 367–383.
Merrison (1962)	A.W. Merrison, 'A Powerful Electron Accelerator for British Physicists', *New Scientist*, **312** (1962), 312–313.
Moorhead & Krischer (1963)	W.G. Moorhead and W. Krischer, 'Future Plans for CERN HAZE', *FSD Conf. 1963*, 125–130.
Multiparameter Analyzers Conf. (1962)	*Conference on the Utilization of Multiparameter Analyzers in Nuclear Physics*, Grossinger, N.Y., 12–15/11/62 (New York: Columbia University, 1963).
MURA staff (1961a)	The MURA staff, 'Design of a 10 BeV FFAG Accelerator', *Accelerator Conf. (1961)*, 57–63.
MURA staff (1961b)	The MURA staff, 'Progress on the MURA Two-Way Electron Accelerator', *Accelerator Conf. (1961)*, 344–350.
MURA 50 MeV (1964)	'MURA 50 MeV Electron Accelerator', *The Review of Scientific Instruments*, **35** (1964), special issue.
Murray (1960)	J.L. Murray, 'Glass Cathodes in Vacuum-Insulated High-Voltage Systems', *Instrumentation Conf. (1960)*, 25–32.
Neal & Panofsky (1956)	R.B. Neal and W.K.H. Panofsky, 'The Standard Mark III Linear Accelerator and Speculations Concerning the Multi BeV Applications of Electron Linear Accelerators', *Accelerator Conf. (1956)*, 530–544.
Nuclear Data Conf. (1964)	*Automatic Acquisition and Reduction of Nuclear Data, Proceedings of a Conference*, Karlsruhe, 13–16/7/64 (Karlsruhe: Gesellschaft für Kernforschung, 1964).
Nuclear Electronics Symp. (1963)	*Nuclear Electronics. Proceedings of the International Symposium Organized by La Société Française des Electroniciens et des Radioélectriciens S.F.E.R.*, Paris, 25–27/11/63 (Paris: OECD, 1964).
Ohkawa (1958)	T. Ohkawa, 'Two-Beam Fixed Field Alternating Gradient Accelerator', *The Review of Scientific Instruments*, **29** (1958), 108–117.
O'Neill (1956a)	G.K. O'Neill, 'Storage Ring Synchrotron: Device for High-Energy Physics Research', *Physical Review*, **102** (1956), 1418–1419.
O'Neill (1956b)	G.K. O'Neill, 'The Storage Ring Synchrotron', *Accelerator Conf. (1956)*, 64–65.
O'Neill (1956c)	G.K. O'Neill, 'Storage Rings for Electrons and Protons', *Accelerator Conf. (1956)*, 125–134.
O'Neill (1959a)	G.K. O'Neill, 'Experimental Methods for Colliding Beams', *Accelerator Conf. (1959)*, 23–26.
O'Neill (1959b)	G.K. O'Neill, 'Storage Rings for Electrons and Protons', *Accelerator Conf. (1959)*, 125–134.
O'Neill (1961a)	G.K. O'Neill, 'Colliding Beam Techniques', 13/7/61, *Berkeley Study (1961)*, 74–80.
O'Neill (1961b)	G.K. O'Neill, 'Component Design and Testing for the Princeton-Stanford Colliding-Beam Experiment', *Accelerator Conf. (1961)*, 247–255.
Pentz (1967)	M.J. Pentz, *Accelerator Research at CERN, 1956–1967* (Geneva: CERN 68-9, 1967).
Pickavance (1965)	T.G. Pickavance, 'The Rutherford High Energy Physics Laboratory', *Cockcroft (1965)*, 215–235.

Pickavance (1968)	T.G. Pickavance, 'Remarks on Some Technical and Organizational Problems of Elementary-Particle Physics', in G. Puppi (ed.), *Old and New Problems in Elementary Particles* (New York: Academic Press, 1968), 240–250.
Pless (1962)	I. Pless, 'High-Energy Physics Applications of Multiparameter Analyzers', *Multiparameter Analyzers Conf. (1962)*, 25–28.
Pless et al. (1964)	I. Pless et al., 'A Precision Encoding and Pattern Recognition System (PEPR)', *HEP Conf. (1964)*, 409–413.
Ramm (1960)	C.A. Ramm, 'Some Features of Beam-Handling Equipment for the CERN Proton Synchrotron', *Instrumentation Conf. (1960)*, 289–298.
Roberts (1963)	A. Roberts, 'Spark Chambers: The State of the Art', *Nuclear Electronics Symp. (1963)*, 21–41.
Robinson (1961)	K.W. Robinson, *A Radiofrequency System for a 300 GeV Proton Synchrotron* (Pasadena: California Institute of Technology, report, CTSL-15, January 1961).
Rosenfeld (1959)	A.H. Rosenfeld, 'Digital Computer Analysis of Data from Hydrogen Bubble Chambers at Berkeley', *Accelerator Conf. (1959)*, 533–541.
Rosenfeld (1962)	A.H. Rosenfeld, 'Current Performance of the Alvarez-Group Data Processing System', *Instrumentation Conf. (1962)*, 422–434.
Rosenfeld & Humphrey (1963)	A. Rosenfeld and W.E. Humphrey, 'Analysis of Bubble Chamber Data', *Annual Review of Nuclear Science*, **13** (1963), 103–144.
Sands (1960)	M. Sands, *A Proton Synchrotron for 300 GeV* (Pasadena: California Institute of Technology, report, SL-10, September 1960).
Sands (1961)	M. Sands, 'Design Concepts for Ultra-High Energy Synchrotrons', *Accelerator Conf. (1961)*, 145–152.
Serber (1961)	R. Serber, 'The Future of High-Energy Physics', *Accelerator Conf. (1961)*, 3–5.
Shutt (1966)	R.P. Shutt, *High Energy Physics with the Brookhaven 80 inch Hydrogen Bubble Chamber* (Brookhaven: BNL-9065, 1966).
Smith (1961)	L. Smith, *The Berkeley Summer Study on High Energy Physics and 100–300 BeV Accelerators, 22/8/61* (Upton: BNL, IA-14, 1961).
Snyder et al. (1964)	J.N. Snyder et al. 'Bubble Chamber Data Analysis Using a Scanning and Measuring Projector (SMP) On-Line to a Digital Computer', *Nuclear Data Conf. (1964)*, 239–248.
Soop (1967)	K. Soop, 'Bubble Chamber Measurements and their Evaluation', *Farley (1967)*, 219–328.
Spark Chamber Meet. (1964)	*Proceedings of the Informal Meeting on Film-less Spark Chamber Techniques and Associated Computer Use,* CERN, 3–6/3/64 (Geneva: CERN 64-30, 1964).
Spark Chamber Symp. (1961)	*Symposium on Spark Chambers,* Argonne National Laboratory, 7/2/61, *The Review of Scientific Instruments,* **32** (1961), 479–531.
Symon (1961)	K.R. Symon, *A Proposal for Accelerating Antiprotons, 10/8/61* (Upton: BNL, IA-10, 1961).
Symon & Tollestrup (1961)	K.R. Symon and A.V. Tollestrup, *Acceleration of Antiprotons, 25/8/61* (Upton: BNL, IA KRS/AVT-1).
Teng (1961)	L.C. Teng, *Matched Long Straight Sections for Alternating Gradient Synchrotron, 19/7/61* (Upton: BNL, IA LCT-4).
Theory Conf. (1961)	*International Conference on Theoretical Aspects of Very High-Energy Phenomena, held at CERN, 5–9/6/61* (Geneva: CERN 61-22, 1961).

816 *Bibliography*

Ticho (1959)	H.K. Ticho, 'The Production, Transport and Separation of Beams', *Accelerator Conf. (1959)*, 387-396.
Tollestrup (1961)	A.V. Tollestrup, *RF Synchronization during Transfer in the Cascade Synchrotron* (Pasadena: California, Institute of Technology, report, CTSL-21, March 1961).
Track Data Meet. (1962)	*Informal Meeting on Track Data Processing*, CERN, 19/7/62 (Geneva: CERN 62-37, 1962).
US Panels (1965)	Joint Committee on Atomic Energy, Congress of the United States, *High Energy Physics Program: Report on National Policy and Background Information* (Washington: US Government Printing Office, 1965).
Van Hove (1961a)	L. Van Hove, 'Summary Lecture', *Theory Conf. (1961)*, 407-423.
Van Hove (1961b)	L. Van Hove, 'The Role of High Energy Accelerators in Particle Physics', *Accelerator Conf. (1961)*, 6-11.
von Dardel et al. (1957)	G. von Dardel, Y. Goldschmidt-Clermont, and F. Iselin, 'The Present State of the Instrument for the Evaluation of Photographs and Future Developments', *Instrumentation Meet. (1957)*, 154-163.
Walker (1961)	R.L. Walker, *Beam Transfer in the Cascade Synchrotron* (Pasadena: California Institute of Technology, report CTSL-16, January 1961).
White et al. (1962)	H.S. White et al., 'Preliminary Operating Experience with Hough-Powell Device Programs', *Instrumentation Conf. (1962)*, 393-400.
White (1965)	H.S. White, 'Status and Future Plans for LRL Flying Spot Digitizer', *Instrumentation Conf. (1965)*, 270-278.

Secondary Literature

400 GeV (1977a)	*Speeches at the Inauguration of the 400 GeV proton synchrotron, 7 May 1977* (Geneva: CERN, 1977).
400 GeV (1977b)	*Le super synchrotron à protons de 400 GeV* (Genève: CERN/SIS-PU 77-13, 1977).
Abragam et al. (1987)	A. Abragam et al., *Final Report of the CERN Review Committee* (Geneva: CERN/1675, 1987).
Allison (1971)	G.T. Allison, *Essence of Decision: Explaining the Cuban Missile Crisis* (Boston: Little, Brown & Co., 1971).
Allison & Morris (1975)	G.T. Allison and F.A. Morris, 'Armaments and Arms Control: Exploring the Determinants of Military Weapons', *Daedalus,* **104** (summer 1975), 99-130.
Alvarez (1970)	L.W. Alvarez, 'Recent developments in particle physics', Conversi (ed.), *Evolution of particle physics* (New York, 1970).
Alvarez (1987)	L.W. Alvarez, *Luis W. Alvarez: Adventures of a Physicist* (New York: Basic Books, 1987).
Amaldi (1977)	E. Amaldi, 'Personal Notes on Neutron Work in Rome in the 30's and Post-war European Collaboration in High Energy Physics', *Scuola Fermi* (1977), 294-351.
Armacost (1969)	M.H. Armacost, *The Politics of Weapons Innovation: The Thor-Jupiter Controversy* (New York: Columbia University Press, 1969).
Armenteros et al. (1978)	R. Armenteros et al. (eds.), *Physics from Friends, Papers Dedicated to Ch. Peyrou on his 60th Birthday,* (Geneva: Multi Office S.A., 1978).

Baracca & Bergia (1975)	A. Baracca and S. Bergia, *La spirale delle alte energie* (Milano: Studi Bompiani, 1975).
Beaver (1986)	D. de B. Beaver, 'Collaboration and Teamwork in Physics', *Czechoslovak Journal of Physics,* **B39** (1986), 14–18.
Belloni & Dilworth (1988)	L. Belloni and C. Dilworth, 'A Story of a European Postwar Collaboration with Nuclear Emulsions', International Conference on the Restructuring of Physical Sciences, 1945–1960, Rome, 19–23 September 1988.
Bernoux (1985)	Ph. Bernoux, *La Sociologie des organisations* (Paris: Seuil, 1985).
Bernstein (1988)	B.J. Bernstein, 'Four Physicists and the Bomb: The Early Years, 1945–1950', *Historical Studies in the Physical and Biological Sciences,* **18:2** (1988), 231–263.
Bianchi-Streit et al. (1984)	M. Bianchi-Streit et al., *Economic Utility Resulting from CERN Contracts, (Second Study)* (Geneva: CERN 84-14, 1984).
Big Science (1988)	*Big Science: The Growth of Large-Scale Research,* August 25–27, 1988, Workshop, Stanford, working drafts.
Bloor (1976)	D. Bloor, *Knowledge and Social Imagery* (London: Routledge & Kegan Paul, 1976).
Braudel (1958)	F. Braudel, 'La longue durée', *Annales ESC,* **4** (1958), 725–753.
Bromberg (1982)	J.L. Bromberg, *Fusion.Science, Politics and the Invention of a New Energy Source* (Cambridge, MIT Press, 1982).
Casimir (1975)	H.B.G. Casimir, 'Big Science and Technological Progess', Speech presented on the 25th Anniversary of CERN, 23/6/79.
Casimir (1983)	H.B.G. Casimir, *Haphazard Reality: Half a Century of Science* (New York: Harper and Row, 1983).
Cohen-Tannoudji & Spiro (1986)	G. Cohen-Tannoudji et M. Spiro, *La Matière-Espace-Temps, La logique des particules élémentaires* (Paris: Fayard, 1986).
Collins (1974)	H.M. Collins, 'The TEA Set: Tacit Knowledge and Scientific Networks', *Science Studies,* **4** (1974), 165–186.
Collins (1985)	H.M. Collins, *Changing Order, Replication and Induction in Scientific Practice* (London: Sage, 1985).
Coulam (1977)	R.F. Coulam, *Illusions of Choice. The F-111 and the Problem of Weapons Acquisition Reform* (Princeton: Princeton University Press, 1977).
Crozier & Friedberg (1977)	M. Crozier and E. Friedberg, *L'acteur et le système* (Paris: Seuil, 1977).
Crozon (1987)	M. Crozon, *La matière première, La rercherche des particules fondamentales et de leurs interactions* (Paris: Seuil, 1987).
Cushing (1982)	J.T. Cushing, 'Models and Methodologies in Current Theoretical High-Energy Physics', *Synthese,* **50** (1982), 5–101.
Darmon (1981)	G. Darmon, *Psychosociologie d'une décision en science lourde, l'Institut franco-allemand de Radio-Astronomie Millimétrique* (Paris: EHESS, Thèse de 3e cycle, 1981).
Darmon & Lemaine (1982)	G. Darmon et G. Lemaine, 'Etude d'une décision en science lourde: le cas de l'Institut de Radio-Astronomie Millimétrique franco-allemand', *Information sur les Sciences Sociales,* **21** (1982), 847–872.
Dufour (1981)	J.M. Dufour, 'Le LEP, avenir de la physique européenne des particules', *Annuaire français de droit international,* **28** (1981), 653–664.
Economic Research (1962)	*The Rate and Direction of Inventive Activity: Economic and Social Factors,* A report of the National Bureau of Economic Research, NY (Princeton: Princeton University Press, 1962).
Edge & Mulkay (1976)	D.O. Edge et M.J. Mulkay, *Astronomy transformed. The emergence of radioastronomy in Britain* (London: Wiley Inter Science, 1976).

Faire de l'Histoire (1974)	J. Le Goff et P. Nora (eds.), *Faire de l'histoire* (Paris: Gallimard, 1974).
Forman (1987)	P. Forman, 'Behind quantum electronics: National security as basis for physical research in the United States, 1940-1960', *Historical Studies in the Physical and Biological Sciences,* **18(1)** (1987), 149-229.
Freeman & Young (1965)	C. Freeman and A. Young, *The Research and Development Effort in Western Europe, North America and the Soviet Union* (Paris: OECD, 1965).
Galison (1985a)	P. Galison, 'Bubble chambers and the experimental workplace', in O. Hannaway and P. Achinstein (eds.), *Observation, Experiment and Hypothesis in Modern Physical Science* (Cambridge: Cambridge University Press, 1985), 309-373.
Galison (1985b)	P. Galison, 'Instruments, Arguments and the Experimental Workplace in High-Energy Physics', *Hist. of Science Conf. (1985),* Session Pn.
Galison (1987a)	P. Galison, 'Bubbles, Sparks and the Post-War Laboratory', *Invited paper, International Symposium on Particle Physics in the 1950's,* May 1-4, 1985, Fermi National Accelerator Laboratory.
Galison (1987b)	P. Galison, *How Experiments End* (Chicago: University of Chicago Press, 1987).
Galison, Hevly & Lowen (1988)	P. Galison, B. Hevly and R. Lowen, 'Controlling the Monster: Stanford and the Growth of Physics Research, 1935-1962', *Big Science (1988),* 41 pp.
Gaston (1973)	J. Gaston, *Originality and Competition in Science, A Study of the British High Energy Physics Community* (Chicago: Chicago University Press, 1973).
Giard (1985)	L. Giard, 'L'Histoire des Sciences, une histoire singulière', *Recherches de Science Religieuse,* **73** (1985), 355-380.
Gibbons (1970)	M. Gibbons, 'The CERN 300 GeV Accelerator: A Case Study in the Application of the Weinberg Criteria', *Minerva,* **8** (1970), 180-191.
Godement (1979)	R. Godement, 'Aux sources du modèle scientifique américain', *La Pensée,* **201** (octobre 1978) 33-69, **203** (février 1979), 95-122, **204** (avril 1979), 86-110.
Goldsmith & Shaw (1977)	M. Goldsmith and E. Shaw, *Europe's Giant Accelerator, The Story of the CERN 400 GeV Proton Synchrotron* (London: Taylor & Francis, 1977).
Greenberg (1971)	D.S. Greenberg, *The Politics of Pure Science* (New York: A Plume Book, 1971).
Gummett (1980)	P. Gummett, *Scientists in Whitehall* (Manchester: Manchester University Press, 1980).
Hagstrom (1964)	W.O. Hagstrom, 'Traditional and Modern Forms of Scientific Teamwork', *Administrative Science Quarterly,* **9** (1964), 241-263.
Hamon & Torres (1987)	M. Hamon and F. Torres, *Mémoires d'avenir, l'histoire dans l'entreprise* (Paris: Economica, 1987).
Heilbron et al. (1981)	J.L. Heilbron, R.W. Seidel and B.R. Wheaton, 'Lawrence and his Laboratory: Nuclear Science at Berkeley', *LBL News Magazine,* **6,** No 3 (1981).
Hertzog (1977)	Y. Hertzog, *Le CERN, Structure et Fonctionnement,* Thèse pour le doctorat d'Etat, 1977, Université de Droit, d'Economie et de Sciences Sociales de Paris (Paris II).
Hine & Taylor (1976)	M.G.N. Hine and C.S. Taylor, 'The Technology of High-Energy Physics', *Physics in Technology,* March 1976, 67-76.
High Tech. Meet. (1974)	*Meeting on Technology Arising from High-Energy Physics,* CERN, 24-26/4/74 (Geneva: CERN 74-9, 1974), 2 vols.

Hist. of Science Conf. (1985)	*XVII International Congress of History of Science,* University of California, Berkeley, 31/7-8/8/85 (Berkeley: Office for History of Science and Technology, 1985).
Histoire Orale (1980)	*Problèmes de méthode en histoire orale, 20/6/80, Table ronde* (Paris: IHTP, 1980).
Histoire Orale (1986)	*Questions à l'histoire orale, 20/6/86, Table ronde* (Paris: Les cahiers de l'IHTP, 4, 1987).
History of CERN (1987)	A. Hermann, J. Krige, U. Mersits and D. Pestre, *History of CERN. Volume I. Launching the European Organization for Nuclear Research* (Amsterdam: North-Holland, 1987).
History of ZGS (1980)	J.S. Day et al. (eds.), *History of the ZGS,* AIP Conference Proceedings (New York: AIP, 1980).
Hoch (1988)	P. Hoch, 'Cristallization of a Strategic Alliance: Big Physics and the Military in the 1940s' *Program, Papers and Abstracts,* Joint Conference (BSHS and AHSS), Manchester, 11-15 July 1988, 366-374.
Hoddeson (1983)	L. Hoddeson, 'Establishing KEK in Japan and Fermilab in the US: Internationalism, Nationalism and High Energy Accelerators', *Social Studies of Science,* **13** (1983), 1-48.
Holton (1972)	G. Holton (ed.), *The Twentieth-Century Sciences* (New York: W.W. Norton, 1972).
Holton (1985)	G. Holton, 'Les hommes de science ont-ils besoin d'une philosophie', *Le Débat,* **35** (1985), 116-138.
Irvine & Martin (1984)	B.R. Martin and J. Irvine, 'CERN: Past Performance and Future Prospects - I. CERN's Position in World High-Energy Physics', *Research Policy,* **13** (1984), 183-210; J. Irvine and B.R. Martin, *'id.* -II. The Scientific Performance of the CERN Accelerators', *id.,* 247-284.
Irvine & Martin (1985)	J. Irvine and B.R. Martin, 'Basic Research in the East and West: a Comparison of the Scientific Performance of High-Energy Physics Accelerators', *Social Studies of Science,* **15** (1985), 293-341.
ISR (1974)	*Anneaux de stockage à intersections* (Genève: CERN/PIO 74-7, 1974).
ISR (1984)	M. Jacob and K. Johnsen, *A Review of Accelerator and Particle Physics at the CERN Intersecting Storage Rings* (Geneva: CERN 74-13, 1984).
Jamous (1969)	H. Jamous, *Sociologie de la décision* (Paris: Editions du CNRS, 1969).
Jungk (1968)	R. Jungk, *The Big Machine* (New York: Scribners, 1968).
Jurdant & Olff-Nathan (1981)	B. Jurdant et J. Olff-Nathan, *Le LEP entre la physique et les physiciens* (Strasbourg: GERSULP, texte ronéotypé, 1981).
Kevles (1978)	D. Kevles, *The Physicists* (New York: A.A. Knopf, 1978).
Kevles (1988)	D. Kevles, 'K_1, S_2: Korea, Science and the State', *Big Science (1988),* 39pp.
Klein (1962)	B.H. Klein, 'The Decision Making Problem in Development', *Economic Research (1962),* 477-507.
Knorr (1981)	K.D. Knorr-Cetina, *The Manufacture of Knowledge: An Essay on the Constructivist and Contextual Nature of Science* (Oxford: Pergamon, 1981).
Knorr & Mulkay (1983)	K.D. Knorr-Cetina and M. Mulkay, *Science Observed: Perspectives on the Social Study of Science* (London: Sage, 1983).
Kowarski (1967)	L. Kowarski, *An Observer's Account of User Relations in the U.S. Accelerator Laboratories* (Geneva: CERN 67-4, 1967).
Kowarski (1977)	L. Kowarski, 'New Forms of Organization in Physical Research after 1945', *Scuola Fermi (1977),* 370-401.

Krige (1988)	J. Krige, 'The Installation of High-Energy Accelerators in Britain after the War. Big Equipment but not «Big Science»', paper presented at the International Conference on the Restructuring of Physical Sciences in Europe and the United States, 1945-1960, Rome, 19-23 September 1988.
Krige (1989)	J. Krige, 'The CERN Beam Transport Programme in the Early 1960s', in F.A.J.L. James (ed.), *Laboratories: The Place of Experiment, Past and Present* (London: Macmillan, 1989).
Krige & Pestre (1986)	J. Krige and D. Pestre, 'The Choice of CERN's First Large Bubble Chambers for the Proton Synchrotron (1957-1958)', *Historical Studies in the Physical Sciences,* **16**:2 (1986), 255-279.
Laqueur (1982)	W. Laqueur, *Europe Since Hitler. The Rebirth of Europe* (Penguin Books, 1982).
Latour (1985)	Bruno Latour, 'Les "vues" de l'esprit', *Culture Technique,* **14** (1985), 4-29.
Latour & Woolgar (1979)	B. Latour and S. Woolgar, *Laboratory Life, The Social Construction of Scientific Facts* (Beverly Hills: Sage, 1979).
Lécuyer (1978)	B.-P. Lécuyer, 'Bilan et perspective de la sociologie de la science dans les pays occidentaux', *Archives Européennes de Sociologie,* **19** (1978), 257-336.
Lemaine et al. (1982)	G. Lemaine, G. Darmon, S. El Nemer, *Noopolis, Les laboratoires de recherche fondamentale: de l'atelier à l'usine* (Paris: CNRS, 1982).
Le Monde	Le Monde, *L'Histoire au jour le jour. Tome I. Les années froides (1944-1954). Tome II. Le temps des ruptures (1955-1962). Tome III. Les printemps éphémères (1963-1973)* (Paris: Le Monde, undated).
Leslie (1987)	S.W. Leslie, 'Playing the Education Game to Win: The Military and Interdisciplinary Research at Stanford', *Historical Studies in the Physical and Biological Sciences,* 18:1 (1987), 55-88.
Levi (1984)	P. Levi, *The Periodic Table* (New York: Schocken Books, 1984).
Lindblom (1959)	C.E. Lindblom, 'The Science of "Muddling Through"', *Public Administration Review,* **19** (1959), 79-88.
MacKenzie & Spinardi (1988)	D. MacKenzie and G. Spinardi, 'The Shaping of Nuclear Weapon System Technology: US Fleet Ballistic Missile Guidance and Navigation', *Social Studies of Science,* **18**(3) (1988), 419-463; **18**(4) (1988), 581-624.
Management (1957)	*The Direction of Research Establishments,* National Physical Laboratory, Symposium held on 26-28/9/56 (London: Her Majesty's Stationery Office, 1957).
Management (1963)	F.E. Kast and J.E. Rosenzweig (ed.), *Science, Technology and Management* (New York: McGraw-Hill, 1963).
March & Olsen (1975)	J.G. March and J.P. Olsen, 'The Uncertainty of the Past: Organizational Learning under Ambiguity', *European Journal of Political Research,* **3** (1975), 147-171.
Marrou (1961)	H.I. Marrou, 'Comment comprendre le métier d'historien', *L'Histoire et ses méthodes* (Paris: Gallimard, Encyclopédie de la Pleiade, 1961), 1465-1540
Marshall & Meckling (1962)	A.W. Marshall and W.H. Meckling, 'Predicability of the Costs, Time, and Success of Development', *Economic Research (1962),* 461-475.
Massey & Robins (1986)	H. Massey and M.O. Robins, *History of British Space Science* (Cambridge: Cambridge University Press, 1986).
Matalon (1986)	B. Matalon, 'Sociologie de la Science et Relativisme', *Revue de Synthèse,* **107**, 4e série (1986), 267-290.
Morris (1985)	P.W.G. Morris, 'The Initiation, Assessment, Securing, and Accomplishment of Major Projects', *Technology in Society,* **7** (1985), 31-45.

Morrison (1978)	D.R.O. Morrison, 'The Sociology of International Scientific Collaborations', *Armenteros et al. (1978),* 351-365.
Mulkay & Edge (1973)	M.J. Mulkay and D.O. Edge, 'Cognitive, Technical and Social Factors in the Growth of Radio Astronomy', *Information sur les Sciences Sociales,* **12** (1973), 25-61.
Needell (1983)	A.A. Needell, 'Nuclear Reactors and the Founding of BNL', *Historical Studies in the Physical Sciences,* **14** (1983), 93-122.
Nimrod (1979)	J. Litt (ed.), *Nimrod. The 7 GeV Proton Synchrotron,* Proceedings of a Nimrod Commemoration Evening, Rutherford Laboratory, 27 June 1978 (Didcot: Science Research Council, 1979).
Nyberg & Zetterberg (1977)	S. Nyberg and K. Zetterberg, *Sweden and CERN, the Decision-Making Process, 1949-1964* (Stockholm: FEK, report 9, 1977).
OECD (1970)	*Inflation. The Present Problem* (Paris: OECD, 1970).
OECD (1971)	*Science Growth and Society. A New Perspective* (Paris: OECD, 1971).
OECD (1984)	*OECD Science and Technology Indicators. Resources Devoted to R & D* (Paris: OECD, 1984).
Pandore (1982)	*La Science telle qu'elle se fait* (Paris: Pandore, 1982).
Pandore (1985)	M. Callon and B. Latour (eds.), *Les scientifiques et leurs alliés* (Paris: Pandore, 1985).
Panofsky (1983)	W.K.H. Panofsky, 'The Evolution of SLAC and its Program' *Physics Today,* **36** (10) (1983), 34-41.
Pestre (1984a)	D. Pestre, *Physique et Physiciens en France, 1918-1940* (Paris: Editions des archives contemporaines, 1984).
Pestre (1984b)	D. Pestre, 'L'Organisation européenne pour la recherche nucléaire (CERN): un succès politique et scientifique?' *Vingtième Siècle, revue d'histoire,* **4** (1984), 65-76.
Pestre (1988a)	D. Pestre, 'Comment se prennent les décisions de très gros équipements dans le laboratoires de science lourde contemporaires' *Revue de Synthèse,* **IV**, 1 (1988), 97-130.
Pestre (1988b)	D. Pestre, 'The Creation of CERN in the Early 50s: Chance or Necessity', paper presented at the International Conference on the Restructuring of Physical Sciences in Europe and the United States, 1945-1960, Rome, 19-23 September, 1988.
Pestre (1989a)	D. Pestre, 'Un changement qualitatif dans les relations scientifiques franco-allemandes', *Revue d'Allemagne,* **XXI** (1) (1989), 415-431.
Pestre (1989b)	D. Pestre, '«Monsters» and Colliders in 1961: the First Debate at CERN on Future Accelerators', in F.A.J.L. James (ed.), *Laboratories: The Place of Experiment, Past and Present* (London: Macmillan, 1989).
Pickering (1984)	A. Pickering, *Constructing Quarks, a Sociological History of Particle Physics* (Edinburgh: Edinburgh University Press, 1984).
Pickering (1985)	A. Pickering, 'From Field Theory to Phenomenology: The History of Dispersion Relations', *Invited Paper, International Symposium on Particle Physics in the 1950's,* May 1-4, 1985, Fermi National Accelerator Laboratory.
Pickering (1989)	A. Pickering, 'Pragmatism in Particle Physics. Scientific and Military Interests in the Postwar United States', in F.A.J.L. James (ed.), *Laboratories: The Place of Experiment, Past and Present* (London: Macmillan, 1989).
Polach (1964)	J.G. Polach, *EURATOM, its Background, Issues and Economic Implications* (New York: Oceana Publications, 1964).

Prost (1988)	A. Prost, 'Les origines de la politique de la recherche en France (1939-1958)', *Cahiers pour l'Histoire du CNRS* (1988-1), 41-62.
Rigden (1987)	J.S. Rigden, *Rabi: American Physicist* (New York: Basic Books, 1987).
Roland (1985)	A. Roland, 'Technology and War. A Bibliographic Essay', in M.R. Smith (ed.), *Military Enterprise and Technical Change* (Cambridge: MIT Press, 1985), 347-379.
Rosenberg (1982)	N. Rosenberg, *Inside the Black Box: Technology and Economics* (Cambridge: Cambridge University Press, 1982).
Rosholt (1966)	R.L. Rosholt, *An Administrative History of NASA, 1958-1963* (Washington: National Aeronautics and Space Administration, 1966).
Rudwick (1985)	M.J. Rudwick, *The Great Devonian Controversy* (Chicago: University of Chicago Press, 1985).
Russo (1986)	A. Russo, 'Science and Industry in Italy between the two World Wars', *Historical Studies in the Physical Sciences,* **16:2** (1986), 281-320.
Salomon (1970)	J.J. Salomon, *Science et Politique* (Paris: Seuil, 1970).
Salomon (1986)	J.J. Salomon, *Le gaulois, le cow-boy et le samourai, la politique française de la technologie* (Paris: Economica, 1986).
Sanford (1976)	J.R. Sanford, 'The Fermi National Accelerator Laboratory', *Annual Review of Nuclear Science,* **26** (1976), 151-198.
Sapolsky (1988)	H.M. Sapolsky, 'Military Support for Academic Research in the United States', paper prepared for the joint meeting of the U.S. and British History of Science Societies, Manchester, England, July 1988.
Schilling (1961)	W.R. Schilling, 'The H-Bomb Decision. How to Decide Without Actually Choosing', *Political Science Quarterly,* **76** (1961), 24-46.
Schmid & Dufour (1976)	F. Schmid and J.M. Dufour, 'Le CERN, exemple de coopération scientifique internationale', *Journal du Droit International,* **1976 (1)**, 46-103.
Schmied (1975)	H. Schmied, *A Study of Economic Utility Resulting from CERN Contracts* (Geneva: CERN 75-5, 1975).
Schopper (1985)	H. Schopper, 'CERN, A Center of Science and High Technology', talk given at 'The Entrepreneurial Imperative', Geneva, 17/6/85.
Schwarz (1979)	M. Schwarz, 'European Policies on Space Science and Technology 1960-1978', *Research Policy,* **8** (1979), 204-243.
Schweber (1985)	S. Schweber, 'Some Reflections on the History of Particle Physics in the 1950's', Invited paper, International Symposium on Particle Physics in the 1950's, May 1-4, 1985, Fermi National Accelerator Laboratory.
Schweber (1986)	Sam Schweber, 'The Empiricist Temper Regnant: Theoretical Physics in the United States, 1920-1950', *Historical Studies in the Physical and Biological Sciences,* **17** (1986), 55-98.
Scuola Fermi (1977)	*Recondiconti della Scuola Internazionale di Fisica Enrico Fermi, LVII Corso* (New York: Academic Press, 1977).
Seidel (1983)	R.W. Seidel, 'Accelerating Science: The Postwar Transformation of the Lawrence Radiation Laboratory', *Historical Studies in the Physical Sciences,* **13** (1988), 375-400.
Seidel (1986)	R.W. Seidel, 'The Political Economy of High Energy Physics, Invited Paper, International Symposium on Particle Physics in the 1950's, May 1-4, 1985, Fermi National Accelerator Laboratory.
Seidel (1988)	R.W. Seidel, 'The Origins of the Lawrence Berkeley Laboratory', *Big Science (1988),* 22pp.

Sfez (1981)	L. Sfez, *Critique de la décision* (Paris: Presses de la Fondation Nationale des Sciences Politiques, 1981).
Six & Artru (1982)	J. Six and S. Artru, 'An Essay Chronology of Particle Physics until 1965', *Journal de Physique,* Colloque C8, supplément au no. 12, **43** (1982), C8-465-C8.496.
Sked & Cook (1979)	A. Sked and C. Cook, *Post-war Britain. A Political History* (Penguin Books, 1979).
Space (1984)	*Europe Two Decades in Space, 1964-1984. Recollections by some of the Principal Pioneers,* Report ESA SP-1060 (Noordwijk: ESA Scientific and Technical Publications Branch, 1984).
Swatez (1970)	G.M. Swatez, 'The Social Organization of a University Laboratory', *Minerva,* **8** (1970), 36-58.
Taubes (1987)	G. Taubes, *Nobel Dreams. Power, Deceit and the Ultimate Experiment* (New York: Random House, 1987).
Traweek (1988)	S. Traweek, *Beamtimes and Lifetimes, the World of High Energy Physicists* (Cambridge: Harward University Press, 1988).
Vogt (1985)	E. Vogt, 'Accelerators—Instruments and Symbols for Power', *IEEE Transactions on Nuclear Science,* **NS-32** (1985), 3834-3837.
Weart (1988)	S. Weart, *Nuclear Fear* (Cambridge: Harvard University Press, 1988).
Weinberg (1972)	A.M. Weinberg, 'Scientific Teams and Scientific Laboratories', *Holton (1972),* 423-442.
Weisskopf (1988)	V.F. Weisskopf, *The Privilege of Being a Physicist* (New York: Freeman, 1988).
Westfall (1988)	C. Westfall, *The First 'Truly National Laboratory': The History of Fermilab,* Ph.D. Thesis, Michigan State University, 1988.
Westfall (1989)	C.L. Westfall, 'The Site Contest for Fermilab', *Physics Today* (1989), 44-52.
Willson (1957)	D.R. Willson, 'Budgets and Administrative Controls', *Management (1957),* 5.p1-5.p13.
Wilson (1972)	R.R. Wilson, 'My Fight Against Team Research', *Holton (1972),* 468-479.
Wise (1985)	George Wise, 'Science and Technology', *Osiris,* **1** (1985), 229-246.
York (1987)	H.F. York, *Making Weapons, Talking Peace* (New York: Basic Books, 1987).
Zilverschoon (1974)	C.J. Zilverschoon, 'Some Aspects of the Realization of High-Energy Projects at CERN', *High Tech. Meet.* (1974), vol. 1, 27-33.

Appendices

APPENDIX 1

Organigramme of CERN, 1955–1960, CERN divisions and their leaders

Proton Synchrotron	Synchro-cyclotron	Theoretical Studies	Scientific and Technical Services	Site and Buildings	Administration
PS	SC	TH	STS	SB	ADM
J.B. Adams (UK), 10/54–	C.J. Bakker (NL), 10/54–6/55 W. Gentner (FRG), 6/55–10/58 G. Bernardini (I), dir. of research, 1/58–10/58 In 1959 and 1960, Bernardini, Gentner and Preiswerk share de facto the direction of SC division	C. Møller (DK) in Copenhagen, 10/54–10/57 B. Ferretti (I) in Geneva, 10/57–4/59 M. Fierz (CH), 4/59–6/60 L. Van Hove (B), 6/60–	L. Kowarski (F), 10/54–	P. Preiswerk (CH), 10/54–10/58 F.A.R. Webb (UK), acting div. leader, 9/59– C. Mallet (F), 9/59–	S.A. ff. Dakin (UK), 10/54–6/55 J. Richemond (F), 6/55–10/58 S.A. ff. Dakin (UK), 10/58–

Organigramme of CERN, 1961–1965, CERN divisions and their leaders

PS Machine	SC Machine	Nuclear Physics (Electronics, emulsions)	Track Chambers	Theory	Nuclear Physics Apparatus
MPS	MSC	NP	TC	TH	NPA

Data Handling	Accelerator Research	Engineering	Site and Buildings	Finance	Administration (61-63) Personnel (64-65)
DD	AR	ENG	SB	FIN	ADM/PE
L. Kowarski (F), 1/61–8/64 G.R. Macleod (UK), acting div. leader, 9/64–6/65 div. leader, 6/65–	A. Schoch (FRG), K. Johnsen (N), C.J. Zilverschoon (NL), (in rotation)	F. Grütter (CH), 1/61–12/63 division abolished in Dec. 1963	C. Mallet (F), 1/61–12/65	C. Tièche (CH), 1/61–	S.A. ff. Dakin (UK), 1/61–9/63 G.H. Hampton (UK), 9/63–12/63 G. Ullmann (FRG), PE, 1/64–

P. Germain (B), 1/61–12/64 P.H. Stanley (UK), 1/65–	P. Lapostolle (F), 1/61–12/63 G. Brianti (I), 1/64–	P. Preiswerk (CH), 1/61– W. Paul (FRG), co-division leader, 9/64–	Ch. Peyrou (F), 1/61–	L. Van Hove (B), 1/61–	C.A. Ramm (UK), 1/61–

APPENDIX 2

CERN Director-Generals and Members of the Directorate

Director-Generals

7/10/1954–31/8/1955	Félix Bloch
1/9/1955–23/4/1960	Cornelis J. Bakker
24/4/1960–31/7/1961	John B. Adams
1/8/1961–31/12/1965	Victor F. Weisskopf
1/1/1966–	Bernard Gregory

Directorate

Director of Research	G. Bernardini (I)	1/1/1961–31/8/1962
	G. Puppi (I)	1/9/1962–31/8/1964
	B. Gregory (F)	1/9/1964–31/12/1965
Director of Applied Physics	M.G.N. Hine (UK)	1/1/1961–31/12/1965
Dirrector of Administration	S.A. ff. Dakin (UK)	1/1/1961–31/8/1963
	G.H. Hampton (UK)	1/9/1963–31/12/1965
Director of Technical Management	P. Germain (B),	1/1/1963–31/12/1965

APPENDIX 3

Member States and their percentage contributions to the CERN basic programme, at three-year intervals

	1955	1958	1961	1964
Austria			1.87	1.96
Belgium	4.88	4.89	4.02	3.85
Denmark	2.48	2.23	1.93	2.09
FR Germany	17.70	18.27	18.92	22.86
France	23.84	22.26	20.57	18.66
Greece	0.97	1.08	1.12	0.60
Italy	10.20	10.61	9.78	10.83
Netherlands	3.68	3.78	3.73	3.94
Norway	1.79	1.72	1.56	1.48
Spain			4.16	1.68
Sweden	4.98	4.85	4.10	4.25
Switzerland	3.71	3.48	3.19	3.20
United Kingdom	23.84	25.00	24.40	24.60
Yugoslavia	1.93	1.83	0.65	

APPENDIX 4

Senior Office Bearers in the Council and its Committees

	Years	Council	Finance Committee	Scientific Policy Committee
President or Chairman	1954–57	Sir Ben Lockspeiser (UK)	J. Willems (B)	W. Heisenberg (FRG)
President or Chairman	1958–60	F. de Rose (F)	J.H. Bannier (NL)	E. Amaldi (I)
Vice-President of Council		Heisenberg, Willems		
President or Chairman	1961–63	J. Willems (B)	A. Hocker (FRG), 1961; G.W. Funke (S), 1962–64	C.F. Powell (UK)
Vice-President of Council		Amaldi, Bannier		
President or Chairman	1964–66	J.H. Bannier (NL)	W. Schulte-Meermann (FRG), 1965–1967	L. Leprince-Ringuet (F)
Vice-President of Council		Sir Harry Melville (UK), Funke (S)		

APPENDIX 5

Members of the SPC, up to 1965

H. Alfvén (S)	1954–1960
G. Bernardini (I)	1954–1957, 1964–
P.M.S. Blackett (UK)	1954–1959
Niels Bohr (DK)	1954–1962
Sir John Cockcroft (UK)	1954–1961
W. Heisenberg (FRG)	1954–1961, Chairman 1954–57
L. Leprince-Ringuet (F)	1954– , Chairman 1964–65
P. Scherrer (CH)	1954–1963
E. Amaldi (I)	1958– , to replace Bernardini, Chairman 1958–60
C.F. Powell (UK)	1959– , to replace Blackett, Chairman 1961–63
C. Møller (DK)	1959–
F. Perrin (F)	1960– , to replace Alfvén
W. Gentner (FRG)	1961– , to replace Heisenberg
J.B. Adams (UK)	1961– , to replace Sir John Cockcroft
W. Jentschke (FRG)	1964–
S.A. Wouthuysen (NL)	1964–

plus the Chairmen of the Emulsion, Track Chamber and Electronic Experiments Committees.

APPENDIX 6

V. Weisskopf's table of some important research results obtained at CERN or elsewhere in Europe in connection with CERN within the period 1960-1965.

Underlined results are those which were either obtained for the first time in Europe, or more actively and thoroughly investigated in Europe than elsewhere.

1. **New particles** (antiparticles, excited states, resonances) discovered and/or analysed. (The first number indicates the mass in MeV, the second the isospin, the third the mechanical spin.)

    (a) Baryons

    Antistates: $\overline{\Xi^-}$ (1321, $1/2$, $1/2$) $\overline{N}^*$ (1238, $3/2$, $3/2$)
    $\overline{Y}^*$ (1385, 1, $3/2$), $\overline{Y}^*$ (1405, 0, $1/2$), $\overline{Y}^*$ (1520, 0, $3/2$)

    Excited states: $\underline{Y^* (1765, 1, 5/2)}$ $\underline{Y^* (1815, 0, 5/2)}$
    $\underline{\Xi^* (1820, 1/2, ?)}$ $\underline{\Xi^* (1935, ?, ?)}$

    (b) Mesons $f (1253, 0, 2)$ $\underline{K^* (1405, 1/2, 2)}$
    $\underline{E (1420, 0, ?)}$ $\underline{D (1286, 0, 1)}$
    $\underline{A_1 (1072, 1, 1)}$ $\underline{A_2 (1324, 1, 2)}$
    $\underline{C (1215, ?, ?)}$ $\underline{(K^+K^+) (1280, ?, ?)}$

    <u>properties of mesons</u>

2. **Strong interactions**

    (a) Theory

    <u>Regge poles</u>
    Diffraction scattering
    Pomeranchuk theorem
    <u>Peripheral model</u>
    (meson exchange)
    General field theory

    (b) Experiment

    <u>Shrinking of diffraction peak</u>
    <u>Quasi-elastic scattering</u>
    Total cross-sections
    <u>Peripheral processes</u>
    <u>Charge exchange collisions</u>

3. **Weak interactions**

    (a) Theory

    <u>Cabibbo angle</u>
    Symmetries
    CP violation
    <u>(5th force)</u>

    (b) Experiment

    Neutrino $\begin{cases} \nu_e \neq \nu_\mu \\ \text{lepton conservation} \\ \underline{M_W > 2 \text{ GeV}} \\ \text{form factors} \end{cases}$

    Σ decay $\begin{cases} \underline{(\Sigma - \Lambda) \text{ parity}} \\ e/\mu \text{ ratio} \\ \Delta S/\Delta Q = 1 \end{cases}$

    K^0 decay $\Delta S/\Delta Q = 1$
    Λ decay C_V/C_A
    Ξ decay $(\Delta I = 1/2)$
    π^+ decay $\underline{(\pi^+ \to \pi^0)}$
    K_2^0 decay <u>(energy dependence)</u>
    μ capture in H
    μ decay (electron polarization)
    Precise measurement of universal weak charge

4. Electromagnetic interactions	
(a) Pure interactions	*(b)* Electric hadron properties
<u>g − 2 experiment</u> (muon magnetic moment)	Λ magnetic moment
γ-ray absorption	<u>π^0 lifetime</u>
e^+ annihilation	$\pi^0 \rightarrow 3\gamma$
μe scattering	<u>Time-like form factor of proton</u>
Theoretical calculations	$\bar{p}p \rightarrow \begin{cases} \bar{e}e \\ \bar{\mu}\mu \end{cases}$

5. Nuclear structure

Double hypernuclei
<u>Muonic X-rays</u> (quadrupole moments)
<u>^{4}He excited states</u>
Pion double charge exchange
Muon capture rates
Theoretical studies.

Who's who in the history of CERN

Armin HERMANN

In addition to our 'Who's who in the foundation of CERN' in Vol. I with its 65 short biographies, the reader will find here another 36 people who played a role for CERN during the fifties and sixties. Our aim is again to give short and precise information on the educational background of these personalities, their career and their functions in or for the Organization. As already indicated in Vol. I, the length of each article depends solely on the information we thought necessary.

A biography which is missing here, may be found in Vol. I, pp. 545-565.

Bernardini, Gilberto (b. 1906) made his Laurea at the University of Pisa, became assistant at the University of Florence in 1930, and went to Berlin for a few months in 1934 to work with Otto Hahn and Lise Meitner. In 1937 he was nominated professor at the University of Camerino, and served as associate professor at the University of Bologna from 1938-41, where he stayed until 1946, finally in the position of full professor and director of the institute for experimental physics.

After the war Bernardini collaborated with Edoardo Amaldi in the reconstruction of the physics department in Rome, and was there nominated professor for spectroscopy in 1947. From 1949 he spent two years as visiting professor at Columbia in New York, and went to the University of Illinois, Urbana, to work with the big betatron of D.W. Kerst.

Bernardini joined CERN on 1 October 1958 to serve as director of the SC division and became a member of the directorate for research on 1 January 1961, a post from which he resigned towards the end of 1962. Bernardini went back to his country and in 1964 became director of the Scuola Normale Superiore in Pisa. He received a great number of distinctions and honorary positions. From 1962-67 he was president of the Società Italiana di Fisica and became the first president of the European Physical Society in 1968.

Burger, Albert (b. 1923) graduated with a Ph.D. in physics at Karlsruhe University in 1955. He came to CERN the same year as a staff number and was sent for $1^1/_2$ years to the Carnegie Institute of Technology, Pittsburgh. After his return he collaborated closely with Charles Peyrou in the development and operation of bubble chambers.

When Peyrou was appointed leader of the Track Chambers Division, Burger became his deputy. Again, when Peyrou was promoted Director of the Physics II Department, Burger became Deputy Department Director. He remained in this position until the end of 1975, except for a year's leave of absence 1972/73. After the reorganization of CERN on 1 January 1976 Burger acted for one year as Deputy Division Leader of the Experimental Facilities Division.

Citron, Anselm (b. 1923) studied physics in Freiburg/Breisgau and Basle and did his Ph.D. in Gentner's institute in 1951 on a problem concerned with cosmic rays.

As a post-doctorate fellow at Cambridge/England, he learnt some fundamentals of accelerator physics and at the same time did the first radiation shielding estimates for the PS machine. In 1953, he joined the PS team at Geneva, where he was responsible for the layout of the PS experimental areas. Later he headed a team that designed and built the muon channel at the SC and used it for a scattering experiment.

In 1963 he spent a year as the 'CERN Ambassador' to BNL and in 1965 he accepted a full professorship at Karlsruhe University and a directorship at the Karlsruhe Nuclear Research Center (KfK). His institute kept working at CERN and also made some technical contributions to it, namely a superconducting separator (1977) and an electron cooling system for LEAR (1988).

Cocconi, Giuseppe (b. 1914) graduated in Milan and had his 'libera docenza' in 1939. He became full professor at the University of Catania in 1942. In 1947 he was invited to Cornell University as Research Associate and was there appointed professor in 1949. He worked first on cosmic rays, then on particle physics with the Brookhaven accelerator. During a sabbatical leave he was at CERN in 1959 and 1960.

In 1962 Cocconi came back to the Organization as a staff member. His main interests were directed towards strong interactions and astrophysics. For three years, from mid-1967 to mid-1970, he took responsibility as director of the Physics I Department, with its two Nuclear Physics (NP) and Synchro-cyclotron Machine (MSC) divisions.

von Dardel, Guy (b. 1919) worked in neutron physics in Stockholm and originated the pulsed neutron diffusion method. He joined CERN in 1954 when Kowarski set up the Scientific and Technical Services division, was transferred to the SC in charge of the electronic workshop in 1957 and set up a first experimental group for the PS in 1958. This group was the first to publish experimental results from the PS on total cross sections in 1960. He served as Coordinator for the PS 1961–62, and Coordinator for the CERN neutrino experiment from 1962–63.

Von Dardel left Geneva in 1965 when he was nominated professor of physics at Lund University. During this period the Scandinavian ISR-collaboration between Lund, Bergen and the Niels Bohr Institute in Copenhagen was organised, which during the life of the ISR made a number of experiments. From 1974–76 he was chairman of ECFA, and as such sponsored the building of PETRA in Hamburg, participated in the discussions on an international Very Big Accelerator, and was responsible for creating a strong European consensus for the building of a large electron-positron storage ring at CERN, later to be known as LEP.

Since he retired from his chair in Lund in 1985 he lives in Geneva and continues to participate in CERN experiments.

Fidecaro, Guiseppe (b. 1926) made his Laurea in physics at the University of Rome in 1947 and joined the staff of the Institute of Physics the same year. He did his research at the 'Centro di Studio per la Fisica Nucleare del C.N.R.' at the University of Rome, at the M.I.T. synchrotron laboratory at Cambridge, Mass., and as deputy director of the Testa Grigia cosmic ray laboratory at Mount Cervinia.

Fidecaro joined the Organization on 1 October 1954 to work for the Theoretical Study Group at the Nuclear Physics Research Laboratory of the University of Liverpool, just when the synchro-cyclotron started operation. He came to Geneva early August 1956 to work for the preparation of experiments. He was one of the authors of the π-eν experiment.

Besides his research position at CERN as senior physicist, he became in 1961 full professor at the University of Trieste. From 1963-70 he was Director of the Trieste Section of the Istituto Nazionale di Fisica Nucleare (INFN) and from 1972-75 Director of the Institute of Physics of the University of Trieste. As associate to the INFN he was member and chairman of several INFN committees, for example for the selection of sites in Italy for the 300 GeV proton synchrotron. In this capacity he played an important role in the Italian approval of the ISR and SPS, which finally led to a considerable larger budget for the Organization.

Fierz, Markus (b. 1912) gained his Ph.D. in 1936 with Gregor Wentzel at Zurich University, and went to Leipzig and Copenhagen to learn directly from Werner Heisenberg and Niels Bohr. His first academic post was as assistant to Wolfgang Pauli at the Federal Institute of Technology, Zurich. In these three and a half years from 1936-40 both men collaborated closely and established a lifelong friendship.

In 1940 Fierz left for Basle, where he became associate professor in 1943 and full professor in 1945. He was guest at the Institute for Advanced Study, Princeton, from 1950-51 and again in 1955, and a visiting professor at Maryland University, College Park.

Fierz came to CERN on 1 April 1959 as head of the Theoretical Physics Division. He left the Organization a year later to become (jointly with Res Jost) Pauli's successor at the Federal Institute of Technology, Zurich.

Germain, Pierre (1922-1988) acquired his 'license' in 1945 and his 'docteur ès sciences mathématiques' in 1952 at Brussels University, the latter for the design and construction of an analogue integrator.

He came to CERN in August 1955 to join the Radiofrequency Group of the PS Division under Christoph Schmelzer. He reconstructed the 'Hall computer' for the frequency programme of the accelerating voltage.

After the completion of the PS he joined the Machine Group, becoming its Leader in 1960. On 1 January 1961, after the internal reorganization, Germain was nominated Leader of the Proton Synchrotron Machine Division, responsible for the operation maintenance and exploitation. On 1 January 1963 he also became a member of the Directorate and served after that, from 1 July 1966 to 31 December 1969, as Department Director of the PS. Until his departure on 30 November 1979 he worked for the Organization as senior physicist.

Gregory, Bernard P. (1919-1977) graduated in 1945 at the Ecole Polytechnique, studied for two years at the Ecole des Mines, and worked under Bruno Rossi at the Massachusetts Institute of Technology on nuclear reactions caused by cosmic rays. He received his Ph.D. in 1950.

Back at the Ecole Polytechnique and working in the laboratory directed by Louis Leprince-Ringuet, he built with Charles Peyrou, André Lagarrigue and Francis Muller a powerful cloud chamber detector system for operation on the Pic du Midi. After a year spent at the Brookhaven National Laboratory he constructed for the CERN-PS the 81 cm hydrogen bubble chamber which provided from 1961–71 over 16 million photographs.

Gregory was nominated professor at the Ecole Polytechnique in 1959 and came to CERN as the leader of a French experimental team. He was appointed directorate member for research in 1964 and Director-General of the Organization from 1 January 1966 until the end of 1970. He then succeeded Leprince-Ringuet as head of the physics laboratory.

He now played a major role in the French scientific community, notably at the CNRS. Links with CERN remained close. From 1971–73 he served as chairman of the ISR committee, French delegate to the Council from 1971 and vice-president from 1 January 1974. He was already elected president of the Council to take over responsibility on 1 January 1978, when he died unexpectedly of a heart attack on 24 December 1977.

Hampton, George Hughan (b. 1920) studied at King's School (Macclesfield), then at Balliol College (Oxford) and received his master's degree in chemistry in 1940. After the war he occupied various positions in the British Civil Aviation Ministry and the Ministry of Transport. Hampton joined the UKAEA Industrial Group (now Production Group) headquarters at Risley, Lancashire, finally holding the post of deputy to the personnel director.

He came to CERN on 1 October 1963 as Member of the Directorate, responsible for the administration, to replace S.A.ff. Dakin, who returned to Britain. When the internal structure of the Organization was changed on 1 July 1966, Hampton became Director of the Administration Department with its three Divisions (Finance, Personnel and Technical Services) and three Groups (Health Physics, General Safety and Central Services). At the end of 1975 Hampton retired from CERN.

Van Hove, Léon C.P. (b. 1924) obtained his Ph.D. in mathematics at the University of Brussels in 1946. While an assistant to Professor Jules Géhéniau at Brussels

University from 1945–52, he spent six months with Niels Bohr in Copenhagen and a whole year at the Institute for Advanced Study. In 1952 he returned to Princeton as a member of the Institute's School of Mathematics, working mainly with George Placzek. He was nominated professor and director of the Institute for Theoretical Physics at the University of Utrecht in 1954.

Van Hove came to CERN in September 1960, taking responsibility for theoretical research, first as division leader of the Theoretical Studies Division and from 1966–68 as director of the Department of Theoretical Physics. From 1972–74 he shared his time between CERN and the Max-Planck-Institute for Physics and Astrophysics, Munich, where he was chairman of the Scientific Directorate.

In 1975 the Council decided to have for the five years' period starting 1 January 1976 two director-generals Léon Van Hove and John Adams. Van Hove was responsible for all research activities of the Organization.

Jentschke, Willibald (b. 1911) received his Ph.D. in his native Vienna in 1935 and went to the United States in 1946. He became assistant professor at the University of Illinois, Urbana, in 1948, associate professor in 1955 and full professor in 1956. He worked there with a low-energy cyclotron on nuclear reactions, and started the transformation into a strong focusing 'spiral ridge' cyclotron.

An offer to take over a full professorship at Hamburg University put Jentschke in the position to propose the construction of an electron accelerator in the GeV range, a project, which finally led to the foundation of the Deutsches Elektronen-Synchrotron (DESY) in 1959 with Jentschke as chairmain of its Board of Directors. In 1964 the 7.5 GeV accelerator came into operation.

From 1964–67 Jentschke was delegate of the Federal Republic of Germany to the Council and member of the Scientific Policy Committee. In 1968 he spent a year at CERN as Guest Professor and took part in the work of the Directorate. In 1969 he was appointed Chairman of the Intersecting Storage Rings Committee. From 1 January 1971 he worked five years as Director-General jointly with John Adams. Jentschke took responsibility for Laboratory I, i.e. all work done except for the SPS under construction.

Johnsen, Kjell (b. 1921) received a degree in electrical engineering (1948) and his doctorate (1954) at the Norwegian Institute of Technology in Trondheim. He joined the Chr. Michelsens Institute in Bergen in 1948. While at this Institute he also spent 15 months with Dennis Gabor at Imperial College. It was in Bergen that he started his work for the PS Group (directed by Odd Dahl), becoming one of the first CERN staff members in July 1952. He came to Geneva at the end of 1953.

In 1957 Johnsen was nominated professor at the Norwegian Institute of Technology and returned to Trondheim. Two years later he came back to CERN to join the Accelerator Research Group. When this Group became a Division on 1 January 1961, a rotation principle made Johnsen its Division Leader for 1962 and

1965. From 1966 to 1973 he served as director for the ISR Construction Department.

Later Johnsen was involved in a diversity of tasks: New collider project studies, 1972–78; creation of the French/German radio-astronomy institute IRAM, 1978–79; The Brookhaven CBA project, 1979–82; CERN Accelerator School Creation, 1983–85; Panel chairman for linear colliders (CLIC), advisor to the CERN Long-Range Planning Committee, 1985–87; HERA Machine Committee chairman, DESY, 1984– ; Part-time professor University in Bergen, 1972–86.

Kummer, Wolfgang (b. 1935) received his degrees in technical physics at the Technische Hochschule in Vienna and came to CERN first as an unpaid visitor (September 1961 to March 1962) and then as a Fellow (January 1963 to September 1964). During the latter period he collaborated with Victor F. Weisskopf in the preparation of Weisskopf's famous 'Seminars for Experimentalists'.

Kummer was from 1966 to 1971 director of the High Energy Physics Institute of the Austrian Academy of Sciences. In 1968 he became full professor for theoretical physics at the Technische Hochschule (now Technische Universität) in Vienna. He spent the year 1972 at the University of Pennsylvania (Philadelphia). During the 70's he stayed for longer periods at different scientific institutions in the U.S. (Princeton University 1975, Brookhaven National Laboratory 1977, etc.).

He worked for CERN as chairman of the Finance Committee from 1968–70, vice-president of the Council from December 1981 to June 1982 and its president from January 1985 to December 1987.

Laporte, Henri Albert (b. 1928) received his education at the Ecole Polytechnique, acquired the diploma in 1951, and finished his studies at the Ecole des Ponts et Chaussées in 1954. He then worked in the Engineering Corps of the French Navy.

Laporte joined CERN on 1 January 1967 as Senior Engineer and Leader of the Technical Services and Buildings Division. He remained in this position until the end of 1975. He then worked as head of the civil engineering group for LEP.

Lock, William Owen (b. 1927) was educated at Cirencester Grammar School and at the University of Bristol, where he graduated with a B.Sc. in Physics in 1948 and a Ph.D. in cosmic ray physics in 1952. The following year he was appointed a Lecturer in Physics at the University of Birmingham.

Lock joined the Organization in 1959 as a Research Associate and in 1960 he became a Staff Member and Joint Group Leader of the Nuclear Emulsion Group, part of the Nuclear Physics Division. From 1965–67 he served as Head of the Fellows and Associates Service, Personnel Division, was Acting Leader of the Personnel Division for the year 1968, Head of Education Services from 1969–77 and concurrently Deputy to the Leader of the Personnel Division from 1969–75 and in 1977.

When John Adams became Executive Director-General he worked closely with W.O. Lock, who was nominated Administrative Assistant to the Director-General in 1978. Lock remained in this position when Herwig Schopper was Director-General until the end of 1988, when he became an Adviser to the new DG, Carlo Rubbia.

Lock founded the CERN School of Physics in 1962 and has been its organiser ever since. With Bernard P. Gregory and a few others he established the Joint CERN-JINR (Dubna) Schools of Physics, and the first course took place in the summer of 1970. He continued to work for European-Soviet-Cooperation in science, while also playing a major role in establishing scientific relations with the People's Republic of China.

In 1978, shortly after its foundation, Lock became Secretary of the International Committee for Future Accelerators (ICFA), and has remained in this position until now.

Macleod, Graham Ross (b. 1929) graduated in physics at Cambridge University and completed his doctorate there in 1955. He joined CERN in the Scientific and Technical Services Division in the same year, working initially in cosmic ray physics.

He then worked on the development of instrumentation and computer programs for the analysis of bubble chamber film, participating in experiments at the SC and PS, and later developing methods for the automatic processing of spark chamber data. From mid-1963 he acted as Division Leader of the Data Handling Division, being appointed Division Leader in June 1965, a position he held until the end of 1975. He initiated the CERN Schools of Computing and from 1970 to 1976 played a leading role in organising the first Schools. From 1977 he was Project Leader for the development of CERN-wide data-communications computer networks for physics experiments and general computing use. He took early retirement, leaving CERN in 1986.

Mallet, Charles (1915-1967) joined, after a splendid examination at the Ecole Polytechnique in 1936, the government public works service as a civil engineer and worked in North Africa from 1939-50. On leave from the government service he then took part in the planning and construction of airports, reservoirs, dams, bridges and irrigation projects.

Mallet was nominated director of the CERN Site and Buildings Division on 1 September 1959. From October 1965 he was a member of the ISR Study Group.

Melville, Sir Harry (b. 1908) was educated at Edinburgh University and at Trinity College, Cambridge. He was Assistant Director of the Colloid Science Laboratory at Cambridge from 1938-40, when he became Professor of Chemistry at the University of Aberdeen.

During the war, Sir Harry was Scientific Adviser to the Chief Superintendent, Chemical Defence, Ministry of Supply, from 1940-43, and then Superintendent of the Radar Research Establishment until 1946. In 1948 he was appointed Mason Professor of Chemistry at Birmingham University, and became Secretary to the Committee of the Privy Council for Scientific and Industrial Research in 1956, a post which he held until 1965.

He served in many British scientific and research councils: Since 1954 he has been a member of the Advisory Council on Research and Development, Ministry of Power, and, since 1956, a member of the governing board of the National Institute for Research in Nuclear Science.

Sir Harry Melville was a delegate of his country in the CERN Council from 1958–65, and was its Vice-President for his last two years 1964–65.

Merrison, Sir Alexander Walter (1924–1989), generally called Sir Alec Merrison, was educated at King's College, London, where he graduated in 1944. He worked for two years on radar for the Ministry of Supply and from 1946–51 on reactor and nuclear physics at Harwell. As Research Fellow and Lecturer at Liverpool University his field of interest shifted to elementary particle physics. He was awarded his Ph.D. in 1957 and came to CERN the same year, working in a team who collected in evidence for the direct disintegration of a π-meson into an electron.

After three years with the Organization he returned to Liverpool as Professor of Experimental Physics. From 1962–69 he was director of the Daresbury Nuclear Physics Laboratory, responsible for the 5 GeV electron synchrotron NINA, which came into operation on 2 December 1966.

From 1969–84 Sir Alec Merrison served as Vice-Chancellor of Bristol University. He became an F.R.S. in 1969 and was appointed chairman of a number of important commissions and committees, for example the Royal Commission on the National Health Service, which issued its report 1979 and, another example, the Government's Advisory Board for the Research Councils, which expressed severe concern at government cuts in scientific research.

Sir Alec Merrison was a good friend of the Organization and worked as a member of the Scientific Policy Committee from 1969–72. He served as President of the Council from 1982–84.

Michaelis, Ernst Gunter (1917–1989) finished secondary school at Leipzig in 1936 and began his University studies there. After emigration to Britain he graduated from London University with a B.Sc. in 1943, B.Sc. Honours (Physics) in 1948 and a Ph.D. in Cosmic Ray Physics in 1953. Afterwards he became a Lecturer in Physics at Birkbeck College, University of London.

Michaelis joined CERN on 15 July 1957 first as a Fellow, then as a Research Associate and became staff member on 19 August 1959, working for the Synchro-cyclotron Machine Division (MSC). He was appointed Deputy Division Leader of the MSC Division on 1 January 1961 and Division Leader on 15 March 1968. This appointment ended with the year 1975.

Paul, Wolfgang (b. 1913) studied for his Ph.D. with Hans Kopfermann at the Technische Hochschule Berlin in 1939, became associate professor in Göttingen (1950) and full professor in Bonn (1952). Here, the first German AG-synchrotron, a 500 MeV electron accelerator, was completed in 1958 under his leadership. From 1964–68 he stimulated again the construction of a 'home made' electron synchrotron, this time a 2.5 GeV machine.

He was deeply involved in the foundation of the Nuclear Research Center Jülich (KFA Jülich) and the Deutsches Elektronen-Synchrotron in Hamburg (DESY). From 1960–62 he served for the KFA as chairman of the Board of Directors. From 1971–73 he was chairman of the Board of Directors and from 1973–75 chairman of the Scientific Council of DESY. In 1979 he became president of the Alexander von Humboldt-Stiftung.

Paul held a visiting professorship at CERN 1958–60, and worked for the Organization as co-leader of the Nuclear Physics Division (first half of 1965) and director of the Physics I Department (July 1965 until December 1967). From 1972–80 he served as member, and from 1975–77 as chairman of the Scientific Policy Committee. He was also the German delegate to the Council from 1968–70 and again in 1978.

Wolfgang Paul shared the 1989 Nobel prize for physics with two other scientists.

Peyrou, Charles (b. 1918) received his scientific education at the Ecole Polytechnique, where he was Maître de conférences from 1949–1955. He worked in the Laboratory of Leprince-Ringuet on strange particles in cosmic rays and spent a year at M.I.T. with Bruno Rossi. He was appointed Extraordinary Professor at Berne University in 1954.

Peyrou became Consultant to CERN in October 1954, and joined the Senior Staff in 1957 with responsibility for the Hydrogen Bubble Chambers. From 1 January 1969 onwards he was leader of the Track Chambers Division (TC).

Peyrou kept his teaching activities at Berne for many years and remained an honorary professor there.

When mid-1966 the internal structures were reorganized, Peyrou became Director of the Physics II Department, which included the Track Chambers Division with Peyrou as its leader. He kept this position until mid-1975. After a sabbatical at SLAC he actively took part in the neutrino-experiment with the SPS and the Big European Bubble Chamber (BEBC) built by the TC Division. He retired from active service in 1983, but kept his interest in CERN's activities.

Prentki, Jacques (b. 1920) received his master's degree at the University of Warsaw in 1946 and his doctorate at the University of Paris in 1952.

He came to CERN in 1955 as one of the first members of the Theory Group and was from 1961–67 deputy leader of the Theoretical Studies Division. When he became full professor at the Collège de France in 1964, he continued his work for

CERN. He served from 1967-70 as leader of the Theoretical Studies Division and again from 1976-82 as leader of the Theoretical Physics Division, as it was then called. He retired in 1985.

Puppi, Gianpietro (b. 1917) learned his physics from Bruno Rossi and Giancarlo Wick at Padua, where he received his diploma in 1939. He resumed his work in 1944 at the universities of Bari, Rome and Padua, and became full professor at the University of Naples in 1950. Two years later Puppi was nominated full professor for advanced physics and director of the Institute of Physics at the University of Bologna, where he has stayed since then, except for extended leaves to the United States and to CERN.

In succession to Gilberto Bernardino he was Member of the Directorate for Research from November 1962 until the end of 1963. From July 1966 to October 1968 Puppi served as chairman of the Scientific Policy Committee, and for the years 1969/70 as its vice-chairman.

Ramm, Colin (b. 1921) began his studies at the University of Western Australia and accepted an invitation of Marcus L.E. Oliphant to be a lecturer at Birmingham University. He built an accelerator tube and spent a few years measuring the energy released in nuclear reactions, for which work he obtained his doctorate in 1951.

He joined the PS Division in 1954, as leader of the magnet group. When the new internal organisation with 12 Divisions instead of the original six came into effect on 1 January 1961, Ramm became leader of the Nuclear Physics Apparatus Division (NPA). He held this position until September 1969.

Richemond, Jean (1896-1987) graduated from Faculté des Sciences and Ecole Supérieure d'Electricité and was Managing Director of various industrial firms until 1930. He finally became Director-General of the Crédit Foncier Colonial Bank. His military service ended as prisoner of war, from which he escaped to join the Free French Forces in London. Here he occupied important administrative functions.

After the liberation of France Richemond became Treasurer-General of the Rassemblement du Peuple Français, Chef de Cabinet, Secrétaire général and Inspecteur général in various ministries and government agencies.

Richemond was appointed Director of Administration on 1 August 1955, and he left this position on 31 August 1958 to return to French government service.

Schoch, Arnold (1911-1967) started his studies at Stuttgart and studied for his Ph.D. at the University of Berlin in 1937. He continued his research on applied and theoretical acoustics during the war years. Schoch qualified for academic teaching (Habilitation) at Göttingen in 1950 and went as 'Dozent' for theoretical physics to Heidelberg.

Early in 1954 he came to CERN to work for the PS Group. When the PS started its operation, Schoch developed new ideas as a member of the Accelerator Research Group. When the Accelerator Research Division was founded on 1 January 1961, a rotating system of leadership was established. Schoch took responsibility in the first year and again from 1964–66.

In 1966 Schoch accepted a position as full professor at Karlsruhe University and department director of the Karlsruhe Nuclear Research Center (KfK). Owing to his untimely death his plans for a 40 GeV proton accelerator were never realized.

Schou-Olsen, Finn (b. 1919) was, after his education in economics, contract officer for Norsk Hydro and Office Manager in a trade firm. He came to CERN in March 1955 as Purchasing Officer, and was nominated Deputy Division Leader in the General Administration in March 1958. Schou-Olsen left the Organization at the end of 1966 to return to Norway.

Schulte-Meermann, Walter (1906–1984) worked, after his law studies, from 1940–44 for the German Foreign Office as expert for economic and financial matters. From its foundation in 1949 he spent ten years with the Federal Ministry for Transport, and went at the end of 1959 to the Federal Ministry for Nuclear Energy (now the Federal Ministry for Research and Technology), where he was head of the subdivision 'International Cooperation in Nuclear Matters'. In this capacity he was also responsible for CERN, especially in representing the Federal Republic in the Council. From 1965–67 he was Chairman of the Finance Committee.

From 1968–72 he worked as administrative manager (Geschäftsführer) for the 'Gesellschaft für Strahlen- und Umweltforschung' at Neuherberge near Munich.

Standley, Peter Hugh (b. 1924) served, after two years at Oxford, as Radar Officer in the Royal Navy from 1944–47. In the following years he worked on the construction and development of the Oxford University 125 MeV electron synchrotron and was awarded his D.Phil. in 1953.

He became a staff member on 1 December 1954, stationed at Harwell, to liaise between the CERN and Harwell linac groups and Metropolitan Vickers, Manchester, who were manufacturing major parts of the 50 MeV proton linac for CERN. Standley came to Meyrin early June 1956 and took charge of the infrastructure, installation and commissioning of the linac. For the years from 1961–64 he was Deputy Division Leader of the MPS Division under Pierre Germain, responsible for the operation and good working status of the PS. The following eight years until the end of 1972 he served as Division Leader of the MPS Division.

Standley then worked, in the first three months of 1973, as Deputy to Günther Plass in the Linac Group, and moved to SC, working with E.G. Michaelis. On 1 January 1976 he became, for a three year period, Deputy Division Leader in the PS Division with responsibility for the SC. In his last years at CERN he worked with J.H.B. Madsen for the LEP injector. He retired on 30 June 1983.

Tièche, Constant (1917–1985), Swiss Civil Servant, joined CERN in 1954. He was the Finance Officer of the Organization and the Leader of the Finance Division (Budget, Accounts and Purchasing) until his retirement in 1982.

Ullmann, Günther (b. 1924) studied, after war service in Africa, law at Frankfurt University. He joined CERN in 1955 as Personnel Officer, was nominated Head of Personnel in 1961, and Leader, Personnel Division, in 1964. In 1978 he became, in addition, member of the directorate for staff policy.

After relinquishing these mandates, he was chosen in 1983 as Administrator (Manager) of the CERN Pension Fund.

Verheyden, André (b. 1913) worked, after a commercial education, in progressively responsible positions in industrial accounting and financial services. He was appointed CERN Chief Accountant on 1 January 1959 and Deputy to the Leader of the Finance Division on 1 January 1961. Verheyden retired at the end of 1976.

Weisskopf, Victor F. (b. 1908) started to study physics in Vienna, his native city, and went, after two years, to Göttingen, where he studied for his Ph.D. with Max Born in 1931. He then worked in other centers of theoretical physics (Leipzig, Berlin, Charkow, Copenhagen, Cambridge) and was from 1933–36 assistant to Wolfgang Pauli at Zurich. Weisskopf left Europe in 1937 and became instructor and then assistant professor at Rochester. From 1943 he worked for the Manhattan Project at Los Alamos, and was appointed in 1945 full professor at M.I.T., a post which he held until his retirement. In 1967 he became chairman of the department and was elected, in the same year, chairman of the commission of high energy physics of the Atomic Energy Commission. In 1975 he became president of the American Academy of Arts and Sciences.

Weisskopf was asked, after the death of Cornelis Bakker, to take responsibility for CERN as Director-General. He served in this position from 1 August 1961 until the end of 1965. When he returned to the MIT the Organization still enjoyed his advice, formally as member of the Scientific Policy Committee from 1966–77 and informally at his regular stays during the summer months.

Zilverschoon, Cornelis J. (b. 1923) started his studies with electrical engineering and turned to physics, when he was able to enter Delft Technical University in 1945. After his degree in physical engineering (1949) he moved to Amsterdam University, where he was awarded his Ph.D. for the construction of an electromagnetic isotope separator early in 1954.

He came to CERN in May 1954 as head of the PS Mechanical Engineering Group. When the PS operated successfully, he transferred to the Accelerator Research Group at the end of 1959. After the formation of the Accelerator Research Division he served as Division Leader in 1963 and Deputy Division Leader for the second half of 1962 and again for two and a half years from 1 January 1964 onwards. Zilverschoon then became Deputy to the Director of the ISR Construction Department and from 1970–75 Director of the Proton Synchrotron Department.

Chronology of events[1]

John KRIGE and Dominique PESTRE

---

1. We have restricted scientific developments in this appendix to a list of the most outstanding items. Generally the date attributed to a scientific item is the date on which the paper making the announcement was received by the scientific journal which first published the result. For a more detailed list see Appendix 6.

Contextual	**CERN**
	1954
October	7/8 - Council, Sir Ben Lockspeiser (UK) President of Council (see appendix 3); Felix Bloch (USA) Director-General; six divisions created (see appendix 1)
1955	
February	24 - Council accepts Bloch's resignation; invites Cornelis Bakker (NL) to accept post of DG; confirms nomination of J.B. Adams (UK), L. Kowarski (F), C. Møller (DK), and P. Preiswerk (CH) as divisional directors
April – Fifth annual Rochester conference 3/8 - Churchill resigns and Eden succeeds him as UK Prime Minister 18 - Alvarez proposes building a very large hydrogen bubble chamber. 50 x 20 x 20 inches is suggested 19 - Death of Albert Einstein	
May 5 - West Germany regains its sovereignty 9 - Adenauer announces the entry of Germany into NATO 14 - Warsaw Pact signed 15 - Last French forces leave North Vietnam 26 - USSR and Yugoslavia reconciled 27 - Conservative Party wins UK general election	
June	10 - The foundation stone is laid at the Meyrin site

Contextual

CERN

10 - Council approves nominations of W. Gentner (FRG) and J. Richemond (F) as directors of SC and ADM divisions (see appendix 1)
11 - Signature of the Headquarters Agreement

July
27 - Austria regains its sovereignty

August
8/20 - First International Conference on the Peaceful Uses of Atomic Energy, Geneva
September

1 - Bakker succeeds Bloch as Director-General

October
10 - European G Stack collaboration reports branching ratios of the K^+
23 - The referendum in Sarre rejects the statutes proposed in the Paris agreements

December
- Nobel prize to W.E. Lamb for 'his discoveries concerning the fine structure of the hydrogen spectrum' and P. Kusch for 'his precise determination of the magnetic moment of the electron'

1956

February
14 - Opening of XXth Communist Party Congress inaugurates the process of 'destalinization'

April
3/7 - Sixth annual Rochester conference
4 - Alvarez increases the size of his envisaged hydrogen bubble chamber to 72 inches

May
23 - Calder Hall, first UK nuclear power station, comes into operation

Contextual

June

11/23 - First international conference on high-energy accelerators, at CERN
13 - 'Let a hundred flowers blossom' campaign launched in Communist China
22 - Hypothesis of parity violation in weak interactions

July
26 - Nasser nationalizes the Suez canal

September
28 - Electricity produced from an atomic pile at Marcoule (F) for the first time in Continental Europe

October

21 - Soviet troops advancing on Warsaw
22 - Statutes of International Atomic Energy Agency adopted in New York
23 - Start of the Hungarian uprising
27 - Franco-German agreement for the political union of the Sarre with the FRG

November
2/5 - Franco-British intervention in Egypt
4 - Soviet troops occupy Budapest
6 - Eisenhower re-elected US President
15 - Two-component theory of the neutrino

1957

January
7 - Start of the 'Battle of Algiers'
10 - Macmillan succeeds Eden as UK Prime Minister

CERN

- Assembly of the double cloud chamber completed by the STS division

- The SC group drastically revises its forward estimates to accommodate an unexpected increase in the costs of research
4 - Finance Committee agrees to award the contract for the PS magnet blocks to Ansaldo San Giorgio, Genoa

Contextual

February

14 - Foundation of the National Institute for Research in Nuclear Science announced in British parliament

March
25 - Common market and EURATOM Treaties signed in Rome by the six (B,F,FRG,I,L,NL)

April
12 - Eighteen German atomic scientists, including Born and Heisenberg, solemnly declare their opposition to nuclear weapons
13 - Norway refuses to stock nuclear weapons on its territory
15/19 - Seventh annual Rochester conference

16 - First UK thermonuclear test, Christmas Islands
26 - British decision to build a 1.5 m hydrogen bubble chamber

June

August

19 - Article by Ohkawa on a two-beam FFAG accelerator

September
- Universal (V-A) weak interaction

CERN

1 - B. Ferretti (I) head of the Theory group in Geneva
28 - First report by an SPC Working Party on the organization of the experimental work (SC machine)

May
1/2 - Finance Committee, the UK delegation calls for a financial policy to control CERN's expenditure

4 - Informal meeting to discuss the PS experimental programme

1 - First full beam on the Synchro-cyclotron

29 - The FC and the DG agree that it may become necessary to impose financial ceilings on CERN

852   *Chronology of events*

Contextual   **CERN**

17 - Alvarez suggests that 'in the high energy field the bubble chamber [would become] the universal detector'
24 - Racial violence at Little Rock, Arkansas

October

1 - Theory division transferred from Copenhagen to Geneva; B. Ferretti (I) divisional director

4 - USSR launches Sputnik 1 satellite

16 - The FC is asked to support a budget for 1958 which exceeds the earlier estimate by 50%

22 - Adenauer re-elected FRG Chancellor
24 - Pomeranchuk theorem

November
7 - Eisenhower announces the appointment of Killian as Presidential Science Adviser

December
- Nobel prize to C.N. Yang and T.D. Lee for 'their penetrating investigation of the so-called parity-laws which have led to important discoveries regarding the elementary particles'

19 - Council, F. de Rose (F) succeeds Lockspeiser as Council President (appendix 4); a ceiling of 100 MSF imposed on CERN's expenditure for 1958 and 1959; the creation of an Accelerator Research Group under Schoch accepted

1958

January

1 - G. Bernardini (I) first Director of Research (SC division)
13 - Internal memorandum: a CERN bubble chamber project 'should be declared urgent and started at once'

24 - Cockcroft announces the progress made on fusion by the ZETA machine at Harwell

Contextual	CERN
Contextual	**CERN**

Contextual

31 - USA launches its first satellite (Explorer 1)

April
15/17 - NATO approves the Norstad plan which allows for 30 divisions equipped with tactical nuclear weapons
17 - King Baudouin opens the World Fair in Brussels

June
1 - The French National Assembly accepts de Gaulle as head of the Government
6 - French decision to copy an American 81cm hydrogen chamber
23 - Eisenhower signs an agreement with EURATOM

30/6-5/7 - Annual conference on high-energy physics, at CERN (Eighth Rochester conference)
30 - US Congress amends the MacMahon act so as to allow the sharing of atomic secrets with its allies

August
20/26 - De Gaulle announces the right to independence of French overseas colonies
22 - French 3 GeV PS 'Saturne' reaches design energy
22/23 - Communist China artillery attack on two Nationalist Chinese islands

September
1/13 - Second International Conference on the Peaceful Uses of Atomic Energy, Geneva.

CERN

29/4-1/5 - Official proposal to build at CERN a 1 m propane chamber and a 2 m hydrogen chamber

May
1 - First successful trial of CERN 10 cm hydrogen chamber

27 - Creation of a CERN Study Group on Fusion after contacts with EURATOM

Contextual

CERN

15/16 - International meeting on instruments for the evaluation of photographs, at CERN
25 - First meeting of the Executive Committee for the PS experimental programme

October
1 - Creation of NASA (National Aeronautics and Space Administration)

- Acceptance tests of Ferranti 'Mercury' computer completed
9 - Council, G. Bernardini (I) director of SC division; W. Gentner (FRG) in charge of research along with P. Preiswerk (CH) who leaves SB division; S.A. ff. Dakin (UK) succeeds J. Richemond (F) as director of Administration (see appendix 1)

November

20–21 - First schematic secondary beam layouts for the PS drawn up

December
- Nobel prize to P.A. Čerenkov, I.M. Frank, I.J. Tamm for 'the discovery and interpretation of the Čerenkov effect'

3 - Council votes the 1959 budget at 55 MSF, and breaks the 1958/59 ceiling; Sweden votes against
20/21 - First meeting of European emulsion specialists at CERN for work with the PS

21 - De Gaulle elected President of the French Republic

1959

January
1 - Entry into force of the Common Market

20 - Start of the negotiations with the British on the use of their hydrogen bubble chamber at CERN

Contextual

February

March
- Alvarez commissions his 72 inch hydrogen bubble chamber

April

May

16 - US Piore Panel report on high-energy accelerator physics published

June
- O'Neil proposal for two tangent storage rings for the electron Mark 3 machine at Stanford
- Sands' first report on a 300 GeV PS during a MURA conference at Madison (Wisconsin)
9 - First nuclear submarine equipped with ballistic missiles launched at Groton, USA

July

12/25 - Ninth (Rochester) annual conference on high-energy physics, at Kiev
18 - Introduction of Regge poles
20 - Spain becomes the 18th member of the OECE

CERN

11 - SPC, Sir John Cockcroft (UK) proposes 'to form mixed teams of CERN staff and visiting scientists' at CERN

21 - Start of negotiations with the French on the use of their chambers at CERN

1 - M. Fierz (CH) succeeds Ferretti as director of Theory division

7 - First successful trial of CERN 30 cm hydrogen chamber

25 - FC approves order for first set of secondary beam magnets
26 - Council, C. Møller (DK) and C.F. Powell (UK) enter the SPC (see appendix 5); a ceiling of 120 MSF for 1959 and 1960 is imposed and it is resolved to stabilize the budgets for 1960-1962 at 65 MSF annually

1 - Austria becomes the 13th member state of CERN

| **Contextual** | **CERN** |

September

14/19 - Second international conference on high-energy accelerators, at CERN
16 - De Gaulle proclaims the right of Algerians to self-determination (after pacification)
22 - Inauguration of first undersea telephone cable between Europe and the United States

1 - C. Mallet (F) director of SB division

November
- McCone-Emilianov meeting: the construction of an international accelerator put on the agenda

19 - Document 'National Participation in Research at CERN' accepted
24 - First full beam on the Proton Synchrotron

December
- Nobel prize to E.G. Segré and O. Chamberlain for their discovery of the antiproton
- MURA 50 MeV electron two-way FFAG model ready (it never worked properly)
1 - Signature of treaty restricting scientific activities in Antarctica to peaceful purposes

2 - Council, creation of an Advisory Committee for the PS experimental programme (it never met)
15 - Hereward, Johnsen, Lapostolle and Zilverschoon join the Accelerator Research Group (see appendix 1)

1960

January

20 - First European meeting held at CERN to discuss the experimental programme for the PS
21 - First meeting of the European bubble chamber specialists for work with the PS

Contextual

February
3 - Macmillan announces that British policy must take account of the 'wind of change' blowing through the African continent

13 - The first French A-bomb explodes in the Sahara

March

April
21 - USSR–China split appears publicly

June

30 - Independence of (Belgian) Congo proclaimed; civil war begins

July

29 - Brookhaven 33 GeV AGS put into operation

August

25/8-1/9 - Tenth (Rochester) annual conference on high-energy physics, at Rochester

September
14 - Creation of OPEC, the Organization of Petroleum Exporting Countries

CERN

6 - Council, session devoted to the future internal structure of CERN

8 - First meeting of the committee for organizing CERN PS bubble chamber experiments

23 - Bakker's death in an air crash; Adams acting DG

14 - Council, L. Van Hove (B) will succeed Fierz as director of the Theory division in September; exchange of scientific personnel with Dubna accepted

11 - Council, Adams Director-General for one year

- Paris Ecole Polytechnique 1 m propane chamber at CERN
18 - 'A Method for Faster Analysis of Bubble Chamber Photographs' (the HPD) reported by Hough and Powell

Contextual

15 - US Congress approves the financing of the 2 mile electron linear accelerator at Stanford (SLAC)

November

8 - J.F. Kennedy elected US President

December
- Nobel prize to D.A. Glaser 'for the invention of the bubble chamber'
- 2nd US Piore Panel report on high-energy accelerator physics published
1 - Intergovernmental conference at CERN sets up the European Preparatory Commission for Space Research

1961

January

2/3 - Break of diplomatic relations between Cuba and the USA

February

CERN

20 - Padua University's offer to loan CERN an electrostatic separator (two elements each 3.5 m long) accepted

- IBM709 computer delivered and put into operation

21 - First meeting of the NPRC, the 'supreme' committee for the experimental programme

8/9 - Council, J. Willems (B) succeeds de Rose as President of the Council, V.F. Weisskopf elected Director-General; official creation of the committee system for the experimental programme; new internal structure of CERN accepted (see appendix 1); budget for 1961 voted against Britain and France

1 - Spain becomes the 14th member state of CERN

17/1-1/2 - Difficulties with the Electronic Experiments Committee

- French 81 cm hydrogen bubble chamber at CERN

Chronology of events 859

Contextual

7 - Spark chamber symposium at Argonne
Feb/Mar - The eightfold way (SU_3) and its use to classify particles

March
- e^+e^- storage ring Ada put into operation at Frascati (I)

April
12 - First manned space flight by Yuri Gagarin (USSR)
16/20 - 'Bay of Pigs' invasion (Cuba)

May
20 - Opening of Evian conference to end the Algerian war
25 - Kennedy announces Apollo moon programme to Congress

June

July

19 - Greece is formally associated with the EEC

August

CERN

29 - The SPC refuses to take a decision on the storage rings; it suggests that the 1962 ceiling be broken by some 13 MSF; the secondary beam situation described as 'very unsatisfactory'

30/31 - First meeting of the European Accelerator Group, at Saclay

2 - Council, observer status granted to Turkey; Yugoslavia officially declares that it will leave CERN; Perrin proposes that the CERN budget, exclusive of any new projects, should grow annually
5/9 - International conference on theoretical aspects of very high-energy phenomena, at CERN

- CERN 1 m heavy liquid chamber takes 120,00 photographs in a neutrino beam

21 - The SPC proposes the study of a high-intensity PS for CERN

1 - Victor Weisskopf takes up his post as Director-General

Contextual

9 - United Kingdom requests membership of the EEC
10 - Denmark requests membership of the EEC
12/13 - Start of the construction of the Berlin wall

September
6/12 - Third international conference on high-energy accelerators, at Brookhaven

October

November

December

- Nobel prize to R. Hofstadter and R.L. Mössbauer
16 - Kennedy decides to increase US forces in Vietnam to 15000

CERN

27/28 - The experiments committee structure rediscussed during the 6th meeting of the NPRC

27 - SPC, serious doubts raised about the adequacy of CERN's data-handling equipment

20 - The UK Foreign Office circulates an Aide-Mémoire suggesting that an upper limit to the CERN budget be agreed directly between governments
25 - Bohr attends an emergency meeting of the SPC
27 - K.Johnsen's report, CERN will study a 300 GeV PS and the ISR
Last week the British halt the shipment of parts of their 1.5 m bubble chamber to CERN

- First 10 m long CERN electrostatic separator runs in a beam

18 - Committee of Council, J.H. Bannier (NL) and O. Obling (DK) condemn the British for trying to by-pass the Council on financial decisions; F. de Rose (F) threatens retaliation in negotiations over ELDO if Britain refuses to support a growing CERN

Contextual

19 - Creation of the French Centre national d'études spatiales (CNES)

1962

February
14 - Project for a 60 GeV PS discussed in France

21 - John Glenn becomes the first American to make an orbital flight around the earth

March
18 - Signature of the Evian peace agreement

19 - Effective cease-fire in Algeria
29 - ELDO Convention signed in London

April
26 - First UK satellite launched from Cape Canaveral

May

June

14 - ESRO Convention signed in Paris
15 - Observation of two kinds of neutrinos at the Brookhaven AGS

CERN

19 - Council, the 1962 budget is voted at 78 MSF against the UK and Sweden; J. H. Bannier (NL) is invited to chair a working party to propose suitable rates of growth for CERN budgets; G. Funke (S) replaces A. Hocker (FRG) as FC Chairman;

31 - Withdrawal of Yugoslavia from CERN

20 - The experiments committee system is reorganized by V. Weisskopf

26/3-6/4 - CERN seminar on experimental physics with proton storage rings

29 - Publication of Bannier Report recommending that CERN should grow at an average of at least 11.5% per annum for the next four years

13 - Council, observer status granted to Yugoslavia; Weisskopf claims that there is an 'equipment gap' at CERN

Contextual

July
- 4 GeV electron synchrotron NINA decided on in Britain
2/8 - Adenauer visits France. The official communiqué explicitly takes note of Franco-German reconciliation
4/11 - Eleventh (Rochester) international conference on high energy physics, at CERN

August
5 - Death of Marilyn Monroe

September

4/9 - De Gaulle visits Germany
30 - Outbreak of bloody race riots in Mississippi

October
11 - Pope John XXIII opens the second Vatican Council
23/28 - Cuban missile crisis

November

18 - Death of Niels Bohr

29 - Decision to build the Anglo-French supersonic aeroplane, Concorde

December
- Nobel prize to L.D. Landau for 'his pioneering theories for condensed matter, especially liquid helium'

CERN

1 - G. Puppi (I) replaces Bernardini as director of research

12 - F. de Rose (F) announces that CERN was to extend onto adjacent French territory

19/30 - Decision by CERN Directorate to set up ECFA (European Committee for Future Accelerators)

3/5 - Fourth and last meeting of the European Accelerator Study Group, at Frascati
12 - The UK delegation inform the DG that they would support rates of growth of 13%-11%-10%-9% for the next four years

Contextual

14 - Mariner-II transmits information on Venus back to earth
17 - Switzerland enters the Council of Europe

1963

January
14 - De Gaulle vetoes Britain's entry into the EEC

21/22 - Adenauer visits Paris; major Franco-German agreement signed

April
6 - USA agrees to furnish UK with Polaris missiles

May

20 - US Ramsey Panel report on high-energy accelerator physics published

June
2 - First photographs taken with the 80 inch hydrogen bubble chamber at BNL

July
15/25 - UK-USA-USSR disarmament conference agrees to halt nuclear testing in the atmosphere, in space, and beneath the sea (The Test Ban Treaty); de Gaulle rejects the agreement on 29 July

CERN

19/20 - Meeting of 'founding fathers' to prepare first ECFA meeting

17/18 - First ECFA meeting at CERN

1/2 - Second ECFA meeting at CERN

29 - Introduction of Cabibbo angle

12 - Fast ejection system extracts full PS beam; neutrino horn ready

7 - Third ECFA meeting recommends the construction of concentric ISR and of a 300 GeV PS
20 - Council, observer status granted to Poland; Council accepts by 7 votes to 4 that the rates of growth for 1965 and 1966 should be provisionally set at 10% and 9%

Contextual

August
21/27 - Fourth international conference on high-energy accelerators, at Dubna
27 - UK 8GeV PS Nimrod achieves its design energy

October

16 - Ludwig Erhard replaces Adenauer as Chancellor (FRG)

November

22 - J.F. Kennedy assassinated

December
- Nobel prize to E.P. Wigner, M. Goeppert-Mayer and J.H.D. Jensen

1964

January

4 - First introduction of quarks

February
- German 6 GeV electron synchrotron DESY accelerates its first beam to full energy
11 - Observation of the Ω^- at BNL
29 - ELDO Convention enters into force

March
20 - ESRO Convention enters into force

CERN

11 - Council, new financial programme to study the ISR and a 300 GeV PS

13 - FC meeting, Scandinavian (and Benelux) countries declare that it is too expensive to build a regional accelerator of 8–10 GeV

17 - Council, J. H. Bannier (NL) succeeds Willems as President of Council (see appendix 4); UK agrees to 11% increase in budget for 1965; provisional estimates for 1966 and 1967 left in abeyance

- UK 1.5 m hydrogen chamber at CERN

- Weinberg's article "Criteria for Scientific Choice"; Weisskopf replies in the CERN Courier

Chronology of events

Contextual	**CERN**
	9 - Internal report by working group on the PS improvement programme
April	
	14 - CERN is given details of BNL's envisaged $50m AGS improvement programme
	22 - Weisskopf presents CERN's PS improvement programme to the FC
28 - De Gaulle withdraws French naval officers from NATO command	
May	
	- First Cathode Ray Tube Flying Spot Digitizer (Luciole) undergoing operating tests at CERN
	6 - Weisskopf proposes to stagger in time the decisions on the ISR and the 300 GeV PS
	12 - Technical study for the ISR published
28/5-2/6 - Creation of PLO (Palestine Liberation Organization)	
June	
12 - First European nuclear-propelled ship launched at Kiel (FRG)	
	18/19 - Council, first move for a decision on new accelerators for CERN
July	
	- Preliminary proposal for developing an isotope separator 'on-line' to the SC (ISOLDE)
10 - Discovery of CP violation at BNL	
July/Aug - SU_6 and its use to classify particles	
August	
2/7 - Incident in the Gulf of Tonkin. President Johnson has Congress accept that America intervene in South-East Asia	
5 - Twelfth (Rochester) international conference on high-energy physics, at Dubna	

Contextual

31/8-9/9 - Third International Conference on the Peaceful Uses of Atomic Energy, Geneva

September

17 - Yugoslavia signs an agreement with Comecon

October

13/14 - Brezhnev replaces Khrushchev as First Secretary of the Soviet Communist Party
14 - Martin Luther King awarded Nobel peace prize
15 - Harold Wilson's Labour Party wins UK general election
16 - Explosion of the first Chinese atomic bomb

November
3 - Johnson elected US President

December

CERN

- First small computer (SDS 920) connected on-line to a physics experiment
1 - B. Gregory (F) replaces Puppi as director of research

- Paris Ecole Polytechnique 1 m propane chamber dismantled
9 - Exceptional ECFA meeting to rediscuss priorities between ISR and 300 GeV PS

11 - Technical study for a 300 GeV PS published
13 - CERN 2m hydrogen bubble chamber commissioned
15/16 - Council, A, F, FRG, NL in favour of the ISR programme; W. Schulte-Mermann (FRG) replaces Funke as FC chairman; W.Paul (FRG) nominated as co-director of NP division (see appendix 1)

Contextual

CERN

1965

January

25 - US Seaborg Panel report on high-energy physics published

- CDC6600 computer delivered
25 - First test photographs with a 10 GeV/c separated K^- beam using an RF separator built at CERN

April
6 - NASA launches Early Bird, the first telecommunications satellite for commercial use

May
12 - Diplomatic relations opened between Israel and the FRG
24 - Britain adopts the metric system
31/5-1/6 - NATO meeting in Paris; France will not join in next round of manoeuvres

June

- UK 1.5 m hydrogen chamber dismantled

8 - Henceforth American forces can participate openly in action on the ground in Vietnam

16/17 - Council, the ISR programme is approved; B. Gregory (F) elected as new Director-General

July
1 - French EEC representative recalled from Brussels; policy of the 'chaise vide' commences (lasts until 29 January 1966)
5 - Accident on 40 inch hydrogen bubble chamber at Cambridge electron accelerator (USA) kills one man and injures six others
16 - Mont Blanc tunnel joining France and Italy by road inaugurated

August
6 - President Johnson signs the law giving blacks the right to vote

September
6 - Outbreak of war between India and Pakistan

Contextual

9/16 - Fifth international conference on high-energy accelerators, at Frascati

October

November
11 - Unilateral Declaration of Independence (of the UK) by Rhodesian Prime Minister Ian Smith
26 - First French satellite launched and placed in orbit
27 - NATO meets in Paris without France being represented

December
- Nobel prize to S.J. Tomonaga, J. Schwinger, and R.P. Feynmann for 'their fundamental work in quantum electrodynamics, with deepploughing consequences for the physics of elementary particles'

19 - De Gaulle re-elected President of the French Republic

1966

January

CERN

13 - Signature of agreement relating to the site leased to CERN by the French government

4/5 - SPC, the new proposals by the physicists to replace the ISR by another machine are defeated

15 - Council, all member states bar Greece agree to contribute to the construction of the ISR; Council votes to spend some 750 MSF on CERN from 1967 to 1969

1 - B. Gregory replaces Weisskopf as Director-General

Name index

Adams, J.B.   66, 76, 149, 151, 219, 231, 292, 347, 440, 459, 525-6, 535, 537, 541, 613, 641, 646, 649, 657, 670
   and new accelerators for CERN   684, 700, 708-9, 715, 722, 736, 752, 757-8
   and the CERN bubble chamber construction programme   519-25
   and the Group on Fusion   418-19, 422, 424, 426
   and the organization of experimental work   504, 511-6
   as British delegate   350, 363, 368, 734, 747
   as Director-General   380-4, 391, 461-4, 465-7, 469
   as PS director   370, 373, 376, 520-1
Alfvén, H.   145, 351
Alston, M.H.   274
Alvager, T.   215
Alvarez, L.W.   83, 142, 156, 157, 271-3, 281, 286, 288, 375, 507, 511, 515, 519, 549, 798, 799, 801
   and his bubble chamber philosophy   516-8
Amaldi, E.   62, 71, 285, 346-9, 351, 367, 384, 441-3, 463, 509, 537, 548, 654-5, 710-1
   and ECFA   610, 734-5, 736-8, 750, 753-4
Amato, G.   289, 292, 310
Amiot, P.   286, 287, 292, 304, 307, 309, 313, 513, 515
Anderson, H.L.   554
Argan, P.E.   274
Armand, L.   418, 420
Armenteros, R.   178, 179, 184
Astbury, P.   197, 201, 211
Asner, A.   317
Auger, P.   70, 81, 346, 371, 384
Bacon, F.   80, 83
Bakker, C.J.   62, 66, 98, 101, 102, 106, 347, 367, 384, 524, 596, 641
   and CERN organization   372, 377, 379-80, 391
   and the Group on Fusion   418-9, 421-2, 426
   and the experimental work   438-9, 440, 443, 459, 577, 580

Ball, T.   300
Ballario   531
Bannier, J.H.   368, 580, 609, 614-5, 734, 752
   as Council delegate and FC member   346-52, 363-6, 582-4, 600
   opposed to setting budget ceilings   591
   and the Bannier Working Party   604-6, 608, 785
Bassi, P.   271, 274
Bernardini, G.   71, 109, 110, 218, 219, 225, 230, 232, 351, 375, 379, 382, 384-5, 435, 454-5, 463, 469, 487, 509, 516, 521, 537, 711, 795
Berthelot, A.   447-8, 450
Blackett, P.   351, 420, 443, 516, 521
Blatt, J.M.   56
Blewett, M.H.   74, 77, 146, 149
Blewett, J.P.   146, 149, 694-5, 715, 723
Blinov, G.A.   278
Bloch, F.   106, 371, 385, 390, 784
Blythe, F.   127
Bøggild, J.K.   349-50, 619
Bohr, N.   55, 60, 76, 351, 371, 791
Bonaudi, F.   102, 108, 127, 731
Bonzano, G.   299, 317
Bradner, H.   273, 546
Bridge, H.   292
Brobeck, W.M.   687-8
Bronca, G.   730
Bruck   723
Budde, R.   169, 285, 292, 302, 322, 475, 509-11
Budker, G.J.   685
Burhop, E.   730, 734, 738
Butler, C.C.   351, 437-9, 459, 473-6, 587, 619, 757
Campiche   582
Casimir, H.B..   384-5
Cassels, J.M.   105, 351, 509, 722, 738
Chamberlain, O.   83, 513
Chavanne, A.   349
Christofilos, N.   692
Churchill, Sir Winston   50

Citron, A. 109, 110, 112-4, 122, 123, 124, 127, 149, 374, 382, 434
    and the organization of experimental work 468-9, 481, 483
Cocconi, G. 191-3, 195, 197, 199, 203, 455, 462-3, 536, 730
Cockcroft, Sir John 150, 347, 351, 372, 384, 437, 444, 710-1
Colonnetti, G. 346
Conte, M. 274
Conversi, M. 78, 113, 125, 208, 436
Cool, R. 470
Cork, B. 798
Courant, E.D. 74
Courtillet 612
Crawford, F. 271
Cronin, J. 798
Cresti, M. 537
Dahl, O. 145, 147, 149-51, 346, 653
Dakin, S. 219, 366, 371, 377-8, 381, 385, 386, 438-9, 445, 641
Dautry, R. 70, 346
de Boer, J. 447
Debrain, P. 104
de Broglie, L. 791
Depommier, P. 122
de Raad, B. 514, 531, 730, 733
de Rose, F. 346-52, 363-6, 384, 588-91, 598-601, 609, 611-2, 734, 737
    and the Group on Fusion 418, 426
de Rougemont, D. 81
Deutschmann, M. 459
de Witt, C. (née Morette) 70, 71
Dodd, C. 274
Drell, S. 733
Dürrenmatt, F. 86
Ehrenfest, P. 55
Eklund 384
Ekspong, A.G. 351, 460, 738
Emilianov, V.S. 693
Faissner, H. 218, 220, 232
Falk-Vairant, P. 197, 201
Farley, F. 112, 120, 121, 208, 215, 466
Feld, B.T. 71
Ferger, F. 700
Fermi, E. 42, 71, 389
Ferretti, B. 371, 512-3
Feynman, R. 70, 83

Fidecaro, G. 110, 112, 116, 118, 119, 127, 166, 179, 190-3, 195, 196, 205, 206, 454
Fierz, M. 349
Fitch, Val. L. 52, 55
Filtuth, H. 459
Fisher, E. 700
Florent, R. 275
Franzinetti, C. 459
Franck, J.V. 547
Fredriksson, O. 98, 100
French, B. 181
Frisch, D.H. 466
Frisch, O. 459
Fry, D.W. 146, 147, 421
Fukui, Shuji 78
Funke, G. 348, 350, 364, 367, 619
Garwin, R. 120, 466
Geibel 483
Gell-Mann, M. 83, 116, 122, 142, 176, 183, 185, 188, 206, 207, 216, 237, 740
Gentner, W. 106, 107, 109, 127, 145, 147, 285, 351, 371, 379, 577
Germain, C. 531, 539
Germain, P. 383, 386
Giacomelli, G. 191, 207, 210
Gibson, W. 110, 531, 535
Gigli, A. 274
Glaser, D.A. 78, 83, 270, 271, 275-9, 506, 515
Goebel, K. 110, 127
Goldhaber, M. 796, 802
Goldschmidt-Clermont, Y. 168, 169, 181, 186, 235, 285, 374, 510, 514, 545, 547, 553, 559
Gonella, R. 274
Gosta, G. 204
Goudsmit, S. 78
Gow, D. 523
Goward, F.K. 145, 147, 149, 151
Gozzini, A. 78
Gregory, B. 163, 275, 351, 386, 538, 548, 560, 738, 741, 751, 755, 757-8, 784
    and bubble chamber experiments 460, 463, 469, 473-5
Grivet, P. 149
Groves, L. 417
Grütter, F. 150, 159, 383, 385, 386, 731
Guéron, J. 418-20
Hagedorn, R. 218
Lord Hailsham 384

Name index

Hampton, G.H. 386
Harting, D. 110, 112, 127, 197, 200, 202, 207
Hedin, B. 98, 100, 127
Heintze 115, 122
Heisenberg, W. 43, 55, 60, 71, 76, 84, 85, 98, 346-7, 351, 361, 475, 734
Helmholtz, H. von 83, 84
Henderson, C. 274
Hereward, H.G. 150, 156, 510, 703-4
Hildebrand, R.H. 271, 272
Hine, M.G.N. 149, 150, 219, 455-6, 467, 510, 513, 533, 553, 559-60, 598, 693
    and forward planning for CERN 605-8, 613-7, 622-3
    and the organisational structure 374, 385, 387, 392
    and the new accelerators for CERN 707-8, 713, 716, 722-3, 725, 734, 748-9
    on the beam transport situation in the early '60s 536-9
Hocker, A. 346-51, 361, 364, 367, 585
Hofstadter, R. 83
Hough, P. 548-9, 551
Hubbard, E.L. 362
Hyams, B. 112, 197, 201, 214
Ibsen 619
Iselin, F. 547
Jacobsen, J.C. 149
Jentschke, W. 349, 351, 723
Johnsen, K. 72, 146, 147, 149, 156, 385, 510, 703-4, 722-4, 730, 732, 734, 738, 756
Jones, L.M. 733
Jost, R. 70
Judd, D.L. 685, 689
Källen, G. 72
Keller, R. 110
Kerst, D.W. 687, 690
Kipphardt, H. 86
Kjellmann, J. 215
Kolomenskij, A.A. 690
Koshiba, M. 78
Kowarski, L. 53, 274, 285, 292, 346, 432, 641
    and CERN's computer policy 554-8
    and the image of CERN 791
    and the PS experimental programme 514, 519
    and the STS division 371, 379, 381-3, 385, 386, 459, 509
    and track chamber picture measuring instruments 545-50

Krienen, F. 100, 102, 127, 219
Kuiper, B. 223
Kusch, P. 83
Lagarrigue, A. 164, 218, 220, 275, 440-1, 459-60, 473
Lamb, W.E. 83
Lapostolle, P. 150, 382, 385, 703
Laslett, L.J. 723
Lawrence, E.O. 389, 792, 802
Lazanski, M. 104, 127
Lee, Tsung-Dao 83
Lederman, L. 219
Leibniz, G.W. von 83
Leighton, R.B. 43
Lenard, Ph. 51
Leprince-Rinquet, L. 62, 219, 275, 348, 351, 367, 439-41, 443, 463, 524, 611, 752, 755
Liebig, J. von 80
Linhart, J.G. 700
Livingston, M.S. 74
Lock, W.O. 82
Lockspeiser, Sir Ben 346-7, 365-6, 574, 578, 585-6, 654
Luders, G. 147
Lundby, A. 110, 112, 127, 191, 197, 200, 202, 382, 454, 465
Lofgren, E. 508, 799
Macleod, G.R. 386, 548, 550, 559
Maglic, B. 199, 204
Maillet, R. 275
Mallet, C. 385, 652, 661
Marshak, R.E. 42, 61, 62
McCone, J.A. 493
McMillan, E.M. 61, 63, 799
Mannelli, I, 271
Meadows, A.J. 55, 60
Melville, Sir Harry 350, 364, 593-4, 599, 611, 756
Mermod, R. 193
Merrison, Sir Alec 110, 112, 116, 118, 119, 127, 190, 191, 360, 521, 723
Middelkoop, W. 191
Mills, F.E. 723, 730
Mitter, H. 72
Miyamoto, Sigenori 78
Møller, Ch. 70, 351, 367, 371, 734
Moneti, G.C. 274
Montanet, L. 169, 548, 550
Moore, M.J. 102, 105, 437-8
Moorhead, G. 551

Morat 584, 589
Morpurgo, M. 100, 315, 316
Morrison, D.R.O. 169, 171, 176, 181, 189, 290, 322
Munday, G.L. 150, 730
Nagle, D. 271, 272
Ne'eman, Y. 142, 183, 206, 237
Neumann, F. 582, 584, 589, 593, 646
Nullens, G. 288, 292
Obling, M.O. 349-50, 364, 582, 584, 590, 598
Okhawa, T. 690-91
O'Neill, G.K. 687-8, 690-2, 697, 724-5, 730
Oppenheimer, J.R. 70, 86
Ørsted, Ch. 83
Ortega y Gasset, J. 54
Overas, H. 109, 123
Panofsky, W. 67, 389, 687, 690, 696, 740, 794, 799, 801
Parmentier, D. 272
Parain, J. 730
Paul, W. 55, 60, 70, 386, 482, 738
Peierls, R. 62
Pentz, M.J. 700
Perkins, D.H. 62
Perrin, F. 275, 346-9, 351, 361, 367, 384, 538, 544, 588-9, 710-11
  and the French accelerator programme 360, 730, 733, 757
  and the French bubble chambers 439, 442-3
  and the growth of CERN 588-9, 597, 733-5, 748, 751
Peters 454
Petrucci, G. 299, 310, 316
Peyrou, Ch. 151, 163, 169, 171, 281, 285, 288, 292, 294, 297, 305, 322, 374, 382-3, 532, 547
  and bubble chamber experiments 458-9, 469, 473
  and the CERN bubble chamber construction programme 510-15, 519-23, 544
Picasso, E. 274
Piccioni, O. 512
Pickavance, T.G. 105, 350, 737
Picot, A. 346
Pizer, H.I. 110
Plass, G. 223
Pontecorvo, B. 217, 218
Powell, B. 548, 550
Powell, C.F. 78, 367, 598-9, 734, 737, 747, 750-2, 790
  and the Emulsion Committee 456-7, 463
  and the SPC 347, 351
Preiswerk, P. 346, 379
  and the CERN building programme 641, 653
  and the NP division 382, 386, 465, 467, 481-3
  and the SB division 371, 651-3
Puppi, G. 351, 386, 435, 734, 738, 757
Quercia, I.F. 723
Rahm, D. 271
Ramm, C.A. 150, 152, 164, 218, 220, 225, 292, 383, 440-1, 473, 523-4, 531-5, 646, 711, 737
Ramsey, N.F. 698, 740
Regenstreif, E. 145, 147
Reinhard, H.-P. 322
Resegotti, L. 292, 514, 523, 531, 730, 733
Richemond, J. 371-2, 579
Riddiford, L. 62
Roaf 459
Roberts, A. 212, 483, 488, 798
Rosenfeld, A. 555
Rossi, B. 71
Rougemont, D. de 81
Rubbia, C. 115, 122, 213, 214
Salam, A. 378, 722
Salmeron, R.A. 219
Salvini, G. 463, 465, 467, 538
Sands, M. 692-4, 696
Schein, M. 78
Scherrer, P. 346, 349, 588
Schlüter 701
Schmeissner, F. 305, 306
Schmelzer, Ch. 146, 147, 149, 510
Schnell, W. 147, 730
Schoch, A. 150, 151, 374, 385, 700-1, 703-4, 706, 730, 734, 757
Schulte-Merman, W. 348-9
Schwartz, M. 217, 219
Schwemin, A.J. 272
Schwinger, J. 83
Scotoni, G. 312
Segré, E. 70, 83, 384
Seitz, F. 277
Serber, R. 77, 719
Shutt, R. 273, 275, 287, 309, 511, 549
Siegbahn, M. 509
Snyder, H.S. 74
Soergel, V. 115, 122
Solla Price, D.J. de 51
Sommerfeld, A. 55, 60

Name index

Stanley, P.H.   386, 390
Steiger, R.   151, 651-4
Steinberger, J.   219, 271, 273, 285, 286, 309
Stevenson, L.   271
Strauss, L.   417
Symon, K.R.   686, 730
Taconis, K.W.   307
Tamm, I.   62
Taylor, A.   192, 197
Tièche, C.   381
Toffolo, G.   365-6
Tomasini, G.   274
Tomonaga, Sin-Itiro   83
Trembley, J.   299, 304
Trumpy, B.   384
Tycho, D.   551
Uexküll, J. von   81
Ullmann, G.   381, 386
Urban, P.   72
Valeur, R.   354
Van der Meer, S.   222, 798
Van Hove, L.   70, 73, 385
    and new accelerators for CERN   708, 712, 714, 725, 734, 737, 751
Veksler, V.I.   62
Verry, H.L.   347, 350, 363, 421, 578-82, 585-6, 654
Vilain, J.   292, 295, 303
Vivargent, M.   193
Vogt-Nielsen, N.   700
von Dardel, G.   190, 193, 195, 210, 220-2, 230, 374, 382, 454, 465, 467, 513-4, 518, 521, 531-5, 547

Walker, R.St.J.   609
Waller, I.   346, 350
Wallin   215
Webb, F.A.R.   652
Weinberg, A.M.   51, 54, 747
Weisskopf, V.F.   50, 56, 70, 82, 84, 130, 219, 221-2, 232, 236, 239, 348-9, 557, 560, 598-9, 607-8, 790
    and CERN's 'lateness' in 1960-62   505, 539-40, 544-5
    and ECFA   734-5, 738, 748-9, 752-5, 756-8
    and the experimental work   467, 468-70
    as Director-General   353, 354, 367, 383-90
White, H.   549-550
Wideröe, R.   145, 147
Wigner, E.P.   54, 80, 82, 83
Willems, J.   346-52, 363-6, 575, 580-4
Wilkinson, Sir Denys   359
Wilson, R.R.   693, 736, 799
Winter, K.   122, 193, 204, 212
Winter   723
Winzeler, H.   460
Wittgenstein, F.   318
Wood, J.   272
Wouthuysen   738
Yang, Chen Ning   83
York, H.   792
Zavattini, E.   112
Zichichi, A.   72, 192, 193, 207, 208
Zilverschoon, C.J.   150, 383, 385, 646, 703-4, 730, 740-2
Zweig, G.   142, 176, 207, 237

Thematic subject index

Accelerators

   AG focusing principle and its problems   146-50
   AGS   218, 219, 230, 504, 533
   American high energy synchrotron   684, 687-8, 690, 692-6
   Bevatron   270, 273, 278, 504, 508, 513
   CERN high energy synchrotron   721-3, 724, 736, 737-9, 740-2, 748-50, 752-3, 756-8
   CERN high intensity synchrotron   714-6, 724
   CERN ISR   707, 709, 712-3, 721, 724, 730-2, 739, 748, 752-3, 756-7, 798
   CERN PS   285, 297, 299
      decision to construct   145-7
      running in   456, 536
   Cosmotron   141, 270, 272, 273, 504, 508
   decision to construct the SC   97-8
   design and construction of CERN SC   98-108
   design and construction of CERN PS   147-62
      magnet   148, 152-4
      RF system   148, 154-6
      linac   148, 156-9
      vacuum system   160-2
      position of the machine   162
      main machine parameters for PS   146, 149-50
      collaboration with the Unites States   146, 149, 151
   DESY   729
   electron colliders   697, 699
   FFAG accelerators   685-6, 689-90, 696-7, 698-9, 701, 703
   Frascati   697, 699
   history of new concepts   684-97
   linear accelerators   687, 695, 730
   Nimrod   164, 274, 275, 359, 521, 581, 690, 729-30
   Nina   360, 729-30, 733
   plasma accelerator   685, 689, 697-8, 701
   proton colliders   686-8, 690, 697, 699, 704, 712
   Saturne   275, 730
   SC improvements   112-4
   the notion of 'monster'   685, 693, 696, 710-1, 714, 716
   ZGS   432, 690

Bubble chambers

10 cm HC, CERN (HBC10)   110, 151, 509, 511
30 cm HC, CERN (HBC30)   112, 151, 163, 236, 476, 510, 511, 522, 536
  experiments   169-70, 176, 177, 182
2 m HC, CERN (HBC200)   152, 163, 164, 236, 476-9, 506, 507, 520-1, 543-5
  experiments   176, 177, 189
1 m HLC, CERN   152, 164-5, 233, 236, 284, 476-9, 522-4, 543
  experiments   185, 218, 220
Gargamelle, HLC   234, 284, 321, 477
Big European HC (BEBC)   284, 321, 477
Camera nazionale italiana   274
1.5 m British National HC   163, 164, 236, 274, 294, 297, 299, 313, 314, 315, 476-9, 521, 543, 600
  experiments   176-8, 188-9
1 m HLC, Ecole Polytechnique (BP3)   164, 236, 275, 297, 476-9, 524, 536, 590
  experiments   170, 174, 176, 183, 218, 220
81 cm HC, Saclay   163, 274, 291, 293, 294, 297, 307, 313, 476-9, 505, 524, 543
  experiments   170-81, 183, 188
  filled with deuterium   163, 173-4, 185
72 inch Alvarez HC   142, 271, 273, 282, 283, 285
20 inch Shutt HC   273
as universal detector   451-3, 516-7
novelty of   509-11, 517
safety of   523-5
see also chapter 6

Construction time   291-4, 299-300, 303, 315, 317-8

and engineers and physicists   521-2, 541, 797-8
and size of bubble chambers   518-22, 798
of electrostatic separators   541-3, 797
of the British National HC   521, 524
of the CERN 2 m HBC   520-1, 528, 543-5

Costs and cost escalation

for research equipment in general   576, 578
in the bubble chamber programme   271, 273, 291, 295, 305-6, 319, 321, 523, 527, 587-9
in the construction of buildings   297, 298, 594
in the SC experimental programme   577
of the PS   574
of the PS experimental programme   589-93
threatening the construction of the SC   114

Decision-making process   694-5, 716-9, 724-5

and temporal processes   726-9, 756-7, 759-62
and the way scientists debate   742-5
and voting procedures   365-7, 578, 602, 787
rational model of   643, 647-8, 759
setting priorities   514-22

Detectors, equipment and measuring devices
    bending and focusing magnets   507, 526, 533-5, 537-41
    Cerenkov counters   533
    cloud chambers   220
    computers/computing   168-9, 199, 238-9, 315, 550, 554-61
    electrostatic separators   165-8, 508, 525-6, 532, 537-43
    emulsions   110-1, 275, 533
    fast ejection system   223-5
    for the East experimental area   525-6, 539
    for the North hall   538, 540
    for the South hall   532-3, 538-40
    HPD   169, 315, 548-52
    Iep   168, 289, 315, 547-8
    internal targets   507
    ISOLDE   116
    muon channel   109, 110, 112, 113, 123-4
    muon storage ring   208
    neutrino horn   222-3
    RF separators   508
    spark chambers   218, 219, 225-6
        sonic spark chambers   199
    Wilson (cloud) chambers   270, 275, 277, 285, 509

Experimental physics
    beam survey   190-2
    bubble chamber experiments using as projectiles
        protons   169, 177-8
        anti-protons   178-82
        pions   169-77
        kaons   182-9
    CP violation   142, 185, 212-3, 238
    detection of the anti-xi   180
    detection of the Ω^- 185-9, 206
    deuterons/anti-deuterons   192, 193
    electron decay of the pion   112, 116-8, 130
    (g-2) experiment   112, 113, 114, 119-21, 208
    lambda-sigma parity   183-4, 206, 237
    lifetime of neutral pion   209-10
    magnetic monopoles   210
    measurement of cross-sections   193-6, 238
    missing mass spectroscopy   204-5
    muon physics   97, 109, 114, 123-6, 130, 214
    neutrino physics   142, 144, 215-29
    nuclear chemistry   96, 111, 115-6
    Panofsky ratio   112, 118-9
    PAPEP/PAPLEP   208-9
    peripheral interactions, OPE model   170-3, 177-8, 181
    pion beta decay   115, 122-3, 130
    physics done at BNL   142, 171, 178-80, 212, 222, 229, 238

physics done at Berkeley   142, 172, 178-9, 182, 187, 195, 212
production mechanism of resonances   187
production of strange particles   170-1, 179
quarks   142, 176, 207, 237
relativity theory   215
scattering experiments   171, 172-3, 181, 197-202, 238
study of neutral kaons   183, 211-3
study of resonant states   142, see chapter 5.3.2 and 5.4.1.4
   diboson resonances   202-4
   hyperon resonances   142, 170, 177, 181, 182, 184-5, 202
   nucleon isobar production   173, 178, 181, 199
   A-mesons   172-4, 176, 204-5, 237
   g-meson   174, 177
   E-meson   179, 181, 237
   C-meson   179, 181, 237
   D-meson   179
   $K^*(890)$ 179, 182
   $K^*(1400)$ 186, 237
   rho-meson   142, 170, 175, 203, 205-6
   w-meson   142, 179, 182, 183, 206
   f-meson   170, 171, 174, 176
   eta-meson   142, 182, 206
SU(3) theory and tests   176, 179, 183, 184, 206, 237
W-boson   219, 222, 228-9, 231
weak interaction, V–A theory   110, 115, 129, 214, 216
$\Delta S = \Delta Q$ selection rule   183, 185

Experimental work

  and CERN failures   505, 548, 607, 795-6
  and CERN unpreparedness in 1960–62   505, 511, 535-45, 795, 798
  as managed by the EEC   465-8, 481-4
  as managed by the NPRC   468-71
  as managed by the TCC   458-61, 538
  as practiced in mixed teams   444-50
  as practiced in truck teams   434-6, 438, 442-3
  in bubble chamber collaborations   458-61, 473-6, 479
  in emulsion collaborations   456-8
  organization around the British bubble chamber   434-39, 442-4, 793
  organization around the CERN chambers   291, 296, 458-60
  organization around the French bubble chambers   439-44, 458-60, 793
  organization around the SC   434-6, 489-90
  PS experimental programme plans, 1958-60   454, 491-2, 512-8, 531-3
  SC experimental programme plans, 1956-57   433-4, 509-11
  with electronic detectors   451-3, 480-9

Industry

  ACEC   318
  Ansaldo-San Giorgio, Genoa (I)   315, 316, 646

Ateliers de Constructions Mécaniques   659-60
Boehler   300
Compagnie des Aciéries et des Forges de la Loire (C.A.F.L.)   301
Fischer   300
Gandini, Guffanti and Vandoni   656-7, 660, 663
Locher, Cuénod and Induni   655
Loos Co.   305
Oerlikon   537
St. Chamond   300
Schott, Mainz   314
Schweizerischer Verein von Dampfkessel Besitzern   320
Societa Italiana Acciaierie Cornigliano   647, 650
Société de Chaudronnerie et de Montage, Tissot et Cie (F)   320
Société des Forges et Ateliers du Creusot (SFAC), Schneider factory   317, 318
Sogenique   553
Spinedi   653
Sulzer Bros.   305, 306
Uddeholm Steel   300
Ugine   299
Zapp   300
Zschokke, Spinedi, and Losinger   656-7, 660, 663

Institutions

ELDO   601, 613, 786
ESA   584
ESRO   601, 613, 622, 638, 786
EURATOM   368, 369, 417-22
IAEA   368
MURA   685-6, 689-90, 692, 696-7, 698-9, 706, 723
NASA   648-9
NATO   368
NIRNS   581

Policy

and national bureaucracies   785-6, 787-8, 792
as formulated by ECFA   619, 736-42, 753-5
for a budget ceiling   534-5, 579-85, 590-2, 598-600
for budget growth   597-601, 604-6, 620-2, 784-5
for awarding contracts   640-3, 665-7, 671
on fusion research   419-20
on standard beam equipment   533-5, 796-7
on the acquisition of computers   554-9
on the distribution of bubble chamber photographs   291, 296, 458-61, 465-6, 475, 476-9, 552-4
on the experiments committee system   461-4, 471-2, 489-94, 795-6, 801-2
on the redeployment of PS engineers   373, 520-4, 541, 705-6
on the setting up of ECFA   733-6
on the use of visitors chambers   436, 442-4, 523-4
vis-à-vis the 'world accelerator'   693, 704, 715-6, 722, 729, 740

Thematic subject index

Social and Political Relations

among
    electronic physicists   488-9
    member states   384, 419-20, 422, 599-601, 620, 622
    physicists from different countries   475-6

between
    CERN Council and the Member States   367-8, 788
    CERN Management and the Council   343, 357-9, 753, 784, 791
    CERN Management and the FC   343, 358, 595-6, 663-4
    CERN Management and the SPC   343, 608
    scientists and the Council   345, 353-6
    scientists and non-scientists in the Council   357-9
    the DG and the Division Leaders   381-3, 387
    the DG and the Senior Staff   373-4, 376, 378, 390-3, 471-2
    the DG and the non-scientific staff   389
    the PS Division and the rest of CERN   375, 376, 379, 388, 419, 535
    electronic and bubble chamber physicists   451-3, 468-71, 516, 518
    experimental and theoretical physicists   515, 517, 716, 718
    engineers/builders and physicists/users   510, 521-2, 707-8, 712-4, 716-9, 749-51, 797, 800-1
    physicists from different countries   475-76
    non CERN scientists and the organization   359-62, 580-1, 785, 793-4
    European and American physicists   550-2, 719-21, 795-6, 799-802
    scientific visitors and CERN staff   446-50, 462-4, 466, 484-9
    European physicists and CERN   359-62, 446-50, 489-93, 618, 754-5, 756-7

CERN and the history project   803-6
CERN as a laboratory in its own right   524, 529
CERN as multinational laboratory   792-3, 801-2
selling strategies to the FC and Council   354-6, 526-30, 544-5, 739-42, 802
the CERN 'lobby'   362-8, 583-4, 586, 595-6, 753, 786-7, 790
the 'corporate spirit' of CERN   388, 789-91

Statistical Data on

    Austria   409, 603
    Belgium   408
    Denmark   409
    France   406-7, 409-10, 476-9, 486-8
    Italy   402-5, 477-9, 487-8
    Germany   407, 477-9, 487-8
    Netherlands   408
    Norway   408
    Spain   409
    Sweden   408, 488-9
    Switzerland   407-8, 409-10, 487-8
    United Kingdom   405-6, 476-9, 487-8
    American visitors at CERN   486, 488
    bubble chamber work   476-9
    CERN divisions   395-7

CERN professional structure   397-9
distribution of competitive and non-competitive industrial contracts   665-7, 669
distribution of industrial contracts by country   668-71
electronic teams   480, 484-9
European economic situation   785
European equipment for measuring bubble chamber photographs   553-4
growth of computer use   561
the output of CERN's track chamber film measuring instruments   548, 551, 554
visitors at CERN   399-401